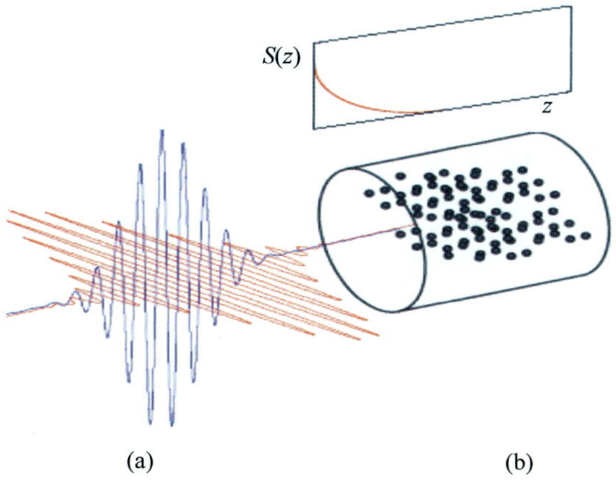

Fig. 1-2 (a) Show the control field (red) and the input single-photon pulse (blue) that are properly timed for absorption of the input pulse in the atomic ensemble. (b) Schem-atically shows the probability of the spin excitation that is decreasing exponentially over the length of the medium.

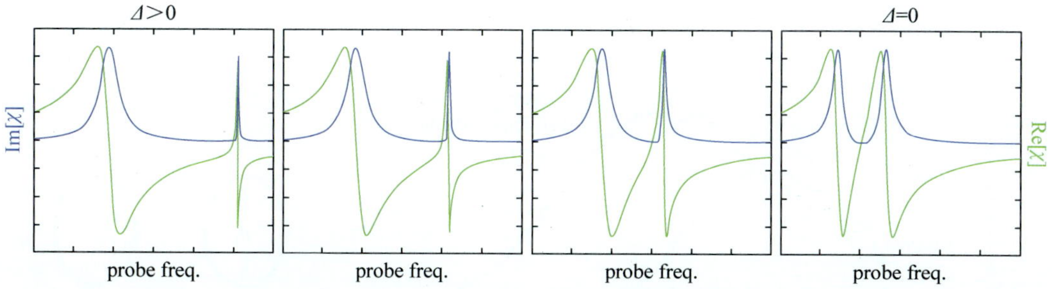

Fig. 1-3 This figure shows the imaginary part (blue) and real part (green) of the susceptibility that is experienced by the probe field $\hat{E}(t)$, see Figure 1-1. In the L configuration in Figure 1-1, by reducing the detuning Δ from left to the right, the scheme approaches the condition for EIT. The plot on the right presents a transparency window that is associated to a reduction in the group velocity that can be determined by the slope of the refernces.

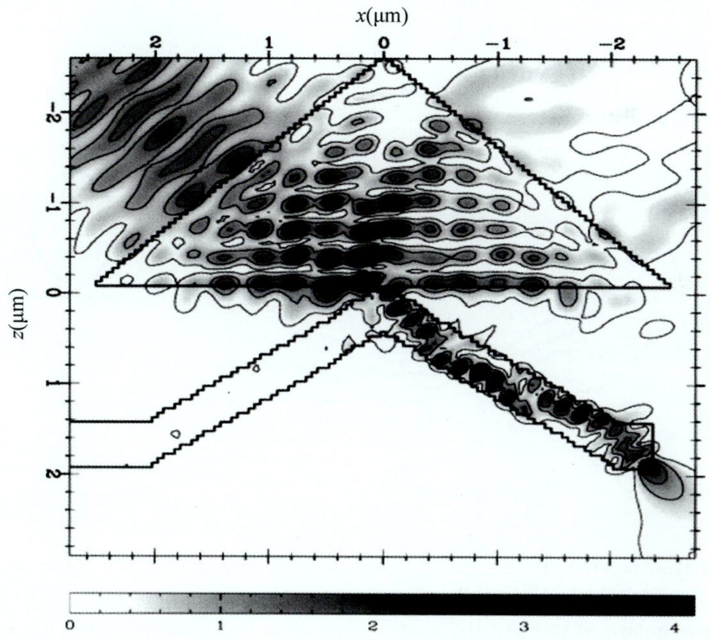

Fig. 4-23 Illustration of the tip-sample optical interaction between a 2D silicon nitride tip and a glass prism. The parameters are described in Figure 4-19. This simulation has been performed with a two-dimensional numerical code built with the Green function technique. In this example the tip touches the surface of the prism.

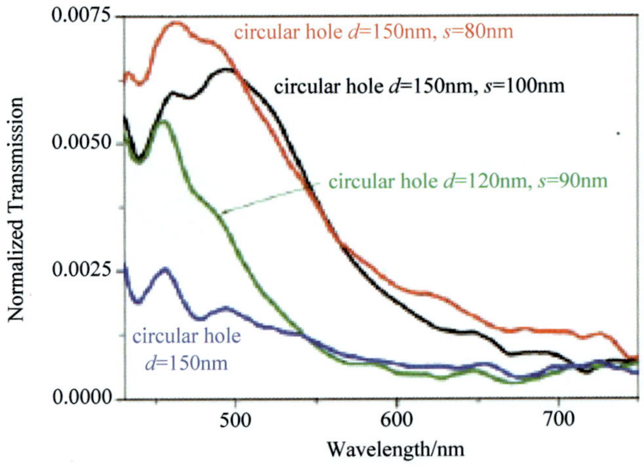

Fig. 4-79 Measured transmission spectra through single and double circular apertures in the suspended 200nm silver film. The white-light source was unpolarized. The spectra are normalized by the transmissivity of a 2μm-diameter aperture milled in the silver film. Blue: single hole, $d = 150$nm. Green: hole pair, $d = 120$nm, $s = 90$nm. Red: hole pair, $d = 150$nm, $s = 60$nm. Black: hole pair, $d = 150$nm, $s = 100$nm.

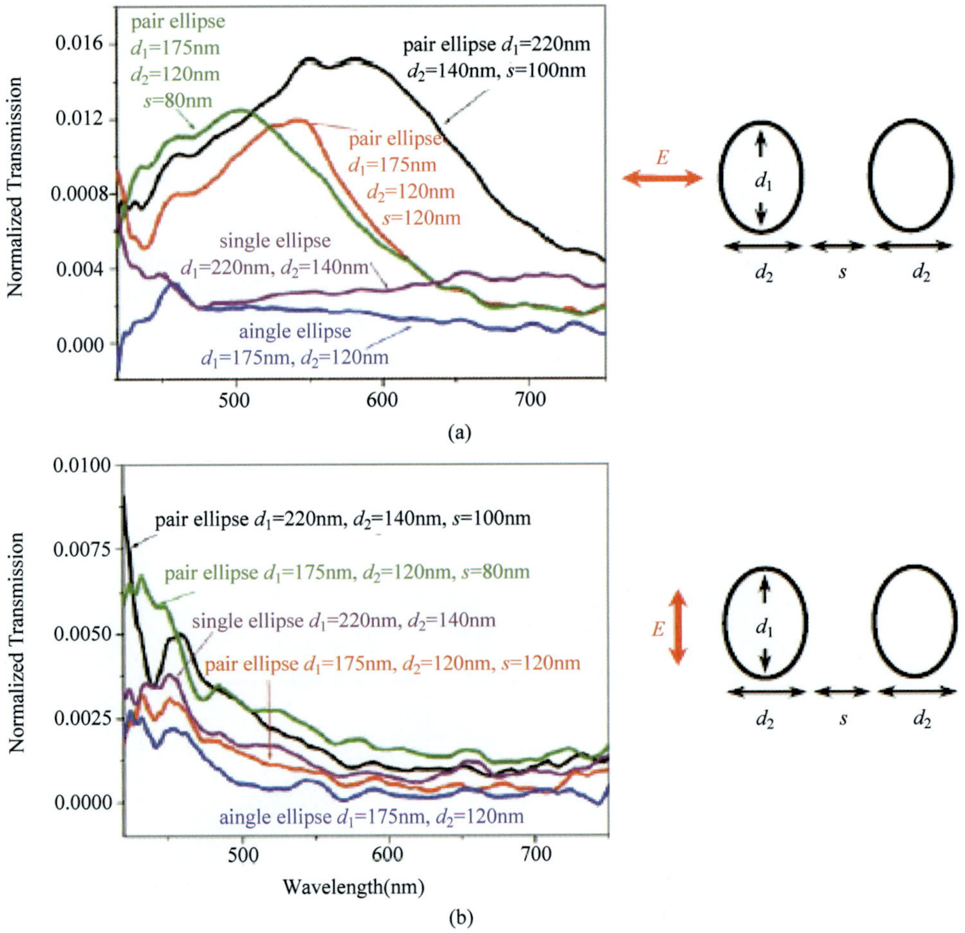

Fig. 4-80 Measured transmission spectra through single and double elliptical nano-holes in the suspended 200nm-thick silver film depicted in Figure 4-77. The white-light source used in these measurements was linearly polarized (a) parallel to the short axes, (b) parallel to the long axes of the elliptical apertures. The spectra, labeled by aperture diameters (d_1, d_2) and pair separation s, are normalized by the transmissivity of a 2μm-diameter aperture milled in the silver film.

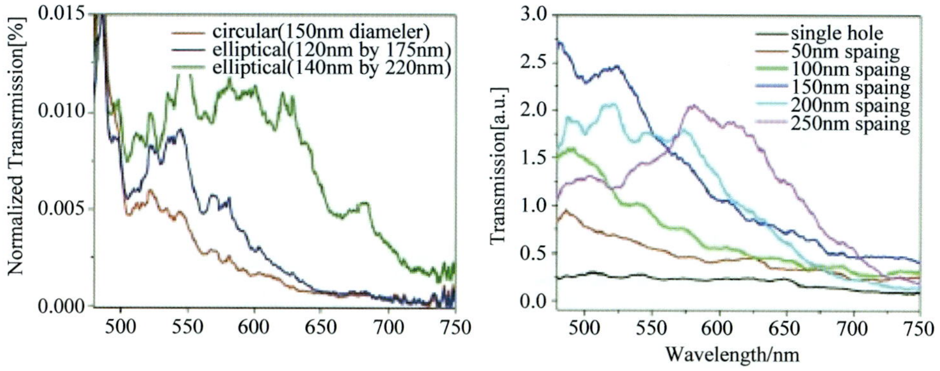

(a) (b)

Fig. 4-81 Transmission spectra through nano-apertures measured with a super-continuum source. (a) Pairs of air-filled circular and elliptical apertures in a suspended 200nm-thick silver film. The circular holes (red) have $d = 150$nm. In the case of elliptical holes, major and minor diameters (d_1, d_2) are (175nm × 120nm) (blue) and (220nm × 140nm) (green). Separation between the apertures is $s = 100$nm in all cases. (b) Single and double circular holes having diameter $d = 100$nm. The 200nm-thick silver film is deposited on a glass substrate. A droplet of index-matching fluid ($n_0 \sim 1.5$) is placed atop the silver surface prior to measurements. Vertical scale is not normalized. While transmission through the single hole (black curve) is relatively weak, double holes exhibit progressively stronger transmission with increasing hole separation. The spectra of double-holes with separation ≥150nm extend as far as $\lambda \sim 700$nm. In both (a) and (b), the various spectra are clearly distinguishable from each other, each representing a unique signature for the corresponding hole pattern.

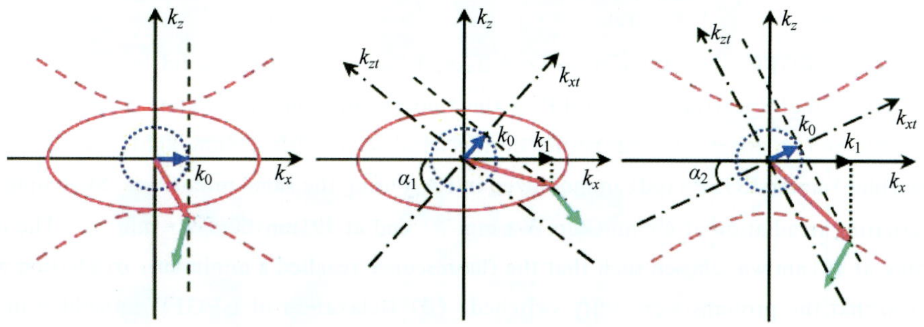

(a) (b) (c)

Fig. 4-87 Equifrequency curves (EFCs) of air (dotted blue circle) and the metamaterials with an elliptic (solid red ellipse) and a hyperbolic (dashed red hyperbola) dispersion, respectively. (a) The interface of the metamaterials is a straight line along the k_x direction; (b) The interface of the meta materials with an elliptic dispersion has an angle α_1 with respect to the k_x axis; (c) The interface of the meta materials with an hyperbolic dispersion has an angle α_2 with respect to the k_x axis. That is, the k_{xt} axis is the interface in both (b) and (c).

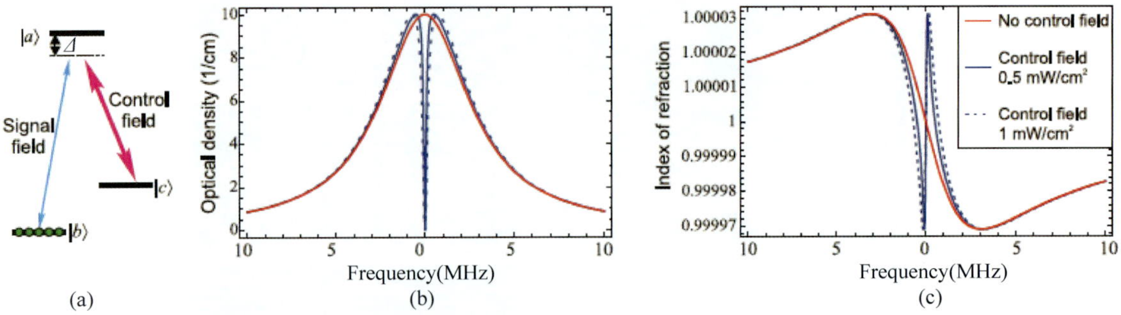

Fig. 5-5 Electromagnetically-induced transparency. (a) Atomic level configuration. Both fields are detuned from the resonance by the same frequency so the two-photon resonance condition is fulfilled. (b) Optical density. (c) index of refraction of an ensemble of atoms in the absence (red) and in the presence (blue) of EIT. In spite of a significant optical depth, the variation of the index of refraction is very small. The atomic parameters used to generate the plots correspond to a cloud of ultracold rubidium atoms.

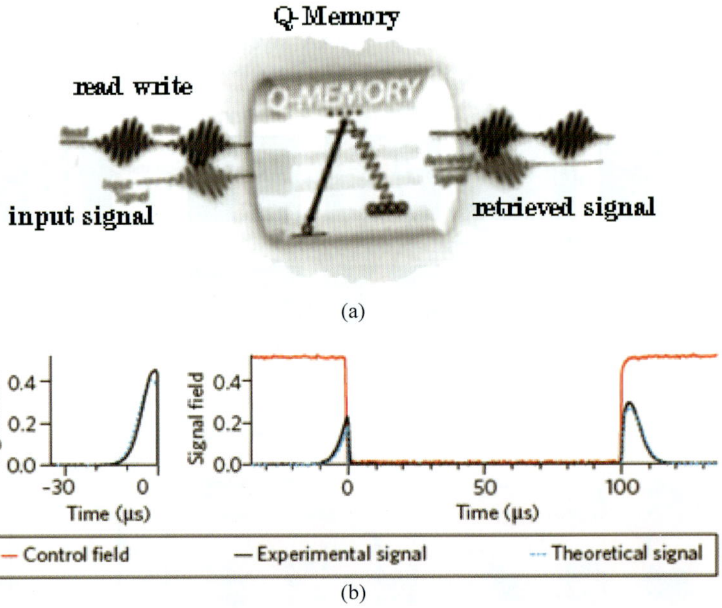

Fig. 5-6 Storage of light by means of electromagnetically induced transparency. (a) Idealized picture. The signal pulse enters the cell under the EIT conditions (with control field on, top image). While the spatially compressed pulse propagates inside the EIT cell, the control field is switched off, so the quantum information carried by the pulse is stored as a collective excitation of the ground states (middle image). When the pulse needs to be retrieved, the control field is switched back on (bottom image). (b) Optimized classical light storage in a rubidium vapor cell with a buffer gas, $\alpha L = 24$. The red curve shows the control field, solid black—experimental signal, dashed blue—theoretical signal. Left: input signal pulse of optimal shape. Right: storage and retrieval.

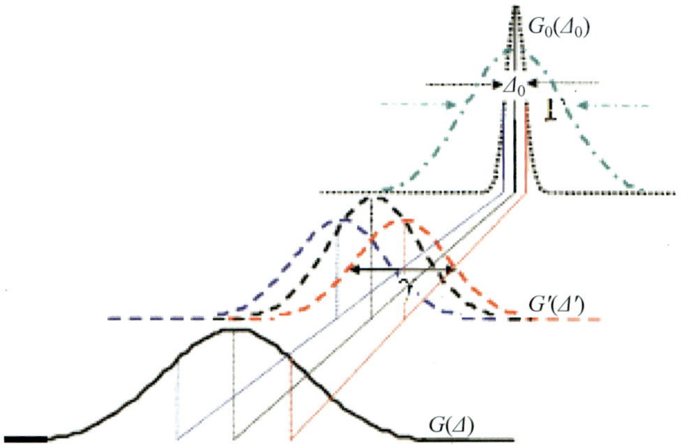

Fig. 5-11 (Color online) Schematic representation of the spectral atomic distribution. The initial distribution $G_0(\Delta_0)$ with characteristic bandwidth 0 is represented as black dotted line. Three spectral components of the initial distribution are considered (blue, black and red vertical lines) and each of them is broadened according to a distribution $G'(\Delta')$ (blue, black and red dashed lines) with bandwidth. The final broadened distribution, called $G(\Delta)$ is the convolution of the initial distribution $G_0(\Delta_0)$ and of the broadened distribution $G'(\Delta')$ associated to a single initial absorption line (Δ_0). The pulse shape with bandwidth γ is represented as green dashed dotted line.

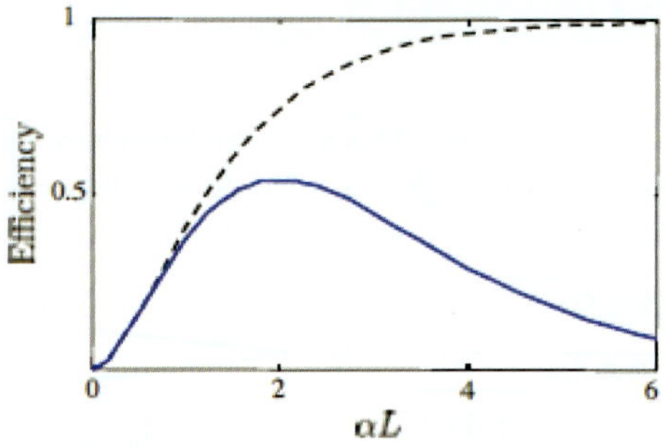

Fig. 5-12 (Color online) Efficiency of the light pulse reemission in forward (full blue line) and backward (dashed black line) direction as a function of the optical depth for a constant broadening. The efficiencies vary as $(\alpha L)^2 e^{-\alpha L}$ for the forward protocol and as $(1-e^{-\alpha L})^2$ for the backward protocol.

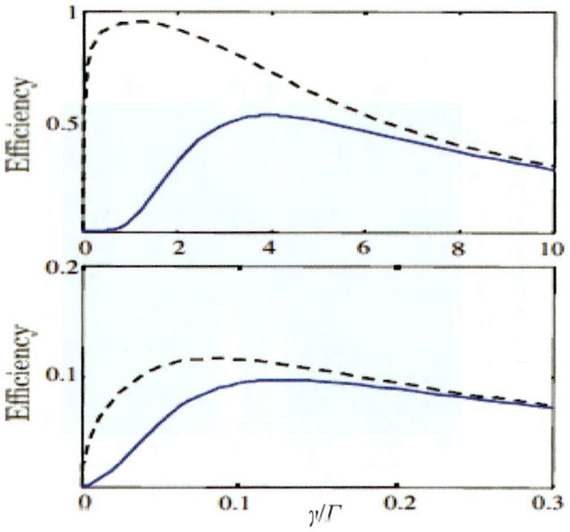

Fig. 5-13 (Color online) Efficiency of the light pulse reemission for forward (full blue line) and backward (dashed black line) protocol as a function of the broadened distribution bandwidth γ (in units of the pulse bandwidth Γ) for effective widths of the initial distribution $v = 2\Gamma$ (top) and $v = 0.05\Gamma$ (bottom).

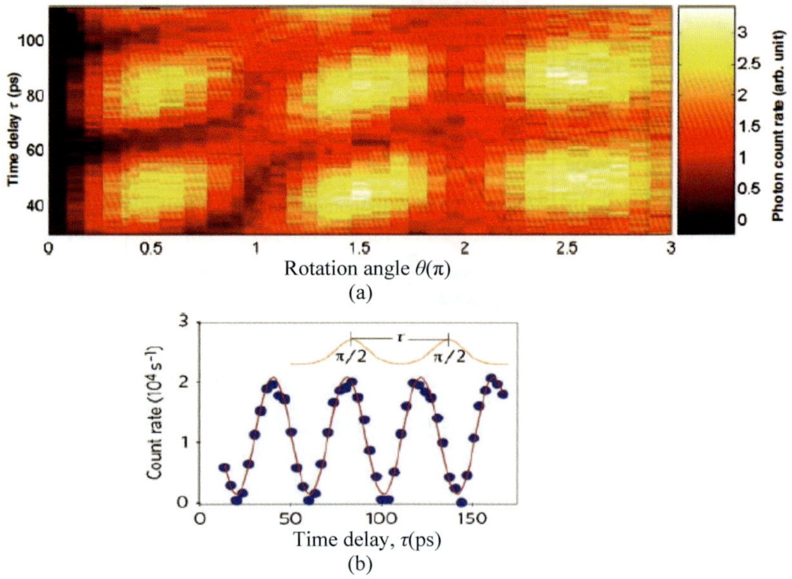

Fig. 5-28 (Color online) Experimental results on Ramsey interference of electron spins obtained using two optical pulses. (a) Dependences of photon count rate and time difference between two pulses on rotation angle due to optical pulse irradiation. (b) Ramsey fringe with respect to time difference between two $\pi/2$ pulses.

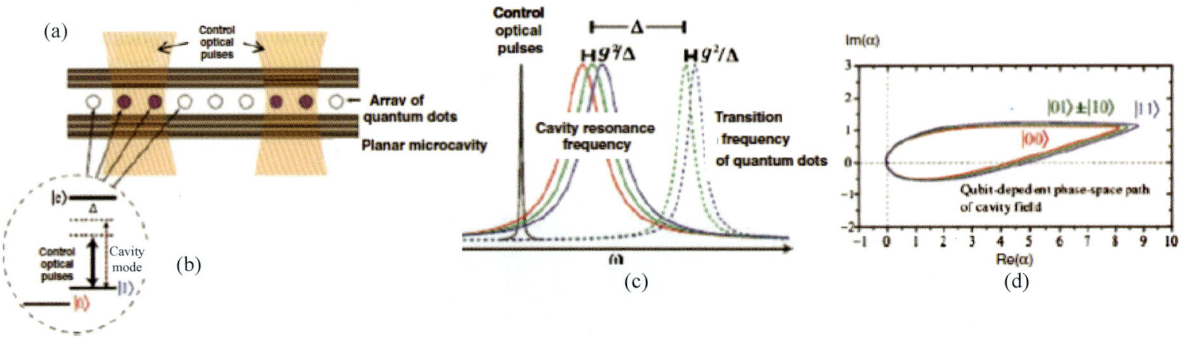

Fig. 5-29 (Color online) (a) Implementation of control phase gate to two adjacent qubits by coherent control optical pulse. (b) Relationships among transition frequency of artificial atom, cavity resonance frequency, and center frequency of control optical pulse. (c) Dependence of shift of cavity resonance frequency on state of two qubits. (d) Qubit-dependent phasespace path with respect to real and imaginary parts of complex amplitude α of coherent optical field inside a cavity.

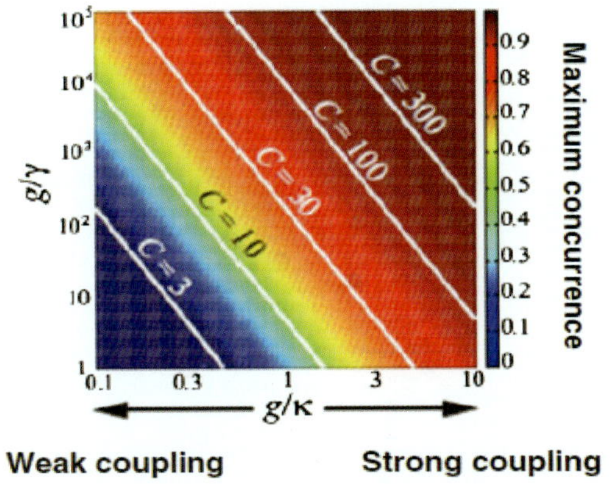

Fig. 5-30 (Color online) Concurrence of entangled quantum states created by two-qubit (control phase) gate. g is the vacuum Rabi frequency, k is the cavity decay rate, and γ is the excitonic spontaneous emission rate.

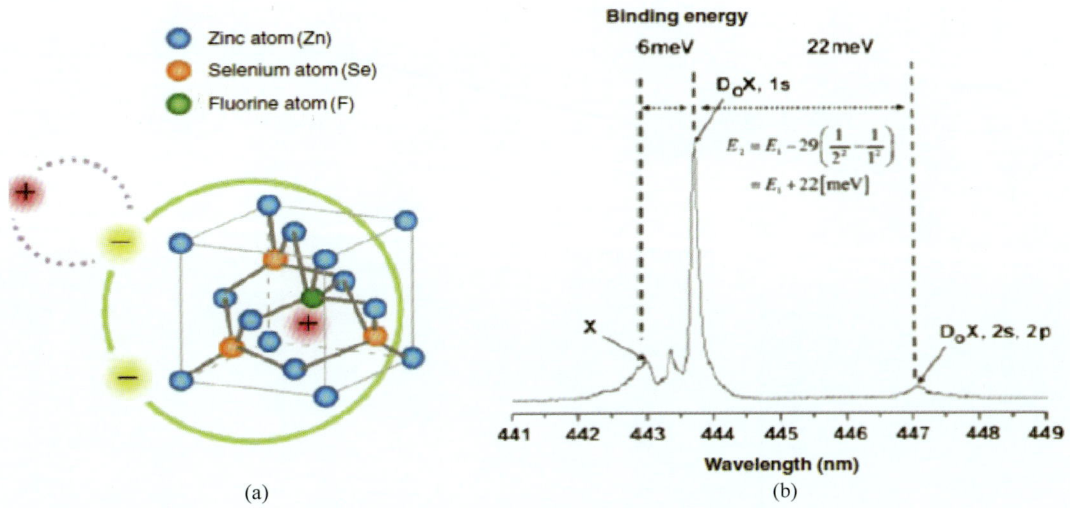

Fig. 5-31 (Color online) (a) Donor impurity (^{19}F:ZnSe), bound electron (D_0), and bound exciton (D_0X). (b) PL spectrum of bound exciton in ^{19}F:ZnSe donor impurity.

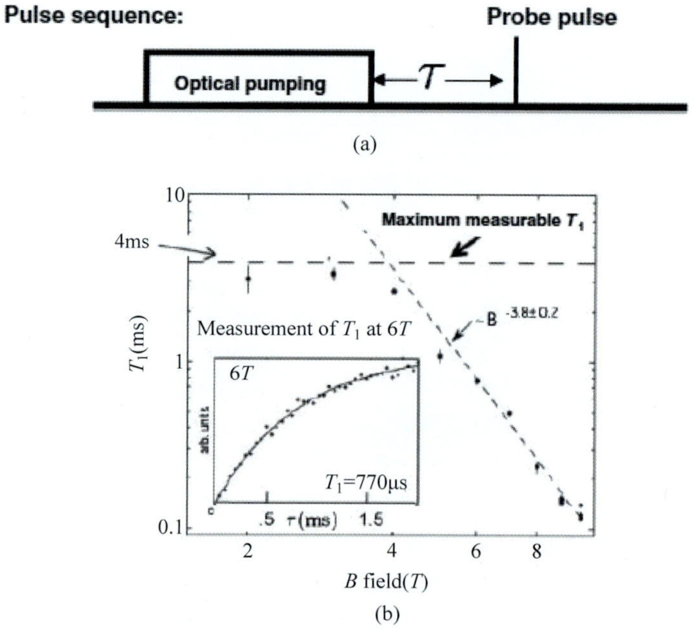

Fig. 5-32 (a) Measurement of T_1 of Si:GaAs electron spins based on initialization by optical pumping. (b) Measured T_1 vs DC magnetic field B_0. T_1 is expected to be longer than its maximum measurement limit (4ms) when $B_0 > 4T$.

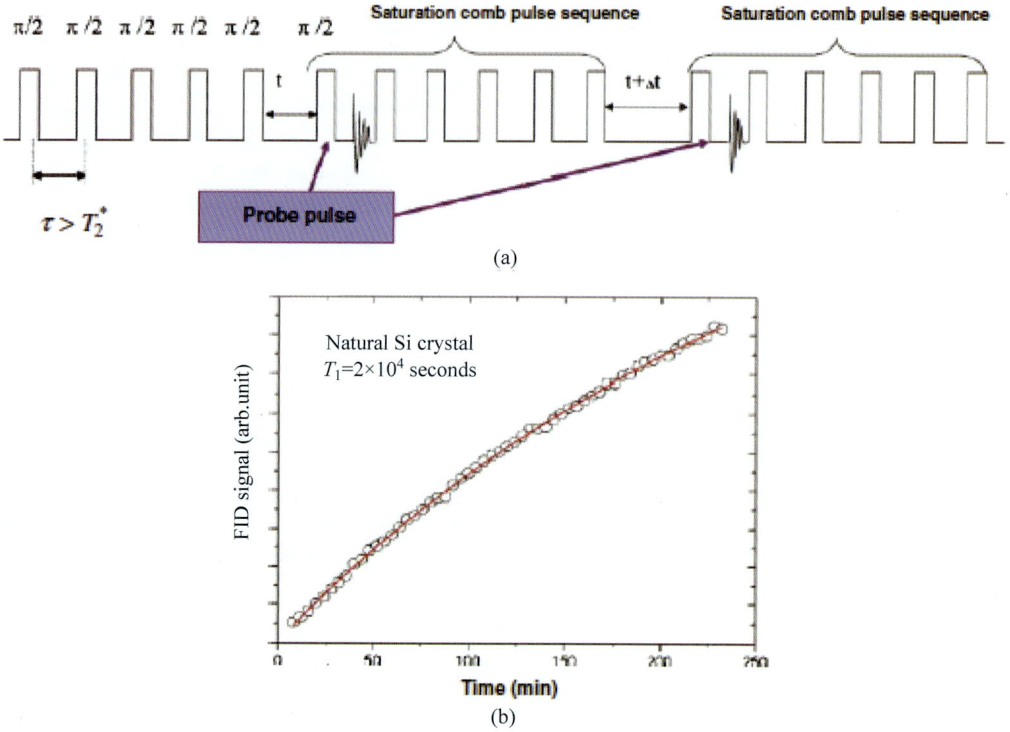

Fig. 5-34 (Color online) (a) Saturation comb pulse sequences used to measure T_1 for nuclear spins. (b) Measurement result indicating that T_1 for Si nuclear spins in natural Si crystal is more than 5h at room temperature.

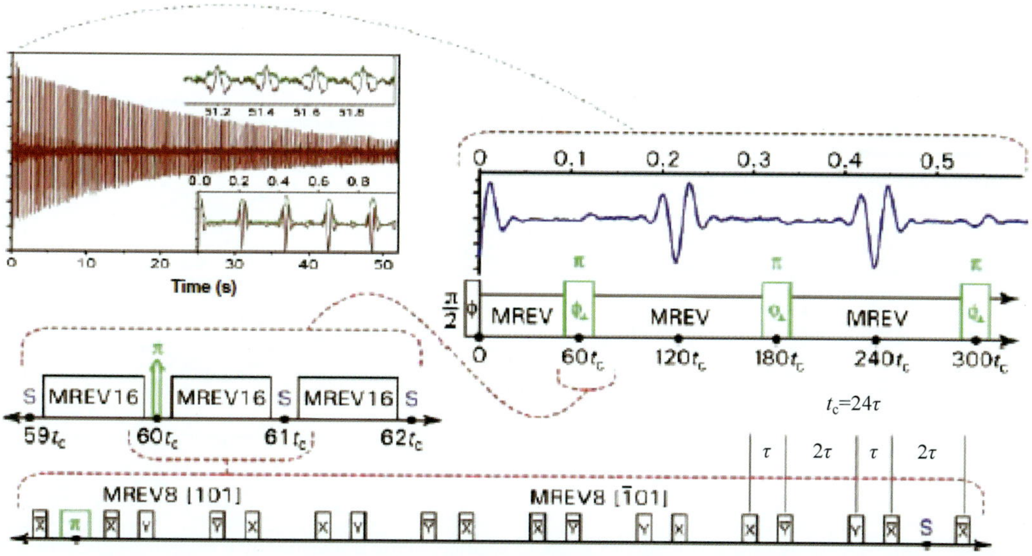

Fig. 5-35 (Color online) Experimental result indicating that T_2 for ^{29}Si nuclear spins in natural Si crystal is 25s at room temperature. CPMG π-pulse sequences and MREV-16 $\pi/2$-pulse sequences are alternately irradiated Time (s).

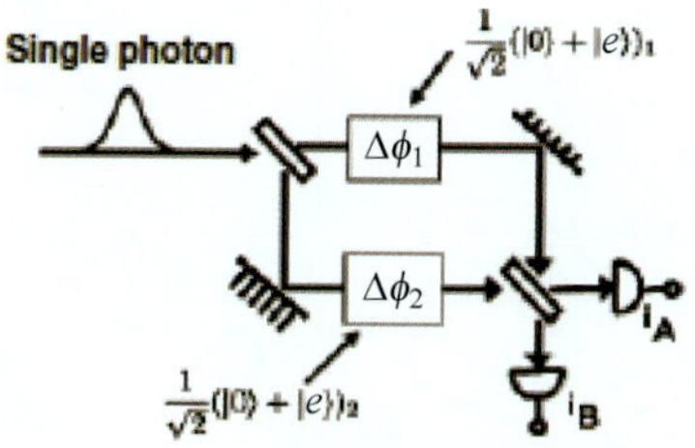

(a)

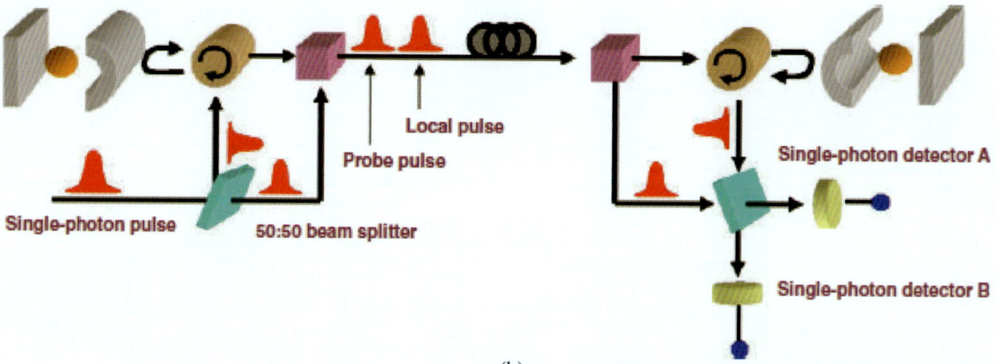

(b)

Fig. 5-36 (a) System for distributing entangled quantum states by driving Mach-Zehnder interferometer equipped with quantum phase modulators using single-photon pulses. (b) Distribution of entangled quantum states based on differential phase shift of single-photon pulses.

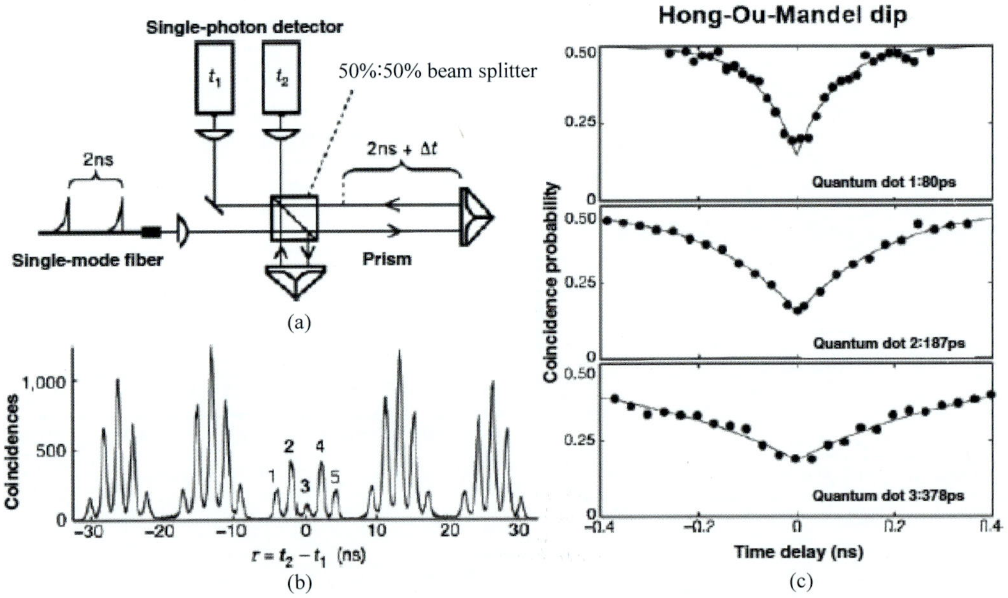

Fig. 5-37 Hong-Ou-Mandel interferometer used to evaluate indistinguishability of two single photons.

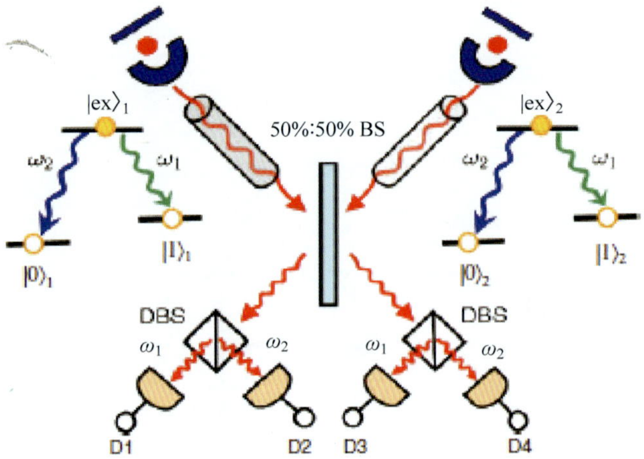

Fig. 5-39 (Color online) Creation of entangled quantum states by coincidence counting of indistinguishable single photons emitted from two quantum memories. Square lattice embedded in a planar microcavity, using optical pulses, as discussed in this section.

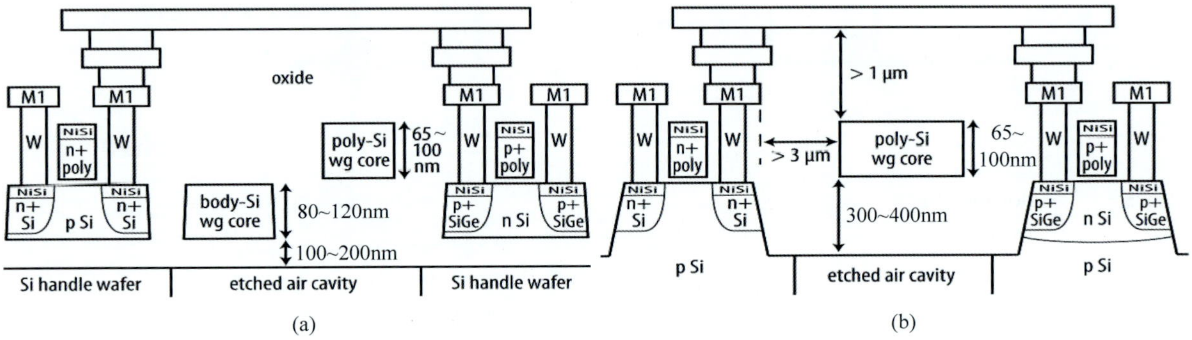

Fig. 5-40 Monolithic EPIC cross-sections integrated within (a) sub-65nm node SOI-CMOS and (b) bulk-CMOS processes.

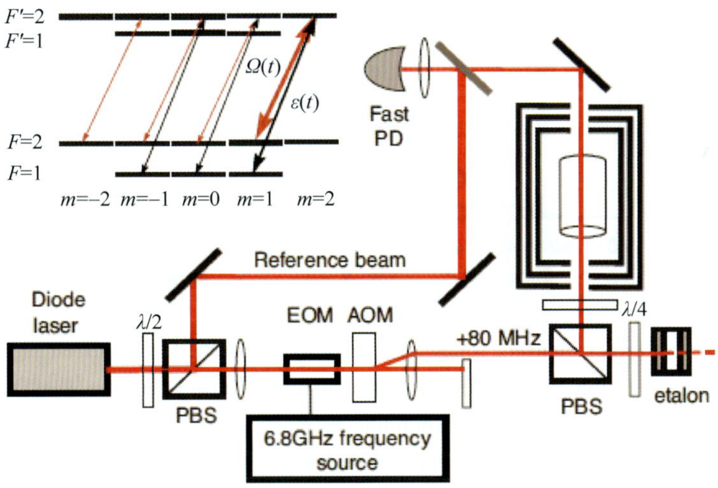

Fig. 5-57 (Color online) Experimental apparatus (see text for abbreviations). Inset: Schematic of the ^{87}Rb D1 line level structure and relevant Λ systems formed by control and signal fields.

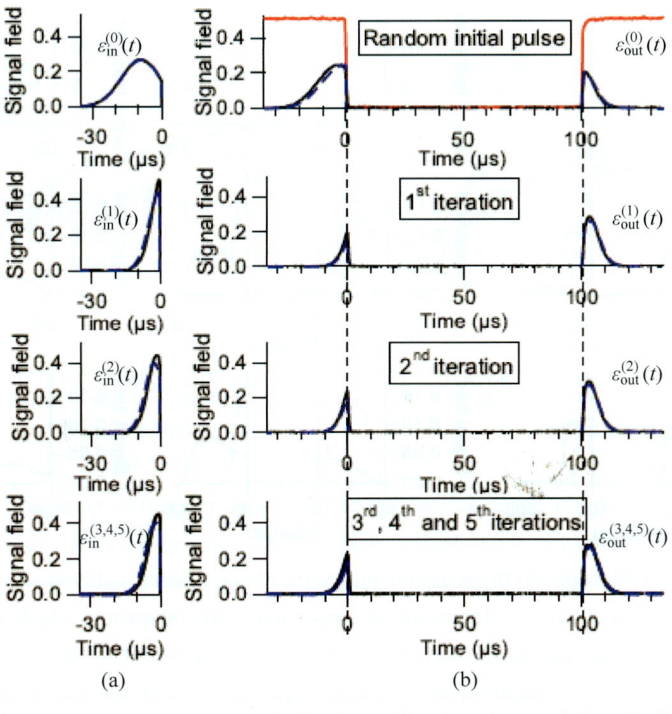

Fig. 5-58 (Color online) Iterative signal pulse optimization. The experimental data (solid black lines) is taken at 60.5℃ ($\alpha L=24$) using 16mW constant control field during writing and retrieval (solid red line in the top panel) with a $\tau=100\mu s$ storage interval. Numerical simulations are shown with blue dashed lines. (a) Input pulses for each iteration. (b) Signal field after the cell, showing leakage of the initial pulse for $t<0$ and the retrieved signal field ε_{out} for $t>100\mu s$.

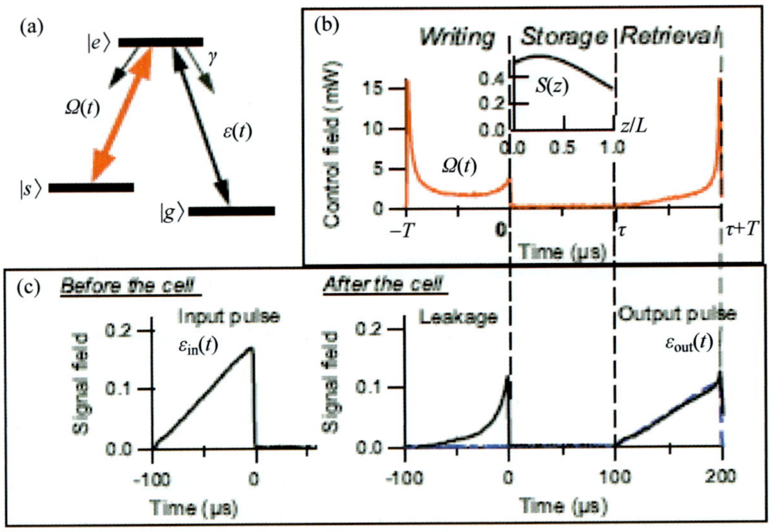

Fig. 5-91 (Color online) (a) Schematic of the three-level Λ interaction scheme. Control (b) and signal (c) fields in pulseshape-preserving storage of a "positive-ramp" pulse using a calculated optimal control field envelope $\Omega(t)$. During the writing stage ($t<0$), the input pulse $\varepsilon_{in}(t)$ is mapped onto the optimal spin-wave $S(z)$ [inset in (b)], while a fraction of the pulse escapes the cell (leakage). After a storage time τ, the spin-wave $S(z)$ is mapped into an output signal pulse $\varepsilon_{out}(t)$ during the retrieval stage. The dashed blue line in (c) shows the target output pulse shape.

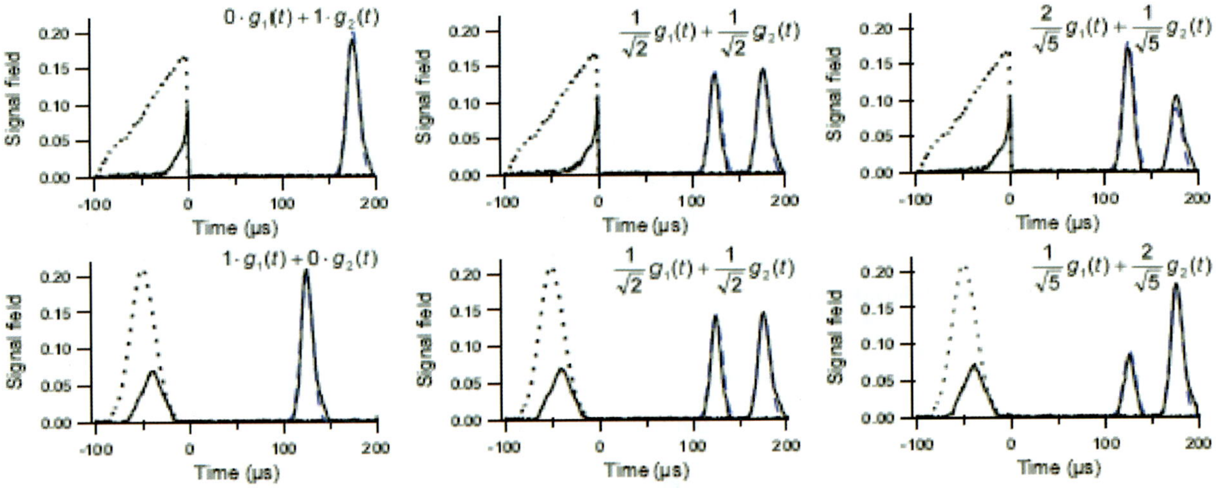

Fig. 5-93 (Color online) Examples of storage of signal input pulses with Gaussian and triangular envelopes, followed by retrieval in a linear combination of two time-resolved Gaussian pulse shapes $g_1(t)$ and $g_2(t)$. Input and output signal fields are shown in dotted and solid black lines, respectively. Dashed blue lines show the target envelopes.

Advanced Quantum Photonics Memory

现代光量子存储

徐端颐 著
Xu Duanyi

清华大学出版社
Tsinghua University Press

北 京
Bei Jing

内容简介

The author Professor Duanyi Xu began to research principles of photonics quantum memory in 1990s early provided the data on which future theories a few of remark for photonics memory. That bear out the current thermal effects optical memory, which will replaced by photonic quantum effects of reaction speed in the future. As quantum photonic memory is a rapidly developing field of research spanning both photonics and materials science. Therefore, he incorporated this research into the national key basic research project under his responsibility. Professor Duanyi Xu collected his research in recent years and up to date literatures in the world, systematic introduces the principles of photonics quantum memory and engineering implementation, examines the impact new technologies and challenges, summary and offers a thought-provoking and entertaining vision about quantum photonic memory to write this monograph *Advanced Quantum Photonics Memory*. This book introduced with more space to present a variety of new principles and techniques for quantum photonic memory, such as nano-photonic nonlinear-optics quantum memories, quantum entanglement memory, photochemistry solid-state memory, optical physics memory and dynamic quantum memory etc.

图书在版编目（CIP）数据

现代光量子存储 = Advanced Quantum Photonics Memory / 徐端颐著. -- 北京 ：清华大学出版社，2024.9.

ISBN 978-7-302-67392-7

Ⅰ. TP333.4

中国国家版本馆 CIP 数据核字第 2024N3N656 号

责任编辑：王一玲　李　晔
封面设计：傅瑞学
责任校对：王勤勤
责任印制：宋　林

出版发行：清华大学出版社
网　　址：https://www.tup.com.cn，https://www.wqxuetang.com
地　　址：北京清华大学学研大厦 A 座　　**邮　　编：**100084
社 总 机：010-83470000　　**邮　　购：**010-62786544
投稿与读者服务：010-62776969，c-service@tup.tsinghua.edu.cn
质量反馈：010-62772015，zhiliang@tup.tsinghua.edu.cn
课件下载：https://www.tup.com.cn,010-83470236
印 装 者：三河市龙大印装有限公司
经　　销：全国新华书店
开　　本：210mm×285mm　**印　张：**53　**插　页：**11　**字　数：**1569 千字
版　　次：2024 年 10 月第 1 版　**印　次：**2024 年 10 月第1 次印刷
定　　价：290.00 元

产品编号：099590-01

Foreword

Professor Duanyi Xu was director and founder of Optical Memory National Engineering Research Center at Tsinghua University, chief scientist of National Basic Research Project and foremost a bands-on experimental application optical memory scientist in China. He led and achieved Chinese National Basic Research Planning on "Super-density and Super-speed Optical Memory" in 2005. Professor Xu began to research principles of photonic quantum memory in 1990s early, introduced using of medium to absorption of photon of different frequencies or intensity to achieve a multi-wavelength and multi-level memory. He provided the data on which future theories a few of remark: "for photonic memory, the current thermal effects will be replaced by photonic quantum effects of reaction speed in the future." (see *High-density Optical Memory*, Tsinghua University Press, 2002 PP-21). Then he utilized linear absorption characteristics of photochromic medium to different wavelength with high sensitivity that achieved experiment of different wavelength photochromic memory and writing process with photochromic fulgide based photon effect for photonic volume memory. There was also his extensive work on photonic quantum memory theory of experimental tests emission and absorption of light fluorescence. He wrote another book on the results of his research *Multi-dimension Optical Memory* published by the Springer Press and Tsinghua University Press in 2017. His another monograph *Photonic Memory* was published by Tsinghua University Press in 2022. As Quantum Photonics Memory is a rapidly developing field of research spanning both photonics and materials science. Therefor Professor Duanyi Xu collected his research in recent years and up to date literatures in the world, systematic introduces the principles of photonics quantum memory and engineering implementation, examines the impact new technologies and challenges, summary and offers a thought-provoking and entertaining vision about quantum photonic memory to write this monograph *Advanced Quantum Photonic Memory and Application*. This book introduced with more space to present a variety of new principles and techniques for quantum photonic memory, such as nano-photonic nonlinear-optics quantum memories, quantum entanglement memory, photochemistry solid-state memory, optical physics memory and dynamic quantum memory etc. Meanwhile devoted to description of the key building principles, experiment and testing technology for research and manufacture the optical solid state memory, including double photon-photochemistry cell, self-assembled ultrathin films, photocyclization switching, dual-mode molecular modulator and nanoscale plasmon devices fabrication etc. This book is the first provide a framework for thinking about the future Photonics Quantum Memory that create a structure for strategic planning and development for exploring potential development paths to the optical memory and should appeal to the readers in universities and industry to understand the fundamental principles of Quantum Photonic Memory. I am confident it will be an important valuable resource for readers and future specialists who engaged upon information science and technology.

President of the Chinese Optical Society
Member of the Chinese Academy of Sciences
Member of International Optical Society
Professor of Department of Electronic Engineering at Tsinghua University

Zhou Bingkun
March 2023

Preface

Information memory is an important means of human civilization transmission and a core link of modern information technology. Quantum photonic memory is an essential basic device in the era from classical information to quantum information. Quantum photonic memory should be able to store various quantum states including with any quantum state. Like classical computers, general-purpose quantum computers require quantum memory for complex computational functions. Depending on the specific computing chip, the memory must store the corresponding quantum information carrier. Usually classical memory measured in bits, and today's classical memory can reach the order of terabytes (2^{40}). So the Optical Memory National Engineering Research Centre (OMNERC) at Tsinghua University has been engaged in optical memory research since the early 1990s. Classical memory a memory unit stores only one bit, so the capacity of the memory is actually the number of classical memory units. Due to the characteristics of quantum coherence, one memory unit of quantum memory can store N qubits at one time. Recent studies have shown that quantum photonic memory can store up to 100 qubits and more than all the classical memory. Therefore, Quantum photonic memory is more important in quantum information than classical memory in classical information because quantum information cannot be copied and amplified. The single photon can be efficiently stored in long-lived spin states and the ability to resist ambient noise in actual system transportation can improve more. With the gradual advancement of the above research, quantum USB disk will be enter the practical link first. Quantum photonic memory is more important in quantum information than classical memory in classical information because quantum information cannot be copied and amplified. There are many research groups in the world including OMNERC at Tsinghua University engaged in quantum memory research at present that all the independent indexes of quantum photonic memory have good results. Application of quantum photonic memory has just become so widely used while the quantum processor evolves. The quantum processor designed mapping between the two systems. The quantum processor then yield information about the target quantum system. Difficult electronic structure problem of a target molecule can mappe onto the qubits of the quantum processor for solving optimization problems: The solution of an optimization problem can encode into the ground state of a Hamiltonian. This ground state can be using an iterative, quantum-classical algorithm illustrated at bottom. The quantum processor is prepared. The energy of the state is measured and can be used the classical computer. A classical optimization algorithm then suggests a new quantum state. This quantum speedup is possible by being able to encode the component vector. Therefor quantum technologies become part of everyday lives in the coming decades. So quantum information science are rapidly developing, including ultraprecise quantum sensors that could propel fundamental science forward by leaps and bounds; powerful quantum computers to tackle insoluble problems in finance and logistics; and quantum communications to connect these machines as part of long-distance networks, quantum computers operate on the 1000-qubit scale. Anticipate millions of qubits are required to solve important problems that are out of reach of today's most powerful supercomputers. There is a global quantum race to develop quantum computers that can help in many important societal challenges from drug discovery to making fertilizer production more energy efficient and solving important problems in nearly every industry, ranging from aeronautics to the financial sector. That works so well and the potential to scale-up by connecting hundreds or even thousands of quantum computing microchips. Towards quantum computers that are robust to errors, suppressing quantum errors by scaling a surface code logical qubit could be the most advanced supercomputer. All experiments validate the unique architecture that the quantum photonic memory been developing—providing an exciting route towards truly large-scale quantum computing. We are still growing our research and teaching in this area, with plans for new teaching programs and appointments. Quantum photonic memory will be pivotal in helping to solve some of the most pressing global issues. And with teams spanning the quantum photonic memory and technology research, OMNERC has both a breadth and a depth of expertise in this. I have been engaged in the research of photonic memory and press published a monograph *Photonic Memory* in 2021, which is very popular with readers. As the world confronted with challenge by exploded increasing amounts of big data. Every day zillions of data generated through the events of the world. I collected and sorted out the new research results of OMRC and at home and abroad in this field in recent years and wrote this monograph, which named *Advanced Quantum Photonic Memory & Application*.

However, the book was a textbook indeed, mainly introducing theories and principles, with little introduction to engineering applications. In order to meet the needs of the development of light quantum technology and the requirements of the vast number of readers to republish the book. The book supplement to introduction applications of photonic memory technology and devices also added some advanced photonics and memory technology to obtain advanced achievements in recent years. Therefore this book summarized the finally efforts of photonic memory with super resolution and capacity, thereby proposed and described systematically adoption of photonics principles and applied implementation technologies to make big data memory devices. That will have higher memory density, capacity, data transfer rate and low power consumption that is one of the most promising next-generation data memory and can be for primary memory, secondary memory and tertiary memory that is photonic RAM, ROM and removable UD. The idea of writing this book was a result of frequent enquiries about the possibility of published a book on Advanced Photonic Memory (APM) in English.

A preliminary survey of the literature showed that numerous researches on almost every aspect of photonics carried out for the past few years, so that the book gives a comprehensive and balanced picture of the field. The book based on quantum physics as quantum entanglement, nano-photonics and photochemistry. From the reversible transfer between a photon and a collective atomic excitation, which in a solid-state device and then accurate expressions. That derived through use of the density matrix equations of motion in detail in order to render this important discussion accessible to general reader a neodymium doped yttrium other-silicate crystal served as quantum memory, with an optical transition with good coherence properties, which employ a thulium-doped lithium niobate waveguide in conjunction with a photon-echo quantum memory protocol. The photons generated in quadratic nonlinear waveguides. that control photon onto nonlinear crystal with entangling, physic-mathematical model of heralded photons in solid-state memory, multimode capability of storing photon pair entanglement, photon nonlinear transport, static model of light-matter entangled state, energy-time entangled photons onto the photonic memory, violation of a bell inequality and dynamic model of entangled photons to photonic memory are discussed in detail. Photochemistry solid state memory presents an introduction to another PM based on the principles of two photon-photochemistry and photo-chromism, include coupled wave equations for different frequency photon, photon nonlinear transport in medium, stereochemistry and isomerisation, preservation of photonic energy during storage, margin analysis based on rigorous modeling, conversion efficiency nano-crystalline film, photochromic dye in amorphous state, electron delocalization valence, error correction and application probabilities. Strong advantages like more performance while less consumption and more ergonomic (less noise, smaller and more flexible cases) stand opposite to disadvantages of more temporary nature (incompatibility and production problems). Photon and light seem to be better than electrons and electric current to carry information. The question how long this will take and the factors influencing it discussed.

The book is organized as follows: Chapter 1 presents an introduction to the latest development in photonic memory including new developments in photonics, Maxwell-Bloch equations, Application of quantum science and technology, Photonic integration solid state memory, Precision of spin-echo-based quantum memories, nanophotonics quantum memories, Raman-type quantum memory and other typical new quantum memory technologies. Moreover, multi-level optical storage. Key principles and technologies for application of photonics, quantum science and technology for memory and relevant technologies in the future. Chapter 2 treats fundamentals of quantum photonics, information. That include theoretical entanglement measures, channel capacity, multiple inputs, quantum probability, dense quantum coding. Theory of quantum entanglement, uncertainty principle in quantum memory, quantum entanglement between photon and spin qubit or quantum entanglement between two crystals, that are basic knowledge for photonic memory and entanglement raman scattering, mapping photonic entanglement in/out of photonic memory. Chapter 3 introduces multi-dimension photonic memory, including mechanism of photonic multi-dimension memory, experiments for multi-wavelength and multi-level storage, crosstalk and non-destructive readout. in photonic memory, modulation coding and error correction, multi-layer photonic memory, application of multi-wavelength and multilevel storage, dynamic and static speckle multiplexing, aberration auto-correction for multi-layer storage. Chapter 4 introduces photonic super-resolution photonic memory that are the important way for increasing density and capacity of photonic storage include basic concepts, essential examples, quantum probability, dense quantum coding, channel capacity and mathematical background. Chapter 5 presented nanophotonics for nano-quantum memory, that analysis the quantum characteristics for photons, photonic quantum memory function. That inclood poton-controlled quantum memory function, photonic integrated circuits and optical networks, solid state light-matter interface at photon, which photon memory in atomic vapour and in atomic media, optical dense atomic memory medium, control field optimization for adiabatic storage, analysis of photon number in quantum memory and typical photon memory devices. By the way to analyze how the mathematical formalism of quantum theory leads to a nonrealist interpretation of the theory in detail. The key technologies as self-assembled ultrathin films, optical-electrical molecular switching, performances detection for photon entanglement states, neodymium doped yttrium ortho-silicate crystal, coherences measuring of the metastable atomic states, testing of electron spin resonance absorption spectra, detected to statistical error of photon pairs, testing of neodymium doped yttrium ortho-silicate crystal are described also. Meanwhile, special testing method, technology and instrument of high efficiency low noise and fast responsibility are decrypted function. Most chapters are essentially independent of each other, providing flexibility in choice of topics to be covered. Each chapter begins with an introduction describing its contents. This introduction for every chapter provides a guide also to what could omitted by a reader familiar with the specific content or by someone who is less inclined to go through details. Each chapter ends with highlights of the content covered in it. The highlights provide the message from the chapter and serve to review the materials learned. For researchers interested in a cursory glimpse of a chapter, the highlights provide an overview of topics covered that the highlights may also be useful in the preparation of lecture notes or PowerPoint presentations.

PQM bring out more expansive development space, as a lot of new research achievement on the fundamental theory of the interaction of light with medium for next generation optical and quantum memory. That are notational very complicated unfortunately, much of notational complication is unavoidable. These results in infinitely many quantum states that a single quantum bit, or "qubit" can take together with another strange property of quantum mechanics entanglement it allows for a much more powerful information platform than is possible with conventional components. As PQM related with quantum information processing (QIP) closely which uses qubits as its basic information units. QIP has many facets, from quantum simulation, to cryptography, to quantum computation, which is expected to solve problems more complex than those within the capabilities of traditional computer. A qubit needs to be both isolated from its environment and tightly controllable, which places stringent requirements on its physical realization. It is need of a scalable architecture and error correction performed in parallel with computation. A number of qubit types have been proposed and experimentally realized that satisfy at least some of these criteria, and tremendous progress has been made over the past decade in improving the figures of merit, such as the coherence time, that focus on the many promising qubit flavours based on spins in semiconductors. The future of QIP appears bright in spite of the many remaining challenges, overcoming these challenges will probably also advance basic research.

This book is a textbook on optical engineering for graduate student originally that intent is to provide most advanced progress and future development to the field of photonics memory. That provid a multidisciplinary training for a future generation of researchers at both undergraduate and graduate levels in the world. A worldwide recognition of this vital need is evident from the growing number of conferences and workshops held on this topic, as well as from the education and training programs offered and contemplated at various institutions. Much of the material covered in this book developed during the teaching of these courses and refined by valuable feedback from these course participants. Of course, the book will serve both as an education and training text and as a reference book for research and development that value to industries and businesses, because the last chapter attempts to provide a critical evaluation of the current nanophotonic-based technologies. Therefore, much gained by creating an environment that includes these disciplines and facilitates their interactions. This book will address all issues that propose to fill the existing void by providing the following features: A unifying, multifaceted description of nanophotonics that includes near-field interactions, nonmaterial, photonic crystals and nanofabrication. A focus on nanoscale optical interactions, nanostructure optical materials and applications of nanophotonics are discussed that is coverage of inorganic, organic materials and biomaterials, as well as their hybrids are introduction. At the same times, it is opportunity for basic research and development of new memory technologies. Due to highly efficient broadband and lightweight rolls, quantum cutters to split vacuum UV photons into two visible photons for new-generation fluorescent principles to guide development photonic materials, novel photonic structures, optical nanoprobes, light-activated nanoparticles and sensors. This book is Available-for multidisciplinary readership with the goal to introduction for a wide range of photonics. A major emphasis placed elucidating concepts with minimal mathematical details, examples provided to illustrate principles and applications. The book can readily enable a newcomer to this field to acquire the minimum necessary background to undertake research and development of PQM. In addition, I hope the book to be helpful to promote recent optical memory developing and accessible to engineers in this field. The soled state photonic memory notational very complicated and much of notational complication is unavoidable unfortunately. The book uses primarily the Gaussian system of units, both to establish a connection with the historical papers of PQM. At several places in the text, tables provided to facilitate conversion to other systems of units. I have attempted to treat the topics that are covered in a reasonably self-contained manner (for each chapter), and consideration perfect generalize up to date research achievements to the greatest extent. Advance at Tsinghua University could enable creation of photonic memory using transnational optical manufacturing equipment with laser array scanning system and precision injection printing to process the photochemistry memory on photonic medium with multi-wavelength and multi-level modulation. These new materials are flexibility for fabrication of components with diverse functionalities and their heterogeneous integration in dealing. Novel synthetic technique and processing of nanomaterials, as new types of molecular nanostructures and supramolecular assemblies with varied nanoarchitectures, self-assembled periodic to induce multifunctionality and cooperative effects will be application.

Finally, I would like to acknowledge and my deep appreciation discussions of the materials in this book with my colleagues and graduate students at Tsinghua University in China. Discussion with numerous professors and graduate students of University of California San Diego, Massachusettes Institute of Technology, New York State University at Buffalo in United States, Yonsei University in Korea, National University of Singapore, Kyoto University in Japan and University of Toronto in Canada, that I am sure that I learned much from them. The assistance of Dr. C. George of Stanford University, Dr. W. Q. David of CalTech, Professor S. Esener of UC San Diego, Professor N. Thomas and Professor K. S. Immink of Eindhoven University in review of this book have to be gratefully acknowledged. In addition, this book is impossible to accomplished and published without the strong support of Prof. W. Imopto and editors of Tsinghua University Press. I think that readers will show deferential regard deeply for them.

Duanyi Xu in Tsinghua University

March 2023

Contents

Chapter 1

The latest development in photonic memory

1.1 New developments in photonics

Optical storage has long-term history if to include photography, which is over 1000 years. Alhazen studied the camera obscura and pinhole camera beginning to the first permanent photograph was an image produced in 1826 by the French inventor Joseph Nicéphore Niépce that experienced more 800 years. But the advanced optical storage especially digital data storage with computer application is less than 50 years. It covers a wide range from optics, photonics, materials science, information science, device physics/chemistry to precision engineering and nanotech applications so that is a typical highly multidisciplinary field. This chapter is focused on summary and exploring the role of optics and photon in storage systems beyond optical data storage per second. For example, future data storage systems may combine optics, electronics and quantum technology to increase storage density, data transfer rates and reliability which include the state-of-the-art in optical data storage, magnetic and hybrid recording, semiconductor phase change media, novel optical storage systems, holographic data storage, photonic quantum memory, long-term archival data storage product and technology roadmaps.

Photonic quantum memory has made significant progress and it has begun to enter the practical. Current trends in photonic quantum computing and emphasize Photonic Quantum communications, for example the use of free-space optical interconnects as a potential solution to alleviate bottlenecks experienced in electronic architectures, including loss of communication efficiency in multiprocessors and difficulty of scaling down the IC technology to nano-meter levels. Light beams can travel very close to each other, and even intersect, without observable or measurable generation of unwanted signals. Therefore, dense arrays of interconnects can be built using optical systems. In addition, risk of noise reduced further, as light is immune to electromagnetic interferences. Finally, as light travels fast and it has extremely large spatial bandwidth and physical channel density, photons are uncharged and do not interact with one another as readily as electrons, light beams may pass through one another in full duplex operation, for example without distorting the information carried. It appears to be an excellent media for information transport and hence can harnessed for data processing. This high bandwidth capability offers a great deal of architectural advantage and flexibility. Based on the technology now available, the systems could have more 1024 smart pixels with each channel clocked at GHz for a chip I/O of 500Gbits per second, giving aggregate data capacity in the parallel optical highway of more that 500Gbits per second, even this could be further increased to Tbits. Free-space photonic techniques are used in scalable crossbar systems also, which allow arbitrary interconnections between a set of inputs and a set of outputs. Photonic sorting and photonic crossbar inter-connects are used in asynchronous transfer modes or packet routing and in shared memory multiprocessor systems. In quantum photonic computing two types of memory are applications. One consists of arrays of one-bit-store elements and the other is mass storage, which is applied three-dimensional array photonic quantum memories or by storage systems integrated by them. This type of memory promises very high capacity and storage density. The primary benefits offered by array photonic quantum memories over current storage technologies include significantly higher storage capacities and faster read-out rates. These researchs expected to lead to compact, high-capacity, rapid-and random-access, radiation-resistant, low-power and low-cost data storage devices necessary for future intelligent spacecraft, as well as to massive-capacity and fast-access terrestrial data archives. As multimedia applications and services become more and more prevalent, entertainment and data storage companies are looking at ways to increase the amount of stored data and reduce the time it takes to get that data out of storage. The RAID and the linear array beam steerer used in array photonic quantum memories also. These devices used to write data into the array photonic quantum memories at more high speed. The analog nature of these devices means that data can be stored at much higher density than data written by conventional devices. Researchers around the world are evaluating a number of inventive ways to store optical data while improving the perfor-free-space photonic techniques used in scalable crossbar systems also, which allow arbitrary interconnections between a set of inputs and a set of outputs. Photonic quantum memory has been used in asynchronous transfer modes or packet routing and in shared memory multiprocessor systems. As quantum photonic memory can be writing and reading at the same times. However, general standard storage systems only record data in a dispatch only, wasting a certain amount of times. Advanced photonic quantum memory records data on three dimensional arrays by neural network channel at the same times. Both input and output and, by choosing spacing interval approximately 1/6 the wavelength of the reading laser beam, and the system can eliminate the crosstrack and crosstalk that would normally be the result of recording data and times. Even conventional photonic quantum memory data from neighboring memory, but this information filtered out for reducing the signal-to-noise ratio. By closely controlling the distance from each other can be maximizing the signal-to-noise ratio that are expected to produce removable photonic quantum memory with capacities to 10GB, which is the same size as a standard Integrated circuit USB also, but that holds 640MB only. Magnetic amplifying magneto-optical control systems with a standard connectors with two or three ports. In general, photonic quantum memory is similar to conventional USB, that when the data are copied from the bottom to the more uppers layer, it is expanded in size, amplifying the signal. According to the advanced technology experimental test results of Optical Memory National Engenering Research Center(OMNERC) at Tsinghua University represents a two-fold increase in storage capacity over 50GB.

The multilayer recording technology could help bridge the gap between solid photonic quantum memory and other optical memory as holographic memories for example. It is called two-photon memory technology the systems under development use a single beam to write the data in either solid photonic quantum memory with up to by hundreds that into more hundreds cubes of solid photonic quantum memory and on every layer of every cube. In operation, the records data on every layers and cubes by choosing location groove depths approximately 1/6 the wavelength of the reading laser light, so the system can eliminate the crosstrack crosstalk that would normally be the result of recording on each cube and layers. Representing data is illuminated by a mode-locked Nd:YAG laser emitting at 1064nm with pulse durations of 35ps. The focal point of the beam intersects a second beam formed by the second harmonic of the same beam at 532nm. The second beam fixes the data spatially and temporally. A third beam from a laser emitting at 543nm reads the data by causing the material to fluoresce. The fluorescence is read out by a charge coupled device (CCD) chip and converted through proprietary algorithms back into data. Newer versions of the system use a Ti:Sapphire laser with 200-fs pulses. The newer and older approaches offer different strengths. The YAG system can deliver higher-power pulses capable of storing megabits of data with a single pulse, but at much lower repetition rates than the Ti:Sapphire laser with its lower-power pulses. Thus, it is a trade-off. OMNERC has demonstrated the system using portable apparatus comprised of a simple stepper-motor-driven stage and 200-microwatt laser in conjunction with a digital camera, that an optimized system could produce static bit error rates (BER) of less than 9×10^{-12}. A final prototype operating at standard data rates would offer BERs that match or slightly exceed conventional optical communication technology. Researchers of OMNERC using active-molecule-doped polymers to store optical data holographically. Their system uses a thin polymer layer of PMMA doped with phenanthrene quinone (PQ). When illuminated with two coherent beams, the subsequent interference pattern causes the PQ molecules to bond to the PMMA host matrix to greater extent in brighter areas and to a lesser extent in areas where the intensity drops due to destructive interference. As a result, a pair of partially offsetting index gratings formed in the PMMA matrix. After writing the hologram into the polymer material, the substrate is baked, which causes the remaining unbounded PQ molecules to diffuse throughout the polymer, removing the offsetting grating and leaving the hologram. A uniform illumination is the final step, bonding the diffuse PQ throughout the matrix and fixing the hologram in the polymer material. The devices based on this method could hold 220GB of data on a solid photonic quantum memory array. More approaches to holographic storage using doped lithium niobate crystal to store pages of data conventional. As a systems developed by Colorado State University use the associative search capabilities of holographic memories. Associative or content-based data access enables the search of the entire memory space in parallel for the presence of a keyword or search argument. Conventional systems use memory addresses to track data and retrieve the data at that location when requested. Several applications can benefit from this mode of operation including management of large multimedia databases, video indexing, image recognition, and data mining.

Different types of data such as formatted and unformatted text, gray scale and binary images, video frames, alphanumeric data tables, and time signals can interleave in the same medium and can search the memory with either data type. The system uses a data and a reference beam to create a hologram. Holographic memory cubes use a spatial light modulator to search the entire memory for a searchable object to be text simultaneously image or something else. This associative memory search process promises significant benefits for database searching and other applications. Associative or content-based data access enables the search of the entire memory space in parallel for the presence of a keyword or search argument. This associative memory search process promises significant benefits for database searching and other applications with plane inside the lithium niobate. By changing the angle of the reference beam, more data can written into the cube just like pages in a book. The current systems have stored up to 1000 pages per spatial location in either VGA or SVGA resolutions. To search the data, a binary or analog pattern that represents the search argument is loaded into a spatial light modulator and modulates a laser beam. The light diffracted by the holographic cube on a CCD generates a signal that indicates the pages that match the sought data. Recent results of Tsinghua University have shown the system can find the correct data 90 percent of the time when using patterns as small as 5 percent of the total page. That level goes up to 95 percent by increasing the amount of data included in the search argument. The multilayer holography system could be PB level storage capacity with page or picture data frame. At the same time a new achievements of quantum technology will take an immensity development space for optical memory, especially for solid photonic quantum memory array with application of nanotechnology and photonic integration.

1.2 Other big data storage technology

Timely information on scientific and engineering developments occurring in laboratories around the world provides a critical input to maintaining the economic and technological strength of the every country. Moreover, sharing this information quickly with other countries eachother can greatly enhance the productivity of scientists and engineers. The information (data) storage could be same important research area with other specific technologies such as communication biotechnology and nanotechnology are studied since its launch just over 55 years ago, its stored density for information on magnetic hard disk drives (HDDs) has increased by a factor of about 5×10^8 times.

Capacity per HDD increasing from 3.75 megabytes increased to 10 terabytes or more, more than a million times larger. Physical volume of HDD decreasing from 2,000 liter (comparable to a large side-by-side refrigerator) decreased to less than 5 milliliter. Weight decreasing from 2,000lbs (~900kg) decreases to 35 grams. Average access time decreasing from over 100 milliseconds decreased to a few milliseconds, a greater than 40-to-1 improvement. Areal density is growing 30% per year average. Price from about US $15,000 per megabyte decreased to less than US $0.000025 per megabyte (US $25/ per terabyte), is greater than 30-billion-to-1 decrease. Market application expanding from mainframe computers of the late 1950s take to most mass storage applications including computers and consumer applications such as storage of entertainment content.

Optical Data Storage has come a long way in the past 42 years. World's First MO Optical Disc Recorder MnBi film coated optically flat disc on air bearing spindle with HeNe laser, E-O modulator and galvo deflector. By the early 1970s analog video disc systems were commercially available. These were closely followed by 12" write-once (WORM) drives and media. In 1982 Sony and Philips announced the 120mm diameter compact disc (CD-DA) followed by the CD-ROM in 1984. In 1995 the DVD-ROM was announced and in 2002 Blu-ray Disc (BD). Each of these technologies increased capacity significantly and mainly supported important consumer electronics applications. Also in 1995, the EIDE/ATAPI standard was promulgated, which allowed these drives to become a standard part of a PC's storage suite. Consequently, sales grew exponentially. Other types of optical storage of various disc diameters and storage mechanisms were extant in 1995. In 2012, nearly 30 years after the introduction of the CD, classical optical data storage has perhaps reached, or even passed, both its technology zenith and market zenith. Solid state flash drives, portable hard drives, and downloading of music and video have begun to erode significantly the optical data storage market share. Moreover, optical data storage technology appears to have reached some fundamental physical limits i. e. laser wavelength at 405nm and numerical aperture at 0.85 that may be optical data storages "superparamagnetic limit." then as magnetic data storage. The utility of optical data storage (ODS) is derived from how small a diffraction-limited laser beam can be focused for writing and reading; in other words, spot size. From basic optical theory to know that optical spot size is proportional to wavelength λ and inversely proportional to the effective numerical aperture (NA) of the optical system. Spot size is proportional to λ/NA and storage densities is proportional to $(\lambda/NA)^2$. With 25GB/storage a surface BD, the 405nm wavelength and 0.85NA pretty much exhaust the basic potential of optical data storage. But like magnetic data storage, optical data storage has several non-conventional means that may permit the technology to reach new capacity plateaus. These range from multi-layer discs and near-field recording (NFR) to UV lasers, negative refraction and plasmaron lenses. There are also several consumer applications that may justify pushing disc capacity to 100GB, or more. The standard is being developed and will require 100GB disc capacity. Super-high Vision is another possibility, which is even more capacity hungry (requires 400GB). The future of optical storage will be analyzed in terms of advanced technologies and meet difficulty of implementation, cost, impact on manufacturing yield and market need. But some related data storage technologies that promise multi-TB capacities are developed. The engineering challenges of these advanced optical read/write methods on lasers, media, optical pickups, servos, and read/write channels will be surmounted or achievable. They could be done if optical data storage is to survive that can confidently predict the future of optical storage capacity will be over many TB. As the Holographic Versatile Disc (HVD) capacity can be to 5TB. HVD alliance hopes to improve this efficiency with capabilities of around 60,000 bits per pulse in an inverted, truncated cone shape that has a 200μm diameter at the bottom and a 500μm diameter at the top. High densities are possible by moving these closer on the tracks: 100GB at 18μm separation, 200GB at 13μm, 500GB at 8μm, and most demonstrated of 5TB for 3μm on a 10cm disc. The system uses a green laser, with an output power of 1 watt which is high power for a consumer device laser. Possible solutions include improving the sensitivity of the polymer used, or developing and commoditizing a laser capable of higher power output while being suitable for a consumer unit. Same companies have heavily invested in optical data storage to the point that now the Japanese industry enjoys a comfortable lead in this area. This lead is not only in the manufacturing and R & D of conventional optical disk media and disk drive systems, but also in the manufacturing of the enabling optoelectronic components such as various wave length lasers and photodetectors. Many big and very big companies in Japan and China. Similar advances have been made with optical disk systems as they have progressed through three generations of products. Blu-Ray and CBHD-China Blue

High-definition Disc. Some universities of U.S. take on an important role in conventional optical storage for both Japan and the United States, since a significant amount of know-how on optical disk heads and metrology exists at these institutions still. In U.S. universities and same U.S. R & D companies on several long-term optical data storage approaches that promise data densities approaching Tb/in^2. These include near field and solid immersion lens approaches, volumetric (multi-layer and holographic) storage, and probe storage techniques. In recent years, under government funding, the United States has gained an advantage on certain potentially enabling technologies such as vertical cavity lasers (VCLs), array optics and MEMS. These powerful technologies may impact or become affected by optical data storage. VCLs and optical arrays may enable high data rate optical drives by exploiting parallelism. Micro and nano machining can find a magnitude application drive industry for solid photonic quantum memory. The United States has lead in research, application and the investment that could lead position quickly. Solid immersion lens based approaches appear promising in the short term volumetric parallel accessible storage systems like holographic and two-photon multi-layer recording techniques appear most promising. exception of certain R & D the recently instituted cooperative university-industry-government cooperative research programs in Japan. In the United States suggest that Japanese companies will assume a larger portion of this future market and certain products in optical recording attempt to control the standard. Opportunities do also exist for the United States to re-enter the optical data storage market via new technologies. The solid photonic quantum memory array based approaches appear promising. Volumetric parallel accessible storage systems like solid photonic quantum memory array and two-photon multi-layer recording techniques appear most promising. The key issue is an inexpensive yet reliable write once material or preferably an erasable volumetric material. With the information explosion on the net, searching for desired data becomes a critical factor. Development of suitable hardware that exploits parallel readout to facilitate content-based data search may point to a potential opportunity. In addition, investing in micro-mechanics for micro-actuators as well as for probe storage and creating a new infrastructure in the United States to support future data storage approaches certainly appears compelling at this time. Chinese companies get great succeed to produce VCD series products from early 1990 and optical disc and drive volume manufacture growth rate could be over 50% in China in deed. Chinese government make a larger research commitment as set up Optical Memory Engineering Research Centre (OMNERC) and National Key Basic Research Project ("973" project in China) to long-term research and should consider working cooperatively with Japan and United States in this area. In addition, various mastering techniques including UV laser, SIL lens, e-beam, and probe mastering are being developed to extend the effective areal density to $50Gb/in^2$. This rate of progress is however now unsustainable without a complete paradigm shift since it will require one bit of information to be stored on considerably fewer than 100 atoms.

Optical data storage research is devoted to researching solutions to the problem of storing and accessing information at densities of 100 atoms per bit and higher. The real power of optical storage has yet to be exploited. It is the only technology that can easily operate in the frequency domain using techniques of a spectroscopic nature. This gives it the power to precisely select very small elements within a larger volume. In principle it can extend down to the atomic level. Current interest is focused on multi-layers of nano-scale silver oxide particles that have been demonstrated able to reversibly store information in the frequency domain, as multi-wavelength memory for example. Simultaneously studying new optoelectronic inertia-less access systems to complete the transformation of optical storage systems are thrown up.

A number of technologies are attempting to surpass the densities of all of these media. This is about the same capacity that perpendicular hard drives are expected to be, and Millipede technology has so-far been losing the density race with hard drives. The latest demonstrator with 2.7Tb/in^2 is seemed promising that the IBM technology, racetrack memory uses an array of many small nanoscopic wires arranged in 3D, each holding numerous bits to improve density. Although exact numbers have not been mentioned, IBM news articles talk of "100 times" increases. Various holographic storage technologies are also attempting to leapfrog existing systems, but they too have been losing the race, and are estimated to offer 1Tb/in^2 as well, molecular polymer storage has been shown to store 10Tb/in^2. By far the densest type of memory storage experimentally to date is electronic quantum holography. By superimposing images of different wavelength into the same hologram, a Stanford research team was able to achieve a bit density of 35bit/electron, that is approximately 3EB/in^2. Of the many potential applications of nanotechnology, one of the most promising ones is in data storage, particularly the hard disk drives. This is because the physical size of the recording bits of hard disk drives is already in the nanometer scope, and continues shrinking due to the ever-increasing demand for higher recording densities. If the pace of areal density increase is maintained at the current level for about ten years, the dimension of the recording bit will reach the sub ~10nm scope. The rapid shrinkage of bit size poses formidable challenges to the read sensors. Its sensitivity must be improved continuously so as to compensate the loss in signal-to-noise ratio due to the decrease in the bit size. The former has to rely heavily on the advance of nanotechnology and spintronics. The combination of two fields has played an important role in advancing the areal density of magnetic recording from 500Gb/in^2 to some Tb/in^2 and capacity will be 100TB or more. In addition to hard disk drives, the technologies developed have also been applied to magnetic random access memories (MRAMs). Further advance in the two fields is the key to realizing Tb/in^2 hard disk drives and gigabit nonvolatile memories within this decade. Spintronics is positioned in the hierarchy of various different types of data storage technologies. Among the magnetic storage devices, the hard disk drive (HDD) is the dominant secondary mass storage device for computers, and very likely also for consumer electronic products in the near future. The ever-increasing demand for higher areal densities has driven the read head evolving from a thin-film inductive head to an anisotropic magnetoresistive head, recently the giant magnetoresistive spin-valve head.

Although the effect on performance is most obvious on rotating media, similar effects come into play even for solid-state media like Flash RAM or DRAM. In this case the performance is generally defined by the time it takes for the electrical signals to travel though the computer bus to the chips, and then through the chips to the individual "cells" used to store data (each cell holds one bit). One defining electrical property is the resistance of the wires inside the chips. As the cell size decreases, through the improvements in semiconductor fabrication that lead to Moore's Law, the resistance is reduced and less power is needed to operate the cells. That less electrical current is needed for operation, and thus less time is needed to send the required amount of electrical charge into the system. In DRAM in particular the amount of charge that needs to be stored in a cell's capacitor also directly affects this time. As fabrication has improved, solid-state memory has improved dramatically in terms of performance. Modern DRAM chips had operational speeds on the order of nanosecond and the minimal line width less 10nm. A obvious effect is that as density improves, the number of DIMMs needed to supply any particular amount of memory decreases, which in turn means less DIMMs overall in any particular computer. Solid-state storage has seen similar dramatic reductions in cost per bit. In this case the primary determinant of cost is yield, the number of working chips produced in a unit time. Chips are produced in batches lithography on the surface of a single large silicon wafer, which is then cut up and non-working examples are discarded. To improve yield, modern fabrication has moved to ever-larger wafers, and made great improvements in the quality of the production environment. Other factors include packaging the resulting wafer, which puts a lower limit on this process of about US $1 per completed chip. The magnetization reversal Fe(100) disk can storage 10 Terabyte would be 100,000 Times the capacity of Blu-Ray, disk drives, or tape drive. That is 1, 000 times any State of the Art hard disk technology with 100 Gigabytes on one disk. Hard drive technology will exceed 10 Terabytes on a disk recent year. Atomic holographic optical image data storage bandwidth is 400,000 times faster than binary bit text processing bandwidths used in other storage technology. Colossal storage of spintronic, polymer molecular and Memory Molecular Image (MMI) technology has been shown to store also. The size of drive could be reduced with MEOMs constantly. The size of a postage stamp, Toshiba introduced a 0.75 inch hard drive for mobile devices and shipped 8GB units, 16GB and 32GB. However, solid state flash memory card have long surpassed 512GB and size smaller than it.

1.3 Photonic quantum for memory

The section contributes to the development of memory protocols and physical systems for implementations of quantum memories. To explore this in more detail, first introduce tools to assess the performance of a quantum memory and introduce the figures of merit for quantum memories. Followed by the discussion of the quantum memory protocols, physical systems, state-of-the-art and applications for quantum memories are discussed. The photonics quantum memory is a device that can faithfully store and re-emit photons. Quantum memories are physical systems that are operating based on quantum memory protocols. The physical system can be comprised of atoms, ions or defects in solids that are able to interact with photons. Some of the properties of a quantum memory, such as storage time, are mainly determined by the coherence properties of the physical system. Other specifications, such as efficiency or multimode capacity of memory as a function of optical depth, are determined by the protocol. Fidelity is a commonly used criterion for assessing the performance of a quantum memory. In order to be able to define a proper measure for fidelity, one needs to understand the application of the quantum memory. Here, I focus on quantum memories for single-photons. For this purpose, conditional fidelity is a proper measure for faithfulness of the quantum memory. The conditional fidelity is the overlap of the state of the re-emitted photon with the state of the input photon that is conditioned on the successful retrieval of the single-photon. Assuming that the quantum state of the single-photon is pure, the overlap of the single-photon wave functions is equivalent to the general definition of the quantum fidelity. It has to be noted that poor fidelity due to a unitary evolution applied by quantum memories may arise in some cases. This does not rule out practicality of the quantum memory in some of applications (e.g. quantum repeaters). The fidelity is mainly determined by the quantum memory protocol, and limitations that come from the physical system and implementation. Here describe these limitations when the physical systems are discussed. Efficiency is a key feature for quantum memories as it can affect performance of possible applications, such as single photon sources that are based on quantum memories. For single photon storage and retrieval, the efficiency can be described based on the single photon probabilities. The storage efficiency could be defined as one minus the probability of having a photon at the output during the storage, if transmission is the only loss channel. Similarly, the retrieval efficiency is the probability of retrieving a photon given that one photon is successfully stored. High efficiency is an essential feature for quantum memories to be used as single photon sources and elements of a long-distance quantum communication. The efficiency depends on specifications of the physical systems that are to be used for implementation of the quantum memory. For quantum memories that are based on an ensemble of atoms, increasing the optical depth enhances the coupling between the photon and the memory. Cavities also can enhance the efficiency. The efficiency with respect to optical depth scales differently in different protocols. This is further discussed with some of the examples, later in the thesis. Storage time is one of the crucial aspects of quantum memories. Quantum memories are developed for synchronization of different events in implementations of quantum information processing. Performance of some of the applications such as quantum repeaters rely on the storage time of quantum memories. In quantum repeaters, quantum memories are crucial to store the entanglement at different links. The minimum storage time is proportional to L_0/c, where L_0 is the length of one link. It has been shown that for storage times of $T_s \ll L/c$, where L is the total communication distance, the rate of entanglement distribution degrades exponentially in $\sqrt{L}$ as opposed to the polynomial scaling with L for quantum memories with infinitely long storage time. In general, the storage time is limited by relevant coherence times of the atomic level configuration that are used for the storage. In solid-state systems, where the optical coherence is stored in spin states, the spin inhomogeneous broadening limits the storage time. In some of the memory schemes, dynamical decoupling approaches have been used to extend the storage time. The precision requirements for one of these schemes were studied in this thesis and are presented. Multimode storage capacity can be defined as the maximum number of modes that can be stored simultaneously in a memory with certain efficiency. Here, the focus is on quantum memories that are based on ensembles of atoms. In addition to the advantage of atomic ensembles in enhancing the photon-memory coupling, ensembles allow one to implement quantum memory protocols with large multimode storage capacity. One way of multimode storage is based on storing photons at different frequencies. This requires large memory bandwidth. Memory bandwidth is the available spectral bandwidth for storage. This parameter is noteworthy as it sets a limit on the rate that the memory can operate. There are other parameters such as the wavelength of the quantum memory. This can be important to match the memory wavelength with an appropriate transmission channel (optical fibers or free-space transmission) in order to optimize the performance of the communication scheme. Furthermore, for quantum memories that are used as single photon sources or in quantum repeaters, it is essential to be compatible with the available parametric down conversion sources. Quantum memories inherit some of their properties from memory protocols as they determine a procedure for the operation of quantum memories. Part of my research contributes to developing new protocols and exploring the connection of these protocols with known schemes. This contribution is briefly introduced in the next sections of this chapter and is presented in detail in the following chapters.

- Off-resonant Raman Coupling Photonic Memory

In 2007, in two separate works, the quantum memory protocol based on the off-resonant Raman interaction has been proposed.

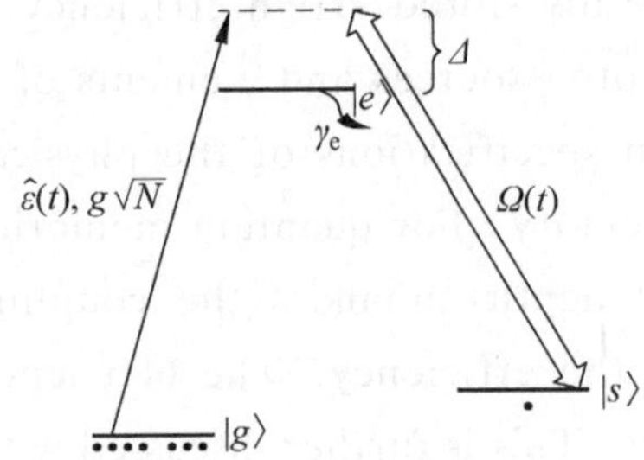

Fig. 1-1 The figure shows the off-resonant Raman coupling that allows the absorption of a single-photon through the creation of a spin excitation. One can retrieve the stored photon by applying the same control field after the storage time.

The scheme is based on an ensemble of 3-level atoms, and it operates by application of a properly shaped control field. To understand the basic principles of this protocol, one can imagine that all atoms are prepared in the ground state g, see Figure 1-1. The purpose is to store and retrieve the single photon pulse that is characterized by $\hat{E}(t)$. Coupling of the s-e transition to the applied control field is determined by $\Omega_c(t)$. Both the input single photon and control field are not in resonant with the g-e and s-e transitions, respectively. However, two-photon resonance is necessary for an efficient coupling, such that $\omega_p - \omega_c = \delta_{gs}$. In order to analyze this system, two paths have to be taken. First, one needs to use Maxwell's wave equation to consider the propagation of the electromagnetic fields in the medium. Second, a Hamiltonian that consists of energy of the levels and the atom-field interaction terms has to be considered. The propagation equation is a second order differential equation. The fact that the bandwidth of the pulse is much smaller than its central frequency leads to simplification of the ave equation. For a more complicated case, the refractive index of the medium is time-dependent. The Hamiltonian can be used in the Heisenberg equation,

$$\dot{\hat{A}} = \frac{i}{\hbar}[\hat{H}, \hat{A}] + \frac{\partial \hat{A}}{\partial t} \tag{1-1}$$

to find the dynamics of any operator in this system. The Heisenberg equation gives a set of equations for the level populations and transition operators. The point that the number of photons is much smaller than the number of atoms allows for a significant simplification, see the appendix of. Finally, Δ much larger than the bandwidth of the input pulse and the excited state broadening leads to the elimination of the excited state from the equations of motion. It has to be noted that sources of decoherence, such as the ground-state spin broadening, can be added to the equations. These considerations provide the tools to study the properties of the Raman quantum memory.

In principle, the fidelity of the Raman quantum memory can approach the ideal fidelity. However, due to the coupling to the control field, excitation of atoms from ground state g generate unwanted spin excitation, and consequently result in noise at the readout. This will limit the fidelity. One can avoid this process, by choosing appropriate levels with the opposite polarization selection rules. The efficiency depends on the effective optical depth. The effective optical depth is determined by the single-photon coupling, control field Rabi frequency, number of atoms, detuning and the ground state spin broadening. The efficiency in the forward direction can be limited, due to re-absorption during the read-out. The backward retrieval is not limited, and the efficiency can reach 100%. High efficiencies may require more control field power. This may lead to a time-dependent phase modulation that is given by

$$\phi(t) = \int_0^t \mathrm{d}t' \frac{|\Omega_c(t')|^2}{\Delta} \tag{1-2}$$

The effect is called the AC Stark shift and can significantly limit the performance of the quantum memory. One can apply a proper phase modulation on the input pulse to cancel the effect of the AC Stark shift, and therefore reach the ideal efficiency. Multimode storage capacity of quantum memories based on atomic ensembles have been studied. For quantum memories without controlled inhomogeneous broadening, including Raman-type quantum memory, the multimode storage capacity does not scale favorably with optical depth. Specifically, the studies show that the number of modes that can be stored simultaneously with efficiency above a certain threshold, in the Raman-type quantum memory, scales with the square-root of the optical depth. It is important to understand that the bandwidth of the Raman quantum memory is determined by the bandwidth of the control field. This allows one to implement a broadband quantum memory based on this protocol in contrast to the limited bandwidth in the electromagnetically induced transparency.

This is the reason for the limited efficiency for the forward retrieval, as the re-emitted pulse is affected by re-absorption in the rest of the medium. The backward retrieval is experimentally more demanding, but it allows to reach 100% efficiency. Finally, it has to be noted that the proposed protocol is equivalent to the Raman quantum memory in a 2-level system and without application of an optical control field.

- Electromagnetically induced transparency quantum memory

Slow light has been one of the most exotic effects in optics. Electromagnetically induced transparency (EIT) is well-known to exhibit slow light effect. As it can be seen in Figure 1-2, the imaginary part of the susceptibility that is responsible for absorption features a transparency window.

Similar to Figure 1-1, the light that propagates in a medium of three-level atoms can be controlled by applying a control field, $\Omega(t)$. In contrast to the previous scheme that is demonstrated in Figure 1-1, when the detuning Δ approaches zero, this transparency window in Figure 1-2 appears. This can be explained by interference between different possibilities for absorption of the probe field. For the detuning $\Delta = 0$ (see Figure 1-3), one can

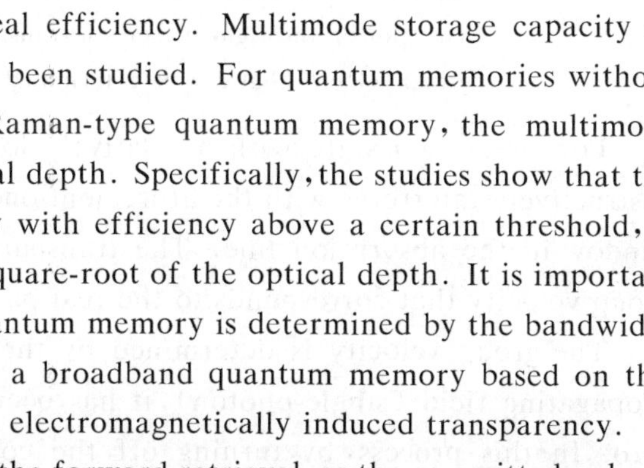

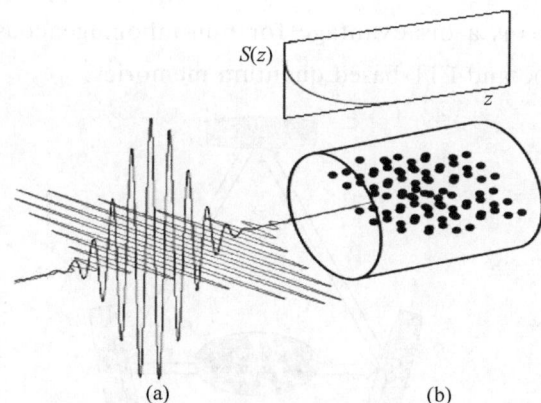

Fig. 1-2 (a) Show the control field (red) and the input single-photon pulse (blue) that are properly timed for absorption of the input pulse in the atomic ensemble. (b) Schematically shows the probability of the spin excitation that is decreasing exponentially over the length of the medium.

imagine that the probe field, $\hat{E}(t)$, can be absorbed through the g-e transition.

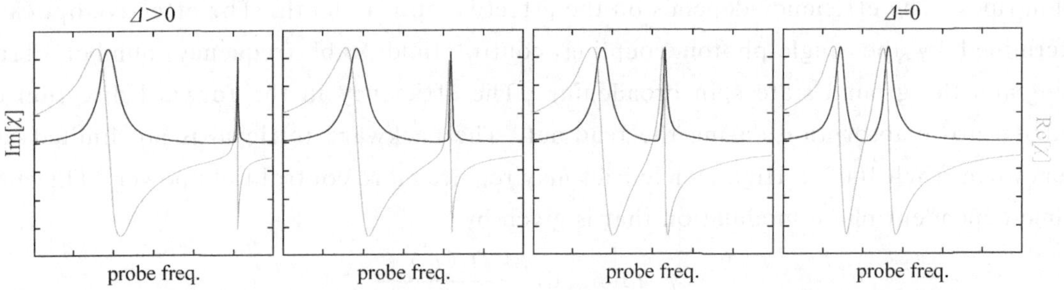

Fig. 1-3 This figure shows the imaginary part (blue) and real part (green) of the susceptibility that is experienced by the probe field $\hat{E}(t)$, see Figure 1-1. In the L configuration in Figure 1-1, by reducing the detuning Δ from left to the right, the scheme approaches the condition for EIT. The plot on the right presents a transparency window that is associated to a reduction in the group velocity that can be determined by the slope of the references.

For an excited state with a relatively long life time, the absorption through the transition destructively interferes with the aforementioned absorption process. This results in a transparency window in the absorption line. The transparency window is associated with a reduction in the group velocity that corresponds to the real part of the susceptibility for the propagating light.

The group velocity is determined by the Rabi frequency of the control field. For a weak propagating field (single-photon), it has been shown that the group velocity can be reduced to zero. In this process, by turning off the control field gradually, the optical excitation in the propagating probe field is converted to a spin excitation and is stored in the ground state of the system. By turning on the control field, the stored excitation can be read out. The optical depth and the ground state coherence time are necessary for higher efficiency and storage time, respectively. Similar to the Raman-type quantum memory, the available bandwidth for storage is determined by specifications of the control field that results in the transparency window, see Figure 1-4. For multimode storage, the capacity scales with the square-root of the optical depth. This is a disadvantage for non-inhomogeneously broadened ensemble memories, such as Raman-type and EIT-based quantum memories.

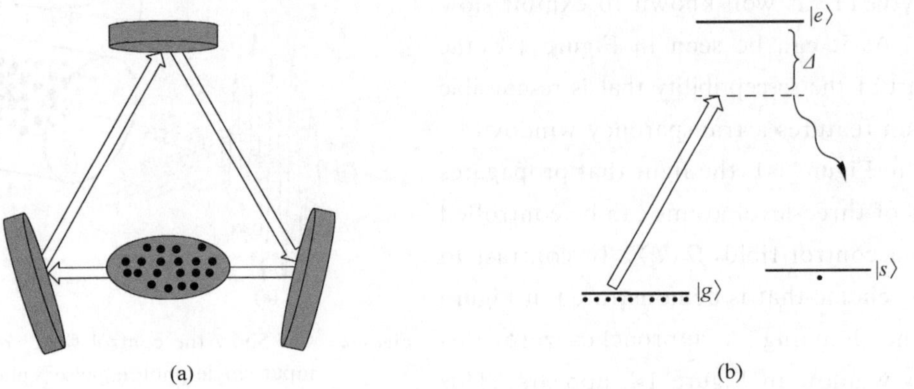

Fig. 1-4 (a) Ensemble of atoms in a ring cavity that is interacting with the cavity field with a well-defined propagation direction. (b) L-level configuration of the atoms. All atoms are prepared in the ground state. The cavity field can result in scattering a photon and generate a collective spin excitation.

• Off-resonant Raman scattering protocol (DLCZ)

As another scheme that is based on an ensemble of three-level atoms, I introduce the off-resonant Raman scattering scheme that is known as the DLCZ protocol. This protocol has been proposed as a part of a scheme for long-distance quantum communication. This scheme is not a quantum memory protocol in the sense that it does not allow for reversible mapping of externally provided single-photons. However, the protocol allows to generate entanglement between the atomic ensemble and a scattered photon. This can be utilized toward implementations of single-photon sources or long-distance quantum communication schemes. Applying a detuned optical pulse to the atomic ensemble scatters one photon and generates a spin excitation. In order to avoid absorption of the pulse, the field has to be detuned by Δ from the $|g\rangle$-$|e\rangle$ transition. However, a greater Δ may reduce the probability of scattering a photon and generating a collective spin excitation. The use of the ring cavity is a way to enhance this probability (coupling).

The spontaneous emission from the excited state might seem to be problematic in this scheme. However, it can be shown that the spontaneous emission distributes excitation over all possible modes (wave vectors). The Raman scattering generates a collective spin excitation, where its wave vector (mode) is determined by the wave vector of the incoming and scattered photon. Therefore, at the read out there is a preferred direction for which the read out emission is collectively enhanced by the number of atoms in the ensemble. This results in a high signal-to-noise ratio (a suppressed noise). This concept is discussed in detail, and used to study the precision requirements for spin-echo quantum memories.

• Controlled-reversible inhomogeneous broadening quantum memory

Controlled-reversible inhomogeneous broadening (CRIB) quantum memory have introduced and analyzed in. As its title suggests, it is based on controlling the inhomogeneous broadening of the relevant transition. Inhomogeneous broadening refers to the variance in energy of the atoms in an ensemble. In this protocol, the photon is stored in a collective excitation of an inhomogeneously broadened atomic ensemble. After absorption of an incoming photon, this collective atomic state can be described by

$$|\psi(t)\rangle = \frac{1}{\sqrt{N}}\sum_{j=1}^{N} e^{i\Delta t} \mid gg\ldots e^{j}g\ldots g\rangle \tag{1-3}$$

where N is the number of atoms in the ensemble and dj is the detuning of the central frequency of the incoming photon from the jth atom. The variance in energy (level splitting) of atoms leads to dephasing of the collective atomic state. The dephasing happens at the rate that is determined by the width of the inhomogeneous broadening.

This prevents any application, including storage that requires coherence times that are longer than the inverse of the width of the broadening. However, a controlled inhomogeneous broadening allows one to rephase the collective atomic state. After some time $t = T$, using this control the inhomogeneous broadening can be reversed. Therefore, the jth atom in the ensemble will acquire a reversed detuning of $i\Delta t$. It can be seen from the collective atomic state $|y\rangle$ that the dephasing can be reversed at the time $t = 2T$, see Figure 1-5. The fidelity of the CRIB quantum memory can in principle approach 100%. Similar to the other protocols, the efficiency depends on the optical depth of the ensemble. As it has been discussed for the Raman-type quantum memory, the reabsorption effect in the medium limits the forward retrieval efficiency. The backward retrieval efficiency approaches 100% for high optical depths. For a modified CRIB protocol that offers a solution to suppress the reabsorption effect in the CRIB protocol.

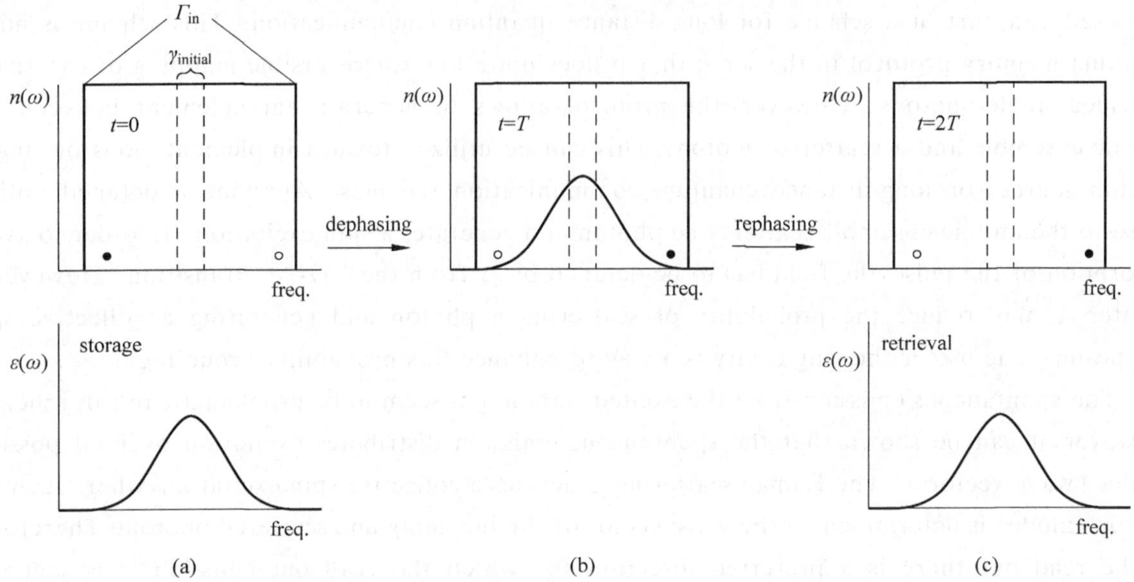

Fig. 1-5 (a) A schematic absorption profile of an atomic ensemble with initial broadening of ginitial is broadened to Γ_{in} by applying an inhomogeneous external (magnetic/electric) field. This broadened ensemble allows to absorb the incoming pulse E (w). (b) The broadened ensemble dephases due to the broadening. After the time $t = T$, the external field is reversed leading to the inversion of the detuning of all absorbers, see the open and solid circles. (c) Reversing the broadening at time $t = T$ allows to get all the atoms rephased at $t = 2T$. Efficient retrieval is expected at $t = 2T$.

The limit on the storage time is determined by properties of the physical system. For a storage time of T_{stor}, one has to reverse the inhomogeneous broadening at $t = T_{stor}/2$ to achieve an efficient recall at $t = T_{stor}$, see Figure 1-5. This property has used for pulse sequencing in. Multimode storage is one of the most important features of quantum memory protocols that are based on inhomogeneous broadening. As opposed to the protocols that are based on unbroadened atomic ensembles, the number of modes that can be stored with a certain efficiency scales linearly with the optical depth. In CRIB, depending on the physical system, the inhomogeneous broadening is produced by an external electric or magnetic field. Increasing the external field strength results in a greater inhomogeneous broadening and consequently a greater memory bandwidth.

This controllable memory bandwidth can be used to store multiple input modes. The gradient echo memory (longitudinal CRIB) is based on the CRIB protocol, where the broadening is produced by a longitudinally (along the propagation direction of the input field) varying external field. This is in contrast to the transverse CRIB, where every slice of the medium along the propagation direction contains all of the frequency components. The most distinguishing feature of the gradient echo memory (GEM) compared to CRIB is that the forward retrieval efficiency is not limited by re-absorption as the atoms are broadened longitudinally in space that prevents re-absorption at the retrieval. The gradient echo memory protocol has led to some of the influential experimental results on quantum memories. A proposal for a novel quantum memory protocol based on a different principle that resembles the gradient echo memory under certain conditions.

• Atomic frequency comb quantum memory

The atomic frequency comb (AFC) quantum memory is another well-known quantum memory protocol that is based on inhomogeneous broadening. The principle of the protocol based on periodicity in the absorption frequency of the atomic ensemble, which see Figure 1-6. An incoming pulse interacting with an ensemble of atoms with a comb-like atomic frequency distribution generates a collective atomic state.

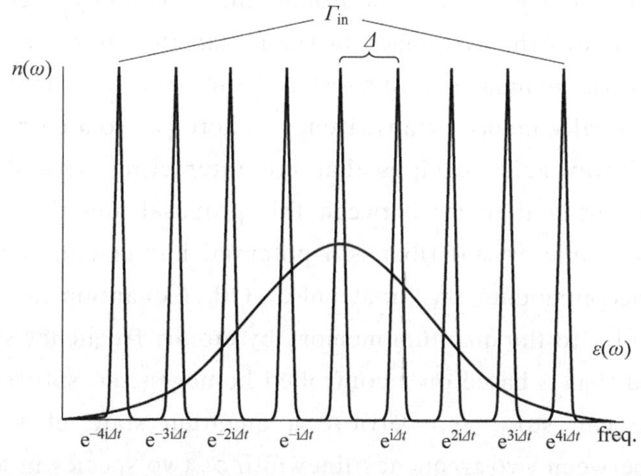

Fig. 1-6 (a) The figure shows the absorption profile of inhomogeneously broadened ensemble that is prepared for the AFC protocol. The periodic absorption peaks are prepared by optical pumping. The incoming field $E\ (w)$ interacts with the ensemble. The collective atomic state dephases as atoms in different absorption peaks precess at different rates. However, the collective atomic state automatically rephases (due to the periodicity) at $t = 2n\pi/\Delta$, where $n = 1.2$.

This collective atomic state dephases as different have different frequencies (detunings). However, due to the comb-like distribution (periodicity in absorption peak frequencies), the collective atomic state rephases at certain times. The rephrasing times are integer multiples of $2\pi/\Delta$, where Δ is the angular frequency difference between the absorption peaks of the AFC. The storage time is determined by this dephasing-rephasing time. Therefore, once the atomic frequency comb is prepared, the storage time is fixed. Transferring the stored excitation to a third level allows one to eliminate the limitation due to the fixed storage time. In addition, for communication purposes, the functionality of the AFC quantum memory can be improved by transferring optically excited state to the collective spin ground state for longer storage times and on-demand retrieval. The fidelity of the AFC has been shown to approach the ideal fidelity. The efficiency in the forward direction is limited by re-absorption. However, the backward retrieval approaches 100% efficiency for high optical depths. AFC provides multimode storage.

Quantum memory based on the refractive index modulation is another novel scheme that utilizes a time-dependent refractive index of a medium to generate an effective longitudinal controlled reversible broadening. The time-dependent refractive index of the medium allows one to effectively modulate the frequency of the propagating pulse as a linear function of the position in the medium. This results in an effective longitudinal position-dependent detuning between the pulse and atoms. A comparison shows that this scheme is equivalent to the gradient echo memory, where the position-dependent detuning is due to an external position-dependent (electric or magnetic) field. That is based on Tm^{3+} ion that are doped in the lithium niobate crystal (medium), in which one can modulate the refractive index of the medium by a time-dependent electric field, such that it provides enough bandwidth to store the input pulse. Reversing the time-dependent electric field allows for retrieving the stored pulse.

Quantum memory by atomic frequency sweeping is an emerging scheme that is based on modulating the transition frequency of an ensemble of two-level atoms. The results show that changing the transition frequency of a narrow atomic line from a large negative to a large positive detuning allows to store pulses that are much broader than the atomic frequency linewidth. The polaritonic description of the dynamics in this system indicates a slow-light effect that is similar to that of the electromagnetically induced transparency. There is also a correspondence between this system and an array of waveguide cavities that are interacting with side cavities. Numerical analysis also sheds light on similarities between this proposal and the gradient echo memory. Atoms in hollow-core photonic crystal fiber is a potential implementation of this scheme as the medium is required to accommodate the input pulse. Other quantum memory protocols are also recently developed. Similar to the quantum memory by atomic frequency sweeping, a protocol has recently been introduced that is based on a controlled homogeneous splitting. The authors of this paper showed that one can store and retrieve a quantum state of light by controlling the homogeneous splitting between two frequency linewidth of two species in an ensemble. They also observe the slow-light effect that is not based on the ground-state coherence (as opposed to EIT), therefore it shares partial similarities with our abovementioned protocol.

- Physical systems

Any stationary quantum system that can be coupled to photons and has a coherence time that is longer than the duration of the photons can be a candidate for the implementation of a quantum memory. A trapped single-atom, or ensembles of atoms in a trap or in a gas cell, or in a crystal are other possibilities. Artificial atoms such as quantum dots and nitrogen vacancy (NV) centers in diamond are also among the possible physical systems. Coherence times vary from about 1ms in the electronic ground states of Rb atoms in a Vapour cell to over 1s in nuclear spin states of some rare-earth ion doped crystals. The efficiency of the atom-photon interface is determined by the optical depth that varies from one implementation to another. Research is focused on the solid-state candidates. In particular, rare-earth ion doped crystals are attractive because of their properties such as relatively long optical coherence times and inhomogeneous broadening that can be engineered to be utilized in some of the protocols. In addition, NV centers in diamond possess various useful properties that propose this system as a promising candidate for photonic quantum information processing tasks. The NV centers in diamond show a significant ground state nuclear spin coherence times even at room temperature. This has provided over a second coherence time that can be used for storage of microwave photons. Below, I provide details about these two solid-state candidates for implementations of quantum memories.

- Rare-earth ion doped crystals

Rare-earth elements are already an inevitable part of the current technology. They became particularly attractive since the invention of the laser in 1960s. In industry, they are used in producing magnets and batteries. In life sciences, their fluorescence properties are used for examining biological fluids and drug research. Rare-earth elements are known to be comprised of scandium, yttrium and 15 other metallic elements that are called Lanthanides. There are four elements that are particularly appealing to the quantum memory community. These four elements are praseodymium (Pr), neodymium (Nd), erbium (Er) and thulium (Tm) as shown in the Figure 1-7. These rare earth elements belong to the 4f block that corresponds to the filling of the 4f electronic shell. The 5s, 5p and 6s shells are filled and have a larger radial distribution, which they partially shield the 4f electrons that results in a narrow homogeneous optical linewidth and reduces the influence of crystal strain and lattice phonons. There are few crystals, such as YAG ($Y_3Al_5O_{12}$), Y_2SiO_5 or $LiNbO_3$ that are often used as the host crystals for the rare-earth ions. The wavelength of the transition is determined by the rare-earth ion. The doped ion replaces yttrium and lithium in the host crystals. The host crystal can offer crystallographically equivalent or inequivalent sites for the dopants. This may have an impact on the response to the external magnetic or electric fields. Homogeneous broadening is limited by the excited state lifetime of the relevant transition. However, depending on the temperature it is determined by phonon interactions (high temperatures) or energy exchange between spins (cryogenic temperatures). Inhomogeneous broadening, which is due to inhomogeneity in the environment of the dopants, is caused through spin-spin interaction and strain that are sources of a spatially varying potential for dopants. Therefore, the density of dopants becomes an important factor that has an impact on the inhomogeneous broadening. This is the reason that one cannot increase the optical depth arbitrarily at a certain frequency by increasing the density of dopants. There is a chance to increase the optical density by using the stoichiometric rare-earth crystals. The proposed implementation of the Controllable-dipole quantum memory is in Tm:YAG. The photon is stored in a two-level configuration that is based on the optical coherence of that transition. It is also noteworthy that one can control the transition dipole moment of the certain transitions of Tm^{3+} ion in Tm:YAG by applying small changes to the external magnetic field. The proposed protocol requires a host that has a variable refractive index under the external electric field. Therefore, $Tm:LiNbO_3$ is an appropriate choice of the rare-earth doped crystals to implement the protocol.

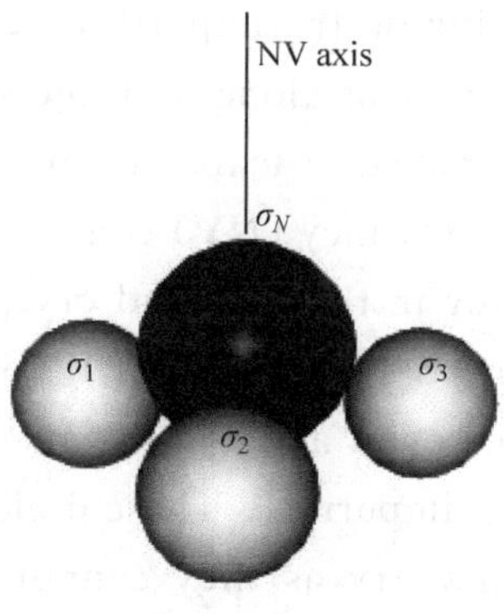

Fig. 1-7 The figure shows the dangling electronic orbitals σ_1, σ_2 and σ_3 that are associated with the electrons shared by three neighboring carbons and σ_N that corresponds to electrons from the nitrogen In the basis of the NV center the z direction is determined by a vector from the nitrogen toward the vacancy. One of the carbon's dangling electrons defines the x direction, and the y direction is correspondingly defined perpendicular to x and z.

- NV centers in diamond

Defects in solids are attractive due to their potential to provide atom-like properties with relatively long coherence times. The defect can be a displaced atom or a vacancy in the crystal structure. In addition to the defects, impurities exist in or can be added to the crystal structure.

Depending on the impurity, it can add an excessive electron or a hole (lack of an electron) to the structure. This along with the symmetry of the defect can determine some of important properties of the impurity-vacancy centers in crystals. Nitrogen is the most common impurity in diamond. A nitrogen vacancy (NV) center in diamond is a nitrogen that replaces a carbon, which is neighbor to a vacancy in the diamond crystal structure. A negatively charged NV center that is the focus of this study is comprised of 6 contributing electrons (3 electrons from 3 carbon atoms, 2 electron from nitrogen and 1 electron from the environment that is possibly from another single atomic nitrogen impurity). These 6 electrons are confined in the defect (vacancy) in the diamond with C3v symmetry as they cannot contribute in any covalent bond with a neighboring site in the lattice. The C3v symmetry denotes symmetry under rotations around a vertical axis and here it is the NV axis, see Figure 1-7 and 3 mirror planes. For comparison this is the same symmetry as for the NH3 molecule. This information can be used to determine the electronic level configuration and energies.

There has been significant interest in NV centers for studying and utilizing their properties for various applications. NV centers have relatively long electronic spin coherence times (up to about 1ms) at room temperature. These coherence times become much longer at low temperatures. The spin state can be read optically by exciting the NV center and measure the polarization of the fluorescent emission. These properties have made NV centers an attractive tool for magnetic sensing and nano-scale nuclear magnetic resonance. Storage of microwave photons in a single NV center or an ensemble of NV centers has been studied theoretically and experimentally. For an NV center that is not coupled to any neighboring nuclear spin the coherence (storage) time is given by the electronic ground state T_2 time. T_2 time of about 1ms has been observed at room temperature. T_2 time of 0.5s has been reported at low temperatures around 77°K. NV centers can couple to their neighboring nuclear spins (of C^{13} or N^{15}) that are within few nm distances. This potentially allows one to take advantage of the even longer coherence time of nuclear spin that has been shown to be over one second. It has also been shown that the parallel electron-nuclear coupling rate exceeds the $1/T_{1e}$, where T_{1e} is the electronic spin lifetime. This means that the flip-flop of the electronic spin that is happening at the time scale of about 10ms leads to dephasing of the nuclear spin. Therefore, the nuclear spin coherence time is limited by the electronic spin lifetime. A dynamical decoupling technique has been used to achieve over one second coherence time at room temperature. Despite all these attractive properties, the storage of optical photons becomes problematic due to the short excited state lifetime of individual NVs and the broad optical linewidth (excited state inhomogeneous broadening) in NV ensembles. A solution to these difficulties is presented, and an NV ensemble coupled to a cavity is shown to be a promising solid-state candidate for micron-scale on-chip optical quantum memories.

Apart from the above-mentioned solid-state systems, there are other physical systems that are being used by several groups around the world. The most commonly used systems are ensemble of trapped cold atoms and hot atomic gases. A hot atomic gas cell of Rb atoms has been used for the implementation of the Raman-type quantum memory and for realizing the gradient echo memory. Trapped cold atoms in a ring cavity have been used to implement DLCZ type protocol that resulted in a combination of high efficiency and storage time. In addition, a cold trapped single atom in a cavity has also been shown to operate as a quantum memory. Atoms trapped in a hollow core fiber provide a relatively large optical depth, which is one of the requirements for an efficient quantum memory. Hollow core photonic crystal fibers as a trap for Cs atoms have been shown to provide even higher optical depths that are promising for the Raman-type scheme. Another candidate is a rare-earth ion doped fiber. Despite their small optical coherence, they can be useful for protocols that are based on inhomogeneous broadening. Very recently, inhomogeneous and homogeneous optical linewidths of rare-earth ion doped transparent ceramics have been studied. The results from this study are comparable to rare-earth ion doped single crystals. This suggests that ceramic materials can be competitive with single crystals for applications in quantum information, including quantum memories.

- State of the art

Quantum memory protocols and their experimental demonstrations progressed rapidly during the past decade. Various quantum memory protocols have been adapted to realize each of the quantum memory criteria in various setups. Even though high efficiency, long storage time and multimode capacity have not been achieved in one single implementation, there have been attempts to address these requirements individually.

- Solid-state systems

Achieving high efficiency has been the focus of many experiments. Since the transition dipole moment in rare-earth doped solids are weak, achieving high optical depth and, consequently, high efficiency is challenging. Higher concentration of dopants could increase the inhomogeneous broadening and prevent one from arbitrarily increasing the optical depth. Isotropically pure, stoichiometric rare-earth crystals promise much higher optical depths compared to the rare-earth doped crystals. In 2010, Pr^{3+} doped in Y_2SiO_5 crystal was used to implement the gradient echo memory (GEM) protocol. In this study, an efficiency of about 69% was achieved and currently constitutes the solid-state quantum memory with the highest efficiency. Very recently, implementing the AFC protocol in the same material within an impedance-matched cavity resulted in about 58% efficiency. It has to be noted that this result is achieved based on a weak absorbing sample. This promises more progress in terms of the efficiency based on impedance-matched

quantum memories. In 2005, Pr^{3+} : Y_2SiO_5 was used to implement EIT-based storage. This experiment demonstrated storage times of over 1s. In this experiment, multiple p-pulses have been used to prevent dephasing due to the spin inhomogeneous broadening. This technique only allowed to store strong pulses, due to a limited signal-to-noise ratio. For evaluation of the error due to the uncertainty in p-pulses. In terms of the bandwidth, the AFC protocol in Tm^{3+} doped lithium niobate waveguide provided 5GHz memory bandwidth. This also provides the potential for multimode capacity. Quantum memory based on AFC protocol in Tm^{3+} : YAG allowed storage of more than 1000 temporal modes. Storage of entangled photons has been demonstrated in Tm^{3+} : $LiNbO_3$ and Nd^{3+} : Y_2SiO_5 based on the AFC protocol. This is important as it is a key component in long-distance quantum communication based on quantum repeaters.

- Atomic gases

Some of the major achievements in developing quantum memories are based on physical realizations in atomic gases. The highest efficiency for quantum memories has been accomplished in hot atomic gas of Rb atoms in a cell. The efficiency of about 87% was achieved based on the

GEM protocol. The same protocol has been employed to use the quantum memory as an optical pulse sequencer to store and re-order multiple pulses in time. In a similar setup, Raman-type quantum memory has been implemented. Despite the limited efficiency, the scheme provided one of the broadband quantum memory implementations with a bandwidth of about 1.5GHz. An ensemble of cold Rb atoms in a magneto-optical trap inside a ring cavity, resulted in the best available combination of storage time and efficiency. The off-resonant Raman scattering (DLCZ) protocol has been employed in this experiment and about 73% efficiency and over 3ms lifetime have been demonstrated. The next step is to combine the required features in one single implementation to take a step toward more practical quantum memories.

- Deterministic single-photon sources

Immediate application of an efficient quantum memory is the implementation of a deterministic single-photon source. A parametric down-conversion source can produce photon pairs in a nondeterministic way. One can combine this source with a quantum memory. Detecting one of the photons at one arm determines the presence of another photon at the other arm to be stored in the quantum memory. Later the stored excitation can be retrieved to construct an on-demand heralded single-photon source. A highly efficient single-photon source is a crucial element in photonic quantum information processing, especially in linear optical quantum computation based on the KLM scheme. The off-resonant Raman scattering approach has been used to realize a deterministic and storable single-photon source. Applying an off-resonant weak laser pulse allows to scatter one photon and create a collective spin excitation in the ground state of an atomic ensemble. Detecting the scattered photon determines the stored spin excitation and its mode (wave vector). Later, the stored excitation can be retrieved to serve as a deterministic and storable single-photon source.

1.4 Controllable-dipole quantum memory

Quantum memories are implemented based on quantum memory protocols. The protocol specifies all the steps that have to be taken to store in and recall from a quantum memory. Taking the quantum memory protocol into account and considering the light-atom interaction allows to analyze the performance of the quantum memory. This chapter presents a new quantum memory protocol for storage in two-level systems that are based on direct control of the transition dipole moment. Mainly, the analysis is focused on the case in which the atomic ensemble is inside a cavity. This enhances the light-atom interaction. The analysis is focused on finding the conditions for maximizing the efficiency. The optimal write process is related to the optimal read process by a reversal of the effective time $t = \int dt g^2(t)/k$, where $g(t)$ is the time-dependent light-atom coupling and k is the cavity decay rate. Then the paper shows results on the free-space case, where Maxwell equation is used for considering the propagation of the photon in the medium. A possible implementation is discussed based on Tm^{3+}: YAG. In Tm^{3+}: YAG, the transition dipole moment can be controlled by changing the direction of the external magnetic field. Finally, the performance of the controllable-dipole quantum memory is compared with that of the controlled reversible inhomogeneous broadening (CRIB) quantum memory. The proposed protocol based on the modulation of the transition dipole moment is shown to be equivalent to the Raman-type quantum memory without the requirement for application of an optical control field.

- Controllable-dipole quantum memory in a cavity

Consider an ensemble of two-level atoms coupled to a cavity mode, see Figure 1-8. The system that we consider is formally equivalent to a Raman memory in a cavity, if the excited state is adiabatically eliminated in the Raman case, and where the two-photon spin transition is replaced by a single-photon optical transition. There is also some similarity, where the light-matter coupling is controlled by tuning a cavity instead of the transition dipole moment.

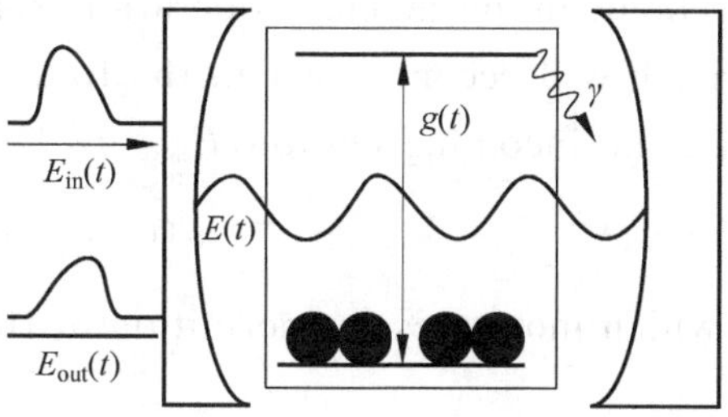

Fig. 1-8 Consider an ensemble of two-level systems inside a one-sided cavity, where the time-dependence of the light-matter coupling $g(t)$ can be controlled. See also Eq. (1-4).

The usual input-output formalism for a single-sided, fairly high-finesse cavity. The basic equations are then

$$\begin{aligned}\dot{\sigma}(t) &= -i\Delta(t)\sigma(t) - \gamma\sigma(t) + \mathrm{i}g(t)E(t)\\ \dot{E}(t) &= \mathrm{i}g(t)\sigma(t) - \kappa E(t) + \sqrt{2\kappa}E_{\mathrm{in}}(t)\\ E_{\mathrm{out}}(t) &= -E_{\mathrm{in}}(t) + \sqrt{2\kappa}E(t)\end{aligned}\tag{1-4}$$

The linearity of the dynamics, s and E can be interpreted as the atomic polarization and cavity fields (in the semi-classical regime), but also as the probability amplitudes corresponding to a single atomic excitation in the ensemble and a single cavity photon respectively; $E_{\mathrm{in}}(t)$ and $E_{\mathrm{out}}(t)$ are the incoming and outgoing fields (photon wave functions); $g(t)$ is the time-dependent light-matter coupling, which is proportional to the transition dipole matrix element between the ground and excited atomic states; k is the cavity decay rate; g is the atomic decay rate; $D(t)$ is a time-dependent detuning, which may arise in practice as a consequence of applying a time-dependent external field in order to control the dipole element and thus $g(t)$; g and $D(t)$ are imperfections that we will neglect at first to keep the discussion simple, but whose effect will be discussed later in the paper.

In the (realistic) situation where the cavity decay defines the shortest relevant timescale. In this case it is well justified to adiabatically eliminate the cavity field, setting $\dot{E}=0$. This gives

$$E(t) = \frac{1}{\kappa}(\mathrm{i}g(t)\sigma(t) + \sqrt{2\kappa}E_{\mathrm{in}}(t))\tag{1-5}$$

and hence

$$\begin{cases}\dot{\sigma}(t) = -\dfrac{g^2(t)}{\kappa}\sigma(t) + \mathrm{i}\sqrt{\dfrac{2}{\kappa}}g(t)E_{\mathrm{in}}(t)\\ E_{\mathrm{out}}(t) = E_{\mathrm{in}}(t) + \mathrm{i}\sqrt{\dfrac{2}{\kappa}}g(t)\sigma(t)\end{cases}\tag{1-6}$$

where have set $\Delta(t) = g = 0$, as mentioned above. It is straightforward to derive the (very intuitive) continuity equation

$$\frac{\mathrm{d}}{\mathrm{d}t}\mid \sigma(t)\mid^2 = \mid E_{\mathrm{in}}(t)\mid^2 - \mid E_{\mathrm{out}}(t)\mid^2 \tag{1-7}$$

- Read process

The quantum memory operation, starting is with the read process. (The motivation for this approach will become clear in the following.) The read process corresponds to a situation where there is no incoming photon, $E_{\mathrm{in}} = 0$. The continuity equation (1-7) implies

$$\mid \sigma(0)\mid^2 = \mid \sigma(t)\mid^2 + \int_0^t \mathrm{d}t' \mid E_{\mathrm{out}}(t')\mid^2 \tag{1-8}$$

which motivates the definition of the read efficiency η_{r} as

$$\eta_{\mathrm{r}} = \frac{\int_0^\infty \mathrm{d}t \mid E_{\mathrm{out}}(t)\mid^2}{\mid \sigma(0)\mid^2} \tag{1-9}$$

Here have defined $t = 0$ as the starting time of the read process. The solution of Eq. (1-6) with $E_{\mathrm{in}} = 0$ is given by

$$\begin{cases} \sigma(t) = \sigma(0)\mathrm{e}^{-\int_0^t \mathrm{d}t' g^2(t')/\kappa} \\ E_{\mathrm{out}}(t) = \mathrm{i}\sqrt{\dfrac{2}{\kappa}}g(t)\sigma(t) \end{cases} \tag{1-10}$$

Using Eqs. (1-9) and (1-10) one finds

$$\eta_{\mathrm{r}} = 1 - \mathrm{e}^{-2\int_0^\infty \mathrm{d}t g^2(t)/\kappa} \tag{1-11}$$

Eq. (1-11) motivates the introduction of the effective time variable

$$\tau = \int_0^t \mathrm{d}t' g^2(t')/\kappa \tag{1-12}$$

giving the simple expression $\eta_{\mathrm{r}} = 1 - \mathrm{e}^{-2\tau_{\mathrm{r}}}$, where $\tau_{\mathrm{r}} = \int_0^\infty \mathrm{d}t g^2(t)/\kappa$ is the total effective time that elapses during the read process. This means that in order to maximize the read efficiency one simply has to maximize τ_{r}. The shape of $g(t)$ has an impact on the form of the output field, but the efficiency only depends on τ_{r}. In order to rewrite the whole dynamics in terms of the effective time variable τ, furthermore introduce effective input, output and cavity fields,

$$\mathscr{E} = \frac{\kappa}{g}E,\ \mathscr{E}_{\mathrm{in}} = \frac{\sqrt{\kappa}}{g}E_{\mathrm{in}},\ \mathscr{E}_{\mathrm{out}} = \frac{\sqrt{\kappa}}{g}E_{\mathrm{out}} \tag{1-13}$$

One then finds the new equations of motion (after adiabatic elimination of E)

$$\begin{cases} \dfrac{\mathrm{d}}{\mathrm{d}\tau}\sigma(\tau) = -\sigma(\tau) + \mathrm{i}\sqrt{2}\,\mathscr{E}_{\mathrm{in}}(\tau) \\ \mathscr{E}_{\mathrm{out}}(\tau) = \mathscr{E}_{\mathrm{in}}(\tau) + \mathrm{i}\sqrt{2}\,\sigma(\tau) \end{cases} \tag{1-14}$$

The read efficiency can be rewritten as

$$\eta_{\mathrm{r}} = \frac{\int_0^{\tau_{\mathrm{r}}} \mathrm{d}\tau \mid \mathscr{E}_{\mathrm{out}}(\tau)\mid^2}{\mid \sigma(0)\mid^2} \tag{1-15}$$

The solution of Eq. (1-10) in the read case ($E_{\mathrm{in}} = 0$) is simply

$$\begin{cases} \sigma(\tau) = \sigma(0)\mathrm{e}^{-\tau} \\ \mathscr{E}_{\mathrm{out}}(\tau) = \mathrm{i}\sqrt{2}\,\sigma(\tau) \end{cases} \tag{1-16}$$

Eq. (1-16) shows that in terms of the effective time (and of the effective fields) the read process is a simple exponential decay —a remarkable simplification considering that the time dependence of $g(t)$ (and hence $E_{out}(t)$) is completely arbitrary.

- Write process

Now ready to discuss the write process. Immediately use the effective variables. Solving Eq. (1-14) for non-zero E_{in} one finds

$$\sigma(0) = i\sqrt{2}\int_{-\tau_w}^{0} d\tau' e^{\tau'} \mathscr{E}_{in}(\tau') \tag{1-17}$$

where τ_w is the total elapsed effective time for the write process and $\sigma(-\tau_w) = 0$. Note that no effective time elapses during times when the transition dipole is zero (i. e. during storage). Define the write efficiency as

$$\eta_w = \frac{|\sigma(0)|^2}{\int_{-\tau_w}^{0} d\tau |\mathscr{E}_{in}(\tau)|^2} \tag{1-18}$$

- Optimal input field

Our goal is to find the form of $E_{in}(\tau)$ that maximizes η_w. Since the solution for σ is linear in E_{in}, maximizing η_w corresponds to maximizing $|\eta_s(\tau_w)|^2$ for a normalized input field satisfying

$$\int_{-\tau_w}^{0} d\tau |\mathscr{E}_{in}(\tau)|^2 = 1 \tag{1-19}$$

Before discussing the formal optimization, let take a step back and try to make a guess for the optimum input field. That when expressed in terms of effective time rather than real time, the read process simply corresponded to an exponential decay, see Eq. (1-16). It is natural to suspect that inverting this decay (in effective time) will give the optimum effective input field. This means that our guess for the optimum solution is $E_{in}(\tau) \propto e^{\tau}$. This can be proved by functional differentiation. The optimum solution has to satisfy

$$\frac{\delta}{\delta \mathscr{E}_{in}^*(\tau)}\left[|\sigma(0)|^2 + \lambda\left(\int_{-\tau_w}^{0} d\tau |\mathscr{E}_{in}(\tau)|^2 - 1\right)\right] = 0 \tag{1-20}$$

where λ is a Lagrange multiplier, and $E_{in}(\tau)$ and $E_{in}^*(\tau)$ are independent variables for each τ. Solving this equation using Eq. (1-17) gives $E_{in}(\tau) \propto e^{\tau}$, confirming the intuitive guess, see also Figure 1-9. For this optimal solution the write efficiency is analogous to the read efficiency,

$$\eta_w = 1 - e^{-2\tau_w} \tag{1-21}$$

The total efficiency (ignoring losses during storage) is then

$$\eta_{tot} = \eta_w \eta_r = (1 - e^{-2\tau_w})(1 - e^{-2\tau_r}) \tag{1-22}$$

which can obviously be simplified further if $\tau_w = \tau_r$. Provided that the optimum input field is chosen for the write process, the efficiency is thus maximized by maximizing τ_w and τ_r. In real time the input field for the write process and the output field for the read process satisfy

$$\begin{cases} E_{in}(t) \propto g_w(t) e^{\int_{-\infty}^{t} dt' g_w^2(t')/\kappa} \\ E_{out}(t) \propto g_r(t) e^{-\int_0^t dt' g_r^2(t')/\kappa} \end{cases} \tag{1-23}$$

where $g_w(t)$ and $g_r(t)$ are the light-matter coupling for the write and read processes respectively, and the proportionality constants are such that

$$\begin{cases}\int_{-\infty}^{0} \mathrm{d}t \mid E_{\mathrm{in}}(t) \mid^2 = 1 \\ \int_{0}^{\infty} \mathrm{d}t \mid E_{\mathrm{out}}(t) \mid^2 = \eta_{\mathrm{tot}}\end{cases} \tag{1-24}$$

shows that if the light-matter couplings are simple square functions in time, then the input and output fields are growing and declining exponentials in real time, respectively. However, there is no general requirement to choose the couplings in this way. On the one hand, one can achieve optimal write efficiency for any form of g_{w}, as long as the input field satisfies the above equation; on the other hand, the form of the output field can be tailored by choosing the form of g_{r}.

- Optimal coupling

This means in particular that memory performance can be optimal even if the input and output fields are not related by time reversal in real time. For example, let us suppose that the input and output fields to have the same temporal shape, $E_{\mathrm{out}}(t) = -\sqrt{\eta_{\mathrm{w}}\eta_{\mathrm{r}}}E_{\mathrm{in}}(t-T)$, where T is the storage time, while still satisfying Eq. (1-24). By inverting Eq. (1-24) one can show that this can be achieved by choosing the following time-dependent couplings for the write and read processes:

$$\begin{aligned} g_{\mathrm{w}}(t) &= \sqrt{\frac{\kappa\eta_{\mathrm{w}} \mid E_{\mathrm{in}}(t) \mid^2}{2\left(1-\eta_{\mathrm{w}}+\eta_{\mathrm{w}}\int_{-\infty}^{t} \mathrm{d}t' \mid E_{\mathrm{in}}(t') \mid^2\right)}} \\ g_{\mathrm{r}}(t) &= \sqrt{\frac{\kappa\eta_{\mathrm{r}} \mid E_{\mathrm{in}}(t-T) \mid^2}{2\left(1-\eta_{\mathrm{r}}\int_{-\infty}^{t} \mathrm{d}t' \mid E_{\mathrm{in}}(t'-T) \mid^2\right)}} \end{aligned} \tag{1-25}$$

This choice of $g_{\mathrm{w}}(t)$ achieves the optimal write efficiency $\eta_{\mathrm{w}} = 1 - \mathrm{e}^{-2\tau_{\mathrm{w}}}$ for any input field $E_{\mathrm{in}}(t)$ and any value of $\tau_{\mathrm{w}} = \int_{-\infty}^{0} \mathrm{d}t g_{\mathrm{w}}^2(t)/\kappa$. On the other hand, the above choice of $g_{\mathrm{r}}(t)$ ensures that the output field is proportional to the input field (shifted in time by T). The read efficiency always satisfies $\eta_{\mathrm{r}} = 1 - \mathrm{e}^{-2\tau_{\mathrm{r}}}$ with $\tau_{\mathrm{r}} = \int_{0}^{\infty} \mathrm{d}t g_{\mathrm{r}}^2(t)/\kappa$. Note that arbitrary output field shapes are possible for appropriately chosen $g_{\mathrm{r}}(t)$, see Figure 1-9 also.

- Imperfections

It has neglected the spontaneous decay rate g. It is not difficult to include in the above approach, but it obviously leads to somewhat lower efficiencies, because its effect is irreversible. The optimum input field can still be found by functional differentiation. To discuss the simplest example, let us consider square coupling pulses of strength $g_{\mathrm{w}}(t)$ and duration $t_{\mathrm{w}}(t)$. Then the optimized input field for writing satisfies $E_{\mathrm{in}}(t) \propto g_{\mathrm{w}} \mathrm{e}^{\frac{g_{\mathrm{w}}^2(t)}{\kappa} + \gamma t}$ and the output field from the read process fulfills $E_{\mathrm{out}}(t) \propto g_{\mathrm{r}} \mathrm{e}^{-\frac{g_{\mathrm{r}}^2(t)}{\kappa} - \gamma t}$, while the efficiencies satisfy

$$\eta_{\mathrm{w}}(r) = \frac{\frac{g_{\mathrm{w}}^2(t)}{\kappa}}{\frac{g_{\mathrm{w}}^2(t)}{\kappa} + \gamma}\left(1 - \mathrm{e}^{-2\left(\frac{g_{\mathrm{w}}^2(t)}{\kappa} + \gamma\right) t_{\mathrm{w}}(t)}\right) \tag{1-26}$$

One can see that for large effective times the efficiencies tend towards $C/(C+1)$, where $C =$

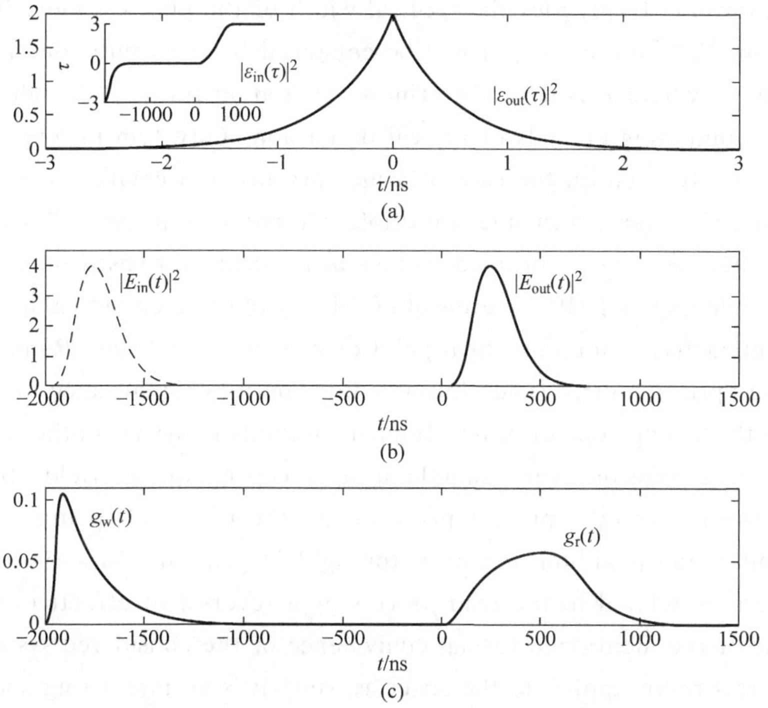

Fig. 1-9 (a) The effective input and output fields $E_{in}(t)$ and $E_{out}(t)$ of Eq. (1-13) in terms of the effective time t of Eq. (1-12). The inset shows the effective time versus real time. It can be seen that the effective time elapses only when the coupling is on. (b) An example for the possible time dependence of the real fields $E_{in}(t)$ and $E_{out}(t)$. (c) The corresponding write and read couplings $g_w(t)$ and $g_r(t)$. Any $E_{in}(t)$ can be absorbed with the optimal efficiency $\eta_w = 1 - e^{-2\tau_w}$ for $g_w(t)$ satisfying Eq. (1-25); and $E_{out}(t)$ can, for example, be chosen to be proportional to $E_{in}(t-T)$ (where T is the storage time) for $g_r(t)$ satisfying Eq. (1-13).

$g^2/k\gamma$, which is essentially the optical depth in the presence of the cavity. High efficiencies require large C. For a given decay rate, C can in principle always be increased by increasing g (which requires increasing the dipole moment or the number of atoms), or by decreasing k (which requires increasing the finesse of the cavity, i.e. the number of roundtrips). The general case also includes a time-dependent detuning $\Delta(t)$. By functional differentiation one finds that the optimum input field has a phase dependence that exactly compensates the detuning. If this is not possible, the achievable efficiencies will again be reduced. However, in analogy to the case of spontaneous decay, the effect will be small as long as the ratio $g^2/k\Delta$ is large.

In certain rare-earth ion doped crystals optical transitions can be switched on and off by changing the applied magnetic field. This is due to the coupling of the electronic Zeeman and hyperfine interactions in the presence of the crystal field. This coupling yields a substantial contribution to the overall nuclear Zeeman effect which is different for the ground and excited states, allowing one to control the branching ratios of optical transitions. For example, in Tm: YAG adding a field of order 80mT transversally to a static applied field of 1 T will turn on a previously forbidden transition to a point where its optical depth d is of order 1/cm. It is possible to control magnetic fields of this order (tens of mT) on ns timescales, making it possible to store light pulses whose duration is on these timescales. See also the appendix for more details on the proposed implementation. In practice the spectral width of the pulses is more likely to be limited by nearby transitions. The optical depth will be enhanced by the cavity, one has $C \approx dF$ for the ratio C defined above, where F is the cavity finesse. Based on Eq. (1-26) high efficiencies should thus be achievable combining crystals of typical dimensions (say 1cm in length) with moderate-finesse cavities. It has focused on the case of a memory inside a cavity. However, good memory performance based on the same principle is possible without a cavity as well, see the appendix.

In particular, that the present protocol outperforms memories based on controlled reversible inhomogeneous broadening (CRIB) in terms of efficiency for a given optical depth. The described memory could be attractive from a practical point of view as a solid-state Ramanlike memory that does not require an optical control field, thus avoiding spurious signal detections (i. e. noise) due to the presence of the strong control beam. Implementations in systems other than rare-earth ion doped crystals may be possible, for example using electric control fields for NV centers in diamond. More conceptually, the present protocol has the potential to provide insight into the basic principles underlying quantum memories for light in general. As a first example, That the optimal write process is related to the read process by a reversal of effective, but not necessarily real, time. Because of the mentioned formal equivalence of the considered system to off-resonant Raman memories, this result applies to the latter as well. It is an interesting question whether the same also holds for other memory protocols for appropriately defined effective variables. Even more generally, the present protocol seems well placed to serve as an "archetype" for quantum memories, because, as discussed above, in all memory protocols the light-matter interaction is controlled in some fashion. Mapping various protocols onto the controllable-dipole memory discussed here may be a good way of analyzing their similarities and differences.

- Photonic quantum memory in two-level ensembles

The section presents studies of a new quantum memory protocol that allows storing light in ensembles of two-level atoms by modulating the refractive index of the host medium. One can imagine an ensemble of atoms in a host, e. g. rare-earth ions doped in a crystal. Results in this paper show that linear modulation of the refractive index of the host medium in time induces a position-dependent detuning between the photon and the atoms. This can be used to store and recall photons. Under certain conditions that are explained in the manuscript, the dynamics of the proposed system is shown to be equivalent to that underlying the gradient echo memory protocol. It has to be noted that the gradient echo memory is based on a position-dependent modulation on the transition energy of the atoms. The experimental implementation is proposed based on the Tm^{3+} ions doped in lithium niobate waveguide. A proper external electric field can modulate the refractive index of the lithium niobate without affecting the transition energy of the thulium ions. Since this scheme is based on the application of a time-dependent electric field, it could introduce simplicity in practice compared to the requirement for a position-dependent magnetic or electric field in the implementation of GEM.

Quantum memory for light is an essential element for the photonic implementation of quantum communication and information processing. In recent years there has been a lot of work both on theoretical proposals and on experimental implementations. Thus far optical control, using relatively strong laser pulses, has been exploited for electromagnetically induced transparency and off-resonant Raman-type storage in ensembles of three-level systems. More recently, a direct control of the transition dipole-moment has been proposed that emulates Raman-type quantum memories in a two-level atomic configuration. In photon-echo based memories, the light-matter coupling is controlled in a more indirect way by exploiting the dephasing and rephasing of inhomogeneously broadened atomic ensembles. This includes the controlled reversible inhomogeneous broadening protocol, the atomic frequency comb protocol, and the gradient echo memory (GEM) protocol. The GEM protocol has allowed the demonstration of the highest memory efficiency (in the quantum regime) so far. Recently proposed a quantum memory protocol based on controlling the refractive index. Considering an ensemble of three-level atoms inside a host medium (e. g. rare-earth ions doped into a crystal), which is located in a circular optical cavity, the authors showed that a continuous change of the refractive index of the host medium during an off-resonant Raman interaction between a single photon, a classical control pulse and the atomic ensemble allowed mapping the state of the single photon into a collective atomic excitation. Consider storing quantum states of light in an ensemble of two-level atoms in a host medium, where the refractive index of the medium can be modulated during the interaction of the light with the atoms. There is no optical control pulse (which is related to the fact that we consider two-level instead of three-level atoms) and no cavity. Interestingly, That under certain conditions the considered system leads to dynamics that are equivalent to those of the GEM protocol; controlling the refractive index of the host medium in time can mimic the effect of the spatial frequency gradient present in GEM.

1.5 Maxwell-Bloch equations

Here we study the propagation of the light and its interaction with two-level atoms inside a host medium whose refractive index varies in time. That in a certain parameter regime the time-dependent refractive index does not play a role in the propagation equation for the light. In contrast, it plays an essential role in the dynamics of the atomic polarization. For simplicity, assume the field is propagating in a certain direction with a fixed linear polarization. (This is well justified for our choice of possible implementation in a waveguide, see below.) The wave equation for the electric field operator is analogous to the classical equation, namely

$$\frac{\partial^2 E}{\partial z^2} = \mu_0 \frac{\partial^2 D}{\partial t^2} = \mu_0 \frac{\partial^2}{\partial t^2}(\varepsilon E + P) \tag{1-27}$$

where E is the electric field, z is the direction of propagation, m_0 is the vacuum permeability, Δ is the electric displacement field, ε is the permittivity of the propagation medium and P is the polarization of the embedded two-level atoms. There are thus two fundamentally different contributions to Δ. The εE term is due to the permittivity of the host medium, whereas P describes the polarization of the two-level atoms that are the actual memory system for the light. Consider the case where E is time-dependent. The permittivity of the medium is related to its refractive index as $\varepsilon(t) = n^2(t)\varepsilon_0$. Consider a medium with a linearly changing refractive index, $n(t) = n_i + \dot{n}t$. Based on this, only the first derivative of the refractive index remains in the Eq. (1-27), giving

$$\left(\frac{\partial^2}{\partial z^2} - \frac{n^2(t)}{c^2}\frac{\partial^2}{\partial t^2}\right)E = \frac{1}{c^2}(2\dot{n}^2 E + 4n(t)\dot{n}\dot{E}) + \mu_0 \ddot{P} \tag{1-28}$$

where c is the speed of light. The slowly varying components of the signal field E and the atomic polarization P, $E = \mathscr{E}\mathrm{e}^{-\mathrm{i}(\omega_0 t - k_0(t)z)}$ and $P = \mathscr{P}\mathrm{e}^{-\mathrm{i}(\omega_0 t - k_0(t)z)}$. Here the wave vector $k_0(t) = k_i + \dot{k}t = (n_i + \dot{n}t)\frac{\omega_0}{c}$ is a function of time and $\dot{k} = \dot{n}\omega_0/c$, where ω_0 is the central frequency of the signal.

The wave equation can be greatly simplified provided that a number of (realistic) conditions are fulfilled. The second-order spatial derivative for E can be dropped provided that the field amplitude changes appreciably over the length of the medium (such that the derivative is comparable to E/L) and $k_0(t) \ll 1/L$. Similarly, the second order time-derivative can be ignored if $\omega_0 \ll 1/\tau$, where τ is the duration of the pulse. The same conditions also allow one to drop the first and second order derivatives of the slowly varying polarization operator. The regime where the extent of the pulse in space (outside the medium), $L = c\tau$, is much greater than the length of the medium, L. This is a realistic condition which is satisfied even in experiments with broadband pulses. This allows one to drop the first-order time derivative of E compared to the first-order spatial derivative. Note that this condition is not crucial; the time derivative can also be eliminated by transforming the equation to the co-propagating frame, provided that $\Delta_n \ll n_i$, where Δ_n is the total change in the refractive index. Finally, by assuming $\Delta_n \ll n_i$ and $\dot{k}L \ll 2c/nL$ one obtains the simplified propagation equation

$$\frac{\partial \mathscr{E}}{\partial z} = \frac{i\mu_0\omega_0^2}{2k_i}\mathscr{P} \tag{1-29}$$

This shows that, under the above conditions, the propagation equation remains unchanged compared to that of systems with constant refractive index ($\dot{k}$ does not play an appreciable role in the propagation). The polarization of the j-th atom is $P^j = \langle g^j | \hat{d} | e^j \rangle \sigma_{ge}^j$, where $\sigma_{ge}^j = | g^j \rangle \langle e^j |$ and $\langle g^j | \hat{d} | e^j \rangle$ is the matrix element of the corresponding dipole moment component between the ground and excited states. The collective atomic polarization at a certain position z is the sum over the individual atoms in a slice of width Δz. The slow component of this collective operator is given by

$$\mathscr{P} = \frac{1}{A\Delta z}\langle g | \hat{d} | e \rangle \sum_{j=1}^{N_z} \sigma_{ge}^j \mathrm{e}^{\mathrm{i}(\omega_0 t - k_0(t) z_j)} \equiv \langle g | \hat{d} | e \rangle \frac{N}{V}\tilde{\sigma}_{ge} \tag{1-30}$$

where assume equivalent dipole moment for all of the atoms; A and V are the cross-section area and volume of the light-atom interface; N is the number of the dopant atoms and $\tilde{\sigma}_{ge} = \frac{1}{N_z}\sum_{j=1}^{N_z}\sigma_{ge}^j \mathrm{e}^{\mathrm{i}(\omega_0 t - k_0(t)z)}$ is the average atomic polarization at position z. The Hamiltonian of the ensemble of the dopant atoms interacting with the light field can be written as

$$\begin{aligned} H &= H_0 + H_{\text{int}} \\ &= \sum_{j=1}^{N} \hbar\Omega\sigma_{ee}^j - \langle e | \hat{d} | g \rangle \sum_{j=1}^{N} \sigma_{eg}^j E(z_j, t) + h.c. \end{aligned} \tag{1-31}$$

where assume uniform excited state energy Ω for all of the atoms. Now derive the dynamics of the slowly varying collective atomic polarization using

$$\frac{\mathrm{d}\tilde{\sigma}_{ge}}{\mathrm{d}t} = -\frac{i}{\hbar}[\tilde{\sigma}_{ge}, H] + \frac{\partial \tilde{\sigma}_{ge}}{\partial t} \tag{1-32}$$

For the next step we assume that all of the atoms are initialized in the ground state and the number of atoms $N \gg 1$. For weak (quantum) signals one can then ignore the change in the excited state population.

Using Eqs. (1-29), (1-30), (1-31), (1-32), the above definition of the slowly varying field, and including the atomic excited state linewidth g, one finds the Maxwell-Bloch equations describing the interaction of the light with the collective atomic polarization in a medium with linearly time-varying refractive index,

$$\begin{cases} \dfrac{d\tilde{\sigma}_{ge}(z,t)}{dt} = -(\gamma + i(\Delta + \dot{k}z))\tilde{\sigma}_{ge}(z,t) + ig\tilde{\mathscr{E}}(z,t) \\ \dfrac{\partial \tilde{\mathscr{E}}(z,t)}{\partial z} = i\dfrac{nNg}{c}\tilde{\sigma}_{ge}(z,t) \end{cases} \tag{1-33}$$

where $\tilde{\mathscr{E}} = \sqrt{\dfrac{\hbar\omega_0}{2\varepsilon_i V}}\mathscr{E}$ and $\Delta = \Omega - \omega_0$ is the detuning. The coupling constant $g = \langle e \mid \hat{d} \mid g \rangle \sqrt{\dfrac{\omega_0}{2\hbar\varepsilon_i V}}$, where ω_0 is the central frequency of the pulse and $\varepsilon_i = n_i^2 e_0$ is the initial refractive index of the medium. Note that the time dependence of the permittivity can be ignored in the definitions of $\tilde{E}$ and g because we are interested in the regime where $\Delta_n \ll n_i$. The above set of equations shows the role of the linearly changing refractive index of the host medium in the regime that have discussed. One sees that the linear change of the refractive index in time results in a space-dependent frequency shift given by the $\dot{k}z$ term in Eq. (1-33), see also Figure 1-10(b). The above Maxwell-Bloch equations are identical to those underlying the GEM quantum memory protocol. In the next section we therefore discuss in detail how the present proposal compares to GEM.

- Comparison with Gradient Echo Memory

The dynamical Eq. (1-11) derived in the previous section (under a number of realistic conditions) are exactly equivalent to those underlying the GEM quantum memory protocol. In the GEM protocol, an initially narrow atomic absorption line is broadened by applying an external (longitudinal) field gradient. This longitudinal broadening allows one to accommodate all frequency components of the incoming pulse, see Figure 1-10(a). Once the pulse is absorbed, the produced collective atomic excitation starts to dephase due to the position-dependent detuning. By inverting the external field that is used to generate the gradient, one can rephase the collective atomic excitation, which leads to the re-emission of the light. At first sight it may seem surprising that a time variation of the refractive index leads to a spatial gradient in the detuning.

This can be understood in the following way. In the definition for the slowvarying field $\mathscr{E}$, $E = e^{-i(\mathscr{E}\omega_0 t - k_0(t)z)}$, the fast-varying phase $-i(\omega_0 t - k_0(t)z)$ has a temporal and a spatial part. One can define an effective frequency for the light by taking the time derivative of the phase, $\omega_{\text{eff}} = \omega_0 - k_z$. One can see that this corresponds to a spatial gradient in the frequency of the light, leading to a spatial gradient in the light-atom detuning, see Figure 1-10(b). A quantum memory can then be realized in analogy with GEM. The light is absorbed while the refractive index is changing linearly in time. Then the refractive index is kept constant for storage. The light can be retrieved by changing the refractive index linearly in time again, but with the opposite sign to before. As for other two-level memory protocols including GEM, the storage time can be increased if necessary

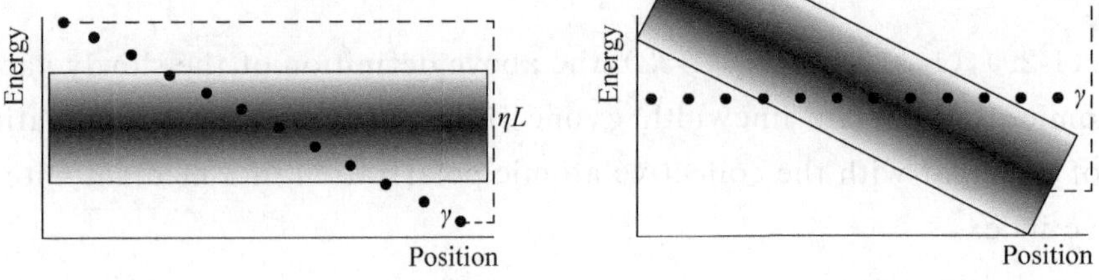

Fig. 1-10 (Color online) (a) In the GEM protocol, a longitudinal energy shift in the atoms (solid dots) allows one to cover all of the frequency components of the incoming light. (b) In the protocol proposed here, due to the linear change of the refractive index in time, the light experiences an effective position-dependent frequency shift k_z. This allows different frequency components of the light to interact on resonance with a spectrally narrow line of atoms.

by transferring the population (after the absorption has been completed) from the excited state to a long-lived third state. The total shift in the effective frequency over the length of the medium is kL. In order to accommodate all frequency components of the signal kL has to be larger than the frequency band-width of the signal, $\Delta\omega \ll 1/\tau$. Therefore, $\dot{k}L$ can be understood as the memory bandwidth. This is equivalent to the role of hL in GEM, see Figure 1-10 and Figure 1-11.

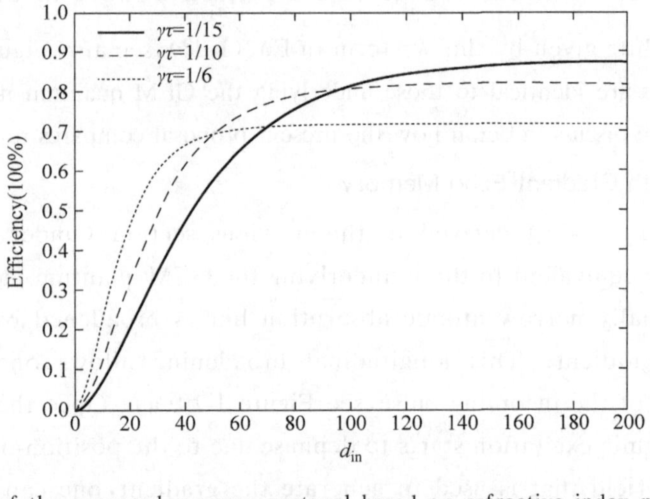

Fig. 1-11 The efficiency of the proposed memory protocol based on refractive index modulation in terms of the initial optical depth din. The efficiency is given by $e^{-2\gamma\pi}(1-\exp(-d_{in}\gamma/kL))^2$. The figure shows the efficiency for different pulse durations t, relative to the excited state line width, g. assume $\dot{k}L_\tau = 2$. Depending on the available optical depth, one can optimize the achievable efficiency by choosing an appropriate pulse duration.

The efficiency of the present quantum memory proposal can be found by analogy to GEM. Converting the equations of motion to the frequency domain, one finds that the transmitted pulse is attenuated by a factor of $\exp(-\beta\pi)$, where $\beta = nN_g^2/ck$. This implies that the optical depth of the system is $d = 2\beta\pi = 2\pi\frac{nN_g^2}{ck} = d_{in}\frac{\gamma}{kL}$. Here d is the optical depth that is associated with the effectively broadened line, with the initial optical depth d_{in} $2\pi(nN_g^2L/c\gamma$. The retrieval efficiency is then given by $(1-\exp(-d))^2 e^{-2\gamma\tau}$, see Figure 1-11. Here assumed that the decay of the excited state only has an effect during absorption and retrieval, but not during storage. As mentioned before, this can be achieved e. g. by transferring the excitation to a third, longer-lived state. Hyperfine ground states in rare-earth doped crystals can have coherence times of many seconds.

- Possible implementation

A potential experimental implementation of our proposed protocol. Using thulium ions doped into a lithium niobate waveguide. This system was used in a recent implementation of an atomic frequency comb memory. Lithium niobate is an attractive host for the present proposal because of its electro-optic properties, see below. The thulium ions interact with near-infrared light at a wavelength of 795nm. The transition is naturally inhomogeneously broadened. To prepare an initial atomic linewidth of $\gamma = 10\text{MHz}$ by optical pumping, which is very realistic. Consider the case where $kL_\tau = 2$. This assures that the memory bandwidth is large enough that it can accommodate the incoming pulse. Assuming $L = 3\text{cm}$ this leads to the requirement $\Delta n \approx 1.5 \times 10^{-5}$. Choose the pulse duration $\tau = 1/6\gamma$. The above values assure that $\omega_0 \ll 1/\tau$, $L \ll \mathcal{L} = c\tau$, $\dot{k}L \ll 2c/nL$, $\Delta n \ll n_i$ and $k_0(t) \ll 1/L$, as required for the derivation. The given length corresponds to an optical depth d_{in} 18 (at the peak) for a doping concentration of $1.35 \times 10^{20}\text{cm}^{-3}$. For the given parameter values one would achieve about 43% efficiency, see Figure 1-11, which would be largely sufficient for a proof of principle experiment. Much larger optical depths, which would allow greater efficiencies (see Figure 1-11), have already been achieved in other crystals doped with rare-earth-ions. Consider the refractive-index modulation of the ordinary optical axis of lithium niobate by a fast varying electric field. The case where the crystal is clamped (spatially confined) and the temperature is a few Kelvin. Under these conditions the refractive index of the ordinary axis $n_o \approx 2.26$. The change in the refractive index through the electrooptical effect is governed by $\Delta\left(\frac{1}{n^2}\right)_{ij} = \sum_k r_{ijk} E_k$, where $i, j = 1$ is associated with the refractive index change of the ordinary axis. This means that a time-dependent external field in a certain direction, E_k, can impose a timedependent refractive index for the ordinary axis if there exists a non-zero linear electro-optical coefficient for that direction, r_{11k}. For lithium niobate $r_{113} \approx 10 \times 10^{-12}\text{m/V}$ and $r_{112} \approx -3 \times 10^{-12}\text{m/V}$. This leads to $\Delta n \approx 1.8 \times 10^{-5}$ under $0.3 \times 10^6 \sim 1.0 \times 10^6\text{V/m}$ electric field, depending on the direction of the field in the 2~3 plane. This is equivalent to applying $3 \times 10\text{V}$ to a system that has $10\mu\text{m}$ thickness, comparable to the waveguide used. The maximum change in the refractive index in lithium niobate is expected to be 10^{-3}, which is limited by the breakdown electric field. Applying the external electric field to change the refractive index is potentially accompanied by level shifts, due to the linear Stark shift, for the atomic ground and excited states. On the other hand, for a certain type of dopant, by having the external electric field orthogonal to the difference between the permanent electric dipole moment of the ground and excited states, one can keep the resonant frequency between these states unchanged. In the proposed system the permanent dipoles are aligned with the 3-axis, therefore the electric field should be applied along the 2-axis in order to avoid level shifts.

1.6 Raman-type optical quantum memory

Future photonic quantum technology requires miniaturized quantum memories that can be integrated with other element from photon sources to the detection. Current quantum memories that are implemented in atomic gas cells have over 10cm length. Rare-earth doped crystals are at about 1cm length. The recent impedance-matched quantum memory has about 2mm length. These sizes are far beyond the required sizes for integrated quantum photonics. Nitrogen vacancy (NV) centers in diamond is an attractive candidate as it can be fabricated in sizes that are comparable with the wavelength of light. Coherence properties of these artificial atoms suggest that NVs could be the next implementations of the quantum memories. A scheme to realize the Raman-type optical quantum memory has been proposed that is based on an ensemble of nitrogen vacancy centers in a diamond that are coupled to a cavity. The scheme allows generating a collective spin excitation in the presence of the excited state inhomogeneous broadening through the off-resonant Raman coupling. As requirements for achieving high efficiency and high fidelity, the manuscript has not been published yet. The work is based on collaboration with several co-authors. I performed the theoretical analysis and used numerical methods to predict the results based on the available realistic parameters.

Quantum memories for light are known to be vital for photonic quantum information processing, specifically in long-distance quantum communications based on quantum repeaters. An efficient quantum memory can be utilized as a deterministic single photon source. The highest available efficiency is >80% that is achieved in a warm Rb vapor cell and is based on gradient echo memory protocol in a L-system. The similar protocol in a two-level atomic configuration allowed about 69% efficiency in a highly absorptive rare earth ion doped crystal. These crystals combined with an impedance-matched cavity have been shown to reach 58% efficiency with much shorter crystal length (about 2mm length). However, none of the present implementations of the quantum memory is suitable to be integrated with other elements for an on-chip quantum information processing architecture with multiple operant quantum memories. In this regard, nitrogen vacancy (NV) centers in diamond are attractive systems to be exploited as quantum memories for photons due to their spin coherence time and efficient light-NV coupling. The electronic spin coherence time of 0.5s has recently been observed in NV ensembles that opens a path toward long storage times in electronic spins of NV ensembles. NV ensembles have been used for storage and retrieval of microwave fields. In contrast to the rare earth ion doped crystals, the large intractable excited state inhomogeneous broadening of the NV ensembles in addition to their short excited state lifetimes prevents storage based on the optical coherence. Electromagnetically induced transparency has been observed in the NV ensemble, where 17% transparency has been achieved. The large inhomogeneous optical linewidth affects the efficiency of this scheme that prevents quantum storage for optical photons. However, off-resonant Raman coupling approach allows to circumvent the excited state inhomogeneous broadening and store optical photons in the collective spin coherence.

Each negatively charged NV consists of 6 contributing electrons in the C3v symmetry that is imposed by the environment (diamond). Figure 1-12 shows the ground and excited states triplets, where the ground state splitting of $m_s = 1$ is due to an external background magnetic field. There are four different orientations for the NV centers in the ensemble. Applying the external magnetic field allows to select a subgroup of NV centers that are aligned with the magnetic field.

Originally, all NV centers are initialized in the $^3A_{20}$ ground state with electronic spin of $m_s = 0$. A preparation microwave π-pulse prepares all NVs in $m_s = -1$ ground state, see below for further discussion. As shown in Figure 1-12, the cavity field and the classical control field are both interacting with NV centers. For storage, the input field is coupled to the cavity and the control field with the perpendicular polarization is simultaneously applied to the ensemble. This requires a cavity bandwidth that is larger than the input field's bandwidth. This allows store the input field in the ground state coherence of NV ensemble. For retrieval, one applies a similar control field to read the stored pulse out. It is crucial to consider selection rules in order to recognize active transitions and construct a proper Hamiltonian for studying the dynamics in this setup. Under the low magnetic field and low strain condition only spin-conserving transitions are allowed. Therefore, transitions from $^3A_{2-}$ ground state to the $E_{x,y}$ excited states are cyclic and do not allow a Λ-level configuration with $^3A_{2-}$. Here, all NVs are prepared in $^3A_{2-}$ ground state and we consider coupling of the $^3A_{2-}$ and the excited states triplets to realize the L configuration. This is similar to the L system that has been employed to generate entanglement between an optical photon and a solid-state spin. For practical reasons we choose to study this system in high strain regime (due to linearity of the polarizations and the increased excited state splitting, see below for more details). In the high strain regime the excited state triplets split into two branches, where each branch consists of 3 states. In the upper branch $A_{1,2}$ excited states are coupled to $^3A_{2-}$ ($^3A_{2+}$) ground states with $\hat{x}$ ($\hat{y}$) linear polarizations, where $\hat{x}$ and $\hat{y}$ are determined by the direction of the strain in the xy-plane of the NV. In the lower branch $E_{1,2}$ excited states show the opposite polarizations of $\hat{y}$($\hat{x}$) to the $^3A_{2-}$ ($^3A_{2+}$) ground states. In the high strain regime, in the lower branch the non-spinpreserving transition appears that led to realizing a L-level configuration that includes the $^3A_{20}$ ground state. In the model, since the population is transferred to the $^3A_{2-}$ ground state, consider the coupling between $^3A_{2-}$ ground states and the $A_{1,2}$ and $E_{1,2}$ excited states in the upper and lower branches. As the excited state inhomogeneous broadening is larger than the splitting between A_1 and A_2 (and also E_1 and E_2) treat the upper and lower branches as two inhomogeneously broadened excited state with the opposite polarization selection rules, which leaves us with a 4-level system. These

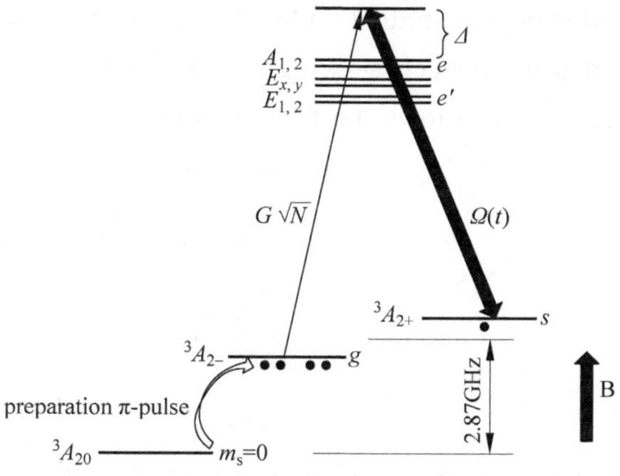

Fig. 1-12 (Color online) The figure shows the ground and excited states triplet for a negatively charged NV center in diamond. Initially, a microwave π-pulse transfers the population to the ground state g. The ground states g and s are coupled to the excited state e, where the couplings are $G\sqrt{N}$ and $\Omega(t)$. The excited state detuning, Δ, allows off-resonant Raman optical storage in the presence of the excited state inhomogeneous broadening.

conditions are studied under the δ_{E1}^{a} strain. The dynamics for this system, the input and the control field are interacting with an ensemble of four-level artificial atoms (NVs). The following Hamiltonian shows the free evolution and the interaction terms for the proposed scheme.

$$\begin{aligned}H &= H_0 + V \\ &= \hbar\sum_{j=1}^{N}\delta_g\hat{\sigma}_{ss}^{j} + (\omega_p - \Delta)\hat{\sigma}_{ee}^{j} + (\omega_p - \Delta - \delta_e)\hat{\sigma}_{e'e'}^{j} - \\ &\quad \hbar\sum_{j=1}^{N}\hat{\mathscr{E}}G\mathrm{e}^{-\mathrm{i}\omega_p(t-z_j/c)}\hat{\sigma}_{eg}^{j} + \hat{\mathscr{E}}G'\mathrm{e}^{-\mathrm{i}\omega_p(t-z_j/c)}\hat{\sigma}_{e's}^{j} + \\ &\quad \Omega(t)\mathrm{e}^{-\mathrm{i}\omega_c(t-z_j/c)}\hat{\sigma}_{es}^{j} + \Omega'(t)\mathrm{e}^{-\mathrm{i}\omega_c(t-z_j/c)}\hat{\sigma}_{e'g}^{j} + h.c.\end{aligned} \tag{1-34}$$

where e, e', s and g represent $A_{1,2}$, $E_{1,2}$, ${}^3A_{2+}$ and ${}^3A_{2-}$ states, respectively, see Eq. (1-34). The excited state inhomogeneous broadening that we consider later is larger than the energy splitting between A_1 and A_2, which justifies to treat these two levels as one that is labeled as e. The same assumption holds for E_1 and E_2. In the above Hamiltonian, ω_p and ω_c are the input and the control field frequencies; δ_g and δ_e are the ground and excited states splittings and Δ is the input and the control field detuning from the $A_{1,2}$ excited states.

$$\begin{cases} G = \langle e \mid \hat{d}.\hat{\varepsilon}_p \mid g\rangle\sqrt{\dfrac{\omega_p}{2\hbar\varepsilon_0 V}} \\ \Omega(t) = \langle e \mid \hat{d}.\hat{\varepsilon}_c \mid g\rangle E_c(t)/2\hbar \end{cases} \tag{1-35}$$

is the Rabi frequency describing coupling of the control field to the e-s transition. The G' and $\Omega'(t)$ are the similar quantities for the unwanted coupling to the e' excited state. Analyzing the collective spin

$$\hat{S} = \frac{1}{\sqrt{N}}\sum_{j=1}^{N}\hat{\sigma}_{gs}^{j}\mathrm{e}^{\mathrm{i}(\omega_p-\omega_c)(t-z_j/c)} \tag{1-36}$$

based on dynamics of the collective polarizations

$$\hat{P} = \frac{1}{\sqrt{N}}\sum_{j=1}^{N}\hat{\sigma}_{ge}^{j}\mathrm{e}^{\mathrm{i}\omega_p(t-z_j/c)} \text{ and } \hat{P}' = \frac{1}{\sqrt{N}}\sum_{j=1}^{N}\hat{\sigma}_{ge'}^{j}\mathrm{e}^{\mathrm{i}\omega_c(t-z_j/c)} \tag{1-37}$$

and coupling to the cavity field $\hat{E}$. Assuming that the number of NVs, N, is much larger than the number of input photons leads to simplification of the dynamics of the level populations. In addition, having Δ $1/\tau$, where τ is the input pulse duration, results in adiabatic elimination of the excited states. These give

$$\begin{aligned}\hat{S} &= -\left(\gamma_s + \frac{G^2\hat{\mathscr{E}}^*\hat{\mathscr{E}}}{\Gamma + i\Delta} + \frac{|\Omega(t)|^2}{\Gamma' + i(\delta_e + \delta_g + \Delta)}\right)\hat{S} + \\ &\quad iG'\hat{\mathscr{E}}^*\left(\frac{iG\hat{\mathscr{E}}\hat{S}}{\gamma' - i(\delta_e - \delta_g + \Delta)} + \frac{i\sqrt{N}\Omega'(t)}{\gamma' - i(\delta_e - \delta_g + \Delta)}\right) + i\Omega^*\left(\frac{iG\sqrt{N}\hat{\mathscr{E}}}{\gamma - i\Delta} + \frac{i\Omega(t)\hat{S}}{\gamma - i\Delta}\right)\end{aligned} \tag{1-38}$$

where γ_s, γ and γ_0 are the ground state spin and the excited states (e, e') inhomogeneous broadening. G and G' are the excited states lifetimes for e and e_0. The cavity field dynamics can be simplified to

$$\hat{\mathscr{E}}(t)=\left(\frac{G'\sqrt{N}\Omega(t)\hat{S}^*}{\Gamma'-i(\delta_g+\delta_e+\Delta)}-\frac{G\sqrt{N}\Omega(t)\hat{S}}{\gamma-i\Delta}+\sqrt{2\kappa}\,\hat{\mathscr{E}}_{in}\right)\times\frac{1}{\kappa+\dfrac{G^2N}{\gamma-i\Delta}} \tag{1-39}$$

The later by assuming that the cavity decay rate k, is faster than the cavity field dynamics that is determined by the input pulse duration, τ. Here, we also replaced $\hat{P}$ that is found based on adiabatic elimination of the excited state dynamics. $\hat{E}_{in}(t)$ represents the input quantum field. The cavity input-output equation, $\hat{E}_{out}(t)=-\hat{E}_{in}(t)+\sqrt{(2\kappa)}\,\hat{E}(t)$, in combination with Eqs. (1-38), (1-39) allows to analyze the proposed memory and study the performance in terms of fidelity and efficiency, for a detailed derivation. Use the above equations to numerically evaluate the single excitation wavefunctions for the spin and the fields (input, output and cavity fields). In a separate evaluation, we consider a three-level configuration that only includes e' as the excited state. Based on this we predict how much spin excitation (during read-in) and consequently optical noise (in read-out) is generated due to the unwanted coupling to $e'(E_{1,2})$. This is justified as the noise that is generated through the above-mentioned process is not coherently interfering with the singlephoton wave function, which is evaluated to predict the signal intensity. Figure 1-13 shows results in which one can compare the retrieved pulse with the input field. The increased excited state splitting (due to strain) the noise is suppressed. The total efficiency is found based on

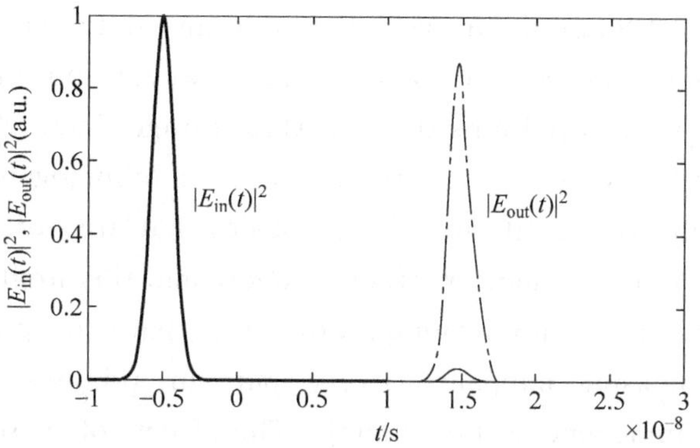

Fig. 1-13 (Color online) The figure shows the input signal pulse (blue solid line) that is being stored and after 20ns is retrieved. The pulse bandwidth D_w is about 1.9GHz, the excited state inhomogeneous broadening, γ_e is 10GHz, the spin inhomogeneous broadening, γ_s, is 200kHz. δ_e is 50GHz and δ_g is chosen to be 1GHz. The control Rabi frequency is about 3GHz, which corresponds to applying of about 0.24 mW power. The detuning from the excited state, Δ = 16GHz. These results in about 88% absorption efficiency and about 84% total efficiency. The red thin line is associated with the predicted noise that contains less than 5% energy of the input light. This results in about 94% fidelity, see below for more details.

$$\eta_{tot}=\frac{\int|\mathscr{E}_{out}(t)|^2\mathrm{d}t}{\int|\mathscr{E}_{in}(t)|^2\mathrm{d}},\quad \text{where } \mathscr{E}_{in/out}(t)=\langle 0|\hat{\mathscr{E}}_{in/out}(t)\mid 1\rangle \tag{1-40}$$

is the single excitation wave function that is defined based on its corresponding operator. One can compare the probabilities for reading out the signal and the noise to estimate the fidelity based on $1-P_{noise}/P_{sig}$. In the following we explain physical requirements for a proper NV ensemble in order to provide a satisfying performance for optical storage in the electronic spin coherence of the NV ensemble.

1.7 Precision of spin-echo-based quantum memories

Electronic or nuclear spin ground states correspondingly provide few or many seconds lifetime that could be used for storage. However, spin inhomogeneous broadening in solid-state ensembles significantly limits the available storage time. Storage of a photon in a spin ensemble generates a collective spin excitation. The spin inhomogeneous broadening results in the dephasing of this collective excitation in time scales that are orders of magnitude shorter than the individual spins coherence time. In order to overcome this limitation, the spin echo technique has to be used. In the spin echo technique, one uses pairs of p-pulses to invert the spin population before the dephasing happens. Every pair of p-pulses allows to prevent dephasing due to the spin inhomogeneous broadening. The effects of pulse imperfections are studied in detail, using both a semi-classical and a fully quantum-mechanical approach. The results show that high efficiencies and low noise-tosignal ratios can be achieved for the quantum memories in the singlephoton regime for realistic levels of the control pulse precision. Errors due to imperfect initial state preparation (optical pumping) are also studied. It can be shown that they are likely to be more influential than control pulse errors in many practical circumstances. These results are crucial for future developments of solid state quantum memories.

Quantum memories for light are key elements of quantum repeaters, which are necessary to distribute entanglement over long distances for future quantum networks. Quantum memories based on atomic ensembles are particularly attractive in practice because the light-matter coupling is enhanced by the large number of atoms and by collective interference effects. In the retrieval process collective interference can strongly enhance the re-emission of the stored light in a well-defined direction, compared to the non-directional background emission. This makes it possible to achieve high retrieval efficiencies and small noise-to-signal ratios. For long-distance applications such as quantum repeaters it is essential for the memories to allow long storage times. This can be achieved by using low-lying atomic states (spin states) for storage. However, spin states are typically affected by inhomogeneous broadening, i.e. different atoms in the ensemble have slightly different energies. For atomic gases this can be due to residual external magnetic fields or intensity-dependent light shifts (for optical dipole traps). In atomic gases it is possible to work with field-insensitive clock transitions to suppress inhomogeneous broadening due to magnetic fields. Solid-state atomic ensembles, such as rare-earth ion doped crystals, are attractive because there are no unwanted effects due to atomic motion and because solidstate systems promise enhanced scalability. However, also have inhomogeneous broadening of the spin transitions. For example, in rare-earth doped crystals the rare-earth ions themselves produce a spatially varying potential due to spin-spin interactions. Inhomogeneous broadening is important because in the absence of control techniques it limits the coherence time of collective memory excitations to the inverse of the inhomogeneous linewidth, which is typically in the tens of microseconds range and is much shorter than the desired storage times. This effect can be compensated using spin-echo techniques, such as the application of a single or a pair of p pulses. The coherence time can be further extended even beyond the singleatom T_2 time by applying chains of p pulses (bang-bang control). In practice the control pulses are never perfect. In a recent experiment the most important imperfection was shown to be an inhomogeneity in the rf intensity across the sample, leading to a variation of about 1% in the total pulse area seen by individual atoms. For successful operation in the quantum regime (i. e. when single atomic excitations are stored) the p pulses would have to be precise to of order $1/N$, where N is the number of atoms. Typically, solid-state ensembles contain of order 10^7 to 10^9 atoms, such a level of precision would thus be completely out of reach. These emissions are non-collective and hence non-directional. As a consequence, good memory operation is achievable with realistic π pulses. There are several different ensemble-based quantum memory protocols. For definiteness, in the following focus on the well-known Duan-Lukin-Cirac-Zoller (DLCZ) protocol. However, with small modifications the results apply to many other protocols, including storage based on electromagnetically induced transparency, off-resonant Raman transitions, controlled reversible inhomogeneous broadening, and atomic frequency combs. In the DLCZ quantum memory protocol, as shown in Figure 1-14(a), an off-resonant write pulse undergoes Raman scattering, leading to the creation of a single photon and a single collective excitation in the state s. This collective excitation dephases due to inhomogeneous broadening of the g-s transition, but the application of a π pulse in the middle of the storage time, see Figure 1-14(b), can prepare a rephased atomic collective excitation at the time of retrieval. A read pulse can now be applied, which leads to the directional emission of the read photon, see Figure 1-14(c). The effects of spin-echo related imperfections in the described protocol. We begin with a semi-classical treatment for uniform errors in the π pulses.

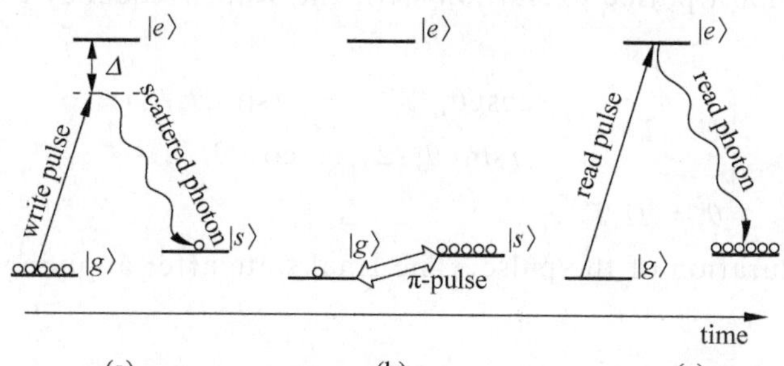

Fig. 1-14 Basic level scheme in the Duan-Lukin-Cirac-Zoller protocol. (a) The far detuned write pulse scatters a write photon and creates a single collective atomic excitation (spin-wave) in the state s. (b) Applying a p pulse on the g-s transition in the middle of the process interchanges the roles of g and s, leading to rephasing at the end of storage period. (c) Shining the read pulse transforms the single collective atomic excitation in g into a read photon.

Then give a fully quantum-mechanical treatment for the (most relevant) case of small p pulse errors, and we show that its results agree with the semi-classical approach. Consider the effects of imperfect optical pumping, i. e. an imperfect initial state. Using the semi-classical approach discuss the application of multiple p pulses (bang-bang control).

- Semi-classical approach

A semi-classical approach, that the collective atomic state is treated as a tensor product of single-atom states. The single-excitation component of this tensor product corresponds to the true quantum state, which is why the two approaches give equivalent results in the singlephoton regime. Consider an ensemble of Λ type three-level atoms with two slightly split ground states g and s, and an excited state e. At the beginning, all of the atoms are ideally pumped into the ground state g.

Applying the write pulse that scatters a (Stokes) photon, transforms the state of the kth atom at the position X_k into,

$$|\psi^{(k)}(t_0)\rangle = |g\rangle - i\xi e^{i\Delta\vec{k}_1\cdot\vec{X}_k}|s\rangle \tag{1-41}$$

where ξ represents the contribution of each atom to the collective excitation, and $\Delta\vec{k}_1 = \vec{k}_w - \vec{k}_s$, where $\vec{k}_w$ is the k-vector of the write pulse and $\vec{k}_s$ that of the scattered (Stokes) photon. For singlephoton storage using an ensemble with N atoms, $N^2 = 1$. For weak light storage this limit is equivalent to limit of large number of atoms, $N \gg 1$. Consider the effect of inhomogeneous broadening and the π pulse. In the semiclassical picture one can easily use unitary evolution of a single atom in absence of the electromagnetic field to represent the dephasing (rephasing) before (after) applying the π pulse. The propagator is given by

$$U^{\Delta_k}(t_f, t_i) = \begin{pmatrix} 1 & 0 \\ 0 & e^{-i\Delta_k(t_f - t_i)} \end{pmatrix} \tag{1-42}$$

where Δ_k is the detuning from the central transition for the kth atom. The Δ_k have an inhomogeneous distribution with a width Γ. Then, the state of an atom after the time interval $\tau_1 = t_1 - t_0$ is $|\psi^{(k)}(t_1)\rangle = U^{\Delta_k}(t_1, t_0)|\psi^{(k)}(t_0)\rangle$. Applying the rf π pulse, which is tuned to the central frequency of the g-s transition, can bring this random phase into a negligible global phase at a certain time. In order to represent the p pulse, let us recall the expression of the propagator of two levels of the atom under a pulsed excitation with the Rabi frequency Ω_i in the rotating wave approximation,

$$\begin{cases} U^{\theta}(T) = \begin{pmatrix} \cos(\theta_i/2) & -i\sin(\theta_i/2) \\ -i\sin(\theta_i/2) & \cos(\theta_i/2) \end{pmatrix} \\ \theta_i = \Omega_i T \end{cases} \tag{1-43}$$

where T is the temporal duration of the pulse. The final state after applying a π pulse at $t = t_1$ and waiting the time interval

$$|\psi^{(k)}(t_2)\rangle = U^{\Delta_k}(t_2, t_1)U^{\theta}(T)U^{\Delta_k}(t_1, t_0)|\psi^{(k)}(t_0)\rangle \tag{1-44}$$

where $\theta = \pi\varepsilon$ and $\theta = \pi$ represents a perfect p pulse and e is the error. One can now retrieve the read photon by applying the read pulse. Hence, consider the spatial phase that comes from the read pulse to find the direction of the read photon. The spatial phase dependence due to the rf pulse can be ignored. Considering this fact, the final state after applying the read pulse is given by

$$\begin{aligned} |\psi_f^{(k)}\rangle = {} & e^{i\vec{k}_r\cdot\vec{X}_k}(\cos(\theta/2) - \xi\sin(\theta/2)e^{-i\Delta_k\tau_1}e^{i\Delta\vec{k}_1\cdot\vec{X}_k})|e\rangle - \\ & ie^{-i\Delta_k\tau_2}(\sin(\theta/2) + \xi e^{-i\Delta_k\tau_1}\cos(\theta/2)e^{i\Delta\vec{k}_1\vec{X}_k})|s\rangle \end{aligned} \tag{1-45}$$

Where $\vec{r}$ is the k-vector of the read pulse.

By transferring the population of the state g into an excited state, the spin coherence transforms into an optical coherence, which leads to emission of the optical echo. The following shows how the atomic polarization would serve as the source of the echo signal,

$$\begin{cases} I_{echo} = I_0 \dfrac{|P_f|^2}{\mu^2} \\ P_f = \sum\limits_{k=1}^{N} \mu\langle e|\psi_f^{(k)}\rangle\langle\psi_f^{(k)}|s\rangle \end{cases} \tag{1-46}$$

where, μ is the electric dipole moment and I0 is the radiation intensity of one isolated atom. It can be seen from Eq. (1-45) that terms with atom-dependent temporal phases appear in the atomic polarization in the Eq. (1-46). Since the emission is governed by the average over single atomic polarizations, only those terms for which these phases are canceled contribute significantly to the echo intensity. Analyzing different terms in $\langle e | \psi_f^{(k)} \rangle \langle \psi_f^{(k)} | s \rangle$ one finds that the term $i\xi\sin^2(\theta/2)e^{i\Delta_k(\tau_2-\tau_1)}e^{i(\vec{\Delta k}_1+\vec{k}_r)\vec{x}_k}$ is the only one for which one can exclude the atom-dependent temporal phase, which is called rephasing. This can take place under the condition $\tau_1 = \tau_2$, which implies that the μ pulse has to be applied at the middle of the process. The dipole moment of each single atom serves as a source of radiation. In the far field approximation and under the rephasing condition ($\tau_1 = \tau_2$), one can show that the amplitude of the readout from the whole ensemble in the direction $\vec{k}_{ro}$ is proportional to $\sum_{j=1}^{N} e^{i(\vec{\Delta k}_1+\vec{\Delta k}_2)\vec{x}_j}\sin^2\theta/2$, where $\vec{\Delta k}_2 = \vec{k}_r - \vec{k}_{ro}$. Hence, for $\vec{\Delta k}_1 + \vec{\Delta k}_2 = 0$ the readout intensity is proportional to N^2.

This corresponds to constructive interference of the re-emission from all of the atoms at a certain direction $\vec{k}_{ro} = \vec{k}_w + \vec{k}_r - \vec{k}_s$. However, even by applying an ideal π pulse the radiation at other directions is not zero. The background re-emission (non-directional) intensity is proportional to $\left\langle \sum_{j,k=1}^{N} e^{i(\vec{\Delta k}_1+\vec{\Delta k}_2)\cdot(\vec{x}_j-\vec{x}_k)} \right\rangle = N$. Accordingly, the intensity ratio of the collectively enhanced re-emission (directional) and the randomly distributed background re-emission (non-directional) is N, see Figure 1-15. For example, for counter-propagation of the write and read pulses the phase-matching condition gives $\vec{k}_{ro} = -\vec{k}_s$. However, the error in the p pulse prevents achieving the highest possible the echo amplitude and presents a source of noise. Efficiency reduction. By considering these points one can calculate the atomic polarization, $P_f = -iN\xi\mu\cos^2(\varepsilon/2)$. Consequently, the intensity of the echo is

$$I_{\text{echo}} = I_0 N^2 \xi^2 \cos^4(\varepsilon/2) \tag{1-47}$$

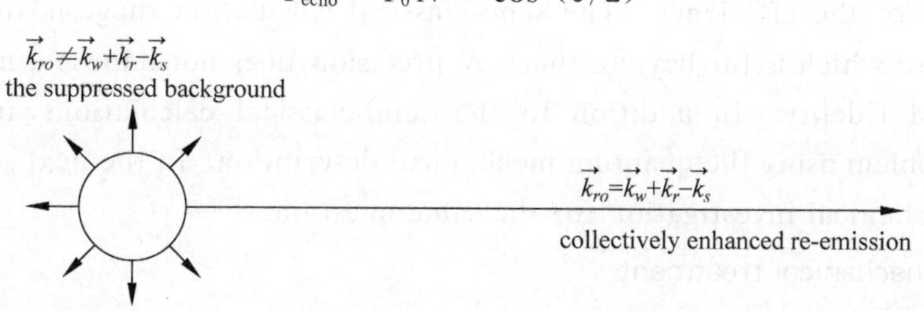

Fig. 1-15 A schematic representation of the collective enhancement at a certain direction that is given by the phase-matching condition. Non-directional re-emission is suppressed by a factor $1/N$, where N is the number of atoms.

where I_0 has the same definition as in Eq. (1-46). Eventually, the total efficiency of any process depends on the optical depth, performance of the experimental facilities and other theoretical and experimental details which are related to the protocol and the experimental setup. However, the present result allows to find the efficiency reduction due to the error in the p pulse that is important in the analysis.

- Noise-to-signal ratio

Due to error in the rephasing pulse there is a noise in the reemissinon. It is important to analyze the intensity of the noise, because for high memory fidelity the noise to echo ratio has to be small. A single π pulse interchanges the states of the atoms from s to g and vice versa. However, an error in the π pulse produces population in the ground state g even without applying write pulse at the beginning. This would lead to fluorescent radiation. Hence, it can be distinguished by the non-zero terms in $|\psi_f^{(k)}\rangle$ for $\xi=0$, which means even with no write pulse the error by itself can produce population in g. Quantitatively, the fluorescent radiation can be specified by

$$|\langle g|\psi_f^{(k)\xi=0}\rangle|^2=\cos^2(\theta/2) \tag{1-48}$$

which is the term in $|\psi_f\rangle$ that does not originate from the read-in process ($\xi=0$), but gives a non-zero projection on $|g\rangle$. Therefore, the intensity of the noise that originates from error in the p pulse can be given by

$$\begin{aligned} I_{\text{noise}} &= I_0\sum_{k=1}^{N}|\langle g|\psi_f^{(k)\xi=0}\rangle|^2 \\ &= I_0 N\sin^2(\varepsilon/2) \end{aligned} \tag{1-49}$$

Obviously, the fluorescent radiation as the source of the noise has the spatial dependence of a single photon radiation. Consequently, the semi-classical picture yields an equally distributed noise (nondirectional) that comes from single atom radiation of ensemble of the atoms, because of error in the rephasing pulse. In order to analyze the effect of an imperfect p pulse on the fidelity, it is important to study the noise-to-signal ratio. The following presents the noise-to-signal ratio,

$$r=\frac{I_{\text{noise}}}{I_{\text{echo}}}=\frac{\sin^2(\varepsilon/2)}{\cos^4(\varepsilon/2)} \tag{1-50}$$

keeping in mind that $N\xi^2=1$ for single-photon storage. The higher the noise-to-signal ratio, the less fidelity we have. The 1% variation has been realized in the intensity of the r_f pulse that causes the same error in the pulse area. Such error gives quite low noise-to-signal ratio of 0.25×10^{-4} and only wastes 0.1% of the efficiency. The semi-classical calculation suggests that the typical 1% error in the p pulse which is far beyond the $1/N$ precision, does not impose a major constraint on the efficiency and fidelity. In addition to the semi-classical calculations, it is interesting to reconsider the problem using the quantum mechanical description. In the next section, we perform fully quantum mechanical investigation for the same question.

- Quantum mechanical treatment

The large detuning of the write laser from g,e leads to scattering of a photon and creating a collective atomic excitation. The theory of light-atom interaction is well established to describe interaction between the field and the atomic dipole moment. In general, one can represent the interaction Hamiltonian as

$$H_{\text{int}}=\sum_{j=1}^{N}G\int d\vec{k}\hat{a}(\vec{k})e^{i\vec{k}\vec{X}_j}\hat{\sigma}_{\rho\nu}^{j}+h.c. \tag{1-51}$$

Where

$$\begin{cases} G=\langle\rho|\hat{\mu}_j\cdot\vec{\varepsilon}|\nu\rangle\sqrt{\dfrac{\hbar\omega}{2\varepsilon_0 V}}, \\ \hat{\sigma}_{\rho\nu}^{j}=|\rho\rangle_{jj}\langle\nu| \end{cases} \tag{1-52}$$

where r and n denote atomic levels g, e and s. This Hamiltonian shows the interaction between a field and the dipole moment of the jth atom, $\hat{\mu}_j$. For simplicity, we assume the dipole-field coupling is identical for all the atoms and have not considered the transverse profile. One can extract the interaction Hamiltonian for the write laser and the scattered photon from Eq. (1-51). By combining the interaction Hamiltonians and considering adiabatic elimination of the excited, one can describe the read-in part of the process as follows,

$$H_{\text{int}}^{\text{eff}} = \sum_{j=1}^{N} G' \int d k_s \hat{a}_s^{\dagger}(\vec{k}_s) \mathrm{e}^{\mathrm{i}(\vec{k}_w - \vec{k}_s)\vec{x}_j} \hat{\sigma}_{sg}^{j} + h.c. \tag{1-53}$$

where

$$\begin{cases} G' = \langle s \mid \hat{\mu}_j, \vec{\varepsilon}_s \mid e\rangle\langle e \mid \hat{\mu}_j, \vec{\varepsilon}_w \mid g\rangle \sqrt{\dfrac{\hbar\omega_j}{2\varepsilon_0 V}} \varepsilon_w(\tau) \\ \Omega_w(\tau) = \langle e \mid \hat{\mu}_j, \vec{\varepsilon}_w \mid g\rangle \varepsilon_w(\tau) \end{cases} \tag{1-54}$$

is the Rabi frequency of the classical write field. The unitary evolution under this effective interaction Hamiltonian describes the creation of a Stokes photon via Raman scattering, which is accompanied by the creation of a collective atomic state. The collective state is

$$|\psi(t_0)\rangle = \frac{1}{\sqrt{N}} \sum_{k=1}^{N} \mathrm{e}^{\mathrm{i}\Delta\vec{k}_1 \cdot \vec{x}_k} \mid g \ldots s^{(k)} \ldots g\rangle \tag{1-55}$$

where $|g \ldots s^{(k)} \ldots g\rangle$ shows all atoms in the ground state and the kth atom in the spin state s and $\Delta\vec{k}_1 = \vec{k}_w - \vec{k}_s$. Here, neglect the terms with more than one excitation, because interested in single photon storage. The interaction Hamiltonian governs the evolution of the atomic and the photonic state of the system. The analysis follows the evolution of the atomic state. However, later in this section, we will refer to the above discussion in order to find the intensity of the readout based on the norm of the final atomic state. One can find the state $|\psi(t_0)\rangle$ in Eq. (1-55) by applying the Schwinger bosonic creation operator or equally weighted superposition of $\sigma_+ = |s\rangle\langle g|$ operators, $J_+(\Delta\vec{k}_1) = \sum_{k=1}^{N} \mathrm{e}^{\mathrm{i}\Delta\vec{k}_1 \cdot \vec{x}_k} \sigma_+^{(k)} \otimes \mathbb{1}$. The inhomogeneous spin broadening implies that each atom has a slightly different energy than the other atom's spin level that indicates any atom will evolve based on its small energy detuning. After the time interval $\tau_1 = t_1 - t_0$ the state $|\psi(t_0)\rangle$ will be evolved to $|\psi(t_1)\rangle = \frac{1}{\sqrt{N}} \sum_{k=1}^{N} \mathrm{e}^{\mathrm{i}\Delta_k \tau_1} \mathrm{e}^{\mathrm{i}\Delta\vec{k}_1 \cdot \vec{x}_k} \mid g \ldots s^{(k)} \ldots g\rangle$. The effect of dephasing as a result of inhomogeneous spin broadening can be described as an atom-dependent phase accumulation of the single collective atomic excitation. The Following operator is appropriate to mathematical modeling of the dephasing after the time t,

$$\mathrm{e}^{\mathrm{i}\hat{\Omega}t} = \bigotimes_{k=1}^{N} \mathrm{e}^{\mathrm{i}\Delta_k t |s\rangle_k\langle s|} \tag{1-56}$$

where $\hat{\Omega} = \sum_{k=1}^{N} \Delta_k \mid s\rangle_k\langle s \mid \otimes \mathbb{1}$, the operator 1 shows the identity operator that acts on the rest of the atoms. Hence, one can represent the dephased state as $|\psi(t_1)\rangle = \frac{1}{\sqrt{N}} \mathrm{e}^{\mathrm{i}\hat{\Omega}\tau_1} J_+(\Delta\vec{k}_1)|_{gg\ldots g_i}$. At first using recall how an ideal p pulse treat the atoms, which it swaps g to s and vice versa. One can describe that as $\mathrm{e}^{\mathrm{i}\pi/2J_x} = \mathrm{e}^{\mathrm{i}\pi/2\sum_{j=1}^{N}\sigma_x^{(j)}\otimes\mathbb{1}}$, where $\mathrm{e}^{\mathrm{i}\pi/2J_x} = \bigotimes_{k=1}^{N} i\sigma_x^{(k)}$.

After applying the p pulse, the operator $e^{i\hat{\Omega}t}$ leads to a rephasing of the collective excitation because for each term in the collective state the previously non-excited atoms now acquire phases whereas the previously excited atom doesn't. Finally, the stored pulse will be retrieved by applying the read field, leading to the emission of the readout photon, see Figure 5.1(c). The readout part of the process can be represented by the operator $J_+(\vec{\Delta k}_2) = \sum_{k=1}^{N} e^{i\vec{\Delta k}_2 \vec{x}_k} \sigma_+^{(k)} \otimes \mathbb{1}$. Eventually, the whole process comprised of creation of collective excitation, dephasing, p pulse, rephasing and the read pulse that leads to the final state can be described as

$$|\psi_f(\vec{\Delta k}_1, \vec{\Delta k}_2)\rangle = \frac{1}{\sqrt{N}} J_+(\vec{\Delta k}_2) e^{i\hat{\Omega}\tau_2} e^{i\pi/2J_x} e^{i\hat{\Omega}\tau_1} J_+(\vec{\Delta k}_1) | gg\ldots g\rangle \tag{1-57}$$

Using Eq. (1-51) and the analogous Hamiltonian for the readout process it is straightforward to include the quantum states of the light field for the Stokes and readout photons into the description. One sees that the emission amplitude for the readout photon can be obtained directly from the norm of the atomic state $|\psi_f(\vec{\Delta k}_1, \vec{\Delta k}_2)\rangle$. Note that there is no preferred direction for the Stokes photon emission. As before, Not really interested in the absolute emission probability, but in how the probability varies as a function of the direction of emission for the read photon, and under the influence of errors in the control pulses. The collective enhancement happens again under the phase-matching condition $\vec{\Delta k}_1 + \vec{\Delta k}_2 = 0$. In the ideal case, considering these conditions one can easily derive that

$$\langle \psi_f(\vec{\Delta k}_1, \vec{\Delta k}_2) | \psi_f(\vec{\Delta k}_1, \vec{\Delta k}_2)\rangle = N \tag{1-58}$$

The later result is based on considering an ideal p pulse in the calculations and it takes place under the phase-matching condition, $\vec{\Delta k}_1 + \vec{\Delta k}_2 = 0$. As we studied in the semiclassical approach, the re-emission in other directions ($\vec{\Delta k}_1 + \vec{\Delta k}_2 \neq 0$) is negligible for the case with large number of atoms in the ensemble. Discussion here was focused on the case of the DLCZ protocol, but the evolution of the atomic state is identical in the other quantum memory protocols mentioned in the introduction. The only difference is that in those protocols the creation of the initial atomic excitation is accompanied by the absorption of a single photon rather than its emission.

- Global error in the rephasing pulse

Now study the effect of imperfect π pulse as one of the sources of error in the spin echo memories. The most general error in the π pulse can be considered as a small ε_k rotation around a random direction $\hat{n}_k$ for each atom that can be represented by the operator

$$e^{\sum_{k=1}^{N} i\varepsilon_k/2\vec{\sigma}^{(k)} \cdot \hat{n}^{(k)} \otimes \mathbb{1}} \tag{1-59}$$

This operator can show the effect of the inhomogeneity in the intensity of the p pulse across the sample. For simplicity, we first consider a global error that can be interpreted as lack of accuracy in pulse shaping. The most general error that affects each atom differently is studied in the appendix. The corresponding operator for such error which is equally distributed over all atoms is

$$e^{i\varepsilon/2J.\hat{n}} = e^{i\varepsilon/2\sum_{k=1}^{N}\vec{\sigma}^{(k)}.\hat{n}\otimes\mathbb{1}} \tag{1-60}$$

Essentially, the question of studying the effect of the error in the π pulse is reduced to the problem of analyzing norm of the state

$$|\psi_f(\vec{\Delta k}_1,\vec{\Delta k}_2,\varepsilon)\rangle = \frac{1}{\sqrt{N}}J_+(\vec{\Delta k}_2)e^{i\hat{\Omega}\tau_2}e^{i\pi/2J_x}e^{i\varepsilon/2J.\hat{n}}e^{i\hat{\Omega}\tau_1}J_+(\vec{\Delta k}_1)|gg...g\rangle \tag{1-61}$$

In order to facilitate analyzing the norm of the state $|\psi_f(\vec{\Delta k}_1,\vec{\Delta k}_2,\varepsilon)\rangle$ now simplify the final state. Since only interested in its norm, any unitary operator can be used to simplify the state. By applying the unitary operator $e^{-i\pi/J_x}$ from the left and adding the identity $e^{i\pi/2J_x}e^{-i\pi/2J_x}$, one can represent the final state up to a unitary and a global phase as

$$\frac{1}{\sqrt{N}}J_-(\vec{\Delta k}_2)e^{-i\hat{\Omega}\tau_2}e^{i\varepsilon/2J.\hat{n}}e^{i\hat{\Omega}\tau_1}J_+(\vec{\Delta k}_1)|gg...g\rangle \tag{1-62}$$

Where

$$J_-(\vec{\Delta k}_2) = \sum_{k=1}^{N}e^{i\vec{\Delta k}_2.\vec{X}_k}\sigma_-^{(k)}\otimes\mathbb{1} \tag{1-63}$$

Under the condition $\tau=\tau_1=\tau_2$ and by conducting some algebra one can derive that

$$e^{-i\hat{\Omega}\tau}e^{i\varepsilon/2J.\hat{n}}e^{i\hat{\Omega}\tau} = e^{\sum_{k=1}^{N}i\varepsilon_k/2\vec{\sigma}^{(k)}.\hat{n}'^{(k)}\otimes\mathbb{1}} \tag{1-64}$$

which shows rotation in a new direction

$$\hat{n}'^{(k)} = (n_x\cos\Delta_k\tau + n_y\sin\Delta_k\tau)\hat{x} + (-n_x\sin\Delta_k\tau + n_y\cos\Delta_k\tau)\hat{y} + n_z\hat{z} \tag{1-65}$$

Finally, by applying the unitary

$$e^{\sum_{k=1}^{N}-i\varepsilon_k/2\vec{\sigma}^{(k)}.\hat{n}'^{(k)}\otimes\mathbb{1}} \tag{1-66}$$

the final state can be simplified to $\hat{\mathcal{O}}J_+|gg...g\rangle$ where the operator $\hat{\mathcal{O}}$ is given by

$$\hat{\mathcal{O}} = \frac{1}{\sqrt{N}}\sum_{k=1}^{N}e^{i\vec{\Delta k}_2.\vec{X}_k}(\alpha\sigma_-^{(k)} + \beta e^{i\Delta_k\tau}\sigma_z^{(k)} + \gamma e^{2i\Delta_k\tau}\sigma_+^{(k)})\otimes\mathbb{1} \tag{1-67}$$

where

$$\begin{cases}\alpha = \cos^2(\varepsilon/2) + 2i\sin(\varepsilon/2)\cos(\varepsilon/2) - n_z^2\sin^2(\varepsilon/2)\\ \beta = -i\sin(\varepsilon/2)\cos(\varepsilon/2)(n_x - in_y) + n_z\sin^2(\varepsilon/2)(n_x - in_y)\\ \gamma = \sin^2(\varepsilon/2)(n_x - in_y)^2\end{cases} \tag{1-68}$$

This simplification allows us to distinguish three different terms in the final state based on the effect of $\hat{\mathcal{O}}J_+$ on the $|gg...g\rangle$. The first term corresponds to $\sigma_-^{(k)}\sigma_+^{(j)}|gg...g\rangle$ which gives $\delta_{jk}|gg...g\rangle$. The other combination $\sigma_z^{(k)}\sigma_+^{(j)}$ in $\hat{\mathcal{O}}J_+$ yields $(-1)^{\delta_{jk}}|g...s^{(j)}...g\rangle$. The later gives N^2 terms that contains one excitation. Finally, the $\sigma_+^{(k)}\sigma_+^{(j)}$ leads to $N(N-1)$ terms with two excitations. One can benefit from these terms to represent the final state as

$$\alpha|\psi_1\rangle + \beta|\psi_2\rangle + \gamma|\psi_3\rangle \tag{1-69}$$

where

$$|\psi_1\rangle = \frac{1}{\sqrt{N}}\sum_{j=1}^{N}e^{i(\vec{\Delta k}_1+\vec{\Delta k}_2).\vec{X}_j}|gg...g\rangle,\quad |\psi_2\rangle = \frac{1}{\sqrt{N}}\sum_{j,k=1}^{N}(-1)^{\delta_{jk}}e^{i\Delta_k\tau}e^{i\vec{\Delta k}_1.\vec{X}_j}e^{i\vec{\Delta k}_2.\vec{X}_k}|g...s^{(j)}...g\rangle$$

$$|\psi_3\rangle = \frac{1}{\sqrt{N}}\sum_{j,k=1,j\neq k}^{N}e^{2i\Delta_k\tau}e^{i\vec{\Delta k}_1.\vec{X}_j}e^{i\vec{\Delta k}_2.\vec{X}_k}|g...s^{(j)}...s^{(k)}...g\rangle \tag{1-70}$$

Obviously, because of different number of excitations, these terms are perpendicular. Thus, one can study the norm of the final state, easily.

As it can be seen from the Eq. (1-69), because of δ_{jk}, the first term corresponds to the directional emission of the readout photon. As discussed in the previous section, the readout is strongly peaked around the direction for which all of the single-atom re-emissions can constructively interfere with each other. The readout intensity at the other directions (the nondirectional background) is suppressed by the ratio $1/N$ that is a quite small number for a typical $N(10^7 \sim 10^9)$. Hence, the a term determines the intensity of the echo. In other words, shining the read laser with $\vec{k}_r = -\vec{k}_w$ result in the emission of a readout photon that its intensity is peaked around $\vec{k}_{ro} = -\vec{k}_s$. The non-directional noise in the re-emission can be attributed to the terms that correspond to β and γ coefficients, because taking average over position of the atoms leads to randomly distributed re-emission from each single atom. It can be shown that the norm of the state is

$$\frac{|\alpha|^2}{N}\left|\sum_{j=1}^{N} e^{i(\Delta\vec{k}_1+\Delta\vec{k}_2)\cdot\vec{X}_j}\right|^2 + \frac{|\beta|^2}{N}\sum_{j=1}^{N}\left|\sum_{k=1}^{N}(-1)^{\delta_{jk}} e^{i\Delta_k\tau} e^{i(\Delta\vec{k}_1)\cdot\vec{X}_j} e^{i(\Delta\vec{k}_2)\cdot\vec{X}_k}\right|^2 + \frac{1}{N}\sum_{j,k=1,j\neq k}^{N}|\gamma|^2 \tag{1-71}$$

Efficiency reduction. In order to study the worst case scenario, upper bound of the noise strength and lower bound of the echo intensity have to be studied separately. Considering the phase-matching condition it can be shown that $I_{\text{echo}}\ N|\alpha|^2$. For small errors, $\varepsilon \ll 1$ and keeping terms up to $\mathcal{O}(\varepsilon^3)$, it can be shown that the following lower bound can be achieved for $n_z = 0$,

$$|\alpha|^2 \leqslant (1-2(\varepsilon/2)^2) \tag{1-72}$$

As in the ideal case the signal intensity is proportional to 1, one can analyze the efficiency reduction. So from quantum mechanical point of view the efficiency reduction factor is given by $1-2(\varepsilon/2)^2$ that gives the same results for the typical 1% error as the semi-classical calculation. Noise. With the aim of studying the intensity of the noise that is proportional to the

$$\frac{1}{N}|\beta|^2\sum_{j=1}^{N}\left|\sum_{k=1}^{N}(-1)^{\delta_{jk}} e^{i\Delta_k\tau} e^{i\Delta\vec{k}_1\cdot\vec{X}_j} e^{i\Delta\vec{k}_2\cdot\vec{X}_k}\right|^2 + \frac{1}{N}\sum_{j,k=1,j\neq k}^{N}|\gamma|^2 \tag{1-73}$$

in the limit of the small errors, $\varepsilon \ll 1$, to discuss the effect of $e^{i\Delta_k\tau}$. For long enough times that τ is comparable with $1/\Gamma$, where Γ is the inhomogeneous linewidth, the $e^{i\Delta_k\tau}$ becomes a completely random phase. It can be shown that for random phases Δ_k,

$$\left|\sum_k e^{i\phi_k} c_k\right|^2 = \left|\sum_{k,l} c_k c_l^* e^{i(\phi_k-\phi_l)}\right| = \sum_k |c_k|^2 \tag{1-74}$$

because $\langle e^{i(\phi_k-\phi_l)}\rangle 1 = \delta_{kl}$. Hence can treat the term that corresponds to β as

$$\frac{1}{N}|\beta|^2\sum_{j,k=1}^{N}|(-1)^{\delta_{jk}} e^{i\Delta\vec{k}_1\cdot\vec{X}_j} e^{i\Delta\vec{k}_2\cdot\vec{X}_k}|^2 = N|\beta|^2 \tag{1-75}$$

Taking these considerations into account shows that the intensity of the noise is proportional to

$$\frac{1}{N}|\beta|^2\sum_{j=1}^{N}\left|\sum_{k=1}^{N}(-1)^{\delta_{jk}} e^{i\Delta_k\tau} e^{i(\Delta\vec{k}_1)\cdot\vec{X}_j} e^{i(\Delta\vec{k}_2)\cdot\vec{X}_k}\right|^2 + \frac{N(N-1)}{N}|\gamma|^2$$

$$\leqslant N(\varepsilon/2)^2\max(n_z^2(n_x^2+n_y^2)(\varepsilon/2)^2 + n_x^2 + n_y^2) + \frac{N^2-N}{N}(\varepsilon/2)^4\max|(n_x - in_y)^2|^2$$

$$\approx N(\varepsilon/2)^2 + \mathcal{O}(\varepsilon^4) \tag{1-76}$$

So the choice of uniformly directed error with $n_z = 0$ gives the upper bound for the noise. Obviously, the second term in the noise intensity is proportional to ε^4 that is negligible for the small errors. These results find the upper bound for the noise-to-signal ratio. Recall the results for the noise and the echo from the semi-classical approach in Eqs. (1-48), (1-49). Obviously, the Taylor expansion of the results obtained in semi-classical treatment and eliminating the terms $\mathcal{O}(\varepsilon^4)$ and higher, demonstrates the agreement between the results of the both approaches in the limit of small errors ($\varepsilon \ll 1$). In the quantum mechanical approach the noise is proportional to the term given in Eq. (1-76). The amplitude of the noise shows a correspondence with the fluorescent radiation in the semiclassical approach. Indeed the spatial phase dependence implies that the direction of the emission of the noise from each atom varies from one to another, leading to a non-directional noise. Thanks to collective enhancement, the non-directional noise will not swamp the echo signal, for realistic control pulse accuracy.

- Imperfect initial state and rephasing pulse

An imperfection in the p pulse is not the only source of inefficiency in the spin echo memories. In our calculations so far we have assumed an ideal situation where all the atoms are initially in the ground state. The quantum mechanical approach allows us to study the effect of an imperfect initial state with n atoms excited to the state s. This can happen in experiments as a result of an imperfect optical pumping in the initialization of the atomic ensemble. Without loss of generality, consider the initial state $|g\ldots gs\ldots s\rangle$ that has n sorted excited atoms instead of randomly positioned excited atoms. Applying the operator $\mathcal{O}J_+$ gives the final state. The directional echo as result of applying $\sum_{j,k=1}^{N} \sigma_-^{(k)}\sigma_+^{(j)}$ on the initial state $|g\ldots gs\ldots s\rangle$. In contrast to the perfect initial case, this will lead to $N-n$ terms with n excitations correspond to $|g\!::\!gs\!::\!s\rangle$, and also $(N-n)n$ terms connected with $|g\ldots s^{(j)}\ldots gs\ldots g^{(k)}\ldots s\rangle$ which has n excitations. By conducting some algebra, one can easily show that only the first case gives δ_{jk} which leads to directional (collectively enhanced) re-emission under the phase-matching condition. This implies that by considering imperfection in both the initial state and the π pulse, intensity of the echo is proportional to

$$I_{\text{echo}} \propto \frac{(N-n)^2}{N}(1-2(\varepsilon/2)^2) \tag{1-77}$$

The imperfection in the initial state reduces the intensity of the echo and also introduces a new source of the non-directional noise. Previously, analyzed the noise that corresponds to $|\psi_2\rangle$ and $|\psi_3\rangle$ by finding the upper bound for b and g. Now, the $(N-n)n$ terms of the form $|g\ldots s^{(j)}\ldots gs\ldots g^{(k)}\ldots s\rangle$ also contribute to the noise. Consequently, as the upper bound, for the term that corresponds to α can be achieved for $n_z = 1$, one can obtain the following upper bound for the noise-to-signal ratio r,

$$r \leqslant \frac{(N-n)n + N^2(\varepsilon/2)^2}{(N-n)^2(1-2(\varepsilon/2)^2)} \tag{1-78}$$

Investigating this equation it can be seen that the two different sources of error compete in producing noise. For the case with small fraction of excitations in the initial state, $n/N \ll 1$, such that $\varepsilon^2 > 4n/N$ then the error in π pulse is the dominant term in the noise. Otherwise, in case of

$4n/N > \varepsilon^2$ the imperfection in the initial state plays an important role in increasing the upper bound of the noise-to-signal ratio. For instance, 2% of the atoms not in the ground state, which is a typical value, will cause a 2% error. The error in the p pulse would have to be as large as $\varepsilon = 0.4$ in order to be comparable in importance.

- Semi-classical treatment of multiple rephasing pulses

So far have studied the application of a single p pulse. Much longer storage times are achievable by applying sequences of p pulses. This is also known as bang-bang control. To analyze this case using the semi-classical approach before, semi-classical and quantum treatment lead to identical conclusions for small errors. The intensity of the signal after applying m pairs of π pulses is given by

$$I_{\text{echo}} \approx I_0 N^2 \xi^2 (1 - (2m^2 - m + 1)\varepsilon^2/2) \tag{1-79}$$

The intensity that is associated to the noise also can be approximated by $I_{\text{noise}} \approx I_0 N m^2 \varepsilon^2/2$. Thus, after applying sequence of the rephasing pulses, the noise-to-signal ratio reads

$$r \approx m^2 \varepsilon^2 / 2 \tag{1-80}$$

For instance, with typical 1% error in the π pulse one can benefit from 30 pairs of rephasing pulses, while still achieving an efficiency factor of 91% and a noise-to-signal ratio of 0.04. If the pulses are 4msec apart, pairs of pulses correspond to a storage time of 240msec. managed to increase the T_2 time to 1sec, by proper alignment of the magnetic field, that is much longer than 80msec. This field configuration would allow one to put the rephasing pulses further apart. For example, one can consider the rephasing pulses 40msec apart, giving a 2.4 sec storage time with an efficiency factor of 91% and a noise-to-signal ratio of 0.04.

- Spatial inhomogeneity in the r_f pulse

The main error in r_f pulses is the variance in the intensity of the r_f pulse that leads to inhomogeneity of the r_f pulse across the sample. Consequently, the error would be different for the atoms at the different positions. Here extend analysis to study such errors to study the spatial inhomogeneity in the r_f pulse by considering $\mathrm{e}^{\sum_{k=1}^{N} \mathrm{i}\varepsilon_k/2\vec{\sigma}^{(k)}\cdot\hat{n}^{(k)}\otimes\mathbb{1}}$ as the operator which represents the error. Therefore, the final state is

$$|\psi_f(\vec{\Delta k}_1, \vec{\Delta k}_2, \varepsilon)\rangle = \frac{1}{\sqrt{N}} J_+(\vec{\Delta k}_2) \mathrm{e}^{\mathrm{i}\hat{\Omega}\tau_2} \mathrm{e}^{\mathrm{i}\pi/2J_x} \mathrm{e}^{\sum_{k=1}^{N} \mathrm{i}\varepsilon_k/2\vec{\sigma}^{(k)}\cdot\hat{n}^{(k)}\otimes\mathbb{1}} \mathrm{e}^{\mathrm{i}\hat{\Omega}\tau_1} J_+(\vec{\Delta k}_1) |gg\ldots g\rangle \tag{1-81}$$

Follow the same approach as we used in the paper to simplify the final state that eases the rest of the calculation. Therefore, the final state $|\psi_f(\vec{\Delta k}_1; \vec{\Delta k}_2; \varepsilon)\rangle$ up to a unitary operator and a global phase is

$$\frac{1}{\sqrt{N}} J_-(\vec{\Delta k}_2) \mathrm{e}^{-\mathrm{i}\hat{\Omega}\tau_2} \mathrm{e}^{\sum_{k=1}^{N} i\xi_k \vec{\sigma}^{(k)}\cdot\hat{n}^{(k)}\otimes\mathbb{1}} \mathrm{e}^{\mathrm{i}\hat{\Omega}\tau_1} J_+(\vec{\Delta k}_1) |gg\ldots g\rangle \tag{1-82}$$

Considering the p pulse at the middle of the process, $t = t_1 = t_2$, and by conducting some algebra one can derive that

$$\mathrm{e}^{-\mathrm{i}\hat{\Omega}\tau} \mathrm{e}^{\sum_{k=1}^{N} i\xi_k \vec{\sigma}^{(k)}\cdot\hat{n}^{(k)}\otimes\mathbb{1}} \mathrm{e}^{\mathrm{i}\hat{\Omega}\tau} = \mathrm{e}^{\sum_{k=1}^{N} i\xi_k \vec{\sigma}^{(k)}\cdot\hat{n}'^{(k)}\otimes\mathbb{1}} \tag{1-83}$$

which shows rotation in a new direction

$$\hat{n}'^{(k)} = (n_x^{(k)}\cos\Delta_k\tau + n_y^{(k)}\sin\Delta_k\tau)\hat{x} + (-n_x^{(k)}\sin\Delta_k\tau + n_y^{(k)}\cos\Delta_k\tau)\hat{y} + n_z^{(k)}\hat{z} \tag{1-84}$$

Finally, by applying the unitary $e^{\sum_{k=1}^{N} i\xi_k \vec{\sigma}^{(k)}\cdot\hat{n}'^{(k)}\otimes\mathbb{1}}$ the final state can be simplified to

$$|\psi_f(\vec{\Delta k}_1, \vec{\Delta k}_2, \varepsilon)\rangle = \hat{O}J_+ |gg\ldots g\rangle \tag{1-85}$$

where the operator $\hat{O}$, which acts on k^{th} atom is given by

$$\hat{O} = \frac{1}{\sqrt{N}}\sum_{k=1}^{N} e^{i\vec{\Delta k}_2\cdot\vec{X}_k}(\alpha^{(k)}\sigma_-^{(k)} + \beta^{(k)}e^{i\Delta_k\tau}\sigma_z^{(k)} + \gamma^{(k)}e^{2i\Delta_k\tau}\sigma_+^{(k)})\otimes\mathbb{1} \tag{1-86}$$

and

$$\begin{cases}\alpha^{(k)} = \cos^2(\varepsilon_k/2) + 2i\sin(\varepsilon_k/2)\cos(\varepsilon_k/2) - n_z^{(k)^2}\sin^2(\varepsilon_k/2), \beta^{(k)} \\ \quad = -i\sin(\varepsilon_k/2)\cos(\varepsilon_k/2)(n_x^{(k)} - in_y^{(k)}) + n_z^{(k)}\sin^2(\varepsilon_k/2)(n_x^{(k)} - in_y^{(k)}) \\ \gamma^{(k)} = \sin^2(\varepsilon_k/2)(n_x^{(k)} - in_y^{(k)})^2\end{cases} \tag{1-87}$$

Finally, the state can be simplified to

$$\begin{aligned}|\psi_f(\vec{\Delta k}_1, \vec{\Delta k}_2, \varepsilon)\rangle = &\frac{1}{\sqrt{N}}\sum_{j=1}^{N}\alpha^{(j)}e^{i(\vec{\Delta k}_1+\vec{\Delta k}_2)\cdot\vec{X}_j}|g\ldots g\rangle + \\ &\frac{1}{\sqrt{N}}\sum_{j,k=1}^{N}(-1)^{\delta_{jk}}\beta^{(k)}e^{i\Delta_k\tau}e^{i(\vec{\Delta k}_1)\cdot\vec{X}_j}e^{i(\vec{\Delta k}_2)\cdot\vec{X}_k}|g\ldots s^{(j)}\ldots g\rangle + \\ &\frac{1}{\sqrt{N}}\left(\sum_{j,k=1,j\neq k}^{N}\gamma^{(k)}e^{2i\Delta_k\tau}e^{i(\vec{\Delta k}_1)\cdot\vec{X}_j}e^{i(\vec{\Delta k}_2)\cdot\vec{X}_k}\right)|g\ldots s^{(j)}\ldots s^{(k)}\ldots g\rangle\end{aligned} \tag{1-88}$$

Fortunately, having different number of excitations in the final state simplifies the calculation of the norm. As the first term contributes are in directional emission of the readout photon. Hence, the $\alpha^{(k)}$ s play role in finding the intensity of the echo. In other words, shining the read laser with $\vec{k}_r = -\vec{k}_w$ result in the emission of a readout photon that is peaked around $\vec{k}_{ro} = -\vec{k}_s$. The non-directional noise in the re-emission can be attributed to the terms corresponding to $\beta^{(k)}$ and $\gamma^{(k)}$ coefficients, because average over the position of the atoms leads to randomly distributed re-emission. It can be shown that the following gives norm of the $|\psi_f\rangle$ in Eq. (1-88).

$$A^2 = \frac{1}{N}\left|\sum_{j=1}^{N}\alpha^{(j)}e^{i(\vec{\Delta k}_1+\vec{\Delta k}_2)\cdot\vec{X}_j}\right|^2 + \frac{1}{N}\sum_{j=1}^{N}\left|\sum_{k=1}^{N}(-1)^{\delta_{jk}}e^{i\Delta_k\tau}\beta^{(k)}e^{i(\vec{\Delta k}_1)\cdot\vec{X}_j}e^{i(\vec{\Delta k}_2)\cdot\vec{X}_k}\right|^2 + \frac{1}{N}\sum_{j,k=1,j\neq k}^{N}|\gamma^{(k)}|^2 \tag{1-89}$$

In order to analyze the efficiency reduction the lower bound of echo intensity have be studied. Considering the phase-matching condition it can be shown that

$$I_{\text{echo}} \propto \frac{1}{N}\left|\sum_{j=1}^{N}\alpha^{(j)}\right|^2 \tag{1-90}$$

For small errors, $\varepsilon_j \ll 1$ and keeping terms to $\mathcal{O}(\varepsilon^3)$, it can be shown that the following lower bound can be achieved for $n_z^{(j)} = 0$

$$\frac{1}{N}\left|\sum_{j=1}^{N}\alpha^{(j)}\right|^2 \leqslant N(1 - 2(\varepsilon_{\max}2/)^2) \tag{1-91}$$

where $\varepsilon_{\max}$ is the largest error and N is number of the atoms. With the aim of studying the intensity of the noise that is proportional to the

$$\frac{1}{N}\sum_{j=1}^{N}\left|\sum_{k=1}^{N}(-1)^{\delta_{jk}}e^{i\Delta_k\tau}\beta^{(k)}e^{i(\vec{\Delta k}_1)\cdot\vec{X}_j}e^{i(\vec{\Delta k}_2)\cdot\vec{X}_k}\right|^2 + \frac{1}{N}\sum_{j,k=1,j\neq k}^{N}|\gamma^{(k)}|^2 \tag{1-92}$$

For small errors $\varepsilon_j \ll 1$, one needs to consider the $e^{i\Delta_k \tau}$ as a random phase, that takes place for long enough times that τ is comparable with the $1/\Gamma$. It can be shown that for random

$$\left| \sum_k e^{i\phi_k} \beta^{(k)} \right|^2 = \left| \sum_{k,l} \beta^{(k)} \beta^{(l)*} e^{i(\phi_k - \phi_l)} \right|^2 = \sum_k \mid \beta^{(k)} \mid^2 \tag{1-93}$$

because

$$\langle e^{i(\phi_k - \phi_l)} \rangle = \delta_{kl} \tag{1-94}$$

Hence, one can conclude that

$$\left| \sum_{k=1}^{N} (-1)^{\delta_{jk}} e^{i\Delta_k \tau} \beta^{(k)} e^{i(\vec{\Delta k}_1) \cdot \vec{X}_j} e^{i(\vec{\Delta k}_2) \cdot \vec{X}_k} \right|^2 \tag{1-95}$$

can be approximated as

$$\sum_{k=1}^{N} \mid (-1)^{\delta_{jk}} \beta^{(k)} e^{i(\vec{\Delta k}_1) \cdot \vec{X}_j} e^{i(\vec{\Delta k}_2) \cdot \vec{X}_k} \mid^2 = \sum_{k=1}^{N} \mid (-1)^{\delta_{jk}} \beta^{(k)} \mid^2 \tag{1-96}$$

By taking these considerations into account and keeping terms up to $\mathcal{O}(\varepsilon^4)$ give the noise intensity upper bound as

$$\begin{aligned} I_{\text{noise}} &\leqslant N \max (n_z^{(k)^2} (n_x^{(k)^2} + n_y^{(k)^2}) (\varepsilon_k / 2)^4 + (n_x^{(k)^2} + n_y^{(k)^2}) (\varepsilon_k / 2)^2) + \\ &\quad \frac{N^2 - N}{N} \max ((\varepsilon_k / 2)^4 \mid (n_x^{(k)} - i n_y^{(k)})^2 \mid^2) \\ &= N (\varepsilon_{\max} / 2)^2 + \mathcal{O}(\varepsilon^4) \end{aligned} \tag{1-97}$$

which implies that a uniformly directed error with $n_z = 0$ gives the upper bound for the noise. Obviously, the second term in the noise intensity is proportional to ε^4 that is negligible for the small errors. Then it leads to $I_{\text{noise}} \approx N(\varepsilon_{\max}/2)^2$. This shows that the fully quantum mechanical treatment for an inhomogeneous error is in good agreement with the results for global errors derived. The emergence of companies that are focused on quantum technologies are an indication of advances of practical applications. The fast progress within the past decade in quantum memories promises significant developments in the near future to realize the first real-world implementations of long-distance quantum communication based on quantum repeaters. For this purpose, developments of quantum memory protocols and physical systems are essential to approach combination of higher efficiencies, multimode capacity and longer storage times. The present thesis contributed to developments of quantum memories by proposing new quantum memory protocols. Furthermore, nitrogen vacancy (NV) centers are studied for a new implementation that can lead to the first micron-scale quantum memory. The precision requirements for the spin echo technique is also studied, due to the importance of long storage time in quantum repeaters. The controllable-dipole quantum memory showed that a scheme equivalent to the Raman-type quantum memory can be implemented without application of an optical control field. The scheme proposed a way to store quantum states of light in a two-level system by direct control of the transition dipole moment and without an engineered inhomogeneous broadening. The proposal has been analyzed for the in-cavity and free-space cases. For the experimental implementation, Tm^{3+} : YAG has been proposed, in which the transition dipole moment can be turned on and off by changes to the angle of the external magnetic field. The quantum memory protocol based on the refractive index modulation has also been proposed in the present thesis. Interestingly, this scheme resembles the gradient echo memory (GEM). As opposed to the GEM protocol, there is no modulation on the energy of the relevant atomic transition. It has been shown that a time-dependent modulation of the refractive index of the host medium implies a position-dependent frequency of the photon along the propagation direction. Lithium niobate is an attractive candidate for refractive index modulation as it is one of the main materials that are used the electro-optical modulators. Tm^{3+} doped lithium niobate waveguide already has been used to implement the AFC protocol. Implementing electrodes along the waveguide allows to implement this proposal. This can be an implementation of a GEM-like protocol without difficulties due to application of a position dependent field. An ensemble of nitrogen vacancy (NV) centers that are coupled to a cavity is examined as a potential physical system for implementation of quantum memories. As the NV ensembles suffer from a large excited state inhomogeneous broadening, the Raman-type scheme can be used to circumvent the broadening for storage of optical photons. This proposal allows to implement micron-scale quantum memories that can be incorporated with sources and detectors for the implementation of an on-chip quantum register. Precision requirements for the spin echo technique, which is essential for extending the storage time in many solid-state quantum memories are studied. It has been shown how the imperfection in the initially prepared atomic state and the error in the applied π-pulses for the spin echo technique can contribute to the noise. Based on the results, for a limited error in the π-pulses, multiple p-pulses can be used to extend the storage time by about 2 orders of magnitude. Limited but effective steps have been taken toward more practical solid-state quantum memories for future applications. However, different aspects of quantum memories are expected to develop in the future. See the next chapter for a brief description of some of the foreseeable directions for future studies.

1.8 Integrated photonics for memory

Integrated photonics is one of key technology for optical memory in the future also that can be defined as the science and technology of harnessing photons as they propagate and interact with micro-or nano-scale structures that are fabricated on a chip. The field is attracting significant R&D interest globally and making an impact on many high technological areas, including green information and communications technology, high-performance computing systems and on-chip biosensing. The photonic-integrated circuits (PICs) is similar to standard integrated circuits (ICs)—with multiple functionalities, high performance and potentially low cost. As a result, many exciting results and proposals in this field are emerging now with lasers and fiber optics. Silicon photonic devices was funded 973 (The 973 Program is Major State Basic Research Development Program in China) project 2011—2015, titled, "Key issues on photonic integrated chips for high-speed and low-power signal processing." The participating institutions include Tsinghua University, the Huazhong University of Science and Technology (HUST), the University of Electronic Science and Technology of China (UESTC), the Beijing University of Post and Telecommunications (BUPT), the Institute of Semiconductors and the Institute of Microelectronics of the Chinese Academy of Sciences (CAS). The project issues involved in peta-bits-per-second communication node processing in backbone networks—a capability that is expected to be implemented in China around 2020. The team proposes five optoelectronic chips on various semiconductor platforms (silicon, indium phosphide) and hybrid material platforms based on their functions in optical networking signal processing—namely (1) integrated chips for optical switching and buffering,

(2) integrated chips for optical clock recovery, (3) 2R/3R regeneration ICs, (4) multi-channel all-optical format conversion chips and (5) optical tunable wavelength conversion chips. Zhejiang University leads another 973 project (2007—2011) titled, "Fundamentals of silicon-based light emission and optical interconnects." Participating institutions include Peking University, Nanjing University and the Institute of Semiconductors of the CAS. The team has proposed and demonstrated various approaches for obtaining light emission from silicon-based materials, including silicon/germanium nanocrystals, rare-earth doping and hybrid technologies. The team developed several devices based on silicon-rich silicon nitride (SiN), nano-silicon multi-quantum-well (MQW) thin films, doped monodispersed nanocrystal silicon and rare-earth-doped silicon-rich silicon nitride/oxide that demonstrated tunable photoluminescence in the visible range based on silicon-rich SiNx films, and shown electroluminescence devices based on such SiNx films. They developed two novel laser structures for silicon photonics, namely a selective-area metal bonding InGaAsP-Si laser (created by Guogang Qin's group of Peking University) and an electrically pumped random zinc oxide laser grown on silicon in a metal-insulator-semiconductor structure as shown in Figure 1-16.

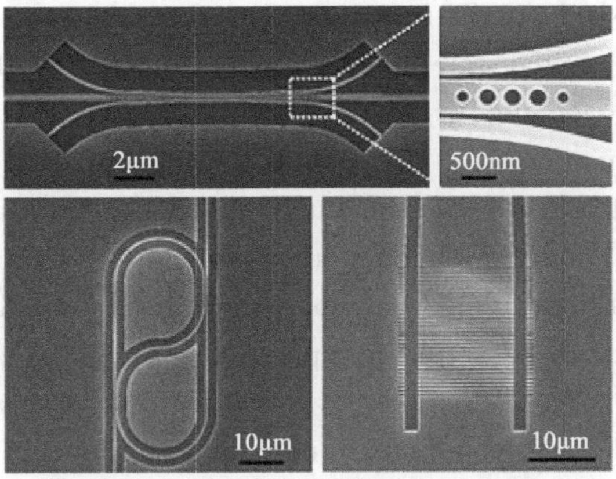

Fig. 1-16 Structure of the metal-insulator-semiconductor photonic devices.

Specifically, for the InGaAsP-Si laser, the InGaAsP MQW distributed feedback (DFB) structure was integrated into a silicon waveguide with the optical coupling and the metal bonding transversely separated. The laser operated with a threshold current density of 1.7kA/cm^2 and a slope efficiency of 0.05W/A under pulsed wave operation. The researchers realized room-temperature continuous lasing with a maximum output power of 0.45mW. Peking University's 973 project (2007—2011): "Mesoscopic optics and novel micro/nano photonic devices." studied a metal-coated microtoroid resonate system in detail that the hybrid system supports a high-Q exterior plasmonic whispering-gallery mode (WGM), a high field locality concentrated in the exterior metal-dielectric interface. These exterior modes are long-lived and enable interaction with external media such as microfluidics that has great potential applications, including highly sensitive biosensors, photonic crystal all-optical switching, ultrafast (picoseconds) and low-threshold-power (100kW/cm^2 order) all-optical switching (80 percent efficiency) in two-dimensional composite photonic crystals. National 863 high-tech projects: "Photonics and monolithic integration research (Pamir)", "Photonic integration technology and its system applications."

The main topics it will address are: Ultra-small and ultra-high-capacity Si-based PIC design and fabrication, Si-based monolithic photonic integrated 100Gb/s coherent receivers. InP monolithic integrated 10 × 10Gb/s transmitters for long-distance transmission, InP monolithic integrated 16×2.5Gb/s transmitters for wavelength-division multiplexing passive optical networks applications, InP monolithic integrated 8×6GHz transmitters for radio-over-fiber applications, and System demonstration and applications. The National Natural Science Foundation of China recently funded a major project: "Basic research on high-speed semiconductor integrated optoelectronic devices." The main topics are:

- Channel InP transmitter PICs capable of a 100Gb/s rate

100Gb/s all-optical wavelength conversion with a semi-conductor optical amplifier integrated with a distributed Bragg reflector laser. Research highlights from Tsinghua University, CAS, Peking University, and Zhejiang University on work related to advanced laser structures, monolithic photonic integration technologies, photodetectors, silicon photonic devices and microsystems. Yi Luo's group at Tsinghua University has filished electroabsorption modulated lasers (EMLs) for high-speed fiber communication systems. EMLs monolithically integrate a DFB laser with an electroabsorption modulator (EAM), and PICs to date to realize wavelength compatibility, the DFB laser wavelength on the longer wavelength side of the excitonic peak of the EAM.

Schematic of a metal-coated silica toroidal microcavity supported by a silicon pillar is shown in Figure 1-17 (Inset), and used to false-color representation of the electric field for an exterior plasmonic WGM. Luo's group solved this problem by adopting the identical epitaxial layer (IEL) structure, in which the DFB laser and the EAM share the MQW layer. The IEL integration scheme greatly simplifies device fabrication, since no additional regrowth step is involved. Luo's group was among the first in the world to propose and independently realize the IEL structure. By adopting the IEL integration technique and suppressing package-induced parasitic crosstalk between the laser and the modulator as well as optical and electrical crosstalk, Luo and his colleagues were the first in China to fabricate 2.5Gb/s, 10Gb/s and 40Gb/s EML transmitter modules in 1998, 2003 and 2007, respectively. Recently, Luo's group demonstrated an EML module exhibiting an over 10dB dynamic extinction ratio upon 43Gb/s non-return-to-zero (NRZ) bit sequence modulation.

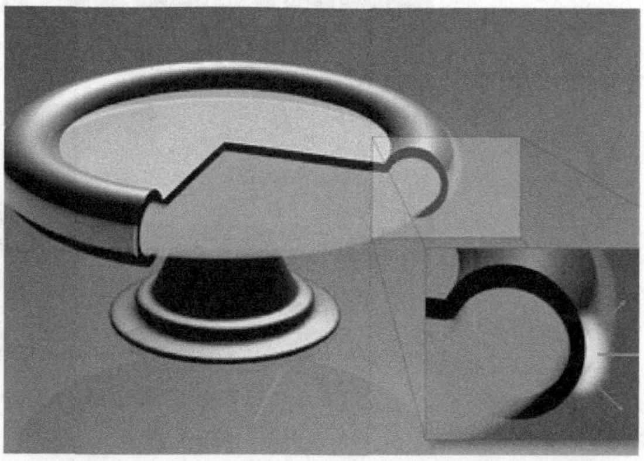

Fig. 1-17 Metal-coated silica toroidal microcavities.

The research group of Wei Wang (an academician of CAS) in the Institute of Semiconductors, CAS, research on InP-based semiconductor lasers and monolithic photonic integration technologies finished a tunable radio frequency source based on a three-section amplified feedback laser (AFL) design. The photonic integration is realized by using the quantum well intermixing technique, and 32～51GHz of continuous tuning, and established 40Gbit/s all-optical clock recovery using this device. The AFL-extracted stable optical clock exhibited a timing jitter of 123.8 fs from a 40Gbit/s return-zero pseudo-random bit stream (RZ-PRBS) pulse train. Main achievements including: the sample-grating-based broadly tunable DFB laser array for high capacity transmission (a national 863 project), array of eight sample grating tunable DFB lasers integrated with one combined output waveguide in n-InP based materials, optical logic gate based on evanescently coupled uni-traveling carrier photodetectors (EC-UTC-PD) integrated with intra-step quantum well EAM, the EC-UTC-PD/EAM optical logic gate operated with an 8.4dB extinction ratio and performed NRZ optical logic—while operating at 5Gb/s.

A monolithically integrated tunable microwave source using an EAM integrated with a DFB laser subject to optical injection. This source is particularly attractive due to its small size and wide frequency tuning range without using any cavity. The monolithically integrated tunable microwave source using an EAM integrated with a DFB laser subject to optical injection. Identical epitaxial layer structure AlGaInAs multiple-quantum-well electroabsorption modulated semiconductor microcavity lasers for PICs as shown in Figure 1-18. Measured 43Gb/s eye diagram at voltage swing of 2 V is shown in Figure 1-19.

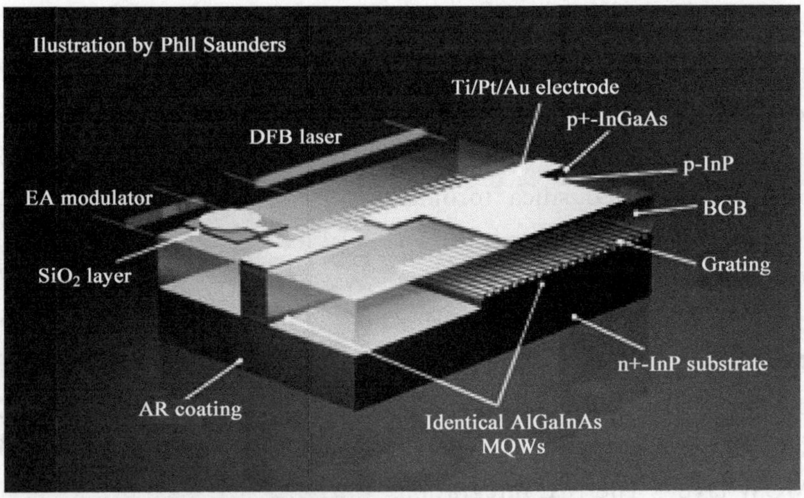

Fig. 1-18 Microcavity AlGaInAs laser chips.

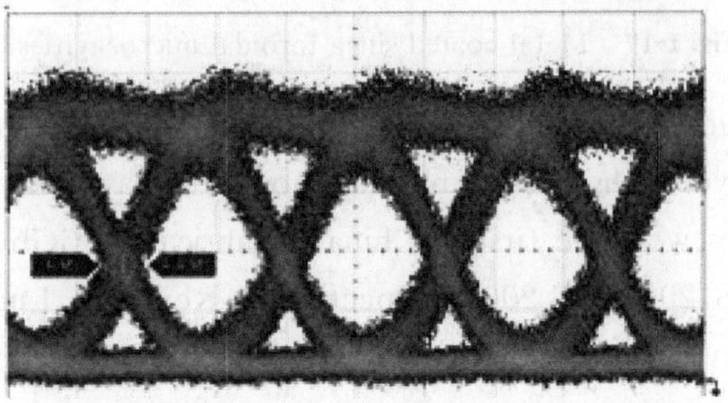

Fig. 1-19 43Gb/s eye diagram at measured voltage swing of 2V.

Unidirectional-emission semiconductor microcavity lasers fabricated by planar technique processes are potential light sources for PICs, and on-chip optical interconnections, showed that integrating a bus-waveguide to an optical microcavity is a potentially simple way to realize unidirectional-emission microlasers and fabricated unidirectional emission microlasers on standard edge-emitting 1,550-nm laser wafers using an integrated bus-waveguide realized room-temperature continuous-wave injection lasing with device sizes down to 10μm. The lowest threshold current demonstrated is 4mA for an AlGaInAs/InP microcylinder laser with the radius of 10μm. Single-mode lasing with an SMSR of 33dB was realized as well. The germanium-on-silicon (Ge-on-Si) materials for silicon-compatible near-infrared photodetectors, Ge-on-Si exhibits high absorption coefficients at the telecommunication wavelengths 1,310nm and 1,550nm, and is compatible with silicon technology. The Ge layers on Si and silicon-on-insulator (SOI) substrates using ultrahigh vacuum chemical vapor deposition based on lower-temperature buffer technology. The Ge layer has a high quality with a threading dislocation density of $1\times10^5/cm^2$ and a root-mean-square roughness of 0.33nm. For Ge p-i-n PD and arrays on SOI, the device with a diameter of 25μm exhibited responsivity of 0.65A/W at 1.31μm and 0.32A/W at 1.55μm and a dark current density of 13.9mA/cm^2 at 1 V reverse bias. The 3-dB bandwidth of the detector with a diameter of 25μm is 13.30GHz at -3V. The Ge/Si avalanche photodetector (APD) separated absorption and charge multiplication

(SACM) structure. In the SACM-APD, the light absorption occurs inside germanium and the carrier multiplication occurs inside silicon. The APD exhibited a responsivity of 4.4A/W at 1,310nm when biased at 90 percent of the breakdown voltage (39.5V at a dark current of 100μA). The APD displayed a gain of 8.8 at 1,310nm at 35.5V and a gain of 40 at 39V. Moreover, Wang's group also explored germanium-tin (GeSn) alloys, which feature direct band gap and the bandgap energy can be adjusted over a wide range in infrared wavelengths. Wang's group fabricated GeSn photodetectors for longer telecommunications wavelengths in the L and U bands, with responsivity at 1,310, 1,540 and 1,640nm as 0.52, 0.23 and 0.12A/W, respectively. The next generation of compactly integrated low-cost optoelectronic systems on the silicon platform: include demonstrating an ultra-compact plasmonic slot filter with band selection and spectrally splitting capabilities and showing perfect vertical fiber coupling by using asymmetric sub-grating structures in which a period consists of two sub-gratings with an identical etched height and a different width. They demonstrated coupling efficiency as high as 69 percent at a wavelength of 1.52μm, and 65 percent at a wavelength of 1.55μm, with 1dB wavelength bandwidth around 80nm and realized a broadband compact polarizing beam splitter (PBS) by using only a single layer of subwavelength multi-subpart profiles grating. The grating PBS demonstrated high diffraction efficiencies (more than 97 percent) over a broad spectrum of 1.53-1.62μm, with an extinction ratio exceeding 16dB and a comparatively wide angular bandwidth (about 8°). Finally, the team investigated waveguide amplifiers based on erbium (Er) silicates with yttrium (Y) and ytterbium (Yb) ions for silicon photonics integration. They observed enhanced Er^{3+} photoluminescence of about 100 times by optimizing the ion ratio of Y(Yb) and Er.

On integrated photonics for optical communications area, optical interconnects and optical sensing, with state-of-the-art equipment in their large cleanroom (including their electron-beam lithography system of the latest Raith 150 Ⅱ model), ultra-small SOI devices and high-speed hybrid Ⅲ-Ⅴ/SOI devices for optical interconnects, ultra-small 4 × 4 arrayed waveguide grating (AWG) with a size of only 40 × 50μm by using their original idea of overlapped free propagation regions of the AWG and experimentally demonstrating an ultra-compact reflective Si-nanowire AWG based on SOI nanowires. Each waveguide in the array has an individual photonic crystal reflector at its end. The total size of the fabricated 8-channel 400GHz-spacing AWG is as small as 134 × 115μm. He's team also proposed two novel types of Si hybrid plasmonic waveguides recently; the waveguides provide a nano-scale confinement of the optical field (e.g., 50 × 5nm at wavelength λ = 1,550nm) and the ability for a relatively long propagation distance (on the order of several tens of λ). In addition, they demonstrated a Mach-Zehnder interferometer (MZI)-coupled microring with a high sensitivity and a large measurement range for assessing the change in refractive index. The MZI-coupled microring sensor demonstrated a sensitivity of 111nm/RIU for measuring the change of the ambient refractive index (from 1.0 to 1.538).

Arrayed waveguide grating multiplexers and double-ring sensor

Fabricated reflective arrayed waveguide grating multiplexer with 8 × 400GHz channels is shown Figure 1-20 (a). Optical microscope image of the double-ring sensor fabricated on a silicon-on-insulator substrate is shown on top of Figure 1-20(b). The two rings have slightly different free spectral ranges. The chip size is only 1.3 × 0.7mm. Normalized output power versus the sample refractive index change when a broadband source with 28nm bandwidth centered at 1,553nm was used as Figure 1-20(b).

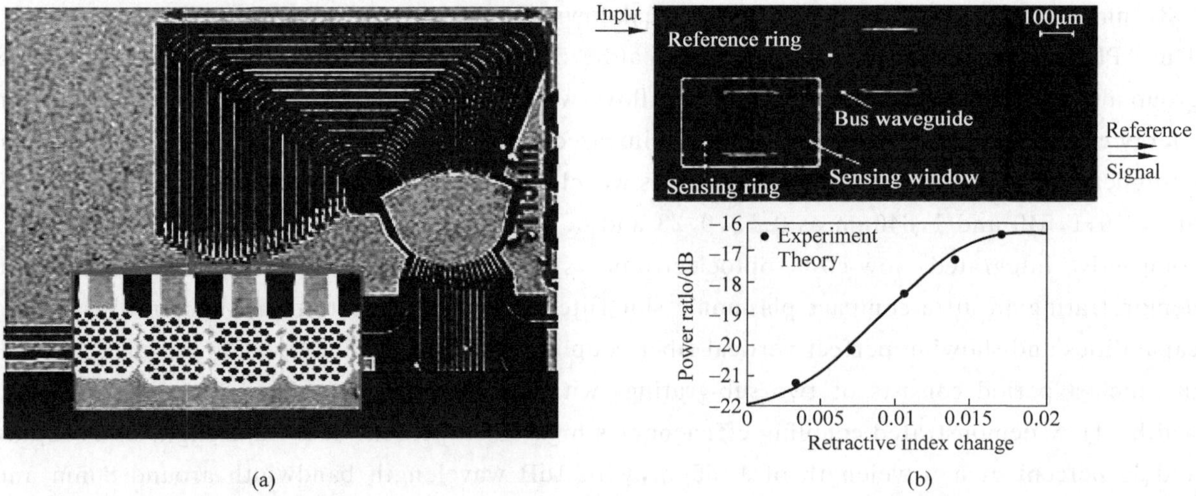

Fig. 1-20 The semiconductor lasers and photonic integrated devices of arrayed waveguide grating multiplexers and double-ring sensors.

The digital wavelength switchable laser based on V-coupled cavities employing the Vernier effect. The fabrication process and chip size are similar to those of a Fabry-Perot laser, without complex gratings and multiple epitaxial growths. Single-electrode-controlled digital wavelength switching over 16-channel with 100GHz spacing has been demonstrated, with a side-mode suppression ratio as high as 40dB. Such a simple and potentially low-cost wavelength-switchable laser has great potential for applications in optical networks and beyond.

A highly sensitive sensor based on two cascaded microring resonators that employ the Vernier effect. By coupling the light from a broadband source such as a light-emitting diode (LED) into the input-port and detecting the power ratio between the signal and reference ports, He's group has demonstrated a highly sensitive intensity-interrogated sensor. The elimination of the requirement for an external bulky and expensive tunable laser or spectrometer will pave the way for low-cost practical applications of planar waveguide ring-resonator biosensors.

1.9 Photonic integration solid state memory

Researchers of California Institute of Technology announced that they have constructed a memory circuit from molecules and nanometer-size wires that is as dense as what manufacturers expect to be building in 2020. The circuit, which stores 0s and 1s by switching clusters of molecules between two states, contains 1.6 × 104bits jammed together at a density of 1011bits per square centimeter. Conventional microchips are at least 10 times less dense. The prototype is not yet as stable or reliable as commercial computer memory, and building it would require manufacturers to learn to harness materials other than silicon, the workhorse of computing technology. But the scale of the device dwarfs any electronic circuit previously constructed using nanotechnology.

The group of California Institute of Technology built the device that major goal here was never to just make a memory circuit to develop a manufacturing technique that could work at molecular dimensions as shown in Figure 1-21. The device is "a true tour de force," says nano researcher Charles Lieber of Harvard University, who was not part of the study. It has pushed far beyond previous limits of integration density and bit numbers realized previously in the field of molecular electronics. Researchers are exploring nano-size electronics systems because silicon circuits cannot be packed with wires at increasing densities—yielding higher numbered Pentium processors—forever. Eventually, electrons will start seeping between wires and lithography techniques for stamping out silicon circuits may reach their physical limit.

Fig. 1-21 Researchers have built an ultradense integrated memory circuit from nanowires and molecules.

The Caltech group combined two approaches: molecular electronics (transistors made of molecules) and nanowire crossbars, which are perpendicular junctions of ultrathin wires. To make their device, the team laid down a tightly packed series of 400 parallel silicon wires (separated by just 33 nanometers) and coated them with a layer of barbell-shaped rotaxane molecules. They created a grid of wires by covering the molecule layer with 400 more platinum wires, resulting in groups of molecules sandwiched between each node formed by the crisscrossed wires.

To switch between 0 and 1 the researchers applied a voltage across a group of Advanced Optical and Quantum Memory molecules at a node, which toggled the molecules between two states. The rotaxane molecules each contain a ring around the "handle" of the barbell. A voltage applied across the molecule caused the ring to slide up or down, changing the electrical conductivity of the molecule. The wires were so crowded that the team could not build conventional electrodes capable of electrifying only two wires at a time (those that define a node); instead they switched the junctions on and off in groups of nine.

One indication that more works needs to be done is that the junctions routinely broke down after being switched more than about 10 times, says Jonathan Green, a physics and chemistry Ph. D. candidate in Heath's lab of Caltech and first author of the report published online in Nature. The molecules spontaneously flipped back to their previous state after nearly an hour, which is another limitation for a memory device. Commercial flash memory is stable for up to years. The molecules were also slow to switch between states that although this time can probably be improved, the speed of such a memory circuit would not come from switching one junction at a time. Instead it would result from switching many junctions at once.

Spintronics storage

Fueled by the ever increasing data density in magnetic storage and the need for a better understanding of the physical properties of magnetic nanostructures there exists a strong demand for high-resolution magnetically sensitive microscopy techniques. The technique with the highest available resolution is spin-polarized scanning tunneling microscopy (SP-STM) which combines the atomic-resolution capability of conventional STMs with spin-sensitivity. Beyond the investigation of ferromagnetic surfaces, thin films, and epitaxial nanostructures with unforeseen precision, it also allows the achievement of a longstanding dream, i. e. the real space imaging of atomic spins in antiferromagnetic surfaces as shown Figure 1-22. The phenomena in surface magnetism in most cases could not be imaged directly before the advent of SPSTM. After starting with a brief introduction to basics of the contrast mechanism, recent major achievements will be presented, like the direct observation of the atomic spin structure of domain walls in antiferromagnets and the visualization of thermally driven switching events in superparamagnetic particles consisting of a few hundreds atoms only. Recently observed complex spin structures containing 15 or more atoms will be presented.

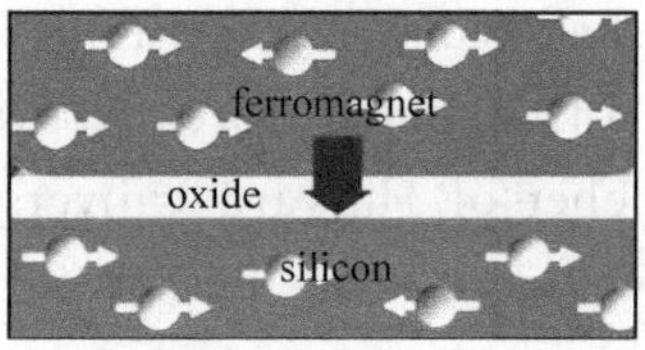

Fig. 1-22 The Storage device: transfers magnetic information into silicon through oxide layer at room temperature is the vital step forward for future spintronics storage.

The first generation optical disk with the wavelength 632nm HeNe laser was used for playing movies. But now every computer must have a BD drive with the wavelength 405nm semiconductor laser. Optical disk is the cheapest storage and distribution of huge amount of data. Optical disk is only 10 cents to print the disk, may even be lower in the future. It is still the cheapest means. A personal library needs only ONE optical disk to storage all textbooks. People could use optical disk to learn, training and entertainment more convenient than internet. Meanwhile optical disc is local and data center need huge amount of power to maintain magnetic disk storage also.

Optical storage will continue to be the best choice for physical media distribution. The likelihood function of Flash memory devices displacing CD, DVD or Blu-ray discs in the near future is minimal. However, Flash and online downloads are major competitors. Optical storage probably has a technical life of 5～10 years and a product life of 10～20 years. Many foreseeable technology as multi-layer and multi-wavelength (two photon), EB mastering and near-field read out, UV laser, plasmonic optical storage etc., will arrive in 10 years. However, many potential technology for optical storage to be competitive in the age of nanotechnology-enabled data storage products, new components and design strategies and significant investments will be required. Future optical storage will probably be based on new principles and technologies. Older concepts, such as 3D holographic memories and Millipede etc. are very difficult to be commercially viable. The plasmatic, negative refraction optical devices, nanotech, molecular self-assembling and photon-entanglement etc. principles will be applied in this area that just like semiconductor memory, magnetic disk, optical storage will keep increasing within a certain time. By the way, the three kinds of storage principles of optics, magnetic and semiconductor will be combined to create new perfects storage devices as optical solid state storage, optics-magnetic storage devices and magnetoresistive random access memory for example and may be becoming a true universal memory. Summary of various potential optical storage technologies are shown in Table 1-1.

Table 1-1 Potential optical storage roadmap to increase density and capacity

Types (feature)	Min. Mark Bit(nm)	Track (nm)	Density (Gb/in^2)	Capacity(GB 120mm disc)	Principles and Technologies
Blu-ray and 4～8 layers disc	112～149	320	18.1	100～200	Multi-layer and Multi-wavelength
Near-Field (NFR) Read only	41.3～58	108	144.8	220	EB Mastering and Near-Field read out

continued

Types (feature)	Min. Mark Bit(nm)	Track (nm)	Density (Gb/in^2)	Capacity(GB 120mm disc)	Principles and Technologies
UV laser diodes and NFR	20~25	75	351	500	EB Mastering and Doubled 405nm to 202.5nm
Plasmonic optical storage	15~20	45	1×1000	1.7×10^3	Plasmonic optical pick-up with negative refraction lense
Nanotech	10~15	30	1.8×10^3	3×10^3	DUV and EB Mastering nano-imprinting and nano-detector
Optical solid state storage	~20	~20	$1\sim2\times10^3$	$1\sim10\times10^3$/unite	Nanotech, molecular self-assembling and photon-entanglement
Holograms recordable storage	—	—	$2.5\sim5.2\times10^3$	$5.4\sim10.7\times10^3$	NA 0.65/λ407nm and medium thickness 1.5mm
Molecular polymer storage	8~12	8~12	$10\sim12\times10^3$	$15\sim20\times10^3$	Self-assembling polymer arrays
X-ray digital holograms storage	—	—	160×10^3	50×10^3	X-ray laser (λ0.5nm),system NA 0.5
Nano-probe storage	1	2~3	645×10^3	1.13×10^6	Dynamic Contact-Electrostatic Force Microscopy (DC-EFM)
Electronic quantum holography	35bit/electron	—	3×10^6	$5\sim10\times10^6$	Written in electron waves and read out with the scanning tunneling microscope

Following the traditional optical disc technology development model to reduce mark size and increase areal density that the CD is 1G (generation),DVD is 2G,BD is 3G that ODS has reached the classical technology end of life with BD discs and drives at $\lambda = 405$nm and NA = 0.85. But extending the definition of classical optics,optical storage used very small part of electro-magnetic spectrum in the 400 ~ 700nm range as shown Figure 1-23. However, extending the meaing of "optical" to include UV,DUV (deep ultraviolet) and X-radiation,opens new frontiers for high-density data storage, 4G and 5G are proven in the lab, 6G and 7G are pure speculation, but illustrate the challenges faced by optical storage to reach 1TB capacity. Multi-layer solutions are feasible. 4-layer,8-layer,12-layer,and 16-layer discs are proven in the lab. NFR is also feasible, but needs to be proven outside the lab. The real potential of nanotech is yet to be determined. Electron Beam Mastering and Near Field Read (5G) 120mm Capacity is 220GB,storage density is 144.8Gb/in^2.

By continuation of the classical optical roadmap,potential future ODS technologies is UV disc that requires UV laser diodes, next X-ray disc as digital holography for example. Atomic/ Molecular data storage means of configuration or quantum state or both. Some enabling means negative refraction (spot sizes less than the diffraction limit),variable focus lenses for multi-layer discs to correct for spherical aberration and nanotech as super high storage densities,self assembly, patterned media etc, nanophotonics as modulators, lasers, gratings implemented in Silicon—plasmonics (spot sizes less than the diffraction limit—photon sieves for far UV and X-ray spot formation).

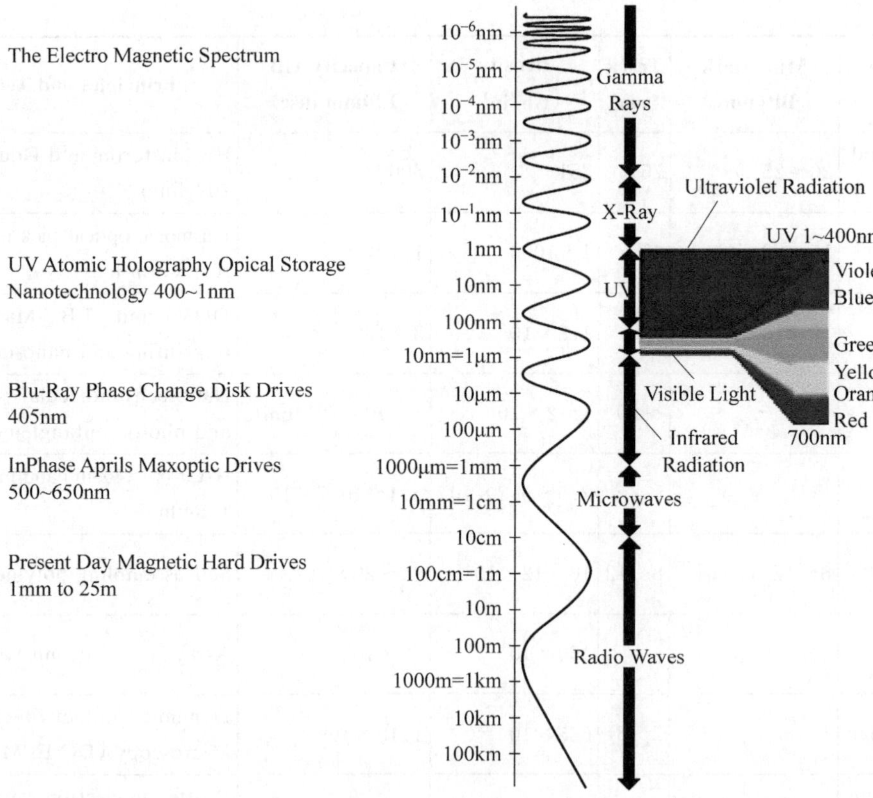

Fig. 1-23 The electro-magnetic spectrum in the 400～700nm range.

1.10 Other new quantum memory technologies

1.10.1 Ultraviolet photonic storage

Ultraviolet optical storage was classical optical data storage end with λ = 405nm when the technology uses near and mid-range ultraviolet (UV). Much of UV optical storage technology will be adapted from semiconductor UV and EUV lithography. UV optical storage will be far more challenging than near-IR and visible optical storage ever was. Front surface recording layer and reflection component OPU (optical pickup unit) required. Using frequency doubled 650nm can get λ = 325nm laser for example that will increases capacity to 39GB/ layer and reach 100GB for 3 layers per disc. For λ = 202.5nm in vacuum UV regime use frequency doubled 405nm that the areal density increase 4× to BD. This increases the capacity per layer to 100GB. However, the technology will probably be abandoned before reaching λ = 202.5nm, owing to cost and complexity. UV optical storage challenges UV laser diodes, including UV optical components for OPU, low noise UV media, mastering and replication processes that not be proportional of cost to complexity for capacity increase. UV laser diodes technology is still immature and only a few commercial products that engineering samples from Nichia is 200 mW CW output 375nm, 340～360nm is current R&D sweet spot, 240～260nm demonstrated in the lab. The diode-pumped solid state (DPSS) lasers, which can be frequency tripled or quadrupled, but must be greatly reduced in size and cost to be candidates. Other options to UV laser diodes and DPSS (for example, KrF or F_2

fiber) have no possibility of meeting size and cost requirements for UV optical storage. Nanotech may hold the key to long-term prospects on structural enhancements and materials improvements. So UV laser diodes are in about the same position as blue lasers in 1995 for solutions need a long term.

1.10.2 Plasmonic optical storage

Plasmonics is a branch of physics in which surface plasmon resonances of metals are used to manipulate light at the sub-wavelength scale. Surface plasmon polaritons (SPPs) are collective oscillations of electron density at an interface of a metal and dielectric. Because SPPs can be excited and strongly coupled with incident light, they have many potential applications in high-resolution optical imaging and storage and lithography. Some metals (gold and silver, for example) exhibit strong SPPs resonance in certain wavelength ranges, and therefore can be used to guide and concentrate light to nanoscale spots less than the classical diffraction limit. Some of the SPPs resonant structures can produce a field irradiance (W/m^2) at the near field that is greater by orders of magnitude than the incident light. Some resonant optical antennas can concentrate laser light into < 25nm FWHM size spots (as defined by gap widths). Plasmonic optical pick-up (OPU) utilise negative refraction variable focus lenses which needed to aid layer-to layer focusing for ML discs) Photon Sieves The above enabling technologies may provide the means to write/read significantly smaller marks. Plasmonic optical pick-up (POPU) utilise negative refraction variable focus lenses which needed to aid layer-to layer focusing for ML discs and similar to NFR that flies are 20nm or thin on the disc surface. The spot sizes can be to 25nm or smaller. This corresponds to an areal density is about $1Tb/in^2$, or a capacity 1.75TB on a 120mm disc to increase 70× for Blu-ray Disc. Plasmonic optical storage will permit multi-channel read/write and integrated POPU per side optical disc drives. Major challenges will be servo control of POPU flying height and tracking with the air bearing surface (ABS) slide. Volume holography storage theoretical maximum volume storage density of $\rho_v \sim 1/\lambda^3$ was possible for binary data, i.e. $\lambda = 500nm$ calculates density about to be $8bits/\mu m^3$ or $131Tb/in^3$ when the NA of optical system is large enough. The corresponding theoretical areal density ρ_A is about $8N bits/\mu m^2$ ($\sim 5.2N\ Gb/in^2$). N is the number of page that can be independently stacked in a common volume of the storage medium. For $N = 1,000$, the areal storage density is $\sim 5.2Tb/in^2$.

1.10.3 X-ray storage

X-ray Storage concept is designed for x-radiation with $\lambda \leqslant 1nm$ 1D or 2D computer-generated Fourier transform holograms Select page size (N or $N \times N$ pixels) and offset angle compute and sample analog interference pattern apply data coding and EDAC modulate and scan write spot to form hologram parallel read by means of holographic reconstruction position read beam over hologram project N or $N \times N$ pixels onto photodetector array and process and format serial data stream. For X-ray storage challenges are safe, inexpensive compact X-ray laser, compact photodetector arrays, all optics must be reflective and new mastering (write) and replication methods. The advantages are needless to page composer (SLM), 3D media and incoherent superposition (stacking) of holograms and can apply method to all media formats that read out servo may be same as today's DVD. Meanwhile, X-ray storage performance potential of digital fourier transform holograms, for a disc of 50mm diameter and a recording area of $1600mm^2$, λ0.5nm and NA 0.5, the storage density ρ = $250Gb/mm^2$ ($160Tb/in^2$), capacity C = 50TB unformatted access time <10ms, read out data rate, which is function of pixels, read power, detector sensitivity and scan speed, could be 50Gbps or higher. State-of-the-art X-ray laser from University of Hamburg, the free-electron laser (FEL) may be suitable for optical storage application. Next generation light sources, based on Free-Electron Lasers (FEL) will be capable of producing X-rays of such extreme brilliance that new possibilities of focusing the radiation emerge. An optical element, based on the simple concept of an array of pinholes (a photon sieve), exploits the monochromaticity and coherence of light from a free-electron laser to focus soft X-rays with unprecedented sharpness (high irradiance levels).

1.10.4 Nano-probe and molecular polymer storage

- Nano-probe storage

The nano-probe storage has higher storage density that bits written on ferroelectric film the marks are roughly on 25nm centers corresponding to a storage density of about $1TB/in^2$. Writing and reading used a voltage nano-probe with switching between two polarization states. Imaging is done with Dynamic Contact-Electrostatic Force Microscopy (DC-EFM), density can be to $645Tb/in^2$, i.e. 1130TB (unformatted) on a 120mm disc with 1nm bit and track pitches. The highest storage density could be atom storage as Hydrogen (H) atom and Fluorine (F) atom level storage that storage density can be to 51.6 Pb/in^2 and represent is about 0s and F atoms represent 1s only.

- Molecular polymer storage

Molecular polymer storage builds on existing approaches by combining the lithography techniques traditionally used to pattern microelectronics with novel self-assembling materials called block copolymers. Molecular polymer storage has been shown to store $10Tbit/in^2$. When added to a lithographically patterned surface, the copolymers' long molecular chains spontaneously assemble into the designated arrangements. There's information encoded in the molecules that results in getting certain size and spacing of features with certain desirable properties. Thermodynamic driving forces make the structures more uniform in size and higher density than traditional materials. The block copolymers pattern the resulting array down to the molecular level, offering a precision unattainable by traditional lithography-based methods alone and even correcting irregularities in the underlying chemical pattern. Such nanoscale control also allows the researchers to create higher-resolution arrays capable of holding more information than those produced. In addition, the self-assembling block copolymers only need one-fourth as much patterning information as traditional materials to form the desired molecular architecture, making the process more efficient. The large potential gains in density offered by patterned media make it one of the most promising new technologies on the horizon for future hard disk drives. In its current form, this method is very suited for designing hard drives and other data-storage devices, which need uniform patterned templates exactly the types of arrangements the block copolymers form most readily. With additional advances, the approach may also be useful for designing more complex patterns such as microchips. These results have profound implications for advancing the performance and capabilities of lithographic materials and processes beyond current limits.

A number of technologies are attempting to surpass the densities of all of these media. IBM's Millipede memory is attempting to commercialize a system at 1Tbit/in^2 in 2007. This is about the same capacity that perpendicular hard drives are expected to "top out" at, and Millipede technology has so-far been losing the density race with hard drives. Development since mid-2006 appears to be moribund, although the latest demonstrator with 2.7Tbit/in^2. A newer IBM technology, racetrack memory, uses an array of many small nanoscopic wires arranged in 3D, each holding numerous bits to improve density. Although exact numbers have not been mentioned, IBM news articles talk of "100 times" increases. Various holographic storage technologies are also attempting to leapfrog existing systems, but they too have been losing the race, and are estimated to offer 1Tbit/in^2 as well, with about 250GB/in^2 being the best demonstrated to date-for non-quantum holography systems.

1.10.5 Electronic quantum holography

The electronic quantum holography is the densest type of memory storage experimentally so far. By superimposing images of different wavelengths into the same hologram, a Stanford research team was able to achieve a bit density of 35bit/electron i. e. approximately 3 Exabytes/in^2. This was demonstrated using electron microscopes and a copper medium as reported in the Stanford Report on January 28, 2009. The initials for Stanford University are written in electron waves on a piece of copper and projected into a tiny hologram. On the two-dimensional surface of the copper, electrons zip around, behaving as both particles and waves, bouncing off the carbon monoxide molecules the way ripples in a shallow pond might interact with stones placed in the water. The ever-moving waves interact with the molecules and with each other to form standing "interference patterns" that vary with the placement of the molecules. In a traditional hologram, laser light is shined on a 2-dimensional image and a ghostly 3D object appears. In the new holography, the 2-dimensional "molecular holograms" are illuminated not by laser light but by the electrons that are already in the copper in great abundance. The resulting "electronic object" can be read with the scanning tunneling microscope. Several images can be stored in the same hologram, each created at a different electron wavelength. The researchers read them separately, like stacked pages of a book. The experience is roughly analogous to an optical hologram that shows one object when illuminated with red light and a different object in green light.

1.10.6 Compositive application of the different principles

The integration all of the different principles and technologies enable to improve the density and capacity of optical storage and make a roadmap in the future to show in Table 1-2. The basic idea in seeking outside the traditional engineering optical technology to improve the resolution of the system, give full play to the advantages of optical storage that expanded to full electromagnetic spectrum, multi-dimensional space and more interaction mechanism of light with medium. Meantime optical storage have to combine with other up to date technologies as nanotech, semiconductor memory and magnetic memory tech to explore new development roadmap.

Table 1-2 Comparison between traditional storage principles/methods and new principles as well as the implementation of technology roadmap to increase optical storage density and capacity.

Traditional methods and technology	Further principles and the implementation
Two dimensional record	Three-dimensional space and multi-dimensional storage as frequency dimension for example
Visible light sources (405～650nm)	UV, EUV, DUV, FEL (Free-Electron Lasers)
Traditional far-field optics	Near-field optics and super-resolution technology
Single parameter changes	Variety of parameter changes: optical interference, changes in reflectivity, refractive index, absorption rate or magnetic declination, while taking advantage of chroma, gray scale, polarized, photochemistry and quantum entanglement efficient etc.
Effects of light or heat to write	Effects of photon quantum and electron transfer
Single disc form	Multi-disc or solid state form
Single-channel timing to write/read	Multi-channel parallel reading and writing
linear encoder	Serial multi-dimensional matrix Coding
Traditional recessional manufacturing(sub-micro)	Nanotech and MEOMs

The basic technical measures to achieve the above ideas and technologies routes can be summarized as follows:

(1) The optical storage from a flat two-dimensional extended to three-dimensional, such as holographic storage and multiple tiers of storage. Then extended to multi-dimensional, such as the use of the frequencies of light, gray scale, the polarization properties of the coordinate system of the optical storage, disc storage capacity from traditional linear function of growth, increase to the growth of the exponential function.

(2) Nonlinear absorption near-field optical recording medium to reduce the effective size of the storage spot, so that the storage density than traditional optical resolution, super-resolution storage.

(3) Research and development of new optical storage materials to replace the traditional dye and phase change measurement as a recording medium, disc storage write speed changed to photothermal effect of the slow response from the current mechanism of photon quantum effects reduce the information written power and energy consumption, improve writing data rate.

(4) A variety of modulation of the optical parameters of principles and methods of optical modulation and coding multi-variable multi-step modulation changed from the traditional modulation and coding, including grayscale, refractive index, color, polarization and other changes to improve the coding efficiency and storage capacity.

- External quantum efficiency(EQE)

Although some progress has been made in the development of UV light sources, the major obstacle remains the extremely low external quantum efficiency (EQE) of the nitride emitters; there is a notable drop in efficiency as emission wavelengths approach deep-UV.[2] The EQE for LEDs is defined as the product of the internal quantum efficiency, carrier injection efficiency, and light-extraction efficiency. The internal quantum efficiency is related to the crystalline quality of the epitaxial layers; fewer defects and dislocations lead to better performance.

If the device structures are grown on sapphire substrates, the lattice-mismatchinduced defects and dislocations (approaching 10^{10}cm^{-2} in some cases) cannot be reduced to a desired level on the order of 104cm^{-2} or lower. Two other issues that affect mainly the carrier-injection efficiency are a difficulty in achieving highly doped p-type AlGaN cladding and barrier layers, and reduced magnitude of the band discontinuities at key heterojunction interfaces. Also, UV light is absorbed at the p-side electrode.

To minimize the impact of lattice-mismatch-induced defects and dislocations, bulk AlN material has recently been proposed and used as a substrate for UV light emitters. The AlN substrate leads to lower lattice and thermal mismatches between it and subsequent AlGaN layers. To reduce dislocations when bulk AlN is used as the substrate, the AlGaN layers grown on top must be sufficiently thin if the Al composition in them is high to guarantee that the structure is pseudomorphic (that is, they maintain the crystalline structure of the substrate). This may represent an unacceptable device design compromise. Although improved performance characteristics in UV LEDs that use this approach has been reported, the EQE for deep-UV LEDs (< 350nm emission wavelength) remains low (<2%).

In a recent paper, argued that in addition to all the known problems that prevent the fabrication of highly efficient UV emitters, another limitation is the amount of residual strain in the device structures. Residual strain modifies the valence-band structure of AlGaN layers used in UV devices in a way that can negatively affect the emitted light polarization properties.

To complicate matters, UV light emitted from an active $Al_xGa_{1-x}N/Al_yGa_{1-y}N$ multi-quantum-well structure can switch its polarization from transverse electric (TE) to transverse magnetic (TM), depending on the amount of residual strain present in the active layer—which further depends on the choice of template/ substrate. When the emitted light is TM-polarized, it is difficult to extract along the c-axis (the direction along which the layers are grown). This leads to decreased light-extraction efficiency, and thus lower overall EQE. Thus, it is important to know in advance the aluminum composition of an $Al_xGa_{1-x}N/Al_yGa_{1-y}N$ quantum-well structure and its relationship to either the bulk AlN substrate used or the $Al_zGa_{1-z}N$ ($x<y<z$) template-on-sapphire substrate. These factors control the amount of residual strain in the active layer.

The ideal situation is where the Al content (and residual strain) does not lead to emitted light with a polarization that makes light extraction difficult. This basically implies that once the operating wavelength for a UV emitter has been selected, the active region and its associated layer structure should dictate the choice of template/substrate. This is not how UV emitters are conventionally designed. The usual approach is to accept the lattice and thermal constraints dictated by the template/substrate—which makes it difficult to decide how much residual strain can be managed.

- Application-oriented nitride

The ultimate solution, we believe, is to use a substrate whose lattice constant closely matches the lattice constant of thick barrier layers in the UV device structure. At best one should strive to design the device structure to be pseudomorphic—with minimal residual strain at the barrier/ template interface. The rest of the layers can be chosen to have some residual strain with the appropriate thicknesses necessary to ensure pseudomorphic growth.

In the absence of bulk binary nitride substrates that could meet the latticematching requirements for photonic-device structures that span the potential spectral band enabled by

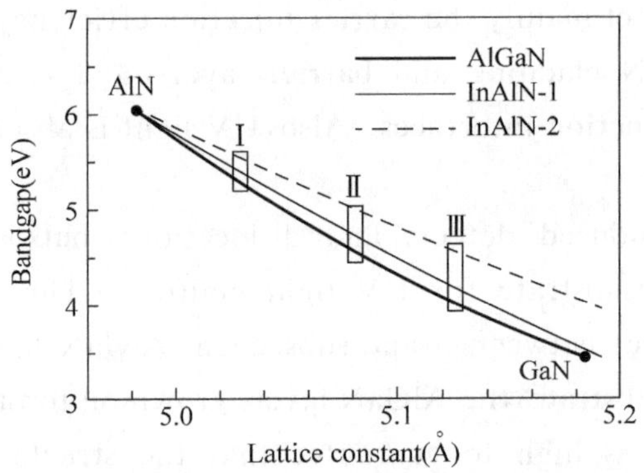

Fig. 1-24 Band gaps of UV nitride ternary alloys (AlGaN with a bowing parameter b = 0.89eV, In AlN-1 with b = 5eV) plotted against the c-axis lattice constant. The dotted red line (InAlN-2) is for a modified bandgap of the InAlN alloy using a low value of b = 2.5eV. Vertical bars denote the range of band gaps for a particular lattice constant accessible via AONS.

nitrides, we have proposed the development of new application-oriented nitride substrates (AONS). In one example, a possible LED structure designed to emit at about 240nm in the deep UV region is based on this concept (see Figure 1-24).

The need for AONS is best explained by considering the variation of the bandgap—and wavelength—of $Al_xGa_{1-x}N$ as a function of lattice constant (see Figure 1-25). Here, the bandgap is plotted as a function of the c-xis lattice constant, c_o; one could just as well have plotted it as a function of the ao lattice constant. For AlN with a bandgap of 6.13eV(x = 1), c_o is 4.9816Å; for GaN, with a bandgap of 3.43eV(x = 0), it is 5.1815Å. This range of bandgaps corresponds to UV wavelengths from about 202nm to about 360nm. For any other value of the Al mole fraction, x, the lattice constant lies between the two extremes. To minimize lattice-mismatch induced defects and dislocations, the $Al_xGa_{1-x}N/Al_yGa_{1-y}N$ active structure should be grown on a substrate whose lattice constant closely matches that of the thickest structural layer. The substrates are necessarily bulk AlGaN ternary alloys; neither the bulk AlN binary alloy nor sapphire possesses the right lattice constant to match that of any ternary ($Al_xGa_{1-x}N$ or $In_xAl_{1-x}N$) or quaternary layers in any canonical nitride light-emitting device structure. As an alternative alloy for active layers, one could use InAlN as shown in Figure 1-25 to cover the same UV spectral region and the other UV wavelengths between 359nm and 400nm.

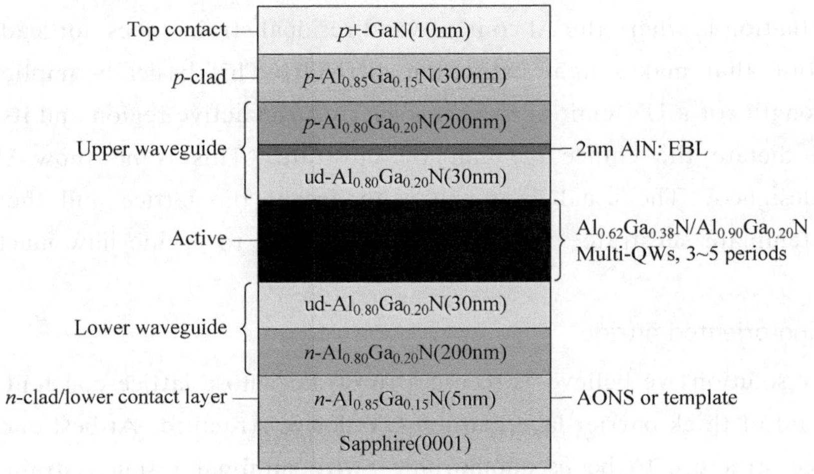

Fig. 1-25 A schematic shows the layer structure of a pseudomorphic LED designed to emit at about 240nm. The device structure includes an application-oriented nitride substrate (AONS), multiple quantum wells (QWs), and an electron-blocking layer (EBL).

Application-oriented nitride substrates are bulk $Al_xGa_{1-x}N$ ternary substrates with Al compositions chosen to lattice-match thick barrier layers. In some cases, an Al composition could be found where a fairly broad spectral region can be covered by a single AONS. The concept can be extended into the visible and to the IR region through use of $In_xGa_{1-x}N$ ternary substrates. No ternary nitride substrates currently exist. The challenges are balanced by the potential benefits for a range of electronic and photonic devices based on nitride compounds.

(1) Two-photon recording medium, information recording process from a single photon effect to the two-photon effect, the space of three-dimensional storage by the principle of the information recorded in the two-photon interchange.

(2) Through a variety of optical parameters at the same time store information, so that the CD signal read from the traditional single channel to multi-channel, in the same CD speed proportional to the increased disc read and write data transfer rate.

(3) To establish a new codec system, the traditional second-order travel-length coding to the multi-wavelength multi-degree tour length coding to achieve the sector guidance, correction and supporting code sharing, reduce redundancy, improve the effective capacity of the CD users utilization.

(4) Various wavelength lasers, photodetectors, and the digital processing circuit hybrid integration of different wavelengths, from the traditional stand-alone unit assembly combination, changed the standard of miniaturization module integrated system, greatly reduce the system size, lower energy consumption, improve storage system reliability, productivity and yield, reduce manufacturing costs.

The alloy semiconductor family, the aluminum gallium nitride AlGaN alloys are the most versatile for design and fabrication of emitters that span the UV spectrum from the UV-A (400~320nm) through the UV-B (320~280nm) to the UV-C (280~200nm) as Figure 1-26. Most devices reported in the literature have been grown on sapphire substrates. Before the active region is grown, one typically deposits an AIN buffer layer on top of the sapphire substrate as a lattice constant bridge that mediates between the sapphire and subsequent AlGaN layers. This is then followed by a silicon-doped $Al_yGa_{1-y}N$ cladding layer on top of which is grown the active region, composed of $Al_xGa_{1-x}N/Al_yGa_{1-y}N$ ($y<x\geqslant0.6$) multiple quantum wells. A top cladding layer made of p-type $Al_yGa_{1-y}N$ is then grown.

Before the top contact layer is grown, one of two alternative schemes is generally used to block the escape of electrons from the active region. The first incorporates a high-aluminum-content $Al_zGa_{1-z}N$ quantum barrier ($z\geqslant0.9$). This barrier is followed by a heavily doped p-type GaN contact layer. In a second form of the electron-blocking scheme, other researchers have used a multi-quantum-barrier structure in place of the single quantum barrier. In the latter case, the implementation employs several periods of an $Al_xGa_{1-x}N/Al_zGa_{1-z}N$ structure that has been reported to yield slightly better operating characteristics.

Optical media in the next generations will be more complex. Required yield, throughput, and quality will be harder to achieve, regardless of the future technology winner(s). The cost and complexity of processes and equipment and the unit cost of media will increase, perhaps significantly in some cases. A major challenge to the industry is to prevent or minimize this. New or modified processes, manufacturing equipment, and quality control methods will be required for

N-layer MLD and NFR media. More sophisticated and complex in-line and off-line test and measurement equipment will be required. The cost of R&D will increase significantly; more materials scientists, chemists, and physicists will be needed.

On the positive side, new opportunities are plentiful, and provide a natural evolutionary path. On the negative side, a finite probability exists that increasing the capacity of optical storage media may well become too expensive (diminishing economic returns).

The market success of optical storage products can be correlated with design to a specific application (notably audio or video). CD, DVD, and Blu-ray Disc (BD) are the most relevant examples. Computer applications were extensions for CD (CD-ROM), inherent, but secondary for DVD and BD. If consumer applications no longer require optical discs, ODS will then depend on computer storage applications. Existing optical storage technologies still have at least a 10-year useful product life cycle. However, classical optical storage will have reached the end of its technology life before then. Future ODS products will primarily be the blue-disc progeny of Blu-ray Disc.

Optical storage 5～10 years from today will be provided mainly by evolved versions of today's proven technologies. Optical storage will continue to dominate the removable-media AV applications sector in consumer electronics for the near future. "HDTV" playback and recording and personal storage applications will remain the dominant applications. Over the 10-year horizon, optical storage will likely be provided by a mixture of today's evolving and future technologies. Displacement technologies cannot be ruled out. To secure its future in the mainstream storage world, ODS must expand its horizons. This will require significant investment and risk, given its many challenges and competitors.

Anyone in the ODS drive or media business, whether OEM or ODM, should accept the certainty that ODS technology is at a tipping point. Optical Media manufacturers must develop equipment that can handle spatial structures of less than 50nm and inline manufacturing of 4～16 layer discs. Future ODS products will have a large percentage of what today are considered esoteric components; most have not been developed to commercial status—R&D must focus on "future" components, if future ODS products are to be evolved.

- Solid-state quantum memory

Zong-Quan Zhou etc. of Key Laboratory of Quantum Information, University of Science and Technology of China realized solid-state quantum memory in 2012 that faithfully storing an unknown quantum light state. It is essential to advanced quantum communication and distributed quantum computation applications. The required quantum memory must have high fidelity to improve the performance of a quantum network. They report the reversible transfer of photonic polarization states into collective atomic excitation in a compact solid-state device. The quantum memory is based on an atomic frequency comb (AFC) in rare-earth ion-doped crystals. They obtain up to 0.999 process fidelity for the storage and retrieval process of single-photon-level coherent pulse. This reliable quantum memory is a crucial step toward quantum networks based on solid-state devices.

Entangled memory

Physicists in the USA are the first to store two entangled quantum states in a memory device and then retrieve the states with their entanglement intact. Their demonstration, which involves "stopping" photons in an ultracold atomic gas, could be an important step towards the practical implementation of quantum computers. The basic unit of information in a quantum computer is the qubit, which can take the value 0, 1 or—unlike a classical bit—a superposition of 0 and 1 together. A photon could be used as a qubit, for example, with its "up" and "down" polarization states representing 0 or 1. If many of these qubits are combined or "entangled" together in a quantum computer, they could be processed simultaneously and allow the device to work exponentially faster than its classical counterpart for certain operations. Entanglement could also play an important role in the secure transmission of information because the act of interception would destroy entanglement and reveal the presence of an eavesdropper.

Micron ships phase-change memory

Memory chip vendor Micron Technology Inc. of UK has backfilled its range of 45-nm phase-change memory (PCM) which shown in Figure 1-26 (construction of section in). They has announced it has been shipping 45-nm 1-Gbit PCM multichip package to Nokia for use in the phones for some time and is now sampling a 512-Mbit PCM plus 512-Mbit LPDDR2 MCP. Nokia is using Micron's PCM solution to enrich the functionality of "select devices" in its portfolio. At the time Micron will ship tens of millions of units of 2-Gbit PCM. Micron's volume comes from a couple of cell phone design wins. Micron's 45nm PCM solution targeted currently for utilization in mobile phones bringing enhanced capabilities, with a future roadmap aimed at addressing smart phones and media tablets. Its PCM can speed up boot time for mobile phones at switch-on, simplify software development and boost performance with its overwrite capability. Micron's 45-nm PCM components targeted at mobile phones with future devices aimed at smart phones and tablet computers. Micron is trusted enhancement a number of feature and performances with the new technology.

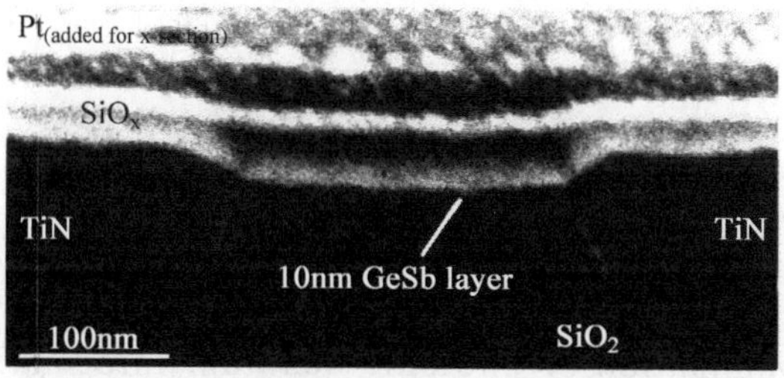

Fig. 1-26 The cross-section micrograph of the phase-change memory bridge.

Chapter 2

Fundamentals of quantum information

2.1 Introduction

The photonic brain inspilded information processing theory and technology formed by the combination of photonic quantum information science, brain information processing mechanism and artificial intelligence algorithm is the development direction of information technology artificial intelligence in the future, which is a combination of brain intelligent computing theory and optical quantum frontier technology. It is also a major challenge facing the scientific community in this century and a frontier science that attracts great attention. Photonic brain information memory and processing should deep integration of photonics, machine intelligence, brain and neuroscience, cognitive science and psychology. It has become the strategic goal of major developed countries and the direction of the future development of artificial intelligence in the world.

The section introduce a self-contained introduction to the conceptional, mathematical foundations and development roadmap of quantum information theory. The first part introduces the basic notions like entanglement, channels, teleportation, etc. and their mathematical description. The second part focuses on a presentation of the quantitative aspects of the theory. Topics discussed in this context include entanglement measures, channel capacities, relations between both, additivity and continuity properties and asymptotic rates of quantum operations. Finally, give an overview on some recent developments and open questions.

Quantum information and quantum computation have recently attracted a lot of interest. The promise of new technologies like safe cryptography and new "super computers", capable of handling otherwise untractable problems, has excited not only researchers from many different fields like physicists, mathematicians and computer scientists, and a large public audience. On a practical level all these new visions are based on the ability to control the quantum states of (a small number of) microsystems individually and to use them for information transmission and processing. From a more fundamental point of view, the crucial point is a reconsideration of the foundations of quantum mechanics in an information theoretical context. The purpose of this work is to follow the second path and to guide physicists into the theoretical foundations of quantum information and some of the most relevant topics of current research. To this end, the outline of this paper is as follows: The rest of this introduction is devoted to a rough and informal overview of the. eld, discussing some of its tasks and experimental realizations. Afterwards, consider the basic formalism, which is necessary to present results that are more detailed. Typical keywords in this context are systems, states, observables, correlations, entanglement and quantum channels. Ten clarify these concepts (in particular, entanglement and channels) with several examples discuss the most important tasks of quantum information in greater detail, and devoted to a more quantitative analysis, where make closer contact to current research, then discuss how entanglement can be measured. The topic channel capacities, i. e. are looking at the amount of information which can maximally be transmitted over a noisy channel and consider state estimation, optimal cloning and related tasks.

Quantum information is a rapidly developing field and the present work can of course only a small part of it. An incomplete list of other general sources the reader should consult is the books of Lo, Gruska, Nielsen and Chuang, Bouwmeester et al, and Alber et al. the lecture notes of Preskill and the collection of references by Cabello, which particularly contains many references to other reviews.

2.1.1 Quantum computing (QC) roadmap

Within these overall goals, different scientific approaches to QC will play a variety of roles: it expected that one or more approaches could emerge that will actually attain these goals. Other approaches may not—but will instead play other vitally important roles, such as offering better scalability potential in the post-years or exploring different ways to implement quantum logic, that will be essential to the desired development of the field as a whole. It was the unanimous opinion of the Technology Experts Panel (TEP) that it is too soon to attempt to identify a smaller number of potential "winners;" the ultimate technology may not even been invented yet. Considerable evolution of and hybridization between approaches has already taken place and should be expected to continue in the future, with existing approaches being superseded by even more promising ones.

A second function of the roadmap is to allow informed decisions about future directions will made by tracking progress and elucidating interrelationships between approaches, which assist researchers to develop synergistic solutions to obstacles within any one approach. To this end, the roadmap presents a "mid-level view" that segments the field into the different scientific approaches and provides a simple graphical representation using a common set of criteria and metrics to capture the promise and characterize progress towards the high-level goals within each approach. A "detailed-level view" incorporates summaries of the state-of-play within each approach, provides a timeline for likely progress, and attempts to capture its role in the overall development of the field. A summary provides some recommendations for moving toward the desired goals. The panel members developed the first version of the QC roadmap from the La Jolla meeting and five follow-up meetings held in conjunction with the annual ARO/ARDA/NSA/NRO Quantum Computing Program Review (QCPR) in Nashville, Tennessee, USA, in August 2002. The present update developed out of further meetings at the August 2003 QCPR; the roadmap will continue updated annually. The quantum computer-science test-bed destination that envision in this roadmap will open up fascinating, powerful new computational capabilities: for evaluating quantum-algorithm performance; allowing quantum simulations performed and for investigating alternative architectures, such as networked quantum subprocessors. The journey to this destination will lead to many new scientific and technological developments with potential societal and economic benefits. Quantum systems of unprecedented complexity created and controlled, that potentially leading to greater fundamental understanding of how classical physics emerges from a quantum world, which is as perplexing and as an important question today as it was when quantum mechanics invented. That can foresee these QC capabilities will lead into an era of "quantum machines" such as atomic clocks with increased precision with benefits to navigation, and "quantum enhanced" sensors. Quantum light sources will be developed that will be enabling technologies for other applications such as secure communications, and single atom doping techniques will be developed that will open up important applications in the semiconductor industry. It could anticipate that there will be considerable synergy with nanotechnology and spintronics also. The journey ahead will be challenging, that will lead to unprecedented advances in both fundamental scientific understanding and practical new technologies.

- Background of quantum computation

The representation of information by classical physical quantities such as the voltage levels in a microprocessor is familiar to everyone. However, quantum information science and technology (QIST) have developed to describe binary information in the form of two-state quantum systems, such as two distinct polarization states of a photon; two energy levels of an atomic electron; or the two spin directions of an electron or atomic nucleus in a magnetic field. A single bit of information in this form has come to known as a "qubit." With two or more qubits, it becomes possible to consider quantum logical-"gate" operations in which a controlled interaction between qubits produces a (coherent) change in the state of one qubit that is contingent upon the state of another. These gate operations are the building blocks of a quantum computer. In principle, a quantum computer is a very much more powerful device than any existing or future classical computer because the superposition principle allows an extraordinarily large number of computations performed simultaneously. For some certain problems, such as integer factorization and the discrete-logarithm problem, which believed to be intractable on any present-day or future conventional computer, this "quantum parallelism" would permit their efficient solution. These are important problems as they form the foundation of nearly all publicly used encryption techniques.

Another example of great potential impact, as first described by Feynman, which is quantum modeling and simulation (e. g., for designing future nanoscale electronic components), that the exact calculations of such systems can only be performed using a quantum computer. This simulation capability has the potential for discovering new phenomenology in mesoscopic/nanoscopic physics, which in turn could lead to new devices and technologies. It not known that if quantum computers will offer computational advantages over conventional computers for generalpurpose computation. To realize this potential will require the engineering and control of quantum-mechanical systems on a scale far beyond anything yet achieved in any physics laboratory.

Many approaches to QC from diverse branches of science may pursued. These present-day QC technologies are some orders of magnitude away in both numbers of qubits and numbers of quantum logic operations that can performed from the sizes that would be required for solving interesting problems. A few experimental approaches are now capable of performing small numbers of quantum operations on small numbers of qubits, with realistic assessments of the challenges for scale-up, while the bulk of the field is at the singlequbit stage with optimistic ideas for producing large-scale systems. There are both fundamental and technical challenges to bridging this gap. A serious obstacle to practical QC is the propensity for qubit superpositions of 0 and 1 to "decohere" into either 0 or 1. (This phenomenon of decoherence is invoked to explain why macroscopic objects are not observed in quantum superposition states.) However, theoretical breakthroughs have been made in generalizing conventional error-correction concepts to correct decoherence in a quantum computer. A single logical bit would encoded as the state of several physical qubits and quantum logic operations used to correct decoherence errors. These quantum error-correction ideas have shown to allow robust or fault-tolerant QC with the encoded logical qubits, at the expense of introducing considerable overhead in the numbers of physical qubits and elementary quantum logic operations on them. For example, one logical qubit may encoded as a state of five physical qubits in one scheme, although the number of physical qubits constituting a logical qubit could well be different for different physical QC and Quantum Information Science and Technology (QIST) roadmap implementations. It has established, under certain assumptions, that if a threshold precision per gate operation achieved, quantum error correction would allow a quantum computer to compute indefinitely.

An essential ingredient of quantum error-correction techniques and QC in general, is the capability to create entangled states of multiple qubits on demand. In these peculiarly quantum mechanical states, the joint properties of several qubits defined uniquely, even though the individual qubits have no definite state. The strength of the correlations between qubits in entangled states is the most prominent feature distinguishing quantum physics from the familiar world of classical physics. The unusual properties of these states, which do not readily exist in nature, underlie the potential new capabilities of QC and other quantum technologies. Although present-day QC experiments are making rapid progress, demonstrations of on demand entanglement are few and the precision of gate operations is quite far from the fault tolerant thresholds. However, experimental capabilities will progress and the fault-tolerant requirements are likely to be relaxed once the underlying assumptions are adapted to specific approaches. The overall purpose of this roadmap is to help achieve these thresholds and to facilitate the progress of QC research towards the quantum computer-science era.

- Purpose and methodology

This roadmap has been formulated and written by the members of a Technology Experts Panel (TEP or the "panel"), whose membership of internationally recognized researchers in QIST held a kick-off meeting in La Jolla, California, USA, in late January 2002 to develop the underlying roadmap methodology.

The TEP held a further five meetings in conjunction with the annual ARO/ARDA/NSA/NRO Quantum Computation Program Review (QCPR) meeting in Nashville, Tennessee, USA, in August 2002. The sheer diversity and rate of evolution of this field, which are two of its significant strengths, made this a particularly challenging exercise. To accommodate the rapid rate of new developments in this field, the roadmap will be a living document and dated annually, and at other times on an ad hoc basis if merited by significant developments. Certain topics will revisited in future versions of the roadmap and additional ones added; it expected that there will significant changes in both content and structure. At the La Jolla meeting, TEP members decided that the overall purpose of the roadmap should be to set as a desired future objective for QC β to develop by 2012 a suite of viable emerging-QC technologies of sufficient complexity to function as quantum computer-science test-beds in which architectural and algorithmic issues can explored.

The roadmap intended to function in several ways to aid this development. It has a prescriptive role by identifying what scientific, technology, skills, organizational, investment, and infrastructure developments will be necessary to achieve the desired goal, while providing options for how to get there. It also performs a descriptive function by capturing the status and likely progress of the field while elucidating the role that each aspect of the field is expected to play toward achieving the desired goal. The roadmap can identify gaps and opportunities, where strategic investments are beneficial. It will provide a framework for coordinating research activities and a venue for experts to provide advice. The roadmap will therefore allow informed decisions about future directions to made, while tracking progress, and elucidating interrelationships between approaches to assist researchers to develop synergistic solutions to obstacles within any one approach. The roadmap intended an aid to researchers and those managing or observing in the field. Underlying the overall objective for the QC roadmap, the panel members decided on a four level structure with a division into "high level goals" "mid-level descriptions" "detailed level summaries" and a summary that includes the panel's recommendations for optimizing the way forward. The panel members decided on specific ambitious, but attainable five- and ten-year high-level technical goals for QC. These technical goals set a path for the field to follow that lead to the desired QC test-bed in 2022. The mid-level roadmap view captures the breadth of approaches to QC on the international scale and uses a graphical format to describe in general terms how the different research approaches are progressing towards these technical goals relative to common sets of criteria and metrics. The panel decided to first segment the field into a few broad categories, with multiple projects grouped together in each category according to their underlying similarities. The panel decided that two types of measures were necessary adequately represent the status of each category: a set of criteria characterizes promises of a class of approaches as a candidate QC technology; whereas a set of metrics captures the status of the approach in terms of technical advances along the way to achieving the high-level goals. The "detailed summaries" provide more information on the essential concept of each approach, the breadth of projects involved, the advantages and challenges of the class of approaches, and a timeline for likely progress according to a common format. These summaries, written by subgroups of the panel members after soliciting input from their respective scientific communities, that intended to provide a brief, readable account that represents the status and potential of the entire approach from a worldwide perspective. The panel has endeavored to provide a complete, balanced, and inclusive picture of each research approach, but with the caveat that it is expected that additional content will need to be added to each summary in future versions of the roadmap, after further input from the scientific community. The panel members decided that it was not appropriate for the roadmap to attempt to describe the relative status of different individual projects within each approach.

The panel members found it was especially challenging represent the status and role of theory in the roadmap adequately. Clearly, theory has been pivotal in the development of QC to its present state, providing often-unanticipated advances that have stimulated experimental investigations.

At the same time, it is difficult to schedule or define meaningful "metrics" for such future breakthroughs. For Version 1.0 of the roadmap the panel decided that the primary focus would be on experimental approaches to QC and limited the description of theory to its historical role. In the present Version 2.0 release, all sections have updated to reflect advances in the 14 months since release of Version 1.0. In addition, new sections on cavity-QED approaches to QC and a full theory section, with coverage of decoherence theory, quantum information theory, quantum algorithms and QC complexity, and quantum computer architectures added. In addition, each detailed summary for the different experimental areas provides an overview of the specific areas in which additional theory work needed.

- Quantum computation roadmap and goals

Although QC is a basic-science endeavor today, it is realistic to predict that within a decade fault-tolerant QC could achieved on a small scale. The overall objective of the roadmap can accomplished by facilitating the development of QC to reach a point from which scalability into the fault-tolerant regime can reliably inferred. It is essential to appreciate that "scalability" has two aspects: the ability to create registers of sufficiently many physical qubits to support logical encoding and the ability to perform qubit operations within the fault-tolerant precision thresholds. The future high-level goals of the roadmap for QC are therefore by the year 2023 or later, Quantum computer is a kind of subversive computing technology, which has advantages and potential that traditional computers cannot match. By deeply understanding the working principle, application and future development trend of quantum computer, we can better understand the value and potential of this emerging technology, and provide reference and inspiration for future research and development. Although quantum computers are still in the early stages of development, they have a wide range of potential applications and broad prospects for future development. With continuous progress of science and technology that have to believe that the future quantum computer will bring us more surprises and breakthroughs, and promote the development of human society to a higher level to encode a single qubit into the state of a logical qubit formed from several physical qubits, perform repetitive error correction of the logical qubit, and transfer the state of the logical qubit into the state of another set of physical qubits with high fidelity to implement a concatenated quantum error-correcting code. With the continuous development of quantum computer technology, there will be more applications of quantum computers in the future. At the same time, with the advancement of algorithms and physical implementation technology, the performance of quantum computers will significantly improved. However, it is also important to note the challenges faced by quantum computers, such as how achieve larger scale quantum entanglement and maintain stability of qubits. In order to solve these problems, continuous innovation and optimization from the hardware and software aspects are required. Meeting these goals will require both experimental and theoretical advances. While remaining within the basic-science regime that requires the achievement of four ingredients that are necessary for fault-tolerant scalability, creating deterministic, on-demand quantum entanglement, encoding quantum information into a logical qubit, extending the lifetime of quantum information; and communicating quantum information coherently from one part of a quantum computer to another. This is a challenging goal requiring something in the order of ten physical qubits and multiple logic operations between them, yet it is within reach of some present-day QC approaches and new approaches that may emerge from synergistic interactions between present approaches. The high-level QC requires of 100 or more physical qubits, exercises multiple logical qubits through the full range of operations required for fault tolerant QC in order to perform a simple instance of a relevant quantum algorithm, and approaches a natural experimental QC benchmark: the limits of full-scale simulation of a quantum computer by a conventional computer. The main goal would be within reach of approaches that attain the 2023 goal. It would extend QC into the quantum computer test-bed regime, in which architectural and algorithmic issues could explored experimentally. Quantum computers of this size would also open up the possibilities of quantum simulation as originally envisioned by Feynman. New ways of using the computational capabilities of these small quantum computers could be explored, such as distributed QC and classically networked arrays ("type Ⅱ" quantum computers), which recent work suggests may be advantageous for partial differential equation simulations, even though in contrast to other potential QC applications no exponential or polynomial speed-up would be possible.

Within these overall goals, different scientific approaches will play a variety of roles. It is expected that one or more approaches will emerge that will actually attain these goals, while others will not, but will instead play vitally important supporting roles by exploring different ways to implement quantum logic, for instance that will be essential to the desired development of the field as a whole. It was the unanimous opinion of the TEP that it is too soon to attempt to identify a smaller number of potential "winners;" the ultimate technology may not have even invented yet. Considerable evolution of and hybridization between the various approaches has already taken place and should be expected to continue in the future, with some existing approaches being superseded even more promising ones.

• Quantum computation roadmap mid-level view

The mid-level roadmap view intended to describe in general terms how the entire field of QC is progressing towards the high-level goals and provides a simple graphical tool to characterize the promise and development status according to common sets of criteria and metrics, respectively. The requirements for quantum computer hardware capable of achieving the high-level goals are simply stated but are very demanding in practice.

(1) A quantum register of multiple qubits must be prepared in an addressable form and isolated from environmental influences, which cause the delicate quantum states to decohere.

(2) Although coupled to the outside world weakly, the qubits must nevertheless strongly coupled to perform logic-gate operations together.

(3) There must be a readout method to determine the state of each qubit at the end of the computation.

Many different routes from diverse fields of science to realizing these requirements pursued. Consequently. In order to represent progress adequately, the TEP decided to segment the field into several broad classes, based on their underlying experimental physics subfields. These subfields are β nuclear magnetic resonance (NMR) quantum computation, β ion trap quantum computation, β neutral atom quantum computation, β cavity quantum electro-dynamic (QED) computation β optical quantum computation, β solid state (spin-based and quantum-dot-based) quantum computation,

β superconducting quantum computation, and β "unique" qubits (e.g., electrons on liquid helium, spectral hole burning, etc.) quantum computation. β the theory subfield, including quantum information theory, architectures, and decoherence challenges.

Each of the different experimental approaches has its own particular strengths as a candidate QC technology. For example, atomic, optical, and NMR approaches build on well-developed experimental capabilities to create and control the quantum properties necessary for QC, whereas the solid-state and superconducting approaches can draw on existing large investments in fabrication technologies and materials studies. However, the different approaches are at different stages of development. Insights from the more developed approaches can incorporated into other usefully, less advanced approaches, which may hold out greater potential for leading to larger-scale quantum computers. The panel decided that to represent this diversity required a set of criteria for the 'promise' of each approach, and a set of metrics for its 'status' state of progress towards the high-level goals adequately. To represent the promise of each approach the panel decided to adopt the "DiVincenzo Criteria." Necessary conditions for any viable QC technology can simply stated as:

(1) A scalable physical system of well-characterized qubits.

(2) Ability to initialize the state of the qubits to a simple fiducial state.

(3) Long (relative) decoherence times, much longer than the gate-operation time.

(4) A universal set of quantum gates.

(5) A qubit-specific measurement capability. Two additional criteria, which are necessary conditions for quantum computer network.

(6) The ability to interconvert stationary and flying qubits and ability to transmit flying qubits between specified locations faithfully.

The physical properties, such as decoherence rates of the two-level quantum systems (qubits) used to represent quantum information must understood well. The physical resource requirements must scale linearly in the number of qubits, not exponentially, if the approach is to be a candidate for a large-scale QC technology. It must be possible to initialize a register of qubits to some state from which QC can performed. The time to perform a quantum logic operation must be much smaller than the time-scales over which the system's quantum information decoheres. There must be a procedure identified for implementing at least one set of universal quantum logic operations. In order to read out the result of a quantum computation there must be a mechanism for measuring the final state of individual qubits in a quantum register. The two networking criteria are necessary if it desired to transfer quantum information from one location to another, between different registers or between different processors in a distributed computing situation. Many different QC architectures are possible within the DiVincenzo framework. For example, architectures based on "clocked" or "ballistic" quantum logic implementations repursued. Some approaches are intrinsically limited to quantum logic gates between nearest neighbor qubits, which would allow parallel operations within a QC, whereas other approaches are capable of performing logic gates between separated qubits widely but are limited to serial operations. To represent visually the DiVincenzo promise criteria of each QC approach, the panel decided to use a simple three-color scheme as shown below (Table 2-1).

Table 2-1 The Mid-level quantum computation roadmap: promise criteria Quantum Computation QC Networkability.

QC Approach	The DiVincenzo Criteria						
	Quantum Computation					QC Networkability	
	#1	#2	#3	#4	#5	#6	#7
NMR							
Trapped Ion							
Neutral Atom							
Cavity QED							
Optical							
Solid State							
Superconducting							
Unique Qubits	This field is so diverse that it is not feasible to label the criteria with "Promise" symbols.						

This field is so diverse that it is not feasible to label the criteria with "Promise" symbols. Legend: as a potentially viable approach has achieved sufficient proof of principle as a potentially viable approach has been proposed, but there has not been sufficient proof of principle no viable approach is known The column numbers correspond to the following QC criteria:

(1) A scalable physical system with well-characterized qubits.

(2) Ability to initialize the state of the qubits to a simple fiducial state.

(3) Long (relative) decoherence times, much longer than the gate-operation time.

(4) A universal set of quantum gates.

(5) A qubit-specific measurement capability.

(6) The ability to interconvert stationary and flying qubits.

(7) The ability to transmit flying qubits faithfully between specified locations.

The values assigned to these criteria constitute a snapshot in time of the panel's opinions on the potential of each approach as a candidate QC technology. Future developments within an approach will lead to these values are updated. To represent the present status of each approach the panel developed a set of metrics that represent relevant steps on the way to the 2007- and 2012-year goals. The panel decided to use a similar color- coding to indicate the status of each approach (Table 2-2). The "development status metrics", which have been augmented somewhat for this version 2.0 on the page facing Table 2-2.

The development status metrics 1 through 4 correspond to steps on the way to achieving the high-level goals for 2007, while development status metrics 5 through 7 correspond to steps leading up to the high-level goal for 2012. For each QC approach, the TEP members have assigned a status code for each of these metrics. These codes could updated in future versions of the roadmap to reflect significant developments within each approach.

(1) Creation of a qubit: Demonstrate preparation and readout of both qubit states.

(2) Single-qubit operations:

① Demonstrate Rabi flops of a qubit.

② Demonstrate decoherence times much longer than the Rabi oscillation period.

③ Demonstrate control of both degrees of freedom on the Bloch sphere.

(3) Two-qubit operations:

① Implement coherent two-qubit quantum logic operations.

Table 2-2　The mid-level QC roadmap development status metrics.

QC Approach	1	1.1	2	2.1	2.2	2.3	3	3.1	3.2	3.3	3.4	3.5	3.6	4	4.1	4.2	4.3	4.4
NMR																		
Trapped Ion																		
Neutral Atom																		
Cavity QED																		
Optical																		
Solid State:																		
Charged or excitonic qubits																		
Spin qubits																		
Superconducting																		

QC Approach	4	4.5	4.6	4.7	4.8	5	5.1	5.2	6	6.1	6.2	6.3	7	7.1	7.2	7.3	7.4	7.5
NMR																		
Trapped Ion																		
Neutral Atom																		
Cavity QED																		
Optical																		
Solid State:																		
Charged or excitonic qubits																		
Spin qubits																		
Superconducting																		

Legend: = sufficient experimental demonstration

= preliminary experimental demonstration, but further experimental work is required

= no experimental demonstration　　and = a change in the development status between Versions 1.0 and. 2.0

② Produce and characterize the Bell entangled states.

③ Demonstrate decoherence times much longer than two-qubit gate times.

④ Demonstrate quantum state and process tomography for two qubits.

⑤ Demonstrate a two-qubit decoherence-free subspace (DFS).

⑥ Demonstrate a two-qubit quantum algorithm.

(4) Operations on 3-10 physical qubits:

① Produce a Greenberger, Horne, and Zeilinger (GHZ) entangled state of three physical qubits.

② Produce maximally entangled states of four or more physical qubits.

③ Quantum state and process tomography.

④ Demonstrate DFSs.

⑤ Demonstrate the transfer of quantum information e. g. teleportation, entanglement swapping, ultiple SWAP operations etc. between physical qubits.

⑥ Demonstrate quantum error-correcting codes.

⑦ Demonstrate simple quantum algorithms e. g. Deutsch-Josza.

⑧ Demonstrate quantum logic operations with faulttolerant precision.

(5) Operations on one logical qubit:

① Create a single logical qubit and keep it alive using repetitive error correction.

② Demonstrate fault-tolerant quantum control of a single logical qubit.

(6) Operations on two logical qubits:

① Implement two-logical-qubit operations.

② Produce two-logical-qubit Bell states.

③ Demonstrate fault-tolerant two-logical-qubit operations.

(7) Operations on 3-10 logical qubits:

① Produce a GHZ-state of three logical qubits.

② Produce maximally entangled states of four or more logical qubits.

③ Demonstrate the transfer of quantum information between logical qubits.

④ Demonstrate simple quantum algorithms (e. g. , Deutsch-Josza) with logical qubits.

⑤ Demonstrate fault-tolerant implementation of simple quantum algorithms with logical qubits.

When interpreting this mid-level graphical part of the roadmap, it is important to appreciate that both the promise criteria and development status metrics need considered also. For example, the "promise criterion" for NMR QC (in the liquid state) indicates that it does not have good scalability potential, but the "development status" metric shows that multiple steps have already achieved in this approach. Although not likely in its current form to be a candidate for a large-scale QC technology, the opportunity to learn how to perform QIP tasks within this approach is of tremendous value to the field in general. Conversely, some approaches are much less far along in their development status metrics, but an inspection of their promise criteria reveals that they offer significantly greater potential for achieving a large-scale QC technology. Intermediate between extremes a few approaches have the essential ingredients for QC under sufficient control that they have started to make the first steps towards developing a scalable architecture. The detailed-level view of the roadmap provides the means more fully understand these subtleties of interpretation.

- Quantum computation roadmap detailed-level

The purpose of the detailed-level roadmap summaries is to provide a short description of each of the experimental approaches, along with explanations of the graphical representation of the metrics in the mid-level view and descriptions of the likely developments over the next decade. A common set of points addressed in each summary:

(1) β is working on this approach.

(2) β location and the size of the group.

(3) β is a brief description of the essential idea of the approach and how far it is developed.

(4) β summary of the approach, which meets the DiVincenzo criteria and their status.

(5) β list of that have accomplished, when it accomplished. For the development status metrics.

(6) β special strengths of this approach.

(7) β the unknowns and weaknesses of this approach.

(8) β 5 year goals for this approach.

(9) β the 10-year goals for this approach, β the necessary achievements to make the 5- and 10-year goals for the approach possible.

(10) β scientific trophies could produced. These defined to be breakthrough-quality results β what developments in other areas of QIST or other areas of science will useful or necessary in this approach. β how will developments within this approach have benefits to others areas of QIST or other areas of science in general. β the role of theory in this approach, and β a timeline that shows the necessary achievements and makes connection to the mid-level development status metrics.

Note: The TEP decided that assessments of individual projects within an approach would not made a part of the roadmap because this is a program-management function. In addition to the theory component of the detailed-level summary for each approach, there is a separate summary for fundamental theory. This summary provides historical background on significant theory contributions to the development of QC and spells out general areas of theoretical work that needed on the way to achieving to 2021-year level goals.

- Detailed quantum computation summaries

The summaries of the different research approaches to QC listed in the table below (Table 2-3). Each of the summaries listed below linked to a file on this web site (click on the summary title below to view/download that document).

Table 2-3 Detailed summaries of quantum computation approaches.

Quantum Computation Approach Summary	Compiled by
6.1 Nuclear magnetic resonance approaches to quantum-information processing and quantum computing	David Cory
6.2 Ion trap approaches to quantum-information processing and quantum computing	David Wineland
6.3 Neutral atom approaches to quantum-information processing and quantum computing	Carlton Caves
6.4 Cavity QED approaches to quantum-information processing and quantum computing	Michael Chapman
6.5 Optical approaches to quantum-information processing and quantum computing	Paul Kwiat and Gerard Milburn
6.6 Solid state approaches to quantum-information processing and quantum computing	David Awschalom, Robert Clark, David DiVincenzo, P. Chris Hammel, Duncan Steel and, Birgitta Whaley

continued

Quantum Computation Approach Summary	Compiled by
6.7 Superconducting approaches to quantum-information processing and quantum computing	Tenry Orlando
6.8 "Unique" qubit approaches to quantum-information processing and quantum computing	P. Chris Hammel and Seth Lloyd
6.9 Theory component of the quantum computing roadmap	David DiVincenzo, Gary Doolen, Seth Llovd, Umesh Vazirani, Brigitta Whaley

- Quantum computation roadmap summary: the way forward

For a successful technology, reality must take precedence over public relations, for Nature cannot fooled. In 1986 Richard P. Feynman take on a basic scientific challenge of the complexity and magnitude of QC, diversity of approaches, persistence, and patience are essential. Major strengths of QC research are the breadth of concepts pursued, the high level of experimental and theoretical innovations, and the quality of the researchers involved. The rate of progress and level of achievements are very encouraging, but breakthroughs in basic science cannot expected to happen to a schedule. Nevertheless, the desired 2012 QC destination and the high-level goals that are set out in this roadmap, although ambitious, are within reach if experimenters and theorists work together, appropriate strategic basic research is pursued, and relevant technological developments from closely related fields, such as nanotechnology and spintronics, are incorporated. In developing the document, TEP members have noted several areas where additional attention, effort more resources would be advantageous.

The emphasis of the quantum computing roadmap out to 2007 is on the experimental development of error-corrected logical qubits. Without this critical building block, plans for further scale-up would be premature; they would not have a firm foundation. Nevertheless, it is important to begin investigations aimed at evaluating key factors associated with scaled architectures at an exploratory design level, for the various implementation approaches.

Such pathway studies, carried out in parallel with the qubit demonstration programs, will require expertise outside of the quantum information science framework. By examining the feasibility of the qubit schemes from a systems perspective, this exercise would define sensible metrics for scale-up, and initiate a closing of the gap between conventional computer systems protocols and quantum information science requirements. It would also encourage a dialogue between quantum information scientists and engineers that will become increasingly important as the field moves toward the logical qubit milestones.

As one looks to the future development of QC should anticipate the need for an increasing industrial involvement as the first steps into the realm of scalability. For example, much learned by trying to develop a few qubit quantum subprocessor that incorporates the quantum ingredients and the classical control and readout in a single device. It will involve a level of applied-science expertise and capability is unlikely found in a university environment. University-industry partnerships would offer an effective route forward. The first steps in this direction are already taking place, e.g. the Australian Centre for Quantum Computing Technology and the panel recommends that further interactions of this type need to encourage and facilitate.

While the intrinsic scalability of qubits is a central issue, it is also important to think in parallel about the more conventional scalability of experimental infrastructure and techniques required to control and readout the qubits, in order to meet the roadmap timeline. At present, single and few-qubit implementations often involve a substantial array of complex, expensive, and highly specialized equipment items. The step-up from few-qubit experiments to high-level goal of encoding quantum information into a logical qubit formed by several physical qubits. The demonstration of fault tolerant control via repetitive error correction goes beyond replicating qubit cells. That will place stringent demands on the overall experimental configuration. In the case of all-electronic solid-state qubits for example, the development of a fast (classical) control chip interfaced to a qubit chip is being pursued to address this issue, where it is instructive to consider the electronics and procedures required to operate a single rf-SET readout element. The control chip in this case may well involve a mix of technologies operating at different temperature levels, such as RSFQ and rf-CMOS, requiring collaboration across traditional boundaries. The drive towards fault-tolerant logical qubit operations separately raises many engineering, as opposed to physics, issues and the early involvement of industry will be important. These issues brought into sharp focus by objectives requiring some 50 physical qubits. β another area in which the TEP members foresee a future need for increased industrial involvement is in the general area of "supporting technology". Efforts made already to ensure that certain critical capabilities are available to researchers in the superconducting QC community, and analogous needs in other areas of QC research anticipated. Examples of relevant areas including materials and device fabrication, electrooptics, and single-photon detectors. The panel intends to amplify on the role of industry in future versions of this roadmap. Theory is an area in which the panel believes that some refocusing or expansion of effort would benefit the development of QC towards the roadmap objectives. Continued research efforts on high-quality, fundamental QC theory remain essential, but additional emphasis on theory and modeling that is directed at specific experimental QC approaches is required if this field is to move forward effectively. For example, further study of the fault-tolerant requirements in the context of the physics of specific approaches to QC is necessary. Closer involvement of theorists with their experimental colleagues is encouraged. The panel also recommends that additional effort directed at QC architectural issues. For example, what architectures are suitable for a scalable system, and how demanding requirements for scalable QC traded-off against each other. The quantum logic units need integrate with data storage, data transmission, and schedulers, some or all of which can benefit from quantum implementation.

Additional efforts within the mathematics and theoretical computer-science communities to better define the classes of problems that are amenable to speed-up on a QC should be encouraged, as should the more mundane but very important analysis of how abstract quantum algorithms mapped onto physical implementations of QC. The desired developments set out in this roadmap cannot happen without an adequate number of highly skilled and trained people to carry them out. The panel notes that graduate-student demand for research opportunities in QC is outstripping resources in many university departments. The panel believes that additional measures should be adopted to ensure that an adequate number of the best physics, mathematics, and computer-science graduate students can find opportunities to enter this field, and to provide a career path for these future researchers. Additional graduate-student fellowships and postdoctoral positions are essential, especially in experimental areas, and there is a need for additional faculty appointments, and the associated start-up investments, in quantum information science. The quantum computer-science test-bed destination that envision in this roadmap will open up fascinating, powerful new computational capabilities, for evaluating quantum algorithm performance, allowing quantum simulations to be performed, and for investigating alternative architectures, such as networked quantum subprocessors. The journey to this destination will lead to many new scientific and technological developments with myriad potential societal and economic benefits. A quantum computer provides the capability to create arbitrary quantum states of its qubits and so could be used as a tool for fundamental science and as an ingredient of quantum technologies that will open up new capabilities utilizing the uniquely quantum.

2.1.2 New quantum computation roadmap

The new Version 2. 0, release, while retaining the majority of the Version 1. 0 content, provides an opportunity to β incorporate advances in the field, that have occurred during the intervening 14 months; β make minor modifications to the roadmap structure to better capture the challenges involved in transitioning from a single qubit to two. β add major sections on topics that could not covered in Version 1.0. β reflect on the purpose, impact, and scope of the roadmap, as well as its future role. Some of the most significant changes in this Version 2.0 of the QC roadmap have been to incorporate the major advances that have occurred since the release of Version 1.0. These include β realization of probabilistic controlled-NOT quantum logic gates in linear optics. β the controlled-NOT quantum logic gates demonstrated in two-ion traps. β the achievement of near single-shot sensitivity for single electron spins in quantum dots, and β excellent coherence times observed in Josephson qubits, which together with the other multiple advances noted in the roadmap.

That indicative of the continued healthy rate of development of this challenging field toward the roadmap goals. In meetings of the roadmap experts panel members at the August 2003 Quantum Computing Program Review in Nashville Tennessee. That decided to increase the number of two qubits development status-metrics in the mid-level roadmap view more accurately reflect the distinct. That challenging scientific steps encountered within each QC approach in moving from one qubit to two. It decided also to relegate coverage of the DiVincenzo "promise criteria" and development status metrics for "unique qubits" from the mid-level view roadmap tables to the appropriate summary section. With these changes and additions, version 2. 0 of the QC roadmap provides a more precise and up-to-date account of the status of the field and its rate of development toward the roadmap desired goal. Perhaps the most unsatisfactory aspect of Version 1. 0 of the QC roadmap was that with its almost exclusive focus on experimental implementations, only a limited coverage of the important role of theory in reaching road map, which desired goals was possible. One of the major additions in Version 2.0 is the expansion of the theory summary section adequately represent the pivotal roles of theory, with sections of quantum algorithms and quantum computational complexity, quantum information theory, quantum computer architectures, and the theory of decoherence. A second major addition in Version 2.0 is a full summary section on cavity-QED approaches to QC. Another significant change in Version 2.0 is in the coverage of solid-state QIST Quantum Computing Roadmap QC, where the summary section are streamlined, and in the roadmap's mid-level view the great diversity of SSQC approaches has been captured into just two categories: "charge or excitonic qubits" and "spin qubits." With these major additions and changes, Version 2.0 of the QC roadmap provides a significantly more comprehensive view of the entire field and the role of each element in working toward the roadmap high-level desired goals. With the benefit of just over one year of experience with the impact of and community response to the first version of the QC roadmap, this Version 2.0 release provides an opportunity to reflect on its structure, scope, and future role. One of the most useful features of the roadmap is that by proposing specific desired development targets and an associated timeline it has focused attention and inspired debate, which are essential for effectively moving forward. The roadmap expert members received considerable input regarding the roadmap's chosen desired high-level goals; the majority of comments characterize these goals as falling into ambitious yet attainable category. Nevertheless, in the light of the recent progress noted in this roadmap update, it is worth asking whether an even more aggressive time line could be envisioned leading to a significantly more advanced development destination for QC (beyond the roadmap's desired quantum computation test-bed era) within the 2012 time horizon.

This question can best considered by comparing the QC roadmap with generally accepted principles of science and technology roadmaps. The research degree of difficulty involved in reaching 2017.

That desired high-level goal is unquestionably very high, but the risk associated with the fundamental scientific challenges involved, that mitigated by pursuing the multiple paths described in the roadmap. Achieving the high-level goals along one or more of these paths will require a sustained and coordinated effort; the uncertainties remain too high today to pick out a more focused development path. An attempt to do so at this time could potentially divert resources away from ultimately more promising research directions. This would increase the risk that QC could fail to reach the quantum computational test-bed era by 2015, beyond the considerable but acceptable levels of the path defined in this roadmap. However, this issue reassessed once the field moves closer to the 2017 desired goal. The roadmap experts panel members believe that the QC roadmap's desired high-level goals and timeline, while remaining consistent with accepted norms of risk within advanced, fundamental science and technology research programs, are sufficiently challenging to effectively stimulate progress. They intend to revisit these important issues in future updates.

Quantum computation (QC) holds out tremendous promise for efficiently solving some of the most difficult problems in computational science, such as integer factorization, discrete logarithms, and quantum simulation and modeling that are intractable on any present or future conventional computer. New concepts for QC implementations, algorithms, and advances in the theoretical understanding of the physics requirements for QC appear almost weekly in the scientific literature. This rapidly evolving field is one of the most active research areas of modern sciences attracting substantial funding that supports research groups at internationally leading academic institutions, national laboratories, and major industrial-research centers. Wellorganized programs are underway in the United States, the European Union and its member nations, Australia, and in other major industrial nations. Start-up quantum-information companies are already in operation. The diverse range of experimental approaches from a variety of scientific disciplines are pursuing different routes to meet the fundamental quantum mechanical challenges involved. Yet experimental achievements in QC, although of unprecedented complexity in basic quantum physics, are only at the proof-of-principle stage in terms of their abilities to perform QC tasks. It will be necessary to develop significantly more complex quantum-information processing (QIP) capabilities before quantum computer-science issues begin experimentally studied. To realize this potential will require the engineering and control of quantum-mechanical systems on a scale far beyond anything yet achieved in any physics laboratory. This required control runs counter to the tendency of the essential quantum properties of quantum systems to degrade with time (decoherence). Yet, it known that it should be possible to reach the quantum computer-science test-bed regime. If challenging requirements for the precision of elementary quantum operations and physical scalability can met. Although a considerable gap exists between these requirements and any of the experimental implementations today, this gap continues to close. To facilitate the progress of QC research towards the quantum computer-science era, a two-day in January 2008 in La Jolla, California, USA. Quantum Information Science and Technology Experts from all over the world attended the meeting. The meeting was attended by with the objective of formulating a QC roadmap. The panel's members decided that a desired future objective for QC should be β to develop by 2018 with a suite of viable emerging-QC technologies of sufficient complexity to function as quantum computer-science test-beds in which architectural and algorithmic issues explored. The panel's members emphasize that although this is a desired outcome, not a prediction, that it is attainable if the momentum in this field maintained with focus on this objective. The intent of this roadmap is to set a path leading to the desired QC test-bed era by 2018 by providing some direction for the field with specific five and ten year technical goals. While remaining within the "basic science" regime, the five-year goal would project QC far enough in terms of the precision of elementary quantum operations and correction of quantum errors that the potential for further scalability could reliably assessed. The ten-year goal would extend QC into the architectural/algorithmic regime, involving a quantum system of such complexity that it is beyond the capability of classical computers to simulate. These high-level goals are ambitious but attainable as a collective effort with cooperative interactions between different experimental approaches and theory.

2.2 Basic concepts

2.2.1 Quantum information

Classical information is speaking roughly, which everything can transmitted from a sender to a receiver with "letters" from a classical alphabet e.g. the two digits 0 and 1 or any other finite set of symbols. In the context of classical information theory, it is completely irrelevant which type of physical system used to perform the transmission. This abstract approach is successful because it is easy to transform information between different types of carriers like electric currents in a wire, laser pulses in an optical fiber, or symbols on a piece of paper without loss of data; and even if there are losses they are well understood and it is known how to deal with them. However, quantum information theory breaks with this point of view. It studies, loosely speaking, that kind of information ("quantum information") which is transmitted by microparticles from a preparation device (sender) to a measuring apparatus (receiver) in a quantum mechanical experiment, the distinction between carriers of classical and quantum information becomes essential. This approach is justified by the observation that a lossless conversion of quantum information into classical information is in the above sense not possible. Therefore, quantum information is a new kind of information. In order to explain why there is no way from quantum to classical information and back, let us discuss how such a conversion would look like. To convert quantum to classical information need a device which takes quantum systems as input and produces classical information as output—this is nothing else than a measuring apparatus. The converse translation from classical to quantum information can be rephrased similarly as "parameter-dependent preparation", i.e. the classical input to such a device is used to control the state (and possibly the type of system) in which the microparticles should be prepared.

A combination of these two elements can done in two ways. Primarily consider the device, which goes from classical to quantum to classical information. This is a possible task and in fact technically realized already. A typical example is the transmission of classical information via an optical ber. The information transmitted through the. The ber carried by microparticles (photons) and is therefore quantum information (in the sense of our preliminary definition). To send classical information we have to prepare first photons in a certain state send them through the channel and measure an appropriate observable at the output side. This is exactly the combination of a classical → quantum with a quantum → classical device just described.

The crucial point is now that the converse composition, which performing the measurement M first and preparation P afterwards (cf. Figure 2-1) is more problematic. Such a process called classical teleportation, if the particles produced by P are "indistinguishable" from the input systems. The impossibility of such a device via a hierarchy of other "impossible machines" which traces the problem back to the fundamental structure of quantum mechanics. This finally will prove the statement that quantum information is a new kind of information. To start with, we have to clarify the precise meaning of "indistinguishable" in this context. This has done in a statistical way, because the only possibility to compare quantum mechanical systems is in terms of statistical experiments. Hence, need an additional preparation device P′ and an additional measuring apparatus M′. Indistinguishable now means that it does not matter whether we perform M′ measurements directly on P′ outputs or whether we switch a teleportation device in between; cf. Figure 2-2. In both cases should get the same distribution of measuring results for a large number of repetitions of the corresponding experiment. This requirement should hold for any preparation P′ and any measurement M′, but for fixed M and P. The latter means that are not allowed to use a priori knowledge about P′ or M′ to adopt the teleportation process. Otherwise can choose in most extreme case always P′ for P and the whole discussion becomes meaningless.

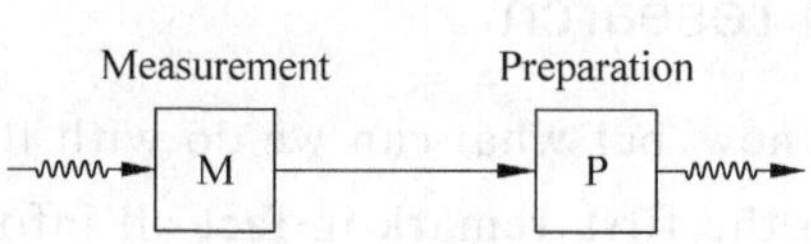

Fig. 2-1 Schematic representation of classical teleportation. Here and in the following diagrams which is curly arrow stands for quantum systems and a straight one for the Jow of classical information.

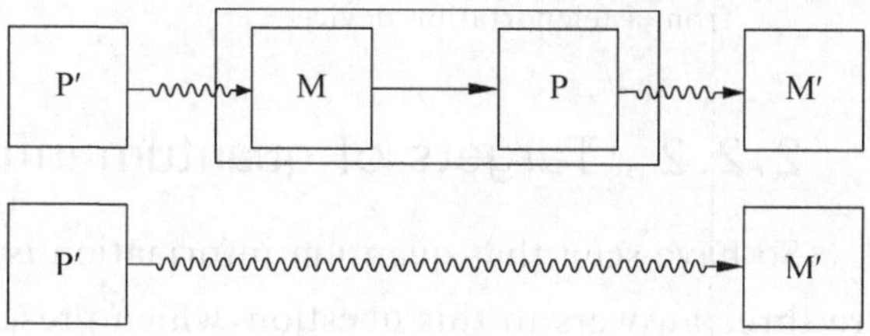

Fig. 2-2 A teleportation process should not affect the results of a statistical experiment with quantum systems. A more precise explanation of the diagram given in the text.

The second impossible machine we have to consider the quantum- copying machine. This is a device C, which takes one quantum system p as input and produces two systems p_1; p_2 of the same type as output. The limiting condition on C is that p_1 and p_2 are indistinguishable from the input, where "indistinguishable" has to be understood in the same way as above: Any statistical experiment performed with one of the output particles (i. e. always with p_1 or always with p_2) yields the same result as applied directly to the input p. To get such a device from teleportation is easy: Jast have to perform an M measurement on p, make two copies of the classical data obtained, and run the preparation P on each of them; cf. Figure 2-3. If teleportation possible copying is possible as well. According to the "no-cloning theorem" of Wootters and Zurek, however, quantum copy machine does not exist and conclude. However, we will give an easy argument for this theorem in terms of the third impossible machine—a joint measuring device MAB for two arbitrary observables A and B. This is a measuring apparatus which produces each time it is invoked a pair (a;b) of classical outputs, where is a possible output of A and b a possible output of B. The crucial requirement for MAB again is of statistical nature: The statistics of the outcomes is the same as for device A, and similarly for B. It known from elementary quantum mechanics that many quantum observables are not jointly measurable in this way. The most famous examples are position and momentum or different components of angular momentum. Nevertheless, a device MAB could constructed for arbitrary A and B from a quantum copy machine C. We simply have to operate with C on the input system p producing two outputs p_1 and p_2 and to perform an A measurement on p_1 and a B measurement on p_2; cf. Figure 2-4. Since the outputs p_1, p_2 are, by assumption, indistinguishable from the input p the overall device constructed this way would give a joint measurement for A and B. Hence, a quantum-copying machine cannot exist, as stated by the no-cloning theorem. This in turn implies that classical teleportation is impossible, and therefore cannot transform quantum information lossless into classical information and back. This concludes the chain of arguments.

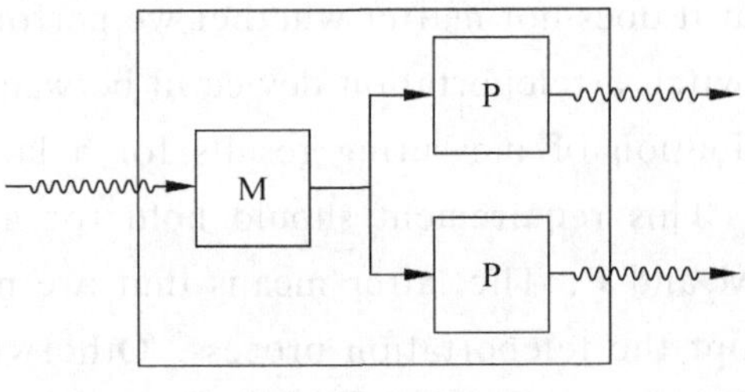

Fig. 2-3 Constructing a quantum copying machine from a teleportation device.

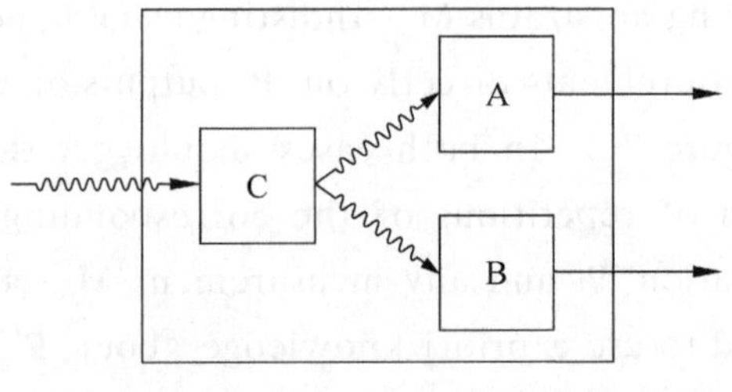

Fig. 2-4 Constructing a joint measurement for the observables A and B from a quantum-copying machine.

2.2.2 Targets of quantum information research

So have seen that quantum information is something new, but what can we do with it? There are three answers to this question, which present here. In the first, remark in fact all information in a modern data processing environment carried by microparticles (e. g. electrons or photons). Hence, quantum information comes automatically into play. Currently, it is safe to ignore this and to use classical information theory to describe all relevant processes. If the size of the structures on a typical circuit decreases below a certain limit, however, this is no longer true and quantum information will become relevant. This leads us to the second answer. Although it is far too early to say which concrete technologies will emerge from quantum information in the future, several interesting proposals show that devices based on quantum information can solve certain practical tasks much better than classical ones. The most well known and exciting one is, without a doubt, quantum computing. The basic idea is, roughly speaking, that a quantum computer can operate not only on one number per register on superpositions of numbers. This possibility leads to an "exponential speedup" for some computations, which makes problems feasible that are considered intractable by any classical algorithm. This demonstrated by Shor's factoring algorithm impressively. A second example which is quite close to a concrete practical realization (i. e. outside the laboratory; see next section) is quantum cryptography. The fact that it is impossible to perform a quantum mechanical measurement without disturbing the state of the measured system is used here for the secure transmission of a cryptographic key i.e. each eavesdropping attempt can be detected with certainty. Together with a subsequent application of a classical encryption method known as the "one-time" pad this leads to a cryptographic scheme with provable security—in contrast to currently used public key systems whose security relies on possibly doubtful assumptions about (pseudo) random number generators and prime numbers. We will come back to both subjects, quantum computing and quantum cryptography. The third answer to the above question of more fundamental nature. The discussion of questions from information theory in the context of quantum mechanics leads to a deeper and in many cases to more quantitative understanding of quantum theory. Maybe the most relevant example for this statement is the study of entanglement, i.e. non-classical correlations between quantum systems, which lead to violations of the Bell inequalities. Entanglement is a fundamental aspect of quantum mechanics and demonstrates the differences between quantum and classical physics in the most drastical way—this can be seen from Bell-type experiments, like the one of Aspect et al. Nevertheless, for a long time it considered as an exotic feature of the foundations of quantum mechanics only, that are not so relevant from a practical point of view. Since quantum information attained broader interest, however, this has changed completely. It has turned out that entanglement is an essential resource whenever classical information processing outperformed by quantum devices. One of the most remarkable examples is the experimental realization of "entanglement enhanced" teleportation, which have argued in Section 1. 1 that classical teleportation, i. e. transmission of quantum information through a classical information channel, is impossible. If sender and receiver share, however, an entangled pair of particles (which can be used as an additional resource) the impossible task becomes, most surprisingly, possible which will discuss this fact in detail in future. Study of entanglement and in particular of the question how it can be quantied is therefore a central topic within quantum information theory. Further examples for fields where quantum information has led to a deeper and in particular more quantitative insights include "capacities" of quantum information channels and "quantum cloning". Finally, let remark that classical information theory benefits in a similar way from the synthesis with quantum mechanics. Beside the just mentioned channel capacities this concerns, for example, the theory of computational complexity which analyzes the scaling behavior of time and space consumed by an algorithm in dependence of the size of the input data. Quantum information challenges here, in particular, the fundamental Church-Turing hypotheses claims that each computation can be simulated "eSciently" on a Turing machine; we come back to this topic.

2.2.3 Experiments

Although this is a theoretical paper, it is of course necessary to say something about experimental realizations of the ideas of quantum information. Consider quantum computing whatever way that need systems prepared very precisely in few distinct states, which need "qubits" and can manipulated afterwards individually (now have to realize "quantum gates") and can nally measured with an appropriate observable, which have to read out the result. One of the most far developed approaches to quantum computing is the ion trap technique. A "quantum register" realized here by a string of ions kept by electromagnetic fields in high vacuum inside a Paul trap, and two long-living states of each ion chosen to represent 0 and 1. A single ion manipulated by laser beams and this allows the implementation of all "one-qubit gates". To get two-qubit gates as well (for a quantum computer need at least one two qubits gate together with all one-qubit operations) the collective motional state of the ions used. A "program" on an ion trap quantum computer starts now with a preparation of the register in an initial state—usually the ground state of the ions. This is optical pumping and laser cooling (which is in fact one of the most diScult parts of the whole procedure, in particular if many ions are involved). Then the "network" of quantum gates applied, in terms of a (complicated) sequence of laser pulses. The readout nally by laser beams illuminate the ions subsequently. The beams tuned to a fast transition affects only one of the qubit states and the Juorescent light detected. Concrete implementations are currently restricted to two qubits; however, there is some hope that we will be able to control up to 10 or 12 qubits in the not too distant future.

A second quite successful technique is NMR quantum computing. NMR stands for "nuclear magnetic resonance" and it is the study of transitions between Zeeman levels of an atomic nucleus in a magnetic field. The qubits are in this case different spin states of the nuclei in an appropriate molecule and quantum gates realized with highfrequency oscillating magnetic. elds in pulses of controlled duration. In contrast to ion traps, however, we do not use one molecule but a whole cup of liquid containing some 1020 of them. This causes a number of problems, concerning in particular the preparation of an initial state, fluctuations in the free time evolution of the molecules and the readout. There are several ways to overcome these difficulties and we refer the reader again for details. Concrete implementations of NMR quantum computers are capable to use up to five qubits. Other realizations include the implementation of several known quantum algorithms on two and three qubits. The fundamental problem of the two methods for quantum computation discussed so far is their lack of scalability. It is realistic to assume that NMR and ion trap quantum computer with up to tens of qubits will exist somewhere in the future but not with thousands of qubits which are necessary for "real-world" applications. There are, however, many other alternative proposals available and some of them might be capable to avoid this problem. The following is a small (not at all exhaustive) list: atoms in optical lattices, semiconductor nanostructures such as quantum dots (there are many works in this area, some recent are) and arrays of Josephson junctions. A second circle of experiments we want to mention here grouped around quantum communication and quantum cryptography (for a more detailed overview). Realizations of quantum cryptography are fairly far developed and it is currently possible to span up to 50km with optical fibers. Potentially greater distances can bridged by "free space cryptography" where the quantum information transmitted through the air. With technology satellites can used as some sort of "relays", thus enabling quantum key distribution over arbitrary distances. In the meantime, there are quite a lot of successful implementations. For a detailed discussion will refer the reader to the review of Gisin et al. and the references therein. Other experiments concern the usage of entanglement in quantum communication. The creation and detection of entangled photons is here a fundamental building block. Nowadays this is no problem and the most famous experiment in this context is the one of aspect et al., where the maximal violation of Bell inequalities demonstrated with polarization-correlated photons. Another spectacular experiment is the creation of entangled photons over a distance of 10km using standard telecommunication optical fibers by the Geneva group. Among the most exciting applications of entanglement is the realization of entanglement based quantum key distribution, the first successful "teleportation" of a photon and the implementation of "dense coding".

2.2.4 Primary concepts

After got a first, rough impression of the basic ideas and most relevant subjects of quantum information theory, start with a more detailed presentation. First, have to introduce the fundamental notions of the theory and their mathematical description. Fortunately, much of the material we should have to present here, like Hilbert spaces, tensor products and density matrices, is known already from quantum mechanics and can focus discussion to concepts, that are less familiar like POV measures, completely positive maps and entangled states.

- Systems, states and effects

As classical probability theory of quantum mechanics is a statistical theory. Hence, its predictions are of probabilistic nature and only be tested if the same experiment is repeated very often and the relative frequencies of the outcomes are calculated. In more operational terms, the means: The experiment has repeated according to the same procedure as it can be set out in a detailed laboratory manual. If consider a somewhat idealized model of such a statistical experiment get, in fact, two different types of procedures: first preparation procedures which prepare a certain kind of physical system in a distinguished state and second registration procedures measuring a particular observable. A mathematical description of such a setup basically consists of two sets S and E and a map $S \times E \ni (\rho A) \to (A) \in [0;1]$. The elements of S describe the states, i. e. preparations, while the $A \in E$ represent all yes = no measurements ($e >$ ects) which can be performed on the system. The probability (i. e. the relative frequency for a large number of repetitions) to get the result "yes", if are measuring the effect A on a system prepared in the state is given by (A). This is a very general scheme applicable not only to quantum mechanics but also to a very broad class of statistical models, containing, in particular, classical probability. In order to make use of it have to specify, of course, the precise structure of the sets S and E and the map $\rho\ (A)$ for the types of systems want to discus.

- Operator algebras

Throughout this paper will encounter three different kinds of systems: Quantum and classical Systems and hybrid systems that are half classical and half quantum. In this subsection will describe a general way to define states and effects which is applicable to all three cases and which therefore provides a handy way to discuss all three cases simultaneously (this will become most useful). The scheme are going to discuss is based on an algebra A of bounded operators acting on a Hilbert space H. More precisely, A is a (closed) linear subspace of $B(H)$, the algebra of bounded operates on H, which contains the identity ($5 \in A$) and is closed under products ($A; B \in A \Rightarrow AB \in A$) and adjoints ($A \in A \Rightarrow A^* \in A$). For simplicity will refer to each such A as an observable algebra. The key observation is now that each type of system will study in the following can be completely characterized by its observable algebra A, i.e. once A is known there is a systematic way to derive the sets S and E and the map $(\rho A) \mapsto \rho(A)$ from it. Frequently make use of this fact by referring to systems in terms of their observable algebra A, or even by identifying them with their algebra and saying that A is the system. Although A and H can be infinite dimensional in general, will consider only finite-dimensional Hilbert spaces, as long as nothing else is explicitly stated. Since most research in quantum information is done up to now for finite-dimensional systems (the only exception in this work is the discussion of Gaussian systems) this is not a too severe loss of generality. Hence can choose $H = C^d$ and $B(H)$ is just the algebra of complex $d \times d$ matrices. Since A is a subalgebra of $B(H)$ it operates naturally on H and it inherits from $B(H)$ the operator norm $\| A \| = \sup_{\|\psi\|=1} \| A\psi \|$ and the operator ordering $A \geqslant B \Leftrightarrow \langle \psi, A\psi \rangle \geqslant \langle \psi, B\psi \rangle \ \forall \psi \in H$. Now can define

$$\mathcal{S}(\mathcal{A}) = \{\rho \in \mathcal{A}^* \mid \rho \geqslant 0, \rho(\mathbb{1}) = 1\} \tag{2-1}$$

where A^* denotes the dual space of A, i.e. the set of all linear functionals on A, and $\rho \geqslant 0$ means $\rho(A) \geqslant 0$; $\forall A \geqslant 0$. Elements of $S(A)$ describe the states of the system in question while effects are given by

$$\mathcal{E}(\mathcal{A}) = \{A \in \mathcal{A} \mid A \geqslant 0, A \leqslant \mathbb{1}\} \tag{2-2}$$

The probability to measure the effect A in the state ρ is $\rho(A)$. More generally, we can look at $\rho(A)$ for an arbitrary A as the expectation value of A in the state ρ. Hence, the idea behind Eq. (2-1) is to define states in terms of their expectation value functionals. Both spaces are convex, i.e. $\rho\sigma \in S(A)$ and $0 \leqslant \lambda \leqslant 1$ implies $\lambda\rho + (1-\lambda)\rho \in S(A)$ and similarly for $E(A)$. The extremal points of $S(A)$, respectively, $E(A)$, i.e. those elements which do not admit a proper convex decomposition($x = \lambda y + (1-\lambda) z \Rightarrow \lambda = 1$ or $\lambda = 0$ or $y = z = x$), play a distinguished role: The extremal points of $S(A)$ are pure states and those of $E(A)$ are the propositions of the system in question. The latter represent those effects register a property with certainty in contrast to non-extremal effects admit some "fuzziness". As a simple example for the latter consider a detector which registers particles not with certainty but only with a probability which is smaller than one. Finally, let us note that the complete discussion of this section can be generalized easily to infinite-dimensional systems, if replace $H = C^d$ by an infinite-dimensional Hilbert space (e.g. $H = L^2(R)$). This would require, however, more material about C^* algebras and measure theory than we want to use in this paper.

- Quantum mechanics

For quantum mechanics have

$$A = B(H) \tag{2-3}$$

where chosen again $H = C^d$. The corresponding systems are called d-level systems or qubits if $d =$

2 holds. To avoid clumsy notations we frequently write $S(H)$ and $E(H)$ instead of $S[B(H)]$ and $E[B(H)]$. An element $P \in E(H)$ is a propositions $i = P$ is a projection operator ($P^2 = P$). States described in quantum mechanics usually by density matrices, i. e. positive and normalized trace class 3 operators. To make contact to the general definition in Eq. (2-1) note first that $B(H)$ is a Hilbert space with the Hilbert-Schmidt scalar product $\langle A; B\rangle = \mathrm{tr}(A^*, B)$. Hence, each linear functional $\rho \in B(H)^*$ can be expressed in terms of a (trace class) operator $\tilde{\rho}$ by $A \mapsto \rho(A) = \mathrm{tr}(\tilde{\rho}A)$. It is obvious that each $\tilde{\rho}$ defines a unique functional ρ. If start on the other hand with can recover the matrix elements of $\tilde{\rho}$ from ρ by $\tilde{\rho}_{kj} = \mathrm{tr}(\tilde{\rho}|j\rangle\langle k|) = (\rho|j\rangle\langle k|)$, where $|j\rangle\langle k|$ denotes the canonical basis of $B(H)$(i.e. $|j\rangle\langle k|_{ab} = \delta_{ja}\delta_{kb}$). More generally, can get for $\psi, \phi \in H$ the relation $(\psi, \tilde{\rho}\psi) = \rho(|\psi\rangle\langle\phi|)$, where $|\psi\rangle\phi|$ now denotes the rank one operator which maps $\eta \in H$ to $\langle\phi, \eta\rangle\psi$. In the following we drop the ~ and use the same symbol for the operator and the functional whenever confusion can be avoided. Due to the same abuse of language will interpret elements of $B(H)^*$ frequently as (trace class) operators instead of linear functionals (and write tr (A) instead of (A)). However, do not identify $B(H)^*$ with $B(H)$ in general, because the two different notations help to keep track of the distinction between spaces of states and spaces of observables. In addition, equip $B^*(H)$ with the trace-norm $\|\rho\|_1 = \mathrm{tr}\,|\rho|$ instead of the operator norm. Positivity of the functional implies positivity of the operator ρ due to $0 \leqslant \rho(|\psi\rangle\langle\psi|) = |\psi\rangle\langle\rho\phi|$ and the same holds for normalization: $1 = \rho(\mathbb{1}) = \mathrm{tr}(\rho)$. Hence, we can identify the state space from Eq. (2-1) with the set of density matrices, as expected for quantum mechanics. Pure states of a quantum system are the one-dimensional projectors. As usual, will frequently identify the density matrix $|\psi\rangle\langle\psi|$ with the wave function ψ and call the latter in abuse of language a state. To get a useful parameterization of the state space consider again the Hilbert-Schmidt scalar product $\langle\rho, \sigma\rangle = \mathrm{tr}(\rho^*\sigma)$, but now on $B^*(H)$. The space of trace free matrices in $B^*(H)$ (alternatively the functionals with $\rho(\mathbb{1}) = 0$) is the corresponding orthocomplement $I^{\perp}$ of the unit operator. If choose a basis $\sigma_1 \cdots\cdots \sigma_{d^2-1}$ with $\langle\sigma_j; \sigma_k\rangle = 2\delta_{jk}$ in $I^{\perp}$ can write each self-adjoint (trace class) operator ρ with $\mathrm{tr}(\rho) = 1$ as

$$\rho = \frac{\mathbb{1}}{d} + \frac{1}{2}\sum_{j=1}^{d^2-1} x_j\sigma_j =: \frac{\mathbb{1}}{d} + \frac{1}{2}\vec{x} \cdot \vec{\sigma} \quad \text{with } \vec{x} \in \mathbb{R}^{d^2-1} \tag{2-4}$$

If $d = 2$ or $d = 3$ holds, it is most natural to choose the Pauli matrices, respectively, the Gell-Mann matrices for the σ_j. In the qubit case it is easy to see that $\rho \geqslant 0$ holds if $|\vec{x}| \leqslant 1$. Hence the state space $S(C^2)$ coincides with the Bloch ball $\{\vec{x} \in \mathbb{R}^3 | \vec{x}| \leqslant 1\}$, and the set of pure states with its boundary, the Bloch sphere $\{\vec{x} \in \mathbb{R}^3 | \vec{x}| \leqslant 1\}$. This shows in a very geometric way that the pure states are the extremal points of the convex set $S(H)$. If it is more generally a pure state of a d-level system can get

$$1 = \mathrm{tr}(\rho^2) = \frac{1}{d} + \frac{1}{2}|\vec{x}|^2 \Rightarrow |\vec{x}| = \sqrt{2(1 - 1/d)} \tag{2-5}$$

This implies that all states are contained in the ball with radius $2^{1/2}(1 - 1/d)^{1/2}$, however, not all operators in this set are positive. A simple example is $d^{-1}\mathbb{1} \pm 2^{1/2}(1 - 1/d)^{1/2}\sigma_j$, which is positive only if $d = 2$ holds.

- Classical probability

Since the difference between classical and quantum systems is an important issue in this work let us reformulate classical probability theory according to the general scheme. The restriction to finite-dimensional observable algebras leads now to the assumption that all systems are considering admit a finite set X of elementary events. Typical examples throwing a dice $X = \{1, 2, \cdots, 6\}$, tossing a coin $X = \{$ "head"; "number" $\}$ or classical bits $X = \{0; 1\}$. To simplify the notations write (as in quantum mechanics) $S(X)$ and $E(X)$ for the spaces of states and effects. The observable algebra A of such a system is the space

$$A = C(X) = \{f : X \rightarrow C\} \tag{2-6}$$

of complex-valued functions on X. To interpret this as an operator algebra acting on a Hilbert space H choose an arbitrary but fixed orthonormal basis $|x\rangle; x \in X$ in H and identify the function $f \in C(X)$ with the operator $f = \sum_x f_x |x\rangle\langle x| \in B(H)$ (use the same symbol for the function and the operator, provided confusion can be avoided). Most frequently have $X = \{1\ 2 \cdots d\}$ and we can choose $H = C^d$ and the canonical basis for $|x\rangle$. Hence, $C(X)$ becomes the algebra of diagonal $d \times d$ matrices. Using Eq. (2-2) we immediately see that $f \in C(X)$ is an effect iff $0 \leqslant f_x \leqslant 1; \forall x \in X$. Physically, can interpret f_x as the probability that the effect f registers the elementary event x. This makes the distinction between propositions and "fuzzy" effects very transparent: $P \in E(X)$ is a proposition iff have either $P_x = 1$ or $P_x = 0$ for all $x \in X$. Hence, the propositions $P \in C(X)$ are in one-to-one correspondence with the subsets $\omega_P = \{x \in X | P_x = 1\} \subset X$ which in turn describe the events of the system. Hence, P registers the event ω_P with certainty, while a fuzzy effect $f < P$ does this only with a probability less than one.

Since $C(X)$ is finite dimensional and admits the distinguished basis $|x\rangle\langle x|; x \in X$ it is naturally isomorphic to its dual $C^*(X)$. As in the quantum case will identify the function with the linear functional and use the same symbol for both, although keep the notation $C^*(X)$ to indicate that are talking about states rather than observables. Positivity of $\rho \in C^*(X)$ is given by $\rho_x \geqslant 0$ for all x and normalization leads to $1 = \rho(\mathbb{1}) = \rho(\sum_x |x\rangle\langle x|) = \sum_x \rho_x$. Hence to be a state $\rho \in C^*(X)$ must be a probability distribution on X and x is the probability that the elementary event x occurs during statistical experiments with systems in the state. More generally $\rho(f) = \sum_j \rho j f j$ is the probability to measure the effect f on systems in the state ρ. If P is in particular, a proposition, $\rho(P)$ gives the probability for the event ω_P. The pure states of the system are the Dirac measures $\delta x; x \in X$; with $\delta x(|y\rangle\langle y|) = \delta_{xy}$.

- Observables

Up to now we have discussed only effects, i. e. yes = no experiments. In this subsection will have a first short look at more general observables. Can think of an observable E taking its values in a finite set X as a map which associates to each possible outcome $x \in X$ the effect $E_x \in E(A)$ (if A is the observable algebra of the system in question) which is true if x is measured and false otherwise. If the measurement is performed on systems in the state ρ get for each $x \in X$ the probability $p_x = \rho(E_x)$ to measure x. Hence, the family of the p_x should be a probability distribution on X, and this implies that E should be a positive operator-valued measure (POV measure) on X.

Definition 2.1 Consider an observable algebra $A \subset B(H)$ and a finite 5 set X. A family $E =$

$(E_x)x \in X$ of effects in A (i.e. $0 \leqslant E_x \leqslant \mathbb{1}$) called a POV measure on X if $\sum_{x \in X} E_x = \mathbb{1}$ holds. If all E_x are projections, E is called projection-valued measure (PV measure). From basic quantum mechanic know that observables described by self-adjoint operators on a Hilbert space H. How does this point of view fit into the previous definition? The answer is given by the spectral theorem: Each self-adjoint operator A on a finite-dimensional Hilbert space H has the form $A = \sum_{\lambda \in \sigma(A)} \lambda P_\lambda$ where $\sigma(A)$ denotes the spectrum of A, i.e. the set of eigenvalues and P_λ denotes the projection onto the corresponding eigenspace. Hence, there is an unique PV measure $P = (P_\lambda)_{\lambda \in \sigma(A)}$ associated to A which is called the spectral measure of A. It is uniquely characterized by the property that the expectation value $\sum_\lambda \lambda \rho(P_\lambda)$ of P in the state ρ is given for any state ρ by $\rho(A) = \mathrm{tr}\,(\rho_A)$; as it is well known from quantum mechanics. Hence, the traditional way to define observables within quantum mechanics perfectly fits into the scheme just outlined, however it only covers the projection-valued case and therefore admits no fuzziness. For this reason POV measures are sometimes called generalized observables. Finally, note that the eigenprojections P_λ of A are elements of an observable algebra $A(A \in \mathscr{A})$. This shows two things: First of all we can consider self-adjoint elements of any $*$-subalgebra A of $B(H)$ as observables of A-systems, and this is precisely the reason why have called A observable algebra. Secondly, why it is essential that A is really a subalgebra of $B(H)$: if it is only a linear subspace of $B(H)$ the relation $A \in \mathscr{A}$ does not imply $P_\lambda \in A$.

- Composite systems and entangled states

Composite systems occur in many places in quantum information theory. A typical example is a register of a quantum computer, which can be regarded as a system consisting of N qubits (if N is the length of the register). The crucial point is that this opens the possibility for correlations and entanglement between subsystems. In particular, entanglement is of great importance, because it is a central resource in many applications of quantum information theory like entanglement enhanced teleportation or quantum computing—already discussed this in Section 1.2 of the Introduction. To explain entanglement detail and introduce some necessary formalism have to complement the scheme developed in the last section by the procedure, which allows to construct states and observables of the composite system from its subsystems. In quantum mechanics this is done, of course, in terms of tensor products, and will review in the following some of the most relevant material.

- Tensor products

Consider two (finite dimensional) Hilbert spaces H and K. To each pair of vectors $\psi_1 \in H$; $\psi_2 \in K$ can associate a bilinear form $\psi_1 \otimes \psi_2$ called the tensor product of ψ_1 and ψ_2 by $\psi_1 \otimes \psi_2(\varphi_1; \varphi_2) = \langle \psi_1, \varphi_1 \rangle \langle \psi_2, \varphi_2 \rangle$. For two product vectors $\psi_1 \otimes \psi_2$ and $\eta_1 \otimes \eta_2$ their scalar product is defined by $\langle \psi_1 \otimes \psi_2; \eta_1 \otimes \eta_2 \rangle = \langle \psi_1; \eta_1 \rangle \langle \psi_2; \eta_2 \rangle$ and it can be shown that this definition extends in a unique way to the span of all $\psi_1 \otimes \psi_2$ which therefore defines the tensor product $H \otimes K$. If have more than two Hilbert spaces $H_j, j = 1, 2, \cdots, N$ their tensor product $H_1 \otimes \cdots \otimes H_N$ can be defined similarly. The tensor product $A_1 \otimes A_2$ of two bounded operators $A_1 \in B(H); A_2 \in B(K)$ is defined first for product vectors $\psi_1 \otimes \psi_2 \in H \otimes K$ by $A_1 \otimes A_2(\psi_1 \otimes \psi_2) = (A_1 \psi_1) \otimes (A_2 \psi_2)$ and then extended by linearity. The space $B(H \otimes K)$ coincides with the span of all $A_1 \otimes A_2$. If $\rho \in B(H \otimes$

K) is not of product form (and of trace class for infinite-dimensional H and K) there is nevertheless a way to define "restrictions" to H, respectively, K called the partial trace of. It is defined by the equation

$$\mathrm{tr}\,[\mathrm{tr}_{\mathscr{K}}(\rho)A] = \mathrm{tr}(\rho A \otimes \mathbb{1}) \quad \forall A \in \mathscr{B}(\mathscr{H}) \tag{2-7}$$

where the trace on the left-hand side is over H and on the right-hand side over $H \otimes K$. If two orthonormal bases $\varphi_1 \varphi_2 \cdots \varphi_n$ and $\psi_1 \psi_2 \cdots \psi_m$ are given in H, respectively, K can consider the product basis $\varphi_1 \otimes \psi_1 \cdots \varphi_n \otimes \psi_m$ in $H \otimes K$, and can expand each "$\psi \in H \otimes K$ as $\psi = \sum_{jk} \psi_{jk} \varphi_j \otimes \psi_k$ with $\psi_{jk} = \langle \varphi_j \otimes \psi_k ; \psi \rangle$. This procedure works for an arbitrary number of tensor factors. However, if we have exactly a twofold tensor product, there is a more economic way to expand", called Schmidt decomposition in which only diagonal terms of the form $\varphi_j \otimes \psi_j$ appear.

Proposition 2.2 For each element of the twofold tensor product $H \otimes K$ there are orthonormal systems φ_j; $j = 1, 2, \cdots, n$ and ψ_k; $k = 1, 2, \cdots, n$ (not necessarily bases; i.e. n can be smaller than dim H and dim K) of H and K; respectively; such that $\psi = \sum_j \sqrt{\lambda_j}, \varphi_j \otimes \psi_j$ holds. The φ_j and ψ_j uniquely determined by ψ. The expansion called Schmidt decomposition and the numbers $\sqrt{\lambda_j}$ are the Schmidt coefficients.

Proof. Consider the partial trace $\rho_1 = \mathrm{tr}_K(|\psi\rangle\langle\psi|)$ of the one-dimensional projector $|\psi\rangle\langle\psi|$ associated to ψ. It can be decomposed in terms of its eigenvectors φ_n and get $\mathrm{tr}_K(|\psi\rangle\langle\psi|) = \rho_1 = \sum_n \lambda_n |\varphi\rangle\langle\varphi|$. Now we can choose an orthonormal basis ψ'_k; $k = 1, 2, \cdots, m$ in K and expand ψ with respect to $\varphi_j \otimes \psi_k$. Carrying out the k summation can get a family of vectors $\psi''_j = \sum_k \langle \Psi, \phi_j \otimes \psi'_k \rangle \psi'_k$ with the property $\Psi = \sum_j \phi_j \otimes \psi''_j$.

Then can calculate the partial trace and get for any $A \in B(H_1)$:

$$\sum_j \lambda_j \langle \phi_j, A\phi_j \rangle = \mathrm{tr}(\rho_1 A) = \langle \Psi, (A \otimes \mathbb{1}) \Psi \rangle = \sum_{j,k} \langle \phi_j, A\phi_k \rangle \langle \psi''_j, \psi''_k \rangle \tag{2-8}$$

Since A is arbitrary we can compare the left- and right-hand side of this equation term by term and can get:

$$\langle \psi''_j, \psi''_k \rangle = \delta_{jk} \lambda_j \tag{2-9}$$

Hence $\psi_j = \lambda_j^{-1/2} \psi''_j$ is the desired orthonormal system. As an immediate application of this result that each mixed state $\rho \in B^*(H)$ (of the quantum system $B(H)$) can be regarded as a pure state on a larger Hilbert space $H \otimes H'$ that just have to consider the eigenvalue expansion $\rho = \sum_j \rho_j |\varphi_j\rangle\langle\varphi_j|$ of ρ and to choose an arbitrary orthonormal system ψ_j; $j = 1, 2, \cdots, n$ in H'. Using Proposition 2.2 can get.

Corollary 2.3 Each state $\rho \in B^*(H)$ can be extended to a pure state ψ on a larger system with Hilbert space $H \otimes H'$ such that $\mathrm{tr}_{H'} |\psi\rangle\langle\psi| = \rho$ holds.

- Compound and hybrid systems

To discuss the composition of two arbitrary (i.e. classical or quantum) systems it is very convenient to use the scheme developed in Section 2.1.1 and to talk about the two subsystems in terms of their observable algebras $A \subset B(H)$ and $B \subset B(K)$. The observable algebra of the composite system is then simply given by the tensor product of A and B, i.e.

$$\mathscr{A} \otimes \mathscr{B} := \text{span}\{A \otimes B \mid A \in \mathscr{A}, B \in \mathscr{B}\} \subset \mathscr{B}(\mathscr{K} \otimes \mathscr{H}) \tag{2-10}$$

The dual of $A \otimes B$ is generated by product states, $(\rho \otimes \sigma)(A \otimes B) = \rho(A)\sigma(B)$ and therefore write $A^* \otimes B^*$ for $(A \otimes B)^*$. The interpretation of the composed system $A \otimes B$ in terms of states and effects straightforward and therefore postponed to the next subsection. Consider first the special cases arising from different choices for A and B. If both systems are quantum ($A = B(H)$ and $B = B(K)$) can get

$$B(H) \otimes B(K) = B(H \otimes K) \tag{2-11}$$

As expected. For two classical systems $A = C(X)$ and $B = C(Y)$ recall that elements of $C(X)$ (respectively, $C(Y)$) are complex-valued functions on X (on Y). Hence, the tensor product $C(X) \otimes C(Y)$ consists of complex-valued functions on $X \times Y$, i.e. $C(X) \otimes C(Y) = C(X \times Y)$. In other words, states and observables of the composite system $C(X) \otimes C(Y)$ are, in accordance with classical probability theory, given by probability distributions and random variables on the Cartesian product $X \times Y$. If only one subsystem is classical and the other is quantum; e.g. a microparticle interacting with a classical measuring device we have a hybrid system. The elements of its observable algebra $C(X) \otimes B(H)$ can be regarded as operator-valued functions on X, i.e. $X \ni x \mapsto A_x \in B(H)$ and A is an effect iff $0 \leqslant A_x \leqslant \mathbb{1}$ holds for all $x \in X$. The elements of the dual $C^*(X) \otimes B^*(H)$ are in a similar way $B^*(X)$-valued functions $X \ni x \mapsto \rho_x \in B^*(H)$ and is a state iff each ρ_x is a positive trace class operator on H and $\sum_x \rho_x = 1$. The probability to measure the effect A in the state ρ is $\sum_x \rho_x(A_x)$.

- Correlations and entanglement

Now consider two effects $A \in \mathscr{A}$ and $B \in \mathscr{B}$ then $A \otimes B$ is an effect of the composite system $A \otimes B$. It interpreted as the joint measurement of A on the first and B on the second subsystem, where the "yes" outcome means: both effects give yes. In particular, $A \otimes \mathbb{1}$ means to measure A on the first subsystem and to ignore the second one completely. If ρ is a state of $A \otimes B$ can define its restrictions by $\rho^A(A) = \rho(A \otimes \mathbb{1})$ and $\rho^B(A) = (\mathbb{1} \otimes A)$. If both systems are quantum the restrictions of are the partial traces, while in the classical case have to sum over the B, respectively A, variables. For two states $\rho_1 \in S(A)$ and $\rho_2 \in S(B)$ there is always a state ρ of $A \otimes B$ such that $\rho_1 = \rho^A$ and $\rho_2 = \rho^B$ holds: Just have to choose the product state $\rho_1 \otimes \rho_2$. However, in general, have $\rho \neq \rho^A \otimes \rho^B$ which means nothing else then ρ also contains correlations between the two subsystems.

Definition 2.4 A state ρ of a bipartite system $A \otimes B$ is called correlated if there are some $A \in A$; $B \in B$ such that $\rho(A \otimes B) \neq \rho^A(A)\rho^B(B)$ holds. Immediately see that $\rho = \rho_1 \otimes \rho_2$ implies $\rho(A \otimes B) = \rho_1(A)\rho_2(B) = \rho^A(A)\rho^B(B)$ hence ρ is not correlated. If on the other hand $\rho(A \otimes B) = \rho^A(A)\rho^B(B)$ holds get $= \rho^A \otimes \rho^B$. Hence, the definition of correlations just given perfectly fits into the intuitive considerations.

An important issue in quantum information theory is the comparison of correlations between quantum systems on the one hand and classical systems on the other. Hence, let have a closer look on the state space of a system consisting of at least one classical subsystem.

Proposition 2.5 Each state ρ of a composite system $A \otimes B$ consisting of a classical ($A = C(X)$) and an arbitrary system (B) has the form

$$\rho = \sum_{j \in X} \lambda_j \rho_j^{\mathscr{A}} \otimes \rho_j^{\mathscr{B}} \tag{2-12}$$

with positive weights $\lambda_j > 0$ and $\rho_j^A \in S(A)$; $\rho_j^B \in S(B)$.

Proof. Since $A = C(X)$ is classical; there is a basis $|j\rangle\langle j| \in A$; $j \in X$ of mutually orthogonal one-dimensional projectors and can write each $A \in \mathscr{A}$ as $\sum_j a_j |j\rangle\langle j|$ (cf. Subsection 2.1.3). For each state $\rho \in S(A \otimes B)$ can now define $\rho_j^A \in S(A)$ with $\rho_j^A = \mathrm{tr}(A|j\rangle\langle j|) = a_j$ and $\rho_j^B \in S(B)$ with $\rho_j^B(B) = \lambda_j^{-1}(|j\rangle\langle j| \otimes B)$ and $\lambda_j = (|j\rangle\langle j| \otimes \mathbb{1})$. Hence get $\rho = \sum_{j\in X} \lambda_j \rho_j^A \otimes \rho_j^B$ with positive λ_j as stated. If A and B are two quantum systems it is still possible for them to be correlated in the way just described. Then can simply prepare them with a classical random generator which triggers two preparation devices to produce systems in the states ρ_j^A; ρ_j^B with probability λ_j. The overall state produced by this setup is obviously the ρ from Eq. (2-11). However, the crucial point is that not all correlations of quantum systems are of this type. This is an immediate consequence of the definition of pure states $\rho = |\psi\rangle\langle\psi| \in S(H)$: Since there is no proper convex decomposition of, it can be written as in Proposition 2.5 iff ψ is a product vector, i.e. $\Psi = \varphi \otimes \psi$ This observation motivates the following definition.

Definition 2.6 A state ρ of the composite system $B(H_1) \otimes B(H_2)$ is called separable or classically correlated if it can be written as

$$\rho = \sum_j \lambda_j \rho_j^{(1)} \otimes \rho_j^{(2)} \tag{2-13}$$

with states $\rho_j^{(k)}$ of $B(H_k)$ and weights $\lambda_j > 0$. Otherwise ρ is called entangled. The set of all separable states is denoted by $D(H_1 \otimes H_2)$ or just D if H_1 and H_2 are understood.

- Bell inequalities

Just seen that it is quite easy for pure states to check whether they entangled or not. In the mixed case however this is a much bigger, and in general unsolved problem. In this subsection will have a short look at the Bell inequalities, which are maybe the oldest criterion entanglement. Today more powerful methods, which most of them based on positivity properties, are available. Postpone the corresponding discussion to the end of the following section, after we have studied (completely) positive maps (cf. Section 2.4). Bell inequalities discussed in the framework of "local hidden variable theories" traditionally. More precisely that a state ρ of a bipartite system $B(H \otimes K)$ admits a hidden variable model, if there is a probability space $(X;\mu)$ and (measurable) response functions $X \ni x \mapsto F_A(x;k); F_B(x;1) \in R$ for all discrete PV measures $A = A_1, A_2, \cdots, A_N \in B(H)$, respectively $B = B_1, B_2, \cdots, B_M \in B(K)$, such that:

$$\int_X F_A(x,k) F_B(x,l) \mu(dx) = \mathrm{tr}(\rho A_k \otimes B_l) \tag{2-14}$$

Hold for all, k; l and A; B. The value of the functions $F_A(x;k)$ is interpreted as the probability to get the value k during an A measurement with known "hidden parameter" x. The set of states admitting a hidden variable model is a convex set and as such can described by an (infinite) hierarchy of correlation inequalities. Any one of these inequalities usually called (generalized) Bell inequality. The most well-known one is those given by Clauser et al. The state ρ satisfies the CHSH-inequality if

$$\rho(A \otimes (B + B') + A' \otimes (B - B')) \leqslant 2 \tag{2-15}$$

holds for all $A; A' \in B(H)$, respectively $B; B' \in B(K)$, with $-\mathbb{1} \leqslant A; A' \leqslant \mathbb{1}$ and $-\mathbb{1} \leqslant B; B' \leqslant \mathbb{1}$. F or the special case of two dichotomic observables the CHSH inequalities are sufficient to characterize the states with a hidden variable model. In the general, CHSH inequalities are a necessary but not a sufficient condition and a complete characterization is not known. It is now easy to see that each separable state

$$\rho = \sum_{j=1}^{n} \lambda_j \rho_j^{(1)} \otimes \rho_j^{(2)} \tag{2-16}$$

If admits a hidden variable model: have to choose $X = 1, 2, \cdots, n$; $\mu(\{j\}) = \lambda_j$; $F_A(x;k) = \rho_x^{(1)}(A_k)$ and F_B analogously. Hence, immediately see that each state of a composite system with at least one classical subsystem satisfies the Bell inequalities (in particular the CHSH version) while this is not the case for pure quantum systems. The most prominent examples are "maximally entangled states" which violate the CHSH inequality (for appropriately chosen A; A'; B; B') with a maximal value of $2\sqrt{2}$. This observation is the starting point for many discussions concerning the interpretation of quantum mechanics, in particular because the maximal violation of $2\sqrt{2}$ observed experimentally by aspect and coworkers. Do not want to follow this path. Interesting is the fact that Bell inequalities, in particular the CHSH case in Eq. (2-15), provide a necessary condition for a state ρ to be separable. However, there exist entangled states admitting a hidden variable model. Hence, Bell inequalities are not sufficient for separability.

- Channels

Assume now that have a number of quantum systems, e. g. a string of ions in a trap. To "process" the quantum information they carry have to perform, in general, many steps of a quite different nature. Typical examples are: free time evolution, controlled time evolution (e. g. the application of a "quantum gate" in a quantum computer), preparations and measurements. The purpose of this section is to provide a united framework for the description of all these different operations. The basic idea is to represent each processing step by a "channel", which converts input systems, described by an observable algebra A into output systems described by a possibly different algebra B. Henceforth will call A the input and B the output algebra. If consider e. g. the free time evolution, need quantum systems of the same type on the input and the output side; hence, in this case have $A = B = B(H)$ with an appropriately chosen Hilbert space H. If on the other hand, want to describe a measurement we have to map quantum systems (the measured system) to classical information (the measuring result). Therefore, need in this example $A = B(H)$ for the input and $B = C(X)$ for the output algebra, where X is the set of possible outcomes of the measurement. The aim is now to get a mathematical object, which can used to describe a channel. To this end consider an effect $A \in B$ of the output system. If invoke first a channel which transforms A systems into B systems, and measure A afterwards on the output systems, end up with a measurement of an effect $T(A)$ on the input systems. Hence, we get a map $T^*: E(B) \to E(A)$. Alternatively, we can look at the states and interpret a channel as a map $T^*: S(A) \to S(B)$ which transforms A systems in the state $\rho \in S(A)$ into B systems in the state $T^*(\rho)$. To distinguish between both maps we can say that T describes the channel in the Heisenberg picture and T^* in the Schrödinger picture. On the level of the statistical interpretation both points of view should coincide of course, i. e. the probabilities $(T^*\rho)(A)$ and $\rho(T_A)$ to get the result "yes" during an A measurement on B systems in the state $T^*\rho$, respectively, a T_A measurement on A systems in the state ρ, should be the same. Since $(T^*\rho)(A)$ is linear in A immediately, that T

must be an affine map, i. e. $T(\lambda_1 A_1 + \lambda_2 A_2) = \lambda_1 T(A_1) + \lambda_2 T(A_2)$ for each convex linear combination $\lambda_1 A_1 + \lambda_2 A_2$ of effects in B, this in turn implies that T can extende naturally to a linear map, which will identify in the following with the channel itself, i. e. that T is the channel.

- Completely positive maps

Let change now slightly the point of view and start with a linear operator $T: A \to B$. To be a channel, T must map effects to effects, i. e. T has to be positive: $T(A) \geqslant 0 \ \forall A \geqslant 0$ and bounded from above by $\mathbb{1}$, i. e. $T(\mathbb{1}) \leqslant \mathbb{1}$. In addition it is natural to require that two channels in parallel are again a channel. More precisely, if two channels $T: A_1 \to B_1$ and $S: A_2 \to B_2$ are given we can consider the map $T \otimes S$ which associates to each $A \otimes B \in A_1 \otimes A_2$ the tensor product $T(A) \otimes S(B) \in B_1 \otimes B_2$. It is natural to assume that $T \otimes S$ is a channel which converts composite systems of type $A_1 \otimes A_2$ into $B_1 \otimes B_2$ systems. Hence $S \otimes T$ should be positive as well.

Definition 2.7 Consider two observable algebras A; B and a linear map $T: A \to B \subset B(H)$.

Consider now the map $T^*: B^* \to A^*$ which is dual to T, i. e. $T^*(A) = T(A)$ for all $\in B^*$ and $A \in \mathscr{A}$. It called the Schrödinger picture representation of the channel T, since it maps states to states provided T is unital. (Complete) positivity can be defined in the Schrödinger picture as in the Heisenberg picture and immediately see that T is (completely) positive $i = T^*$ is. It is natural to ask whether the distinction between positivity and complete positivity really necessary, i. e. whether there are positive maps which are not completely positive. If at least one of the algebras A or B is classical, answer is no: each positive map is completely positive in this case. If both algebras are quantum, however, complete positivity not implied by positivity alone.

If item 2 holds only for a fixed $n \in N$ the map T is called n-positive. This is obviously a weaker condition than complete positivity. Consider now the question whether a channel should be unital or not. That have already mentioned that $T(\mathbb{1}) \leqslant \mathbb{1}$ must hold since effects should be mapped to effects. If $T(\mathrm{J})$ is not equal to $\mathbb{1}$ can get $\rho(T\mathbb{1}) = T^* \rho(\mathbb{1}) < 1$ for the probability to measure the effect $\mathbb{1}$ on systems in the state $T^* \rho$, but this is impossible for channels which produce an output with certainty, because 5 is the effect which is always true. In other words: If a cp map is not unital it describes a channel which sometimes produces no output at all and $T(\mathbb{1})$ is the effect which measures whether have got an output. Assume in the future that channels are unital if nothing else is explicitly stated.

- The Stinespring theorem

Consider now channels between quantum systems, i. e. $A = B(H_1)$ and $B = B(H_2)$. A fairly simple example (not necessarily unital) is given in terms of an operator $V: H_1 \to H_2$ by $B(H_1) \ni A \mapsto VAV^* \in B(H_2)$. A second example is the restriction to a subsystem, which is given in the Heisenberg picture by $B(H) \ni A \mapsto A \otimes 5K \in B(H \otimes K)$. Finally, the composition $S \circ T = ST$ of two channels is again a channel. The following theorem, which is the most fundamental structural result about cp maps, that each channel can be represented as a composition of these two examples.

Theorem 2.8 (Stinespring dilation theorem). Every completely positive map $T: B(H_1) \to B(H_2)$ has the form

$$T(A) = V^*(A \otimes 5K)V \tag{2-17}$$

with an additional Hilbert space K and an operator $V: H_2 \to H_1 \otimes K$. Both (i. e. K and V) can be chosen such that the span of all $(A \otimes \mathbb{1})V\varphi$ with $A \in B(H_1)$ and $\varphi \in H_2$ is dense in $H_1 \otimes K$. This particular decomposition is unique (up to unitary equivalence) and called the minimal decomposition. If $\dim H_1 = d_1$ and $\dim H_2 = d_2$ the minimal K satisfies $\dim K \leqslant d_1^2 d_2$. By introducing a family $|X_j\rangle\langle X_j|$ of one-dimensional projectors with $\sum_j |X_j\rangle\langle X_j| = \mathbb{1}$ can define the "Kraus operators" $\langle \psi V_j \varphi\rangle = \langle \psi \otimes + j, V\varphi\rangle$. In terms of them we can rewrite Eq. (2-17) in the following form:

Corollary 2.9 (Kraus form). Every completely positive map $T: B(H_1) \to B(H_2)$ can be written in the form

$$T(A) = \sum_{j=1}^{N} V_j^* A V_j \tag{2-18}$$

with operators $V_j: H_2 \to H_1$ and $N \leqslant \dim(H_1)\dim(H_2)$.

- The duality lemma

Consider a fundamental relation between positive maps and bipartite systems, which will allow us later on to translate properties of entangled states to properties of channels and vice versa. The basic idea originates from elementary linear algebra: A bilinear form φ on a d-dimensional vector space V can be represented by a $d \times d$-matrix, just as an operator on V. Hence, we can transform φ into an operator simply by reinterpreting the matrix elements. In the situation things are more difficult, because the positivity constraints for states and channels should match up in the right way. Nevertheless, have the following theorem.

Theorem 2.10 Let ρ be a density operator on $H \otimes H_1$. Then there is a Hilbert space K a pure state σ on $H \otimes K$ and a channel $T: B(H_1) \to B(K)$ with:

$$\rho = (\mathrm{Id} \otimes T^*)\sigma \tag{2-19}$$

Where Id denotes the identity map on $B^*(H)$. The pure state σ can be chosen such that $\mathrm{tr}H(\sigma)$ has no zero eigenvalue. In this case T and σ are uniquely determined (up to unitary equivalence) by Eq. (2-17); i.e. if $\tilde{\sigma}$; $\tilde{T}$ with $\rho = (\mathrm{Id} \otimes \tilde{T}^*)\tilde{\sigma}$ are given; have $\tilde{\sigma} = (\mathbb{1} \otimes U) * (\mathbb{1} \otimes U)$ with an appropriate unitary operator U.

Proof. The state σ is obviously the purification of $\mathrm{tr}H_1(\rho)$. Hence if λ_j and j are eigenvalues and eigenvectors of $\mathrm{tr}H_1(\rho)$ can set $\sigma = |\Psi\rangle\langle\Psi|$ with $\psi = \sum_j \sqrt{\lambda j_j}\psi_j \otimes \varphi_j$ where φ_j is an (arbitrary) orthonormal basis in K. It is clear that σ is uniquely determined up to a unitary. Hence; only have to show that a unique T exists if ψ is given. To satisfy Eq. (2-17) we must have

$$\rho(|\psi_j \otimes \eta_k\rangle\langle\psi_l \otimes \eta_l|) = \langle\Psi, (\mathrm{Id} \otimes T)(|\psi_j \otimes \eta_k\rangle\langle\psi_l \otimes \eta_l|)\Psi\rangle \tag{2-20}$$

$$= \langle\Psi, |\psi_j\rangle\langle\psi_l| \otimes T(|\eta_k\rangle\langle\eta_p|)\Psi\rangle \tag{2-21}$$

$$= \sqrt{\lambda_j\lambda_l}\langle\phi_j, T(|\eta_k\rangle\langle\eta_p|)\phi_l\rangle \tag{2-22}$$

Where η_k is an (arbitrary) orthonormal basis in H_1. As T determined uniquely by ρ in terms of its matrix elements. Only have to check complete positivity. To the end is useful to note that the map $\rho \mapsto T$ is linear if the λ_j is fixed. Hence, it is sufficient to consider the case $\rho = |X\rangle\langle X|$. Inserting this into Eq. (2-22) immediately see that:

$$T(A) = V^* AV \text{ with } \langle V\phi_j, \eta_k\rangle = \lambda_i^{-1/2}\langle\psi_j \otimes \eta_k, \chi\rangle \tag{2-23}$$

Therefore, T is completely positive. Since normalization $T(\mathbb{1}) = \mathbb{1}$ follows from the choice of the λ_j the theorem is proved.

2.2.5 Separability criteria and positive maps

Although no such map represents a valid quantum operation, but they are great importance in quantum information theory, due to their deep relations to entanglement properties. Hence, this section is a continuation of the study of separability criteria, which have started. In contrast to the rest of this section, all maps considered in the Schrödinger rather than in the Heisenberg picture.

- Positivity

Consider now an arbitrary positive, but not necessarily completely positive map $T^*: B^*(H) \rightarrow B^*(K)$. If Id again denotes the identity map, it is easy to see that $(\mathrm{Id} \otimes T^*)(\sigma_2 \otimes \sigma_2) = \sigma_1 \otimes T^*(\sigma_2) \geqslant 0$ holds for each product state $\sigma_1 \otimes \sigma_2 \in S(H \otimes K)$. Hence $((\mathrm{Id} \otimes T^*) \geqslant 0$ for each positive T^* is a necessary condition for ρ to be separable. The following theorem proved in shows that sufficiency holds as well.

Theorem 2.11 A state $\rho \in B^*(H \otimes K)$ is for any positive map $T^*: B^*(K) \rightarrow B^*(H)$ the operator $(\mathrm{Id} \otimes T^*)\rho$ is positive.

Proof. It will only give a sketch of the proof for details. The condition is obviously necessary since $(\mathrm{Id} \otimes T^*)\rho_1 \otimes \rho_2 \geqslant 0$ holds for any product state provided T^* is positive. The proof of sufficiency relies on the fact that it is always possible to separate a point ρ (an entangled state) from a convex set D (the set of separable states) by a hyperplane. A precise formulation of this idea leads to the following proposition.

Proposition 2.12 For any entangled state $\rho \in S(H \otimes K)$ there is an operator A on $H \otimes K$ called entanglement witness for ρ; with the property $\rho(A) \geqslant 0$ and $\rho(A) \geqslant 0$ for all separable $\sigma \in S(H \otimes K)$.

Proof. Since $D \subset B^*(H \otimes K)$ is a closed convex set; for each $\rho \in S \subset B^*(H \otimes K)$ with $\rho \notin D$ there exists a linear functional α on $B^*(H \otimes K)$; such that $\alpha(\rho) < \gamma \leqslant \alpha(\rho)$ for each $\sigma \in D$ with a constant γ. This holds as well in infinite-dimensional Banach spaces and is a consequence of the Hahn-Banach theorem. Without loss of generality; can assume that $\gamma = 0$ holds. Otherwise, just have to replace α by α-γ tr. Hence; the result follows from the fact that each linear functional on $B^*(H \otimes K)$ has the form $\alpha(\sigma) = \mathrm{tr}(A\sigma)$ with $A \in B(H \otimes K)$. To continue the proof of Theorem 2.11 associate now to any operator $A \in B(H \otimes K)$ the map $T_A^*: B^*(K) \rightarrow B^*(H)$ with

$$\mathrm{tr}(A\rho_1 \otimes \rho_2) = \mathrm{tr}(\rho_1^{\mathrm{T}} T_A^*(\rho_2)) \tag{2-24}$$

Where $(\cdot)^{\mathrm{T}}$ denotes the transposition in an arbitrary. It is easy to see that T_A^* is positive if $\mathrm{tr}(A\rho_1 \otimes \rho_2) \geqslant 0$ for all product states $\rho_1 \otimes \rho_2 \in S(H \otimes K)$. A straightforward calculation shows in addition that

$$\mathrm{tr}(A\rho) = \mathrm{tr}(|\Psi\rangle\langle\Psi| (\mathrm{Id} \otimes T_A^*)(\rho)) \tag{2-25}$$

Holds, where $\psi = d^{-1/2} \sum_j |j\rangle \otimes |j\rangle$. Assume now that $(\mathrm{Id} \otimes T^*) \geqslant 0$ for all positive T^* Since T_A^* is positive this implies that the left-hand side of Eq. (2-25) is positive; hence $\mathrm{tr}(A\rho) \geqslant 0$ provided $\mathrm{tr}(A\sigma) \geqslant 0$ holds for all separable σ, and the statement follows.

- The partial transpose

The most typical example for a positive non-cp map is the transposition $\theta A = A^{\mathrm{T}}$ of $d \times d$ matrices, which have just used in the proof of Theorem. θ is obviously a positive map, but the partial transpose

$$\mathscr{B}^*(\mathscr{H} \otimes \mathscr{K}) \ni \rho \mapsto (\mathrm{Id} \otimes \Theta)(\rho) \in \mathscr{B}^*(\mathscr{H} \otimes \mathscr{K}) \tag{2-26}$$

is not. The latter can be easily checked with the maximally entangled state.

$$\Psi = \frac{1}{\sqrt{d}} \sum_j |j\rangle \otimes |j\rangle \tag{2-27}$$

where $|j\rangle \in C^d$; $j = 1, 2, \cdots, d$ denote the canonical basis vectors. In low dimensions of transposition are positive basial map only. Due to results of Sthrmer and Woronowicz have dim $H=2$ and dim $K=2;3$ imply that each positive map $T^*: B^*(H) \to B^*(K)$ has the form $T^* = T_1^* + T^*2/$ with two cp maps T_1^*, T_2^* and the transposition on $B(H)$. This immediately implies that positivity of the partial transpose is necessary and sufficient for separability of a state $\rho \in S(H \otimes K)$.

Consider a bipartite system $B(H \otimes K)$ with dim $H=2$ and dim $K=2,3$. A state $\rho \in S(H \otimes K)$ is separable iff its partial transpose is positive. To use positivity of the partial transpose as a separability criterion proposed for the first time by Peres, and he conjectured that it is a necessary and sufficient condition in arbitrary finite dimension. Although it has turned out in the meantime that this conjecture is wrong in general, partial transposition has become a crucial tool within entanglement theory and define.

Definition 2.13 A state $\rho \in B^*(H \otimes K)$ of a bipartite quantum system is called ppt-state if $(\mathrm{Id} \otimes \theta)\rho \geqslant 0$ holds and npt-state otherwise (ppt = "positive partial transpose" and npt = "negative partial transpose").

- The reduction criterion

Another frequently used example of a non-cp but positive map is $B^*(H) \ni \rho \mapsto T^*(\rho) = (\mathrm{tr}\,\rho)\mathbb{1} - \rho \in B^*(H)$. The eigenvalues of $T^*(\rho)$ are given by $\mathrm{tr}\rho - \lambda_i$, where λ_i are the eigenvalues of ρ. If $\rho \geqslant 0$ have $\lambda_i \geqslant 0$ and therefore $\sum_j \lambda_j - \lambda_k \geqslant 0$. Hence T^* is positive. That T^* is not completely positive follows if consider again the example $|\Psi\rangle\langle\Psi|$ from Eq. (2-24); hence can get

$$\begin{cases} \mathbb{1} \otimes \mathrm{tr}_2(\rho) - \rho \geqslant 0 \\ \mathrm{tr}_1(\rho) \otimes \mathbb{1} - \rho \geqslant 0 \end{cases} \tag{2-28}$$

for any separable state $\rho \in B^*(H \otimes K)$. These equations are another non-trivial separability criterion, which called the reduction criterion. It closely related to the ppt criterion, due to the following proposition. Each ppt-state $\rho \in S(H \otimes K)$ satisfies the reduction criterion. If dim $H=2$ and dim $K=2;3$ both criteria are equivalent. Hence that a state ρ in 2×2 or 2×3 dimensions is separable iff it satisfies the reduction criterion.

2.3 Basic concepts

After the somewhat abstract discussion in the last section will become more concrete now. In the following, will present a number of examples, which help on the one hand to understand the structures just introduced, and which are of fundamental importance within quantum information on the other. Although entanglement definition of entanglement is applicable in arbitrary dimensions, detailed knowledge about entangled states is available only for low-dimensional systems or for states with very special properties. In this section will discuss some of the most basic examples.

2.3.1 Maximally entangled states

Let start with a look on pure states of composite systems $A \otimes B$ and their possible correlations.

If one subsystem is classical, i. e. $A = C(\{1, 2, \cdots, d\})$, the state space is given according $S(B)^d$ and $\rho \in S(B)^d$ is pure iff $\rho = (\delta_{j1}\tau, \delta_{j2}\tau, \cdots, \delta_{jd}\tau)$ with $j = 1, 2, \cdots, d$ and a pure state τ of the B system. Hence, the restrictions of ρ to A, respectively, B are the Dirac measure $\delta_j \in S(X)$ or $\tau \in S(B)$, in other words both restrictions are pure. This is completely different if A and B are quantum, i.e. $A \otimes B = B(H \otimes K)$: Consider $\rho = |\psi\rangle\langle\psi|$ with $\psi \in H \otimes K$ and Schmidt decomposition (Proposition 2.2) $\psi = \sum_j \lambda_j^{1/2} \varphi_j \otimes \psi_j$. Calculating the A restriction, i.e. the partial trace over K can get

$$\mathrm{tr}\,[\mathrm{tr}_{\mathscr{K}}(\rho)A] = \mathrm{tr}\,[|\Psi\rangle\langle\Psi|A \otimes \mathbb{1}] = \sum_{jk} \lambda_j^{1/2}\lambda_k^{1/2}\langle\phi_j, A\phi_k\rangle\delta_{jk} \tag{2-29}$$

Hence $\mathrm{tr}_K(\rho) = \sum_j \lambda_j |\varphi_j\rangle\langle\varphi_j|$ is mixed iff ψ is entangled. The most extreme case arises if $H = K = C^d$ and $\mathrm{tr}_K(\rho)$ is maximally mixed, i.e. $\mathrm{tr}_K(\rho) = \mathbb{1}/d$. to get for ψ

$$\Psi = \frac{1}{\sqrt{d}} \sum_{j=1}^{d} \phi_j \otimes \psi_j \tag{2-30}$$

With two orthonormal bases $\varphi_1\varphi_2\cdots\varphi_d$ and $\psi_1\psi_2\cdots\psi_d$. In $2n \times 2n$ dimensions these states violate maximally the CHSH inequalities, with appropriately chosen operators A, A', B, B'. Such states are therefore called maximally entangled. The most prominent examples of maximally entangled states are the four "Bell states" for two qubit systems, i. e. $H = K = C^2$; $|1\rangle$; $|0\rangle$ denotes the canonical basis and

$$\Phi_0 = \frac{1}{\sqrt{2}}(|11\rangle + |00\rangle), \quad \Phi_j = i(\mathbb{1} \otimes \sigma_j)\Phi_0, \quad j = 1,2,3 \tag{2-31}$$

Where have used the shorthand notation $|jk\rangle$ for $|j\rangle \otimes |k\rangle$ and the σ_j denote the Pauli matrices. A mixture of them, i.e. a density matrix $\rho \in S(C^2 \otimes C^2)$ with eigenvectors φ_j and eigenvalues $0 \leqslant \lambda_j \leqslant 1$; $\sum_j \lambda_j = 1$, is called a Bell diagonal state. It can be shown that ρ is entangled iff $\max|\lambda\rangle > 1/2$ holds. Let come back to the general case now and consider an arbitrary $\rho \in S(H \otimes H)$. Using maximally entangled states, we can introduce another separability criterion in terms of the maximally entangled fraction

$$\mathscr{F}(\rho) = \sup_{\psi\ \mathrm{max.\,ent.}} \langle\Psi, \rho\Psi\rangle \tag{2-32}$$

If ρ separable the reduction criterion (2-25) implies $\langle\psi; [\mathrm{tr}(\rho) \otimes \mathbb{1} - \rho] - \psi\rangle \geqslant 0$ for any maximally entangled state. Since the partial trace of $|\psi\rangle\langle\psi|$ is $d^{-1}\mathbb{1}$ can get

$$d^{-1} = \langle\Psi, \mathrm{tr}_1(\rho) \otimes \mathbb{1}\Psi\rangle\rangle\langle\Psi, \rho\Psi\rangle \tag{2-33}$$

hence $F(\rho) \leqslant 1/d$. This condition is not very sharp however. Using the ppt criterion it can be shown that $\rho = \lambda|\varphi_1\rangle\langle\varphi_1| + (1-\lambda)|00\rangle\langle 00|$ (with the Bell state φ_1) is entangled for all $0 < \lambda \leqslant 1$ but a straightforward calculation shows that $F(\rho) \leqslant 1/2$ holds for $\lambda \leqslant 1/2$.

Finally, have to mention here a very useful parameterization of the set of pure states on $H \otimes H$ in terms of maximally entangled states: If ψ is an arbitrary but fixed maximally entangled state, each $\varphi \in H \otimes H$ admits (uniquely determined) operators X_1; X_2 such that

$$\phi = (X_1 \otimes \mathbb{1})\Psi = (\mathbb{1} \otimes X_2)\Psi \tag{2-34}$$

holds. This can be easily checked in a product basis.

- Werner states

If we consider entanglement of mixed states rather than pure ones, the analysis becomes quite difficult, even if the dimensions of the underlying Hilbert spaces are low. The reason is that the state space $S(H_1 \otimes H_2)$ of a two-partite system with dim $H_i = d_i$ is a geometric object in a $(d_1^2 d_2^2 - 1)$-dimensional space. Hence even in the simplest non-trivial case (two qubits) the dimension of the state space becomes very high (15 dimensions) and naive geometric intuition can be misleading. Therefore, it is often useful to look at special classes of model states, which can characterized by only few parameters. A quite powerful tool is the study of symmetry properties; i.e. to investigate the set of states, which is invariant under a group of local unitaries. A general discussion of this scheme can found that present three of the most prominent examples. Consider first a state $\rho \in S(H \otimes H)$ (with $H = C^d$) which is invariant under the group of all $U \otimes U$ with a unitary U on H; i.e. $[U \otimes U; \rho] = 0$ for all U.

Theorem 2.14 Each operator A on the N-fold tensor product $H \otimes N$ of the (finite dimensional) Hilbert space H which commutes with all unitaries of the form $U \otimes N$ is a linear combination of permutation operators; i.e. $A = \sum_{\pi} \lambda_{\pi} V_{\pi}$; where the sum is taken over all permutations π of N elements; $\lambda_{\pi} \in C$ and V_{π} is defined by

$$V_{\pi} \phi_1 \otimes \cdots \otimes \phi_N = \phi_{\pi^{-1}(1)} \otimes \cdots \otimes \phi_{\pi^{-1}(N)} \tag{2-35}$$

Hence $\rho = a\,\mathbb{1} + bF$ with appropriate coefficients $a; b$. Since ρ is a density matrix, which a and b are not independent. To get a transparent way to express these constraints, it is reasonable to consider the

Eigen projections $P_{\pm}$ of F rather than $\mathbb{1}$ and F; i.e. $FP_{\pm}\psi = \pm P_{\pm}\psi$ and $P_{\pm} = (\mathbb{1} \pm F)/2$. The $P_{\pm}$ are the projections on the subspaces $H_{\pm}^{\otimes 2} \subset H \otimes H$ of symmetric, respectively antisymmetric, tensor products (Bose-, respectively, Fermi-subspace). If write $d_{\pm} = d(d \pm 1)/2$ for the dimensions of $H_{\pm}^{\otimes 2}$ can get for each Werner state ρ

$$\rho = \frac{\lambda}{d_+} P_+ + \frac{(1-\lambda)}{d_-} P_-, \quad \lambda \in [0,1] \tag{2-36}$$

On the other hand, it is obvious that each state of this form is $U \otimes U$ invariant, hence a Werner state. If ρ is given, it is very easy to calculate the parameter λ from the expectation value of ρ and the flip $\mathrm{tr}(\rho F) = 2\lambda - 1 \in [-1;1]$. Therefore, can write for an arbitrary state $\rho \in S(H \otimes H)$

$$P_{UU}(\sigma) = \frac{\mathrm{tr}(\sigma F) + 1}{2d_+} P_+ + \frac{(1 - \mathrm{tr}\sigma F)}{2d_-} P_- \tag{2-37}$$

This defines a projection from the full state space to the set of Werner states which is called the twirl operation. In many cases, it is quite useful that it can written alternatively as a group average of the form

$$P_{UU}(\sigma) = \int_{U(d)} (U \otimes U)\sigma(U^* \otimes U^*)\,\mathrm{d}U \tag{2-38}$$

Where $\mathrm{d}U$ denotes the normalized, left invariant Haar measure on $U(d)$. To check this identity note first that its right-hand side is indeed $U \otimes U$ invariant, due to the invariance of the volume element $\mathrm{d}U$. Hence, have to check only that the trace of F times the integral coincides with $\mathrm{tr}(F\sigma)$:

$$\mathrm{tr}\left[F \int_{U(d)} (U \otimes U)\sigma(U^* \otimes U^*)\,\mathrm{d}U\right] = \int_{U(d)} \mathrm{tr}\,[F(U \otimes U)\sigma(U^* \otimes U^*)]\,\mathrm{d}U \tag{2-39}$$

$$= \mathrm{tr}(F\sigma) \int_{U(d)} \mathrm{d}U = \mathrm{tr}(F\sigma) \tag{2-40}$$

where have used the fact that F commutes with $U \otimes U$ and the normalization of $\mathrm{d}U$ can apply P_{UU} obviously to arbitrary operators $A \in B(H \otimes H)$ and, as an integral over unitarily implemented operations, get a channel. Substituting $U \to U^*$ in (2-35) and cycling the trace $\mathrm{tr}(AP_{UU}(\rho))$ find $\mathrm{tr}(P_{UU}(A)) = \mathrm{tr}(AP_{UU}(\rho))$, hence P_{UU} has the same form in the Heisenberg and the Schrödinger picture (i.e. $P^*_{UU} = P_{UU}$). If $\sigma \in S(H \otimes H)$ is a separable state the integrand of $P_{UU}(\sigma)$ in Eq. (2-35) consists entirely of separable states, which hence $P_{UU}(\sigma)$ is separable. Since each Werner state ρ is the twirl of itself, that is separable iff it is the twirl $P_{UU}(\sigma)$ of a separable state $\sigma \in S(H \otimes H)$. To determine the set of separable Werner states therefore have to calculate only the set of all $\mathrm{tr}(F\sigma) \in [-1;1]$ with separable σ. Since each such σ admits a convex decomposition into pure product states it is sufficient to look at

$$\langle \psi \otimes \phi, F\psi \otimes \phi \rangle = |\langle \psi, \phi \rangle|^2 \tag{2-41}$$

Which ranges from 0 to 1. Hence ρ from Eq. (2-33) is separable iff $1/2 \leqslant \lambda \leqslant 1$ and entangled otherwise (due to $\lambda = (\mathrm{tr}(F\rho) + 1)/2$). If $H = C^2$ holds, each Werner state is Bell diagonal and recover the result (separable if highest eigenvalue less or equal than 1/2).

- Isotropic states

To derive a second class of states consider the partial transpose $(\mathrm{Id} \otimes \theta)\rho$ (with respect to a distinguished base $|j\rangle \in H, j = 1, 2, \cdots, d$) of a Werner state ρ. Since ρ is, by definition, $U \otimes U$ invariant, it is easy to see that $(\mathrm{Id} \otimes \theta)\rho$ is $U \otimes \bar{U}$ invariant, where $\bar{U}$ denotes componentwise complex conjugation in the base $|j\rangle$ (just have to use that $U^* = \bar{U}^{\mathrm{T}}$ holds). Each state 1 with this kind of symmetry is called an isotropic state, and our previous discussion shows that 1 is a linear combination of $\mathbb{1}$ and the partial transpose of the flip, which is the rank one operator

$$\widetilde{F} = (\mathrm{Id} \otimes \Theta)F = |\Psi\rangle\langle\Psi| = \sum_{jk=1}^{a} |jj\rangle\langle kk| \tag{2-42}$$

Where $\psi = \sum_j |jj\rangle$ is, up to normalization a maximally entangled state. Hence, each isotropic τ written as

$$\tau = \frac{1}{d}\left(\lambda \frac{\mathbb{1}}{d} + (1-\lambda)\widetilde{F}\right), \quad \lambda \in \left[0, \frac{d^2}{d^2-1}\right] \tag{2-43}$$

Where the bounds on λ follow from normalization and positivity. As above can determine the parameter λ from the expectation value

$$\mathrm{tr}(\widetilde{F}\tau) = \frac{1-d^2}{d}\lambda + d \tag{2-44}$$

which ranges from 0 to d and this again leads to a twirl operation: For an arbitrary state $\sigma \in (H \otimes H)$ can define:

$$P_{U\bar{U}}(\sigma) = \frac{1}{d(1-d^2)}([\mathrm{tr}(\widetilde{F}\sigma) - d]\mathbb{1} + [1 - d\,\mathrm{tr}(\widetilde{F}\sigma)]\widetilde{F}) \tag{2-45}$$

As for Werner states $P_{U\bar{U}}$ can be rewritten in terms of a group average:

$$P_{U\bar{U}}(\sigma) = \int_{U(d)} (U \otimes \bar{U})\sigma(U^* \otimes \bar{U}^*)\mathrm{d}U \tag{2-46}$$

Now can proceed in the same way as above: $P_{U\bar{U}}$ is a channel with $P^*_{U\bar{U}} = P_{U\bar{U}}$, its fixed points $P_{U\bar{U}}(\tau) = \tau$ are exactly the isotropic states, and the image of the set of separable states under $P_{U\bar{U}}$ coincides with the set of separable isotropic states. To determine the latter have to consider the expectation values (cf. Eq. (2-38))

$$\langle \psi \otimes \phi, \widetilde{F}\psi \otimes \phi \rangle = \left|\sum_{j=1}^{d} \psi_j \phi_j\right| = |\langle \psi, \bar{\phi} \rangle|^2 \in [0, 1] \tag{2-47}$$

This implies that τ is separable iff

$$\frac{d(d-1)}{d^2-1} \leqslant \lambda \leqslant \frac{d^2}{d^2-1} \tag{2-48}$$

Holds and entangled otherwise. For $\lambda = 0$ we recover the maximally entangled state. For $d = 2$, again recover again the special case of Bell diagonal states encountered already in the last subsection.

• OO-invariant states

Combine now Werner states with isotropic states, i. e. look for density matrices which can written as $\rho = a\,\mathbb{I} + bF + c\widetilde{F}$, or, if introduce the three mutually orthogonal projection operators

$$p_0 = \frac{1}{d}\widetilde{F}, \quad p_1 = \frac{1}{2}(\mathbb{1} - F), \quad \frac{1}{2}(\mathbb{1} + F) - \frac{1}{d}\widetilde{F} \tag{2-49}$$

as a convex linear combination of $\operatorname{tr}(p_j)^{-1}p_j, j = 0,1,2$:

$$\rho = (1 - \lambda_1 - \lambda_2)p_0 + \lambda_1 \frac{p_1}{\operatorname{tr}(p_1)} + \lambda_2 \frac{p_2}{\operatorname{tr}(p_2)}, \quad \lambda_1, \lambda_2 \geqslant 0, \lambda_1 + \lambda_2 \leqslant 1 \tag{2-50}$$

Each such operator is invariant under all transformations of the form $U \otimes U$ if U is a unitary with $U = \bar{U}$, in other words: U should be a real orthogonal matrix. A little bit representation theory of the orthogonal group shows that in fact all operators with this invariance property have the form given in (2-47). The corresponding states are therefore called OO-invariant, and can apply basically the same machinery if replace the unitary group $U(d)$ by the orthogonal group $O(d)$. This includes, in particular, the definition of a twirl operation as an average over $O(d)$ (for an arbitrary $\rho \in S(H \otimes H)$):

$$P_{\mathrm{OO}}(\rho) = \int_{O(d)} U \otimes U\rho U \otimes U^* \mathrm{d}U \tag{2-51}$$

Which can express alternatively in terms of the expectation values $\operatorname{tr}(F)$, $\operatorname{tr}(\widetilde{F})$ by

$$P_{\mathrm{OO}}(\rho) = \frac{\operatorname{tr}(\widetilde{F}\rho)}{d} p_0 + \frac{1 - \operatorname{tr}(F\rho)}{2\operatorname{tr}(p_1)} p_1 + \left(\frac{1 + \operatorname{tr}(F\rho)}{2} - \frac{\operatorname{tr}(\widetilde{F}\rho)}{d}\right) \frac{p_2}{\operatorname{tr}(p_2)} \tag{2-52}$$

The range of allowed values for $\operatorname{tr}(F\rho)$, $\operatorname{tr}(\widetilde{F}\rho)$ is given by

$$\begin{cases} -1 \leqslant \operatorname{tr}(F\rho) \leqslant 1 \\ 0 \leqslant \operatorname{tr}(\widetilde{F}\rho) \leqslant d \\ \operatorname{tr}(F\rho) \geqslant \dfrac{2\operatorname{tr}(\widetilde{F}\rho)}{d} - 1 \end{cases} \tag{2-53}$$

For $d = 3$ this is the upper triangle in Figure 2-5. The values in the lower (dotted) triangle belong to partial transpositions of OO-invariant states. The intersection of both, i. e. the gray-shaded square $Q = [0;1] \times [0;1]$, represents therefore the set of OO-invariant ppt states, and at the same time the set of separable states, since each OO-invariant ppt state is separable. To see the latter note that separable OO-invariant states form a convex subset of Q. Hence, we only have to show that the corners of Q are separable. To do this note that (1) $P_{\mathrm{OO}}()$ is separable whenever is and

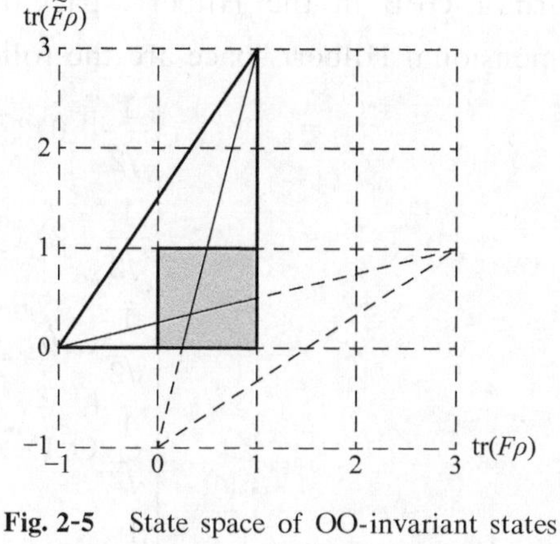

Fig. 2-5 State space of OO-invariant states (upper triangle) and its partial transpose (lower triangle) for $d = 3$. The special cases of isotropic and Werner states drawn as thin lines.

(2) that $\mathrm{tr}(FP_{OO}(\rho)) = \mathrm{tr}(F)$ and $\mathrm{tr}(\widetilde{F}P_{OO}(\rho)) = \mathrm{tr}(F)$ holds (cf. Eq. (2-40)). Consider pure product states $|\varphi\otimes\psi\rangle\langle\varphi\otimes\psi|$ for ρ and get $(|\langle\varphi,\psi\rangle|^2; |\langle\varphi,\bar{\psi}\rangle|^2)$ for the tuple $(\mathrm{tr}(F\rho); \mathrm{tr}(\widetilde{F}\rho))$. Now the point (1;1) in Q is obtained if $\psi=\varphi$ is real, the point (0;0) is obtained for real and orthogonal φ,ψ; and the point (1;0) belongs to the case $\psi=\varphi$ and $\langle\varphi,\bar{\varphi}\rangle=0$. Symmetrically can get (0;1) with the same φ and $\psi=\bar{\varphi}$.

- PPT states

Another class of examples with this property are OO-invariant states just studied. Nevertheless, separability and a positive partial transpose are not equivalent. An easy way to produce such examples of states, which entangled and ppt is given in terms of unextendible product bases. An orthonormal family $\varphi_j\in H_1\otimes H_2, j=1,2,\cdots,N<d_1 d_2$ (with $d_k=\dim H_k$) is called an unextendible product basis 9 (UPB) iff (1) all φ_j are product vectors and (2) there is no product vector orthogonal to all φ_j. Let denote the projector to the span of all φ_j by E, its orthocomplement by $E^{\perp}$, i. e. $E^{\perp}=\mathbb{1}-E$, and define the state $\rho=(d_1d_2-N)^{-1}E^{\perp}$. It is entangled because there is by construction no product vector in the support of ρ, and it is ppt. The latter can be seen as follows: The projector E is a sum of the one-dimensional projectors $|\varphi_j\rangle\langle\varphi_j|$, $j=1,2,\cdots,N$. Since all φ_j are product vectors the partial transposes of the $|\varphi_j\rangle\langle\varphi_j|$, are of the form $|\widetilde{\varphi}_j\rangle\langle\widetilde{\varphi}_j|$, with another UPB $\widetilde{\varphi}_j, j=1,2,\cdots,N$ and the partial transpose $(\mathbb{1}\otimes\theta)E$ of E is the sum of the $|\widetilde{\varphi}_j\rangle\langle\widetilde{\varphi}_j|$. Hence $(\mathbb{1}\otimes\theta)E^{\perp}=\mathbb{1}-(\mathbb{1}\otimes\theta)E$ is a projector and therefore positive. To construct entangled ppt states have to find UPBs. The following two examples taken from. Consider first the five vectors

$$\phi_j = N(\cos(2\pi j/5), \sin(2\pi j/5), h),\quad j=0,1,\cdots,4 \tag{2-54}$$

with $N=2/\sqrt{5+\sqrt{5}}$ and $h=1/2\sqrt{1+\sqrt{5}}$. They form the apex of a regular pentagonal pyramid with height h. The latter chosen such that non-adjacent vectors are orthogonal. It is now easy to show that the five vectors

$$\Psi_j = \phi_j\otimes\phi_{2j \bmod 5},\quad j=0,1,\cdots,4 \tag{2-55}$$

Form a UPB in the Hilbert space $H\otimes H$, $\dim H=3$. A second example, again in (3×3)-dimensional Hilbert space are the following five vectors:

$$\begin{cases} \frac{1}{\sqrt{2}}|0\rangle\otimes(|0\rangle-|1\rangle) \\ \frac{1}{\sqrt{2}}|2\rangle\otimes(|1\rangle-|2\rangle) \\ \frac{1}{\sqrt{2}}(|0\rangle-|1\rangle)\otimes|2\rangle \\ \frac{1}{\sqrt{2}}(|1\rangle-|2\rangle)\otimes|0\rangle \\ \frac{1}{3}(|0\rangle+|1\rangle+|2\rangle)\otimes(|0\rangle+|1\rangle+|2\rangle) \end{cases} \tag{2-56}$$

where $|k\rangle, k=0,1,2$ denotes the standard basis in $H=C^3$.

- Multipartite states

In many applications of quantum information rather big systems, consisting of a large number of subsystems, occur (e.g. a quantum register of a quantum computer) and it is necessary to study the corresponding correlation and entanglement properties. Since this is a fairly difficult task, there is not much known about—much less as in the two-partite case, which is mainly consider in this section. Nevertheless, in this subsection we will give a rough outline of some of the most relevant aspects. At the level of pure state of the most signi. Which is the lack of an analog of the Schmidt decomposition. More precisely, there are elements in an N-fold tensor product $H^{(1)} \otimes \cdots \otimes H^{(N)}$ (with $N>2$) which cannot be written as

$$\Psi = \sum_{j=1}^{d} \lambda_j \phi_j^{(1)} \otimes \cdots \otimes \phi_j^{(N)} \tag{2-57}$$

With N orthonormal bases $\varphi_1^{(k)} \cdots \varphi_d^{(k)}$ of $H^{(k)}, k = 1, 2, \cdots, N$. To get examples for such states in the tri-partite case, notefirst that any partial trace of $|\psi\rangle\langle\psi|$ with ψ from Eq. (2-57) has separable eigenvectors. Hence, each purification of an entangled, two-partite, mixed state with inseparable eigenvectors (e.g. a Bell diagonal state) does not admit a Schmidt decomposition. This implies on the one hand that there are interesting new properties discovered, but on the other see that many techniques developed for bipartite pure states can generalized in a straightforward way only for states, which are Schmidt decomposable in the sense of Eq. (2-57). The most well known representative of this class for a tripartite qubit system is the GHZ state

$$\Psi = \frac{1}{\sqrt{2}}(|000\rangle + |111\rangle) \tag{2-58}$$

Which has the special property that contradictions between local hidden variable theories and quantum mechanics occur even for non-statistical predictions. A second new aspect arising in the discussion of multiparty entanglement is the fact that several different notions of separability occur. A state ρ of an N-partite system $B(H_1) \otimes \cdots \otimes B(H_N)$ is called N-separable if

$$\rho = \sum_J \lambda_J \rho_{j_1} \otimes \cdots \otimes \rho_{j_N} \tag{2-59}$$

with states $\rho_{jk} \in B^*(H_k)$ and multiindices $J = (j_1, j_2, \cdots, j_k)$. Alternatively, however, we can decompose $B(H_1) \otimes \cdots \otimes B(H_N)$ into two subsystems (or even into M subsystems if $M < N$) and call ρ biseparable if it is separable with respect to this decomposition. It is obvious that N-separability implies biseparability with respect to all possible decompositions. The converse is—not very surprisingly—not true. One way to construct a corresponding counterexample is to use an unextendable product base. In it is shown that the tripartite qubit state complementary to the UPB

$$|0,1,+\rangle, |1,+,0\rangle, |+,0,1\rangle, |-,-,-\rangle \text{ with } |\pm\rangle = \frac{1}{\sqrt{2}}(|0\rangle \pm |1\rangle) \tag{2-60}$$

is entangled (i.e. tri-inseparable) but biseparable with respect to any decomposition into two subsystems. Another, maybe more systematic, way to find examples for multipartite states with interesting properties is the generalization of the methods used for Werner states, i.e. to look for density matrices $\rho \in B^*(H^{\otimes N})$ which commute with all unitaries of the form $U \otimes N$. Applying again Theorem 2.14, each such ρ is a linear combination of permutation unitaries. Hence, the structure of the set of all $U^{\otimes N}$ invariant states can derived from representation theory of the symmetric group (which can be tedious for large N!). For $N = 3$ this program is carried out and it turns out that the corresponding set of invariant states is a five-dimensional (real) manifold. Skip the details here and refer to instead.

2.3.2 Channels

In Section 2.3 introduced channels as very general objects transforming arbitrary types of information (i.e. classical, quantum and mixtures of them) into one another. In the following, will consider some of the most important special cases.

- Quantum channels

Many tasks of quantum information theory require the transmission of quantum information over Long distances, using devices like optical fibers or storing quantum information in some sort of memory. Ideally, prefer those channels which do not affect the information at all, i.e. $T = \mathbb{1}$, or, as the next best choice, a T whose action can be undone by a physical device, i.e. T should be invertible and $T-1$ is again a channel.

The Stinespring Theorem immediately shows that this implies $T^* \rho = U\rho U^*$ with a unitary U; in other words, the systems carrying the information do not interact with the environment. Such a kind of channel an ideal channel. In real situations, however, interaction with the environment, i.e. additional, unobservable degrees of freedom, cannot avoided. The general structure of such a noisy channel is given by

$$T^*(\rho) = \mathrm{tr}K\ (U(\rho \otimes \rho_0)U^*) \tag{2-61}$$

Where $U: H \otimes K \to H \otimes K$ is a unitary operator describing the common evolution of the system (Hilbert space H) and the environment (Hilbert space K) and $0\rho \in S(K)$ is the initial state of the environment. It is obvious that the quantum information originally stored in $\rho \in S(H)$ cannot be completely recovered from $T^*(\rho)$ if only one system is available. It is an easy consequence of the Stinepspring theorem, which each channel can expressed in this form corollary. Assume that $T: B(H) \to B(H)$ is a channel. Then there is a Hilbert space K; a pure state 0 and a unitary map $U: H \otimes K \to H \otimes K$ such that Eq. (2-61) holds. It is always possible; to choose K such that dim $(K) = \dim (H)^3$ holds.

Proof. Consider the Stinespring form $T(A) = V^*(A \otimes 5)V$ with $V: H \to H \otimes K$ of T and choose a vector $\psi \in K$ such that $U(\varphi \otimes \psi) = V(\varphi)$ can be extended to a unitary map $U: H \otimes K \to H \otimes K$ (this is always possible since T is unital and V therefore isometric). If $e_j \in H; j = 1, 2, \cdots, d_1$ and $f_k \in K; k = 1, 2, \cdots, d_2$ are orthonormal bases with $f_1 = \psi$ can get:

$$\mathrm{tr}\,[T(A)\rho] = \mathrm{tr}\,[\rho V^*(A \otimes \mathbb{1})V] = \sum_i \langle V\rho e_j, (A \otimes \mathbb{1})Ve_j \rangle \tag{2-62}$$

$$= \sum_{jk} \langle U(\rho \otimes |\psi\rangle\langle\psi|)(e_j \otimes f_k), (A \otimes \mathbb{1})U(e_j \otimes f_k) \rangle \tag{2-63}$$

$$= \mathrm{tr}\,[\mathrm{tr}K[U(\rho \otimes |\psi\rangle\langle\psi|)U^*]A] \tag{2-64}$$

The proves the statement.

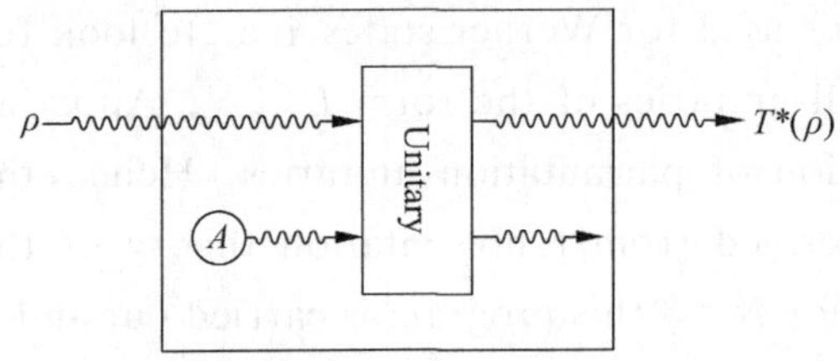

Fig. 2-6 Noisy channel.

The most prominent example for a noisy channel is the depolarizing channel for d-level systems (i.e. $H = C^d$):

$$\mathcal{S}(H) \ni \rho \mapsto \vartheta\rho + (1-\vartheta)\frac{\mathbb{1}}{d} \in \mathcal{S}(H), \quad 0 \leqslant \vartheta \leqslant 1 \tag{2-65}$$

or in the Heisenberg picture

$$\mathscr{B}(H) \ni A \mapsto \vartheta A + (1-\vartheta)\frac{\operatorname{tr}(A)}{d}\mathbb{1} \in \mathscr{B}(H) \tag{2-66}$$

A Stinespring dilation of T (not the minimal one—this can checked by counting dimensions) is given by $K = H \otimes H \oplus C$ and $V: H \to H \otimes K = H^{\otimes 3} \oplus H$ with

$$|j\rangle \mapsto V|j\rangle = \left[\sqrt{\frac{1-\vartheta}{d}}\sum_{k=1}^{d}|k\rangle \otimes |k\rangle \otimes |j\rangle\right] \oplus [\sqrt{\vartheta}\,|j\rangle] \tag{2-67}$$

where $|k\rangle, k = 1, 2, \cdots, d$ denotes again the canonical basis in H. An ancilla form of T with the same K is given by the (pure) environment state

$$\psi = \left[\sqrt{\frac{1-\vartheta}{d}}\sum_{k=1}^{d}|k\rangle \otimes |k\rangle\right] \oplus [\sqrt{\vartheta}\,|0\rangle] \in K \tag{2-68}$$

and the unitary operator $U: H \otimes K \to H \otimes K$ with

$$U(\phi_1 \otimes \phi_2 \otimes \phi_3 \oplus \chi) = \phi_2 \otimes \phi_3 \otimes \phi_1 \oplus \chi \tag{2-69}$$

i.e. U is the direct sum of a permutation unitary and the identity.

- Channels under symmetry

To be more precise, consider a group G and two unitary representations 31; 32 on the Hilbert spaces H_1 and H_2, respectively. A channel $T: B(H_1) \to B(H_2)$ is called covariant (with respect to π_1 and π_2) if

$$T[\pi_1(U) A \pi_1(U)^*] = \pi_2(U) T[A] \pi_2(U)^*, \quad \forall A \in \mathscr{B}(H_1), \forall U \in G \tag{2-70}$$

The general structure of covariant channels governed by a powerful variant fairly of Stinespring's theorem which will state below (and which will be very useful for the study of the cloning problem). Before do this, have a short look on a particular class of examples which is closely related to OO-invariant states.

Hence consider a channel $T: B(H) \to B(H)$ which is covariant with respect to the orthogonal group, i.e. $T(UAU^* = UT(A)U^*$ for all unitaries U on H with $\bar{U} = U$ in a distinguished basis $|j\rangle, j = 1, 2, \cdots$, e maximally entangled state $\varphi = d^{-1/2}\sum_j |jj\rangle$ is OO-invariant, i.e. $U \otimes U\psi = \psi$ for all these U. Therefore, each state $\rho = (\mathrm{Id} \otimes T^* |\psi\rangle\langle\psi|$ is OO-invariant as well and by the duality lemma T and ψ are uniquely determined (up to unitary equivalence) by ρ. This means can use the structure of OO-invariant states derived to characterize all orthogonal covariant channels. As a fist step consider the linear maps $X_1(A) = d\ \operatorname{tr}(A)5$, $X_2(A) = \mathrm{d}A^{\mathrm{T}}$ and $X_3(A) = \mathrm{d}A$. They are not channels (they are not unital and X_2 is not cp) but they have the correct covariance property and it is easy to see that they correspond to the operators $\mathbb{1}; F; \widetilde{F} \in B(H \otimes H)$, i.e.

$$(\mathrm{Id} \otimes X_1)|\psi\rangle\langle\psi| = \mathbb{1}, \quad (\mathrm{Id} \otimes X_2)|\psi\rangle\langle\psi| = F, \quad (\mathrm{Id} \otimes X_3)|\psi\rangle\langle\psi| = \widetilde{F} \tag{2-71}$$

Using Eq. (2-49), can determine therefore the channels which belong to the three extremal OO-invariant states (the corners of the upper triangle in Figure 2-7):

$$T_0(A) = A, \quad T_1(A) = \frac{\operatorname{tr}(A)\mathbb{1} - A^{\mathrm{T}}}{d-1} \tag{2-72}$$

$$T_2(A) = \frac{2}{d(d+1)-2}\left[\frac{d}{2}(\operatorname{tr}(A)\mathbb{1} + A^{\mathrm{T}}) - A\right] \tag{2-73}$$

Each OO-invariant channel is a convex linear combination of these three. Special cases are the channels corresponding to Werner and isotropic states. The latter leads to depolarizing channels

$T(A) = \theta A + (1 - \theta) d - 1\mathrm{tr}(A)\mathbb{1}$ with $\theta \in [0; d^2 = (d^2 - 1)]$; cf. Eq. (2-40), while Werner states correspond to

$$T(A) = \frac{\vartheta}{d+1}[\mathrm{tr}(A)\,\mathbb{1} + A^{\mathrm{T}}] + \frac{1-\vartheta}{d-1}[\mathrm{tr}(A)\,\mathbb{1} - A^{\mathrm{T}}], \vartheta \in [0,1] \tag{2-74}$$

cf. Eq. (2-33).

Let come back now to the general case, here the covariant version of the Stinespring theorem (see proof). The basic idea is that all covariant channels parameterized by representations on the dilation space.

Theorem 2.15 Let G be a group with finite-dimensional unitary representations $\pi_j: G \to U(H_j)$ and $T: B(H_1) \to B(H_2)$ a $\pi_1; \pi_2$-covariant channel. Then there is a finite-dimensional unitary representation $\tilde{\pi}: G \to U(K)$ and an operator $V: H_2 \to H_1 \otimes K$ with $V\pi_2(U) = \pi_1(U) \otimes \tilde{\pi}(U)$ and $T(A) = V * A \otimes \mathbb{1} V$.

To get an explicit example consider the dilation of a depolarizing channel given in Eq. (2-64). In this case have $\pi_1(U) = \pi_2(U) = U$ and $\tilde{\pi}(U) = (U \otimes \bar{U}) \oplus \mathbb{1}$. The check that the map V has indeed the intertwining property $V\pi_2(U) = \pi_1(U) \otimes \tilde{\pi}(U)$ stated in the theorem is left as an exercise to the reader.

- Classical channels

The classical analog to a quantum operation is a channel $T: C(X) \to C(Y)$ which describes the transmission or manipulation of classical information. As have mentioned already in Section 2.3.1 positivity and complete positivity are equivalent in this case. Hence, have to assume only that T is positive and unital. Obviously, T is characterized by its matrix elements $T_{xy} = \delta_y(T|x\rangle\langle x|)$, where $\delta_y \in C^*(X)$ denotes the Dirac measure at $\delta_y \in Y$ and $|x\rangle\langle x| \in C(X)$ is the canonical basis in $C(X)$. Positivity and normalization of T imply that $0 \leqslant T_{xy} \leqslant 1$ and

$$1 = \delta_y(\mathbb{1}) = \delta_y(T(\mathbb{1})) = \delta_y\left[T\left(\sum_x |x\rangle\langle x|\right)\right] = \sum_x T_{xy} \tag{2-75}$$

Hence, the family $(T_{xy})_{x \in X}$ is a probability distribution on X and T_{xy} is therefore the probability to get the information $x \in X$ at the output side of the channel if $y \in Y$ was send. Each classical channel is uniquely determined by its matrix of transition probabilities. F or $X = Y$ that the information is transmitted without error $i = T_{xy} = \delta_{xy}$, i.e. T is an ideal channel if $T = \mathrm{Id}$ holds and noisy otherwise.

2.3.3 Observables and preparations

Consider now a channel which transforms quantum information $B(H)$ into classical information $C(X)$. Since positivity and complete positivity are again equivalent, just have to look at a positive and unital map $E: C(X) \to B(H)$. With the canonical basis $|x\rangle\langle x|, x \in X$ of $C(X)$ can get a family $E_x = E(|x\rangle\langle x|), x \in X$ of positive operators $E_x \in B(H)$ with $\sum_{x \in X} E_x = \mathbb{1}$. Because the E_x form a POV measure, i. e. an observable. If on the other hand a POV measure $E_x \in B(H), x \in X$ is given can define a quantum to classical channel $E: C(X) \to B(H)$ by $E(f) = \sum_x f(x)\ E_x$ This shows that the observable E_x; $x \in X$ and the channel E can be identified and say E is the observable. Keeping this interpretation in mind it is possible to have a short look at continuous observables without the need of abstract measure theory: Only have to define the classical algebra $C(X)$ for a set X which is not finite or discrete. For simplicity, assume that $X = R$ holds; however, the generalization to other locally compact spaces is forward straightly. Choose for $C(R)$ the space of continuous, complex-valued functions vanishing at infinity, i. e. $|f(x) < \varepsilon$ for each $\varepsilon < 0$ provided $|x|$ is large enough. $C(R)$ can be equipped with the sup-norm and becomes an Abelian C^*-algebra. To interpret it as an operator algebra as assumed have to identify $f \in C(R)$ with the corresponding multiplication operator on $L_2(R)$. An observable taking arbitrary real values can now be defined as a positive map $E: C(R) \to B(H)$. The probability to get a result in the interval $[a;b] \subset R$ during an E measurement on systems in the state ρ is

$$\mu([a,b]) = \sup\{\operatorname{tr}(E(f)\rho) \mid f \in \mathscr{C}(r), 0 \leqslant f \leqslant \mathbb{1}, \operatorname{supp} f \subset [a,b]\} \tag{2-76}$$

Where supp denotes the support of f. The most well known example for R valued observables are of course position Q and momentum P of a free particle in one dimension. In this case have $H = L_2(r)$ and the channels corresponding to Q and P are (in position representation) given by $C(R) \ni f \mapsto E_Q(f) \in B(H)$ with $E_Q(f)\psi = f\psi$, respectively, $C(R) \ni f \mapsto E_P(f) \in B(H)$ with $E_P(f)\psi = (f\hat{\psi})\vee$ where $\wedge$ and $\vee$ denote the Fourier transform and its inverse. Let return now to a finite set X and exchange the role of $C(X)$ and $B(H)$; in other words let consider a channel $R: B(H) \to C(X)$ with a classical input and a quantum output algebra. In the Schrödinger picture can get a family of density matrices $\rho_x := R^*(\delta_x) \in B^*(H), x \in X$, where $\delta_x \in C^*(X)$ again denote the Dirac measures. Hence, can get a parameter-dependent preparation which can be used to encode the classical information $x \in X$ into the quantum information $\rho_x \in B^*(H)$.

- Instruments and parameter-dependent operations

An observable describes only the statistics of measuring results, but does not contain information about the state of the system after the measurement. Following Davies will call such an object an instrument. From T can derive the subchannel

$$\mathscr{C}(X) \ni f \mapsto T(\mathbb{1} \otimes f) \in \mathscr{B}(\mathscr{K}) \tag{2-77}$$

Which is the observable measured by T, i. e. $\operatorname{tr}[T(\mathbb{1} \otimes |x\rangle\langle x|)]$ is the probability to measure $x \in X$ on systems in the state ρ. On the other hand, can get for each $x \in X$ a quantum channel (which is not unital)

$$\mathscr{B}(\mathscr{H}) \ni A \mapsto T_x(A) = T(A \otimes |x\rangle\langle x|) \in \mathscr{B}(\mathscr{K}) \tag{2-78}$$

It describes the operation performed by the instrument T if $x \in X$ was measured. More precisely if a measurement on systems in the state ρ gives the result $x \in X$ can get (up to normalization) the state $T^* x(\rho)$ after the measurement (cf. Figure 2-7), while

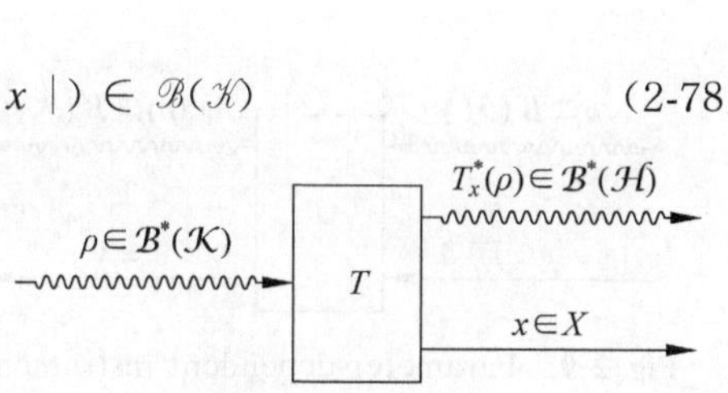

Fig. 2-7 Instrument.

$$\operatorname{tr}(T_x^*(\rho)) = \operatorname{tr}(T_x^*(\rho)\mathbb{1}) = \operatorname{tr}(\rho T(\mathbb{1} \otimes |x\rangle\langle x|)) \tag{2-79}$$

is (again) the probability to measure $x \in X$ on ρ. The instrument T can be expressed in terms of the operations T_x by

$$T(A \otimes f) = \sum_x f(x) T_x(A) \tag{2-80}$$

hence, can identify T with the family $T_x, x \in X$. Finally, can consider the second marginal of T

$$\mathscr{B}(\mathscr{H}) \ni A \mapsto T(A \otimes \mathbb{1}) = \sum_{x \in X} T_x(A) \in \mathscr{B}(\mathscr{K}) \tag{2-81}$$

It describes the operation can get if the outcome of the measurement is ignored. The most well-known example of an instrument is a von Neumann measurement associated to a PV measure given by family of projections E_x, $x = 1, 2, \cdots, d$; e.g. the eigenprojections of a self-adjoint operator $A \in B(H)$. It is defined as the channel

$$T: \mathcal{B}(\mathcal{H}) \otimes \mathcal{C}(X) \to \mathcal{B}(\mathcal{H}) \quad \text{with } X = \{1, 2, \cdots, d\} \quad \text{and} \quad T_x(A) = E_x A E_x \tag{2-82}$$

Hence, can get the final state $\mathrm{tr}(E_x \rho)^{-1} E_x \rho E_x$ if measure the value $x \in X$ on systems initially in the state ρ—this is well known from quantum mechanics.

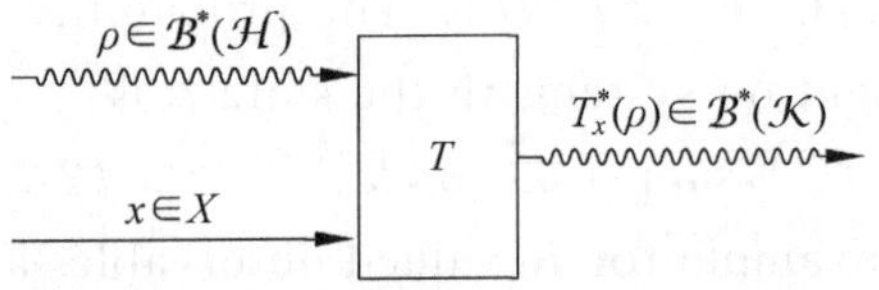

Fig. 2-8 Parameter-dependent operation.

Let change now the role of $B(H) \otimes C(X)$ and $B(K)$; in other words, consider a channel $T: B(K) \to B(H) \otimes C(X)$ with hybrid input and quantum output. It describes a device, which changes the state of a system depending on additional classical information. As for an instrument, Tdecomposes into a family of (unital!) channels $T_x: B(K) \to B(H)$ such that get $T^*(\rho \otimes p) = \sum_x p_x T_x^*(\rho)$ in the Schrödinger picture. Physically T describes a parameter-dependent operation: depending on the classical information $x \in X$ the quantum information $\rho \in B(K)$ is transformed by the operation T_x (cf. Figure 2-8).

Finally, Consider a channel $T: B(H) \otimes C(X) \to B(K) \otimes C(Y)$ with hybrid input and output to get a parameter-dependent instrument (cf. Figure 2-9): Similar to the discussion in the last paragraph we can define a family of instruments $T_y: B(H) \otimes C(X) \to B(K)$, $y \in Y$ by the equation $T^*(\rho \otimes p) = \sum_y p_y T^* y(\rho)$. Physically, T describes the following device: It receives the classical information $y \in Y$ and a quantum system in the state $\rho \in B^*(K)$ as input. Depending on y a measurement with the instrument T_y is performed, which in turn produces the measuring value $x \in X$ and leaves the quantum system in the state (up to normalization) $T_{y,x}^*(\rho)$; with $T_{y,x}$ given as in Eq. (2-75) by $T_{y,x}(A) = T_y(A \otimes |x\rangle\langle x|)$.

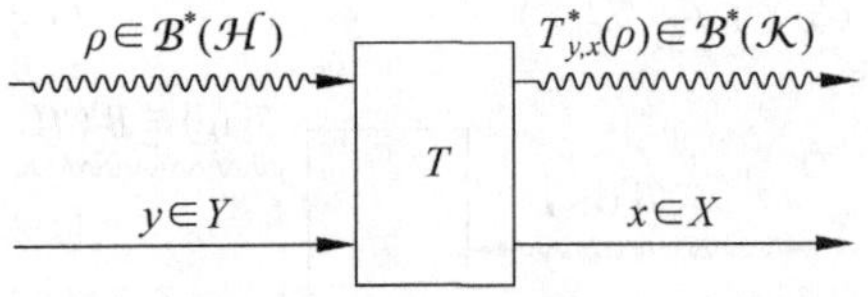

Fig. 2-9 Parameter-dependent instrument.

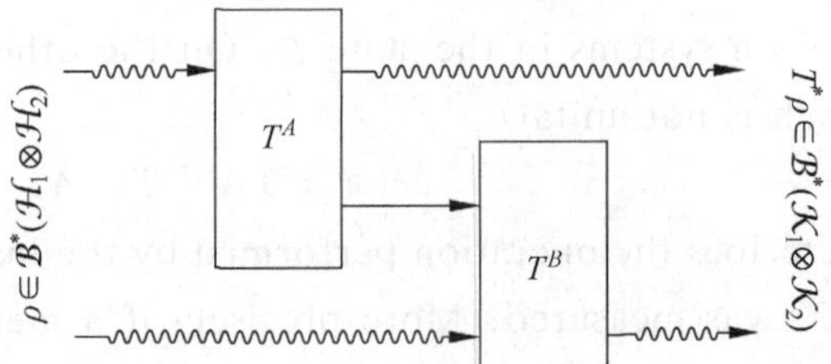

Fig. 2-10 One-way LOCC operation

- LOCC and separable channels

Consider channels acting on. nite-dimensional bipartite systems: $T: B(H_1 \otimes K_2) \to B(K_1 \otimes K_2)$. In this case can ask the question whether a channel preserves separability. Simple examples are local operations (LO$_s$), i.e. $T = T^A \otimes T^B$ with two channels $T^{A,B}: B(H_j) \to B(K_j)$. Physically, think of such a T in terms of two physicists Alice and Bob both performing operations on their own particle but without information transmission neither classical nor quantum. The next difficult step are LOs with one-way classical communications (one way is LOCC). This means Alice operates on her system with an instrument, communicates the classical measuring result $j \in X =$

$\{1,2,\cdots,N\}$ to Bob and he selects an operation depending on these data. can write such a channel as a composition $T=(T^A\otimes \mathrm{Id})(\mathrm{Id}\otimes T^B)$ of the instrument $T^A: B(H_1)\otimes C(X_1)\rightarrow B(K_1)$ and the parameter-dependent operation $T^B: B(H_2)\rightarrow C(X_1)\otimes B(K_2)$

$$\mathscr{B}(\mathscr{H}_1\otimes\mathscr{H}_2)\xrightarrow{\mathrm{Id}\otimes T^B}\mathscr{B}(\mathscr{H}_1)\otimes\mathscr{C}(X)\otimes\mathscr{B}(\mathscr{K}_2)\xrightarrow{T^A\otimes \mathrm{Id}}\mathscr{B}(\mathscr{K}_1\otimes\mathscr{K}_2) \tag{2-83}$$

It is of course possible to continue the chain in Eq. (2-80), i.e. instead of just operating on his system, Bob can invoke a parameter-dependent instrument depending on Alice's data $j_1\in X_1$, send the corresponding measuring results $j_2\in X_2$ to Alice and so on. To write down the corresponding chain of maps (as in Eq. (2-80)) is simple but not very illuminating and therefore omitted; cf. Figure 2-11 instead. If we allow Alice and Bob to drop some of their particles, i. e. the operations they perform need not unital, that get an LOCC channel ("local operations and classical communications"). It represents the most general physical process, which can performed on a two partite system if only classical communication (in both directions) is available. The LOCC channels play a significant role in entanglement theory, but they are difficult to handle. Fortunately, it is often possible to replace them by closely related operations with a more simple structure: A not necessarily unital channel $T: B(H_1\otimes K_2)\rightarrow B(K_1\otimes K_2)$ is called separable, if it is a sum of (in general non-unital) local operations, i.e.

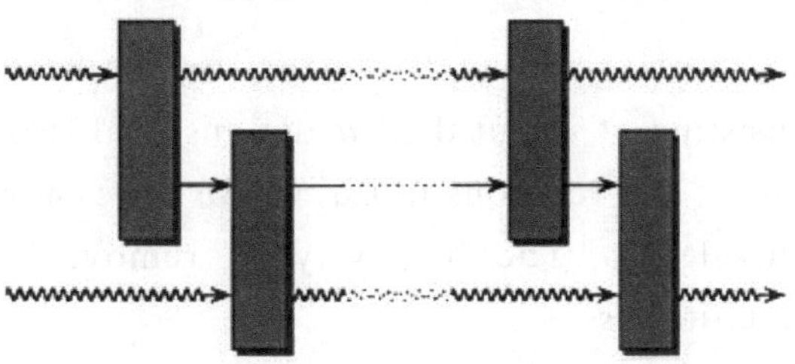

Fig. 2-11 LOCC operation. The upper and lower curly arrows represent Alice's respectively Bob's, quantum system, while the straight arrows in the middle stand for the classical information Alice and Bob exchange. The boxes symbolize the channels applied by Alice and Bob. It is easy to see that a separable T maps separable states to separable states (up to normalization) and that each LOCC channel is separable. The converse, however, is (somewhat surprisingly) not true: there are separable channels, which are not LOCC, for a concrete example.

$$T=\sum_{j=1}^{N}T_j^A\otimes T_j^B \tag{2-84}$$

2.3.4 Quantum mechanics in phase space

Up to now have considered only finite-dimensional systems and even in this extremely idealized situation it is not easy to get non-trivial results. At a first, the discussion of continuous quantum systems seems therefore to be hopeless. If restrict our attention however to small classes of states and channels, with sufficiently simple structure, many problems become tractable. Before start, let add some remarks to discussion, which have restricted to finite-dimensional Hilbert spaces. The material considered there can generalized in a straight forward way basically, as long as topological issues like continuity and convergence arguments are treated carefully enough. There are of course some caveats; however, they do not lead to problems in the framework are going to discuss and can therefore be ignored.

- Weyl operators and the CCR

The kinematical structure of a quantum system with d degrees of freedom is usually described by a separable Hilbert space H and $2d$ self-adjoint operators $Q_1 Q_2\cdots Q_d; P_1 P_2\cdots P_d$ satisfying the canonical commutation relations $[Q_j; Q_k] = 0, [P_j; P_k] = 0, [Q_j; P_k] = i\delta_{jk} \mathbb{1}$. The latter can rewritten in a more compact form as

$$R_{2j-1} = Q_j, R_{2j} = P_j, \quad j = 1, 2, \cdots, d, [R_j, R_k] = -\mathrm{i}\sigma_{jk} \tag{2-85}$$

Here σ denotes the symplectic matrix

$$\sigma = \mathrm{diag}\,(J, J, \cdots, J), \quad J = \begin{bmatrix} 0 & 1 \\ -1 & 0 \end{bmatrix} \tag{2-86}$$

Which plays a crucial role for the geometry of classical mechanics. And call the pair $(V; \sigma)$ consisting of σ and the $2d$-dimensional real vector space $V = R^{2d}$ henceforth the classical phase space. The relations in Eq. (2-85) are, however, not scient to x the operators R_j up to unitary equivalence. The best way to remove the remaining physical ambiguities is the study of the unitaries

$$W(x) = \exp\,(\mathrm{i}x \cdot \sigma \cdot R), \quad x \in V, \quad x \cdot \sigma \cdot R = \sum_{jk=1}^{2d} x_j \sigma_{jk} R_k \tag{2-87}$$

instead of the R_j directly. If the family $W(x), x \in V$ is irreducible (i.e. $[W(x); A] = 0, \forall x \in V$ implies $A = \lambda\,\mathbb{1}$ with $\lambda \in C$) and satisfies

$$W(x)W(x') = \exp\left(-\frac{i}{2} x \cdot \sigma \cdot x'\right) W(x + x') \tag{2-88}$$

it is called an (irreducible) representation of the Weyl relations (on $(V; \sigma)$) and the operators $W(x)$ are called Weyl operators. By the well-known Stone-von Neumann uniqueness theorem all these representations are mutually unitarily equivalent, i.e. if have two of them $W_1(x); W_2(x)$, there is a unitary operator U with $UW_1(x)U^* = W_2(x)\ \forall x \in V$. This implies that it does not matter from a physical point of view which representation we use. The most well-known one is of course the Schrödinger representation where $H = L^2(R^d)$ and Q_j, P_k are the usual position and momentum operators.

- Gaussian states

A density operator $\rho \in S(H)$ has finite second moments if the expectation values $\mathrm{tr}(\rho Q_j^2)$ and $\mathrm{tr}\,(\rho P_j^2)$ are finite for all $j = 1, 2 \cdots, d$. In this case we can de. ne the mean $m \in R^{2d}$ and the correlation matrix α by

$$m_j = \mathrm{tr}(\rho R_j), \quad \alpha_{jk} + \mathrm{i}\sigma_{jk} = 2\mathrm{tr}\,[(R_j - m_j)\rho(R_k - m_k)] \tag{2-89}$$

The mean m can be arbitrary, but the correlation matrix α must be real and symmetric and the positivity condition

$$\alpha + \mathrm{i}\sigma \geqslant 0 \tag{2-90}$$

must hold (this is an easy consequence of the canonical commutation relations (2-85)). The aim is now to distinguish exactly one state among all others with the same mean and correlation matrix. This is the point where the operators come into play. Each state $\rho \in S(H)$ can be characterized uniquely by its quantum characteristic function $X \ni x \mapsto \mathrm{tr}[W(x)\rho] \in C$ which should be regarded as the quantum Fourier transform of ρ and is in fact the Fourier transform of the Wigner function of ρ and call ρ Gaussian if

$$\mathrm{tr}\,[W(x)\rho] = \exp\left(\mathrm{i}m \cdot x - \frac{1}{4} x \cdot \alpha \cdot x\right) \tag{2-91}$$

By differentiation, it is easy to check that ρ indeed mean m and covariance matrix α. The most prominent examples for Gaussian states are ground state ρ_0 of a system of d harmonic oscillators, where the mean is 0 and α is given by the corresponding classical Hamiltonian and its phase space translates $\rho_m = W(m)\rho W(-m)$, which are known from quantum optics as coherent states. ρ_0 and ρ_m are pure states and it can be shown that a Gaussian state is pure $i = \sigma^{-1}\alpha = -\mathbb{1}$ holds. Examples for mixed Gaussians are temperature states of harmonic oscillators. In one degree of freedom this is

$$\rho_N = \frac{1}{N+1}\sum_{n=0}^{\infty}\left(\frac{N}{N+1}\right)^n \mid n\rangle\langle n \mid \tag{2-92}$$

where $|n\rangle\langle n|$ denotes the number basis and N is the mean photon number. The characteristic function of ρ_N is

$$\mathrm{tr}\left[W(x)\rho_N\right] = \exp\left[-\frac{1}{2}\left(N+\frac{1}{2}\right)\mid x \mid^2\right] \tag{2-93}$$

Its correlation matrix is simply $\alpha = 2(N+1/2)\mathbb{1}$.

- Entangled Gaussians

Let us now consider bipartite systems. Hence the phase space $(V;\sigma)$ decomposes into a direct sum $V = V_A \oplus V_B$ (where A stands for "Alice" and B for "Bob") and the symplectic matrix $\sigma = \sigma_A \oplus \sigma_B$ is block diagonal with respect to this decomposition. If $W_A(x)$, respectively $W_B(y)$, denote Weyl operators, acting on the Hilbert spaces H_A, H_B, and corresponding to the phase spaces V_A and V_B, it is easy to see that the tensor product $W_A(x)\otimes W_B(y)$ satisfies the Weyl relations with respect to $(V;\sigma)$. Hence by the Stone-von Neumann uniqueness theorem we can identify $W(x\otimes y)$, $x\otimes y \in V_A \oplus V_B = V$ with $W_A(x)\otimes W_A(y)$. This immediately shows that a state ρ on $H = H_A \otimes H_B$ is a product state iff its characteristic function factorizes. Separability characterized as follows.

Theorem 2.16 A Gaussian state with covariance matrix—is separable $i>$ there are covariance matrices $\alpha_A;\alpha_B$ such that holds:

$$\alpha \geqslant \begin{bmatrix} \alpha_A & 0 \\ 0 & \alpha_B \end{bmatrix} \tag{2-94}$$

It provides a useful criterion as long as abstract considerations are concerned, but not for explicit calculations. In contrast to finite-dimensional in systems, however, separability of Gaussian states can decided by an operational criterion in terms of nonlinear maps between matrices. To state it have to introduce some terminology first. The key tool is a sequence of $2n+2m \times 2n+2m$ matrices α_N, $N \in \mathbb{N}$, written in block matrix notation as

$$\alpha_N = \begin{bmatrix} A_N & C_N \\ C_N^{\mathrm{T}} & B_N \end{bmatrix} \tag{2-95}$$

Given α_0 the other α_N are recursively defined by

$$\begin{cases} A_{N+1} = B_{N+1} = A_N - \mathrm{Re}\,(X_N) \\ C_{N+1} = -\mathrm{Im}\,(X_N) \end{cases} \tag{2-96}$$

if $\alpha N - \mathrm{i}\sigma \geqslant 0$ and $\alpha_{N+1} = 0$ otherwise. Here set $X_N = C_N(B_N - i\sigma_B)^{-1}C_N^{\mathrm{T}}$ and the inverse denotes the pseudo-inverse if $B_N - \mathrm{i}\sigma_B$ is not invertible. Now can state the following theorem.

Theorem 2.17 Consider a Gaussian state ρ of a bipartite system with correlation matrix α_0 and the sequence α_N; $N \in \mathbb{N}$ just defined.

(1) If r some $N \in \mathbb{N}$ we have $A_N - \mathrm{i}\sigma_A \not\geqslant 0$ then ρ is not separable.

(2) If there is; on the other hand an $N \in \mathbb{N}$ such that $A_N - \| C_N \| \mathbb{1} - i\sigma_A \geqslant 0$; then the state ρ is separable ($\| C_N \|$ denotes the operator norm of C_N). To check whether a Gaussian state ρ is separable or not, that have to iterate through the sequence α_N until either condition 1 or 2 holds. In the first case, know that ρ entangled and separable in the second. Hence, only the question remains whether the whole procedure terminates after a finite number of iterations. This problem is treated and it turns out that the set of ρ for which separability is decidable after a finite number of steps is the complement of a measure zero set (in the set of all separable states). Numerical calculations indicate in addition that the method converges usually very fast (typically less than five iterations). To consider ppt states first have to characterize the transpose for infinite-dimensional systems. There are different ways to do that. That will use the fact that the adjoint of a matrix can be regarded as transposition followed by componentwise complex conjugation. Hence, define for any (possibly unbounded) operator $A^{\mathrm{T}} = CA * C$, where $C: H \rightarrow H$ denotes complex conjugation of the wave function in position representation. This implies $Q_j^{\mathrm{T}} = Q_j$ for position and $P_j^{\mathrm{T}} = -P_j$ for momentum operators. If insert the partial transpose of a bipartite state ρ into Eq. (2-89) that the correlation matrix $\tilde{\alpha}_{jk}$ of T picks up a minus sign whenever one of the indices belongs to one of Alice's momentum operators. To be a state $\tilde{\alpha}$ should satisfy $\tilde{\alpha} + i\sigma \geqslant 0$, but this is equivalent to $\alpha + i\tilde{\sigma} \geqslant 0$, where in $\tilde{\sigma}$ the corresponding components are reversed i. e. $\tilde{\sigma} = (-\sigma_A) \oplus \sigma_B$. A Gaussian state is ppt iff its correlation matrix α satisfies

$$\alpha + i\tilde{\sigma} \geqslant 0 \quad \text{with } \tilde{\sigma} = \begin{bmatrix} -\sigma_A & 0 \\ 0 & \sigma_B \end{bmatrix} \tag{2-97}$$

The interesting question is now whether the ppt criterion is (for a given number of degrees of freedom) equivalent to separability or not. The following theorem for 1×1 systems and in $1 \times d$ case gives a complete answer.

Theorem 2.18 A Gaussian state of a quantum system with $1 \times d$ degrees of freedom (i. e. dim $X_A = 2$ and dim $X_B = 2d$) is separable iff it is ppt. For other kinds of system the ppt criterion may fail which means that there are entangled Gaussian states which are ppt. A systematic way to construct such states can be found in Roughly speaking, it is based on the idea to go to the boundary of the set of ppt covariance matrices, i. e. —has to satisfy Eqs. (2-87) and (2-94) and it has to be a minimal matrix with this property. Using the method explicit examples for ppt and entangled Gaussians are constructed for 2×2 degrees of freedom.

- Gaussian channels

Finally, as want to give a short review on a special class of channels for infinite-dimensional quantum systems. To explain the basic idea firstly note that each finite set of Weyl operators $(W(x_j), j = 1, 2, \cdots, N, x_j \neq x_k$ for $j \neq k)$ is linear independent. This can checked easily using expectation values of $\sum_j \lambda_j W(x_j)$ in Gaussian states. Hence, linear maps on the space of finite linear combinations of Weyl operators can be defined by $T[W(x)] = f(x) W(A_x)$ where f is a complex-valued function on V and A is a $2d \times 2d$ matrix. If choose A and f carefully enough, such that some continuity properties match T can be extended in a unique way to a linear map on $B(H)$-which is, however, in general not completely positive. This means have to consider special choices for A and f. The most easy case arises if $f \equiv 1$ and A is a symplectic isomorphism, i. e.

$A^{T}\sigma A=\sigma$. If this holds the map $V \ni x \mapsto W(Ax)$ is a representation of the Weyl relations and therefore unitarily equivalent to the representation have started with. In other words, there is a unitary operator U with $T[W(x)]= W(A_x)=UW(x)U^*$ i.e. T is unitarily implemented, hence completely positive and, in fact, well known as Bogolubov transformation. If A does not preserve the symplectic matrix, $f\equiv 1$ is no option. Instead, we have to choose f such that the matrices

$$M_{jk}=f(x_j-x_k)\exp\left(-\frac{i}{2}x_j\cdot\sigma x_k+\frac{i}{2}Ax_j\cdot\sigma Ax_k\right) \tag{2-98}$$

That are positive. Complete positivity of the corresponding T is then a standard result of abstract C^*-algebra theory. If the factor f is in addition a Gaussian, i. e. $f(x)=\exp(-1/2x\beta x)$ for a positive definite matrix β the cp-map T is called a Gaussian channel. A simple way to construct a Gaussian channel is in terms of an ancilla representation. More precisely, if $A: V \to \mathcal{V}$ *is an arbitrary linear map can extend it to a symplectic map* $V \ni x \mapsto A_x \oplus A'x \in V \oplus V'$, where the symplectic vector space $(V';\sigma')$ now refers to the environment. Consider now the Weyl operator $W(x)\otimes W'(x)=W(x;x')$ on the Hilbert space $H\otimes H'$ associated to the phase space element $x\oplus x' \in V\oplus V'$. Since $A\oplus A'$ is symplectic it admits a unitary Bogolubov transformation $U: H\otimes H' \to H\otimes H'$ with $U^* W(x;x')U=W(Ax;A'x)$. If ρ' denotes now a Gaussian density matrix on H' describing the initial state of the environment can get a Gaussian channel by

$$\mathrm{tr}\,[T^*(\rho)W(x)]=\mathrm{tr}\,[\rho\otimes\rho' U^* W(x,x')U]=\mathrm{tr}\,[\rho W(Ax)]\mathrm{tr}\,[\rho' W(A'x)] \tag{2-99}$$

Hence $T[W(x)]=f(x)W(Ax)$ with $f(x)=\mathrm{tr}[\rho' W(A'x)]$. Particular examples for Gaussian channels in the case of one degree of freedom are attenuation and amplification channels. They are given in terms of a real parameter $k\neq 1$ by $\mathbb{R}^2 \ni x \mapsto Ax=kx\in\mathbb{R}^2$

$$\mathbb{R}^2 \ni x \mapsto A'x=\sqrt{1-k^2}\,x\in\mathbb{R}^2<1 \tag{2-100}$$

for $k<1$ and

$$\mathbb{R}^2 \ni (q,p)\mapsto A'(q,p)=(\kappa q,-\kappa p)\in\mathbb{R}^2 \quad \text{with } \kappa=\sqrt{k^2-1} \tag{2-101}$$

for $k>1$. If the environment is initially in a thermal state $\rho_{\tilde{N}}$ this leads to

$$T[W(x)]=\exp\left[\frac{1}{2}\left(\frac{|k^2-1|}{2}+N_c\right)x^2\right]W(kx) \tag{2-102}$$

If start initially with a thermal state N it is mapped by T again to a thermal state $\rho_{N'}$ with mean photon number N' given by

$$N'=k^2N+\max\{0,k^2-1\}+N_c \tag{2-103}$$

If $N_c=0$ this means that T amplifies $(k>1)$ or damps $(k<1)$ the mean photon number, while $N_c>0$ leads to additional classical, Gaussian noise. Reconsider this channel will be more detail in future.

2.4 Micro-aperture laser for photonic memory

After discussed the conceptual foundations of quantum information consider some of its study tasks. The spectrum ranges here from elementary processes, like teleportation or error correction, which are building blocks for more complex applications, up to possible future technologies like quantum cryptography and quantum computing.

2.4.1 Teleportation and dense coding

Maybe the most striking feature of entanglement is the fact that otherwise impossible machines become possible if entangled states used as an additional resource. The most prominent examples are teleportation and dense coding which want to discuss latter. Impossible machines revisited: classical teleportation Already pointed out in the introduction that classical teleportation, i. e. transmission of quantum information over a classical information channel is impossible. With the material introduced in the last two chapters it is now possible to reconsider this subject in a slightly more mathematical way, which makes the following treatment of entanglement' enhanced teleportation more transparent. To "teleport" the state $\rho \in B^*(H)$ Alice performs a measurement (described by a POV measure $E_1 \cdots E_N \in B(H)$) on her system and gets a value $x \in X = \{1, 2, \cdots, N\}$ with probability $p_x = \mathrm{tr}(E_x)$. These data she communicates to Bob and he prepares a $B(H)$ system in the state x. Hence, the overall state Bob gets if the experiment is repeated many times is $\tilde{\rho} = \sum_{x \in X} \mathrm{tr}(E_x \rho) \rho_x$. The latter can be rewritten as the composition

$$\mathscr{B}^*(\mathscr{H}) \xrightarrow{E^*} \mathscr{C}(X)^* \xrightarrow{D^*} \mathscr{B}^*(\mathscr{H})^* \tag{2-104}$$

of the channels

$$\mathscr{C}(X) \ni f \mapsto E(f) = \sum_{x \in X} f(x) E_x \in \mathscr{B}(\mathscr{H}) \tag{2-105}$$

and

$$\mathscr{C}^*(X) \ni p \mapsto D^*(p) = \sum_{x \in X} p_x \rho_x \in \mathscr{B}^*(\mathscr{H}) \tag{2-106}$$

i.e. $\tilde{\rho} = D^* E^*(\rho)$ and this equation makes sense even if X is not finite. The teleportation is successful if the output state $\tilde{\rho}$ cannot be distinguished from the input state ρ by any statistical experiment, i. e. if $D^* E^*(\rho) = \rho$. Hence the impossibility of classical teleportation can be rephrased simply as $ED \neq \mathrm{Id}$ for all observables E and all preparations D.

2.4.2 Entanglement enhanced teleportation

Assume that Alice wants to send a quantum state $\rho \in B^* H^*(\rho)$ to Bob and that she shares an entangled state $\sigma \in B^*(K \otimes K)$ and an ideal classical communication channel $C(X) \to C(X)$ with him. Alice can perform a measurement E: $C(X) \to B(H \otimes K)$ on the composite system $B(H \otimes K)$ consisting of the particle to teleport ($B(H)$) and her part of the entangled system ($B(K)$). Then she communicates the classical data $x \in X$ to Bob and he operates with the parameter—dependent operation $D: B(H) \to B(K) \otimes C(X)$ appropriately on his particle (cf. Figure 2-12). Hence, the overall procedure can be described by the channel $T = (E \otimes \mathrm{Id}) D$, or in analogy to (2-101)

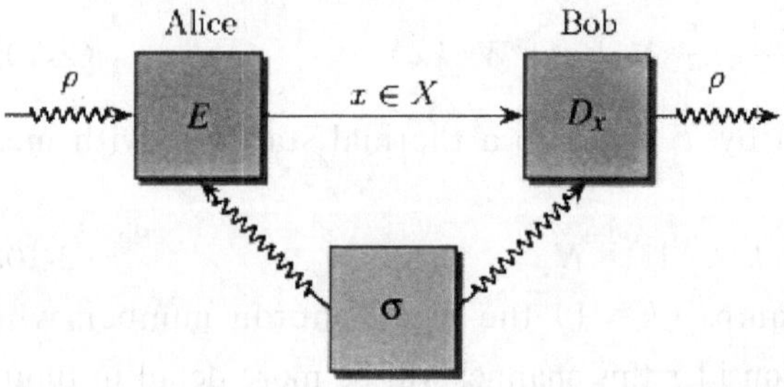

Fig. 2-12 Entanglement enhanced teleportation.

$$\mathscr{B}^*(\mathscr{H} \otimes \mathscr{K}^{\otimes 2}) \xrightarrow{E^* \otimes \mathrm{Id}} \mathscr{C}^*(X) \otimes \mathscr{B}^*(\mathscr{K}) \xrightarrow{D^*} \mathscr{B}^*(\mathscr{H}) \tag{2-107}$$

The teleportation of ρ is successful if

$$T^*(\rho \otimes \sigma) := D^*((E^* \otimes \mathrm{Id})(\rho \otimes \sigma)) = \rho \tag{2-108}$$

Holds, in other words if there is no statistical measurement which can distinguish the final state $T^*(\rho \otimes \sigma)$ of Bob's particle from the initial state ρ of Alice's input system. The two channels E and

D and the entangled state σ form a teleportation scheme if Eq. (2-105) holds for all states ρ of the $B(H)$ system, i. e. if each state of a $B(H)$ system can be teleported without loss of quantum information. Assume now that $H = K = C^d$ and $X = \{0, 1, \cdots, d^2 - 1\}$ holds. In this case can define a teleportation scheme as follows: The entangled state shared by Alice and Bob is a maximally entangled state $\rho = |\Omega\rangle\langle\Omega|$ and Alice performs a measurement, which is given by the one-dimensional projections $E_j = |\Phi_j\rangle\langle\Phi_j|$, where $\Phi_j \in H \otimes H$, $j = 0, 1, \cdots, d^2 - 1$ is a basis of maximally entangled vectors. If her result is $j = 0, 1, \cdots, d^2 - 1$ Bob has to apply the operation $\tau \mapsto U_{j\tau}^* U_j$ on his partner of the entangled pair, where the $U_j \in B(H)$, $j = 0, 1, \cdots, d^2 - 1$ are an orthonormal family of unitary operators, i. e. $\operatorname{tr}(U_j^* U_k) = d\delta_{jk}$. Hence, the parameter-dependent operation D has the form (in the Schrödinger picture):

$$\mathscr{C}^*(X) \otimes \mathscr{B}^*(\mathscr{H}) \ni (p, \tau) \mapsto D^*(p, \tau) = \sum_{j=0}^{d^2-1} p_j U_j^* \tau U_j \in \mathscr{B}^*(\mathscr{H}) \tag{2-109}$$

Therefore, can get for $T^*(\rho \otimes \sigma)$ from Eq. (2-105)

$$\operatorname{tr}[T^*(\rho \otimes \sigma) A] = \operatorname{tr}[(E \otimes \mathrm{Id}) * (\rho \otimes \sigma)$$

$$= \operatorname{tr}\left[\sum_{j=0}^{d^2-1} \operatorname{tr}_{12}[|\Phi_j\rangle\langle\Phi_j|(\rho \otimes \sigma)] U_j^* A U_j\right] \tag{2-110}$$

$$= \sum_{j=0}^{d^2-1} \operatorname{tr}[(\rho \otimes \sigma)|\Phi_j\rangle\langle\Phi_j| \otimes (U_j^* A U_j)] \tag{2-111}$$

Here tr_{12} denotes the partial trace over the first two tensor factors (= Alice's qubits). If Ω, the Φ_j and the U_j related by the equation

$$\Phi_j = (U_j \otimes \mathbb{1})\Omega \tag{2-112}$$

it is a straightforward calculation to show that $T^*(\rho \otimes \sigma) = \rho$ holds as expected. If $d = 2$ there basically a unique choice: the Φ_j, $j = 0, 1, \cdots, 3$ are the four Bell states, $\Omega = \Phi_0$ and the U_j are the identity and the three Pauli matrices. In this way, recover the standard example for teleportation.

2.4.3 Dense coding

Just shown how quantum information can transmitted via a classical channel, if entanglement is available as an additional resource. Looking at the dual procedure: transmission of classical information over a quantum channel. To send the classical information $x \in X = \{1, 2, \cdots, n\}$ to Bob, Alice can prepare a d-level quantum system in the state $\rho^x \in B^*(H)$, sends it to Bob and measures an observable given by positive operators $E_1 \cdots E_m$. The probability for Bob to receive the signal $y \in X$ if Alice has sent $x \in X$ is $\operatorname{tr}(\rho_x E_y)$ and this defines a classical information channel by

$$\mathscr{C}^*(X) \ni p \mapsto \left(\sum_{x \in X} p(x) \operatorname{tr}(\rho_x E_1), \cdots, \sum_{x \in X} p(x) \operatorname{tr}(\rho_x E_m)\right) \in \mathscr{C}^*(X) \tag{2-113}$$

To get an ideal channel we just have to choose mutually orthogonal pure states $\rho_x = |\psi_x\rangle\langle\psi_x|$, $x = 1, 2, \cdots, d$ on Alice's side and the corresponding one-dimensional projections $E_y = |\psi_y\rangle\langle\psi_y|$ $y = 1, 2, \cdots, d$ on Bob's. If $d = 2$ and $H = C^2$ it is possible to send one bit classical information via one qubit quantum information. The crucial point is now that the amount of classical information can be increased (doubled in the qubit case) if Alice shares an entangled state $\sigma \in S(H \otimes H)$ with Bob. To send the classical information $x \in X = \{1, 2, \cdots, n\}$ to Bob, Alice operates on her particle with an operation $D_x : B(H) \to B(H)$, sends it through an (ideal) quantum channel to Bob and he performs a measurement $E_1 \cdots E_n \in B(H \otimes H)$ on both particles. The probability for Bob to measure $y \in X$ if Alice has send $x \in X$ is given by

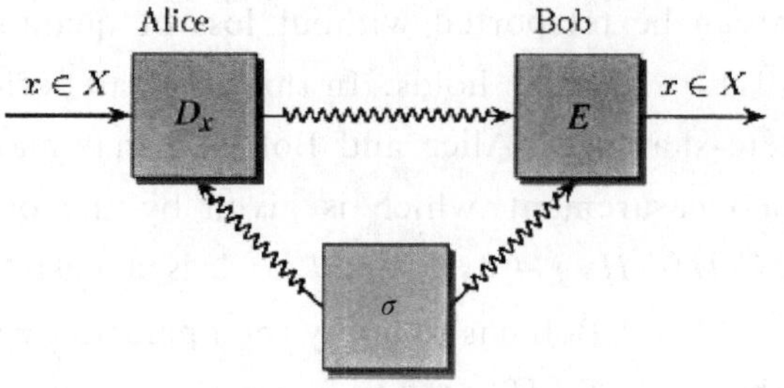

Fig. 2-13 Dense coding.

$$\mathrm{tr}\left[(D_x \otimes \mathrm{Id})^*(\sigma)E_y\right] \tag{2-114}$$

And this defines the transition matrix of a classical communication channel T. If T is an ideal channel, i.e. if the transition matrix (2-114) is the identity, we will call E, D and σ a dense coding scheme (cf. Figure 2-13). In analogy to Eq. (2-104) can rewrite the channel T defined by (2-114) in terms of the composition

$$\mathscr{C}^*(X) \otimes \mathscr{B}^*(\mathscr{H}) \otimes \mathscr{B}^*(\mathscr{H}) \xrightarrow{D^* \otimes \mathrm{Id}} \mathscr{B}^*(\mathscr{H}) \otimes \mathscr{B}^*(\mathscr{H}) \xrightarrow{E^*} \mathscr{C}^*(X) \tag{2-115}$$

of the parameter-dependent operation

$$D: \mathscr{C}^*(X) \otimes \mathscr{B}^*(\mathscr{H}) \to \mathscr{B}^*(\mathscr{H}), \quad p \otimes \tau \mapsto \sum_{j=1}^{n} p_j D_j(\tau) \tag{2-116}$$

and the observable

$$E: \mathscr{C}(X) \to \mathscr{B}(\mathscr{H} \otimes \mathscr{H}), \quad p \mapsto \sum_{j=1}^{n} p_j E_j \tag{2-117}$$

i.e. $T^*(p) = E^* \circ (D^* \otimes \mathrm{Id})(p \otimes \sigma)$. The advantage of this point of view is that it works as well for infinite-dimensional Hilbert spaces and continuous observables. Finally, let again consider the case where $H = C^d$ and $X = \{1, 2, \cdots, d^2\}$. If choose as in the last paragraph a maximally entangled vector $\Omega \in H \otimes H$, an orthonormal base $\Phi_x \in H \otimes H$, $x = 1, 2, \cdots, d^2$ of maximally entangled vectors and an orthonormal family $U_x \in B(H \otimes H)$, $x = 1, 2, \cdots, d^2$ of unitary operators, we can construct a dense coding scheme as follows: $E_x = |\Phi_x\rangle\langle\Phi_x|$, $D_x(A) = U_x^* A U_x$ and $\sigma = |\Omega\rangle\langle\Omega|$. If Ω, the Φ_x and the U_x are related by Eq. (2-110) it is easy to see that we really get a dense coding scheme. If $d = 2$ holds, have to set again the Bell basis for the Φ_x, $\Omega = \Phi_0$ and the identity and the Pauli matrices for the U_x. Recover in this case the standard example of dense coding proposed and that can transfer two bits via one qubit, as stated above.

2.4.4 Estimating and copying

The impossibility of classical teleportation can rephrased as follows: It is impossible to get complete information about the state ρ of a quantum system by one measurement on one system. However, if have many systems, say N, all prepared in the same state ρ it should be possible to get (with a clever measuring strategy) as much information on ρ as possible, provided N is large enough. In this way, can circumvent the impossibility of devices to like classical teleportation or quantum copying at least in an approximate way.

- Quantum state estimation

To discuss this idea in a more detailed way consider a number N of d-level quantum systems, all of them prepared in the same (unknown) state $\rho \in B^*(H)$. Our aim is to estimate the state ρ by measurements on the compound system $\rho^{\otimes N}$. This is described in terms of an observable E^N: $C(X_N) \to B(H^{\otimes N})$ with values in a finite subset $X_N \subset S(H)$ of the quantum state space $S(H)$. According to each such E^N is given in terms of a tuple E_σ^N, $\sigma \in X_N$, by $E(f) = \sum_\sigma f(\sigma) E_\sigma^N$; hence, can get for the expectation value of an E_N measurement on systems in the state $\rho^{\otimes N}$ the density matrix $\hat{\rho}_N \in S(H)$ with matrix elements

$$\langle \phi, \hat{\rho}_N \psi \rangle = \sum_{x \in X_N} \langle \phi, \sigma\psi \rangle E_\sigma^N \tag{2-118}$$

Call the channel E^N an estimator and the criterion for a good estimator E^N is that for any one-particle density operator ρ, the value measured on a state $\rho^{\otimes N}$ is likely to be close to ρ, i.e. that the probability

$$K^N(\omega) := \mathrm{tr}(E^N(\omega)\rho^{\otimes N}) \quad \text{with} \quad E^N(\omega) = \sum_{\sigma \in X_N \cap \omega} E_\sigma^N \tag{2-119}$$

is small if $\omega \subset S(H)$ is the complement of a small ball around ρ. Of course, look at this problem for large N. So the task is to find a whole sequence of observables $E^N, N = 1, 2\cdots$, making error probabilities like Eq. (2-119) go to zero as $N \to \infty$. The most direct way to get a family $E^N, N \in \mathbb{N}$ of estimators with this property is to perform a sequence of measurements on each of the N input systems separately. A finite set of observables, which leads to a successful estimation strategy is usually called a "quorum". E.g. for $d = 2$ can perform alternating measurements of the three spin components. If $\rho = 1/2(\mathbb{1} + \vec{x}\vec{\sigma})$ is the Bloch representation of ρ that the expectation values of these measurements are given by $1/2(1 + x_j)$. Hence an arbitrarily good estimate if N is large enough. A similar procedure is possible for arbitrary d if consider the generalized Bloch representation for ρ. There are however more efficient strategies based on "entangled" measurements (i.e. the $E^N(\sigma)$ cannot be decomposed into pure tensor products) on the whole input system $\rho^{\otimes N}$. Somewhat in between are "adaptive schemes" consisting of separate measurements but the jth measurement depend on the results of $(j-1)$ th. Reconsider this circle of questions in a quantitative way.

- Approximate cloning

By virtue of the no-cloning theorem, it is impossible to produce M perfect copies of a d-level quantum system if $N < M$ input systems in the common (unknown) state $\rho^{\otimes N}$ are given. More precisely there is no channel $T_{MN}: B(H^{\otimes M}) \to B(H^{\otimes N})$ such that $T_{MN}^*(\rho^{\otimes N}) = \rho^{\otimes M}$ holds for all $\rho \in S(H)$. Using state estimation, however, it is easy to find a device T_{MN} which produces at least approximate copies which become exact in the $\lim_{N;M\to\infty}$: If $\rho^{\otimes N}$ is given, we measure the observable E^N and get the classical data $\sigma \in X_N \subset S(H)$, which we use subsequently to prepare M systems in the state $\sigma^{\otimes M}$. In other words, T_{MN} has the form

$$\mathcal{B}^*(\mathcal{H}^{\otimes N}) \ni \tau \mapsto \sum_{\sigma \in X_N} \mathrm{tr}(E_\sigma^N \tau)\sigma^{\otimes M} \in \mathcal{B}^*(\mathcal{H}^{\otimes M}) \tag{2-120}$$

Immediately that the probability get wrong copies coincides exactly with the error probability of the estimator given in Eq. (2-119). This shows first that we get exact copies in the limit $N \to \infty$ and second that the quality of the copies does not depend on the number M of output systems, i.e. the asymptotic rate $\lim_{N;M\to\infty} M/N$ of output systems per input system can be arbitrary large. The fact that can get classical data at an intermediate step allows a further generalization of this scheme. Instead of just preparing M systems in the state σ detected by the estimator, can apply first an arbitrary transformation $F: S(H) \to S(H)$ on the density matrix σ and prepare $F(\sigma)^{\otimes M}$ instead of $\sigma^{\otimes M}$. In this way, can get the channel (cf. Figure 2-14)

$$\mathcal{B}^*(\mathcal{H}^{\otimes N}) \ni \tau \mapsto \sum_{\sigma \in X_N} \mathrm{tr}(E_\sigma^N \tau)F(\sigma)^{\otimes M} \in \mathcal{B}^*(\mathcal{H}^{\otimes M}) \tag{2-121}$$

i.e. a physically realizable device which approximates the impossible machine F. The probability to get a bad approximation of the state $F(\rho)^{\otimes M}$ (if the input state was $\rho^{\otimes N}$) is again given by the error probability of the estimator and get a perfect realization of F at arbitrary rate as $M;N \to \infty$.

There are in particular two interesting tasks, which become possible this way: The first is the "universal not gate" which associates to each pure state of a qubit the unique pure state orthogonal to it. This is a special example of a antiunitarily implemented symmetry operation and therefore not completely positive. The second example is the purification of states. Here it is assumed that the input states were once pure but have passed later on a depolarizing channel $|\Phi\rangle\langle\Phi| \mapsto \theta|\Phi\rangle\langle\Phi| + (1-\theta)\mathbb{1}/d$. If $\theta > 0$ this map is invertible but its inverse does not describe an allowed quantum operation because it maps some density operators to operators with negative eigenvalues. Hence the reversal of noise is not possible with a one-shot operation but can be done with high accuracy if enough input systems are available.

2.4.5 Distillation of entanglement

Now return to entanglement, have seen that maximally entangled states play a crucial role for processes like teleportation and dense coding. In practice however entanglement is a rather fragile property: If Alice produces a pair of particles in a maximally entangled state $|\Omega\rangle\langle\Omega| \in S(H_A \otimes H_B)$ and distributes one of them over a great distance to Bob, both end up with a mixed state ρ which contains much less entanglement then the original and which cannot be used any longer for teleportation. The latter can be seen quite easily if try to apply the qubit teleportation scheme with a non-maximally entangled isotropic state instead of Ω.

Hence the question arises, whether it is possible to recover $|\Omega\rangle\langle\Omega|$ from ρ, or, following the reasoning from the last section, at least a small number of (almost) maximally entangled states from a large number N of copies of ρ. However, since the distance between Alice and Bob is big (and quantum communication therefore impossible) only LOCC operations are available for this task (Alice and Bob can only operate on their respective particles, drop some of them and communicate classically with one another). This excludes procedures like the puri. cation scheme just sketched, because would need "entangled" measurements to get an asymptotically exact estimate for the state ρ. Hence, need a sequence of LOCC channels

$$T_N : \mathcal{B}(\mathbb{C}^{d_N} \otimes \mathbb{C}^{d_N}) \to \mathcal{B}(\mathcal{H}_A^{\otimes N} \otimes \mathcal{H}_B^{\otimes N}) \tag{2-122}$$

such that

$$\| T_N^*(\rho^{\otimes N}) - |\Omega_N\rangle\langle\Omega_N| \|_1 \to 0 \quad \text{for} \quad N \to \infty \tag{2-123}$$

holds, with a sequence of maximally entangled vectors $\Omega_N \in \Omega^{d_N} \otimes \Omega^{d_N}$. Note that have to use here the natural isomorphism $H_A^{\otimes N} \otimes H_B^{\otimes N} \approx (H_A \otimes H_B)^{\otimes N}$, i.e. have to reshuffle $\rho^{\otimes N}$ that the fit N tensor factors belong to Alice (H_A) and the last N to Bob (H_B). If confusion can avoided will use this isomorphism in the following without a further note. Call a sequence of LOCC channels, T_N satisfying (2-121) with a state $\rho \in S(H_A \otimes H_B)$ a distillation scheme for ρ and ρ is called distillable if it admits a distillation scheme. The asymptotic rate with which maximally entangled states can distilled with a given protocol is

$$\liminf_{n \to \infty} \log_2(d_N)/N \tag{2-124}$$

This quantity will become relevant in the framework of entanglement measures.

• Distillation of pairs of qubits

Concrete distillation protocols are in general rather complicated procedures. Sketch in the following how any pair of entangled qubits can distilled. The first step is a scheme proposed for the first time by Bennett et al. It can applied if the maximally entangled fraction F (cf. Eq. (2-29)) is greater than 1/2. As indicated above, we assume that Alice and Bob share a large amount of pairs in the state ρ, so that the total state is $\rho^{\otimes N}$. To obtain a smaller number of pairs with a higher F they proceed as follows:

(1) First they take two pairs (let call them pairs 1 and 2), i. e. $\rho\otimes\rho$ and apply to each of them the twirl operation $P_{U\bar{U}}$ associated to isotropic states. This can be done by LOCC operations in the following way: Alice selects at random a unitary operator U applies it to her qubits and sends to Bob which transformation she has chosen; then he applies $\bar{U}$ to his particles. They end up with two isotropic states $\tilde{\rho}\otimes\tilde{\rho}$ with the same maximally entangled fraction as ρ.

(2) Each party performs the unitary transformation

$$U_{\mathrm{XOR}}: |a\rangle\otimes|b\rangle\mapsto|a\rangle\otimes|a+b \bmod 2\rangle \tag{2-125}$$

on his/her members of the pairs.

(3) Finally, Alice and Bob perform local measurements in the basis $|0\rangle;|1\rangle$ on pair 1 and discards it afterwards. If the measurements agree, pair 2 is kept and has a higher F. Otherwise pair 2 is discarded as well. If this procedure is repeated over again, it is possible to get states with an arbitrarily high F, but have to sacrifice more and more pairs and the asymptotic rate is zero. To overcome this problem can apply the scheme above until $F(\rho)$ is high enough such that $1+\mathrm{tr}(\rho\ln\rho)\geqslant 0$ holds and then continue with another scheme called hashing which leads to a non-vanishing rate. If finally $F(\rho)\leqslant 1/2$ but ρ is entangled, Alice and Bob can increase F for some of their particles by filtering operations. The basic idea is that Alice applies an instrument $T: C(X)\otimes B(H)\to B(H)$ with two possible outcomes ($X=\{1;2\}$) to her particles. Hence, the state becomes $\rho\mapsto p_x^{-1}(T_x\otimes\mathrm{Id})*(\rho), x=1;2$ with probability $p_x=\mathrm{tr}[T_x^*(\rho)]$ (in particular Eq. (2-75) for the definition of T_x). Alice communicates her measuring result x to Bob and if $x=1$ they keep the particle otherwise ($x=2$) they discard it. If the instrument T chosen Alice and Bob end up with a state $\tilde{\rho}$ with higher maximally entangled fraction correctly. To find an appropriate T firstly note that there are $\psi\in H\otimes H$ with $\langle\psi(\mathrm{Id}\otimes\theta)\rho\psi\rangle\leqslant 0$ and second that can write each vector $\psi\in H\otimes H$ as $(X_\psi\otimes\mathbb{1})\Phi_0$ with the Bell state Φ_0 and an appropriately chosen operator X. Now can define T in terms of the two operations $T_1;T_2$ (cf. Eq. (2-80)) with

$$\begin{cases} T_1(A)=X_\psi^* A X_\psi^{-1} \\ \mathrm{Id}-T_1=T_2 \end{cases} \tag{2-126}$$

It is straight forward to check that end up with

$$\tilde{\rho}=\frac{(T_x\otimes\mathrm{Id})^*(\rho)}{\mathrm{tr}\left[(T_x\otimes\mathrm{Id})^*(\rho)\right]} \tag{2-127}$$

such that $F(\tilde{\rho})>1/2$ holds and can continue with the scheme described in the previous paragraph.

• Distillation of isotropic states

Consider now an entangled isotropic state ρ in d dimensions, i. e. we have $H=C^d$ and $0\leqslant\mathrm{tr}(\tilde{F}\rho)\leqslant 1$ (with the operator $\tilde{F}$). Each such state is distillable via the following scheme: First,

Alice and Bob apply a filter operation $TC(X)\otimes B(H)\rightarrow B(H)$ on their respective particle given by $T_1(A)=PAP, T_2=1-T_1$ where P is the projection onto a two-dimensional subspace. If both measure the value 1 they get a qubit pair in the state $\tilde{\rho}=(T_1\otimes T_1)(\rho)$. Other wise they discard their particles (this requires classical communication). Obviously, the state $\tilde{\rho}$ is entangled (this can be easily checked), hence they can proceed as in the previous subsection. The scheme just proposed can be used to show that each state ρ, which violates the reduction criterion distilled. The basic idea is to project ρ with the twirl $P^{U\bar{U}}$ (which is LOCC as have seen above.) to an isotropic state $P^{U\bar{U}}(\rho)$ and to apply the procedure from the last paragraph afterwards. Only have to guarantee that $P^{U\bar{U}}(\rho)$ entangled. To this end use a vector $\psi\in H\otimes H$ with $\langle\psi(\mathbb{1}\otimes \mathrm{tr}\ 1(\rho)-\rho)\psi\rangle<0$ (which exists by assumption since ρ violates the reduction criterion) and to apply the filter operation given by ψ via Eq. (2-126).

- Bound entangled states

It is obvious that separable states are not distillable, because an LOCC operation map separable states to separable states. However, is each entangled state distillable. The answer, maybe somewhat surprising, is no and an entangled state which is not distillable is called bound entangled (distillable states are sometimes called free entangled, in analogy to thermodynamics). Examples of bound entangled states are all ppt entangled states: This is an easy consequence of the fact that each separable channel (and therefore each LOCC channel as well) maps ppt states to ppt states (this is easy to check), but a maximally entangled state is never ppt. It is not yet known, whether bound entangled npt states exists, however, there are at least some partial results:

(1) It is sufficient to solve this question for Werner states, i. e. if can show that each npt Werner state is distillable it follows that all npt states are distillable.

(2) Each npt Gaussian state is distillable.

(3) For each $N\in\mathbb{N}$ there is an npt Werner state ρ which is not " N-copy distillable", i. e. $\langle\psi, \rho\otimes N\psi\rangle\geqslant 0$ holds for each pure state ψ with exactly two Schmidt summands. This gives some evidence for the existence of bound entangled npt states because ρ is distillable iff it is N-copy distillability for some N. Since bound entangled states cannot distilled, they cannot used for teleportation. Never the less bound entanglement can produce a non-classical effect, called "activation of bound entanglement". To explain the basic idea, assume that Alice and Bob share one pair of particles in a distillable state ρ_f and many particles in a bound entangled state ρ_b. Assume in addition that ρ_f cannot be used for teleportation, or, in other words if ρ_f is used for teleportation the particle Bob receives is in a state σ' which differs from the state σ Alice has send. This problem cannot be solved by distillation, since Alice and Bob share only one pair of particles in the state ρ_f. Nevertheless, they can try to apply an appropriate filter operation on ρ to get with a certain probability a new state which leads to a better quality of the teleportation (or, if the filtering fails, to get nothing at all).

It can be shown, however, that there are states ρ_f such that the error occurring in this process (e.g. measured by the trace norm distance of σ and σ') is always above a certain threshold. This is the point where the bound entangled states ρ_b come into play: If Alice and Bob operate with an appropriate protocol on ρ_f and many copies of ρ_b the distance between σ and σ' can be made arbitrarily small (although the probability to be successful goes to zero). Another example for an activation of bound entanglement is related to distillability of npt states: If Alice and Bob share a certain ppt-entangled state as additional resource each npt state ρ becomes distillable (even if is bound entangled). For a more detailed survey of the role of bound entanglement and further references.

2.4.6 Quantum error correction

If try to distribute quantum information over large distances or store it for a long time in some sort of "quantum memory" always have to deal with "decoherence effects", i. e. unavoidable interactions with the environment. This results in a significant information loss, which is particularly bad for the functioning of a quantum computer. Similar problems arise as well in a classical computer, but the methods used there to circumvent the problems cannot transferred to the quantum regime. E. g. the most simple strategy to protect classical information against noise is redundancy: instead of storing the information once we make three copies and decide during readout by a majority vote which bit to take. It is easy to see that this reduces the probability of an error from order ε to ε^2. Quantum mechanically however such a procedure forbidden by the no cloning theorem. Nevertheless, quantum error correction is possible although have to do it in a more subtle way than just copying; this was observed for the first time independently. Consider first the general scheme and assume that $T: B(K) \to B(K)$ is a noisy quantum channel. To send quantum systems of type $B(H)$ undisturbed through T we need an encoding channel E: $B(K) \to B(H)$ and a decoding channel $D: B(H) \to B(K)$ such that $ETD = \mathrm{Id}$ holds, respectively $D^* T^* E^* = \mathrm{Id}$, in the Schrödinger picture; cf. Figure 2-14.

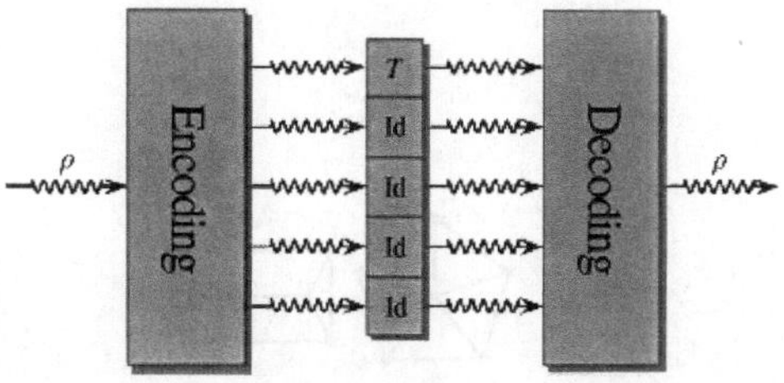

Fig. 2-14 Five-bit quantum code: encoding one qubit into five and correcting one error.

A powerful error correction scheme should not be restricted to one particular type of error, i.e. one particular noisy channel T. Assume instead that $E \subset B(K)$ is a linear subspace of "error operators" and T is any channel given by

$$T_*(\rho) = \sum_j F_j \rho F_j^*, \quad F_j \in \mathfrak{E} \tag{2-128}$$

An isometry $V: H \to K$ is called an error correcting code for E if for each T of form (2-126) there is a decoding channel $D: B(H) \to B(K)$ with $D^*(T(VV^*)) = \rho$ for all $\rho \in S(H)$. By the theory of Knill and La Jamme this is equivalent to the factorization condition

$$\langle V\psi, F_j^* F_k V\phi \rangle = \omega(F_j^* F_k)\langle \psi, \phi \rangle \tag{2-129}$$

where $\omega(F_j^* F_k)$ is a factor which does not depend on the arbitrary vectors $\psi, \Phi \in H$. The most relevant examples of error correcting codes, which generalize the classical idea of sending multiple copies in a certain sense. This means encode a small number N of d-level systems into a big number $M \gg N$ of systems of the same type, which then transmitted and decoded back into N systems afterwards. During the transmission $K < M$ arbitrary errors are allowed. Hence, have $H = H_1^{\otimes N}$, $K = H_1^{\otimes M}$ with $H_1 = C_d$ and T is an arbitrary tensor product of K noisy channels S_j, $j = 1, 2, \cdots, K$ and $M - K$ ideal channels Id. To define the corresponding error space E consider the finite sets $X = \{1, 2, \cdots, N\}$ and $Y = \{1 + N, \cdots, M + N\}$ and define first for each subset $Z \subset Y$:

$$\mathfrak{E}(Z) = \text{span}\{A_1 \otimes \cdots \otimes A_M \in \mathcal{B}(\mathcal{H}) \mid$$
$$A_j \in \mathcal{B}(\mathcal{K}_1)\text{arbitrary for } j + N \in Z, \quad A_j = \mathbb{1} \text{ otherwise}\} \tag{2-130}$$

E is now the span of all $E(Z)$ with $|Z| \ll K$ (i.e. the length of Z is less or equal to K). That an error correcting code for this particular E corrects K errors. There are several ways to construct error correcting codes. Most of these methods are somewhat involved however and require knowledge from classical error correction, which want to skip. Therefore, only present the scheme proposed, which is quite easy to describe and admits a simple way to check the error correction condition. Sketch first the general scheme. Start with an undirected graph Γ with two kinds of vertices: A set of input vertices, labeled by X and a set of output vertices labeled by Y. The links of the graph are given by the adjacency matrix, i.e. an $N + M \times N + M$ matrix Γ with $\Gamma_{jk} = 1$ if node k and j are linked and $\Gamma_{jk} = 0$ otherwise. With respect to Γ can define now an isometry VF: $H \otimes N_1 \rightarrow H \otimes M_1$ by

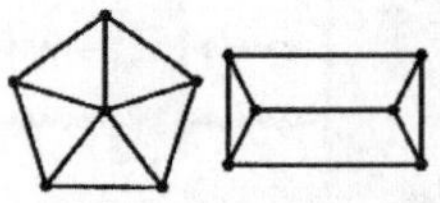

Fig. 2-15 Two graphs belonging to (equivalent) five bit codes. The input node can be chosen in both cases arbitrarily.

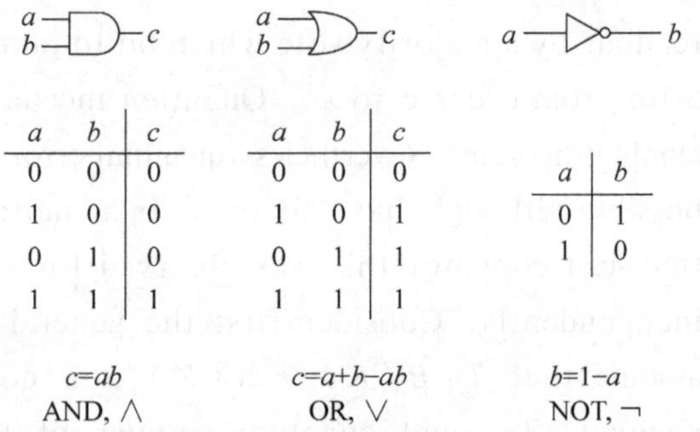

Fig. 2-16 Symbols and definition for the three elementary gates AND, OR and NOT.

Linked and $\Gamma_{jk} = 0$ otherwise. With respect to F can define now an isometry $V_\Gamma : H_1^{\otimes N} \rightarrow H_1^{\otimes M}$ by

$$\langle j_{N+1} \cdots j_{N+M} \mid V_\Gamma \mid j_1 \cdots j_N \rangle = \exp\left(\frac{1\pi}{d}\vec{j} \cdot \Gamma\vec{j}\right) \tag{2-131}$$

With $\vec{j} = (j_1, j_2, \cdots, j_{N+M}) \in Z_d^{N+M}$ (where Z_d denotes the cyclic group with d elements). An easy condition under which V_Γ is an error correcting code. To write it down need the following additional terminology: That an error correcting code $VH_1^{\otimes N} \rightarrow H_1^{\otimes M}$ detects the error configuration $Z \subset Y$ if

$$\langle V\psi, FV\phi \rangle = \omega(F)\langle \psi, \phi \rangle \quad \forall F \in \mathfrak{E}(Z) \tag{2-132}$$

holds. With Eq. (2-127) it is easy to see that V corrects K errors iff it detects all error configurations of length $2K$ or less. Now we have the following theorem:

Theorem 2.19 The quantum code V_Γ defined in Eq. (2-129) detects the error configuration $Z \subset Y$ if the system of equations

$$\sum_{l \in X \cup Z} \Gamma_{kl} g_l = 0, \quad k \in Y \backslash E, \quad g_l \in \mathbb{Z}_d \tag{2-133}$$

implies that holds.

$$g_l = 0, \quad l \in X \quad \text{and} \quad \sum_{l \in Z} \Gamma_{kl} g_l = 0, \quad k \in X \tag{2-134}$$

Omit the proof, instead. Two particular examples (which are equivalent!) given in Figure 2-17. In both cases have $N = 1, M = 5$ and $K = 1$ i.e. one input node, which can be chosen arbitrarily, five output nodes and the corresponding codes correct one error. For a more detailed survey on quantum error correction, in particular for more examples.

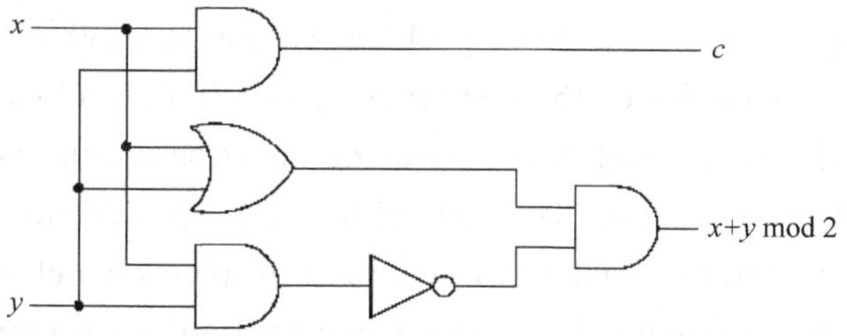

Fig. 2-17 Half-adder circuit as an example for a Boolean network.

2.4.7 Quantum computing

Quantum computing is without a doubt the most prominent and most far reaching application of quantum information theory, since it promises on the one hand, "exponential speedup" for some problems which are "hard to solve" with a classical computer, and gives completely new insights into classical computing and complexity theory on the other. Unfortunately, an exhaustive discussion would require its own review article. Hence, are only able to give a short overview (for a more complete presentation and for further references).

• The network model of classical computing

Start with a brief (and very informal) introduction to classical computing, what need first is a mathematical model for computation. There are, in fact, several different choices and the Turing machine is the most prominent one. More appropriate for our purposes is, however, the so-called network model, since it allows an easier generalization to the quantum case. The basic idea is to interpret a classical (deterministic) computation as the evaluation of a map $f: B^N \rightarrow B^M$ (where $B = \{0;1\}$ denotes the field with two elements) which maps N input bits to M output bits. If $M = 1$ holds f is called a Boolean function and it is for many purposes sufficient to consider this special case—each general f is in fact a Cartesian product of Boolean functions. Particular examples are the three elementary gates AND, OR and NOT defined in Figure 2-17 and arbitrary algebraic expressions constructed from them: e.g. the XOR gate $(x;y) \mapsto x + y \bmod 2$ which can be written as $(x \vee y) \wedge (x \wedge y)$. It is now a standard result of Boolean algebra that each Boolean function can represented in this way and there are in general many possibilities to do this. A special case is the disjunctive normal form of f. To write such an expression down in form of equations is, however, somewhat confusing. f is therefore expressed most conveniently in graphical form as a circuit or network, i. e. a graph C with nodes representing elementary gates and edges ("wires") which determine how the gates should be composed. A classical computation can defined as a circuit applied to a specified string of input bits. Variants of this model arise if replace AND, OR and NOT by another (finite) set G of elementary gates. Only have to guarantee that each function f can expressed as a composition of elements from G. A typical example for G is the set which contains only the NAND gate $(x;y) \mapsto x \uparrow y = (x \wedge y)$. Since AND, OR and NOT can be rewritten in terms of NAND (e. g. $x = x \uparrow x$) can calculate each Boolean function by a circuit of NAND gates.

• Computational complexity

One of the most relevant questions within classical computing, and the central subject of computational complexity, is whether a given problem is easy to solve or not, where "easy" is de. ned in terms of the scaling behavior of the resources needed in dependence of the size of the input data. In the following will give a rough survey over the most basic aspects of this field, while refer the reader for a detailed presentation. To start with, let us specify the basic question in detail. First of all the problems we want to analyze are decision problems which only give the two possible values "yes" and "no". They mathematically described by Boolean functions acting on bit strings of arbitrary size. A well-known example is the factoring problem given by the function fac with fac $(m;l)=1$ if m (more precisely the natural number represented by m) has a divisor less then 1 and fac $(m;l)=0$ otherwise. Note that many tasks of classical computation can reformulated this way, so that we do not get a severe loss of generality. The second crucial point have to clarify which is the question what exactly are the resources have mentioned above and how we have to quantify them. A natural physical quantity which come into mind immediately is the time needed to perform the computation (space is another candidate, which do not discuss here, however). Hence the question have to discuss is how the computation time t depends on the size L of the input data x (i.e. the length L of the smallest register needed to represent x as a bit string). However, a precise definition of "computation time" is still model dependent. For a Turing machine can take simply the number of head movements needed to solve the problem, and in the network model choose the number of steps needed to execute the whole circuit, if gates which operate on different bits are allowed to work simultaneously. Even with a fixed type of model the functional behavior of t depends on the set of elementary operations choose, e.g. the set of elementary gates in the network model. It is therefore useful to divide computational problems into complexity classes whose definitions do not suer under model-dependent aspects. The most fundamental one is the class P contains all problems, which can computed in "polynomial time", i.e. t is, as a function of L, bounded from above by a polynomial. The model independence of this class basically the content of the strong Church Turing hypotheses which states, roughly speaking, that each model of computation can be simulated in polynomial time on a probabilistic Turing machine. Problems of class P are considered "easy", everything else is "hard". However, even if a (decision) problem is hard the situation is not hopeless. E.g. consider the factoring problem fac described above. It is believed (although not proved) generally that this problem is not in class P. But if somebody gives us a divisor p of m it is easy to check whether p is really a factor, and if the answer is true we have computed fac $(m;l)$. This example motivates the following definition: A decision problem f is in class NP ("non-deterministic polynomial time") if there is a Boolean function f' in class P such that $f'(x;y)=1$ for some y implies $f(x)$. In the example fac$'$ is obviously defined by fac$'(m;l;p)=1 \Leftrightarrow p<1$ and p is a devisor of m. It is obvious that P is a subset of NP the other inclusion however is rather non-trivial. The conjecture is that $P \neq$ NP holds and great parts of complexity theory based on it. Its proof (or disproof), however, represents one of the biggest open questions of theoretical informatics. To introduce a third complexity class have to generalize the point of view slightly. Instead of a function $f: B^N \rightarrow B^M$ can look at a noisy classical T which sends the input value $x \in$ BN to a probability distribution T_{xy}, $y \in$ BM on BM (i.

e. T_{xy} is the transition matrix of the classical channel T). Roughly speaking, we can interpret such a channel as a probabilistic computation, which can be realized as a circuit consisting of "probabilistic gates".

This means there are several different ways to proceed at each step and use a classical random number generator to decide which of them we have to choose. If we run our device several times on the same input data x can get different results y with probability T_{xy}. The crucial point is now that can allow some of the outcomes to be wrong as long as there is an easy way (i.e. a class P algorithm) to check the validity of the results. Hence, define BPP ("bounded error probabilistic polynomial time") as the class of all decision problems which admit a polynomial time probabilistic algorithm with error probability less than $1/2-\varepsilon$ (for fixed ε). It is obvious that $P\subset$ BPP holds but the relation between BPP and NP is not known.

- Reversible computing

In the last subsection have discussed the time needed to perform a certain computation. Other physical quantities seem to be important are space and energy. Space can treated in a similar way as time and there are in fact space-related complexity classes e.g. PSPACE stands for "polynomial space". However, the energy different, because it turns surprisingly out that it is possible to do any calculation without expending any energy! One source of energy consumption in a usual computer is the intrinsic irreversibility of the basic operations. E.g. a basic gate like AND maps two input bits to one output bit, which obviously implies that the input cannot be reconstructed from the output. In other words, one bit of information is erased during the operation of the AND gate; hence a small amount of energy is dissipated to the environment. If want to avoid this kind of energy dissipation we are restricted to reversible processes, i.e. it should be possible to reconstruct the input data from the output data. This is called reversible computation and it is performed in terms of reversible gates, which in turn can be described by invertible functions $f: B^N \to B^M$. This does not restrict the class of problems which can be solved however: can repackage a non-invertible function $f: B^N \to B^M$ into an invertible one $f': B^{N+M} \to B^{N+M}$ simply by $f'(x;0)=(x;f(x))$ and an appropriate extension to the rest of B^{N+M}. It can even shown that a reversible computer performs as good as a usual one, i.e. an "irreversible" network can simulated in polynomial time by a reversible one. This will be of particular importance for quantum computing, because a reversible computer is, as will see soon, a special case of a quantum computer.

- The network model of a quantum computer

Now are ready to introduce a mathematical model for quantum computation. To this end will generalize the network model discussed to the network model of quantum computation.

A classical computer operates by a network of gates on a finite number of classical bits. A quantum computer operates on a finite number of qubits in terms of a network of quantum gates—this is the initial concept. To be more precise consider the Hilbert space $H^{\otimes N}$ with $H=C^2$ which describes a quantum register consisting of N qubits. In H there is a preferred set $|0\rangle; |1\rangle$ of orthogonal states, describing the two values a classical bit can have. Hence, can describe each possible value x of a classical register of length N in terms of the computational basis $|x\rangle = |x_1\rangle \otimes \cdots \otimes |x_N\rangle$, $x \in B^N$. A quantum gate is now nothing else but a unitary operator acting on a small number of qubits (referably 1 or 2) and a quantum network is a graph representing the composition of elementary gates taken from a small set G of unitaries. A quantum computation can now be defined as the application of such a network to an input state ψ of the quantum register (cf. Figure 2-18 for an example). Similar to the classical case the set G should be universal; i. e. each unitary operator on a quantum register of arbitrary length can represented as a composition of elements from G. Since the group of unitaries on a Hilbert space is continuous, it is not possible to do this with a finite set G. However, can find at least suitably small sets which have the chance to be realizable technically (e.g. in an ion-trap) somehow in the future. Particular examples are on the one hand the controlled U operations and the set consisting of CNOT and all one-qubit gates on the other (cf. Figure 2-19; for a proof of universality).

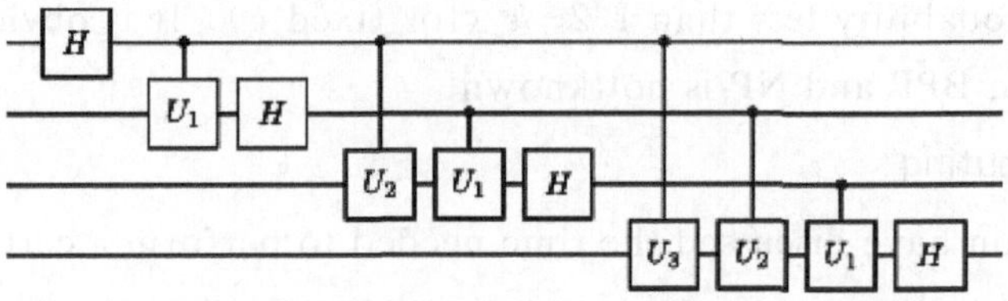

$$H = \frac{1}{\sqrt{2}}\begin{bmatrix} 1 & 1 \\ 1 & -1 \end{bmatrix} \quad U_k = \begin{bmatrix} 1 & 0 \\ 0 & e^{2^{-k}\pi} \end{bmatrix}$$

Fig. 2-18 Quantum circuit for the discrete Fourier transform on a 4-qubit register.

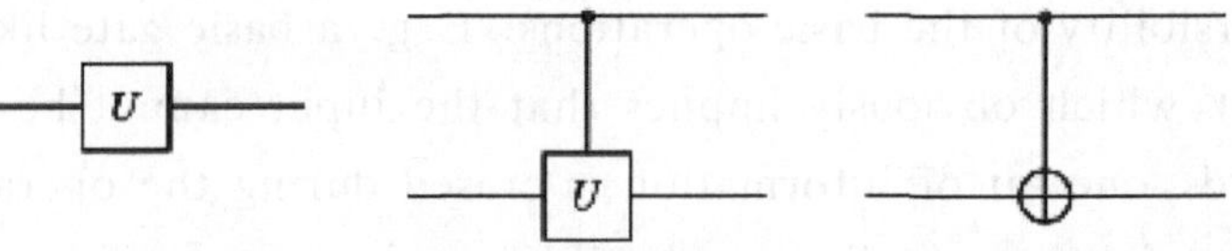

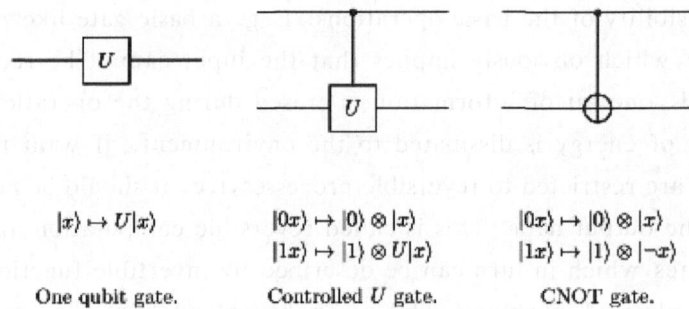

$|x\rangle \mapsto U|x\rangle$

$|0x\rangle \mapsto |0\rangle \otimes |x\rangle$
$|1x\rangle \mapsto |1\rangle \otimes U|x\rangle$

$|0x\rangle \mapsto |0\rangle \otimes |x\rangle$
$|1x\rangle \mapsto |1\rangle \otimes |\neg x\rangle$

One qubit gate. **Controlled U gate.** **CNOT gate.**

Fig. 2-19 Universal sets of quantum gates.

Basically, could have considered arbitrary quantum operations instead of only unitaries as gates. However that can implement each operation unitarily if add an ancilla to the systems. Hence, the kind of generalization already covered by the model. (As long as non-unitarily implemented operations are a desired feature. Decoherence effect due to unavoidable interaction with the environment are a completely different story; we come back to this point at the end of the subsection.) The same holds for measurements at intermediate steps and subsequent conditioned operations. In the case get basically the same result with a different network where all measurements are postponed to the end. (Often it is however very useful to allow measurements at inter mediate steps as will see in the next subsection.) Having a mathematical model of quantum computers in mind we are now ready to discuss how it would work in principle.

(1) The first step is in most cases preprocessing of the input data on a classical computer. E. g. the Shor algorithm for the factoring problem does not work if the input number m is a pure prime power. However, in this case there is an efficient classical algorithm. Hence, we have to check first whether m is of this particular form and use this classical algorithm where appropriate.

(2) In the next step we have to prepare the quantum register based on these preprocessed

data. This means in the most simple case to write classical data, i. e. to prepare the state $|x\rangle \in H^{\otimes N}$ if the (classical) input is $x \in B^N$. In many cases, however, it might be more intelligent to use a superposition of several $|x\rangle$, e. g. the state

$$\Psi = \frac{1}{\sqrt{2^N}} \sum_{x \in B^N} |x\rangle \tag{2-135}$$

which represents actually the superposition of all numbers the registers can represent—this is indeed the crucial point of quantum computing and come back to it below.

(3) Now we can apply the quantum circuit C to the input state ψ and after the calculation get the output state $U\psi$, where U is the unitary represented by C.

(4) To read out the data after the calculation perform a von Neumann measurement in the computational basis, i. e. measure the observable given by the one-dimensional projectors $|x', x|$, $x \in B^N$. Hence, get $x \in B^N$ with probability $P_N = |\langle \psi | x \rangle|^2$.

(5) Finally, have to postprocess the measured value x on a classical computer to end up with the final result x'. If, however, the output state $U\psi$ is a proper superposition of basis vectors $|x\rangle$. Hence, have to check the validity of the results (with a class P algorithm on a classical computer) and if they are wrong have to go back to step (2). So, why is quantum computing potentially useful? First of all, a quantum computer can perform at least as good as a classical computer. This follows immediately from our discussion of reversible computing and the fact that any invertible function $f: B^N \to B^N$ defines a unitary by $U_f: |x\rangle \mapsto |f(x)\rangle$. But, there is on the other hand strong evidence which indicates that a quantum computer can solve problems in polynomial time which a classical computer cannot. The most striking example for this fact is the Shor algorithm, which provides a way to solve the factoring problem (which is most probably not in class P) in polynomial time. If we introduce the new complexity class BQP of decision problems which can be solved with high probability and in polynomial time with a quantum computer, which can express this conjecture as BPP$\neq$BQP.

The mechanism which gives a quantum computer its potential power is the ability to operate not just on one value $x \in B^N$, but on whole superpositions of values, as already mentioned in step (2) above. E. g. consider a, not necessarily invertible, map $f: B^N \to B^M$ and the unitary operator U_f

$$\mathscr{H}^{\otimes N} \otimes \mathscr{H}^{\otimes M} \ni |x\rangle \otimes |0\rangle \mapsto U_f |x\rangle \otimes |0\rangle = |x\rangle \otimes |f(x)\rangle \in \mathscr{H}^{\otimes N} \otimes \mathscr{H}^{\otimes M} \tag{2-136}$$

Where network all measurements are postponed to the end. Often it is however very useful to allow measurements at intermediate steps, see in the next subsection.

If let act U_f on a register in the state $\psi \otimes |0\rangle$ can get the result

$$U_f(\Psi \otimes |0\rangle) = \frac{1}{\sqrt{2^N}} \sum_{x \in B^N} |x\rangle \otimes |f(x)\rangle \tag{2-137}$$

Hence, a quantum computer can evaluate the function f on all possible arguments $x \in B^N$ at the same time. To benefit from this feature—usually called quantum parallelism—is, however, not as easy as it looks like. If perform a measurement on $U_f(\psi \otimes |0\rangle)$ in the computational basis get the value of f for exactly one argument and the rest of the information originally contained in $U_f(\psi \otimes |0\rangle)$ is destroyed. In other words it is not possible to read out all pairs $(x; f(x))$ from $U_f(\psi \otimes |0\rangle)$ and to fill a (classical) lookup table with them. To take advantage from quantum parallelism have to use a clever algorithm within the quantum computation step (step (3) above). In the next section will consider a particular example for this. Before we come to this point, let give some additional comments, which link this section to other parts of quantum information. The first point concerns entanglement. The state $U_f(\psi\otimes|0\rangle)$ is highly entangled (although ψ is separable since $\psi=[2^{-1/2}(|0\rangle+|1\rangle)]^{\otimes N}$), and this fact is essential for the "exponential speedup" of computations we could gain in a quantum computer. In other words, to outperform a classical computer, entanglement is the most crucial resource—this will become more transparent in the next section. The second remark concerns error correction. Up to now implicitly assumed that all components of a quantum computer work perfectly without any error. In reality, however, decoherence effects make it impossible to realize unitarily implemented operations, and have to deal with noisy channels. Fortunately, it is possible within quantum information to correct at least a certain amount of errors. Hence, unlike an analog computer a quantum computer can be designed fault tolerant, i.e. it can work with imperfectly manufactured components.

- Simons problem

Consider now a particular problem (known as Simons problem) which shows explicitly how a quantum computer can speed up a problem which is hard to solve with a classical computer. The term "oracle" indicates here that are not interested in the time the black box needs to perform the calculation but only in the number of times we have to access it. Hence, this example does not prove the conjecture BPP $\neq$ BQP stated above. Other quantum algorithms, which have not room discuss. Include the Deutsch and Deutsch-Josza problem, the Grover search algorithm and of course Shor's factoring algorithm. Hence, assume that our black box calculates the unitary U_f from Eq. (2-135) with a map $f: B^N \to B^N$ which is two to one and has period a, i.e. $f(x)=f(y)$ iff $y=x+a \bmod 2$. The task is to find a. Classically, this problem is hard, i.e. we have to query the oracle exponentially often. To see this note first that have to find a pair $(x;y)$ with $f(x)=f(y)$ and the probability to get it with two random queries is 2^{-N} (since there is for each x exactly one $y\neq x$ with $f(x)=f(y)$). If use the box $2^{N/4}$ times, get less than $2^{N/2}$ different pairs. Hence, the probability to get the correct solution is $2^{-N/2}$, i.e. arbitrarily small even with exponentially many queries. Assume now that let the box act on a quantum register $H^{\otimes N}\otimes H^{\otimes N}$ in the state $\psi\otimes|0\rangle$ with ψ from Eq. (2-133) to get $U_f(\psi\otimes|0\rangle)$ from (2-135). Now we measure the second register. The outcome is one of 2^{N-1} possible values (say $f(x_0)$), each of which occurs equiprobable. Hence, after the measurement the first register is the state $2^{-1/2}(|x\rangle+|x+a\rangle)$. Now let a Hadamard gate H (cf. Figure 2-19) act on each qubit of the first register and the result is (this follows with a short calculation)

$$\frac{1}{\sqrt{2}}H^{\otimes N}(|x\rangle+|x+a\rangle)=\frac{1}{\sqrt{2^{N-1}}}\sum_{a\cdot y=0}(-1)^{x\cdot y}|y\rangle \tag{2-138}$$

where the dot denotes the (B-valued) scalar product in the vector space B^N. Now perform a measurement on the first register (in computational basis) and we get a $y\in B^N$ with the property $y\,a=0$. If we repeat this procedure N times and if get N linear-independent values y_j can determine a as a solution of the system of equations $y_1a=0=y_2a=\cdots y_Na=0$. The probability to appear as an outcome of the second measurement is for each y with $ya=0$ given by 2^{1-N}. Therefore, the success probability can be made arbitrarily big while the number of times we have to access the box is linear in N.

2.4.8 Quantum cryptography

Finally, want to have a short look on quantum cryptography—another more practical application of quantum information, which has the potential to emerge into technology in the not so distant future for some experimental realizations and for a more detailed overview.

Hence, let assume that Alice has a message $x \in B^N$ which she wants to send secretly to Bob over a public communication channels. One way to do this is the so-called "one-time pad": Alice generates randomly a second bit-string $y \in B^N$ of the same length as x sends $x + y$ instead of x. Without knowledge of the key y it is completely impossible to recover the message x from $x + y$. Hence, this is a perfectly secure method to transmit secret data. Unfortunately, it is completely useless without a secure way to transmit the key y to Bob, because Bob needs y to decrypt the message $x + y$ (simply by adding y again). What makes the situation even worse is the fact that the key can used only once (therefore the name one-time pad). If two messages x_1, x_2 are encrypted with the same key can use x_1 as a key to decrypt x_2 and vice versa: $(x_1 + y) + (x_2 + y) = x_1 + x_2$, hence both messages are partly compromised. Due to these problems completely different approaches, namely "public key systems" like DSA and RSA used today for cryptography. The idea is to use two keys instead of one: a private key which used for decryption and only known to its owner and a public key used for encryption, which is publicly available (we do not discuss the algorithms needed for key generation, encryption and decryption here and the references therein instead). To use this method, Bob generates a key pair $(z; y)$, keeps his private key (y) at a secure place and sends the public one (z) to Alice over a public channel. Alice encrypts her message with z sends the result to Bob and he can decrypt it with y. The security of this scheme relies on the assumption that the factoring problem is computationally hard, i. e. not in class P, because to calculate y from z requires the factorization of large integers. Since the latter is tractable on quantum computers via Shor's algorithm, the security of public key systems breaks down if quantum computers become available in the future. Another problem of more fundamental nature is the unproven status of the conjecture that factorization is not solvable in polynomial time. Consequently, security of public key systems not proven either. The crucial point is now that quantum information provides a way to distribute a cryptographic key y in a secure way, such that y can be used as a one-time pad afterwards. The basic idea is to use the no cloning theorem to detect possible eavesdropping attempts. To make this more transparent, let us consider a particular example here, namely the probably most prominent protocol proposed by Benett and Brassard.

(1) Assume that Alice wants to transmit bits from the (randomly generated) key $y \in B^N$ through an ideal quantum channel to Bob. Before they start they settle upon two orthonormal bases $e_0; e_1 \in H$, respectively $f_0; f_1 \in H$, which are mutually non-orthogonal, i. e. $|\langle e_j; f_k \rangle| \geqslant \varepsilon \geqslant 0$ for each $j; k = 0; 1$. If photons used as information carrier a typical choice linearly polarized photons with polarization direction rotated by 45° against each other.

(2) To send one bit $j \in B$ Alice selects now at random one of the two bases, say $e_0; e_1$ and then she sends a qubit in the state $|e_j\rangle\langle e_j|$ through the channel. Note that neither Bob nor a potential eavesdropper knows which bases she has chosen.

(3) When Bob receives the qubit he selects, as Alice before, at random a base and performs the corresponding von Neumann measurement to get one classical bit $k \in B$, which he records together with the measurement method.

(4) Both repeat this procedure until the whole string $y \in B^N$ is transmitted and then Bob tells Alice (through a classical, public communication channel) bit for bit which base he has used for the measurement (but not the result of the measurement). If he has used the same base as Alice both keep the corresponding bit otherwise they discard it. They end up with a bit-string $y' \in B^M$ of a reduced length M. If this is not suse Scient they have to continue sending random bits until the key is long enough. For large N the rate of successfully transmitted bits per bits sended is obviously 1/2. Hence, Alice has to send approximately twice as many bits as they need. Why this procedure is secure, assume now that the eavesdropper Eve can listen and modify the information sent through the quantum channel and that she can listen on the classical channel but cannot modify it (come back to this restriction in a minute). Hence, Eve can intercept the qubits sent by Alice and make two copies of it. One she forwards to Bob and the other she keeps for later analysis. Due to the no cloning theorem, however, she has produced errors in both copies and the quality of her own decreases if she tries to make the error in Bob's as small as possible. Even if Eve knows about the two bases $e_0 ; e_1$ and $f_0 ; f_1$ she does not know which one Alice uses to send a particular qubit. Hence, Eve has to decide randomly which base to choose (as Bob). If $e_0 ; e_1$ and $f_0 ; f_1$ are chosen optimal, i.e. $|\langle e| ; |k\rangle|^2 = 0.5$ it is easy to see that the error rate Eve necessarily produces if she randomly measures in one of the bases is 1/4 for large N. To detect this error Alice and Bob simply have to sacrify portions of the generated key and compare randomly selected bits using their classical channel. If the error rate they detect is too big they can decide to drop the whole key and restart from the beginning. So let us discuss finally a situation where Eve is able to intercept the quantum and the classical channel. This would imply that she play Bob's part for Alice and Alice's for Bob. As a result, she shares a key with Alice and one with Bob. Hence, she can decode all secret data Alice sends to Bob, read it, and encode it finally again to forward it to Bob. To secure against such a "woman in the middle attack", Alice and Bob can use classical authentication protocols, which ensure that the correct person is at the other end of the line. This implies that they need a small amount of initial secret material, which can renewed, however, from the new key they have generated through quantum communication.

2.5 Entanglement measures

In last paragraph can see that entanglement is an essential resource for many tasks of quantum information theory, like teleportation or quantum computation. This means that entangled states needed for the functioning of many processes and that they consumed during operation. It is therefore necessary to have measures, which tell whether the entanglement contained in a number of quantum systems is sufficient to perform a certain task. What makes this subject difficult is the fact that cannot restrict the discussion to systems in a maximally or at least highly entangled pure state. Due to unavoid able decoherence effects realistic applications have to deal with imperfect systems in mixed states, and exactly in this situation the question for the amount of available entanglement is interesting.

2.5.1 General properties and definitions

The difficulties arising if we try to quantify entanglement can be divided, roughly speaking, into two parts: Firstly, have to find a reasonable quantity, which describes exactly those properties, which are interested in, and secondly have to calculate it for a given state. In this section will discuss the first problem and consider several different possibilities to define entanglement measures.

• Axiomatics

First of all, collect some general properties which a reasonable entanglement measure should have. To quantify entanglement, means nothing else but to associate a positive real number to each state of (finite dimensional) two-partite systems.

Axiom E0 An entanglement measure is a function E which assigns to each state ρ of a finite dimensional bipartite system a positive real number $E(\rho)\in\mathbb{R}^+$. Note that have glanced over some mathematical subtleties here, because E is not just defined on the state space of $B(H\otimes K)$ systems for particularly chosen Hilbert spaces H and K – E is defined on any state space for arbitrary finite dimensional H and K. This is expressed mathematically most conveniently by a family of functions which behaves naturally under restrictions (i. e. the restriction to a subspace $H'\otimes K'$ coincides with the function belonging to $H'\otimes K'$). However, see soon that can safely ignore this problem. The next point concerns the range of E. If ρ is unentangled $E(\rho)$ should be zero of course and it should be maximal on maximally entangled states. But what happens if we allow the dimensions of H and K to grow? To get an answer consider first a pair of qubits in a maximally entangled state ρ. It should contain exactly one-bit entanglement, i.e. $E(\rho)=1$ and N pairs in the state $\rho^{\otimes N}$ hould contain N bits. If interpret $\rho^{\otimes N}$ as a maximally entangled state of a $H\otimes H$ system with $H=C^N$ get $E(\rho^{\otimes N})=\log_2(\dim(H))=N$, where have to reshuffle in $\rho^{\otimes N}$ the tensor factors such that $(C^2\otimes C^2)\otimes N$ becomes $(C^2)^{\otimes N}\otimes(C^2)^{\otimes N}$ (i.e. "all Alice particles to the left and all Bob particles to the right".) This observation motivates the following.

Axiom E1 (Normalization). E vanishes on separable and takes its maximum on maximally entangled states. More precisely; this means that $E(\sigma)\leqslant E(\rho)=\log_2(d)$ for $\rho\sigma\in S(H\otimes H)$ and maximally entangled.

One thing an entanglement measure tell that is how much quantum information can be maximally teleported with a certain amount of entanglement, where this maximum is taken over all possible teleportation schemes and distillation protocols. Hence it cannot be increased further by additional LOCC operations on the entangled systems in question. This consideration motivates the following Axiom.

Axiom E2 (LOCC monotonicity). E cannot increase under LOCC operation; i.e. $E[T(\rho)]\leqslant E(\rho)$ for all states ρ and all LOCC channels T. A special case of LOCC operations are, of course, local unitary operations $U\otimes V$.

Axiom E2 implies now that $E(U\otimes V\rho U^*\otimes V^*)\leqslant E(\tilde{\rho})$ and on the other hand $E(U^*\otimes V^*\tilde{\rho}U\otimes V)\leqslant E(\tilde{\rho})$ hence with $\tilde{\rho}=U\otimes V\rho U^*\otimes V$ get $E(\rho)\leqslant(U\otimes V\rho V^*\otimes U^*)$ therefore $E(\rho)=E(U\otimes V\rho U^*\otimes V^*)$. This axiom shows why do not have to bother about families of functions as mentioned above. If E is defined on $S(H\otimes H)$ it is automatically defined on $S(H_1\otimes H_2)$ for all Hilbert spaces H_k with $\dim(H_k)\leqslant\dim(H)$, because can embed $H_1\otimes H_2$ under this condition unitarily into $H \otimes H$. Consider now a convex linear combination $\lambda\rho + (1-\lambda)\sigma$ with $0 \leqslant \lambda \leqslant 1$. Entanglement cannot be "generated" by mixing two states, i.e. $E(\lambda\rho + (1-\lambda)\sigma) \leqslant \lambda E(\rho) + (1-\lambda)E(\sigma)$.

Axiom E3 (Convexity). E is a convex function; i.e. $E(\lambda\rho + (1-\lambda)\sigma) \leqslant \lambda E(\rho) + (1-\lambda)E(\sigma)$ for two states ρ, σ and $0 \leqslant \lambda \leqslant 1$. The next property concerns the continuity of E, i.e. if perturb ρ slightly the change of $E(\rho)$ should be small. This can be expressed most conveniently as continuity of E in the trace norm. At this point, however, it is not quite clear, how have to handle the fact that E is defined for arbitrary Hilbert spaces. The following version is motivated basically by the fact that it is a crucial assumption.

Axiom E4 (Continuity). Consider a sequence of Hilbert spaces $H_N; N \in \mathbb{N}$ and two sequences of states $\rho_N; \sigma_N \in S(H_N \otimes H_N)$ with $\lim \| \rho_N - \sigma_N \|_1 = 0$. Then have

$$\lim_{N \to \infty} \frac{E(\rho_N) - E(\sigma_N)}{1 + \log_2(\dim \mathscr{H}_N)} = 0 \tag{2-139}$$

The last point have to consider here are additivity properties: Since we are looking at entanglement as a resource, it is natural to assume that we can do with two pairs in the state ρ twice as much as with one ρ, or more precisely $E(\rho \otimes \rho) = 2E(\rho)$ (in $\rho \otimes \rho$ have to reshuffle tensor factors again; see above).

Axiom E5 (Additivity). For any pair of two-partite states ; $\rho\sigma \in S(H \otimes K)$ have $E(\sigma \otimes \rho) = E(\sigma) + E(\rho)$. Unfortunately, this rather natural looking axiom seems to be too strong (it excludes reasonable candidates). It should be however, always true that entanglement cannot increase if put two pairs together. For any pair of states $\rho\sigma$ have $E(\rho \otimes \sigma) \leqslant E(\rho) + E(\sigma)$. There are further modifications of additivity available in the literature. Most frequently used is the following, which restricts Axiom E5 to the case $\rho = \sigma$. For any state ρ of a bipartite system have $N^{-1}E(\rho^{\otimes N}) = E(\rho)$.

Finally, the weakest version of additivity only deals with the behavior of E for large tensor products, i.e. $\rho^{\otimes N}$ for $N \to \infty$. Existence of a regularization. For each state ρ the limit exists.

$$E^{\infty}(\rho) = \lim_{N \to \infty} \frac{E(\rho^{\otimes N})}{N} \tag{2-140}$$

- Pure states

Consider now a pure state $\rho = |\psi\rangle\langle\psi| \in S(H \otimes K)$. If it is entangled its partial trace $\sigma = \mathrm{tr}_H |\psi\rangle\langle\psi| = \mathrm{tr}_K |\psi\rangle\langle\psi|$ is mixed and for a maximally entangled state it is maximally mixed. This suggests to use the von Neumann entropy of ρ, which measures how much a state is mixed, as an entanglement measure for pure states, i.e. define

$$E_{vN}(\rho) = -\mathrm{tr}\left[\mathrm{tr}_{\mathscr{K}} \rho \ln (\mathrm{tr}_{\mathscr{K}} \rho)\right] \tag{2-141}$$

It is easy to deduce from the properties of the von Neumann entropy that E_{vN} satisfies Axioms E0, E1, E3 and E5. Somewhat more difficult is only Axiom E2, which follows, however, from a nice theorem of Nielsen, which relates LOCC operations (on pure states) to the theory of majorization. To state it here we need first some terminology. Consider two probability distributions $\lambda = (\lambda_1, \lambda_2, \cdots, \lambda_M)$ and $\mu = (\mu_1, \mu_2, \cdots, \mu_N)$ both given in decreasing order (i.e. $\lambda_1 \geqslant \lambda_2 \geqslant \cdots \geqslant \lambda_M$ and $\mu_1 \geqslant \mu_2 \geqslant \cdots \geqslant \mu_N$). That λ is majorized by μ, in symbols $\lambda \prec \mu$, if

$$\sum_{j=1}^{k} \lambda_j \leqslant \sum_{j=1}^{k} \mu_j \quad \forall\, k = 1, 2, \cdots, \min M, N \tag{2-142}$$

holds. Now have the following result.

Theorem 2.20 A pure state

$$\psi = \sum_j \lambda_j^{1/2} e_j \otimes e'_j \in \mathscr{H} \otimes \mathscr{K} \tag{2-143}$$

It can transformed into another pure state

$$\phi = \sum_j \mu_j^{1/2} f_j \otimes f'_j \in \mathscr{H} \otimes \mathscr{K} \tag{2-144}$$

via an LOCC operation; ψ the Schmidt coefficients of ψ are majorized by those of Φ; i.e. $\lambda \prec \mu$. The von Neumann entropy of the restriction $\operatorname{tr}_H |\psi\rangle\langle\psi|$ can be immediately calculated from the Schmidt coefficients λ of ψ by $E_{vN}(|\psi\rangle\langle\psi|) = -\sum_j \lambda_j \ln(\lambda_j)$ Axiom E2 follows therefore from the fact that the entropy $S(\lambda) = -\sum_j \lambda_j \ln(\lambda_j)$ of a probability distribution λ is a Shur concave function, i.e. $\lambda \prec \mu$ implies $S(\lambda) \geqslant (\mu)$. Hence, so far that E_{vN} is one possible candidate for an entanglement measure on pure states. In the following will see that it is in fact the only candidate which is physically reasonable. There are basically two reasons for this. The first one deals with distillation of entanglement. It was shown by Bennett et al. that each state $\psi \in H \otimes K$ of a bipartite system can be prepared out of (a possibly large number of) systems in an arbitrary entangled state Φ by LOCC operations. To be more precise, we can find a sequence of LOCC operations

$$T_N : \mathscr{B}[(\mathscr{H} \otimes \mathscr{K})^{\otimes M(N)}] \to \mathscr{B}[(\mathscr{H} \otimes \mathscr{K})^{\otimes N}] \tag{2-145}$$

such that

$$\lim_{N\to\infty} \| T_N^*(|\phi\rangle\langle\phi|^{\otimes N}) - |\psi\rangle\langle\psi| \|_1 = 0 \tag{2-146}$$

Holds with a non-vanishing rate $r = \lim_{N\to\infty} M(N) = N$. This is done either by distillation ($r < 1$ if ψ is higher entangled then Φ) or by "diluting" entanglement, i.e. creating many less entangled states from few highly entangled ones ($r > 1$). All this can be performed in a reversible way: can start with some maximally entangled qubits, dilute them to get many less entangled states which can be distilled afterwards to get the original states back (again only in an asymptotic sense). The crucial point is that the asymptotic rate r of these processes is given in terms of E_{vN} by $r = E_{vN}(|\Phi\rangle\langle\Phi|)/E_{vN}(|\psi\rangle\langle\psi|)$. Hence, we can say, roughly speaking, that $E_{vN}(|\psi\rangle\langle\psi|)$ describes exactly the amount of maximally entangled qubits which is contained in $|\psi\rangle\langle\psi|$. A second somewhat more formal reason is that E_{vN} is the only entanglement measure on the set of pure states which satisfies the axioms formulated above. In other words the following "uniqueness theorem for entanglement measures" holds.

Theorem 2.21 The reduced von Neumann entropy E_{vN} is the only entanglement measure on pure states, which satisfies Axioms E0~E5.

- Entanglement measures for mixed states

To find reasonable entanglement measures for mixed states is much more difficult. There are in fact many possibilities and we want to present therefore only four of the most reasonable candidates. Among those measures, which do not discuss here are negativity quantities and the references therein, the "best separable approximation", the base norm associated with the set of separable states and ppt-distillation rates. The first measure we want to present is oriented along the discussion of pure states: To define, roughly speaking, the asymptotic rate with which maximally entangled qubits can be distilled at most out of a state $\rho \in S(H \otimes K)$ as the entanglement of distillation $E_D(\rho)$. To be more precise consider all possible distillation protocols for ρ, i.e. all sequences of LOCC channels:

$$T_N : \mathscr{B}(\mathbb{C}^{d_N} \otimes \mathbb{C}^{d_N}) \to \mathscr{B}(\mathscr{H}^{\otimes N} \otimes \mathscr{K}^{\otimes N}) \tag{2-147}$$

such that:

$$\lim_{N\to\infty} \| T_N^*(\rho^{\otimes N}) - | \Omega_N \rangle\langle \Omega_N | \|_1 = 0 \tag{2-148}$$

holds with a sequence of maximally entangled states $\Omega_N \in \mathbb{C}^{d_N}$. Now can define

$$E_D(\rho) = \sup_{(T_N)_{N\in\mathbb{N}}} \limsup_{N\to\infty} \frac{\log_2(d_N)}{N} \tag{4-149}$$

where the supremum is taken over all possible distillation protocols $(T_N)_{N\in\mathbf{N}}$. It is not very difficult to see that E_D satisfies Axioms E0, E1, E2 and E5. It is not known whether continuity (Axiom E4) and convexity (Axiom E3) holds. It can be shown, however, that E_D is not convex (and not additive; Axiom E5) if npt bound entangled states exist. For pure states have discussed beside distillation the "dilution" of entanglement and can use, similar to E_D, the asymptotic rate with which bipartite systems in a given state ρ can prepared out of maximally entangled singlets. Hence, consider again a sequence of LOCC channels:

$$T_N : \mathscr{B}(\mathscr{H}^{\otimes N} \otimes \mathscr{K}^{\otimes N}) \to \mathscr{B}(\mathbb{C}^{d_N} \otimes \mathbb{C}^{d_N}) \tag{2-150}$$

and a sequence of maximally entangled states $\Omega_N \in \mathbb{C}^{d_N}$, $N \in \mathbb{N}$, but now with the property:

$$\lim_{N\to\infty} \| \rho^{\otimes N} - T_N^*(| \Omega_N \rangle\langle \Omega_N |) \|_1 = 0 \tag{2-151}$$

Then can define the entanglement cost $E_{C(\rho)}$ of ρ as:

$$E_C(\rho) = \inf_{(S_N)N\in\mathbf{N}} \liminf_{N\to\infty} \frac{\log_2(d_N)}{N} \tag{2-152}$$

where the infimum is taken over all dilution protocols S_N, $N \in \mathbb{N}$. It is again easy to see that E_C satisfies Axioms E0, E1, E2 and E5. In contrast to ED however it can shown that E_C is convex (Axiom E3), while it is not known, whether E_C is continuous (Axiom E4). E_D and E_C directly based on operational concepts. The remaining two measures we want to discuss here are defined in a more abstract way. The first can characterized as the minimal convex extension of E_{vN} to mixed states: Can define the entanglement of formation E_F of ρ as

$$E_F(\rho) = \inf_{\rho = \sum_j p_j |\psi_j\rangle\langle\psi_j|} \sum p_j E_{vN}(| \psi_j \rangle\langle \psi_j |) \tag{2-153}$$

where the infimum is taken over all decompositions of ρ into a convex sum of pure states. E_F satisfies E0~E4 and E5 (the rest follows directly from the definition). Whether E_F is (weakly) additive is not known. Furthermore, it is conjectured that E_F coincides with E_C. However, proven is only the identity $E_F^\infty = E_C$, where the existence of the regularization E_F^∞ of E_F follows directly from subadditivity. Another idea to quantify entanglement is to measure the "distance" of the (entangled) from the set of separable states D. It hat turned out that among all possible distance functions the relative entropy is physically most reasonable. Hence, we define the relative entropy of entanglement as

$$E_R(\rho) = \inf_{\sigma\in\mathscr{D}} S(\rho \mid \sigma), \quad S(\rho \mid \sigma) = [\mathrm{tr}(\rho \log_2 \rho - \rho \log_2 \sigma)] \tag{2-154}$$

where the infimum is taken over all separable states. It can shown that E_R satisfies, as E_F the Axioms E0~E4 and E5, where E1 and E2 are shown and the rest follows directly from the definition. It is shown that E_R does not satisfy E5. Hence, the regularization E_R^∞ of E_R differs from E_R. Finally, give now some comments on the relation between the measures just introduced. On pure states all measures just discussed, coincide with the reduced von Neumann entropy—this follows from Theorem 2.21 and the properties stated in the last subsection. For mixed states, the situation is more difficult. It can shown however that $E_D \leqslant E_C$ holds and that all "reasonable" entanglement measures lie in between.

Theorem 2.22. For each entanglement measure E satisfying E0, E1, E2 and E5 and each state $\rho \in S(H \otimes K)$ have $E_D(\rho) \leqslant E_E(\rho) \leqslant E_C(\rho)$. Unfortunately, no measure have discussed in the last subsection satis. es all the assumptions of the theorem. It is possible, however, to get a similar statement for the regularization E^∞ with weaker assumptions on E itself (in particular, without assuming additivity).

2.5.2 Two qubits

Even more difficult than finding reasonable entanglement measures are explicit calculations. All measures have discussed above involve optimization processes over spaces which grow exponentially with the dimension of the Hilbert space. A direct numerical calculation for a general state ρ is therefore hopeless. There are, however, some attempts get either some bounds on entanglement measures or to get explicit calculations for special classes of states. That will concentrate this discussion to some relevant special cases. On the one hand, and will concentrate on E_F and E_R and on the other will look at two special classes of states where explicit calculations are possible: Two qubit systems in this section and states with symmetry properties in the next one are given.

- Pure states

Assume for the rest of this section that $H = C^2$ holds and consider first a pure state $\psi \in H \otimes H$. To calculate $E_{vN}(\psi)$ is of course not difficult and it is straight forward to see that for all material of this

$$E_{vN}(\psi) = H\left[\frac{1}{2}(1 + \sqrt{1 - C(\psi)^2})\right] \tag{2-155}$$

holds, with

$$H(x) = -x\log_2(x) - (1 - x)\log_2(1 - x) \tag{2-156}$$

and the concurrence $C(\psi)$ of ψ which is defined by

$$C(\psi) = \left|\sum_{j=0}^{3} \alpha_j^2\right| \quad \text{with} \quad \psi = \sum_{j=0}^{3} \alpha_j \Phi_j \tag{2-157}$$

Where Φ_j, $j = 0, 1, 2, 3$ denotes the Bell basis. Since C becomes rather important in the following let reexpress it as $C(\psi) = |\psi L \Sigma \psi|$, where $\psi \mapsto \Sigma\psi$ denotes complex conjugation in the Bell basis. Hence, Σ is an antiunitary operator and it can be written as the tensor product $\Sigma = \xi \otimes \xi$ of the map $H \ni \Phi \mapsto \sigma_2 \bar{\Phi}$, where $\bar{\Phi}$ denotes complex conjugation in the canonical basis and σ_2 is the second Pauli matrix. Hence, local unitaries (i.e. those of the form $U_1 \otimes U_2$) commute with L and it can shown that this is not only a necessary but also a sufficient condition for a unitary to be local. Can see from Eqs. (2-153) and (2-155) that $C(\psi)$ ranges from 0 to 1 and that $E_{vN}(\psi)$ is a monotone function in $C(\psi)$. The latter can considered therefore as an entanglement quantity in its own right. For a Bell state we get in particular $C(\Phi_j) = 1$ while a separable state $\Phi_1 \otimes \Phi_2$ leads to $C(\Phi_1 \otimes \Phi_2) = 0$; this can be seen easily with the factorization $\Sigma = \xi \otimes \xi$. Assume now that one of the α_j say α_0 satisfies $|\alpha_0|^2 > 1/2$. This implies that $C(\psi)$ cannot be zero since

$$\left|\sum_{j=1}^{3}\alpha_j^2\right| \leqslant 1-|\alpha_0|^2 \tag{2-158}$$

must hold. Hence, $C(\psi)$ is at least $1-2|\alpha_0|^2$ and this implies for E_{vN} and arbitrary ψ

$$E_{vN}(\psi) \geqslant h(|\langle\Phi_0,\psi\rangle|^2) \quad \text{with} \quad h(x)=\begin{cases} H\left[\frac{1}{2}+\sqrt{x(1-x)}\right], & x\geqslant\frac{1}{2} \\ 0, & x<\frac{1}{2} \end{cases} \tag{2-159}$$

This inequality remains valid if replace Φ_0 by any other maximally entangled state $\Phi\in H\otimes H$. To see this note that two maximally entangled states $\Phi,\Phi'\in H\otimes H$ are related (up to a phase) by a local unitary transformation $U\otimes U_2$ (this follows immediately from their Schmidt decomposition). Hence, if replace the Bell basis in Eq. (2-159) by $\Phi'_j=U_1\otimes U_2\Phi_j$, $j=0,1,2,3$ can get for the corresponding C' the equation $C'(\psi)=U_1^*\otimes U_2^*\psi,\Sigma U_1^*\otimes U_2^*\psi=C(\psi)$ since Σ commutes with local unitaries. Can even replace $|\Phi_0;\psi|^2$ with the supremum over all maximally entangled states and therefore get

$$E_{vN}(\psi)\geqslant h[\mathcal{F}(|\psi\rangle\langle\psi|)] \tag{2-160}$$

Where $F(|\psi\rangle\langle\psi|)$ is the maximally entangled fraction of $|\psi\rangle\langle\psi|$ which have introduced. That even equality holds in Eq. (2-160) note first that it is sufficient to consider the case $\psi=a|00\rangle+b|11\rangle$ with $a;b\geqslant0$, $a^2+b^2=1$, since each pure state ψ can be brought into this form (this follows again from the Schmidt decomposition) by a local unitary transformation which on the other hand does not change E_{vN}. The maximally entangled state which maximizes $|\langle\psi,\Phi\rangle|^2$ is in this case Φ_0 and get $F(|\psi\rangle\langle\psi|)=(a+b)^2/2=/1/2+ab$. Straightforward calculations now show that $h[F(|\psi\rangle\langle\psi|)]=h(1/2+ab)=E_{vN}(\psi)$ holds as stated.

- EOF for Bell diagonal states

It is easy to extend inequality (2-160) to mixed states if use the convexity of E_F and the fact that E_F coincides with E_{vN} on pure states. Hence, Eq. (2-160) becomes

$$E_F(\rho)\geqslant h[\mathcal{F}(\rho)] \tag{2-161}$$

For general two-qubit states this bound is not achieved however. This can be seen with the example $\rho=1/2(|\Phi_1\rangle\langle\Phi_1|+|00\rangle\langle00|)$, which have already considered in the last paragraph. It is easy to see that $F(\rho)=12$ holds hence $h[F(\rho)]=0$ but ρ is entangled. To prove this statement have to find a convex decomposition $\rho=\sum_j\mu_j|\psi\rangle\langle\psi|$ of such a ρ into pure states $|\psi_j\rangle\langle\psi_j|$ such that $h[F(\rho)]=\sum_j\mu_j(E_{vN}|\psi\rangle\langle\psi|)$ holds. Since $E_F(\rho)$ cannot be smaller than $h[F(\rho)]$ due to inequality (2-161) this decomposition must be optimal and equality is proven. To find such ψ_j assume first that the biggest eigenvalue of ρ is greater than 1/2, and let, without loss of generality, λ_1 be this eigenvalue. A good choice for the ψ_j are then the eight pure states

$$\sqrt{\lambda_0}\Phi_0+\mathrm{i}\left(\sum_{j=1}^{3}(\pm\sqrt{\lambda_j})\Phi_j\right) \tag{2-162}$$

The reduced von Neumann entropy of all these states equals $h(\lambda_1)$, hence $\sum_j\mu_jE_{vN}(|\psi_j\rangle\langle\psi_j|)=h(\lambda_1)$ and therefore $E_F(\rho)=h(\lambda_1)$. Since the maximally entangled fraction of ρ is obviously λ_1 can see that Eq. (2-161) holds with equality. Assume now that the highest eigenvalue is less than 1/2. Then we can find phase factors $\exp(\mathrm{i}\Phi_j)$ such that $\sum_{j=0}^{3}\exp(\Phi_{ij})\lambda_j=0$ holds and ρ can be expressed as a convex linear combination of the states

$$e^{i\phi_0/2}\sqrt{\lambda_0}\Phi_0 + i\left(\sum_{j=1}^{3}(\pm e^{i\phi_j/2}\sqrt{\lambda_j})\Phi_j\right) \tag{2-163}$$

The concurrence C of all these states is 0 hence their entanglement is 0 by Eq.(2-157), which in turn implies $E_F(\rho)=0$. Again, that equality achieved in Eq.(2-161) since the maximally entangled fraction is less than 1/2. Summarizing this discussion have shown in Figure 2-20. A Bell diagonal state ρ is entangled λ its highest eigenvalue λ is greater than 1/2. In this case the entanglement of formation of ρ is given by

$$E_F(\rho) = H\left[\frac{1}{2} + \sqrt{\lambda(1-\lambda)}\right] \tag{2-164}$$

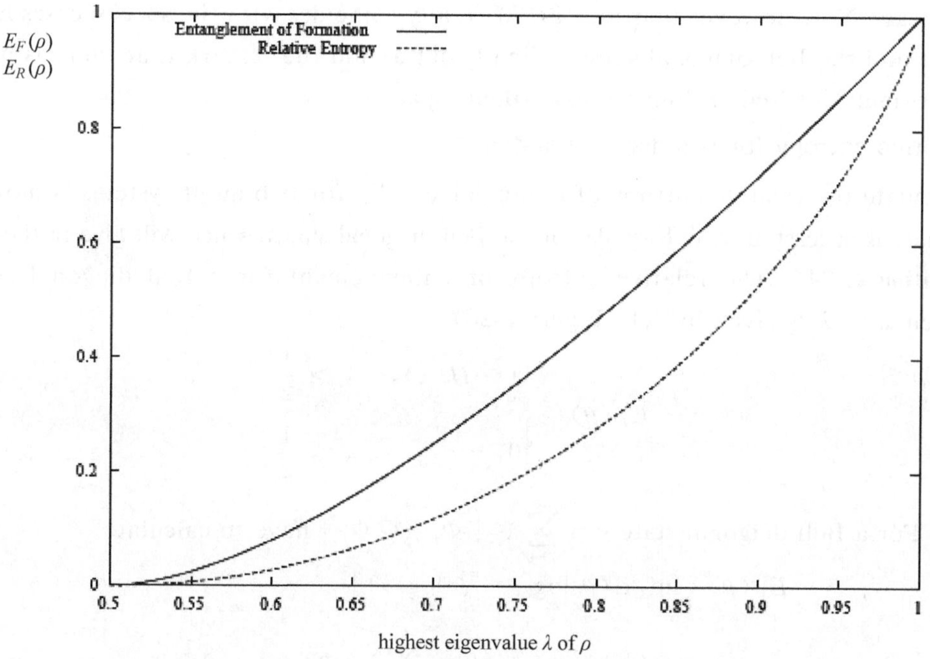

Fig. 2-20 Entanglement of formation and relative entropy of entanglement for the Bell diagonal states, plotted as a function of the highest eigenvalue λ of ρ.

- Wootters formula

If have a general two-qubit state ρ there is a formula of Wootters which allows an easy calculation of E_F. It based on a generalization of the concurrence C to mixed states. To motivate it rewrite $C^2(\psi)=|\langle\psi,\Sigma\psi\rangle|$ as

$$C^2(\psi) = \mathrm{tr}(|\psi\rangle\langle\psi \| \Xi\psi\rangle\langle\Xi\psi|) = \mathrm{tr}(\rho\Xi\rho\Xi) = \mathrm{tr}(R^2) \tag{2-165}$$

with

$$R = \sqrt{\sqrt{\rho}\Xi\rho\Xi\sqrt{\rho}} \tag{2-166}$$

Here have set $\rho = |\psi\rangle\langle\psi|$. The definition of the Hermitian matrix R however makes sense for arbitrary ρ as well. If we write $\lambda_j; j=1,2,3,4$ for the eigenvalues of R and λ_1 is without loss of generality, the biggest one can define the concurrence of an arbitrary two-qubit state ρ as

$$C(\rho) = \max(0, 2\lambda_1 - \mathrm{tr}(R)) = \max(0, \lambda_1 - \lambda_2 - \lambda_3 - \lambda_4) \tag{2-167}$$

It is easy to see that $C(|\psi\rangle\langle\psi|)$ coincides with $C(\psi)$ from (2-157). The crucial point is now that Eq.(2-153) holds for $E_F(\rho)$ if we insert $C(\rho)$ instead of $C(\psi)$:

Theorem 2.23 (Wootters formula). The entanglement of formation of a two-qubit system in a state ρ given by

$$E_F(\rho) = H\left[\frac{1}{2}(1 + \sqrt{1 - C(\rho)^2})\right] \tag{2-168}$$

Where the concurrence of ρ is given in Eq. (2-159) and H denotes the binary entropy from Eq. (2-158). To prove this theorem firstly have to find a convex decomposition $\rho = \sum_j \mu_j \mid \psi_j\rangle\langle\psi_j \mid$ of ρ into pure states ψ_j such that the average reduced von Neumann entropy $\sum_j \mu_j E_{vN}(\psi_j)$ coincides with the right-hand side of Eq. (2-170). Secondly, have to show that have really found the minimal decomposition. Since this is much more involved than the simple case discussed omit the proof and refer to instead. Note however that Eq. (2-170) really coincides with the special cases have derived for the pure and the Bell diagonal states. Finally, let us add the remark that there is no analog of Wootters' formula for higher dimensional Hilbert spaces.

- Relative entropy for Bell diagonal states

To calculate the relative entropy of entanglement E_R for two-qubit systems is more difficult. However, there is at least an easy formula for the Bell diagonal states, which will give in the following:

Proposition 2.24 The relative entropy of entanglement for a Bell diagonal state ρ with highest eigenvalue λ is given by (cf. Figure 2-20)

$$E_R(\rho) = \begin{cases} 1 - H(\lambda), & \lambda > \frac{1}{2} \\ 0, & \lambda \leqslant \frac{1}{2} \end{cases} \tag{2-169}$$

Proof. For a Bell diagonal state $\rho = \sum_{j=0}^{3} \lambda_j \mid \Phi_j\rangle\langle\Phi_j \Phi_j \mid$ have to calculate

$$E_R(\rho) = \inf_{\sigma\in\mathcal{D}}[\mathrm{tr}(\rho \log_2 \rho - \rho \log_2 \sigma)] \tag{2-170}$$

$$= \mathrm{tr}(\rho \log_2 \rho) + \inf_{\sigma\in\mathcal{D}}\left[-\sum_{j=0}^{3} \lambda_j \langle\Phi_j, \log_2(\sigma)\Phi_j\rangle\right] \tag{2-171}$$

Since log is a concave function have $-\log_2\langle\Phi_j; \lambda\Phi_j\rangle \leqslant \langle\Phi_j - \log_2(\sigma)\Phi_j\rangle$ and therefore

$$E_R(\rho) \geqslant \mathrm{tr}(\rho \log_2 \rho) + \inf_{\sigma\in\mathcal{D}}\left[-\sum_{j=0}^{3} \lambda_j \log_2\langle\Phi_j, \sigma\Phi_j\rangle\right] \tag{2-172}$$

Hence, only the diagonal elements of σ in the Bell basis enter the minimization on the right-hand side of this inequality and this implies that we can restrict the infimum to the set of separable Bell diagonal state. Since a Bell diagonal state is separable iff all its eigenvalues are less than 1/2 can get

$$E_R(\rho) \geqslant \mathrm{tr}(\rho \log_2 \rho) + \inf_{p_j\in[0,1/2]}\left[-\sum_{j=0}^{3} \lambda_j \log_2 p_j\right] \quad \text{with} \quad \sum_{j=0}^{3} p_j = 1 \tag{2-173}$$

This is an optimization problem (with constraints) over only four real parameters and easy to solve.

If the highest eigenvalue of ρ is greater than 1/2 get $p_1 = 1/2$ and $p_j = \lambda_j/(2 - 2\lambda)$; where have chosen without loss of generality $\lambda = \lambda_1$. Can get a lower bound on $E_R(\rho)$ which is achieved if insert the corresponding σ in Eq. (2-169). Hence; have proven the statement for $\lambda > 1/2$. which completes the proof; since have already seen that $\lambda \leqslant 1/2$ implies that ρ is separable.

2.5.3 Entanglement measures under symmetry

The problems occurring if we try to calculate quantities like E_R or E_F for general density matrices arise from the fact that we have to solve optimization problems over very high dimensional spaces. One possible strategy to get explicit results is therefore parameter reduction by symmetry arguments. This can done if the state in question admits some invariance properties like Werner, isotropic or OO-invariant states. In the following, will give some particular examples for such calculations, while a detailed discussion of the general idea (together with much more examples and further references) found.

- Entanglement of formation

Consider a compact group of unitaries $G \subset B(H \otimes H)$ (where H is again arbitrary finite dimensional), the set of G-invariant states, i. e. all ρ with $[V;\rho]=0$ for all $V \in G$ and the corresponding twirl operation $P_{G\sigma} = \int G(V\sigma V * dV)$. Particular examples are looking at are:

(1) Werner states where G consists of all unitaries $U \otimes U$.

(2) isotropic states where each $V \in G$ has the form $V = U \otimes \bar{U}$.

(3) OO-invariant states where G consists of unitaries $U \otimes U$ with real matrix elements ($U = \bar{U}$).

One way to calculate E_F for a G-invariant state ρ consists now of the following steps:

(1) Determine the set M of pure states 2 such that $P_G |\Phi\rangle\langle\Phi| = \rho$ holds.

(2) Calculate the function

$$P_G \mathcal{S} \ni \rho \mapsto \varepsilon_G(\rho) = \inf\{E_{vN}(\sigma) \mid \sigma \in M_\rho\} \in \mathbb{R} \tag{2-174}$$

Where have denoted the set of G-invariant states with P_{GS}.

(3) Determine $E_F(\rho)$ then in terms of the convex hull of ε i.e.

$$E_F(\rho) = \inf\{\sum_j \lambda_j \varepsilon(\sigma_j) \mid \sigma_j \in P_G \mathcal{S}, 0 \leqslant \lambda_j \leqslant 1, \rho = \sum_j \lambda_j \sigma_j, \sum_j \lambda_j = 1\} \tag{2-175}$$

The equality in the last equation is of course a non-trivial statement which has to be proved. Skip this point. The advantage of this scheme relies on the fact that spaces of G invariant states are in general very low dimensional (if G is not too small). Hence, the optimization problem contained in step (3) has a much bigger chance to be tractable than the one have to solve for the original definition of E_F. There is of course no guarantee that any of this three steps can be carried out in a concrete situation. For the three examples mentioned above, however, there are results available, which we will present in the following.

- Werner states

Let start with Werner states. In this case ρ is uniquely determined by its Jip expectation value $\mathrm{tr}(\rho F)$. To determine $\Phi \in H \otimes H$ such that $P_{UU} |\Phi\rangle\langle\Phi| = \rho$ holds, have to solve therefore the equation

$$\langle \Phi, F\Phi \rangle = \sum_{jk} \Phi_{jk} \overline{\Phi_{kj}} = \mathrm{tr}(F\rho) \tag{2-176}$$

where Φ_{jk} denote components of Φ in the canonical basis. On the other hand, the reduced density matrix $\rho = \mathrm{tr}_1 |\Phi\rangle\langle\Phi|$ has the matrix elements $\rho_{jk} = \sum_l \Phi_{jl} \bar{\Phi}_{kl}$. By exploiting $U \otimes U$ invariance can assume without loss of generality that ρ is diagonal. Hence, to get the function ε_{UU} have to minimize

$$E_{vN}(|\Phi\rangle\langle\Phi|) = \sum_j S\left[\sum_k |\Phi_{jk}|^2\right] \tag{2-177}$$

under constraint (2-174), where $S(x) = -x\log_2(x)$ denotes the von Neumann entropy. Skip these calculations here and state the results only. For $\mathrm{tr}(F\rho) \geqslant 0$ can get $\varepsilon(\rho) = 0$ (as expected since ρ is separable in this case) and with H from (2-158)

$$\varepsilon_{UU}(\rho) = H\left[\frac{1}{2}(1 - \sqrt{1 - \mathrm{tr}(F\rho)^2})\right] \tag{2-178}$$

for $\mathrm{tr}(F\rho) < 0$. The minima are taken for Φ where all Φ_{jk} except one diagonal element are zero in the case $\mathrm{tr}(F\rho) \geqslant 0$ and for Φ with only two (non-diagonal) coefficients Φ_{jk}; Φ_{kj}, $j \neq k$ non-zero if $\mathrm{tr}(\rho F) < 0$. For any Werner state ρ the entanglement of formation is given by (cf. Figure 2-21)

$$E_F(\rho) = \begin{cases} H\left[\frac{1}{2}(1 - \sqrt{1 - \mathrm{tr}(F\rho)^2})\right], & \mathrm{tr}(F\rho) < 0 \\ 0, & \mathrm{tr}(F\rho) \geqslant 0 \end{cases} \tag{2-179}$$

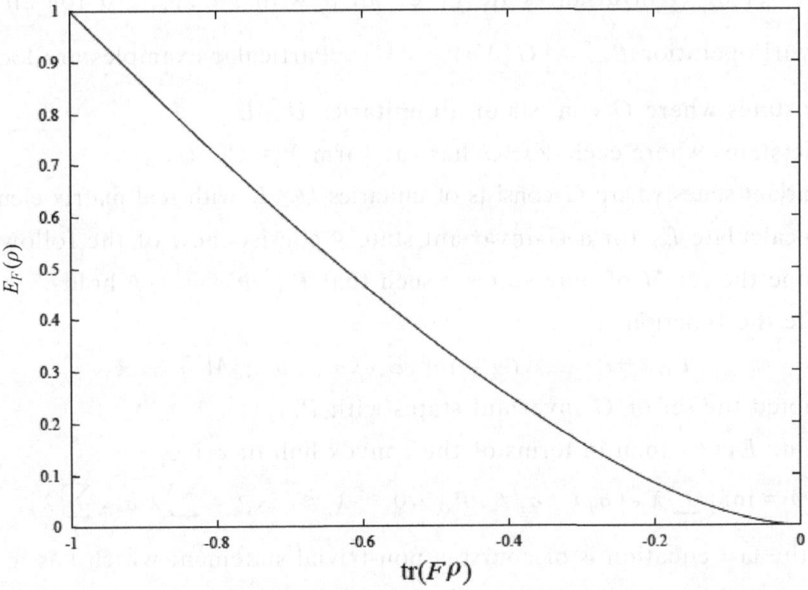

Fig. 2-21 Entanglement of formation for Werner states plotted as function of the Jip expectation.

- Isotropic states

Now consider isotropic, i.e. $U \otimes \bar{U}$ invariant states. They are determined by the expectation value $\mathrm{tr}(\rho\widetilde{F})$ with $\widetilde{F}$ from Eq. (2-128). Hence, have to look first for pure states Φ with $\langle \Phi \widetilde{F} \Phi \rangle = \mathrm{tr}(\rho\widetilde{F})$ (since this determines, as for Werner states above, those Φ with $P_{U\bar{U}}(|\Phi\rangle\langle\Phi|) = \rho$). To this end assume that Φ has the Schmidt decomposition

$$\Phi = \sum_j \lambda_j f_j \otimes f'_j = U_1 \otimes U_2 \sum_j \lambda_j e_j \otimes e_j \tag{2-180}$$

With appropriate unitary matrices U_1; U_2 and the canonical basis e_j, $j = 1, 2, \cdots, d$. Exploiting the $U \otimes \bar{U}$ invariance of ρ can get:

$$\mathrm{tr}(\rho\widetilde{F}) = \left\langle (\mathbb{1} \otimes V) \sum_j \lambda_j e_j \otimes e_j, \widetilde{F}(\mathbb{1} \otimes V) \sum_k \lambda_k e_k \otimes e_k \right\rangle \tag{2-181}$$

$$= \sum_{j,k,l,m} \lambda_j \lambda_k \langle e_j \otimes V e_j, e_l \otimes e_l \rangle \langle e_m \otimes e_m, e_k \otimes V e_k \rangle \tag{2-182}$$

$$= \left| \sum_j \lambda_j \langle e_j, V e_j \rangle \right|^2 \tag{2-183}$$

with $V = U_1^{\mathrm{T}} U_2$ and after inserting the definition of $\widetilde{F}$. Following our general scheme, have to minimize $E_{vN}(|\Phi\rangle\langle\Phi|)$ under the constraint given in Eq. (2-180). This is explicitly done that will only state the result here, which leads to the function

$$\varepsilon_{U\bar{U}}(\rho)=\begin{cases}H(\gamma)+(1-\gamma)\log_2(d-1), & \mathrm{tr}(\rho\widetilde{F})\geqslant\frac{1}{d}\\ 0, & \mathrm{tr}(\rho\widetilde{F})<0\end{cases} \tag{2-184}$$

with

$$\gamma=\frac{1}{d^2}\left(\sqrt{\mathrm{tr}(\rho\widetilde{F})}+\sqrt{[d-1][d-\mathrm{tr}(\rho\widetilde{F})]}\right)^2 \tag{2-185}$$

For $d\geqslant 3$ this function is not convex (cf. Figure 2-22). For any isotropic state the entanglement of formation is given as the convex hull

$$E_F(\rho)=\inf\left\{\sum_j\lambda_j\varepsilon_{U\bar{U}}(\sigma_j)\,\middle|\,\rho=\sum_j\lambda_j\sigma_j,P_{U\bar{U}}\sigma=\sigma\right\} \tag{2-186}$$

of the function $\varepsilon_{U\bar{U}}$ in Eq. (2-183).

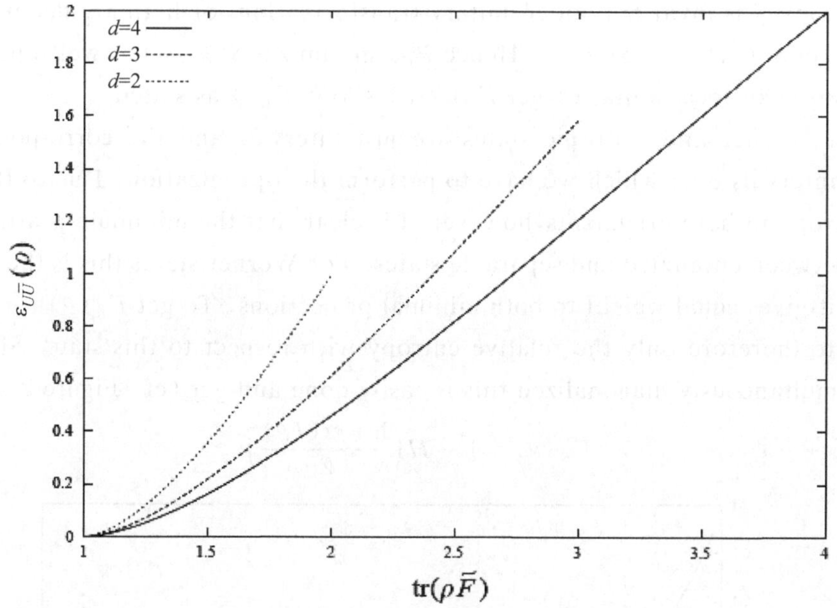

Fig. 2-22 ε-function for isotopic states plotted as a function of the Jip expectation. For $d>2$ it is not convex near the right endpoint.

- OO-invariant states

The results derived for isotropic and Werner states can be extended now to a large part of the set of OO-invariant states without solving new minimization problems. This is possible, because the definition of E_F in Eq. (2-149) allows under some conditions an easy extension to a suitable set of non-symmetric states. If more precisely a non-trivial, minimizing decomposition $\rho=\sum_j p_j|\psi_j\rangle\langle\psi_j|$ of ρ is known, all states ρ' which are a convex linear combination of the same $|\psi_j\rangle\langle\psi_j|$ but arbitrary p_j' have the same E_F as ρ (for proof of the statement). For the general scheme have presented this implies the following: If know the pure states $\sigma\in M$ which solve the minimization problem for $\varepsilon(\rho)$ in Eq. (2-176) can get a minimizing decomposition of ρ in terms of $U\in G$ translated copies of σ. This follows from the fact that ρ is by definition of $M\rho$ the twirl of σ. Hence any convex linear combination of pure states $U_\rho U^*$ with $U\in G$ has the same E_F as ρ.

A detailed analysis of the corresponding optimization problems in the case of Werner and

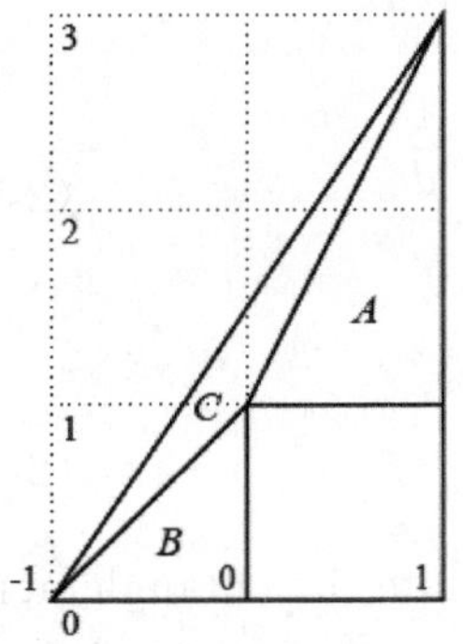

Fig. 2-23 State space of OO-invariant states.

isotropic states leads therefore to the following results about OO-invariant states: The space of OO-invariant states decomposes into four regions: The separable square and three triangles A; B;C; cf. Figure 2-23.

This implies in particular that E_F depends in A only on tr(ρF) and in B only on tr($\rho\widetilde{F}$) and the dimension.

- Relative entropy of entanglement

To calculate $E_R(\rho)$ for a symmetric state ρ is even easier as the treatment of $E_F(\rho)$, because can restrict the minimization in the definition of $E_R(\rho)$ in Eq. (2-152) to G-invariant separable states, provided G is a group of local unitaries. To see this assume that $\sigma \in D$ minimizes $S(\rho|\sigma)$ for a G-invariant state ρ. Then get $S(\rho|U\sigma U^*) = S(\rho|\sigma)$ for all $U \in G$ since the relative entropy S is invariant under unitary transformations of both arguments and due to its convexity even get $S(\rho|P_G\sigma) \leqslant S(\rho|\sigma)$. Hence $P_G\sigma$ minimizes $S(\rho|\cdot)$ as well, and since $P_G\sigma \in D$ holds for a group G of local unitaries, get $E_R(\sigma;\rho) = S(\rho|P_G\sigma)$ as stated.

The sets of Werner and isotropic states are just intervals and the corresponding separable states form subintervals over which we have to perform the optimization. Due to the convexity of the relative entropy in both arguments, however, it is clear that the minimum is attained exactly at the boundary between entangled and separable states. For Werner states this is the state σ_0 with tr($F\sigma_0$) = 0, i.e. it gives equal weight to both minimal projections. To get $E_R(\rho)$ for a Werner state have to calculate therefore only the relative entropy with respect to this state. Since all Werner states can be simultaneously diagonalized this is easily done and get (cf. Figure 2-24)

$$E_R(\rho) = 1 - H\left(\frac{1 + \mathrm{tr}(F\rho)}{2}\right) \tag{2-187}$$

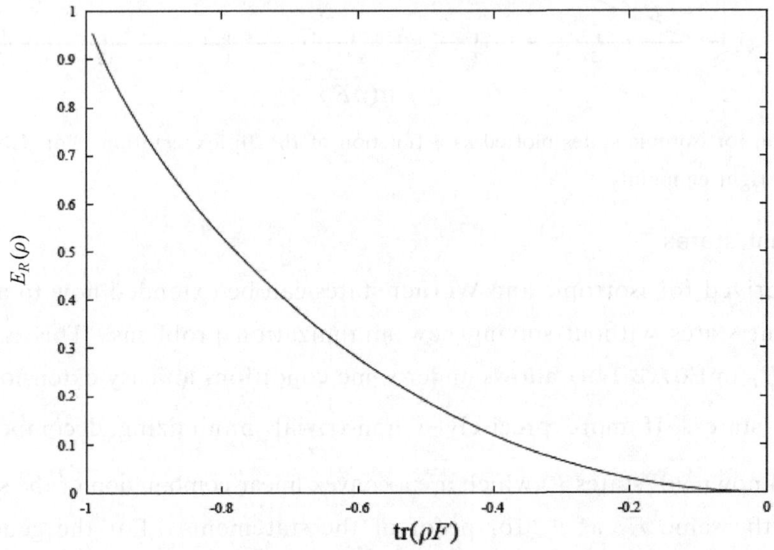

Fig. 2-24 Relative entropy of entanglement for Werner states, plotted as a function of the Jip expectation.

Similarly, the boundary point σ_1 for isotropic states is given by tr($\widetilde{F}\sigma_1$) = 1 which leads to

$$E_R(\rho) = \log_2 d - \left(1 - \frac{\mathrm{tr}(\widetilde{F}\rho)}{d}\right)\log_2(d-1) - S\left(\frac{\mathrm{tr}(\widetilde{F}\rho)}{d}, \frac{1 - \mathrm{tr}(\widetilde{F}\rho)}{d}\right) \tag{2-188}$$

for each entangled isotropic state ρ, and 0 if ρ is separable. ($S(p_1;p_2)$ denotes here the entropy of the probability vector $(p_1;p_2)$.

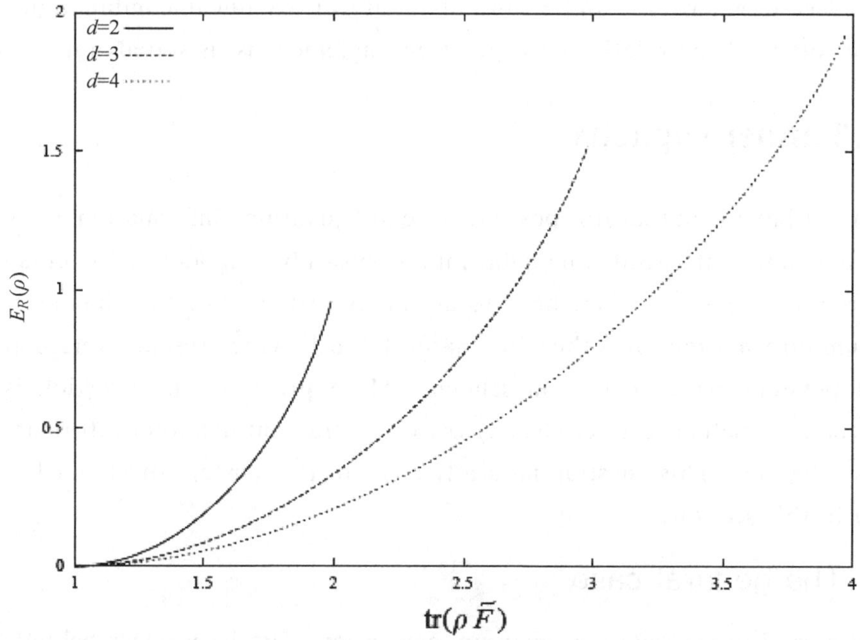

Fig. 2-25 Relative entropy of entanglement for isotropic states and $d=2,3,4$, plotted as a function of $\mathrm{tr}(\widetilde{F})$.

Now consider OO-invariant states. As for EOF divide the state space into the separable square and the three triangles A, B, C; cf. Figure 2-23. The state at the coordinates $(1,d)$ is a maximally entangled state and all separable states on the line connecting $(0,1)$ with $(1,1)$ minimize the relative entropy for this state. Hence consider a particular state σ on this line. The convexity property of the relative entropy immediately shows that σ is a minimizer for all states on the line connecting σ with the state at $(1,d)$. In this way, it is easy to calculate $E_R(\rho)$ for all ρ in A. In a similar way can treat the triangle B: Just have to draw a line from ρ to the state at $(-1,0)$ and find the minimizer for ρ at the intersection with the separable border between $(0,0)$ and $(0,1)$. For all states in the triangle C the relative entropy is minimized by the separable state at $(0,1)$. An application of the scheme just reviewed is a proof that E_R is not additive. To see this consider the state $\rho=\mathrm{tr}(P_-)^{-1}P_-$ where P_- denotes the projector on the antisymmetric subspace. It is a Werner state with Jip expectation -1 (i.e. it corresponds to the point $(-1,0)$ in Figure 2-23). According to the discussion above $S(\rho|\cdot)$ is minimized in this case by the separable state σ_0 and get $E_R(\rho)=1$ independently of the dimension d. The tensor product $\rho^{\otimes 2}$ can regarded as a state in $S(H^{\otimes 2}\otimes H^{\otimes 2})$ with $U\otimes U\otimes V\otimes V$ symmetry, where U, V are unitaries on H. Note that the corresponding state space of $UUVV$ invariant states can be parameterized by the expectation of the three operators $F\otimes\mathbb{1}$, $\mathbb{1}\otimes F$ and $F\otimes F$ and can apply the machinery just described to get the minimizer $\tilde{\sigma}$ of $S(\rho|)$. If $d>2$ holds it turns out that

$$\tilde{\sigma}=\frac{d+1}{2d\,\mathrm{tr}(P_+)^2}P_+\otimes P_+ + \frac{d-1}{2d\,\mathrm{tr}(P_-)^2}P_-\otimes P_- \tag{2-189}$$

holds (where $P_\pm$ denote the projections onto the symmetric and antisymmetric subspaces of $\otimes H$) and not $\tilde{\sigma}=\sigma_0\otimes\sigma_0$ as one would expect. As a consequence get the inequality:

$$E_R(\rho^{\otimes 2}) = 2 - \log_2\left(\frac{2d-1}{d}\right) < 2 = S(\rho^{\otimes 2} \mid \sigma_0^{\otimes 2}) = 2E_R(\rho) \tag{2-190}$$

$d=2$ is a special case, where $\sigma_0^{\otimes 2}$ and $\tilde{\sigma}$ (and all their convex linear combination) give the same value 2. Hence for $d>2$ the relative entropy of entanglement is, as stated, not additive.

2.6 Channel capacity

In Section 2.4 have seen that it is possible to send (quantum) information undisturbed through a noisy quantum channel, if encode one qubit into a (possibly long and highly entangled) string of qubits. This process is wasteful, since have to use many instances of the channel to send just one qubit of quantum information. It is therefore natural to ask, which resources need at least if we are using the best possible error correction scheme. More precisely the question is: With which maximal rate, i.e. information sent per channel usage, can transmit quantum information undisturbed through a noisy channel. This question naturally leads to the concept of channel capacities which we will review in this section.

2.6.1 The general case

Mainly interested in classical and quantum capacities. The basic ideas behind both situations are however quite similar. In this section we will consider therefore a general definition of capacity which applies to arbitrary channels and both kinds of information.

- The definition

Hence consider two observable algebras A_1, A_2 and an arbitrary channel $T: A_1 \to A_2$. To send systems described by a third observable algebra B undisturbed through T need an encoding channel $E: A_2 \to B$ and a decoding channel $D: B \to A_1$ such that ETD equals the ideal channel $B \to B$, i.e. the identity on B. Note that the algebra B describing the systems to send. The input respectively output, algebra of T need not to be the same type, e.g. B is classical while A_1, A_2 are quantum or vice versa. In general (i.e. for arbitrary T and B) it is of course impossible to find such a pair E and D. In this case are interested at least in encodings and decodings that make the error produced during the transmission as small as possible. To make this statement precise we need a measure for this error and there are in fact many good choices for such a quantity (all of them leading to equivalent results). It will use in the following the "cb-norm difference" $\| \text{ETD-Id} \|_{\text{cb}}$, where Id is the identity (i.e. ideal) channel on B and $\| \cdot \|_{\text{cb}}$ denotes the norm of complete boundedness ("cb-norm" for short)

$$\| T \|_{\text{cb}} = \sup \| T \otimes \text{Id}_n \|, \quad \text{Id}_n : \mathscr{B}(\mathbb{C}^n) \to \mathscr{B}(\mathbb{C}^n) \tag{2-191}$$

The cb-norm improves the sometimes annoying property of the usual operator norm that quantities like $\| T \otimes \text{Id}_{B(C^d)} \|$ may increase with the dimension d. On infinite-dimensional observable algebras $\| T \|_{\text{cb}}$ can be infinite although each term in the supremum is finite. A particular example for a map with such a behavior is the transposition on an infinite-dimensional Hilbert space. A map with finite cb-norm therefore called completely bounded. In a finite-dimensional setups each linear map is completely bounded. For the transposition θ on C^d have in particular $\| \theta \|_{\text{cb}} = d$. The cb-norm has some nice features which we will use frequently; this includes its multiplicativity

$\| T_1 \otimes T_2 \|_{cb} = \| T_1 \|_{cb} \| T_2 \|_{cb}$ and the fact that $\| T \|_{cb} = 1$ holds for each (unital) channel. Another useful relation is $\| T \|_{cb} = \| T \otimes \mathrm{Id}_{B(H)} \|$, which holds if T is a map $B(H) \to B(H)$. For more properties of the cb-norm. Now can define the quantity

$$\Delta(T, \mathscr{B}) = \inf_{E,D} \| ETD - \mathrm{Id}_{\mathscr{B}} \|_{cb} \tag{2-192}$$

where the infimum is taken over all channels $E: A_2 \to B$ and $D: B \to A_1$ and Id_B is again the ideal B-channel. N describes, as indicated above, the smallest possible error have to take into account if try to transmit one B system through one copy of the channel T using any encoding E and decoding D. However, have seen that can reduce the error if take M copies of the channel instead of just one. More generally are interested in the transmission of "codewords of length" N, i.e. $B^{\otimes N}$ systems using M copies of the channel T. Encodings and decodings are in this case channels of the form $E: A \otimes M_2 \to B^{\otimes N}$ respectively $D: B^{\otimes N} \to A^{\otimes M_1}$. If increase the number M of channels the error $N(T^{\otimes M}; B^{\otimes N}(M))$ decreases provided the rate with which N grows as a function of M is not too large. A more precise formulation of this idea leads to the following definition.

Definition 2.25 Let T be a channel and B an observable algebra. A number $c \geqslant 0$ is called achievable rate for T with respect to B; if for any pair of sequences $M_j; N_j; j \in N$ with $M_{j \to \infty}$ and $\limsup_{j \to \infty} N_j / M_j < c$ then have

$$\lim_{j \to \infty} \Delta(T^{\otimes M_j}, \mathscr{B}^{\otimes N_j}) = 0 \tag{2-193}$$

The supremum of all achievable rates is called the capacity of T with respect to B and denoted by $C(T; B)$. Note that by definition $c = 0$ is an achievable rate hence $C(T; B) \geqslant 0$. If on the other hand each $c > 0$ is achievable write $C(T; B) = \infty$. At a first look it seems cumbersome to check all pairs of sequences with given upper ratio when testing c. Due to some monotonicity properties of N, however, it can be shown that it is sufficient to check only one sequence provided the M_j satisfy the additional condition $M_j / (M_{j+1}) \to 1$.

- Simple calculations

There are in fact many different capacities of a given channel depending on the type of information want to transmit. However, there are only two different cases are interested in: B can be either classical or quantum. To discuss both special cases in great detail in the next two sections. Before do this, however, will have a short look on some simple calculations, which can be done in the general case. To this end it is convenient to introduce the notations

$$\mathscr{M}_d = \mathscr{B}(\mathbb{C}^d) \quad \text{and} \quad \mathscr{C}_d = \mathscr{C}(\{1, 2, \cdots, d\}) \tag{2-194}$$

as shorthand notations for $B(C^d)$ and $C(\{1, 2, \cdots, d\})$ since some notations become otherwise a little bit clumsy. First of all, have a look on capacities of ideal channels. If Id_{M_f} and Id_{C_f} denote the identity channels on the quantum algebra M_f, respectively the classical algebra C_f, get

$$C(\mathrm{Id}_{\mathscr{C}_f}, \mathscr{M}_d) = 0, \quad C(\mathrm{Id}_{\mathscr{C}_f}, \mathscr{C}_d) = C(\mathrm{Id}_{\mathscr{M}_f}, \mathscr{M}_d) = C(\mathrm{Id}_{M_f}, \mathscr{C}_d) = \frac{\log_2 f}{\log_2 d} \tag{2-195}$$

The first equation is the channel capacity version of the no-teleportation theorem: It is impossible to transfer quantum information through a classical channel. The other equations follow simply by counting dimensions. For the next relation it is convenient to associate to a pair of channels T, S the quantity $C(T, S)$ which arises if replace in definition Eq. (2-187) the ideal channel Id_B by an arbitrary channel S. Hence $C(T, S)$ is a slight generalization of the channel capacity which describes with which asymptotic rate the channel S can be approximated by T (and appropriate encodings and decodings). These generalized capacities satisfy the two-step coding inequality, i.e. for the three channels T_1, T_2, T_3 have

$$C(T_3, T_1) \geqslant C(T_2, T_1)C(T_3, T_2) \tag{2-196}$$

To prove it consider the relations

$$\| T_1^{\otimes N} - E_1 E_2 T_3^{\otimes K} D_2 D_1 \|_{cb}$$

$$= \| T_1^{\otimes N} - E_1 T_2^{\otimes M} D_1 + E_1 T_2^{\otimes M} D_1 - E_1 E_2 T_3^{\otimes K} D_2 D_1 \|_{cb} \tag{2-197}$$

$$\leqslant \| T_1^{\otimes N} - E_1 T_2^{\otimes M} D_1 \|_{cb} + \| E_1 \|_{cb} \| T_2^{\otimes M} - E_2 T_3^{\otimes K} D_2 \|_{cb} \| D_1 \|_{cb} \tag{2-198}$$

$$\leqslant \| T_1^{\otimes N} - E_1 T_2^{\otimes M} D_1 \|_{cb} + \| T_2^{\otimes M} - E_2 T_3^{\otimes K} D_2 \|_{cb} \tag{2-199}$$

where used for the last inequality the fact that the cb-norm of a channel is one. If c_1 is an achievable rate of T_1 with respect to T_2 such that $\lim\sup_{j\to\infty} M_j/N_j < c_1$ and c_2 is an achievable rate of T_2 with respect to T_3 such that $\lim\sup_{j\to\infty} N_j/K_j < c_2$ can see that

$$\limsup_{j\to\infty} \frac{M_j}{K_j} = \limsup_{j\to\infty} \frac{M_j}{N_j}\frac{N_j}{K_j} \leqslant \limsup_{j\to\infty} \frac{M_j}{N_j} \limsup_{k\to\infty} \frac{N_k}{K_k} \tag{2-200}$$

If choose the sequences M_j, N_j and K_j clever enough this implies that $c_1 c_2$ is an achievable rate for T_1 with respect to T_3 and this proves Eq. (2-196). As a first application of (2-196), can relate all capacities $C(T, M_d)$ (and $C(T, C_d)$) for different d to one another. If choose $T_3 = T$, $T_1 = \mathrm{Id}_{M_d}$ and $T_2 = \mathrm{Id}_{M_f}$ get with (2-197) $C(T, M_d) \leqslant (\log_2 f/\log_2 d) C(T, M_f)$, and exchanging d with f shows that even equality holds. A similar relation can be shown for $C(T, C_d)$. Hence, the dimension of the observable algebra B describing the type of information to be transmitted, enters only via a multiplicative constant, i.e. it is only a choice of units and define the classical capacity $C_c(T)$ and the quantum capacity $C_q(T)$ of a channel T as

$$C_c(T) = C(T, \mathscr{C}_2), \quad C_q(T) = C(T, \mathscr{M}_2) \tag{2-201}$$

A second application of Eq. (2-191) is a relation between the classical and the quantum capacity of a channel. Setting $T_3 = T$, $T_1 = \mathrm{Id}_{C_2}$ and $T_2 = \mathrm{Id}_{M_2}$ we get again with (2-201),

$$C_q(T) \leqslant C_c(T) \tag{2-202}$$

Note that it is now not possible to interchange the roles of C_2 and M_2. Hence equality does not hold here. Another useful relation concerns concatenated channels: Transmit information of type B First through a channel T_1 and then through a second channel T_2. It is reasonable to assume that the capacity of the composition $T_2 T_1$ cannot be bigger than capacity of the channel with the smallest bandwidth. This conjecture is indeed true and known as the "Bottleneck inequality":

$$C(T_2 T_1, \mathscr{B}) \leqslant \min\{C(T_1, \mathscr{B}), C(T_2, \mathscr{B})\} \tag{2-203}$$

This consider an encoding and a decoding channel E, respectively D, for $(T_2 T_1)^{\otimes M}$, i.e. in the definition of $C(T_2 T_1; B)$ look at

$$\| \mathrm{Id}_{\mathscr{B}}^{\otimes N} - E(T_2 T_1)^{\otimes M} D \|_{cb} = \| \mathrm{Id}_{\mathscr{B}}^{\otimes N} - (E T_2^{\otimes M}) T_1^{\otimes M} D \|_{cb} \tag{2-204}$$

This implies that $ET^{\otimes M_2}$ and D are an encoding and a decoding channel for T_1. Something similar holds for D and $T^{\otimes M_1} D$ with respect to T_2. Hence each achievable rate for $T_2 T_1$ is also an achievable rate for T_2 and T_1, and this proves Eq. (2-203). Finally, want to consider two channels T_1, T_2 in parallel, i.e. we consider the tensor product $T_1 \otimes T_2$. If $E_j, D_j, j = 1, 2$ are encoding, respectively decoding, channels for $T^{\otimes M_1}$ and $T^{\otimes M_2}$ such that $\| \mathrm{Id}_B^{\otimes N_j} - E_j T^{\otimes M_j} D_j \|_{cb} < \varepsilon$ holds, can get

$$\| \mathrm{Id} - \mathrm{Id} \otimes (E_2 T^{\otimes M} D_2) + \mathrm{Id} \otimes (E_2 T^{\otimes M} D_2) - E_1 \otimes E_2 (T_1 \otimes T_2)^{\otimes M} D_1 \otimes D_2 \|_{cb} \tag{2-205}$$

$$\leqslant \| \mathrm{Id} \otimes (\mathrm{Id} - E_2 T^{\otimes M} D_2) \|_{cb} + \| (\mathrm{Id} - E_1 T_1^{\otimes M} D_1) \otimes E_2 T^{\otimes M} D_2 \|_{cb} \tag{2-206}$$

$$\leqslant \| \mathrm{Id} - E_2 T^{\otimes M} D_2 \|_{\mathrm{cb}} + \| \mathrm{Id} - E_1 T_1^{\otimes M} D_1 \|_{\mathrm{cb}} \leqslant 2\varepsilon \tag{2-207}$$

Hence $c_1 + c_2$ is achievable for $T_1 \otimes T_2$ if c_j is achievable for T_j. This implies the inequality

$$C(T_1 \otimes T_2, \mathscr{B}) \geqslant C(T_1, \mathscr{B}) + C(T_2, \mathscr{B}) \tag{2-208}$$

When all channels are ideal, or when all systems involved are classical even equality holds, i. e. channel capacities are additive in this case. However, if quantum channels are considered, it is one of the big open problems of the field, to decide under which conditions additivity holds.

2.6.2 The classical capacity

In this section will discuss the classical capacity $C_c(T)$ of a channel T. There are in fact three different cases to consider: T can be either classical or quantum and in the quantum case can use either ordinary encodings and decodings or a dense coding scheme.

- Classical channels

Consider first a classical to classical channel $T: C(Y) \to C(X)$. This is basically the situation of classical information theory and we will only have a short look here—mainly to show how this (well known) situation fits into the general scheme described in the last section. As stated T is completely determined by its transition probabilities T_{xy}, $(x, y) \in X \times Y$ describing the probability to receive $x \in X$ when $y \in Y$ was sent. Since the cb-norm for a classical algebra coincides with the ordinary norm can get (have set $X = Y$ for this calculation)

$$\| \mathrm{Id} - T \|_{\mathrm{cb}} = \| \mathrm{Id} - T \| = \sup_{x, \delta} \left| \sum_y (\delta_{xy} - T_{xy}) f_y \right| \tag{2-209}$$

$$= 2 \sup_x (1 - T_{xx})$$

where the supremum in the first equation is taken over all $f \in C(X)$ with $\| f \| = \sup_y | f_y | < 1$. That the quantity in Eq. (2-209) is exactly twice the maximal error probability, i. e. the maximal probability of sending x and getting anything different. Inserting this quantity for N applied to a classical channel T and the "bit-algebra" $B = C_2$, get exactly the Shannons classical definition of the capacity of a discrete memoryless channel. Hence can apply the Shannons noisy channel coding theorem to calculate $C_c(T)$ for a classical channel. To state it we have to introduce first some terminology. Consider therefore a state $p \in C^*(X)$ of the classical input algebra $C(X)$ and its image $q = T^*(p) \in C^*(Y)$ under the channel. p and q are probability distributions on X, respectively Y, and p_x can be interpreted as the probability that the "letter" $x \in X$ was send. Similarly $q_y = \sum_x T_{xy} p_x$ is the probability that $y \in Y$ was received and $P_{xy} = T_{xy} p_x$ is the probability that $x \in X$ was sent and $y \in Y$ was received. The family of all P_{xy} can be interpreted as a probability distribution P on $X \times Y$ and the T_{xy} can be regarded as conditional probability of P under the condition x. Now can introduce the mutual information

$$I(p, T) = S(p) + S(q) - S(P) = \sum_{(x, y) \in X \times Y} P_{xy} \log_2 \left(\frac{P_{xy}}{p_x q_y} \right) \tag{2-210}$$

where $S(p)$, $S(q)$ and $S(P)$ denote the entropies of p, q and P. The mutual information describes, roughly speaking, the information that p and q contain about each other. E.g. if p and q are completely uncorrelated (i.e. $P_{xy} = p_x q_y$) we get $I(p; T) = 0$. If T is on the other hand an ideal bit-channel and p equally distributed we have $I(p; T) = 1$. Now we can state the Shannons Theorem which expresses the classical capacity of T in terms of mutual informations: The classical capacity of $C_c(T)$ of a classical communication channel $T: C(Y) \rightarrow C(X)$ is given by

$$C_c(T) = \sup_p I(p, T) \tag{2-211}$$

where the supremum is taken over all states $p \in C^*(X)$.

- Quantum channels

If transmit classical data through a quantum channel $T: B(H) \rightarrow B(H)$ the encoding $E: B(H) \rightarrow C_2$ is a parameter-dependent preparation and the decoding D: $C_2 \rightarrow B(H)$ is an observable. Hence, the composition ETD is a channel $C_2 \rightarrow C_2$, i.e. a purely classical channel and can calculate its capacity in terms of the Shannons Theorem. This observation leads to the definition of the "one-shot" classical capacity of T:

$$C_{c,1}(T) = \sup_{E,D} C_c(\text{ETD}) \tag{2-212}$$

where the supremum is taken over all encodings and decodings of classical bits. The term"one-shot" in this definition arises from the fact that we need apparently only one invocation of the channel T. However, many uses of the channel hidden in the definition of the classical capacity on the right-hand side. Hence, $C_{c,1}(T)$ can be de. ned alternatively in the same way as $C_c(T)$ except that no entanglement is allowed during encoding and decoding, or more precisely consider only encodings $E: B(K)^{\otimes M} \rightarrow C^{\otimes N_2}$ which prepare separable states and only decodings $D: C^{\otimes N_2} \rightarrow B(H)^{\otimes M}$ which lead to separable observables. It is not yet known, whether entangled codings can help to increase the transmission rate. Therefore, know that

$$C_{c,1}(T) \leqslant C_c(T) = \sup_{M \in \mathbb{N}} \frac{1}{M} C_{c,1}(T^{\otimes M}) \tag{2-213}$$

holds. One reason why $C_{c,1}(T)$ is an interesting quantity relies on the fact that we have, due to the following theorem by Holevo, a computable expression for it. The one-shot classical capacity $C_{c;1}(T)$ of a quantum channel $T: B(H) \rightarrow B(H)$ is given by

$$C_{c,1}(T) = \sup_{p_j, \rho_j} \left[S\left(\sum_j p_j T^*[\rho_j]\right) - \sum_j p_j S(T^*[\rho_j]) \right] \tag{2-214}$$

where the supremum is taken over all probability distributions p_j and collections of density operators ρ_j.

- Entanglement assisted capacity

Another classical capacity of a quantum channel arises, if we use dense coding schemes instead of simple encodings and decodings to transmit the data through the channel T. In other words we can define the entanglement enhanced classical capacity $C_e(T)$ in the same way as $C_c(T)$ but by replacing the encoding and decoding channels in Definition 2.25 and Eq. (2-189) by dense coding protocols. Note that this implies that the sender Alice and the receiver Bob share an (arbitrary) amount of (maximally) entangled states prior to the transmission. For this quantity a coding theorem was recently proven by Bennett and others which want to state in the following. To this end assume that are transmitting systems in the state $\rho \in B^*(H)$ through the channel and that ρ has the purification $\psi \in H \otimes H$, i. e. $\rho = \text{tr}_1 |\psi\rangle\langle\psi| = \text{tr}_2 |\psi\rangle\langle\psi|$. Then we can de. ne the entropy exchange

$$S(\rho T) = S[(T \otimes \text{Id})(|\psi\rangle\langle\psi|)] \tag{2-215}$$

The density operator $(T \otimes \text{Id})(|\psi\rangle\langle\psi|)$ has the output state $T^*(\rho)$ and the input state ρ as its partial traces. It can regarded therefore as the quantum analog of the input = output probability distribution T_{xy} defined. Another way to look at $S(\rho T)$ is in terms of an ancilla representation of T: If $T^*(\rho) = \text{tr}_K(U_{\rho\otimes K}U^*)$ with a unitary $U:H\otimes K$ and a pure environment state K it can be shown that $S(\rho T) = S[T_K^*]$ where T_K is the channel describing the information transfer into the environment, i.e. $T^* K(\rho) = \text{tr}_H(U\rho\otimes\rho_K U^*)$, in other words $S(\rho, T)$ is the final entropy of the environment. Now can define

$$I(\rho, T) = S(\rho) + S(T^*\rho) - S(\rho, T) \tag{2-216}$$

which is the quantum analog of the mutual information given in Eq. (2-210). It has a number of nice properties, in particular positivity, concavity with respect to the input state and additivity and its maximum with respect to ρ coincides actually with $C_e(T)$. The entanglement assisted capacity $C_e(T)$ of a quantum channel $T: B(H) \to B(H)$ is given by

$$C_e(T) = \sup_\rho I(\rho, T) \tag{2-217}$$

where the supremum is taken over all input states $\rho \in B^*(H)$. Due to the nice additivity properties of the quantum mutual information $I(\rho, T)$ the capacity $C_e(T)$ is known to be additive as well. This implies that it coincides with the corresponding "one-shot" capacity, and this is an essential simplification compared to the classical capacity $C_c(T)$.

- Examples

As a first example will consider the "quantum erasure channel" which transmits with probability $1-\theta$ the d-dimensional input state intact while it is replaced with probability θ by an "erasure symbol", i. e. a $(d+1)$ th pure state e which is orthogonal to all others. In the Schrödinger picture this is

$$\mathscr{B}^*(\mathbb{C}^d) \ni \rho \mapsto T^*(\rho) = (1-\vartheta)\rho + \vartheta \text{tr}(\rho) \mid \psi_e\rangle\langle\psi_e \mid \in \mathscr{B}^*(\mathbb{C}^{d+1}) \tag{2-218}$$

This example is very unusual, because all capacities discussed up to now (including the quantum capacity) can be calculated explicitly: Can get $C_{c,1}(T) = C_c(T) = (1-\theta)\log_2(d)$ for the classical and $C_e(T) = 2C_c(T)$ for the entanglement enhanced classical capacity. Hence the gain by entanglement assistance is exactly a factor two; cf. Figure 2-26.

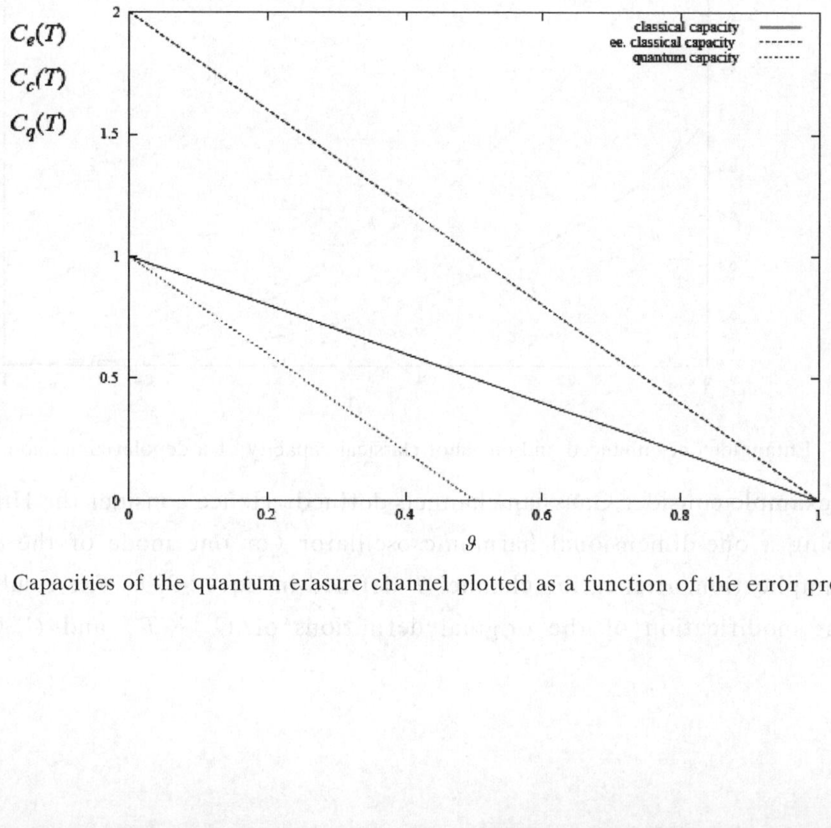

Fig. 2-26 Capacities of the quantum erasure channel plotted as a function of the error probability.

The next example is the depolarizing channel

$$\mathscr{B}^*(\mathbb{C}^d) \ni \rho \mapsto T^*(\rho) = (1-\vartheta)\rho + \vartheta \operatorname{tr}(\rho)\frac{\mathbb{1}}{d} \in \mathscr{B}^*(\mathbb{C}^d) \tag{2-219}$$

already discussed. It is more interesting and more difficult to study. It is in particular not known whether C_c and $C_{c,1}$ coincide in this case (i. e. the value of C_c is not known). Therefore can compare $C_e(T)$ only with $C_{c,1}$. Using the unitary covariance of T. First that $I(U_\rho U^*; T) = I(\rho; T)$ holds for all unitaries U (to calculate $S(U_\rho U^*; T)$ note that $U \otimes U\psi$ is a purification of $U_\rho U^*$ if ψ is a purification of ρ). Due to the concavity of $I(\rho; T)$ in the first argument can average over all unitaries and see that the maximum in Eq. (2-217) is achieved on the maximally mixed state. Straightforward calculation therefore shows that

$$C_e(T) = \log_2(d^2) + \left(1 - \vartheta\frac{d^2-1}{d^2}\right)\log_2\left(1 - \vartheta\frac{d^2-1}{d^2}\right) + \vartheta\frac{d^2-1}{d^2}\log_2\frac{\vartheta}{d^2} \tag{2-220}$$

holds, while have

$$C_{c,1}(T) = \log_2(d) + \left(1 - \vartheta\frac{d-1}{d}\right)\log_2\left(1 - \vartheta\frac{d-1}{d}\right) + \vartheta\frac{d-1}{d}\log_2\frac{\vartheta}{d} \tag{2-221}$$

where the maximum in Eq. (2-214) is achieved for an ensemble of equiprobable pure states taken from an orthonormal basis in H. This is plausible since the first term under the sup in Eq. (2-214) becomes maximal and the second becomes minimal: $\sum_j p_j T^* \rho_j$ is maximally mixed in this case and its entropy is therefore maximal. The entropies of the $T^* \rho_j$ are on the other hand minimal if the ρ_j are pure. In Figure 2-27 we have plotted both capacities as a function of the noise parameter θ and in Figure 2-28 have plotted the quotient $C_e(T)/C_{c,1}(T)$ which gives an upper bound on the gain we get from entanglement assistance.

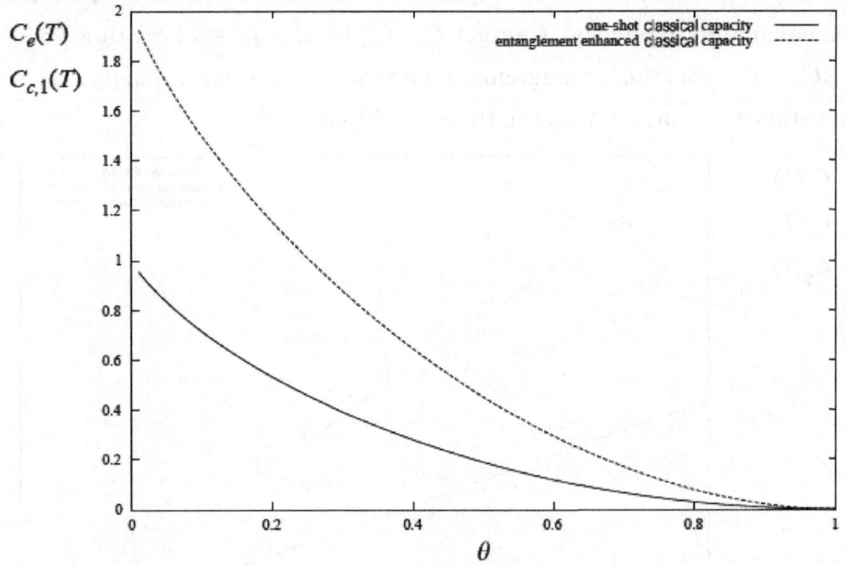

Fig. 2-27 Entanglement enhanced and one-shot classical capacity of a depolarizing qubit channel.

As a third example consider Gaussian channels defined. Hence consider the Hilbert space $H = L^2(R)$ describing a one-dimensional harmonic oscillator (or one mode of the electromagnetic field) and the amplication/attenuation channel T defined in Eq. (2-99). The results want to state concern a slight modification of the original definitions of $C_{c,1}(T)$ and $C_e(T)$: Consider

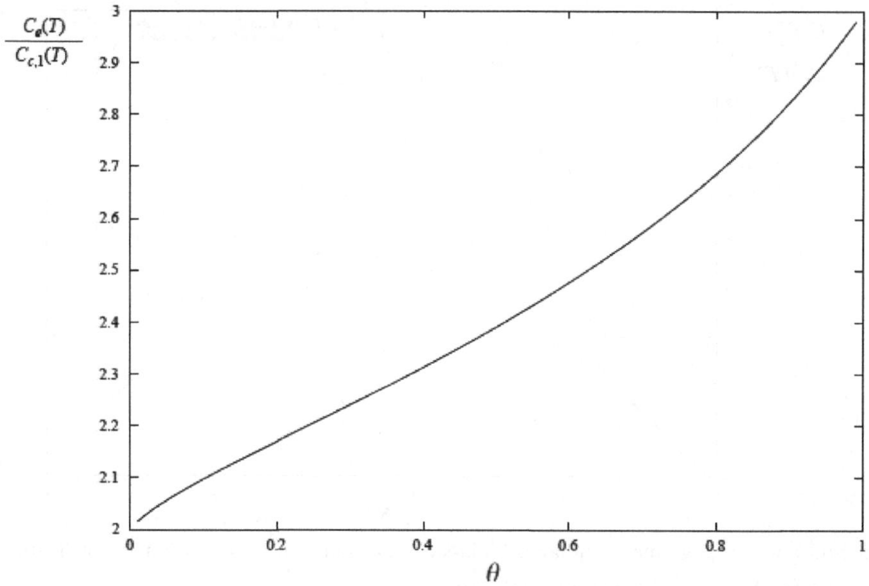

Fig. 2-28 Gain of using entanglement assisted versus unassisted classical capacity for a depolarizing qubit channel.

capacities for channels with constraint input. This means that only a restricted class of states ρ on the input Hilbert space of the channel allowed for encoding. In the case this means that consider the constraint $\mathrm{tr}(\rho aa^*) \leqslant N$ for a positive real number $N > 0$ and with the usual creation and annihilation operators a^*, a. This can be rewritten as an energy constraint for a quadratic Hamiltonian; hence this is a physically realistic restriction.

For the entanglement enhanced capacity it can be shown now that the maximum in Eq. (2-217) is taken on Gaussian states. To get $C_e(T)$ it is sufficient therefore to calculate the quantum mutual information $I(T;\rho)$ for the Gaussian state ρ_N from Eq. (2-89). The details can be found, will only state the results here. With the abbreviation

$$g(x) = (x+1)\log_2(x+1) - x\log_2 x \tag{2-222}$$

can get $S(\rho N) = g(N)$ and $S(T[\rho N']) = g(N')$ with $N' = k^2 N + \max\{0; k^2 - 1\} + N_c$ (cf. Eq. (2-100)) for the entropies of input and output states and

$$S(\rho, T) = g\left(\frac{D + N' - N - 1}{2}\right) + g\left(\frac{D - N' + N - 1}{2}\right) \tag{2-223}$$

with

$$D = \sqrt{(N + N' + 1)^2 - 4k^2 N(N+1)} \tag{2-224}$$

for the entropy exchange.

The sum of all three terms gives $C_e(T)$ which have plotted in Figure 2-29 as a function of k. To calculate the one-shot capacity $C_{c,1}(T)$ the optimization in Eq. (2-214) has to be calculated over probability distributions p_j and collections of density operators ρ_j such that $\sum_j p_j \mathrm{tr}(aa^* \rho_j) \leqslant N$ holds. It is conjectured but not yet proven that the maximum is achieved on coherent states with Gaussian probability distribution $p(x) = (3N)^{-1}\exp(-|x|^2/N)$. If this is true can get

$$C_{c,1}(T) = g(N') - g(N'_0) \quad \text{with} \quad N'_0 = \max\{0, k^2 - 1\} + N_c \tag{2-225}$$

The result is plotted as a function of k in Figure 2-29 and the ratio $G = C_e/C_1$ in Figure 2-30. G gives an upper bound on the gain of using entanglement assisted versus unassisted classical capacity.

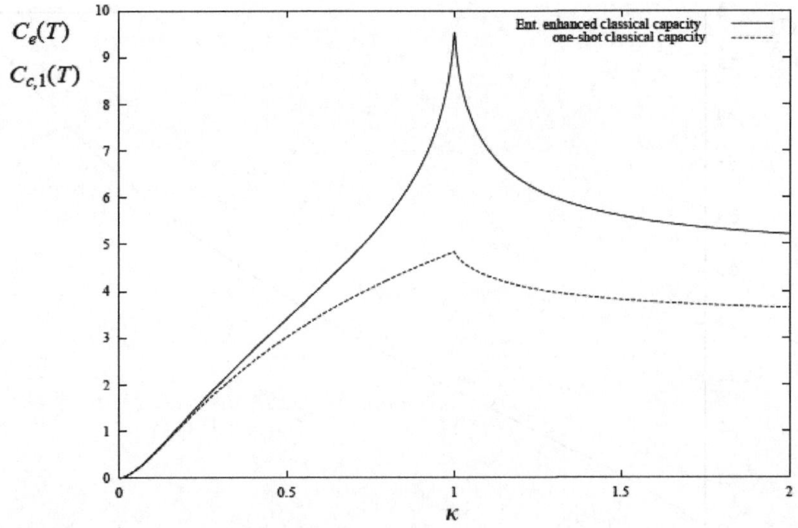

Fig. 2-29 One-shot and entanglement enhanced classical capacity of a Gaussian amplification = attenuation channel with $N_c = 0$ and input noise $N = 10$.

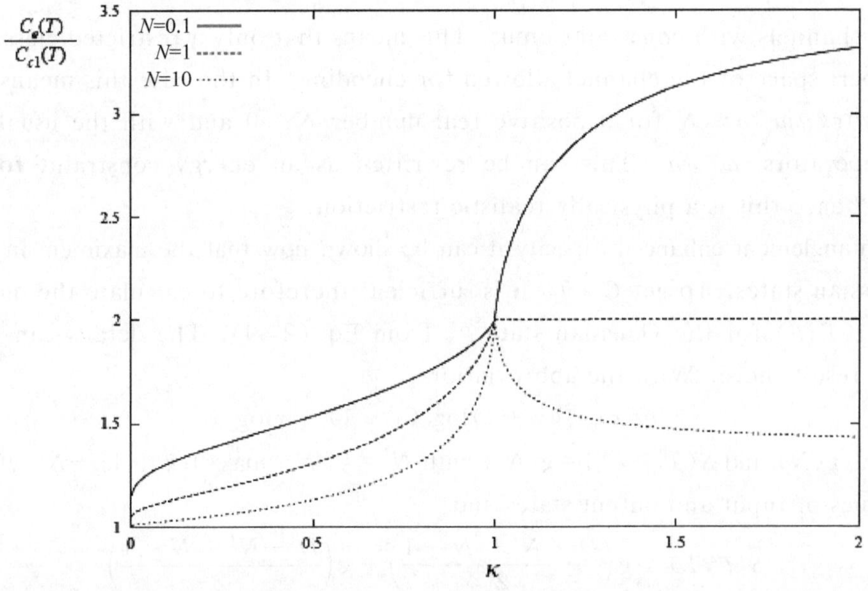

Fig. 2-30 Gain of using entanglement assisted versus unassisted classical capacity for a Gaussian amplification = attenuation channel with $N_c = 0$ and input noise $N = 0.1;1;10$.

2.6.3 The quantum capacity

The quantum capacity of a quantum channel $T:B(H)\rightarrow B(H)$ is more difficult to treat than the classical capacities discussed in the last section. There is, in particular, no coding theorem available which would allow explicit calculations. Nevertheless, there are partial results available, which will review in the following.

- Alternative definitions

Let start with two alternative definitions of $C_q(T)$. The first one proposed by Bennett differs only in the error quantity which should go to zero. Instead of the cb-norm the minimal fidelity is used. For a channel $T:B(H)\rightarrow B(H)$ and a subspace $H'\subset H$ it is defined as

$$\mathcal{F}_p(\mathcal{H}', T) = \inf_{\psi \in \mathcal{H}'} \langle \psi, T[|\psi\rangle\langle\psi|]\psi\rangle \tag{2-226}$$

and if $H' = H$ holds we simply write $F_p(T)$. Hence a number c is an achievable rate if

$$\lim_{j\to\infty} \mathcal{F}_p(E_j T^{\otimes M_j} D_j) = 1 \tag{2-227}$$

holds for sequences

$$E_j : \mathcal{B}(\mathcal{H})^{\otimes M_j} \to \mathcal{M}_2^{\otimes N_j}, \quad \mathcal{D}_j : \mathcal{M}_2^{\otimes N_j} \to \mathcal{B}(\mathcal{H})^{\otimes M_j}, \quad j \in \mathbb{N} \tag{2-228}$$

of encodings and decodings and sequences of integers $M_j, N_j, j \in \mathbb{N}$ satisfying the same constraints (in particular $\lim_{j\to\infty} N_j/M_j < c$). The equivalence to the version of $C_q(T)$ follows now from the estimates.

$$\| T - \mathrm{Id} \| \leqslant \| T - \mathrm{Id} \|_{\mathrm{cb}} \leqslant 4\sqrt{\| T - \mathrm{Id} \|} \tag{2-229}$$

$$\| T - \mathrm{Id} \| \leqslant 4\sqrt{1 - \mathcal{F}_p(T)} \leqslant 4\sqrt{\| T - \mathrm{Id} \|} \tag{2-230}$$

A second version of $C_q(T)$ is given. To state it let us define first a quantum source as a sequence $\rho_N; N \in \mathbb{N}$ of density operators $\rho_N \in B^*(K \otimes N)$ (with an appropriate Hilbert space K) and the entropy rate of this source as $\limsup_{N\to\infty} S(\rho_N)/N$. In addition need the entanglement fidelity of a state ρ (with respect to a channel T)

$$\mathcal{F}_e(\rho, T) = \langle \Psi, (T \otimes \mathrm{Id})[|\Psi\rangle\langle\Psi|]\Psi\rangle \tag{2-231}$$

where ψ is the purification of ρ. Now we define $c \geqslant 0$ to be achievable if there is a quantum source $\rho_N, N \in \mathbb{N}$ with entropy rate c such that

$$\lim_{n\to\infty} \mathcal{F}_e(\rho_N, E'_N T^{\otimes N} D'_N) = 1 \tag{2-232}$$

holds with encodings and decodings

$$E'_N : \mathcal{B}(\mathcal{H})^{\otimes N} \to \mathcal{B}(\mathcal{K}^{\otimes N}), \quad \mathcal{D}'_N : \mathcal{B}(\mathcal{K}^{\otimes N}) \to \mathcal{B}(\mathcal{H})^{\otimes N}, \quad j \in \mathbb{N} \tag{2-233}$$

Note that these E'_N, D'_N play a slightly different role than the E_j, D_j in Eq. (2-228), because the number of tensor factors of the input and the output algebra is always identical, while in Eq. (2-228) the quotients of these numbers lead to the achievable rate. To relate both definitions we have to derive an appropriately chosen family of subspaces $H'N \subset K \otimes N$ from the N such that the minimal fidelities $F_p(H'_N; E'_N T^{\otimes N'} D'_N)$ of these subspaces go to 1 as $N \to \infty$. If identify the H'_N with tensor products of C^2 and the E_j, D_j of Eq. (2-228) with restrictions of E'_N, D'_N to these tensor products recover Eq. (2-227). A precise implementation of this rough idea can be found and it shows that both definitions just discussed are indeed equivalent.

- Upper bounds and achievable rates

Although there is no coding theorem for the quantum capacity $C_q(T)$, there is a fairly good candidate which is related to the coherent information

$$J(\rho T) = S(T^* \rho) - S(\rho T) \tag{2-234}$$

Here $S(T^* \rho)$ is the entropy of the output state and $S(\rho T)$ is the entropy exchange defined in Eq. (2-215). It is argued that $J(\rho T)$ plays a role in quantum information theory which is an alogousto that of the (classical) mutual information (2-210) in classical information theory. $J(\rho T)$ has some nasty properties, however: it is negative and it is known not additive. To relate it to $C_q(T)$ it is therefore not sufficient to consider a one-shot capacity as in the Shannons Theorem. Instead, have to define

$$C_s(T) = \sup_N \frac{1}{N} C_{s,1}(T^{\otimes N}) \quad \text{with} \quad C_{s,1}(T) = \sup_\rho J(\rho, T) \tag{2-235}$$

That $C_s(T)$ is an upper bound on $C_q(T)$. Equality, however, is conjectured but not yet proven, although there are good heuristic arguments. A second interesting quantity which provides an upper bound on the quantum capacity uses the transposition operation θ on the output systems. More precisely it is shown in that

$$C_q(T) \leqslant C_\theta(T) = \log_2 \| T\Theta \|_{cb} \tag{2-236}$$

holds for any channel. In contrast to many other calculations in this field it is particular easy to derive this relation from properties of the cb-norm. Hence we are able to give a proof here. Start with the fact that $\| \theta \|_{cb} = d$ if d is the dimension of the Hilbert space on which θ operates. Assume that $N_j = M_j \to c \leqslant C_q(T)$ and j large enough such that $\| \mathrm{Id}_2^{N_j} - E_j T^{\otimes M_j} D_j \| \leqslant \varepsilon$ with appropriate encodings and decodings E_j, D_j, get

$$\begin{aligned} 2^{N_j} &= \| \mathrm{Id}_2^{N_j} \Theta \|_{cb} \leqslant \| \Theta(\mathrm{Id}_2^{N_j} - E_j T^{\otimes M_j} D_j) \|_{cb} + \| \Theta E_j T^{\otimes M_j} D_j \|_{cb} \\ &\leqslant 2^{N_j} \| \mathrm{Id}_2^{N_j} - E_j T^{\otimes M_j} D_j \|_{cb} + \| \Theta E_j \Theta (\Theta T)^{\otimes M_j} D_j \|_{cb} \\ &\leqslant 2^{N_j} \varepsilon + \| \Theta T \|_{cb}^{M_j} \end{aligned} \tag{2-237}$$

where have used for the last equation the fact that D_j and $\theta E_j \theta$ are channels and that the cb-norm is multiplicative. Taking logarithms on both sides get

$$\frac{N_j}{M_j} + \frac{\log_2(1-\varepsilon)}{M_j} \leqslant \log_2 \| \Theta T \|_{cb} \tag{2-238}$$

In the limit $j \to \infty$ this implies $c \leqslant \log_2 \| \theta T \|$ and therefore $C_q(T) \leqslant \log_2 \| \theta T \|_{cb} = C_\theta(T)$ as stated. Since $C_\theta(T)$ is an upper bound on $C_q(T)$ it is particularly useful to check whether the quantum capacity for a particular channel is zero. If, e. g., T is classical have $\theta T = T$ since the transposition coincides on a classical algebra C_d with the identity (elements of C_d are just diagonal matrices). This implies $C_\theta(T) = \log_2 \| \theta T \|_{cb} = \log_2 \| T \|_{cb} = 0$, because the cb-norm of a channel is 1. Therefore that the quantum capacity of a classical channel is 0—this is just another proof of the no-teleportation theorem. A slightly more general result concerns channels $T = RS$ which are the composition of a preparation $R: M_d \to C_f$ and a subsequent measurement $S: C_f \to M_d$. It is easy to see that $\theta T = \theta RS$ is a channel, because $\theta R \theta$ is a channel and θ is the identity on C_f, hence $\theta R \theta = \theta R$ and $\theta R \theta S = \theta RS = \theta T$. Again get $C_\theta(T) = 0$. Consider now some examples. The most simple case is again the quantum erasure channel from Eq. (2-218). As for the classical capacities its quantum capacity can be explicitly calculated and have $C_q(T) = \max(0; (1-2\theta)\log_2(d))$.

For the depolarizing channel (2-217) precise calculations of $C_q(T)$ are not available. Hence consider first the coherent information. $J(T;\rho)$ inherits from T its unitary covariance, i.e. have $J(U\rho U^*; T) = J(\rho T)$. In contrast to the mutual information, however, it does not have nice concavity properties, which makes the optimization over all input states more difficult to solve. Nevertheless, the calculation of $J(\rho T)$ is straightforward and get in the qubit case (if θ is the noise parameter of T and λ is the highest eigenvalue of ρ):

$$\begin{aligned} J(\rho, T) = {} & S\left(\lambda(1-\vartheta) + \frac{\vartheta}{2}\right) - S\left(\frac{1-\vartheta/2+A}{2}\right) - S\left(\frac{1-\vartheta/2-A}{2}\right) \\ & - S\left(\frac{\lambda\vartheta}{2}\right) - S\left(\frac{(1-\lambda)\vartheta}{2}\right) \end{aligned} \tag{2-239}$$

where $S(x) = -x\log_2(x)$ denotes again the entropy function and

$$A = \sqrt{(2\lambda-1)^2(1-\vartheta/2)^2 + 4\lambda(1-\lambda)(1-\vartheta)^2} \tag{2-240}$$

Optimization over λ can be performed at least numerically (the maximum is attained at the left boundary ($\lambda = 1/2$) if J is positive there, and the right boundary otherwise). The result is plotted together with $C_\theta(T)$ in Figure 2-31 as a function of θ. The quantity $C_\theta(T)$ is much easier to compute and can get

$$C_\theta(T) = \max\left\{0, \log_2\left(2 - \frac{3}{2}\theta\right)\right\} \tag{2-241}$$

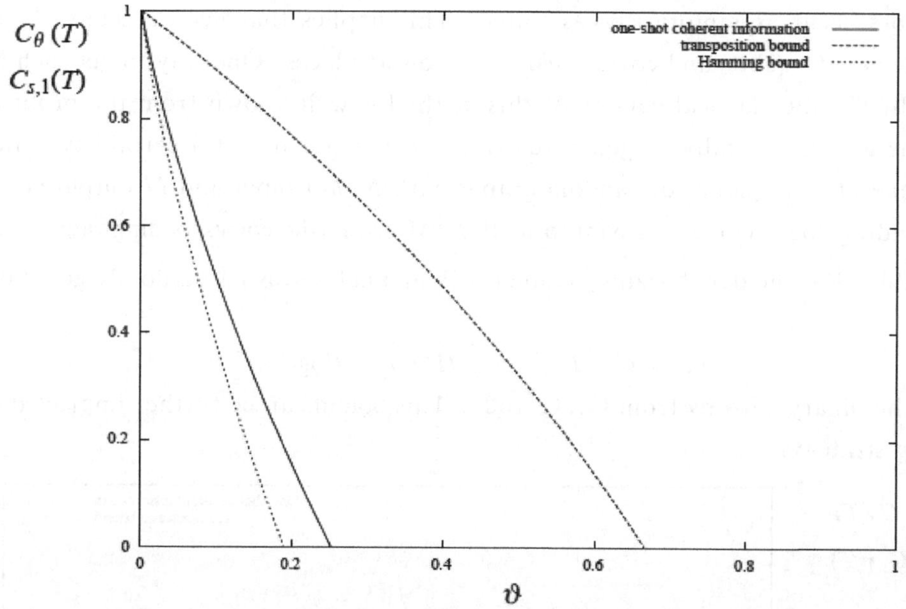

Fig. 2-31 $C_q(T)$, $C_s(T)$ and the Hamming bound of a depolarizing qubit channel plotted as function of the noise parameter θ.

To get a lower bound on $C_q(T)$ we have to show that a certain rate $r \leqslant C_q(T)$ can be achieved with an appropriate sequence

$$E_M : \mathcal{M}_d^{\otimes M} \to \mathcal{M}_2^{\otimes N(M)}, \quad M, N(M) \in \mathbb{N} \tag{2-242}$$

of error correcting codes and corresponding decodings D_M. I.e. need

$$\lim_{j\to\infty} N(M)/M = r \quad \text{and} \quad \lim_{j\to\infty} \| E_M T^{\otimes M} D_M - \mathrm{Id} \|_{\mathrm{cb}} = 0 \tag{2-243}$$

To find such a sequence note first that can look at the depolarizing channel as a device which produces an error with probability θ and leaves the quantum information intact otherwise. If more and more copies of T are used in parallel, i.e. if M goes to infinity, the number of errors approaches therefore θM. In other words, the probability to have more than θM errors vanishes asymptotically. To see this consider

$$T^{\otimes M} = ((\vartheta - 1)\mathrm{Id} + \vartheta d^{-1}\mathrm{tr}(\cdot)\mathbb{1})^{\otimes M} = \sum_{K=1}^{M} (1-\vartheta)^K \vartheta^{N-K} T_K^{(M)} \tag{2-244}$$

where $T_K^{(M)}$ denotes the sum of all M-fold tensor products with $d^{-1}\mathrm{tr}(\cdot)\mathbb{1}$ on N places and Id on the $N-K$ remaining—i.e. $T_K^{(M)}$ is a channel which produces exactly K errors on M transmitted systems. Now have

$$\left\| T^{\otimes M} - \sum_{K \leqslant \vartheta M} (1-\vartheta)^K \vartheta^{N-K} T_K^{(M)} \right\|_{\mathrm{cb}} \tag{2-245}$$

$$= \left\| \sum_{K > \vartheta M} (1-\vartheta)^K \vartheta^{N-K} T_K^{(M)} \right\|_{\mathrm{cb}} \tag{2-246}$$

$$\leqslant \sum_{K>\vartheta M}^{M} (1-\vartheta)^{K} \vartheta^{N-K} \| T_{K}^{(M)} \|_{cb} \tag{2-247}$$

$$\leqslant \sum_{K>\vartheta M}^{M} \binom{M}{K} (1-\vartheta)^{K} \vartheta^{N-K} = R \tag{2-248}$$

The quantity R is the tail a of binomial series and vanishes therefore in the limit $M\to\infty$. This shows that for $M\to\infty$ only terms $T_{K}^{(M)}$ with $K\leqslant\theta M$ are relevant in Eq. (2-242)—in other words at most θM errors occur asymptotically, as stated. This implies that need a sequence of codes EM which encode $N(M)$ qubits and correct θM errors on M places. One way to get such a sequence is "random coding"—the classical version of this method is well known from the proof of Shannons theorem. The idea is, basically, to generate error correcting codes of a certain type randomly. E. g. we can generate a sequence of random graphs with $N(M)$ input and M output vertices. If that the corresponding codes correct (asymptotically) θM errors, the corresponding rate $r=\lim_{M\to\infty} N(M)=M$ is achievable. For the depolarizing channel such an analysis, using randomly generated stabilizer codes shows

$$C_{q}(T)\leqslant 1 - H(\vartheta) - \vartheta\log_{2}3 \tag{2-249}$$

where H is the binary entropy from Eq. (2-152). This bound can be further improved using a more clever coding strategy.

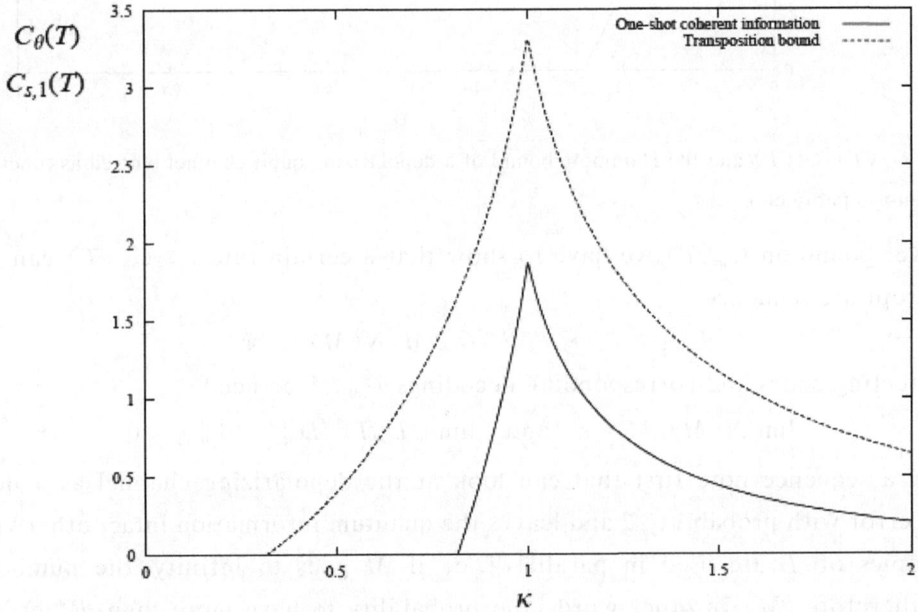

Fig. 2-32 $C_\theta(T)$ and $C_s(T)$ of a Gaussian amplification/attenuation channel as a function of amplification parameter k.

As a third example let us consider again the Gaussian channel studied already. For $C_q(T)$ have (the corresponding calculation is not trivial and uses properties of Gaussian channels which we have not discussed.)

$$C_{\theta}(T) = \max\{0, \log_2(k^2+1) - \log_2(|k^2-1| + 2N_c)\} \tag{2-250}$$

and see that $C_\theta(T)$ and therefore $C_q(T)$ become zero if N_c is large enough (i. e. $N_c \geqslant \max\{1; k^2\}$). The coherent information for the Gaussian state ρ_N from Eq. (2-64) has the form

$$J(\rho_N, T) = g(N') - g\left(\frac{D+N'-N-1}{2}\right) - g\left(\frac{D-N'+N-1}{2}\right) \tag{2-251}$$

with N', D and g that increases with N and we can calculate therefore the maximum over all Gaussian states (which might di= er from $C_s(T)$) as

$$C_G(T) = \lim_{N\to\infty} J(\rho_N, T) = \log_2 k^2 - \log_2 |k^2 - 1| - g\left(\frac{N_c}{k^2 - 1}\right) \tag{2-252}$$

Have plotted both quantities in Figure 2-33 as a function of k. Finally short look on the special case $k = 1$, i. e. T describes in this case only the influence of classical Gaussian noise on the transmitted qubits. If set $k = 1$ in Eq. (2-251) and take the limit $N \to \infty$ get $C_G(T) = -\log_2(N_c e)$ and $C_\theta(T)$ becomes $C_\theta(T) = \max\{0; -\log_2(N_c)\}$; both quantities are plotted in Figure 2-32. This special case is interesting because the one-shot coherent information $C_G(T)$ is achievable, provided the noise parameter N_c satisfies certain conditions. Hence there is strong evidence that the quantum capacity lies between the two lines in Figure 2-33.

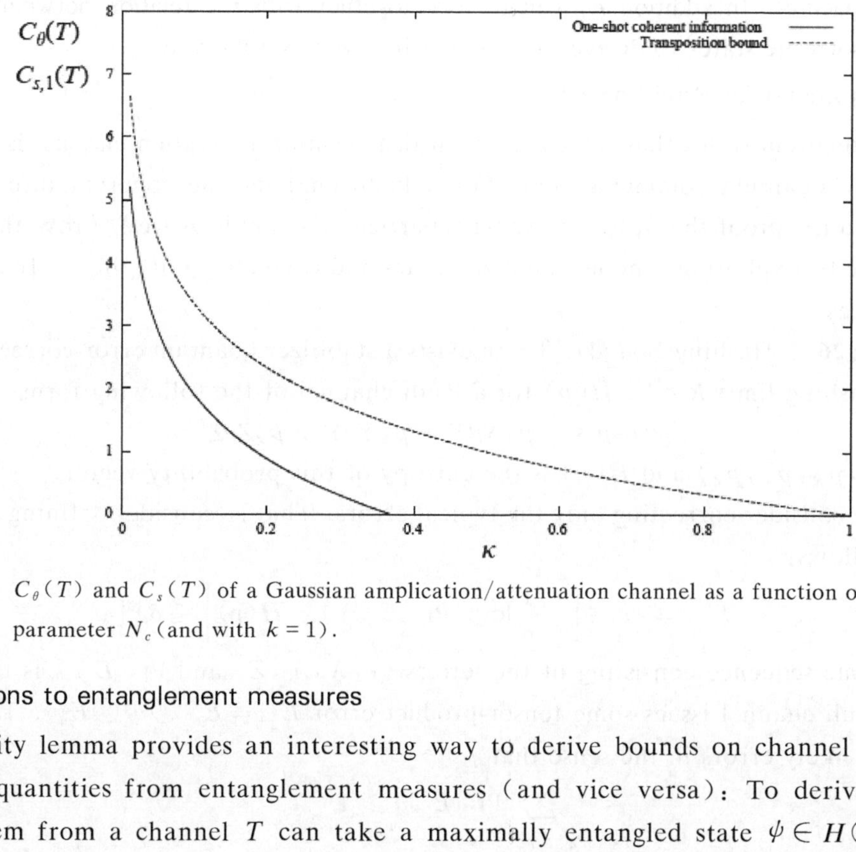

Fig. 2-33 $C_\theta(T)$ and $C_s(T)$ of a Gaussian amplication/attenuation channel as a function of the noise parameter N_c (and with $k = 1$).

- Relations to entanglement measures

The duality lemma provides an interesting way to derive bounds on channel capacities and capacity-like quantities from entanglement measures (and vice versa): To derive a state of a bipartite system from a channel T can take a maximally entangled state $\psi \in H \otimes H$, send one particle through T and get a less entangled pair in the state $T = (\mathrm{Id}\ \rho \otimes T^*) |\psi\rangle\langle\psi|$. If on the other hand an entangled state $\rho \in S(H \otimes H)$ is given, can use it as a resource for teleportation and get a channel T. The two maps $\rho \mapsto T$ and $T\rho \mapsto \rho T$ are, however, not inverse to one another. This can be seen easily from the duality lemma: For each state $\rho \in S(H \otimes H)$ there is a channel T and a pure state $\Phi \in H \otimes H$ such that $\rho = (\mathrm{Id} \otimes T^*) |\Phi\rangle\langle\Phi|$ holds; but Φ is in general not maximally entangled (and uniquely determined by ρ). Nevertheless, there are special cases in which the state derived from T_ρ coincides with ρ: A particular class of examples is given by teleportation channels derived from a Bell-diagonal state. On ρ_T can evaluate an entanglement measure $E(\rho_T)$ and get in this way a quantity which is related to the capacity of T. A particularly interesting candidate for E is the "one-way LOCC" distillation rate. It is defined in the same way as the entanglement of distillation E_D, except that only one-way LOCC operation are allowed in Eq. (2-144). According to ED $\rightarrow$ is related to C_q by the inequalities $E_D;\rightarrow(\rho) \geqslant C_q(T_\rho)$ and $E_D;\rightarrow(T_\rho) \leqslant C_q(T_\rho)$. Hence if $\rho T_\rho = \rho$ can calculate $E_D;\rightarrow(\rho)$ in terms of $C_q(T)$ and vice versa. A second interesting example is the transposition bound $C_\theta(T)$ introduced in the last subsection. It is related to the logarithmic negativity

$$E_\theta(\rho_T) = \log_2 \| (\mathrm{Id} \otimes \Theta)\rho_T \|_1 \tag{2-253}$$

which measures the degree with which the partial transpose of fails to be positive. E_Q can be regarded as entanglement measure although it has some drawbacks: it is not LOCC monotone (Axiom E2), it is not convex (Axiom E3) and most severe: It does not coincide with the reduced von Neumann entropy on pure states, which we have considered as "the" entanglement measure for pure states. On the other hand, it is easy to calculate and it gives bounds on distillation rates and teleportation capacities. In addition E_Q can be used together with the relation between depolarizing channels and isotropic states to derive Eq. (2-241) in a very simple way.

- Hashing Bound for Pauli Channels

The LSD theorem states that the coherent information of a quantum channel is an achievable rate for reliable quantum communication. For a Pauli channel, the coherent information has a simple form and the proof that it is achievable is particularly simple as well. Prove the theorem for this special case by exploiting random stabilizer codes and correcting only the likely errors that the channel produces.

Theorem 2.26 (Hashing Bound). There exists a stabilizer quantum error-correcting code that achieves the hashing limit $R = 1 - H(p)$ for a Pauli channel of the following form:

$$\rho \mapsto p_I\rho + p_X X\rho X + p_Y Y\rho Y + p_Z Z\rho Z \tag{2-254}$$

where $p = (p_I, p_X, p_Y, p_Z)$ and $H(p)$ is the entropy of this probability vector.

Proof. We consider correcting only the typical errors. That is, consider defining the typical set of errors as follows:

$$T_\delta^{p^n} \equiv \left\{ a^n : \left| -\frac{1}{n}\log_2(\Pr\{E_{a^n}\}) - H(\mathrm{p}) \right| \leqslant \delta \right\} \tag{2-255}$$

where a^n is some sequence consisting of the letters $\{I, X, Y, Z\}$ and $\Pr\{E_{a^n}\}$ is the probability that an IID Pauli channel issues some tensor-product error $E_{a^n} \equiv E_{a_1} \otimes \cdots \otimes E_{a_n}$. This typical set consists of the likely errors in the sense that

$$\sum_{a^n \in T_\delta^{p^n}} \Pr\{E_{a^n}\} \geqslant 1 - \varepsilon \tag{2-256}$$

for all $\varepsilon > 0$ and sufficiently large n. The error-correcting conditions for a stabilizer code $\mathcal{S}$ in this case are that $\{E_{a^n} : a^n \in T_\delta^{p^n}\}$ is a correctable set of errors if:

$$E_{a^n}^\dagger E_{b^n} \notin N(\mathcal{S}) \backslash \mathcal{S} \tag{2-257}$$

for all error pairs E_{a^n} and E_{b^n} such that $a^n, b^n \in T_\delta^{p^n}$ where $N(\mathcal{S})$ is the normalizer of $\mathcal{S}$. Also, we consider the expectation of the error probability under a random choice of a stabilizer code. We proceed as follows:

$$E_S\{p_e\} = E_S\left\{ \sum_{a^n} \Pr\{E_{a^n}\} \mathcal{I}(E_{a^n} \text{ is uncorrectable under } \mathcal{S}) \right\}$$

$$\leqslant E_S\left\{ \sum_{a^n \in T_\delta^{p^n}} \Pr\{E_{a^n}\} \mathcal{I}(E_{a^n} \text{ is uncorrectable under } \mathcal{S}) \right\} + \varepsilon$$

$$= \sum_{a^n \in T_\delta^{p^n}} \Pr\{E_{a^n}\} \mathbb{E}_{\mathcal{S}}\{\mathcal{I}(E_{a^n} \text{ is uncorrectable under } \mathcal{S})\} + \varepsilon$$

$$= \sum_{a^n \in T_\delta^{p^n}} \Pr\{E_{a^n}\} \Pr_{\mathcal{S}}\{E_{a^n} \text{ is uncorrectable under } \mathcal{S}\} + \varepsilon \tag{2-258}$$

The first equality follows by definition—$\mathcal{I}$ is an indicator function equal to one if E_{a^n} is uncorrectable under $\mathcal{S}$ and equal to zero otherwise. The first inequality follows, since we correct only the typical errors because the atypical error set has negligible probability mass. The second equality follows by exchanging the expectation and the sum. The third equality follows because the expectation of an indicator function is the probability that the event it selects occurs. Continuing, have

$$= \sum_{a^n \in T_\delta^{p^n}} \Pr\{E_{a^n}\} \Pr_{\mathcal{S}}\{\exists E_{b^n} : b^n \in T_\delta^{p^n}, b^n \neq a^n, E_{a^n}^\dagger E_{b^n} \in N(\mathcal{S}) \setminus \mathcal{S}\}$$

$$\leqslant \sum_{a^n \in T_\delta^{A^n}} \Pr\{E_{a^n}\} \Pr_{\mathcal{S}}\{\exists E_{b^n} : b^n \in T_\delta^{p^n}, b^n \neq a^n, E_{a^n}^\dagger E_{b^n} \in N(\mathcal{S})\}$$

$$= \sum_{a^n \in T_\delta^{p^n}} \Pr\{E_{a^n}\} \Pr_{\mathcal{S}}\left\{ \bigcup_{b^n \in T_b^{p^n}, b^n \neq a^n} E_{a^n}^\dagger E_{b^n} \in N(\mathcal{S}) \right\}$$

$$\leqslant \sum_{a^n, b^n \in T_\delta^{p^n}, b^n \neq a^n} \Pr\{E_{a^n}\} \Pr_{\mathcal{S}}\{E_{a^n}^\dagger E_{b^n} \in N(\mathcal{S})\}$$

$$\leqslant \sum_{a^n, b^n \in T_\delta^{p^n}, b^n \neq a^n} \Pr\{E_{a^n}\} 2^{-(n-k)}$$

$$\leqslant 2^{2n[H(p)+\delta]} 2^{-n[H(p)+\delta]} 2^{-(n-k)}$$

$$= 2^{-n[1-H(p)-k/n-3\delta]} \tag{2-259}$$

The first equality follows from the error-correcting conditions for a quantum stabilizer code, where $N(\mathcal{S})$ is the normalizer of $\mathcal{S}$. The first inequality follows by ignoring any potential degeneracy in the code—we consider an error uncorrectable if it lies in the normalizer $N(\mathcal{S})$ and the probability can only be larger because $N(\mathcal{S}) \setminus \mathcal{S} \in N(\mathcal{S})$. The second equality follows by realizing that the probabilities for the existence criterion and the union of events are equivalent. The second inequality follows by applying the union bound. The third inequality follows from the fact that the probability for a fixed operator $E_{a^n}^\dagger E_{b^n}$ not equal to the identity commuting with the stabilizer operators of a random stabilizer can be upper bounded as follows:

$$\Pr_{\mathcal{S}}\{E_{a^n}^\dagger E_{b^n} \in N(\mathcal{S})\} = \frac{2^{n+k} - 1}{2^{2n} - 1} \leqslant 2^{-(n-k)} \tag{2-260}$$

The reasoning here is that the random choice of a stabilizer code is equivalent to fixing operators $z_1, \cdots, z_{n-k}$ and performing a uniformly random Clifford unitary. The probability that a fixed operator commutes with $\bar{z}_1, \cdots, \bar{z}_{n-k}$ is then just the number of non-identity operators in the normalizer $(2^{n+k} - 1)$ divided by the total number of non-identity operators $(2^{2n} - 1)$. After applying the above bound, then exploit the following typicality bounds:

$$\forall a^n \in T_\delta^{p^n} : \Pr\{E_{a^n}\} \leqslant 2^{-n[H(p)+\delta]} \tag{2-261}$$

$$| T_\delta^{p^n} | \leqslant 2^{n[H(p)+\delta]} \tag{2-262}$$

To conclude that as long as the rate $k/n = 1 - H(p) - 4\delta$, the expectation of the error probability becomes arbitrarily small, so that there exists at least one choice of a stabilizer code with the same bound on the error probability.

2.7 Multiple inputs

In Section 2.4 that many tasks of quantum information which are impossible with one-shot operations can be approximated by channels which operate on a large number of equally prepared inputs. Typical examples are approximate cloning, undoing noise and distillation of entanglement.

There are basically two questions which are interesting for a quantitative analysis: First, can search for the optimal solutions for a fixed number N of input systems and second can ask for the asymptotic behavior in the limit $N \to \infty$. In the latter case the asymptotic rate, i.e. the number of outputs (of a certain quality) per input system is of particular interest.

Both types of questions just mentioned can be treated (up to certain degree) independently from the (impossible) task we are dealing with. In the following we will study the corresponding general scheme. Hence consider a channel $T: B(H^{\otimes M}) \to B(H^{\otimes N})$ which operates on N input systems and produces M outputs of the same type. The aim is to optimize a "figure of merit" $F(T)$ which measures the deviation of $T^*(\rho^{\otimes N})$ from the target functional we want to approximate. The particular type of device are considering is mainly fixed by the choice of $F(T)$ and will discuss in the following the most relevant examples. (Note that have considered them already on a qualitative level in Section 2.4.)

- Figures of merit

Let start with pure state cloning, i.e. for each (unknown) pure input state $\sigma = |\psi\rangle\langle\psi|, \psi \in H$ the M clones $T^*(\sigma^{\otimes N})$ produced by the channel T should approximate M copies of the input in the common state $\sigma^{\otimes M}$ as good as possible. There are in fact two different possibilities to measure the distance of $T^*(\sigma^{\otimes N})$ to $\sigma^{\otimes M}$. Can either check the quality of each clone separately or we can test in addition the correlations between output systems. With the notation

$$\sigma^{(j)} = \mathbb{1}^{\otimes(j-1)} \otimes \sigma \otimes \mathbb{1}^{\otimes(M-j)} \in \mathcal{B}(\mathcal{H}^{\otimes M}) \tag{2-263}$$

a figure of merit for the first case is given by

$$\mathcal{F}_{c,1}(T) = \inf_{j=1,\cdots,N} \inf_{\sigma \text{ pure}} \operatorname{tr}(\sigma^{(j)} T^*(\sigma^{\otimes N})) \tag{2-264}$$

It measures the worst one-particle fidelity of the output state $T^*(\rho^{\otimes N})$. If are interested in correlations too, have to choose

$$\mathcal{F}_{c,\text{all}}(T) = \inf_{\sigma \text{ pure}} \operatorname{tr}(\sigma^{\otimes M} T^*(\sigma^{\otimes N})) \tag{2-265}$$

which is again a "worst case" fidelity, but now of the full output with respect to M uncorrelated copies of the input σ.

Instead of fidelities we can consider other error quantities like trace-norm distances or relative entropies. In general, however, we do not get significantly different results from such alternative choices; hence, can safely ignore them. Real variants arise if consider instead of the infima over all pure states quantities which prefer a (possibly discrete or even finite) class of states. Such a choice leads to "state-dependent cloning", because the corresponding optimal devices perform better as "universal" ones (i.e. those described by the figures of merit above) on some states but much worse on the rest. Ignore state-dependent cloning in this work, because the universal case is physically more relevant and technically more challenging. Other cases which do not discuss either include "asymmetric cloning", which arises if trade in Eq. (2-264) the quality of one particular output system against the rest, and cloning of mixed states. The latter is much more difficult than the pure state case and even for classical systems, where it is related to the so-called "bootstrap" technique, non-trivial. Closely related to cloning is puri. cation, i. e. undoing noise. This means consider N systems originally prepared in the same (unknown) pure state ρ but which have passed a depolarizing channel

$$R^{*}\sigma=\vartheta\sigma+(1-\vartheta)\mathbb{1}/d \tag{2-266}$$

after wards. The task is now to find a device T acting on N of the decohered systems such that $T^{*}(R^{*}\rho)$ is as close as possible to the original pure state. The same basic choices for a figure of merit as in the cloning problem. Hence, define

$$\mathscr{F}_{R,1}(T)=\inf_{j=1,\cdots,N}\inf_{\sigma \text{ pure}}\operatorname{tr}(\sigma^{(j)}T^{*}[(R^{*}\sigma)^{\otimes N}]) \tag{2-267}$$

and

$$\mathscr{F}_{R,\text{all}}(T)=\inf_{\sigma \text{ pure}}\operatorname{tr}(\sigma^{\otimes M}T^{*}[(R^{*}\sigma)^{\otimes N}]) \tag{2-268}$$

These quantities can be regarded as generalizations of $F_{c,1}$ and F_c, all which we recover if R^{*} is the identity.

Another task consider is the approximation of a map θ which is positive but not completely positive, like the transposition. An explicit example is the universal not gate (UNOT) which maps each pure qubit state σ to its orthocomplement $\sigma^{\perp}$. It is given the anti-unitary operator

$$\psi=\alpha\mid 0\rangle+\beta\mid 1\rangle\mapsto\Theta\psi=\bar{\alpha}\mid 0\rangle-\bar{\beta}\mid 1\rangle \tag{2-269}$$

Since $\theta\sigma$ is a state if σ is, we can ask again for a channel T such that $T^{*}(\sigma^{\otimes N})$ approximates $(\theta\sigma)^{\otimes M}$. As in the two previous examples have the choice to allow arbitrary correlations in the output or not and get the following figures of merit:

$$\mathscr{F}_{\theta,1}(T)=\inf_{j=1,\cdots,N}\inf_{\sigma \text{ pure}}\operatorname{tr}((\Theta\sigma)^{(j)}T^{*}(\sigma^{\otimes N})) \tag{2-270}$$

and

$$\mathscr{F}_{\theta,\text{all}}(T)=\inf_{\sigma \text{ pure}}\operatorname{tr}((\Theta\sigma)^{\otimes M}T^{*}(\sigma^{\otimes N})) \tag{2-271}$$

Note that, plug in for θ basically any functional which maps states to states. In addition combine Eqs. (2-267), (2-268), (2-270) and (2-271) on the other. As result would get a measure for devices which undo an operation R and approximate an impossible machine θ at the same time.

- Covariant operations

All the functionals just defined give rise to optimization problems which will study in greater detail in the next sections. This means are interested in two things: First of all the maximal value of $F_{\#,\#}$ (with $\#=c,R,Q$ and $\#=1,\text{all}$) given by

$$\mathscr{F}_{\#,\#}(N,M)=\inf_{T}\mathscr{F}_{\#,\#}(T) \tag{2-272}$$

where the supremum is taken over all channels $T:B(H^{\otimes M})\to B(H^{\otimes N})$, and second the particular channel $\hat{T}$ where the optimum is attained. At a first look a complete solution of these problems seems to be impossible, due to the large dimension of the space of all T, which scales exponentially in M and N. Fortunately, all $F_{\#,\#}(T)$ admit a large symmetry group which allows in many cases the explicit calculation of the optimal values $F_{\#,\#}(N,M)$ and the determination of optimizers $\hat{T}$ with a certain covariance behavior. Note that this is an immediate consequence of decision to restrict the discussion to "universal" procedures, which do not prefer any particular input state. Consider permutations of the input systems first: If $p \in S_N$ is a permutation on N places and V_p the corresponding unitary on $H^{\otimes N}$ get obviously $T^*(V_p \rho^{\otimes N} V_p^*) = T^*(\rho^{\otimes N})$, hence

$$\mathscr{F}_{\#,\#}[\alpha_p(T)] = \mathscr{F}_{\#,\#}(T) \quad \forall\, p \in S_N \quad \text{with } [\alpha_p(T)](A) = V_p^* T(A) V_p \tag{2-273}$$

In other words: $F_{\#,\#}(T)$ is invariant under permutations of the input systems. Similarly, can show that $F_{\#,\#}(T)$ is invariant under permutations of the output systems:

$$\mathscr{F}_{\#,\#}[\beta_p(T)] = \mathscr{F}(T) \quad \forall\, p \in S_M \quad \text{with } [\beta_p(T)](A) = T(V_p^* A V_p) \tag{2-274}$$

To see this consider e.g. for $\# = c$ and $\# = $ all

$$\operatorname{tr}[\sigma^{\otimes M} V_p T^*(\rho^{\otimes N}) V_p^*] = \operatorname{tr}[V_p \sigma^{\otimes M} V_p^* T^*(\rho^{\otimes N})] = \operatorname{tr}[\sigma^{\otimes M} T^*(\rho^{\otimes N})] \tag{2-275}$$

For the other cases similar calculations apply. Finally, none of the $F_{\#,\#}(T)$ singles out a preferred direction in the one-particle Hilbert space H. This implies that we can rotate T by local unitaries of the form $U \otimes N$, respectively $U \otimes M$, without changing $F_{\#,\#}(T)$. More precisely have

$$\mathscr{F}_{\#,\#}[\gamma_U(T)] = \mathscr{F}_{\#,\#}(T) \quad \forall\, U \in U(d) \tag{2-276}$$

with

$$[\gamma_U(T)](A) = U^{*\otimes N} T(U^{\otimes M} A U^{*\otimes M}) U^{\otimes N} \tag{2-277}$$

The validity of Eq. (2-276) can be proven in the same way as (2-273) and (2-274). The details are therefore left to the reader. Now average over the groups S_N, S_M and $U(d)$. Instead of the operation T consider

$$\overline{T} = \frac{1}{N!M!} \sum_{p \in S_N} \sum_{q \in S_M} \int_G \alpha_p \beta_q \gamma_U(T) \mathrm{d}U \tag{2-278}$$

where $\mathrm{d}U$ denotes the normalized, left invariant Haar measure on $U(d)$. Immediately that $\overline{T}$ has the following symmetry properties:

$$\alpha_p(\overline{T}) = \overline{T}, \quad \beta_q(\overline{T}) = \overline{T}, \quad \gamma_U(\overline{T}) = \overline{T}, \quad \forall\, p \in S_N, \quad \forall\, q \in S_M, \quad \forall\, U \in U(d) \tag{2-279}$$

and call each operation T fully symmetric, if it satis. es this equation. The concavity of $F_{\#,\#}$ implies immediately that it cannot decrease if replace T by $\overline{T}$:

$$\mathscr{F}_{\#,\#}(T) = \mathscr{F}_{\#,\#}\left(\frac{1}{N!M!} \sum_{p \in S_N} \sum_{q \in S_M} \int_G \alpha_p \beta_q \gamma_U(T) \mathrm{d}U\right) \tag{2-280}$$

$$\geqslant \frac{1}{N!M!} \sum_{p \in S_N} \sum_{q \in S_M} \int_G \mathscr{F}_{\#,\#}[\alpha_p \beta_q \gamma_U(T)] \mathrm{d}U = \mathscr{F}_{\#,\#}(T) \tag{2-281}$$

To calculate the optimal value $F_{\#,\#}(N,M)$ it is therefore completely sufficient to search a maximize for $F_{\#,\#}(T)$ only among fully symmetric T and to evaluate $F_{\#,\#}(T)$ for this particular operation. This simplifies the problem significantly because the size of the parameter space is extremely reduced. Of course, do not know from this argument whether the optimum is attained on non-symmetric operations, however this information is in general less important (and for some problems like optimal cloning a uniqueness result is available).

- Group representations

To get an idea how this parameter reduction can be exploited practically, which reconsider Theorem 2.14: The two representations $U \mapsto U^{\otimes N}$ and $p \mapsto V_p$ of $U(d)$, respectively S_N, on $H^{\otimes N}$ are "commutants" of each other, i.e., any operator on $H^{\otimes N}$ commuting with all $U^{\otimes N}$ is a linear combination of the V_p, and conversely. This knowledge can be used to decompose the representation $U^{\otimes N}$ and V_p as well into irreducible components. To reduce the group theoretic overhead, will discuss this procedure first for qubits only and come back to the general case afterwards. Hence assume that $H = C^2$ holds. Then $H^{\otimes N}$ is the Hilbert space of N distinguishable spin $-1/2$ particles and it can be decomposed into terms of eigenspaces of total angular momentum. More precisely consider

$$L_k = \frac{1}{2}\sum_j \sigma_k^{(j)}, \quad k = 1,2,3 \tag{2-282}$$

The k-component of total angular momentum (i.e. σ_k is the kth Pauli matrix and $\sigma^{(j)} \in B(H^{\otimes N})$ defined according to Eq. (2-263)) and $\vec{L}^2 = \sum_k L_k^2$. The eigenvalue expansion of $\vec{L}^2$ is well known to be

$$\vec{L} = \sum_j s(s+1)P_s \quad \text{with } s = \begin{cases} 0,1,\cdots,N/2, & N \text{ even} \\ 1/2,3/2,\cdots,N/2, & N \text{ odd} \end{cases} \tag{2-283}$$

where the P_s denote the projections to the eigenspaces of $\widetilde{L}^2$. It is easy to see that both representations $U \mapsto U^{\otimes N}$ and $p \mapsto V_p$ commute with $\widetilde{L}$. Hence the eigenspaces $P_s H^{\otimes N}$ of $\widetilde{L}^2$ are invariant subspaces of $U \otimes N$ and V_p and this implies that the restriction of $U \otimes N$ and V_p to them are representations of $S_U(2)$, respectively S_N. Since $\widetilde{L}^2$ is constant on $P_s H^{\otimes N}$ the $S_U(2)$ representation get in this way must be (naturally isomorphic to) a multiple of the irreducible spins representation 3s. It is defined by

$$\pi_s\left[\exp\left(\frac{\mathrm{i}}{2}\sigma_k\right)\right] = \exp(\mathrm{i}L_k^{(s)}) \quad \text{with } L_k^{(s)} = \frac{1}{2}\sum_{j=1}^{2s}\sigma_k^{(j)} \tag{2-284}$$

on the representation space

$$\mathscr{H}_s = \mathscr{H}_+^{\otimes 2s} \tag{2-285}$$

(the Bose-subspace of $H^{\otimes 2s}$). Hence get

$$P_s\mathscr{H}^{\otimes N} \cong \mathscr{H}_s \otimes \mathscr{K}_{N,s}, \quad U^{\otimes N}\psi = (\pi_s(U) \otimes \mathbb{1})\psi \quad \forall \psi \in P_s\mathscr{H}^{\otimes N} \tag{2-286}$$

Since V_p and $U^{\otimes N}$ commute the Hilbert space $K_{N,s}$ carries a representation $\hat{\pi}_{Ns}(p)$ of S_N which is irreducible as well. Note that $K_{N,s}$ depends in contrast to H_s on the number N of tensor factors and its dimension is

$$\dim \mathscr{K}_{N,s} = \frac{2s+1}{N/2+s+1}\binom{N}{N/2-s} \tag{2-287}$$

Summarizing the discussion get

$$\mathscr{H}^{\otimes N} \cong \bigoplus_s \mathscr{H}_s \otimes \mathscr{K}_{N,s}, \quad U^{\otimes N} \cong \bigoplus_s \pi_s(U) \otimes \mathbb{1}, \quad V_p \cong \bigoplus_s \mathbb{1} \otimes \hat{\pi}(p) \tag{2-288}$$

Consider now a fully symmetric operation T. Permutation invariance ($\alpha_p(T) = T$ and $\beta_p(T) = T$) implies together with Eq. (2-288) that

$$T(A_j \otimes B_j) = \bigoplus_s \left[\frac{\mathrm{tr}(B_j)}{\dim \mathscr{K}_{N,j}} T_{sj}(A_j) \otimes \mathbb{1}\right] \quad \text{with } T_{sj}: \mathscr{B}(\mathscr{H}_j) \to \mathscr{B}(\mathscr{H}_s) \tag{2-289}$$

holds if $A_j \otimes B_j \in B(H_j \otimes K_{N,j})$. The operations T_{sj} are unital and have, according to $\gamma_U(T) = T$ the following covariance properties:

$$\pi_s(U)T(A_j)\pi_s(U^*) = T[\pi_j(U)A_j\pi_j(U^*)], \quad \forall U \in S_U(2) \tag{2-290}$$

The classification of all fully symmetric channels T is reduced therefore to the study of all these T_{sj}. Can apply now the covariant version of Stinespring's theorem to find that

$$T_{sj}(A_j) = V^*(A_j \otimes \mathbb{1})V, \quad V: \mathscr{H}_s \to \mathscr{H}_j \otimes \widetilde{\mathscr{H}}, \quad V\pi_s(U) = \pi_j(U) \otimes \widetilde{\pi}(U)V \tag{2-291}$$

where $\tilde{\pi}$ is a representation of $S_U(2)$ on $\tilde{H}$. If π is irreducible with total angular momentum l the "intertwining operator" V is well known: Its components in a particularly chosen basis coincide with certain Clebsh-Gordon coefficients. Hence, the corresponding operation is uniquely determined (up to unitary equivalence) and write

$$T_{sjl}(A_j) = [V_l(A_j \otimes \mathbb{1})V_l], \quad V_l\pi_s(U) = \pi_j(U) \otimes \pi_l(U)V_l \tag{2-292}$$

where l can range from $|j-s|$ to $j+s$. Since in a general representation $\tilde{\pi}$ can be decomposed into irreducible components that each covariant T_{sj} is a convex linear combination of the T_{sjl} and get with Eq. (2-289)

$$T(A_j \otimes B_j) = \bigoplus_s \left[\sum_l c_{jl}[T_{sjl}(A_j) \otimes (\mathrm{tr}(B_j)\mathbb{1})]\right] \tag{2-293}$$

where the c_{jl} are constrained by $c_{jl} > 0$ and $\sum_j cjl = (\dim K_{Nj})^{-1}$. In this way we have parameterized the set of fully symmetric operations completely in terms of group theoretical data and can rewrite $F_{\#,\#}(T)$ accordingly. This leads to an optimization problem for a quantity depending only on s, j and l, which is at least in some cases solvable. To generalize the scheme just presented to the case $H = C^d$ with arbitrary d only have to find a replacement for the decomposition in Eq. (2-288). This, however, is well known from group theory:

$$\mathcal{H}^{\otimes N} \cong \bigoplus_Y \mathcal{H}_Y \otimes \mathcal{K}_Y, \quad U^{\otimes N} \cong \bigoplus_Y \pi_Y(U) \otimes \mathbb{1}, \quad V_p \cong \bigoplus_Y \mathbb{1} \otimes \hat{\pi}_Y(p) \tag{2-294}$$

where $\pi_y: U(d) \rightarrow B(H_Y)$ and $\hat{\pi}_Y: S_N \rightarrow B(K_Y)$ are irreducible representations. The summation index Y runs over all Young frames with d rows and N boxes, i. e. by the arrangements of N boxes into d rows of lengths $Y_1 \geqslant Y_2 \geqslant \cdots \geqslant Y_d \geqslant 0$ with $\sum_k Y_k = N$. The relation to total angular momentum s used as the parameter for $d = 2$ is given by $Y_1 - Y_2 = 2s$, which determines Y together with $Y_1 + Y_2 = N$ completely. The rest of the arguments applies without signi. cant changes, this is in particular the case for Eq. (2-293) which holds for general d if replace s, j and l by Young frames. However, the representation theory of $U(d)$ becomes much more difficult. The generalization of results available for qubits ($d = 2$) to $d > 2$ is therefore not straightforward. Finally, let give a short comment on Gaussian states here. Obviously, the methods just described do not apply in this case. However, consider instead of $U^{\otimes N}$-covariance, covariance with respect to phase-space translations. Following this idea some results concerning optimal cloning of Gaussian states are obtained, but the corresponding general theory is not as far developed as in the. nite-dimensional case.

- Distillation of entanglement

Finally, have another look at distillation of entanglement. The basic idea is quite the same as for optimal cloning: Use multiple inputs to approximate a task, which is impossible with one-shot operations. From a more technical point of view, however, it does not. t into the general scheme proposed up to now. Nevertheless, some of the arguments can adopted in an easy way. First of all we have to replace the "one-particle" Hilbert space H with a twofold tensor product $H_A \otimes H_B$ and the channels we have to look at are LOCC operations

$$T: \mathcal{B}(\mathcal{H}_A^{\otimes M} \otimes \mathcal{H}_B^{\otimes M}) \rightarrow \mathcal{B}(\mathcal{H}_A^{\otimes N} \otimes \mathcal{H}_B^{\otimes N}) \tag{2-295}$$

The aim is to determine T such that $T^*(\rho^{\otimes N})$ is for each distillable (mixed) state $\rho \in B^*(H_A \otimes H_B)$, close to the M-fold tensor product $|\psi\rangle\langle\psi|^{\otimes M}$ of a maximally entangled state $\psi \in H_A \otimes H_B$.

A figure of merit with a similar structure as the $F_{\#,\text{all}}$ studied above can be derived directly from the definition of the entanglement measure E_D: define (replacing the trace-norm distance with a fidelity)

$$\mathscr{F}_D(T) = \inf_{\rho} \inf_{\Psi} \langle \Psi^{\otimes M}, T^*(\rho^{\otimes N}) \Psi^{\otimes M} \rangle \tag{2-296}$$

where the infima are taken over all maximally entangled states Ψ and all distillable states ρ Alternatively, can look at state-dependent measures, which seem to be particularly important if try to calculate $E_D(\rho)$ for some state ρ. In this case simply get

$$\mathscr{F}_{D,\rho}(T) = \inf_{\Psi} \langle \Psi^{\otimes M}, T^*(\rho^{\otimes N}) \Psi^{\otimes M} \rangle \tag{2-297}$$

To translate the group theoretical analysis of the last two subsections is somewhat more difficult. As in the case of $F_{\#,\#}$ can restrict the search for optimizers to permutation invariant operations, i.e. $\alpha_p(T) = T$ and $\beta_p(T) = T$ in the terminology. Unitary covariance

$$U^{\otimes N} T(A) U^{*\otimes N} = T(U^{\otimes M} A U^{*\otimes M}) \tag{2-298}$$

however, cannot be assumed for all unitaries U of $H_A \otimes H_B$, but only for local ones ($U = U_A \otimes U_B$) in the case of F_D or only for local U which leave ρ invariant for $F_{D,\rho}$. This makes the analog of the decomposition scheme more difficult and such a study is (up to my knowledge) not yet done. A related subproblem arises if consider $F_{D,\rho}$ from Eq. (2-297) for a state ρ with special symmetry properties; e.g. an OO-invariant state. The corresponding optimization might be simpler and a solution would be relevant for the calculation of E_D.

- Optimal devices

Now consider the optimization problems associated to the figures of merit discussed in the last section. This means that is searching for those devices which approximate the impossible tasks in question in the best possible way. As pointed out at the beginning of this Section this can be done for finite N and in the limit $N \to \infty$. The latter is postponed to the next section.

- Optimal cloning

The quality of an optimal, pure state cloner is defined by the figures of merit $F_{c,\#}$ in Eqs. (2-264) and (2-265) and the group theoretic ideas sketched allow the complete solution of this problem. That will demonstrate some of the basic ideas in the qubit case first and state the final result afterwards in full generality. The solvability of this problem relies in part on the special structure. If consider e.g. $F_{c,1}(T)$ (the other case works similarly) get

$$\mathscr{F}_{c,1}(T) = \inf_{j=1,\cdots,N} \inf_{\sigma \text{ pure}} \operatorname{tr}(\sigma^{(j)} T^*(\sigma^{\otimes N})) \tag{2-299}$$

$$= \inf_{j=1,\cdots,N} \inf_{\sigma \text{ pure}} \operatorname{tr}(T(\sigma^{(j)}) \sigma^{\otimes N})) \tag{2-300}$$

$$= \inf_{j=1,\cdots,N} \inf_{\psi} \langle \psi^{\otimes N}, T(\sigma^{(j)}) \psi^{\otimes N} \rangle \tag{2-301}$$

Hence $F_{c,\#}$ only depends on the $B(H_+^{\otimes N})$ component (where $H_+^{\otimes N}$ denotes again the Bose-subspace of $H^{\otimes N}$) of T and can assume without loss of generality that T is of the form:

$$T: \mathscr{B}(\mathscr{H}^{\otimes M}) \to \mathscr{B}(\mathscr{H}_+^{\otimes N}) \tag{2-302}$$

The restriction of $U^{\otimes N}$ to $H_+^{\otimes N}$ is an irreducible representation (for any d) and in the qubit case ($d = 2$) have $U^{\otimes N}\psi = \pi_s(U)\psi$ with $s = N/2$ for all $\psi \in H_+^{\otimes N}$. The decomposition of T from Eq. (2-289) contains therefore only those summands with $s = N/2$. This simplifies the optimization problem significantly, since the number of variables needed to parametrize all relevant cloning maps according to Eq. (2-293) is reduced from 3 to 2. A more detailed (and non-trivial) analysis shows that the maximum for $F_{c,1}$ and $F_{c,\text{all}}$ is attained if all terms in (2-293) except the one with $s = N/2$; $j = N/2$ and $l = (M - N)/2$ vanish. The precise result is stated in the following theorem.

Theorem 2. 27 For each $H = C^d$ both figures of merit $F_{c,l}$ and F_c; all are maximized by the cloner

$$\hat{T}^*(\rho) = \frac{d[N]}{d[M]} S_M(\rho \otimes \mathbb{1}) S_M \tag{2-303}$$

where $d[N]$, $d[M]$ denote the dimensions of the symmetric tensor products $H_+^{\otimes N}$; respectively $H_+^{\otimes M}$; and S_M is the projection from $H^{\otimes M}$ to $H_+^{\otimes M}$. This implies for the optimal fidelities

$$\mathscr{F}_{c,1}(N, M) = \frac{d-1}{d} \frac{N}{N+d} \frac{M+d}{M} \tag{2-304}$$

and

$$\mathscr{F}_{c,\text{all}}(N, M) = \frac{d[N]}{d[M]} \tag{2-305}$$

$\hat{T}$ is the unique solution for both optimization problems; i.e. there is no other operation T of form (2-302) which maximizes $F_{c,l}$ or $F_{c,\text{all}}$. There are two aspects of this result which deserve special attention. One is the relation to state estimation which is postponed. The second concerns the role of correlations: It does not matter whether we are looking for the quality of each single clone ($F_{c,l}$) only, or whether correlations are taken into account ($F_{c,\text{all}}$). In both cases get the same optimal solution. This is a special feature of pure states, however. Although there are no concrete results for quantum systems, it can be checked quite easily in the classical case that considering correlations changes the optimal cloner for arbitrary mixed states drastically.

• Purification

To find an optimal purification device, i. e. maximizing $F_{R,\#}$, is more difficult than the cloning problem, because the simplification from Eq. (2-302) does not apply. Hence we have to consider all the summands in the direct sum decomposition of T from Eq. (2-293) and solutions are available only for qubits. Therefore will assume for the rest of this subsection that $H = C^2$ holds. The $S_U(2)$ symmetry of the problem allows us to assume without loss of generality that the pure initial state ψ coincides with one of the basis vectors. Hence get for the (noisy) input states of the

$$\rho(\beta) = \frac{1}{2\cosh(\beta)} \exp\left(2\beta \frac{\sigma_3}{2}\right) = \frac{1}{e^{\beta} + e^{-\beta}} \begin{pmatrix} e^{\beta} & 0 \\ 0 & e^{-\beta} \end{pmatrix} \tag{2-306}$$

$$= \tanh(\beta) \mid \psi\rangle\langle\psi \mid + (1 - \tanh(\beta)) \frac{1}{2} \mathbb{1}, \quad \psi = \mid 0\rangle \tag{2-307}$$

The parameterization of ρ in terms of the "pseudo-temperature" β is chosen here, because it simplifies some calculations significantly (as will see soon). The relation to the form of $\rho = R * \sigma$ initially given in Eq. (2-266) is obviously $\theta = \tanh(\beta)$. To state the main result of this subsection have to decompose the product state $\rho(\beta)^{\otimes N}$ into spin-s components. This can be done in terms of Eq. (2-288). $\rho(\beta)$ is not unitary of course. However, can apply Eq. (2-288) by analytic continuation, i.e. we treat $\rho(\beta)$ in the same way as would $\exp(i\beta\rho_3)$. It is then straightforward to get

$$\rho(\beta)^{\otimes N} = \bigoplus_s w_N(s) \rho_s(\beta) \otimes \frac{\mathbb{1}}{\dim \mathscr{K}_{N,s}} \tag{2-308}$$

with

$$w_N(s) = \frac{\sinh((2s+1)\beta)}{\sinh(\beta)(2\cosh(\beta))^N} \dim \mathscr{K}_{N,s} \tag{2-309a}$$

and

$$\rho_s(\beta) = \frac{\sinh(\beta)}{\sinh((2s+1)\beta)} \exp(2\beta L_3^{(s)}) \tag{2-309b}$$

where $L_3^{(s)}$ is the three-component of angular momentum in the spin-s representation and the dimension of $K_{N,s}$ is given in Eq. (2-287). By Eq. (2-285) the representation space of π_s coincides with the symmetric tensor product H_+^{2s}. Hence we can interpret $\rho_s(\beta)$ as a state of $2s$ (indistinguishable) particles. In other words the decomposition of $\rho(\beta)^{\otimes N}$ leads in a natural way to a family of operations

$$Q_s : \mathscr{B}(\mathscr{H}_+^{\otimes 2s}) \rightarrow \mathscr{B}(\mathscr{H}^{\otimes N}) \quad \text{with } Q_s^*[\rho(\beta)^{\otimes N}] = \rho_s(\beta) \tag{2-310}$$

Think of the family Q_s, of operations as an instrument Q which measures the number of output systems and transforms $\rho(\beta)^{\otimes N}$ to the appropriate $\rho_s(\beta)$. The crucial point is now that the purity of $\rho_s(\beta)$, measured in terms of fidelities with respect to increases provided $s > 1/2$ holds. Hence, think of Q as a purifier which arises naturally by reduction to irreducible spin components. Unfortunately, Q does not produce a fixed number of output systems. The most obvious way to construct a device which produces always the same number M of outputs is to run the optimal $2s \rightarrow M$ cloner $\hat{T}_{2s\rightarrow M}$ if $2s < M$ or to drop $2s \rightarrow M$ particles if $M \leqslant 2s$ holds. More precisely can define $\hat{Q}: B(H^{\otimes M}) \rightarrow B(H^{\otimes N})$ by

$$\hat{Q}^*[\rho(\beta)^{\otimes N}] = \sum_s w_N(s) \hat{T}_{2s\rightarrow M}^*[\rho_s(\beta)] \tag{2-311}$$

with

$$\hat{T}_{2s\rightarrow M}^*(\rho) = \begin{cases} \dfrac{d[2s]}{d[M]} S_M(\rho \otimes \mathbb{1}) S_M, & \text{for } M > 2s \\ \operatorname{tr}_{2s-M} \rho & \text{for } M \leqslant 2s \end{cases} \tag{2-312}$$

tr_{2s-M} denotes here the partial trace over the $2s - M$ first tensor factors. Applying the general scheme shows that this is the best way to get exactly M purified qubits:

Theorem 2.28 The operation $\hat{Q}$ defined in Eq. (2-311) maximizes $F_{R,1}$ and $F_{R,\text{all}}$. It is called therefore the optimal purifier. The maximal values for $F_{R,1}$ and $F_{R,\text{all}}$ are given by

$$\mathscr{F}_{R,1}(N,M) = \sum_s w_N(s) f_1(M,\beta,s), \quad \mathscr{F}_{R,\text{all}}(N,M) = \sum_s w_N(s) f_{\text{all}}(M,\beta,s) \tag{2-313}$$

with

$$2f_1(M,\beta,s) - 1 = \begin{cases} \dfrac{2s+1}{2s}\coth((2s+1)\beta) - \dfrac{1}{2s}\coth\beta, & 2s > M \\ \dfrac{1}{2s+2}\dfrac{M+2}{M}((2s+1)\coth((2s+1)\beta) - \coth\beta), & 2s \leqslant M \end{cases} \tag{2-314}$$

and

$$f_{\text{all}}(M,\beta,s) = \begin{cases} \dfrac{2s+1}{M+1}\dfrac{1-e^{-2\beta}}{1-e^{-(4s+2)\beta}}, & M \leqslant 2s \\ \dfrac{1-e^{-2\beta}}{1-e^{-(4s+2)\beta}}\binom{2s}{M}^{-1}\sum_K \binom{K}{M} e^{2\beta(K-s)}, & M > 2s \end{cases} \tag{2-315}$$

The expression for the optimal fidelities given here look rather complicated and are not very illuminating. That have plotted there both quantities as a function of θ (Figure 2-34) of N (Figure 2-35) and M (Figure 2-36). While the first two plots looks quite similar the functional behavior in dependence of M seems to be very different. The study of the asymptotic behavior in the next section will give a precise analysis of this observation.

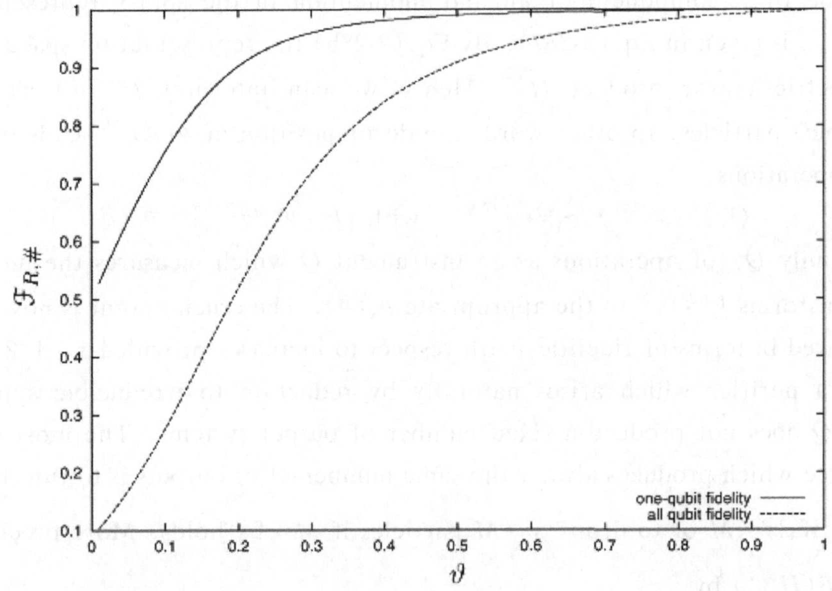

Fig. 2-34 One-and all-qubit fidelities of the optimal purifier for $N = 100$ and $M = 10$. Plotted as a function of the noise parameter θ.

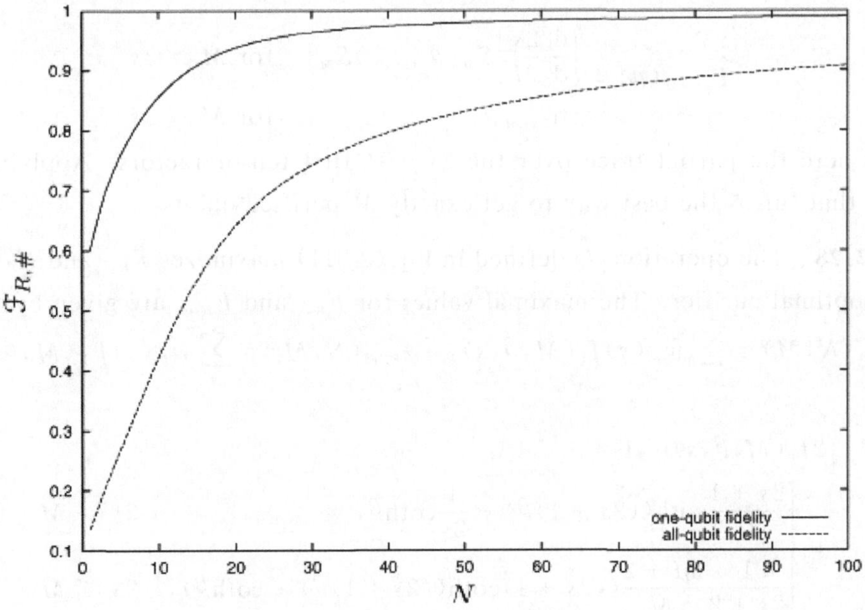

Fig. 2-35 One-and all-qubit fidelities of the optimal purifier for $\theta = 0{:}5$ and $M = 10$. Plotted as a function of N.

- Estimating pure states

Already seen that the cloning problem and state estimation are closely related, because we can construct an approximate cloner T from an estimator E simply by running.

E on the N input states, and preparing M systems according to the attained classical information. In this section we want to go the other way round and show that the optimal cloner derived in Theorem 2.27 leads immediately to an optimal pure state estimator. To this end let us assume that E has the form

$$\mathscr{C}(X) \ni f \mapsto E(f) = \sum_{\sigma \in X} f(\sigma) E_\sigma \in \mathscr{B}(\mathscr{H}^{\otimes N}) \tag{2-316}$$

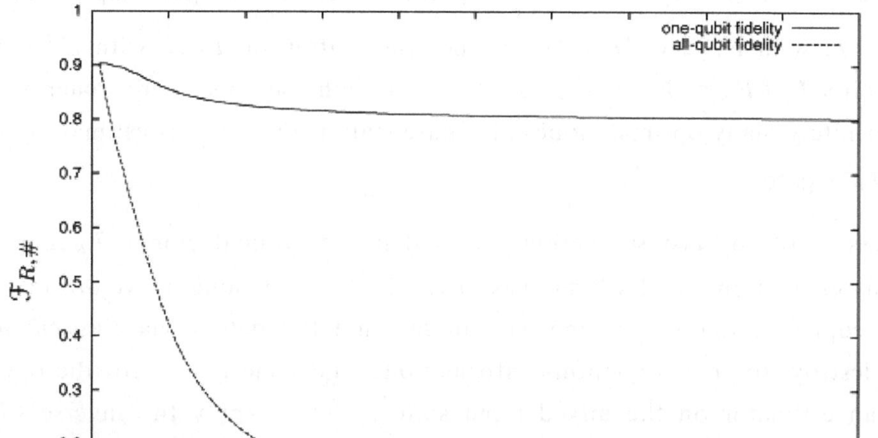

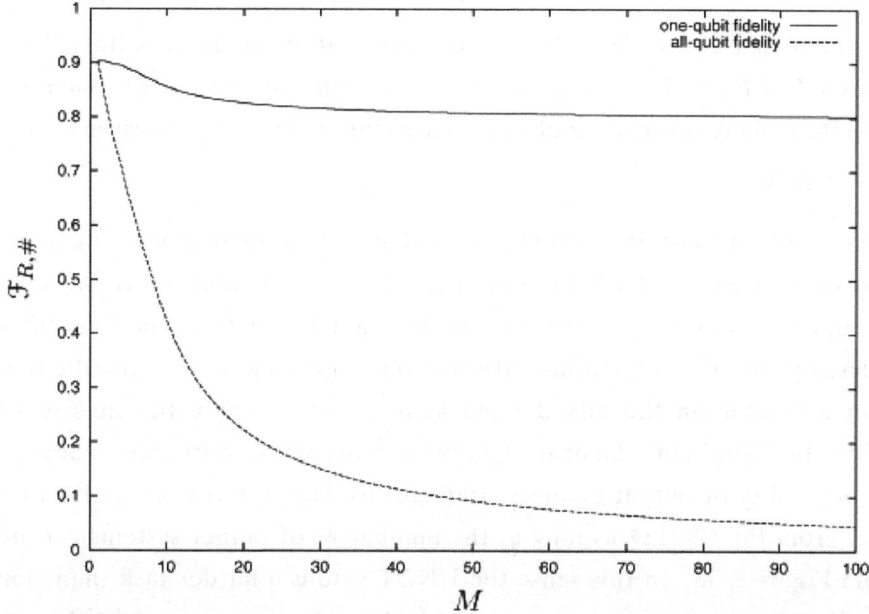

Fig. 2-36 One-and all-qubit fidelities of the optimal purifier for $\theta = 0{:}5$ and $N = 10$. Plotted as a function of M.

where $X \subset B^*(H)$ is a finite set of pure states. The quality of E can be measured in analogy by a fidelity-like quantity:

$$\mathscr{F}_s(E) = \inf_{\psi \in \mathscr{H}} \langle \psi, \rho_\psi \psi \rangle = \inf_{\psi \in \mathscr{H}} \sum_{\sigma \in X} \langle \psi^{\otimes N}, E_\sigma \psi^{\otimes N} \rangle \langle \psi, \sigma \psi \rangle \tag{2-317}$$

where $\rho_\psi = \sum_\sigma \langle \psi \otimes N; E_\sigma \psi \otimes n \rangle \sigma$ is the (density matrix valued) expectation value of E and the infimum is taken over all pure states ψ. Hence $F_s(E)$ measures the worst fidelity of ρ_ψ with respect to the input state ψ. If construct now a cloner T_E from E by

$$T_E^*(|\psi\rangle\langle\psi|^{\otimes N}) = \sum_\sigma \langle \psi^{\otimes N}, E_\sigma \psi^{\otimes n} \rangle \sigma^{\otimes M} \tag{2-318}$$

its one-particle fidelity $F_{c,1}(T_E)$ coincides obviously with $F_s(E)$. Since can produce in this way arbitrary many clones of the same quality we see that $F_s(E)$ is smaller than $F_{c,1}(N, M)$ for all M and therefore

$$\mathscr{F}_s(E) \leqslant \mathscr{F}_{c,1}(N, \infty) = \lim_{M \to \infty} \mathscr{F}_{c,1}(N, M) = \frac{d-1}{d} \frac{N}{N+d} \tag{2-319}$$

where can look at $F_{c,1}(N, \infty)$ as the optimal quality of a cloner which produces arbitrary many outputs from N input systems. To see that this bound can be saturated consider an asymptotically exact family

$$\mathscr{C}(X_M) \ni f \mapsto E^M(f) = \sum_{\sigma \in X} f(\sigma) E_\sigma^M \in \mathscr{B}(\mathscr{H}^{\otimes M}), \quad X_M \subset \mathscr{S}(\mathscr{H}) \tag{2-320}$$

of estimators, i.e. the error probabilities vanish in the limit $N \to \infty$. If the $E_\sigma^M \in B(H^{\otimes M})$ are pure tensor products (i.e. the E^M are realized by a "quorum" of observables as described) they cannot distinguish between the output state $\hat{T}^*(\rho^{\otimes N})$ (which is highly correlated) and the pure product state $\tilde{\rho}^{\otimes M}$ where $\tilde{\rho} \in B^*(H)$ denotes the partial trace over $M-1$ tensor factors (due to permutation invariance it does not matter which factors we trace away here). Hence if apply E^M to the output of the optimal N to M cloner $\hat{T}_{N\to M}$ get an estimate for $\tilde{\rho}$ and in the limit $M\to\infty$ this estimate is exact. The fidelity $\langle \psi, \tilde{\rho}\psi \rangle$ of $\tilde{\rho}$ with respect to the pure input state ψ of $\hat{T}_{N\to M}$ coincides however with $F_{c,1}(N,M)$. Hence the composition of $\hat{T}_{N\to M}$ with E^M converges to an estimator E with $F_e(E) = F_{c,1}(N;\infty)$. We can rephrase this result roughly in the from: "producing infinitely many optimal clones of a pure state is the same as estimating optimally".

- The UNOT gate

The discussion of the last subsection shows that the optimal cloner $\hat{T}_{N\to M}$ produces better clones than any estimation-based scheme (as in Eq. (2-318)), as long as we are interested only in finitely many copies. Loosely speaking we can say that the detour via classical information is wasteful and destroys too much quantum information. The same is true for the optimal purifier: can first run an estimator on the mixed input state $\rho(\beta)^{\otimes N}$, apply the inverse $(R^*)^{-1}$ of the channel map to the attained classical data and reprepare arbitrarily many purified qubits accordingly. The quality of output systems attained this way is, however, worse than those of the optimal purifier from Eq. (2-311) as long as the number M of output systems is finite; this can be seen easily from Figure 2-36. In this sense the UNOT gate is a harder task than cloning and puri. cation, because there is no quantum operation which performs better than the estimation-based strategy. The following theorem can be proved again with the group theoretical scheme. Let $H = C^2$. Among all channels $T: B(H) \to B(H_+^{\otimes N})$ the estimation-based scheme just described attains the biggest possible value for the fidelity $F_{\theta,\#}$; namely:

$$\mathscr{F}_{\theta,1}(N,1) = \mathscr{F}_{\theta,\mathrm{all}}(N,1) = 1 - \frac{1}{N+2} \tag{2-321}$$

The dependence on the number M of outputs is not interesting here, because the optimal device produces arbitrarily many copies of the same quality.

- Asymptotic behaviour

If a device, such as the optimal cloner, is given which produces M output system from N inputs it is interesting to ask for the maximal rate, i.e. the maximal ratio $M(N) = N$ in the limit $N\to\infty$ such that the asymptotic fidelity $\lim_{N\to\infty} F(N, M(N))$ is above a certain threshold (preferably equal to one). Note that this type of question was very important as well for distillation of entanglement and channel capacities, but almost not computable in there. In the current context this type of question is somewhat easier to answer.

- Estimating mixed state

If do not know a priori that the input systems are in a pure state much less is known about estimating and cloning. It is, in particular, almost impossible to say anything about optimality for finitely many input systems (only if N is very small e.g.). Nevertheless, some strong results are available for the behavior in the limit $N\to\infty$ and will give here a short review of some of them. One quantity, interesting to be analyzed for a family of estimators E^N in the limit $N\to\infty$ is the variance of the E^N. To state some results in this context it is convenient to parameterize the state space $S(H)$ or parts of it in terms of n real parameters $x=(x_1,x_2,\cdots,x_n)=\Sigma\subset R^n$ and to write $\rho(x)$ as the corresponding state. If want to cover all states, one particular parameterization is e.g. the generalized Bloch ball. An estimator taking N input systems is now a (discrete) observable $E_x^N\in B(H^{\otimes N})$; $x\in X_N$ with values in a (finite) subset X_N of Σ. The expectation value of E^N in the state $\rho(x)^{\otimes N}$ is therefore the vector $\langle E^N\rangle x$ with components $\langle E^N\rangle_{x,j}$; $j=1,2,\cdots,N$ given by

$$\langle E^N\rangle_{x,j}=\sum_{y\in X_N} y_j \operatorname{tr}(E_y^N\rho(x)^{\otimes N}) \tag{2-322}$$

and the mean quadratic error is described by the matrix

$$V_{jk}^N(x)=\sum_{y\in X_N}(\langle E_N\rangle_{x,j}-y_j)(\langle E_N\rangle_{x,k}-y_k)\operatorname{tr}(E_y^N\rho(x)^{\otimes N}) \tag{2-323}$$

For a good estimation strategy expect that $V_{jk}(x)$ decreases as $1/N$, i.e.

$$V_{jk}^N(x)\simeq\frac{W_{jk}(x)}{N} \tag{2-324}$$

where the scaled mean quadratic error matrix $W_{jk}(x)$ does not depend on N. The task is now to find bounds on this matrix. State here one result taken from. To this end need the quantum information matrix

$$H_{jk}(x)=\operatorname{tr}\left[\rho(x)\frac{\lambda j(x)\lambda_k(x)-\lambda_k(x)\lambda_j(x)}{2}\right] \tag{2-325}$$

which is defined in terms of symmetric logarithmic derivatives λ_j, which in turn are implicitly given

$$\frac{\partial\rho(x)}{\partial x_j}=\frac{\lambda_j(x)\rho(x)+\rho(x)\lambda_j(x)}{2} \tag{2-326}$$

Now have the following theorem:

Theorem 2.29 Consider a family of estimators E^N; $N\in\mathbb{N}$ as described above such that the following conditions hold:

(1) The scaled mean quadratic error matrix $NV_{jk}^N(x)$ converges uniformly in x to $W_{jk}(x)$ as $N\to\infty$.

(2) $W_{jk}(x)$ is continuous at a point $x_0=x$.

(3) $H_{jk}(x)$ and its derivatives are bounded in a neighborhood of x_0.

Then can have:

$$\operatorname{tr}[H^{-1}(x_0)W^{-1}(x_0)]\leqslant(d-1) \tag{2-327}$$

For qubits this bound can be attained by a particular estimation strategy which measures on each qubit separately. A second quantity interesting to study in the limit $N\to\infty$ is the error probability defined in Eq. (2-117). For a good estimation strategy it should go to zero of course, an additional question, however, concerns the rate with which this happens. Review here a result from which concerns the subproblem of estimating the spectrum. Hence are looking now at a family of observables $E^N:C(X_N)\to B(H^{\otimes N})$; $N\in\mathbb{N}$ taking their values in a finite subset X_N of the set:

$$\Sigma=\left\{(x_1,\cdots,x_d)\in\mathbb{R}^d\mid x_1\geqslant\cdots\geqslant x_d\geqslant 0,\sum_j x_j=1\right\} \tag{2-328}$$

of ordered spectra of density operators on $H=C^d$. The aim is to determine the behavior of the error probabilities

$$K_N(\Delta)=\sum_{x\in\Delta\cap X_N}\operatorname{tr}(E_x^N\rho^{\otimes N}) \tag{2-329}$$

in the limit $N\to\infty$. Following the general arguments can restrict our attention here to covariant observables, i. e. can assume without loss of cloning quality that the $E^N x$ commute with all permutation unitaries $V_p; p\in S_N$ and all local unitaries $U\otimes N; U\in U(d)$. If restrict attention in addition to projection-valued measures, which is suggestive for ruling out unnecessary fuzziness, that each E_x^N must coincide with a (sum of) projections P_Y from $H\otimes N$ onto the $U(d)$, respectively V_p, invariant subspace $H_Y\otimes K_Y$, which is de. ned in Eq. (2-294), where $Y=(Y_1, Y_2, \cdots, Y_d)$ refers here to Young frames with d rows and N boxes. The only remaining freedom for the E^N is the assignment $x(Y)\in\Sigma$ of Young frames (and therefore projections E^N) to points in Σ. Since the Young frames themselves have up to normalization the same structure as the elements of Σ, one possibility for $s(Y)$ is just $s(Y)=Y/N$. Written as quantum to classical channel this is

$$\mathscr{C}(X_N)\ni f\mapsto\sum_Y f(Y/N)P_Y\in\mathscr{B}(\mathscr{H}^{\otimes N}) \tag{2-330}$$

where $X_N\subset\Sigma$ is the set of normalized Young frames, i.e. all Y/N if Y has d rows and N boxes. It turns out, somewhat surprisingly that this choice leads indeed to an asymptotically exact estimation strategy with exponentially decaying error probability (2-329). The following theorem can be proven with methods from the theory of large deviations:

Theorem 2.30 The family of estimators $E^N; N\in\mathbb{N}$ given in Eq. (2-330) is asymptotically exact; i.e. the error probabilities $K_N(\Delta)$ vanish in the limit $N\to\infty$ if Δ is a complement of a ball around the spectrum $r\in\Sigma$ of ρ. If Δ is a set (possibly containing r) whose interior is dense in its closure have the asymptotic estimate for $K_N(\Delta)$:

$$\lim_{N\to\infty}\frac{1}{N}\ln K_N(\Delta)=\inf_{s\in\Delta} I(s) \tag{2-331}$$

where the "rate function" $I: U\to R$ is just the relative entropy between the two probability vectors s and r

$$I(s)=\sum_j s_j(\ln s_j-\ln r_j) \tag{2-332}$$

To make this statement more transparent, note that can rephrase Eq. (2-331) as

$$K_N(\Delta)\approx\exp\left(-N\inf_{s\in\Delta} I(s)\right) \tag{2-333}$$

Since the rate function I vanishes only for $s=r$ that the probability measures K_N converge (weakly) to a point measure concentrated at $r\in\Sigma$. The rate of this convergence is exponential and measured exactly by the function I.

- Purification and cloning

Let come back now to the discussion of purification started (consequently have $H=C^2$ again). The aim is now to calculate the fidelities $F_{R,\#}(N, M(N))$ in the limit $N\to\infty$ for a sequence $M(N); N\in\mathbb{N}$ such that $M(N)/N$ converges to a value $c\in R$. The crucial step to do this is the application of Theorem 2.30. The density matrices $\rho_s(\beta)$ from Eq. (2-308) can be defined alternatively by

$$\begin{cases}\rho_s(\beta)\otimes\dfrac{\mathbb{1}}{\dim\mathscr{K}_{N,s}}=w_N(s)^{-1}P_s\rho(\beta)^{\otimes N}P_s\\ w_N(s)=\operatorname{tr}(\rho(\beta)^{\otimes N}P_s)\end{cases} \tag{2-334}$$

where P_s is the projection from $H^{\otimes N}$ to $H_s\otimes K_{N,s}$. In other words P_s is equal to P_Y from Eq. (2-330) if apply the reparametrization

$$(Y_1, Y_2) \mapsto (s, N) = ((Y_1 - Y_2)/2, Y_1 + Y_2) \tag{2-335}$$

In a similar way can rewrite the set of ordered spectra by $\Sigma \ni (x_1, x_2) \mapsto x_1 - x_2 \in [0,1]$ and $K_N(\Delta)$ becomes a measure on $[0,1]$ (i.e. $\Delta \subset [0,1]$):

$$K_N(\Delta) = \sum_{2s/N \in \Delta} \operatorname{tr}(\rho(\beta)^{\otimes N} P_s) = \sum_{2s/N \in \Delta} w_N(s) \tag{2-336}$$

and the sum

$$\mathscr{F}_{R,\#}(N, M(N)) = \sum_s w_N(s) f_\#(M(N), \beta, s) \tag{2-337}$$

can be rephrased as the integral of a function $[0,1] \ni x \mapsto \tilde{f}_\#(N, \beta, x) \in R$ with respect to this measure, provided $\tilde{f}_\#$ is related to $f_\#$ by $\tilde{f}_\#(N, \beta, 2s/N) = f_\#(M(N), \beta, s)$. According to Theorem 2.30 the K_N converge to a point measure concentrated at the ordered spectrum of $\rho(\beta)$; but the latter corresponds, according to the reparametrization above, to the noise parameter $\theta = \tanh\beta$.

Hence, if the sequence of functions $\tilde{f}_\#(N, \beta \cdot)$ converges for $N \to \infty$ uniformly (or at least uniformly on a neighborhood of #) to $\tilde{f}_\#(\beta)$ get

$$\lim_{N \to \infty} \mathscr{F}(N, M(N)) = \lim_{N \to \infty} \sum_s \tilde{f}_\#(N, \beta, s) = \tilde{f}_\#(\beta, \vartheta) \tag{2-338}$$

for the limit of the fidelities. A precise formulation of this idea leads to the following theorem.

Theorem 2.31 The two purification fidelities $F_{R,\#}$ have the following limits:

$$\lim_{N \to \infty} \lim_{M \to \infty} \mathscr{F}_{R,1}(N, M) = 1 \tag{2-339}$$

and

$$\Phi(\mu) = \lim_{\substack{N \to \infty \\ M/N \to \mu}} \mathscr{F}_{R,\text{all}}(N, M) = \begin{cases} \dfrac{2\vartheta^2}{2\vartheta^2 + \mu(1 - \vartheta)}, & \mu \leqslant \vartheta \\ \dfrac{2\vartheta^2}{\mu(1 + \vartheta)}, & \mu \geqslant \vartheta \end{cases} \tag{2-340}$$

If only interested in the quality of each qubit separately we can produce arbitrarily good purified qubits at any rate. If on the other hand the correlations between the output systems should vanish in the limit the rate is always zero. This can be seen from the function Φ, which is the asymptotic all-qubit fidelity which can be reached by a given rate μ. It is plotted it in Figure 2-37. Note finally that the results just stated contain the rates of optimal cloning machines as a special case; only have to set $\theta = 1$.

This course will present the quantum analog of Shannon's information theory. This area has seen an explosion of interest and a correspondingly rapid technical advance over the past ten years, largely in response to the development of quantum-mechanically based cryptographic protocols and Shor's famous algorithm for factoring integers. The unavoidable presence of noise in any quantum-mechanical information processing device means that error-correction techniques will play a crucial role in any practical application of quantum cryptography or computing. This course will focus on asymptotic protocols for compression, communication, error correction and state distillation, identifying the absolute limits placed on those tasks by quantum mechanics.

Familiarity with quantum mechanics is recommended. The course content is very mathematical, but elementary. Students should be comfortable with basic probability theory, linear algebra and real analysis. The material will be covered through a combination of lectures and student presentations.

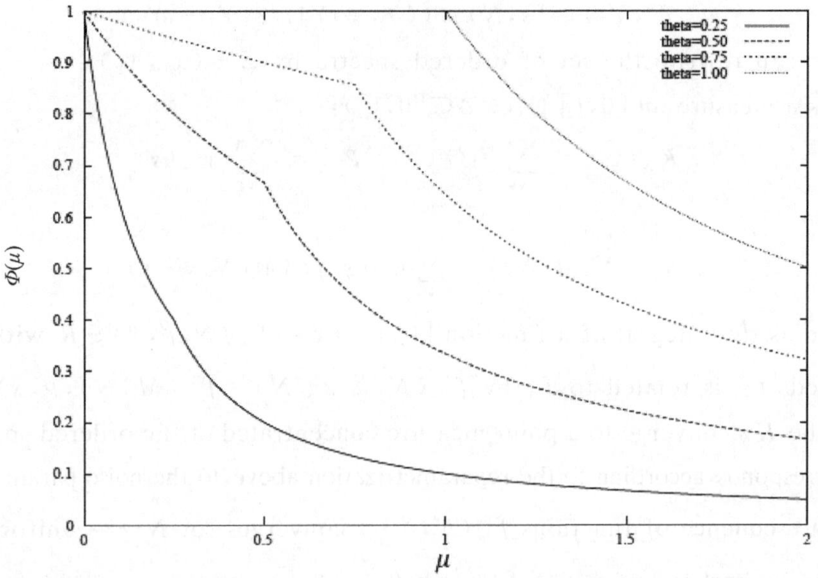

Fig. 2-37 Asymptotic all-qubit fidelity $\Phi(\mu)$ plotted as function of the rate μ.

2.8 Quantum probability

According to these "realistic" intuitions all things have their definite place and sharply determined qualities, such as speed, color, and weight. Quantum theory, however, refuses to precisely pinpoint them. With respect to this apparent shortcoming of the theory different points of view can be taken. It could be suspected that quantum theory is incomplete, that it gives a coarse description of a reality that is actually more refined. This is the viewpoint once taken by Einstein, and it still has adherents today. It calls for a search for finer mathematical models of physical reality, often referred to as "hidden variable models". One such attempt is Bohm's theory of non-relativistic quantum mechanics. However, the work of John Bell in the 60's and of Alain Aspect in the 80's strongly favors the opposite point of view: their work has made clear that such models with an underlying realistic structure are necessarily affected with a certain weakness: they must at least allow action at a distance. This regard as a bad property for a theory which aims to describe a physical world where no signals have been observed to travel faster than light. And apart from that, the hidden variable theories which have been found so far are highly artificial and do not predict any new phenomena. It is for these reasons that we decide to accept quantum theory with its inherent strangeness. Note, in passing, that that there is no paradox in quantum mechanics: the theory does not contradict itself. It is only at variance with our intuition, which we think must be adapted.

2.8.1 Review of quantum probability

- Quantum probability

So quantum mechanics does not predict the results of physical experiments with certainty, but calculates probabilities for their possible outcomes. Now, the mathematical theory of probability obtained a unified formulation, when Kolmogorov introduced his axioms, defining the universal structure $(\Omega;\Sigma,P)$ of a probability space. For a long time this theory of probability (dealing with probability distributions, stochastic processes, Markov chains, martingales, etc.) remained completely separate from the mathematical development of quantum mechanics (involving vectors in a Hilbert space, hermitian operators, unitary transformations, and such like). It consists of ordinary Hilbert space quantum theory, with the emphasis moved towards operators on Hilbert space, and the algebras which they generate. The main objective of this course is to sketch the outlines of this framework, and show its usefulness for information theory.

• Quantum Information

In Shannon's (classical) information theory, a single unit, the bit, serves to quantify all forms of information, be it in print, in computer memory, CDROM or strings of DNA. The physical states of quantum systems, however, cannot be copied into such "classical" information, but can be converted into each other. This leads to a new unit of information: the qubit. Quantum Information theory studies the handling of this new form of information by information-carrying channels. The basic properties of these channels, and some impossibilities as well as new possibilities connected with quantum information.

• Quantum Computing

It was Richard Feynman who first thought of actually employing the strangeness of quantum mechanics to do things that would be impossible in a classical world. The idea was developed in the 1980's and 1990's by David Deutsch, Peter Shor, and many others into a ourishing branch of science called "quantum computing": how to make quantum mechanical systems perform calculations more efficiently than ordinary computers can do. This research is still in a predominantly theoretical stage: the quantum computers actually built are as yet extremely primitive and can by no means compete with even the simplest pocket calculator, but expectations are high.

2.8.2 Why classical probability does not suffice

• An Experiment with Polarizers

To start with, we consider a simple experiment. In a beam of light of a fixed color we put a pair of polarizing filters, each of which can be rotated around the axis formed by the beam. As is well known, the light falling through both filters changes in intensity when the filters rotated relative to each other. Starting from the orientation where the resulting intensity is maximal, and rotating one of the filters through an angle α, the light intensity decreases with α, vanishing for $\alpha = \pi/2$. If call the intensity of the beam before the filters I_0, after the first I_1, and after the second I_2, then $I_1 = 1/2 I_0$, (assume the original beam to be unpolarized), and

$$I_2 = I_1 \cos^2 \alpha \tag{2-341}$$

So far the phenomenon is described well by classical physics. During the last century, however, it has been observed that for very low intensities (monochromatic) light comes in small packages, which were called photons, whose energy depends on the color, but not on the total intensity.

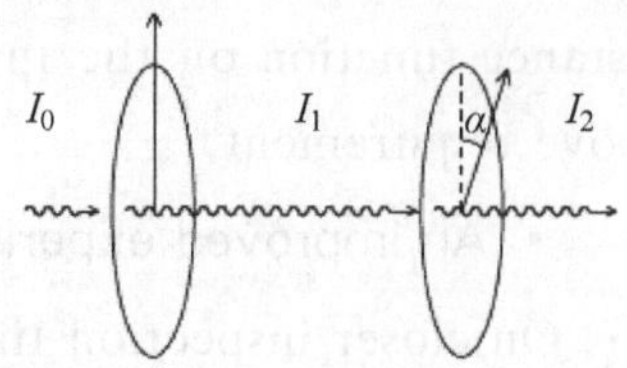

Fig. 2-38 Two polarizers in conjunction.

Thinking along the lines of classical probability, may associate to a polarization filter in the direction α a random variable P_α, taking the value $P_\alpha(\omega) = 0$ if the photon ω is absorbed by the filter and $P_\alpha(\omega) = 1$ if it passes through. For two filters in the directions α and

β these random variables then should be correlated as follows:

$$\mathbb{E}(P_\alpha P_\beta) = \mathbb{P}[P_\alpha = 1 \text{ and } P_\beta = 1] = \frac{1}{2}\cos^2(\alpha - \beta) \tag{2-342}$$

Here hit on a difficulty: the function on the right hand side is not a possible correlation function! This can be seen as follows. Take three polarizing filters, having polarization directions α_1, α_2 and α_3 respectively. Put them on the optical bench in pairs. They should give rise to random variables P_1, P_2 and P_3 satisfying

$$\mathbb{E}(P_i P_j) = \frac{1}{2}\cos^2(\alpha_i - \alpha_j) \tag{2-343}$$

Proposition 2.32 (Bell's 3 variable inequality) For any three 0-1-valued random variables P_1, P_2, and P_3 on a probability space (ΩP) the following inequality holds:

$$\mathbb{P}[P_1 = 1, P_3 = 0] \leqslant \mathbb{P}[P_1 = 1, P_2 = 0] + \mathbb{P}[P_2 = 1, P_3 = 0] \tag{2-344}$$

Proof.

$$\begin{aligned}\mathbb{P}[P_1 = 1, P_3 = 0] &= \mathbb{P}[P_1 = 1, P_2 = 0, P_3 = 0] + \mathbb{P}[P_1 = 1, P_2 = 1, P_3 = 0] \\ &\leqslant \mathbb{P}[P_1 = 1, P_2 = 0] + \mathbb{P}[P_2 = 1, P_3 = 0]\end{aligned} \tag{2-345}$$

In the example have

$$\begin{aligned}\mathbb{P}[P_i = 1, P_j = 0] &= \mathbb{P}[P_i = 1] - \mathbb{P}[P_i = 1, P_j = 1] \\ &= \frac{1}{2} - \frac{1}{2}\cos^2(\alpha_i - \alpha_j) = \frac{1}{2}\sin^2(\alpha_i - \alpha_j)\end{aligned} \tag{2-346}$$

Bell's inequality thus reads

$$\frac{1}{2}\sin^2(\alpha_1 - \alpha_3) \leqslant \frac{1}{2}\sin^2(\alpha_1 - \alpha_2) + \frac{1}{2}\sin^2(\alpha_2 - \alpha_3) \tag{2-347}$$

which is clearly violated for the choices $\alpha_1 = 0$; $\alpha_2 = \pi/6$ and $\alpha_3 = \pi/3$.

This example suggests that classical probability cannot even describe this simple experiment!

Remark.

The above calculation could be summarized as follows: we are in fact looking for a family of 0-1-valued random variables

$$(P_\alpha)_{0\leqslant\alpha<\pi} \quad \text{with } \mathbb{P}[P_\alpha = 1] = \frac{1}{2} \tag{2-348}$$

satisfying the requirement that

$$\mathbb{P}[P_\alpha \neq P_\beta] = \sin^2(\alpha - \beta) \tag{2-349}$$

Now, on the space of 0-1-valued random variables on a probability space the function $(X;Y) \mapsto \mathbb{P}[X \neq Y]$ equals the L^1-distance of X and Y:

$$\mathbb{P}[X \neq Y] = \int_\Omega |X(\omega) - Y(\omega)| \mathbb{P}(d\omega) = \| X - Y \|_1 \tag{2-350}$$

On the other hand, the function $(\alpha;\beta) \mapsto \sin^2(\alpha\beta)$ does not satisfy the triangle inequality for a distance function on the interval $[0,\pi)$. Therefore no family $(P_\alpha)_{0\leqslant\alpha<\pi}$ exists which meets the above requirement.

- An improved experiment

On closer inspection the above example is not very convincing. Indeed, when two polarizers are arranged on the optical bench, why should not the random variable for the second polarizer depend on the first angle. The correlation would then read

$$\mathbb{E}(P_\alpha P_{\alpha,\beta}) = \mathbb{P}[P_\alpha = 1 \text{ and } P_{\alpha,\beta} = 1] = \frac{1}{2}\cos^2(\alpha - \beta) \tag{2-351}$$

which can easily be satisfied, so that the whole reasoning collapses. So should do a better experiment. Must let the filters act on the photons without influence on each other. Maybe can separate them spatially.

Here a clever technique from quantum optics comes to the aid. It is possible to build a device that produces pairs of photons, such that the members of each pair move in opposite directions and show opposite behavior towards parallel polarization filters: if one passes the filter, then the other is surely absorbed. The device contains Calcium atoms, which are excited by a laser to a state they can only leave under emission of such a pair. With these photon pairs, the very same experiment can be performed, but this time the polarizers are far apart, each one acting on its own photon.

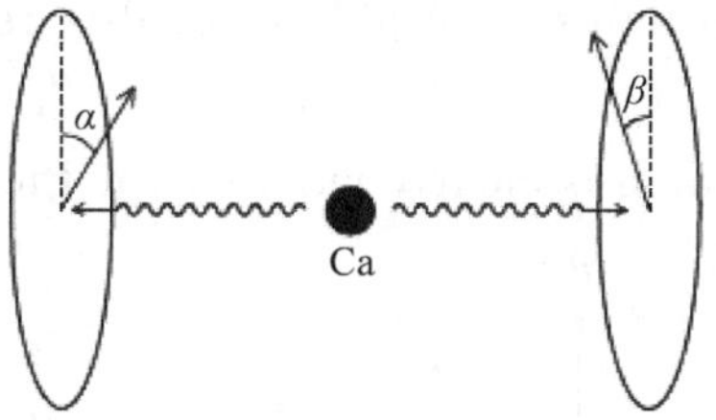

Fig. 2-39 Photon pair production.

The same correlations are measured, say first between $P_{\alpha 1}$ on the left and $P_{\alpha 2}$ on the right, then between $P_{\alpha 1}$ on the left and $P_{\alpha 3}$ on the right, and finally between $P_{\alpha 2}$ on the left and $P_{\alpha 3}$ on the right. The same outcomes are found, violating Bell's three variable inequality, thus strengthening the case against classical probability.

- The decisive experiment

Advocates of classical probability could still find serious fault with the argument given so far. Indeed, do really have to assume that are measuring the same random variable $P_{\alpha 2}$ on the right as later on the left. Is it really true that the polarizations in these pairs are exactly opposite? There could exist a probabilistic explanation of the phenomena without this assumption. So the argument has to be tightened still further. This brings us to the experiment which was actually performed by *A*. Aspect in Orsay (near Paris). In this experiment a random choice out of two different polarization measurements was performed on each side of the pair-producing device, say in the direction α_1 or α_2 on the left and in the direction β_1 or β_2 on the right, giving rise to four random variables $P_1 := P(\alpha_1)$, $P_2 := P(\alpha_2)$ and $Q_1 := Q(\beta_1)$, $Q_2 := Q(\beta_2)$, two of which are measured and compared at each trial.

Proposition 2.33 (Bell's 4 variable inequality) For any quadruple P_1, P_2, Q_1, and Q_2 of 0-1-valued random variables on (ΩP) the following inequality holds:

$$\mathbb{P}[P_1 = Q_1] \leqslant \mathbb{P}[P_1 = Q_2] + \mathbb{P}[Q_2 = P_2] + \mathbb{P}[P_2 = Q_1] \tag{2-352}$$

(In fact, by symmetry, neither of these four probabilities is larger than the sum of the other three.)

Proof. It is easy to see that for all ω:

$$P_1(\omega) = Q_1(\omega) \Rightarrow P_1(\omega) = Q_2(\omega) \text{ or } Q_2(\omega) = P_2(\omega) \text{ or } P_2(\omega) = Q_1(\omega) \tag{2-353}$$

Bell's 4-variable inequality can be viewed as a "quadrangle inequality" with respect to the metric

$$(X, Y) \mapsto \| X - Y \|_1$$

On the other hand, quantum mechanics predicts, and the experiment of Aspect showed, that one has

$$\mathbb{P}[P(\alpha) = Q(\beta) = 1] = \frac{1}{2}\sin^2(\alpha - \beta) \tag{2-354}$$

Similarly

$$\mathbb{P}[P(\alpha) = Q(\beta) = 0] = \frac{1}{2}\sin^2(\alpha - \beta) \tag{2-355}$$

Hence

$$\mathbb{P}[P(\alpha) = Q(\beta)] = \sin^2(\alpha - \beta) \tag{2-356}$$

So Bell's 4 variable inequality reads in this example:

$$\sin^2(\alpha_1 - \beta_1) \leqslant \sin^2(\alpha_1 - \beta_2) + \sin^2(\alpha_2 - \beta_1) + \sin^2(\alpha_2 - \beta_2) \tag{2-357}$$

which is clearly violated for the choices $\alpha_1 = 0, \alpha_2 = \pi/3, \beta_1 = \pi/2$, and $\beta_2 = \pi/6$.

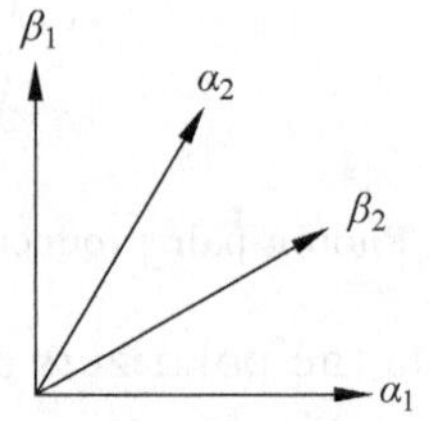

Fig. 2-40 Directions violating Bell's inequality.

Remarks:

(1) A crucial assumption that goes into Bell's inequality, is that it makes sense to compare the (possibly random) reactions which a given photon would show to different filters, including those it does not actually meet. This assumption is called realism; it is made in all classical probabilistic physical theories, but is abandoned in quantum mechanics.

(2) A second important assumption, necessary for the validity of Bell's inequality, was mentioned before: the outcome on the right (described by $Q(\beta)$ for some β) should not depend on the angle α of the polarizer on the left. This assumption is called "locality". In order to justify this assumption, Aspect has made considerable efforts. In his (third) experiment, the choice of what to measure on the left (α_1 or α_2) and on the right (β_1 or β_2) was made during the right of the photons, so that any influence which each of these choices might have on the outcome on the opposite end would have to travel faster than light. By the causality principle of Relativity Theory such influences are excluded.

(3) The Orsay experiment refutes all imaginable physical theories which are both local and realistic. Quantum mechanics is local, but not realistic. Its great successes lead to believe that realism is false in nature. Some prefer to adhere to realism, and so they must give up locality, and hence Einstein causality.

(4) In the opinion, the phrase "quantum non-locality", which is often heard in the context of Bell's inequalities, signals a misconception. Quantum mechanics is local. But it describes phenomena which in a classical theory could only be explained using some action at a distance.

- The Orsay experiment as a card game

To illustrate the above refutation of local realism more vividly, we shall present the experiment in the form of a card game. Nature can win this game.

Two players, P and Q, are sitting at a table. They are cooperating to achieve a single goal. There is an arbiter present to deal cards and to count points. On the table there is a board consisting of four squares as drawn in Figure 2-41. There are dice and an ordinary deck of playing cards. The deck of cards is shuffled well. (In fact we shall assume that the deck of cards is an infinite sequence of independent cards, chosen fully at random.) First the players are given some time to make agreements on the strategy they are going to follow. Then the game starts, and from this moment on they are no longer allowed to communicate. The following sequence of actions is then repeated many times.

(1) The dealer hands a card to P and one to Q. Both look at their own card, but not at the other one's. (The only feature of the card that matters is its colour: red or black.)

(2) The dice are thrown.

(3) P and Q simultaneously say "yes" or "no", according to their own choice.

They are free to make their answer depend on any information they possess, such as the color of their own card, the agreements made in advance, the numbers shown by the dice, the weather, the time, et cetera.

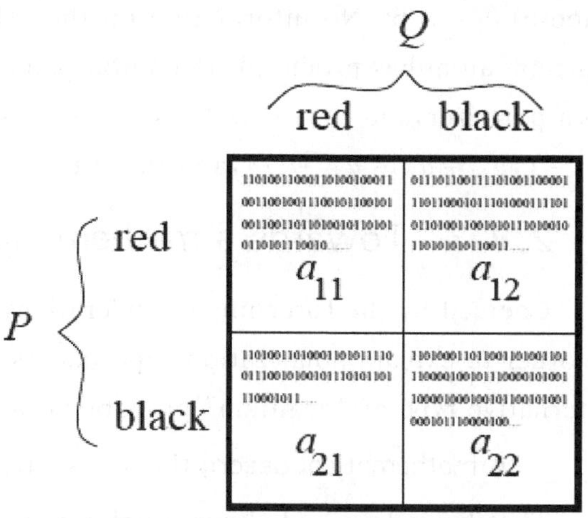

Fig. 2-41 Board for the Bell game.

(4) The cards are laid out on the table. The pair of colors of the cards determines one of the four squares on the board: these are labeled (red, red), (red, black), (black, red) and (black, black).

(5) In the square so determined a 0 or a 1 is written: a 0 when the answers of P and Q have been different, a 1 if they have been the same. In the course of time, the squares on the board get filled with 0's and 1's. The arbiter keeps track of the percentage of 1's in proportion to the total number of bits in each square; call the time limits of these percentages as the game proceeds: a_{11}, a_{12}, a_{21}, and a_{22}. The aim of the game, for both P and Q, is to get a_{11} larger than the sum of the other three limiting percentages.

So P and Q must try to give identical answers as often as they can when both their cards are red, but different answers otherwise.

Proposition 2.34 (Bell's inequality for the game) P and Q cannot win the game by classical means, namely:

$$a_{11} \leqslant a_{12} + a_{21} + a_{22} \tag{2-358}$$

Proof.

The best P and Q can do, in order to win the game, is to agree upon some (possibly random) strategy for each turn. For instance, they may agree that P will always say "yes" (i.e., $P_{\text{red}} = P_{\text{black}} =$ "yes") and that Q will answer the question "Is my card red" (i.e., $Q_{\text{red}} =$ "yes" and $Q_{\text{black}} =$ "no"). This will lead to a 1 in the (red, red) square or the (black, red) square or to a 0 in one of the other two. So if the players repeat this strategy indefinitely, on the long run they would get $a_{11} = a_{12} = 1$ and $a_{21} = a_{22} = 0$, disappointingly satisfying Bell's inequality. The above example is an extremal strategy. There are many (in fact, sixteen) strategies like this. By the pointwise of Bell's 4-variable inequality, none of these sixteen extremal strategies wins the game. Inclusion of the randomness coming from the dice yields a full polytope of random strategies, having the above sixteen as its extremal points. But since the inequalities are linear, this averaging procedure does not help.

Strangely enough, however, Nature does provide us with a strategy to win the game, solely based on the $\cos^2$ law (2) for photon absorption. Instead of the dice, put a Calcium atom on the table. When the cards have been dealt, P and Q put their polarizers in the direction indicated by their cards. If P has a red card, then he chooses the direction $\alpha_1 = 0$. If his card is black, then he chooses $\alpha_2 = \pi/3$. If Q has a red card, then he chooses $\beta_1 = \pi/2$. If his card is black, then he chooses $\beta_2 = \pi/6$. No information on the colours of the cards needs to be exchanged. When the Calcium atom has produced its photon pair, each player looks whether his own photon passes his own polarizer, and then says "yes" if it does, "no" if it does not. On the long run they will get $a_{11} = 1$, $a_{12} = a_{21} = a_{22} = 1/4$, and thus they win the game.

2.8.3 Towards a mathematical model

Coerced by the foregoing considerations, give up trying to make a classical probabilistic model in order to explain polarization experiments. Instead, take these experiments as a paradigm for an alternative type of "quantum" probability, to be developed now.

- A mathematical description of polarization

That have discussed (linear) polarization of a light beam. This is completely characterized by a direction in the plane perpendicular to the light beam. So simply describe states of polarization by different directions in a twodimensional real plane R^2, or equivalently by unit vectors $\psi \in R^2$, $\| \psi \| = 1$, pointing in this direction. Actually, since we cannot distinguish between two states which differ by a rotation of π, that shall describe states of polarization by one-dimensional subspaces of R^2. Given two directions of polarization with an angle α between them, spanned by two unit vectors $\psi; \theta \in R^2$, the probability to find polarization θ when a photon is in the state ψ, can be expressed as

$$\cos^2 \alpha = \langle \psi, \theta \rangle^2 \tag{2-359}$$

where $\langle \psi, \theta \rangle$ denotes the scalar product between ψ and θ.

In the mathematical model we should distinguish between the physical state of polarization of a photon on the one hand and the filter on the other hand, i.e., the 0-1-valued random variable which asks, whether a photon is polarized in a certain direction. This can be done by identifying the random variable with the orthogonal projection P onto the one-dimensional subspace.

Then can write

$$\cos^2 \alpha = \langle \psi, \theta \rangle^2 = \langle \psi, P\psi \rangle \tag{2-360}$$

Since P is 0-1-valued, (a photon passes or is absorbed), this probability is equal to the expectation of this random variable:

$$\langle \psi, P\psi \rangle = \mathbb{E}(P) \tag{2-361}$$

- The full truth about polarization: the qubit

In the foregoing description of polarization things were presented somewhat simpler than they are: we considered only linear polarization, thus disregarding circular polarization. The full description of polarization leads to the quantum mechanics of a 2-level system or qubit:

(1) State of polarization of a photon $\hat{=}$ one-dimensional subspace of C^2, described by a unit vector ψ spanning this subspace (and determined only up to a phase).

(2) Polarization filter or generalized 0-1-valued random variable $\hat{=}$ orthogonal projection P onto a complex one-dimensional subspace. (Also for left- or right-circular polarization there exist physical filters.)

(3) Probability for a photon, described $\hat{=} \langle \psi; P\psi \rangle$ by ψ, to pass through a filter, described by P.

The set of all states is conveniently parametrized by the unit vectors of the form

$$(\cos\alpha, e^{i\phi}\sin\alpha) \in \mathbb{C}^2, \quad \frac{-\pi}{2} \leqslant \alpha \leqslant \frac{\pi}{2}, \quad 0 \leqslant \phi \leqslant \pi \tag{2-362}$$

- Finite dimensional models

The mathematical model that used by quantum mechanics is the straightforward generalization of the above description. In order to keep things simple, in this course we restrict ourselves to the quantum mechanics in finite dimension. This generalizes the probability theory of systems with only finitely many states. As in classical probability, the generalization to systems with a countably infinite number of states or a continuum of states is analytically more involved. The model is as follows: States correspond to one-dimensional subspaces of C^n, where the dimension n determined by the model. Again, a state described conveniently by some unit vector spanning this subspace.

0-1-valued random variables or events described by orthogonal projections onto linear subspaces of C^n. Here also projections onto higher dimensional subspaces make sense. The probability that a measurement of a random variable P on a system in a state ψ gives the value 1 is given by $\langle \psi; P\psi \rangle$. Note that do not assume that every unit vector $\psi \in C^n$ describes a state of the system, nor that every orthogonal projection corresponds to a meaningful random variable. Specializing these two sets is part of the description of the mathematical model for a given system. In a truly quantum mechanical situation, typically all possible vectors and projections are used. In contrast to this, a model from classical probability incorporated into this description as follows.

- Finite classical models

A finite probability space is usually described by a finite set $\Omega = \{\omega_1, \omega_2, \cdots, \omega_n\}$ and a probability distribution $(p_1, p_2, \cdots, p_n)$, $0 \leqslant p_i \leqslant 1$, $\sum_i P_i = 1$, such that the probability for ω_i is p_i. A 0-1-valued random variable is a 0-1-valued function on Ω, i.e., a characteristic function X_A of some subset $A \subseteq \Omega$. In order to describe such a system in our model, think of C^n as the space of complex valued functions on Ω, and use the functions δ_i with $\delta_i(\omega_j) = \delta_{i,j}$ as basis. The states of the system, i.e., the points ω_i of Ω, are now represented by the unit vectors δ_i, $1 \leqslant i \leqslant n$. The random variable X_A is identified with the orthogonal projection P_A onto the linear span of the vectors $\{\delta_i : \omega_i \in A\}$. In our basis X_A becomes a diagonal matrix with a 1 at the ith place of the diagonal $\omega_i \in A$, and a 0 otherwise. It is obvious that $\omega_i \in A$ if and only if $X_A(\omega_i) = 1$ if and only if $\langle \delta_i; P_A \delta_i \rangle = 1$.

Conversely, any set of pairwise commuting projections on C^n can be diagonalized simultaneously and thus have an interpretation as a set of classical 0-1-valued random variables. Therefore: Classical probability corresponds to sets of pairwise commuting projections.

- Mixed states

In the above sketch of quantum probability an important point is still missing: How can we describe a situation where a photon has one polarization with some probability q and in another with probability $1-q$? Indeed, since states must play the role of probability distributions, this combination should expressed as a single state of the photon.

In general, if P is any 0-1-valued (quantum) random variable and $\psi_1, \psi_2, \cdots, \psi_k$ are arbitrary P quantum states, each occuring with a probability p_i, $1 \leqslant i \leqslant k$ $\sum_i p_i = 1$, then the probability that a measurement of P gives 1 is clearly given by

$$\sum_i p_i \langle \psi_i, P\psi_i \rangle \tag{2-363}$$

A more convenient description of mixed states is obtained as follows. For a unit vector $\psi \in C^n$ denote by Φ_ψ the orthogonal projection onto the one-dimensional subspace generated by Φ_ψ. In the physics literature, Φ_ψ is frequently denoted by $|\psi\rangle\langle\psi|$. By tr denote the trace on the $n \times n$-matrices, summing up the diagonal entries of such a matrix. Then one obtains

$$\langle \psi, P\psi \rangle = \mathrm{tr}(\Phi_\psi \cdot P) \tag{2-364}$$

Hence

$$\sum_i p_i \langle \psi_i, P\psi_i \rangle = \mathrm{tr}\left(\sum_i p_i \Phi_{\psi_i} \cdot P\right) = \mathrm{tr}(\Phi \cdot P) \tag{2-365}$$

Where

$$\Phi := \sum_i p_i \Phi_{\psi_i} \tag{2-366}$$

Being a convex combination of 1 -dimensional projections, Φ obviously is a positive (i. e., self-adjoint positive semidefinite) $n \times n$-matrix with $\mathrm{tr}(\Phi) = 1$. Conversely, from diagonalizing positive matrices it is clear that any such positive matrix Φ with $\mathrm{tr}(\Phi) = 1$ can be written as a convex combination of 1-dimensional projections. The set of these matrices forms a closed (even compact) convex set, and its extreme points are precisely the 1 -dimensional projections, which in turn correspond to pure states, represented also by unit vectors. Therefore it is precisely this class of matrices which represents mixed states. These matrices are frequently called density matrices.

Thus, a general mixed state is described by a density matrix Φ and the probability for an observation of P to yield the value 1 is given by $\mathrm{tr}(\Phi \cdot P)$.

- Remarks

(1) The decomposition of a density matrix Φ into a convex combination of 1-dimensional projections is by no means unique. So the compact convex set of density matrices is not a simplex at all. Indeed, on C^2 it can be a finely identified with a full ball in R^3, by taking in R^3 the convex hull of the sphere that was described above.

(2) In classical probability the convex set of mixed states is the simplex of all probability distributions. In our picture, if insist on decomposing a mixed state given by $\Phi = \sum_i P_i P_{\delta i}$ into a convex combination of pure states (within the convex hull of $\{P_{\delta i} : 1 \leqslant i \leqslant n\}$ which is a simplex), then it becomes unique.

(3) Physically, a state Φ is completely described by all of its values $\mathrm{tr}(\Phi \cdot P)$, where P runs through the random variables of the model. Thus, if consider only subsets of projections, then two different density matrices can represent the same physical state of the system. As a drastic example, consider the classical system $\Omega = \{\omega_1, \omega_2, \cdots, \omega_n\}$ with equidistribution, i. e., $p_i(\omega_i) = 1/n$, leading to the density matrix $\Phi = \sum_i P_{\delta i}/n = 1/n \cdot \mathbb{1}$. On the other hand, with the unit vector

$$\psi = \left(\frac{1}{\sqrt{n}}, \cdots, \frac{1}{\sqrt{n}}\right) \in \mathbb{C}^n \tag{2-367}$$

can obtain for any subset

$$A \subseteq \Omega : \mathrm{tr}(\Phi \cdot P_A) = \frac{1}{n} \cdot |A| = \langle \psi, P_A \psi \rangle \tag{2-368}$$

Therefore, on the random variables

$$\{P_A : A \subseteq \Omega\} \tag{2-369}$$

the rank-one-density matrix P_ψ represents the same state as the density matrix $\mathbb{1}/n$. Note, however, that P_ψ is not in the convex hull of $\{P_{\delta i} : 1 \leqslant i \leqslant n\}$.

- The mathematical model of Aspect's experiment

As an illustration, we shall now explain the photon correlation in the Orsay experiment, given by the $\cos^2$-law. Note that here cannot simply refer to the basic $\cos^2$-law of quantum probability, since the filters are acting on two different photons. The polarization of a pair of photons is described by a unit vector in the tensor product $C^2 \otimes C^2 = C^4$, where use the basis

$$\begin{cases} (1,0,0,0) = e_1 \otimes e_1 =: e_{11} \\ (0,1,0,0) = e_1 \otimes e_2 =: e_{12} \\ (0,0,1,0) = e_2 \otimes e_1 =: e_{21} \\ (0,0,0,1) = e_2 \otimes e_2 =: e_{22} \end{cases} \tag{2-370}$$

with $e_1 = (1,0) \in C^2$ and $e_2 = (0,1) \in C^2$. For example, in the pure state e_{12} the left-hand photon is vertically polarized and the right-hand photon horizontally. As it turns out, the state of the pair of photons as produced by the Calcium atom is described by the state

$$\psi = \frac{1}{\sqrt{2}}(e_{12} - e_{21}) \tag{2-371}$$

Now, the filters $P(\alpha)$ on the left and $Q(\beta)$ on the right, introduced, are represented by two-dimensional projection operators on C^4, which are the "2-right amplification" and the "2-left-amplification" of the polarization matrix

$$\begin{pmatrix} \cos^2\alpha & \cos\alpha\sin\alpha \\ \cos\alpha\sin\alpha & \sin^2\alpha \end{pmatrix} \tag{2-372}$$

namely

$$P(\alpha) = \begin{pmatrix} \cos^2\alpha & \cos\alpha\sin\alpha \\ \cos\alpha\sin\alpha & \sin^2\alpha \end{pmatrix} \otimes \begin{pmatrix} 1 & 0 \\ 0 & 1 \end{pmatrix}$$

$$= \begin{bmatrix} \cos^2\alpha & 0 & \cos\alpha\sin\alpha & 0 \\ 0 & \cos^2\alpha & 0 & \cos\alpha\sin\alpha \\ \cos\alpha\sin\alpha & 0 & \sin^2\alpha & 0 \\ 0 & \cos\alpha\sin\alpha & 0 & \sin^2\alpha \end{bmatrix} \tag{2-373}$$

$$Q(\beta) = \begin{pmatrix} 1 & 0 \\ 0 & 1 \end{pmatrix} \otimes \begin{pmatrix} \cos^2\beta & \cos\beta\sin\beta \\ \cos\beta\sin\beta & \sin^2\beta \end{pmatrix}$$

$$= \begin{bmatrix} \cos^2\beta & \cos\beta\sin\beta & 0 & 0 \\ \cos\beta\sin\beta & \sin^2\beta & 0 & 0 \\ 0 & 0 & \cos^2\beta & \cos\beta\sin\beta \\ 0 & 0 & \cos\beta\sin\beta & \sin^2\beta \end{bmatrix} \tag{2-374}$$

Note that $P(\alpha)$ and $Q(\beta)$ are commuting projections for fixed α and β. It follows that $P(\alpha)Q(\beta)$ is again a projection, as well as the products $P(\alpha)(\mathbb{1} - Q(\beta))$, $(\mathbb{1} - P(\alpha))Q(\beta)$, and $(\mathbb{1} - P(\alpha))(\mathbb{1} - Q(\beta))$. So obtain the description of a classical probability space with four states, to be interpreted as "left photon passes, right photon passes"), ("left photon passes", "right photon is absorbed");

("left photon is absorbed","right photon passes"),("left photon is absorbed","right photon is absorbed"). The probabilities of these four events are found by the actions on $\psi = 1/\sqrt{2}(e_{12} - e_{21}) = 1/2(0,1-1,0)$ of the four projections. In particular, the probability that both photons pass is given by:

$$\langle \psi, P(\alpha)Q(\beta)\psi \rangle = \frac{1}{2}(0,1,-1,0) \times$$

$$\begin{pmatrix} \cos^2\alpha\cos^2\beta & \cos^2\alpha\cos\beta\sin\beta & \cos\alpha\sin\alpha\cos^2\beta & \cos\alpha\sin\alpha\cos\beta\sin\beta \\ \cos^2\alpha\cos\beta\sin\beta & \cos^2\alpha\sin^2\beta & \cos\alpha\sin\alpha\cos\beta\sin\beta & \cos\alpha\sin\alpha\sin^2\beta \\ \cos\alpha\sin\alpha\cos^2\beta & \cos\alpha\sin\alpha\cos\beta\sin\beta & \sin^2\alpha\cos^2\beta & \sin^2\alpha\cos\beta\sin\beta \\ \cos\alpha\sin\alpha\cos\beta\sin\beta & \cos\alpha\sin\alpha\sin^2\beta & \sin^2\alpha\cos\beta\sin\beta & \sin^2\alpha\sin^2\beta \end{pmatrix}$$

$$\begin{pmatrix} 0 \\ 1 \\ -1 \\ 0 \end{pmatrix} = \frac{1}{2}(\cos^2\alpha\sin^2\beta + \sin^2\alpha\cos^2\beta - 2\cos\alpha\sin\alpha\cos\beta\sin\beta)$$

$$= \frac{1}{2}(\cos\alpha\sin\beta - \sin\alpha\cos\beta)^2$$

$$= \frac{1}{2}\sin^2(\alpha - \beta) \tag{2-375}$$

2.8.4 Quantum probability

In classical probability a model-or probability space-is determined by giving a set Ω of outcomes ω, by specifying what subsets $S \subset \Omega$ are to be considered as events, and by associating a probability $P(S)$ to each of these events. Requirements: the events must form a σ-algebra, the probability measure IP must be σ-additive, and normalized, i.e. $P(\Omega) = 1$.

In quantum probability must loosen this scheme somewhat. Must give up the set Ω of sample points: a point $\omega \in \Omega$ in a classical model decides about the occurrence or non-occurrence of all events simultaneously, and this abandon. Following our polarization example we take as events certain closed subspaces of a Hilbert space, or, equivalently, a set of projections. To all these projections we associate probabilities.

Requirements:

(1) The set of ε of all events of a quantum model must be the set of projections in some $*$-algebra A of operators on H.

(2) The probability function $P: \varepsilon \to [0,1]$ must be σ-additive. According to a theorem of Gleason, for dim $(H) \geqslant 3$ this implies that the probabilities are given by a state φ on A:

$$\mathbb{P}(E) = \varphi(E), \quad (E \in \mathcal{A} \text{ a projection}) \tag{2-376}$$

In this section we shall work out the above notions in some detail.

- $*$-algebras of operators and states

A Hilbert space is a complex linear space H with a sesquilinear function

$$\mathcal{H} \times \mathcal{H} \to \mathbb{C}: \quad (\psi, \chi) \mapsto \langle \psi, \chi \rangle \tag{2-377}$$

the inner product. For the defining properties of the inner product and the main facts about Hilbert spaces refer to the contribution of Defienes Petz to this volume. Let H be a finite-dimensional Hilbert space. By an operator on H mean a linear map $A: H \to H$. Operators can be added and multiplied in the natural way. By the adjoint of an operator A mean the unique operator A^* on H satisfying

$$\forall_{\psi,\vartheta\in\mathcal{H}}:\langle A^*\psi,\vartheta\rangle=\langle\psi,A\vartheta\rangle \tag{2-378}$$

The norm of an operator A is defined by

$$\|A\|:=\sup\{\|A\psi\| \mid \psi\in\mathcal{H},\|\psi\|=1\} \tag{2-379}$$

It has the property

$$\|A^*A\|=\|A\|^2 \tag{2-380}$$

Exercise: Prove this!

By a (unital) $*$-algebra of operators on H we mean a subspace A of the space of all linear maps $A: H\to H$ such that $\mathbb{1}\in A$ and

$$A,B\in\mathcal{A}\Rightarrow\lambda A,A+B,A\cdot B,A^*\in\mathcal{A} \tag{2-381}$$

By a state on A we mean a linear functional $\varphi:A\to C$ satisfying

(1) $\forall_{A\in A}: \varphi(A^*A)\geqslant 0$,

(2) $\varphi(\mathbb{1})=1$. (2-382)

Call a pair $(A,')$ of the above kind a quantum probability space.

Examples

(1) Let $P_1,P_2,\cdots,P_k$ be mutually orthogonal projections on H with sum $\mathbb{1}$. Then their linear span

$$\mathcal{A}:=\left\{\sum_{j=1}^{k}\lambda_jP_j \mid \lambda_1,\cdots,\lambda_k\in\mathbb{C}\right\} \tag{2-383}$$

forms a unital $*$-algebra of operators on H. If ψ is some vector in H of unit length, it determines a state φ by

$$\varphi(A):=\langle\psi,A\psi\rangle \tag{2-384}$$

The probabilities of this classical model are

$$p_j:=\varphi(P_j)=\|P_j\psi\|^2 \tag{2-385}$$

Note that there are many ψ's, and even more density matrices Φ determining the same state φ on A.

(2) Let A be the $*$-algebra M_n of all complex $n\times n$ matrices. Let $\varphi(A):=\mathrm{tr}(\Phi A)$ with $\Phi\geqslant 0$ and $\mathrm{tr}(\Phi)=1$, as introduced. The state φ' is called a pure state if $\Phi=|\psi\rangle\langle\psi|$ for some unit vector $\psi\in H$. The qubit corresponds to the case $n=2$. The most general way of representing M_n on a (finite dimensional) Hilbert space is

$$\begin{cases}\mathcal{H}=\mathbb{C}^m\otimes\mathbb{C}^n\ (m\geqslant 1)\\ \mathcal{A}=\{\mathbb{1}\otimes A \mid A\in M_n\}\end{cases} \tag{2-386}$$

(3) Let $k,n_1,n_2,\cdots,n_k,m_1,m_2,\cdots,m_k$ be natural numbers, and let the Hilbert space H be given by

$$\mathcal{H}:=(\mathbb{C}^{m_1}\otimes\mathbb{C}^{n_1})\oplus(\mathbb{C}^{m_2}\otimes\mathbb{C}^{n_2})\oplus\cdots\oplus(\mathbb{C}^{m_k}\otimes\mathbb{C}^{n_k}) \tag{2-387}$$

Let A be the $*$-algebra given by

$$\mathcal{A}:=\{(\mathbb{1}\otimes A_1)\oplus\cdots\oplus(\mathbb{1}\otimes A_k) \mid A_j\in M_{n_j}\ \text{for}\ j=1,2,\cdots,k\} \tag{2-388}$$

Let $\psi=\psi_1\oplus\psi_2\oplus\cdots\oplus\psi_k$ be a unit vector in H and

$$\varphi(A):=\langle\psi,A\psi\rangle=\sum_{j=1}^{k}\langle\psi_j,A_j\psi_j\rangle \tag{2-389}$$

If $m_j\geqslant n_j\ \forall_j$ then every state on A is of the above form. Otherwise, density matrices may be needed. In finite dimension Example (1) is the only commutative possibility, Example (2) is the "purely quantum mechanical" situation, and Example (3) is the most general case.

Theorem 2.35 Every commutative $*$-algebra of operators on a finite-dimensional Hilbert space is isomorphic to $C(\Omega)$ for some finite Ω.

Proof. Since the operators in A all commute, there exists an orthonormal basis $e_1, e_2, \cdots, e_n$ in H on which they are all represented by diagonal matrices. Then the states $\omega_j : A \mapsto \langle e_j ; Ae_j \rangle$ are multiplicative:

$$\omega_j(AB) = \langle e_j, ABe_j \rangle = \sum_{i=1}^{n} \langle e_j, Ae_i \rangle \langle e_i, Be_j \rangle = \langle e_j, Ae_j \rangle \langle e_j, Be_j \rangle = \omega_j(A)\omega_j(B) \quad (2\text{-}390)$$

These These states need not all be different; let $\Omega := (\omega_{j1}, \omega_{j2}, \cdots, \omega_{jk})$ be a maximal set of different ones. Then the map

$$\iota : \mathcal{A} \rightarrow \mathcal{C}(\Omega) : \iota(A)(\omega) := \omega(A) \quad (2\text{-}391)$$

is an isomorphism. The projections of Example (1) are found back as the operators $P\omega := l^{-1}(\delta\omega)$.

Exercise: Check that the map l defined above is indeed an isomorphism of $*$-algebras.

Definition 2.36 By the commutant of a set S of operators on H mean the $*$-algebra

$$\mathcal{S}' := \{B : \mathcal{H} \rightarrow \mathcal{H} \text{ linear} \mid \forall_{A \in S} : AB = BA\} \quad (2\text{-}392)$$

The algebra generated by $\mathbb{1}$ and S denote by alg (S). The center of a $*$-algebra A is the (commutative) $*$-algebra Z given by

$$\mathcal{Z} := \mathcal{A} \cap \mathcal{A}' \quad (2\text{-}393)$$

Exercise: Find the center of A in each of the examples (1), (2) and (3) above.

Theorem 2.37 (Double commutant theorem.) Let S be a set of operators on a finite dimensional Hilbert space H, such that $X \in S \Rightarrow X^* \in S$. Then

$$\text{alg}(\mathcal{S}) = \mathcal{S}'' \quad (2\text{-}394)$$

Proof. Clearly $S \subset S''$, and since S'' is a $*$-algebra, have alg $(S) \subset S''$ Now prove the converse inclusion. Let $B \in S00$, and let $A := \text{alg}(S)$. must show that $B \in A$.

Step 1: Choose $\psi \in H$, and let P be the orthogonal projection onto $A\psi$. Then for all $X \in S$ and $A \in \mathcal{A}$:

$$XPA\psi = XA\psi \in \mathcal{A}\psi \Rightarrow XPA\psi = PXA\psi \quad (2\text{-}395)$$

So XP and PX coincide on the space $A\psi$. But if $\theta \perp A\psi$, then $P\theta = 0$ and for all $A \in \mathcal{A}$:

$$\langle X\vartheta, A\psi \rangle = \langle \vartheta, X^* A\psi \rangle = 0 \quad (2\text{-}396)$$

so $X\theta \perp A\psi$ as well. Hence $PX\theta = 0 = XP\theta$, and the operators XP and PX also coincide on the orthogonal complement of $A\psi$. Conclude that $XP = PX$, i.e. $P \in S'$. But then we also have $BP = PB$, since $B \in S''$, So

$$B\psi = BP\psi = PB\psi \in \mathcal{A}\psi \quad (2\text{-}397)$$

and $B\psi$ is of the form $A\psi$ for some $A \in \mathcal{A}$.

Step 2: But this is not sufficient: that $B\psi = A\psi$ for all ψ in a basis for H. So choose a basis 1, 2, $\cdots$, n of H. can define:

$$\begin{cases} \widetilde{\mathcal{H}} := \mathcal{H} \oplus \mathcal{H} \oplus \cdots \oplus \mathcal{H} = \mathbb{C}^n \otimes \mathcal{H} \\ \widetilde{\mathcal{A}} := \{A \oplus A \oplus \cdots \oplus A \mid A \in \mathcal{A}\} = \mathcal{A} \otimes \mathbb{1} \\ \widetilde{\psi} := \psi_1 \oplus \psi_2 \oplus \cdots \oplus \psi_n \end{cases} \quad (2\text{-}398)$$

Then $(\widetilde{\mathcal{A}})' = (\mathcal{A} \otimes \mathbb{1})' = \mathcal{A}' \otimes M_n$ and $(\widetilde{\mathcal{A}})'' = (\mathcal{A}' \otimes M_n)' = \mathcal{A}'' \otimes \mathbb{1}$. So $B \otimes \mathbb{1} \in (\mathcal{A})''$. By step 1 find an element $\widetilde{A}$ of $\widetilde{\mathcal{A}}$, such that

$$\widetilde{A}\widetilde{\psi} = (B \otimes \mathbb{1})\widetilde{\psi} \tag{2-399}$$

But $\widetilde{A} \in \widetilde{\mathcal{A}}$ must be of the form $A \otimes \mathbb{1}$ with $A \in \mathcal{A}A$, so

$$A\psi_1 \oplus \cdots \oplus A\psi_n = B\psi_1 \oplus \cdots \oplus B\psi_n \tag{2-400}$$

This implies that $A = B$, hence $B \in \mathcal{A}$.

Exercise: Find the algebra generated by $\mathbb{1}$ and the matrix

$$\begin{pmatrix} 0 & 1 & 0 \\ 1 & 0 & 0 \\ 0 & 0 & 0 \end{pmatrix} \tag{2-401}$$

Give the following proposition without proof. It characterizes the situation of Example 2.

Proposition 2.38 If the center of $\mathcal{A}$ contains only multiples of $\mathbb{1}$, then $\mathcal{H}$ and $\mathcal{A}$ must be of the form

$$\mathcal{H} = \mathbb{C}^m \otimes \mathbb{C}^n, \quad \text{with} \quad \mathcal{A} = \{\mathbb{1} \otimes A \mid A \in M_n\} \tag{2-402}$$

Proposition 2.39 Let $\mathcal{H}$ be a finite-dimensional Hilbert space. Then every $*$-algebra of operators on $\mathcal{H}$ can be written in the form of Example (3) above.

Proof. The center $\mathcal{A} \cap \mathcal{A}'$ is an abelian $*$-algebra, so Theorem 2.35 applies, giving a set of projections $P_j, j = 1, 2, \cdots, k$. Then it is not difficult to show that the unital $*$-algebras $P_j \mathcal{A} P_j$ on the Hilbert subspaces $P_j \mathcal{H}$ satisfy the condition of Proposition 2.38. The statement follows.

- The qubit

The simplest non-commutative $*$-algebra is M_2, the algebra of all 2×2 matrices with complex entries. The events in this probability space are the orthogonal projections in M_2: the complex 2×2 matrices E satisfying

$$E^2 = E = E^* \tag{2-403}$$

What these projections look like. Since E is self-adjoint, it must have two real eigenvalues, and since $E^2 = E$ these must both be 0 or 1. So have three possibilities.

- Both are 0 ; i.e. $E = 0$.
- One of them is 0 and the other is 1.
- Both are 1 ; i.e. $E = 1$.

In the second case, E is a one-dimensional projection satisfying

$$\text{tr}\, E = 0 + 1 = 1 \text{ and } \det E = 0 \cdot 1 = 0 \tag{2-404}$$

As $E^* = E$ and $\text{tr}E = 1$ may write

$$E = E(x, y, z) = \frac{1}{2}\begin{pmatrix} 1 + z & x - iy \\ x + iy & 1 - z \end{pmatrix} \tag{2-405}$$

Then $\det E = 0$ implies that

$$\frac{1}{4}((1 - z^2) - (x^2 + y^2)) = 0 \Longrightarrow x^2 + y^2 + z^2 = 1 \tag{2-406}$$

So the one-dimensional projections in M_2 are parametrised by the unit sphere S_2.

Notation: For $a = (a_1, a_2, a_3) \in R^3$ can write

$$\sigma(a) := \begin{pmatrix} a_3 & a_1 - ia_2 \\ a_1 + ia_2 & -a_3 \end{pmatrix} = a_1\sigma_1 + a_2\sigma_2 + a_3\sigma_3 \tag{2-407}$$

where σ_1, σ_2 and σ_3 are the Pauli matrices

$$\begin{cases} \sigma_1 := \begin{pmatrix} 0 & 1 \\ 1 & 0 \end{pmatrix} \\ \sigma_2 := \begin{pmatrix} 0 & -\mathrm{i} \\ \mathrm{i} & 0 \end{pmatrix} \\ \sigma_3 := \begin{pmatrix} 1 & 0 \\ 0 & -1 \end{pmatrix} \end{cases} \tag{2-408}$$

Note that for all $a, b \in R^3$ have

$$\sigma(a)\sigma(b) = \langle a, b \rangle \cdot \mathbb{1} + \mathrm{i}\sigma(a \times b) \tag{2-409}$$

Now write (2-409) as

$$E(a) := \frac{1}{2}(\mathbb{1} + \sigma(a)), \quad \| a \| = 1 \tag{2-410}$$

In the same way the possible states on M_2 can be calculated. Can find that

$$\varphi(A) = \mathrm{tr}(\rho A) \quad \text{where} \quad \rho = \rho(a) := \frac{1}{2}(\mathbb{1} + \sigma(a)), \quad \| a \| \leqslant 1 \tag{2-411}$$

The probability of the event $E(\alpha)$ in the state $\rho(b)$ is given by

$$\mathrm{tr}(\rho(b)E(a)) = \frac{1}{2}(1 + \langle a, b \rangle) \tag{2-412}$$

The events $E(a)$ and $E(b)$ are compatible if and only if $a = \pm b$. Moreover have for all $a \in S_2$:

$$\begin{cases} E(a) + E(-a) = \mathbb{1} \\ E(a)E(-a) = 0 \end{cases} \tag{2-413}$$

Interpretation: The state of the qubit is given by a vector b in the three dimensional unit ball. For every a on the unit sphere can say with probability one that of the two events $E(a)$ and $E(-a)$ exactly one will occur, $E(a)$ having probability $1/2(1 + ha; b_i)$. So have a classical coin toss with probability for heads equal to $1/2(1 + \langle a; b \rangle)$ for every direction in R^3. The coin tosses in different directions are incompatible. (See Figure 2-42) The quantum coin toss is realised in nature: apart from photon polarization, the spin direction of a particle with total spin 1/2 behaves in this way.

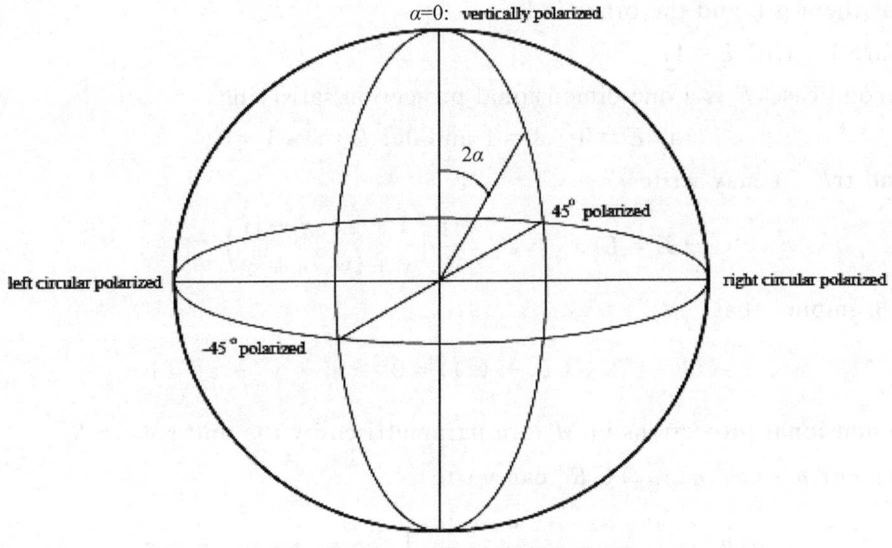

Fig. 2-42 Bloch sphere of the qubit.

• Photons

There is a second natural way to parametrize the one-dimensional projections in M_2, which is closer to the description of polarization of photons, as treated. The projection onto the one-dimensional subspace spanned by the unit vector $(\cos\alpha, e^{i\varphi}\sin\alpha)$ mentioned in Eq. (2-409) of that section is given by

$$F(\alpha,\varphi)=\begin{pmatrix}\cos^2\alpha & e^{-i\varphi}\cos\alpha\sin\alpha\\ e^{i\varphi}\cos\alpha\sin\alpha & \sin^2\alpha\end{pmatrix}\tag{2-414}$$

Equating this projection to $E(x,y,z)$ in (2-405) obtain the relations

$$\begin{cases}x=\sin2\alpha\cos\varphi\\ y=\sin2\alpha\sin\varphi\\ z=\cos2\alpha\end{cases}$$

which define a mapping between the polarization states of a photon and the points of the unit sphere in R^3, called the Bloch sphere in this context. In particular, the projection $F(\alpha,0)$ onto the line in C^2 with real slope $\tan\alpha$ with $\alpha\in[-\pi/2;\pi/2)$ is given by

$$F(\alpha,0)=\begin{pmatrix}\cos^2\alpha & \cos\alpha\sin\alpha\\ \cos\alpha\sin\alpha & \sin^2\alpha\end{pmatrix}=E(\sin2\alpha,0,\cos2\alpha)\tag{2-415}$$

Finally, any atomic or molecular system, only two energy levels of which are of importance in the experiment, can be described by some $(M_2,')$.

Exercise: Let $f: CU\{\infty\}\to S_2$ be given by

$$\begin{cases}f(0):=(0,0,1)\\ f(\infty):=(0,0,-1)\\ f(re^{i\varphi}):=(\sin\vartheta\cos\varphi,\sin\vartheta\sin\varphi,\cos\vartheta)\\ \qquad\text{with}\quad \vartheta=2\arctan r,\quad r\in(0,\infty),\varphi\in[0,\pi)\end{cases}\tag{2-416}$$

Show that $E(f(z))$ is the one-dimensional projection onto the line in C^2 with slope $z\in C$.

2.8.5 Operations on probability spaces

Our main objects of study will be operations on probability spaces. This means that we shall focus attention on the input-output aspect of probabilistic systems.

• Operations on classical probability spaces

It could be maintained that operations are already the core of classical probability. Start with a definition on the level of points.

Definition 2.40 By an operation from a finite classical probability space Ω to a finite classical probability space Ω' mean an $\Omega\times\Omega'$ transition matrix, i. e. a matrix $(t_{\omega\omega'})$ of nonnegative numbers satisfying

$$\forall_{\omega\in\Omega}:\sum_{\omega'\in\Omega'}t_{\omega\omega'}=1\tag{2-417}$$

Examples:

(1) Let τ be a bijection $\Omega\to\Omega'$. may think of rearranging a deck of cards, ($\Omega=\Omega'=\{\text{cards}\}$), or the time evolution of a mechanical system ($\Omega=\Omega'=$ phase space), or the shift on sequences of letters, or just some relabeling of the outcomes of a statistical experiment. The associated matrix is

$$t_{\omega\omega'}:=\begin{cases}1, & \text{if } \omega'=\tau(\omega)\\ 0, & \text{otherwise}\end{cases}\tag{2-418}$$

(2) Let $X: \Omega \to \Omega'$ be surjective. Think of X as an Ω'-valued random variable, where Ω' is usually some subset of R or R^n or so. The associated operation is that of "measuring X" or "forgetting everything about ω except the value of X". The associated matrix is again

$$t_{\omega\omega'} := \begin{cases} 1, & \text{if } \omega' = X(\omega) \\ 0, & \text{otherwise} \end{cases} \tag{2-419}$$

(3) An inverse to the operation of Example (2) is given by

$$t_{\omega'\omega} := \begin{cases} \dfrac{\pi(\{\omega\})}{\pi(X^{-1}(\{\omega'\}))}, & \text{if } \omega' = X(\omega) \\ 0, & \text{otherwise} \end{cases} \tag{2-420}$$

Here π is some probability distribution, which assume to be everywhere nonzero. This operation describes the immersion of a system Ω into the larger system Ω. It can be shown that every transition matrix can be decomposed as a product of matrices of the types 3, 1 and 2. So every operation can be decomposed as an immersion, followed by a rearrangement and a restriction. Such a decomposition is called a dilation of the operation in question.

- Quantum Operations

If $\mathcal{A}$ is a unital $*$-algebra describing a quantum system, then denote by $\mathcal{A}^*$ the dual of $\mathcal{A}$, and by $\mathcal{A}^*_{+,1}$ the positive normalized functionals, i.e. the states on $\mathcal{A}$. By $M_n(\mathcal{A})$ we denote the unital $*$-algebra of all $n \times n$-matrices with entries in $\mathcal{A}$. Now suppose that perform a physical operation which takes as input a state on the system A, and yields as its output a state on the system B. Which maps $f: \mathcal{A}^*_{+,1} \to B^*_{+,1}$ can occur as descriptions of such an operation? Formulate three natural requirements.

(1) f must be an affine map. This means that for all ρ, θ $\mathcal{A}^*_{+,1}$ and all $\lambda \in [0;1]$:

$$\lambda f(\rho) + (1-\lambda) f(\vartheta) = f(\lambda\rho + (1-\lambda)\vartheta) \tag{2-421}$$

This requirement is a consequence of the stochastic equivalence principle which states that a system which is in state ρ with probability λ and in state θ with probability $1-\lambda$ can not be distinguished from a system in the state $\lambda\rho + (1-\lambda)\theta$.

A map f satisfying this condition can be extended to a unique linear map $\mathcal{A}^* \to \mathcal{B}^*$, since every element of $\mathcal{A}^*$ can be written as a linear combination of (at most four) states on $\mathcal{A}$. So f must be the adjoint of some linear map $T: B \to \mathcal{A}$. We shall henceforth write T^* instead of f.

(2) Of course, $f = T^*$ must still map $\mathcal{A}^*_{+,1}$ to $B^*_{+,1}$: for all $\rho \in \mathcal{A}^*$,

$$\begin{cases} \text{tr}(T^*\rho) = \text{tr}(\rho) \\ T^*\rho \geqslant 0, \quad \rho \geqslant 0 \end{cases} \tag{2-422}$$

(3) It would seem at first sight that nothing more can be said a priori about T^*. However, it was realised in the early that the positivity property has to be strengthened in quantum mechanics: if the system under consideration is in a combined state with some other system, then after performing the operation T^* on the former system, the whole combination must still be in some (positive) state. Surprisingly, this is not automatic in the quantum situation, where "entanglement", as treated, can occur between the two systems. Therefore this stronger form of positivity must be added as a requirement: For all $n \in N$: $id_n \otimes T^*$ maps states on $M_n \otimes \mathcal{A}$ to states on $M_n \otimes B$:

Requirement (3) is called complete positivity of the map T^* (or T for that matter).

Summarizing arrive at the following definition, which we shall formulate in the contravariant,

"Heisenberg" picture.

Definition 2.41 A linear map $T: \mathcal{B} \to \mathcal{A}$ is called an operation (from $\mathcal{A}$ to $\mathcal{B}'$) if the following conditions hold:

(1) $T(\mathbb{1}\mathcal{B}) = \mathbb{1}\mathcal{A}$;

(2) T is completely positive, i.e. id $n \otimes T$ is positive $M_n(\mathcal{B}) \to M_n(\mathcal{A})$ for all $n \in N$.

Here $M_n(\mathcal{A})$ stands for the algebra of $n \times n$ matrices with entries in A. This algebra is isomorphic to $M_n \otimes \mathcal{A}$.

Example. A map which is positive, but not completely positive: Let $\mathcal{A} := M_2$ and let

$$T^* : \mathcal{A}^* \to \mathcal{A}^* : \begin{pmatrix} a & b \\ c & d \end{pmatrix} \mapsto \begin{pmatrix} a & c \\ b & d \end{pmatrix} \tag{2-423}$$

be the transposition map. Then T^* is linear, positive, and preserves the trace. However, T^* is not completely positive since

$$\mathrm{id}_2 \otimes T^* : \frac{1}{2}\begin{pmatrix} 1 & 0 & 0 & 1 \\ 0 & 0 & 0 & 0 \\ 0 & 0 & 0 & 0 \\ 1 & 0 & 0 & 1 \end{pmatrix} \mapsto \frac{1}{2}\begin{pmatrix} 1 & 0 & 0 & 0 \\ 0 & 0 & 1 & 0 \\ 0 & 1 & 0 & 0 \\ 0 & 0 & 0 & 1 \end{pmatrix} \tag{2-424}$$

The matrix on the left is a projection whereas the matrix on the left has eigenvalues 1/2; 1/2; 1/2 and $-1/2$, hence is not a valid density matrix. However, if $\mathcal{A}$ or B is abelian, then any positive operator $T: \mathcal{A} \to B$ is automatically completely positive.

2.8.6 Examples of quantum operations

- Let $U \in M_n$ be unitary. Then the automorphism $T: M_n \to M_n : A \mapsto U^* AU$ is an operation. (See Lemma 2.42 below.)
- The $*$-homomorphism $j: M_k \to M_l \otimes M_k : A \mapsto \mathbb{1} \otimes A$ is an operation. (See Lemma 2.42 below.)
- Let φ be a state on M_k. Then the map $E: M_l \otimes M_k \to M_k \otimes B \otimes A \mapsto \varphi(B)A$ is an operation.

The above examples are to be compared with those. To prove their validity in two Lemmas.

Lemma 2.42 If $\mathcal{A} \subset M_k$ and $T: \mathcal{A} \to B \subset M_l$ is a $*$-homomorphism, i.e. if for all $A, B \in \mathcal{A}$ have $T(AB) = T(A)T(B)$ and $T(A^*) = T(A)^*$, then T is completely positive.

Proof. That for all $n \in N$ the map

$$\mathrm{id}_n \otimes T: (A_{ij})_{i,j=1}^n \mapsto (T(A_{ij}))_{i,j=1}^n \tag{2-425}$$

is positive. Indeed, for all $\psi = (\psi_1, \psi_2, \cdots, \psi_n) \in (C^l)^n$, putting $A = X^* X$ with $X \in M_n(\mathcal{A})$:

$$\begin{aligned} \langle \psi, (\mathrm{id}_n \otimes T)(X^* X)\psi \rangle &= \sum_{i,i'=1} \langle \psi_i, T((X^* X)_{ii'})\psi_{i'} \rangle \\ &= \sum_{i,i'=1}^{l} \sum_{j=1}^{n} \langle \psi_i, T(X_{ji}^* X_{ji'})\psi_{i'} \rangle \end{aligned} \tag{2-426}$$

$$\begin{aligned} &= \sum_{i,i'=1}^{l} \sum_{j=1}^{n} \langle \psi_i, T(X_{ji})^* T(X_{ji'})\psi_{i'} \rangle \\ &= \sum_{j=1}^{n} \left\| \sum_{i=1}^{l} T(X_{ji})\psi_i \right\|^2 \geqslant 0 \end{aligned} \tag{2-427}$$

Lemma 2.43 Let $A \subset M_k$, $B \subset M_l$ and let V be a linear map $C^l \to C^k$. Then

$$T: \mathcal{A} \to \mathcal{B}: A \mapsto V^* AV \tag{2-428}$$

is completely positive.

Proof.

If $(A_{ij})_{i,j=1}^n \in M_n(A)$ is positive, then for all $(\psi_1, \psi_2, \cdots, \psi_n) \in (C^l)^n = C^n \otimes C^l$ have

$$\begin{aligned}\langle \psi, (\mathrm{id}_n \otimes T)(A)\psi \rangle &= \sum_{i,j=1}^{n} \langle \psi_i, T(A_{ij})\psi_j \rangle \\ &= \sum_{i,j}^{n} \langle \psi_i, V^* A_{ij} V\psi_j \rangle \\ &= \sum_{i,j}^{n} \langle V\psi_i, A_{ij} V\psi_j \rangle \geqslant 0 \end{aligned} \tag{2-429}$$

Lemma 2.44 covers the third case in Example (3) above since can be decomposed into pure states as $\varphi = \sum_i \langle \psi, \psi \rangle$ and

$$\varphi(B)A = \sum_{i=1}^{l} \lambda_i \langle \psi_i, B\psi_i \rangle A = \sum_{i=1}^{l} \lambda_i V_i^* (B \otimes A) V_i \tag{2-430}$$

where $V_i: C^k \to C^l \otimes C^k: \theta \mapsto \psi_i \otimes \theta$.

- Unraveling quantum operations

Theorem 2.45 Let T be a linear map $M_k \to M_l$. Then T is completely positive if and only if there exist $m \in N$ and operators $V_1, V_2, \cdots, V_m: C^l \to C^k$ such that for all $A \in M_k$:

$$T(A) = \sum_{i=1}^{m} V_i^* A V_i \tag{2-431}$$

To give a proof based on a physical argument. The system is put in an entangled state with a second system, which for convenience describe by the opposite algebra. Then act on the main system with our operation T, and by complete positivity get a new state on the pair. Surprisingly, this state fully characterizes the operation T. By decomposing the state into vector states we shall obtain the unraveling wanted. First introduce some notation. If H is a (finite dimensional) Hilbert space, let H' denote its dual, the space of all linear functionals $H \to C$. The elements of H' are of the form $\bar{\theta}: X \mapsto \langle \theta, X \rangle$; in Dirac notation θ is denoted as $\langle \theta |$. This dual H' is actually isomorphic to H itself, but it is convenient to maintain the distinction, as see below. In particular, if $H = C^n$, then there is a natural action on H' of the algebra M_n^t, the opposite algebra of M_n, which has the multiplication reected: $A^t B^t = (BA)^t$. The operator At acts on $\bar{X}$ as $A^t \bar{X} = X \otimes A$. Consider the tensor product $H_{k1} := C^k \otimes (C^l)')$ of the Hilbert space C^k and the dual of C^l. By identifying the vector $\psi \otimes \bar{\theta} \in H_{kl}$ with the operator $|\psi\rangle, \langle \theta |: X \mapsto \langle \theta, X \rangle \psi$, the Hilbert space H_{kl} can alternatively be viewed as the space of all operators $C^l \to C^k$. On this Hilbert space the algebra $M_k \otimes M_l^t$ acts naturally as follows:

$$A \otimes B^t: \psi \otimes \bar{\vartheta} \mapsto A\psi \otimes B^t \bar{\vartheta} [\approx A \mid \psi \rangle \langle \vartheta \mid B] \tag{2-432}$$

The space H_{ll} has a rotation invariant vector (the so-called fully entangled state on $M_l \otimes M_l^t$), given by

$$\Omega := \frac{1}{\sqrt{l}} \sum_{i=1}^{l} e_i \otimes \bar{e}_i \left[\approx \frac{1}{\sqrt{l}} \sum_{i=1}^{l} \mid e_i \rangle \langle e_i \mid = \mathbb{1}_l / \sqrt{l} \right] \tag{2-433}$$

for any orthonormal basis $e_1, e_2, \cdots, e_l$ of C^l. This vector has the property that

$$\langle \Omega, (A \otimes B')\Omega \rangle = \frac{1}{l}\sum_{i=1}^{l}\sum_{j=1}^{l}\langle e_i \otimes \overline{e_i}, (A \otimes B')e_j \otimes \overline{e_j}\rangle$$
$$= \frac{1}{l}\sum_{i=1}^{l}\sum_{j=1}^{l}\langle e_i, Ae_j\rangle\langle \overline{e_i}, B'\overline{e_j}\rangle$$
$$= \frac{1}{l}\sum_{i=1}^{l}\sum_{j=1}^{l}\langle e_i, Ae_j\rangle\langle e_j, Be_i\rangle = \frac{1}{l}\mathrm{tr}(AB) \tag{2-434}$$

For the "only if" part, assume that $T: M_k \to M_l$ is completely positive. Let $H_{lU} := C^l \otimes (C^l)'$ as above, and let ω denote the state

$$\omega(X) := \langle \Omega, X\Omega\rangle \tag{2-435}$$

on $B(H_u) \approx M_l M_l'$. Since T is completely positive, the functional ω_T on $B(H_{kl}) \approx M_k \otimes M_l'$, given by

$$\omega_T(A \otimes B') := \omega(T(A) \otimes B') \tag{2-436}$$

is also a state. Decompose ω_T into pure states given by vectors $v_1, v_2, \cdots, v_m \in H_{kl}$:

$$\omega_T(X) = \sum_{i=1}^{m}\langle v_i, Xv_i\rangle \tag{2-437}$$

Now, as noted above, $v_i \in H_{kl}$ can be considered as an operator $V_i: C^1 \to C^k$. these operators satisfy the requirement (2-431) of the theorem. Indeed, for all $\psi; \theta \in C^1$:

$$\sum_{i=1}^{m}\langle \psi, V_i^* AV_i\theta\rangle = \sum_{i=1}^{m}\langle V_i\psi, AV_i\theta\rangle$$
$$= \sum_{i=1}^{m}\langle v_i, (A \otimes (|\bar{\psi}\rangle\langle\bar{\theta}|))v_i\rangle \mathcal{H}_{kl}$$
$$= \omega_T(A \otimes (|\bar{\psi}\rangle\langle\bar{\theta}|))$$
$$= \omega(T(A) \otimes (|\bar{\psi}\rangle\langle\bar{\theta}|))$$
$$= \mathrm{tr}(T(A)(|\theta\rangle\langle\psi|))$$
$$= \langle\psi, T(A)\theta\rangle \tag{2-438}$$

The second step is verified by substituting

$$V_i = \sum_j |\alpha_j^i\rangle\langle\overline{\beta_j^i}| \text{ with } \alpha_j^i \in \mathbb{C}^k, \beta_j^i \in \mathbb{C}^l \tag{2-439}$$

and realizing that

$$v_i = \sum_j \alpha_j^i \otimes \overline{\beta_j^i} \tag{2-440}$$

- Uniqueness of unravelings

(This section elaborates on a remark by Mark Fannes during the Summer School. It can skipped in a first reading.) The unraveling (2-432) is not unique. If the matrices $V_1, V_2, \cdots, V_m$ are linearly independent, then they are determined by the completely positive map T up to a transformation of the form

$$V_i' := \sum_{j=1}^{m} u_{ij}V_j \tag{2-441}$$

where u is a unitary $m \times m$-matrix of complex numbers. In this independent case the number m of terms in the unraveling takes its minimal value, which call the rank of the operation T. In general, any number m of terms, also larger than the rank, can occur in the unraveling of T. But in that case the operators V_i are not linearly independent. In fact, the space D of dependencies, given by

$$\mathcal{D}:=\left\{\lambda\in\mathbb{C}^m\;\middle|\;\sum_{i=1}^{m}\bar{\lambda}_i V_i=0\right\}\tag{2-442}$$

has dimension m-rank (T), and the matrix u of (2-442) is a partial isometry with initial space $D^{\perp}$ and final space $(D')^{\perp}$, where D' denotes the space of dependencies of the V_i'. Now prove these statements in the context of the decomposition of states.

Proposition 2.46 Let φ be a state on $A:=M_k$, and let two decompositions states be given:

$$\varphi(A)=\sum_{i=1}^{m}\langle\psi_i,A\psi_i\rangle=\sum_{j=1}^{n}\langle\theta_j,A\theta_j\rangle\tag{2-443}$$

Let $D\subset C^m$ and $D'\subset C^n$ denote the dependency spaces of $\psi=(\psi_1,\psi_2,\cdots,\psi_m)$ and $\theta=(\theta_1,\theta_2,\cdots,\theta_n)$ respectively. Then ψ and θ are connected by a transfor-mation of the form

$$\theta_j=\sum_{i=1}^{m}u_{ji}\psi_i\tag{2-444}$$

where the $n\times m$ matrix u describes a partial isometry $C^m\to C^n$ with initial space $D^{\perp}$ and final space $(D')^{\perp}$. In particular, if the m-tuple $(\psi_1,\psi_2,\cdots,\psi_m)$ and the n-tuple $(\theta_1,\theta_2,\cdots,\theta_n)$ are both sequences of independent vectors, then $n=m$ and u is unitary.

Proof.

Consider ψ and θ as vectors in $H:=(C^k)m=C^m C^k$ and $H':=(C^k)^n=C^n\otimes C^k$ respectively. Then Eq. (2-444) can be written in the form

$$\varphi(A)=\langle\psi,(\mathbb{1}_m\otimes A)\psi\rangle=\langle\theta,(\mathbb{1}_n\otimes A)\theta\rangle\tag{2-445}$$

Let $L\subset H$ and $L'\subset H'$ be the subspaces consisting of the vectors $(\mathbb{1}_m\otimes A)\psi$ and $(\mathbb{1}_n\otimes A)\theta$ respectively, where A runs through the matrix algebra $A\to M_k$. Let $U:L\to L'$ be given by

$$U(\mathbb{1}_m\otimes A)\psi:=(\mathbb{1}_n\otimes A)\theta\tag{2-446}$$

Then U is well-defined, isometric, and onto since

$$\begin{aligned}\|(\mathbb{1}_n\otimes A)\theta\|^2&=\langle(\mathbb{1}_n\otimes A)\theta,(\mathbb{1}_n\otimes A)\theta\rangle=\langle\theta,(\mathbb{1}_n\otimes A^*A)\theta\rangle\\&=\varphi(A^*A)=\|(\mathbb{1}_m\otimes A)\psi\|^2\end{aligned}\tag{2-447}$$

Extend U to a map $H\to H'$ by putting $U_X=0$ for all $X\in H$ which are orthogonal to L. Next, let us show that U is actually of the form $u\otimes\mathbb{1}_k$ for some partial isometry $u:C^m\to C^n$. This is equivalent to the statement that for all $A\in M_k$:

$$U(\mathbb{1}_m\otimes A)=(\mathbb{1}_n\otimes A)U\tag{2-448}$$

which is true since $(\mathbb{1}_m\otimes A)$ leaves $L^{\perp}$ invariant, so that both sides vanish on $L^{\perp}$. And for $X\in L$, i.e. for $X=(\mathbb{1}_m\in X)\psi$ with $X_2\in M_k$, have

$$\begin{aligned}U(\mathbb{1}_m\otimes A)\chi&=U(\mathbb{1}_m\otimes A)(\mathbb{1}_m\otimes X)\psi=U(\mathbb{1}_m\otimes AX)\psi=(\mathbb{1}_n\otimes AX)\theta\\&=(\mathbb{1}_n\otimes A)(\mathbb{1}_n\otimes X)\theta=(\mathbb{1}_n\otimes A)U(\mathbb{1}_m\otimes X)\psi=(\mathbb{1}_n\otimes A)U_X\end{aligned}\tag{2-449}$$

It remains to be shown that

$$\mathcal{L}^{\perp}=\mathcal{D}\otimes\mathbb{C}^k\tag{2-450}$$

(and analogously $(L')^{\perp}=D'\otimes\mathbb{C}^k$). Clearly, for all $\lambda\in\mathbb{C}^m$ and $\mu\in\mathbb{C}^k$,

$$\langle\lambda\otimes\mu,(\mathbb{1}\otimes A)\psi\rangle=\sum_{i=1}^{m}\bar{\lambda}_i\langle\mu,A\psi_i\rangle=\left\langle A^*\mu,\left(\sum_{i=1}^{m}\bar{\lambda}_i\psi_i\right)\right\rangle\tag{2-451}$$

It follows that for $\lambda\in D$ the vector $\lambda\in\mu$ is orthogonal to L, so have $D\otimes\mathbb{C}^k\subset L^{\perp}$. To prove the converse inclusion, first note that the orthogonal projection onto L is $U^*U=u^*u\otimes\mathbb{1}_k$, hence $L=\varepsilon\otimes\mathbb{C}_k$ for some subspace ε of $\mathbb{C}^m$. That $\varepsilon^{\perp}\subset D$. So suppose that $\lambda\perp\varepsilon$, so that $\lambda\otimes\mu\perp L$ for all $\mu \in C^k$ Putting $A = \mathbb{1}$ in (2-452) can find that the left hand side, and hence the right hand side, is 0 for all μ, so

$$\sum_{i=1}^{m} \bar{\lambda}_i \psi_i = 0 \text{ and } \lambda \in \mathcal{D} \tag{2-452}$$

- Properties of quantum operations

When A and B are operators on a Hilbert space, we mean by $A \geqslant B$ that the difference $A - B$ is a positive operator. The following is an extremely useful inequality for operations.

Proposition 2.47 (Cauchy-Schwarz for operations) Let A and B be $*$-algebras of operators on Hilbert spaces H and K, and let $T: \mathcal{A} \to \mathcal{B}$ be an operation. Then we have for all $A \in \mathcal{A}$:

$$T(A^* A) \geqslant T(A)^* T(A) \tag{2-453}$$

Proof.

The operator $X \in M_2 \otimes \mathcal{A}$ given by

$$X := \begin{pmatrix} A^* A & -A^* \\ -A & \mathbb{1} \end{pmatrix} = \begin{pmatrix} A & -\mathbb{1} \\ 0 & 0 \end{pmatrix} \begin{pmatrix} A & -\mathbb{1} \\ 0 & 0 \end{pmatrix} \tag{2-454}$$

is positive. Since T is completely positive and $T(\mathbb{1}) = \mathbb{1}$, it follows that also

$$(\mathrm{id} \otimes T)(X) = \begin{pmatrix} T(A^* A) & -T(A)^* \\ -T(A) & \mathbb{1} \end{pmatrix} \tag{2-455}$$

is a positive operator. Putting $\xi := \psi \oplus T(A)\psi$ can find that

$$\langle \xi, (\mathrm{id} \otimes T) X \xi \rangle = \langle \psi, (T(A^* A) - T(A)^* T(A)) \psi \rangle \tag{2-456}$$

is positive for all $\psi \in H$.

Theorem 2.48 (Multiplication Theorem) If $T: \mathcal{A} \to \mathcal{B}$ is an operation and $T(A^* A) = T(A)^* T(A)$ for some $A \in \mathcal{A}$, then $T(A^* B) = T(A)^* T(B)$ and $T(B^* A) = T(B)^* T(A)$.

Proof.

Take any $B \in \mathcal{A}$ and $\lambda \in R$. Then

$$T((A^* + \lambda B^*)(A + \lambda B)) = T(A)^* T(A) + \lambda T(A^* B + B^* A) + \lambda^2 T(B^* B) \tag{2-457}$$

while by Cauchy-Schwartz

$$T((A^* + \lambda B^*)(A + \lambda B))$$
$$\geqslant T(A)^* T(A) + \lambda (T(A)^* T(B) + T(B)^* T(A)) + \lambda^2 T(B)^* T(B)) \tag{2-458}$$

This inequality holds for all $\lambda \in R$ which implies

$$T(A^* B + B^* A) \geqslant T(A)^* T(B) + T(B)^* T(A) \tag{2-459}$$

Replacing A by iA and B by $-iB$ shows that the opposite inequality also holds, so have equality. Finally replacing only B by iB shows that $T(A^* B) = T(A)^* T(B)$ and $T(B^* A) = T(B)^* T(A)$.

In particular, if a Cauchy-Schwarz equality holds for an operation T then T is a $*$-homomorphism.

Theorem 2.49 (Embedding theorem) Let $(A; \varphi)$ and $(B; \psi)$ be nondegenerate quantum probability spaces, and let $j: \mathcal{A} \to \mathcal{B}$, $E: \mathcal{B} \to \mathcal{A}$ be operations which preserve the states. If

$$E \circ j = \mathrm{id}_{\mathcal{A}} \tag{2-460}$$

then j is an injective $*$-homomorphism and $P := j \circ E$ is a conditional expectation, i.e.,

$$P(C_1 B C_2) = C_1 P(B) C_2 \tag{2-461}$$

for all $C_1, C_2 \in j(\mathcal{A})$ and all $B \in \mathcal{B}$.

Following the language used that call j a random variable and P the conditional expectation with respect to ψ, given j.

Proof. For any $A \in \mathcal{A}$ have by Cauchy-Schwartz

$$A^* A = E \circ j(A^* A) \geqslant E(j(A)^* j(A)) \geqslant E \circ j(A)^* E \circ j(A) = A^* A \tag{2-462}$$

so have equalities here. In particular

$$\psi(j(A^* A) - j(A)^* j(A)) = \varphi \circ E(j(A^* A) - j(A)^* j(A)) = 0 \tag{2-463}$$

and as (B, ψ) is non-degenerate, $j(A^* A) = j(A)^* j(A)$, i. e. j is a $*$-homomorphism. j is injective since it has the left-inverse E.

But also from (2-462) have

$$E(j(A)^* j(A)) = E \circ j(A)^* E \circ j(A) \tag{2-464}$$

The Multiplication Theorem 2.46 then implies that for all $B \in \mathcal{B}$ and $A_1 \in \mathcal{A}$,

$$E(j(A_1)^* B) = E \circ j(A_1)^* E(B) = A_1^* E(B) \tag{2-465}$$

and similarly, with $A_2 \in \mathcal{A}$:

$$E(j(A_1)^* Bj(A_2)) = E(j(A_1)^* B)E \circ j(A_2) = A_1^* E(B) A_2 \tag{2-466}$$

Applying j to both sides find (2-461).

2.8.7 Quantum impossibilities

The result of any physical operation applied on a probabilistic system (quantum or not) is described by a completely positive identity preserving map from the state space of that system to the state space of the resulting system. This imposes strong restrictions on what can be done. Some of these are well-known quantum principles, such as the Heisenberg principle (no measurement without disturbance), some are surprising and relatively recent discoveries (no cloning), but all of them obtain quite neat formulations in the language of quantum probability.

- No cloning

In its original formulation the "No Cloning Theorem" dealt with the reproduction of nonorthogonal vector states. Here give an algebraic version, which distinguishes clearly between the classical and the quantum case. "Cloning", or—more mundanely—copying a stochastic object is an operation which takes as input an object in some state ρ and yields as its output a pair of objects with identical state spaces, such that, if throw away one of them, are left with a single object in the state ρ. (see Figure 2-43, which is actually not complete: the same equality should hold with the other output line blocked.)

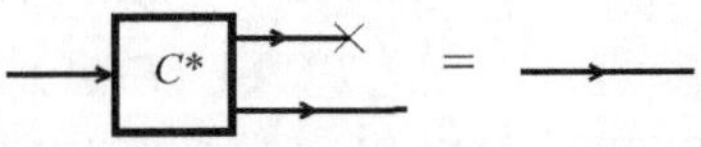

Fig. 2-43 Definition of a copier.

In a formula: for all $\rho \in A^*_{+,1}$:

$$(\mathrm{tr} \otimes \mathrm{id}) \circ C^*(\rho) = (\mathrm{id} \otimes \mathrm{tr}) \circ C^*(\rho) = \rho \tag{2-467}$$

Reformulated in the Heisenberg picture: Call an operation $C: \mathcal{A} \otimes \mathcal{A} \to \mathcal{A}$ a copying operation or copier if for all $A \in \mathcal{A}$:

$$C(\mathbb{1} \otimes A) = C(A \otimes \mathbb{1}) = A \tag{2-468}$$

As is well known, copying presents no problem in classical physics, or classical probability. Here is an example of a classical copying operation. For simplicity, let us think of the operation of copying n bits. Let Ω denote the space $\{0,1\}^n$ of all strings of n bits, and let γ be the "copying" map

$$\Omega \to \Omega \times \Omega : \omega \mapsto (\omega, \omega) \tag{2-469}$$

This map induces an operation

$$C: \mathcal{C}(\Omega) \times \mathcal{C}(\Omega) \to \mathcal{C}(\Omega): Cf(\omega) := f \circ \gamma(\omega) = f(\omega, \omega) \tag{2-470}$$

Clearly, for all $f \in C(\Omega)$:

$$C(\mathbb{1} \otimes f)(\omega) = (\mathbb{1} \otimes f)(\omega, \omega) = f(\omega) \tag{2-471}$$

and the same holds for $C(f \otimes \mathbb{1})$. In the Schrödinger picture the operation looks as follows: for any probability distribution π on Ω,

$$(C^* \pi)(\nu, \omega) = \delta_{\nu\omega} \pi(\omega) \tag{2-472}$$

and can see that:

$$(\mathrm{tr} \otimes \mathrm{id}) \circ C^*(\pi)(\omega) = \sum_{\nu \in \Omega} \delta_{\nu\omega} \pi(\omega) = \pi(\omega) \tag{2-473}$$

The following theorem says that this construction is only possible in the abelian (i.e. commutative) case.

Theorem 2.50 (No cloning) Let $\mathcal{A}$ be a $*$-algebra of operators on a (finite dimensional) Hilbert space. Then A admits a copying operation if and only if $\mathcal{A}$ is abelian.

Proof. If $\mathcal{A}$ is abelian, by Gel'fands Theorem (Theorem 2.35), $\mathcal{A}$ is isomorphic to $C(\Omega)$ for some finite set Ω, and the above construction of a copier applies. Conversely, suppose that $C: \mathcal{A} \otimes \mathcal{A} \to \mathcal{A}$ is a copying operation. Then for all $A \in \mathcal{A}$:

$$C((\mathbb{1} \otimes A)^* (\mathbb{1} \otimes A)) = C(\mathbb{1} \otimes A^* A) = A^* A = C(\mathbb{1} \otimes A)^* C(\mathbb{1} \otimes A) \tag{2-474}$$

Then it follows from the Multiplication Theorem 2.46 that for all $A, B \in \mathcal{A}$:

$$\begin{aligned} AB &= C(A \otimes \mathbb{1}) C(\mathbb{1} \otimes B) = C((A \otimes \mathbb{1})(\mathbb{1} \otimes B)) \\ &= C((\mathbb{1} \otimes B)(A \otimes \mathbb{1})) = C(\mathbb{1} \otimes B) C(A \otimes \mathbb{1}) = BA \end{aligned} \tag{2-475}$$

- No classical coding

Closely related to the above is the rule that "quantum information cannot be classically coded": It is not possible to operate on a quantum system, extracting some information from it, and then from this information reconstruct the quantum system in its original state:

$$\rho \in \mathcal{A}^* \overset{C^*}{\mapsto} \pi \in \mathcal{B}^* \overset{D^*}{\mapsto} \rho \in \mathcal{A}^* \tag{2-476}$$

Formulate this theorem in the contravariant (Heisenberg) picture:

Theorem 2.51 Let $\mathcal{A}$ and $\mathcal{B}$ be $*$-algebras, and let $C: \mathcal{B} \to \mathcal{A}$ and $D: \mathcal{A} \to \mathcal{B}$ be operations, ("Coding" and "Decoding"), such that $C * D = \mathrm{id}_{\mathcal{A}}$. Then if $\mathcal{B}$ is abelian, so is $\mathcal{A}$.

Proof. Have for all $A \in \mathcal{A}$:

$$\begin{cases} A^* A = C \circ D(A^* A) \geqslant C(D(A)^* D(A)) \geqslant A^* A \\ AA^* = C \circ D(AA^*) \geqslant C(D(A) D(A)^*) \geqslant AA^* \end{cases} \tag{2-477}$$

so that again have equality everywhere. If $\mathcal{B}$ is abelian, have $D(A)^* D(A) = D(A) D(A)^*$, so that $A^* A = AA^*$.

Exercise: Prove that, if $A^* A = AA^*$ for all $A \in \mathcal{A}$, then $\mathcal{A}$ is abelian.

- The Heisenberg principle

The Heisenberg principle states—roughly speaking—that no information on a quantum system can be obtained without changing its state. In this form, the statement is not so interesting: if realise that the state of the system expresses the expectations of its observables, given the information have on it, it is no wonder that this state changes once gain information! A more precise formulation is the following: If extract information from a system whose algebra $\mathcal{A}$ is a factor (i.e. $\mathcal{A} \cap \mathcal{A}' = \mathbb{C}\,\mathbb{1}$), and if throw away (disregard) this information, then still it can not be avoided that some initial states are altered. Let work towards a mathematical formulation.

A measurement is an operation performed on a physical system, which results in the extraction of information from that system, while possibly changing its state. So a measurement is an operation

$$M^*: \mathcal{A}^* \to \mathcal{A}^* \otimes \mathcal{B}^* \tag{2-478}$$

where $\mathcal{A}$ describes the physical system, and $\mathcal{B}$ the output part of a measurement apparatus which couple to it. $\mathcal{A}^*$ consists of states and $\mathcal{B}^*$ of probability distributions on the outcomes. So $\mathcal{B}$ will be commutative, but do not need this property here. Now suppose that no initial state is altered by the measurement:

$$(\mathrm{id} \otimes \mathrm{tr}) M^*(\rho) = \rho \quad \forall_{\rho \in \mathcal{A}^*} \tag{2-479}$$

Suppose also that $\mathcal{A}$ is a factor claim that no information can be obtained on ρ:

$$(\mathrm{tr} \otimes \mathrm{id}) M^*(\rho) = \vartheta \tag{2-480}$$

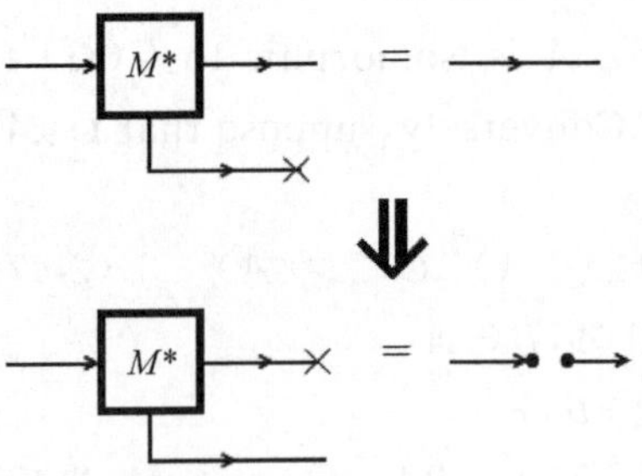

Fig. 2-44 The Heisenberg Principle.

where θ does not depend on ρ. The diagram of Figure 2-44 symbolically expresses this fact.

Again formulate and prove the theorem in the contravariant picture:

Theorem 2. 52 (Heisenberg's Principle) Let M be an operation $\mathcal{A} \otimes \mathcal{B} \to \mathcal{A}$ such that for all $A \in \mathcal{A}$,

$$M(A \otimes \mathbb{1}) = A \tag{2-481}$$

Then:

$$M(\mathbb{1} \otimes B) \in \mathcal{A} \cap \mathcal{A}' \tag{2-482}$$

In particular, if $\mathcal{A}$ is a factor, then for some fixed state θ on B:

$$M(\mathbb{1} \otimes B) = \vartheta(B) \cdot \mathbb{1}_{\mathcal{A}} \tag{2-483}$$

Note that (2-483) implies (2-480), since for all ρ on $\mathcal{A}$ and all $B \in \mathcal{B}$:

$$((\mathrm{tr} \otimes \mathrm{id}) M^* \rho)(B) = \rho(M(\mathbb{1} \otimes B)) = \rho(\vartheta(B) \mathbb{1}_{\mathcal{A}}) = \vartheta(B) \tag{2-484}$$

Proof.

As in the proof of the "no cloning" theorem we have by the multiplication theorem for all $A \in \mathcal{A}, B \in \mathcal{B}$:

$$M(\mathbb{1} \otimes B) \cdot A = M(\mathbb{1} \otimes B) M(A \otimes \mathbb{1}) = M(A \otimes B) \tag{2-485}$$

But also,

$$A \cdot M(\mathbb{1} \otimes B) = M(A \otimes \mathbb{1}) M(\mathbb{1} \otimes B) = M(A \otimes B) \tag{2-486}$$

So $M(\mathbb{1} \otimes B)$ lies in the center of A. If A is a factor, then $B \mapsto M(\mathbb{1} \otimes B)$ is an operation from B to $C \cdot \mathbb{1}_A$, i.e. a state on B times $\mathbb{1}_A$.

- Random variables and von Neumann measurements

Define a random variable to be a $*$-homomorphism from one algebra B to a (larger) algebra A:

$$\mathcal{A} \xleftarrow{j} \mathcal{B} \tag{2-487}$$

In the covariant (Schrödinger) picture this describes the operation j^* of restriction to the subsystem B:

$$\mathcal{A}^* \xrightarrow{j^*} \mathcal{B}^* \tag{2-488}$$

An important case is when $B = C(\Omega)$ for some finite set: then j is to be viewed as an Ω-valued random variable. Let $\Omega = \{x_1, x_2, \cdots, x_n\}$. Then $j(\mathbb{1}_{\{x_i\}})$ is a projection, P_i say, in $\mathcal{A}$, with the properties that

$$\sum_{i=1}^{n} P_i = \sum_{i=1}^{n} j(1_{\{x_i\}}) = j(\mathbb{1}_B) = \mathbb{1}_{\mathcal{A}} \tag{2-489}$$

and for $i \neq j$,

$$P_i P_k = j(1_{\{x_i\}}) j(1_{\{x_k\}}) = j(1_{\{x_i\}} \cdot 1_{\{x_k\}}) = 0 \tag{2-490}$$

Interpret P_i as the event "the random variable described by j takes the value x_i". Note that j can be written as

$$j(f) = j\left(\sum_{i=1}^{n} f(x_i) 1_{\{x_i\}}\right) = \sum_{i=1}^{n} f(x_i) P_i \tag{2-491}$$

In particular, if $\Omega \subset R$, then j defines a hermitian operator

$$j(\mathrm{id}) = \sum_{i=1}^{n} x_i P_i =: X \tag{2-492}$$

which completely determines j.

Proposition 2.53 Let $\mathcal{A}$ be a finite-dimensional $*$-algebra with unit. Then there is a one-to-one correspondence between injective $*$-homomorphisms $j: C(\Omega) \to \mathcal{A}$ for some finite $\Omega \subset R$ and self-adjoint operators $X \in \mathcal{A}$, given by

$$j(\mathrm{id}) = X \tag{2-493}$$

Proof.

If j is a $*$-homomorphism $C(\{x_1, x_2, \cdots, x_n\}) \mathscr{A}$ with $x_1, x_2, \cdots, x_n$ real, then

$$X := j(\mathrm{id}) = \sum_{i=1}^{n} x_i j(1_{\{x_i\}}) =: \sum_{i=1}^{n} x_i P_i \tag{2-494}$$

is a hermitian element of $\mathcal{A}$. Conversely, if $X \in \mathcal{A}$ is hermitian, then let $x_1, x_2, \cdots, x_n$ be its eigenvalues. Let $p: C \to C$ denote the polynomial

$$p(x) := (x - x_1)(x - x_2) \cdots (x - x_n) \tag{2-495}$$

and let, for $i = 1, 2, \cdots, n$, the (Lagrange interpolation) polynomial p_i be given by

$$p_i(x) := \frac{p(x)}{(x - x_i) p(x_i)} \tag{2-496}$$

Then $p_i(x_k) = \rho_{ikpk}$, so have on the spectrum $\{x_1, x_2, \cdots, x_n\}$ of X:

$$\sum_{i=1}^{n} p_i = 1 \quad \text{and} \quad p_i \cdot p_k = \delta_{ik} p_k \tag{2-497}$$

It follows that the projections $P_i := p_i(X)$, with $i = 1, 2, \cdots, n$, lie in the algebra $\mathcal{A}$ and satisfy

$$\sum_{i=1}^{n} P_i = \mathbb{1} \quad \text{and} \quad P_i P_k = \delta_{ik} P_k \tag{2-498}$$

Hence, if define

$$j(f) := \sum_{i=1}^{n} f(x_i) P_i \tag{2-499}$$

then j is a $*$-homomorphism with the property that $j(\mathrm{id}) = X$. Clearly, different X's correspond to different j's.

- The joint measurement apparatus

Let X and Y be self-adjoint elements of the $*$-algebra $\mathcal{A}$. Consider X and Y as random variables taking values in the spectra sp (X) and sp (Y). By a joint measurement M^* of these random variables mean an operation that takes a state ρ on $\mathcal{A}$ as input, and yields a probability distribution π on sp $(X) \times$ sp (Y) as output, in such a way that for all functions f on sp (X), gon sp (Y):

$$\begin{cases}\rho(f(X)) = \sum_{x\in \mathrm{sp}\,(X)} \sum_{y\in \mathrm{sp}\,(Y)} \pi(x,y)f(x) \\ \rho(g(X)) = \sum_{x\in \mathrm{sp}\,(X)} \sum_{y\in \mathrm{sp}\,(Y)} \pi(x,y)g(x)\end{cases} \tag{2-500}$$

A contravariant formulation of these requirements is

$$\begin{cases}M(f \otimes \mathbb{1}) = f(X) \\ M(\mathbb{1} \otimes g) = g(Y)\end{cases} \tag{2-501}$$

Theorem 2.54 If two random variables X and Y allow a joint measurement operation, then they commute.

Proof.

Let denote by x the identity function on sp (X), and by y that on sp (Y). Apply the multiplication theorem on the measurement operation M, which is supposed to exist. Since

$$M((x \otimes \mathbb{1})^* (x \otimes \mathbb{1})) = M(x^2 \otimes \mathbb{1}) = X^2 = M(x \otimes \mathbb{1})^* M(x \otimes \mathbb{1}) \tag{2-502}$$

and have

$$M((x \otimes \mathbb{1})^* (\mathbb{1} \otimes y)) = M(x \otimes \mathbb{1})^* M(\mathbb{1} \otimes y) = XY \tag{2-503}$$

and

$$M((\mathbb{1} \otimes y)^* (x \otimes \mathbb{1})) = M(\mathbb{1} \otimes y)^* M(x \otimes \mathbb{1}) = YX$$

as

$$(x \otimes \mathbb{1})^* (\mathbb{1} \otimes y) = x \otimes y = (\mathbb{1} \otimes y)^* (x \otimes \mathbb{1}), \text{we have } XY = YX \tag{2-504}$$

2.8.8 Quantum novelties

In the previous section certain strange limitations that quantum operations are subject to. Let now look at the other side of the coin: some surprising possibilities.

Leave treatment of the really sensational features to other contributions in this volume, such as very fast computation and secure cryptography. Here treat "teleportation" of quantum states and "superdense coding".

- Teleportation of quantum states

Suppose that Alice wishes to send to Bob the quantum state ρ of a qubit over a (classical) telephone line. In no classical coding have seen that, without any further tools, this is impossible. If Alice were to perform measurements on the qubit, and tell the results to Bob over the telephone, these would not enable Bob to reconstruct the state ρ. However, suppose that Alice and Bob have been together in the past, and that at that time they have created an entangled pair of qubits, as introduced, each taking one qubit with them. Wootters, Peres and others, that by making use of this shared entanglement, Alice is indeed able to transfer her qubit to Bob. Of course, she cannot avoid destroying the original state ρ in the process; otherwise Alice and Bob would have copied the state ρ. It is for this reason that the procedure is called "teleportation". Illustrate the procedure in a picture as:

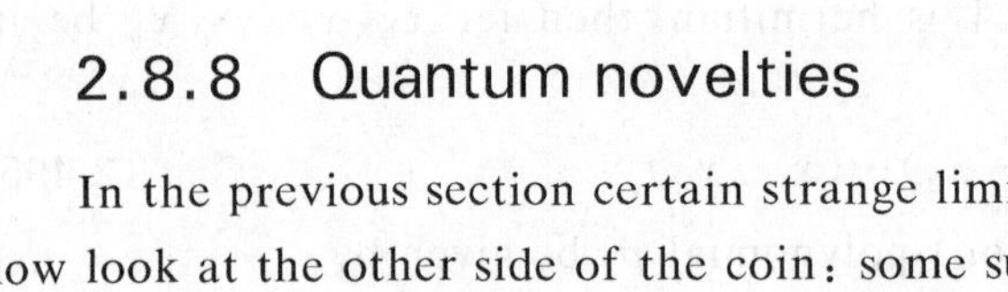

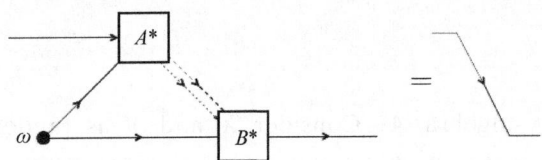

Fig. 2-45 Teleportation based on shared entanglement.

Here ω is the fully entangled state $X \mapsto \langle \Omega,$

$X\Omega\rangle$ on $M2$, The procedure runs as follows. Alice possesses two qubits, one from the entangled pair, and one which she wishes to send to Bob. She performs a von Neumann measurement on these two qubits along the four Bell projections

$$\begin{cases} Q_{00} := \frac{1}{2}\begin{bmatrix} 1 & 0 & 0 & 1 \\ 0 & 0 & 0 & 0 \\ 0 & 0 & 0 & 0 \\ 1 & 0 & 0 & 1 \end{bmatrix} \\ Q_{01} := \frac{1}{2}\begin{bmatrix} 1 & 0 & 0 & -1 \\ 0 & 0 & 0 & 0 \\ 0 & 0 & 0 & 0 \\ -1 & 0 & 0 & 1 \end{bmatrix} \\ Q_{10} := \frac{1}{2}\begin{bmatrix} 0 & 0 & 0 & 0 \\ 0 & 1 & 1 & 0 \\ 0 & 1 & 1 & 0 \\ 0 & 0 & 0 & 0 \end{bmatrix} \\ Q_{11} := \frac{1}{2}\begin{bmatrix} 0 & 0 & 0 & 0 \\ 0 & 1 & -1 & 0 \\ 0 & -1 & 1 & 0 \\ 0 & 0 & 0 & 0 \end{bmatrix} \end{cases} \tag{2-505}$$

The operation performed by Alice has the contravariant description:

$$A: C_2 \otimes C_2 \to M_2 \otimes M_2: \quad A(e_i \otimes e_j) := Q_{ij}$$

He then takes his own qubit from the entangled pair, and if $j = 1$ performs the "phase flip" operation

$$Z: \begin{pmatrix} \rho_{00} & \rho_{01} \\ \rho_{10} & \rho_{11} \end{pmatrix} \mapsto \begin{pmatrix} \rho_{00} & -\rho_{01} \\ -\rho_{10} & \rho_{11} \end{pmatrix} = \begin{pmatrix} 1 & 0 \\ 0 & -1 \end{pmatrix} \begin{pmatrix} \rho_{00} & \rho_{01} \\ \rho_{10} & \rho_{11} \end{pmatrix} \begin{pmatrix} 1 & 0 \\ 0 & -1 \end{pmatrix} \tag{2-506}$$

and if $j = 0$ he does nothing. Then, if $i = 1$ he performs the "quantum not" operation

$$X: \begin{pmatrix} \rho_{00} & \rho_{01} \\ \rho_{10} & \rho_{11} \end{pmatrix} \mapsto \begin{pmatrix} \rho_{11} & \rho_{10} \\ \rho_{01} & \rho_{00} \end{pmatrix} = \begin{pmatrix} 0 & 1 \\ 1 & 0 \end{pmatrix} \begin{pmatrix} \rho_{00} & \rho_{01} \\ \rho_{10} & \rho_{11} \end{pmatrix} \begin{pmatrix} 0 & 1 \\ 1 & 0 \end{pmatrix} \tag{2-507}$$

and if $i = 0$ he does nothing. In the Heisenberg picture, the result of Bob's actions is the operation

$$B: M_2 \to C_2 \otimes C_2 \otimes M_2: \quad M \mapsto M \oplus \sigma_3 M \sigma_3 \oplus \sigma_1 M \sigma_1 \oplus \sigma_2 M \sigma_2 \tag{2-508}$$

where

$$\begin{cases} \sigma_1 := \begin{pmatrix} 0 & 1 \\ 1 & 0 \end{pmatrix} \\ \sigma_2 := \begin{pmatrix} 0 & -i \\ i & 0 \end{pmatrix} \\ \sigma_3 := \begin{pmatrix} 1 & 0 \\ 0 & -1 \end{pmatrix} \end{cases} \tag{2-509}$$

are Pauli's spin matrices. Bob ends up with a qubit in exactly the same state as Alice wanted to send. Formulate this result in the Heisenberg picture.

Proposition 2.55 The state and the operations A and B described above satisfy

$$(\mathrm{id}_{M_2} \otimes \omega) \circ (A \otimes \mathrm{id}_{M_2}) \circ B = \mathrm{id}_{M_2} \tag{2-510}$$

Proof.

just calculate for $M \in M_2$:

$$M \overset{B}{\longmapsto} M \oplus \sigma_3 M \sigma_3 \oplus \sigma_1 M \sigma_1 \oplus \sigma_2 M \sigma_2$$

$$\overset{A\otimes \mathrm{id}}{\longmapsto} (Q_{00} \otimes M) + (Q_{01} \otimes \sigma_3 M \sigma_3) + (Q_{10} \otimes \sigma_1 M \sigma_1) + (Q_{11} \otimes \sigma_2 M \sigma_2)$$

$$= \frac{1}{2}\begin{pmatrix} M + \sigma_3 M \sigma_3 & 0 & 0 & M - \sigma_3 M \sigma_3 \\ 0 & \sigma_1 M \sigma_1 + \sigma_2 M \sigma_2 & \sigma_1 M \sigma_1 - \sigma_2 M \sigma_2 & 0 \\ 0 & \sigma_1 M \sigma_1 - \sigma_2 M \sigma_2 & \sigma_1 M \sigma_1 + \sigma_2 M \sigma_2 & 0 \\ M - \sigma_3 M \sigma_3 & 0 & 0 & M + \sigma_3 M \sigma_3 \end{pmatrix}$$

$$= \left(\begin{array}{cccc|cccc} m_{00} & 0 & 0 & 0 & 0 & 0 & 0 & m_{01} \\ 0 & m_{11} & 0 & 0 & 0 & 0 & m_{10} & 0 \\ 0 & 0 & m_{11} & 0 & 0 & m_{01} & 0 & 0 \\ 0 & 0 & 0 & m_{00} & m_{01} & 0 & 0 & 0 \\ \hline 0 & 0 & 0 & m_{10} & m_{11} & 0 & 0 & 0 \\ 0 & 0 & m_{01} & 0 & 0 & m_{00} & 0 & 0 \\ 0 & m_{01} & 0 & 0 & 0 & 0 & m_{00} & 0 \\ m_{10} & 0 & 0 & 0 & 0 & 0 & 0 & m_{11} \end{array}\right)$$

$$\overset{\mathrm{id}\otimes\omega}{\longmapsto} \begin{pmatrix} m_{00} & m_{01} \\ m_{10} & m_{11} \end{pmatrix} = M \tag{2-511}$$

Teleportation has been carried out succesfully in the lab by Zeilinger et al. using polarized photons, and by other experimenters using different techniques later. For the sake of such experiments explicit operations have been developed that form the "building blocks" of the diversity of quantum operations needed. For example the operation performed by Alice to prepare the teleportation of a qubit can be decomposed into an interaction and a measurement. Let j be the ordinary measurement operation of a qubit:

$$j: \mathcal{C}_2 \to M_2: \quad (f_0, f_1) \mapsto \begin{pmatrix} f_0 & 0 \\ 0 & f_1 \end{pmatrix} \tag{2-512}$$

Let H denote the Hadamard gate, which acts on states or observables by multiplication on the left and on the right by the Hadamard matrix

$$\frac{1}{\sqrt{2}}\begin{pmatrix} 1 & 1 \\ 1 & -1 \end{pmatrix} \tag{2-513}$$

and let C denote the controlled not gate $M_2 \otimes M_2 \to M_2 \otimes M_2$ which sandwiches a matrix with

$$\begin{pmatrix} 1 & 0 & 0 & 0 \\ 0 & 0 & 0 & 1 \\ 0 & 0 & 1 & 0 \\ 0 & 1 & 0 & 0 \end{pmatrix} \tag{2-514}$$

The operation C performs a not operation on the first qubit provided that the second is a 1. In diagrams:

Check that, using the above building blocks, the procedure of quantum teleportation can be charted as in Figure 2-47.

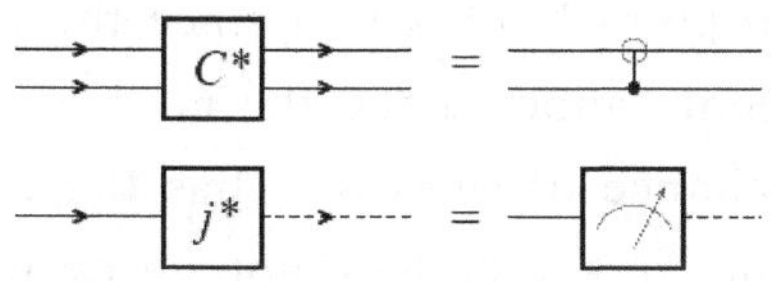

Fig. 2-46 Conventional signs used for the C and j operations.

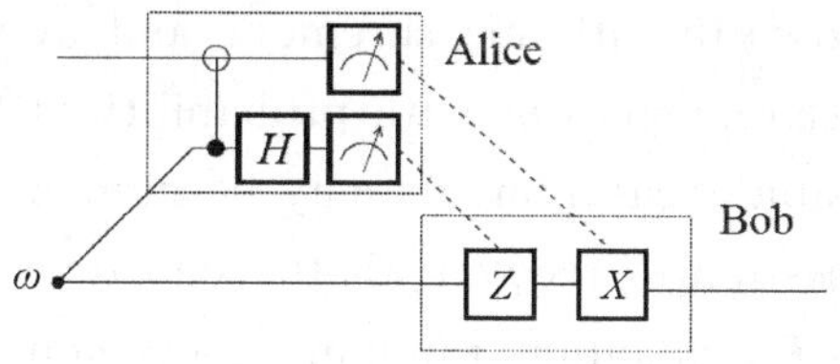

Fig. 2-47 More detailed scheme of teleportation.

- Superdense coding

That Alice can "teleport" a qubit using two classical bits, given a pre-entangled qubit pair. A kind of converse is also possible: Bob can communicate two classical bits to Alice by sending her a single qubit, again given a shared pre-entangled qubit pair.

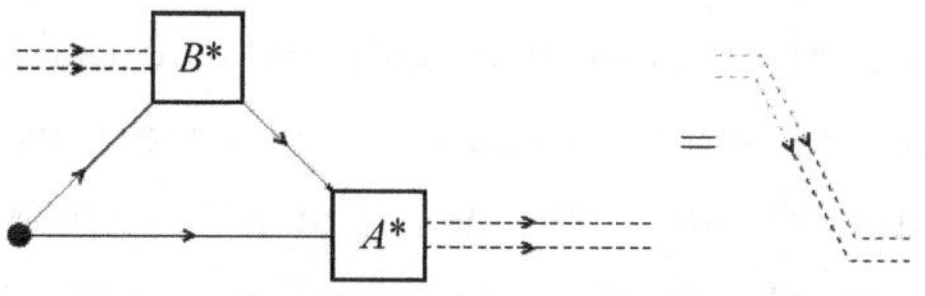

Fig. 2-48 Superdense coding: two bits in a single photon.

Proposition 2.56 Taking ω, A and B as in Proposition 2.53, have

$$(\mathrm{id}_{C_2\otimes C_2} \otimes \omega) \circ (B \otimes \mathrm{id}_{M_2}) \circ A = \mathrm{id}_{C_2\otimes C_2} \tag{2-515}$$

2.9 Dense quantum coding and quantum finite automata

The tremendous information processing capabilities of quantum mechanical systems may be attributed to the fact that the state of an n quantum bit (qubit) system is given by a unit vector in a $2n$ dimensional complex vector space. This suggests the possibility that classical information might be encoded and transmitted with exponentially fewer qubits. Yet, according to a fundamental result in quantum information theory, Holevo's theorem, no more than n classical bits of information can faithfully be transmitted by transferring n quantum bits from one party to another. In view of this result, it is tempting to conclude that the exponentially many degrees of freedom latent in the description of a quantum system must necessarily stay hidden or inaccessible. However, the situation is more subtle since the recipient of the n-qubit quantum state has a choice of measurement he or she can make to extract information about their state. In general, these measurements do not commute. Thus making a particular measurement will disturb the system, thereby destroying some or all the information that would have been revealed by another possible measurement. This opens up the possibility of quantum random access codes, which encode classical bits into many fewer qubits, such that the recipient can choose which bit of classical information he or she would like to extract out of the encoding. We might think of this as a disposable quantum phone book, where the contents of an entire telephone directory are compressed into a few quantum bits such that the recipient of these qubits can, via a suitably chosen measurement, look up any single telephone number of his or her choice. Such quantum codes, if possible, would serve as a powerful primitive in quantum communication.

To formalize this, say we wish to encodem bits $b_1 b_2 \cdots b_m$ into n qubits ($m \gg n$). Then a quantum random access encoding with parameters m, n, p (or simply an $m \mapsto pn$ encoding) consists of an encoding map from $\{0,1\}$m to mixed states with support in $C_2 n$, together with a sequence of m possible measurements for the recipient. The measurements are such that if the recipient chooses the ith measurement and applies it to the encoding of $b_1 b_2 \cdots b_m$, the result of the measurement is b_i with probability at least p. The main point here is that since the m different possible measurements may be noncommuting, the recipient cannot make the m measurements in succession to recover all the encoded bits with a good chance of success. Thus the existence of $m \mapsto PN$ quantum random access codes with $m \gg n$ and $p > 1/2$ does not necessarily violate Holevo's bound. Furthermore, even though C_k can accommodate only k mutually orthogonal unit vectors, it can accommodate a_k almost mutually orthogonal unit vectors (i.e., vectors such that the inner product of any two has an absolute value less than, say, 1/10) for some $a>1$. Indeed, there is no a priori reason to rule out the existence of codes that represent an classical bits in n quantum bits for some constant $a>1$. Start by showing that quantum encodings are more powerful than classical ones. We describe a $2 \mapsto 0.851$ quantum encoding, and prove that there is no $2 \mapsto P_1$ classical encoding for any $p > 1/2$. The quantum encoding may generalized to a $3 \mapsto 0.781$ encoding, as was shown by Chuang, and to encodings of more bits into one quantum bit. The main result in this paper is that (despite the potential of quantum encoding shown by the arguments and results presented above) quantum encoding does not provide much compression. We prove that any $m \mapsto PN$ quantum encoding satisfies $n \geqslant (1 - H(p))m$, where $H(p) = -\log p - (1-p)\log(1-p)$ is the binary entropy function. The main technique in the proof is the use of the entropy coalescence lemma, which quantifies the increase in entropy when we take a convex combination of mixed states. This lemma obtained by viewing Holevo's theorem from a new perspective. We turn to upper bounds on compression next, and show that the lower bound is asymptotically tight up to an additive logarithmic term, and can be achieved even with classical encoding. For any $p > 1/2$, we give a construction for $m \mapsto PN$ classical codes with $n = (1 - H(p))m + O(\log m)$. Thus, even though quantum random access codes can be more succinct as compared to classical codes, they may be only a logarithmic number of qubits shorter. In many of the existing quantum computing implementations, the complexity of implementing the system grows tremendously as the number of qubits increases.

Moreover, even discarding one qubit and replacing it by a new qubit initialized to $|0\rangle$ (often called a clean qubit) while keeping the total number of qubits the same might be difficult or impossible (as in NMR quantum computing). This has motivated a huge body of work on one-way quantum finite automata (QFAs), which are devices that model computers with a small finite memory. During the computation of a QFA, no clean qubits are allowed, and in addition intermediate measurements are allowed, except to decide whether to accept or reject the input. To define generalized one-way quantum finite automata (GQFAs) that capture the most general quantum computation that can be carried out with restricted memory and no extra clean qubits. In particular, the model allows arbitrary measurements upon the state space of the automaton as long as the measurements can carried out without clean qubits. The model accurately incorporates the capabilities of today's implementations of quantum computing.

Kondacs and Watrous was shown that not every language recognized by a classical deterministic finite automaton (DFA) is recognized by a QFA. On the other hand, there are languages that are recognized byQFAs with sizes exponentially smaller than those of corresponding classical automata. It remained open whether for any language that can recognized by a one-way finite automaton both classically and quantum-mechanically, a classical automaton can be efficiently simulated by a QFA with no extra clean qubits. We answer this question in the negative. Can apply the entropy coalescence lemma in a computational setting to give a lower bound on the size of (GQFAs). There is a sequence of languages for which the minimal GQFA has exponentially more states than the minimal DFA. It may be surprising that despite their quantum power (and irreversible computation, thanks to the intermediate measurements) GQFAs are exponentially less powerful for certain languages than classical DFAs. This lower bound highlights the need for clean qubits for efficient computation.

- Quantum systems

Just as a bit is a fundamental unit of classical information, a qubit is the fundamental unit of quantum information. A qubit is described by a unit vector in the two-dimensional Hilbert space C_2. Let $|0\rangle$ and $|1\rangle$ be an orthonormal basis for this space. In general, the state of the qubit is a linear superposition of the form $\alpha|0\rangle+\beta|1\rangle$. The state of n qubits described by a unit vector in the n-fold tensor product $C_2 \otimes C_2 \otimes \cdots \otimes C_2$. An orthonormal basis for this space is now given by the $2n$ vectors $|x\rangle$, where $x \in \{0,1\} n$. This often referred to as the computational basis. In general, the state of n qubits is a linear superposition of the $2n$ computational basis states. Thus the description of an n qubit system requires $2n$ complex numbers. This is arguably the source of the astounding information processing capabilities of quantum computers. The information in a set of qubits may be "read out" by measuring it in an orthonormal basis, such as the computational basis. When a state $\sum_x \alpha_x |x\rangle$ is measured in the computational basis, can get the outcome x with probability $|\alpha_x|2$. More generally, a (von Neumann) measurement on a Hilbert space His defined by a set of orthogonal projection operators $\{P_i\}$. When a state $|\phi\rangle$ is measured according to this set of projection operators, can get outcome i with probability $\|P_i|\phi\rangle\|^2$. Moreover, the state of the qubits "collapses" to (i.e., becomes) $P_i|\phi\rangle / \|P_i|\phi\rangle\|$, when the outcome i is observed. In order to retrieve information from an unknown quantum state $|\phi\rangle$, it is sometimes advantageous to augment the state with some ancillary qubits, so that the combined state is now $|\phi\rangle \otimes |\bar{0}\rangle$ before measuring them jointly according to a set of operators $\{P_i\}$ as above. This is the most general form of quantum measurement, and called a positive operator valued measurement (POVM).

- Density matrices

In general, a quantum system may be in a mixed state—a probability distribution over superpositions. For example, such a mixed state may result from the measurement of a pure state $|\phi\rangle$. Consider the mixed state $\{p_i, |\phi_i\rangle\}$, where the superposition $|\phi_i\rangle$ occurs with probability p_i. The behavior of this mixed state is completely characterized by its density matrix $\rho=\sum_i P_i |\phi_i\rangle\langle\phi_i|$. (The "bra" notation $\langle\phi|$ here is used to denote the conjugate transpose of the superposition (column vector) $|\phi\rangle$. Thus $|\phi\rangle\langle\phi|$ denotes the outer product of the vector with itself.) For example, under a unitary transformation U, the mixed state $\{p_i, |\phi_i\rangle\}$ evolves as $\{p_i, |\phi_i\rangle\}$ so that the resulting density matrix is $U\rho U^\dagger$. When measured according to the projection operators $\{P_j\}$, the probability q_j of getting outcome j is $q_j=\sum_i P_i \|P_j\langle\phi_i|\|^2=\mathrm{tr}(P_j\rho P_j)$ and the residual density matrix is $P_j\rho P_j/q_j$. Thus, two mixed states with the same density matrix have the same behavior under any physical operation. That will therefore identify a mixed state with its density matrix.

The following properties of density matrices follow from the definition. For any density matrix ρ,

(1) ρ is Hermitian, that is, $\rho = \rho^{\dagger}$.

(2) ρ has unit trace, that is $\mathrm{tr}(\rho) = \sum_i \rho(i,i) = 1$.

(3) ρ is positive semidefinite, that is, $\langle\psi|\rho|\psi\rangle \geqslant 0, 0$ for all $|\psi\rangle$.

Thus, every density matrix is unitarily diagonalizable and has nonnegative real eigenvalues that sum up to 1. Recall that the amount of randomness (or the uncertainty) in a classical probability distribution may be quantified by its Shannon entropy. Doing the same for a mixed state is tricky because all mixed states consistent with a given density matrix are physically indistinguishable, and therefore contain the same amount of "entropy." Before we do this, we recall the classical definitions.

- Classical entropy and mutual information

The Shannon entropy $S(X)$ of a classical random variable X that takes values x in some finite set with probability p_x is defined as

$$S(X) = -\sum_x p_x \log p_x \tag{2-516}$$

The mutual information $I(X:Y)$ of a pair of random variables X, Y is defined by

$$I(X:Y) = S(X) + S(Y) - S(XY) \tag{2-517}$$

where XY denotes the joint random variable with marginals X and Y. It quantifies the amount of correlation between the random variables X and Y. Fano's inequality asserts that if Y can predict X well, then X and Y have large mutual information. We use a simple form of Fano's inequality, referring only to Boolean variables X and Y.

Fact 2.57 (Fano's Inequality). Let X be a uniformly distributed boolean random variable, and let Y be a boolean random variable such that $\Pr(X = Y) = p$. Then $I(X:Y) \geqslant 1 - H(p)$.

- Von neumann entropy

Consider the mixed state $X = \{p_i, |\phi_i\rangle\}$, where the superposition $|\phi_i\rangle$ occurs with probability p_i. Since the constituent states $|\phi_i\rangle$ of the mixture are not perfectly distinguishable in general, cannot define the entropy of this mixture to be the Shannon entropy of $\{p_i\}$. Another way to see this is that this mixture is equivalent to any other mixture with the same density matrix, and so should have the same entropy as that mixture. Indeed, a special such equivalent mixture can be obtained by diagonalizing the density matrix—the constituent states of this mixture are orthogonal, and therefore perfectly distinguishable. Now, the entropy of the density matrix can be defined to be the Shannon entropy of these probabilities. To formalize this, recall that every density matrix ρ is unitarily diagonalizable:

$$\rho = \sum_j \lambda_j |\psi_j\rangle\langle\psi_j| \tag{2-518}$$

and has nonnegative real eigenvalues, $\lambda_j \geqslant 0$ that sum up to 1, and the corresponding eigenvectors $|\psi_j\rangle$ i are all orthonormal. The von Neumann entropy $S(\rho)$ of the density matrix ρ is then defined as $S(\rho) = \sum_i \lambda_i \log\lambda_i$. In other words, $S(\rho)$ is the Shannon entropy of the distribution induced by the eigenvalues of ρ on the corresponding eigenvectors. Summarize some basic properties of von Neumann entropy below. For a comprehensive introduction to this concept and its properties, see, for instance, Preskill and Wehrl.

If the constituent states of a mixture lie in a Hilbert space H, then the corresponding density matrix is said to have support in H. A density matrix with support in a Hilbert space of dimension d, has d eigenvalues, and hence the entropy of any such distribution is at most log d. i.e.,

Fact 2.58 If ρ is a density matrix with support in a Hilbert space of dimension d, then $0 \leqslant S(\rho) \leqslant \log d$.

Quantum mechanics requires that the evolution of the state of an isolated system be unitary, and therefore reversible. This implies that information cannot be erased and entropy is invariant under unitary operations:

For any density matrix ρ and unitary operator U, $S(U\rho U^{\dagger}) = S(\rho)$. This is easy to see since the eigenvalues of the resulting matrix $U\rho U^{\dagger}$ are the same as those of ρ. In the classical world observing a value does not disturb its state, and as a result measurements (or observations) do not change entropy. In the quantum world, however, measurements usually disturb the system, introducing new uncertainties. Thus, the entropy increases. Consider for example a system of one qubit that is with probability 1 in the pure state $1/\sqrt{2}(|0\rangle + |1\rangle)$, and thus has 0 entropy. Suppose measure it in the $|0\rangle; |1\rangle$ bas is. That get each result with equal probability, and the resulting mixed state of the qubit, disregarding the outcome of the measurement is $\left\{\left(\frac{1}{2}, |0\rangle\right), \left(\frac{1}{2}, |1\rangle\right)\right\}$ which has entropy 1.

Let ρ be the density matrix of a mixed state in a Hilbert space H and let the set of orthogonal projections $\{P_j\}$ define a von Neumann measurement in H. If $\rho' = \sum_j P_j \rho P_j$ is the density matrix resulting from a measurement of the mixed state with respect to these projections (disregarding the measurement outcome), then $S(\rho') \geqslant S(\rho)$.

2.9.1 Holevo's theorem and the entropy coalescence lemma

Consider two parties Alice and Bob communicating over a quantum channel, where Alice wishes to transmit some classical information, given by a random variable X, to Bob by encoding it into some number of qubits and sending these qubits to Bob. Holevo's theorem bounds the amount of information Bob can extract from the quantum encoding.

Let $x \mapsto \rho_x$ be any quantum encoding of bit strings into density matrices. Let X be a random variable with a distribution given by $\Pr(X = x) = p_x$, and let $\rho = \sum_x p_x \rho_x$ be the state corresponding to the encoding of the random variable X. If Y is any random variable obtained by performing a measurement on the encoding, then

$$I(X:Y) \leqslant S(\rho) - \sum_x p_x S(\rho_x) \tag{2-519}$$

Viewing Holevo's theorem from a different perspective is the key to the lowerbound results in this paper. When consider the scenario where a mixture ρ is obtained as the convex combination of two mixtures $\rho_0; \rho_1$ of equal entropy. When can say that the combination results in a mixture of higher entropy? This is the content of the entropy coalescence lemma below. This lemma can quite easily be generalized to the case where ρ is obtained from a more general mixture of density matrices.

Let ρ_0 and ρ_1 be two density matrices, and let $\rho = 1/2(\rho_0 + \rho_1)$ be a uniformly random mixture of these matrices. If O is a measurement with outcome 0 or 1 such that making the measurement on ρ_b yields the bit b with probability at least p, then

$$S(\rho) \geqslant \frac{1}{2}[S(\rho_0) + S(\rho_1)] + (1 - H(p)) \tag{2-520}$$

Proof.

To view ρ_b as an encoding of the bit b. If X is an unbiased random variable over $\{0;1\}$, then ρ represents the encoding of X. Let Y be the outcome of the measurement of this encoding according to O. By the hypothesis of the lemma, $\Pr(Y = X) \geqslant p$. Thus, by Fano's inequality—Fact 2.55:

$$I(X:Y) \geqslant 1 - H(p) \tag{2-521}$$

Also, by Holevo's Theorem:

$$I(X:Y) \leqslant S(\rho) - \frac{1}{2}[S(\rho_0) + S(\rho_1)] \tag{2-522}$$

Rearranging,

$$S(\rho) \geqslant \frac{1}{\rho}[S(\rho_0) + S(\rho_1)] + (1 - H(p)) \tag{2-523}$$

as desired.

- Random access encodings

First define random access encodings.

Definition 2.59 A $m \mapsto p_n$ quantum random access encoding is a function $f: \{0;1\} m \times R \mapsto C^{2^n}$ (here R is the set of random choices in the encoding) such that for every $1 \leqslant i \leqslant m$, there is a measurement $\mathcal{O}_i$ that returns 0 or 1 and has the property that

$$\forall\, b \in \{0,1\}^m : \Pr_r(\mathcal{O}_i \mid f(b,r)\rangle = b_i) \geqslant p \tag{2-524}$$

Call f the encoding function, and $\mathcal{O}_i$ the decoding function. The encoding is classical if f is a mapping into $\{0;1\}\ n$.

- A quantum encoding with no classical counterpart

When begin by constructing a random access encoding of two classical bits into one qubit. This encoding first used by Bennett et al. in the context of quantum cryptography and independently rediscovered by the authors of this paper in the context of coding.

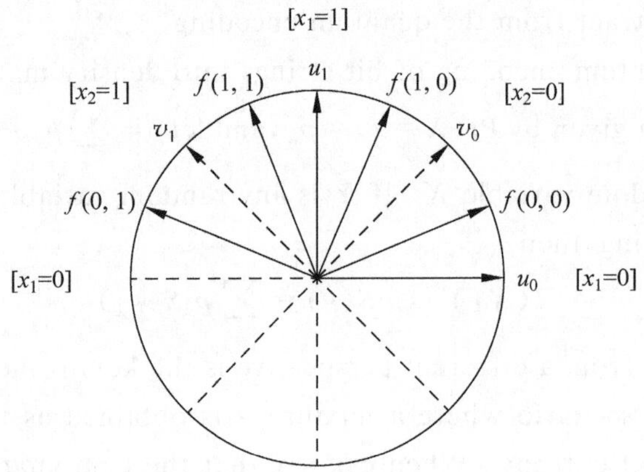

Fig. 2-49 A two-into-one quantum encoding with probability of success ≈0.85.

Lemma 2.60 There is a 2 07:85 1 quantum encoding.

Proof.

Let $|u_0\rangle = |0\rangle$, $|u_1\rangle = |1\rangle$, and $|v_0\rangle = 1/2(|1\rangle + |0\rangle)$, $|v_1\rangle = 1/2(|1\rangle - |0\rangle)$. Define $f(x_1; x_2)$, the encoding of the string $(x_1; x_2)$, to be $|u_{x1}\rangle + |v_{x2}\rangle$ normalized, unless $x_1 x_2 = 01$, in which case it is $-|u_0\rangle + |v_1\rangle$ normalized. The four vectors $f(0,0)$, $f(0,1)$, $f(1,0)$, $\cdots$, $f(1,1)$ appear in Figure 2-49. The decoding functions are defined as follows: for the first bit x_1, measure the message qubit according to the u basis and associate $|u_0\rangle$ with $x_1 = 0$ and $|u_1\rangle$ with $x_1 = 1$. Similarly, for the second bit, measure according to the v basis, and associate $|v_0\rangle$ with $x_2 = 0$ and $|v_1\rangle$ with $x_2 = 1$, as see Figure 2-49. For all four codewords, and for any $i = 1,2$, the angle between the codeword and the correct subspace is $\pi/8$. Hence the success probability is $\cos^2(\pi/8) \approx 0.85$. This example was further refined into a $3 \mapsto 0.78$ 1 quantum encoding by Chuang. The next lemma shows that such classical codes are not possible.

Lemma 2.61 No $2 \mapsto p1$ classical encoding exists for any $p > 1/2$.

Proof. Let there be a classical $2 \mapsto p1$ encoding for some p. Let $f: \{0;1\}^2 \times R' \mapsto \{0;1\}$ be the corresponding probabilistic encoding function and $V_i: \{0;1\} \times R' \mapsto \{0;1\}$ the probabilistic decoding functions. First give a geometric characterization of the decoding functions. Each V_i depends only on the encoding, which is either 0 or 1. Define the point P_j (for $j = 0,1$) in the unit square $[0;1]^2$ as $P^j = (a_1^j; a_2^j)$, where $a_i^j = \Pr_r 0(V_i(j, r') = 1)$. The point P^0 characterizes the decoding functions when the encoding is 0, and P^1 characterizes the decoding functions when the encoding is 1. For example, $P^1 = (1;1)$ means that given the encoding 1, the decoding functions return $y_1 D_1$ and $y_2 D_1$ with certainty, and $P^0 = (0; 1/4)$ means that given the encoding 0, the decoding functions return $y_1 = 1$ with probability zero and $y_2 = 1$ with probability 1/4.

Now fix the decoding functions V_1, V_2. They define two points P^0 and P^1 in $[0;1]^2$ and the line connecting them:

$$P(q) = (1 - q)P^0 + qP^1 \tag{2-525}$$

To divide $[0;1]^2$ to four quadrants, and we associate each quadrant with its corner $(x_1; x_2) \in \{0;1\}^2$ (see Figure 2-50). The connecting line $P(q)$ cannot strictly pass through all four quadrants. To see that, let us assume without loss of generality that the point $P(1/2)$ is at or above the center $(1/2; 1/2)$. If the line is monotone increasing, then the line must miss the bottom right quadrant, while if it is monotone decreasing, it must miss the left bottom quadrant. If the line misses the quadrant associated with $(x_1; x_2)$, the decoding functions miss $(x_1; x_2)$. Now look at the encoding. Can

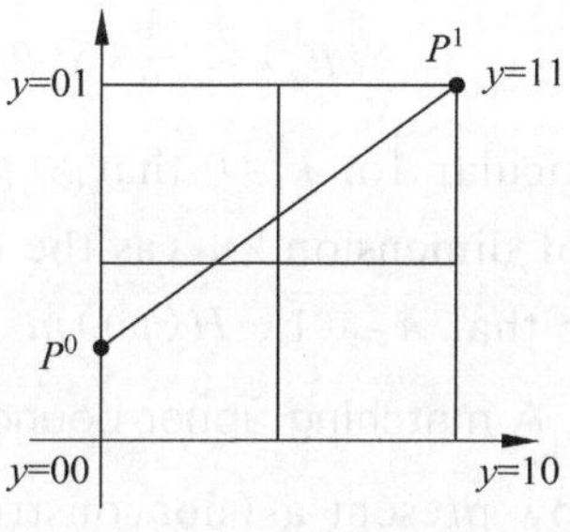

Fig. 2-50 A geometric characterization of the probabilistic decoding functions of Lemma 2.58.

know the decoding functions miss some $(x_1; x_2)$ and without loss of generality that they miss (1; 0). Given the input $x = (1;0)$, the encoder can choose (based on r) whether to encode x as 0 or 1. That he or she encodes x as 1 with probability q_x. Denote by $P^x = (a_1(x); a_2(x))$ the point with $a_i(x) = \Pr'_{r,r}(V_i(f(x;r); r'))$.

$$P^x = (1 - q_x)P^0 + q_x P^1 = P(q_x) \tag{2-526}$$

In particular it lies on the line connecting P^0 and P^1 and therefore it is not in the interior of the bottom right quadrant. Thus, either $a_1(x)$ is at most 1/2 or $a_2(x)$ is at least 1/2. It follows that either the first bit ($x_1 = 1$) or the second bit ($x_2 = 0$) is decoded correctly with probability $p \leqslant 1/2$.

2.9.2 The asymptotic of random access codes

- The lower bound

We now prove a lower bound on the number of qubits required for quantum random access codes.

Theorem 2.62 Let $1/2 < p \leqslant 1$. Any quantum (and hence any classical) $mp \mapsto N$ encoding satisfies $n \geqslant (1 - H(p))m$.

Proof.

Let ρ_x denote the density matrix corresponding to the encoding of the m-bit string x, and let ρ be the density matrix corresponding to picking x uniformly from $\{0;1\}m$ and encoding it. Then

$$\rho = \frac{1}{2^m} \sum_x \rho_x \tag{2-527}$$

Furthermore, for any $y \in \{0;1\}^k$, where $0 \leqslant k \leqslant m$, let

$$\rho_y = \frac{1}{2^{m-k}} \sum_{z \in (0,1)^{m-k}} \rho_{zy} \tag{2-528}$$

be the mixture corresponding to picking x uniformly from all strings in $\{0;1\}m$ with postfix y and encoding it. That prove by downward induction on k that $S(py) \geqslant (1 - H(p))(m - k)$. Base case. Assume $k = m$ and $y \in \{0;1\}^m$. Need to prove that $S(\rho y) \geqslant 0$. Indeed, the von Neumann entropy of any mixed state is nonnegative. Induction step. Suppose the claim is true for $k + 1$. Have $\rho_y = 1/2(\rho_0 y + \rho_1 y)$. By hypothesis,

$$S(\rho_{by}) \geqslant (1 - H(p))(m - k - 1) \tag{2-529}$$

For $b = 1, 0$. Moreover, ρ by is a mixture arising for encoding strings with b in the $(m - k)$ th bit. In particular, the measurement Om $- k$ when applied to the density matrix ρ by returns b with probability at least p. Thus, by the entropy coalescence lemma, can get

$$S(\rho_y) \geqslant \frac{1}{2}(S(\rho_{0y}) + S(\rho_{1y})) + (1 - H(p)) \geqslant (1 - H(p))(m - k) \tag{2-530}$$

In particular, for $k = 0$ that $S(\rho) \geqslant (1 - H(p))\ m$. On the other hand, ρ is defined over a Hilbert space of dimension $2n$ (as the encoding uses only n qubits) and implies that $S(\rho) \leqslant n$. Together can see that $n \geqslant (1 - H(p))m$ as desired.

- A matching upper bound

Now present a (nonconstructive) classical encoding scheme that asymptotically matches the lower bound derived in the previous section.

Theorem 2.63 For any $p > 1/2$ there is a classical m $p \mapsto n$ encoding with $n = (1 - H(p))m + O(\log m)$.

Proof. If $p > 1 - 1/m$, $H(p) \leqslant (\log m + 2)/m$ and there is a trivial encoding—the identity map. So turn to the case where $p \leqslant 1 - 1/m$. That use a code $S \subseteq \{0;1\}m$ such that, for every $x \in \{0;1\}m$, there is a $y \in S$ within Hamming distance $(1 - p - 1/m)m$. It is known (see, e.g., Cohen et al., Theorem that there is such a code S, called a covering code, of size

$$|S| \leqslant 2^{\left(1-H\left(p+\frac{1}{m}\right)\right)m+2\log m} \leqslant 2^{(1-H(p))m+4\log m} \tag{2-531}$$

For explicit constructions of covering codes, refer the reader to Cohen et al. (The explicit constructions, however, do not achieve the bound seek.) Let $S(x)$ denote the code word in the covering code S as above closest to x. One possibility is to encode a string x by $S(x)$. This would give us an encoding of the right size. Further, for every x, at least $(p+1/m)m$ out of the m bits would be correct.

This means that the probability (over all bits i) that $x_i = S(x)_i$ is at least $p+1/m$. However, for the encoding to need this probability to be at least p for every bit, not just on average over all bits. So introduce the following modification: Let r be an m-bit string, and π be a permutation of $\{1,2,\cdots,m\}$. For a string $x \in \{0;1\}^m$, let $\pi(x)$ denote the string $x_{\pi(1)} X_{\pi(2)} \cdots X_{\pi(m)}$.

Consider the encoding $S_{\pi r}$ defined by $S_{\pi r(x)} = \pi - 1(S(\pi(x+r))) + r$. That if π and r are chosen uniformly at random, then for any x and any index i, the probability that the ith bit in the encoding is different from x_i is at most $1-p-1/m$. First, note that if i is also chosen uniformly at random, then this probability is bounded by $1-p-1/m$. So all need to do is to show that this probability is independent of i. If π and r are uniformly random, then $\pi(x+r)$ is uniformly random as well. Furthermore, for a fixed $y=\pi(x+r)$, there is exactly one r corresponding to any permutation π that gives $y=\pi(x+r)$. Hence, if condition on $y=\pi(x+r)$, all π (and, hence, all $\pi^{-1}(i)$) are equally likely. This means that the probability that $x_i \neq S_{\pi r}(x)_i$ (or, equivalently, that $\pi(x+r)\pi-1(i) \neq (S(\pi(x+r))_{\pi-1}(i)$ for random π and r is just the probability of $y_j \neq S(y)_j$ for random y and j. This is independent of i (and x). Finally, that there is a small set of permutation-string pairs such that the desired property continues to hold if choose $\pi; r$ uniformly at random from this set, rather than the entire space of permutations and strings. Employ the probabilistic method to prove the existence of such a small set of permutationstring pairs. Let $l = m^3$, and let the strings $r_1, r_2, \cdots, r_1 \in \{0;1\}\, m$ and permutations $\pi_1, \pi_2, \cdots, \pi_l$ be chosen independently and uniformly at random. Fix $x \in \{0;1\}\, m$ and $i \in \{1,2,\cdots,m\}$. Let X_j be 1 if $x_i \neq S_{\pi_j r_j}(x)_i$ and 0 otherwise. Then $\sum_{j=1}^{i} X_j$ is a sum of independent Bernoulli random variables, the mean of which is at most $(1-p-1/m)l$. Note that $1/1 \sum_{j=1}^{l} X_j$ is the probability of encoding the ith bit of x erroneously when the permutation-string pair is chosen uniformly at random from the set $\{(\pi_1; r_1)\cdots\cdots(\pi_1; \pi,)\}$. By the Chernoff bound, the probability that the sum $1/1 \sum_{j=1}^{l} X_j$ is at least $(1-p-1/m)l+m^2$ (i.e., that the error probability $1\Big/ \sum_{j=1}^{l} X_j$ jmentioned above is at least $(1-p)$ is bounded by $e^{-2m^4} = e^{-2m}$. Thus, there is a combination of strings $r_1, r_2, \cdots, r_l$ and permutations $\pi_1, \pi_2, \cdots, \pi_l$ with the property seek, such a set of strings and permutations. Now can define the random access code as follows: To encode x, select $j \in \{1,2,\cdots,l\}$ uniformly at random and compute $y = S_{\pi_j r_j}(x)$. This is the encoding of x. To decode the ith bit, just take y_i. For this scheme, need $\log(l|S|) \leqslant \log l + \log|S| = (1-H(p))m + 7\log m$ bits. This completes the proof of the theorem.

2.9.3 One-way quantum finite automata

In this section, define generalized one-way quantum finite automata, and use the techniques developed above to prove size lower bounds on GQFAs. First introduce the model.

- The abstract model

A one-way quantum finite automaton is a theoretical model for a quantum computer with finite work space. QFAs were first considered by Moore and Crutchfield. These models do not allow intermediate measurements, except to decide whether to accept or reject the input. The model describe below allows the full range of operations permitted by the laws of quantum physics, subject to a space constraint. In particular, allow any orthogonal (or von Neumann) measurement as a valid intermediate computational step. The model may be seen as a finite memory version of the mixed-state quantum computers defined in Aharonov et al. That has to take care to formulate the model to properly account for all the qubits that are used in the computation. Thus any clean qubits must accounted for explicitly in the finite memory of the automaton. For example, performing a general "positive operator valued measurement" on the state of the automaton would require a joint measurement of the state with a fresh set of ancilla qubits. Once these ancillary qubits are explicitly included in the accounting, the same effect can achieved by a von Neumann measurement.

In abstract terms, may define a GQFA as follows: A GQFA has a finite set of basis states Q, which consists of three parts: accepting states, rejecting states, and nonhalting states. The sets of accepting, rejecting and nonhalting basis states denoted by Q_{acc}; Q_{rej}, and Q_{non}, respectively. One of the states, q_0, distinguished as the starting state. Inputs to a GQFA are words over a finite alphabet. Also use the symbols "¢" and "\$" that do not belong to Σ to denote the left and the right end marker, respectively. The set $\Gamma = \sum \cup \{ ¢ ; \$ \}$ denotes the working alphabet of the GQFA. For each symbol $\sigma \in \Gamma$, a GQFA has a corresponding "superoperator" U_σ that is given by a composition of a finite sequence of unitary transformations and von Neumann measurements on the space CQ. A GQFA is thus defined by describing Q; Q_{acc}; Q_{rej}; Q_{non}; q_0; Σ, and U_σ for all $\sigma \in \Gamma$. At any time, the state of a GQFA described by a density matrix with support in C^Q. The computation starts in the state $|q_0\rangle\langle q_0|$. A transformation corresponding to a symbol $\sigma \in \Gamma$ consists of two steps:

(1) First, U_σ is applied to ρ, the current state of the automaton, to obtain the new state ρ'.

(2) Then, ρ' is measured with respect to the operators $\{P_{\text{acc}}; P_{\text{rej}}; P_{\text{non}}\}$, where the P_i are orthogonal projections on the spaces E_i defined as follows: $E_{\text{acc}} = \text{span}\ \{|q\rangle \mid q \in Q_{\text{acc}}\}$, $E_{\text{rej}} = \text{span}\ \{|q\rangle \mid q \in Q_{\text{rej}}\}$, and $E_{\text{non}} = \text{span}\ \{|q\rangle \mid q \in Q_{\text{non}}\}$. The probability of observing $i \in \{\text{acc}; \text{rej}; \text{non}\}$ is equal to $\text{Tr}\ (P_i\rho')$. If observe acc (or rej), the input is accepted (or rejected). Otherwise, the computation continues (with the state $P_{\text{non}}\ \rho'/P_{\text{non}} = \text{Tr}\ (P_{\text{non}}\ \rho')$), and the next transformation, if any, is applied. Regard these two steps together as reading the symbol σ. A GQFA M is said to accept (or recognize) a language L with probability $p > 1/2$ if it accepts every word in L with probability at least p, and rejects every word not in L with probability at least p.

A reversible finite automaton (RFA) is a GQFA such that, for any $\sigma \in \Gamma$ and $q \in Q$, $U_\sigma |q\rangle\langle q| = |q'\rangle\langle q'|$ for some distinct $q' \in Q$. In other words, the operator U_σ is a permutation over the basis states. The size of a finite automaton is defined as the number of (basis) states in it. The "space used by the automaton" refers to the number of (qu)bits required to represent an arbitrary automaton state.

- GQFA for checking evenness

Theorem 2.64 Let

$$L_m = \{w_0 \mid w \in \{0,1\}^*, |w| \leqslant m\}, \quad m \geqslant 1 \tag{2-532}$$

define a family of regular languages. Then,

(1) L_m is recognized by a one-way deterministic automaton of size $O(m)$,

(2) L_m is recognized by a one-way quantum finite automaton, and

(3) any generalized one-way quantum automaton recognizing L_m with some constant probability greater than 1/2 has $2\Omega(m)$ states.

Theorem 2.65 compares classical and quantum automata for checking if a given input is a small even number (an even number less than $2m+1$). The proof of the first two parts of Theorem 2.61 is easy. Figure 2-51 shows a DFA with $2m+3$ states for the language L_m. Also, since each L_m is a finite language, there is a one-way reversible finite automaton, and hence a one-way QFA that accepts it. What then remains to be shown is the lower bound on the size of a one-way GQFA accepting the language. Define an r-restricted one-way GQFA for a language L as a one-way GQFA that recognizes the language with probability $p>1/2$, and which halts with nonzero probability before seeing the right end marker only after it has read r letters of the input. First show a lower bound on the size of m-restricted GQFAs that accept L_m. Let M be any m-restricted GQFA accepting L_m with constant probability $p>1/2$. So, at the end of reading the entire m-bit input string, the state of M can be shown to have entropy of at least $(1-H(p))m$. However, this entropy is bounded by $\log|Q|$ by Fact 2.56, where Q is the set of basis states of M. This gives the claimed bound, as explained in detail below.

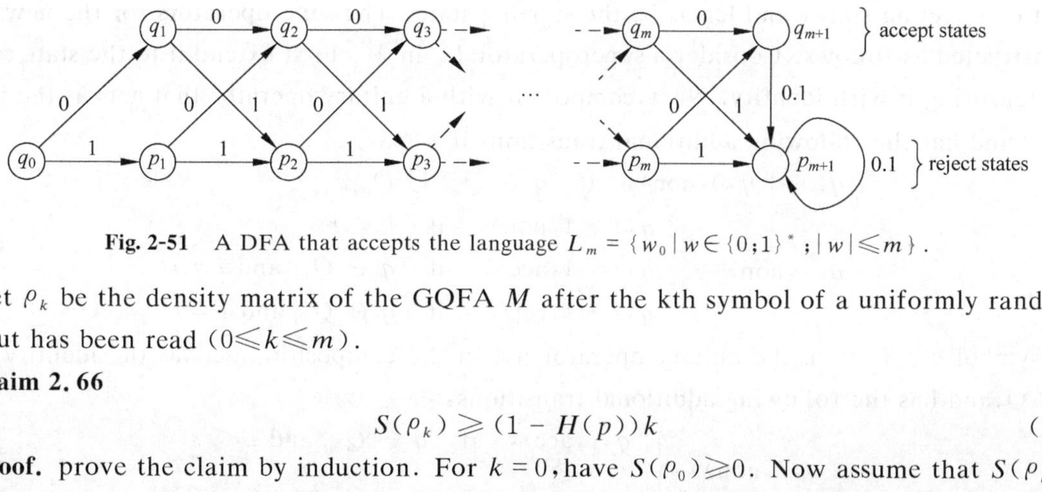

Fig. 2-51 A DFA that accepts the language $L_m=\{w_0 \mid w\in\{0;1\}^*;|w|\leqslant m\}$.

Let ρ_k be the density matrix of the GQFA M after the kth symbol of a uniformly random m-bit input has been read $(0\leqslant k\leqslant m)$.

Claim 2.66

$$S(\rho_k)\geqslant(1-H(p))k \tag{2-533}$$

Proof. prove the claim by induction. For $k=0$, have $S(\rho_0)\geqslant 0$. Now assume that $S(\rho_{k-1})\geqslant(1-H(p))(k-1)$. After the kth random input symbol is read, and the unitary transformation U_a is applied, the state of M becomes $\rho_k=1/2(U_0\rho_{k-1}+U_1\rho_{k-1})$. By the definition of M, if to see the right end marker, can learn the value of the last bit b: there is a local measurement O that yields b with probability at least $p>1/2$. So, have

$$S(\rho_k)\geqslant\frac{1}{2}(S(\mathcal{U}_0\rho_{k-1})+S(\mathcal{U}_1\rho_{k-1}))+(1-H(p)) \tag{2-534}$$

But the entropy of a mixed state is preserved by unitary transformations, and may not decrease when subjected to a von Neumann measurement, so

$$S(\mathcal{U}_b\rho_{k-1})\geqslant S(\rho_{k-1})\geqslant(1-H(p))(k-1) \tag{2-535}$$

Inequality (2-534) now gives the claimed bound. It only remains to show that the lower bound on the size of restricted GQFAs obtained above implies a lower bound on the size of arbitrary GQFAs accepting L_m. This by showing that we can convert any one-way GQFA to an r-restricted one-way

GQFA which is only $O(r)$ times as large as the original GQFA.

The idea behind the construction of a restricted GQFA, which given an arbitrary GQFA, as follows. Carry the halting parts of the state of the original automaton as "distinguished" nonhalting parts of the state of the new automaton till at least r more symbols of the input read since the halting part was generated, or until the right end marker is encountered. Then map the distinguished parts of the state to accepting or rejecting subspaces appropriately.

Lemma 2. 67 Let M be a one-way GQFA with S states recognizing a language L with probability p. Then there is an r-restricted one-way GQFA M0 with $O(rS)$ states, that recognizes L with probability p.

Proof. Let M be a GQFA with Q as the set of basis states, Q_{acc} as the set of accepting states, Q_{rej} as the set of rejecting states, and q_0 as the starting state. Let M_0 be the automaton with basis state set

$$Q \cup (Q_{acc} \times \{0,1,\cdots,r+1\} \times \{acc,non\}) \\ \cup (Q_{rej} \times \{0,1,\cdots,r+1\} \times \{rej,non\}) \tag{2-536}$$

Let:

$$Q_{acc} \cup (Q_{acc} \times \{0,1,\cdots,r+1\} \times \{acc\}) \tag{2-537}$$

its set of accepting states, let

$$Q_{rej} \cup (Q_{rej} \times \{0,1,\cdots,r+1\} \times \{rej\}) \tag{2-538}$$

The set of rejecting states, and let q_0 be the starting state. The superoperators for the new GQFA M' constructed as follows. Consider a superoperator U_σ in M. First extend it to the state space of M_0 by tensoring it with identity. Next, compose it with a unitary operator that acts as the identity on C^{Qnon} and has the following additional transitions if $\sigma \neq \$$:

$$|q\rangle \mapsto |q,0,non\rangle, \quad \text{if} \quad q \in Q_{acc} \cup Q_{rej}$$

$$|q,i,non\rangle \mapsto \begin{cases} |q,i+1,non\rangle, & \text{if} \quad i < r \\ |q,i+1,acc\rangle, & \text{if} \quad q \in Q_{acc} \text{ and } i = r \\ |q,i+1,rej\rangle, & \text{if} \quad q \in Q_{rej} \text{ and } i = r \end{cases} \tag{2-539}$$

If the symbol $\sigma = \$$, then the unitary operator use in the composition acts as the identity on the space CQ, and has the following additional transitions:

$$|q,i,non\rangle \mapsto \begin{cases} |q,i,acc\rangle, & \text{if} \quad q \in Q_{acc} \text{ and } i \leqslant r \\ |q,i,rej\rangle, & \text{if} \quad q \in Q_{rej} \text{ and } i \leqslant r \end{cases} \tag{2-540}$$

This gives the superoperator for the symbol σ in the new GQFA M'. It is not difficult to verify that M' is an r-restricted one-way QFA (of size $O(rS)$) accepting the same language as M, and with the same probability. The bounds were slightly generalized (to the case of interactive communication with prior entanglement) in Nayak. Buhrman and deWolf observed that our results imply an $\Omega(m)$ lower bound for the single-round communication complexity of determining whether two subsets of $\{1,2,\cdots,m\}$ are disjoint. There is an $O(\sqrt{m}\log m)$ qubit protocol with $O(pm)$ rounds of communication for this problem, so we see that greater interaction leads to a decrease in the communication required to solve certain problems.

As noted in Nayak, the results imply a stronger dependence of communication complexity on the number of rounds. Suppose there are two players Alice and Bob. Alice holds an m bit string $x \in \{0; 1\}m$ and Bob holds $i \in \{1,2,\cdots,m\}$. Bob would like to know the value x_i. If allow two rounds of interaction, Bob can send i to Alice, who can respond with the value x_i, and the overall communication cost is log $(m)+1$. On the other hand, if the players are limited to sending one message, then the result shows that $\Omega(m)$ qubits of communication are necessary. This was further extended in Klauck et al. showing an exponential separation between quantum communication complexity with k and $k+1$ rounds of message exchange, for any constant k.

2.9.4 Quantum advantage for dense coding

Entanglement plays a central role in quantum information theory, especially in quantum communication. It is a physical resource exploited in tasks like teleportation and dense coding. In the last communication problem, the sender, Alice, and the receiver, Bob, share a pair of two-level systems, or qubits, in a maximally entangled state:

$$|\psi_0\rangle = \frac{|00\rangle + |11\rangle}{\sqrt{2}} \tag{2-541}$$

Alice can transmit two classical bits of information sending her qubit to Bob, i.e. with a exchange of just one qubit. To achieve this result, Alice sends her qubit after having applied an appropriate unitary rotation, corresponding for example to the identity σ_0 and the Pauli matrices σ_i, $i=1,2,3$. The resulting states $|\psi_\mu\rangle = (\sigma\mu \otimes 1)|\psi_0\rangle$ (Bell states) are orthogonal, so that when Bob has received Alice's subsystem they can be unambiguously distinguished.

The previous result is possible because the two parties share initially an entangled state. Indeed, the Holevo bound implies that one qubit may carry at most one classical bit of information, if no pre-established (quantum) correlations between the parties exist. Unfortunately, in real world applications that have to deal with imperfect knowledge and noisy operations, therefore the resulting (shared) quantum states, which are mixed and described in terms of density matrices. From the point of view of quantum information, this is most relevant in the distant-labs paradigm. In this case two (or more) parties may want to share a maximally entangled system, but their actions are limited to local operations and classical communication (LOCC), so that they cannot create shared entangled states, but only process by LOCC pre-existing (imperfect) quantum correlated resources. One possible way out entanglement distillation, which can realized by LOCC. Therefore, it is interesting to study coding protocols for Alice to send classical information to Bob, directly acting on copies of a shared mixed state ρ_{AB}. The problem was further generalized, where the notion of distributed quantum dense coding was introduced: in that picture, many senders, called Alices, share states with many receivers, called Bobs. In the encoding is purely unitary, as in the standard pure two-parties setting, i.e. a letter in an alphabet is associated to a unitary operation. With this protocol, it was shown that, for a given network scheme—i.e., for a given choice of which Alice sends her subsystem to which Bob—the possibility of dense coding does not depend on the allowed operations among the senders, but does depend on the allowed operations among the receivers, i.e. on the allowed decoding processes. In the case global operations are allowed among the Bob's, i.e. there is essentially only a single receiver, dense coding is possible if and only if the coherent information between the Alices and the Bobs,

$$I(A\rangle B) = S(\rho_B) - S(\rho_{AB}) \tag{2-542}$$

with $S(\sigma) = -\mathrm{Tr}\,\rho\log\rho$ the von Neumann entropy, is strictly positive. Both $I(A\rangle B)$ and $I(B\rangle A)$ are less or equal to zero for separable or bound entangled states. Indeed, if $I(A\rangle B)$ or $I(B\rangle A)$ is strictly positive, that know that the state ρ_B is distillable thanks to the hashing inequality. Therefore neither separable nor bound entangled states (along the cut $A:B$) are useful for dense coding. In the present paper we consider a general encoding scheme: each letter in an alphabet is associated to a completely positive trace preserving (CPTP) map. Such a scheme was already presented, in the one-sender-one-receiver context, where it was found that the optimal encoding is still unitary, but only after a "pre-processing" operation which optimizes the coherent information $I(A\rangle B)$ between the parties and is independent from the letter of the alphabet to be sent. Indeed, while no CPTP operation by Bob can increase $I(A\rangle B)$, an operation by Alice can. The simplest example was given the sender discards a noisy subsystem of hers, which is factorized with respect to the remnant state and as such can not increase the capacity, i. e., the maximal rate of transmission. As consequence, in the case of multisenders, the capacity depends on the allowed operations among the senders because this may restrict the pre-processing operation to be non-optimal (with respect to the global operation, i. e. to the one-sender setting). This may understood considering the case where the noise to be traced out to increase the coherent information, is not concentrated in some factorized subsystem, but spread over many subsystems; it may be possible to concentrate and discard such noise with some global (on Alice's side) operation, but not with local ones. A classification of quantum states according to their usefulness for distributed dense coding with respect to the allowed operation on the receiver's side depicted. Considering pre-processing, a similar classification can made with respect to the allowed operations on the sender's side. Provide concrete examples of states which are useful for dense coding only if, e. g., global operations are allowed among the senders—i. e., there is only one sender—but not if they are limited to LOCC (similarly, there are states that are useful only if LOCC are allowed, but not if the senders are limited to local operations). Remark that the constraints on the usefulness of a given multipartite state, based on the allowed operations, persist even in the most general scenario of preprocessing on many copies. As a general observation about dense coding, further show that, in the basic one-sender-one-receiver scenario, what call the quantum advantage of dense coding satisfies a monogamy relation with the so-called entanglement of purification. The section organized as follows. Define the terms of the problem and present a formula for the capacity of two-party dense coding with pre-processing. The many-senders-one-receiver setting, specifying the various classes of operations that will allow among the senders described in the section. Then analyze the many-senders-onereceiver setting, giving examples of how the dense-codeability of multipartite states depends on the allowed pre-processing operations among the senders, particular providing sufficient conditions for non-dense-codeability with restricted operations. Latter discuss how dense-codeability related to distillability and the concept of symmetric extensions. Briefly consider the asymptotic setting, and argue that the limits to dense-codeability presented in the previous sections remain valid, and elaborate on the monogamy relation between what call the quantum advantage of dense coding and a measure of correlations known as entanglement of purification.

- Two-party dense coding with pre-processing

Rederive here some known results for the two-party case, i. e. allow global operations on both sides (sender's and receiver's). Such results will be the backbone of our further discussion. Moreover, we discuss some subtle points of dense coding which arise already at the level of the two-party setting, and stress the importance of a quantity defined in terms of a maximization of coherent information, that we call quantum advantage of dense coding. Alice and Bob share a mixed state ρ^{AB} in dimension $d_A \otimes d_B$, i. e. $\rho^{AB} \in M(C^{d_A}) \otimes M(C^{d_B})$. The protocol want to optimize is the following.

(1) Alice performs a local CPTP map $\Lambda_i : M(C^{d_A}) \to M(C^{d'_A})$ (note that the output dimension d'_A is different than the input one d_A) with a priori probability p_i on her part of ρ^{AB}. She therefore transforms ρ^{AB} into the ensemble $\{(p_i, \rho_i^{AB})\}$, with $\rho_i^{AB} = (\Lambda_i \otimes \mathrm{id})[\rho^{AB}]$.

(2) Alice sends her part of the ensemble state to Bob.

(3) Bob, having at disposal the ensemble $\{(p_i, \rho_i^{AB})\}$, extracts the maximal possible information about the index i. Notice that, for the moment being, allow only one-copy actions, i. e. Alice acts partially only on one copy of ρ^{AB} a time. On the other hand, analyze the asymptotic regime where long sequences sent. Moreover, notice that stated as it is, the protocol requires a perfect quantum channel of dimension equal to the output dimension d'_A, i. e. the ability to send perfectly a quantum system characterized by the mentioned dimension. Now prove that the capacity (rate of information transmission per shared state used) for this protocol corresponds to

$$C^{d'_A}(\rho^{AB}) = \log d'_A + S(\rho^B) - \min_{\Lambda_A}((\Lambda_A \otimes \mathrm{id}_B)[\rho^{AB}])$$
$$= \log d'_A + \max_{\Lambda_A} I'(A\rangle B) \tag{2-543}$$

where $I'(A_i B)$ is the coherent information of the transformed state $(\Lambda_A \otimes \mathrm{id}_B)[\rho^{AB}]$. Note that the quantity depends both on the shared state ρ^{AB} and on the output dimension d'_A of the maps $\{\Lambda_i\}$, not only through the first logarithmic term, but also because the minimum runs over such maps. Reproduce here essentially the same proof used, but our attention will ultimately focus on the rate of communication per copy of the state used, not on the capacity of a perfect quantum channel of a given dimension assisted by an unlimited amount of noisy entanglement. In this sense, the approach is strictly related to the one pursued, as will discuss in the following. That there will be an important difference: there is no optimization over the output dimension of Λ_A (i. e. of the quantum channel), when it comes to consider the rate per copy of the state, with one use of the channel per copy. The capacity (attained in the asymptotic limit where Alice sends long strings of states ρ_i^{AB}) of the dense coding protocol depicted above is given by the Holevo quantity:

$$C^{d'_A}(\rho^{AB}) = \max_{\{(p_i, \Lambda_i)\}} \left(S\left(\sum_i \rho_i^{AB}\right) - \sum_i p_i S(\rho_i^{AB})\right) \tag{2-544}$$

Bound it from above considering an optimal set $\{(\hat{p}_i, \hat{\Lambda}_i)\}$. Since the entropy is subadditive and no operation by Alice can change the reduce state ρ^B, so have:

$$C^{d'_A}(\rho^{AB}) \leqslant S\left(\sum_i \hat{\rho}_i^A\right) + S(\rho^B) - \sum_i p_i S(\hat{\rho}_i^{AB})$$
$$\leqslant \log d'_A + S(\rho^B) - \min_{\Lambda_A} S((\Lambda_A \otimes \mathrm{id}_B)[\rho_{AB}]) \tag{2-545}$$

The quantity in the last line of the previous inequality, corresponding to (2-543), can be actually achieved by an encoding with:

$$p_i = 1/d'^2_A \quad \text{and} \quad \Lambda_i[X] = U_i \Lambda_A[X] U_i^\dagger$$

The unitaries $\{U_i\}_{i=1}^{d'^2_A}$ are orthogonal, $\mathrm{Tr}(U_i^\dagger U_j) = d'_A \delta_{ij}$, and satisfy $1/d'_A \sum_i U_i X U_i^\dagger = \mathrm{Tr}(X)\,\mathbb{1}$, for all $X \in \mathcal{M}(\mathbb{C}^{d'_A})$ while the CPTP map Λ_A corresponds to the pre-processing operation. Indeed, in this case

$$\sum_i \rho_i^{AB} = \mathbb{1}/d'_A \otimes \rho^B$$

so that

$$S(\sum_i \rho_i^{AB}) = \log d'_A + S(\rho_B)$$

and

$$\sum_i p_i S(\rho_i^{AB}) = S((\Lambda_A \otimes \mathrm{id}_B)[\rho_{AB}]) \tag{2-546}$$

Notice in particular that Alice may always choose to substitute her part of the shared state with a fresh ancilla in a pure state. This corresponds to a preprocessing $\Lambda_{\mathrm{sub}}[X] = \mathrm{Tr}\,(X)\,|\psi\rangle\langle\psi|$ and gives a rate $\log d'_A$, corresponding to classical transmission of information with a d'_A-long alphabet, i.e. without a quantum advantage. A quantum effect is present if $X > \log d'_A$, i.e. if a local operation on Alice side is able to reduce the entropy of the global state strictly below the local entropy of Bob, or if $I(A \rangle B) > 0$ from the very beginning. An almost trivial case where preprocessing has an important role is

$$\rho^{AA'B} = \rho^{AB} \otimes \rho^{A'}$$

with $S(\rho^{A\bar{B}}) < S(\rho^{\bar{B}})$ but

$$S(\rho^{AB}) + S(\rho^{A'}) \geqslant S(\rho^B) \tag{2-547}$$

Here we consider AA' as a composite system Alice can globally act on. Then a possible preprocessing operation is

$$\mathrm{id}_A \otimes \Lambda_{\mathrm{sub}}^{A'}[\rho^{AA'B}] = \rho^{AB} \otimes |\psi\rangle_{A'}\langle\psi|$$

(which can be realized acting on A' only).

Since the $\log d'_A$ contribution in (2-543) can be considered purely classical, choose a different way of counting the rate of transmission: indeed, it appears natural to subtract the logarithmic contribution in order to define the quantum advantage of dense coding as

$$\Delta(A\rangle B) \equiv S(\rho^B) - \inf_{\Lambda_A} S((\Lambda_A \otimes \mathrm{id}_B)[\rho_{AB}])$$
$$= \sup_{\Lambda_A} I'(A\rangle B) \tag{2-548}$$

where it makes sense to consider the infimum (or the supremum) over all maps Λ_A, with whatever output dimension. Since a possible map is Λ_{sub}, we have $\Delta(A;B) \geqslant 0$. That a state is Dense-Codeable (DC) if $\Delta(A;B)$ is strictly positive. It may be that Δ is not additive, hence to ensure that the state is not useful at all for dense coding, which one has to consider its regularization. Remark that the classification of states in terms of their dense-codeability for different classes of encoding operations, which will not depend on such redefinition. Moreover, the redefined quantity appears more information-theoretical and depends only on the state. Now recall the analysis of a similar optimization problem, which occurs in the study of entanglement of purification. Since von Neumann entropy is concave, it is sufficient to consider extremal maps. The input of the map is an operator acting on a d_A-dimensional system, thus, according to Λ_A is extremal it can be written by means of at most d_A Kraus operators, i.e., as

$$\Lambda_A[X] = \sum_{i=1}^{d_A} A_i X A_i^{\dagger} \tag{2-549}$$

The range of the operator $\Lambda_A[X]$ given by all the columns of Kraus operators A_i, each operator A_i has d_A(the input dimension) columns. Therefore, the optimal output dimension $d_{A'}$ can taken to be d_A^2, and the infimum in (2-548) is actually a minimum. It is possible to relate the quantum advantage of dense coding with entanglement of purification further. Exploiting the convexity of entropy, it is immediate to find the following upper bound for $\Delta(A\rangle B)$:

$$\Delta(A\rangle B) \leqslant S(\rho^B) - \min_A S(\rho^{AB}(A))$$

$$\rho^{AB}(A) = \frac{A \otimes \mathbb{1}\, \rho^{AB} A^\dagger \otimes \mathbb{1}}{\mathrm{Tr}(A \otimes \mathbb{1}\, \rho^{AB} A^\dagger \otimes \mathbb{1})} \tag{2-550}$$

Where, according to the reasoning of the previous paragraph, A can be taken as a $d_A \times d_A$ square matrix. Remark that this is only an upper bound: local filtering not allowed in the framework, because it requires classical communication. Moreover, with a true local filtering, the reduced density matrix ρ^B changes, while keep it fixed in Eq. (2-550). Example 1. In Eq. (2-551) the dense-codeability by unitaries of the Werner state equivalent to

$$\rho_p = p \mid \psi_0\rangle\langle\psi_0 \mid + (1-p)\frac{\mathbb{1}}{4} \tag{2-551}$$

with ψ_0 defined in (2-541), was studied. The state ρ_p is entangled for $p > 1/3$, but unitarily-DC only for $p > p_{\text{U-DC}} = 0.7476$. One ask different questions: (2-541). Is the state ρ_p DC for some $p < p_{\text{U-DC}}$ if allow general encoding operations. i. e pre-processing (2-542) Can $\Delta(A\rangle B)$ be greater that $I(A > B)$ when the latter is strictly positive Question (2-542) addresses. the problem of deciding whether a state that is not useful with some restricted encoding, which instead useful if allow more general operations. Question (2-542) addresses instead the problem of a "greater usefulness" by general encoding. In numerical evidence was found that no pre-processing Λ_A: $M(C^2) \to M(C^2)$ can enhance $I(A;B)$ in the case of a shared two-qubit state. However, note that, as previously discussed, optimal pre-processing maps are in principle of the form $\Lambda_A: M(C^2) \to M(C^4)$, i.e. with a larger output. Here concentrate instead on the bound (2-550), for which, as discussed, we can consider the matrix A as $A: C^2 \to C^2$. Observe that, thanks to the $U \otimes U^*$ symmetry of the state, and to the invariance under unitaries of both the entropy and the trace, we can take A to be diagonal in Alice's Schmidt basis for ψ_0, i.e. to be of the form:

$$A = \begin{pmatrix} r & 0 \\ 0 & 1-r \end{pmatrix}$$

with $0 \leqslant r \leqslant 1$. It is possible to compute analytically the entropy $S(\rho^{AB}(A))$. One finds that the optimal choice is $r = 0, 1$, and the bound is $1 - H_2((1+p)/2)$, where $H_2(x) = -x\log x - (1-x)\log(1-x)$ is the binary entropy. Thus, in this example we see that the bound is far from being tight, since it is strictly positive for every $p > 0$, i.e. even for separable states. Therefore, it is not possible to use it to conclude something as regards question (2-549). Anyway, it constitutes a non-trivial limit on $\Delta(A\rangle B)$, and provides some information about question (2-550).

- Preprocessing with many senders

If the there are many senders, it may happen that the operations among them are restricted, for example no communication may be allowed, or they may collaborate only trough LOCC. Both latter situations do not affect the unitary encoding part of the dense coding. Indeed, it is possible to realize the optimal unitary operations locally: each Alice acts with unitaries satisfying the optimal condition on her subsystem. Notice though that in the case of many senders and many receivers, even considering just unitary encoding, it happens that in certain cases local unitary encoding is not enough to take advantage of quantum correlations. This may be due, for example, to the fact that with restricted operations at the senders' and at the receivers', the question of who sends what to whom is really important. A very simple case where this is evident, is that of two senders A and A', and two receivers B and B', with A and B' (A' and B) sharing an EPR pair, A (A') sending her quantum system to B (B'), and both the senders and the receivers restricted to act by local operations. It is then clear that though the senders and the receivers share two EPR pairs, these are useless for the sake of dense coding, because they happen to pertain to the wrong pairs of senders and receivers. It is also obvious that the availability of global operations—or even just of the swap operation—on the senders' side would make dense coding possible, by using fully the quantum correlations existing between the set of senders and the set of receivers. Considering the case of many senders and one receiver, but allowing encoding by general operations, one realizes that it might not be possible to apply the optimal pre-processing map. Indeed, just repeating the considerations which brought to (2-543), it is clear that, in the case of many senders and one receiver, the dense coding capacity may be expressed as

$$\chi_O^{d'_A} = \log d'_A + S(\rho^B) - \min_{\Lambda \in O}((\Lambda \otimes \mathrm{id})[\rho^{AB}]) \tag{2-552}$$

where O is the set of allowed operations on the senders' side, for example global (G), LOCC, or LO. Obviously,

$$\chi_G^{d'_A} \geqslant \chi_{\mathrm{LOCC}}^{d'_A} \geqslant \chi_{\mathrm{LO}}^{d'_A} \geqslant \log d'_A \tag{2-553}$$

The capacity corresponds at least to the classical one with many senders and one receiver, because it is always possible for the Alices to apply locally the substitution map $\Lambda_{\mathrm{sub}}^{A} = \Lambda_{\mathrm{sub}}^{A1} \otimes \Lambda_{\mathrm{sub}}^{A2} \cdots \otimes \Lambda_{\mathrm{sub}}^{AN}$. The only subtle point is the compatibility of the choice of the target output dimension d'_A: suppose it is always of the factorized form $d'_A = d'_{A1} d'_{A2} \cdots d'_{AN}$, so that it can be achieved exactly by an optimal local unitary encoding. It corresponds to the case where there is only one sender. Thus, can define the corresponding (non-negative) quantum advantages

$$\Delta_G \geqslant \Delta_{\mathrm{LOCC}} \geqslant \Delta_{\mathrm{LO}} \geqslant 0 \tag{2-554}$$

Can obtain an upper bound for Δ_{LOCC}—and therefore valid also for Δ_{LO}—similar to the presented in (2-550):

$$\Delta_{\mathrm{LOCC}}(A \rangle B) \leqslant S(\rho^B) - \min_{A_{\mathrm{prod}}} S\left(\frac{A_{\mathrm{prod}} \otimes \mathbb{1}\, \rho^{AB} A_{\mathrm{prod}}^{\dagger} \otimes \mathbb{1}}{\mathrm{Tr}(A_{\mathrm{prod}} \otimes \mathbb{1}\, \rho^{AB} A_{\mathrm{prod}}^{\dagger} \otimes \mathbb{1})}\right) \tag{2-555}$$

Where Aprod $= A_1 \otimes A_2 \otimes \cdots \otimes A_N$, with each A_i a $d_{Ai} \times d_{Ai}$ square matrix.

- Examples of the hierarchy of capacities for multi-senders

To provide examples of the hierarchy (2-553), more precisely of shared states that are not DC for certain classes of allowed operations among senders, but are DC for more general operations.

First analyze the case where the state is not LO-DC but it is LOCC-DC: $\Delta_{\mathrm{LO}} = 0$ while $\Delta_{\mathrm{LOCC}} > 0$. which need the following lemma:

Lemma 2.68 Consider a tripartite state $\rho_{AA'B}$ such that (2-556) it is separable under the A': AB cut, and (2-557) its reduction ρ_{AB} is also separable. Then, after any bilocal operation $\Lambda_{AA'} = \Lambda_A \otimes \Lambda_{A'}$ of parties A and A', we have $I'(AA'_i B) \leqslant 0$.

Proof. For separable states coherent information is always non-positive. For any state separable under A': AB cut then have:

$$S(A'AB) \geqslant S(AB) \tag{2-556}$$

Now, if the state ρ_{AB} is also separable, then:

$$S(AB) \geqslant S(B) \tag{2-557}$$

Thus, for a tripartite state separable along the A': AB cut and such that its AB reduction is separable, $I(AA'_{;}B) \leqslant 0$. Moreover, after any bilocal operation $\Lambda_A \otimes \Lambda_{A'}$ the state still satisfies the above separability features so that $I'(AA'_{;}B) \leqslant 0$. Note that the separability properties used in the previous lemma may not preserved when the parties AA' can communicate classically.

Consider the state:

$$\rho_{AA'B} = \frac{1}{2}(|\phi_0\rangle\langle\phi_0| + |\phi_1\rangle\langle\phi_1|) \tag{2-558}$$

where

$$|\phi_0\rangle = |0\rangle_{A'} \otimes \frac{1}{\sqrt{2}}(|00\rangle_{AB} + |11\rangle_{AB}) \tag{2-559}$$

$$|\phi_1\rangle = |1\rangle_{A'} \otimes \frac{1}{\sqrt{2}}(|01\rangle_{AB} + |10\rangle_{AB}) \tag{2-560}$$

The initial entropies $S(AA'B) = S(B) = 1$ and $I(AA'_{;}B) = 0$. The state is explicitly separable with respect to the A': AB cut. The trace over A' gives an equal mixture of two qubit orthogonal maximally entangled states, which hence it is separable. Thus, according to Lemma 2.64, parties A and A' cannot locally decrease the total entropy below $S(B)$. Namely, A' can measure in the $\{0,1\}$ basis, communicate the result to A, and further substitute the subsystem A' with one in a pure state. Then, after a suitable local unitary rotation, A will share a maximally entangled state with B and $S'(AA'B) = 0$, so that $I'(AA'_{;}B) = 1$. The previous example is the simplest possible one that illustrates how some "spread" noise, which conflicts with (unitary) dense coding can be undone only allowing operations among the senders that are more general than local operations. Indeed, in the system AA' one can single out a virtual qubit, carrying the whole noise. The noisy qubit encoded non-locally into the system AA', so that the senders do not have local access to it. Effective tracing out of the unwanted noise (prior to unitary encoding) is possible only if A and A' communicate. Indeed, one can go from $\rho_{AA'B}$ to $\rho_{\tilde{A}'\tilde{A}'B} = \mathbb{1}_{\tilde{A}'/2} \otimes \varphi^{\tilde{A}B}$ by an invertible A: A'-LOCC operation, but not by an A: A'-LO operation. The previous tripartite (two senders, one receiver) case can be generalized straightforwardly. Following the definition of multipartite mutual information, one can define a quantity, which is not an entanglement measure, but may be useful:

$$D(B:A_1:\cdots:A_n) \equiv E(B:A_1) + E(BA_1:A_2) + \cdots + E(BA_1A_2\cdots A_{n-1}:A_n) \tag{2-561}$$

where E is any entanglement parameter, i.e. it is positive and $E(X:Y) = 0$ if and only if the state $\rho_{X:Y}$ is separable. Similarly as in tripartite case, can obtain that if D is zero, then parties $A_1A_2\cdots A_n$ cannot make the global entropy be less than $S(B)$ by LO (but not necessarily by LOCC), so that the state is useless for superdense coding from $A_1A_2\cdots A_n$ to B. Consider the state

$$\rho_{A_1\cdots A_nB} = \frac{1}{2^{n-1}} \sum_{i_2,\cdots,i_n=0}^{1} |i_2\rangle\langle i_2| \otimes \cdots \otimes |i_n\rangle\langle i_n| \otimes \left(\left(\sigma^{A_1}_{\oplus_{j=2}^{n} i_j} \otimes \mathbb{1}^B\right) P^0_{A_1B} \left(\sigma^{A_1}_{\oplus_{j=2}^{n} i_j} \otimes \mathbb{1}^B\right)\right) \tag{2-562}$$

where σ_0 and σ_1 are the identity and the flip operator, respectively, P_0 is the projector onto the maximally entangled state $|\psi_0\rangle$, and $\oplus$ corresponds to addition modulo 2. It easily checked that $\rho_{A1}\cdots A_nB$ satisfies $D=0$. The unitary rotation σ_a applied to the A_1 part of the maximally entangled state depends on the "parity" of the state of the other Alices. It is correctly identified if all $A_2,\cdots,A_n$ measure their qubits in the computational basis, and communicate their results to A_1, which can then share a singlet with B.

- G-DC but not LOCC-DC

To have an example of a state for which $\Delta_{LOCC}=0$ while $\Delta_G>0$, consider the Smolin state

$$\rho_{A_1BA_2A_3}=\sum_{\mu=0}^{3}|\psi_\mu\rangle_{A_1B}\langle\psi_\mu|\otimes|\psi_\mu\rangle_{A_2A_3}\langle\psi_\mu| \tag{2-563}$$

Where ψ_μ are Bell states. Note that states ψ_μ are indistinguishable by LOCC. Hence it seems reasonable, that the state cannot be used for super dense coding, even if the parties A_1, A_2 and A_3 can use LOCC. For example, the parties A_2 A_3 cannot distinguish which Bell state they have, hence cannot tell A_1 what rotation to apply, in order to share singlet with B. Let us now prove that this is true. The state is A_1 A_2: BA_3 separable (from Eq.(2-563) it is explicitly $A_1B:A_2A_3$ separable, however it is permutationally invariant). After any LOCC operation this will not change. Thus the output state of systems A_1B will be separable, hence $S'(A_1B)\geqslant S'(B)=S(B)$. Moreover the total output state will remain $A_1B:A_2A_3$ separable, which implies $S'(A_1BA_2A_3)\geqslant S'(A_1B)$. Combining the two inequalities can get

$$S'(A_1BA_2A_3)\geqslant S(B) \tag{2-564}$$

Of course, if for example A_2 and A_3 could meet and perform global operations, the state would become useful for dense coding, as they could help A_1 to share a singlet with B.

- Limits on pre-processing from one-way distillability and symmetric extensions

The possibility of global pre-processing makes non-trivial the identification of states which, although $A_1,A_2,\cdots,A_n$: B is entangled, are not G-DC ($\Delta G\leqslant 0$). One has to exclude that the coherent information can made strictly positive by any action on the side of Alices. It will now see how this may related to one-way distillation and the concept of symmetric extension. Since we will focus on global operations, may as well consider a bipartite setting. Loosely speaking, entanglement distillation consists of the process of obtaining m copies of the highly entangled pure states, starting from n copies of a mixed entangled state, by means of a restricted class of operations that cannot create entanglement. The optimal rate, i.e. the optimal ratio m/n, of the conversion for n that goes to infinity, is the distillable entanglement under the given constraint on operations. The class may be chosen to be LOCC operations—in such case speak simply of distillable entanglement—or, more restrictively, one-way LOCC operations, for which classical communication is allowed only from one party to the other, and not in both directions. In the latter speak of one-way distillable entanglement. If suppose that the communication goes from Alice to Bob, it has been showed that the one-way distillable entanglement ED ($A\rangle B$) of a state ρ^{AB} satisfies the hashing inequality

$$\mathrm{ED}(A\rangle B)\geqslant I(A\rangle B) \tag{2-565}$$

hence

$$\mathrm{ED}(A\rangle B)\geqslant \Delta(A\rangle B) \tag{2-566}$$

i.e., it is greater than the quantum advantage of dense coding. It follows that any DC state is not only distillable, which even one-way distillable. In turn, if a state is not one-way distillable, then it cannot be DC. There are entangled states for which know ED $(A\rangle B)=0$: states which admit B-symmetric extensions. A state ρ^{AB} admits a B-symmetric extension if there exists a state $\sigma^{ABB'}$ such that its reductions satisfy $\sigma^{AB}=\sigma^{AB'}=\rho^{AB}$. Suppose ρ^{AB} has a tripartite symmetric extension $\sigma^{ABB'}$ and is at the same time one-way distillable. A one-way distillation protocol consists of an Alice operation whose result—the index of the Kraus operator in (2-549)—communicated to the other party. Bob can then perform an operation depending on the result received; no further action of Alice is required. The communication involved is classical, so it can freely sent to many parties. If, having at disposal $\rho^{ABB'}$, run the oneway LOCC protocol, which by hypothesis allows distillation, in parallel between A and B, and A and B', end up with a subsystem A which is at the same time maximally entangled both with B and B$'$. However, this is impossible, because of monogamy of entanglement. Conclude that a one-way distillable state does not admit a symmetric B-extension. As regards the case of the two-qubit Werner state, it proved that it admits a symmetric extension for $p\leqslant 2/3$.

- Limits on many-copies processing

The examples of the classification discussed depend only on relations among entropies, which rely on separability properties. As such, the action on many copies of the state at disposal not help. Indeed, following, one can define the quantum advantage per copy when the encoding allowed on n-copies of the state at the same time:

$$\Delta^{(n)}(A\rangle B)=\frac{1}{n}\Delta(A\rangle B)_{\rho_{AB}^{\otimes n}}=S(\rho^{B})-\frac{1}{n}\min_{\Lambda_A^{(n)}}S((\Lambda_A^{(n)}\otimes \mathrm{id}_B)[\rho_{AB}^{\otimes n}]) \tag{2-567}$$

where now $\Lambda_A^{(n)}$ acts on (C_A^{dn}), and the asymptotic quantum advantage per copy:

$$\Delta^{\infty}(A\rangle B)=\lim_{n\to\infty}\Delta^{(n)}(A\rangle B) \tag{2-568}$$

Correspondingly, one has the multipartite quantum advantages $\Delta_O^{(n)}(A\rangle B)$ and $\Delta_O^{\infty}(A\rangle B)$ where the Alices are restricted to the class of operations O. It is clear that $\Delta_{\mathrm{LO}}^{(n)}(A\rangle B)=\Delta_{\mathrm{LO}}^{\infty}(A\rangle B)=0$ and $0<\Delta_{\mathrm{LOCC}}^{(n)}(A\rangle B)\leqslant\Delta_{\mathrm{LOCC}}^{\infty}(A\rangle B)$ for the state (2-558), while $\Delta_{\mathrm{LOCC}}^{(n)}(A\rangle B)=\Delta_{\mathrm{LOCC}}^{\infty}(A\rangle B)=0$ and $0<\Delta_G^{(n)}(A\rangle B)\leqslant\Delta_G^{\infty}(A\rangle B)$ for the state (2-558) for the Smolin state (2-563).

- Monogamy relation between entanglement of purification and the advantage of dense coding

There are similarities in the calculation of the advantage of dense coding and in that of entanglement of purification. In this paragraph will see that this relation is more than a coincidence: there is in fact a monogamy relation between the advantage of dense coding and the entanglement of purification, that does not seem to have already been reported in literature. Start by recalling the definition of entanglement of purification for a bipartite state $\rho_{AB}\in M(\mathbb{C}^{d_A})\otimes M(\mathbb{C}^{d_B})$:

$$E_p(\rho_{AB})=E_p(A:B)=\min_{\psi:\mathrm{Tr}_{A'B'}(\psi)=\rho_{AB}}S(\psi_{AA'}) \tag{2-569}$$

where the minimum runs over all purifications:

$$\psi=|\psi\rangle\langle\psi|_{AA'BB'},\ |\psi\rangle\in\mathbb{C}^{d_A}\otimes\mathbb{C}^{d_{A'}}\otimes\mathbb{C}^{d_B}\otimes\mathbb{C}^{d_{B'}} \tag{2-570}$$

Such that $\mathrm{Tr}_{A'B'}(\psi)=\rho_{AB}$. Entanglement of purification is a measure of total correlations, where all correlations—even those of separable states somehow thought as being due to entanglement. Indeed, in the bipartite pure-state case, the entropy of one subsystem is an entanglement measure. For $\{\lambda_i, |\lambda_i\rangle\langle\lambda_i|\}$ the spectral ensemble of ρ_{AB}, consider its purification

$$|\psi\rangle = \sum_i \sqrt{\lambda_i}\ |\lambda_i\rangle_{AB}\ |i\rangle_{A'}\ |0\rangle_{B'} = |\psi\rangle_{AA'B}\ |0\rangle_{B'} \tag{2-571}$$

Then any other purification can be obtained from $|\psi\rangle$ by means of an isometry $U_{A'B'}$ as $|\psi\rangle = U_{AB} \otimes U_{A'B'|}\ |\psi\rangle$ Following, can find

$$\begin{aligned}
\psi_{AA'} &= \mathrm{Tr}_{BB'}(\psi_{AA'BB'}) \\
&= \mathrm{Tr}_{BB'}(U_{A'B'}\bar{\psi}_{AA'B} \otimes |0\rangle\langle 0|_{B'} U^{\dagger}_{A'B'}) \\
&= (\Lambda_{A'} \otimes \mathrm{id}_A)[\mathrm{Tr}_B(\bar{\psi}_{AA'B})] \\
&= (\Lambda_{A'} \otimes \mathrm{id}_A)[(\bar{\psi}_{AA'})]
\end{aligned} \tag{2-572}$$

Where

$$\Lambda_{A'}[X_{A'}] = \mathrm{Tr}_{B'}(U_{A'B'}X_{A'} \otimes |0\rangle\langle 0|_{B'} U^{\dagger}_{A'B'}), \text{for all } X_{A'} \in C(\mathbb{C}^{d_{A'}}) \tag{2-573}$$

By varying $U_{A'B'}$ that is the purification ψ vary $\Lambda_{A'}$. Thus,

$$E_p(\rho_{AB}) \equiv \min_{\Lambda_{A'}} S((\Lambda_{A'} \otimes \mathrm{id}_A)[\bar{\psi}_{AA'}]) \tag{2-574}$$

we then conclude that, given a pure tripartite state ψ_{ABC}, one has

$$S(B) = \Delta(A\rangle B) + E_p(B:C) \tag{2-575}$$

For fixed entropy $S(B)$, this means that the more B is correlated with C, the less dense coding is advantageous from A to B. For a tripartite mixed state ρ_{ABC}, following may consider a purification ψ_{ABCD}, and apply equation (2-575) to the three parties (AD), B and C to find

$$S(B) \geqslant \Delta(A\rangle B) + E_p(B:C) \tag{2-576}$$

Indeed, from the definition of advantage it is easy to check that $\Delta(AD\rangle B) \geqslant \Delta(A\rangle B)$ for all tripartite states ρ_{ABD}, in particular for the ABD reduction of ψ_{ABCD}. Following again, may consider the asymptotic case, applying the just found relations to $\varphi_{ABC}^{\otimes n}$ ($\rho_{ABC}^{\otimes n}$), using the additivity of von Neumann entropy, dividing by n, and taking the limit $n \to \infty$, to find

$$S(B) = \Delta^{\infty}(A\rangle B) + E_{\mathrm{LO}q}(B:C) \tag{2-577}$$

and

$$S(B) \geqslant \Delta^{\infty}(A\rangle B) + E_{\mathrm{LO}q}(B:C) \tag{2-578}$$

for the case of pure and mixed states, respectively. Here:

$E_{\mathrm{LO}q}(A:B) = \lim_n \frac{1}{n} E_p(\rho_{AB}^{\otimes n})$ is the cost—in singlets—to create ρ^{AB} in the asymptotic regime, allowing approximation, from an initial supply of EPR-pairs by means of local operations and asymptotically vanishing communication. For the pure state case it is fascinating to put together the results of Theorem and the present ones, to find relations between different notions of correlations and entanglement measures/parameters:

$$I_{HV}(A\rangle B) - \Delta(A\rangle B) = E_p(B:C) - E_F(B:C) \tag{2-579}$$

$$C_D(A\rangle B) - \Delta^{\infty}(A\rangle B) = E_{\mathrm{LO}q}(B:C) - E_C(B:C) \tag{2-580}$$

where, for a bipartite state ρ^{AB}: I_{HV} is the measure of correlations defined in as

$$I_{\mathrm{HV}}(A\rangle B) = \max_{\{M_x\}}[S(\rho_B) - \sum_x p_x S(\rho_B^x)] \tag{2-581}$$

where the maximum taken over all the POVMs $\{M_x\}$, hich applied on system, A, $p_x \equiv \mathrm{Tr}((M_x \otimes I)\rho_{AB})$ is the probability of the outcome x, $\rho_x B \equiv \mathrm{Tr}A((M_x \otimes I)\rho_{AB})/p_x$ is the conditional state on B given the outcome x on A. here $p_x \rho_x B = \mathrm{Tr}\ A(\rho_{AB})$. CD is the common randomness distillable by means of one-way classical communication from A to B, that is the net amount of correlated classical bits that A and B can asymptotically share starting from an initial supply of copies of ρ_{AB}. Iit is equal to $CD(A\rangle B) = -\lim n \ln I_{\mathrm{HV}}(A\rangle B)\rho_{AB}^{\otimes n}$; E_F is the entanglement of formation:

$$E_F(A:B) = \min_{\{(p_i,\psi_i^{AB})\}} \sum_i p_i S(\psi_i^A) \tag{2-582}$$

Where the minimum runs over all pure ensembles such that:

$$\sum_i p_i \psi_i^{AB} = \rho_{AB} \tag{2-583}$$

E_C is the entanglement cost, that is the cost—in singlets—to create ρ_{BC} in the asymptotic regime, allowing approximation, from an initial supply of EPR-pairs by means of local operations and classical communication; it is equal to $E_C(A:B) = \lim n 1/EF(A:B)\rho_{AB}^{\otimes n}$. Note that the differences appearing in (2-579) and (2-580) are positive.

The difference with the results presented is two-fold. Firstly, in defining the quantum advantage of dense coding, immediately consider a maximum over maps without restricting the dimension of the output. This means that focus on the property of the state, rather than of a couple state + channel. Secondly, exactly for the same reason, do not distinguish between many uses of the state and many uses of the channel: the rate always defined in terms of the number of copies of the state used, even when allow encoding on many copies. The two facts make the quantities Δ and Δ^{∞} different from all the ones presented. In particular, we claim that the quantity Δ^{∞} is more information theoretical than the quantity:

$$\overline{DC^{(\infty)}}(\rho) = 1 + \sup_n \sup_{\Lambda_A} \frac{nS(\rho^B) - S((\Lambda_A \otimes \mathrm{id}^{\otimes n})[\rho^{\otimes n}])}{S(\rho_A^{\otimes n})} \tag{2-584}$$

which, according to correspond to the rate of classical communication per qubit sent, i.e. per use of a two dimensional quantum channel. Indeed, in the latter case one considers the use of whatever number of copies of the shared state per use of the channel. In particular, remark that for pure states one has $DC^{(\infty)}(\psi^{AB}) = 2$ as soon as the state AB is entangled—whatever the degree of its entanglement—while $\Delta(\psi^{AB}) = S(\rho^B)$. Further, notice that the distinction of usefulness of states for dense coding according to the allowed encoding operations, holds also for the quantities presented, as it is evident. Considered the transmission of classical information by exploiting (many copies of) a shared quantum state, both in the bipartite and in the multipartite—more specifically, in the many-to-one—setting. Discussed fundamental limits on the usefulness of states for multipartite dense coding, for given constraints on the operations allowed among senders. Such limits are not removed even if allow the most general encoding under whatever number of copies of the shared state. Such analysis leads to a non-trivial classification of quantum states, parallel to the one suggested, where constraints on the operations allowed on the receivers side (in a many-to-many communication setting) considered. Indeed, one can depict a subdivision of multipartite states into classes of states that are many-to-one dense-codeable if certain operations, for example LOCC, are allowed among the senders, but not if the senders are restricted to local operations. Finally, focussing on general properties of dense-codeability of states, observed that there exist a monogamy relation between the quantum advantage of dense coding and the entanglement of purification. Such a relation puts in quantitative terms the fact that the quantum advantage of dense coding is (or can be) large (only) if the disorder—as quantified by the von Neumann entropy—of the receiver is due to correlations with the sender, rather than with a third party.

2.10 Quantum data compression

Consider a long message consisting of n letters, where each letter is chosen at random from the ensemble of pure states:

$$\{|\varphi_x\rangle, p_x\} \tag{2-585}$$

and the $|\varphi_x\rangle$'s are not necessarily mutually orthogonal. (For example, each $|\varphi_x\rangle$ might be the polarization state of a single photon.) Thus, each letter is described by the density matrix

$$\rho = \sum_x p_x |\varphi_x\rangle\langle\varphi_x| \tag{2-586}$$

and the entire message has the density matrix

$$\rho^n = \rho \otimes \cdots \otimes \rho \tag{2-587}$$

How redundant is this quantum information. To devise a quantum code that enables us to compress the message to a smaller Hilbert space, but without compromising the fidelity of the message. For example, perhaps have a quantum memory device (the hard disk of a quantum computer), and we know the statistical properties of the recorded data (i.e., know ρ). When want to conserve space on the device by compressing the data. The optimal compression can found by Ben Schumacher. The best possible compression compatible with arbitrarily good fidelity as $n\to\infty$ is compression to a Hilbert space H with

$$\log(\dim \mathcal{H}) = nS(\rho) \tag{2-588}$$

In this sense, the Von Neumann entropy is the number of qubits of quantum information carried per letter of the message. For example, if the message consists of n photon polarization states, we can compress the message to $m = nS(\rho)$ photons—compression is always possible unless $\rho = 1/2\,\mathbb{I}$. It can't compress random qubits just as we can't compress random bits. Once Shannon's results known and understood, the proof of Schumacher's theorem is not difficult. Schumacher's important contribution was to ask the right question, and so to establish for the first time a precise (quantum) information theoretic interpretation of Von Neumann entropy.

2.10.1 Quantum data compression: an example

Before discussing Schumacher's protocol of quantum data compression in full generality, it is helpful to consider a simple example. So suppose that the letters are single qubits drawn from the ensemble

$$\begin{cases} |\uparrow_z\rangle = \begin{pmatrix}1\\0\end{pmatrix}, & p = \dfrac{1}{2} \\ |\uparrow_x\rangle = \begin{pmatrix}1/\sqrt{2}\\1/\sqrt{2}\end{pmatrix}, & p = \dfrac{1}{2} \end{cases} \tag{2-589}$$

So that the density matrix of each letter is

$$\rho = \frac{1}{2}|\uparrow_z\rangle\langle\uparrow_z| + \frac{1}{2}|\uparrow_x\rangle\langle\uparrow_x|$$

$$= \frac{1}{2}\begin{pmatrix}1 & 0\\0 & 0\end{pmatrix} + \frac{1}{2}\begin{pmatrix}\frac{1}{2} & \frac{1}{2}\\ \frac{1}{2} & \frac{1}{2}\end{pmatrix} = \begin{pmatrix}\frac{3}{4} & \frac{1}{4}\\ \frac{1}{4} & \frac{1}{4}\end{pmatrix} \tag{2-590}$$

As It is obvious from symmetry, the eigenstates of ρ are qubits oriented up and down along the axis $\hat{n}=1/\sqrt{2}(\hat{x}+\hat{z})$,

$$\begin{cases} |0'\rangle \equiv |\uparrow_{\hat{n}}\rangle = \begin{pmatrix} \cos\dfrac{\pi}{8} \\ \sin\dfrac{\pi}{8} \end{pmatrix} \\ |1'\rangle \equiv |\downarrow_{\hat{n}}\rangle = \begin{pmatrix} \sin\dfrac{\pi}{8} \\ -\cos\dfrac{\pi}{8} \end{pmatrix} \end{cases} \tag{2-591}$$

The eigenvalues are

$$\begin{cases} \lambda(0') = \dfrac{1}{2} + \dfrac{1}{2\sqrt{2}} = \cos^2\dfrac{\pi}{8} \\ \lambda(1') = \dfrac{1}{2} - \dfrac{1}{2\sqrt{2}} = \sin^2\dfrac{\pi}{8} \end{cases} \tag{2-592}$$

Evidently $\lambda(0')+\lambda(1')=1$ and $\lambda(0')\lambda(1')=1/8=\det(\lambda)$. The eigenstate $|0'\rangle$ has equal (and elatively large) overlap with both signal states

$$|\langle 0' | \uparrow_z\rangle|^2 = |\langle 0' | \uparrow_x\rangle|^2 = \cos^2\frac{\pi}{8} \approx 0.8535 \tag{2-593}$$

while $|1'\rangle$ has equal (and relatively small) overlap with both

$$|\langle 1' | \uparrow_z\rangle|^2 = |\langle 1' | \uparrow_x\rangle|^2 = \sin^2\frac{\pi}{8} \approx 0.1465 \tag{2-594}$$

Thus if don't know whether $|\uparrow_z\rangle$ or $|\uparrow_x\rangle$ was sent, the best guess can make is $|\psi\rangle=|0'\rangle$. This guess has the maximal fidelity:

$$F = \frac{1}{2}|\langle\uparrow_z|\psi\rangle|^2 + \frac{1}{2}|\langle\uparrow_x|\psi\rangle|^2 \tag{2-595}$$

among all possible qubit states $|\psi\rangle$ ($F=0.8535$).

Now imagine that Alice needs to send three letters to Bob. But she can afford to send only two qubits (quantum channels are very expensive!). Still she wants Bob to reconstruct her state with the highest possible fidelity. She could send Bob two of her three letters, and ask Bob to guess $|0'\rangle$ for the third. Then Bob receives the two letters with $F=1$, and he has $F=0.8535$ for the third; hence $F=0.8535$ overall. But is there a more clever procedure that achieves higher fidelity?

There is a better procedure. By diagonalizing ρ, decomposed the Hilbert space of a single qubit into a "likely" one-dimensional subspace (spanned by $|0'\rangle$ and an "unlikely" one-dimensional subspace (spanned by $|1'\rangle$). In a similar way we can decompose the Hilbert space of three qubits into likely and unlikely subspaces. If $|\psi\rangle=|\psi_1\rangle|\psi_2\rangle|\psi_3\rangle$ is any signal state (with each of three qubits in either the $|\uparrow_z\rangle$ or $|\uparrow_x\rangle$ state), have

$$|\langle 0'0'0' | \psi\rangle|^2 = \cos^6\left(\frac{\pi}{8}\right) \approx 0.6219$$

$$|\langle 0'0'1' | \psi\rangle|^2 = |\langle 0'1'0' | \psi\rangle|^2 = |\langle 1'0'0' | \psi\rangle|^2 = \cos^4\left(\frac{\pi}{8}\right)\sin^2\left(\frac{\pi}{8}\right) \approx 0.1067$$

$$|\langle 0'1'1' | \psi\rangle|^2 = |\langle 1'0'1' | \psi\rangle|^2 = |\langle 1'1'0' | \psi\rangle|^2 = \cos^2\left(\frac{\pi}{8}\right)\sin^4\left(\frac{\pi}{8}\right) \approx 0.0183$$

$$|\langle 1'1'1' | \psi\rangle|^2 = \sin^6\left(\frac{\pi}{8}\right) \approx 0.0031 \tag{2-596}$$

Thus, may decompose the space into the likely subspace Λ spanned by $\{|0'0'0'\rangle, |0'0'1'\rangle, |0'1'0'\rangle, |1'0'0'\rangle\}$, and its orthogonal complement $\Lambda^{\perp}$. If make a ("fuzzy") measurement that projects a signal state onto Λ or $\Lambda^{\perp}$, the probability of projecting onto the likely subspace is

$$P_{\text{likely}} = 0.6219 + 3 \times 0.1067 \approx 0.942 \tag{2-597}$$

while the probability of projecting onto the unlikely subspace is

$$\text{Punlikely} = 3 \times 0.0183 + 0.0031 = 0.0581 \tag{2-598}$$

To perform this fuzzy measurement, Alice could, for example first apply a unitary transformation U that rotates the four high-probability basis states to:

$$|\cdot\rangle\ |\cdot|\ 0\rangle \tag{2-599}$$

The four low-probability basis states to:

$$|\cdot\rangle\ |\cdot\rangle\ |\ 1\rangle \tag{2-600}$$

Then Alice measures the third qubit to complete the fuzzy measurement. If the outcome is $|0i$, then Alice's input state has been projected (in effect) onto Λ. She sends the remaining two (unmeasured) qubits to Bob. When Bob receives this (compressed) two-qubit state $|$ compi, he decompresses it by appending $|0\rangle$ and applying U^{-1}, obtaining

$$|\psi'\rangle = U^{-1}(|\psi_{\text{comp}}\rangle\ |\ 0\rangle) \tag{2-601}$$

If Alice's measurement of the third qubit yields $|1\rangle$, she has projected her input state onto the low-probability subspace $\Lambda^{\perp}$. In this event, the best thing she can do is send the state that Bob will decompress to the most likely state $|0'0'0'\rangle$—that is, she sends the state $|\Psi_{\text{comp}}\rangle$ such that

$$|\psi'\rangle = U^{-1}(|\psi_{\text{comp}}\rangle\ |\ 0\rangle) = |0'0'0'\rangle \tag{2-602}$$

Thus, if Alice encodes the three-qubit signal state $|\psi\rangle$, sends two qubits to Bob, and Bob decodes as just described, then Bob obtains the state ρ'

$$|\psi\rangle\langle\psi| \to \rho' = E\ |\psi\rangle\langle\psi|\ E + |0'0'0'\rangle\langle\psi|\ (1-E)\ |\psi\rangle\langle 0'0'0'| \tag{2-603}$$

where E is the projection onto Λ. The fidelity achieved by this procedure is

$$F = \langle\psi|\rho'|\psi\rangle = (\langle\psi|E|\psi\rangle)^2 + (\langle\psi|(1-E)|\psi\rangle)(\langle\psi|0'0'0'\rangle)^2$$
$$= (0.9419)^2 + (0.0581)(0.6219) = 0.9234 \tag{2-604}$$

This is indeed better than the naive procedure of sending two of the three qubits each with perfect fidelity. As consider longer messages with more letters, the fidelity of the compression improves. The Von-Neumann entropy of the one-qubit ensemble is

$$S(\rho) = H\left(\cos^2\frac{\pi}{8}\right) = 0.60088\cdots \tag{2-605}$$

Therefore, according to Schumacher's theorem, can shorten a long message by the factor 0.6009, and still achieve very good fidelity.

2.10.2 Schumacher encoding in general

The key to Shannon's noiseless coding theorem is that can code the typical sequences and ignore the rest, without much loss of fidelity. To quantify the compressibility of quantum information, we promote the notion of a typical sequence to that of a typical subspace. The key to Schumacher's noiseless quantum coding theorem is that we can code the typical subspace and ignore its orthogonal complement, without much loss of fidelity. Consider a message of n letters where each letter is a pure quantum state drawn from the ensemble $\{|\varphi_x\rangle, p_x\}$, so that the density matrix of a single letter is

$$\rho = \sum_x p_x |\varphi_x\rangle\langle\varphi_x| \tag{2-606}$$

Furthermore, the letters are drawn independently, so that the density matrix of the entire message is

$$\rho^n \equiv \rho \otimes \cdots \otimes \rho \tag{2-607}$$

That, for n large, this density matrix has nearly all of its support on a subspace of the full Hilbert space of the messages, where the dimension of this subspace asymptotically approaches $2^{nS(\rho)}$. This conclusion follows directly from the corresponding classical statement, f we consider the orthonormal basis in which ρ is diagonal. Working in this basis, may regard our quantum information source as an effectively classical source, producing messages that are strings of ρ eigenstates, each with a probability given by the product of the corresponding eigenvalues. For a specified n and δ, define the typical subspace Λ as the space spanned by the eigenvectors of ρ^n with eigenvalues λ satisfying

$$2^{-n(S-\delta)} \geqslant \lambda \geqslant \mathrm{e}^{-n(S+\delta)} \tag{2-608}$$

Borrowing directly from Shannon, conclude that for any $\delta, \varepsilon > 0$ and n sufficiently large, the sum of the eigenvalues of ρ^n that obey this condition satisfies

$$\mathrm{tr}(\rho^n E) > 1 - \varepsilon \tag{2-609}$$

Where E denotes the projection onto the typical subspace and the number dim of such eigenvalues satisfies

$$2^{n(S+\delta)} \geqslant \dim(\Lambda) \geqslant (1-\varepsilon)2^{n(S-\delta)} \tag{2-610}$$

The coding strategy is to send states in the typical subspace faithfully. For example, can make a fuzzy measurement that projects the input message onto either Λ or $\Lambda^{\perp}$; the outcome will be Λ with probability $P_\Lambda = \mathrm{tr}(\rho^n E) > 1 - \varepsilon$. In that event, the projected state coded and sent. Asymptotically, the probability of the other outcome becomes negligible, so it matters little what we do in that case. The coding of the projected state merely packages it so it carried by a minimal number of qubits. For example, we apply a unitary change of basis U that takes each state $|\psi\rangle_{\mathrm{typ}}$ in Λ to a state of the form

$$U|\psi_{\mathrm{typ}}\rangle = |\psi_{\mathrm{comp}}\rangle|0_{\mathrm{rest}}\rangle \tag{2-611}$$

where $|\psi\rangle_{\mathrm{comp}}$ is a state of $n(S+\delta)$ qubits, and $|0_{\mathrm{rest}}\rangle$ denotes the state $|0\rangle\cdots|0\rangle$ of the remaining qubits. Alice sends $|\psi\rangle_{\mathrm{comp}}$ to Bob, who decodes by appending $|0_{\mathrm{rest}}\rangle$ and applying U^{-1}. Suppose that

$$|\varphi_i\rangle = |\varphi_{x_1(i)}\rangle\cdots|\varphi_{x_n(i)}\rangle \tag{2-612}$$

Denotes any one of the n-letter pure state messages that might sent. After coding, transmission and decoding that carried out as just described, Bob has reconstructed a state

$$|\varphi_i\rangle\langle\varphi_i| \to \rho_i' = E|\varphi_i\rangle\langle\varphi_i|E + \rho_{i,\mathrm{Junk}}\langle\varphi_i|(1-E)|\varphi_i\rangle \tag{2-613}$$

where $\rho_{i,\mathrm{Junk}}$ is the state choose to send if the fuzzy measurement yields the outcome $\Lambda^{\perp}$. What can say about the fidelity of this procedure? The fidelity varies from message to message (in contrast to the example discussed above), so consider the fidelity averaged over the ensemble of possible messages:

$$\begin{aligned} F &= \sum_i p_i \langle\varphi_i|\rho_i'|\varphi_i\rangle \\ &= \sum_i p_i \langle\varphi_i|E|\varphi_i\rangle\langle\varphi_i|E|\varphi_i\rangle + \sum_i p_i \langle\varphi_i|\rho_{i,\mathrm{Junk}}|\varphi_i\rangle\langle\varphi_i|1-E|\varphi_i\rangle \\ &\geqslant \sum_i p_i \| E|\varphi_i\rangle \|^4 \end{aligned} \tag{2-614}$$

where the last inequality holds because the "junk" term is nonnegative. Since any real number satisfies

$$(x-1)^2 \geqslant 0, \quad \text{or} \quad x^2 \geqslant 2x-1 \tag{2-615}$$

Here have (setting $x = \| E|\rho_i\rangle \|^2$)

$$\| E \mid \varphi_i\rangle \|^4 \geqslant 2\| E \mid \varphi_i\rangle \|^2 - 1 = 2\langle \varphi_i \mid E \mid \varphi_i\rangle - 1 \tag{2-616}$$

hence

$$\begin{aligned} F &\geqslant \sum_i p_i(2\langle \varphi_i \mid E \mid \varphi_i\rangle - 1) \\ &= 2\mathrm{tr}(\rho^n E) - 1 > 2(1-\varepsilon) - 1 = 1 - 2\varepsilon \end{aligned} \tag{2-617}$$

That it is possible to compress the message to fewer than $n(S+\delta)$ qubits, while achieving an average fidelity that becomes arbitrarily good a n gets large. So have established that the message compressed, with insignificant loss of fidelity, to $S+\delta$ qubits per letter. Is further compression possible? Suppose that Bob will decode the message $\rho_{\mathrm{comp},i}$ that he receives by appending qubits and applying a unitary transformation U^{-1}, obtaining:

$$\rho'_i = U^{-1}(\rho_{\mathrm{comp},i} \otimes |0\rangle\langle 0|)U \tag{2-618}$$

Unitary Decoding: Suppose that ρ_{comp} has been compressed to $n(S-\delta)$ qubits. Then, no matter how the input message have been encoded, the decoded messages are all contained in a subspace Λ' of Bob's Hilbert space of dimension $2^{n(S-\delta)}$. (We are not assuming now that Λ' has anything to do with the typical subspace.) If the input message is $|\rho_i\rangle$, then the message reconstructed by Bob is ρ'_i which can be diagonalized as

$$\rho'_i = \sum_{a_i} | a_i\rangle \lambda_{a_i} \langle a_i | \tag{2-619}$$

where the $|a_i\rangle$'s are mutually orthogonal states in Λ'. The fidelity of the reconstructed message is

$$\begin{aligned} F_i &= \langle \varphi_i \mid \rho'_i \mid \varphi_i\rangle \\ &= \sum_{a_i} \lambda_{a_i} \langle \varphi_i \mid a_i\rangle\langle a_i \mid \varphi_i\rangle \\ &\leqslant \sum_{a_i} \langle \varphi_i \mid a_i\rangle\langle a_i \mid \varphi_i\rangle \leqslant \langle \varphi_i \mid E' \mid \varphi_i\rangle \end{aligned} \tag{2-620}$$

where E' denotes the orthogonal projection onto the subspace Λ'. The average fidelity therefore obeys

$$F = \sum_i p_i F_i \leqslant \sum_i p_i \langle \varphi_i \mid E' \mid \varphi_i\rangle = \mathrm{tr}(\rho^n E') \tag{2-621}$$

But since E' projects onto a space of dimension $2^{n(S-\delta)}$, $\mathrm{tr}^{(\rho^n}E')$ can be no larger than the sum of the $2^{n(S-\delta)}$ largest eigenvalues of ρ^n. It follows from the properties of typical subspaces that this sum becomes as small as please; for n large enough

$$F \leqslant \mathrm{tr}(\rho^n E') < \varepsilon \tag{2-622}$$

Thus have shown that, if attempt to compress to $S-\delta$ qubits per letter, then the fidelity inevitably becomes poor for n sufficiently large. Conclude then, that $S(\rho)$ qubits per letter is the optimal compression of the quantum information that can be attained if we are to obtain good fidelity as n goes to infinity. This is Schumacher's noiseless quantum coding theorem. The above argument applies to any conceivable encoding scheme, but only to a restricted class of decoding schemes (unitary decodings). A more general decoding scheme can certainly contemplated, described by a superoperator. More technology is then required to prove that better compression than S qubits per letter is not possible. But the conclusion is the same. The point is that $n(S-\delta)$ qubits are not sufficient to distinguish all of the typical states. To summarize, there is a close analogy between

Shannon's noiseless coding theorem and Schumacher's noiseless quantum coding theorem. In the classical case, nearly all long messages are typical sequences, so can code only these and still have a small probability of error. In the quantum case, nearly all long messages have nearly unit overlap with the typical subspace, so can code only the typical subspace and still achieve good fidelity. In fact, Alice could send effectively classical information to Bob—the string $x_1 x_2 \cdots x_n$ encoded in mutually orthogonal quantum states—and Bob could then follow these classical instructions to reconstruct Alice's state.

By this means, they could achieve high-fidelity compression to $H(X)$ bits—or qubits—per letter. But if the letters are drawn from an ensemble of nonorthogonal pure states, this amount of compression is not optimal; some of the classical information about the preparation of the state has become redundant, because the nonorthogonal states cannot be perfectly distinguished. Thus Schumacher coding can go further, achieving optimal compression to $S(\rho)$ qubits per letter. The information has been packaged more efficiently, but at a price—Bob has received what Alice intended, but Bob can't know what he has. In contrast to the classical case, Bob can't make any measurement that is certain to decipher Alice's message correctly. An attempt to read the message

2.10.3 Mixed-state coding: Holevo information

The Schumacher theorem characterizes the compressibility of an ensemble of pure states. But what if the letters are drawn from an ensemble of mixed states? The compressibility in that case is not firmly established, and is the subject of current research. It is easy to see that $S(\rho)$ won't be the answer for mixed states. To give a trivial example, suppose that a particular mixed state ρ_0 with $S(\rho_0) \neq 0$ is chosen with probability $p_0 = 1$. Then the message is always $\rho_0 \otimes \rho_0 \otimes \cdots \otimes \rho_0$ and it carries no information; Bob can reconstruct the message perfectly without receiving anything from Alice. Therefore, the message can be compressed to zero qubits per letters, which is less than $S(\rho) > 0$. To construct a slightly less trivial example, recall that for an ensemble of mutually orthogonal pure states, the Shannon entropy of the ensemble equals the Von Neumann entropy

$$H(X) = S(\rho) \tag{2-623}$$

So that the classical and quantum compressibility coincide. This makes sense, since the orthogonal states are perfectly distinguishable. In fact, if Alice wants to send the message

$$|\varphi_{x_1}\rangle \varphi_{x_2}\rangle \cdots |\varphi_{x_n}\rangle \tag{2-624}$$

to Bob, she can send the classical message $x_1, x_2, \cdots, x_n$ to Bob, who can reconstruct the state with perfect fidelity. But now suppose that the letters are drawn from an ensemble of mutually orthogonal mixed states $\{\rho_x, p_x\}$ is

$$\operatorname{tr} \rho_x \rho_y = 0 \text{ for } x \neq y \tag{2-625}$$

The ρ_x and ρ_y have support on mutually orthogonal subspaces of the Hilbert space. These mixed states are also perfectly distinguishable, so again the messages are essentially classical, and therefore can be compressed to $H(X)$ qubits per letter. For example, can extend the Hilbert space H_A of the letters to the larger space $H_A \otimes H_B$, and choose a purification of each ρ_x, a pure state $|\rho_x\rangle_{AB} \in H_A \otimes H_B$ such that

$$\operatorname{tr}_B(|\varphi_x\rangle_{AB\,AB}\langle\varphi_x|) = (\rho_x)_A \tag{2-626}$$

These pure states are mutually orthogonal, and the ensemble $\{|\rho_x\rangle_{AB}, p_x\}$ has Von Neumann entropy $H(X)$; hence may Schumacher compress a message

$$|\varphi_{x_1}\rangle_{AB}\cdots|\varphi_{x_n}\rangle_{AB} \tag{2-627}$$

to $H(X)$ qubits per letter (asymptotically). Upon decompressing this state, Bob can perform the partial trace by "throwing away" subsystem B, and so reconstruct Alice's message. To make a reasonable guess about what expression characterizes the compressibility of a message constructed from a mixed state alphabet, we might seek a formula that reduces to $S(\rho)$ for an ensemble of pure states, and to $H(X)$ for an ensemble of mutually orthogonal mixed states. Choosing a basis in which

$$\rho = \sum_x p_x \rho_x \tag{2-628}$$

is block diagonalized, see that

$$\begin{aligned} S(\rho) &= -\operatorname{tr} \rho \log \rho = -\sum_x \operatorname{tr}(p_x \rho_x) \log(p_x \rho_x) \\ &= -\sum_x p_x \log p_x - \sum_x p_x \operatorname{tr} \rho_x \log \rho_x \\ &= H(X) + \sum_x p_x S(\rho_x) \end{aligned} \tag{2-629}$$

Here recalling that $\operatorname{tr}\rho_x = 1$ for each x. Therefore we may write the Shannon entropy as

$$H(X) = S(\rho) - \sum_x p_x S(\rho_x) \equiv \chi(\mathcal{E}) \tag{2-630}$$

The quantity $X(\varepsilon)$ is called the Holevo information of the ensemble $\varepsilon = \{\rho_x, p_x\}$. Evidently, it depends not just on the density matrix ρ, but also on the particular way that ρ is realized as an ensemble of mixed states. Now have found that, for either an ensemble of pure states, or for an ensemble of mutually orthogonal mixed states, the Holevo information $X(\varepsilon)$ is the optimal number of qubits per letter that can be attained if to compress the messages while retaining good fidelity for large n. The Holevo information can be regarded as a generalization of Von Neumann entropy, reducing to $S(\rho)$ for an ensemble of pure states. It also bears a close resemblance to the mutual information of classical information theory:

$$I(Y;X) = H(Y) - H(Y \mid X) \tag{2-631}$$

How much, on the average, the Shannon entropy of Y is reduced once we learn the value of X; similarly,

$$\chi(\mathcal{E}) = S(\rho) - \sum_x p_x S(\rho_x) \tag{2-632}$$

how much, on the average, the Von Neumann entropy of an ensemble is reduced when we know which preparation was chosen. Like the classical mutual information, the Holevo information is always nonnegative, as follows from the concavity property of $S(\rho)$ as

$$S\left(\sum p_x \rho_x\right) \geqslant \sum_x p_x S(\rho_x) \tag{2-633}$$

Now wish to explore the connection between the Holevo information and the compressibility of messages constructed from an alphabet of nonorthog-onal mixed states. In fact, it can be shown that, in general, high-fidelity compression to less than X qubits per letter is not possible. To establish this result we use a "monotonicity" property of X that was proved by Lindblad and by Uhlmann: A superoperator cannot increase the Holevo information. That is, if $\$$ is any superoperator, let it act on an ensemble of mixed states according to

$$\$: \mathcal{E} = \{\rho_x, p_x\} \rightarrow \mathcal{E}' = \{\$(\rho_x), p_x\} \tag{2-634}$$

then

$$\chi(\mathcal{E}') \leqslant \chi(\mathcal{E}) \tag{2-635}$$

Lindblad-Uhlmann monotonicity is closely related to the strong subadditivity of the Von Neumann entropy, as you will show in a homework exercise. The monotonicity of X provides a further indication that X quantifies an amount of information encoded in a quantum system. The decoherence described by a superoperator can only retain or reduce this quantity of information—it can never increase it. Note that, in contrast, the Von Neumann entropy is not monotonic. A superoperator might take an initial pure state to a mixed state, increasing $S(\rho)$. But another superoperator takes every mixed state to the "ground state" $|0\rangle\langle 0|$, and so reduces the entropy of an initial mixed state to zero. It would be misleading to interpret this reduction of S as an "information gain," in that our ability to distinguish the different possible preparations has been completely destroyed. Correspondingly, decay to the ground state reduces the Holevo information to zero, reflecting that have lost the ability to reconstruct the initial state. Now consider messages of n letters, each drawn independently from the ensemble $\varepsilon = \{\rho_x, p_x\}$; the ensemble of all such input messages is denoted $\varepsilon^{(n)}$. A code is constructed that compresses the messages so that they all occupy a Hilbert space $\widetilde{H}^{(n)}$; the ensemble of compressed messages is denoted $\widetilde{\varepsilon}^{(n)}$. Then decompression is performed with a superoperator $\$$,

$$\$: \widetilde{\varepsilon}^{(n)} \to \varepsilon^{(n)} \tag{2-636}$$

to obtain an ensemble $E_0(n)$ of output messages.

Now suppose that this coding scheme has high fidelity. To minimize technicalities, let not specify in detail how the fidelity of $\varepsilon'^{(n)}$ relative to $E^{(n)}$ should be quantified. Just accept that if $E'^{(n)}$ has high fidelity, then for any δ and n sufficiently large

$$\frac{1}{n}\chi(\varepsilon^{(n)}) - \delta \leqslant \frac{1}{n}\chi(\varepsilon'^{(n)}) \leqslant \frac{1}{n}\chi(\varepsilon^{(n)}) + \delta \tag{2-637}$$

the Holevo information per letter of the output approaches that of the input. Since the input messages are product states, it follows from the additivity of $S(\rho)$ that

$$\chi(\varepsilon^{(n)}) = n\chi(\varepsilon) \tag{2-638}$$

and also know from Lindblad-Uhlmann monotonicity that

$$\chi(\varepsilon'^{(n)}) \leqslant \chi(\widetilde{\varepsilon}^{(n)}) \tag{2-639}$$

By combining Eqs. (2-637)~(2-639), find that

$$\frac{1}{n}\chi(\widetilde{\varepsilon}^{(n)}) \geqslant \chi(\varepsilon) - \delta \tag{2-640}$$

Finally, $X(\widetilde{\varepsilon}^{(n)})$ is bounded above by $S(\widetilde{\rho}^{(n)})$, which is in turn bounded above by log dim $\widetilde{H}^{(n)}$. Since δ may be as small, conclude that, asymptotically as $n \to \infty$,

$$\frac{1}{n}\log(\dim \widetilde{\mathcal{H}}^{(n)}) \geqslant \chi(\varepsilon) \tag{2-641}$$

high-fidelity compression to fewer than $X(\varepsilon)$ qubits per letter is not possible. One is sorely tempted to conjecture that compression to $X(\varepsilon)$ qubits per letter is asymptotically attainable.

2.10.4 Accessible information

The close analogy between the Holevo information $X(\varepsilon)$ and the classical mutual information $I(X;Y)$, as well as the monotonicity of X, suggest that X is related to the amount of classical information that can be stored in and recovered from a quantum system. In this section, will make this connection precise. The previous section was devoted to quantifying the quantum information content—measured in qubits—of messages constructed from an alphabet of quantum states. But turn to a quite different topic. Quantify the classical information content—measured in bits—that can be extracted from such messages, particularly in the case where the alphabet includes letters that are not mutually orthogonal.

Now, why would be so foolish as to store classical information in nonorthogonal quantum states that cannot be perfectly distinguished? Storing information this way should surely be avoided as it will degrade the classical signal. But perhaps we can't help it. For example, maybe I am a communications engineer, and I am interested in the intrinsic physical limitations on the classical capacity of a high bandwidth optical fiber. Clearly, to achieve a higher throughout of classical information per unit power, choose to encode information in single photons, and to attain a high rate, we should increase the number of photons transmitted per second. But if squeeze photon wavepackets together tightly, the wavepackets will overlap, and so will not be perfectly distinguishable. How do maximize the classical information transmitted in that case? As another important example, maybe an experimental physicist, and want to use a delicate quantum system to construct a very sensitive instrument that measures a classical force acting on the system. Can model the force as a free parameter x in the system's Hamiltonian $H(x)$. Depending on the value of x, the state of the system will evolve to various possible final (nonorthogonal) states ρ_x. How much information about x can our apparatus acquire? While physically this is a much different issue than the compressibility of quantum information, mathematically the two questions are related. Now can find that the Von Neumann entropy and its generalization the Holevo information will play a central role in the discussion. Suppose, for example, that Alice prepares a pure quantum state drawn from the ensemble $\varepsilon = \{|\rho_x\rangle, p_x\}$. Bob knows the ensemble, but not the particular state that Alice chose. He wants to acquire as much information as possible about x.

Bob collects his information by performing a generalized measurement, the POVM $\{F_y\}$. If Alice chose preparation x, Bob will obtain the measure ment outcome y with conditional probability

$$p(y \mid x) = \langle \varphi_x \mid F_y \mid \varphi_x \rangle \tag{2-642}$$

These conditional probabilities, together with the ensemble X, determine the amount of information that Bob gains on the average, the mutual information $I(X;Y)$ of preparation and measurement outcome. Bob is free to perform the measurement of his choice. The "best" possible measurement, that which maximizes his information gain, is called the optimal measurement determined by the ensemble. The maximal information gain is

$$\mathrm{Acc}\,(\varepsilon) = \underset{\{F_y\}}{\mathrm{Max}}\, I(X;Y) \tag{2-643}$$

where the Max is over all POVM's. This quantity is called the accessible information of the ensemble ε. Of course, if the states $|\varphi_x\rangle$ are mutually orthogonal, then they are perfectly distinguishable. The orthogonal measurement:

$$E_y = |\varphi_y\rangle\langle\varphi_y| \tag{2-644}$$

has conditional probability

$$p(y \mid x) = \delta_{y,x} \tag{2-645}$$

So that $H(X \mid Y) = 0$ and $I(X;Y) = H(X)$. This measurement is clearly optimal—the preparation is completely determined—so that

$$\mathrm{Acc}\,(\varepsilon) = H(X) \tag{2-646}$$

for an ensemble of mutually orthogonal (pure or mixed) states. But the problem is much more interesting when the signal states are nonorthogonal pure states. In this case, no useful general formula for Acc $(\mathcal{E})$ is known, but there is an upper bound

$$\text{Acc}\ (\mathcal{E}) \leqslant S(\rho) \tag{2-647}$$

That this bound is saturated in the case of orthogonal signal states, where $S(\rho) = H(X)$. In general, we know from classical information theory that $I(X;Y) \leqslant H(X)$; but for nonorthogonal states have $S(\rho) < H(X)$, so that Eq. (2-647) is a better bound. Even so, this bound is not tight; in many cases Acc $(\mathcal{E})$ is strictly less than $S(\rho)$. Can obtain a sharper relation between Acc $(\mathcal{E})$ and $S(\rho)$ if consider the accessible information per letter in a message containing n letters. Now Bob has more flexibility—he can choose to perform a collective measurement on all n letters, and thereby collect more information than if he were restricted to measuring only one letter at a time. Furthermore, Alice can choose to prepare, rather than arbitrary messages with each letter drawn from the ensemble $\mathcal{E}$, an ensemble of special messages (a code) designed to be maximally distinguishable. Then see that Alice and Bob can find a code such that the marginal ensemble for each letter is $\mathcal{E}$, and the accessible information per letter asymptotically approaches $S(\rho)$ as $n \to \infty$. In this sense, $S(\rho)$ characterizes the accessible information of an ensemble of pure quantum states. Furthermore, these results generalize to ensembles of mixed quantum states, with the Holevo information replacing the Von Neumann entropy. The accessible information of an ensemble of mixed states $\{\rho_x, p_x\}$ satisfies

$$\text{Acc}\ (\varepsilon) \leqslant X(\mathcal{E}) \tag{2-648}$$

a result known as the Holevo bound. This bound is not tight in general (though it is saturated for ensembles of mutually orthogonal mixed states). However, if Alice and Bob choose an n-letter code, where the marginal ensemble for each letter is E, and Bob performs an optimal POVM on all n letters collectively, then the best attainable accessible information per letter is $X(\mathcal{E})$—if all code words are required to be product states. In this sense, $X(\mathcal{E})$ characterizes the accessible information of an ensemble of mixed quantum states. One way that an alphabet of mixed quantum states might arise is that Alice might try to send pure quantum states to Bob through a noisy quantum channel. Due to decoherence in the channel, Bob receives mixed states that he must decode. In this case, then, $X(\mathcal{E})$ characterizes the maximal amount of classical information that can be transmitted to Bob through the noisy quantum channel. For example, Alice might send to Bob n photons in certain polarization states. If we suppose that the noise acts on each photon independently, and that Alice sends unentangled states of the photons, then $X(\mathcal{E})$ is the maximal amount of information that Bob can acquire per photon. Since

$$X(\mathcal{E}) \leqslant S(\rho) \leqslant 1 \tag{2-649}$$

it follows in particular that a single (unentangled) photon can carry at most one bit of classical information.

- The Holevo bound

The Holevo bound on the accessible information is not an easy theorem, but like many good things in quantum information theory, it follows easily once the strong subadditivity of Von Neumann entropy is established. Here assume strong subadditivity and show that the Holevo bound follows. Recall the setting: Alice prepares a quantum state drawn from the ensemble $\mathcal{E}=$

$\{\rho_x, p_x\}$, and then Bob performs the POVM $\{F_y\}$. The joint probability distribution governing Alice's preparation x and Bob's outcome y is

$$p(x,y) = p_x \operatorname{tr}\{F_y \rho_x\} \tag{2-650}$$

can show that

$$I(X;Y) \leqslant X(\varepsilon) \tag{2-651}$$

Since strong subadditivity is a property of three subsystems, which will need to identify three systems to apply it to. The strategy will be to prepare an input system X that stores a classical record of what preparation was chosen and an output system Y whose classical correlations with x are governed by the joint probability distribution $p(x,y)$. Then applying strong subadditivity to X, Y, and the quantum system Q, will be able to relate $I(X;Y)$ to $X(\varepsilon)$. Suppose that the initial state of the system XQY is

$$\rho_{XQY} = \sum_x p_x \mid x\rangle\langle x \mid \otimes \rho_x \otimes \mid 0\rangle\langle 0 \mid \tag{2-652}$$

where the $|x\rangle$'s are mutually orthogonal pure states of the input system X, and $|0\rangle$ is a particular pure state of the output system Y. By performing partial traces, can see that

$$\begin{cases} \rho_X = \sum_x p_x \mid x\rangle\langle x \mid \rightarrow S(\rho_X) = H(X) \\ \rho_Q = \sum_x p_x \rho_x \equiv \rho \rightarrow S(\rho_{QY}) = S(\rho_Q) = S(\rho) \end{cases} \tag{2-653}$$

and since the $|x\rangle$'s are mutually orthogonal, we also have

$$\begin{aligned} S(\rho_{XQY}) = S(\rho_{XQ}) &= \sum_x - \operatorname{tr}(p_x \rho_x \log p_x \rho_x) \\ &= H(X) + \sum_x p_x S(\rho_x) \end{aligned} \tag{2-654}$$

Now will perform a unitary transformation that "imprints" Bob's measurement result in the output system Y. To suppose, for now, that Bob performs an orthogonal measurement $\{E_y\}$, where

$$E_y E_{y'} = \delta_{y,y'} E_y \tag{2-655}$$

(consider more general POVM's shortly). The unitary transformation U_{QY} acts on QY according to

$$U_{QY}: \mid \varphi\rangle_Q \otimes \mid 0\rangle_Y = \sum_y E_y \mid \varphi\rangle_Q \otimes \mid y\rangle_Y \tag{2-656}$$

(where the $|y\rangle$'s are mutually orthogonal), and so transforms ρ_{XQY} as

$$U_{QY}: \rho_{XQY} \rightarrow \rho'_{XQY} = \sum_{x,y,y'} p_x \mid x\rangle\langle x \mid \otimes E_y \rho_x E_{y'} \otimes \mid y\rangle\langle y' \mid \tag{2-657}$$

Since Von Neumann entropy is invariant under a unitary change of basis, have

$$\begin{cases} S(\rho'_{XQY}) = S(\rho_{XQY}) = H(x) + \sum_x p_x S(\rho_x) \\ S(\rho'_{QY}) = S(\rho_{QY}) = S(\rho) \end{cases} \tag{2-658}$$

and taking a partial trace of Eq. (2-657) find

$$\begin{aligned} \rho'_{XY} &= \sum_{x,y} p_x \operatorname{tr}(E_y \rho_x) \mid x\rangle\langle x \mid \otimes \mid y\rangle\langle y \mid \\ &= \sum_{x,y} p(x,y) \mid x,y\rangle\langle x,y \mid \rightarrow S(\rho'_{XY}) = H(X,Y) \end{aligned} \tag{2-659}$$

Evidently it follows that

$$\rho'_Y = \sum_y p(y) \mid y\rangle\langle y \mid \rightarrow S(\rho'_Y) = H(Y) \tag{2-660}$$

Now invoke strong subadditivity in the form

$$S(\rho'_{XQY}) + S(\rho'_Y) \leqslant S(\rho'_{XY}) + S(\rho'_{QY}) \tag{2-661}$$

which becomes

$$H(X) + \sum_x p_x S(\rho_x) + H(Y) \leqslant H(X,Y) + S(\rho) \tag{2-662}$$

or

$$I(X;Y) = H(X) + H(Y) - H(X,Y) \leqslant S(\rho) - \sum_x p_x S(\rho_x) = \chi(\mathcal{E}) \tag{2-663}$$

This is the Holevo bound. One way to treat more general POVM's is to enlarge the system by appending one more subsystem Z. Then construct a unitary $UQY\,Z$ acting as

$$U_{QYZ}: |\varphi\rangle_Q \otimes |0\rangle_Y \otimes |0\rangle_Z = \sum_y \sqrt{F_y}\, |\varphi\rangle_A \otimes |y\rangle_Y \otimes |y\rangle_Z \tag{2-664}$$

so that

$$\rho'_{XQYZ} = \sum_{x,y,y'} p_x\, |x\rangle\langle x| \otimes \sqrt{F_y}\rho_x\sqrt{F_{y'}} \otimes |y\rangle\langle y'| \otimes |y\rangle\langle y'| \tag{2-665}$$

Then the partial trace over Z yields

$$\rho'_{XQY} = \sum_{x,y} p_x\, |x\rangle\langle x| \otimes \sqrt{F_y}\rho_x\sqrt{F_y} \otimes |y\rangle\langle y| \tag{2-666}$$

and

$$\begin{aligned} \rho'_{XY} &= \sum_{x,y} p_x \operatorname{tr}(F_y\rho_x)\, |x\rangle\langle x| \otimes |y\rangle\langle y| \\ &= \sum_{x,y} p(x,y)\, |x,y\rangle\langle x,y| \\ &\to S(\rho'_{XY}) = H(X,Y) \end{aligned} \tag{2-667}$$

The rest of the argument then runs as before.

- Improving distinguishability: the Peres-Wootters method

To better acquaint ourselves with the concept of accessible information, let's consider a single-qubit example. Alice prepares one of the three possible pure states

$$\begin{cases} |\varphi_1\rangle = |\uparrow_{\hat{n}_1}\rangle = \begin{pmatrix}1\\0\end{pmatrix} \\ |\varphi_2\rangle = |\uparrow_{\hat{n}_2}\rangle = \begin{pmatrix}\frac{1}{2}\\ \frac{\sqrt{3}}{2}\end{pmatrix} \\ |\varphi_3\rangle = |\uparrow_{\hat{n}_3}\rangle = \begin{pmatrix}-\frac{1}{2}\\ -\frac{\sqrt{3}}{2}\end{pmatrix} \end{cases} \tag{2-668}$$

a spin- 1/2 object points in one of three directions that are symmetrically distributed in the xz-plane. Each state has a priori probability 1/3. Evidently, Alice's "signal states" are nonorthogonal:

$$\langle\varphi_1|\varphi_2\rangle = \langle\varphi_1|\varphi_3\rangle = \langle\varphi_2|\varphi_3\rangle = -\frac{1}{2} \tag{2-669}$$

Bob's task is to find out as much as he can about what Alice prepared by making a suitable measurement. The density matrix of Alice's ensemble is

$$\rho = \frac{1}{3}(|\varphi_1\rangle\langle\varphi_1| + |\varphi_2\rangle\langle\varphi_3| + |\varphi_3\rangle\langle\varphi_3|) \tag{2-670}$$

which has $S(\rho)=1$. Therefore, the Holevo bound tells us that the mutual information of Alice's preparation and Bob's measurement outcome cannot exceed 1 bit. In fact, though, the accessible information is considerably less than the one bit allowed by the Holevo bound. In this case, Alice's ensemble has enough symmetry that it is not hard to guess the optimal measurement. Bob may choose a POVM with three outcomes, where

$$F_{\bar{a}} = \frac{2}{3}(1 - |\varphi_a\rangle\langle\varphi_a|), \quad a = 1,2,3 \tag{2-671}$$

that

$$p(a \mid b) = \langle\varphi_b \mid F_{\bar{a}} \mid \varphi_b\rangle = \begin{cases} 0, & a = b \\ \dfrac{1}{2}, & a \neq b \end{cases} \tag{2-672}$$

Therefore, the measurement outcome a excludes the possibility that Alice prepared a, but leaves equal a posteriori probabilities ($p=1/2$) for the other two states. Bob's information gain is

$$I = H(X) - H(X \mid Y) = \log_2 3 - 1 = 0.58496 \tag{2-673}$$

To show that this measurement is really optimal, may appeal to a variation on a theorem of Davies, which assures us that an optimal POVM can be chosen with three F_a's that share the same three-fold symmetry as the three states in the input ensemble. This result restricts the possible POVM's enough so that can check that Eq. (2-671) is optimal with an explicit calculation. Hence we have found that the ensemble $\varepsilon = \{|\varphi_a\rangle, p_a = 1/3\}$ has accessible information.

$$\text{Acc}\ (\mathcal{E}) = \log_2\left(\frac{3}{2}\right) = 0.58496 \tag{2-674}$$

The Holevo bound is not saturated.

Now suppose that Alice has enough cash so that she can afford to send two qubits to Bob, where again each qubit is drawn from the ensemble ε. The obvious thing for Alice to do is prepare one of the nine states

$$|\varphi_a\rangle|\varphi_b\rangle, \quad a, b = 1,2,3 \tag{2-675}$$

each with $p_{ab} = 1/9$. Then Bob's best strategy is to perform the POVM Eq. (2-671) on each of the two qubits, achieving a mutual information of 0.58496 bits per qubit, as before. But Alice and Bob are determined to do better. After discussing the problem with A. Peres and W. Wootters, they decide on a different strategy. Alice will prepare one of three two-qubit states

$$|\Phi_a\rangle = |\varphi_a\rangle|\varphi_a\rangle, \quad a = 1,2,3 \tag{2-676}$$

each occurring with a priori probability $p_a = 1/2$. Considered one-qubit at a time, Alice's choice is governed by the ensemble E, but now her two qubits have (classical) correlations—both are prepared the same way. The three $|\Phi_a\rangle$'s are linearly independent, and so span a three-dimensional subspace of the four-dimensional two-qubit Hilbert space. In a homework exercise, you will show that the density matrix:

$$\rho = \frac{1}{3}\left(\sum_{a=1}^{3} |\Phi_a\rangle\langle\Phi_a|\right) \tag{2-677}$$

has the nonzero eigenvalues 1/2, 1/4, 1/4, so that

$$S(\rho) = -\frac{1}{2}\log\frac{1}{2} - 2\left(\frac{1}{4}\log\frac{1}{4}\right) = \frac{3}{2} \tag{2-678}$$

The Holevo bound requires that the accessible information per qubit is less than 3/4 bit. This would at least be consistent with the possibility that can exceed the 0.58496 bits per qubit attained by the nine-state method. Naively, it may seem that Alice won't be able to convey as much classical information to Bob, if she chooses to send one of only three possible states instead of nine. But on further reflection, this conclusion is not obvious. True, Alice has fewer signals to choose from, but the signals are more distinguishable; have

$$\langle \Phi_a \mid \Phi_b \rangle = \frac{1}{4}, \quad a \neq b \tag{2-679}$$

instead of Eq. (2-669). It is up to Bob to exploit this improved distinguishability in his choice of measurement. In particular, Bob will find it advantageous to perform collective measurements on the two qubits instead of measuring them one at a time. It is no longer obvious what Bob's optimal measurement will be. But Bob can invoke a general procedure that, while not guaranteed optimal, is usually at least pretty good. Call the POVM constructed by this procedure a "pretty good measurement" (or PGM). Consider some collection of vectors $|\tilde{\Phi}_a\rangle$ that are not assumed to be orthogonal or normalized. Now want to devise a POVM that can distinguish these vectors reasonably well. First construct

$$G = \sum_a |\tilde{\Phi}_a\rangle\langle\tilde{\Phi}_a| \tag{2-680}$$

This is a positive operator on the space spanned by the $|\tilde{\Phi}_a\rangle$'s. Therefore, on that subspace, G has an inverse, G^{-1} and that inverse has a positive square root $G^{-1/2}$. Now we define

$$F_a = G^{-1/2} |\tilde{\Phi}_a\rangle\langle\tilde{\Phi}_a| G^{-1/2} \tag{2-681}$$

and that

$$\begin{aligned}\sum_a F_a &= G^{-1/2}\Big(\sum_a |\tilde{\Phi}_a\rangle\langle\tilde{\Phi}_a|\Big) G^{-1/2} \\ &= G^{-1/2} G G^{-1/2} = 1\end{aligned} \tag{2-682}$$

on the span of the $|\tilde{\Phi}_a\rangle$'s. If necessary, can augment these F_a's with one more positive operator, the projection F_0 onto the orthogonal complement of the span of the $|\tilde{\Phi}_a\rangle$'s, and so construct a POVM. This POVM is the PGM associated with the vectors $|\tilde{\Phi}_a\rangle$. In the special case where the $|\tilde{\Phi}_a\rangle$'s are orthogonal,

$$|\tilde{\Phi}_a\rangle = \sqrt{\lambda_a}\, |\Phi_a\rangle \tag{2-683}$$

(where the $|\Phi_a\rangle$'s are orthonormal), have

$$\begin{aligned}F_a &= \sum_{a,b,c} (|\Phi_b\rangle\lambda_b^{-1/2}\langle\Phi_b|)(\lambda_a |\Phi_a\rangle\langle\Phi_a|)(|\Phi_c\rangle\lambda_c^{-1/2}\langle\Phi_c|) \\ &= |\Phi_a\rangle\langle\Phi_a|\end{aligned} \tag{2-684}$$

this is the orthogonal measurement that perfectly distinguishes the $|\tilde{\Phi}_a\rangle$'s and so clearly is optimal. If the $|\tilde{\Phi}_a\rangle$'s are linearly independent but not orthogonal, then the PGM is again an orthogonal measurement (because n one-dimensional operators in an n-dimensional space can constitute a POVM only if mutually orthogonal), but in that case the measurement may not be optimal.

In the homework, you'll construct the PGM for the vectors $|\tilde{\Phi}_a\rangle$ in Eq. (2-683), and you'll show that

$$p(a \mid a)=\langle \Phi_a \mid F_a \mid \Phi_a \rangle=\frac{1}{3}\left(1+\frac{1}{\sqrt{2}}\right)^2=0.971405$$

$$p(b \mid a)=\langle \Phi_a \mid F_b \mid \Phi_a \rangle=\frac{1}{6}\left(1-\frac{1}{\sqrt{2}}\right)^2=0.0142977 \tag{2-685}$$

(for $b \neq a$). It follows that the conditional entropy of the input is

$$H(X \mid Y)=0.215893 \tag{2-686}$$

and since $H(X)=\log_2 3=1.58496$, the information gain is

$$I=H(X)-H(X \mid Y)=1.36907 \tag{2-687}$$

a mutual information of 0.684535 bits per qubit. Thus, the improved distinguishability of Alice's signals has indeed paid off—have exceeded the 0.58496 bits that can be extracted from a single qubit. Still didn't saturate the Holevo bound ($I<1.5$ in this case), but we came a lot closer than before. This example, first described by Peres and Wootters, teaches some useful lessons. First, Alice is able to convey more information to Bob by "pruning" her set of codewords. She is better off choosing among fewer signals that are more distinguishable than more signals that are less distinguishable. An alphabet of three letters encodes more than an alphabet of nine letters.

Second, Bob is able to read more of the information if he performs a collective measurement instead of measuring each qubit separately. His optimal orthogonal measurement projects Alice's signal onto a basis of entangled states. The PGM described here is "optimal" in the sense that it gives the best information gain of any known measurement. Most likely, this is really the highest I that can be achieved with any measurement, but have not proved it.

- Attaining Holevo: pure states

With these lessons in mind, can proceed to show that, given an ensemble of pure states, we can construct n-letter codewords that asymptotically attain an accessible information of $S(\rho)$ per letter. Must select a code, the ensemble of codewords that Alice can prepare, and a "decoding observable," the POVM that Bob will use to try to distinguish the codewords. The task is to show that Alice can choose $2^{n(S-\delta)}$ codewords, such that Bob can determine which one was sent, with negligible probability of error as $n \to \infty$. That won't go through all the details of the argument, but will be content to understand why the result is highly plausible. The main idea, of course, is to invoke random coding. Alice chooses product signal states

$$\mid \varphi_{x_1} \rangle \mid \varphi_{x_2} \rangle \cdots \mid \varphi_{x_n} \rangle \tag{2-688}$$

by drawing each letter at random from the ensemble $\varepsilon=\{\mid \varphi_x \rangle, p_x\}$. As have seen, for a typical code each typical codeword has a large overlap with a typical subspace $\Lambda^{(n)}$ that has dimension dim $\Lambda^{(n)}>2^{n(S(\rho)-\delta)}$. Furthermore, for a typical code, the marginal ensemble governing each letter is close to E. Because the typical subspace is very large for n large, Alice can choose many codewords, yet be assured that the typical overlap of two typical code words is very small. Heuristically, the typical codewords are randomly distributed in the typical subspace, and on average, two random unit vectors in a space of dimension D have overlap $1/D$. Therefore if $\mid u \rangle$ and $\mid w \rangle$ are two codewords

$$\langle \mid \langle u \mid w \rangle \mid^2 \rangle_\Lambda < 2^{-n(S-\delta)} \tag{2-689}$$

Here $\langle \cdot \rangle_\Lambda$ denotes an average over random typical codewords. Can convince yourself that the typical codewords really are uniformly distributed in the typical subspace as follows: Averaged over the ensemble, the overlap of random codewords

$$|\varphi_{x_1}\rangle\cdots|\varphi_{x_n}\rangle \text{ and } |\varphi_{y_1}\rangle\cdots|\varphi_{y_n}\rangle = \sum p_{x_1}\cdots p_{x_n}p_{y_1}\cdots p_{y_n}(|\langle\varphi_{x_1}|\varphi_{y_1}\rangle|^2\cdots|\langle\varphi_{x_n}|\varphi_{y_n}\rangle|^2)$$

$$= \mathrm{tr}(\rho\otimes\cdots\otimes\rho)^2 \tag{2-690}$$

Now suppose we restrict the trace to the typical subspace $\Lambda^{(n)}$; this space has dim $\Lambda^{(n)} < 2^{n(S+\delta)}$ and the eigenvalues of $\rho^{(n)} = \rho\otimes\cdots\otimes\rho$ restricted to $\Lambda^{(n)}$ satisfy $\lambda < 2^{-n(S-\delta)}$. Therefore

$$\langle|\langle u|w\rangle|^2\rangle_\Lambda = \mathrm{tr}_\Lambda[\rho^{(n)}]^2 < 2^{n(S+\delta)}[2^{-n(S-\delta)}]^2 = 2^{-n(S-3\delta)} \tag{2-691}$$

where tr Λ denotes the trace in the typical subspace. Now suppose that $2^{n(S-\delta)}$ random codewords $\{|u_i\rangle\}$ are selected. Then if $|u_j\rangle$ is any fixed codeword

$$\sum_{i\neq j}\langle|\langle u_i|u_j\rangle|^2\rangle < 2^{n(S-\delta)}2^{-n(S-\delta')} + \varepsilon = 2^{-n(\delta-\delta')} + \varepsilon \tag{2-692}$$

Here the sum is over all codewords, and the average is no longer restricted to the typical codewords—the ε on the right-hand side arises from the atypical case. Now for any fixed δ, can choose δ' and ε as small as for n sufficiently large; conclude that when average over both codes and codewords within a code, the codewords become highly distinguishable as $n\to\infty$. Now invoke some standard Shannonisms: Since Eq. (2-692) holds when average over codes, it also holds for a particular code. (Furthermore, since nearly all codes have the property that the marginal ensemble for each letter is close to ε, there is a code with this property satisfying eq. (2-692).) Now eq. (2-692) holds when average over the particular codeword $|u_j\rangle$. But by throwing away at most half of the codewords, can ensure that each and every codeword is highly distinguishable from all the others. That Alice can choose $2^{n(S-\delta)}$ highly distinguishable codewords, which become mutually orthogonal as $n\to\infty$. Bob can perform a PGM at finite n that approaches an optimal orthogonal measurement as $n\to\infty$. Therefore the accessible information per letter:

$$\frac{1}{n}\mathrm{Acc}(\tilde{\varepsilon}^{(n)}) = S(\rho) - \delta \tag{2-693}$$

It is attainable, where $\tilde{\varepsilon}^{(n)}$ denotes Alice's ensemble of n-letter codewords. Of course, for any finite n, Bob's POVM will be a complicated collective measurement performed on all n letters. To give an honest proof of attainability, analyze the POVM carefully, and bound its probability of error. This has been done by Hausladen et al. The handwaving argument here at least indicates why their conclusion is not surprising. It also follows from the Holevo bound and the subadditivity of the entropy that the accessible information per letter cannot exceed $S(\rho)$ asymptotically. The Holevo bound tells that:

$$\mathrm{Acc}(\tilde{\varepsilon}^{(n)}) \leqslant S(\tilde{\rho}^{(n)}) \tag{2-694}$$

where $\tilde{\rho}^{(n)}$ denotes the density matrix of the codewords, and subadditivity implies that

$$S(\tilde{\rho}^{(n)}) \leqslant \sum_{i=1}^{n} S(\tilde{\rho}_i) \tag{2-695}$$

where $\tilde{\rho}_i$ is the reduced density matrix of the ith letter. Since each $\tilde{\rho}_i$ approaches ρ asymptotically, have

$$\lim_{n\to\infty}\frac{1}{n}\mathrm{Acc}(\tilde{\varepsilon}^{(n)}) \leqslant \lim_{n\to\infty}\frac{1}{n}S(\tilde{\rho}^{(n)}) \leqslant S(\rho) \tag{2-696}$$

To derive this bound, did not assume anything about the code, except that the marginal ensemble for each letter asymptotically approaches ε. In particular the bound applies even if the codewords are entangled states rather than product states. Therefore that have shown that $S(\rho)$ is the optimal accessible information per letter. Can define a kind of channel capacity associated with a specified alphabet of pure quantum states, the "fixed-alphabet capacity." Suppose that Alice is equipped with a source of quantum states. She can produce any one of the states $|\rho_x\rangle$, but it is up to her to choose the a priori probabilities of these states. The fixed-alphabet capacity C_{f_a} is the maximum accessible information per letter she can achieve with the best possible distribution $\{p_x\}$. have found that

$$C_{f_a} = \underset{\{p_x\}}{\text{Max}} S(\rho) \tag{2-697}$$

C_{f_a} is the optimal number of classical bits can encode per letter (asymptotically), given the specified quantum-state alphabet of the source.

- Attaining Holevo: mixed states

Now would like to extend the above reasoning to a more general context. Consider n-letter messages, where the marginal ensemble for each letter is the ensemble of mixed quantum states:

$$\varepsilon = \{\rho_x, p_x\} \tag{2-698}$$

To argue that it is possible (asymptotically as $n \to \infty$) to convey $X(\varepsilon)$ bits of classical information per letter. Again, our task is to:

(1) specify a code that Alice and Bob can use, where the ensemble of codewords yields the ensemble ε letter by letter (at least asymptotically).

(2) Specify Bob's decoding observable, the POVM he will use to attempt to distinguish the codewords.

(3) Show that Bob's probability of error approaches zero as $n \to \infty$. As in our discussion of the pure-state case, Not exhibit the complete proof, see Holevo and Schumacher and Westmoreland. Instead, that offer an argument with even more handwaving than before, if that's possible indicating that the conclusion is reasonable.

As always, demonstrate attainability by a random coding argument. Alice will select mixed-state codewords, with each letter drawn from the ensemble E. That is, the codeword

$$\rho_{x_1} \otimes \rho_{x_2} \otimes \cdots \otimes \rho_{x_n} \tag{2-699}$$

It is chosen with probability $p_{x_1} p_{x_2} \cdots p_{x_n}$. The idea is that each typical codeword can regarded as an ensemble of pure states, with nearly all of its support on a certain typical subspace. If the typical subspaces of the various codewords have little overlap, then Bob will be able to perform a POVM that identifies the typical subspace characteristic of Alice's message, with small probability of error. What is the dimension of the typical subspace of a typical codeword? If we average over the codewords, the mean entropy of a codeword is

$$\langle S^{(n)} \rangle = \sum_{x_1 \cdots x_n} p_{x_1} p_{x_2} \cdots p_{x_n} S(\rho_{x_1} \otimes \rho_{x_2} \otimes \cdots \otimes \rho_{x_n}) \tag{2-700}$$

Using additivity of the entropy of a product state, and $\sum_x p_x = 1$, can obtain:

$$\langle S^{(n)} \rangle = n \sum_x p_x S(\rho_x) \equiv n \langle S \rangle \tag{2-701}$$

For n large, the entropy of a codeword is, with high probability, close to this mean, and furthermore, the high probability eigenvalues of $\rho_{x_1} \otimes \cdots \otimes \rho_{x_2}$ are close to $2^{-n\langle S \rangle}$. In other words a typical $\rho_{x_1} \otimes \cdots \otimes \rho_{x_n}$ has its support on a typical subspace of dimension $2^{n\langle S \rangle}$. This statement is closely analogous to the observation (crucial to the proof of Shannon's noisy channel coding theorem) that the number of typical messages received when a typical message is sent through a noisy classical channel is $2^{nH(Y|X)}$.

Now the argument follows a familiar road. For each typical message $x_1 x_2 \cdots x_n$, Bob can construct a "decoding subspace" of dimension $2^{n(\langle S \rangle + \delta)}$, with assurance that Alice's message is highly likely to have nearly all its support on this subspace. His POVM will designed to determine in which decoding subspace Alice's message lies. Decoding errors will be unlikely if typical decoding subspaces have little overlap.

Although Bob is interested only in the value of the decoding subspace and hence $x_1 x_2 \cdots x_n$, suppose that he performs the complete PGM determined by all the vectors that span all the typical subspaces of Alice's codewords. The PGM will approach an orthogonal measurement for large n, as long as the number of codewords is not too large. Bob obtains a particular result which is likely to be in the typical subspace of dimension $2^{nS(\delta)}$ determined by the source $\rho \otimes \rho \otimes \cdots \otimes \rho$, and furthermore, is likely to be in the decoding subspace of the message that Alice actually sent. Since Bob's measurement results are uniformly distributed in a space on dimension 2^{nS}, and the pure-state ensemble determined by a particular decoding subspace has dimension $2^{n(\langle S \rangle + \delta)}$, the average overlap of the vector determined by Bob's result with a typical decoding subspace is

$$\frac{2^{n(\langle S \rangle + \delta)}}{2^{nS}} = 2^{-n(S - \langle S \rangle - \delta)} = 2^{-n(\chi - \delta)} \tag{2-702}$$

If Alice chooses 2^{nR} codewords, the average probability of a decoding error will be:

$$2^{nR} 2^{-n(\chi - \delta)} = 2^{-n(\chi - R - \delta)} \tag{2-703}$$

Choose any R less than X, and this error probability will get very small as $n \to \infty$. This argument shows that the probability of error is small, averaged over both random codes and codewords. As usual, we can choose a particular code, and throw away some codewords to achieve a small probability of error for every codeword. Furthermore, the particular code may chosen to be typical, so that the marginal ensemble for each codeword approaches ε as $n \to \infty$. Conclude that an accessible information of X per letter is asymptotically attainable.

The structure of the argument closely follows that for the corresponding classical coding theorem. In particular, the quantity X arose much as I does in Shannon's theorem. While 2^{-nI} is the probability that a particular typical sequence lies in a specified decoding sphere, 2^{-nX} is the overlap of a particular typical state with a specified decoding subspace.

- Channel capacity

Combining the Holevo bound with the conclusion that X bits per letter is attainable, we obtain an expression for the classical capacity of a quantum channel (But with a caveat: we are assured that this "capacity" cannot be exceeded only if we disallow entangled codewords.)

Alice will prepare n-letter messages and send them through a noisy quantum channel to Bob. The channel is described by a superoperator, and will assume that the same superoperator $\$$ acts on each letter independently (memoryless quantum channel). Bob performs the POVM that optimizes his information going about what Alice prepared.

It will turn out, in fact, that Alice is best off preparing pure-state messages (this follows from the subadditivity of the entropy). If a particular letter is prepared as the pure state $|\varphi_x\rangle$, Bob will receive

$$|\varphi_x\rangle\langle\varphi_x| \to \$(|\varphi_x\rangle\langle\varphi_x|) \equiv \rho_x \tag{2-704}$$

And if Alice sends the pure state $|\varphi_{x_1}\rangle \cdots |\varphi_{x_n}\rangle$, Bob receives the mixed state $\rho_{x_1} \cdots \rho_{x_n}$. Thus, the ensemble of Alice's codewords determines as ensemble $\tilde{\varepsilon}^{(n)}$ of mixed states received by Bob. Hence Bob's optimal information gain is by definition Acc $(\tilde{\varepsilon}^{(n)})$, which satisfies the Holevo bound

$$\mathrm{Acc}\,(\tilde{\varepsilon}^{(n)}) \leqslant \chi(\tilde{\varepsilon}^{(n)}) \tag{2-705}$$

Now Bob's ensemble is

$$\{\rho_{x_1} \otimes \cdots \otimes \rho_{x_n}, p(x_1, x_2, \cdots, x_n)\} \tag{2-706}$$

where $p(x_1, x_2, \cdots, x_n)$ is a completely arbitrary probability distribution on Alice's codewords. Calculate X for this ensemble, note that

$$\begin{aligned} &\sum_{x_1 \cdots x_n} p(x_1, x_2, \cdots, x_n) S(\rho_{x_1} \otimes \cdots \otimes \rho_{x_n}) \\ &= \sum_{x_1 \cdots x_n} p(x_1, x_2, \cdots, x_n)[S(\rho_{x_1}) + S(\rho_{x_2}) + \cdots + S(\rho_{x_n})] \\ &= \sum_{x_1} p_1(x_1) S(\rho_{x_1}) + \sum_{x_2} p_2(x_2) S(\rho_{x_2}) + \cdots + \sum_{x_n} p_n(x_n) S(\rho_{x_n}) \end{aligned} \tag{2-707}$$

where, e. g., $p_1(x_1) = \sum_{x_2 \cdots x_n} p(x_1, x_2, \cdots, x_n)$ is the marginal probability distribution for the first letter. Furthermore, from subadditivity we have

$$S(\rho^{(n)}) \leqslant S(\rho_1) + S(\rho_2) + \cdots + S(\rho_n) \tag{2-708}$$

where ρ_i is the reduced density matrix for the ith letter. Combining Eq. (2-707) and Eq. (2-708) can find that

$$\chi(\tilde{\varepsilon}^{(n)}) \leqslant \chi(\tilde{\varepsilon}_1) + \cdots + \chi(\tilde{\varepsilon}_n) \tag{2-709}$$

where $\tilde{\varepsilon}_i$ is the marginal ensemble governing the ith letter that Bob receives. Eq. (2-709) applies to any ensemble of product states. Now, for the channel described by the superoperator $\$$, we define the product-state channel capacity

$$C(\$) = \max_{\varepsilon} \chi(\$(\varepsilon)) \tag{2-710}$$

Therefore, $X(\tilde{\varepsilon}_i) \leqslant C$ for each term in Eq. (2-709) and can obtain

$$\chi(\tilde{\varepsilon}^{(n)}) \leqslant nC \tag{2-711}$$

where $\tilde{\varepsilon}^{(n)}$ is any ensemble of product states. In particular, infer from the Holevo bound that Bob's information gain is bounded above by nC. But we have seen that $X(\$(\varepsilon))$ bits per letter can be attained asymptotically for any ε, with the right choice of code and decoding observable. Therefore, C is the optimal number of bits per letter that can be sent through the noisy channel with negligible error probability, if the messages that Alice prepares are required to be product states. That have left open the possibility that the product-state capacity $C(\$)$ might be exceeded if Alice is permitted to prepare entangled states of her n letters. It is not known whether there are quantum channels for which a higher rate can be attained by using entangled messages. This is one of the many interesting open questions in quantum information theory.

2.11 Photonic technologies for quantum information

It generally realized now that quantum-mechanical phenomena can enable significant fundamentally, and in some cases, tremendous, improvement for a variety of tasks important to emergent technologies. Building on decades of successes in the experimental demonstration of such fundamental phenomena, it is not surprising that photonics is playing a preeminent role in this nascent endeavor. Many of the objectives of quantum information processing are inherently suited to optics (e.g., quantum cryptography and optical metrology), while others may have a strong optical component (e.g., distributed quantum computing). In addition, it known that, at least in principle, one can realize *scalable* linear optics quantum computing (LOQC). For these applications to attain their full potential, various photonic technologies are needed, including high fidelity sources of single and entangled photons, and high efficiency photon-counting detectors, both at visible and telecommunication wavelengths. Much progress made on the development of these, though they are still not up to the demanding requirements of LOQC. Nevertheless, even at their present stage they have direct application to initial experiments. Moreover, they may find use in various "adjacent" technologies, such as biomedical and astronomical imaging, and low-power classical telecommunications. Here describe a number of the leading schemes for implementing approximations of sources of single photons on-demand and entangled photons, followed by a review of methods for detecting individual photons.

2.11.1 Single-photon sources

Photon-based quantum cryptography, communication, and computation schemes have increased the need for light sources that produce individual photons. A single-photon source would produce completely characterized single photons on demand ideally. When surveying attempts to create such sources, however, it is important to realize that there never has been and will never be such an ideal source. All of the currently available sources fall significantly short of this ideal. While other factors (such as rate, robustness, and complexity) certainly do matter, two of the most important parameters for quantifying how close a "single-photon source" approaches the ideal are the fraction of the time the device delivers light in response to a request, and the fraction of time that light is just a single photon.

In general single-photon sources fall into two categories—isolated quantum systems, or two-photon emitters. The first type relies on the fact that a single isolated quantum system can emit only one photon each time it is excited. The trick here is obtaining efficient excitation, output collection, and good isolation of individual systems. The second type uses light sources that emit two photons at a time. Here the detection of one photon indicates the existence of the second photon. That knowledge allows the second photon manipulated and delivered to where it needed.

- Quantum dot single-photon sources

A quantum dot is essentially an artificial atom that is easily isolated, so it is an obvious choice as the basis of a single-photon source. Single photons on-demand have generated by a combination of pulsed excitation of a single self-assembled semiconductor quantum dot and spectral filtering. When such a quantum dot is excited, either with a short (e.g., 3ps) laser pulse, or with an electrical pulse, electron-hole pairs created. For laser excitation, this can occur either within the dot itself, when the laser frequency tuned to a resonant transition between confined states of the dot, or in the surrounding semiconductor matrix, when the laser frequency is tuned above the semiconductor band gap. In the latter case, carriers diffuse toward the dot, where they relax to the lowest confined states. Created carriers recombine in a radiative cascade, leading to the generation of several photons for each laser pulse; all of these photons have slightly different frequencies, resulting from the Coulomb interaction among carriers. The last emitted photon for each pulse has a unique frequency, and can be spectrally isolated.

If the dots grown in a *bulk* semiconductor material, the out-coupling efficiency is poor, since the majority of emitted photons are lost in the semiconductor substrate. To increase the efficiency, an optical microcavity can fabricated around a quantum dot. An additional advantage is that the duration of photon pulses emitted from semiconductor quantum dots reduced, due to an enhancement of the spontaneous emission rate. This enhancement, also known as the Purcell factor, is proportional to the ratio of the mode quality factor to the mode volume. In addition, the spontaneous emission becomes directional; the photons emitted into the nicely shaped cavity mode can more easily coupled into downstream optical components.

By embedding InGaAs/GaAs quantum dots inside micropost microcavities with quality (Q)-factors of around 1300 and Purcell factors around five, the properties of a single-photon source have been significantly improved see Figure 2-52. The probability of generating two photons for the same laser pulse [estimated from the zero-time correlation parameter $g^{(2)}(0)$] can be as small as 2% compared to a Poisson-distributed source (i. e., an attenuated laser) of the same mean photon rate. The duration of single-photon pulses is below 200ps, and the sources emit identical (indistinguishable) photons, as confirmed by two-photon interference in a Hong-Ou-Mandel type experiment. Such sources have employed to realize the BB84 QKD protocol, and to generate post-selected polarization-entangled photons.

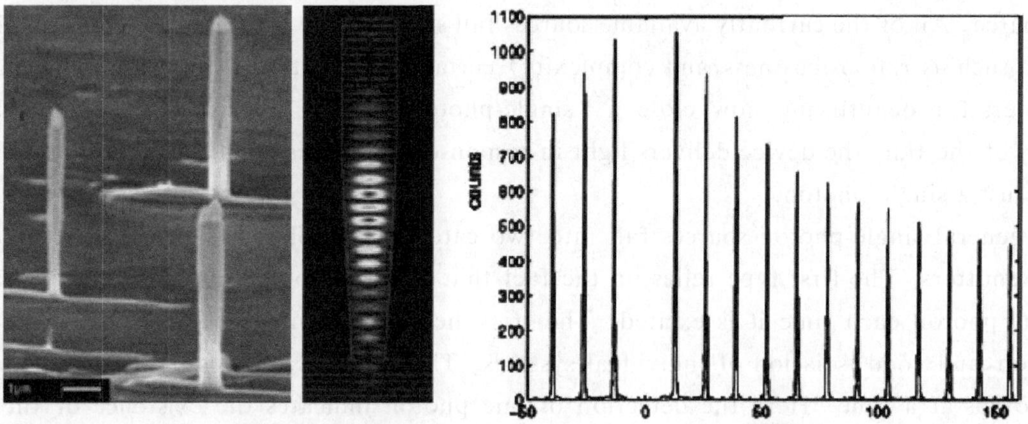

(a) (b) (c)

Fig. 2-52 (a) Scanning electron micrograph showing a fabricated array of GaAs/AlAs microposts ($\sim 0.3\mu m$ diameters, $5\mu m$ heights), with InAs/GaAs quantum dots embedded at the cavity center. (b) Electric field magnitude of the fundamental HE11 mode in a micropost microcavity with a realistic wall profile. (c) Photon correlation histogram for a single quantum dot embedded inside a micropost and on resonance with the cavity, under pulsed, resonant excitation. The histogram is generated using a Hanbury Brown and Twiss-type setup—the vanishing central peak (at $\tau = 0$) indicates a large suppression of two-photon pulses (to $\sim 2\%$ compared to a Poisson-distributed source), e. g., an attenuated laser, of the same intensity. The 13ns peak-to-peak separation corresponds to the repetition period of excitation pulses.

These sources still face several great challenges, however. They require cryogenic cooling ($<10K$), the output wavelengths are not yet readily tunable (present operation is around 900nm), the out-coupling efficiency into a single-mode traveling wave is still rather low ($<40\%$), and excitation of quantum dots in microcavities presently requires optical pumping (electrical pumping would be more desirable and efforts in that direction are underway). In the future, photonic-crystal microcavities may lead to much higher ratios of the quality factor to mode volumes, and therefore, much stronger cavity QED effects should be possible. This would enable an increase in the efficiency and speed of the single-photon devices, and thus open the possibility for building integrated quantum information systems. The spontaneous emission lifetime could reduced further to the order of several picoseconds, which would allow the generation of single photons at a rate higher than 10GHz. Moreover, the Purcell effect would also help in bringing the emitted photons closer to being Fourier-transform limited in bandwidth. Finally, photonic-crystal based cavities could even enable the realization of the strong coupling regime with a single quantum dot exciton, opening the possibility for the generation of indistinguishable single photons by coherent excitation schemes.

- Other single-emitter approaches

Other isolated quantum system approaches to producing single photons include isolated single fluorescence molecules and isolated nitrogen vacancies in diamond. Two significant deficiencies of these sources for many applications are that it is not easy to efficiently out-couple the photons, and that the spectral spread of the light is typically quite large (~120nm), though widths as low as 12nm have been seen in new results. This spectral width is non-optimal for applications relying on two-photon interference effects, which also for quantum cryptographic applications (where one typically desires fairly narrow bandwidths to exclude background light). More recently, single atoms coupled to a high-finesse optical cavity have demonstrated features of single-photon operation. Despite the technological challenges, this approach does offer the large potential advantage that the photons are emitted preferentially into the cavity modes, which are easier to couple out of, with couplings of 40%-70% already achieved. Also, the frequency of the photons is necessarily matched to a strong atomic transition, which may allow for efficient quantum *communication* using photons, while other quantum information processing tasks, such as memory or state readout, are carried out in the atomic system.

- Downconversion single-photon sources

Another effort toward single-photon sources relies on producing photons in pairs, typically via the process of optical parametric down conversion (PDC). The PDC process effectively takes an input photon from a pump beam and converts it into output pairs in a crystal possessing a $\chi^{(2)}$ nonlinearity. Thus the detection of one photon can be used to indicate (or herald) the existence of the second photon, which is available for further use. This second photon is, at low photon rates, left in an excellent approximation to a single-photon number state. It has been demonstrated how these photons may then be converted into completely arbitrary quantum states with fidelities of 99.9%. Recent efforts have focused on improving the collection of those pairs and improving the "single-photon accuracy," e.g., the value of $g^{(2)}(0)$.

The physics of the PDC process guarantees that the output pairs will possess certain energy and momentum constraints, so that under appropriate conditions the detected location of the herald photon tightly defines the location of its twin, a significant advantage over other single-photon schemes. There have been many mode-engineering efforts to improve this collection into a *single* mode, but the current best collection efficiency is still only 58%, including 15% optical-transmittance losses. (Contrast this to the required single-photon efficiency of over 99% for LOQC.) One example of a promising method to improve this is to directly modify the spatial emission profile of the photon pairs (which are usually emitted along cones) so that the photons are emitted preferentially into "beacon"-like beams, which couple more naturally into single-mode optical fibers. Another approach yet are explored is the use of adaptive optics to tailor the output modes. It should noted that for not all quantum information-processing applications require single-mode performance; for example, free-space quantum key distribution is likely to work nearly as well with a small number of modes. Because the conversion of pump photons into pairs via PDC is a random process, these sources suffer from the same problem that afflicts faint laser sources—one cannot guarantee that one and only one photon pair is created at a time (i.e., $g^{(2)}(0) \neq 0$). Multiplexing and storage schemes have proposed to deal with this. They work by similar principles which one scheme is based on space multiplexing and the other is based on temporal multiplexing photons are created at relatively low rates where the probability of simultaneous multi-pair production is low; contingent on the detection of a herald photon. The twin are stored, which to be emitted in a controlled fashion at some later desired time. The overall emission rate reduced, but the rate of producing one and only one photon at regular intervals is improved. By operating an array of simultaneously pumped PDC sources at low photon production rates and optically switching the output of one of the PDC sources that did produce a photon to the single output channel, it is possible to increase the single-photon rate, while maintaining a low rate of unwanted multiphoton pulses

2.11.2 Entangled-photon sources

Entangled states known to be a critical resource for realizing many quantum information protocols, such as teleportation and quantum networking. An on-demand source of entangled photons would also greatly aid the realization of all-optical quantum computing.

- Down-conversion schemes

At present, by far the most prevalent source of entangled photon pairs is parametric down conversion based on crystals with a $\chi^{(2)}$ non-linearity. As discussed above, it is precisely the temporal and spatial correlations between the photon pairs which make them very promising for the realization of an on-demand source of single photons. Much of the effort in studying these sources has been devoted to the generation of *polarization*-entangled photon pairs, an area which has seen tremendous growth—more than a million-fold improvement in the detected rates of polarization—entangled photons has been achieved in the past two decades (see Figure 2-53).

There are now several ways to realize polarization entanglement using the PDC process. One method uses a single nonlinear crystal, cut for "type-Ⅱ" phase matching, and selecting out a particular pair of output directions. Although initially these sources used large gas lasers for pumping, the recent availability of ultraviolet diode lasers has led to Figure 2-53. The apparent "Moore's Law" for entanglement. Shown are the reported detection rates of (polarization-) entangled photon pairs (from down conversion), as a function of year.

The solid line—drawn to guide the eye—indicates the ×100 gain every 5 years. The primary limiting factor has now become the lack of single-photon counting detectors with saturation rates above 10MHz. much more compact sources. A potentially important disadvantage, in addition to

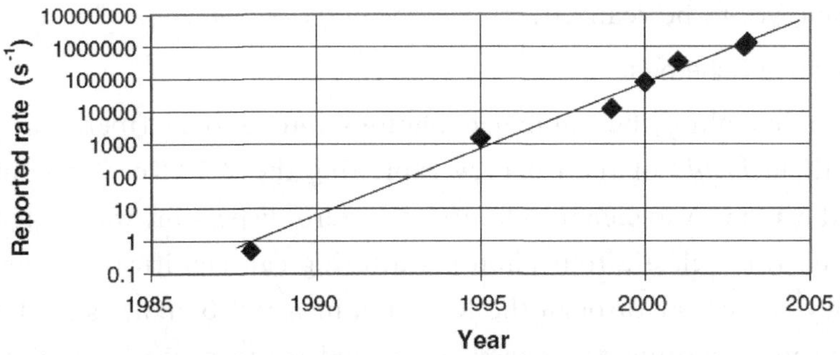

Fig. 2-53 The apparent "Moore's Law" for entanglement. Shown are the reported detection rates of (polarization-) entangled photon pairs (from down conversion), as a function of year. The solid line—drawn to guide the eye—indicates the ×100 gain every 5 years. The primary limiting factor has now become the lack of single-photon counting detectors with saturation rates above 10MHz.

the need to compensate the birefringent walk-off with this scheme, is that the entanglement is present only over a particular pair of modes (corresponding to the intersection of two cones). One method to eliminate this disadvantage is to pump the crystal from two different directions, or to allow the PDC occur in two crystals, the outputs of which are superposed directly or using a beam splitter. By proper alignment, nearly all of the output modes can display polarization entanglement, which moreover is completely tunable. Nearly perfect entanglement (within statistical uncertainty) has observed with such sources. Results with short-pulse pumps are encouraging, but the quality of the entanglement is typically not as high, a problem that will need addressed for future applications. One disadvantage of all of these techniques is that the output spectral bandwidth is still quite wide (typically 1～10nm) for possible coupling to atomic states. Research is underway to circumvent this problem by placing the nonlinear crystals inside high finesse optical cavities, which significantly increases the probability of downconversion into a narrow spectral bandwidth. As discussed above, there are a number of approaches for improving the coupling efficiency into single spatial modes. Improving conversion efficiency by finding higher non-linearity bulk crystals is limited by the choice of available crystals (with BBO and $LiIO_3$ being two of the better ones). Engineering crystals by processes such as periodic poling allows one to take advantage of crystals (e.g., Lithium Niobate) with somewhat higher nonlinearities. The conversion efficiency into a specific mode can further enhanced by some 1-2 orders of magnitude by creating waveguides in these crystals. Because the waveguide is small, possibly even single mode, it can be much easier to collect the output light. However, the net outcoupling efficiencies achieved to date (10%～20%) still require substantial improvement. Finally, by using a buildup cavity to recycle the unconverted pump photons, the effective conversion efficiency may increased (at the expense of a more complicated setup). Entanglement in non-polarization degrees of freedom, such as energy/time-bin and orbital angular momentum, has also realized recently. These may present some advantages over the polarization case, e.g., they allow implementation of higher-order quantum structures, such as qu-trits (3-level systems), and timing entanglement is more robust for transmission through optical fibers. One problem plaguing all of these sources is that the production of pairs is a random process. By using a short-pulsed pump, it is possible to define the times when no photon pairs will produced, but there is still no way to guarantee production of exactly one photon-pair during any given pulse. At least one theoretical scheme has proposed to circumvent this problem, but practical implementations have yet to be realized.

- $\chi^{(3)}$-Nonlinearity Schemes

The difficulty of coupling the entangled photons into optical fibers has been overcome by directly producing them *inside* of the fiber, by exploiting the $\chi^{(3)}$ (Kerr) nonlinearity of the fiber itself. By placing the pump wavelength close to the zero-dispersion wavelength of the fiber, the probability amplitude for inelastic four-photon scattering can significantly enhanced. Two pump photons at frequency ω_p scatter through the Kerr nonlinearity to create simultaneous energy-time-entangled signal and idler photons at frequencies ω_s and ω_i, respectively, such that $2\omega_p = \omega_s + \omega_i$. Because of the isotropic nature of the Kerr nonlinearity in fused-silica-glass fibers, the correlated scattered photons co-polarized with the pump photons predominantly. Two such correlated down-conversion events from temporally multiplexed orthogonally polarized pumps can configured to create polarization entanglement as well. In this way, all four polarization-entangled Bell states have recently been prepared, violating Bell inequalities by up to ten standard deviations of measurement uncertainty. One drawback is the existence of Raman scattering in standard optical fibers due to coupling of the pump photons with optical phonons in the fiber. However, for small pump-signal detunings the imaginary part of $\chi^{(3)}$ in standard fibers is small enough that a 10-fold higher probability of creating a correlated photon-pair in a suitable detection window can be obtained than the probability of two uncorrelated Raman-scattered photons in the same detection window. Further work to quantify Raman scattering at the single-photon level need.

- Quantum dot entangled-photon sources

A biexcitonic cascade from a semiconductor quantum dot might also allow the generation of polarization-entangled photon pairs on demand, since the selection rules should translate the anticorrelation of electron and hole spins in the biexcitonic state into polarization anticorrelation of photons. However, this requires that the two decay paths from the biexcitonic state are indistinguishable; therefore, the effects such as dot anisotropy, strain, piezoelectric effects, and dephasing processes need minimized. To accomplish this, one needs optimize growth condition of quantum dot and employ novel high-Q photonic crystal microcavities, which would increase the radiative recombination rate over the dephasing rate.

2.11.3 Single-photon detectors

As noted in the introduction, photon-based quantum information processing applications require that single photons, or more generally, the photon number in a multiphoton state, which detected with efficiency approaching unity. To end much progress has been made in recent years towards developing high efficiency, low noise, and high count-rate detectors, which can reliably distinguish the photon number in an incident quantum state.

- Avalanche devices

Detection of single photons with avalanche photodiodes (APDs) biased above the breakdown voltage is convenient (no cryogenic temperatures are needed) and relatively efficient. When one or more photons are absorbed, the generated carriers that undergo avalanche gain may cause a detectable macroscopic breakdown of the diode p-n junction. APD photon counters suffer both from dark counts, where thermally generated charge carriers cause a detection event, and from after-pulses, where carriers from a previous avalanche cause subsequent detection events when the APD reactivated. The best counters at visible wavelengths have made with silicon APDs. These work well because of both the material system's ability to provide very low-noise avalanche gain and the availability of silicon of nearly perfect quality. For example, the single-photon counting modules (SPCMs), made by Perkin-Elmer (SPCM-AQR-16), can have 50% ~ 70% quantum efficiency near 700-nm wavelength, dark-count rate <25/s, and can count at rates up to 10-15 MHz. The dark-count rate is low enough for the SPCMs to be operated continuously except for a 50-ns avalanche quench time, although heating effects limit the CW counting rate to about 5MHz. After-pulsing is less than 0.5%. The quantum efficiency of the SPCMs drops at longer wavelengths (2% at 1μm). Attempts to resolve multiple photons by splitting a multi-photon pulse into several time bins (e.g., with a storage loop) have been made, but they are limited by losses in the device switching photons into and out of the loop, and by the non-unity detector efficiencies. The Visible Light Photon Counter (VLPC) and Solid State Photomultiplier (SSPM) are modified Si devices, which operate using a spatially localized avalanche from an impurity band to the conduction band. They possess high quantum efficiency (estimated to be 95%) with low multiplication noise. The localized nature of the avalanche allows high efficiency photon-number discrimination, which is not possible with conventional APDs. Using this capability, the non-classical nature of PDC has investigated and violations of classical statistics demonstrated. Unfortunately, these detectors require cooling to 6K for optimal performance, and even they display dark count rates in excess of $10^4\ s^{-1}$. In the infrared, 1~1.6μm, the best results to date have come from APDs having InGaAs as the absorption region that is separate from a multiplication layer of InP; see Figure 2-54. This has proven to be a better solution than germanium APDs. To suppress the high dark count rate in these devices, at best thousands of times worse than in silicon APDs, cooled InGaAs/InP APDs usually activated for only ~1-10ns duration to coincide with the arrival of the photon detected. The reported quantum efficiencies are typically between 10%~30%, and the APDs are usually operated at a count rate of 100kHz in order to alleviate after-pulsing caused by carriers trapped between the InGaAs and InP layers.

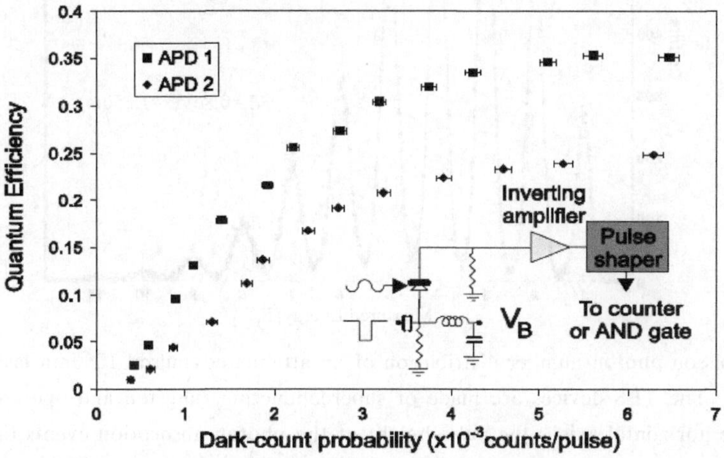

Fig. 2-54 Quantum efficiency versus dark-count probability for two InGaAs APDs operated in gated Geiger mode near 1537nm wavelength. In the gated Geiger mode, the APD is biased below breakdown and a short electrical pulse (~1ns), coincident with the incident light pulse containing the photon to be detected, brings it into the breakdown region momentarily. The inset shows a schematic of the electronic circuit used with the APDs.

• Superconducting devices

Superconducting devices offer the potential to achieve levels of performance that exceed those of conventional semiconductor APDs. Although there are many types of superconducting detectors, only three have used to observe single optical photons. The transition-edge sensor (TES), the superconducting tunnel junction (STJ), and the superconducting single-photon detector (SSPD). Both the TES and the STJ detectors have been able to detect single photons and count the number of photons absorbed by the detector. The TES detector uses the steep slope of the resistance as a function of temperature at the superconducting transition as a very sensitive thermometer. This thermometer is able to measure the temperature change in an absorber when one or more photons are absorbed (see Figure 2-55). The TES detectors are slow, capable of count rates at most up to 100kHz, but essentially have no dark counts. The reported detection efficiency currently varies from 20% to 40% in the telecom to optical band, although significant improvements in detection efficiency and speed are being realized with better detector designs (e.g., anti-reflection coatings) and research into new superconducting materials. In an STJ detector, excitations of the superconductor generated when a photon is absorbed. The excited quasiparticles can create an enhanced tunneling current which is proportional to the energy of the photon (or the number of photons absorbed). These detectors are similar in speed to the TES, which have no dark counts. The detection efficiency demonstrated to date is roughly 40% for visible photons, which could improved with AR coatings. The SSPD detectors are extremely fast detectors (~100ps total pulse duration) that have single photon sensitivities. In an SSPD, the detector is a narrow superconducting current path on a substrate. This path is current-biased at a point just below the superconducting critical current. A local hot spot is formed where a photon(s) is absorbed, locally destroying the superconductivity.

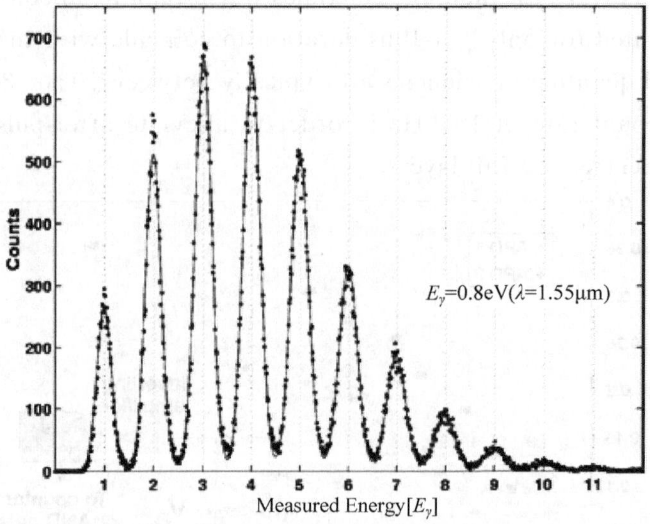

Fig. 2-55 Measured Poisson photon-number distribution of an attenuated, pulsed 1550nm laser, repeatedly measured using a TES. The TES devices are made of superconducting tungsten and operated at a temperature of 100mK. The horizontal axis is the pulse height of the photon absorption events in units of the energy of one 1550nm photon, 0.8eV. The inset shows a photograph of four fiber-coupled devices prepared cooled to 100mK. For the conducting path. Much improvement in device fabrication and design is needed to improve the quantum efficiencies of these devices beyond the current values of ~20%; the detection efficiency is lower still, due to the area effect mentioned above.

This forces the current to flow around the hot spot causing the current density around the hot spot to exceed the critical current density. As a result, the device develops a resistance, causing a voltage to appear across the device. These detectors are single-photon-threshold devices and are not able to resolve the photon number in multiphoton pulses. Typical implementations use meandering paths to increase the sensitive area, which is otherwise very small due to the narrowness required.

- Frequency upconversion

Detection techniques based on frequency upconversion allow IR photons converted into the visible where single photon detection is more efficient and convenient. Frequency upconversion uses sum-frequency generation in a non-linear optical crystal to mix a weak input signal at ω in with a strong pump at ω_p to yield a higher-frequency output field at $\omega_{\text{out}} = \omega_{\text{in}} + \omega_p$. With sufficient pump power, which upconversion can occur with near unity efficiency even for weak light fields at the single-photon level for LOQC and quantum key distribution applications, telecommunication-wavelength photons at 1.55μm can then be efficiently detected with low-noise, high quantum-efficiency Si APDs. Recently, upconversion of single photons from 1.55 to 0.63μm in bulk periodically poled lithium niobate (PPLN) has been demonstrated with an efficiency of 90% limited only by the available continuous wave (CW) pump power at 1.06μm, see Figure 2-56.

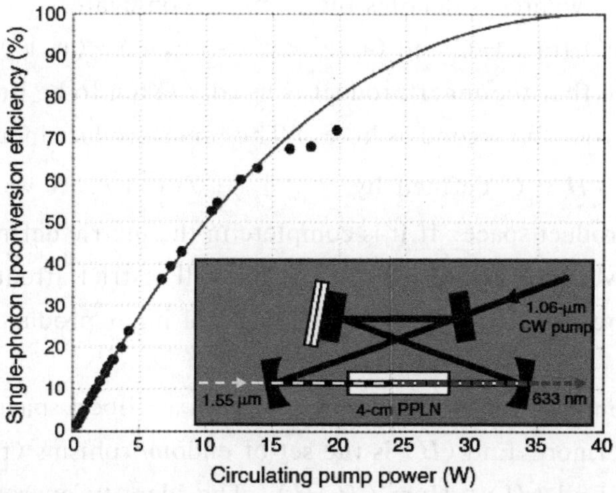

Fig. 2-56 CW single-photon up conversion efficiency versus circulating pump power in the pump enhancement ring cavity (inset). Solid line is a theoretical fit to data. At high pump powers lower than expected efficiencies are due to heating in PPLN that caused thermal instability in the ring cavity lock for results with improved cavity lock. Such a system has enabled single-photon conversion efficiencies of ~80% and backgrounds less than 10^{-3} per pulse. The pump power requirement can be relaxed by using a waveguide PPLN crystal, but the effect of waveguide losses must addressed to achieve the required near-unity net upconversion efficiency. The next step is to demonstrate frequency upconversion of a quantum state, i.e., high fidelity frequency translation of a single photon in an arbitrary quantum polarization state. This will allow a modular approach to developing LOQC technologies. For example, the photonic qubits and ancilla photons can be prepared at wavelengths with the most convenient and efficient methods, and then converted with near-unity efficiency to wavelengths that are optimal for photonic logic gates employing quantum interference. Similarly, tunable quantum frequency upconversion can used to match the required wavelengths to the resonant transitions in various atomic systems, for applications such as quantum repeaters. As another example, there have also been proposals to couple the photons to an atomic vapor system—the excitation of a single atom can made very probable by having many atoms, and that excitation can read out with very high efficiency by using a cycling transition. Such schemes could potentially yield efficiencies in excess of 99.9%. However, there are critical noise issues, which must still addressed.

The bulk PPLN crystal embedded inside a pump enhancement cavity that also imposes a well-defined spatial mode for the single-pass input photons. One approach to eliminate the need for a stabilized buildup cavity is to use a bright pulse escort beam, which is temporally mode-matched to the input photon.

For reasons noted in the introduction, there is intense current interest in creating robust, high-precision sources and detectors of single photons. In the last year alone, two special issues have appeared in the literature focusing just on these topics. Though tremendous progress has achieved, more development is clearly necessary to bring these technologies to the level of operation needed for LOQC. Nevertheless, already they have shown promise, enabling the realization of simple quantum gates, and improved quantum key distribution protocols. That further improvements over the next few years will continue to make optical qubits an attractive system, though it remains to be seen whether the extremely demanding LOQC requirements can be met.

2.11.4 Mathematical background

- Hilbert spaces and operators on them

Consider a vector space H, for concreteness over the the field of complex numbers C. An inner product on H is a bilinear function with the properties that

(1) $(v, v') = (v', v)^*$ where $*$ denotes the complex conjugate.

(2) $(v, \alpha v') = \alpha(v, v')$ for $\alpha \in C$ and $(v, v' + v'') = (v, v') + (v, v'')$.

(3) $(v, v) \geqslant 0$. (Note that the inner product is usually taken to be linear in the first argument in mathematics literature, not the second as here.) The inner product induces a norm on the vector space, a function $\|\cdot\|: H \to C$ defined by $\|v\| := \sqrt{(v, v)}$. A vector space with an inner product called an inner product space. If it is complete in the metric defined by the norm, meaning all Cauchy sequences converge, it called a Hilbert space will restrict attention to finite-dimensional spaces, where the completeness condition always holds and inner product spaces are equivalent to Hilbert spaces.

Denote the set of homomorphisms (linear maps) from a Hilbert space H to a Hilbert space H' by Hom (H, H'). Furthermore, End (H) is the set of endomorphisms (the homomorphisms from a space to itself) on H: End (H) = Hom (H, H). The identity operator $v \mapsto v$ that maps any vector $v \in H$ to itself is denoted by id. The adjoint of a homomorphism $S \in$ Hom (H, H'), denoted S^*, is the unique operator in Hom (H', H) such that:

$$(v', Sv) = (S^* v', v) \tag{2-712}$$

for any $v \in H$ and $v' \in H'$. In particular, have $(S^*)^* = S$. If S is represented as a matrix, then the adjoint operation can be thought of as the conjugate transpose. Here list some properties of endomorphisms $S \in$ End (H):

◇ S is normal if $SS^* = S * S$, unitary if $SS^* = S * S =$ id, and self-adjoint if $S^* = S$.

◇ S is positive if $(v, Sv) \geqslant 0$ for all $v \in H$. Positive operators are always self-adjoint. Sometimes write $S \geqslant 0$ to express that S is positive.

◇ S is a projector if $SS = S$. Projectors are always positive.

Given an orthonormal basis $\{b_i\}_i$ of H, that S is diagonal with respect to $\{b_i\}_i$ if the matrix $(S_{i,j})$ defined by the elements $S_{i,j} = (b_i, Sb_j)$ is diagonal. A map $U \in$ Hom (H, H') with dim $(H') \geqslant$ dim (H) will be called an isometry if $U^* U = \mathrm{id}_H$. It can be understood as an embedding of

H into H', since all inner products between vectors are preserved: $(\Phi', \psi') = (U\Phi, U\psi) = (\Phi, U^* U\psi) = (\Phi, \psi)$.

- The bra-ket notation

In this script will make extensive use of a variant of Dirac's bra-ket notation, where vectors interpreted as operators. More precisely, we can associate any vector $v \in H$ with an endomorphism $Iv\rangle \in \mathrm{Hom}\ (C, H)$, called ket and defined as

$$|v\rangle: \gamma \leftrightarrow \gamma v \tag{2-713}$$

for any $\gamma \in C$. it often regard $|v\rangle$ as the vector itself, a misuse of notation which enables a lot of simplification. The adjoint $|v\rangle^*$ of this mapping is called bra and denoted by $|v\rangle$. It is easy to see that $|v\rangle$ is an element of the dual space $H^* := \mathrm{Hom}\ (H, C)$, namely the linear functional defined by

$$\langle v|: \quad u \mapsto (v, u) \tag{2-714}$$

for any $u \in H$. Note, however, that bras and kets are not quite on equal footing, as the label of a bra is an element of H, not H^*. The reason we can do this is the Riesz representation theorem, which states that every element of the dual space is of the form given in Eq. (2-714). It follows immediately from the above definitions that, for any $u, v \in H$,

$$\langle u| \circ |v\rangle \equiv (u, v) \tag{2-715}$$

Conversely, the concatenation $|v\rangle\langle u|$ is an element of End (H) (or, more generally, of Hom (H, H') if $u \in H$ and $v \in H'$ are defined on different spaces). In fact, any endomorphism $S \in$ End (H) can written as a linear combination of such concatenations:

$$S = \sum_i |u_i\rangle\langle v_i| \tag{2-716}$$

for some families of vectors $\{u\}_i$ and $\{v\}_i$. For example, the identity id $\in$ End (H) can be written as

$$\mathrm{id} = \sum_i |b_i\rangle\langle b_i| \tag{2-717}$$

for any orthonormal basis $\{b_i\}$ of H. This is often called the completeness relation of the basis vectors.

- Representations of operators by matrices

Given an orthonormal basis $\{|b_k\rangle\}k \overset{d}{=} 1$, we can associate a matrix with any operator $S \in$ End (H),

$$S \rightarrow S_{jk} = \langle b_j | S | b_k\rangle \tag{2-718}$$

Here "overloading" the notation a bit, and referring to both the matrix components as well as the matrix itself as S_{jk}. In the study of relativity, this is referred to as abstract index notation or slot-naming index notation. Have chosen j to be the row index and k the column index, so that a product of operators like ST corresponds to the product of the corresponding matrices, but the other choice could have been made.

It is important to realize that the representation of an operator by a matrix is not unique, but depends on the choice of basis. One way to see this is to use the completeness relation, to write:

$$S = \mathrm{id}\ S\ \mathrm{id} \tag{2-719}$$

$$= \sum_{j,k} |b_j\rangle\langle b_j | S | b_k\rangle\langle b_k| \tag{2-720}$$

$$= \sum_{j,k} S_{j,k} |b_j'\rangle\langle b_k| \tag{2-721}$$

Now the basis dependence is plain to see. Matrix representations can be given for more general operators $S \in \mathrm{Hom}(H, H_0)$ by the same technique:

$$S = \mathrm{id}_{\mathscr{H}'} S \, \mathrm{id}_{\mathscr{H}} \tag{2-722}$$

$$= \sum_{j,k} | b'_j \rangle \langle b'_j | S | b_k \rangle \langle b_k | \tag{2-723}$$

$$= \sum_{j,k} S_{j,k} | b'_j \rangle \langle b_k | \tag{2-724}$$

In the version of Dirac notation, $| v \rangle$ is itself an operator, so we can apply the above method to this case. Now, however, the input space is one-dimensional, so drop the associated basis vector and simply write:

$$| v \rangle = \sum_j v_j | b_j \rangle \tag{2-725}$$

According to the above convention, the representation of $| v \rangle$ is automatically a column vector, as it is the column index (which would take only one value) that has been omitted. Following our use of abstract index notation, the (vector) representative of $| v \rangle$ is called v_j, not $\vec{v}$ or similar. In terms of matrix representatives, the inner product of two vectors u and v is given by $u_j^* v_j$, since the inner product is linear in the second argument, but antilinear in the first. Expect the representation of the adjoint of an operator to be the conjugate transpose of the matrix, but let us verify that this is indeed the case. The defining property of the adjoint is Dirac notation:

$$\langle u | Sv \rangle = \langle S^* u | v \rangle \tag{2-726}$$

In terms of matrix representatives, reading the above from right to left have:

$$(S^* u)_j^* \cdot v_j = u_j^* \cdot (Sv)_j \tag{2-727}$$

$$= \sum_{jk} u_j^* S_{jk} v_k \tag{2-728}$$

$$= \sum_{jk} ([S_{jk}]^* u_j)^* v_k \tag{2-729}$$

$$= \sum_{jk} ([S_{jk}]^\dagger u_k)^* v_j \tag{2-730}$$

Here † denotes the conjugate transpose of a matrix. Comparing the first expression with the last, it must be that $[S^*]_{jk} = [S_{jk}]^\dagger$, as we suspected.

- Tensor products

Given vectors u and v from two Hilbert spaces H_A and H_B, may formally define their product $u \times v$, which is an element of the Cartesian product $H_A \times H_B$. However, the Cartesian product does not respect the linearity of the underlying spaces. That is, while we may formally add $u \times v$ and $u' : \times v$, the result is not $(u + u') \times v$; it is just $u' v + u' \times v$. The idea behind the tensor product is to enforce this sort of linearity on $H_A \times H_B$. There are four combinations of vectors which expect to vanish by linearity:

$$\begin{cases} u \times v + u' \times v - (u + u') \times v \\ u \times v + u \times v' - u \times (v + v') \\ \alpha (u \times v) - (\alpha u) \times v \\ \alpha (u \times v) - u \times (\alpha v) \end{cases} \tag{2-731}$$

for any $\alpha \in C$. These vectors define an equivalence relation on $H_A \times H_B$ in that can consider two elements of that space to be equivalent if they differ by some vector of the form in Eq. (2-731). These equivalence classes themselves form a vector space, and the resulting vector space is precisely the tensor product $H_A \otimes H_B$.

Since the construction enforces linearity of the products of vectors, may consider the tensor product to be the space spanned by products of basis elements of each space. Furthermore, the inner product of $H_A \otimes H_B$ is defined by the linear extension of

$$(u \otimes v, u' \otimes v') = \langle u \mid u' \rangle \langle v \mid v' \rangle \tag{2-732}$$

For two homomorphisms $S \in \text{Hom}(H_A, H'_A)$ and $T \in \text{Hom}(H_B, H'_B)$, the tensor product $S \otimes T$ is defined as

$$(S \otimes T)(u \otimes v) := (Su) \otimes (Tv) \tag{2-733}$$

For any $u \in H_A$ and $v \in H_B$. The space spanned by the products $S \otimes T$ can be canonically identified with the tensor product of the spaces of the homomorphisms, i.e.

$$\text{Hom}(\mathcal{H}_A, \mathcal{H}'_A) \otimes \text{Hom}(\mathcal{H}_B, \mathcal{H}'_B) \simeq \text{Hom}(\mathcal{H}_A \otimes \mathcal{H}_B, \mathcal{H}'_A \otimes \mathcal{H}'_B) \tag{2-734}$$

That is, the mapping defined by Eq. (2-733) is an isomorphism between these two vector spaces. This identification allows us to write, for instance,

$$|u\rangle \otimes |v\rangle = |u \otimes v\rangle \tag{2-735}$$

for any $u \in H_A$ and $v \in H_B$.

Trace and partial trace: The trace of an endomorphism $S \in \text{End}(H)$ over a Hilbert space H is defined by

$$\text{tr}(S) := \sum_i \langle b_i \mid S \mid b_i \rangle \tag{2-736}$$

• Decompositions of operators and vectors

where $\{|b_i\rangle\}_i$ is any orthonormal basis of H. The trace is well defined because the above expression is independent of the choice of the basis, as one can easily verify. The trace operation is obviously linear,

$$\text{tr}(\alpha S + \beta T) = \alpha \text{tr}(S) + \beta \text{tr}(T) \tag{2-737}$$

for any $S, T \in \text{End}(H)$ and $\alpha, \beta \in C$. It also commutes with the operation of taking the adjoint,

$$\text{tr}(S^*) = \text{tr}(S)^* \tag{2-738}$$

Since the adjoint of a complex number $\gamma \in C$ is simply its complex conjugate. Furthermore, the trace is cyclic,

$$\text{tr}(ST) = \text{tr}(TS) \tag{2-739}$$

Also, it is easy to verify using the spectral decomposition that the trace $\text{tr}(S)$ of a positive operator $S \geqslant 0$ is positive. More generally

$$(S \geqslant 0) \wedge (T \geqslant 0) \Rightarrow \text{tr}(ST) \geqslant 0 \tag{2-740}$$

The partial trace tr_B is a mapping from the endomorphisms $\text{End}(H_A \otimes H_B)$ on a product space $H_A \otimes H_B$ onto the endomorphisms $\text{End}(H_A)$ on H_A. (Here and in the following, that will use subscripts to indicate the space on which an operator acts.) It defined by the linear extension of the mapping.

$$\text{tr}_B: \quad S \otimes T \mapsto \text{tr}(T) S \tag{2-741}$$

for any $S \in \text{End}(H_A)$ and $T \in \text{End}(H_B)$.

Similarly to the trace operation, the partial trace tr_B is linear and commutes with the operation of taking the adjoint. Furthermore, it commutes with the left and right multiplication with an operator of the form $T_A \otimes \text{id}_B$ where $T_A \in \text{End}(H_A)$. That is, for any operator $S_{AB} \in \text{End}(H_A \otimes H_B)$,

$$\text{tr}_B(S_{AB}(T_A \otimes \text{id}_B)) = \text{tr}_B(S_{AB}) T_A \tag{2-742}$$

and

$$\text{tr}_B((T_A \otimes \text{id}_B) S_{AB}) = T_A \text{tr}_B(S_{AB}) \tag{2-743}$$

The will also make use of the property that the trace on a bipartite system can be decomposed into partial traces on the individual subsystems. That is,

$$\operatorname{tr}(S_{AB}) = \operatorname{tr}(\operatorname{tr}_B(S_{AB})) \tag{2-744}$$

or, more generally, for an operator $S_{ABC} \in \operatorname{End}(H_A \otimes H_B \otimes H_C)$,

$$\operatorname{tr}_{AB}(S_{ABC}) = \operatorname{tr}_A(\operatorname{tr}_B(S_{ABC})) \tag{2-745}$$

- Decompositions of operators and vectors

Singular value decomposition. Let $S \in \operatorname{Hom}(H, H')$ and let $\{b_i\}_i(\{b'_i\}_i)$ be an orthonormal basis of H. Then there exist unitaries $U, V \in \operatorname{End}(H)$ and an operator $D \in \operatorname{End}(H)$ which is diagonal with respect to $\{e_i\}_i$ such that

$$S = UDV^* \tag{2-746}$$

Polar decomposition. Let $S \in \operatorname{End}(H)$. Then there exists a unitary $U \in \operatorname{End}(H)$ such that

$$S = \sqrt{SS^*}\, U \tag{2-747}$$

and

$$S = U\sqrt{S^* S} \tag{2-748}$$

Spectral decomposition. Let $S \in \operatorname{End}(H)$ be normal and let $\{|b_i\rangle\}_i$ be an orthonormal basis of H. Then there exists a unitary $U \in \operatorname{End}(H)$ and an operator $D \in \operatorname{End}(H)$ which is diagonal with respect to $\{|b_i\rangle\}_i$ such that

$$S = UDU^* \tag{2-749}$$

The spectral decomposition implies that, for any normal $S \in \operatorname{End}(H)$, there exists a basis $\{|b_i\rangle\}_i$ of H with respect to which S is diagonal. That is, S can be written as

$$S = \sum_i \alpha_i |b_i\rangle\langle b_i| \tag{2-750}$$

where $\alpha_i \in C$ are the eigenvalues of S.

Eq. (2-750) can be used to give a meaning to a complex function $f: C \to C$ applied to a normal operator S. to define $f(S)$ by

$$f(S) := \sum_i f(\alpha_i) |b_i\rangle\langle b_i| \tag{2-751}$$

- Operator norms and the Hilbert-Schmidt inner product

The Hilbert-Schmidt inner product between two operators $S, T \in \operatorname{End}(H)$ is defined by

$$(S, T) := \operatorname{tr}(S^* T) \tag{2-752}$$

The induced norm $\|S\|_2 := \sqrt{(S,S)}$ is called Hilbert-Schmidt norm. If S is normal with spectral decomposition $S = \sum_i \alpha_i |b_i\rangle\langle b_i|$ then

$$\|S\|_2 = \sqrt{\sum_i |\alpha_i|^2} \tag{2-753}$$

An important property of the Hilbert-Schmidt inner product (S, T) is that it is positive whenever S and T are positive.

Lemma 2.69 Let $S, T \in \operatorname{End}(H)$. If $S \geqslant 0$ and $T \geqslant 0$ then

$$\operatorname{tr}(ST) \geqslant 0 \tag{2-754}$$

- Operator norms and the Hilbert-Schmidt inner product

Proof. If S is positive have $S = \sqrt{S^2}$ and $T = \sqrt{T^2}$. Hence, using the cyclicity of the trace, have

$$\operatorname{tr}(ST) = \operatorname{tr}(V^* V) \tag{2-755}$$

where $V=\sqrt{S}\sqrt{T}$. Because the trace of a positive operator is positive, it suffices to show that $V*V\geqslant 0$. This, however, follows from the fact that, for any $\Phi\in H$,

$$\langle\phi\mid V^{*}V\mid\phi\rangle=\|V\phi\|^{2}\geqslant 0 \tag{2-756}$$

The trace norm of S is defined by

$$\|S\|_{1}:=\operatorname{tr}\mid S\mid \tag{2-757}$$

where

$$\mid S\mid:=\sqrt{S^{*}S} \tag{2-758}$$

If S is normal with spectral decomposition $S=\sum_{i}\alpha_{i}\mid e_{i}\rangle\langle e_{i}\mid$ then

$$\|S\|_{1}=\sum_{i}\mid\alpha_{i}\mid \tag{2-759}$$

The following lemma provides a useful characterization of the trace norm.

Lemma 2.70 For any $S\in\operatorname{End}(H)$,

$$\|S\|_{1}=\max_{U}\mid\operatorname{tr}(US)\mid \tag{2-760}$$

where U ranges over all unitaries on H.

Proof. For any unitary U,

$$\mid\operatorname{tr}(US)\mid\leqslant\operatorname{tr}\mid S\mid \tag{2-761}$$

with equality for some appropriately chosen U.

Let $S=V|S|$ be the polar decomposition of S. Then, using the Cauchy-Schwarz inequality

$$\mid\operatorname{tr}(Q^{*}R)\mid\leqslant\|Q\|_{2}\|R\|_{2} \tag{2-762}$$

with $Q:=\sqrt{|S|}V^{*}U^{*}$ and $R:=\sqrt{|S|}$ can find

$$\begin{aligned}\mid\operatorname{tr}(US)\mid&=\mid\operatorname{tr}(UV\mid S\mid)\mid=\mid\operatorname{tr}(UV\sqrt{|S|}\sqrt{|S|})\mid\\&\leqslant\sqrt{\operatorname{tr}(UV\mid S\mid V^{*}U^{*})\operatorname{tr}(\mid S\mid)}=\operatorname{tr}(\mid S\mid)\end{aligned} \tag{2-763}$$

Which proves (2-761). Finally, it is easy to see that equality holds for $U:=V^{*}$.

- The vector space of Hermitian operators

The set of Hermitian operators on a Hilbert space H, in the following denoted Herm (H), forms a real vector space. Furthermore, equipped with the Hilbert-Schmidt inner product defined in the previous section, Herm (H) is an inner product space.

If $\{e_i\}_i$ is an orthonormal basis of H then the set of operators $E_{i,j}$ defined by

$$E_{i,j}:=\begin{cases}\dfrac{1}{\sqrt{2}}\mid e_{i}\rangle\langle e_{j}\mid+\dfrac{1}{\sqrt{2}}\mid e_{j}\rangle\langle e_{i}\mid, & \text{if } i<j\\ \dfrac{i}{\sqrt{2}}\mid e_{i}\rangle\langle e_{j}\mid-\dfrac{i}{\sqrt{2}}\mid e_{j}\rangle\langle e_{i}\mid, & \text{if } i>j\\ \mid e_{i}\rangle\langle e_{i}\mid, & \text{otherwise}\end{cases} \tag{2-764}$$

Forms an orthonormal basis of Herm (H). To conclude from this that:

$$\dim\operatorname{Herm}(\mathscr{H})=(\dim\mathscr{H})^{2} \tag{2-765}$$

For two Hilbert spaces H_A and H_B, have in analogy to:

$$\operatorname{Herm}(\mathscr{H}_{A})\otimes\operatorname{Herm}(\mathscr{H}_{B})\cong\operatorname{Herm}(\mathscr{H}_{A}\otimes\mathscr{H}_{B}) \tag{2-766}$$

Consider the canonical mapping from Herm $(H_A)\otimes$Herm (H_B) to Herm $(H_A\otimes H_B)$ defined by (2-733). It is easy to verify that this mapping is injective. Furthermore, because by Eq. (2-765) the dimension of both spaces equals dim $(H_A)^2$dim $(H_B)^2$, it is a bijection, which proves Eq. (2-766).

Chapter 3

Multi-dimension Photonic Memory

3.1 Mechanism of photochromic multi-dimension memory

The traditional technology roadmap to improve the optical storage density is to reduce the size of the recording spot that is, by reducing the laser wavelength and increase the number of the value of numerical aperture of the objective lens and reduce the wavelength of the recording laser ray to increase optical storage density and capacity. Continue to rely on traditional optical technology will be increasingly difficult to improve the optical storage density. The first is to continue to reduce the laser wavelength will face a variety of technical barriers is not only difficult to obtain a shorter wavelength semiconductor laser as a light source, and when the wavelength of the ultraviolet or DUV, most of the current optical materials, including CD-ROM disk based materials will have a strong absorption etc series of technical problems. Continue to increase the numerical aperture is also facing the same problem. Now, the numerical aperture has been 0.85, so potential improvement is very small. If it continues to increase the numerical aperture that the disc must be removed the layer of protection, the disc will loss of a variety of advantages, cannot run in the general environment and long-term preservation.

A team of Boffins at the Swinburne University of Technology in the Queendom of Down under have tested a new type of "five-dimensional" optical storage medium that they estimated might hold up to 2,000 times more data than a conventional DVD. Nanomaterials, it seems, are photoreactive and adjust their shape according to different colours of the visible spectrum, which were illuminated by lasers in this case. The team then followed up by applying multiple polarizations to the same physical disc space, effectively writing the data at different angles in the same place.

This means that data—usually written in a typical three dimensional (x, y, z) fashion—acquired two more dimensions. So far this has already resulted in an optical disc sample capable of storing 1.6TB of data, but as development continues, researchers Min Gu, Peter Zijlstra and James Won expect storage capacity to reach a whopping 10TB. None of these techniques are actually new, just the fact that they were all applied at the same time. This brings about at least one major problem that the technology has to contend with, that is, recording speed. Current prototypes record about as fast as a glyph-carver in ancient Egypt, the researchers have implied.

Another problem such high-capacity media are going to have to confront is tied up with several related terms like robustness, reliability and longevity. At least initially, most people will want to have such large capacity physical media offer some assurance that they won't self-destruct within merely ten or twenty years as most presently available CD and DVD discs are very easy to do. But scientists are working on first things, thinking about how to do this before working out how to make people believe it's worth entrusting lots of valuable data to it. The research, in the meantime, has been hoovered up by the storage giant Samsung in Korea, which now seems destined to manufacture the media that records every bit of stored data on the planet. The technology should be ready within the next ten years, but it's also possible to see it application for long-term preservation information of the future.

This chapter proposed the super-resolution multi-wavelength photon bistable storage. The basic idea is not to continue, but to use the increase in information storage space to improve the optical storage density by reducing the laser wavelength and increase the number of the value of the objective lens aperture the basic implementation of the technology roadmap is as follows:

(1) CD storage from a flat two-dimensional extended to three-dimensional or even multidimensional (e.g. frequency dimension), by increasing the coordinate system of the optical storage disc storage capacity from traditional linear function of growth, increase to the growth of the exponential function.

(2) The nonlinear absorption of the recording medium and changing beam propagation direction and phase changes to reduce the effective size of the storage spot, storage density than conventional optical resolution, in order to achieve super-resolution storage.

(3) The photochromic material to replace the traditional dye and phase change measurement as a recording medium, disc storage write mechanism speed changed to photothermal effect of the slow response from the current light-quantum effects, in order to improve the data write/read rate.

(4) Principles and methods of light stress is adopted, optical modulation and coding from the traditional binary system changed to including grayscale, refractive index, color and its intensity changing modulation and coding that great to improve the coding efficiency and storage capacity.

(5) Two-photon recording medium, information recording process from a single photon effect to the two-photon effect, only in two-photon interchanges can record information in order to achieve the spatial three-dimensional storage.

(6) Using a variety of optical parameters to storage information, reading signal of this kind CD from the traditional single channel change to multi-channel that can improve CD reading/writing data at the same rotative speed, with synchronous transfer rate is many times to the traditional CD.

(7) The new codec system, the traditional second-order travel-length coding changed to multi-wavelength, multi-level tour length coding to achieve the sector guidance, correction and supporting code sharing, reduce redundancy, improve the effective capacity of the CD and read out rate.

(8) Bring forward principles and methods of synthesis design and integration for multi-wavelength laser resources, modulation and detection system that is the base for the realization of multi-wavelength, multi-level and multi-layer drives and discs of the different capacity and function.

Apparently the principles of the multi-wavelength and multi-level optical storage are using the multilist spectral absorption characteristics of a photosensitive material or combinations materials on a recording layer. Because these materials can also record a lot of different color (wavelength) and different intensity (gray scale), thereby the capacity of the record information is enhanced that is very similar to color photos. If extract analysis characteristics of record and read out information for this spectral and photometric CD, it can be called luminosity compact disc indeed. In order to facilitate the promotion of research and future applications, the light sources of this CD using the CD industry-scale production of various wavelengths of semiconductor lasers, the hardware structure of modular combinations for use in traditional CD technological innovation, a combination of various size storage systems. For example, the concept of the order of storage can also be used to record information in the "pit" ordinary CD, through the depth or size to change the "pit", "Grayscale" information, to increase storage capacity to a certain extent, the use of the multi-level optical disc recorder can make the traditional CD-R discs recorded this technology to increase capacity three times. In addition, the chapter also introduces the principles of phase shift mask and optical system using super-resolution further reduce the record spot size, and technology to improve the optical storage density and capacity. Experimental study of the basic principles of the multi-dimensional optical information storage has to be based on the necessary experimental devices. This chapter will discuss and analysis of the light-induced photochromic multi-wavelength and multi-level experiment principles, detailing experimental setup as well as high-resolution storage experimental system design, affective factors, including read/write signal crosstalk, non-destructive readout method, the design and experimental parameters of the synthesis of multi-wavelength optical system design model, basic parameters, apochromatic objective lens, memory encoding and error correction principles of multi-wavelength bands, mathematical model, error-correcting capability assessment and the electric field fixed non-destructive readout principles. Super-resolution storage technology can improve the multi-wavelength optical storage density observably. A nonlinear absorption super-resolution storage experiment based on near-field optical principles and the experimental results are discussed. Same important correlative technology, as changing the space phase and intensity distribution of the focused beam to improve the optical system transfer function to raise the resolution of optical system design and development of the experimental apparatus and experimental results are introduced in this section also. The last one of this chapter introduces the multi-wavelength and multi-level optical disc replication and testing technology, which is an important extension for the multi-wavelength, multi-level optical disc storage, the recording medium with the traditional CD is never different. The CD-ROM of multi-wavelength and multi-level disc copy has to used a new kind of mastering disc replication processes only. There are a series of new issues, that will be introduced in this section, including structural principles of multi-wavelength multi-level optical disc mastering system designed, developed and experimental results. This section also introduces the principles of the optical method to copy a CD, the lithography system and the use of this principle to copy the disc read signal. Addition, same consummations of multi-level and multi-wavelength optical storage experimental study would support other high-speed CD-ROM to read out the signal testing system design, and blue light super-resolution storage experimental research. So this chapter provides a brief introduction also.

3.1.1 Photochromic reaction

Photochromic reaction or photochromism does not have a rigorous definition, but is usually used to describe compounds that undergo a reversible photochemical reaction where an absorption band in the visible part of the electromagnetic spectrum changes dramatically in strength or wavelength. The photochromic phenomenon first was discovered in the late 1880s, including work by Markwald. He labeled this phenomenon "phototropy", and this name was used until the 1950s when Yehuda Hirshberg, of the Weizmann Institute of Science in Israel proposed the term "photochromism". Photochromism can take place in both organic and inorganic compounds, and also has its place in biological systems trahydronaphthalene. A phenomenon was known as the photoinduced thermodynamic reversible color change each other (phototropy). Yehuda Hirshberg researched this phenomenon more thorough, clearly put forward the concept of photochromic reaction (photochromism) in 1956, and establishment of the fineness and photobleaching cycle may constitute chemical memory model for the photochemical information storage. In particular, organic light-induced reaction of photochromic materials for laser signals constitute a new generation of optical information storage materials was put forward to the research agenda. Since nearly half a century, the photochromic material to carry out a lot of research work, the photochromic materials specific to give such compounds to bring a broad and important applications. When the photochromic reaction of a compound or complex *A* with certain wavelength of light irradiation $h\nu_1$, the formation of the structure was became to another compound *B*. The compound *B* went back to the original structure *A*, when it was irradiation or heating by other wavelength light $h\nu_2$ or Δ simultaneity, the phenomenon as follows:

$$A \underset{h\nu_2 \text{ or } A}{\overset{h\nu_1}{\rightleftarrows}} B$$

Where, the h is the Planck constant ν_1, ν_2 is the optical frequency Δ said heat. As shown in Figure 3-1, a photochromic materiala has two completely different absorption spectrum of *A* and *B* when it absorbs different laser light. The two kinds of steady state of photochromic materiala that can be represented the number 0 and 1 and rewritable, that is the basis using this photochemical phenomoena to record digital data.

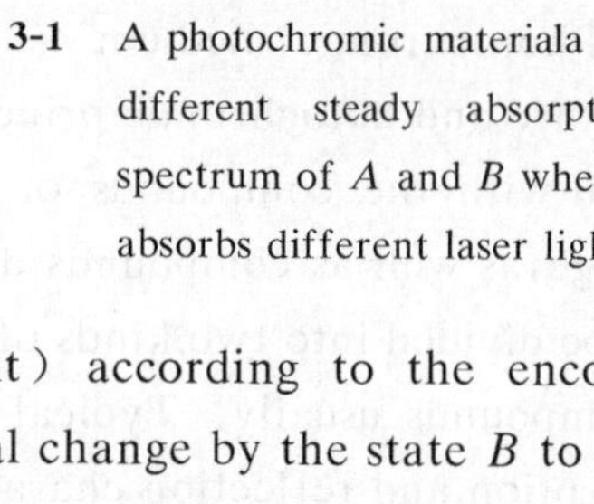

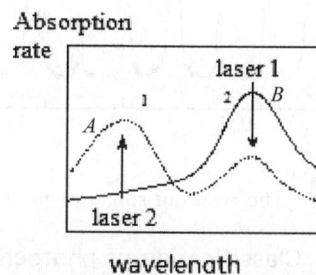

Fig. 3-1 A photochromic materiala has different steady absorption spectrum of *A* and *B* when it absorbs different laser light.

Actually photochromic materials for recording media to store information, the first wavelength λ_1 light (known as the erase light) irradiation, storage media transition from state *A* to state *B*. Record with a wavelength λ_2 is light (write light) according to the encoding information is written, λ_2 is light shines on the area of the material change by the state *B* to state *A*, the binary coded signal representative 1. Not been λ_2 light exposure within the region material is still state *B* corresponds to the binary-coded 0. For information read out, can use the test medium absorption changing rate or the refractive index change, that can get a readout signal also. To measure the change of the material absorption rate (transmission rate) is using weak wavelength λ_2 light often and shown coded of 0 and very small to λ_2 light absorption and transmittance. When the light λ_2 shines, it is shown coded 1 (state *A*), and absorption and transmission rate obtain maximum. In addition, the change of refractive index can also record information, but need to accurately measure the refractive index of the state *A* and state *B*. In addition, due to medium sensitivity is difference so requiring different laser recording powers. Using absorption peaks were 650nm/532nm/405nm recording materials, with same written conditions to experiment that results is shown in Figure 3-2 (a), (b) and (c) for example. Obviously the sensitivity of (a) for 650nm laser is highest and (c) lowest for 405nm.

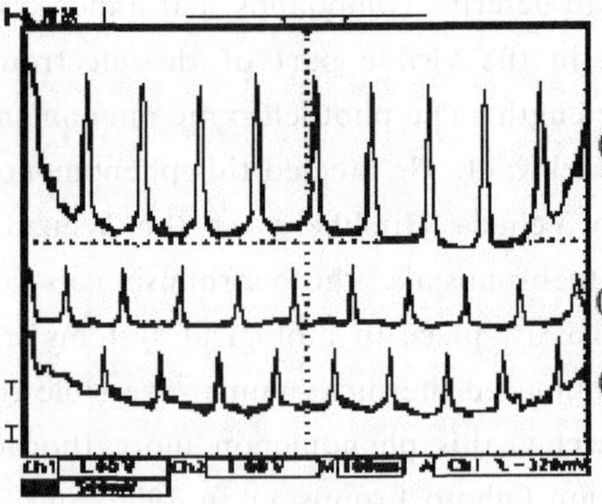

Fig. 3-2 Experiment readout signals of sensitivity wavelength for (a) 650nm, (b) 532nm and (c) 405nm photochromic materials.

The more sensitive photochromic material is diarylethylene which absorption peak is 650nm. The multilevel storage experiment is shown in Figure 3-3 using diarylethylene easily. It can be written by controlling the light intensity to get four-level storage. If the medium contains three kinds of materials for different wavelength absorption peaks and mixed on a recording layer that can achieve the three wavelength and four-level multilevel storage.

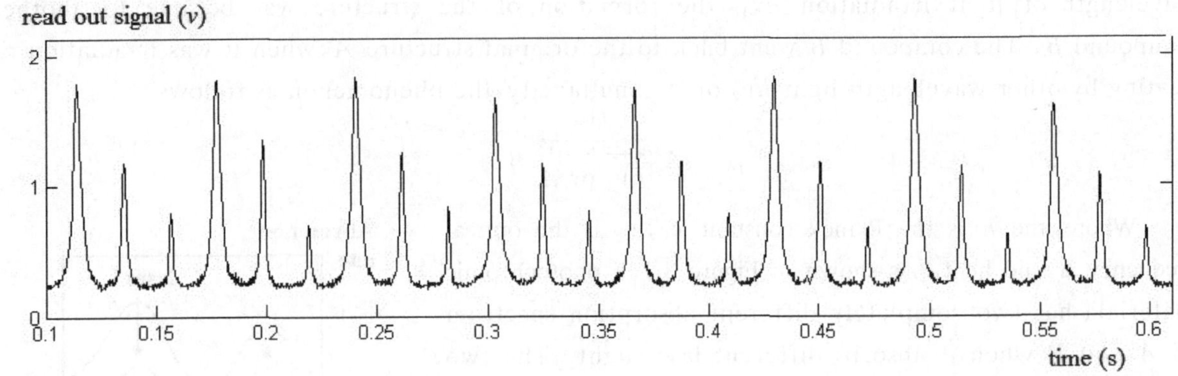

Fig. 3-3 The read out single of multilevel recording experiment on the sensitivity wavelength for 650nm material.

- Classification of photochromic materials

Photochromic medium is including both inorganic and organic materials that molecular structure and hotochromic principles are different absolutely. Inorganic photochromic medium are added with the compounds of metals (mainly transition period of heavy metals) ion valence change, as well as compounds decompose and re-combined to achieve photochromic reaction that can be divided into two kinds of the decomposition of the metal ions of variable valency and halide decompounds usually. Typical inorganic material is tellurium, that the material has a strong absorption and reflection characteristics of a variety of specific wavelength laser. But it is easy to produce the scaly cracking by prolonged exposure of light irradiation, and the control technology of thickness and uniformity are more complex in production, so that not continue to be used. Organic photochromic materials generally rely on organic compounds bond rupture (including the uniform crack and different crack), the key to the restructuring and heterogeneous cause photochromic or discoloration reaction. Research focus began to organic optical storage materials from the 1980s in the world that was used in digital optical storage of organic photochromic compounds commonly and have the following categories:

- Spiropyran/Looxazine compounds

Photochromic reaction of this kind of compounds are because that CS bond (or CO) of spiro compounds (closed-loop body) was cleavaged and generated open-loop body at the short wavelength UV irradiation. It has a strong absorption in the visible region and photochromic reaction. Under visible light irradiation or in the case of dark, open-loop body to re-restore into a closed loop body that has good reversible reaction. Spiropyran has better photochromic properties also, the open-loop maximum absorption wavelength of less than 600nm, its main drawback is that the photochromic body of the open-loop is unstable. Looxazine thermal stability and fatigue resistance are better in the all, and its open-loop maximum absorption wavelength is great than the spiropyran with potential applications.

- Fulgide compounds

Such compounds are used valence bond tautomerism intramolecular pericyclic reactions to be photochromic reaction. Its stability and fatigue resistance are better, but the absorption peak wavelength does not match with the commercial semiconductor lasers, that needs to be adjusted. The representative structure of the fulgide moleculecompounds is shown in Figure 3-4.

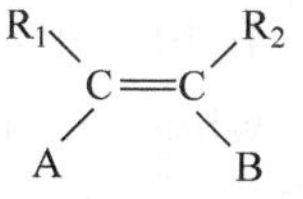

Fig. 3-4 Typical structure of the fulgide.

- Diarylethene compounds

Diarylethene compounds belonging to the cis-trans isomerization of photochromic compounds, due to the absorption spectra before and after the photochromism is smaller, the absorption spectrum will have a greater overlap that is not sutable for multi-wavelength storage. Experiments show that these compounds not only have good photochromic properties and the overlap of the absorption spectrum is also smaller, may be used for CD storage. Study also showed that these compounds thermal stability is good, and the open-loop body at 300℃ does not produce the thermochromic phenomenon. During destructive testing, the closed-loop body may be stable at 80℃ in more than 3 months, and fatigue resistance is also very great. Such compounds have a general form that used the valence tautomerism occurrence of intramolecular pericyclic reactions. Typical diarylethene molecular structure of photochromic reaction is shown in Figure 3-5.

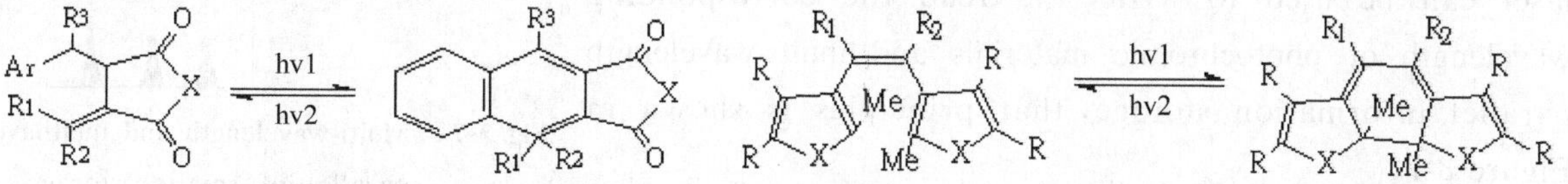

Fig. 3-5 Typical molecular structures of diarylethene fulgide and photochromic reaction.

In addition, the performance of this compounds with substituents of different and will change greatly, that the feature can be used to adjust design of the absorption peak as shown in Figure 3-5, which will be great development for application.

- The azo compound

This kind of compounds is utilized the cis-trans isomerization of key. Its peak of the absorption can be adjust when the electronic grant or electronic receiver body are introduced to different parts of the molecule. The maximum absorption wavelength of some azo dyes can cover 700nm. A typical molecular structure of the azo compound and its photochromic reaction is shown in Figure 3-6.

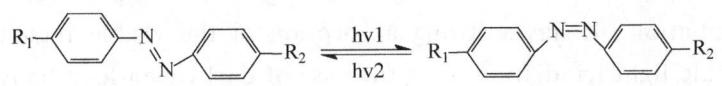

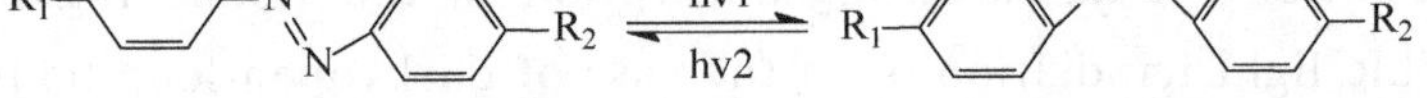

Fig. 3-6 Typical molecular structures of the azo compound and its photochromic reaction.

The wavelength of the absorption spectra of azo compounds is shorter and that is changed very small before and after photochromic reaction, and thermal stability is poor. In other hand the sensitive peak of azo dyes can not match with the laser of current commercialized production, so it is in the experimental study of optical storage only.

- Other organic photochromic compounds

For light-induced photochromic storage materials are still much more indeed, that as long as the following conditions can meet photochromic optical storage. Where high sensitivity and good anti-fatigue performance, can repeatedly write, erase, or stable performance of the material can be used for write-once optical information storage. Best sensitive wavelength match with the industrial production of semiconductor lasers, good solubility, ease of production of the recording layer using the spin coating method. For commercialize application products require materials with good thermal stability and long-term preservation also. Study of photochromic storage materials, including dedicated research of organic photochromic compounds, molecular assembly (such as MTCNQ and MTNAP norbornene, sulfur, indigo, etc.) and inorganic materials with good performance of the photochromic compounds have great potential development.

3.1.2 Multi-wavelength photochromic storage process

With new progress made by the synthetic study of the photochromic material, the absorption peak region can choose and control within a relatively narrow range that greatly reduces the absorption of the material on non-sensitive areas wavelength laser. Different wavelengths of laser can be used to write and read the corresponding wavelength of photochromic materials and multi-wavelength parallel information storage, that principles is shown in Figure 3-7.

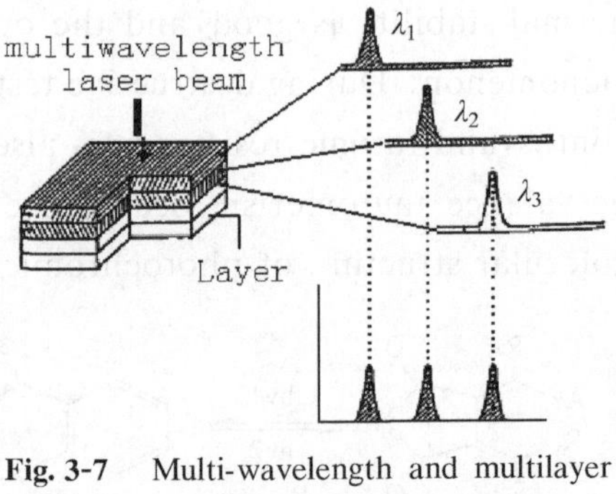

Fig. 3-7 Multi-wavelength and multilayer parallel information storage.

This is a multi-layer and multi-wavelength storage program, the absorption peak of wavelength, respectively, three kinds of wavelength photochromic materials, three-layer coating in experiments on the CD, the absorption spectrum of the selected material requirements is relatively narrow, its width should be less than or close to the difference between the neighboring recording laser wavelength, the absorption spectrum of the material should not generate overlap. Since each recording layer contains only a photochromic materials, each recording layer is only sensitive to a record-wavelength laser, while the other wavelength lasers are transparent. For example, recording laser λ_1 in Figure 3-7, when the laser λ_1 writes the multi-wavelength photochromic plate, only absorption by this photochromic material that is sensitive to the wavelength λ_1, but other recording layers can not be absorbed. When three wavelength synthesis coaxial lasers write the multi-wavelength photochromic plate at the same point of the disc, each wavelength laser will be absorbed by corresponding recorded layer, that all recording layer material in the point are photochromic at the same times and response to record multiple information. Experiment of this system selectes recording lasers wavelength are 780nm, 650nm, 532nm and 405nm. These wavelengths of laser are typical lasers application for CD, DVD or BD and HD-DVD. Therefore, these lasers have good compatibility with existing CD industry and industrial scale production except the laser of wavelengths 532nm semiconductor laser, but it was researched and developed at Tsinghua University in 1996. Interval of these wavelengths of laser is more balanced, control of write, read and crosstalk is easement. The research and development of corresponding record materials is more mature.

Between various photochromic materials do not chemical reaction each other, so that that disc can be used traditional spin-coating process to manufacturing. In addition, to simplify the process, three materials can be mixing dissolution by selection of efficient solvent, and spin-coating a single-layer mixed record layer, to obtain the same effect with the three-layer multi-wavelength CD in Figure 3-8.

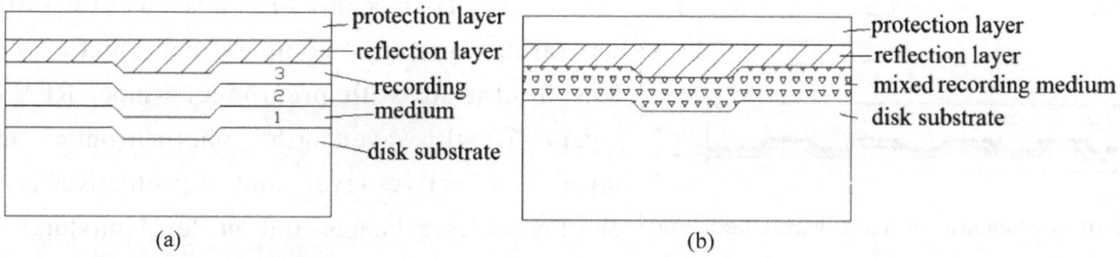

Fig. 3-8 The structure of multi-wavelength photochromic disk (a) Multi-wavelength and multilayer. (b) Photochromic material mixed spin coating to a single recording layer.

The structure of this multi-wavelength photochromic disc is similar to the traditional current CD-R/DVD-R, the CD/DVD-RW disc, that is composed of substrate, recording layer, reflective layer and protective layer. It has pre-molded channel also, that the channel is divided into multiple sectors, each sector the beginning of the sector format code used to identify the sector address, the format code for the data area for storing data. The difference is that the recording layer with variety of wavelength-sensitive to different photochromic materials, can be a recording layer or multi-layer structure also.

An obvious defect of the multi-wavelength photochromic storage system is the size of focus is different. Therefore utilization super-RENS technology to control and reduce the size of focus, that the structure of this disc with super-RENS mask layer is shown in Figure 3-9. The material of mask layer is not transparent to sensitive wavelength laser of recording layer, but it can be made a hole using another wavelength laser with nonlinear effect on mask layer, then recording wavelength lasers are enter recording layer through the openings hole on the mask layer. Therefore solved the problem that the focus spot size is difference of chromatic aberration of multiple wavelengths lasers. At the same time such that the uniform spot size on the super-RENS mask layer is reduced by 20 to 30 percent, can further increase the total storage capacity.

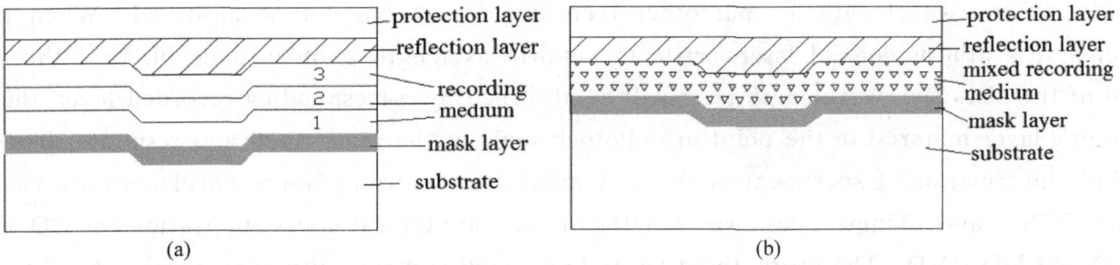

Fig. 3-9 The typical multi-wavelength multi-layer storage disc with super resolution photochromic mask layer: (a) for multilayer recording medium (b) for mixed recording medium.

The manufacturing process of the multi-wavelength disc with super resolution mask is: substract use poly carbon acid resin material and molding injection, then coating mask layer, mixed recording layer, reflection layer and protection layer with spin coating method successively.

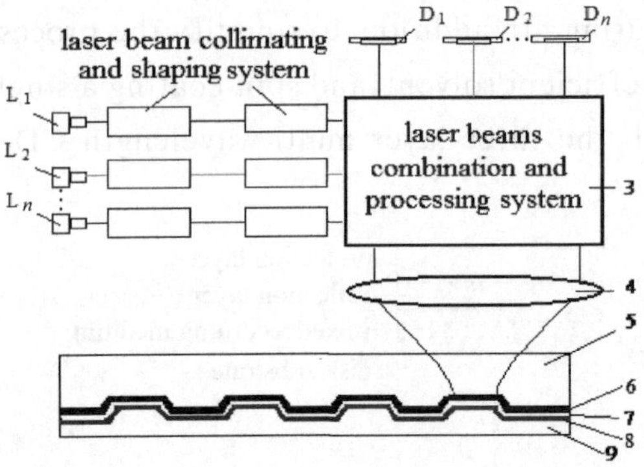

Fig. 3-10 Schematic of multi-wavelength and multilevel storage experiment system.

The storage experiment system of multi-wavelength and multi-level is shown in Figure 3-10. The figure shows $L_1 L_2 \cdots L_n$ are wavelengths lasers and beams collimating/sgaping system, $D_1 D_2 \cdots D_n$ are photodetectors to $L_1 L_2 \cdots L_n$ laser beam, 3-laser beams integrating and separating system, 4-complex achromatic objective lens, 5-substrate with pregroove, 6-super-RENS mask layer, 7-multi-wavelength photochromic medium layer, 8-reflective layer and 9-protective layer. If employ n laser beams and m level modulation and coding to record experiment the storage capacity will be increased by N times ($N = n\ln m$) theoretically.

The key component of the optical system is the laser beam coupling/splitting prism (LCSP). The Figure 3-11 is a three beam LCSP for example that is composed of four cube corner prisms which adhere to each other. Two interfaces are coated with film series by which lights are coupling or splitting. The transmissivities of the two interface are shown in Figure 3-12 and Figure 3-13 respectively. The transmissivity of the Interface 1 is 82.06% for 405nm laser and 11.13% for 650nm laser respectively with the entrance angle of 45°. The transmissivity of the Interface 2 is 89.35% for 405nm laser and 95.11% for 650nm laser, as well as 8.99% for 532nm laser with the entrance angle of 45°.

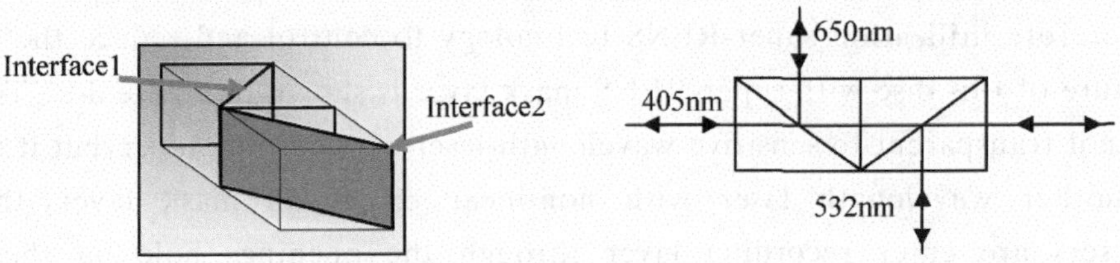

Fig. 3-11 The mechanism of optical coupling/splitting prism (LCSP): Interface 1 transparent to the light of 405nm but reflecting the light of 650nm. Interface 2 transparent to the lights of 405nm and 650nm but reflecting the light to 532nm.

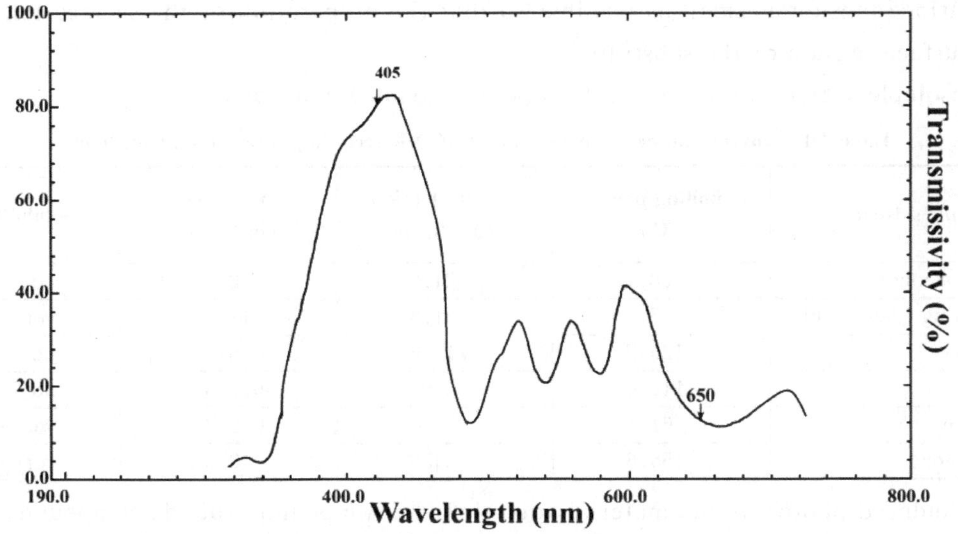

Fig. 3-12 Transmissivity of the Interface 1 with the entrance angle of 45°.

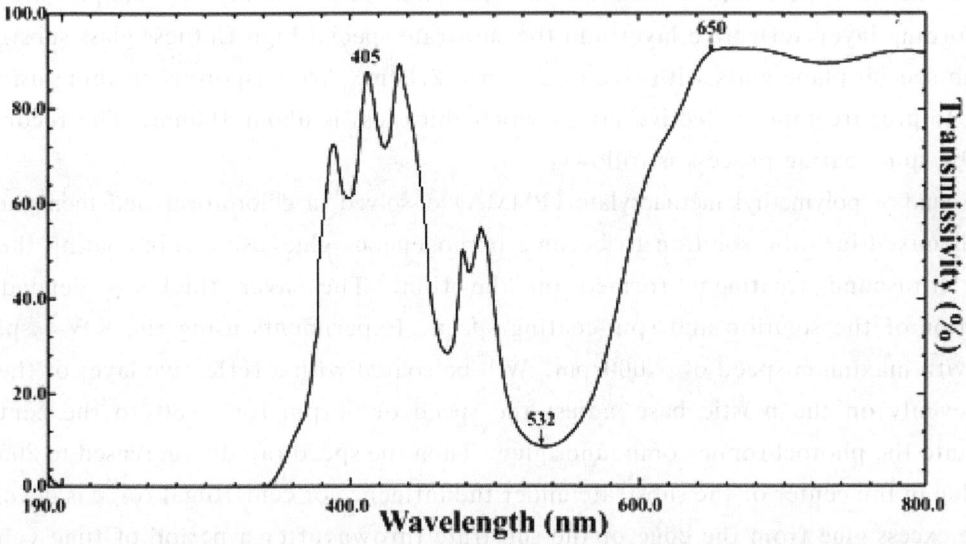

Fig. 3-13 Transmissivity for 405nm, 532nm and 650nm lasers on the interface-2 with the entrance angle of 45°.

In fact, the disc of cphotochromic materials can be manufactured with vacuum sputtering or solvent spin coating. Thickness and uniformity of film are relatively easy to control, but it is high manufacturing costs and lower production efficiency, usually only used for the experimental study. The spin-coating is the best to preparation of uniform organic thin film method, that the cphotochromic materials are directly dissolved in an organic solvent, spin coating to film. The advantage is low cost, thin film smooth, uniform, easy to control the thickness of the medium and widely used in optical disc manufacturing industry. Spin-coating process is generally the first film-forming material soluble in a moderately volatile solvent, the formation of a certain viscosity liquid (or colloidal) spin-coating liquid, so the selection of solvent is very important. Main requirements for organic solvents are shown in Table 3-1, that selection of solvent should meet the following conditions:

(1) good solubility of the dyes and polymers.

(2) volatile moderate boiling point of about 90～120℃.

(3) surface tension as small as possible, so that the preparation of the surface tension is less than the surface tension of the substrate.

(4) insoluble substrate (the substrate is polycarbonate), non-toxic.

Table 3-1 solvent and performances for the CD-R recording layer organic medium.

Organic solvents	Boiling point (℃)	Surface tension (10^{-5} N/cm)	Viscosity (mPa · s)	Solubility to PC
diacetone alcohol	168.1	31.0	2.9	no
ethyl-melting fiber agent	156.3	31.8	1.03	very small
cyclooctane	125.7	21.76	0.547	no
toluene	110.6	30.92	0.773	no
chloroform	61.1	27.14	0.563	very small
cyclohexanone	155.6	3450	2.2	very small

Light-induced photochromic materials, diarylethene compounds, fulgide compounds have good photochromic properties is more suitable for optical storage. Light-induced photochromic multi-wavelength storage experiment is used diarylethene compounds mainly. Its samples are composed of the recording layer, reflective layer and the substrate special high flatness glass substrate which is a 1.1mm special plane glass with size of 2.5cm × 2.5cm. After rigorous cleaning using vacuum sputtering to prepare total reflective layer, which thickness is about 100nm. The recording layer prepared by spin-coating process is follows:

Ultrasound of polymethyl methacrylate (PMMA) dissolved in chloroform, and then photochromic compounds mixed into the solution to become homogeneous glue, using spin-coating the reflector layer, by ultrasound treatment formed on the film. The layer thickness depends on the concentration of the solution and spin-coating speed. Experiments using the KW-4 spin coating machine with maximum speed of 6,000rpm. Will be coated with a reflective layer of the substrate is placed evenly on the plastic base, adjust the speed of 50rpm for ~80, to the center of the substrate into the photochromic compounds glue. Then the speed rapidly increased to 2000rpm, the drop of glue in the center of the substrate under the influence of centrifugal force is developed into a film, the excess glue from the edge of the substrate thrown, after a period of time (about 40s), the solvent is volatilized in the all to obtain the film with thickness of about 600nm. Also using the binary mixed solvent for example, by a certain percentage of mixed-butyl ether solvent and 2,6-dimethyl the heptanone auxiliary solvent, can guarantee the good volatility and effectively prevent the aggregation of the film when the crystallization phenomena. In particular, it can help to balance the appropriate viscosity glue to fill the film base to pre-carved slot for with pre-groove experiments disc. To join in the production of light-induced discoloration of multi-wavelength storage experiment samples of PMMA, in order to prevent the aggregation and crystallization of the photochromic compounds, to ensure a smooth and uniform of the recording films layer.

3.1.3 Model of data writing

In data writing process with photochromic materials to achieve multi-wavelength and multi-level storage, the photochromic recording medium appear photochromic reaction that is relation closely with exposure energe (time, speed and intensity), material sensitivity, absorption rate, absorption spectra, crosstalk and other factors. So it is need to establish a model of photochromic recording process to quantify analysis for various affect factors. Photochromic reaction caused discoloration with exposure that the writing process is an achromatic process. Actually the light-induced photochromic reaction is the discoloration of storage material in the exposure of certain wavelength that is from the closed-loop state transformation to the open-loop state. The transformation process caused to decrease number of the closed-loop state molecules in photochromic materials, that resulting change of light absorption rate in the recording medium for specific wavelength, and manifested as change of reflectivity in measurement. For discussion quantitative relationship between the reflectivity changes and exposure prerequisite in different materials to sensitive wavelength, first have to analyze the absorption process of the photochemical reactions. According to Beer's law, the absorbed amount of radiation is proportional to the number of absorb molecules in the medium, that meaning is only proportional to concentration C of the photochemical material in the medium. Meanwhile by Lambert's law: the absorbed percentage of incident light in the transparent medium is not any relationship to the incident light intensity and luminance. A basic model can be established by the Beer-Lambert principles, as follows:

$$\frac{\mathrm{d}I}{I} = -\alpha_v C \mathrm{d}l \tag{3-1}$$

Where I is the incident light intensity, C is the solute concentration (in mol/L), l is the thickness (in cm), α_v is the constant of proportionality (for 103cm^2/mol). The $C\mathrm{d}l$ is the solution quality per unit area within $\mathrm{d}l$ absorption layer. The $\mathrm{d}I$ is change of the light intensity by the absorption of the $\mathrm{d}l$ layer. For a cell of certain length, the boundary conditions are: on the incident surface, as $l = 0$ so incident light intensity is I_0, but on the exit surface, ie $l = 1$, the light intensity is I therefore:

$$\int_{I_0}^{I} \frac{\mathrm{d}I}{I} = -\int_0^l \alpha_v C \mathrm{d}l \tag{3-2}$$

Thus

$$\ln \frac{I}{I_0} = -\alpha_v Cl \tag{3-3}$$

Where α_v is the absorption coefficient, it is a function of radiation frequency and wavelength also.

Order $\varepsilon = \alpha_v/2.3$, known as the Molar extinction coefficient, its unit is 103cm^2/mol. It is the absorption of 1mol medium to certain wavelengths light under fixed parameters conditions (temperature, concentration and optical path length). Thus available to the integral expression of the Beer-Lambert law:

$$\frac{I}{I_0} = \mathrm{e}^{-2.3\varepsilon Cl} \tag{3-4}$$

Where I_0 is the initial incident light, I is the light after the photochromic materials. The equation describes the remaining intensity of a parallel light beam of certain wavelength through the media which concentration is C and the optical path length is l. $D = \varepsilon Cl$ is defined extinction coefficient of the medium, also known as optical density.

In order to simplify the process of the medium manufacturing, various storage medium is mixed uniformly, and dissolved in PMMA solvent, then spin coating on the glass substrate with aluminum reflective layer of reflector reflectivity of R_f. So it is a single layer disc without cover layer. Because of the role of the reflective layer, the writing light will be twice through recording layer, the writing process can be simplified the structure of a model as shown in Figure 3-14.

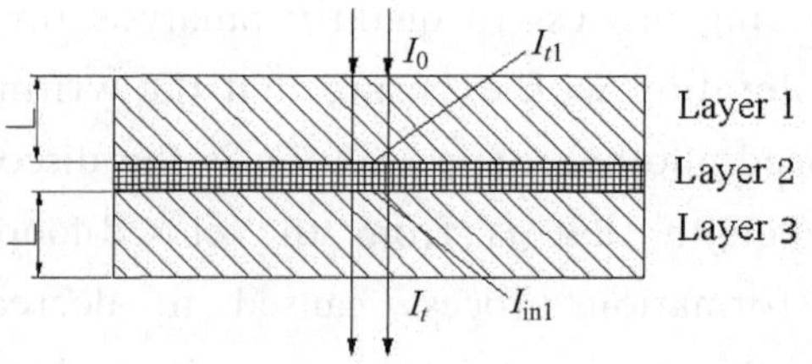

Fig. 3-14 The simplifying structure of photochromic medium samples for multi-wavelength experiment analysis.

In the model, the layer 1 and layer 3 are recording medium and same thickness L. Layer 2 is an absorption layer with transmission rate is R_f but thickness is zero. Layer 1 I_0 is the incident light intensity, I_{t1} is the light intensity through the recording Layer 1 and out to Layer 2. I_{in1} is incident light intensity through the absorption layer Layer 2 to recording Layer 3. I_t is exit light intensity through the recording Layer 3, i.e., reflected light or read out light intensity to medium sample. In the writing process experiments, if the sample has been fully written that the final reflectivity will be defined "maximum reflectivity" $R_{\max}$. The maximum reflectivity $R_{\max}$ is a very important concept to evaluate the performances of recording medium. The $R_{\max}$ of the final reflectivity is a constant for certain wavelength light, that the property of medium has to be experiment. Clearly, the most important influence factor to $R_{\max}$ is the transmittance of the photochromic materials recording layer to a certain wavelength light and was defined T.

Introduction transmittance T, according to Eq. (3-4), the out light intensity through Layer 1 can express to:

$$I_{t1} = I_0 T e^{-2.3D} \tag{3-5}$$

The out light intensity through absorbance layer Layer 2 is

$$I_{\text{in1}} = I_{t1} R_f \tag{3-6}$$

Finally the out light intensity I_t through the absorbing layer Layer 3 is

$$I_t = I_{\text{in1}} T e^{-2.3D} \tag{3-7}$$

Eqs. (3-5), (3-6) and (3-7) to get the light intensity through the simplifying structure sample (in Figure 3-14), i.e., reflected light from really experiment sample is

$$I_t = I_0 T^2 R_f e^{-4.6D} \tag{3-8}$$

The proportion R of transmitted light intensity I_t and the initial light intensity I_0 ultimately is

$$R = \frac{I_t}{I_0} = T^2 R_f e^{-4.6D} \tag{3-9}$$

Taking into the $R_{\max}$ definition, when full writing $R_f = 1$ and molecular concentration of recording photochromic materials is 0, i.e. $D = 0$, thus

$$R_{\max} = T^2 \tag{3-10}$$

Expression of the reflectivity can be obtained:

$$R = \frac{I_t}{I_0} = R_{\max} R_f e^{-4.6D} \tag{3-11}$$

For a mixed variety of photochromic materials for multi-wavelength storage, the recording layer contains a variety of materials, consider the most complex cases, the absorption spectra of various materials of multi-wavelength are absorbed completely. Total of light transmittance of the sample should be equal to the product of the light transmittance of each component separately in the mixed variety of photochromic materials, the total optical density is equal to the sum of all optical density of components and:

$$D_i = \sum_{j=1}^{n} \varepsilon_{j(\lambda_i)} C_j l \tag{3-12}$$

Where, D_i is the total optical density for this system at the wavelength λ_i. The $\varepsilon_j(\lambda_i)$ is Molar extinction coefficient with j-th component in the wavelength λ_i light radiation. C_j is the concentration of material j. l is optical path, ie film thickness of photochromic materials and n is the number of wavelengths. So the reflectivity R_i to wavelength λ_i is:

$$R_i = \frac{I_{t(\lambda_i)}}{I_{0(\lambda_i)}} = R_{\max(\lambda_i)} R_f \mathrm{e}^{D_i} = R_{\max(\lambda_i)} R_f \mathrm{e}^{-4.6\sum_{j=1}^{n} \varepsilon_{j(\lambda_i)} c_j l} \tag{3-13}$$

Where $I_0(\lambda_i)$ is the incident light intensity to wavelength λ_i. $I_t(\lambda_i)$ is reflective light intensity to wavelength λ_i. $R_{\max}(\lambda_i)$ is maximum reflectivity of the samples to wavelength of λ_i. R_f is reflectivity of the reflective layer. The light intensity $I_{ab(\lambda_j)1}$ after through the Layer 1 of wavelength λ_j material is

$$I_{ab(\lambda_j)1} = I_{0(\lambda_j)} \left[1 - \left(\frac{R_j}{R_{\max(\lambda_j)} R_f}\right)^{1/2} \sqrt{R_{\max(\lambda_j)}} - \left(1 - \sqrt{R_{\max(\lambda_j)}}\right)\right] \tag{3-14}$$

The incident light intensity $I_{\mathrm{in}1(\lambda_j)}$ after through Layer 1 and Layer 2 to reach the Layer 3 surface of wavelength λ_j material is

$$I_{\mathrm{in}1(\lambda_j)} = I_{0(\lambda_j)} \left(\frac{R_j}{R_{\max(\lambda_j)} R_f}\right)^{1/2} \sqrt{R_{\max(\lambda_j)}} R_f \tag{3-15}$$

The light intensity $I_{ab(\lambda_j)}$ through Layer 3 with absorption wavelength λ_j is

$$I_{ab(\lambda_j)} = I_{0(\lambda_j)} \left(\frac{R_j}{R_{\max(\lambda_j)} R_f}\right)^{1/2} \sqrt{R_{\max(\lambda_j)}} R_f \left[\left(1 - \left(\frac{R_j}{R_{\max(\lambda_j)} R_f}\right)^{1/2} \sqrt{R_{\max(\lambda_j)}}\right) - \left(1 - \sqrt{R_{\max(\lambda_j)}}\right)\right] \tag{3-16}$$

Thus, the light intensity $I_{ab(\lambda_j)}$ after absorption of photochromic materials to wavelength λ_j can be written:

$$I_{ab(\lambda_j)} = I_{ab(\lambda_j)1} + I_{ab(\lambda_j)3} = I_{0(\lambda_j)} \left(\sqrt{R_{\max(\lambda_j)}} - \sqrt{\frac{R_j}{R_f}}\right)\left(1 + \sqrt{R_f R_j}\right) \tag{3-17}$$

Most of the light-induced discoloration of the material achromatic response can be regarded as a primary photochemical reaction. The multi-wavelength recording layer materials absorb different wavelengths of light and react with an overall response rate for each material to correspondence wavelength. The sum of the reaction rate of all materials is

$$-\frac{\mathrm{d}C_i}{\mathrm{d}t} = \sum_{j=1}^{n}\left(-\frac{\mathrm{d}C_i}{\mathrm{d}t}\right)_{\lambda_j} \tag{3-18}$$

The reaction rate of the material i to wavelength of λ_j is

$$\left(-\frac{\mathrm{d}C_i}{\mathrm{d}t}\right)_{\lambda_j} = \Phi_{i(\lambda_j)} \frac{\lambda_j}{Nhc} K_{i(\lambda_j)} \tag{3-19}$$

Where $\Phi_i(\lambda_j)$ is the quantum yield of material i in the light irradiation of wavelength λ_j. $K_i(\lambda_j)$ is the absorption rate of material i to wavelength λ_j. N is Avogadro constant which equal $6.02214 \times 10^{23}\,\mathrm{mol}^{-1}$. The h is the Planck constant which equal $6.62607 \times 10^{-34}\,\mathrm{Js}$. The c is velocity of light in vacuum which equal $3 \times 10^8\,\mathrm{m/s}$. The generated efficiency of reactant of photochemical reaction materials can be used the quantum yield to characterize. Quantum yield Φ is photochemical reaction efficiency directly which is defined as $\Phi = N_r / N_m$. Here N_r is quantity of molecular of reactant, N_m is quantity of absorption photon of photochemical material.

The quantum yield of photochemical reaction is relation with the photochemical material and the radiation wavelength, that is a constant to the certain photochemical material and certain wavelength of radiation.

Take Eq. (3-19) into Eq. (3-18) can obtain the total reaction rate:

$$-\frac{dC_i}{dt} = \sum_{j=1}^{n} \Phi_{i(\lambda_j)} \frac{\lambda_j}{Nhc} K_{i(\lambda_j)} \tag{3-20}$$

Absorption rate per unit volume absorbed light energy, i. e.,

$$K = \frac{E}{Stl} = \frac{I}{l} \tag{3-21}$$

Where E is the radiant energy of certain wavelengths, S is the radiation area with unit of m^2, l is the optical path.

Material i absorption rate to the wavelength λ_j is

$$K_{i(\lambda_j)} = I_{ab(\lambda_j)} \frac{\varepsilon_{i(\lambda_j)} C_i}{\sum_{k=1}^{n} \varepsilon_{k(\lambda_j)} C_k} \cdot \frac{1}{l} \tag{3-22}$$

Combined with Eqs. (3-17), (3-20) and (3-22) can get the function of the concentration C_i change of multi-wavelength light-induced discoloration material i to time t:

$$-\frac{dC_i}{dt} = \sum_{j=1}^{n} \Phi_{i(\lambda_j)} \frac{\lambda_j}{Nhc} I_{0(\lambda_j)} \left(\sqrt{R_{\max(\lambda_j)}} - \sqrt{\frac{R_j}{R_f}}\right)(1 + \sqrt{R_f R_j}) \cdot \frac{\varepsilon_{i(\lambda_j)} C_i}{\sum_{k=1}^{n} \varepsilon_{k(\lambda_j)} C_k} \cdot \frac{1}{l} \tag{3-23}$$

In multi-wavelength recording process, the reflection rate is decided by optical density of medium of various wavelengths. So the reflection rate is decided by concentration of all recorded material when the Molar extinction coefficient and the thickness of the recording medium are given value. Therefore concentration changes of the recording material can determine reflectivity changes of writing process and aggregate of Eqs. (3-13) and (3-23), can accomplish description of the writing process of the multi-wavelength storage.

Above model has taken into most complex situation in which the absorption spectrum of each material to cover all recorded wavelength. In the actual study on the material absorption spectrum as narrow as possible, the best absorption spectrum of each material is only covered with a record wavelength. In this case, the storage of multi-wavelength response model can be further simplified. Each component absorbed only with its corresponding wavelength of light, but little or not absorb other light absorption, ie $\varepsilon_i(\lambda_j) \approx 0 (j \neq i)$. By Eqs. (3-13) and (3-21) are available:

$$R_i = R_{\max(\lambda_i)} R_f e^{-4.6\varepsilon_{i(\lambda_i)} C_i l} \tag{3-24}$$

$$-\frac{dC_i}{dt} = \Phi_{i(\lambda_i)} \frac{\lambda_i}{Nhc} I_{0(\lambda_i)} \left(\sqrt{R_{\max(\lambda_i)}} - \sqrt{\frac{R_i}{R_f}}\right)(1 + \sqrt{R_i R_f}) \cdot \frac{1}{l} \tag{3-25}$$

And from Eq. (3-22) can get the concentration of material i:

$$C_i = -\frac{1}{4.6\varepsilon_i l} \ln \frac{R_i}{R_{\max(\lambda_i)} R_f} \tag{3-26}$$

Derivative of both sides of the above equation have:

$$\frac{dC_i}{dt} = -\frac{1}{4.6\varepsilon_{i(\lambda_i)} l} \cdot \frac{1}{R_i} \cdot \frac{dR_i}{dt} \tag{3-27}$$

Recording the laser beams are uniform of intensity and parallel beams, namely

$$I = \frac{P}{S} \tag{3-28}$$

Where P is the laser power with unit of W, S is light irradiation area, in units of m^2.

Incorporated Eqs. (3-25), (3-27) and (3-28), can get the relationship between transmission rate and the irradiation time:

$$\frac{dR_i}{dt} = k_i\left(\sqrt{R_{\max(\lambda_i)}} - \sqrt{\frac{R_i}{R_f}}\right)(1 + \sqrt{R_i R_f})R_i \tag{3-29}$$

Where k_i is called to constant of writing time, it is a constant in writing process with units s^{-1}. The constant characterizes the writing speed of wavelength λ_i laser on of the sample, its value is

$$k_i = 4.6\varepsilon_{i(\lambda_i)} \Phi_{i(\lambda_i)} \frac{\lambda_i}{Nhc} \frac{P_i}{S_i} \tag{3-30}$$

Thus available:

$$t = \int_{R_{0(i)}}^{R_{f(i)}} \frac{1}{k_i} \cdot \frac{dR_i}{\left(\sqrt{R_{\max(\lambda_i)}} - \sqrt{\frac{R_i}{R_f}}\right)(1 + \sqrt{R_i R_f})R_i} \tag{3-31}$$

Where $R_{0(i)}$ and the $R_{f(i)}$ are initial transmission rate and the final reflectivity. The k_i is a constant to a recording medium and radiation conditions. Clearly, the changes of reflectivity to wavelength are independent over time, that do not affect each other in the simplified model. Of course, the simplified model can be used to description of the writing process to a single wavelength also. Above mathematical models based on the photochemical reaction principles of multi-wavelength photochromic storage process can be used to analyze crosstalk mechanism of light-induced, abatement crosstalk, optimization of write strategy to restrain crosstalk in the storage process of photochromic multi-wavelength recording, and then control crosstalks within less than 5% for multi-wavelength and multi-level signal.

3.2 Experiments for multi-wavelength and multi-level storage

The crosstalk between the different wavelengths is an important problem for multi-wavelength and multilevel storage. Crosstalk is emerged due to cross of absorption spectrum of the materials in the writing process mainly. As there are many special technical requirements for variety of materials. The characteristics of the recording medium achieved perfect entirely not crosstalk that is impossible almost. The absorption spectrum of the material whether cross and its cross extent, is an important factor to determine the material can be a practical or no. Regardless the cross between the absorption spectrum of materials is controlled in any extents. It has to be calculated with establishment of systems analysis and quantitative evaluation the degree of cross-interference of the materials. In this section first analysis of the material absorption spectra do not cross or very small situation. Representative 1, 2 twofold (2-methyl-5-n-butyl-3-thienyl) perfluorinated cyclopentene and 1, 2 twofold (2-methyl-5-(4-N, N-dimethyl-phenyl)-thiophene thiophene-3-yl) perfluorinated cyclopentene two kinds of materials, for example. Sensitive spectra of these two materials were 532nm and 650nm, uniform mixing of the two kinds of materials soluble in the recording layer by spin coating PMMA solution, the thickness of about 200nm. Their absorption spectra and molecular structure are shown in Figure 3-15 and Figure 3-16. Before the test experiments of the material, the required reflectivity of 99.9% of the standard total reflection film with variety of laser readout signal power calibration, as read out a benchmark.

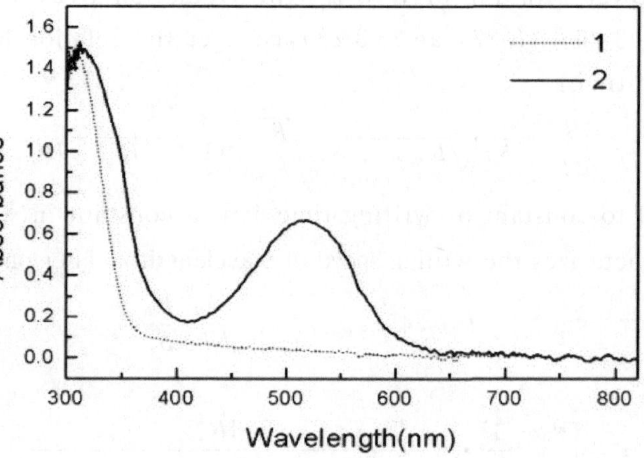

(a)

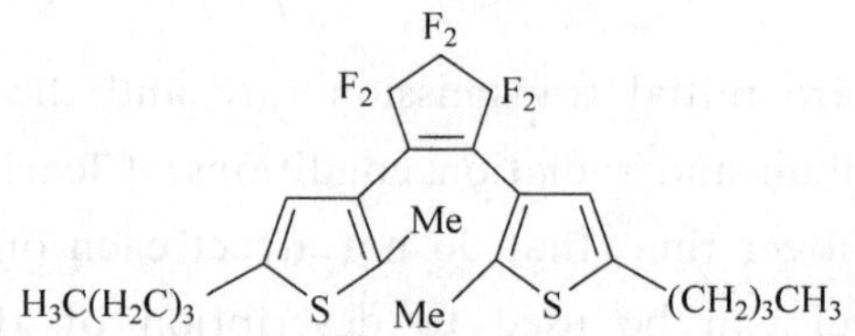

(b)

Fig. 3-15 Absorption spectra (a) and molecular structure (b) of the 1, 2-bis (2-methyl-5-n-butyl-3-thienyl) perfluorinated cyclopentene.

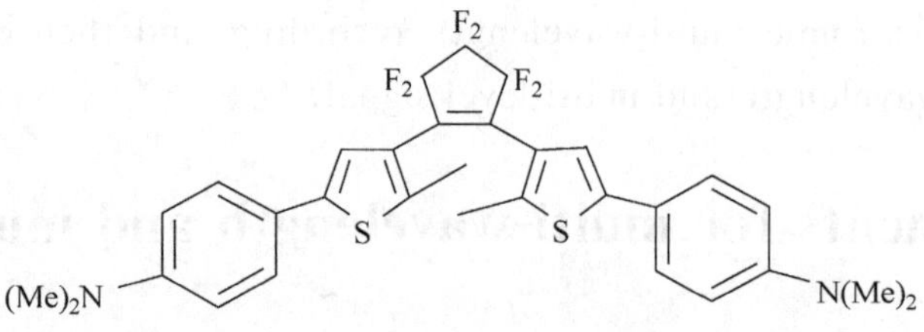

(a) (b)

Fig. 3-16 Absorption spectra (a) and molecular structure (b) of 1, 2-bis (2-methyl-5-(4-N, N-dimethyl-phenyl)-thiophenethiophene-3-yl) perfluorinated cyclopentene.

In addition, the photochromic reaction of the photochromic material and write laser power and time off, and depending on the product of both. Such as 1, 2-bis (2-methyl-5-(4-N, N-methyl-phenyl)-thiophene-3-yl) perfluorinated cyclopentene media write different power 650nm laser experiment, the reflection rate of 0.8, to obtain the power-time curve is shown in Figure 3-17. It can be seen using different laser power and exposure time combination to get the same reaction. This feature material for a certain cross-spectral interference was used to improve or inhibit cross-interference.

The experiments also demonstrated that the need of exposure energy to chang reflectivity of a photochromic material is determine. As using different writing power to reached reflectivity 0.8

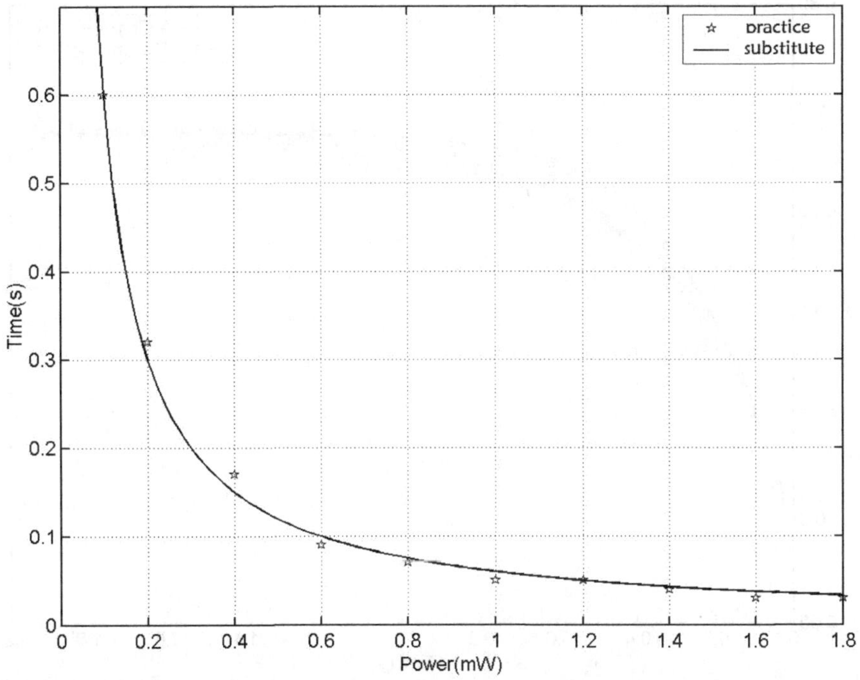

Fig. 3-17 Experiment power-time curve of thiophene-3-yl) perfluorinated cyclopentene media are written with different power and exposure time of 650nm laser to obtain the reflection rate of 0.8.

with a photochromic material for example. The product when multiply writing power by exposure time (equivalent to the exposure energy) is remained unchanging that is about 0.06mJ, ie Pt = 0.06. It can be seen, when a photochromic material was written to reach a certain reflectivity, if the writing power is double, writing time will be reduced by half. Increasing the writing power can accelerate proportional to shorten the writing time. So materials without crossover of absorption spectrum, can choose the shortest exposure time to write. Above-described two kinds of materials without crossover of absorption spectrum, their reflectivity changing are independent, that the writing model can bee simplified. Using 532nm, 650nm wavelength lasers write at the same sample written, the experiment results reflectivity curve are shown in Figure 3-18 and Figure 3-19. The reflectivity of reflective layer of the sample is total reflection, $R_f = 1$, the initial reflectivity of 532nm laser R_{ini} is 0.589, initial reflectivity of 650nm laser R_{ini2} is 0.38. The reflectivity to 532nm laser is $R_{max1} = 0.788$, to 650nm laser is $R_{max2} = 0.86$ after full writing experiment. Based on the above parameters and the experimental data, using Eq. (3-31) can calculate writing time constant of perfluorinated cyclopentene material 1,2-bis (2-methyl-5-n-butyl-3-thiophene thiophene base) $k_1 = 5.61s^{-1}$, and writing time constant of perfluorinated cyclopentene 1,2-bis (2-methyl-5-(4-N, N-dimethyl-phenyl)-thiophene-3-yl) material $k_2 = 5.34s^{-1}$. The experimental and theoretical calculation results of 532nm and 650nm lasers actually writing are compared respectively, as shown in Figure 3-18 and Figure 3-19.

The experimental results and theoretical calculations show that the simplified model can accurately describe the writing process in the multi-wavelength absorption spectra of a variety of materials without cross. The k is an important parameter to characterize the write speed of light-induced discoloration of storage medium. The initial reflectance R_{ini} and maximum reflectance R_{max} are determined by the sample preparation process, and measured by the experiment.

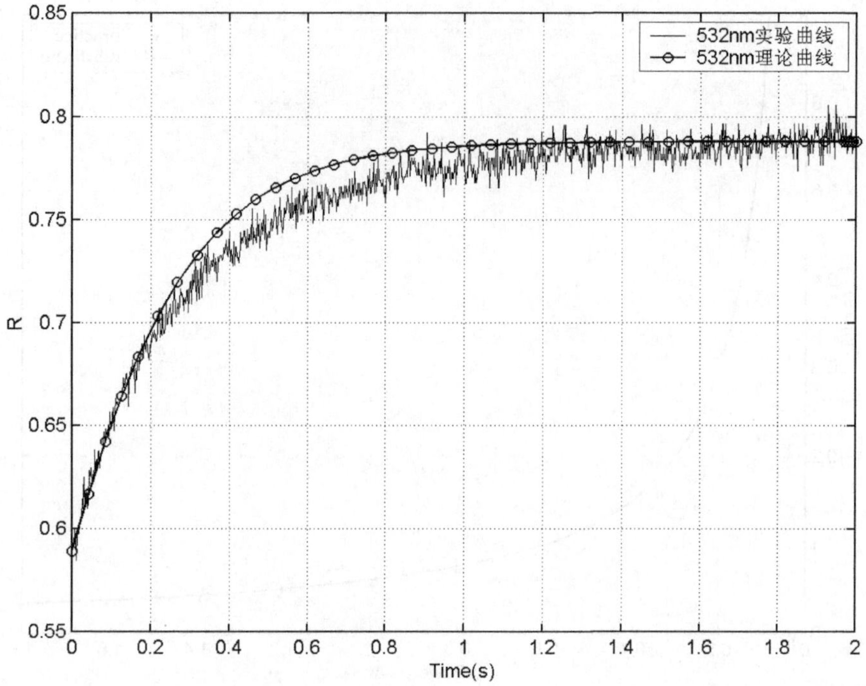

Fig. 3-18 The writing experimental and theoretical calculation results of 1,2-bis (2-methyl-5-n-butyl-3-thiophene thiophene base) perfluorinated cyclopentene.

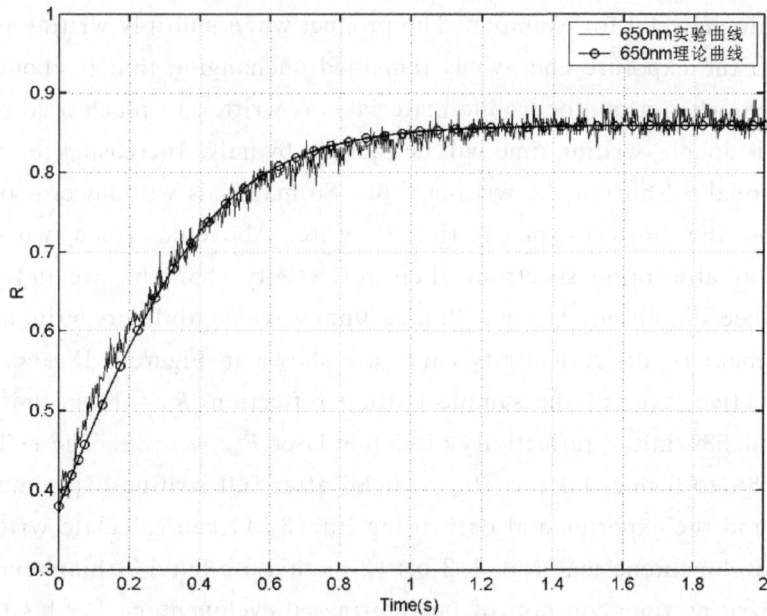

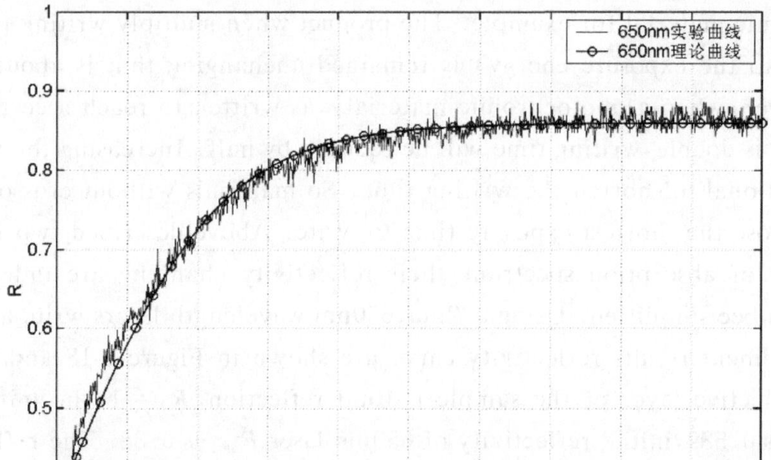

Fig. 3-19 The writing experimental and theoretical calculation results of 1,2-bis (2-methyl-5-(4-N,N-dimethyl-phenyl)-thiophene-3-yl) perfluorinated cyclopentene.

There are a lot of materials can be used to multi-wavelength and multi-level storage that is a great advantage to this technology in the future, but most of the material has some spectral cross often. For prevention of crosstalk of multi-wavelength and multi-level storage, of course, first of all should be choice the material without cross of absorption spectra, but sometimes in order to take into account other characteristics of the material, such as sensitivity, stability and process and so on, have to choose some of the existence of certain cross-interference materials. Facts have proved that as long as choice of their parameters is right that can get the desired results also. So it is necessary to discuss the existence of some cross-interference materials and impact each other. Following introduction an absorption spectra covering the whole fluorine cyclopentene materials for example, that the molecular structure is shown in Figure 3-20. Can be seen from Figure 3-20 in absorption spectrum of the material, the main income of the material the peak is wavelength of 650nm, but has some absorption at wavelength of 532nm also. Therefore it is most representative if using whole fluorine cyclopentene materials to research crosstalk of absorption at 650nm and 532nm wavelength.

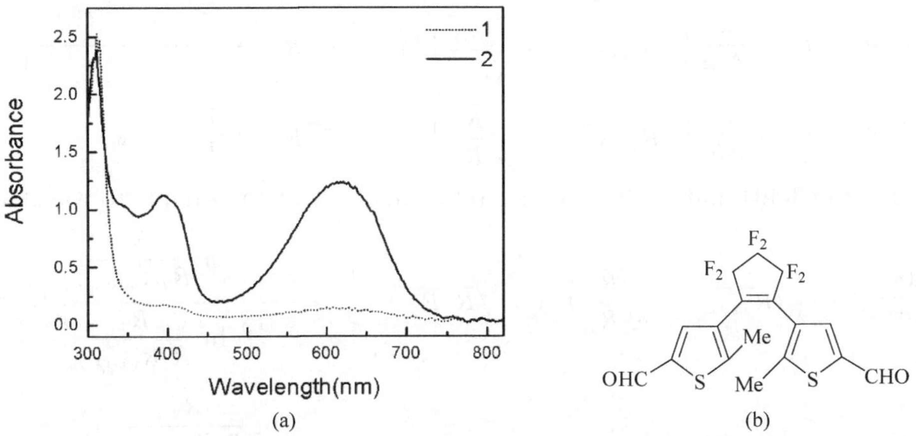

Fig. 3-20 (a) absorption spectra covering the whole fluorine cyclopentene materials(1-open loop absorption spectra, 2-closed loop); (b) the molecular structure of whole fluorine cyclopentene materials.

As the material absorption spectrum cross it cannot description using simplify model, and must employ the Eqs. (3-13) and (3-23), more exact model for light-induced photochromic multi-wavelength storage. If the absorption wavelength of the materials-1 λ_1 is 532nm only but the absorption wavelength λ_2 of materials-2 is 532nm and 650nm. The reflectivity of the sample to wavelength λ_1 is available present with the Eq. (3-13):

$$R_1 = R_f R_{\max(\lambda_1)} \mathrm{e}^{-4.6\sum_{j=1}^{2}\varepsilon_{j(\lambda_1)} C_j l} = R_f R_{\max(\lambda_1)} \mathrm{e}^{-4.6 l(\varepsilon_{1(\lambda_1)} C_1 + \varepsilon_{2(\lambda_1)} C_2)} \tag{3-32}$$

As material-1 absorbs wavelength λ_1 light only, its Molar extinction coefficient approximate to 0 to wavelength λ_2, the reflectivity of the sample to wavelength λ_2 is

$$R_2 = R_{\max(\lambda_2)} R_f \mathrm{e}^{-4.6\sum_{j=1}^{2}\varepsilon_{j(\lambda_2)} C_j l} = R_{\max(\lambda_2)} R_f \mathrm{e}^{-4.6\varepsilon_{2(\lambda_2)} C_2 l} \tag{3-33}$$

According to Eq. (3-32)

$$\varepsilon_{1(\lambda_1)} C_1 + \varepsilon_{2(\lambda_1)} C_2 = -\frac{1}{4.6l}\ln\frac{R_1}{R_{\max(\lambda_1)} R_f} \tag{3-34}$$

$$\varepsilon_{1(\lambda_1)}\frac{\mathrm{d}C_1}{\mathrm{d}t} + \varepsilon_{2(\lambda_1)}\frac{\mathrm{d}C_2}{\mathrm{d}t} = -\frac{1}{4.6l}\cdot\frac{1}{R_1}\cdot\frac{\mathrm{d}R_1}{\mathrm{d}t} \tag{3-35}$$

from Eq. (3-33) can get:

$$C_2 = -\frac{1}{4.6\varepsilon_{2(\lambda_2)} l}\ln\frac{R_2}{R_{\max(\lambda_2)} R_f} \tag{3-36}$$

$$\frac{\mathrm{d}C_2}{\mathrm{d}t} = -\frac{1}{4.6\varepsilon_{2(\lambda_2)} l}\cdot\frac{1}{R_2}\cdot\frac{\mathrm{d}R_2}{\mathrm{d}t} \tag{3-37}$$

Based on the above four equations are available:

$$C_1 = -\frac{1}{4.6\varepsilon_{1(\lambda_1)}l}\ln\frac{R_1}{R_fR_{\max(\lambda_1)}} + \frac{\varepsilon_{2(\lambda_1)}}{\varepsilon_{1(\lambda_1)}}\frac{1}{4.6\varepsilon_{2(\lambda_2)}l}\ln\frac{R_2}{R_fR_{\max(\lambda_2)}} \tag{3-38}$$

Similarly, according to Eq. (3-23), the available material concentration

$$-\frac{dC_1}{dt} = \Phi_{1(\lambda_1)}I_{0(\lambda_1)}\frac{\lambda_1}{Nhc}\left(\sqrt{R_{\max(\lambda_1)}} - \sqrt{\frac{R_1}{R_f}}\right)(1+\sqrt{R_1R_f})\cdot\frac{\varepsilon_{1(\lambda_1)}C_1}{\varepsilon_{1(\lambda_1)}C_1+\varepsilon_{2(\lambda_1)}C_2}\cdot\frac{1}{l} \tag{3-39}$$

Concentration of material-2 is

$$-\frac{dC_2}{dt} = \Phi_{2(\lambda_1)}I_{0(\lambda_1)}\frac{\lambda_1}{Nhc}\left(\sqrt{R_{\max}(\lambda_1)} - \sqrt{\frac{R_1}{R_f}}\right)(1+\sqrt{R_1R_f})\cdot\frac{\varepsilon_{2(\lambda_1)}C_2}{\varepsilon_{1(\lambda_1)}C_1+\varepsilon_{2(\lambda_1)}C_2}\cdot\frac{1}{l} + \Phi_{2(\lambda_2)}I_{0(\lambda_2)}\frac{\lambda_2}{Nhc}\left(\sqrt{R_{\max}(\lambda_2)} - \sqrt{\frac{R_2}{R_f}}\right)(1+\sqrt{R_2R_f})\cdot\frac{1}{l} \tag{3-40}$$

Sum Eqs. (3-34) to (3-40) and (3-28), can get reflectance R_1 of the sample to wavelength λ_1:

$$\frac{dR_1}{dt} = k_1\left(\sqrt{R_{\max(\lambda_1)}} - \sqrt{\frac{R_1}{R_f}}\right)(1+\sqrt{R_1R_f})\cdot\left[1-\frac{\varepsilon_{2(\lambda_1)}}{\varepsilon_{2(\lambda_2)}}\frac{\ln\frac{R_2}{R_fR_{\max(\lambda_2)}}}{\ln\frac{R_1}{R_fR_{\max(\lambda_1)}}}\right]R_1 + \frac{\varepsilon_{2(\lambda_1)}}{\varepsilon_{2(\lambda_2)}}\left[k_2\left(\sqrt{R_{\max(\lambda_1)}} - \sqrt{\frac{R_1}{R_f}}\right)(1+\sqrt{R_1R_f})\cdot\frac{\ln\frac{R_2}{R_fR_{\max(\lambda_2)}}}{\ln\frac{R_1}{R_fR_{\max(\lambda_1)}}} + k_3\left(\sqrt{R_{\max(\lambda_2)}} - \sqrt{\frac{R_2}{R_f}}\right)(1+\sqrt{R_2R_f})\right]R_1 \tag{3-41}$$

And the reflectivity R_2 of this sample to wavelength λ_2:

$$\frac{dR_2}{dt} = \left[k_2\left(\sqrt{R_{\max(\lambda_1)}} - \sqrt{\frac{R_1}{R_f}}\right)(1+\sqrt{R_1R_f})\cdot\frac{\ln\frac{R_2}{R_fR_{\max(\lambda_2)}}}{\ln\frac{R_1}{R_fR_{\max(\lambda_1)}}} + k_3\left(\sqrt{R_{\max(\lambda_2)}} - \sqrt{\frac{R_2}{R_f}}\right)(1+\sqrt{R_2R_f})\right]R_2 \tag{3-42}$$

Where

$$\begin{cases} k_1 = 4.6\varepsilon_{1(\lambda_1)}\Phi_{1(\lambda_1)}\dfrac{\lambda_1}{Nhc}\cdot\dfrac{P_1}{S_1} \\ k_2 = 4.6\varepsilon_{2(\lambda_1)}\Phi_{2(\lambda_1)}\dfrac{\lambda_1}{Nhc}\cdot\dfrac{P_1}{S_1} \\ k_3 = 4.6\varepsilon_{2(\lambda_2)}\Phi_{2(\lambda_2)}\dfrac{\lambda_2}{Nhc}\cdot\dfrac{P_2}{S_2} \end{cases} \tag{3-43}$$

Here k_n is writing time constant, k_2 is the writing time constant with cross of absorption spectrum of material. It can be seen that time constant is relation with the writing power only when the writing wavelength and materials are identified.

In order to correctness of the analysis of above calculation, make of two experimental samples with absorption spectrum of the 532nm and 650nm material respectively, and to write experiment alone. The writing power to reach surface of sample of 532nm wavelength is 0.07mW, and 0.1mW to sample of 650nm wavelength. The using simplified model to calculate the time constant are $k_1 = 5.61\text{s}^{-1}$, $k_2 = 1.5\text{s}^{-1}$, $k_3 = 4.51\text{s}^{-1}$ respectively. Meanwhile, measured initial reflectance $R_{\text{ini1}} = 0.2$in for 532nm light, and the initial reflectivity $R_{\text{ini2}} = 0.03$ for 650nm light at the same experimental conditions. The reflectance of reflective layer of the experimental samples $R_f = 1$. The experimental curves using 532nm and 650nm light to write this sample at the same experimental conditions is shown in Figure 3-18.

According to the experimental curve samples with two wavelength while fully writing, maximum reflectivity of 532nm $R_{\max1} = 0.8$, the maximum reflectivity of 650nm light $R_{\max2} = 0.71$. In addition, the ratio of molar extinction coefficient of 650nm materials in the wavelength of 532nm and 650nm is $\varepsilon_2(\lambda_1)/\varepsilon_2(\lambda_2) = 0.5$. Put above parameters into Eqs. (3-41) and (3-42), available to get theoretical writing curve of mixed-material samples with spectrum cross-coefficient of two materials. The results of theoretical calculations compare with the experimental curve in Figure 3-21, that shows the theoretical curve and experimental curve is basically consistent, and can accurately describe the absorption of a variety of materials in the multi-wavelength photochromic storage with spectrum cross case. Can be seen, written to the time constant of the writing process speed, while the trend of the initial reflectivity, maximum reflectivity and reflectivity of reflective layer will affect the writing process. These differ relationships have great significance for development of reading and writing strategies in the future, the following will be discussed separately.

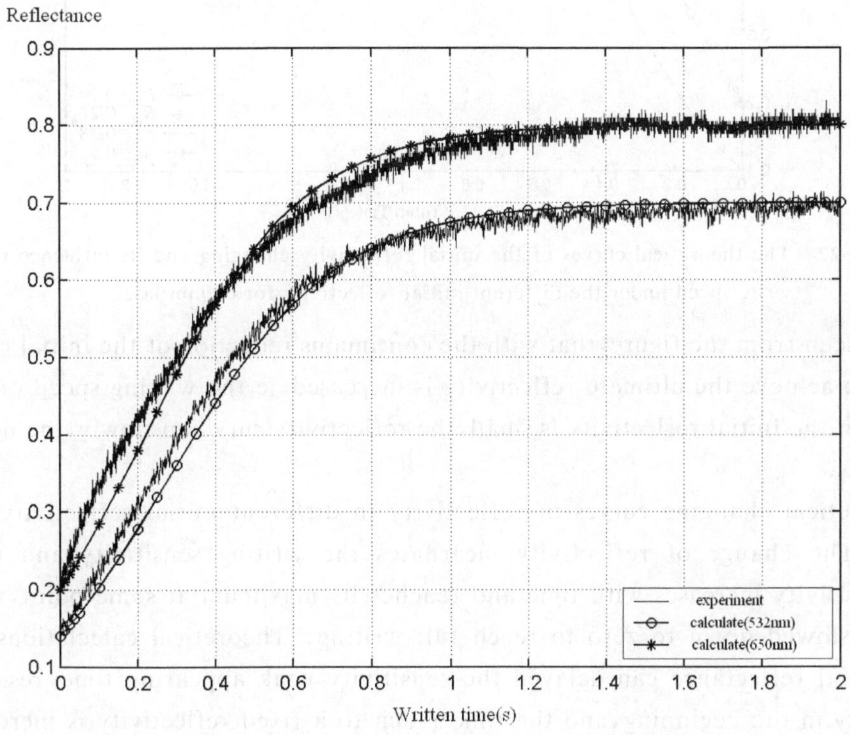

Fig. 3-21 Experimental and theoretical calculation curves of whole fluorine cyclopentene materials are written with 532nm and 650nm rays respectively.

3.2.1 The influence of initial reflectivity to writing speed

The initial reflectivity is determined by the initial optical density or concentration of materials. But concentration of the recording material can not maintain in processes of the recording. Therefore, the initial reflectivity directly affected by the manufacturing processes in the disc production. Of course, it can be controled by adjusting the preparation process parameters also. For disc of multi-materials without cross of absorption spectra, its initial reflection is determined only by the concentration of the material. So change of initial reflectivity of a certain material does not affect other reflectivity of materials. Here analyzation one of the writing process of the 650nm material for example. Keeping the other parameters the same conditions, changing the initial reflectivity of 650nm light only, that the curves of theoretical reflectivity under the different initial reflectivity are shown in Figure 3-22.

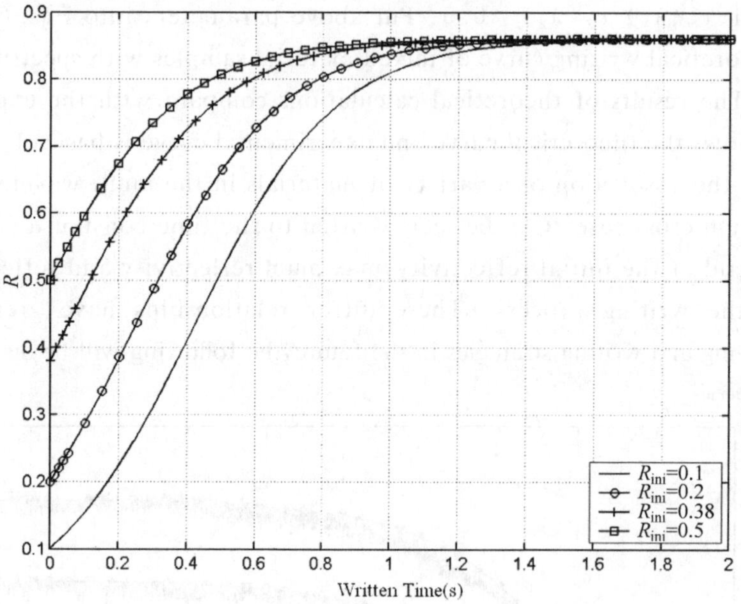

Fig. 3-22 The theoretical curves of the initial reflectivity changing and its influence to the write speed under the different initial reflectivity for 650nm laser.

It can be seen from the figure, that with the continuous reduction of the initial reflectivity, full writing time to achieve the ultimate reflectivity is increased, ie the writing speed of the process is reduced. When the initial reflectivity is small, the reflectivity curve up slowly, the nonlinear effect is more obvious.

The theoretical changing curves of reflectivity in different initial reflectivity are shown in Figure 3-23. The change of reflectivity incarnates the writing sensitivity and non-linear. In beginning, sensitivity increases with time and reaches its maximum at some point, then decreased gradually and slowed down to zero to reach full writing. Theoretical calculations show that to reduce the initial reflectance can delayed the sensitivity peak appearing time, resulting in lower write sensitivity in the beginning, and the time reach to a fixed reflectivity is increased, that will help improve read read out signal contrast and working life. The high initial reflectivity has higher written sensitivity, and soon reached the peak of fully written can reduce and help to improve the writing speed.

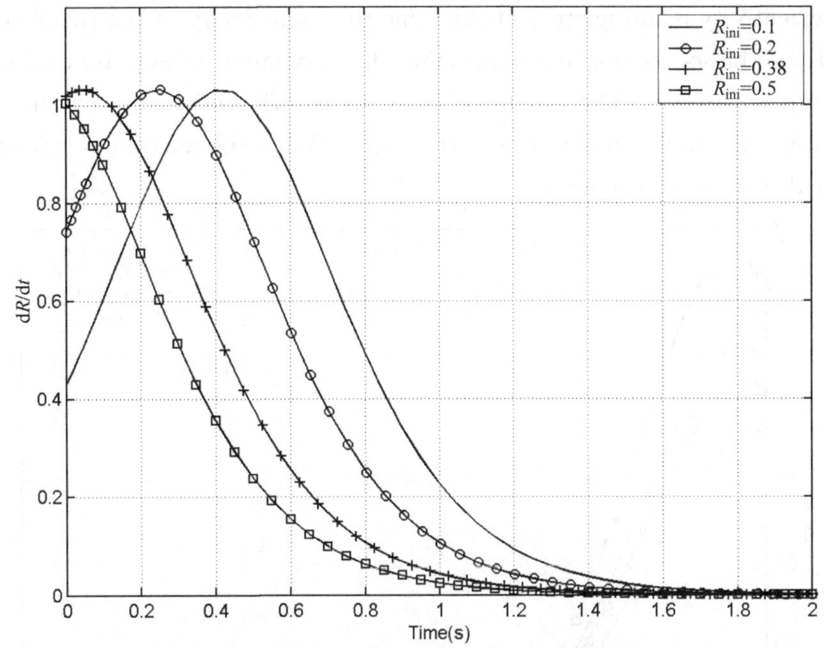

Fig. 3-23 The theoretical curves of the reflectivity changing and influence to sensitivity peak in different initial reflectivity.

3.2.2 The influence of the maximum reflectivity to writing process

Due to exist of impurities and partial loss of the photochromic molecules in medium production process, even after full write, the final reflectance of the samples cannot reach the original reflectivity of the reflectivity. The maximum reflectivity theoretical curves of writing process are shown in Figure 3-24. Can be seen from the figure, the maximum reflectivity changes do not affect the time to reach full written, but influence of the final value of the reflectivity to the writing wavelength after full writing.

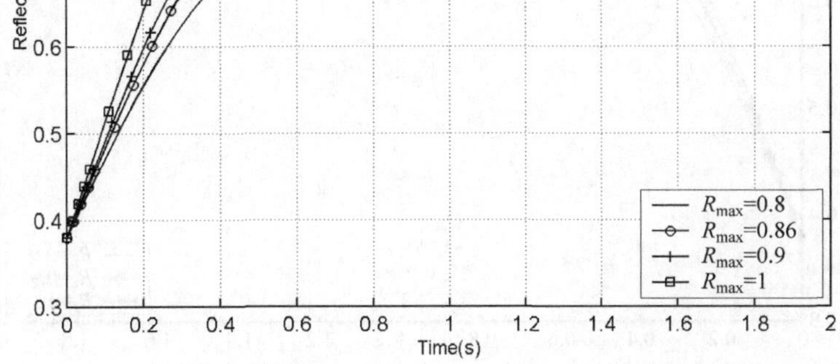

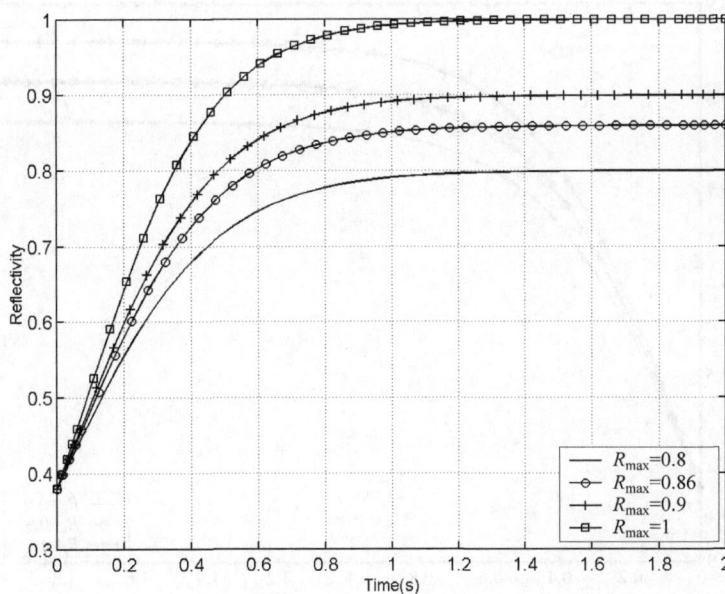

Fig. 3-24 The theoretical curves of the maximum reflectivity to influence to writing speed.

Maximum reflectivity is not written at the same time sensitivity of the theoretical curve shown in Figure 3-25. It can be seen from this figure, that the maximum reflectivity of the writing process is influence writing sensitivity. The greater of maximum reflectivity, the higher of the sensitivity and reduced writing time to reach a certain reflectivity. When the maximum reflectivity increases, its reflectivity peak will be postponed.

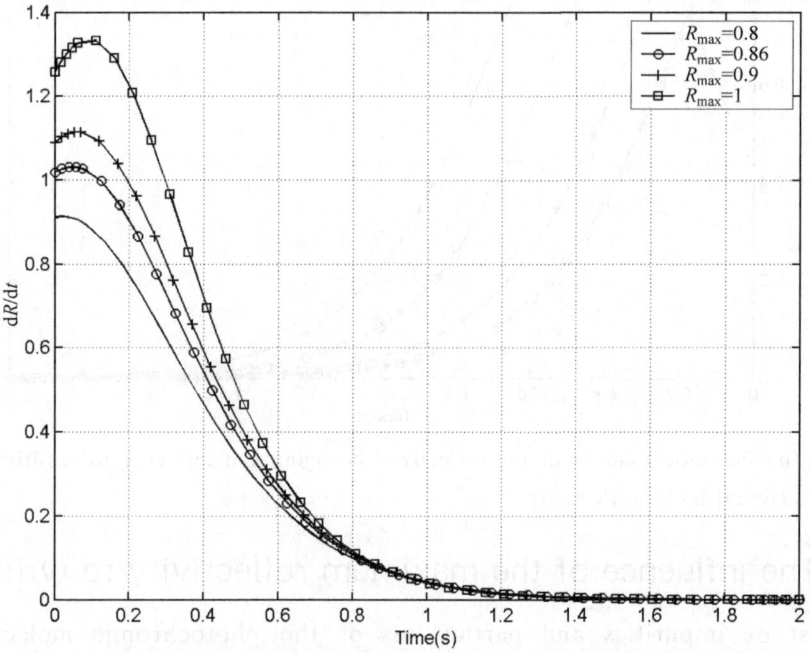

Fig. 3-25 The theoretical curves of the reflectivity changing and influence to sensitivity of different maximum reflectivity.

The theoretical curves of the writing process are shown in Figure 3-26 when the reflectivity of reflective layer is changed. The figure shows that the change of reflectivity will affect the final reflectivity of the full write only, but dos not effect on the time to fully write.

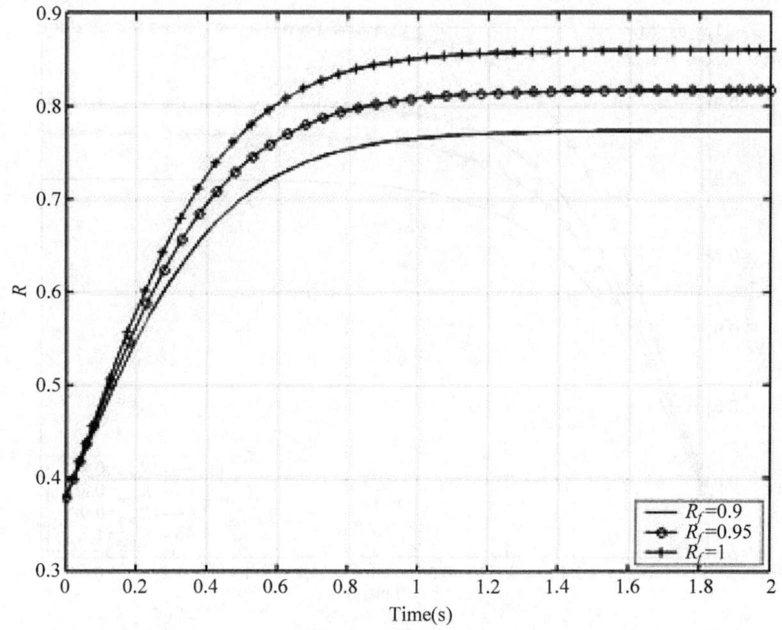

Fig. 3-26 The theoretical curves between reflectivity and writing time on medium with different reflectivity of reflective layer.

3.2.3 Written time constant *k*

Written time constant k is the characterization of the photochromic material reaction speed, that is a constant when parameters of medium and write conditions are certain. The time constant k to a certain writing wavelength is determined by the material properties and the write power. So changing the writing power can change the time constant k for a certain material. The same material is written with different time constant k that the theoretical curves are shown in Figure 3-27.

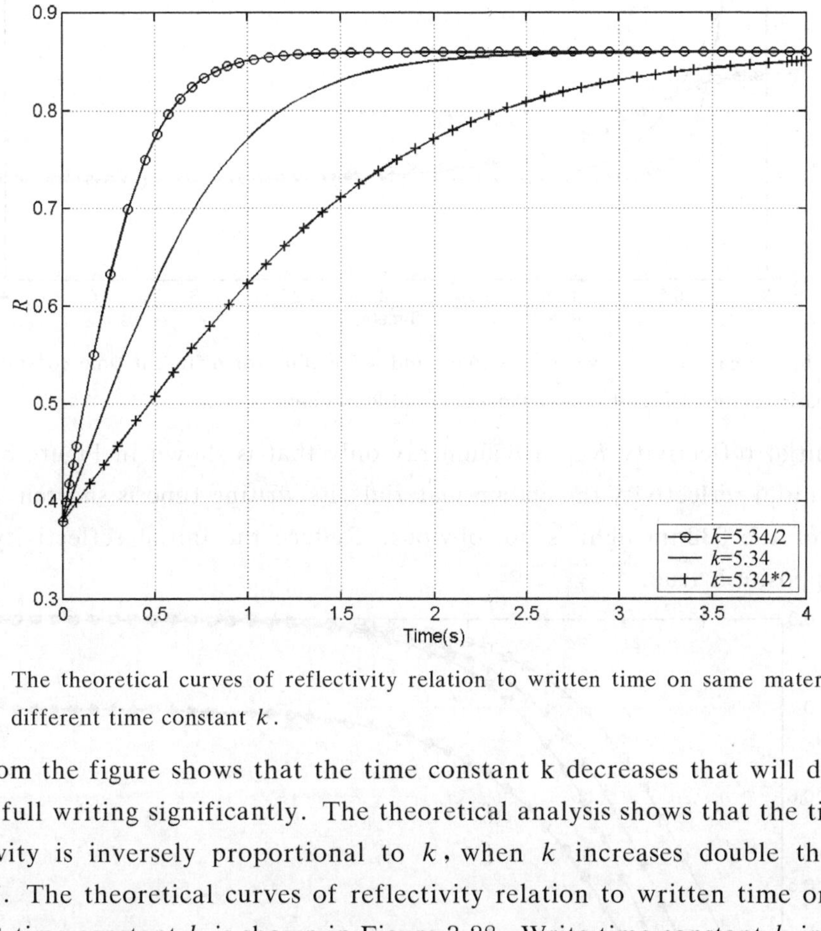

Fig. 3-27 The theoretical curves of reflectivity relation to written time on same material with a different time constant k.

Obvious from the figure shows that the time constant k decreases that will delay completion the process for full writing significantly. The theoretical analysis shows that the time to achieve a certain reflectivity is inversely proportional to k, when k increases double the write time is reduced by half. The theoretical curves of reflectivity relation to written time on same material with a different time constant k is shown in Figure 3-28. Write time constant k increases that can improve the sensitivity of the recording process. When the write time k is very small, the sensitivity will be lower and the writing speed is reduced greatly.

The parameters of the molecular concentration of photochromic materials are influence each other that is very complicated. Using the theoretical model (3-41) and (3-42) analyze and calculate a material with absorption spectra of cross of 532nm and 650nm for example. The reflectivity of the mixed material to the 532nm or 650nm alone is affected by common of 532nm and 650nm light. The initial reflectivity of mixed materials of absorption spectra mutual cross will be decided by both Molar extinction constant and concentration of two materials. Therefore, if changing an initial reflectance of a wavelength but another initial reflectivity of wavelength unchanged that have to improve the ratio of the initial concentration of two materials. The theoretical curves of reflectivity in the writing process to keep the other parameters unchanged but

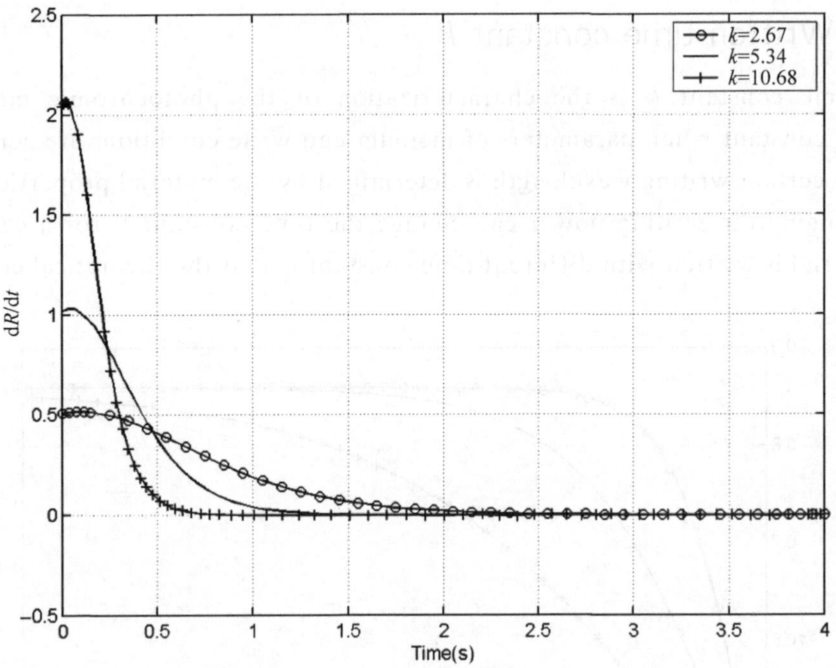

Fig. 3-28 Theoretical curves between sensitivity and write time for different time constant k that is basis for the using of low-power non-volatile readout.

to change the initial reflectivity R_{ini} of 650nm ray only that is shown in Figure 3-29. Evidently, when heighten initial reflectivity the 650nm materials, its writing time is shorten, but influence to writing process of the 532nm light is not obvious. Reduce the initial reflectivity of the 650nm light, that the opposite is true.

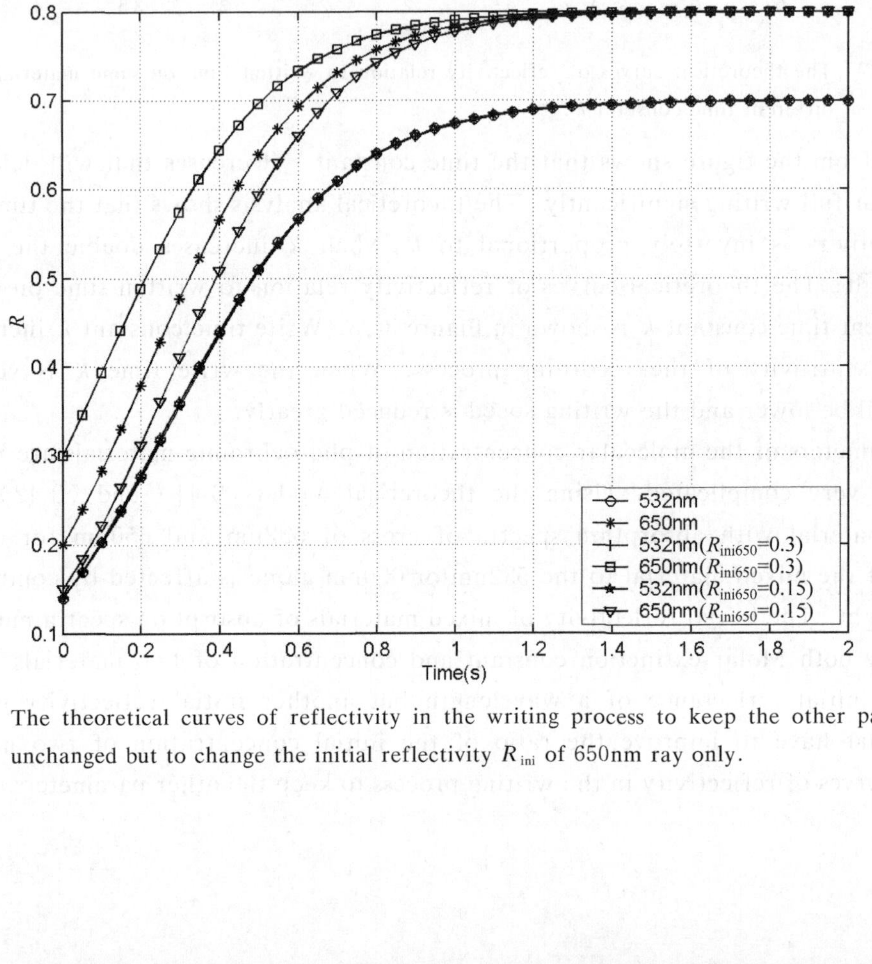

Fig. 3-29 The theoretical curves of reflectivity in the writing process to keep the other parameters unchanged but to change the initial reflectivity R_{ini} of 650nm ray only.

Figure 3-30 shows theoretical curves of writing sensitivity of 650nm when initial reflectivity is changed. The peak of writing sensitivity of 650nm moves to left when its initial reflectivity increases to 0.3, the changing of sensitivity was faster (0.2s) within beginning region of the 0~0.18s, after then the change of sensitivity is slow, and closed to 0 earlier. At the same times the write sensitivity of 532nm light is increased overall. When the initial reflectivity of 650nm is reduced to 0.15, the situation of writing sensitivity is the opposite. Lower writing sensitivity of the medium, exposure time of reaching a certain sensitivity growths that can be used to improve the amount time of non-destructive read out. The theoretical curves of reflectivity in the writing process to keep the other parameters unchanged but to change the initial reflectivity R_{ini} of 532nm ray only is shown in Figure 3-31. Such as in the diagram, the initial reflectivity of the 532nm light changes, influence the full writing time to 532nm light, but that is no effect to 650nm light almost.

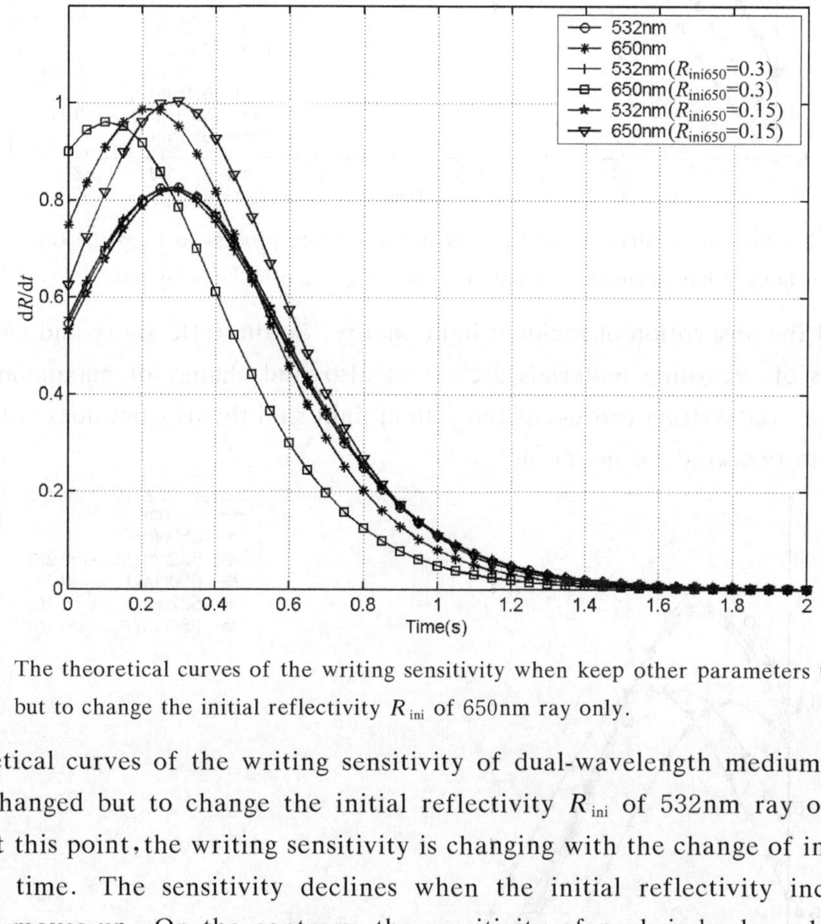

Fig. 3-30 The theoretical curves of the writing sensitivity when keep other parameters unchanged but to change the initial reflectivity R_{ini} of 650nm ray only.

The theoretical curves of the writing sensitivity of dual-wavelength medium keep the other parameters unchanged but to change the initial reflectivity R_{ini} of 532nm ray only is shown in Figure 3-32. At this point, the writing sensitivity is changing with the change of initial reflectivity to 532nm over time. The sensitivity declines when the initial reflectivity increasing, but its sensitivity peak moves up. On the contrary, the sensitivity of peak is backward extension. The sensitivity of 650nm declines overall and the sensitivity peak is premise slightly, when the reflectivity of 532nm increases. Therefore, change of the initial reflectivity of the medium, will affect writing process of both two lasers, but the magnitude of effect is depended on the sensitivity of the materials. For example, when maximum reflectivity are 0.6 and 0.9 respectively, the theoretical curves of writing sensitivity of the dual-wavelength sample in the writing process is shown in Figure 3-33 and Figure 3-34. It can be seen from Figure 3-30, changes of maximum reflectivity to 532nm light is influence the final reflectivity of 532nm for full writing, but does not affect the final reflectivity of 650nm light. Maximum reflectivity of recording layer of photochromic

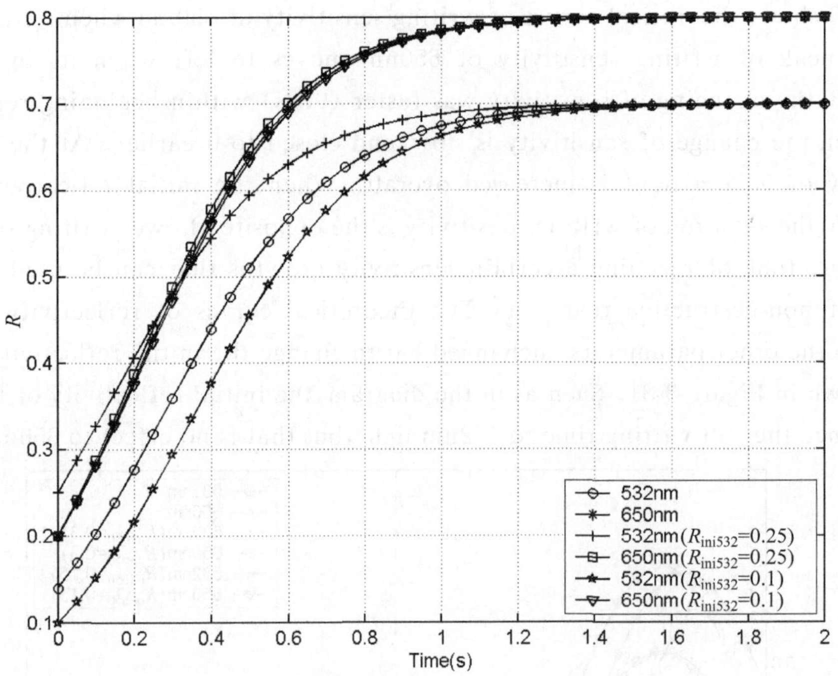

Fig. 3-31 The theoretical curves of reflectivity in the writing process to keep the other parameters unchanged but to change the initial reflectivity R_{ini} of 532nm ray only.

material decided the absorption of incident light energy. So the reflectivity and energy absorption of the two kinds of recording materials are effect also, and change of maximum reflectivity of 532nm is influence the writing process of the 650nm light slightly also, but does not affect the time of fully written to two kinds of materials.

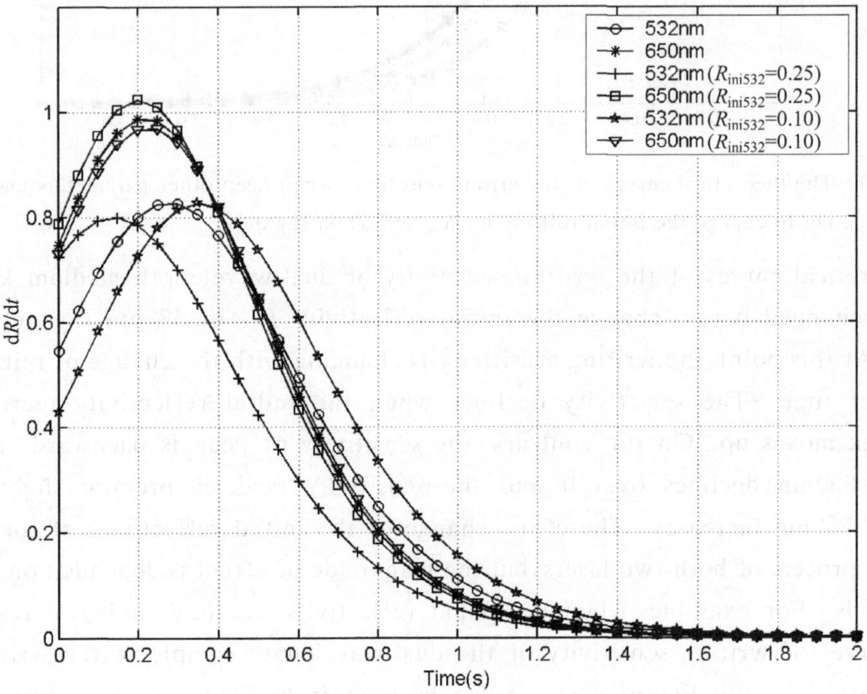

Fig. 3-32 The theoretical curves of the writing sensitivity keep the other parameters unchanged but to change the initial reflectivity R_{ini} of 532nm ray only.

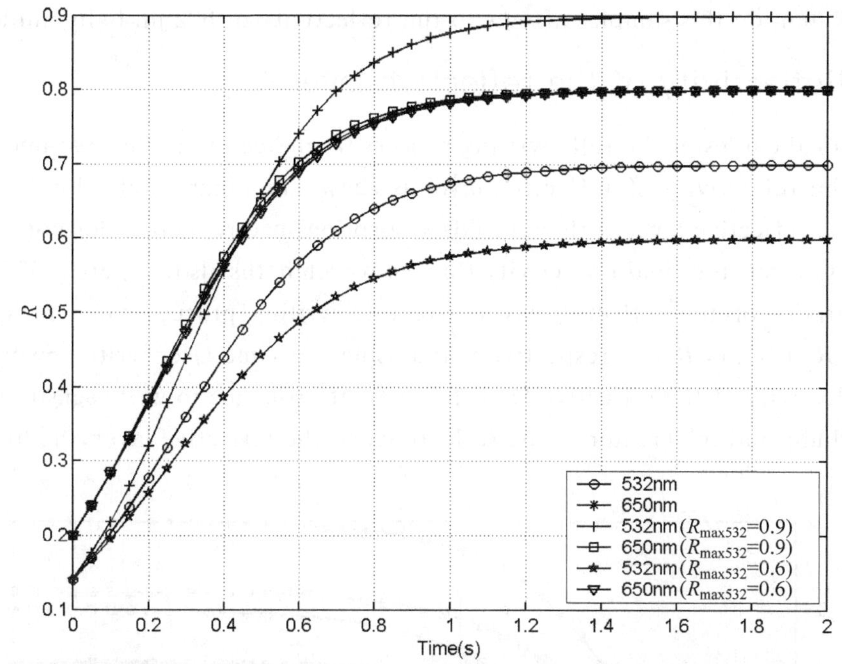

Fig. 3-33 The theoretical curves of the writing process when keep the other parameters unchanged but to change the maximum reflectivity R_{ini} to 650nm and 532nm ray.

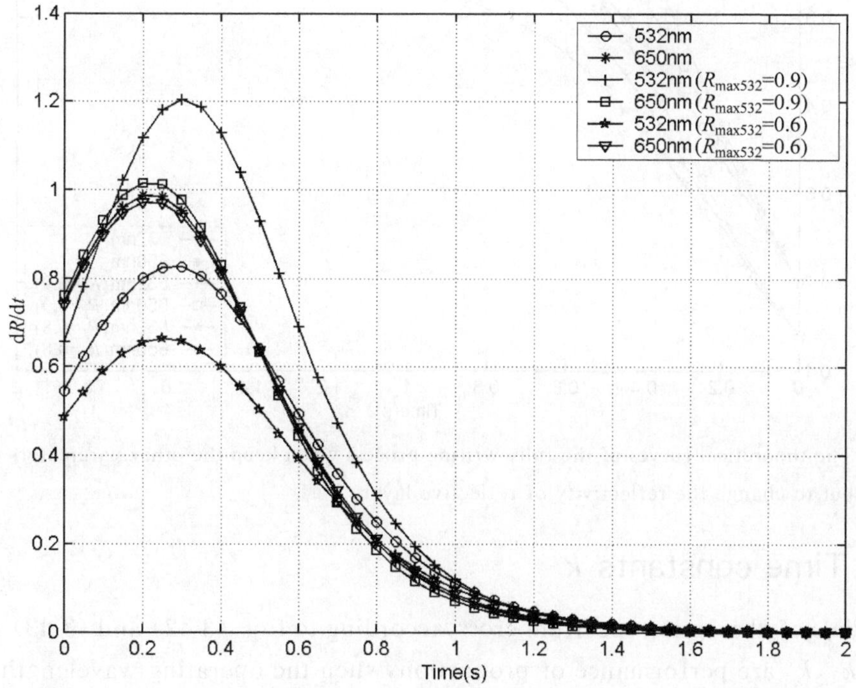

Fig. 3-34 The theoretical curves of the writing sensitivity keep the other parameters unchanged but to change the maximum reflectivity R_{ini} of 650nm and 532nm ray.

From Figure 3-34 can see that improving the maximum reflectivity of 532nm will affect the writing sensitivity of two wavelengths also. Obviously, the larger of the maximum reflectivity of 532nm light, the higher writing sensitivity, that the time to achieve certain reflectivity is shorter, but influence to writing sensitivity of 650nm light slightly. Of course, changing the maximum reflectivity of 650nm on this samples affects to the reflectivity of 532nm light similarly.

3.2.4 Reflectivity of the reflective layer

The theoretical curves of the fully writing process when keep the other parameters unchanged but to change the reflectivity of reflective layer is shown in Figure 3-35. Be seen, changing of reflectivity of the reflective layer with materials absorption spectrum cross dos not change the fully writing time, but affects the final reflectivity for two wavelengths also. Figure 3-36 is experimental curves of reflectivity of the reflective layer affect to writing process when reflectivity of two wavelength are $R_f 0.9$ and $R_f 0.8$ respectively and using the same laser write power to write. Can be seen, the reflectivity of the reflective layer has a great influence to write sensitivity of the dual-wavelength medium, and the greater of the reflectivity of the reflective layer, the higher of writing sensitivity.

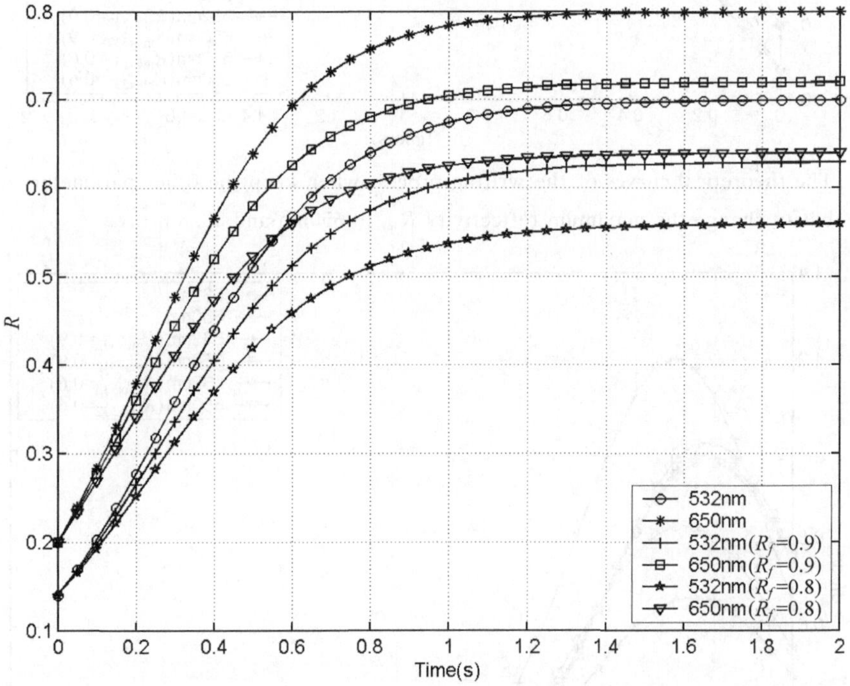

Fig. 3-35 The theoretical curves of the fully writing process when keep the other parameters unchanged but to change the reflectivity of reflective layer.

3.2.5 Time constants *k*

For materials of absorption spectrum cross, according to Eqs. (3-42) and (3-43) shows that the time constant k_1, k_2 are performance of proportion, when the operating wavelength and materials have been identified. For example, increase or reduce the writing power of the 532nm laser, will also change the time constants, k_1 and k_2, as shown in Figure 3-37, k_1, k_2, increase and decrease the time doubled in the writing process. Theoretical calculations show that write the time constants k_1 and k_2 change will directly affect the reflectivity of the two wavelengths of light over time, as reflectivity of any one wavelength of the two materials absorption spectrum cross are influence by writing energy of the two wavelengths at the same time. Improve 532nm writing power, writing time constants k_1 and k_2 increases and accelerate completion of the two wavelengths of

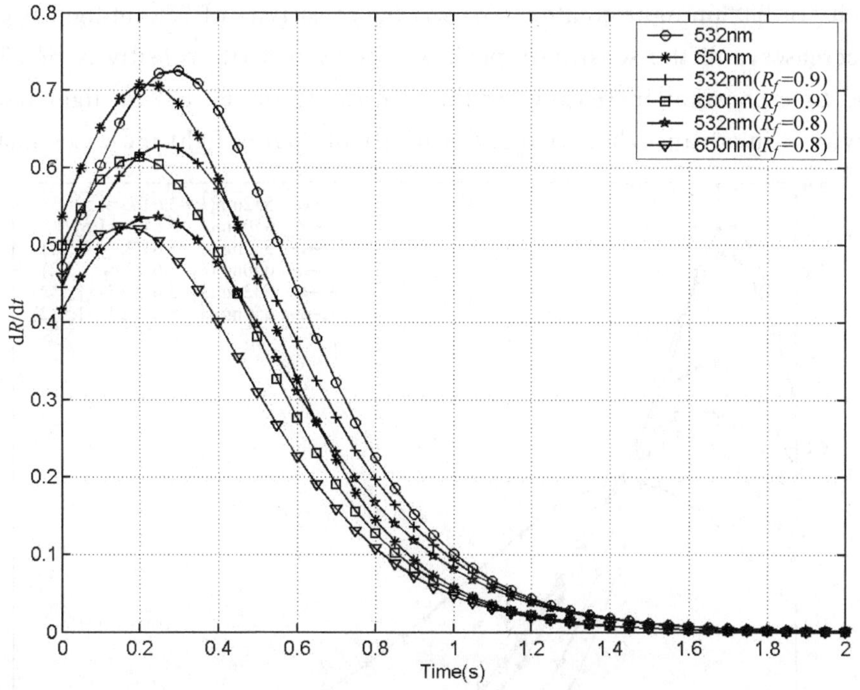

Fig. 3-36 Experimental curves between sensitivity and write time for different reflectivity of reflective layer.

light writing process, in turn will reduce the writing speed. But changed time constant k_1 or k_2 individually, writing speed not strictly inversely proportional to time constant k_n.

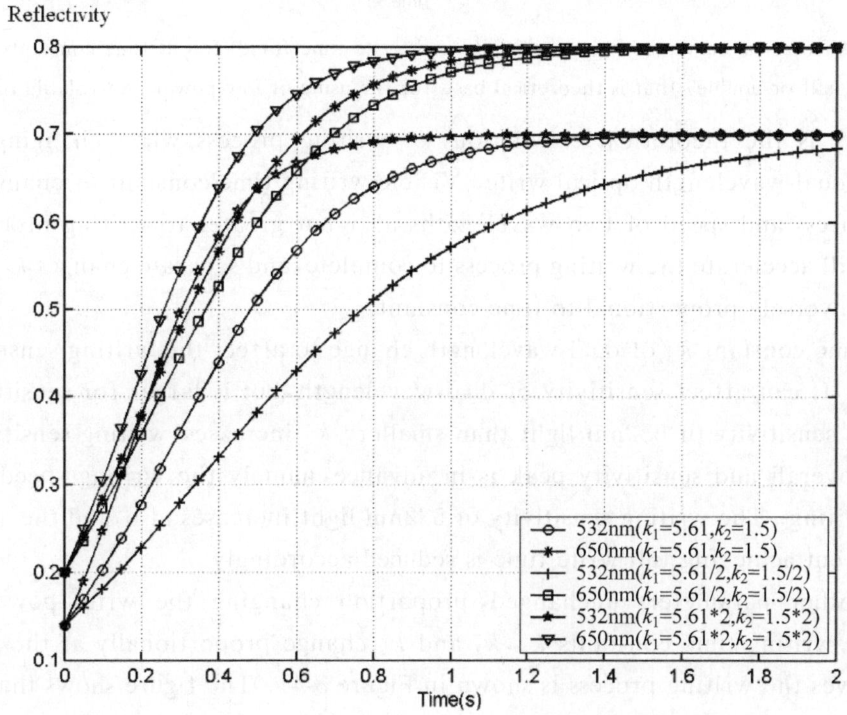

Fig. 3-37 Theoretical curves of the writing process when maintain other parameters but to change the time constants k_1 and k_2 (half or double).

Change time constants of k_1 and k_2 will affect writing sensitivity, the experiments are shown in Figure 3-38. As the write power can change the time constant, so changing of the k_1, k_2 affect writing sensitivity of 532nm light greater. The writing sensitivity of 532nm light is greatly reduced when k_1, k_2 decreases, and the sensitivity peak was delayed, write sensitivity of 650nm light fell also. When the time constant increasing, writing sensitivity of the 532nm light increases, and its peak of sensitivity is in advance, but writing sensitivity of 650nm light increases smaller.

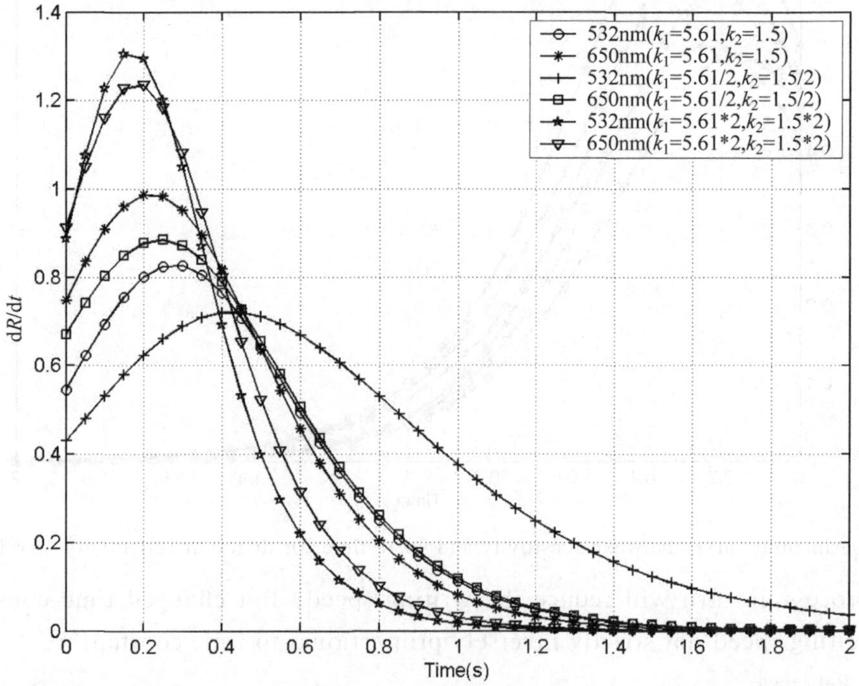

Fig. 3-38 Theoretical curves between sensitivity and write time for different time constants k_1 and k_2 (half or double) that is theoretical basis for the using of low-power non-volatile readout.

Figure 3-39 is the theoretical calculations of writing process when changing writing time constant k_3 to dual-wavelength optical write. Visible writing time constant k_3 changing will affect the writing process and speed of two wavelengths also, but greater to writing process of 650nm. Increased k_3 will accelerate the writing process to complete, and separate changes k_3, writing speed is not strictly inversely proportional to time constant.

Writing time constant k_3 of dual-wavelength change to affect the writing sensitivity is shown in Figure 3-40. It can affect sensitivity of dual-wavelength, but is larger for sensitivity of 650nm light and write sensitivity of 532nm light than smaller. k_3 increases, writing sensitivity of 650nm light increase overall and sensitivity peak is in advance, namely the writing speed up quickly to achieve full writing. The writing sensitivity of 532nm light increases also, and the sensitivity peak slightly ahead, but achieving full write time is reduced accordingly.

Keeping other parameters unchanged, proportion changing the write power of the two wavelengths, ie writing time constants k_1, k_2 and k_3 change proportionally at the same time, the theoretical curves the writing process is shown in Figure 3-41. The figure shows that with a larger of the three time constants proportional, due to the growth of writing power simultaneous, the relationship of distribution of the absorption of light energy to two materials is not affected, but the writing time was reduced proportionately. However, if a separate change any one writing power, ie k_1, k_2 and k_3 is proportional change, the writing time are no longer proportional change.

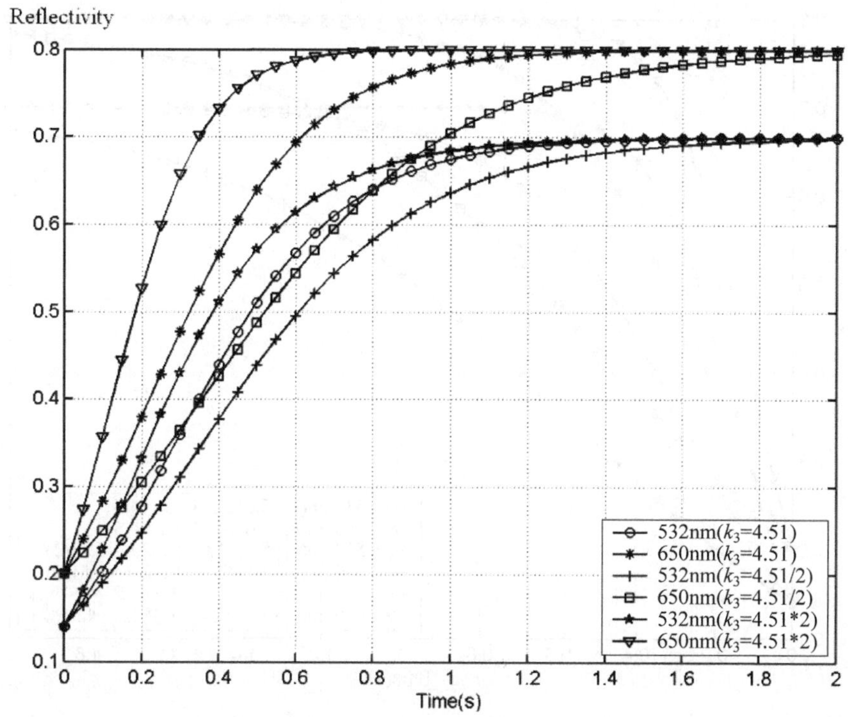

Fig. 3-39 Theoretical curves of the writing process when keep the other parameters unchanged but to change the time constants k_3 (half or double).

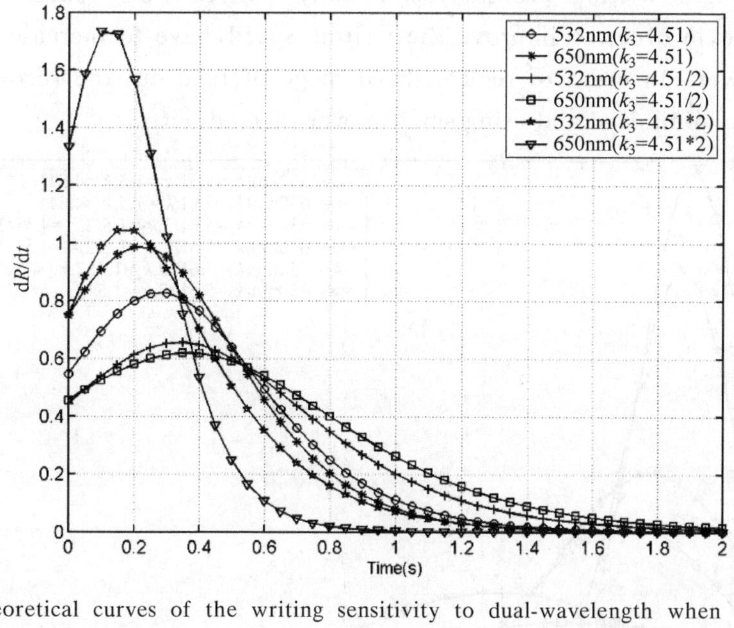

Fig. 3-40 Theoretical curves of the writing sensitivity to dual-wavelength when keep the other parameters unchanged but to change the time constants k_3 (half or double).

Writing sensitivity for different time constants k_1, k_2 and k_3 is shown in Figure 3-42. When the time constant increases at same time, the writing sensitivity of the two wavelengths of light also much increases. Whether changing anyone of the writing time constants, writing sensitivity of the two wavelengths of light are affected for dual-wavelengths materials with absorption spectrum cross. Reducing the time constants of the dual-wavelength writing process slowing down, fully written time required for growth, and writing sensitivity decreases. Increasing the time constants

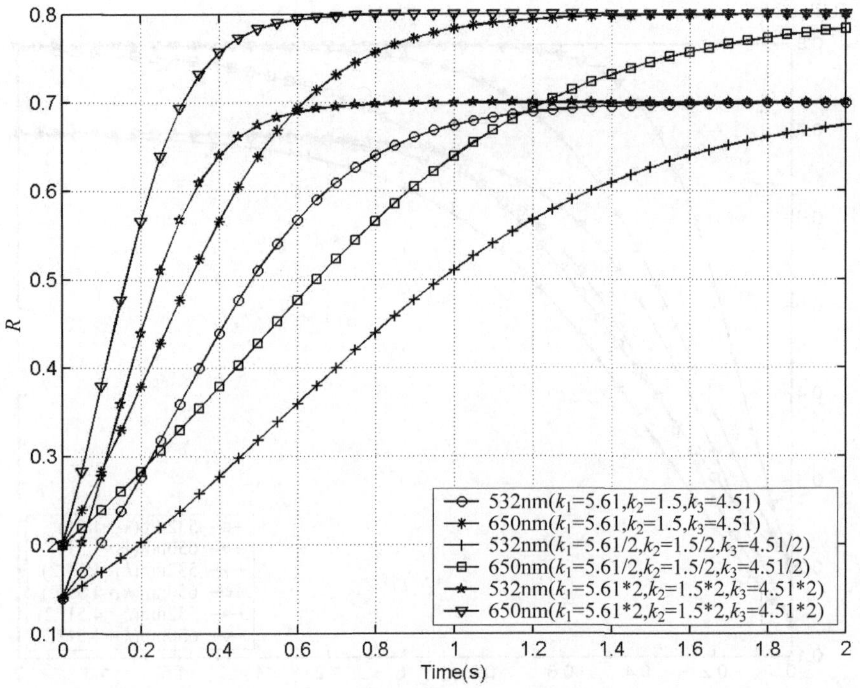

Fig. 3-41 Theoretical curves of the write process when keep the other parameters unchanged but to change the time constants k_1, k_2 and k_3 (half or double).

of the dual-wavelength writing process faster, fully written time reduce, writing sensitivity increases. Therefore, in order to improve the writing speed, have to increase the time constants and the writing power. In order to reduce the damage of read out the recorded data, the time constants have to be minimized and using small power of read out.

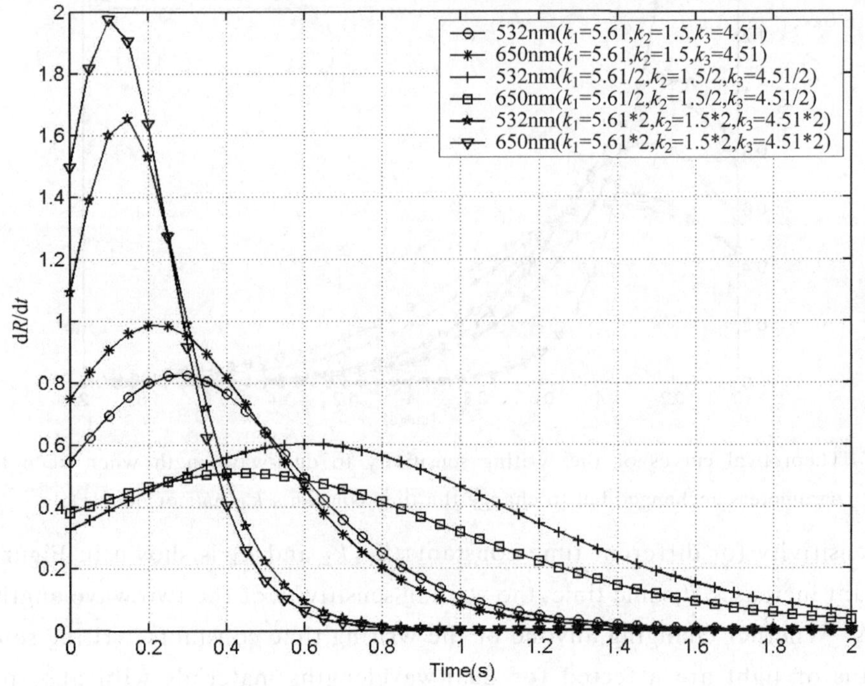

Fig. 3-42 Theoretical curves of the writing sensitivity when keep the other parameters unchanged but to change the time constants k_1, k_2 and k_3 (half or double).

3.3 Crosstalk in multi-wavelength and multi-level storage

3.3.1 Emerging of crosstalk

Light-induced discoloration of multi-wavelength storage photochromic principles and using photochromic materials for the memory medium are established different structures of molecular have different absorption of light. The cross of absorption spectrum of the recording medium to the recorded signal will bring crosstalk. Only through a systematic experimental study on the multi-wavelength photochromic storage process, analysis material absorption spectrum cross-write process and quantified describes with mathematical model of the writing process to photochromic multi-wavelength storage that could bring the theoretical basis to eliminate light-induced crosstalk in photochromic multi-wavelength storage.

The production of light-induced discoloration of the crosstalk of multi-wavelength storage is due to a variety of recording materials absorption spectrum of cross. In the multi-wavelength storage, each wavelength corresponding to the recording material of the absorption spectrum has to be as narrow as possible, ie each recording materials absorb a wavelength of light only. In fact a lot of organic photochromic materials absorption spectrum are wider, but other characteristics of these materials are extremely advantageous and may be adopted. In addition, requirement of more multi-wavelength photochromic recording materials compliance working together, elimination the crosstalk are more difficult. Under normal circumstances, there is always part of the material to cover two or more write-read wavelength, the crosstalk problem is difficult to avoid. The crosstalk seriously affect the multi-wavelength readout signal, the recorded data will be leaded to great error by crosstalk.

A typical recording experiments of photochromic multi-wavelength storage is shown in Figure 3-43, and the experimental parameters are shown in Table 3-2. In this experiment, take six locations to write, where the point 1 and 4 was recorded by 650nm laser and 532 laser at the same time, point 2

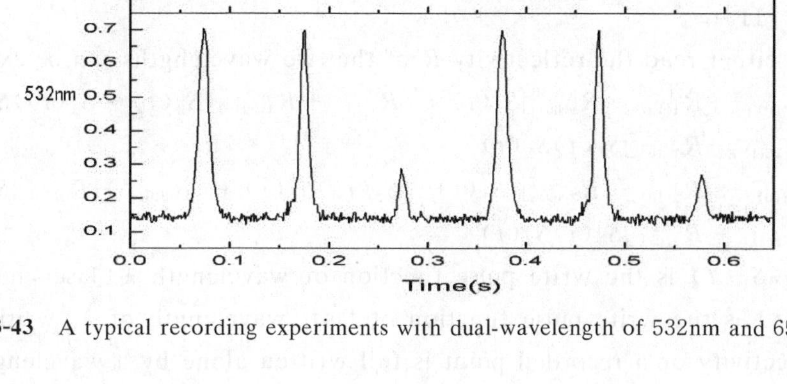

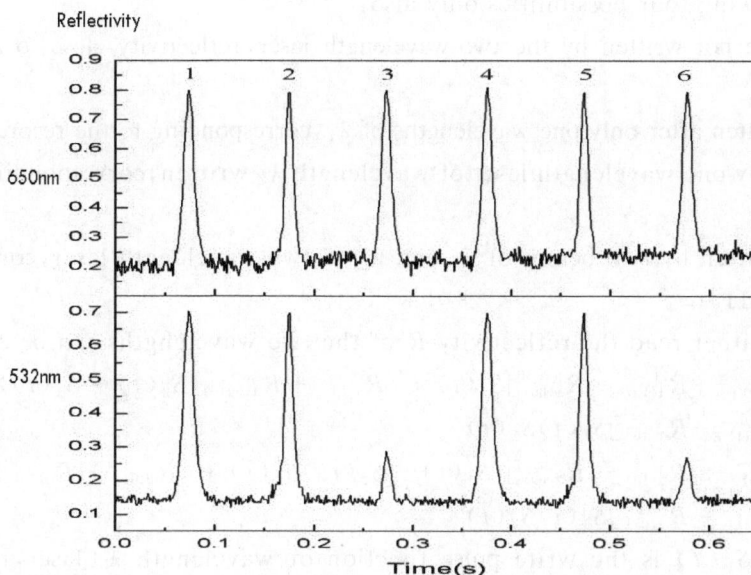

Fig. 3-43 A typical recording experiments with dual-wavelength of 532nm and 650nm.

and 5 was recorded by 532nm wavelength laser alone, but point 3 and 6 was recorded the by 650nm laser alone. The read out power are 0.07mW for 532nm laser and 0.1mW for 650nm light simultaneously, readout speed is 0.1mm/s. From the experimental results can be seen that the points 2,5 are written with 532nm laser only, but the read out reflectivity of 650nm are close to the 650nm laser alone to write at the same points, ie there are serious crosstalk at the two points.

Table 3-2 Experimental writing parameters

Written position	Writing power (mW)		Writing time (ms)	
	532nm	650nm	532nm	650nm
1,4	0.07	0.1	2	2
2,5	0.07	0	5	0
3,6	0	0.1	0	4

Note point 1 and 4 are full written by 532nm laser and 650nm laser, the molecule of material 650nm absorb 532nm light cannot be discovery. But the point 3 and 6 are written by 650nm laser only that is almost no absorption by 532nm laser material, ie the crosstalk is small. Read out signal from above dual-wavelength recording experiment can also see, read out reflectivity with 650nm laser of six point are about 0.8, i. e. six points are written by 650nm laser. But in fact point 2,5 are not written, that is a intolerable errors. Obviously the performance of photochromic materials affects the crosstalk in multi-wavelength storage fundamentally. In addition to solve this problem have to identify the optical absorption cross of materials, and the profiles of storage material absorption of certain cross, that according to the trend of crosstalk select the appropriate write strategy, can reduce or inhibition its impact to a certain degree.

3.3.2 The calculations of crosstalk

Still above storage experiments of two wavelengths recording for example, the experimental arrangements can be recording [00],[01] and [11] at each point recording data only that the read out signal are following four possibilities only also:

(1) $R[00]$, are not written by the two-wavelength laser, reflectivity close to zero, corresponding recorded data [00];

(2) $R[10]$, written after only one wavelengths of λ_1, corresponding to the recorded data [10];

(3) $R[01]$, only one wavelength laser of wavelength λ_2 written, corresponding to the recorded data [01];

(4) $R[11]$, written by two beams of λ_1 and λ_2 of two-wavelength laser, corresponding to the recorded data for [11];

Therefore, in either read the reflectivity R of the two wavelengths can be expressed as:

$$R_1 = R_{1(00)} + [R_{1(10)} - R_{1(00)}]S_1(t) + [R_{1(01)} - R_{1(00)}](S_2(t) - S_1(t))S_2(t) + [R_{1(11)} - R_{1(10)}]S_1(t)S_2(t) \tag{3-44}$$

$$R_2 = R_{2(00)} + [R_{2(10)} - R_{2(00)}](S_1(t) - S_2(t))S_1(t) + [R_{2(01)} - R_{2(00)}]S_2(t) + [R_{2(11)} - R_{2(01)}]S_1(t)S_2(t) \tag{3-45}$$

Among them, $S_1(t)$ is the write pulse function of wavelength λ_1 laser, namely λ_1 written information; $S_2(t)$ is the write pulse function of laser wavelength of λ_2 writing information. Definition of reflectivity of a recorded point is full written alone by a wavelength and read out reflectivity by the same wavelength of light, ie recorded data[1]. From equations (3-44) and (3-45) can see that the read out the reflectivity R_1 to wavelength λ_1 is determined by equation (3-44), where R_1 contains all crosstalk in $S_2(t)$. Recorded different data, crosstalk can be divided into the crosstalk 0 and crosstalk 1 categories. The so-called 1 crosstalk refers the crosstalk of read out signal that the point is recorded by λ_1 and other wavelengths to participate to write. The reflectivity of read out by the λ_1 light at this time will be greater than or equal to the effective reflectivity. Definition the crosstalk is the difference of read out reflectivity and effective reflectivity. The so-called [0] cross-talk is introduction of crosstalk by other wavelengths recording without λ_1 wavelength, ie all other wavelength light except λ_1 wavelength introduced crosstalk. It is defined the difference between read out reflectivity and the initial reflectivity. The readout signals of dual-wavelength recording at the same time have [1] cross-talk for λ_1 and λ_2 also that the [1] cross-talk of λ_1 of read out of signal wavelength is available $R_1[11]$-$R_1[01]$. If the λ_2 wavelength laser alone write on the sample, the wavelength of λ_1 will bring 0 crosstalk to the readout signal, and it is available $R_{1(01)}-R_{1(00)}$. Obviously, the property and impact of the two kinds of cross-talk is different, that affect read out signal [1] smaller, but involve read out signal [0] even more. So selection of the writing strategy is difference for the two types of crosstalk.

(1) The readout signal crosstalk of wavelength λ_1:

When λ_2 laser writing only, the introduction of [0] crosstalk is $R_{1(01)}-R_{1(00)}$;

When two-wavelength lasers writing at the same time, the introduction of the cross-talk is $R_{1(11)}-R_{1(10)}$.

(2) The readout signal crosstalk of wavelength λ_2 respectively:

When only λ_1 laser to write, the introduction of [0] crosstalk is $R_{2(10)}-R_{2(00)}$;

When two-wavelength laser write at the same time, the introduction of the cross-talk is $R_{2(11)}-R_{2(01)}$.

The situation is more complicated in the three materials for multi-wavelength storage. The requirements of performance of materials are higher, especially for the absorption spectra is need of as narrow as possible. Here take the synthesis of 1, 2-bis (2-methyl-5-n-butyl-3-thienyl), perfluorinated cyclopentene; 1, 2-bis (2-methyl-5-(4-N, N-dimethyl) phenyl-thiophene-3-yl), perfluorinated cyclopentene and {1-[2-methyl-3-2-(1,3-dithiolane benzo thienyl)]}; 2-{2-methyl-5-[4-(2, 2-two cyano-vinyl phenyl)]}, thiophene thiophene 3 perfluorinated cyclopentene three materials for example. The absorption peaks of three kinds of materials are 532nm, 650nm and 780nm. The absorption spectra and molecular formula of the materials of 532nm and 650nm have been introduced in last section. The molecular formula and absorption spectra of material 780nm are shown in Figure 3-44.

Performances of these materials are better that the absorption spectra of the three materials is narrow relatively, and does not absorb each other and is not crosstalk generally. The three materials have been used to the three-wavelength storage experiment and achieved the same results.

The experimental parameters of three-wavelength mixed recording media is written and read out (powers are actual power on the focal plane) is shown in Table 3-3. The three-wavelength storage experiment results of readout signals (reflectivity) which is fully written in the same channel at the same time and same position synchronization with three wavelength of 532nm, 650nm and 780nm lasers are shown in Figure 3-45.

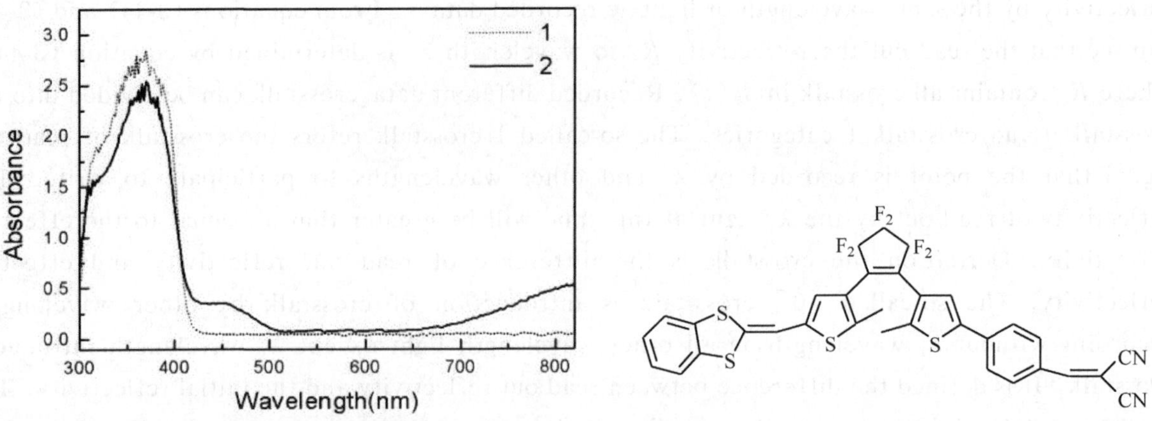

Fig. 3-44 The absorption spectra and molecular formula of 2-{2-methyl-5-[4-(2,2-two cyano-vinyl phenyl)]}, thiophene thiophene-3 perfluorinated cyclopentene for 780nm laser.

Table 3-3 Experimental parameters of three-wavelength mixed recording media is written and read out.

Parameters	780nm	650nm	532nm
Writing power (mW)	2	2.5	2.2
Writing time (s)	0.05	0.05	0.05
Readout power (mW)	0.1	0.1	0.07
Scan speed (m/s)	0.1		

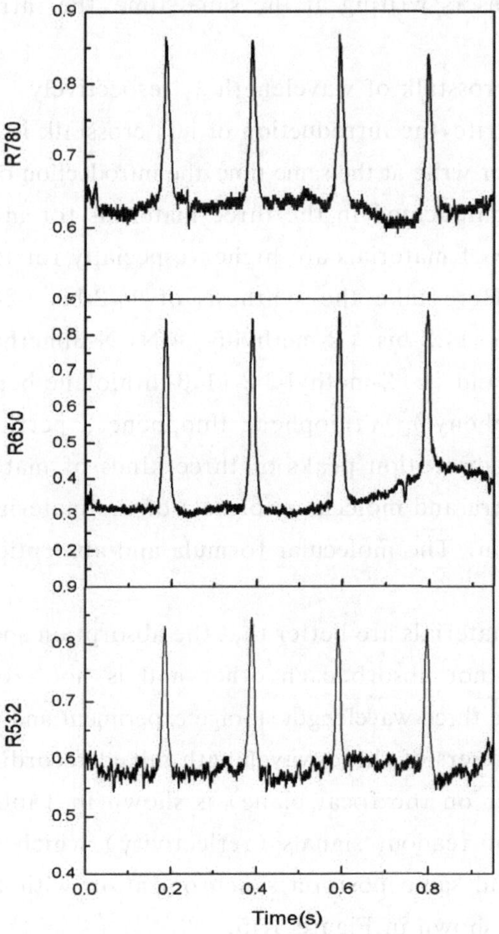

Fig. 3-45 Readout signals (reflectivity) which is fully written in the same channel at the same time and same position synchronization with three wavelength of 532nm, 650nm and 780nm lasers.

It can be seen from Figure 3-45 that in a three-wavelength readout on a point at the same time has significant reflectivity peaks, indicating that the molecules of three photochromic recording materials are transformed completely basic, and cannot absorb light of various wavelengths. The reflectivity of the three wavelengths on the experimental samples is maximum, corresponding to "1" in the digital storage. Experimental recording films on the same location have three "1", ie achieve to record 3bit at the same point. Low reflectivity is unrecorded state, corresponding digital data "0". Some fluctuations of the signal in the Figure 3-45 are cause of uneven of initial reflectivity and the recording layer according to the test experiment.

Based on the models established by photochemical Lambert-Beer law, read out reflectivity of each wavelength is determined by

$$R_i = R_f R_{\max(\lambda_i)} e^{-4.6 \sum_{j=1}^{3} \varepsilon_{j(\lambda_i)} C_j l} \tag{3-46}$$

Where $\varepsilon_j(\lambda_i)$ is the molar extinction coefficient of material j for the wavelength λ_i. C_j is concentration of material j. l is the optical path. R_f is reflectivity of reflective layer. $R_{\max}(\lambda_i)$ is maximum reflectivity for wavelength λ_i of the sample. Three materials mixed in a layer, so the optical path is same also.

From Eq. (3-46) can know, the initial reflectivity to certain wavelength on the samples is determined by the concentration of material molecule and its molar extinction coefficient. As absorption spectra of the materials is not cross each other almost, thus the initial reflectivity of a wavelength only determined by the concentration of material and molar extinction coefficient in above experiment.

To further analyze the crosstalk of various materials and the impact on signal quality, take another experiment to write and read under the same conditions. Modulation of three laser power, along the same channel in the experimental samples in parallel records, write parameters are the same with Table 3-3, but at every point write with the three-wavelength laser staggered on a points. The read out signal is shown in Figure 3-46, that each record of the three materials can be accurately read out and is not fringes on location of all 532nm laser, 650nm and 780nm laser recording and read out. Confirmed 532nm laser writing did not make the 650nm materials and 780nm materials achromatic response to introduction of crosstalk, and the 650nm laser and 780nm laser writing do not introduce crosstalk to other wavelengths. So long as the absorption spectrum of the recording material is not present cross, that the mixture of three materials or any more materials at the same time recording and read out cannot present crosstalk. Fluctuations of the signal are due to uneven of concentration of recording materials that indication control physical parameters of the recording layer materials and uniformity are very important for multi-wavelength photochromic optical storage. But the uniformity of concentration of the mixed materials are relationship with uniformity of solvent, spin coating process and the stability of the material itself and many other factors, so must be resolved in the future.

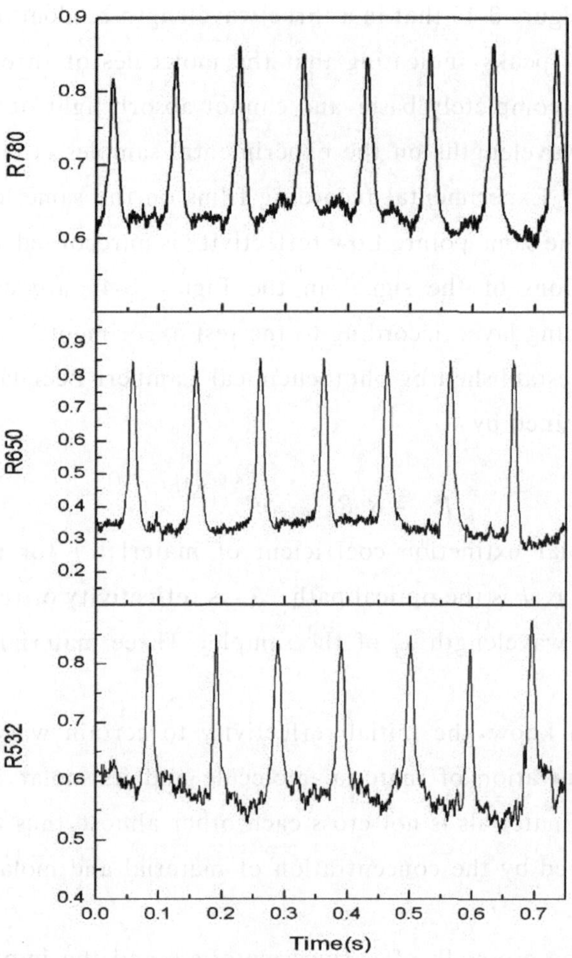

Fig. 3-46 Readout signal (reflectivity) which is fully written in the different channel at same position with three wavelength lasers (532nm, 650nm and 780nm).

3.4 Non-destructive readout

Reaction of photochromic compounds is reversible usually. The photochromic materials of optical storage are used the colored state of molecule to be the recording state and initialized with coloring method. Using an achromatic light write and a very weak light read out. The colored state molecule is achromatic after writing and dos not absorbe wring light again that the transmittance of this point is enhanced. When all materials are achromatic by writing, the transmittance is highest in this point. As read out employs weak light, some molecules of medium will be achromatic in the non-writing area. So the contrast of the reflectivity between writing point and not written point is decreased gradually. After repeatedly read out much more, colored state molecule on all not written area will be achromatic and became colorless state molecule, resulting in the destruction of the recorded data. The achromatic response of some materials has a certain threshold, low-light response is insensitive, but experiments show that most photochromic materials does not exist response threshold to laser power, even if relatively weak laser irradiation, there will be a certain percentage of photochromic molecules react, resulting in the destruction of the information written on the original. In order to explore non-destructive readout methord in photochromic recording to achieve non-destructive read out. Study of non-destructive read out has two scenarios mainly as following. First is using other wavelength laser to read out, which cannot be absorbed by all materials in recording layer, so can achieve non-destructive readout. Another way is utilization the photochromic compounds which can be control by external trigger, such as light, electricity, magnetism, heat or chemical factors stimulate the photochromic reaction. In addition to achieve non-destructive readout, some way so-called gated reactions when no zero quantum yield wavelength readout, temperature threshold control of dual-wavelength readout and ultra low power read out and other programs. This section will introduce an electric locking non-destructive readout method belongs to only a specific external excitation (electric field) occurs only when the photochromic reaction. Fumio MATSUI etc. proposed zero quantum yield wavelength readout method, using change the quantum yield of readout wavelength control achromatic response in the indole fulgide photochromic storage experiments achieved non-destructive read out. As indole fulgide is not sensitive to 780nm laser, so the achromatic quantum yield is almost zero to read out, but there are still large enough to absorb for the detection. Therefore, in case that does not destroy the recorded information read out times as wish. The disadvantage of this method is that have to use another wavelength laser, increasing the complexity of the read-write device especially for multi-wavelength storage systems. In addition, a temperature threshold method non-destructive readout proposed earliest by Fumio Tatazono et al. In this method, recording medium has a characteristics above room temperature threshold T_c, that using λ_1 laser of higher power (to get higher threshold T_c temperature) to write, read out using low power λ_2 (lower temperature) and at the same time using λ_1 and λ_2 two wavelength lasers erases. Because written location has stronger absorption on to λ_2, but dos not absorb without written, so can read out information no any damaging, only when the temperature is higher than the writing threshold T_c (using higher power λ_1 exposure) λ_2 can erase the media. The research team used series of diarylethene photochromic compounds, its threshold temperature is about 85℃, 458nm Ar^+ laser (power 1.6mW, the pulse width 10μs). In the experiments medium is recorded by Ar^+ laser (form of open-loop into a closed loop), and using 633nm He-Ne laser (0.5mW) read out and erase. But erase have to use both 633nm laser and 458nm laser at the same time. According to reports, this method was achieved non-destructive readout more than 10^6. The disadvantage of this method is the thermal effect that the responsing speed is lower and loss characteristics of high sensitivity and high-speed of the photon effect. In addition, this temperature threshold is not too high to the environmental temperature, storage time cannot be too long. The ultra-low power non-destructive read out the experiments carried out earlier in Tsinghua University and has accumulated more data, that is the main method currently employ. Tsuyoshi Tsujioka etc. also carry out experimental research in Japan. Although this method every time readout can make part of the recording medium molecular restore to before writing state, but it is very limited due to the readout power is very low, so can read a certain number of still maintain the required signal to noise ratio. Some theoretical calculations and experimental results demonstrate that effectively read out number can be to 10^6 when the laser power of read out under order level of 10nW. The greatest advantages of this approach is the structure of the system is relatively simple and ease implementation, the difficulty is due to the readout power is very low, weak signal, signal to noise ratio is very small need of special signal processing systems. In addition to above non-destructive readout scheme, use of fluorescent luminescence properties read out, the nature of intramolecular locking, mid-infrared laser to read out etc non-destructive readout method are during the study and exploration.

Tsinghua University starts study of the electric lock non-destructive read out early the program in the implementation is relatively easy to achieve, focusing on the electric lock non-destructive readout of photochromic multi-wavelength storage program of experimental research, this section will introduce the theory and experimental results of the non-destructive read. Electric lock is non-destructive readout refers electrochromic properties of photochromic molecules, that if plus the positive voltage to the written data can be locked, and locked after the recording medium to read information will not damage to achieve security repeatable read out. Negative voltage is applied, the data will be locked and lock re-implement write. This method is theoretically capable of doing the three-state optical storage on the single molecule may be unlimited writing or erasable. Such as solution of diarylethene molecules has the property of the electric lock that can reversibly between the three kinds of state (state A, state B and state C as in Figure 3-47) conversion, and these three states are stable states also, and the absorption spectrum is not same. Conversion relationship between the three states is shown Figure 3-47 in detail.

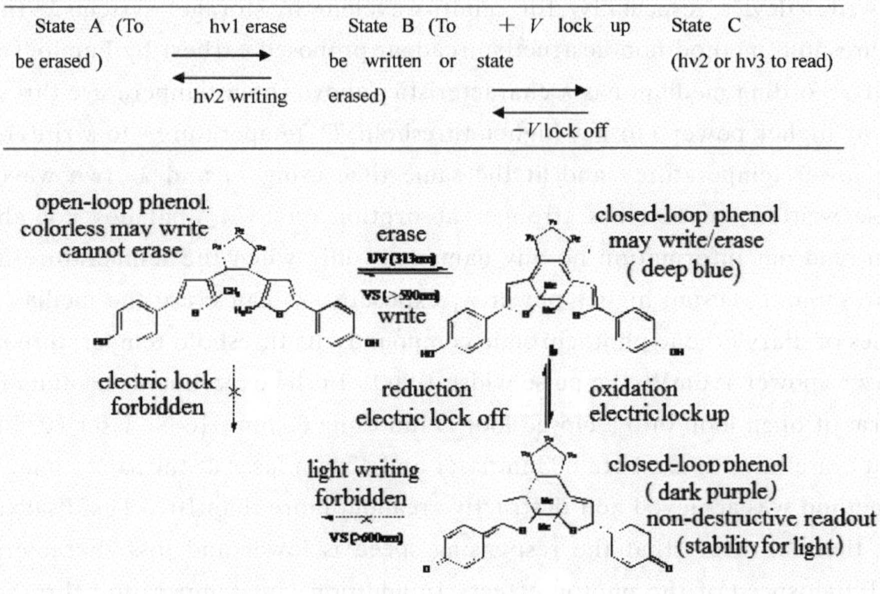

Fig. 3-47 Changing of molecular structure in solution of diarylethene at three states with light action of electric lock.

The three-state molecular structure of this material and state—state to transfer its exchange principles is shown in Figure 3-47. When UV (313~365nm) irradiation, the colorless open-loop solution of the compound state is rapidly becoming dark blue, and generates closed-loop phenol B and corresponds to absorption band of 592nm and 342nm. Closed-loop state of blue solution can return to open-loop state A and the color disappeared under visible light irradiation of wavelength larger than 510nm. A and B state of the UV—visible spectrum is shown in Figure 3-48. In order to add in the electric field, the experimental solution was mixed in the acetonitrile electrolyte. The mixing solution is put in between two transparent plate of electrode with electrode voltage of 1.5V. When solution of state A was UV radiation into the state B, the maximum absorption peak moved to 588nm and 362nm, and become to dark blue. When voltage of 1.5V add to between two transparent plate of electrode, the solution changes from blue to purple C state with maximum absorption of 548nm and 380nm, and the absorption intensity has increased as shown in Figure 3-49. The purple state C solution was long time strong exposure by wavelength is greater than 540nm, experiment result indicate that the state is very stable, the absorption spectrum does not change. Illustrate the photochromic compounds generated very high stability state C with a voltage lock, and can use greater than 510nm according to non-destructive read out.

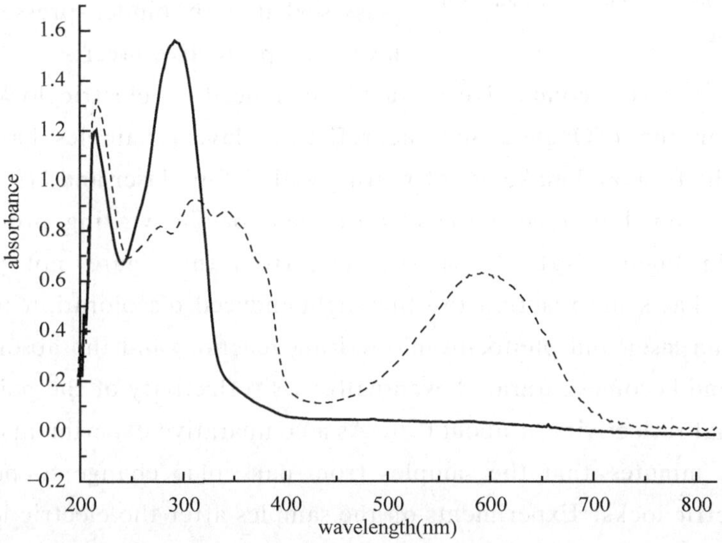

Fig. 3-48 Absorption spectra of two-state (state A: solid line, state B: dotted line) of acetonitrile.

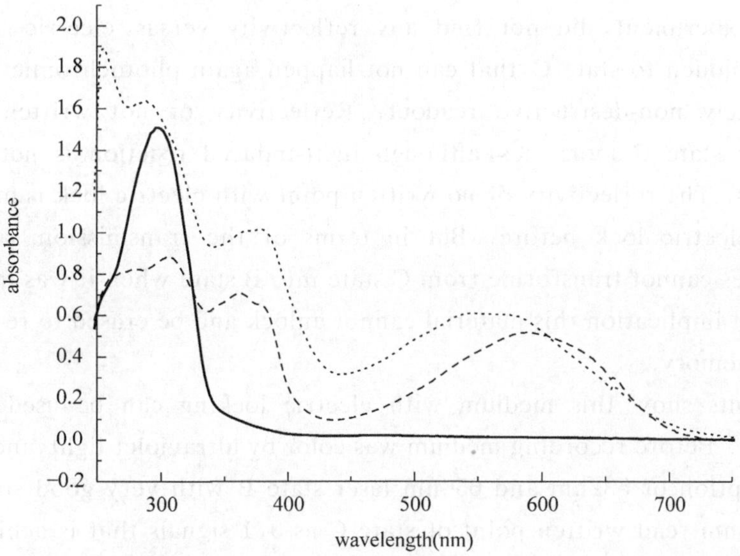

Fig. 3-49 Absorption spectra of three-state (state A: solid line, state B: dotted line and state C: broken painted line) of acetonitrile NBu_4Br in electrolyte.

Experiments in solution only verify that the principles and materials of the electric lock with non-destructive read out the possibility to really be used in the recording layer, and have to in-depth study of the materials in thin film form of electric locking performance.

A reflective store experiment samples was made of electric locks and photochromic

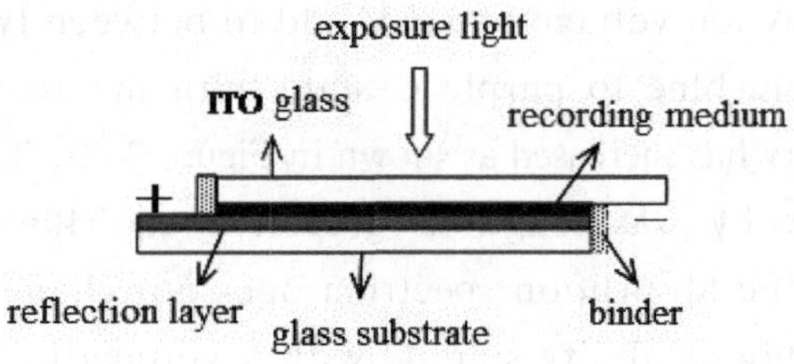

Fig. 3-50 The experiment samples of recording photochromic compounds medium with electric locks.

compounds, that the structure is shown in Figure 3-50. It is easy to write on photochromic storage experimental system experiment. The plated aluminum reflective film, special glass substrate instead of ITO conductive glass as a film base, and even spin-coated on the aluminum reflective layer containing of diarylethene the gel as the record membrane and ITO conductive glass sealed with binder pressed on the recording layer as a protective layer.

The reflective layer is a conductive surface, when need of electric lock that the electrode voltage is applied on the ITO glass and the reflective layer. Samples have to be color with ultraviolet light in the first, and make use of writing with 532nm laser and effective power 0.1mw.

Oscilloscope is applied to record the electric field of the writing process, the reflectivity changes is shown in Figure 3-51. It can be seen from the figure, not power lock before, experimental sample has same reaction to other light induced discoloration recording samples, ie the role of the 532nm laser and photochromic writing reaction, and the absorption of the 532nm laser is decreasing and becomes saturated eventually. Its reflectivity of the point changes from the initial 0.4 to the final reflectivity of about 0.9. As a comparative experiment, samples add lockout voltage 3.2V for 3 minutes, that the samples from dark blue change to purple to appear the phenomenon of electric locks. Experiments on the samples after the electric locks, using the same power wavelength of 532nm laser write on the sample another area, that the changing of reflectivity is shown in Figure 3-52. Can also be seen from Figure 3-52, the samples after electric lock for writing experiment, did not find any reflectivity versus, electric lock has taken the molecules into forbidden to state C, that can not happen again photochromic reaction, namely to achieve a completely non-destructive readout. Reflectivity of not written point was 0.41, indicating that the state C molecules, although light-induced reaction is not sensitive, but still absorb 532nm light. The reflectivity of no written point with electric lock is much lower than the final reflectivity electric lock before. But in terms of the transmission sample or reflective samples, the samples cannot transforme from C state into B state when it was imposed anti-voltage to electric lock that implication this material cannot unlock and be erased to re-write, so only to be a write one-time memory.

The experiments show this medium with electric locking can be used to non-destructive readout completely. Before recording medium was color by ultraviolet light, and into the recording state for the absorption of 532nm and 650nm laser state B with very good stability. Therefore, 532nm laser or 650nm read written point of state C as 0, 1 signals that is achieved accurate non-destructive readout. State C molecules cannot get the H atom in the film, and cannot return to the state B, as the electric lock the recording medium and cannot unlock. The absorption spectrum of photochromic compounds with the nature of the electric lock cover a wide, relatively easy to achieve matching of the semiconductor laser wavelength and industrialization, the selection of the photochromic compounds with absorption spectrum are narrow use to multi-wavelength storage more easy.

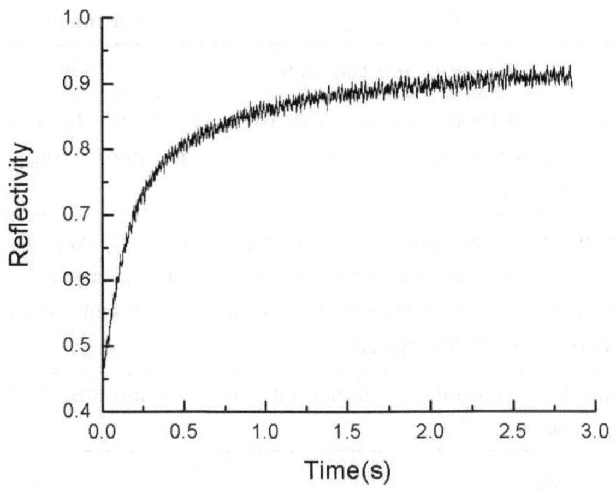

Fig. 3-51 Reflectivity of the recording photochromic compounds medium with electric lock for writing process when electric lock is off.

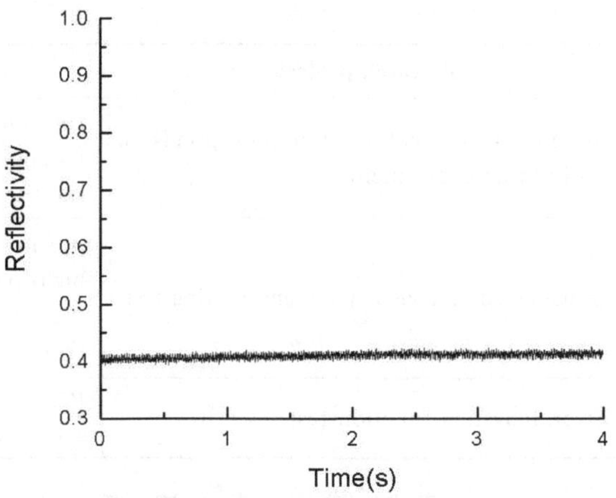

Fig. 3-52 Reflectivity of the recording photochromic compounds medium with electric lock for writing when electric lock is on.

3.5 Multi-wavelength and multi-level storage system

3.5.1 System architecture

Multi-wavelength and multi-level optical storage experimental system is a key equipment for reasearch and development of the mechanism, medium and drive design. A reasearch group of Tsinghua University has developed a dynamic experimental system is shown in Figure 3-53 (a). Functions of the experimental system are shown in Table 3-4. A static read and write system as in Figure 3-53(b) has a three-dimensional precision stage of positioning accuracy better than 50nm and three independent modulation of laser light source. It is not only application to study of multi-wavelength storage process principles, verify the theoretical model of the storage mechanism, and can be used for the test of characteristic, parameters of the recording medium, and improve the recording performance of the recording film in further engineering research and the light-induced photochromic multi-wavelength multi-degree storage practical infrastructure. The angle positioning accuracy of flotation rotary table of the dynamic experimental system is better than 0.2 seconds and displacement positioning accuracy reaches deep sub-micron to track pitch control.

Table 3-4 Main functions of the three wavelength storage experimental system.

Research projects	Features and functions
Read and write laser coupling and separation experiments	modular multi-channel beam combiner, the spectrophotometric device, the optical path adjustment mechanism
Data-parallel read and write flexible interface	coding and data written to the read out control system in parallel
Multilevel writing experiment	laser power control, the write pulse width control, and multilevel writing strategy control
Record material properties testing	read and write laser power calibration, readout of photo-digital converter, the cyclic fatigue experiments control and test data processing software
experiments of system photo-electronic specification and parameters	CCD detection and signal processing systems, read and write of the test light intensity

continued

Research projects	Features and functions
System mechanical motion and precision positioning experiment	four-dimensional closed-loop control precision work platforms: mobile mechanical positioning accuracy of ±0.1 micron angle positioning accuracy of ±0.2 seconds
Dynamic continuous reading and writing test	programmable laser pulse modulation cycle coding experimental control, high-speed read and write data processing systems and high precision of the experimental piece or the wait-free complex Road CD repeatability of positioning servo complex control system
Overall system function test	PC interface, the D/A module, a dedicated control computer and system analysis software

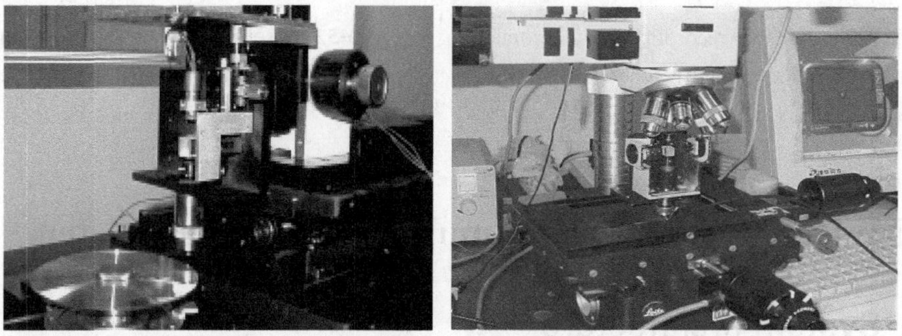

(a) (b)

Fig. 3-53 The dynamic experimental system with air bearing high speed spinning table (a) and (b) three-wavelength multi-level storage static experimental system.

Therefore, the experiment system can be used to multi-wavelength, multi-order continuous data memory write and read out the experiment on without servo track disc. On these devices to carry out the experimental study of a series of read/write signal of the channel characteristics, and selective absorption of different wavelength, mechanism different light intensity nonlinear absorption and multi-wavelength optical system design and signal processing methods for multi-wavelength and multi-level storage. Its optical system is shown in Figure 3-54 where: 1-semiconductor laser, 2-collimator, 3-formation aperture, 4-polarization beam splitter, 5-1/4 wave plate, 6-collimating lens, 7-cylindrical lens, 8-narrowband filter, 9-detector, 10-beam combiner/splitter, 11-transflective mirror-imaging lens, 13-detectors of CCD, 14-achromatic objective lens, 15-the objective lens of pickup, 16-disc or experimental samples. The static experimental system employs the 532nm, 650nm, and 780nm lasers. But the dynamic experimental system uses 405nm, 532nm and 650nm lasers.

Both optical systems adopt the new division and optical components and multilayer dielectric optical thin-film structure, that not only has a highly efficient of multi-wavelength laser beam modulation, synthesis, collimating, shaping and beam functions, while achieving the different wavelengths of light laser. The system is equipped with wide-band apochromatic objective lens that the first order Bessel function of the number of curves is shown in Figure 3-55.

Light-induced photochromic multi-wavelength and multi-level storage experiment control system block diagram is shown in Figure 3-56. Wavelengths are 650nm, 532nm and 405nm semiconductor laser (L_1, L_2, L_3), issued the read, write laser beam coupling system combined into

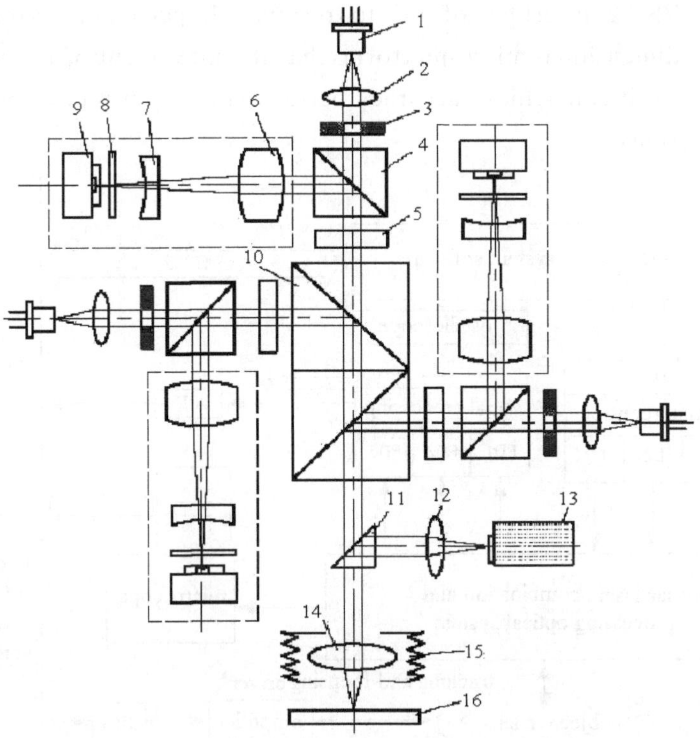

Fig. 3-54 The optical system of the multi-wavelength optical storage experimental system.

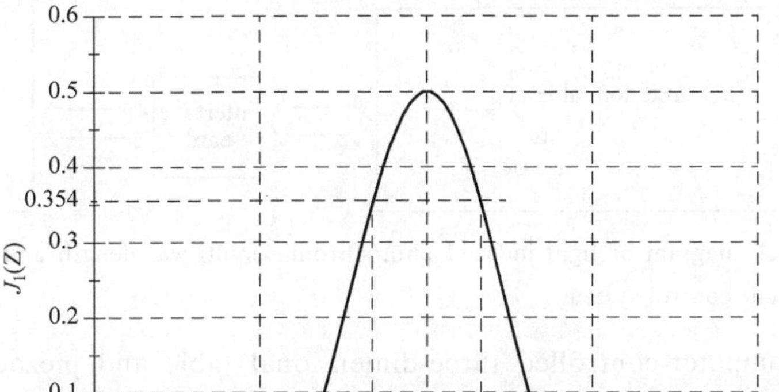

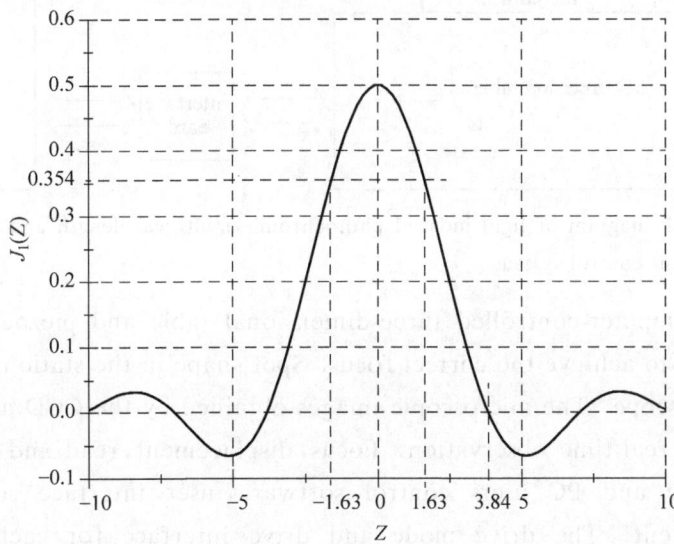

Fig. 3-55 The first order Bessel function (J_1) curves of the apochromatic objective lens for the multi-wavelength optical storage experimental system.

a bunch of coaxial multi-wavelength beam collimation, into the achromatic objective lens focus on the experimental samples or disc. Laer beams is reflected by sample reflective layer and back into the multi-wavelength coupling of the laser lens and beam splitter system, that are separated into three independent readout beam to photodetector PD_1 of PD_2 and PD_3, then the read out signals are converted into electrical signals and complete readout signal extraction. Write and read lasers power control and time control are completed by the write control module. The module consists of the 89C51 microcontroller through the D/A interface control. Microcontroller gets read and write commands from the RS232 interface of PC to receive. Experimental samples on the loading platform of the three-dimensional micro-platform, that the movement of micro platform by the PC via a control card driver, it can achieve accuracy better than 0.1μm repeatability of positioning on the X, Y, and Z directions.

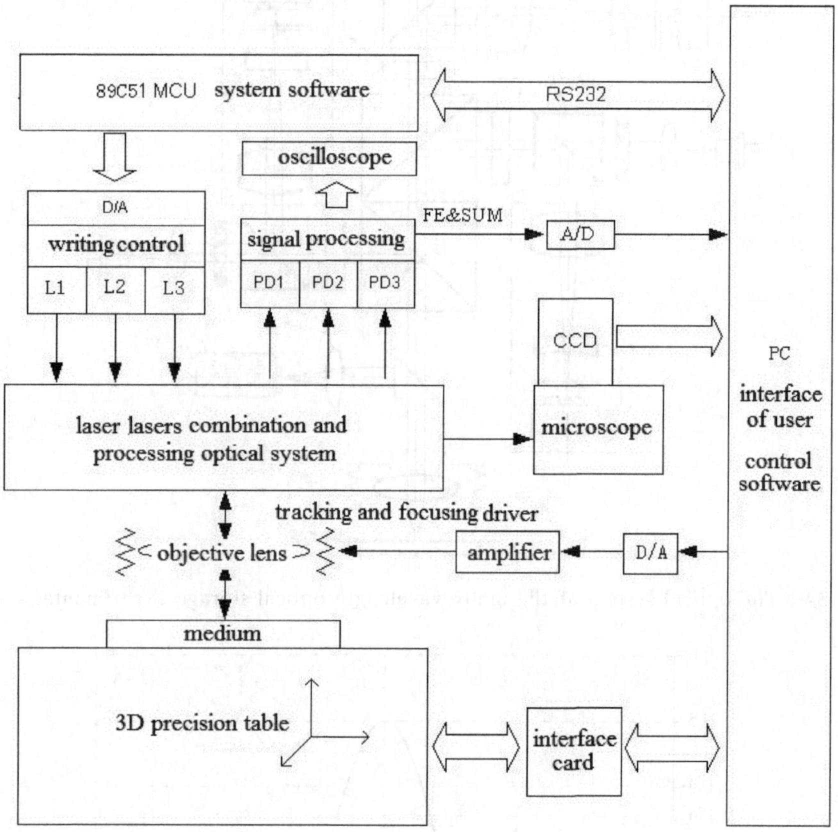

Fig. 3-56 The block diagram of light-induced photochromic multi-wavelength and multi-level storage experiment control system.

Focus control, computer-controlled three-dimensional table and piezoelectric ceramic drive head-driven objective, to achieve the correct focus. Spot shape in the static read-write experiment read out by the microscope. The microscopic images obtained by the CCD and stored by a frame grabber of the PC in real-time observation. Focus, displacement, read and write control by the SCM system software and PC users control software, user interface software, unified and coordinated management. The drive mode and drive interface for each laser with highest modulation frequency that is need of different performances and the external circuit control to finalize the parameters and system functions. The highly integrated semiconductor laser driver AD9660 and AD9661 were employed for laser read/write power control with computer simple D/A interface. The AD9660 is suitable to the driver of the N-type semiconductor laser, but the AD9661 is available for the P-type laser. The following introduce the specific characteristics and working principles of the AD9660 driver chip as an example.

The output of the AD9660 chip parameters such as shown in Figure 3-54, its internal structure and function is shown in block diagram (cf. Figure 3-57). There are three different current output, write current (IWRITE), bias current (IBIAS) and the compensation current (IOFFSET) of AD9660, by the logic control of their combination of output. The first two circuits are closed-loop control, the latter is open loop.

Table 3-5 The output parameters of the AD9660 chip.

List	Parameters	List	Parameters
Up time	⩾1.5ns	Bias current	0~90mA
Down time	⩾2.0ns	Compensation current	0~30mA
Output current	0~180mA	Switching rate	0~200MHz

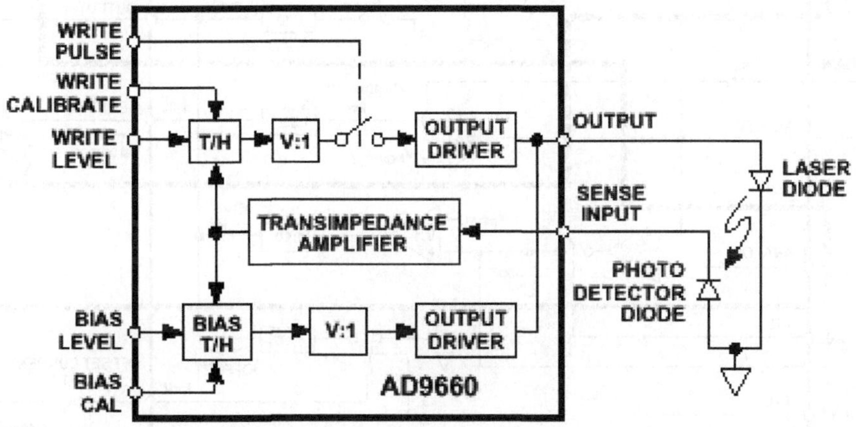

Fig. 3-57 The internal structure and function of the AD9660 in block.

Laser constant power control is achieved by periodic calibration, the timing diagram of the calibration process is shown in Figure 3-58. The power feedback signal of AD9960 is from an external photodetector, and used to control the output current of write circuit, in order to achieve the most stable power output. Pulse-driven semiconductor laser with constant power drive are very different. Semiconductor lasers are damaged very easy by working electronic current. Its control current is shown in Figure 3-59 that the drive current from 0 to instantly rise to the operating current may cause permanent damage to the semiconductor laser.

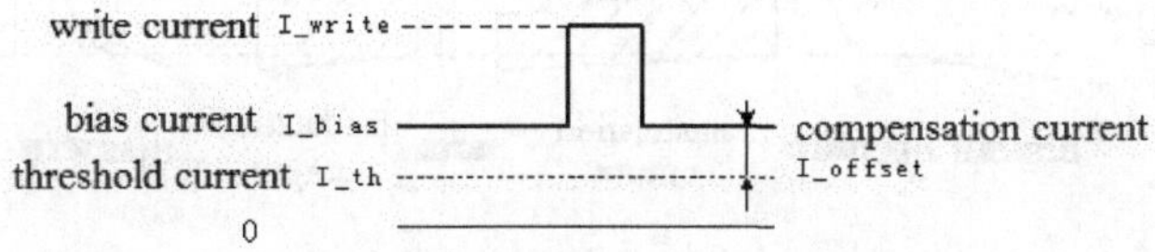

Fig. 3-58 The timing diagram of the calibration process for laser constant power control.

For continuous and single-pulse pulse frequency laser driving, requirement of slower start driving current, then rising to bias current I_{BIAS} that is greater than the threshold current I_{th} of the laser to set it into the lasing state as in Figure 3-60. The semiconductor lasers can be very steep output power of less than 1ns rise and falling edge based on the I_{BIAS}. It is appropriate bias current, write current, and the compensation current that is the key to achieve high-quality laser pulse control. The inner working principles of the main function of the AD9660 chip is described in Figure 3-59, the write-chip bias circuit (bias loop) and loop (write loop) to control the amplitude of the output current I_{out}, the detector (Photo detector) from the SENSE IN pin output of the feedback current $I_{MONITOR}$ is proportional to the input analog voltage of BIAS LEVEL and WRITE LEVEL pin.

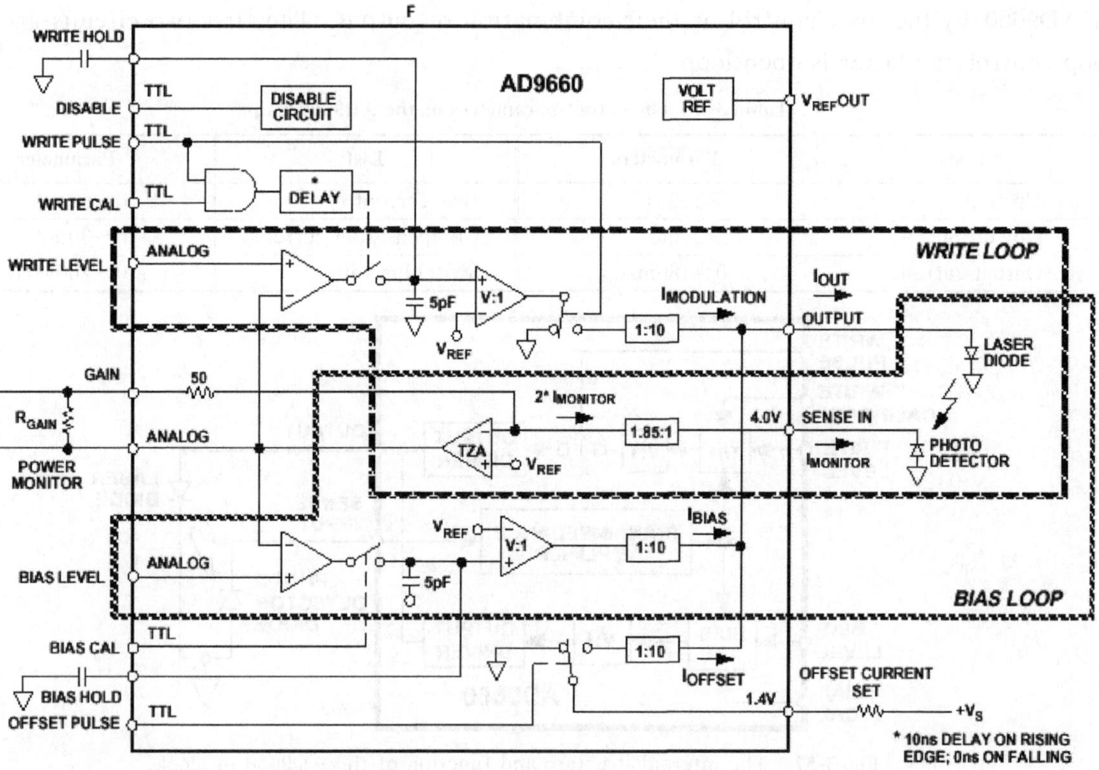

Fig. 3-59 Inner workings principles and of the main function of the AD9660 chip.

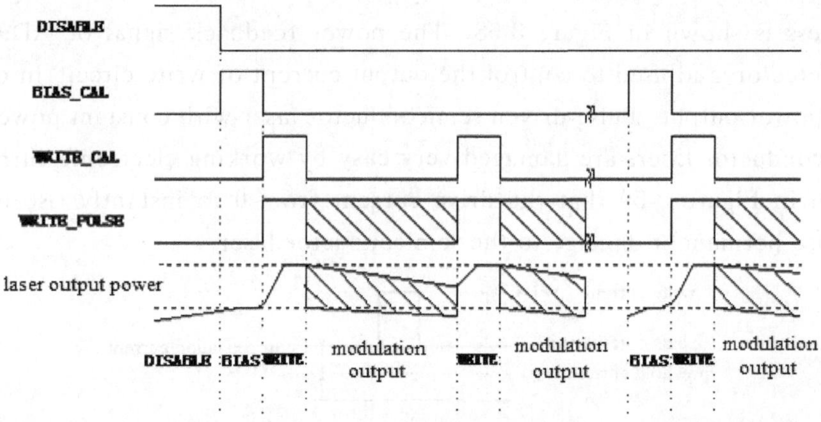

Fig. 3-60 Timing calibration process of AD9660 carried out periodic calibration.

On the other hand, the photodetector output current of the laser (LD) is proportional to the output power, therefore by changing the BIAS LEVEL, WRITE LEVEL input voltage can easily adjust the laser output power. The entire control system is periodic calibration with routing control logic to prevent the power drop caused by the laser aging. Compensation current open loop output current I_{OFFSET}, I_{OFFSET} the size can be adjusted through an external resistor. The AD9660 can achieve continuous laser output, the output of the single pulse and continuous modulation of the pulse output function.

Continuous laser output power can be adjusted to set the BIAS LEVEL pin input voltage. By a WRITE PULSE pin input, TTL or CMOS compatible pulse signal, and can output a single pulse and continuous modulation of the pulse laser, the laser pulse width modulation mode by the microcontroller through IO port control.

In order to ensure the stability of output power, the AD9660 in each start and course of their work should be carried out periodic calibration. Timing calibration process is shown in Figure 3-60, when the DISABLE signal goes low, the first calibration of the bias current is about to bias circuit is closed, the light emitted by the laser LD power detector PD detection, the current-voltage converter a feedback voltage, compared to the reference voltage VREF, the control BIAS_HOLD pin voltage reaches a predetermined value, actual calibration accuracy to be achieved and maintained.

Write current calibration methods and the bias current calibration is similar to the two loops are open loop, the actual output of laser modulation. When the modulation frequency is higher, due to time lag, closed-loop control cannot be good to play a role, so periodic calibration, both to improve accuracy without compromising speed. The AD9660's calibration is divided into two, one is to initialize the calibration (the Initial Calibration), which is a repeat calibration (Recalibration) longer than the former need the latter. Calibration and laser waveform is controlled by TTL level, and therefore it is easy through the computer I/O ports to achieve. The size of the write power and bias power the through the pin WRITE_LEVEL and BIAS_LEVEL are controled by analog voltage compensation power and by adjusting the resistance value on the OFFSET_CURRENT_SET pin. Microcontroller through the I/O ports control the D/A conversion device to generate a certain level of analog voltage, this voltage through the circuit after sent to the AD9660 input, thus completing the write, and the laser bias power control. Compensation power control is adjusted by the potentiometer. The AD9660 digital control signal is the time to read and write control, unified and coordinated by pulse width control circuit and the microcontroller. Read out the signal processing module with four-quadrant photodiode (PD) as a detecting element PD signal detected by the amplification processing to generate read signals and focus error signal.

Therefore, to read out the signal processing module is to ensure the normal read and focus key to accurate. Photodetector under reverse bias conditions, as shown in Figure 3-61, the PN junction reverse bias, the P-type region is equivalent to the base region, the N-type region is equivalent to the collector district, from the light photoinduced carrier caused by the current of the N region changes the output signal.

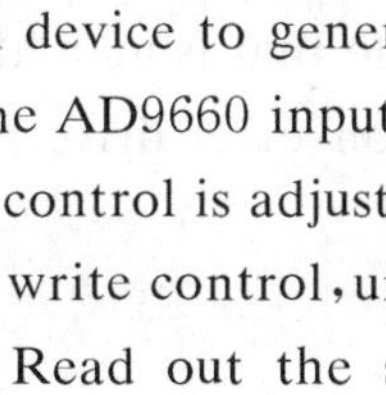

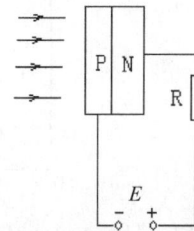

Fig. 3-61 Photodetector under reverse bias conditions.

This photodiode diode PN junction photovoltaic effect of the optical signals into electrical signals multiple devices, the working principle is similar to the transistor, which is made of silicon or germanium. The dark current of the silicon material, the temperature coefficient are small, and the control process is easy, so the most frequently devices employ the silicon photodiode.

The relative curves between sensitivity S and the incident wavelength λ of the germanium and silicon photodiodes is shown in Figure 3-62. Germanium photodiode spectral range is between about $0.4 \sim 1.8\mu m$, the peak wavelength is about $1.4 \sim 1.5\mu m$. Silicon photodiode spectral range is between $0.4 \sim 1.2\mu m$ approximately, its peak was wavelength of $0.8 \sim 0.9\mu m$. This experimental system to detect the laser wavelength range of $532 \sim 780$nm, selected the most appropriate silicon photodiode as the detector elements. The property of volta-current of the photodiode is shown in Figure 3-63. Its output characteristics are very similar to transistor under the conditions of different conversion base current I_0, but using illuminance or luminous flux to replace base current. In addition, when the bias voltage is 0 the current is not normalized to 0, but is the corresponding short-circuit current under photovoltaic effect by light. However, similar to the transistor, the bias voltage, photodiode work in the linear region always. The four-quadrant photodetector D quadrant photoelectric signal amplifying circuit is shown in Figure 3-64 that the differencing combination of the four signals quadrants (A, B, C, D) of photoelectric detector is available use to the auto-focusing signal, i. e. UA + UB + UC + UD is read out signal SUM, but (UA + UC) − (UB + UD) is focusing signal FE. The focusing signal utilizes a differential amplifier, that can overcome the shortcomings of adjust gain may cause the changing of circuit symmetry. The system using the apochromatic objective, its depth of focus is less than 1μm. Auto-focusing control accuracy of the system should adapt to $\pm 0.5\mu$m. The 650nm, 532nm and 405nm laser beam is focused on the recording layer by the achromatic objective lens, so taken the centered 532nm laser to be automatic focusing control signal. The structure of lens of the system is more complex, that itself quality is up to 120 grams, the existing standard torque cannot drive it, adopt piezoelectric ceramic nano-displacement drive P720 with maximum displacement 100μm and resolution of 10nm, and the whole trip repeat accuracy of $\pm$20nm. The drive voltage of P720 is between −20～120V, the computer cannot directly control. Therefore, adopt the E503 amplifier module that is matching with control voltage −2～12V of computer, through the computer and the DA converter drive piezoelectric ceramic.

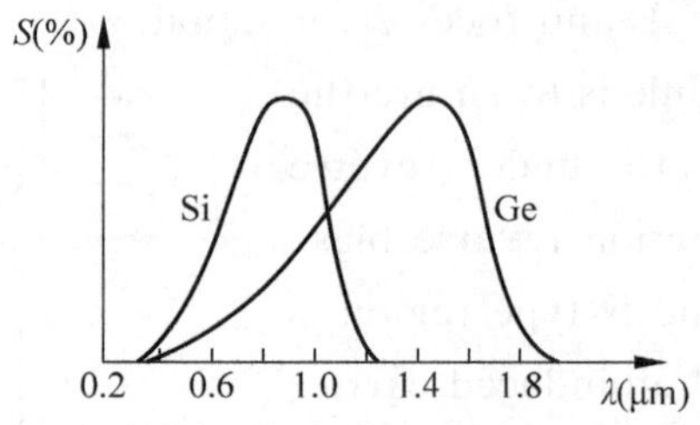

Fig. 3-62 The curves of sensitivity S and the incident wavelength λ of germanium and silicon photodiodes.

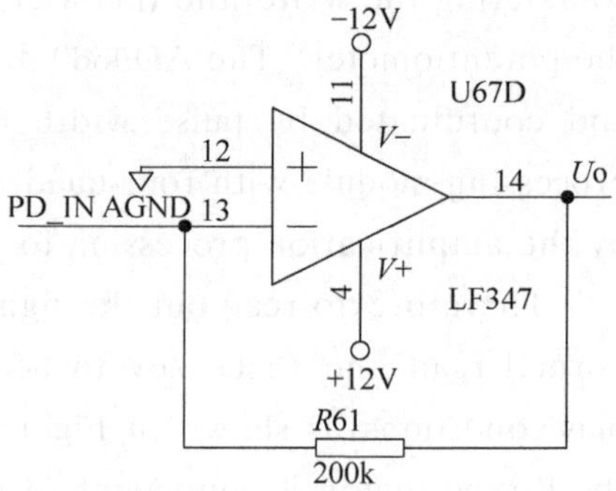

Fig. 3-63 The property of volta-current of the photodiode.

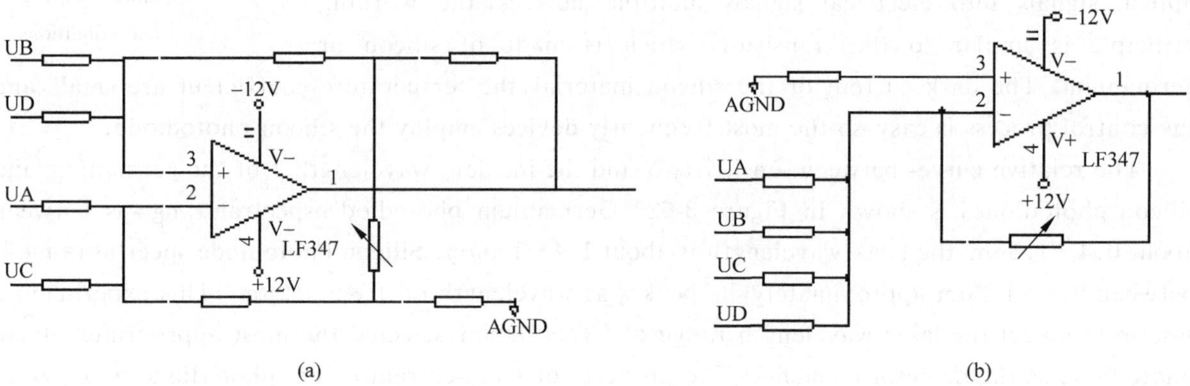

Fig. 3-64 The signal amplifying and process circuit of the four-quadrant photodetector D (a) for focusing control, (b) for read out.

The focusing of the system employ the Z-axis scanning focusing method, focusing principles is shown in Figure 3-55. Piezoelectric ceramics with apochromatic objective head is mounted on the Z axis in three-dimensional precision stage. The focusing at the beginning, by the computer control table along Z axis stepper movement 20μm, computer acquisition AD signal from the 4-quadrant photodetector and process to the minimum pressing. Computer through the DA converter determine the location of objective lens to control the piezoelectric ceramic drive stepping down within 0.01μm to keep the move close to the focus on the linear region until the focus error signal to 0. On the contrary, the objective lens to move in opposite directions to ensure that the objective has always been precisely focused on the recording layer of the experimental film. Z direction of worktable moving control frequency is about 160Hz, all the control process with computer, simply modify the program to adjust the focus accuracy and speed. This principle can be very high focusing accuracy, but the reaction rate and the working stroke are relatively small, only suitable for static or quasi-dynamic experiments. Computers can strictly control the moving distance after each focus, due to the high bench precision flatness 1μm/100mm used in specially designed glass substrate that the flatness is less than 1μm/10mm, so the system is relatively easy to implement and to ensure that the focus accuracy and the need for high speed.

- Lasers source and drive

In this experimental system, the laser driver module consists of the 89C51 microcontroller through IO port unified control. SCM can be achieved through a simple interface circuit and D/A converter links the following functions:

(1) laser power control: three-channel D/A module, respectively, the bias of the three wavelengths of the read-write laser power, write power, read power setting. And completed the periodic calibration of the read-write laser power.

(2) read and write pulse: according to the pre-set, from programming to read and write signals of different frequency and duty cycle pulse sequence for continuous reading and writing test.

(3) S232 serial communication: MCU and a PC computer through RS232 serial interface associated SCM various functional modules realized by the control PC via RS232 interface. Microcontroller interface circuit and interface protocols to ensure correct communication.

Using modular assembly language microcontroller program to write to achieve the function of each program were from a PC via RS232 interface, scalable to add capabilities as needed. The control computer using Visual C++ programming and has the following features:

(1) Through the input interface of the wavelength of laser write power, the write pulse width, to achieve control of the light source. Through the user interface intuitive set of data values of the various functional modules of the system and send various control commands.

(2) AD/DA data acquisition card programming focus error and readout and signal acquisition, control the movement of the workbench to the bench control card programming. DA channel data acquisition card to control the PZT controller to drive the piezoelectric pottery drive heads move the objective lens to achieve precise focus.

(3) The generated laser control parameters and control commands, and control of the user input data into control data for the microcontroller.

(4) Serial communication with microcontroller, and pass the control data and commands;

RS232 serial communication between PC and MCU interface protocol to send and receive data and instructions.

(5) DLL platform control, programming and working platform for all kinds of sports mode, the platform motion mode is set through a unified user interface.

The experiment system was achieved precision: the XY worktable resolution of 25nm, the turntable resolution 0.1arc-sec, the numerical aperture of apochromatic objective lens is 0.95, focusing accuracy $\pm 0.1\mu m$, CCD detector system optical magnification of 3,000X, the total magnification of 25,000X.

3.5.2 Optical channel characteristics and crosstalk analysis

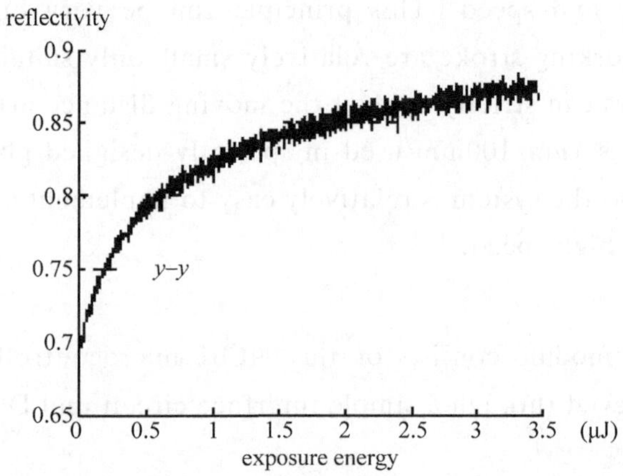

Fig. 3-65 The relationship between reflectivity and exposure energy of light-induced discoloration medium.

Photochromic medium grayscale reaction can be controlled by laser exposure and readout its signal of recording symbol by distribution light field energy and the pupil plane. According to the characteristics of the linear region of the photochromic reaction, using the superposition method and numerical calculation can obtain the optical channel readout signals of multi-level photochromic optical disc and theoretical value of its crosstalk between channels. The measuring experiment of reflectivity to the photochromic reaction process illustrated that have linear and saturation tow regions as shown in Figure 3-65. In the photochromic reaction of initial region (reflectivity under 0.8), the reflectivity is a linear process with exposure energy which was used to multi-level photochromic storage that can be considered linear superposition of the amplitude reflectivity in this region. Therefore the Fourier transform and convolution of linear superposition can be used to overlay analysis to optical channel and crosstalk analysis of photochromic multi-level optical disc.

According to linear superposition reflectivity of the various elements on light-induced discoloration disc surface overlay can create:

$$U_\sigma(u',v') = U_F(u',v') + U_{M1}(u',v') + U_{M2}(u',v') \tag{3-47}$$

Where U_σ is amplitude distribution function of reflectivity on disc surface, U_F on is CD amplitude distribution function of reflectivity with pre-groove, U_{M1} is amplitude distribution function of reflectivity on all current recorded place, U_{M2} is amplitude distribution function of reflectivity on all adjacent tracks of the disc surface. After Fourier transform of Eq. (3-47) the amplitude reflectivity of the different elements on in the pupil plane reflected field of the disc as follows:

$$\widetilde{U}_\sigma(x',y') = \widetilde{U}_F(x',y') + \widetilde{U}_{M1}(x',y') + \widetilde{U}_{M2}(x',y') \tag{3-48}$$

Where the x', y' spherical pupil normalized coordinates. U'_{M1} and U'_{M2} represent the linear superposition of all photochromic recording spots go back to the pupil. According to read out spot center offset on the $U'V'$ plane, introduction a displacement factor for each record spot. The amplitude reflectivity of the blank disc surface with servo track pre-proove can be expressed as

$$U_F(u',v')=\begin{cases} r, & |v'|\leqslant \dfrac{\tau}{2} \\ r\mathrm{e}^{-\mathrm{j}\alpha}, & \dfrac{\tau}{2}\leqslant |v'|\leqslant \dfrac{q}{2} \end{cases} \tag{3-49}$$

Where τ is the width of normalized servo pre-proove, α is differed phase of the shore and pre-proove. Consider the case $\Delta v'$ of the tracking error, the amplitude distribution on the pupil as follows:

$$S_F(x',y')=\sum_n \widetilde{U}_0\left(x',y'-\frac{n}{q}\right)\widetilde{U}_F(0,n)\mathrm{e}^{-\mathrm{j}\frac{2n\pi}{q}\Delta v'} \tag{3-50}$$

Where $U_F(m,n)$ is Fourier series of $U_F(u',v')$, $U_0(x',y')$ is amplitude distribution of readout light on pupil, and can be approximated as

$$\widetilde{U}_0(x',y')=\begin{cases} 0, & \sqrt{(x')^2+(y')^2}\geqslant 1 \\ A\mathrm{e}^{-\frac{(x')^2+(y')^2}{2\sigma}}, & \text{other} \end{cases} \tag{3-51}$$

If consider $n=0$ and $n=\pm 1$ three level diffraction superimposed only. The amplitude reflectivity of single photochromic recording spot ignoring initial reflectivity can be expressed as the following:

$$U_i(u',v')=a\Delta t\cdot U_1(u',v')U_1^*(u',v') \tag{3-52}$$

Where a is material sensitivity, Δt is exposure time, $U_1(u',v')$ is amplitude distribution on the spot of focusing surface that can be obtained by inverse Fourier transform of amplitude distribution on pupil $\widetilde{U}_1(x',y')$. Eq. (3-52) after Fourier transform becomes:

$$\widetilde{U}_i(x',y')=a\Delta t\cdot \widetilde{U}_1(x',y')\otimes \widetilde{U}_1^*(-x',-y') \tag{3-53}$$

The amplitude distribution of return to the pupil $S_i(x',y')$ equal to the convolution between single record spots in the pupil plane amplitude reflectivity $\widetilde{U}_i(x',y')$ and readout light intensity amplitude distribution $\widetilde{U}_0(x',y')$. When the recorded point center has displacement $(\Delta u',\Delta v')$ to the optical center and consider the amplitude distribution of the circular symmetry, the amplitude distribution of return to the pupil can be written:

$$S_i(x',y')=a\Delta t\cdot \widetilde{U}_0(x',y')\otimes\{[\widetilde{U}_1(x',y')\otimes \widetilde{U}_1(x',y')]\,\mathrm{e}^{-\mathrm{j}2\pi(x'\Delta u'+y'\Delta v')}\} \tag{3-54}$$

According to the superposition principles, when the recorded spot 1 and closes recorded spot 2, intensity distribution of the readout signal on the pupil is:

$$\begin{aligned} I_\sigma \propto |S_\sigma|^2 &= |S_F+S_1+S_2|^2 \\ &= (|S_F|^2+|S_1|^2+2|S_FS_1|\cos\Delta\phi_{F1})+ \\ &\quad (|S_2|^2+2|S_FS_2|\cos\Delta\phi_{F2}+2|S_1S_2|\cos\Delta\phi_{12}) \end{aligned} \tag{3-55}$$

Where S_1 and S_2 represent the complex amplitude of the recorded spot 1 and recorded spot 2. Expression in the first parenthesis is the contribution of the light intensity distribution on the record spot 1 on pupil, and expression in the second parenthesis is the crosstalk introduced by the record spot 2, $\Delta\phi_{F1}$, $\Delta\phi_{F2}$ and $\Delta\phi_{12}$ represent the phase differ between the complex amplitude. It can be seen that the composition of the crosstalk can actually be divided into the recorded spot 2 readout signal in a certain offset $(|S_2|^2+2|S_FS_2|\cos\Delta\phi_{F2})$ and the crosstalk signal between recorded spots $(2|S_1S_2|\cos\Delta\phi_{12})$.

Above theoretical analysis of reflected field of the recorded spot is suitable for the gray-scale modulation multi-level storage and photochromic multi-level optical disc to utilize for the optimization design of the different kind of multi-level storage method.

So as to multi-gray modulation storage for example, the theoretical module of pupil function is follows:

$$\overline{T}(m,n) = \delta(m,n) + \frac{2(\sqrt{g}-1)\pi R^2}{pq}E\left(2\pi R\sqrt{\left(\frac{m}{p}\right)^2+\left(\frac{n}{q}\right)^2}\right) \tag{3-56}$$

Where $E(x) = J_1(x)/x$ and $T(m,n)$ square are proportional to the intensity in the pupil, it can describe g modulation of the reflected light, i.e. the multi-level modulated signal by reflectivity. In optical disc storage system, only 0 and ±1 order diffraction spot can be returned to the pupil, so m and n are possible combination for:

$$(m=0, n=0), (m=\pm 1, n=0) \text{ and } (m=0, n=\pm 1) \tag{3-57}$$

Analysis of comparison of 532nm and 650nm photochromic materials with 8 level writing and read out experimental signal are shown in Figure 3-66 and Figure 3-67, that confirm the pupil function of modulation theoretical value and the experimental results to be correspond. As comparison, Figure 3-68 is the theoretical calculation result of effective readout signal and its components. Figure 3-69 is the theoretical calculation result of crosstalk and its components of the readout signal.

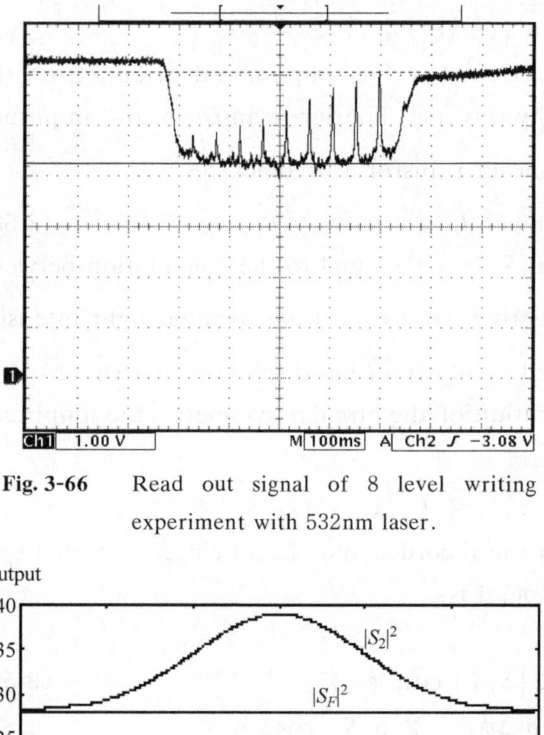

Fig. 3-66 Read out signal of 8 level writing experiment with 532nm laser.

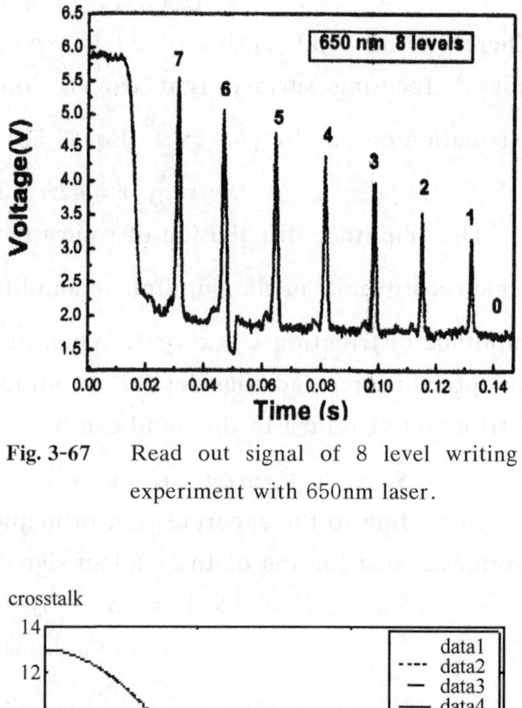

Fig. 3-67 Read out signal of 8 level writing experiment with 650nm laser.

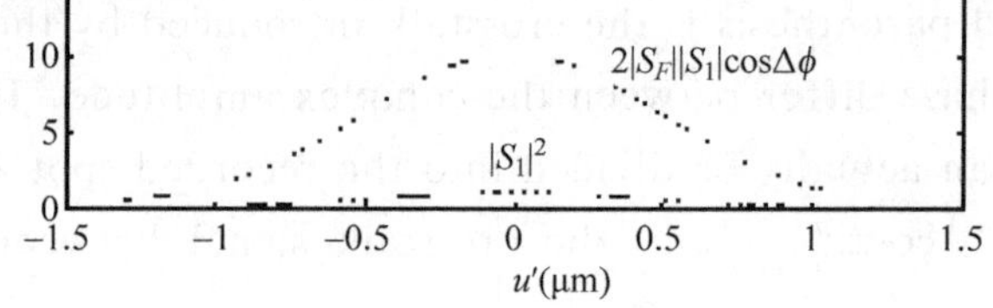

Fig. 3-68 The theoretical calculation result of effective readout signal and its components.

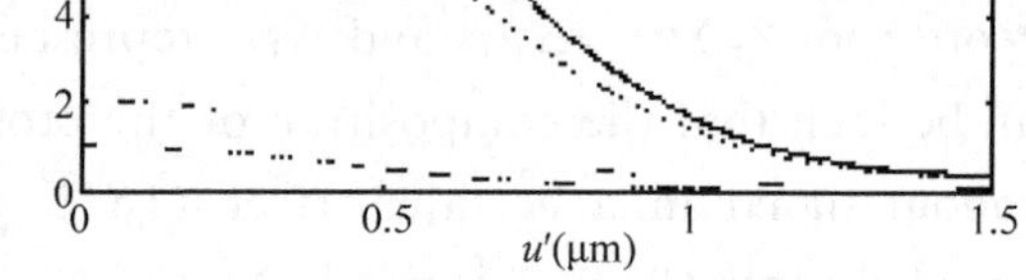

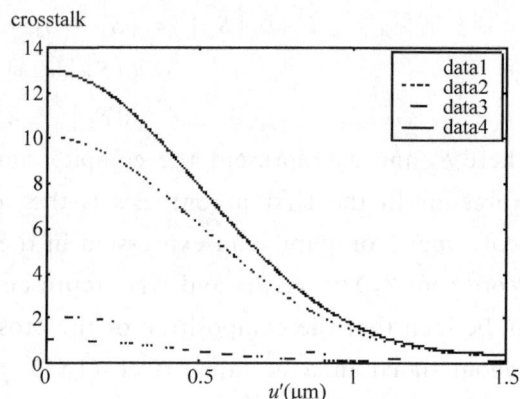

Fig. 3-69 The theoretical calculation result of crosstalk and its components of the readout signal.

• Equalization and compensation of optical channel signal

From the point of view of the channel in the actual CD-ROM storage, the same light induced photochromic multi-level optical disc optical channel compensation and balanced use of photochromic recording medium order storage experiment the compensation and balanced approach as shown in Figure 3-70.

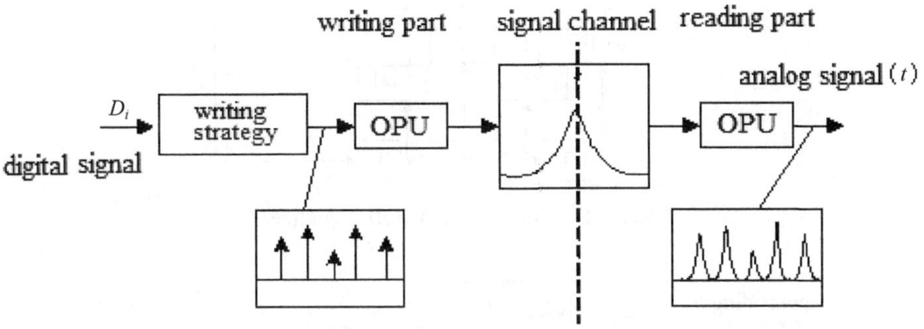

Fig. 3-70 Photochromic multi-level memory read/write channels.

The impulse response function $h(t)$ of the disc channel can be described the read out the signal returned from the optical pickup and to be expressed as

$$f(t) = a\sum_{i=-\infty}^{+\infty} D_i h(t - iT_b) \tag{3-58}$$

The establishment of normalized exposure energy of record spot 1 and spot 2 with gap T_b are P_1 and P_2 respectively, the read out signal can be described as

$$f(t) = h_{P_1}(t) + h_{P_2}(t - T_b) + P_1 P_2 \cdot H_3(T_b) \tag{3-59}$$

Where $H_3(T_b)$ is all crosstalk components with the varies spacing of the two spots except the linear superposition. In order to equalize the channel, the nonlinear factors of the channels have to be reduced to within the reception range. In optical disc storage, the quality of readout signal is often related to the recorded spot and two adjacent recorded spots only, other crosstalk of signal can be ignored, the sequence of compensation to exposure energy of the channel nonlinear $\{P_i,\}$ have to meet the Eq. (3-60):

$$f(n) = \sum_{i=n-1}^{n+1} P_i^2 H_1(nT_b - iT_b) + \sum_{i=n-1}^{n+1} P_i H_2(nT_b - iT_b) + (P_n P_{n-1} + P_n P_{n+1}) H_3(T_b)$$

$$P_i \in [P_{\min}, P_{\max}] \tag{3-60}$$

The sequence $\{P_i\}$ satisfies the given conditions of the upper and lower bounds, $f(n)$ and $f(n)$ have minimum variance, therefore, using minimum variance of the state tree write policy search method can achieve the optimized design to writing strategy. According to ensure the crosstalk of light-induced photochromic multi-level recording channel after the equalizer by the equalizer is reduced to zero, to employ zero crosstalk equalizer as shown in Figure 3-71 and partial response channel equalizer as shown in Figure 3-72.

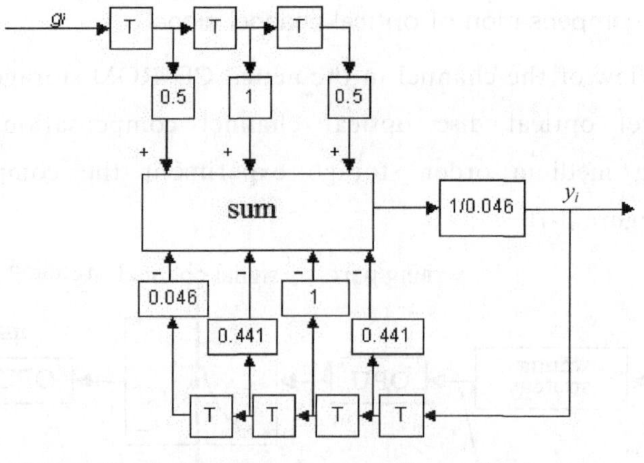

Fig. 3-71 The zero crosstalk equalizer.

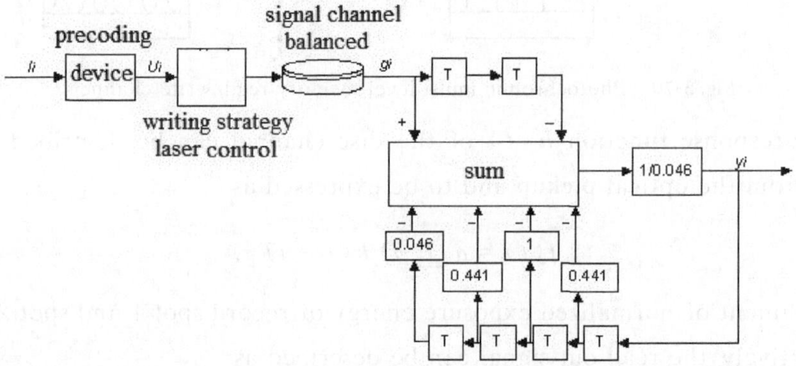

Fig. 3-72 The structure principle of partial response channel equalizer for photochromic disc.

3.6 Modulation coding and error correction

The reflectivity curve of photochromic materials has a wide linear region that is more suitable for amplitude modulation for multi-level storage. The analytical calculations theory of based on Fourier optics and optical channel diffraction theory, as well as the fast Fourier transform algorithm for traditional optical disc storage systems are suitable multi-level optical disc storage also. Korpe and Milsterl etc. proposed overlay analysis method used to physical analysis and calculation of the process directly in the optical disc system for read out, that have confirmed these analytical calculations and important reference value for the modulation, coding and error correction analysis applications in multi-level optical disc storage research.

3.6.1 Modulation coding

This section in order to analyze light-induced discoloration of the multi-level optical disc optical channel response characteristics, is still Fourier optics and overlay analysis method based on the establishment of the coordinate system using the optical principle of superposition and Fourier readout signal model of a single photochromic materials record spots as the basis of the analysis of the optical channel characteristics.

The new coordinate system as shown in Figure 3-73 shows, the set objective of the object space and space-like refractive index and the numerical aperture and the object plane and image plane coordinates are (ξ, η) (ξ', η') and transformation to Eq. (3-61) as following:

$$\begin{cases} u = (n\sin\alpha/\lambda)\xi, \quad v = (n\sin\alpha)/\eta \\ u' = (n'\sin\alpha'/\lambda)\xi', \quad v' = (n'\sin\alpha')/\eta' \end{cases} \tag{3-61}$$

This analysis of the entrance pupil and exit pupil is still using the rectangular coordinates (X, Y) and (X', Y') of position, and the distance between the optical axis and the edge light of pupil h and h' normalized coordinates that can be obtained by

$$\begin{aligned} x &= X/h, \quad y = Y/h \\ x' &= X'/h', \quad y' = Y'/h' \end{aligned} \tag{3-62}$$

In accordance above coordinate system, the coordinates of any point of object plane $Q(u, v)$ has a correspondence coordinates $Q'(u' = u, v' = v)$ on image plane. If the optical disk system to satisfy the sine condition, the point on entrance pupil $B(x, y)$ will be conjugate to point $B'(x' = x, y' = y)$ on exit pupil. Therefore can omit the symbol "′" in mathematics, the physical meaning is not difference between the entrance pupil and exit pupil.

As the laser beam is a Gaussian distribution, the complex amplitude of pupil in normalized coordinate system can be expressed as

$$u_e(x', y') = Ae^{-\frac{(x')^2+(y')^2}{2\sigma}} \mathrm{Circ}\left(\sqrt{(x')^2 + (y')^2}\right)$$

$$\mathrm{Circ}(r') = \begin{cases} 1, & r' \leqslant 1 \\ 0, & r' > 1 \end{cases} \tag{3-63}$$

where x' and y' are the coordinates on pupil sphere, for the ideal objective lens, the light is a same phase spherical wave on exit pupils as in Figure 3-73.

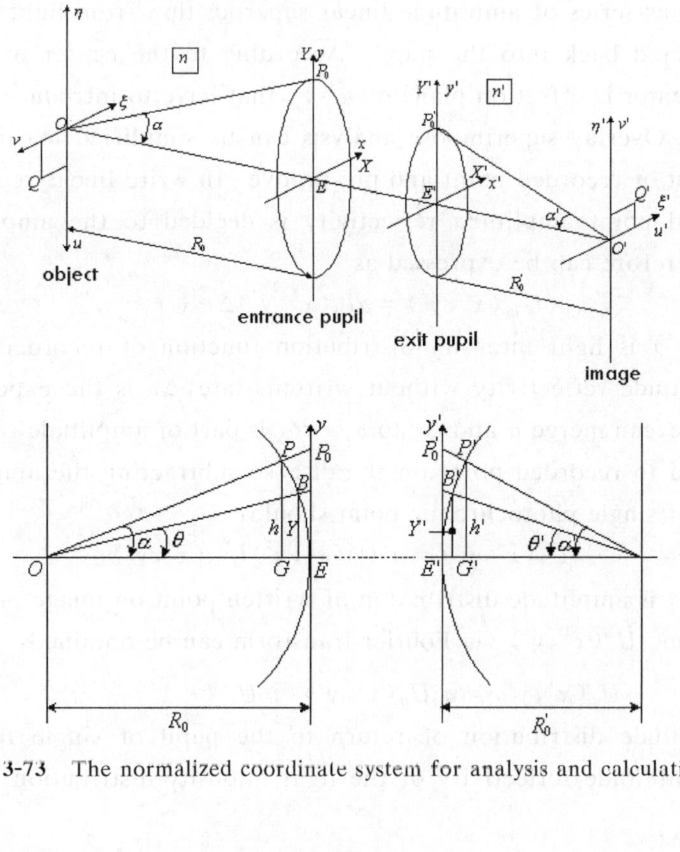

Fig. 3-73 The normalized coordinate system for analysis and calculation.

The normalized coordinate system in Figure 3-73 can be used to superposition analysis conveniently. Line overlay analysis method based on Fourier transform and convolution have a linear superposition of features, but requires the record spot must not be coincidence region merely. For the photochromic multi-level storage, recorded medium is in the linear region of the photochromic reaction, i. e. the light-induced color reaction amplitude reflectivity is a linear process with the exposure energy. Therefore, the recorded points overlap are conform the requirements of linear superposition, i. e. is not limiting the recorded point overlap for light-induced photochromic storage.

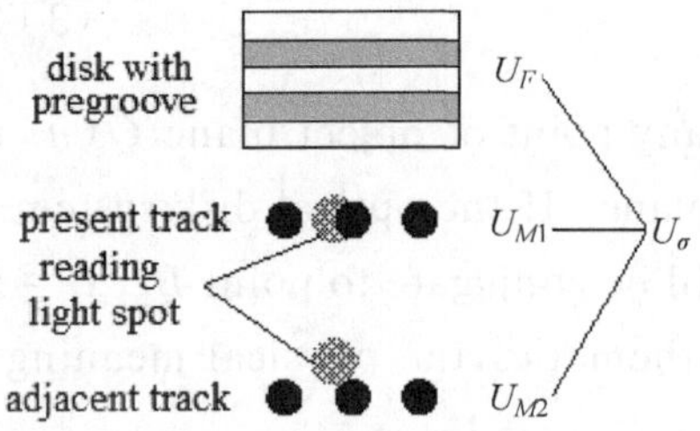

Fig. 3-74 Decomposition of the elements on disc surface.

Photochromic optical disc can be decomposed into the current track and adjacent track or pre-groove is shown in Figure 3-74. According to the amplitude reflectivity is formation superposition on the recording surface of the disc overlay can establish the equation $U_\sigma = U_F + U_{M1} + U_{M2}$. Where the U_σ is whole reflectivity distribution function on the surface of the disc, U_F is reflectivity distribution function on the disc surface with pre-proove, U_{M1} is amplitude reflectivity distribution function currently being reading recorded points in the all. Meanwhile U_{M1}, U_{M2} can be expressed the combination of reflectivity of series of photochromic recording point U_i which depends on the exposure energy U_i (integral of exposure energy in this disc surface).

The reflectivity in $U_\sigma = U_F + U_{M1} + U_{M2}$ via the Fourier transform can obtaine the amplitude reflectivity function on pupil plane of all elements of the disc $\widetilde{U}_\sigma = \widetilde{U}_F + \widetilde{U}_{M1} + \widetilde{U}_{M2}$. But $\widetilde{U}_{M1}$ and $\widetilde{U}_{M2}$ can be expressed as series of amplitude linear superposition from light induced discoloration recorded points reflected back into the pupil. According to the center of each recorded point relative to read out center is offset on plane of u'-v', that have to introduce a weighting factor to superimpose analysis. Overlay superimpose analysis can be simplified to single point diffraction study with some offset of recorded point and pre-proove. In write linear region, the light-induced discoloration recorded point amplitude reflectivity is decided to the amplitude distribution of written point, and therefore can be expressed as

$$U_{i0}(x', y') = aI_0(x', y')\Delta t + r_L \tag{3-64}$$

Where $I_0(x', y')$ is light intensity distribution function of recorded point, α is material sensitivity, r_1 is amplitude reflectivity without written state, Δt is the exposure time. If all the exposed time are same, can merge α and Δt to α_t. R_L is part of amplitude reflectivity U_F actually which is superimposed to recorded point on the disc, so subtracting the amplitude reflectivity of disc, the reflectivity of single photochromic point should:

$$U_i(x', y') = a_t U_0(x', y') U_0^*(x', y') \tag{3-65}$$

Where $U_0(x', y')$ is amplitude distribution of written point on image planet, and $U_0^*(x', y')$ as the conjugated form. $U_i(x', y')$ via Fourier transform can be obtained:

$$\widetilde{U}_i(x', y') = a_t \widetilde{U}_0(x', y') \otimes \widetilde{U}_0^*(-x', -y') \tag{3-66}$$

The light amplitude distribution of return to the pupil of single recorded point in the convolution of the amplitude reflectivity of the light intensity distribution of the surface of the pupil:

$$S_i(x,y) = b\int_{-\infty}^{+\infty}\int_{-\infty}^{+\infty}\widetilde{U}_0(x-x',y-y')\widetilde{U}_i(x',y')\mathrm{d}x'\mathrm{d}y' \tag{3-67}$$

Where b is the ratio of the readout light amplitude to writing light amplitude, according to circular symmetry of $\widetilde{U}_0(x',y')$, and therefore:

$$S_i(x,y) = C\widetilde{U}_0(x,y)\otimes\widetilde{U}_0(x,y)\otimes\widetilde{U}_0(x,y) \tag{3-68}$$

Where C is constant. The calculation of the readout signal of single recorded point was transformed to the convolution of light amplitude distribution on three pupils. The calculation of single recorded point read out signal can be simplified when use of fast Fourier transform.

The above analysis assumed the read out center coincides with the recorded point center, when the recorded point center has a displacement $(\Delta x',\Delta y')$ relative to the optical axis center, the calculation equation have to introduce a phase shift, namely:

$$S_i(x,y) = C\widetilde{U}_0(x,y)\otimes\{[\widetilde{U}_0(x,y)\otimes\widetilde{U}_0(x,y)]\mathrm{e}^{-\mathrm{j}2\pi(x\Delta x+y\Delta y)}\} \tag{3-69}$$

For disc with servo pre-groove, calculation of reflected field is shown in Figure 3-72, that is a blank disc with a servo pre-proove, its surface amplitude reflectivity can be expressed as

$$U_F(u',v') = \begin{cases} r_L, & |v'|\leqslant\dfrac{\tau}{2} \\ r_L\mathrm{e}^{-\mathrm{j}\alpha}, & \dfrac{\tau}{2}\leqslant|v'|\leqslant\dfrac{q}{2} \end{cases} \tag{3-70}$$

Where τ the normalized shore width, q is the normalized the space between pre-prooves, α is phase difference of shore and pre-prooves. Via Fourier transform can find its Fourier series:

$$\overline{U}_F(m,n) = \delta(m)\frac{r_L\sin\left(\dfrac{n\pi\tau}{q}\right)}{n\pi}(1-\mathrm{e}^{-\mathrm{j}\alpha}),\quad n\neq 0 \tag{3-71}$$

When $m=n=0$:

$$\overline{U}_F(0,0) = \frac{r_L\tau}{q}(1-\mathrm{e}^{-\mathrm{j}\alpha})+r_L\mathrm{e}^{-\mathrm{j}\alpha} \tag{3-72}$$

The light amplitude distribution on pupil:

$$S_F(x,y) = b\sum_n U_0\left(x,y-\frac{n}{q}\right)\overline{U}_F(0,n) \tag{3-73}$$

Where (x,y) is a circle which radius is (x,y) within center of $(0,0)$. $U_0(x,y)$ is the light amplitude distribution of disc image plane, b is the ratio of light amplitude between read out and writing. According to the Hopkins analysis, the diffraction spots of order n are greater than $2q$ cannot be considered. Therefore, for blank optical disc, the amplitude on the pupil is the superposition of 0 and ± 1 three diffraction spots. Such as traditional blank optical disc it is 0.9231.

If tracking error exists, and has a displacement $\Delta y'$ in the vertical direction of track that the diffraction spots on the pupil will have corresponding phase shift, and the equation of light amplitude distribution on pupil will be became:

$$S_F(x,y) = \sum_n U_0\left(x,y-\frac{n}{q}\right)\overline{U}_F(0,n)\mathrm{e}^{-\mathrm{j}\frac{2n\pi}{q}\Delta y'} \tag{3-74}$$

According to above analysis method, the readout signal along the track direction of the disc is a function of optical channel impulse response. Ideally, the light intensity distribution on the pupil of a blank light-induced discoloration disc can be expressed as

$$\begin{aligned} I_\sigma \propto |S_\sigma|^2 &= |S_F + S_i|^2 \\ &= |S_F|^2 + |S_i|^2 + 2|S_F S_i|\cos\Delta\phi \end{aligned} \tag{3-75}$$

Read out the signal can be decomposed into three components: $|S_F|^2$ is read out signal of blank disc with servo track, it is a constant always when without channel tracking error. $|S_i|^2$ is readout signal of a single light-induced discoloration recorded point that is proportional to the square of writing power. $2|S_F S_i|\cos\Delta\phi$ is the main components of read out single that is superimposed of blank disc and recorded point and modulo, as well as proportional to writing power, where $\Delta\phi$ is their phase difference. $|S_F|$ is larger than $|S_i|$ several times and almost over 10 times $|S_i|$, so it affects readout signal slightly.

As using substrate of the traditional CD disc in this experiment, so its initial reflectivity is 40%, laser wavelength is 780nm, numerical aperture of objective lens is 0.45, emitted light energy distribution coefficient on pupil σ is selected 1000, interval of track is 1.6μm, shore width is 1μm. The total readout signal and various components are calculated as shown in Figure 3-75 that is normalized with the maximum value.

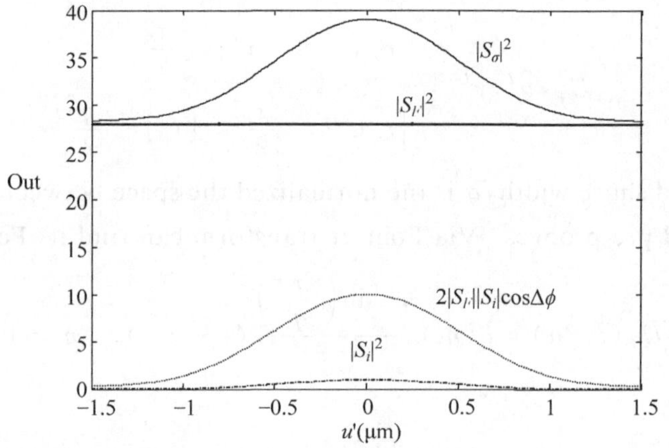

Fig. 3-75 The calculated results of the total readout signal and its various components.

In addition, the shape of $|S_i|^2$ and $2|S_F S_i|\cos\Delta\phi$ is very similar as in Figure 3-76, and similar to typical Gaussian distribution. Therefore, in analysis of the readout signal can consider that the exposure energy is approximated to read out. This calculation is static photochromic recording, but can also be equivalent to that is written with very short exposure time of dynamic recording, i.e. the writing pulse is an impulse function $\delta(t)$. In read out process, disc rotation with constant line velocity v, read out signal $|S_\sigma|^2$ is changing to $h(t)$, so it is an optical channel impulse response function also.

A comparison experiment to the theoretical analysis is following: the medium is fulgide photochromic material, exposure laser is wavelength 650nm and writing power is 12mW, 10mW and 8mW for four level recording experiment. The disc with constant linear velocity rotation, and read power is 0.5mW. The read out signal to 4 level recording are shown in Figure 3-77, as in Figure (a) is written with the 12mW, (b) for 8mW write and (c) for 8mW. Take diagrams of three typical signal (the waveform in the virtual box) compared with the corresponding theoretical calculations of the impulse response curve with its peak normalized (the direct-current component of the measured impulse response curve has been deducted) is shown in Figure 3-78.

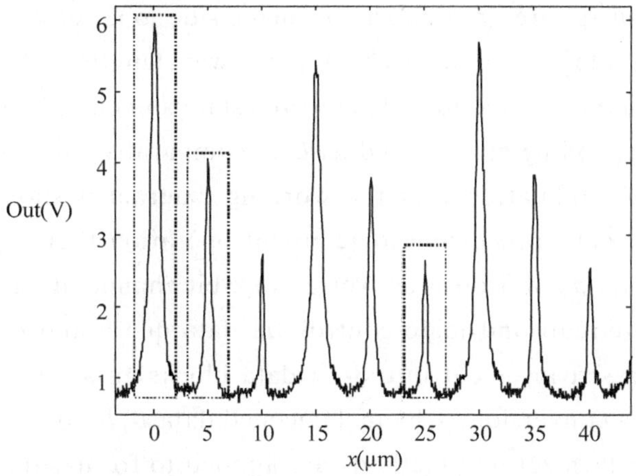

Fig. 3-76 The read out signal of 4 level recording.

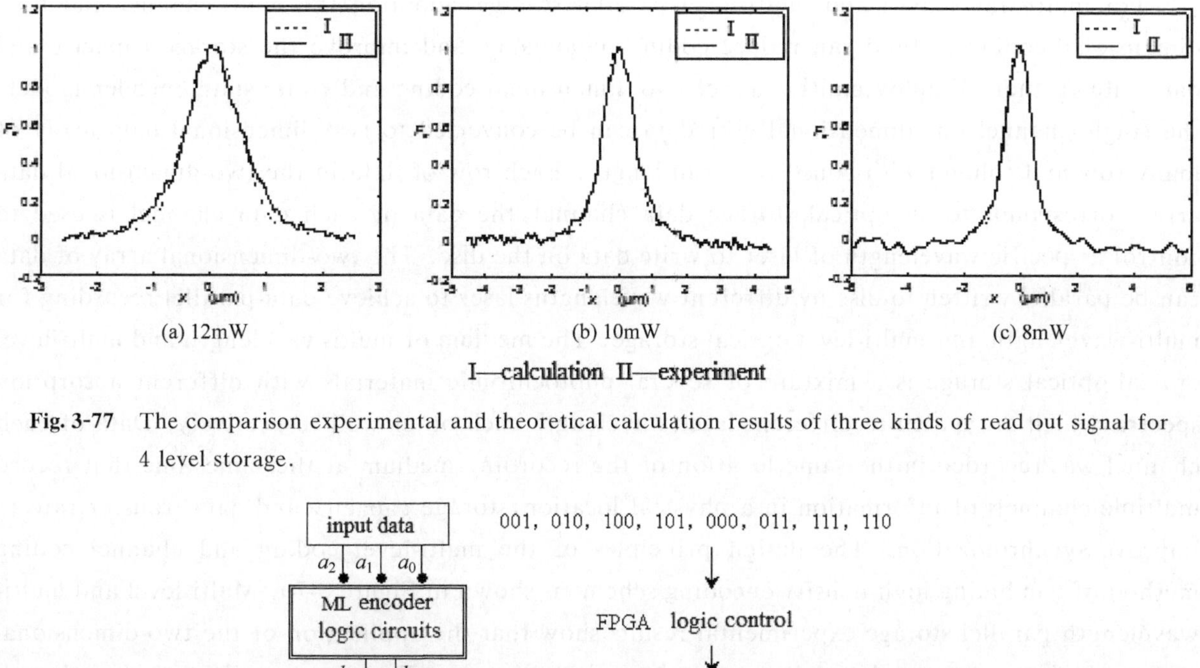

Fig. 3-77 The comparison experimental and theoretical calculation results of three kinds of read out signal for 4 level storage.

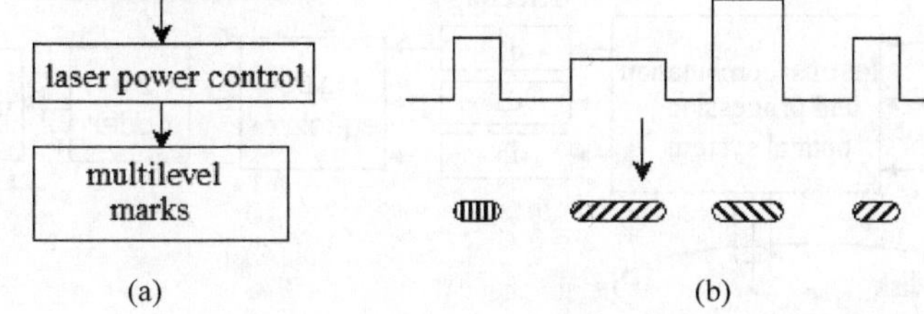

Fig. 3-78 The program to achieve the multi-level encoding and recording.

It can be seen from Figure 3-77(b) and (c) the two curves of read out signal are good agreement between experimental and theoretical calculations results. But for the high-power recording (a), the side lobe of the experimental curve is slightly higher than the theoretical value. The reason is that the larger exposure energy record, the material on recorded center is saturation, in accordance with the central area normalized to make side lobe relative enlarge.

One of the key technologies to research of multi-wavelength and multi-level storage is the modulation, coding and channel detection. Therefore establishment of a bit rate of 2/3 and 8 level (1,2) run-length limited encoding method and base on comparative analysis of run-length encoding and multi-level amplitude modulation that the working diagram is shown in Figure 3-75: multi-level encoder accepts user data, processing logic circuit operation then the output of the encoder. The encoding user data $a_2 a_1 a_0$ of 3-bit is converted to 2-bit channel data $b_1 b_0$ with 3/2 bit rate. A FPGA encoder is used to circuit and logic control via data query and conversion of coding table mapping. User input data stream is cut into 3-bit data blocks, in accordance the established code table mapping to encoding conversion. Channel encoded data $b_1 b_0$ after waveform transformation (utilized non-antagonistic turn zeroing method) are applied to format the laser power control with pulse intensity and width, that the multi-level data is written to the photochromic disc.

For multi-wavelength and multi-level parallel storage, adoption the multi-channel and two-dimensional coding method can reduce coding redundancy and improve the storage capacity and data rate further. Employ multi-channel two-dimensional coding and finite-state encoder model, the single-channel one-dimensional user data can be converted to two-dimensional data array of many row and column with constraints run length. Each row of data in the two-dimensional data array corresponds to an optical storage data channel, the data of each data channel is used to control a specific wavelength of laser to write data on the disc. The two-dimensional array of data can be parallel written to disc by different wavelengths laser to achieve data-parallel recording for multi-wavelength and multi-level optical storage. The medium of multi-wavelength and multi-level optical optical storage is a mixture of several photochromic materials with different absorption spectra, which are match to different laser wavelength each other well respectively. Data of each channel was recorded in the same location of the recording medium at the same time that record multiple channels of information in a physical location, storage capacity and data transfer rate to improve synchronization. The design principles of the multi-level coding and channel coding method of combining high-density encoding scheme is shown in Figure 3-79. Multi-level and multi-wavelength parallel storage experimental results show that the application of the two-dimensional channel detection technology, can overcome to some crosstalk between different wavelength recording materials, effective data storage capacity and recorded data reliability.

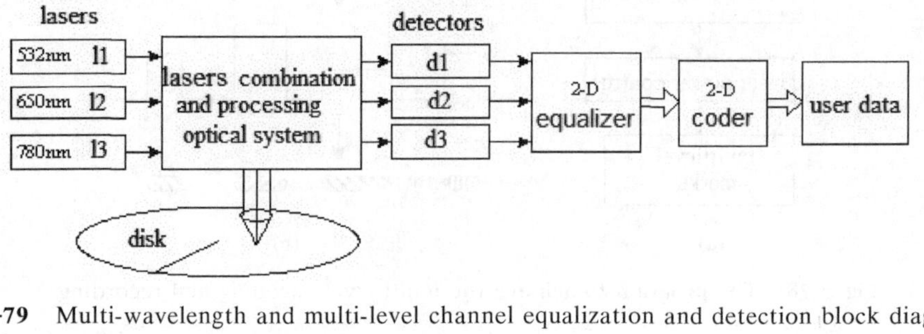

Fig. 3-79 Multi-wavelength and multi-level channel equalization and detection block diagram.

3.6.2 The error correction coding

For the computer system memory, actual error rate has to be $10^{-12} \sim 10^{-13}$. The original error rate of multi-wavelength and multi-level is about $10^{-4} \sim 10^{-5}$. Must be studied and designed for photochromic multi-wavelength and multi-level storage error correction coding scheme, error correction analysis and assessment systems. This section first discusses the principles and methods to reduce the error rate through the detection of defects and to establish a effective error correction code (ECC).

- A single-wavelength multi-level error correction coding

According to the principles of the RS encoder, for the RS code without correction, its error adjoint polynomial as

$$S(x) = S_1 + S_2 x + S_3 x^2 + \cdots + S_{n-k} x^{n-k-1} = \sum_{j=0}^{n-k-1} S_{j+1} x^j \tag{3-76}$$

Where $S_j = \sum_{i \in \varepsilon} E_i \alpha^{ij} = R(\alpha^j) = E(\alpha^j)$.

Constitute the error location polynomial:

$$\sigma(x) = \prod_{i \in \varepsilon} (x - \alpha^{-i}) \tag{3-77}$$

Error evaluation polynomial:

$$\omega(x) = \sum_{i \in \varepsilon} E_i \prod_{\substack{i \in \varepsilon \\ l \neq i}} (x - \alpha^{-l}) \tag{3-78}$$

As RS (n,k) up to only correct the $(nk)/2$ errors so that can be set $|\varepsilon| = \deg(\sigma) \leqslant (n-k)/2$, and $\deg(\omega) \leqslant |\varepsilon| - 1$ must, exist $\mu(x)$ and to meet:

$$\sigma(x) S(x) = \omega(x) + u(x) x^{n-k} \tag{3-79}$$

And can write in the following form:

$$\sigma(x) S(x) \equiv \omega(x) (\mathrm{mod}\, x^{n-k}) \tag{3-80}$$

Errors are:

$$E_i = \frac{\omega(\alpha^{-i})}{\sigma^{l}(\alpha^{-i})} \tag{3-81}$$

$$\sigma^{l}(x) = \sum_{i \in \varepsilon} \prod_{\substack{l \in \varepsilon \\ l \neq i}} (x - \alpha^{-l}) \tag{3-82}$$

$$\sigma^{l}(\alpha^{-j}) = \prod_{\substack{l \in \varepsilon \\ l \neq j}} (\alpha^{-j} - \alpha^{-l}) \tag{3-83}$$

If there are deleted error t, random error s and $s < (nkt)/2$, can use above method. The location of error is alike to an error and to be correction as set to 0. Meanwhile improvement S decoding, ε_1 to be random error domain, ε_2 to be deleted domain and ε is all error domain, i.e. $\varepsilon = \varepsilon_1 \cup \varepsilon_2$.

The random error location polynomial can write:

$$\sigma_1(x) = \prod_{i \in \varepsilon_1} (x - \alpha^{-i}) \tag{3-84}$$

Remove error location polynomial:

$$\sigma_2(x) = \prod_{i \in \varepsilon_2} (x - \alpha^{-i}) \tag{3-85}$$

All error location polynomial:

$$\sigma(x) = \prod_{i \in \varepsilon} (x - \alpha^{-i}) \tag{3-86}$$

Error valuator polynomial as follows:

$$\omega(x) = \sum_{i \in \varepsilon} E_i \prod_{\substack{i \in \varepsilon \\ l \neq i}} (x - \alpha^{-l}) \tag{3-87}$$

When $\mu(x)$ exists and to meet the

$$\sigma(x)S(x) = -\omega(x) + u(x)x^{n-k} \tag{3-88}$$

As $\omega(x)$ is obtained, $\sigma_2(x)$ and $\omega(x)$ can be known soon.

Define a maximum number is $n-k-|\varepsilon_2|$ adjoint polynomial:

$$\hat{S}(x) = \sigma_2(x)S(x) = \Big(\prod_{i \in \varepsilon_2} (x - \alpha^{-i})\Big)S(x) \tag{3-89}$$

The key Eq. (3-79) can be amended as follows:

$$\sigma_1(x)\hat{S}(x) = -\omega(x) + u(x)x^{n-k} \tag{3-90}$$

Or write:

$$\sigma_1(x)\hat{S}(x) = -\omega(x)(\mathrm{mod}x^{n-k})$$

Using the Berlekamp-Massey (BM) iterative algorithm to solving $\sigma(x)$ and $\sigma_1(x)$, then access to Eq. (3-91) can calculate the error:

$$E_i = \frac{\omega(\alpha^{-i})}{\sigma^l(\alpha^{-i})} \tag{3-91}$$

For the RS (n, k) code, when the maximum distance of $n-k+1$ code, can correct $[nk]/2$ burst errors. However, if confined to correct the error only, it can correct nk errors. To need of correct burst errors and delete errors at the same time, it can correct $2t+e \leqslant d-1$ error, where t is the burst error, e is delete errors. So using RS correct is more efficient with burst and delete error.

In the multi-wavelength and multi-level storage, as each wavelength is independent coding, so it can be considered as a M layers disc and coding error correction will appear in corresponding layer. Therefore, as long as to know the wrong location of this layer, it can be deleted by other layer, to reduce the redundancy of other layer. For example, first layer using RS(n, k) coding can correct $[nk]/2$ burst errors, but other layer of RS $(n, k+[nk]/2)$ coding can correct $[nk]/2$ errors at least. When decoding, the wrong location of first layer can be gotten in the first, that the wrong location can be corrected by other layer decoding, simply remove the incorrect decoding can do to these locations. This approach not only improves the coding efficiency and decoding speed. The M layer encoded as shown in Table 3-6.

Table 3-6 Encoding single for M wavelength.

$B_{1,1}$	$B_{1,2}$	…	$B_{1,k}$	$B_{1,k+1}$	…	$B_{1,k+[n-k]/2}$	$B_{1,k+[n-k]/2+1}$	…	$B_{1,n}$
$B_{2,1}$	$B_{2,2}$	…	$B_{2,k}$	$B_{2,k+1}$	…	$B_{2,k+[n-k]/2}$	$B_{2,k+[n-k]/2+1}$	…	$B_{2,n}$
⋮	⋮	…	⋮	⋮	…	⋮	⋮	…	⋮
$B_{M,1}$	$B_{M,2}$	…	$B_{M,k}$	$B_{M,k+1}$	…	$B_{M,k+[n-k]/2}$	$B_{M,k+[n-k]/2+1}$	…	$B_{M,n}$

In the encoding process on each respective independent encode the first layer of the RS (n, k) code, nk-byte checksum. The first $i(2 \leqslant i \leqslant M)$ layer is made of RS $(n, k+[nk]/2)$ encoding, $nk-[nk]/2$-byte checksum. Than each layer uses the RS (n, k) code, save the $M \times [nk]/2$ bytes of space, can increase the effective data capacity. And the error correction performance on each of RS (n, k) encoding is the same as. This decoding process and steps used in this experimental system is shown in Figure 3-80. But the coding method logic is very original, and suitable to basic research for multi-wavelength and multi-level coding only. For example, in the completion of the first layer decoding to get the wrong position, the layers can be used stand-alone decoder of the data at the same time to correct the error to delete the decoding, so that the decoding speed is greatly improved.

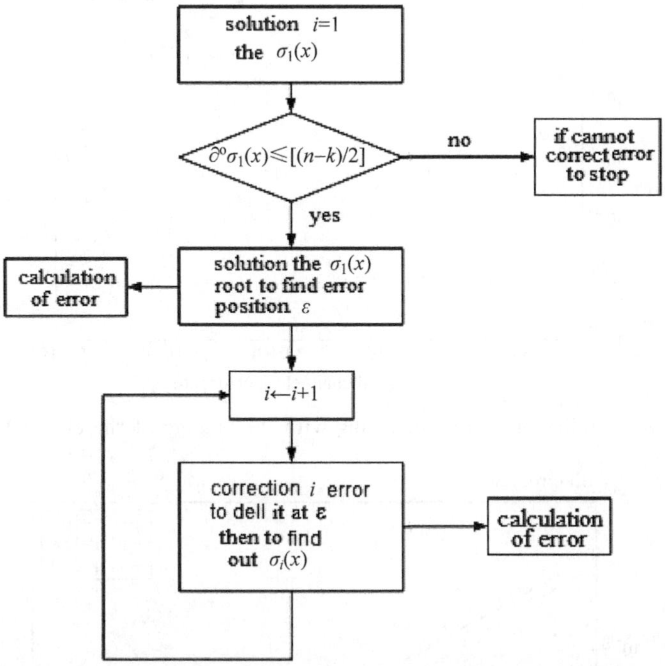

Fig. 3-80 Multi-layer decoding logic diagram.

Actually the probability of wrong coding P_{wc} includes undetected error probability P_{ud}, decoding failure probability P_{df} and error probability of decoding error probability P_{de} of three kinds. If definition decoding error probability P_{ef} is the sum of decoding failure probability P_{df} and decoding error probability P_{de}, $P_{ef} = P_{df} + P_{de}$. Undetected error probability P_{ud} refers to the error of transmission of code, ei the probability of cannot be found error by decoder. The P_{df} is that can detect error, but the number of errors exceed the correction ability of decoder. The P_{de} refers the probability of error of translated of codes after decoder received it. Therefore, the main consideration P_{ef}.

The typical formula of correct decoding probability is:

$$P_{wc} = \sum_{i=0}^{t} \binom{n}{i} p_e^i (1 - p_e)^{n-i} \tag{3-92}$$

Where p_e is the channel bit error rate, so decoding error probability:

$$P_{ef} = 1 - P_{wc} \tag{3-93}$$

For a given n, k, the probability of correct decoding with the change of the channel bit error rate is shown in Figure 3-81. The relationship of decoding error probability and $n - k$ is shown in Figure 3-82. It can be seen from this figure, for n fixed, as nk increases, the error correction performance will soon upgrade. When $n = 188$, $n - k = 16$, $P_{ef} = 6.54962\text{e} - 22$, $n - k = 10$, $P_{ef} = 5.56989\text{e} - 14$. When $n = 200$, $n - k = 10$, $P_{ef} = 8.10498\text{e} - 14$. If need of the decoding error rate is limited to a certain value that can be adjusted by selecting the n and k. For example, when $n = 200$, choose $nk \leqslant 10^{-12}$ bit error rate can be achieve.

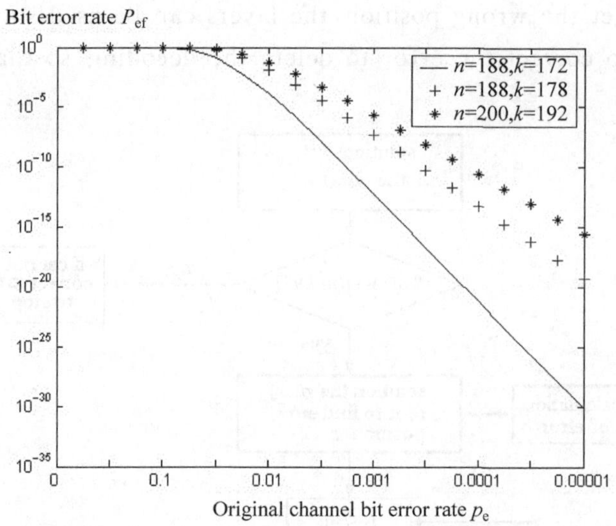

Fig. 3-81 The probability of correct decoding with the change of the channel bit error rate.

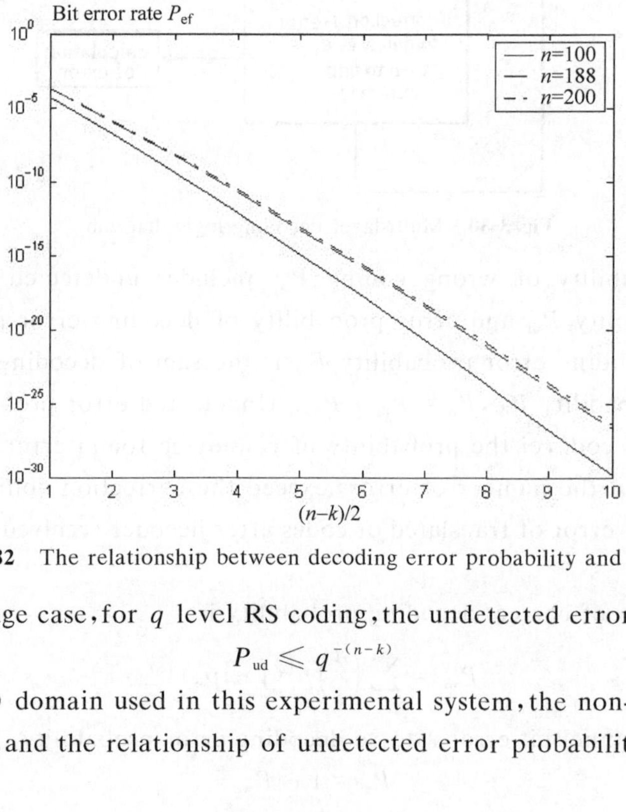

Fig. 3-82 The relationship between decoding error probability and $n-k$.

In multi-level storage case, for q level RS coding, the undetected error probability is

$$P_{ud} \leqslant q^{-(n-k)} \tag{3-94}$$

RS code in GF(28) domain used in this experimental system, the non-detection probability is indicated in Eq. (3-95), and the relationship of undetected error probability and $n-k$ indicate is shown in Figure 3-83.

$$P_{ud} \leqslant 256^{-(n-k)} \tag{3-95}$$

It can be seen from Figure 3-84, undetected error probability of rapid decline when $n-k$ increasing, i.e. $n-k=8$ undetected error probability is 5.4210e − 20, $n-k=10$ undetected error probability is 8.2718e − 25, $n-k=16$ undetected error probability is 2.9387e − 39. Undetected error probability is very small when $n-k$ is greater than a certain value.

For P_{de}, there are:

$$P_{de} = \sum_{j=0}^{n} A_j P_{de}^{j} \tag{3-96}$$

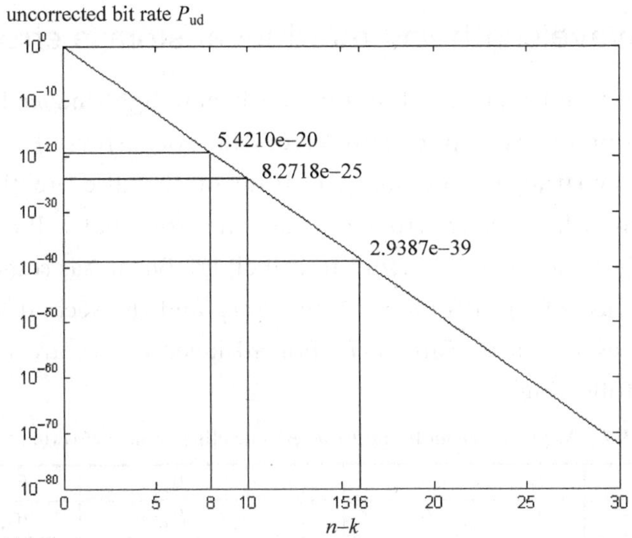

Fig. 3-83 The relationship between undetected error probability and $n-k$ in multi-level coding.

where j is number of the code:

$$A_j = \binom{n}{j}(Q-1)\sum_{i=0}^{j-(n-k+1)}(-1)^i\binom{j-1}{i}Q^{j-(n-k+1)-i} \tag{3-97}$$

As $Q=2^8=256$, for $j<n-k+1$, $A_j=0$, when the received codes are into the A_j domain, the error probability of decoder P_{de}^j is

$$P_{de}^j = \sum_{v=0}^{t}\sum_{w=0}^{t-v}\binom{n-j}{v}\binom{j}{w}(Q-1)^{w-j}\left(1-\frac{p_e}{Q-1}\right)^w(1-p_e)^{n-j-v}p_e^{j+v-w} \tag{3-98}$$

For $n=28$, $k=24$, and the bit error rate $P_{ef}\approx P_{df}$, $P_{de}\ll P_{ef}$ as shown in Figure 3-83, so just principal discussion P_{ef}.

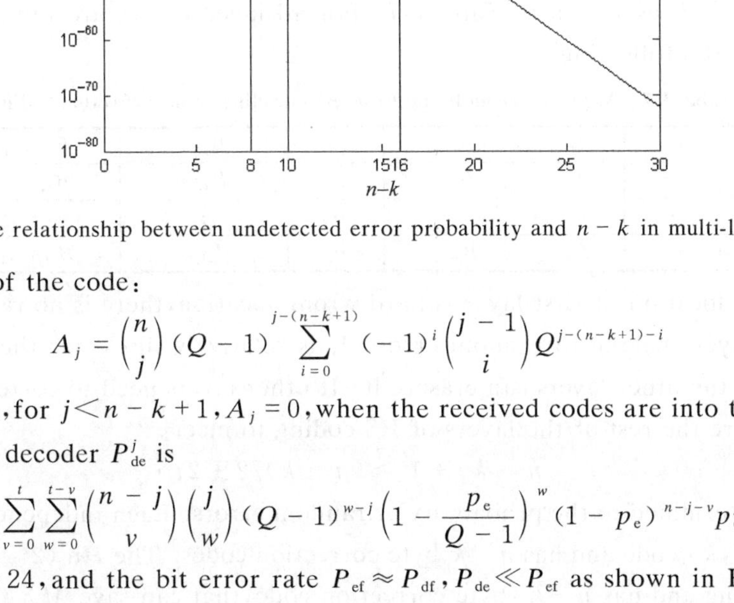

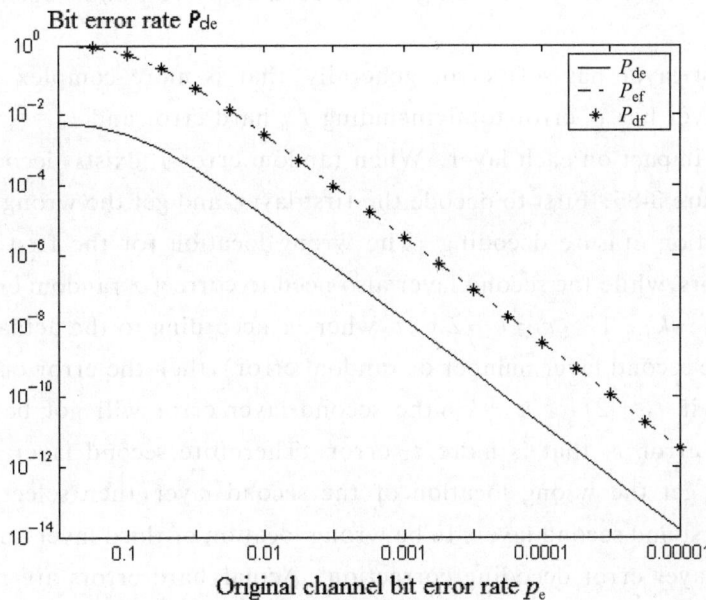

Fig. 3-84 The relationship between decoding error probability P_{de} and the bit error rate p_e.

3.6.3 Multi-wavelength and multi-level storage error code correction

For multi-wavelength and multi-level storage with new light-induced photochromic medium that is need of introduction the concept of hard errors and soft errors. Hard errors include defects of materials of medium, coating process and substrate of the disc etc that appear at the same position of the disc on the whole. Soft errors are random errors that will occur in each wavelength which is no effect to other wavelength (layer) in principle. But it has some problems also. As the first layer (wavelength) use of the RS ($n-k$) encoding, and the second layer use of RS (n, k_1) encoding, which, $k_1 > k$ as shown in Table 3-7. For all layers, there are random errors that have two possibilities as the following:

Table 3-7 Multi-wavelength (multi-level) encoding error correction coding.

$B_{1,1}$	$B_{1,2}$	…	$B_{1,k}$	$B_{1,k+1}$	…	$B_{1,k1}$	$B_{1,k1+1}$	…	$B_{1,n}$
$B_{2,1}$	$B_{2,2}$	…	$B_{2,k}$	$B_{2,k+1}$	…	$B_{2,k1}$	$B_{2,k1+1}$	…	$B_{2,n}$
⋮	⋮	…	⋮	⋮	…	⋮	⋮	…	⋮
$B_{M,1}$	$B_{M,2}$	…	$B_{M,k}$	$B_{M,k+1}$	…	$B_{M,k1}$	$B_{M,k1+1}$	…	$B_{M,n}$

(1) The error location of first layer is hard wrong location, there is no random error, and dos not affect other layers belong to random error. It is relatively disc, that the first layer gets the wrong location, so the other layers can erasure it. If other layers need to correct t random errors, which would require the rest of the layers of RS coding to meet:

$$n - k_1 + 1 \geqslant (n - k)/2 + 2t \tag{3-99}$$

Where t is determined to the probability of random errors. Each independent coding, the first layer use of RS (n, k) code and has $n-k$ byte correction codes. The ith ($2 \leqslant i \leqslant M$) layer employ RS ($n, k_1$) encoding and has $n-k_1$ byte correction code, that can save $M \times (k_1 - k)$ bytes space than each layer uses the RS (n, k) encoding too. Decoding process and steps is shown in Figure 3-85.

(2) In fact, first layer has soft error generally, that is more complex than previous case. Assuming the first layer has t_1 error total, including t_{1h} hard error and $t_s = t_1 - t_{1h}$ random error. The hard error t_{1h} is impact on each layer. When random error t_s exists, decoding process and the steps is shown in Figure 3-85. First to decode the first layer, and get the wrong position, the second layer of error correction erasure decoding. The wrong location for the first layer as the second layer of deletion errors, while the second layer also need to correct t random errors, and requeste k and k_1 to meet the $n - k_1 + 1 \geqslant [n-k]/2 + 2t$, where t according to the actual to choose. If $t_1 + 2t_2 \leqslant n - k_1$ (t_2 is the second layer number of random error), then the error of second layer can be corrected. However, if $t_1 + 2t_2 > n - k_1$, the second layer error will not be corrected. Because second layer correct error t_1 that is more t_s error. Therefore second layer has to be corrected without deleted, and get the wrong location of the second layer, then select the intersection of wrong location on first and second layers to be wrong location of third layer and correction until to completion of Mth layer error decoding correction. Actual, hard errors are much more than the soft error, that this probability is happening less. Therefore, according to the actual situation to choice t appropriately, this probability can be droped to very small. From above analysis shows that this correction decoding speed is considerably limited. So may consideration to correction of first layer error get its wrong location, then using deletion the location of error correct the rest

errors of M_{-1} layer. If the M_i layer error cannot be smooth correction, but make error correction without, and take the intersection of its error location and the first error location to be the new location of the deleted, that can increase in decoding speed appropriately.

Finally the encoding method and capability of error correction are assessed as follows:

For soft and hard errors exist, each of the error will affect the correctness of the decoding. Assume that each layer of the error rate is the same as the first layer $P_{ud}^1 \leqslant 256^{-(n-k)}$. Then undetected error probability for the other layers is $P_{ud}^2 \leqslant 256^{-(n-k_1)} > P_{ud}^1$, undetected error probability for non-inspection of the entire coding error probability is $P_{ud} \leqslant 256^{-(n-k_1)}$, that should be shown in Figure 3-85.

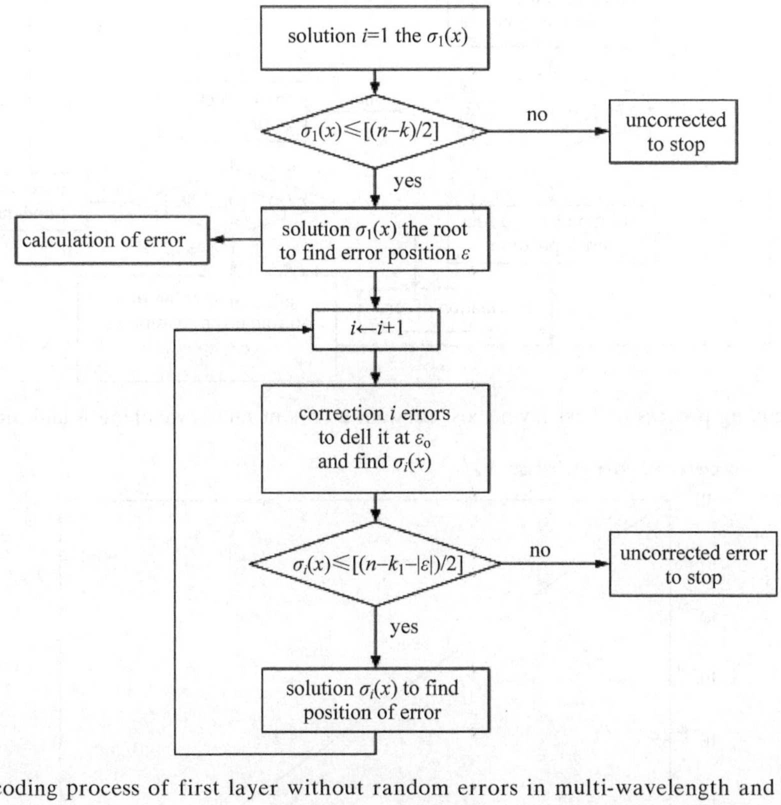

Fig. 3-85 The decoding process of first layer without random errors in multi-wavelength and multi-level storage.

If the probability of correct decoding in the first layer P_{wc}^1 is follows:

$$P_{wc}^1 = \sum_{i=0}^{t_1} \binom{n}{i} p_e^i (1 - p_e)^{n-i} \tag{3-100}$$

Where p_e is the channel error rate, and $t_1 = [(n-k)/2]$. The error decoding probability P_{ef}^1 of first layer is $P_{ef}^1 = 1 - P_{wc}^1$.

For the second layer, if using the wrong location as the deleted location, in addition to correct t random errors, the correct decoding probability $P_{wc}^{2\ 1}$ formula need to re-establish as the following:

$$P_{wc}^{21} = \frac{\Pr(0)}{\Pr} \times P_{wc}^1 + \frac{\Pr(1)}{\Pr} \times P_{wc}^1 \times \sum_{i=0}^{1} \binom{n}{i} p_e^i (1 - p_e)^{n-i} + \cdots + \frac{\Pr(2t_2 - t_1)}{\Pr} \times$$

$$P_{wc}^1 \times \sum_{i=0}^{2t_2 - t_1} \binom{n}{i} p_e^i (1 - p_e)^{n-i}$$

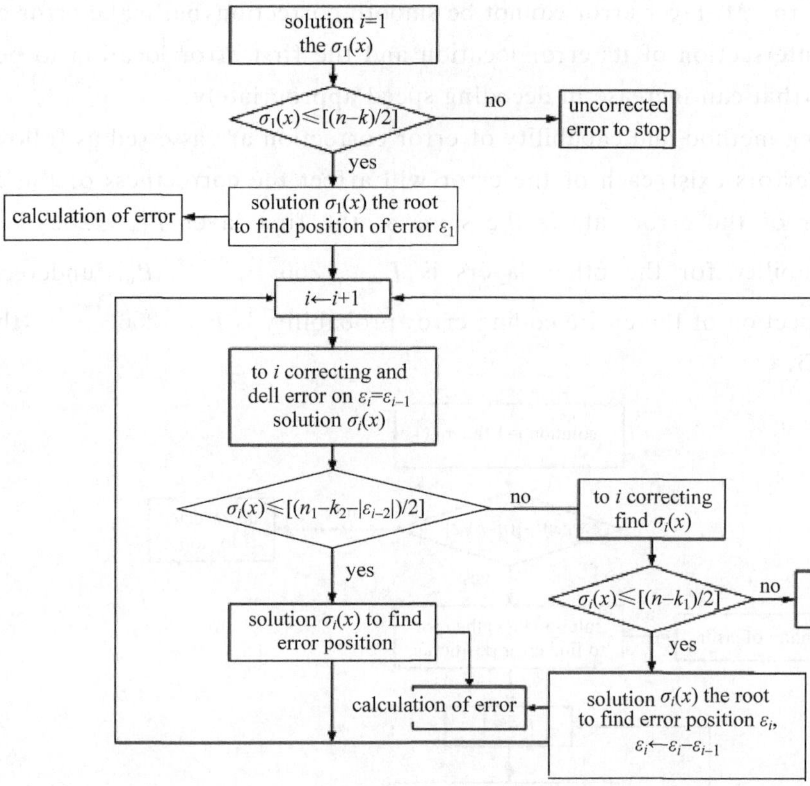

Fig. 3-86 The decoding process of first layer exist random errors in multi-wavelength and multi-level storage.

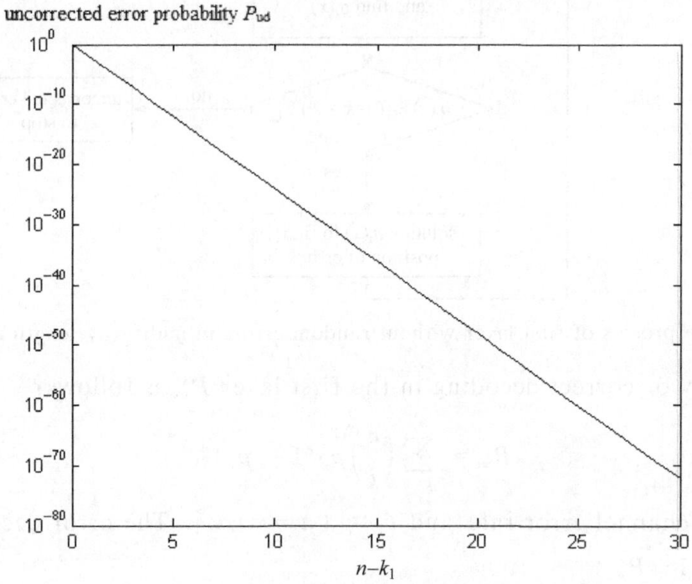

Fig. 3-87 The undetected error probability for error rate is the same of every layer.

$$P_{\mathrm{wc}}^{2\ 1} = \sum_{i=0}^{(2t_2-t_1)/2} \frac{\Pr(i)}{\Pr} P_{\mathrm{wc}}^{1} \sum_{j=0}^{i} \binom{n}{i} p_{\mathrm{e}}^{j}(1-p_{\mathrm{e}})^{n-j} \tag{3-101}$$

Where $\sum_{j=0}^{i}\binom{n}{i} p_{\mathrm{e}}^{j}(1-p_{\mathrm{e}})^{n-j}=1$ is the probability of i soft error. Pr (i) is the total probability of soft errors and $\Pr=\sum_{i=0}^{n}\Pr(i)$. If the second layer using error location of first layer as the deleted position error correction failure, the correct decoding probability is

$$P_{\mathrm{wc}}^{2\ 2} = \sum_{i=0}^{t_2} \binom{n}{i} p_{\mathrm{e}}^{i} (1 - p_{\mathrm{e}})^{n-i} \tag{3-102}$$

Therefore, the error decoding probability of the second layer:

$$P_{\mathrm{ef}}^{2} = 1 - P_{\mathrm{wc}}^{2\ 1} - (1 - P_{\mathrm{wc}}^{2\ 1}) P_{\mathrm{wc}}^{2\ 2} \tag{3-103}$$

The third layer is more complex than the second layer, and can still be divided into the following two situations:

(1) Third layer from second layer get the wrong location to delete errors and completion of the correction, the correct decoding $P_{\mathrm{wc}}^{3\ 1}$ should be:

$$P_{\mathrm{wc}}^{3\ 1} = \begin{cases} \displaystyle\sum_{i=0}^{(2t_2 - t_2')/2} \frac{\Pr(i)}{\Pr} P_{\mathrm{wc}}^{2\ 1} \sum_{j=0}^{i} \binom{n}{i} p_{\mathrm{e}}^{j} (1 - p_{\mathrm{e}})^{n-j}, & t_2' = t_1' \\ \displaystyle\sum_{i=0}^{(2t_2 - t_2')/2} \frac{\Pr(i)}{\Pr} P_{\mathrm{wc}}^{2\ 2} \sum_{j=0}^{i} \binom{n}{i} p_{\mathrm{e}}^{j} (1 - p_{\mathrm{e}})^{n-j}, & t_2' \neq t_1' \end{cases} \tag{3-104}$$

Where t_2' is the wrong location after error correction decoding of second layer. If $t_2' = t_1'$, said second layer using wrong location of first layer to be the deleted position decoding successful, its probability is $P_{\mathrm{wc}}^{2\ 2}$. If $t_2' \neq t_1'$ said second layer using wrong location of first layer to be the deleted position decoding failure, but it is decoding successful independently, the probability is $P_{\mathrm{wc}}^{2\ 2}$.

(2) If the three-story using error location of above layer for error correction fails, but successful correct decoding independently also, its probability $P_{\mathrm{wc}}^{2\ 2}$ is

$$P_{\mathrm{wc}}^{3\ 2} = \sum_{i=0}^{t_2} \binom{n}{i} p_{\mathrm{e}}^{i} (1 - p_{\mathrm{e}})^{n-i} \tag{3-105}$$

The third layer decoding error probability P_{ef}^{3}:

$$P_{\mathrm{ef}}^{3} = 1 - P_{\mathrm{wc}}^{3\ 1} - (1 - P_{\mathrm{wc}}^{3\ 1}) P_{\mathrm{wc}}^{3\ 2} \tag{3-106}$$

For L layer is similar to third layer, can be divided into two cases:

(1) No. L layer using error location of $L-1$ layer as the deleted error correction decoding is successful, the probability of correct decoding P_{wc}^{L-1} is

$$P_{\mathrm{wc}}^{L-1} = \begin{cases} \displaystyle\sum_{i=0}^{(2t_2 - t_{L-1}')/2} \frac{\Pr(i)}{\Pr} P_{\mathrm{wc}}^{L-11} \sum_{j=0}^{i} \binom{n}{i} p_{\mathrm{e}}^{j} (1 - p_{\mathrm{e}})^{n-j}, & t_{L-1}' = t_{L-2}' \\ \displaystyle\sum_{i=0}^{(2t_2 - t_{L-1}')/2} \frac{\Pr(i)}{\Pr} P_{\mathrm{wc}}^{L-12} \sum_{j=0}^{i} \binom{n}{i} p_{\mathrm{e}}^{j} (1 - p_{\mathrm{e}})^{n-j}, & t_{L-1}' \neq t_{L-2}' \end{cases} \tag{3-107}$$

Where t_{L-1}' is the wrong location obtained from first $L-1$ layer decoded. If $t_{L-1}' = t_{L-2}'$ indicated $L-1$ layer using wrong location of $L-2$ layer as delete position decoding is successful, the probability P_{wc}^{L-11}. If the $L-1$ layer using wrong location of $L-2$ layer to delete position decoding failed, but successful correct decoding independently, its probability is P_{wc}^{L-12}.

(2) If the L layer using wrong location of $L-1$ layer to delete position decoding failed, but self-correcting successfully. The correct decoding probability of the L layer P_{wc}^{L-2} should be:

$$P_{\mathrm{wc}}^{L-2} = \sum_{i=0}^{t_2} \binom{n}{i} p_{\mathrm{e}}^{i} (1 - p_{\mathrm{e}})^{n-i} \tag{3-108}$$

L layer decoding error probability P_{ef}^{L}:

$$P_{\mathrm{ef}}^{L} = 1 - P_{\mathrm{wc}}^{L-1} - (1 - P_{\mathrm{wc}}^{L-1}) P_{\mathrm{wc}}^{L-2} \tag{3-109}$$

Decoding error probability with increasing number of layers is more complicated, because the changing of wrong location t'_{L-1} are lager than delete on each layer often. So need of calculation in the decoding process, but can be estimated decoding error probability of upper and lower limits. If success of the first layer decoding, other layers are used the wrong location as the deleted position decoding success also, that is best case and the probability of this time may be $P_{wc}^{L}{}'$. This method is very convenient, and decoding error probability is the lowest. If success of the first layer decoding, but other layers use the wrong location of last layer to be delete wrong location fail, and need to carry out their own error correction decoding, which is the worst-case scenario and probability of decoding error P_{ef}^{L} is largest.

When the the correct decoding probability of first layer is follows:

$$P_{wc}^{1}{}' = \sum_{i=0}^{t_1} \binom{n}{i} p_e^i (1 - p_e)^{n-i} \tag{3-110}$$

If the second layer sues the wrong location decoding of the the first layer success, the probability of correct decoding is

$$P_{wc}^{2}{}' = \sum_{i=0}^{(2t_2-t_1)/2} \frac{\Pr(i)}{\Pr} P_{wc}^{1}{}' \sum_{j=0}^{i} \binom{n}{i} p_e^j (1 - p_e)^{n-j} \tag{3-111}$$

For L layer, all with the error location of before layer to decode success, the probability of correct decoding:

$$P_{wc}^{L}{}' = \sum_{i=0}^{(2t_2-t_1)/2} \frac{\Pr(i)}{\Pr} P_{wc}^{L-1}{}' \sum_{j=0}^{i} \binom{n}{i} p_e^j (1 - p_e)^{n-j} \tag{3-112}$$

As each layer using the the correction position decoding of before layer to decode success, its error correction capability is equivalent to the first layer of error correction capability completely. The lower limit of the decoding error rate $P_{ef}^{L}{}'$ can be estimated by the equation:

$$P_{ef}^{L}{}' = 1 - \prod_{i=1}^{L} P_{wc}^{1}{}' \tag{3-113}$$

If each floor has to sue own error correction decoding, which is the worst case, and the maximum limit $P_{ef}^{L}{}''$ is

$$P_{ef}^{L}{}'' = 1 - \prod_{i=1}^{L} P_{wc}^{i}{}^{2} \tag{3-114}$$

Where the $P_{wc}^{i}{}^{2}$ is correct probability of error correction decoding for the i layer, and which:

$$P_{wc}^{1}{}^{2} = P_{wc}^{1} \tag{3-115}$$

General multi-wavelength (layer) disc channel original bit error rate is $p_e \approx 10^{-4}$, if set $n = 188$, $k = 172$, $k_1 = 178$, then t_1, $= 8$. The probability of correct decoding is shown in Figure 3-88. It can be seen, when Pr (0) + Pr (1) is same, the probability $P_{wc}^{L}{}'$ of correct decoding is same almost. Therefore the number of soft errors the 0 and 1 in n bytes is determined to the probability of correct decoding probability $P_{wc}^{L}{}'$.

Simplify the treatment of $P_{wc}^{L}{}'$, available from Eq. (3-112):

$$P_{wc}^{L}{}' = \sum_{i=0}^{(2t_2-t_1)/2} \frac{\Pr(i)}{\Pr} P_{wc}^{L-1}{}' \sum_{j=0}^{i} \binom{n}{i} p_e^j (1 - p_e)^{n-j} \tag{3-116}$$

Normally, $2t_2 - t_1$ is very small, as $\sum_{j=0}^{i} \binom{n}{i} p_e^j (1 - p_e)^{n-j} \approx 1$ so $P_{wc}^{L}{}'$ so can approximate that:

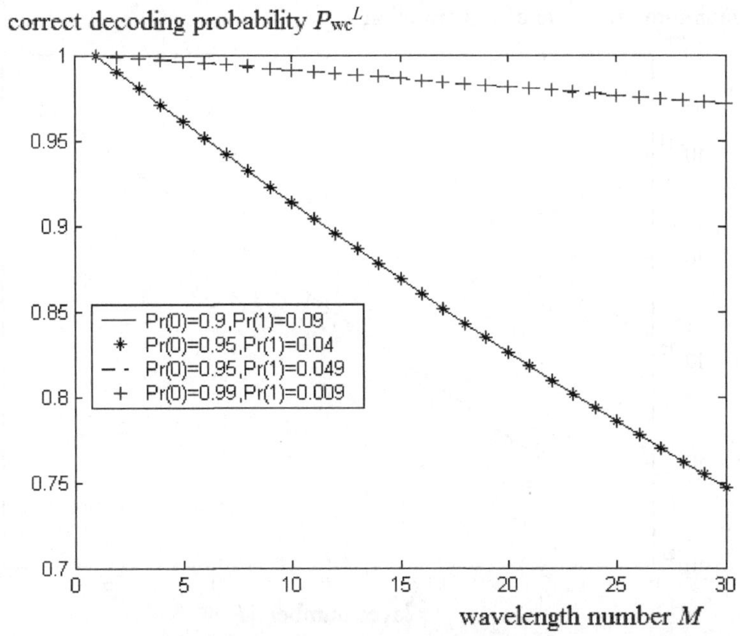

Fig. 3-88 The channel bit error rate of the multi-wavelength (layer) multi-level optical disc.

$$P_{wc}^{L\,\prime} \approx \sum_{i=0}^{(2t_2-t_1)/2} \frac{\Pr(i)}{\Pr} P_{wc}^{L-1\,\prime} = \frac{\sum_{i=0}^{(2t_2-t_1)/2} \Pr(i)}{\Pr} \times P_{wc}^{L-1\,\prime} \tag{3-117}$$

Can be seen that the sum of $(2t_2 - t_1)/2$ of the probability of soft errors determine the probability of correct decoding. When the sum of $(2t_2 - t_1)/2$ of the probability of soft errors is greater than 0.999, the probability of correct decoding is already very close to 1, as shown in Figure 3-89.

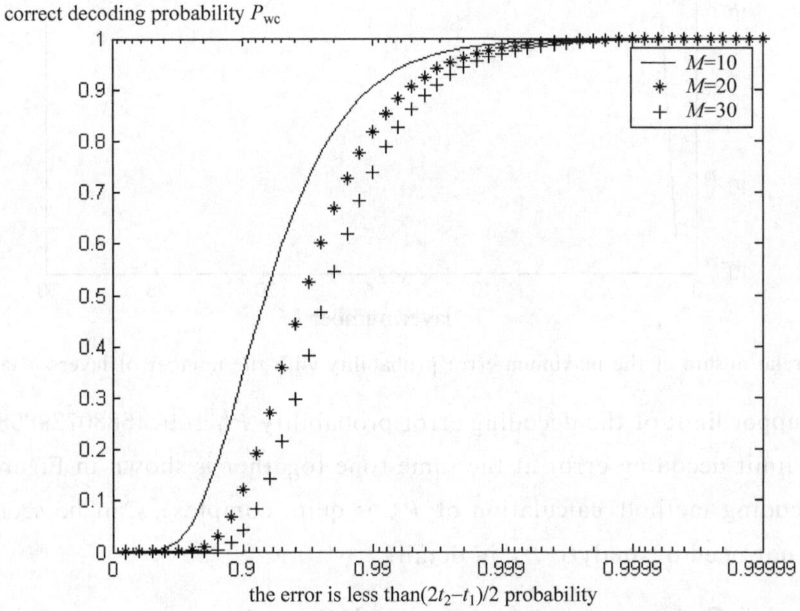

Fig. 3-89 The probability of correct decoding when sum of probability of soft errors is not greater than 1.

Decoding error probability changes with the number of layers M is shown in Figure 3-90. The curve of the maximum error probability and number of layers M is in Figure 3-91.

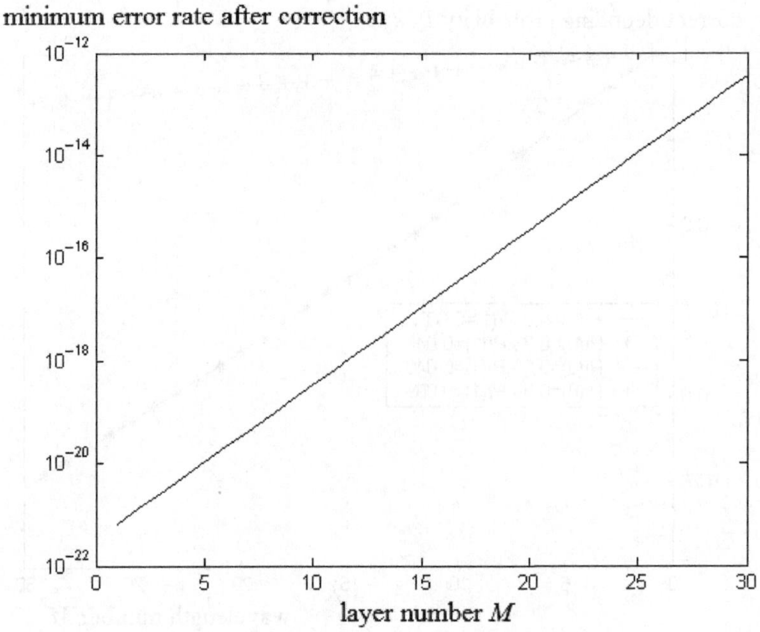

Fig. 3-90 The curve of relationship between decoding error probability and number of layers (wavelength) M.

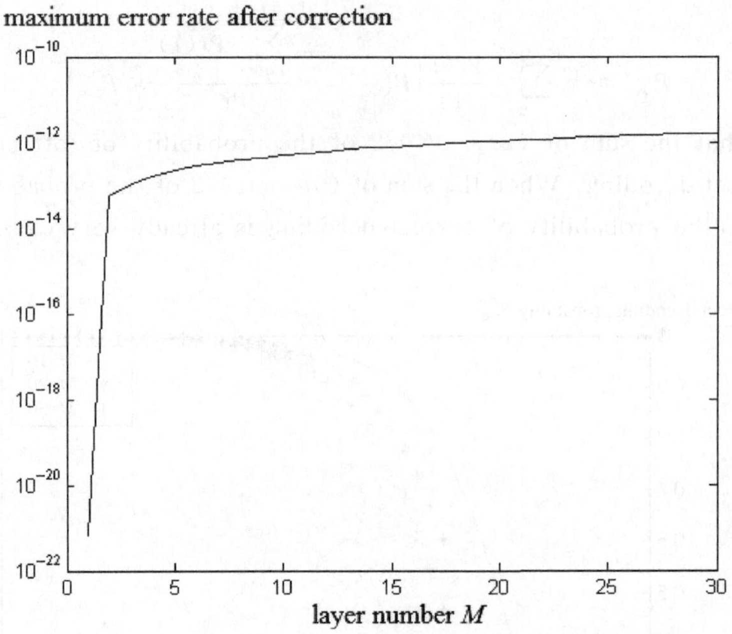

Fig. 3-91 The relationship of the maximum error probability with the number of layers M after correction.

$M=18$, the upper limit of the decoding error probability $P_{\mathrm{ef}}^{L''}=9.468807298581023\times10^{-13}$, the upper and lower limit decoding error at the same time together is shown in Figure 3-92.

In above decoding method, calculation of P_{de} is quite complex. Can be seen that $P_{\mathrm{de}}\ll P_{\mathrm{ef}}$, $P_{\mathrm{ef}}\approx P_{\mathrm{df}}$, so it is not need of analyze P_{de} in detail.

3.6.4 Reed-Solomon error-correcting code

According to the above assessment shows that this codec in the face of a continued burst error, if process of decoding stops that the data on the disc will not be able to read out. In order to

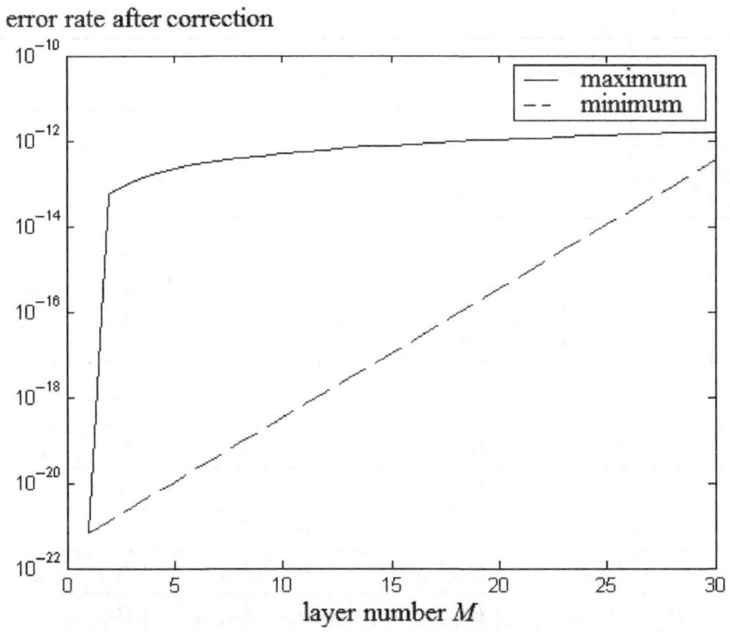

Fig. 3-92 The decoding error probability maximum and minimum when $M = 18$.

improve the performance of anti-burst error, is introduction of the Reed-Solomon error correcting code. It can correct a wide range of continuous burst error in the optical disc widely, that principles are shown in Figure 3-93. In accordance with the first data recording, data error detection code is arranged in a $(n_2 \times M) \times n_1$ ECC matrix. In order to preserve the characteristics of the error between the layers, should be in the decoding first row decoder and then decoding the column. Therefore, the column is encoded as an internal code PI in the first, then to encode the row and to be outer-parity check code PO (Outer-parity Redd-Solomon code). Column encoding of M layer has higher probability of error, in order to avoid the error number exceeds the error correction capability in the columns of error correction, using a staggered column coding, i. e. for each j take the $j + M \times i(0 \leqslant i \leqslant M - 1;\ 0 \leqslant i \leqslant n_2 - 1)$, total n_2 using RS (n_2, k_3) to be encoded. The n_2 columns are corresponding data on the same layer in different regions. Therefore, in the column encoding, on the first layer of the k_1 column of the RS $(n_2 k_3)$ encoding k_2 columns of the RS, while the remaining layers (n_2, k_3) encoding. Column coding is done after the formation of a mixed array, as shown in Table 3-9. Finally, the line of $1 + M \times i\ (0 \leqslant i \leqslant n_2)$ RS (n_1, k_1) encoding, for each $j, j + M \times i(0 \leqslant j \leqslant M;\ 0 \leqslant i \leqslant n_2)$ RS (n_1, k_2) coding gain is shown in Table 3-8 $(n_2 \times M)$ to $\times n_1$ coding matrix.

Table 3-8 The $(n_2 \times M)$ to$\times n_1$ coding matrix.

$B_{1,1,1}$	$B_{1,1,2}$	…	$B_{1,1,k1}$			
$B_{2,1,1}$	$B_{2,1,2}$	…	$B_{2,1,k1}$	$B_{2,1,k1+1}$	…	$B_{2,1,k2}$
⋮	⋮	…	⋮	⋮	…	⋮
$B_{M,1,1}$	$B_{M,1,2}$	…	$B_{M,1,k1}$	$B_{M,1,k1+1}$	…	$B_{M,1,k2}$
$B_{1,2,1}$	$B_{1,2,2}$	…	$B_{1,2,k1}$			
$B_{2,2,1}$	$B_{2,2,2}$	…	$B_{2,2,k1}$	$B_{2,2,k1+1}$	…	$B_{2,2,k2}$
⋮	⋮	…	⋮	⋮	…	⋮
$B_{M,2,1}$	$B_{M,3,2}$	…	$B_{M,2,k1}$	$B_{M,2,k1+1}$	…	$B_{M,2,k2}$
⋮	⋮	…	⋮	⋮	…	⋮

continued

$B_{1,k3,1}$	$B_{1,k3,2}$	⋯	$B_{1,k3,k1}$			
$B_{2,k3,1}$	$B_{2,k3,2}$	⋯	$B_{2,k3,k1}$	$B_{2,k3,k1+1}$	⋯	$B_{2,k3,k2}$
⋮	⋮	⋯	⋮	⋮	⋯	⋮
$B_{M,k3,1}$	$B_{M,k3,2}$	⋯	$B_{M,k3,k1}$	$B_{M,k3,k1+1}$	⋯	$B_{M,k3,k2}$
$B_{1,k3+1,1}$	$B_{1,k3+1,2}$	⋯	$B_{1,k3+1,k1}$			
⋮	⋮	⋯	⋮	⋮	⋯	⋮
$B_{M,k3+1,1}$	$B_{M,k3+1,2}$	⋯	$B_{M,k3+1,k1}$	$B_{M,k3+1,k1+1}$	⋯	$B_{M,k3+1,k2}$
⋮	⋮	⋯	⋮	⋮	⋯	⋮
$B_{1,n2,1}$	$B_{1,n2,2}$	⋯	$B_{1,n2,k1}$			
$B_{M,n2,0}$	$B_{M,n2,1}$	⋯	$B_{M,n2,k1}$	$B_{M,n2,k1+1}$	⋯	$B_{M,n2,k2}$

$B_{1,1,1}$	$B_{1,1,2}$	⋯	$B_{1,1,k1}$	$B_{1,1,k1+1}$	⋯	$B_{1,1,k2}$	$B_{1,1,k2+1}$	⋯	$B_{1,1,n1}$
$B_{2,1,1}$	$B_{2,1,2}$	⋯	$B_{2,1,k1}$	$B_{2,1,k1+1}$	⋯	$B_{2,1,k2}$	$B_{2,1,k2+1}$	⋯	$B_{2,1,n1}$
⋮	⋮	⋯	⋮	⋮	⋯	⋮	⋮	⋯	⋮
$B_{M,1,1}$	$B_{M,1,2}$	⋯	$B_{M,1,k1}$	$B_{M,1,k1+1}$	⋯	$B_{M,1,k2}$	$B_{M,1,k2+1}$	⋯	$B_{M,1,n1}$
$B_{1,2,1}$	$B_{1,2,2}$	⋯	$B_{1,2,k1}$	$B_{1,2,k1+1}$	⋯	$B_{1,2,k2}$	$B_{1,2,k2+1}$	⋯	$B_{1,2,n1}$
$B_{2,2,1}$	$B_{2,2,2}$	⋯	$B_{2,2,k1}$	$B_{2,2,k1+1}$	⋯	$B_{2,2,k2}$	$B_{2,2,k2+1}$	⋯	$B_{2,2,n1}$
⋮	⋮	⋯	⋮	⋮	⋯	⋮	⋮	⋯	⋮
$B_{M,2,1}$	$B_{M,2,2}$	⋯	$B_{M,2,k1}$	$B_{M,2,k1+1}$	⋯	$B_{M,2,k2}$	$B_{M,2,k2+1}$	⋯	$B_{M,2,n1}$
⋮	⋮	⋯	⋮	⋮	⋯	⋮	⋮	⋯	⋮
$B_{1,k3,1}$	$B_{1,k3,2}$	⋯	$B_{1,k3,k1}$	$B_{1,k3,k1+1}$	⋯	$B_{1,k3,k2}$	$B_{1,k3,k2+1}$	⋯	$B_{1,k3,n1}$
$B_{2,k3,1}$	$B_{2,k3,2}$	⋯	$B_{2,k3,k1}$	$B_{2,k3,k1+1}$	⋯	$B_{2,k3,k2}$	$B_{2,k3,k2+1}$	⋯	$B_{2,k3,n1}$
⋮	⋮	⋯	⋮	⋮	⋯	⋮	⋮	⋯	⋮
$B_{M,k3,1}$	$B_{M,k3,2}$	⋯	$B_{M,k3,k1}$	$B_{M,k3,k1+1}$	⋯	$B_{M,k3,k2}$	$B_{M,k3,k2+1}$	⋯	$B_{M,k3,n1}$
$B_{1,k3+1,1}$	$B_{1,k3+1,2}$	⋯	$B_{1,k3+1,k1}$	$B_{1,k3+1,k1+1}$	⋯	$B_{1,k3+1,k2}$	$B_{1,k3+1,k2+1}$	⋯	$B_{1,k3+1,n1}$
⋮	⋮	⋯	⋮	⋮	⋯	⋮	⋮	⋯	⋮
$B_{M,k3+1,1}$	$B_{M,k3+1,2}$	⋯	$B_{M,k3+1,k1}$	$B_{M,k3+1,k1+1}$	⋯	$B_{M,k3+1,k2}$	$B_{M,k3+1,k2+1}$	⋯	$B_{M,k3+1,n1}$
⋮	⋮	⋯	⋮	⋮	⋯	⋮	⋮	⋯	⋮
$B_{1,n2,1}$	$B_{1,n2,2}$	⋯	$B_{1,n2,k1}$	$B_{1,n2,k1+1}$	⋯	$B_{1,n2,k2}$	$B_{1,n2,k2+1}$	⋯	$B_{1,n2,n1}$
$B_{M,n2,0}$	$B_{M,n2,1}$	⋯	$B_{M,n2,k1}$	$B_{M,n2,k1+1}$	⋯	$B_{M,n2,k2}$	$B_{M,n2,k2+1}$	⋯	$B_{M,n2,n1}$

For multi-wavelength and multi-level optical storage, this encoding and decoding algorithms and processing are need to be redesigned. The new algorithm and process of encode and decode is shown in Figure 3-93.

After the above computation each row of M layer to n^2 group decoding, the outer code PO decoding process is completed. If a row cannot be decoded correctly, then all the data of the row are tag. This tag in relation to error code will be deemed to delete, when inner code PI is decoding. The decoding process of inner code PI is shown in Figure 3-94.

In previous section, while correcting coding scheme were estimated to error correction capability of soft and hard errors. After the introduction of the Reed-Solomon code encoding method, on the column direction with the RS encoder at the same time, that the corrected ability is improved greatly.

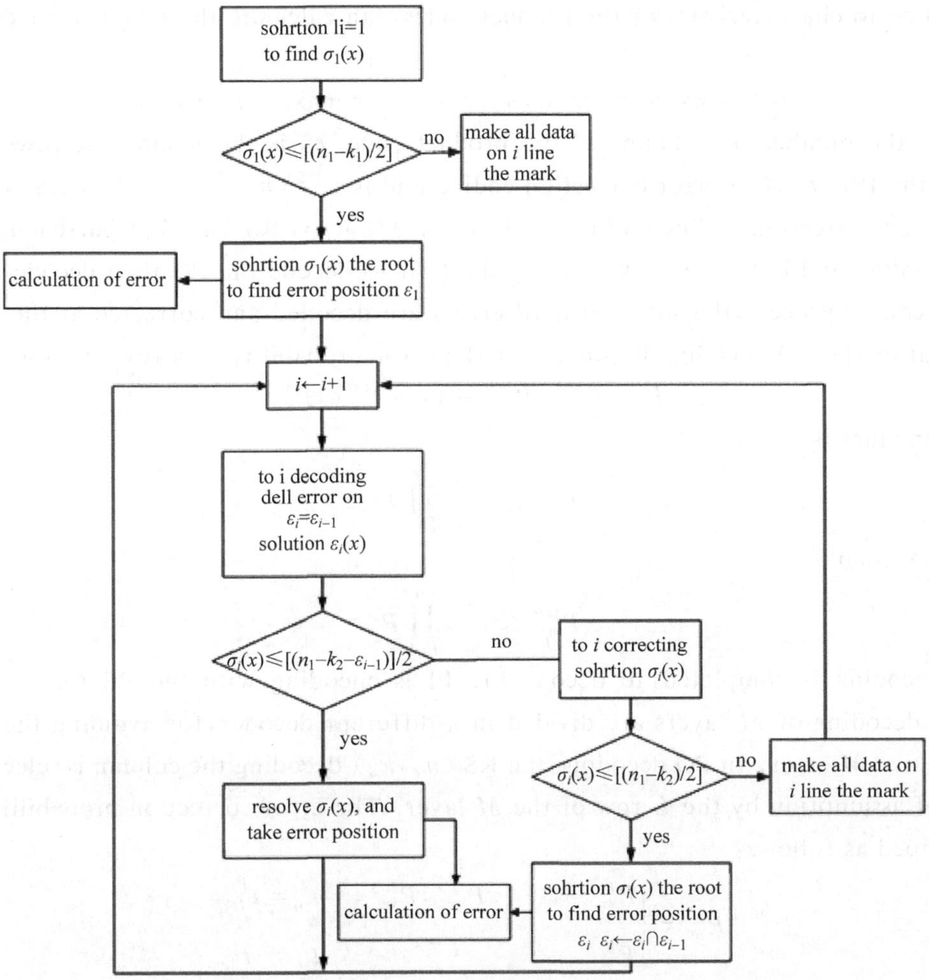

Fig. 3-93 Decoding algorithm and steps of Reed-Solomon codes for multi-wavelength q and multi-level optical storage.

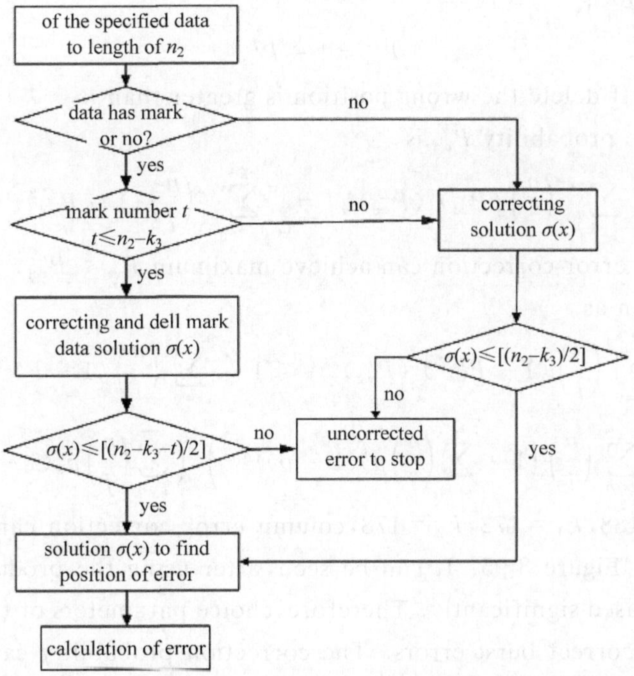

Fig. 3-94 The decoding process logic diagram for inner code.

According to characteristics of the product codes, can calculate the length b of correct burst errors is

$$b \leqslant \max(n_1 t_3, n_1 t_3, n_2 t_1, n_2 t_2) = \max(n_1 t_3, n_2 t_1) \tag{3-118}$$

Where n_1 is the number of columns of the product code, n_2 is the number of rows of product codes, t_1 is the RS (n_1, k_1) error correction coding and is $t_1 = (n_1 - k_1)/2$, t_2 is RS(n_1, k_2) burst length of error correction coding and is $t_2 = (n_1 - k_2)/2$, t_3 is RS (n_2, k_3) burst length of error correction coding and is $t_3 = (n_2 - k_3)/2$. In the first to decode the PO then decode the product code. In decoding process, the soft and hard errors are decoded and corrected at the same time. Therefore, after the PO decoding is completed, the error probability of a row of data is:

$$P_{\mathrm{ef}}^{M} \leqslant 1 - P_{\mathrm{wc}}^{M\,1} - (1 - P_{\mathrm{wc}}^{M\,1}) P_{\mathrm{wc}}^{M\,2} \tag{3-119}$$

The minimum is

$$P_{\mathrm{ef}}^{M\,\prime} \leqslant 1 - \prod_{i=1}^{M} P_{\mathrm{wc}}^{1\ \prime} \tag{3-120}$$

The maximum is

$$P_{\mathrm{ef}}^{M\,\prime\prime} \leqslant 1 - \prod_{i=1}^{M} P_{\mathrm{wc}}^{i\ 2} \tag{3-121}$$

After PO decoding is completed, to decode PI. PI is encoding with the RS (n_2, k_3), and for consecutive decoding of M layers are divided in a different decoder, for avoiding the hard error and consecutive M error. In PO decoding, the RS (n_2, k_3) decoding the column is selected n_2 rows consisting of assumption by the L row of the M layer. The error correction probability of L row P_{wc}^{L} is described as follows:

$$P_{\mathrm{wc}}^{L} = \begin{cases} P_{\mathrm{wc}}^{L\,1} + (1 - P_{\mathrm{wc}}^{L\,1}) P_{\mathrm{wc}}^{L\,2}, & t'_L = t'_{L-1} \\ P_{\mathrm{wc}}^{L\,2}, & t'_L \neq t'_{L-1} \end{cases} \tag{3-122}$$

If the L row decoding fails, plus a label on each error on the row, as the column correction to delete, the probability P_{cd}^{L} is

$$P_{\mathrm{cd}}^{L} = 1 - P_{\mathrm{wc}}^{L\,2} \tag{3-123}$$

It can be seen that if delete the wrong position is greater than $n_2 - k_3$ in the election n_2 rows, and can not correct, the probability P'_{cef} is

$$P'_{\mathrm{cef}} \leqslant \sum_{i=n_2-k_3-1}^{n} \binom{n}{i} (P_{\mathrm{cd}}^{L})^{i} (P_{\mathrm{wc}}^{L})^{n-i} = \sum_{i=n_2-k_3-1}^{n} \binom{n}{i} (1 - P_{\mathrm{wc}}^{L\,2})^{i} (P_{\mathrm{wc}}^{L})^{n-i} \tag{3-124}$$

This is the column error correction can achieve maximum $P_{\mathrm{cef}} < P'_{\mathrm{cef}}$. The Eq. (3-124) can be simplified, and rewritten as

$$P'_{\mathrm{cef}} = 1 - \sum_{i=0}^{n_2-k_3} \binom{n}{i} (1 - P_{\mathrm{wc}}^{L\,2})^{i} (P_{\mathrm{wc}}^{L})^{n-i} \leqslant 1 - \sum_{i=0}^{n_2-k_3} \binom{n}{i} (1 - P_{\mathrm{wc}}^{L\,2})^{i} (P_{\mathrm{wc}}^{L\,2})^{n-i} \tag{3-125}$$

$$P'_{\mathrm{cef}} \leqslant 1 - \sum_{i=0}^{n_2-k_3} \binom{n}{i} \left(1 - \sum_{j=0}^{t_2} \binom{n}{j} p_{\mathrm{e}}^{j} (1 - p_{\mathrm{e}})^{n-j}\right)^{i} \left(\sum_{j=0}^{t_2} \binom{n}{j} p_{\mathrm{e}}^{j} (1 - p_{\mathrm{e}})^{n-j}\right)^{n-i} \tag{3-126}$$

When given $n_1 = 188$, $k_1 = 172$, $k_2 = 178$, column error correction can achieve the maximum probability is shown in Figure 3-95. It can be seen, after using the product code correction, the error probability decreased significantly. Therefore, choice parameters of the product code have to consider the length of correct burst errors. The correction probability can be guaranteed within length of burst error.

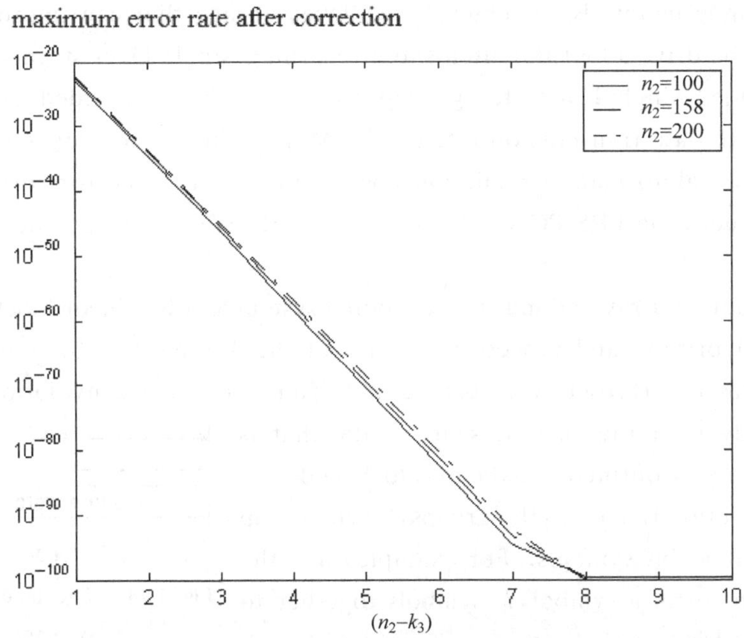

Fig. 3-95 The column correction can reach the maximum error correction probability.

3.7 Application of multi-wavelength and multi-level storage

3.7.1 Multi-level blu-ray disc drive

According to the requirements of the National Key Basic Research Development Program in China, the blu-ray multi-level storage systems are researched and developed to improve the data capacity and transfer rate, as an alternative techniques of the next generation optical disc. The experimental prototype is shown in Figure 3-96. The optical head (pickup) is shown in Figure 3-97. It was use of wavelength 405nm blue semiconductor laser and the numerical aperture 0.65 objective lens, that the spot of focusing diameter can be reduced to 260nm. The two technologies can increase storage density of 2.56x and 1.17x, respectively. The thickness of substrate of disc is 0.6mm, it can be able to use the manufacturing process of DVD. Its main structure of the hardware was design to compatible with DVD drive. The system can use dye, phase change medium

Fig. 3-96 Multi-level experimental prototype of the Blu-ray Disc.

Fig. 3-97 The multi-order Blu-ray Disc optical head.

and new photochromic materials to achieve 4～8 level storage that can improve the capacity of 1.5x to 3x times. The drive adopt the ditch-shore recording, single-layer storage capacity is 13GB, multi-level single-sided single-layer storage capacity is 20GB, single-sided double-layer storage capacity is 38GB with data transmission rate of 35.2Mbps. This section focuses on the modulation and coding and physical formats, modulation code with 2/3 (1, 7), error-correcting codes using Reed-Solomon product code (RS-PC (208, 192, 17) * (182, 172, 11)) and the corresponding new data format.

The system adopts Blu-ray and multi-level coding system and RS linear block codes. Therefore the format, codec algorithms and processing circuits of the disc have to be re-designed in the all. The new block codes is different from before and after data code convolutional code, the code word generated only with the current source data, that is shown in Figure 3-98. Continuous data of block code bit stream is divided into fixed-length groups, each group further split the unit m-bit symbols. For example, take the 3-bit or 8-bit data to form a symbol, k symbols together to form the source word is encoded into a codeword of length n, referred to as m-bit symbols (n, k) block codes. Error-correcting codes using linear coding process are linear transformations, and through the matrix transformation. In this linear space of all possible m-bit source can transform coding, nothing to do with the m-bit data.

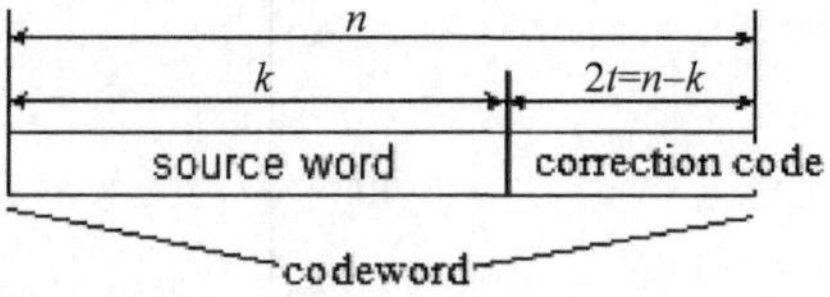

Fig. 3-98 The block code structure in system.

Source word and transform from the check sum code contains the source word is placed in the first half of part of the code, checksum attached in the second half. RS code in part by the parameters of m, n and k three 8-bit symbol RS (204, 188) code, i.e. DVB yards. The difference between the n and k (usually referred to as $2t$) is length of code symbol. RS code can correct no more than $(nk)/2$ errors, that is up to can correct t errors. DVB code, source data is divided into 188 symbols of a group, after the transformation of the coding, the codeword length of 204 symbols. Length of 16 symbol check character, to ensure the correct codeword up to eight errors. RS code is defined in a special limited domain, ie the Galois domain GF (2^m). The nature of the Galois domain with the integer domain, also a plus, subtract, multiply, with the exception of matrix operations, polynomial operations, such as, but the operation was established in 2^m. Codeword on a GF (2^m) has a corresponding polynomial. The code in accordance with the order from high to low, the data bits as a polynomial corresponding coefficient. For example, in GF (2^4), ie code "1010" can be expressed in polynomial $x^3 + x$. Therefore, certain operations to the source word to get the checksum and combined into a code word process, and available polynomial operator. RS code to select a suitable polynomial $g(x)$ (generator polynomial), and makes the codeword polynomial calculated for each source word are $g(x)$ times, the codeword polynomial divided by the generated polynomial from the remainder to 0. If the received codeword polynomial is divided by the generator polynomial remainder is not 0, that there is an error in the received codeword in need of correction. Further calculation shows that error to be corrected up to $t = (nk)/2$. RS code generator polynomial can be according to Eq. (3-127) to select:

$$g(x) = (x - \alpha)(x - \alpha^2)\cdots(x - \alpha^{2t}) = \prod_{i=1}^{2t}(x - \alpha^i) \tag{3-127}$$

If $d(x)$ represents the source word polynomial can constructed as follows codeword polynomial $c(x)$ $x^{n-k} \cdot d(x)/g(x) = h(x) \cdot g(x) + r(x)$. First calculate the quotient $h(x)$ and $r(x)$: the remainder $r(x)$ as a check word, so that the source word is placed in the first half of the codeword, the checksum $c(x) = x^{n-k} \cdot d(x) + r(x)$ is placed in the code word the second half, that can get:

$$\begin{aligned} c(x)/g(x) &= x^{n-k} \cdot d(x)/g(x) + r(x)/g(x) \\ &= h(x) \cdot g(x) + r(x) + r(x) \\ &= h(x) \cdot g(x) \end{aligned} \tag{3-128}$$

Where $r(x) + r(x)$ when modulo 2 is addition, the result is 0, so the codeword polynomial $c(x)$ must be generating polynomial $g(x)$ is divisible. The receiver detects the remainder is not 0, that can determine the received code word errors. In this system, each user sector includes 2048 bytes of user data, coupled with the former in the data sector header of 12 bytes (including 4-byte sectors marked code ID, 2-byte ID error detection code IED and 6 bytes of copy protection information CPR) and 4-byte sector tail (including 4-byte error detection code EDC), a total of 2064 bytes. These 2064 bytes form a 12 for 172 data blocks 2048 user data block and then through the scrambling process, the resulting block of data is the data sector, as shown in Figure 3-99.

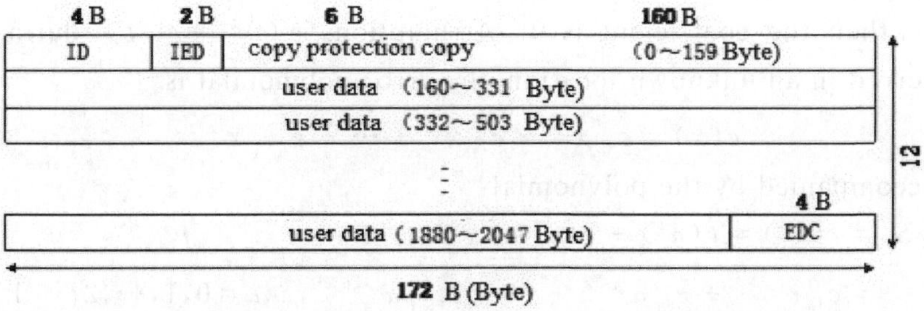

Fig. 3-99 The data sector structure ($172 \times 12 = 2064$ bytes).

The system will be 16 consecutive data sector (total of $16 \times 2064 = 33024$bytes) combined to form a 192 OK 172 ECC blocks, the RS-PC code of the ECC block. By the above formula for each column 172 has a 16-byte external checksum PO. PO data will be attached to the end of the corresponding column, add 16 rows of data blocks. This 208-row (192 row of the original data plus PO data row 16) of each row can calculate the available length of 10 bytes of internal checksum PI. PI data is attached at the end of the corresponding row of data blocks added 10, and finally can get a 208 OK 182 ECC block is shown in Table 3-9. The two generated polynomial factor for the aforementioned factors, despite the difference, but the final effect is still the same in GF (2^m) operations.

This ECC block 192 row per 12 row was splited into 16 block, for 16 row(PO data). Record sector after the modulation and coding can be converted into the physical sector recorded on the disc. This system, the sector size of each record is 2366 bytes, 2048 bytes is the really useful data, in addition to the 10 bytes as a sector marked and copy the information protection, the other 308 bytes for the error detection and correction, including word of the 6-byte sector data error detection and 302 bytes of the RS code checksum (line 12 to the end of each row of 120 bytes, 172 at the end of each column of a total of 172 bytes and the last line and the last 10 cross-regional total of 10 bytes). Therefore, in order to achieve the required level of error correction, to arise about data redundancy.

Table 3-9 10 ECC blocks (a total of 182×208=37856 bytes)

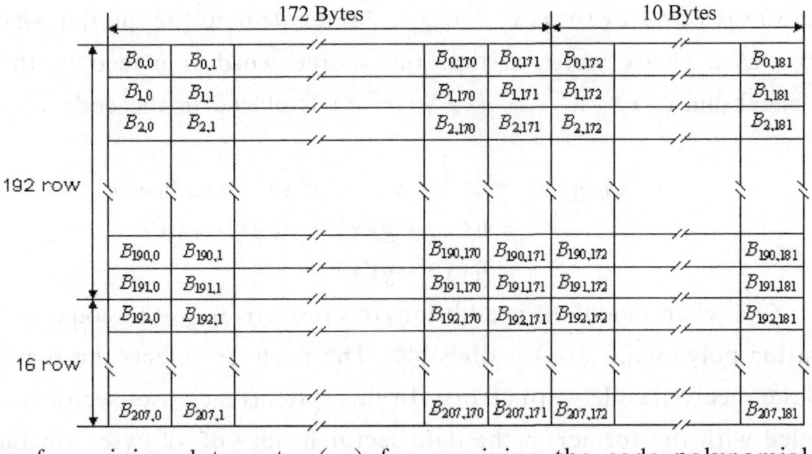
172 Bytes
10 Bytes
192 row
16 row
$B_{0,0}$ $B_{0,1}$ $B_{0,170}$ $B_{0,171}$ $B_{0,172}$ $B_{0,181}$
$B_{1,0}$ $B_{1,1}$ $B_{1,170}$ $B_{1,171}$ $B_{1,172}$ $B_{1,181}$
$B_{2,0}$ $B_{2,1}$ $B_{2,170}$ $B_{2,171}$ $B_{2,172}$ $B_{2,181}$
$B_{190,0}$ $B_{190,1}$ $B_{190,170}$ $B_{190,171}$ $B_{190,172}$ $B_{190,181}$
$B_{191,0}$ $B_{191,1}$ $B_{191,170}$ $B_{191,171}$ $B_{191,172}$ $B_{191,181}$
$B_{192,0}$ $B_{192,1}$ $B_{192,170}$ $B_{192,171}$ $B_{192,172}$ $B_{192,181}$
$B_{207,0}$ $B_{207,1}$ $B_{207,170}$ $B_{207,171}$ $B_{207,172}$ $B_{207,181}$

In the process of receiving data, set $r(x)$ for receiving the code polynomial, $v(x)$ is the original code polynomial, $e(x)$ is polynomial for the error codes generated in the transmission process:

$$r(x) = v(x) + e(x) \tag{3-129}$$

Where $v(x) = r(x) - e(x)$, if the transmission error $e(x)$ all coefficients are 0. The i to produce an error, then the coefficient is 0. Assumptions $c\ (0 \leqslant c \leqslant t)$, during transmission a mistake, and occurred in an unknown location, the error polynomial is

$$e(x) = e_{i_0} x^{i_0} + e_{i_1} x^{i_1} + \cdots + e_{i_{c-1}} x^{i_{c-1}} \tag{3-130}$$

Otherwise accompanied by the polynomial:

$$\begin{aligned} S_i &= r(\alpha^i) = v(\alpha^i) + e(\alpha^i) = e(\alpha^i) \\ &= e_{i_0} \alpha^{i_0 * i} + e_{i_1} \alpha^{i_1 * i} + \cdots + e_{i_{c-1}} \alpha^{i_{c-1} * i}, \quad i = 0, 1, \cdots, 2t - 1 \end{aligned} \tag{3-131}$$

If the error number is less than t, solving the equations of the Si composition may get the error location and error value, and to restore the original data. Si solving is more complex, especially the introduction of an error location polynomial are as follows:

$$\begin{aligned} \Lambda(x) &= (1 - xX_1)(1 - xX_2)\cdots(1 - xX_c) \\ &= \Lambda_c X^c + \Lambda_{c-1} x^{c-1} + \cdots + \Lambda_1 x + 1 \end{aligned} \tag{3-132}$$

Its roots $X_1, X_2, \cdots X_c$ is the inverse of the number of the error position, that the root of the error location polynomial can be obtained by Chien algorithm. So as get the wrong location number, the error value can be obtained by Forney's rule. The format parameters of Blu-ray multi-level read only disc and rewritable disc are shown in Table 3-10 and Table 3-11.

Table 3-10 The format parameters of Blu-ray multi-level read-only.

Parameters	single-side single-layer	double-sided double-layer
User data capacity	20GB/ side	38GB/side
Laser wavelength	405nm	
Objective lens numerical aperture	0.65	
Data length	(a) 0.306μm	
	(b) 0.153μm	
Channel length	(a) 0.204μm	
	(b) 0.102μm	

continued

Minimum record character length	(a) 0.408μm
	(b) 0.204μm
Maximum record character length	(a) 2.652μm
	(b) 1.326μm
Track spacing	(a) 0.680μm
	(b) 0.400μm
Disc diameter	120mm
Disc thickness	1.20(= 0.6 * 2)mm
Disc center hole diameter Disc	15.0mm
Inner diameter of information area	24.1mm
Outer diameter of information area	116.0mm
User data sector size	2048 bytes
Error-correcting code	RS-PC (208,192,17) * (182,172,11)
RS-to-PC Correction region	32 physical sectors
Modulation code	8/12modulation, RLL(1,10)
Burst error correction length	7.1mm
Standard speed	6.61m/s
Channel bit rate (standard rate)	(a) 32.40Mbps
	(b) 64.80Mbps
User bit rate (at standard speed)	(a) 18.28Mbps
	(b) 36.55Mbps

Table 3-11 Rewritable format parameters of Blu-ray multi-level disc

Parameters	single-side single-layer
User data capacity	20GB/side
Laser wavelength	405nm
Objective lens numerical aperture	0.65
Data length	(a) 0.306μm
	(b) 0.130～0.140μm
Channel length	(a) 0.204μm
	(b) 0.087～0.093μm
Minimum length of the record symbol(2T)	(a) 0.408μm
	(b) 0.173～0.187μm
Maximum length of the record symbol (13T)	(a) 2.652μm
	(b) 1.126～1.213μm
Track spacing	(a) 0.680μm
	(b) 0.340μm
Physical address	WAP(Wobble Address in Periodic position)
Disc diameter	120mm
Disc thickness	1.20(= 0.6 * 2)mm
Disc center hole diameter	15.0mm
Disc information area inner diameter	24.1mm

continued

Disc information area outer diameter	115.78mm
User data sector size	2048 bytes
Error-correcting code	RS-PC (208,192,17) * (182,172,11)
Correction region	32 physical sectors
Modulation code	8/12modulation, RLL(1,10)
Burst error correction length	(a) 7.1mm
	(b) 6.0mm
Standard speed	(a) 6.61m/s
	(b) 5.64～6.03m/s
Channel bit rate (standard rate)	(a) 32.40Mbps
	(b) 64.80Mbps
User bit rate (standard speed)	(a) 18.28Mbps
	(b) 36.55Mbps

- Blu-ray super-resolution optical head

The above-described Blu-ray drive used objective lens of numerical aperture 0.65, in order to further improve the Blu-ray multi-level optical disc storage capacity, as well as to be optical head of Blu-ray multi-level mastering system. In the optical system is supplemented the super-resolution technology to reduce the recording symbol size, to improve the optical disc storage density and can be used to Blu-ray multi-level mastering system further. The Blu-ray optical pickup with super-resolution and larger numerical aperture is shown in Figure 3-100. The system is based on the characteristics of the halo ball to design a numerical aperture NA of 0.95, and after accurate calibration for the Blu-ray aberration. Using a special optical phase shift aperture and two collimating lenses are convergence and intercept the laser beam, that reduces the high-frequency diffraction light, and significant improvements in the energy utilization of the laser beam. The Blu-ray lager numerical aperture super-resolution optical pich up is shown in Figure 3-101.

Fig. 3-100 large numerical aperture of Blu-ray optical pickup.

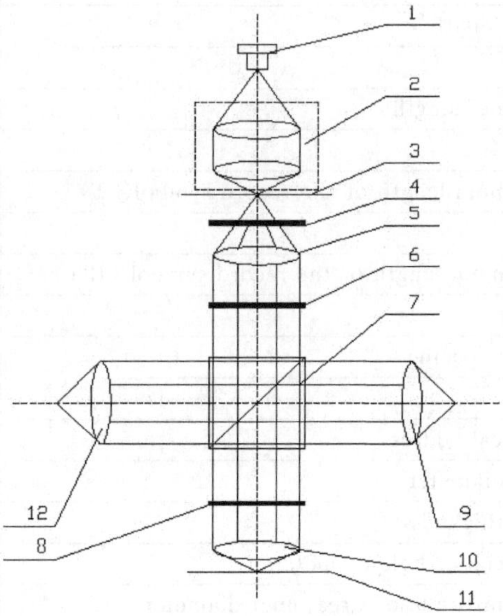

Fig. 3-101 Blu-ray high numerical aperture super-resolution storage optical system.

The optical system of Blu-ray and lager numerical aperture super-resolution storage is shown in Figure 3-101: 1—405nm semiconductor optical devices, 2—collimation and uniform lens group, 3—in hole phase shift aperture, 4—uniform plane, 5—collimator, 6—light apodized, 7—prism, 8—1/4 wave plate, 9—extract lens for readout signal, 10—objective lens, 11—medium, 12—to extract lens for the laser energy control signal. As introduces the pinhole phase shift aperture, uniform plane and phase shift of apodized, the space phase and intensity distribution of the focused laser beam are changed to improve the transfer function of the optical system that is shown in Figure 3-102. Energy distribution of the focusing laser beam shows that the 80% energy is focusing on range of radius of 110nm.

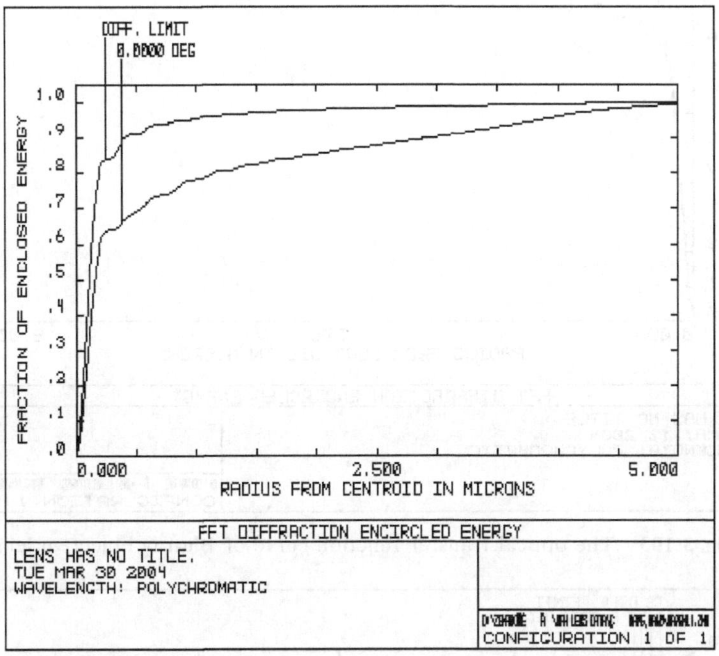

Fig. 3-102 The transfer function distribution of the super-resolution optical system.

The traditional energy distribution of an optical system is described as:

$$
\begin{aligned}
I_i(x_i, y_i) &= f\lfloor\lambda, \mathrm{NA}, \sigma, t_0(x_o, y_o), H(x_s - f_x, y_s - f_y), I_{\mathrm{eff}}(x_s, y_s)\rfloor \\
&= \iint_{\sigma} I_{\mathrm{eff}}(x_s, y_s)\left|\iint U(f_x, f_y) H(x_s - f_x, y_s - f_y)\exp[\mathrm{j}2\pi(f_x x_i + f_y y_i)]\mathrm{d}f_x \mathrm{d}f_y\right|^2 \mathrm{d}x_s \mathrm{d}y_s
\end{aligned}
\tag{3-133}
$$

As the optical system adopted phase shift masks, optical spatial filtering and apodized technology, the light energy distribution of the optical system was changed to the form:

$$
\begin{aligned}
I_{\mathrm{opt}}(x_i, y_i) &= f\lfloor\lambda, \mathrm{NA}, o, t_{\mathrm{opt}}(x_o, y_o), H(x_s - f_x, y_s - f_y), I_{\mathrm{eff}}(x_s, y_s)\rfloor \\
&\approx I_{\mathrm{ideal}}(x_i, y_i)
\end{aligned}
\tag{3-134}
$$

Where $I_{\mathrm{ideal}}(x_i, y_i)$ is light intensity distribution of the ideal image, $t_{\mathrm{opt}}(x_o, y_o)$ is optimized complex amplitude transmittance function, $I_{\mathrm{opt}}(x_i, y_i)$ is the optimized of the space light intensity distribution function. Since then, if reasonable to adjust the complex amplitude distribution of the light apodized to optimize the spatial frequency spectrum distribution with the phase shift mask, that can improve the light intensity distribution of the image plane, to improve the system resolution. Fully consider the features faint ball design, and use of super-resolution mask technology, to improve the transfer function of the optical system. Actual transfer function up to get the curve is shown in Figure 3-103, and the modulation transfer function is shown in Figure 3-104. From analysis of the diffraction energy distribution curve, 80% of the energy distribution is close to the diffraction limit. That the distribution corresponds to 80% energy on field of radius of less than 110nm, the modulation transfer function curve of the cut-off frequency corresponding to a resolution of 210nm.

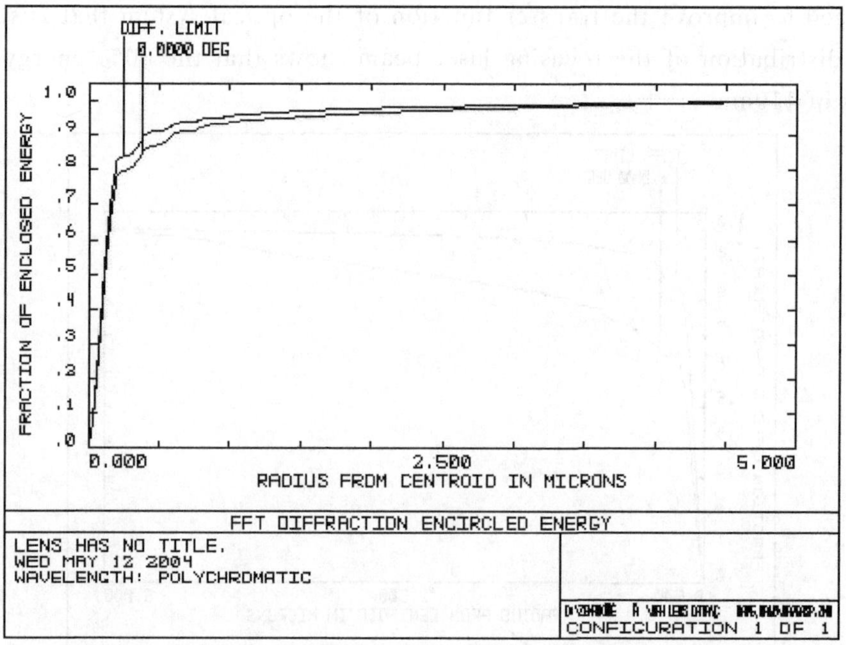

Fig. 3-103 The optical transfer function curve of Blu-ray objective lens.

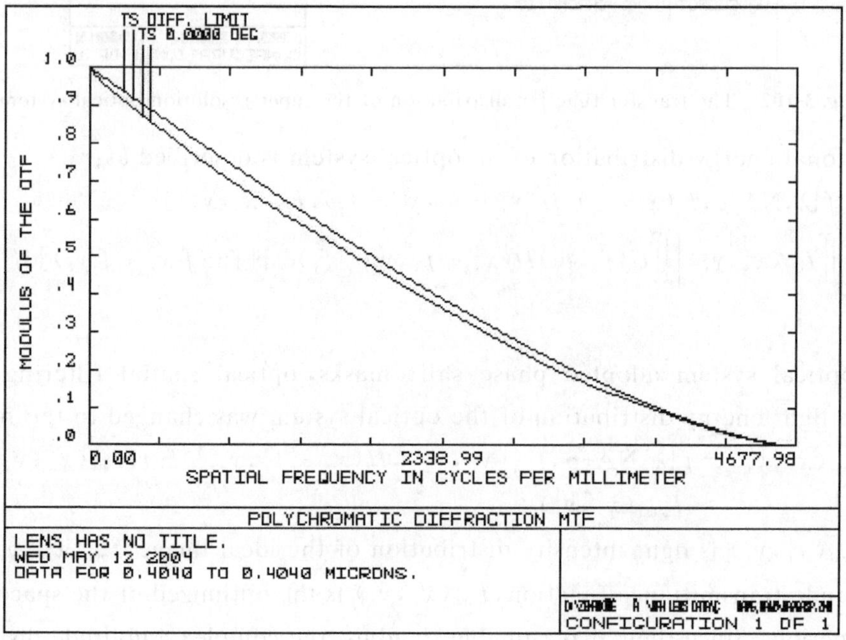

Fig. 3-104 The modulation transfer function curve of the Blu-ray objective lens.

In the development of Blu-ray objective lens, the focusing spot size is smaller than the resolution of optical microscopy in general. In order to precise determination of the actual optical properties of the super-resolution of the Blu-ray high numerical aperture optical system, using the knife-edge scanning method is specifically designed and developed by the diameter of the micro-spot detection system is shown in Figure 3-105. In this system, the knife-edge scanning micro-displacement measuring device, which measurement accuracy reaches ± 10nm sensitivity of 5nm.

Fig. 3-105 The testing system for diameter of the micro-spot of Blu-ray super-resolution optical pickup.

The system measured the diameter of the super-resolution focusing spot is 203nm actually. Above measurement results were compared to a standard DVD disc, that the results are shown in Figure 3-106.

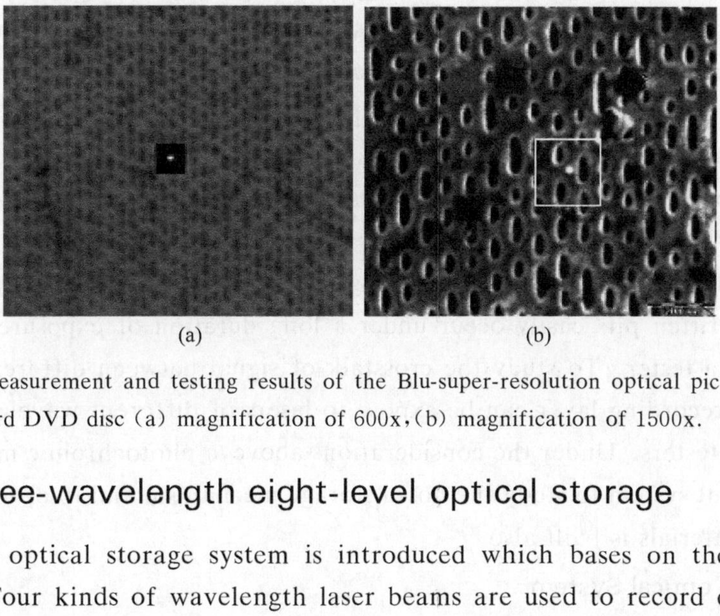

(a) (b)

Fig. 3-106 The measurement and testing results of the Blu-super-resolution optical pickup comparison to standard DVD disc (a) magnification of 600x, (b) magnification of 1500x.

3.7.2 Three-wavelength eight-level optical storage

A high-density optical storage system is introduced which bases on the multicolor photo-chromic medium. Four kinds of wavelength laser beams are used to record digital data on four layers photo-chromic materials independently, which are different spectrum sensitivity suitable for different wavelength laser beams at the same time. The principles of the system construction and the experiment results are described.

To meet requirements, the research of optical storage technology has focused on developing storage methods of high density and high access speed. To break through the bottleneck of the storage capacity, three-dimensional storage method has been introduced. There are three main kinds of three-dimensional storage method. First, the volume holographic storage, using the angular multiplexing technology, appears a feasible high capacity storage method. Second, two-photon storage, based on the two-photon phenomenon, realizes recording information on any points within a cubic medium. Third, multilevel recording in phase-change media technology realized the multilayer and multilevel storage. To increase the access speed, many methods has been studied: multilevel storage method by which several bits can be read or written in exact one pit, multi-beam light source storage method by which several pits can be read or written in the same time. OMNERC has studied a new storage method—multicolor multi-level storage, i. e. sharing some characters of the storage methods mentioned above. And a specific testing system has been developed for this novel storage method.

The disc is composed of *n* different recording medium. Each one contains different photochromic material with different absorbance spectrum. Before the writing process, disc has been exposed in UV light. Under the exposure of UV light, each recording materials have been converted into a colored state absorbing different spectrums. In the writing process, light beam containing lasers with different wavelength are focused on the recording layer by an achromatic objective. Because the thickness of recording medium is less than the focus depth, so the medium can be effectively written. When specific photons in laser beam are enough to make the corresponding recording material to react, the state of recording material is converted into a colorless state, and named written state. In the reading process, a laser beam with relative smaller power lasers is focused on the recording layers. According to the different absorbance of these states, photo detectors of the system can distinguish the written state from the unwritten state. Therefore, the data are recorded by these two states. There are three main advantages in this storage method. First, multilayer structure makes high-capacity storage possible. Second, multi-wavelength parallel recording realizes the high-speed access. Third, under different light power, data can be recorded in multilevel pits, which has different absorbance of the corresponding laser.

To study the novel storage method, a testing system is necessary. On the other hand, it is need of a material tester to optimize the parameters of photochromic materials. The photochromic material is one kind of photon-reaction recording materials that have no power threshold value. Therefore, wrong written pits easily occur under a long duration of exposure to a focusing light that is necessary to a tester. To study the crosstalk of signal between different recording layers, that one pit at the recording layer can be expose to lasers of different wavelengths that is hardly satisfied by current testers. Under the considerations above, a photochromic multi-wavelength and multilevel experiment system to study the novel storage mechanism and to optimize the parameters of photochromic materials is built also.

- Structure of optical System

As the Figure 3-54 and Figure 3-56 illustrated, the system is controlled by a computer. PC gives writing or reading command to the DSP controller and sends locating signals to the 4-freedom precise platform. Then DSP produces a corresponding modulating signals to control the LDs (laser dioxides) group to send out laser beams. Through the ensuing beam coupler, laser beams with different wavelength are coupled into one collimated light beam. It is focused by the achromatic objective, the light beam focus on the recording layers of the disc. Detectors gather the light reflected from the disc gets focusing and tracking error signals which are finally sent to the DSP controller. According to the error signals, DSP controls the actuator which moves the achromatic objective to focus the beam on the track of the recording layers. After accurately locating the focus on the right point, DSP retrieves the signals from the detectors group, that the signals are passed to the PC.

The optical system has been shown in Figure 1-36 (in Chapter 1), that 405nm, 532nm and 650nm three lasers from (1) are collimated by the collimating component (2). Then laser beams are polarized by the polarizing splitting prism(3). After that, the lasers are converted from the linearly polarized light to the circularly polarized light by the 1/4 wavelength plate (4). Following the wavelength plate, the light beam is totally reflected by the light splitting film of the light coupling/splitting prism (7). Reflected by the reflecting prism (8), the light gets to the achromatic objective lens, which focus the light onto the surface of the multicolor multi-level disc mounted on the 4-freedom precise platform. After the light is reflected by the reflecting layer of the disc, the light will return in the negative direction. Then the light gets to the 1/4 wavelength plate (4) again, the 1/4 wavelength plate converts the light from a circularly polarized light to a linearly polarized light, and the polarization direction of the light is normal to the original polarization direction. So the light is reflected by the polarizing splitting prism. Through the lens component (6), the light is imaged on the detector (5) which produces the focus error signals, tracking error signals and RF signals.

The light path represented above is just a branch of the optical system. As the Figure 1-36 and Figure 3-54 illustrates, another two similar branches are mounted in the optical system. The two branches share the same operating principle with the branch formulated in the last paragraph. The difference among the three branches is the wavelength of their laser. They produce light of different wavelength and processing their signals independently, which makes the parallel storage possible. The key component is the light coupling/splitting prism (LCSP). The LCSP. LCSP is composed of four cube corner prisms which adhere to each other. Two interfaces are coated with film series by which lights are coupling or splitting. The transmissivities of the two interface and respectively: The transmissivity of the interface 1 is 81.39% and 12.82% at the wavelength of 780nm and 650nm respectively with the entrance angle of 45°. The transmissivity of the interface2 is 92.14%, 92.18% and 8.99% at the wavelength of 780nm, 650nm and 532nm with the entrance angle of 45°.

- The precision mechanism and control system

The 4-freedom precise platform of the three-wavelength eight-level optical storage is assembled by three linear inching platform and one precise rotating platform. This 4-freedom precise platform realizes all the necessary movement involved in the optical storage. The minimum displacement of the linear inching platform is 0.1μm. The minimum angle of rotation is 5μrad. The maximum speed of rotation is 1r/min. Under the control of a computer, the platform can locate any points of a cubic space precisely. The platform can cooperate with the modulation of the laser dioxides, therefore material testing method can be specially designed to reduce the possibility of writing-when-reading, meeting the specific requirements of the photochromic storage materials.

The controller system of the experiment multi-wavelength and multi-level device includes the signal pre-processing circuits (SPPC), DSP controller and controller of the 4-freedom precise platform.

The front-end circuit receives and amplifies the signals from different quadrants of photodetectors. Processed by the following circuits (filtering the inter track crosstalk and the interlayer crosstalk), multilevel signals, focus error and tracking error signals are produced.

The DSP controller is used to modulate the pulse of laser dioxides, control the focusing process and control the tracking process. To modulate the pulse of the laser dioxides, a computer sends commands to the DSP controller which control the dioxides of 780nm, 650nm and 532nm to emit laser pulse. The maxim power of the dioxides of 780nm, 650nm and 532nm are 30mW, 20mW and 50mW respectively. The minimum impulse width is 100ns. To focus and track, the DSP controller receives the focus error signals and tracking error signals from the signal pre-processing circuits. Then the DSP controller produces the controlling signals to control the actuator to make the corresponding movement.

- Application example

Experimental samples of medium use a slide coated with a reflecting layer and a fulgide layer which reacts under the irradiation of the light of 780nm. The diagram of the reflectance of the samples is under the irradiation of the light of 780nm with a constant power.

In order to reduce the influence of the focus light, we firstly focused the light at a point of the samples. Then the position of the actuator was locked. After that, the 4-freedom precise platform moved the sample for 5μm. Under the movement, the shifting of the focus was neglectable. At the same time, the laser dioxide emitted a laser of 780nm of a required power. Then signals from the detector were recorded. Figure 3-107 and Figure 3-108 are the diagrams of the reflectance of the samples under the irradiation of the light of 780nm with different constant powers. Figure 3-108 shows the reflectance of the samples under the irradiation of the light of 780nm with power of 2.5mW. Figure 3-108 shows the reflectance of the samples under the irradiation of the light of 780nm with power of 1.8mW. From the Figure 3-107 and Figure 3-108, can found that the reflectance of the samples increases under the irradiation of the light of 780nm. And the rate of the increment of the reflectance decreases.

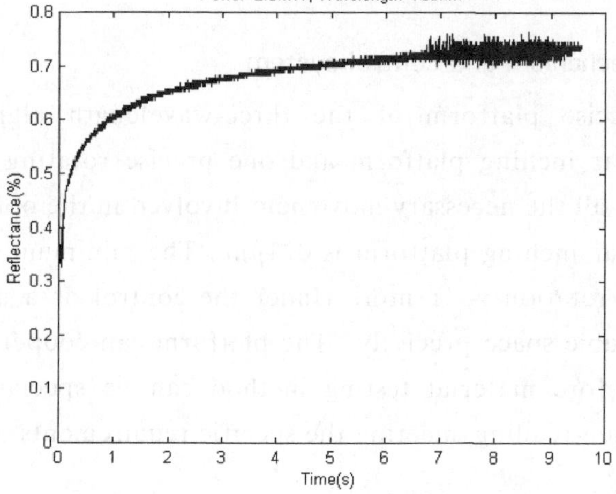

Fig. 3-107 Diagram of the reflectance of the samples under the irradiation of the light of 780nm with power of 2.5mW.

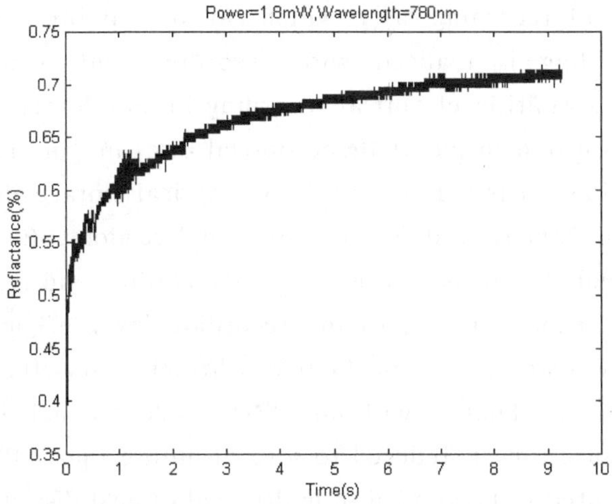

Fig. 3-108 Diagram of the reflectance of the samples under the irradiation of the light of 780nm with power of 1.8mW.

3.7.3 Multi-level photochromic medium

Using the photon-mode multi-level optical storage drive for photochromic diarylethene optical disc is investigated. The reflectivity of photochromic materials varies nonlinearly with the exposure energy, which is used for multi-level optical storage. Eight-level optical storage of amplitude modulation is experimentally realized in a diarylethene optical disc. The 650nm laser beam with power of 0.1mW is used for recording and readout, and multi-level signals with high signal-to-noise ratio (SNR) are detected. Photochromic diarylethene molecules show the considerable potential to be used as multi-level optical disc storage medium. The recording density and data transfer rate can be significantly increased in multi-level optical storage system with coded modulation.

With the increasing requirements of the huge information storage, higher density and higher data transfer rate become the trends of optical data storage. Current commercial recordable and rewritable optical storage systems are heat-mode recording systems, which limits the size of recorded pits and writing speed. Photon-mode optical storage using photochromic materials is a proposed method, which has advantages over heat-mode recording in terms of resolution, speed of data access, and multiplex recording capability. Photochromism is defined as a reversible transformation between two forms with different absorption spectra. Diarylethenes are newcomers to the photochromic field with excellent properties such as good chemical and thermal stability, and remarkable fatigue resistance. These properties enable diarylethenes to have the most applicable potential for future high-density optical storage. However, among them there are few publications that contribute to multi-level optical storage.

Multi-level or nonbinary optical storage is a promising approach to increase the recording density and data transfer rate significantly without changing optical parameters of current optical storage system. Generally speaking, M-level storage allows M statuses in each recorded mark, which increases the recording density by about $\log_2 M$ times of the traditional binary-coded storage systems. There are few papers discussing multi-level optical storage with photon-mode materials. From the experimental study, can find the reflectivity of photochromic materials varies nonlinearly with the exposure energy of recording laser, which can be suitably used for multi-level optical storage. Four-level recording is realized with recording and readout laser of 780nm in photochromic fulgide film. Multi-level optical recording in diarylethene PMMA film has also be studied, but it is not investigated in diarylethene optical disc, and the recording time is too long about 100ms. In this section is carried out multi-level optical storage in the diarylethene optical disc. The laser beam of 650nm is used for recording and readout. By controlling the exposure energy of laser beam delivered to of 650nm is used for recording and readout. By controlling the exposure energy of laser beam delivered to the recording layer of pure diarylethene, that the reflectivity difference between each form of the photochromic diarylethene is considerable. Eight-level recording of amplitude modulation with high SNR readout signals is realized in diarylethene optical disc, and the exposure time is reduced to several milliseconds. Photochromic diarylethene molecules show great potential to be applied in multi-level optical disc storage system.

The photochromic diarylethene used in out experiments is synthesized, and the photoisomerization of diarylethene is shown in Figure 3-109. The a means open form and the b of Figure 3-109 means closed form.

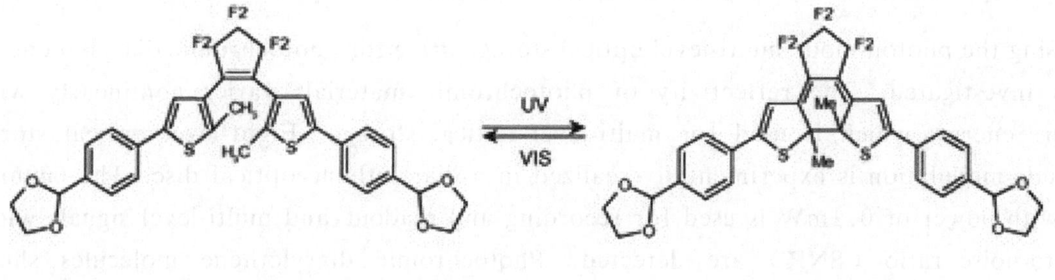

(a) open-form (b) closed-form

Fig. 3-109 The photochromic reaction of diarylethenes.

The structure of diarylethene optical disc is shown in Figure 3-110. It is alike with the standard CD-R, but the thickness of the recording layer is only 60nm. The reflective layer is aluminum, and the substrate is polycarbonate (PC). The photochromic diarylethene is directly evaporated onto the optical disc under vacuum condition. The recording layer of diarylethene optical disc is pure of photochromic material, which is obviously different from that in PMMA film. In the diarylethene optical disc, the recording process is more sensitive and the recording rate is evidently improved, which is suitable for multi-level optical storage.

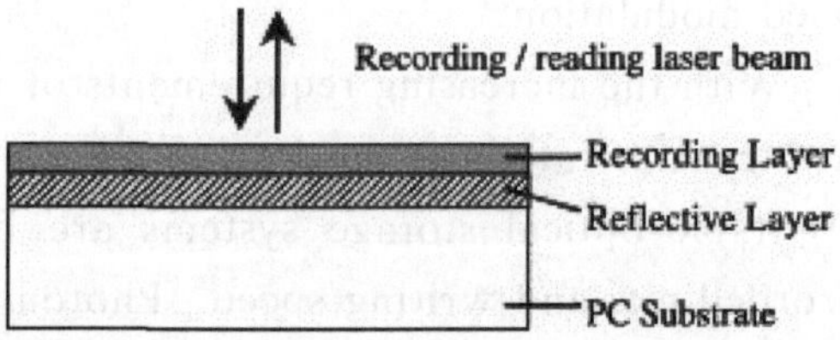

Fig. 3-110 Structure of diarylethene optical disc.

Before experimental recording, we use a UV lamp to initiate the diarylethene optical disc for 5min, making sure that the open form 1a is entirely converted to the closed form 1*b*. Under the UV irradiation, the colorless recording layer turned blue. Multi-level recording optical pick up (OPU) is shown in Figure 3-111. The laser beam of 650nm is used for multi-level recording, whose power and pulse width can be adjusted. The numerical aperture (NA) of the objective lens is 0.6. It should be noted that the laser of 650nm and NA of 0.65.

The shape of the recorded dot in diarylethene recording layer is not like that of CD-R or CD-RW pit, and the heat effect does not contribute to the recording process, so the power of the recording laser can be very low in the photon-type recording. In the experiments of multi-level optical storage, the power of the laser is kept constant to be 0.1mW. In the readout process, we use the scanning speed of 480mm/s, which is limited by the rotating-stage of the experimental system. A photosensitive detector detects the reflected light intensity, and signal outputs are transformed into voltage displayed in the digital oscillograph.

- Absorption spectra of diarylethene

The absorption spectra of diarylethene recording layer of optical disc before and after irradiation are shown in Figure 3-112. The absorption band at about 300nm, corresponding to the open form, disappears along with increase a new absorption band at about 600nm, which attributes to the closed form.

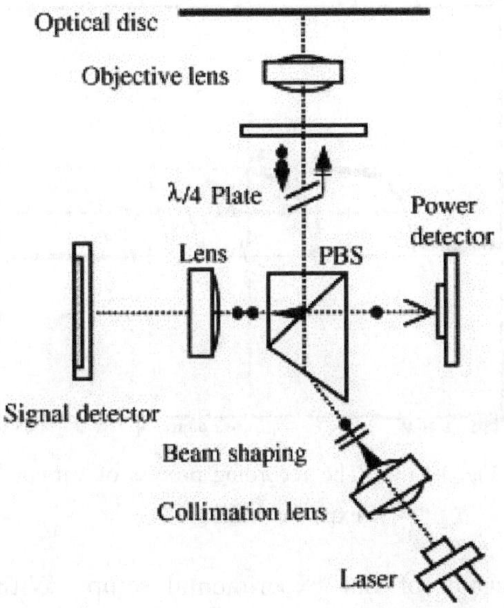

Fig. 3-111 The optical system of multi-level materials to testing experimental system.

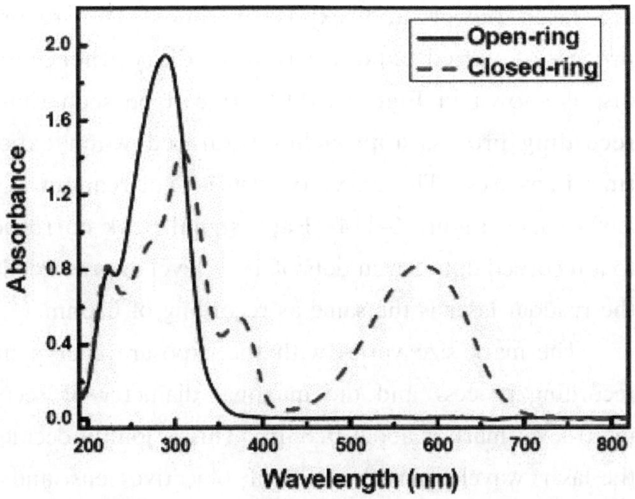

Fig. 3-112 The absorption spectra of diarylethene before (—) and after (---) irradiation.

In multi-level diarylethene optical disc storage, the closed form 1b is regarded as unrecorded state, and the open form 1a is recorded state. In the recording process, the unrecorded state is inverted to the recorded state under the visible irradiation. As shown in Figure 3-113, the diarylethene recording layer is sensitive to the laser of about 600nm, thus laser beam of 650nm can be used for recording and readout.

In the recording process, the reflectivity R of the diarylethene recording layer increases nonlinearly with the exposure energy. This characteristic of different reflectivity corresponding to different exposure energy can be used for multi-level recording. In our experimental study, the power of 650nm laser is kept constant to be 0.1mW, and the exposure time t varies to achieve different exposure energy. The recording process of diarylethene optical disc can be modeled investigated. The relationship between the reflectivity R and exposure time t can be expressed as the following equation:

$$t = \frac{1}{k}\int_{R_{\min}}^{R} \frac{dR}{R(R_{\max} - R)(1 + R)} \tag{3-135}$$

where the expression of recording constant k is

$$k = \frac{2.303\varepsilon I_0 \phi}{N_a hc/\lambda} \tag{3-136}$$

where ε is the mole absorption coefficient; I_0 is the intensity of the incident light; ϕ is the quantum yield of the ring-open reaction; N_a is the Avogadro constant; hc/λ is the energy of one single photon. If know R_{max} and R_{min} from the recording curve, the recording constant k can be calculated and then the exposure time t corresponding to some reflectivity R is decided by the definite integral of Eq. (3-130). The different exposure time t can be used for multi-level photochromic recording in the diarylethene optical disc.

- Readout of multi-level recording

The optimal recording and readout conditions are attempted for several times. The most important parameters are laser power and exposure time. The recording process of varied exposure time in diarylethene optical disc is shown in Figure 3-113. It can be seen that the recording process approaches saturated when exposure time increases. The curve of multi-level readout signals is shown in Figure 3-114. Each signal peak corresponds to a recorded dot. Seven dots of 1~7 level are recorded, and the readout laser is the same as recording of 650nm.

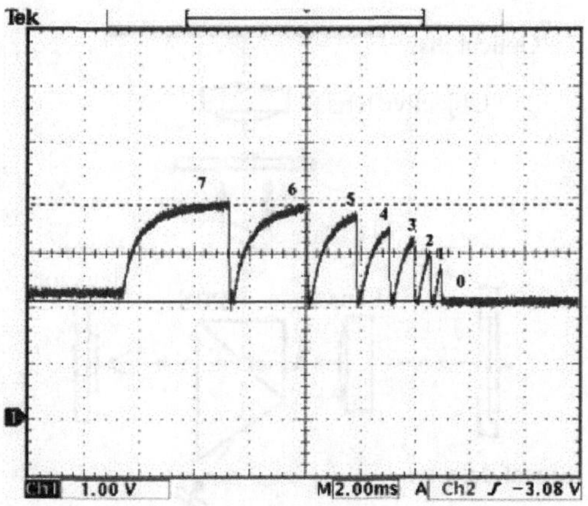

Fig. 3-113 The recording process of varied exposure time.

The mark size varies with the exposure energy in the recording process, and the maximal diameter of recorded multi-level mark is about 0.8μm. This is jointly decided by the laser wavelength, the NA of objective lens, and the precision of our experimental setup. With practicable improvements on the optical system in the experimental setup, we will decrease the maximal mark size to be about 0.5μm, which is more practical for high-density optical storage.

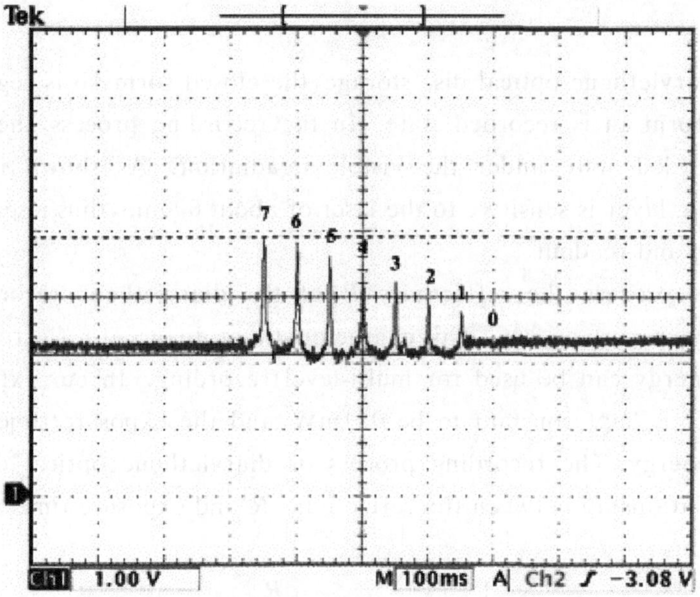

Fig. 3-114 The curve of multi-level readout signals.

In the readout process, the reflectivity of recording layer is transformed into voltage signal. The exposure time and readout signals of multi-level recording are shown in Table 3-12. It is obvious that the amplitudes of multi-level readout signals increase with exposure time. The SNR and contrast of multi-level readout signals are both high enough for subsequent signal processing. Because the process of diarylethene material is photon-mode and very sensitivity to exposure energy, multi-level optical storage can be easily realized in diarylethene optical disc by accurately controlling exposure energy.

Table 3-12 Eight-level recording conditions and readout signals

Signal level	Exposure time (ms)	Signal amplitude (V)
0	0.00	0.00
1	0.34	0.65
2	0.58	0.86
3	0.89	1.12
4	1.22	1.28
5	1.68	1.45
6	2.73	1.61
7	3.84	1.76

In the multi-level storage experiments, can find the response time of the material is on the level of milliseconds, which should be decreased to satisfy the requirements of real application. The response time is related to the process of disc preparation and the property of photochromic material in optical disc structure, which can be improved in further research. According to theoretical analysis of photochromic material, the reaction time of some nanoseconds can be obtained. Thus in principle, our experimental results demonstrate the feasibility that photon-mode diarylethene can be used as a novel rewritable media for multi-level optical storage.

3.7.4 Multi-level amplitude modulation

Multi-level amplitude modulation is used in the experiments of multi-level diarylethene optical disc storage. The multi-level amplitudes of readout signals are used to carry the information recorded in optical disc. The recording density of multi-level storage system can be increased significantly by employing multi-level modulation codes.

As shown in Figure 3-115, a trellis code is used for multi-level amplitude modulation, which includes convolutional coding and signal mapping. The code rate is 8/9, in which one parity bit is added to increase the error correction ability. Then the encoded 9 bits are mapped to three 8-level channel symbols. The recording density is 2.67 bits per minimum mark, which is 78% higher than that of traditional EFM+ modulation code used in DVD systems. With the reduced length of data mark, the storage capacity of diarylethene 8-level optical disc can be increased to be about 10GB.

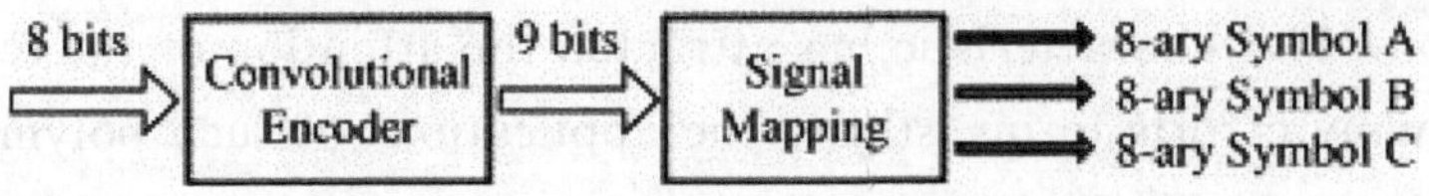

Fig. 3-115 The block of multi-level coding.

In further research, multi-level run-length limited (*M*-ary RLL) modulation will be considered. *M*-ary RLL modulation uses both the mark length and signal amplitude for information storage. But in the readout process, adequate SNR is needed for reliably detection and signal processing. Although much more complex than amplitude modulation, it can be used to achieve higher coding density. For example, an 8-ary (2, 10) code can achieve the recording density of 3.62 bits per minimum mark. Then the storage capacity of 8-level optical disc can be increased to be about 13.5GB. Combined with two-wavelength parallel recording technology, the multi-level and multi-wavelength optical disc with diarylethene media shows great potential to realize the storage capacity of 27GB.

Photon-mode multi-level optical storage in the diarylethene optical disc is carried out. The reflectivity of diarylethene recording layer varies with different exposure energy, which is used for multi-level optical storage. Eight-level recording of amplitude modulation is firstly realized in diarylethene optical disc with good results, and photochromic diarylethene shows great potential to be used as a novel material for multi-level optical storage. The recording and readout processes belong to photon-mode storage, which has many advantages over current heat-mode optical storage technologies. The mark size and response time of photochromic diarylethene optical disc should be improved in the further research. Multi-level diarylethene optical disc storage with coded modulation can significantly increase the recording density and data transfer rate. Retardancy multiplex multilevel optical recording Multi-valued phase patterns were multiply recorded by retardagraphy in order to improve recording density and data transfer rate. In the experiment, the phase pattern consists of four values were recorded on a polarization-sensitive medium by focusing the recording beam, and three patterns were multiply recorded by shifting the focal point. The recorded patterns could be independently reconstructed.

- Optical sub-diffraction multi-level encoding

By exploiting photo-induced reorientation in azo-polymer thin films, that demonstrate all-opticalpolarization-encoded information storage with a scanning near-field optical microscope. In the writing routine, 5-level bits are created by associating different bit values to different birefringence directions, induced in the polymer after illumination with linearly polarized light. The reading routine is then performed by implementing polarization-modulation techniques on the same near-field microscope, inorder to measure the encoded birefringence direction. Since the beginning of the computer era, data storage has been a fundamental issue in the scientific world. During the last decades, considerable efforts have been devoted to find solutions other than magnetic storage, allowing for cheaper substrates, smaller bit definition and lower power consumption during encoding and decoding. In this frame, polymeric materials have been extensively exploited, especially those, such as azo-polymers, that present photo-induced motion and reorientation. When illuminated with an appropriate linearly-polarized source, in fact, these polymers undergo isomerization cycles which eventually orient them with respect to the polarization direction. This characteristic has attracted the attention of the scientific community, which has devoted many efforts to investigate the applications of such polymers to far-field data encoding, optical holography, multipolar nonlinear optical memories, 3D information storage, and near-field binary encoding. With an increasing exposure energy, azo-polymers also present a polarization dependent macroscopic translational movement. This characteristic has been exploited for applications in nano-movement, nanofabrication, visualization of local field enhancements due to plasmonic resonances, and mapping of field components. Initially, azo-polymers were used in the liquid crystal form, but Nathanson et al. have demonstrated that liquid crystallinity is neither a necessary nor a best suited condition for optical storage. Liquid crystal azo-polymers, in fact, present glass transition temperatures comparable with room temperature, leading to a lower storage stability since the ordering tends to smear out in ambient conditions. We demonstrate the possibility to exploit optically-induced orientation of azo-polymers for all-optical, high-density multilevel data encoding and we define a new operative method for the development of organic-based information storage technology. We optimize photon energies and fluencies for the writing and reading routines and we implement near-field scanning optical microscopy (NSOM) with polarization modulation and analysis in order to overcome the diffraction limit for optical resolution. In this respect, the polarization preserving feature of the setup is a key point to reliably control information encoding and decoding. We thus achieve efficient writing/reading/rewriting of 5-level bits (pentabits), with a lateral size of about 250nm.

Use a commercially available azo-moiety Disperse Red-1 blended in a poly(methylmethacrylate)(PMMA) matrix with a molar concentration of 25%. This blend presents a rather high glass transitiontemperature of 102℃. A sample thickness of about 200nm is chosen since, for this film thickness, the optical response (i. e. the microscopic reorientation) of the polymers reaches its asymptotical value, asreported in the literature. For near-field encoding/decoding, use a commercially available NSOM (AlphaSNOM, WITec GmbH) modified in order to accommodate the optical elements presented in Figure 3-116.

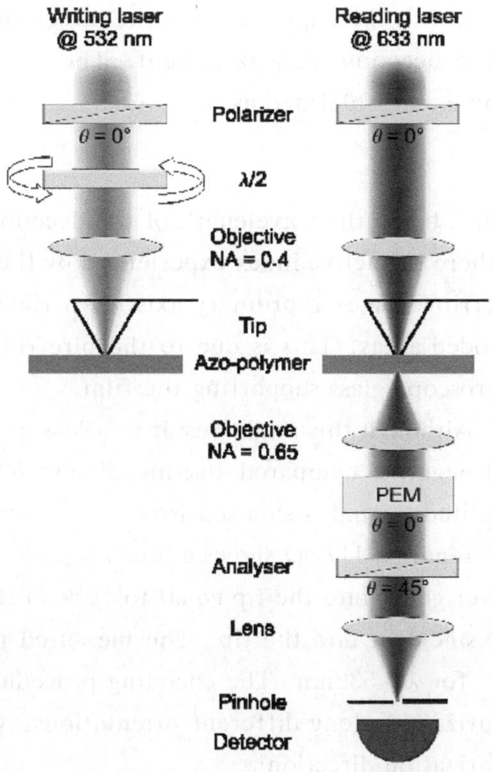

Fig. 3-116 (a) Encoding set up: the birefringence axis induced in the azo-polymer film is determined by the orientation of the half-wave plate. (b) Decoding set up to measure the induced birefringence. The polarizers, the photo-elastic modu-lator (PEM), and the analyzer have their optical axes at fixed angles with respect to the linearly polarized electric field of the light.

The instrument employs a hollow-pyramid probe with an optical resolution comparable with the tip aperture nominal diameter, equal to 120nm. As recently demonstrated, this type of near-field probesensure preservation of the incoming light linear polarization. The encoding routine is performed at a light wavelength $\lambda = 532$nm (corresponding to the emission of a frequency-doubled Nd:YAG laser), at which the film absorbance is relatively high (about 12%). Onthe other hand, a low absorbance is needed during the decoding routine, otherwise the stored information might be corrupted. In this case we employ a wavelength equal to 633nm (supplied by a He-Ne laser), for which the measured absorbance is less than 1%. The information is encoded by optically inducing birefringence over small sub-wavelength volumes. This mechanism involves photochemical excitation of the azobenzene group, activated with linearly polarized light. In this excitation process, the azobenzene unit under goes a trans-cis-trans photo-isomerization cycle, which usually takes about 10^{-9}s. Photo-isomerisation is a reversible process in which the single azo-unit is switched between a ground state (trans) and a more energetic one (cis). Due to thermal relaxation, themolecules then relax back to the ground state, in random orientation. However, those relaxed molecules that fall perpendicular to the incoming light polarization are no longer able to absorb energy, and remain fixed.

Operationally, the pentabit-encoding routine is performed by illuminating the sample surface with an appropriate polarization, keeping the tip in close proximity to the sample at the writing position. In the decoding routine, instead, the tip is raster scanned on the encoded pentabit. The light transmitted through the film is collected by a microscope objective, crosses a photo-elastic modulator and is measured with a lock-in amplifier connected to the detector, yielding an amplitude

$$A \propto \delta \sin(2\phi) \tag{3-137}$$

where Φ is the angle between the direction of the induced birefringence axis and the electric field of the decoding polarized light. The parameter δ stands for the birefringence induced in the sample and is defined as

$$\delta = 2\pi \frac{d}{\lambda} \Delta n \tag{3-138}$$

with λ being the wavelength of the decoding light, d the sample thickness, and Δn the difference in there refractive index experienced by the two light components parallel and perpendicular to the birefringence extraordinary axis. Experimentally verified that, on the average, $A \neq 0$ even in non-encoded areas. This is due to the birefringence introduced by the optical system and/or by the microscope glass supporting the film.

Although this might result in a loss of sensitivity, that can define levels that are both positive and negative compared to this offset. This is possible since the f-dependence shown by the amplitude signal is sign sensitive.

Figure 3-117(a) shows a birefringence map where four different bits have been encoded with a power going into the tip equal to 1μW. From now on, the reported powers are intended to be the one sincident into the tip. The measured far-field throughput of our NSOM tips is typically 5×10^{-3} for $\lambda = 532$nm. The encoding procedure has been performed with light presenting to a linear polarization along different orientations: $\Phi = \pm 45^\circ$ and $\Phi = \pm 15^\circ$ with respect to the decoding polarization direction.

Choose such angles to have equally-spaced 5-level encoding, the 0 (zero) level corresponding to no encoding or, equivalently, to $\Phi = 0$. The differences between the pentabits are not due to a different exposure energy (defined as input power × exposure time) but strictly to the Φ-dependence of this encoding routine. Figure 3-117 therefore demonstrates the possibility offered by the polarization maintaining setup to reliably write and retrieve different units of information in a 5-levels system of encoding by controlling the polarization of the near-field component of the light. The encoded pentabits show a long lasting persistence, which verified by measuring the sample after one month storage at ambient conditions, revealing no appreciable degradation of the stored information. In Figure 3-117 the exposure time is 30s. Chose such a long exposure time since it represents a good compromise between encoding speed and lateral definition of the

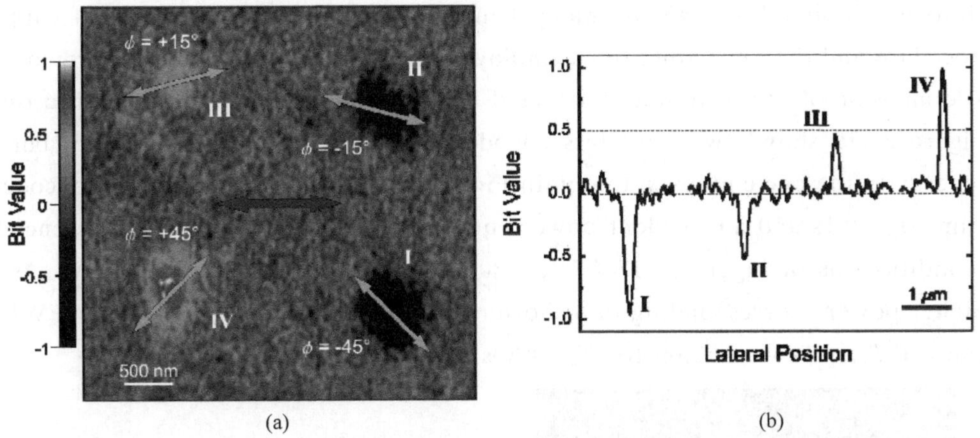

Fig. 3-117 (a) Birefringence map of a sample region containing four different pentabits. A constant background has been subtracted. The grey arrows represent the different orientations of the birefringence ax is associated to each pentabit, while the red arrow indicates the decoding polarization direction. The values of the angle Φ between the green and red arrows are also indicated. (b) Line profile of the pentabits: four well-defined levels (1, 0.5, −0.5, −1) are visible. The fifth level (0) is represented by the signal associated to non encoded areas. The FWHM of each pentabit is 250±15nm.

pentabits. High fluencies, in fact, allow writing the information in much shorter times, but also tend to increase the temperature in the exposed area. This leads to the thermally activated motion of the azo-moieties, resulting in larger bit areas and a worse signal-to-noise ratio, as discussed below.

The decoding routine has been performed on an area of $4\times 4\mu m^2$, with a point integration time of 10ms and an incident laser power of about 50μW. These values have shown to be a good compromise between reading speed and signal-to-noise ratio. The topography and phase signals (not shown) collected during the decoding step are flat within the instrument noise, demonstrating that, with such alow encoding exposure energy, no macroscopic movement is activated. The phase of the decoded signal turns is out to be dominated by the residual birefringence of the optical elements since no contribution is observed in connection with the different pentabits.

All four levels present a lateral dimension with FWHM = 250±15nm. We would like to stress that the size of each pentabit results from the convolution between the lateral size of the encoded area and the resolution associated with the decoding step. Note that this size is below the FWHM lowest limit one could expect by combining diffraction-limited encoding and decoding routines, which can be estimated around 310nm (assuming a numerical aperture equal to 1.4). With a lateral optical resolution associated to the decoding step equal to 120nm, we can estimate that the polymer area reoriented during the encoding step has a FWHM diameter of about 220nm. This area is larger than the one directly exposed to the tip aperture during the encoding routine. The broadening effect is likely due to the fact that each azo-moiety is coupled to the neighbouring ones because of short range interactions.

The system for data encoding also provides a reliable overwriting procedure. In our case, changing the information associated to a pentabit consists in directly writing a new pentabit in the same position, with the advantage that the re-orientation of the azo-moieties does not present any hysteretic behaviour, at variance with magnetic storage supports. Figure 3-118 shows the overwriting of

$a-1$ bit into $a+1$ bit ($\Phi=-45^{\circ}$ is changed into $\Phi=+45^{\circ}$). This is obtained with the same writing procedure and the same conditions employed to obtain Figure 3-117. As previously stated, the lateral dimension of the pentabits is affected by parameters such as the exposure time and/or energy. Figure 3-119 shows two pentabits encoded with different exposure times but the same exposure energy as the one employed to obtain Figure 3-117. The first pentabit is encoded with an exposure time $T_{exp}=1s$ and an incident power equal to $30\mu W$, while the second is encoded under the same conditions as in Figure 3-117 ($T_{exp}=30s$), incident power equal to $1\mu W$. As it can be noted, a higher power (corresponding to a shorter exposure time) leads to a higher FWHM lateral size (300nm for $T_{exp}=1s$ vs. 250nm for $T_{exp}=30s$).

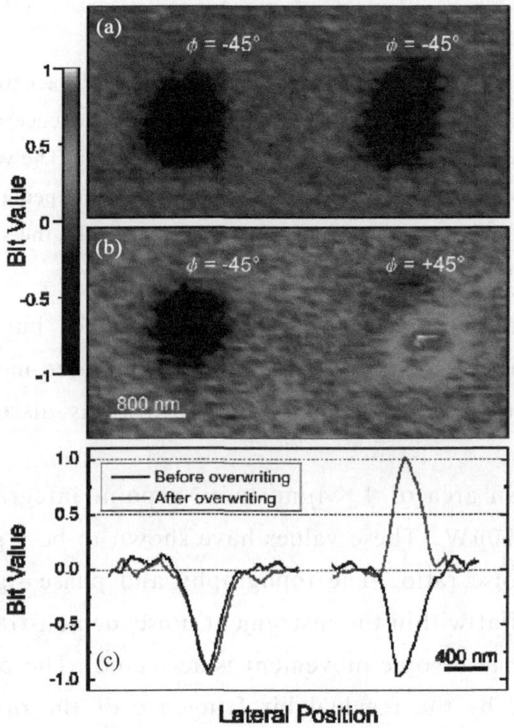

Fig. 3-118 Overwriting the information: the pentabit on the right in (a) is re-encoded with the opposite value in (b), while the left one is maintained as a reference; Φ is the angle between the encodedbirefringence axis and the decoding polarization direction. (c) Pentabit line profile of the pentabits before and after the overwriting procedure.

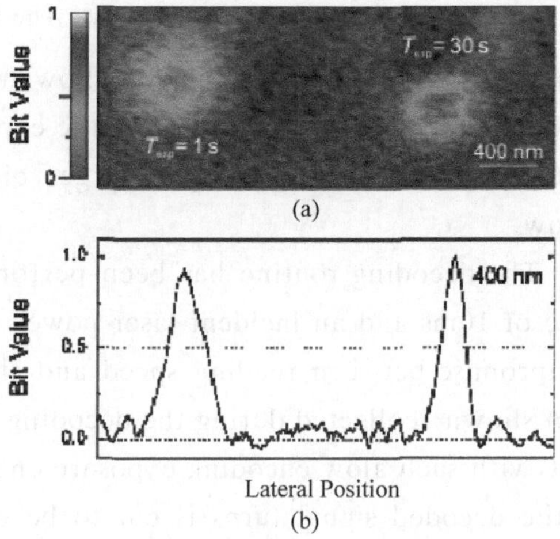

Fig. 3-119 (a) Amplitude signals of two nominally equally pentabits ($\Phi=+45$℃ for both pentabits) encoded with the same total exposure energy ($30\mu W\times s$), but different exposition times T_{exp}. (b) Line profile of the two pentabits. The FWHM of the pentabits is 300nm (left) and 250nm (right), respectively.

A final issue, shown in Figure 3-120 is the importance of the total exposure energy: too low an energy induces only a partial reorientation of the azo-moieties, leading to blurry signals, as shown in Figure 3-120(a). Increasing the exposure energy (from left to right in Figure 3-120(a), (b)), a more uniform and well defined orientation is induced in the exposed area and the associated signal-to-noise ratio improves in the decoded amplitude. On the other hand, too high an exposure can lead to the macroscopic movement of the azo-chain. The result is the presence of undesired artefacts in the decoding maps, due to the interaction of the tip near field with the inhomogeneities caused by the macroscopic movement itself as penta bits in Figure 3-120(a),(c). These effects are even more evident in the couples from Ⅶ to Ⅸ, which show "butterfly shapes" both in the amplitude and in the topographic maps (Figure 3-120(b) and Figure 3-120(d), respectively) when excessive exposure energies are used. In this case, even the topographyremarkably retains a polarization dependence. The demodulated amplitude signal A presents two winglike lobes divided by a central line, due to the dip formed by the macroscopic motion of the azo-moieties involved in the photo chemical excitation process. Upon increasing the exposure energy, more material is accumulated into the side lobes, while a depletion up to 40nm is formed in the middle (see Figure 3-120(f)).

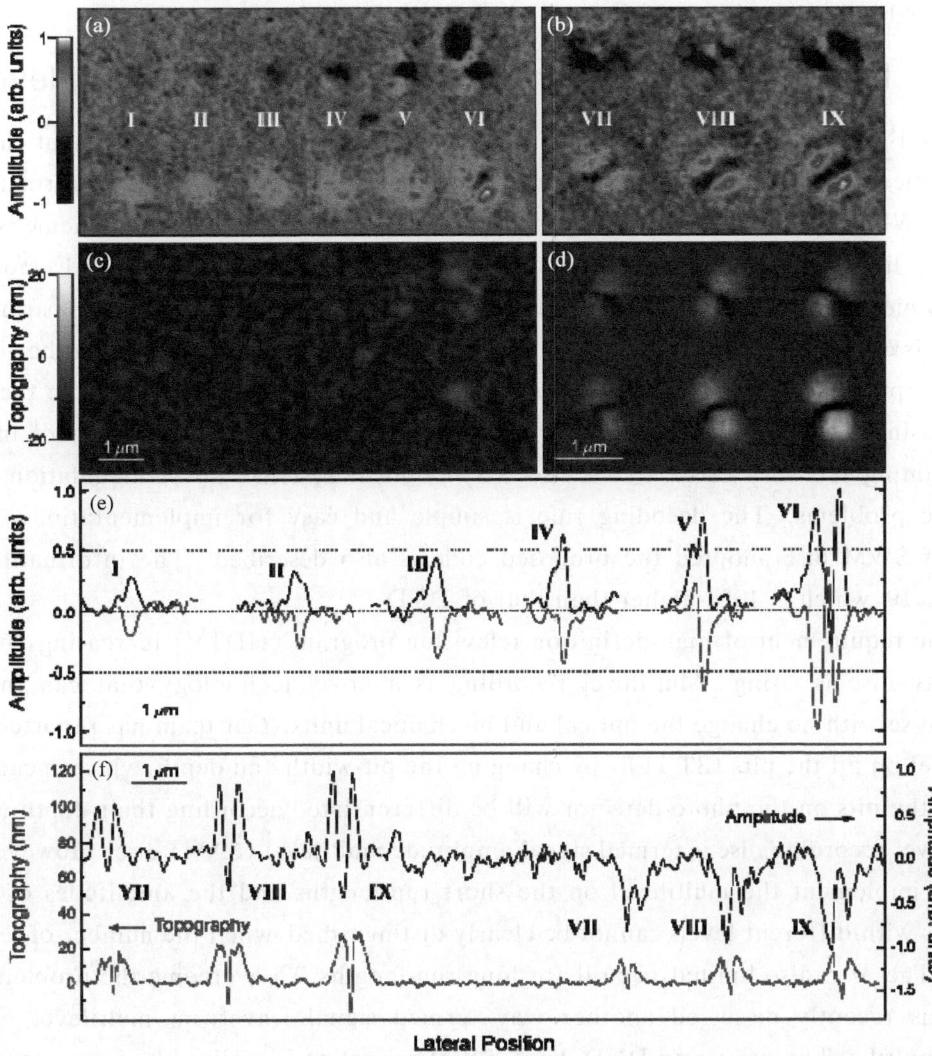

Fig. 3-120 Effects of the total exposure energy on the azo-polymer film. (a),(c) Birefringence map and corresponding topographic signal for couples of pentabits ($\Phi = \pm 45°$) as a function of the exposure time for a fixed incident power equal to 15μW. From Ⅰ to Ⅳ the exposure time is doubled at each step, starting with 1s for Ⅰ. The corrensponding amplitude line profile is presented in (e). (b),(d) Birefringence map and corresponding topographic signal for couples of pentabits ($\Phi = \pm 45°$) as a function of the exposure time for a fixed incident power equal to 1.5mW. The exposure time is 1s, 2s, and 4s for Ⅶ, Ⅷ and Ⅸ, respectively. (f) Topography and amplitude profile of the pentabits in (b),(d).

In conclusion, have demonstrated long lasting all-optical multilevel information storage with sub-diffraction resolution, based on polarization encoding and decoding in azo-polymer thin films. Pentabits with a diameter of roughly 250nm have been written, read and re-written by a proper choice of photon energies, fluencies and polarization. In perspective, encoding on more than 5 levels is possible, the only limitation being the sensitivity of the decoding set up. Nano-patterning the surface of the films to form a grid with squares having dimensions comparable with the tip diameter could possibly be an efficient solution to avoid the smearing of the encoded information over an area larger than the tip aperture. In such a way the resolution and density of the bits could possibly be highly increased.

3.7.5 Rate 7/8 run-length and level modulation for multi-level ROM

Multilevel recording technology is used to improve the recording density without changing the optical and mechanical units. A new modulation code scheme for signal waveform modulation multilevel (SWM) read-only optical disc has been implemented. The proposed scheme is composed of run-length limited (RLL) modulation and level modulation two steps. RLL modulation is employed to meet the requirements of channel. To acquire higher code rate, the parameter d of RLL (d, k) is decreased to 0, which makes the presented scheme difference from other modulation codes of the optical storage systems. Increasing the number of k also contributes to the high code rate. Decreasing d and increasing k will respectively introduce more inter-symbol interference (ISI) and timing recovery error (TRE) to SWM optical system. Level modulation is used to resolve these problems. The decoding rule is simple and easy for implementation. The signal waveform of SWM disc adopted the proposed code is also described. The information bits per 400nm are 2.19, which is 46% higher than that of DVD.

With the requirement of high-definition television program (HDTV) increasing, the need for mass capacity disc is rising. Multilevel recording is a novel technology that can increase the capacity of disc with no change the optical and mechanical units. Our team has reported to realize the multilevel on all the pits (3T 11T) by changing the pit-width and depth, which means the light intensity of the pits on the photo-detector will be differentiated according their depth and width. This multilevel recording disc is termed signal amplitude multilevel (SAM) disc. However, find the difficulty to implement the multilevel on the short run-lengths and the amplitudes of read back signal of pits with different levels cannot be clearly distinguished, when the number of level (M) is increasing. The M is also limited, even if for long run-lengths. To overcome the problem of SAM, our team has recently discussed another way termed signal waveform multilevel (SWM) to implement multilevel recording on DVD discs. SWM recording is realized by inserting a sub-land/sub-pit into an original recording pit/land, which leads to the waveforms of readback signal of pits and lands differentiate according to the position and size of the sub-lands and sub-pits. In the SWM disc, there is no or less multilevel on the short run lengths, which successively avoid the problem of SAM. The more multilevel is implemented, the longer the run-length is. Figure 3-121 is the atomic force microscopy (AFM) image of SWM disc. Figure 3-121(b) describes the profile shapes of the four recording symbols in Figure 3-121(a). The distances between three pairs lines are respectively 894, 1245, 910nm, which means the three recording symbols are $5T$ (land), $7T$ (pit), $5T$ (pit). The variation of size and position of subpit/sub-land, which is the level information of run-length. Modulation code is one of the key technologies to improve the storage capacity of optical disc. The common modulation code used in currently optical recording disc is run-length limited (RLL) code, which is described as RLL (d,k). The number of 0's between two neighboring 1's is at least d and at most k. The modulation code in compact disc (CD), digital versatile disc (DVD) and blu-ray disc are respectively 8/17 RLL (2,10), 8/16RLL (2,10) and 2/3RLL(1,7). There are some papers published for the modulation code of SAM disc. The modulation code for SAM is called M-ary RLL(d,k). M is the level number of every run-length. As the difference between SWM and SAM, the M-ary RLL(d,k) code can not be directly used in SWM disc.

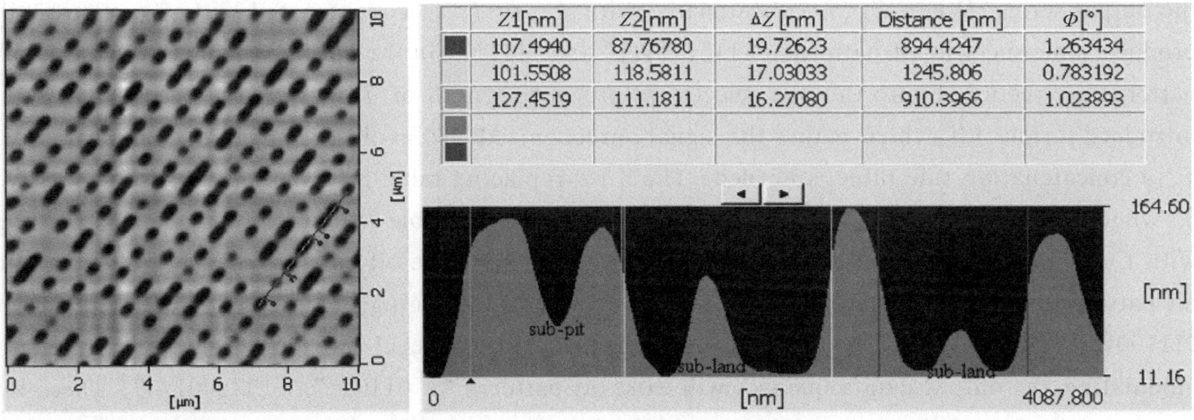

Fig. 3-121 The AFM image of SWM.

In this paragraph the modulation-coding scheme defined as run-length and level modulation (RLM) code is presented and the encoding and decoding rules are discussed detailedly. RLM is carried out through run-length modulating and level modulating two steps. Firstly, use run-length-limited (RLL) code to scheme out a RLL (0,15) code with code rate $R = 7/8$. To avoid the appearance of two or more successive T, a substitution table is needed. Secondly, to alleviate the inter-symbol interference (ISI) and mitigate the timing recovery error, level modulation is employed to eliminate the existence of T, continuous $2T$ and the long run-lengths (more than $10T$).

3.7.6 7/8 run-length and level modulation code

The writing channel for SWM optical disc is described in Figure 3-122, which including ECC encoding, run-length modulation coding, level modulation coding and writing strategy. The presented modulation code scheme of this paper includes run-length modulation and level modulation process. The aim of run-length modulation process is to achieve high code rate $m/(m+1)$ RLL (d,k) code. The level modulation code is used to resolve the problems of ISI and TRE, which are introduced in the run-length modulation process.

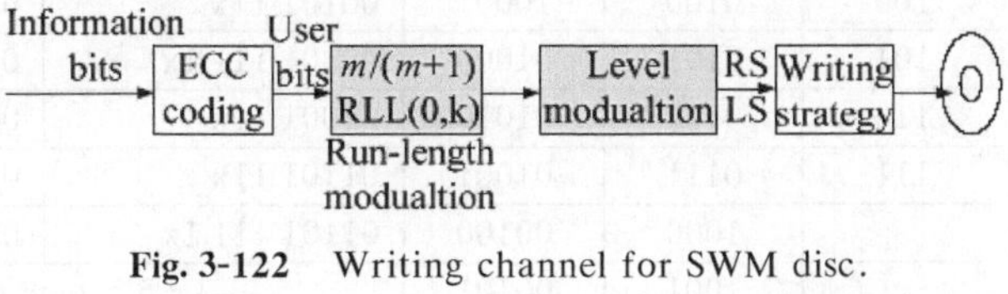

Fig. 3-122 Writing channel for SWM disc.

The higher is the capacity, the higher the code rate is. K. A. S. Immink described the relationship between the capacity of (d,k) constrained channel ($C(d,k)$) and the parameters d and k. If d decrease or the k increase, the $C(d,k)$ will improve. To achieve high code rate $m/$

$(m+1)$, the parameter d of RLL (d,k) is selected as 0. If $d=1$, the capacity of RLL (d,k) is limited to 0.6942 and the code rate $R=7/8$ cannot be achieved. The number of kis set as large as possible. Table 3-13 gives the detailed encoding processing. The encoding table consists of main code table and table of concatenation rule. The main table is employed to convert information bits to satisfy the constraint of RLL (0,15). Every 7 information bits is separated into two parts. The first 4 bits termed $I_{i,4}$ is called control code and the last 3 bits is called basic code ($I_{i,3}$). Basic code means that these bits are unchanged in the modulation processing. Control code $I_{i,4}$ is converted to 5 bits ($C_{i,5}$) according the control code table. Then, $C_{i,5}$ is concatenated with $I_{i,3}$ as the output series. When $C_{i,5}$ is concatenated with $I_{i,3}$ or $I_{i-1,3}$ concatenated with $C_{i,5}$, there will produce some specific bit-patterns, "111", "1111" and "11111" in the output sequences. If these bit-patterns are recorded onto the disc, successive pits with length of T will exist. Successive T_s will introduce serious ISI, which makes the signal unrecognized. To avoid the appearance of successive T_s, a concatenation rule table is needed. The 5+3 replacing table resolves the cases of successive T_s when $C_{i,5}$ is concatenated with $I_{i,3}$. The 3+5 replacing table is used for $I_{i-1,3}$ concatenating with $C_{i,5}$. Tables of 111 replacing are employed to eliminate the bit-patterns of "111". Even if the bit-pattern of "111" is not completely eliminated, all the remaining "111" exist as the patterns of "111001", which is treated as a whole body termed "5n" in level modulation. After run-length modulating, the output data sequences will exist bit patterns "...0110...", "...111001...". These bit patterns will introduce serious ISI to readout signals of SWM disc. With the number of k increasing, it will bring serious timing recovery error to readout RF signal of SWM disc. To decrease TRE, the longest run-lengths, whose length is more than 10, should not exist in the run-length series.

Table 3-13 Encoder of RLL (0,15) main code table of concatenation rule.

Main code table			Table of concatenation rule			
Basic code	Control code table		5+3 replacing table		111 replacing table1	
	Input	Output	Input	Output	Input	Output
001	0001	10100	10101 11x	10011 01x	111 01010	010 00000
010	0010	10101	10101 111 1x	10011 100 1x	111 01001	011 00000
011	0011	10010	01001 11x	01011 00x	111 00100	111 00100
100	0100	10001	00101 11x	01011 01x	111 00101	111 00101
101	0101	01000	00101 111 1x	01011 100 1x	111 00101 11x	100 11100 10x
110	0110	01010	00001 11x	00011 00x	111 00010	011 10010
111	0111	01001	01101 11x	00011 01x	111 00001	100 00000
	1000	00100	01101 111 1x	00011 100 1x	111 01100	101 00000
	1001	00101	3+5 replacing table		111 00110	110 00000
	1010	00010	x11 10000	x00 11000	111 01101	001 11001
	1011	00001	x11 10100	x00 11010	111 01101 1x	010 11011 0x
	1100	01100	x11 10101	x00 11001	111 replacing table 2	
	1101	00110	x11 10101 1x	x00 11011 0x	00110 111 00100	00001 110 01000
	1110	10110	x11 10010	x10 11000	10110 111 00100	10001 110 01000
	1111	01101	x11 10001	x10 11010	00110 111 00101	00001 110 01010
			x11 10110	x10 11001	10110 111 00101	10001 110 01010
			0111 0110 11x	000 11100 10		
			1111 0110 11x	101 11001 00		

If T and successive T_s exist in the actual system, the amplitude of the readback signal of T will be every lower and misidentify the slicer. Moreover, T with neighboring run-lengths will produce ISI and long run-lengths will bring timing recovery error. To lighten the ISI and mitigate the timing recovery error, level modulation is needed.

3.7.7 Level modulation process

The aim of level modulation is to combine T or $5n$ with the latter run-lengths to form longer run-lengths and separate the longest run-lengths to relatively short run-lengths. Run-length-level transition table is the processing of level modulation. The inputs of level modulation are outputs of run-length modulation and the outputs of level modulation are two vectors: run-length series (RS) and level series (LS). The two vectors are used to control the write strategy to record information data onto discs. The symbol "m" in Table 3-14 represents the input series includes bit-pattern "11". For instance, the series of "11001", "110100001", "110110001" are expressed as $4m$, $3m+5T$, $3m+5m$ respectively. Alphabets of A ~ F represent the levels of run-lengths after level modulation. Table 3-14 gives the detailed description of run-length-level transition process. The original run-lengths are the input ones, while the output sequences are the output run-lengths and levels. Run-lengths of $2T$~$10T$ remain unchangeable. Run-lengths of $11T$~$16T$ are cut into two parts to mitigate TRE. One is level 5 of $9T$, the other is level 0 of ($2T$~$7T$). $4m$~$10m$ are encoded as level 1 of ($4T$~$10T$). $11m$~$15m$ is separated into level 6 of $9T$ and level 0 of ($2T$~$7T$). To eliminate the patterns "1101" ($3m$), $3m$ is combined with next run-lengths to form multilevel of longer run-lengths. $3m$ and ($2T$~$7T$) are combined as level 2 of ($5T$~$10T$). $3m$ and ($3m$~$7m$) are united as level 3 of ($6T$~$10T$). To remove the bit-pattern "111001" ($5n$), it is also combined with neighboring run-lengths to constitute multilevel of longer run-lengths. $5n$ and ($2T$~$5T$) are incorporated to level 4 of ($7T$~$10T$). Assumed the output series of run-length modulation is "10001101000101110010001000110110001001...". The input of level modulation: $4T$ $3m+4T$ $2T$ $5n+4T$ $4T$ $3m+5m$ $4T$. Where, "+" means the two run-lengths should be combined as a pit or land. The vectors RS and LS are RS = [$4T$ $7T$ $2T$ $9T$ $4T$ $8T$ $4T$...] and LS = [0 2 0 4 0 3 0 ...]. The elements of RS are used to control the length of pits/lands and that of LS are employed to control the size and position of the sub-pits.

Table 3-14 Run-length-level transition table

Original run-lengths					Output sequences				
$2T$					$2T$				
$3T$					$3T$				
$4T$	$4m$				$4T$	$4A$			
$5T$	$5m$	$3m2$			$5T$	$5A$	$5B$		
$6T$	$6m$	$3m3$	$3m3m$		$6T$	$6A$	$6B$	$6C$	
$7T$	$7m$	$3m4$	$3m4m$	$5n2$	$7T$	$7A$	$7B$	$7C$	$7D$
$8T$	$8m$	$3m5$	$3m5m$	$5n3$	$8T$	$8A$	$8B$	$8C$	$8D$
$9T$	$9m$	$3m6$	$3m6m$	$5n4$	$9T$	$9A$	$9B$	$9C$	$9D$
$10T$	$10m$	$3m7$	$3m7m$	$5n5$	$10T$	$10A$	$10B$	$10C$	$10D$
$11T$	$11m$	$3m8$	$3m8m$	$5n6$	$9E2$	$9F2$	$8E3$	$8F3$	$10F1T$
$12T$	$12m$	$3m9$	$3m9m$	$5n7$	$9E3$	$9F3$	$8E4$	$8F4$	$8E4A$

continued

Original run-lengths					Output sequences				
$13T$	$13m$	$3m10$	$3m10m$	$5n8$	$9E4$	$9F4$	$8E5$	$8F5$	$8E5A$
$14T$	$14m$	$3m11$	$3m11m$	$5n9$	$9E5$	$9F5$	$8E6$	$8F6$	$8E6A$
$15T$	$15m$	$3m12$	$3m12m$	$5n10$	$9E6$	$9F6$	$8E7$	$8F7$	$8E7A$
$16T$		$3m13$	$3m13m$	$5n11$	$9E7$	$9F7$	$8E8$	$8F8$	$8E8A$
		$3m14$	$3m14m$	$5n12$			$SE9$	$8F9$	$8E9A$
		$3m15$					$8E10$		
22222					$10E$				

Table 3-15 gives the realizable level number of each run-length after level modulation. The number of level of different length run-lengths is different.

Table 3-15 Realizable level number of run-lengths.

Run-length	$2T$	$3T$	$4T$	$5T$	$6T$	$7T$	$8T$	$9T$	$10T$
Level number	1	1	2	3	4	5	7	7	7

Figure 3-123 shows the signal processing and the complete decoding scheme for SWM system. When run-length is less than $4T$, the data is directly demodulated and ECC decoded. If the run-length is more than $4T$, the level of readback signal is detected with adaptive level detecting method. The data is also directly demodulated for level $=0$, while the levelrunlengthreverse transition process is applied if level >0. If the level is $5(E)$ or $6(F)$, the reversed data must be combined with next input data before demodulated. Some examples are given to explain SWM in Figure 3-124, the readout signal series represent level 2 of $8T(P)$, level 1 of $10T(L)$, level 1 of $9T(P)$, level 0 of $5T(L)$, level 5 of $10T(P)$, level 0 of $4T(L)$ and level 3 of $8T(P)$. P means the recording pit and L is land. Although the first pit and the last pit are both $8T$, the variation of waveforms is clearly and used to differentiate the level.

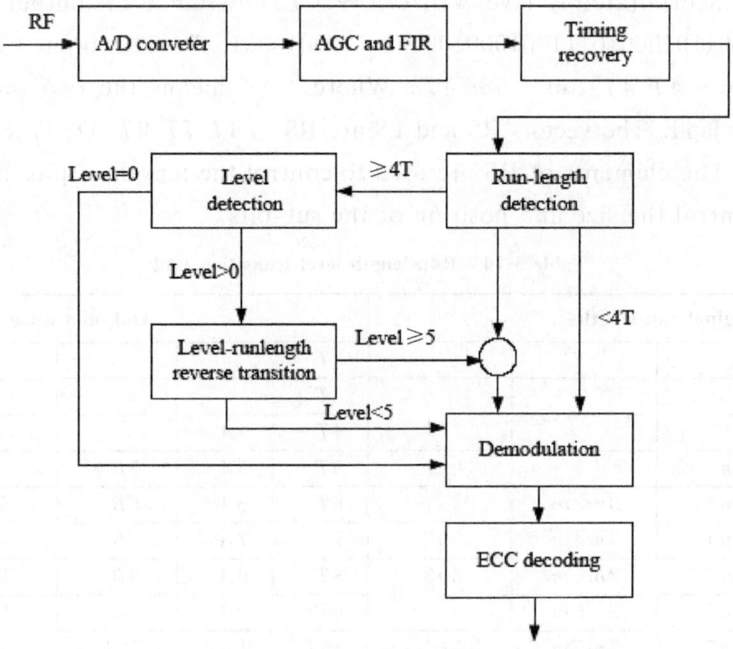

Fig. 3-123 SWM decoding system.

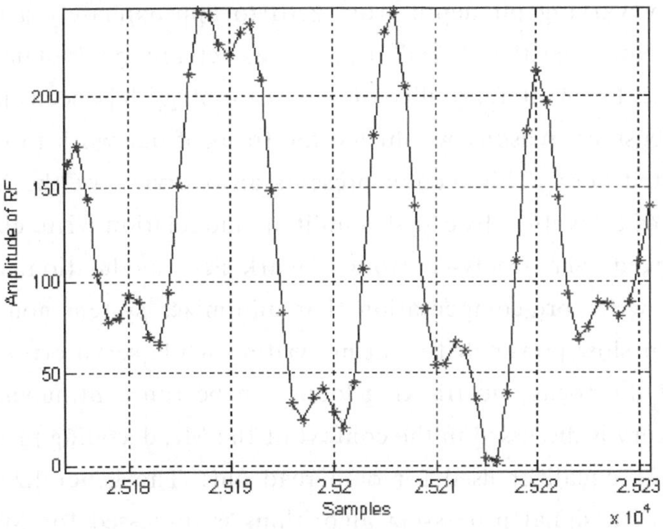

Fig. 3-124 Readback signal of multilevel pits and lands.

- Performance analysis of the proposed code

The density ration (DR) of this code is calculated by DR = $(1 + d) \times R$ = 1.75 bits/Tmin, where Tmin is the minimum recording mark. DR of EFM plus code adopted in DVD is 1.5. They cannot be compared directly, because Tmin in SWM and DVD is different. Tmin is $3T$ and the physical length is 400nm in DVD, while Tmin is $2T$ and 320nm in SWM disc. Hence, can define a new index termed information bits per 400nm (IBP) to characterize the relationship between the code and the capacity of disc. The IBP of SWM is 2.19, which is 46% higher than that of DVD. With the proposed code the capacity of SWM disc is improved 46%. The new modulation code for SWM-ROM is implemented and a capacity increase of 46% can be obtained. The encoding and decoding rules are detailed discussed and the hardware complexity is simple. Thus, the proposed scheme is a promising modulation method for SWM disc.

3.7.8 Multi-level amplitude-modulation

Multi-level (ML) encoding has been demonstrated to significantly increase the linear densities achieved with standard methods of binary encoding in optical data storage systems. An overview of the channel is provided from write encoder and precompensation, to read out with a matched adaptive equalizer and Viterbi decoder. The multilevel channel has been implemented in silicon and designed to operate in parallel with the standard binary channel, thus maintaining full backward functionality of the underlying drive that developed an efficient prototyping and integration process to accomplish final system implementation with minimal impact to the host system. Here discuss servo and firmware functions that ease integration of the multilevel LSI into existing circuit board architecture, firmware, and software bases. Additionally, the same LSI can increase DVD rewritable disc capacity to 7GB when combined with a 0.65NA DVD RW base drivel to 20GB if combined with a DVD-like system using a blue laser. HDML can also provide 25GB on a DVD-base with the addition of a blue laser and 50GB per layer in combination with a blue laser and such technologies that add higher NA optics and thinner disc covers. The initial work on ML was with pit-depth modulation (PDM) on CD-ROM systems to achieve a multilevel reflectivity response by varying pit depth from zero to approximately a quarter of wavelength depth in the medium. Note that the "PDM" process was actually pit "volume" modulation, as later TEM pictures revealed. Proof of feasibility of this technology has been demonstrated on DVD-ROM as well, but for business reasons, we shifted the focus of our work to writable and rewritable forms of ML as reported here. This section presents an overview of the ML system and discuss data encoding process of 8-level trellis coded, amplitude-modulation write channel. The 8 levels of reflectivity are achieved via finely-controlled mark-size modulation. Writing is done in combination with ML write pre-compensation that minimizes system non-linearity. ML power-control compensates for slow power drifts during writing while servo error-signals during writing are normalized so that the focus and tracking loops can be run continuously as in standard read mode. ML clock recovery is discussed in the context of the ML decoding process that is matched to an adaptive zero-forcing equalizer used for data read out. The generalized approach to system development, starting with signal processing algorithms is discussed for ML media research and development, then moving to FPGA prototyping and final system integration and testing of the ML Endec mixed-signal LSI. We conclude with a discussion of ML system implementation issues related to the LSI and drive circuit-board combination and the associated firmware and software. As with multi-amplitude and multi-phase signaling in modem or portable phone technology, ML optical technology transmits more information over a fixed bandwidth channel by using the available SNR more efficiently. Calimetrics has also worked with media manufacturers to slightly boost the SNR of phase-change and dye-based media by fine tuning the optical layer structure and composition for ML use. By combining these two pathways and utilizing an efficient ECC, Calimetrics has demonstrated the ability to manufacture a triple-density ML-R/RW disc and drive without changes to the optical pick-up head (PUH) or drive except for the addition of a chip. The ML physical specifications used in our system are shown in Table 3-16. The ML system can be understood by tracing the processing steps as data is encoded before being written on the disc and as data is decoded after being read from the disc (Figure 3-125).

Table 3-16 ML physical specifications compared to standard CD.

Parameter	Formula	CD	CD+ML
Cover Thickness (mm)	t	1.2	1.2
Laser Diode λ (nm)	λ	780	780
Objective Lens	NA	0.50	0.50
Track Pitch (μm)	p	1.6	1.55
Min. Mark Length or ML Data Cell Length (μm)	MML	0.833	0.600
Code Rate	$r = \frac{\text{data bits}}{\text{ch bits}}$	8/17	5/6
Channel Bit Length (μm)	$c = r/b \times \text{MML}$	0.278	0.200
Density (μm^2/ch bit)	$d = p \times c$	0.445	0.310
Data Bits per Min. Mark	b	1.41	2.50
Data Bit Length (μm)	MML/b	0.591	0.240
Linear Velocity (m/s)	v	14.4	14.4
Channel Bit Rate (MHz)	$f = v/c$	52	72
User Data Rate (Mbps)	$f \times E$	14	41

continued

Parameter	Formula	CD	CD+ML
Encoding Efficiency	$E = \frac{\text{User bits}}{\text{ch bits}}$	27%	57%
Total Efficiency	E/r	57%	69%
Program Area (mm^2)	A	8606	8653
User Data Capacity (GB)	$A/d \times E$	0.65	2.0

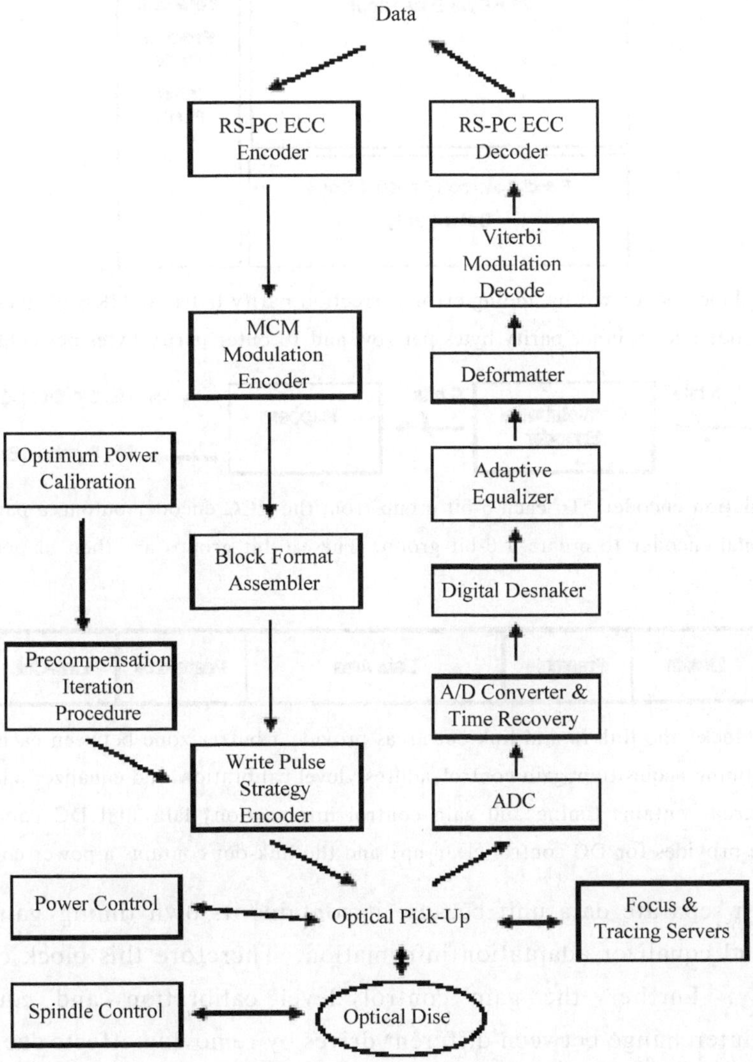

Fig. 3-125 Data flow in the multilevel encoding and decoding system. Data is ECC-, modulation-, and format-encoded before being translated into laser pulses that write data on an ML disc using the laser in the optical pick-up. For decoding ML signals, the all-sum signal (RF signal) is processed by setting the gain, converting from analog to digital, recovering the timing, adjusting for amplitude and offset, equalizing, and finally decoding for format, modulation, and ECC. Spindle, tracking, focusing, and laser control systems are all similar to existing optical disc systems.

On the encoding side, data is error correction encoded using a Reed-Solomon Product Code (RS-PC) (Figure 3-126). Unlike CD encoding, this ECC block is not interleaved with other blocks. Each block is stored on the disc as an independent unit. The modulation encoder processes the ECC block bytes (Figure 3-127) and provides additional error correcting capabilities that allows more effective use of the SNR of the optical data storage system. After modulation encoding, the stream of ML symbols is placed into a complete block structure that includes linking areas, a preamble, a data area, and a postamble (Figure 3-128).

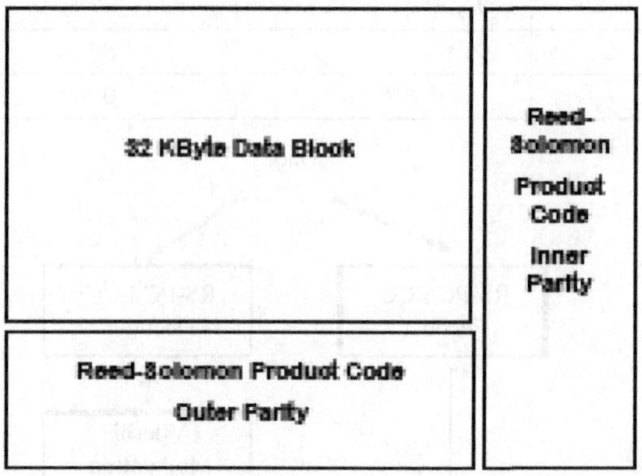

Fig. 3-126 The ECC block is created by adding error correction parity bytes to 32KB of data. The Reed-Solomon Product Code adds 5 inner parity bytes per row and 16 outer parity bytes per column to the data.

Fig. 3-127 The modulation encoder. To each 5-bit group from the ECC encoder output, a parity bit is added by an on volutional encoder to create a 6-bit group. These 6-bit groups are then mapped to two 8-level ML symbols.

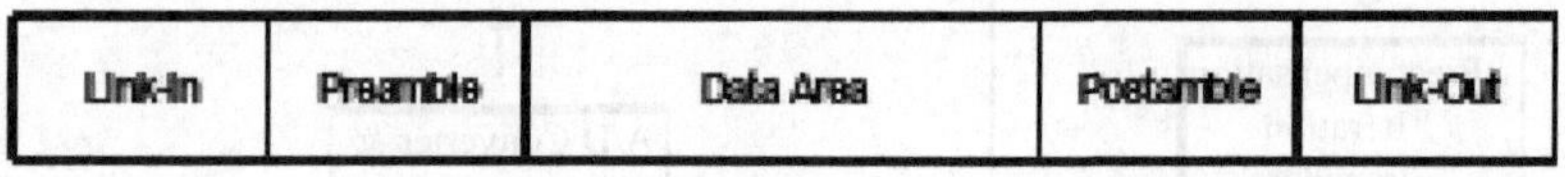

Fig. 3-128 ML data block: the link-in and link-out areas provide a buffer zone between each block; the preamble contains timing acquisition, gain control, address, level calibration, and equalizer adaptation information; the data area contains timing and gain control information, data, and DC control information; the postamble provides for DC control clean-up; and the link-out contains a power control pattern.

Each block is a separate data unit because it contains its own timing, gain control, address, level calibration, and equalizer adaptation information. Therefore this block can be written and read independently. Further, the gain control, level calibration, and equalizer adaptation subsystems enable interchange between different drives by removing effects due to mechanical and optical drive differences. Issues arising from disc defects and consumer abuse are also addressed by these subsystems. There are also subsystems to synchronize within the data block and to control the DC content of the ML signal.

- Writing calibration

The ML symbols are converted by the write strategy to laser pulses that actually write the ML marks on the disc as Figure 3-129. The write strategy is developed by an ML write calibration procedure that occurs when the first write command is issued. This calibration system begins with ML Optimum Power Control (OPC). ML OPC finds the optimum power(s) forwriting by writing a pattern with a pre-selected range of different power levels, reading back this pattern, and measuring distinctive metrics for this pattern (similar to CD's asymmetry or beta). The OPC pattern whose metrics are closest to the target value was written at the optimal powers. Next, the calibration system develops a write strategy with the ML Pre-compensation Iteration Procedure (PIPTM). MLPIP also writes out a test pattern, reads the pattern back, and performs measurements of the writing distortion due to the neighboring marks (Figure 3-129). PIP improves the ML writing strategy iteratively until the nearest-neighbor nonlinear writing distortion is below a threshold. The 8-level writing strategy has 512 different pulse definitions, one for each ML level with each of the combinations of neighboring marks as Figure 3-130. Equalized ML eye pattern showing an other perspective of above level distributions after PIP process is shown in Figure 3-131.

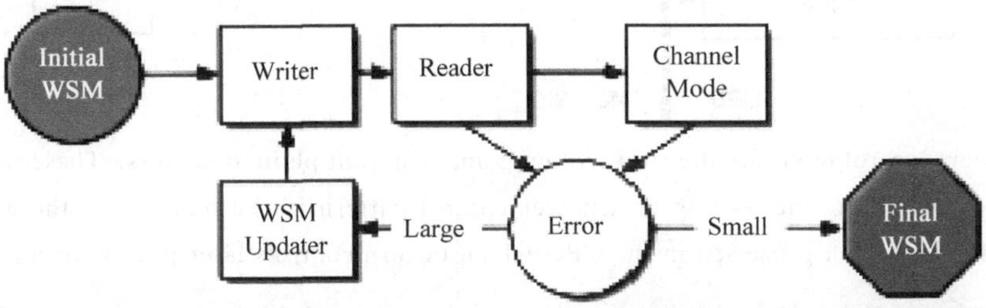

Fig. 3-129 ML PIP writes a pattern on the disc using an initial write strategy matrix (WSM).

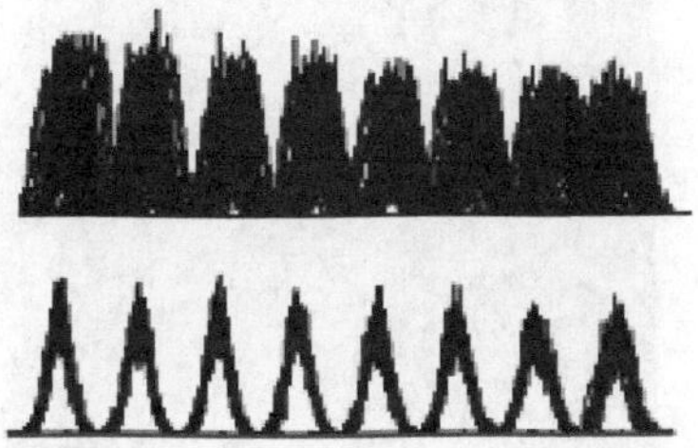

Fig. 3-130 Recovered reflectivity levels from random ML data. Top: before PIP; bottom: after PIP. Each of the 8 distributions contains 64 different traces resultant from every possible combination of the central level with all of its nearest neighboring data cells.

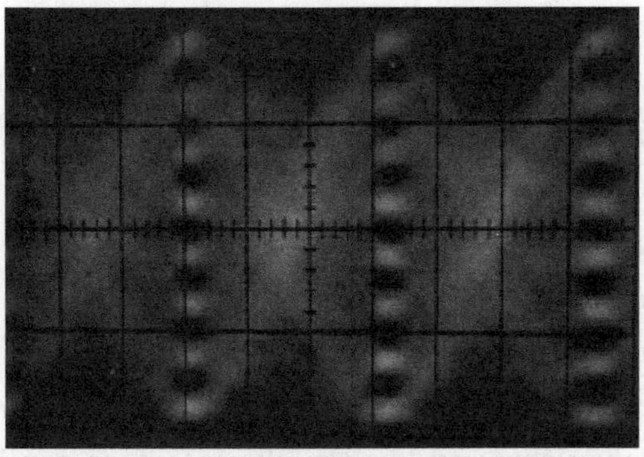

Fig. 3-131 Equalized ML eye pattern showing another perspective of above level distributions after PIP process.

Because the writing is tuned for a particular disc that has been inserted into a given drive, this adaptive write-strategy can correct for drive and/or disc variations. When writing ML-R, laser power is constantly updated against a target power value: back-reflected light from the disc and laser light at the forward-sense diode are measured during writing of a power-calibration pattern in the link-out of each block. The back-reflected signal is divided by a write-power control signal generated from the forward-sense signal and then compared to, and adjusted against, a target value. For ML-RW, constant power is maintained by sampling only the signal from the laser's forward-sense diode (Figure 3-132). For both ML-R and ML-RW, sampling of the photodetector output is synchronized with the writing of the power calibration pattern in the link-out area of the block. The power-calibration pattern contains a repeated sequence of long periods with different constant laser powers.

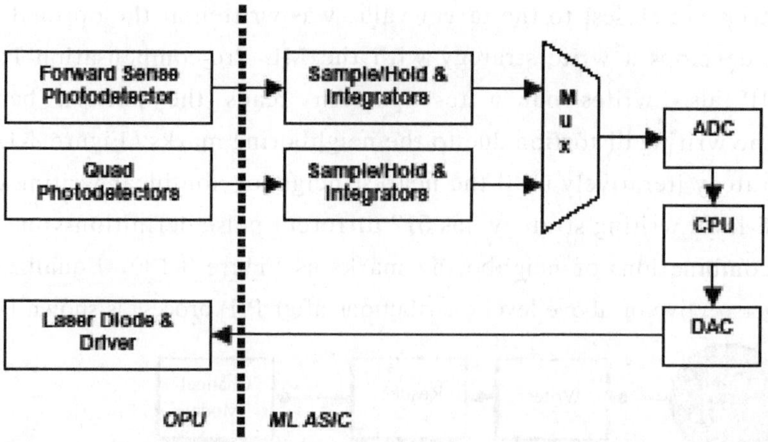

Fig. 3-132 ML Power Control monitors the forward sense and quadrant photo detectors. These signals are sampled at specific times during the writing of a power control pattern in the link out of the ML data block. The sampled data is then processed in the CPU which in turn controls laser power through DACs.

This pattern is read back and compared to a linear model of the channel. If the errors are large, the WSM is updated and the process repeated until the error falls below a threshold and achieves the final WSM.

An example showing the results of the fine control over the laser pulse writing strategy is shown in Figure 3-133. It is important to note that laser writing power levels for the ML-R and ML-RW materials are very similar to the conventional CD-R and CD-RW power levels, so no change is required to the laser. Details about the fundamental media mechanisms and head-media interface can be found in other recent presentations.

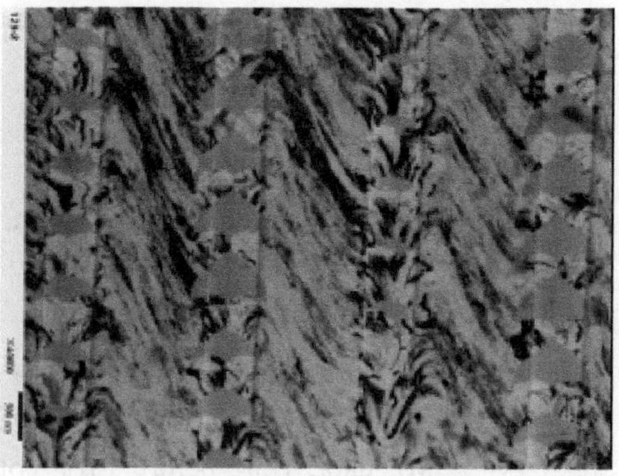

Fig. 3-133 TEM of ML marks in phase change media on CD dimension substrate. ML data cell length (mark edge to edge) is 600nm. Shape and extent of ML mark can be finely controlled with the write-pulse strategy.

There is no servo modification required for reading ML data, but a small change is needed for ML writing. In standard CD/DVD writing, the different lengths of the RLL modulation code are written on the disc using alternating marks and spaces of different lengths. Because of this signaling method, there is always a space following each mark and a mark following each space. Many CD/DVD focus-and tracking-error signal-generators sample the PUH servo-signals during the spaces in the modulation code because during these times there is a stable and constant power level: the reading power for R media and the erasing power for RW media. However, because ML does not have these long spaces during writing, these types of focus-and tracking-error signal generation are not possible. Rather, the ML system separately normalizes the main quadrant photodetector signals and outrigger photodetector signals from the PUH (Figure 3-134). These signals are then used to generate continuous focus-and tracking-error signals.

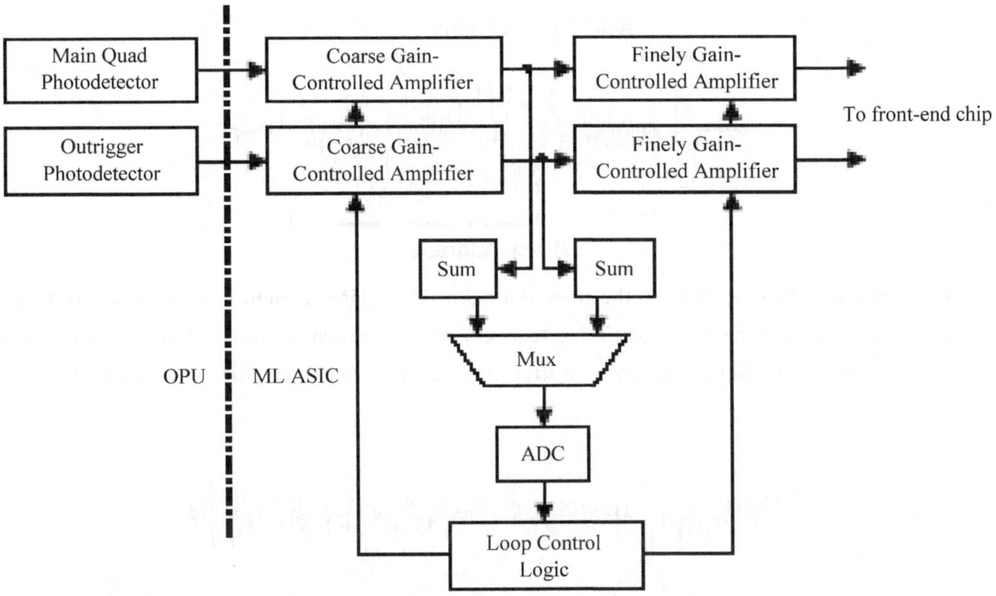

Fig. 3-134 During writing, the main quad and outrigger photodetector signals need to be processed to remove the effects of the writing pulses. This is achieved by normalizing these signals before they go to the front-end chip. The outrigger photodetector signals are separately summed and used to adjust the gain-controlled amplifiers. Once adjusted, the main and outrigger signals are passed to the front-end chip, which uses these signals to calculate the focus and tracking error signals.

For ML spindle control, speed can be adjusted either from an error signal generated from the media's wobble groove or from the recovered data clock. Note that similar to the CD system, the wobble groove also contains addressing and other system information including target media metric values. Refer to the wobble format as Address-in-Pre-groove (AIP).

- Decoding

On the decoding side, the disc is read using the all-sum signal from a standard PUH. After this signal passes through again-controlled amplifier and digitized, timing is recovered using marks in the preamble (Figure 3-135). This timing information is used to sample the all-sum signal to produce a digital stream at twice the ML symbol rate. Timing and gain/offset are maintained by using a series of marks that provide both a clean signal-edge for clock recovery as well as minimum and maximum signal levels for envelope monitoring to provide for AGC. Before the entire block is decoded, the block address is examined to ensure that the correct block is being read (Figure 3-136). Once the block address is confirmed, the digital processing of the multilevel signal begins with a fine adjustment of the gain and offset using measurements of the envelope. An example of an unequalized data signal is shown in Figure 3-137.

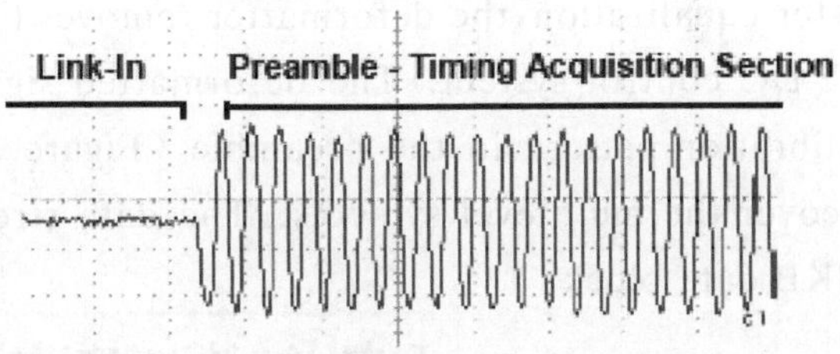

Fig. 3-135 PUH all-sum signal showing the transition from the link-in to the timing acquisition section of the preamble.

The ML data signal is then equalized by an 11-tap fractionally-spaced equalizer. The taps are trained at the beginning of each block using an equalizer adaptation pattern. Equalization of the signal removes the intersymbol interference caused by the interaction of the readout spot on the

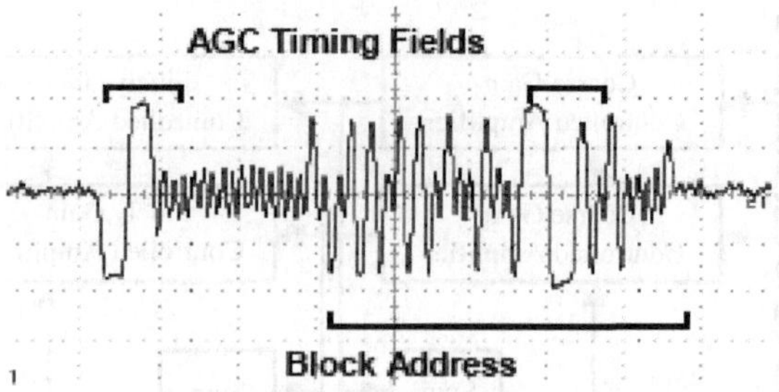

Fig. 3-136 Block address section of the preamble. An AGC/Timing field is seen on the left. This field contains an edge for timing recovery and maximum/minimum levels for gain/offset control. The block address is written in an easy to read code for error resilience.

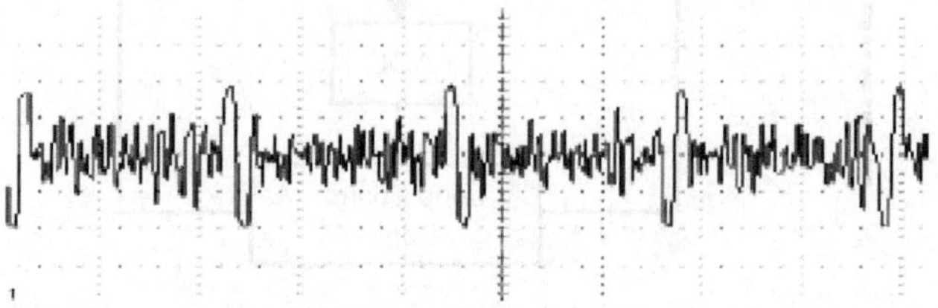

Fig. 3-137 Unequalized random ML data signal. This analog signal is from after the desnaker, but before the equalizer.

disc. Figure 3-138 shows the equalized output of a typical data set. For this data, the standard deviation of each signal level expressed as a percentage of the dynamic range is ~3%. If a hard decision level were placed between each level, it would give a symbol error rate of ~ 10^{-2}. However, because of the advanced coding system design, this data has an ~10^{-5} error rate after the Viterbi decoder and no errors after error correction. The low error rate after the Viterbi decoder is achieved not only because the convolutional coding provides a degree of error correction, but also because of the level calibration system and the adaptive equalization system. These two systems help to compensate for interchange effects that arise from differences in discs and drives. After equalization, the deformatter removes the non-data marks and adjusts the signal according to the DC control system. The deformatted signal and level information, measured from the level calibration pattern in the preamble (Figure 3-139), is used by the 256-state Viterbi decoder to recover the multilevel symbols. The data stream is then RS-PC decoded to produce the original 32KB data block.

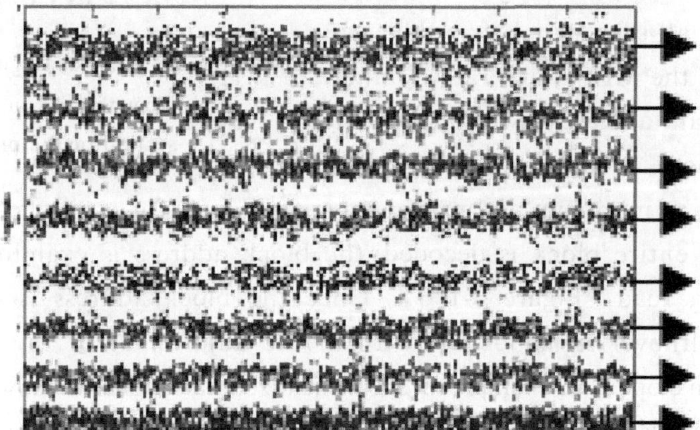

Fig. 3-138 Equalized random data. The 8 levels of the ML signal are indicated by the arrows to the right of the graph.

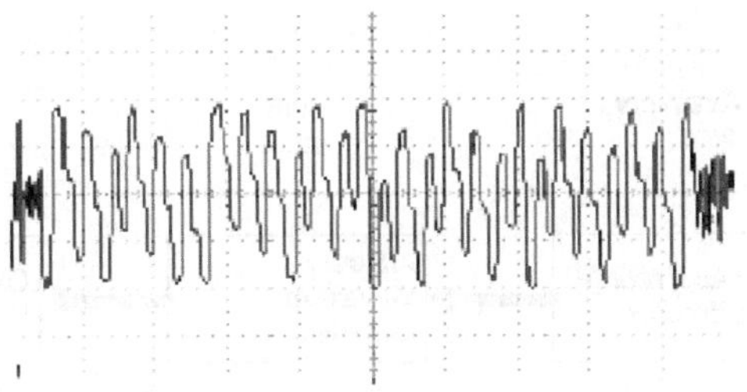

Fig. 3-139 Example of a level calibration section of the preamble. The 8 levels of the ML signal can be measured to counteract effects due to inter change.

Calimetrics has written ML marks with many commercially available OPUs; all OPUs tried to date have been successful. Those familiar with Calimetrics' media metrics will be familiar with SDR of the unequalized signal, which should be less than 2%. The 3% reported here is the equalized SDR, which is typically larger than the unequalized SDR.

ML technology involves several key components that are researched, developed and implemented in parallel, including our proprietary mixed-signal processing LSI, our proprietary firmware, commercial and proprietary media, base-drive integration, and integration into existing commercial software. In this section, we discuss the methodology used to organize a parallel development effort in terms of four phases: Phase A—Signal processing algorithms and media research; Phase B—Signal processing algorithms and media development; Phase C—Systems prototyping; and Phase D—Systems integration.

- Signal processing algorithms and medium

Algorithm and media research begins with a commercial media tester and the MATLAB rapid-prototyping language. Using commercial lab equipment to capture and generate ML waveforms, can rapidly develop optimal, floating-point algorithms for write pre-compensation, writing, and reading. Included full scripting capability in all of our "System A" components used in Phase A is shown in Figure 3-140) such that full margin testing is available with the commercial media tester, including radial and tangential tilt, tracking and focus offsets, and combinations thereof. System A provides a fast and accurate proof-of-concept evaluation of ML technology's effectiveness and robustness for a given base drive or PUH.

Past the proof-of-concept evaluation, the ML technology for a given base drive or PUH can be successfully implemented in a commercial product. This is the purpose of "System B" is shown in Figure 3-141. By building a custom media tester optimized for ML technology and using modular, swappable production-level Opto-Mechanic Assemblies (OMA), that can get an accurate estimation of the integration challenges involved with various researchers.

The signal processing algorithms are partitioned between the analog and digital domain. Analog circuitry is designed and implemented in custom, rack-mount systems. The digital signal processing uses a workstation with a custom C++ software emulator to ensure bit-exact and cycle-exact functionality for the eventual mixed-signal LSI. The emulator serves multiple purposes: it operates precisely as the LSI for system demonstration, development and testing; it provides the LSI designers with a functional specification; and it allows a powerful method of keeping test benches and vectors consistent throughout the entire design process.

Results from media research on system A are passed forward to System B, and the same

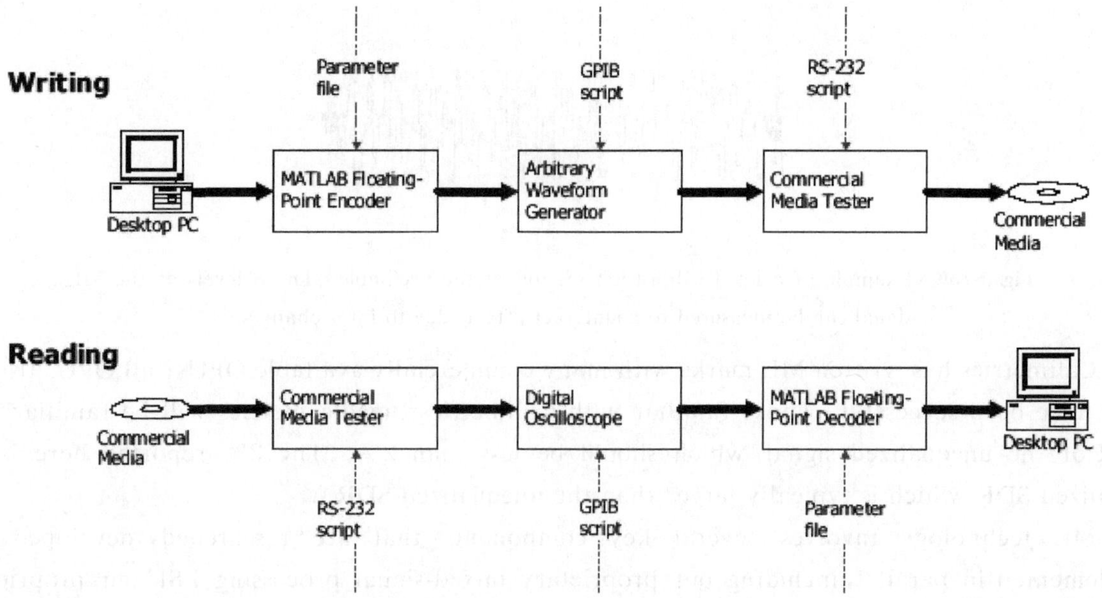

Fig. 3-140 Algorithms and media research: The form-factor of the system is quite large and the read-signal processing rate is about 1/820 of CD 1x read speed, but since the automation level is quite high, it allows the system to be left operating overnight.

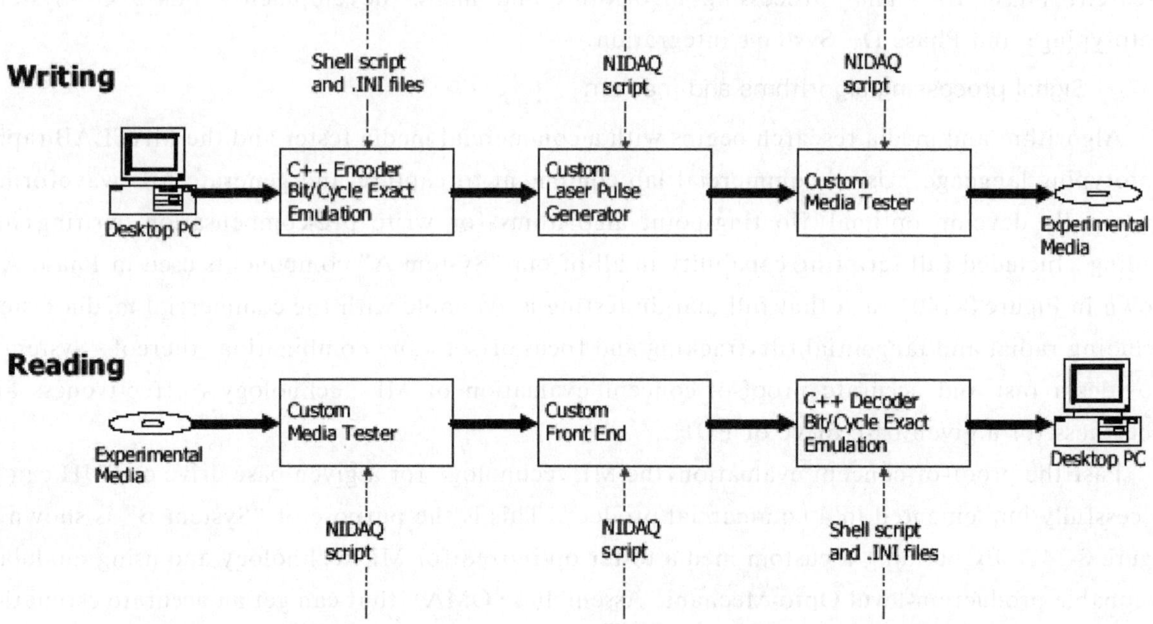

Fig. 3-141 Algorithms and Media Development: The form-factor of the system is now reduced to rack-size, the read-signal processing rate is about 1/140 of CD 1x read speed, and the automation level remains quite high.

battery of speed, capacity, manufacturability, and robustness tests are carried out to evaluate various dye formulations, phase-change optical stack designs, and groove geometries. Commercial media can also be used during this phase depending on the write-speed requirements. ML can provide the same capacity increases on commercial media, albeit at a slower speed than media optimized for ML writing. All hardware is given full script capability to provide automation for extensive testing. Phase B completes the feasibility study for ML technology's effectiveness and robustness for a given optical system.

- Prototype of the systems

Full system prototype is shown in Figure 3-142. An out-of-form-factor board ("Big Board") is used to verify interfaces and functionality, while remaining flexible for rework and debug. In implementing "basic functions" to help with low-level system debug. The same custom analog hardware from Figure 3-141 is used, but now with an FPGA emulation of the digital signal processing in the LSI. The FPGA emulation allows flexible validation of the LSI Hardware Descriptor Language (HDL) coding in two ways. First, the FPGA allows for connecting to internal wires that are not usually brought out to pins. Second, the FPGA can be modified easily during debug to try various fixes. Finally, the FPGA system allows for commencement of firmware design, coding, and testing. Based on past experience, we can get a fully-functional system working at about a third of the target LSI speed. For higher speed tests, System A and System B are used, albeit with less ATAPI functionality than the Big Board.

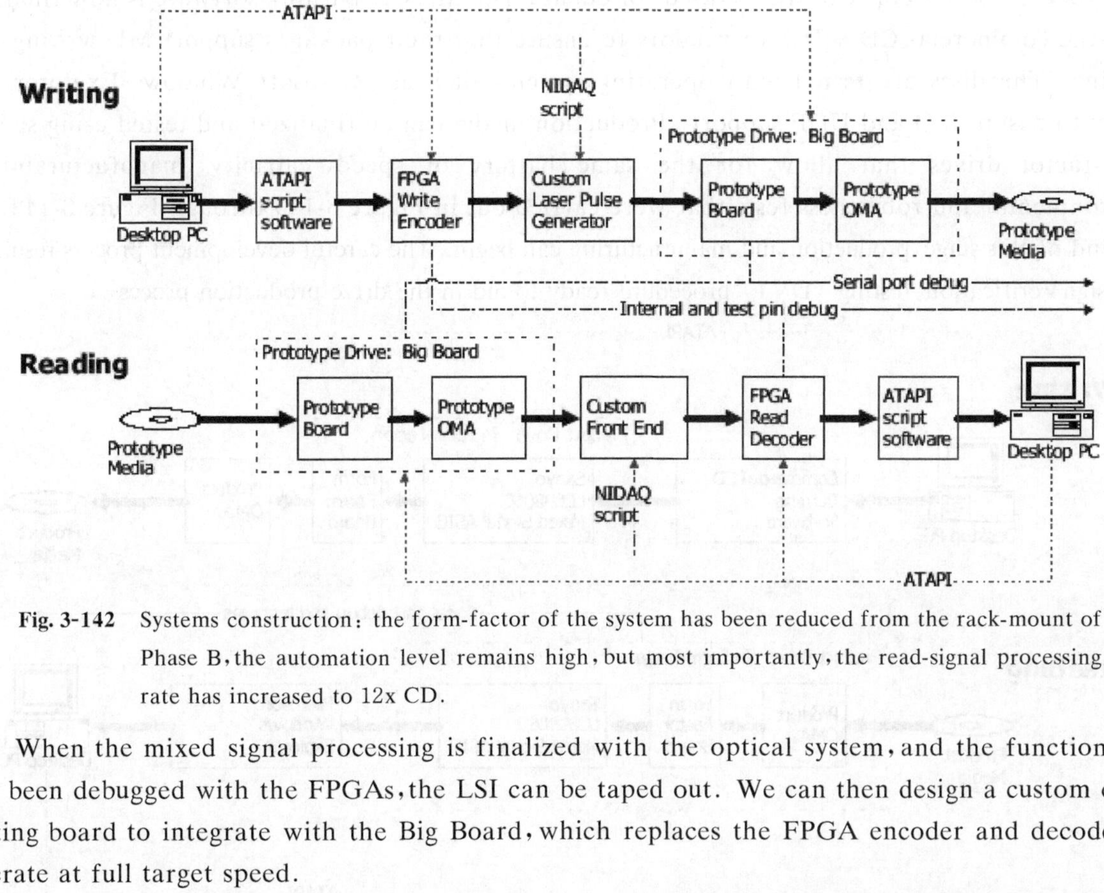

Fig. 3-142 Systems construction: the form-factor of the system has been reduced from the rack-mount of Phase B, the automation level remains high, but most importantly, the read-signal processing rate has increased to 12x CD.

When the mixed signal processing is finalized with the optical system, and the functionality has been debugged with the FPGAs, the LSI can be taped out. We can then design a custom chip-testing board to integrate with the Big Board, which replaces the FPGA encoder and decoder to operate at full target speed.

From the start, ML firmware is carefully crafted to allow debugging, both through the use of in-circuit emulators and using serial port output. Additionally, we have designed a general-purpose, scriptable, low-level ATAPI package to test firmware in an automated fashion. Because the drive is an ATAPI device, all drive functionality (e.g., servo, spindle and seek) can be scripted for extensive testing. To allow commercial CD-burning software to provide ML functionality, our software and firmware teams can provide an ATAPI command-set to vendors to allow for easy integration. The commercial CD-burning packages can begin debug after finalization of related firmware during Phase C.

At this point in the program, media can be tested with the vendor's OMA. Minor modifications can be made to fine-tune media-servo interactions or to provide more robust margin performance if necessary. The Big Board can be used for the same battery of speed, capacity, manufacturability and robustness tests as done as in Figure 3-140 and Figure 3-141. Modifications can also be made to test environmental conditions such as temperature, light fastness, and media defects such as scratches and fingerprints. At the end of Figure 3-142, all complex details with regards to hardware, firmware, and media can be determined and final systems integration can begin.

3.7.9 Systems integration

With all design details set and tested, the final stage of the ML integration process can begin (Figure 3-143). Using an inform-factor drive base, the mixed-signal LSI is integrated on the circuit board. Firmware requires minor modifications going from the Big Board base to the form-factor drive, but the serial port can still be used for debug. The ML CD-burning software is now finalized with the commercial CD software vendors to ensure that their packages support ML writing and reading. The discs are tested with operating systems such as Microsoft Windows Explorer and Linux to ensure ISO and UDF support. Production media can be finalized and tested using special form-factor drives that allow for the same battery of speed, capacity, manufacturability, environmental and robustness tests that were carried out in Figure 3-140 through Figure 3-142. At the end of this stage, production and manufacturing can begin. The careful development process results in a DesignVerification Testing (DVT) procedure ready to aid in the drive production process.

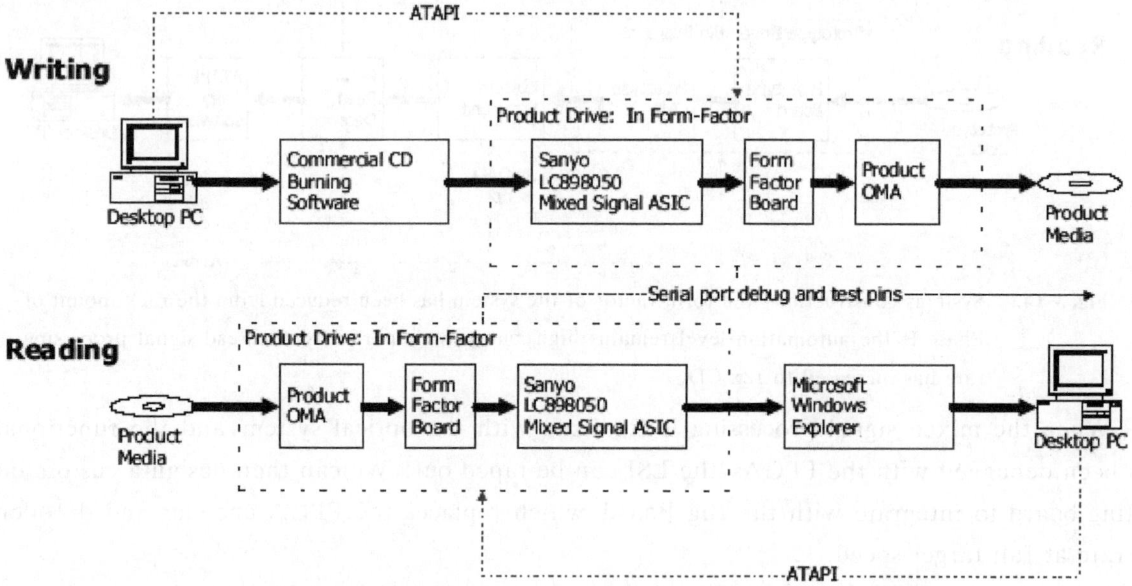

Fig. 3-143 Systems Integration: The form-factor of the system is now that of a standard 120mm disc drive bay, automation still very high, and the read-signal processing rate is 36x CD across the whole disc due to underlying CLV technology.

A key point of ML technology is the ease with which it integrates with existing drive technologies. From inception, the ML LSI and firmware model are designed to simplify integration with a conventional optical drive. In this section, the 2GB CD product and the Sanyo LC898050 LSI are used as a case study. The Sanyo LC898050 and ML firmware library have been successfully integrated with two vendors' drives to date. In this section, discuss the ease of implementing ML technology from three perspectives: (1) LSI and drive circuit board, (2) firmware, and (3) software.

The first-generation ML LSI, the Sanyo LC898050, is an easy-to-integrate, stand-alone ML Endec that is fully self-contained and requires little modification to integrate into existing drive systems. Additionally, we have designed internal interfaces to take advantage of future "cost downs" that can come with LSI core integration.

Part of the strength of ML technology lies in the PIP process described before. To ensure that the PIP procedure is fast, automatic, and compensates for a given drive-media environment. By its nature, PIP doesnot require the time-consuming characterization for every media type to be supported as in the conventional case. PIP, in combination with the adaptive equalization, timing recovery, and level recovery process described in paragraph, ensures a self-calibrating, adaptive, closed-loop system that minimizes the development needs for our system.

By keeping both the digital and analog signal processing within the ML LSI, the interface and signal integrity issues are minimized. In addition, the ML system has been designed to be programmable with many user options corresponding to different environments. This openness also extends to test ports and interrupts, to allow system diagnostic information be accessible for drive bring-up and media analysis. The ML Endec LSI makes the integration virtually a drop-in. Note that the standard drive components and signal paths of the underlying non-ML drive remain unchanged. The OMA, the analog front end, and the servo power drivers are identical to a conventional drive. A minor ML interface is added to the controller chip that remains otherwise unchanged.

The controller's ML interface does three things:

(1) provides a local data input and output bus for the ML LSI;

(2) provides a demodulated wobble signal, shown in Figure 3-144 as the Bi-Clock signal and Bi-Data signal;

(3) accepts a spindle control output interface. The demodulated wobble signal and spindle control interface allow the ML LSI to provide addressing information and speed control when ML media is inserted.

For Sanyo, adding the ML interface to their existing controller was a minor modification that took three months from beginning to end, at which point the modified controller was ready to interface with the ML Endec. Since then, Sanyo has enabled other controllers that provide "combi" drives with both DVD decoder and CD encoder/decoder functionality. A future controller is also planned that integrates a "super-combi" controller that not only includes both DVD and CD encoder/decoders, but also integrates the ML Endec functions as a core, eliminating the need for a separate ML LSI. To reduce cost further, the ML Endec has been designed to easily integrate with the analog front end chip and the controller chip (Figure 3-145).

- Firmware support for ML technology

Adding firmware support for multilevel (ML) technology is accomplished in 4 steps:

(1) develop drop-in ML-Library helper functions;

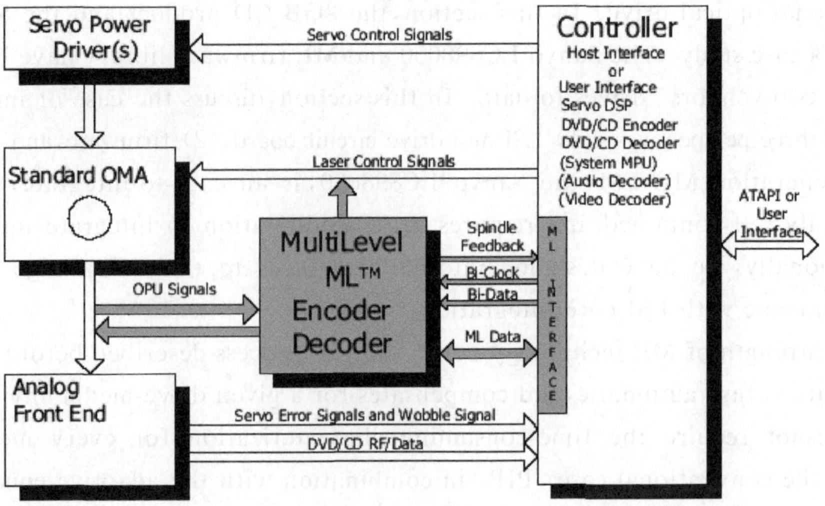

Fig. 3-144 An ML-enabled drive architecture. The shaded region of the drawing shows the necessary ML additions or connections required, while the remainder is a typical drive architecture. The system microprocessor, memory and their associated connections are not included for clarity. At the board level, multiplexers are needed at two signal paths in the integration with the current ML chip. One of the signal paths is the laser control path to maintain backward compatibility with conventional EFM recording. The other signal path needed comes from the PUH photodetector outputs since ML recording uses servo-normalization during writing.

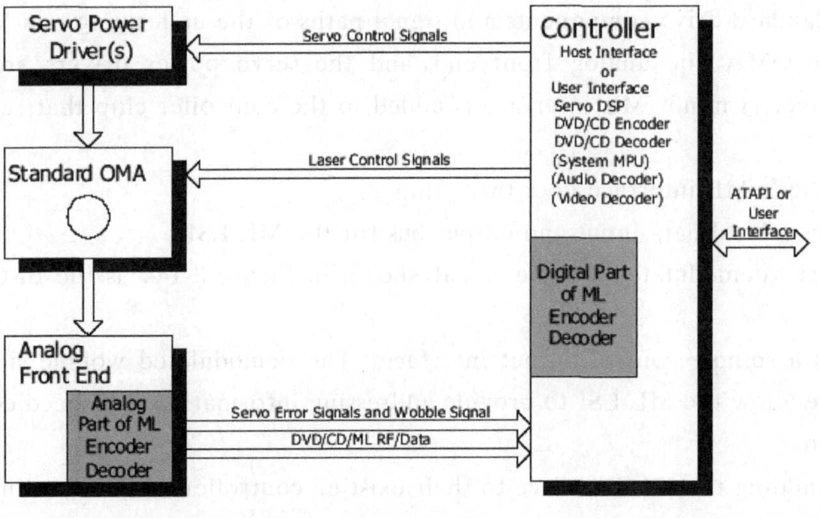

Fig. 3-145 An integrated ML-enabled drive architecture. ML chip is integrated into the controller and the analog front end. The chip number count on board is the same as a non-ML-enabled drive.

(2) provide support for ML-media in disc initialization;

(3) adjust the ATAPI command execution routines to account for ML-media;

(4) provide some simple "callback" routines. Designed the ML-code Library set (ML-Lib) to be easily added to pre-existing firmware sets or incorporated into the design of a new firmware set. ML-Lib is designed as a library of functions to be invoked by vendor firmware code when using the ML LSI. These functions have 2 purposes: to provide a clean "wrapper code" interface to the MLLSI and to provide calculations to aid in the use of ML functionality. ML-Lib code will not take control of the drive.

Rather, it is designed to extend the vendor firmware capability to handle the ML case. The library is a set of "helper functions". ML-Lib code consists of C-language routines that fall into three main groups:

(1) those functions to be called directly by vendor firmware;

(2) those to be called only by other ML-Lib functions;

(3) an interrupt routine. ML-Lib functions are called code in the process of carrying out a large-scale task, such as reading blocks. ML-Lib provides a suggested interrupt service routine to be invoked by the interrupt from the ML LSI. When an ML disc is inserted, standard CD vendor firmware will not recognize the type of disc. So we have provided ML functions that read the ML AIP information in the wobble-groove and thus identify the disc as an ML variety. The library also supplies several routines to calibrate the analog portion of the ML LSI during startup.

In order to make ML-Lib as general, complete, and logically consistent as possible, we define firmware functions that are anticipated from the vendor's code that will also be made available to ML-Lib functions. These are referred to as "callback" functions. Functions in this category are used to set laser power, tracking and focus offsets, etc. Table 3-17 shows the general sequence of events for reading data from ML-media. Each of the called functions sets up a few registers in the ML LSI and returns. The overall operation of the read process and its timing are entirely under control of the host firmware. The write command is similar, containing one additional step of producing a write strategy matrix for the disc and drive combination if necessary.

Table 3-17 General read command sequence.

Read Commands	
Host Firmware	ML-Lib Routines
Receive command	
Call → FindBlock0	Locate Block
Seek to Block	
Call → MLPrepRead0	Prepare chip for reading
Arrive at Block	
Set up buffer pointers	
Call → MLInitRead0	Initialization for reading
Call → MLRead 0	Begin reading. Enable interrupts
Count block received, send to host	Continue reading
Count block received, send to host	Continue reading
Count block received, send to host	Continue reading
⋮	⋮
When all blocks rcvd, Call → MLRdStop0	Stop read process. Disable interrupts

The three main factors responsible for capacity improvement are highlighted: reduction of data cell length, increasing the number of bits per data cell, and increasing the encoding and total efficiency of the system.

• Implementing ML technology from software

The communication interface between the PC host software and the ML drive is across the IDE bus using the ATAPI protocol. The ML ATAPI interface is almost identical to the "Mt. Fuji" (v4) specification. Here have modified the responses to some of the commands and created a mode page to handle responses that do not fit within existing structures. It is straight forward to adapt existing software interfaces to CD/DVD drives to accommodate ML. In fact, it is worked with ahead software to interface the current product, NERO Burning ROM, with the ML optical data storage system. Currently, we use NERO Burning ROM to write all of our ISO-formatted ML discs. These ISO formatted ML discs are recognizable from Windows, and the files can be opened and read by simply double-clicking. ML methodology has been demonstrated on CD/DVD-ROM, CD-R/RW, DVD-R/RW, and blue laser, high-NA systems. From this variety of experience, ML technology can be applied to virtually any optical media system available. With over eight years of experience in optical data storage systems, Calimetrics retains intellectual property, software and hardware tools to easily apply this methodology to any new optical system quickly.

- Multi-valued phase patterns recording

Multi-valued phase patterns were multiply recorded by retardagraphy in order to improve recording density and data transfer rate. In the experiment, the phase pattern consists of four values were recorded on a polarization-sensitive medium by focusing the recording beam, and three patterns were multiply recorded by shifting the focal point. The recorded patterns could be independently reconstructed. Optical recording technique utilizing property of vector wave is expected for optical data storage with high density, large capacity and high data transfer rate. The vector wave recording techniques such as retardagraphy and polarization holography have been investigated. In retardagraphy, optical information consisted of the phase retardation between two orthogonal polarization components is record on a polarization-sensitive medium using a single beam. The advantages of this technique are robustness for vibration and simplicity for optical system. Additionally, this technique can be regarded as a kind of in-line polarization holography, thus it is possible to the holographic multiplex recording cause of Bragg selectivity in the case using a thick recording medium. Then, the optical information is recorded as a three-dimensional birefringence pattern on the polarization-sensitive medium. Thus it has good shift selectivity and can increase more crossover of an adjacent recording region. In our previous work, the improving method of recording density was investigated only the multiplex recording using binary pattern, but now we propose a method that uses the multiplex recording using multi-valued phase pattern to further improve recording density. In this paragraph, the method of improving recording density in retardagraphy is investigated using multi-valued phase data.

The phase of two orthogonal polarization components included in a recording beam can be independently modulated using a spatial retarder. In retardagraphy, the phase retardation between two orthogonal polarization components as signal and reference components is modulated. It adequately converted into the polarization basis and recorded on polarization-sensitive medium. Most of polarization-sensitive media have sensitivity for polarization azimuth. Then, an optical anisotropy such as birefringence and dichroism are induced in the media. The principal axis of the optical anisotropy is dependent on the polarization azimuth. Thus, the phase retardation is correctly recordable by converting so that it may correspond to a polarization azimuth. The principle of optical recording by retardagraphy is shown in Figure 3-146. Here, it is assumed that the spatial phase retardation distribution Φ between p-and s-polarization components is given by using a spatial retarder.

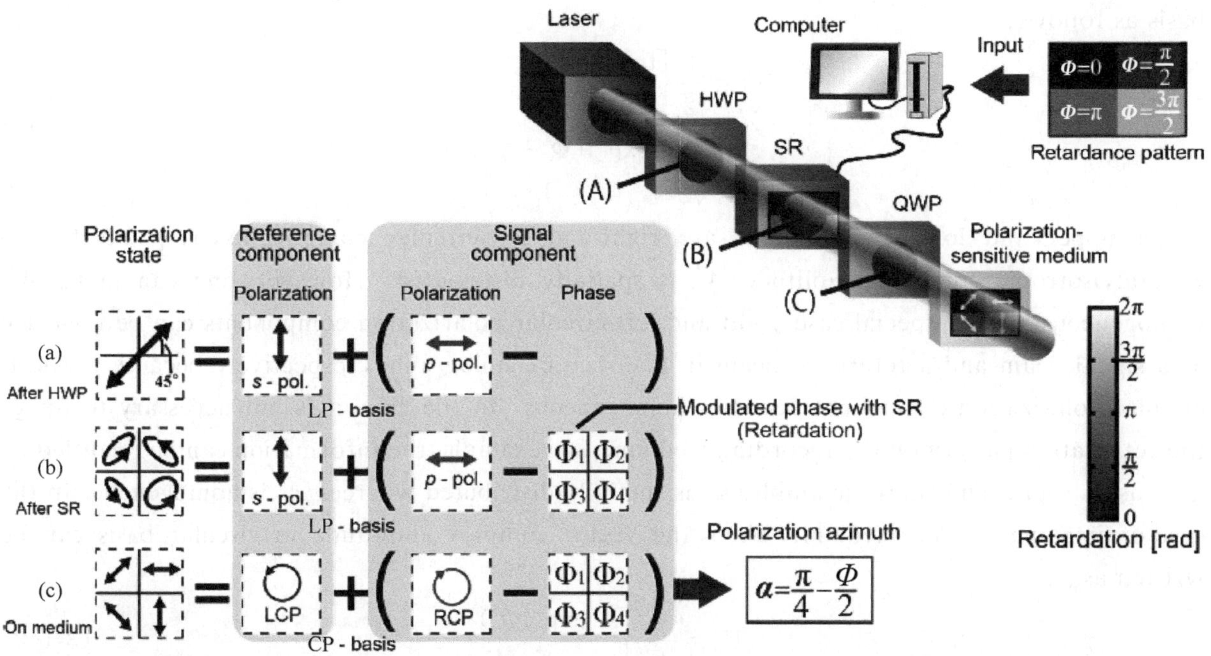

Fig. 3-146 Schematic of retardagraphy: HWP, SR, QWP are half-wave plate, spatial retarder, quarter-wave plate, respectively.

When 45-deg linear polarized light is illuminated onto the spatial retarder, the polarization distribution becomes elliptical and represented by

$$
\begin{aligned}
U_1 &= T_{SR}\begin{bmatrix}\exp(\mathrm{i}\Phi/2) & 0 \\ 0 & \exp(-\mathrm{i}\Phi/2)\end{bmatrix}\frac{1}{\sqrt{2}}\begin{bmatrix}1\\1\end{bmatrix} \\
&= \frac{T_{SR}}{\sqrt{2}}\begin{bmatrix}\cos(\pi/4) & -\sin(\pi/4)\\ \sin(\pi/4) & \cos(\pi/4)\end{bmatrix}\begin{bmatrix}\cos(-\Phi/2)\\ \mathrm{i}\sin(-\Phi/2)\end{bmatrix}
\end{aligned} \tag{3-139}
$$

where TSR is the isotropic amplitude transmissivity of the spatial retarder. The polarization state is elliptical with the azimuth of 45 of -45 degrees, and the ellipticity angle becomes $-\Phi/2$. Polarization-sensitive medium for polarization azimuth is sensitive to not only polarization azimuth but also polarization ellipticity. However, the polarization-sensitive medium cannot recognize the handedness of the elliptical polarization. In order to precisely record the optical information, it is necessary to convert the polarization state corresponding to the properties of the recording medium. Using a quarter-wave plate with a fast axis of 45 degrees, the polarization state becomes as follows:

$$
\begin{aligned}
U_2 &= Q(\pi/4)U_1 \\
&= A_2\begin{bmatrix}\cos(\Phi/2-\pi/4) & \sin(\Phi/2-\pi/4)\\ -\sin(\Phi/2-\pi/4) & \cos(\Phi/2-\pi/4)\end{bmatrix}\begin{bmatrix}1\\0\end{bmatrix}
\end{aligned} \tag{3-140}
$$

where $Q(\pi/4)$ and A_2 are a Jones matrix of the quarter-wave plate with a fast axis of 45 degrees and isotropic complex amplitude, respectively. Thus, the polarization state becomes linear polarization distribution with the azimuth of $-\Phi/2+\pi/4$. When the polarization distribution is imaged onto the polarization-sensitive medium, birefringence distribution is induced and the optical information can be recorded as the principal axis of the birefringence. The polarization basis of Jones vector is linear. Here, the linear polarization basis is converted into circular polarization basis as follows:

$$\begin{aligned} U'_2 &= \frac{1}{\sqrt{2}}\begin{bmatrix} 1 & -\mathrm{i} \\ \mathrm{i} & 1 \end{bmatrix} U_2 \\ &= \frac{A'_2}{\sqrt{2}}\begin{bmatrix} \exp[\mathrm{i}(\Phi - \pi/2)] \\ 1 \end{bmatrix} \end{aligned} \tag{3-141}$$

where upper and lower components are right-and left-circular components, respectively. In general, isotropic complex amplitude A'_2 is spatially distributed. However, one can make A'_2 homogeneous. In the special case, right-and left-circular polarization components can be regarded as a signal beam and a reference beam in polarization holography, respectively because the left-circular polarization component becomes homogeneous. In the case, it is not necessary to image the retardation pattern on the recording medium. For example, the information can be recorded by a focused beam. The isotropic amplitude is spatially distributed whereas A'_2 is homogeneous in the case of imaging on the medium. Then, the vector complex amplitude in circular basis can be written as

$$U_3 = \frac{A'_3}{\sqrt{2}}\begin{bmatrix} A'\exp(\mathrm{i}\Phi') \\ 1 \end{bmatrix} \tag{3-142}$$

where A'_2, A', and Φ' are the distributed isotropic amplitude, the amplitude ratio to the reference component, and the phase difference between the signal and the reference components, respectively. Then, the amplitude transmissivity tensor induced in the medium is expressed by

$$H = T_H R(-\Phi'/2) M R(\Phi'/2) \tag{3-143}$$

$$M = \begin{bmatrix} \exp(\mathrm{i}\Delta\phi/2) & 0 \\ 0 & \exp(-\mathrm{i}\Delta\phi/2) \end{bmatrix} \tag{3-144}$$

$$R(\varphi) = \begin{bmatrix} \cos\varphi & \sin\varphi \\ -\sin\varphi & \cos\varphi \end{bmatrix} \tag{3-145}$$

where T_H and $\Delta\phi$ are isotropic amplitude transmissivity and retardance induced in the medium, respectively. The photoinduced retardance is dependent on the intensity and the ellipticity of the recording polarized beam, that is, isotropic amplitude and the amplitude ratio of right-and left-circular polarization components. In reconstruction, a left circularly polarized beam is illuminated onto the medium. Then, the complex amplitude vector is expressed by

$$U_4 = H\frac{1}{\sqrt{2}}\begin{bmatrix} 1 \\ -\mathrm{i} \end{bmatrix} = \frac{1}{\sqrt{2}}\left\{\cos(\Delta\phi/2)\begin{bmatrix} 1 \\ -\mathrm{i} \end{bmatrix} + \mathrm{i}\sin(\Delta\phi/2)\exp(\mathrm{i}\Phi')\begin{bmatrix} 1 \\ \mathrm{i} \end{bmatrix}\right\} \tag{3-146}$$

The recorded retardation pattern is included in the second term of the equation. When sin $(\Delta\Phi/2)$ is regarded to be proportional to the intensity of the recording signal component, the signal is completely reconstructed. The information in the signal component is extracted by imaging polarimetry. As shown in Figure 3-147, when the recording pattern recorded on a polarization-sensitive medium by focusing the recording beam, the signal and reference components are changed Fourier pattern and spherical wave like the spherical wave shift multiplex in a holography, respectively. Then, the optical information is recorded as a three-dimensional birefringence pattern on the polarization-sensitive medium. Similarly, the Bragg selectivity was formed on a recording medium. Thus it has good shift selectivity and can increase more crossover of an adjacent recording region.

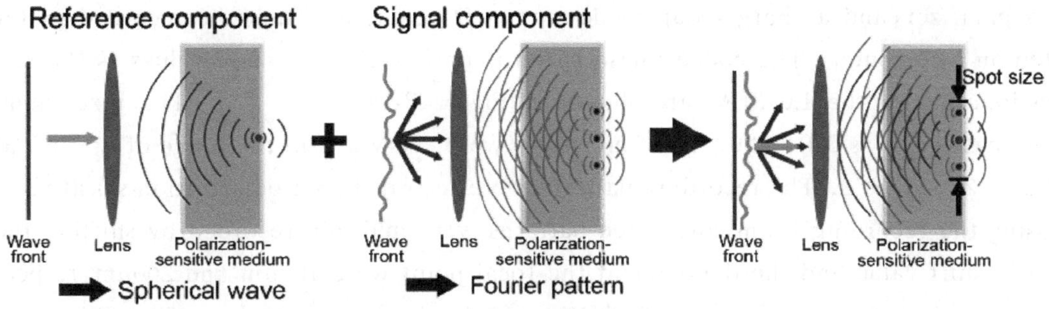

Fig. 3-147 Schematic of Fourier-transform retardagraphy: When the recording pattern recorded on a polarization-sensitive medium by focusing the recording beam, the signal and reference components are changed Fourier pattern and spherical wave. Thus, a phase retardation of two components is difference in every point on a medium.

Figure 3-148 shows the experimental setup for Fourier-transform retardagraphy. In this experiment, a diode laser (406nm) was used recording and reconstruction. The polarization azimuth was adjusted using a half-wave plate so that the polarization state was elliptical with an azimuth of 45 degrees or-45 degrees.

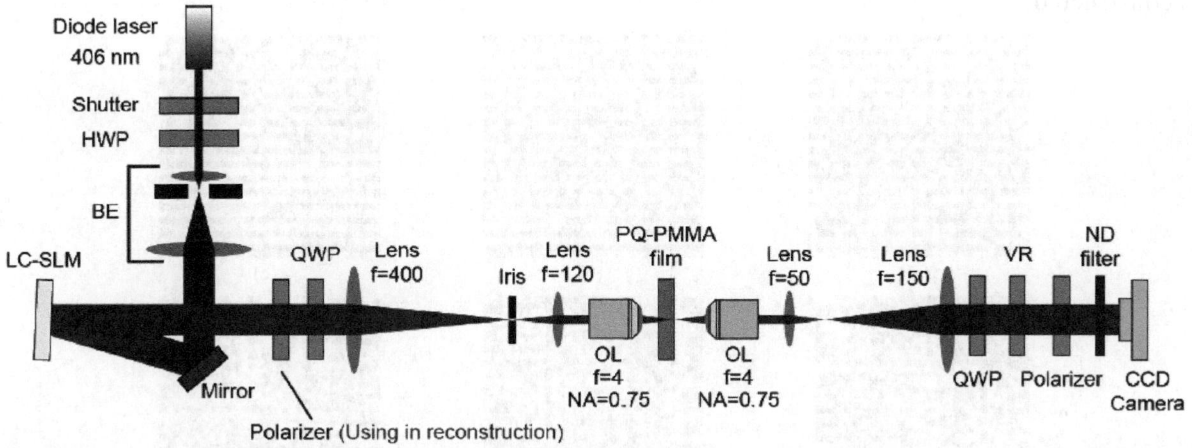

Fig. 3-148 Experimental setup for Fourier-transform retardagraphy: BE, HWP, QWP, OL, LC-SLM, VR, and CCD camera are beam expender, half-wave plate, quarter-wave plate, objective lens, liquid crystal spatial light modulator, variable retarder, and charge coupled device camera, respectively. The polarizer after LC-SLM is used only in reconstruction.

The signal component included in are cording beam was independently modulated by a parallel aligned liquid crystal spatial light modulator (LC-SLM) as a spatial retarder. A recording beam was converted into linear polarization with the azimuth corresponding to the retardation a using quarter-wave plate (QWP). A phenanthrenequinone doped poly-methylmethacrylate (PQ-PMMA) film was used as a polarization-sensitive medium. The thickness was about 1mm. The recording pattern was recorded on a polarization-sensitive medium by focusing the recording beam. In reconstruction, a homogeneous image was displayed on the LC-SLM in order to obtain a homogeneous polarization pattern. The polarization state of the beam was adjusted to the polarization state of reference component in recording using a polarizer and a quarter-wave plate. The reconstructed beam transmitted through the PQ-PMMA film was analyzed by the 8-step phase-shifting method using an imaging polarimetric system that consists of a QWP, a variable retarder

(VR), a polarizer, and a charge-coupled device (CCD) camera. Then the recorded pattern was extracted on a computer. The coded phase pattern consists of four phase values as the recording patterns inputted to the LC-SLM are shown in Figure 3-149 (a) ~ (c). The pixel number of recording patterns was 320 × 320. The four values were every $\pi/2$ in the range of $0 \sim 2\pi$ radian, i.e. $0, \pi/2, \pi/2$, and $3\pi/2$. The recording patterns were recorded on a polarization-sensitive medium by focusing the recording beam, and three patterns were multiply recorded by shifting the focal point. The shift value and the diameter at the focal point were 100μm and 200μm, respectively. And the recording beam power was 2.05W/cm^2, the exposure time was 33 milliseconds. The reconstructed images are shown in Figure 3-149(d) ~ (i). The images (d), (e), (f) are extracted raw data using an imaging polarimetric system, and the images (g), (h), (i) are reconstructed four-valued phase retardation patterns from the images (d), (e), (f), respectively. These images are a pattern in each recorded point, i.e. 0μm, 100μm, and 200μm, respectively. In result, bit error rates of three patterns were obtained 12.3%, 5.7%, and 13.8%, respectively. It is predicted that these errors can be corrected by an error correction technique. As shown in Figure 3-149, it was verified that three images consist of multi-valued phase pattern could be independently recorded and reconstructed.

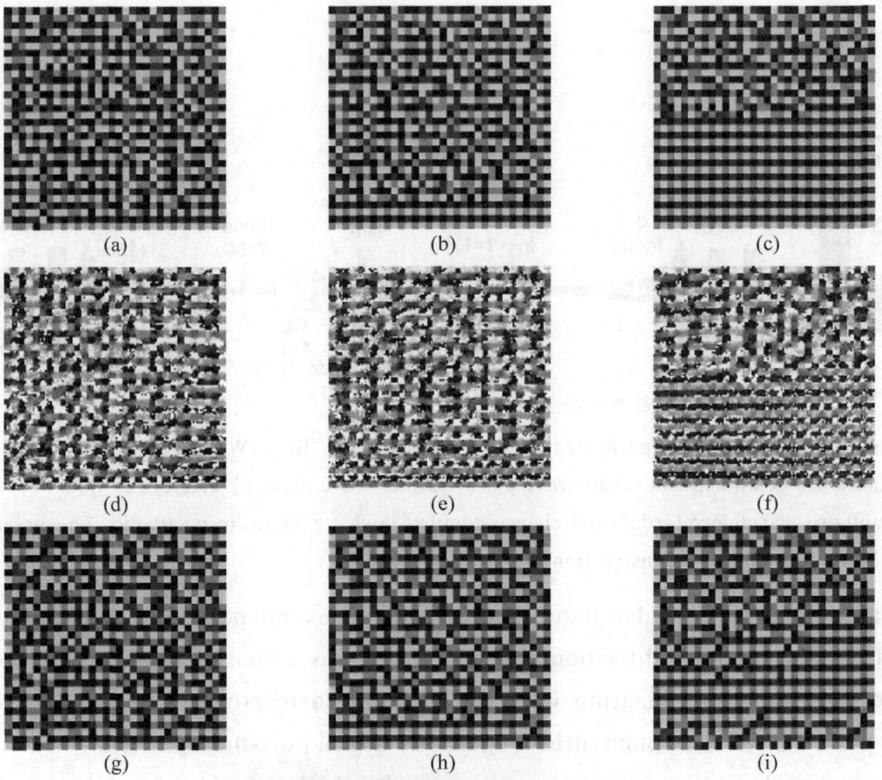

Fig. 3-149 Experimental results of shift multiple recording: The images (a), (b), (c) are four-valued phase patterns on the LC-SLM. The images (d), (e), (f) are extracted raw data using an imaging polarimetric system. The images (g), (h), (i) are reconstructed four-valued phase retardation patterns. The images are a pattern in each recorded point. The shift value is 100μm.

Multilevel phase patterns consists of four values were recorded were multiply recorded on a polarization sensitive medium by Fourier-transform retardagraphy. The combined multiplex recording technique with the multilevel recording can effectively improve recording density and data transfer rate.

3.7.10 Multi-level run-length-limited (ML-RLL) modulation

Writing and reading of 3-level ML-RLL modulation signals on two different write-once phase change materials is discussed. These recordings represent a linear storage density enhancement of 50% Vs conventional (2-level) RLL modulation. Multi-level (ML) recording has recently been utilized to increase optical data recording density. In the ML systems demonstrated to date, recording is accomplished by writing variable size marks within a sequence of fixed size "data cells" that form a spiral data track on the disc. From 8 to 12 distinct amplitude levels are discriminated when such data cells are read. Using 2-stage trellis-coded modulation (2-TCM) and convolutional coding, the raw (no overhead dedicated error-correction redundancy, amplitude reference and synchronization marks) linear storage density achieved is in the range from 2.5 to 3.08 bits per data cell. With this type of ML recording, the signal to noise ratio (SNR) becomes critical for accurate discrimination of the correct signal level and a complex write strategy involving data-dependent pre-emphasis is required when small data cells are used. The ML modulation technique just described as "half-tone 2-TCM" recording.

Different ML recording technique utilizes marks that (1) produce more than two distinct signal levels when read and (2) have several distinct lengths along the direction of the data track. This is very similar to conventional run-length limited (RLL) recording which employs two different types of marks (e.g., pits and lands) which have several distinct lengths; RLL playback signals are comprised of pulses having one of two possible amplitude levels. We refer to the recording scheme considered here as multi-level run-length-limited (ML-RLL) modulation. If the recorded variable-length marks produce a read signal comprised of pulses that have 3(or 4) distinct amplitude levels, then the recording is a 3L-RLL (or 4L-RLL) scheme; conventional RLL modulation is a 2L-RLL scheme. ML-RLL modulation has been previously discussed in the technical literature. The 2L-RLL modulation used in the DVD system (EFM+) achieves a linear storage density of 1.5 user bits per shortest mark, i.e., along the length of data track occupied by the shortest recorded pit or land. EFM+ is a {2,10} RLL modulation scheme since it produces shortest (and longest) recorded marks that contain exactly 3 (and 11) channel bits, or data bit clock windows. A similar 3L-RLL modulation scheme (i.e., 3-level {2,10} RLL) will yield 2.25 user bits per shortest mark; this linear density is increased to more than 2.5 user bits per shortest mark by adding one additional level (i.e., 4-level {2,10} RLL). Moreover, due to the way in which ML-RLL modulation assigns amplitude levels to individual recorded marks, during playback one only needs to detect the change in amplitude level of each pulse relative to the amplitude level exhibited by its previous neighbor pulse in order to demodulate the playback signal. Thus, the amplitude discrimination process is self-referencing and additional marks used to provide reference amplitude levels need not be recorded. This means that ML-RLL modulation with only 3 levels yields approximately the same linear storage density obtained with the 8-level half-tone 2-TCM recording scheme discussed in the previous paragraph, assuming that equivalent minimum recorded mark lengths are used.

Previously, a numerical optical recording model predicted that recording and playback of ML-RLL signals in $Ge_2Sb_2Te_5$ phase change (PC) materials would be problematic. In this presentation we report the experimental recording and playback of 3-level ML-RLL signals on discs that employ two different types of write-once PC materials. The main objective in this research is to develop mark writing strategies that produced useful results, i. e., which yield 3L-RLL playback signals having excellent timing jitter and amplitude level discrimination.

- Recording and readout experiment

The optical head used in this experiment has a 675nm laser and a NA = 0. 5 write/read objective. One of the write once PC recording media one of these shows evidence (via e-beam micrographs and x-ray diffraction) of partially crystallized marks in an amorphous thin film (PC-A) while the other exhibits fully crystallized marks in an amorphous film (PC-B). Only use a single write laser power, i. e., the write laser is switched between a low bias power (read power) and high power (write power) to effect the recording. The reflectance level and length of the written marks are varied by changing the width, duty cycle and number of short sub-pulses used to write each individual mark Vs the mark pattern. Different write strategies were found to optimize the writing of 3L-RLL signals on each medium. We shall discuss recording and playback of two different 3L-RLL waveform patterns on each type of PC medium. The three distinct recorded mark types will produce playback signal amplitude levels that will be referred to as levels 0, 1, and 2. A Nomarski differential interference photo-micrograph of the first 3L-RLL pattern recorded on the PC-A material is shown as Figure 3-150(a). Of the 7 recorded tracks in this photomicrograph, the 3rd to 5th tracks (counting from left to right) represent a repetition of the following three pulses: a $3T$ pulse at level 0; followed by $7T$ pulse at level 1 ; followed by $4T$ pulse at level 2 (where T is the length of a single channel bit). The other tracks are recordings of a 50% duty cycle binary pulse signal which serves as a reference markers. The inter tracks separation is 5 microns. Figure 3-150(b) shows the playback signal obtained from one of the 3L-RLL tracks depicted in Figure 3-150(a). Optimum 3L-RLL playback is obtained when amplitude level 1 is midway between amplitude levels 0 and 2.

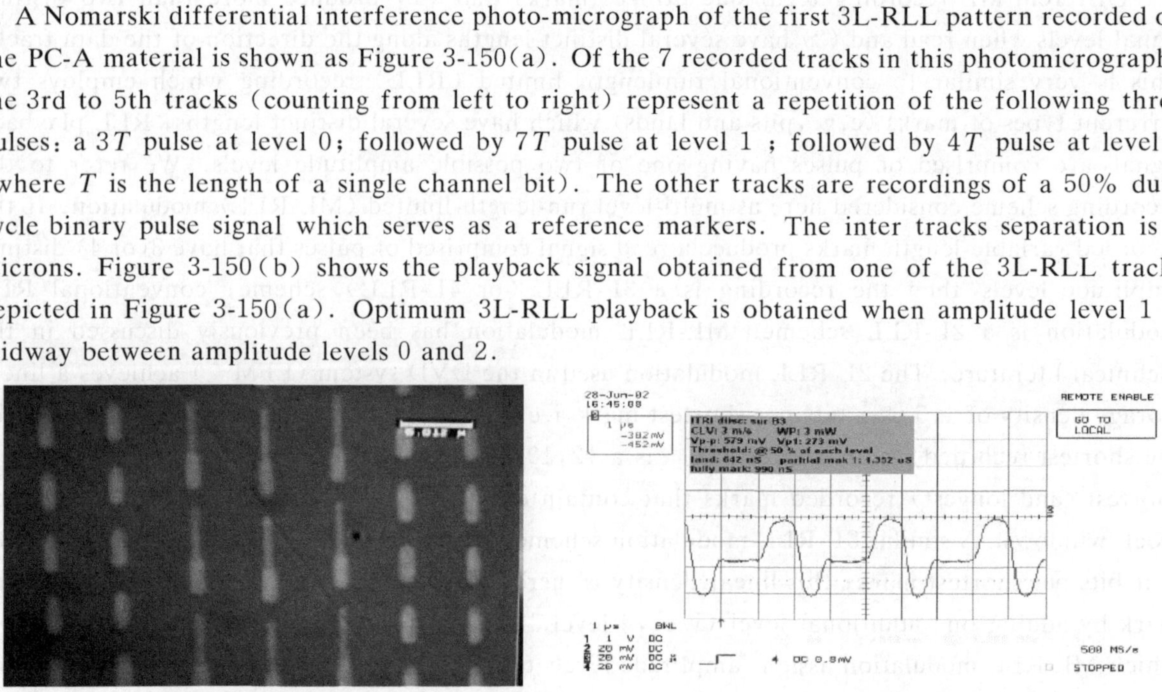

(a) (b)

Fig. 3-150 (a) Photo micrograph showing data tracks that store the first 3L-RLL waveform pattern on the PC-A medium; (b) Playback signal obtained when reading one of the recorded three central tracks shown in (a).

The 3rd to 5th tracks in the photomicrograph shown in Figure 3-151(a) are recordings of the second 3L-RLL pattern which consists of a repetition of the following: a $3T$ pulse at level 0; followed by a $7T$ pulse at level 1; followed by a $4T$ pulse at level 2; followed by a $7T$ pulse at level

1. The corresponding playback signal is shown in Figure 3-151(b). The playback signals shown in Figures 3-150(b) and 3-151(b) are not equalized, but a low pass filter was employed to reduce noise.

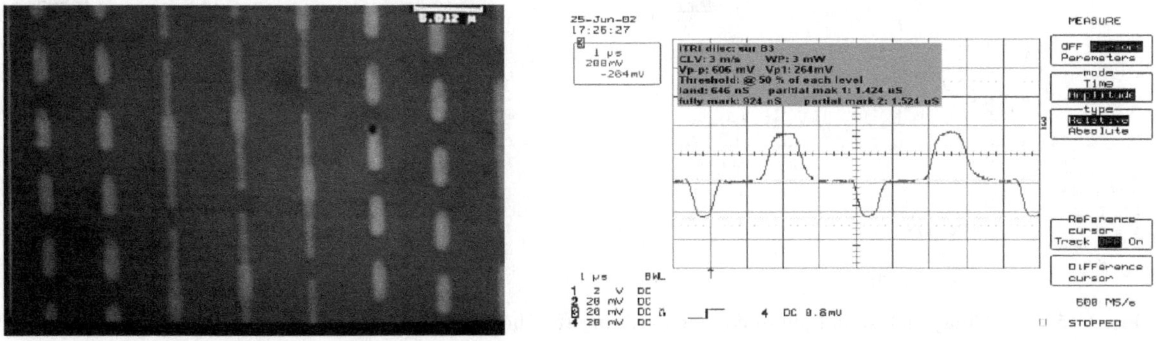

(a) (b)

Fig. 3-151 (a) Photomicrograph showing data tracks that store the second 3L-RLL waveform pattern on the PC-A medium; (b) Playback signal obtained when reading one of the recorded three central tracks shown in (a).

Figure 3-152(a) shows the results of using a different write strategy to control the playback amplitude of level 1 marks (relative to the other playback amplitudes) while maintaining correct mark lengths. Now we also try to record these same waveform patterns on PC-B medium. A photomicrograph of 3L-RLL recording in the PC-B medium is shown in Figure 3-152(b). The first waveform pattern is on the 3rd to 5th tracks (counting from bottom to top); the second waveform pattern is on the 8th to 10th tracks. The recorded marks are fully crystallized and the level 1 marks are not uniformly shaped. Figure 3-153(a) and (b) respectively show the playback signals obtained from an optimum recording of the first and second 3L-RLL patterns on the PCB medium. The level 1 and level 2 pulses are noisy (distorted). It is quite difficult to control the three relative pulse amplitudes in the playback signal when recording on the PC-B medium.

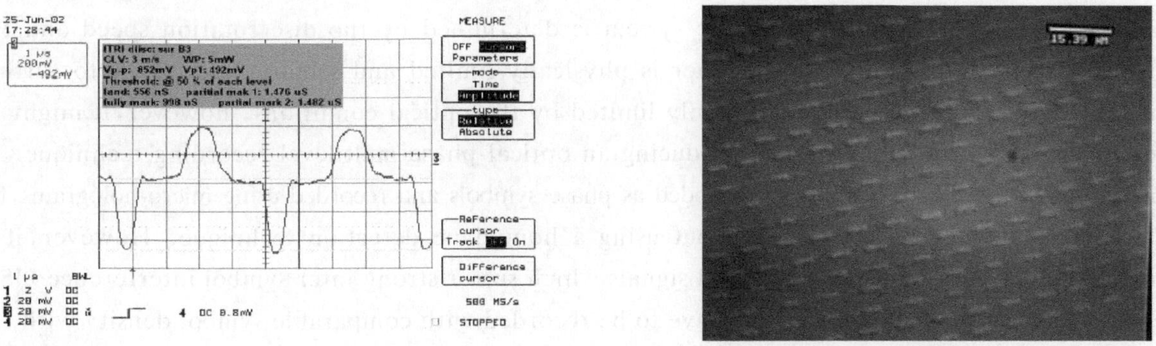

(a) (b)

Fig. 3-152 (a) Playback signal obtained when reading the second 3L-RLL waveform pattern when a non-optimum write strategy used to record the pattern; (b) Photomicrograph showing data tracks that store the second 3L-RLL waveform pattern recorded on the PC-B medium.

Viable recording of 3L-RLL modulated signals was demonstrated on a write-once PC medium (PC-A) that supports partially crystallized written marks. Good playback signals were not obtained when the same signals were recorded on a write-once PC medium that produces fully crystalline marks. These results were obtained using only a single write power level, i.e., the write

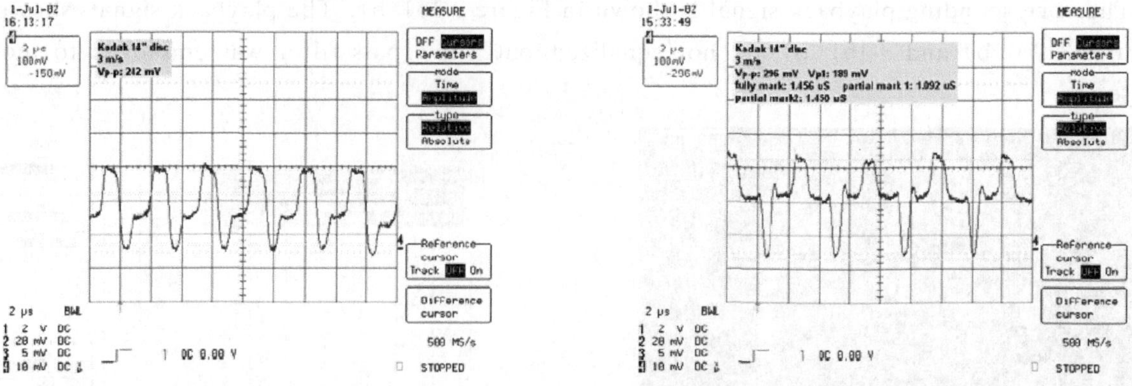

(a) (b)

Fig. 3-153 (a) Playback signal obtained when reading the the first 3L-RLL waveform pattern from the PC-B medium; (b) Playback signal obtained when reading second 3L-RLL waveform pattern from the PC-B medium.

laser was toggled between its write and read levels. The good 3L-RLL playback signals obtained from the recordings on the PC-A medium appear to result from a combination of recorded mark width and mark reflectance variations, the latter resulting from partially crystallized recorded marks. Quantitative measurements of the quality of the 3L-RLL recordings (e. g. mark edge jitter and amplitude level margins) will be presented at the meeting.

- Data rate estimation of phase multilevel recording

Decoding performance of multilevel signals recorded using optical phase modulation was investigated by computer simulations. It was found that 4-ary phase modulated signals can be satisfactorily decoded by applying PR (1,2,1) ML provided that homodyne detector output signal-to-noise ratio (SNR) is equivalent to current optical drives. It was also found that 8-ary phase modulated signals require SNR to be 6.1 to 7.0dB higher. Therefore, it can be concluded that the data transfer rate can be more than doubled by using optical phase recording technology. The data transfer rate (DTR) in an optical disc system is determined by the disc rotation speed and the linear bit recording density. The former is physically limited and seemingly difficult to further improve significantly. The later is mainly limited by the optical conditions; however, it might be increased over a factor of two by introducing an optical phase multilevel recording technique. In this technique, multilevel symbols are encoded as phase symbols and recorded using micro-holograms, for example. Then their phases are read out using a homodyne detection technique. However, it is required to decode the phase modulated signals which suffer strong inter symbol interference (ISI) because the multilevel phase symbols have to be recorded with comparable symbol density with the current optical disc system.

It is readily imagined that the partial response most-likely (PRML) method will solve the ISI problem. However, negative side effects are also likely to occur due to increased numbers of the decoder inner states and branches. Therefore, read performance of an optical phase multilevel recording system has been investigated by computer simulation, and the feasibility of read DTR enhancement has been considered.

- PRML for phase multilevel signal

Figure 3-154 illustrates how ISI is reflected in phase modulated signal while the optical spot moves from a region with phase 0 to that of phase of π. Here, phase reference is the reference light used in the homodyne detector. The phase detected by using the phase-diversity homodyne detector is equivalent to the value obtained by weighted average of the phase within the optical spot. The range of the phase φ is determined as $0 \leqslant \varphi < 2\pi$. Naturally, the phase will pass through $\pi/2$ not $(3/2)\pi$. Figure 3-155 shows the phase plots (constellations) of 4-ary phase modulated signals (QPSK): Figure 3-155(a) without ISI and Figure 3-155(b) with ISI. The horizontal and the vertical axes respectively represent amplitudes of the in-phase and quadrature-phase components of the homodyne detector outputs. Thus, the argument represents the phase. Four two bit long symbols ("00", "01", "10", and "11") are allocated at phases of $0, \pi/2, \pi$, and $(3/2)\pi$. If there is no ISI, the signal phase changesinstantaneously, thus the phase plots appear as four distinct spots like in Figure 3-155(a). Contrarily, if the ISI exists, temporal phase transition becomes gradual, thus the corresponding constellation will appear like in Figure 3-155(b).

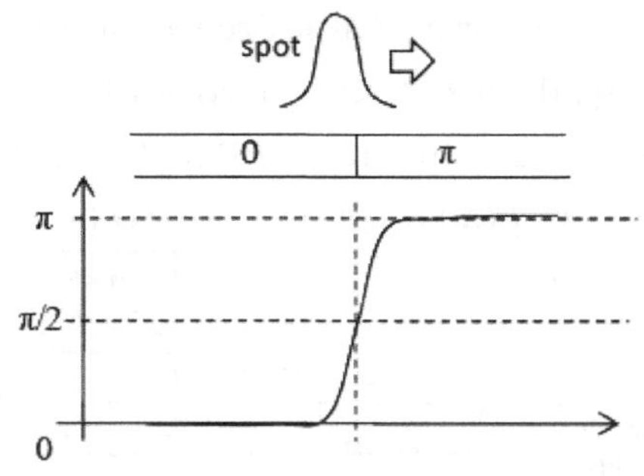

Fig. 3-154 ISI observed at phase transition.

The PRML decoding technique can be extended for multilevel signal decoding in a straight forward manner. If the run-length is not limited, the number of inner states of a PRML decoder for M-ary signal with constraint length of L is increased to ML − 1. The number branches exiting from each state increases to M, thus the total number of the branches becomes ML. Shown in Figure 3-155(c) and Figure 3-155(d) are examples of trellis diagrams for systems with a constraint length of 3: Figure 3-155(c) binary and Figure 3-155(d) 4-ary signals.

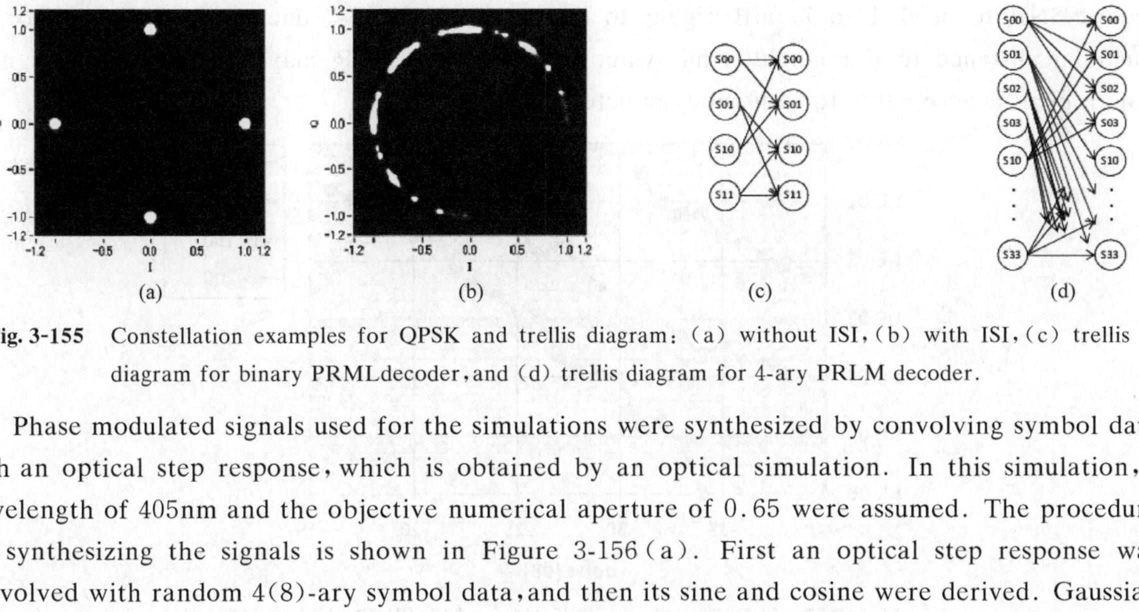

Fig. 3-155 Constellation examples for QPSK and trellis diagram: (a) without ISI, (b) with ISI, (c) trellis diagram for binary PRMLdecoder, and (d) trellis diagram for 4-ary PRLM decoder.

Phase modulated signals used for the simulations were synthesized by convolving symbol data with an optical step response, which is obtained by an optical simulation. In this simulation, a wavelength of 405nm and the objective numerical aperture of 0.65 were assumed. The procedure for synthesizing the signals is shown in Figure 3-156(a). First an optical step response was convolved with random 4(8)-ary symbol data, and then its sine and cosine were derived. Gaussian noises of equal power from different sources were superimposed over each of them, and they were regarded as I and Q output of a homodyne detector. The decoding procedure is illustrated in Figure 3-156(b). The phase signal is derived by calculating the instantaneous arguments from I and Q values. Then, it is equalized by using a 15-tap adaptive equalizer whose tap coefficients were obtained using the least squared error (LSE) algorithm. The target signal used for the adaptive equalization was synthesized from the same random symbol source used for the signal synthesize. The equalizer output is decoded by using the 4 or 8 leveled PRML decoder. The partial response classes used were either PR(1,2,1)ML or PR (1,2,2,1)ML.

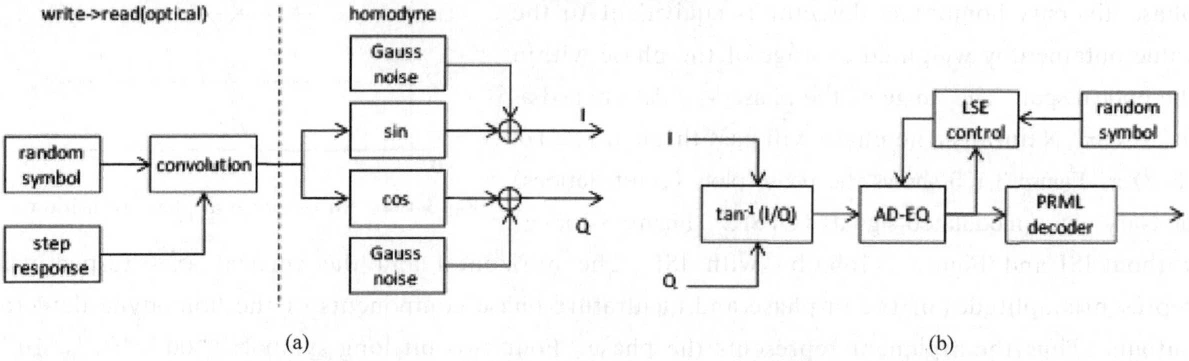

Fig. 3-156 Signal processes: (a) test signal synthesize and (b) decoding.

Simulation results and discussions Figure 3-157(a) shows the symbol error rate (SER) curves for a series of symbol lengths relative to the noise amplitude obtained for QPSK signal. The numerals in the legend represent the minimum symbol length in units of nanometers, and "HD15" refers to the bit error curve obtained for HD DVD (15GB/ layer, channel bit length: 102nm), which is meant for comparison with a binary recording system. It can be seen that the noise levels where curves cross SER of 10^{-5} are comparable to binary recording when symbol length is above 180nm. Thus, it can be said that read DTR may be doubled when PR(1,2,1)ML is applied to a QPSK signal.

The cases for 8-PSK signal are shown in Figure 3-157(b). It is apparent that 8-PSK signals require SNR to be 6.1 to 7.0dB higher to achieve SER of 10^{-5} due to the shrinkage of the Euclidian distance to the neighboring symbols. Thus, read DTR may be trebled if this noise requirement is acceptable for a homodyne detector.

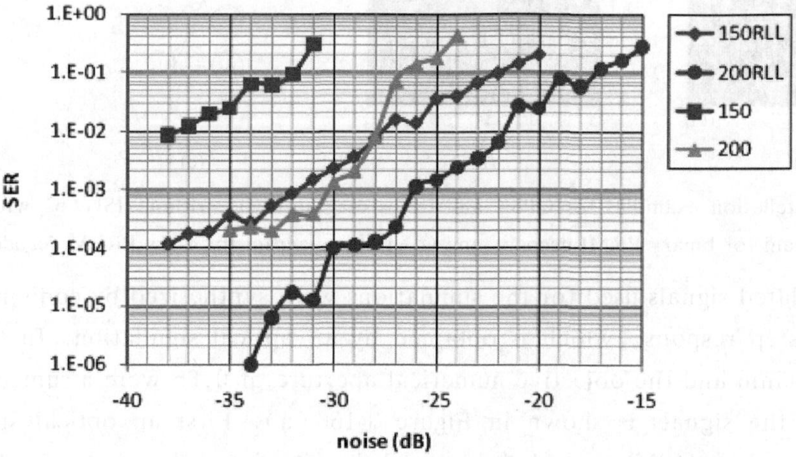

Fig. 3-157 SER curves for 8-PSK obtained by PR (1,2,2,1) ML.

Figure 3-158 shows the SER curves for 8-PSK signals obtained using PR(1,2,2,1) ML, which shows higher performance against ISI when applied to binary recording. When the curves "150" or "200" are compared with corresponding results in Figure 3-159(b), extreme deterioration of the decoding performance is apparent. This is caused by the shrinkage of the Euclidian distance between the neighboring phase targets, which increases the possibility of path miss selection due to noise during the "add-compare-select" process.

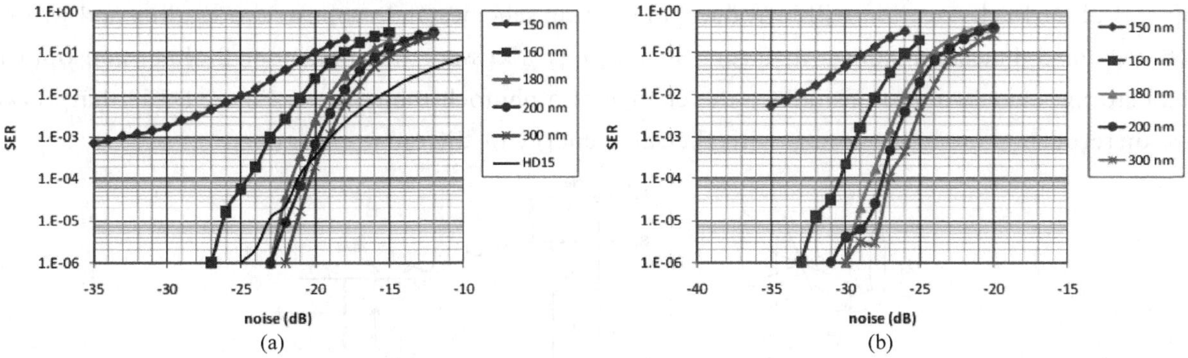

Fig. 3-158 SER curves obtained by PR(1,2,1)ML relative to noise amplitude: (a) QPSK, (b) 8-PSK.

The curves denoted with "RLL" are the cases in which (1,7) RLL modulation for 8-ary symbols were used. With the (1,7) RLL modulation, the number of the states was reduced to 120 from 512, and the total number of the branches was reduced to 568 from 4096 (minimum symbol lengths were adjusted to 150 and 200nm). The effects of these reductions are trivial, but significant SER improvements are valid at all noise amplitudes if compared with the cases without RLL. However, it is still difficult to state that PR(1,2,2,1)ML is superior to PR (1,2,1)ML, especially when considering the increased complexity of the system. Also, the physical nature of the medium should be considered in deciding whether to apply RLL modulation to the optical phase recording.

(1) Read data transfer rate (DTR) may be increased by a factor of at least two compared to conventional system, if PR (1, 2, 1) ML decoding is applied to quadrature phase-shift keying (QPSK).

(2) Read DTR may be further improved by a factor of three by 8-PSK recording and PR (1, 2,1)ML if decoder input signal-to noise ratio (SNR) is acceptable.

(3) Longer constraint length does not necessarily lead to higher decoding performance due to an increased number of the decoder inner states.

3.7.11 Three wavelength and multi-level storage with mask

The mechanism of three dimensions digital data storage, especially for multi-layer optical disk, seeking method, principle and method of reading/writing is described in this section. With the technology of multi-layer, multi-color, longitudinal coding and masking, the new scheme can overcome the bottle neck of 3-D digital storage for tracking, focusing, data rate and do most of the advantages of 3-dimension optical storage for higher density and larger capacity.

Considering diffraction and the wavelength of photo-diodes, 2-dimension optical storage has its own limitation. In general, the limitation is $\lambda/2NA$, where λ is wavelength and NA is numeral aperture. Therefore, multi-layer or 3-Dimension (solid state memory) has been a promising approach to reach higher density and high data rate storage. After the discovery of two-photon absorption materials, a kind of photochromic dye, more and more researchers employ the characteristics of the material to design 3-dimension storage. However, schemes of multi-layer optical disc face a key problem of reading and writing. Due to that is impractical to make pre-grooves on every layer and reflective layers cannot be plated for each layer, not only tracking but also focusing is more difficult. How to solve the problem is the critical point to realize multi-layer optical disk. In this section will introduce an existing schemes and propose two reading and writing schemes for 3-dimension storage in Figure 3-159. It is necessary to implement 3-dimension optical data storage that two photons induce the change of a photochromic dye as Figure 3-159 that λ_1, λ_2 are different wavelength photons with the same energy or different energy.

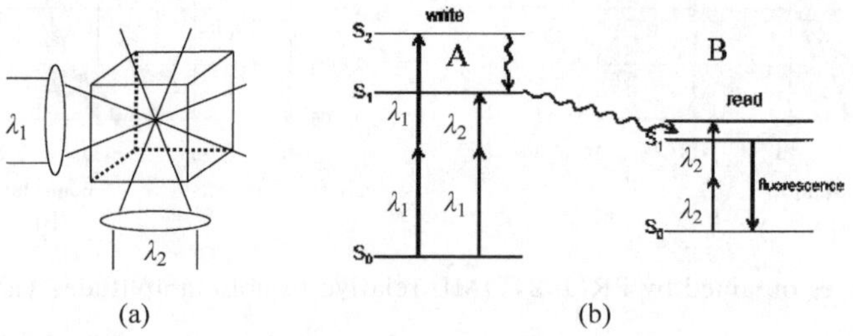

Fig. 3-159 Principles of 3-D storage with double wavelength.

Several photochromic compounds have previously been investigated as two-photon photochromic data storage media. The pioneering studies for 3D optical memory using photochromic compounds as the photochromic spiropyrans etc. for example. The molecules of photochromic compounds exist in two isomeric forms of A and B as shown in Figure 3-159 which are a colorless cyclic form and a colored open form, respectively. Irradiation of light λ_1 to the colorless form A converts to the colored from B which exhibits fluoresce upon photoexcitation. Orthogonal two beams system was employed for the reading and writing, where the molecule was excited at the intersection of the two beams by simultaneous. For writing information, the excitation was performed by two-photon irradiation of λ_1 and λ_2 photon. Then isomer A at the intersection photo isomerized to isomer B. The energy level diagram is changed along with the molecular structures. For reading data, only the λ_2 beam was used for irradiating the media and isomer B can be excited and emit fluorescence.

The media absorbed double photons of λ_1 that will change to status of S_2 as erased. Only when the two photons (converge at a same point, it can be absorbed and the photochromism was induced. In this way, the recording of information is implemented. According to different media, the refractive index, absorbance, fluorescent light and electrical properties can be adopted for reading. However, focusing and tracking are difficult in this mode, since there is not a reflective layer on the recording plane. Two typical schemes of 3-dimension optical storage now are shown in Figure 3-160. Figure 3-160(a) shows that 1—cylindrical lens, 2—cuneal beam λ_2 (as a plane light) for writing, 3—objective lens for writing and reading, 4—λ_1 laser beam for writing and reading, 5—read out objective lens, 6—read out light, 7—two-photon absorption material for recording.

A plane light is adopted to select the layer need to read and write. Another laser beam lights on the layer, thus finishing reading and writing. In this way, the data was stored as an image, which is stored and read concurrently. It can reach a high-intensity and high-speed storage. However, the scheme needs a precision X-Y stage and the storage volume, read and write speed will be limited. But the traditional schemes of 3-dimension optical storage are not practical with current technology. A more practical scheme with advanced nanotechnology is shown in Figure 3-160(b).

This is a 3D integrated optical memory that is based on rectangular waveguides as a solid state memory.

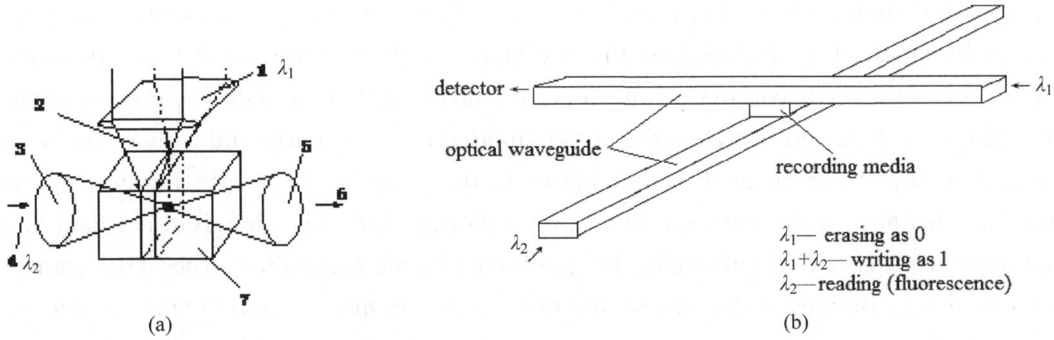

Fig. 3-160 Sketch of the two-photon 3-dimension optical storage scheme.

In order to overcome the shortages of the scheme above, A multi-layer scheme of 3-dimension optical storage is brought forward as shown in Figure 3-161.

In this scheme, the recording media is made the same shape as the general disc. The sensitive materials are coated as layers, whose spacing is greater than the focus depth, in order to avoid the crosstalk of different layers. The lights of the two wavelengths should focus at the same point but binary optical components. When a disc rotates, the focus of the laser beam scans vertically. The pre-grooves are pressed on the top and bottom layers. In that case, the focusing, seeking and tracking of a pick-up head can be implemented.

Because the 2-photon absorption photo-chromical material would not react, unless two wavelength laser beams reach certain level, the smaller one of the two spots, as area 3 shown in Figure 3-162, in fact determines the size of the effective spot. Accurate and effective control of the short wavelength laser beam can gain smaller spot. Though the scheme needs very stable workstation and more accurate actuator of pick-up head to realize focus and track, the scheme is still more practical than the existing one of 3-D optical storage.

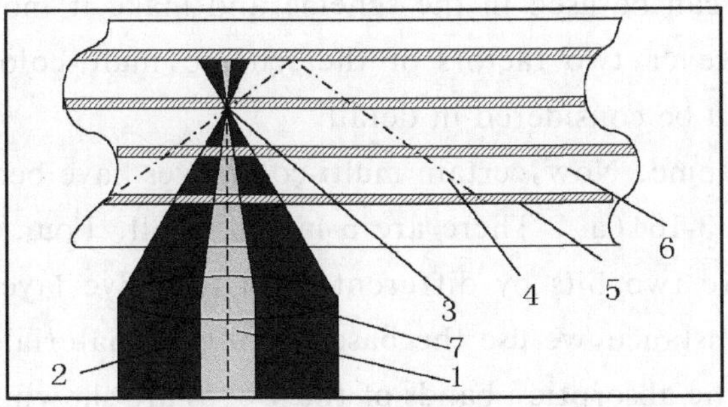

Fig. 3-161 Two wavelength multi-layer storage: 1—Laser beam λ_1, 2—Laser beam λ_2, 3—Focus of two beams, 4—track of focus in recording, 5—recording medium layer, 6—transparent separate layer, 7—objective lens.

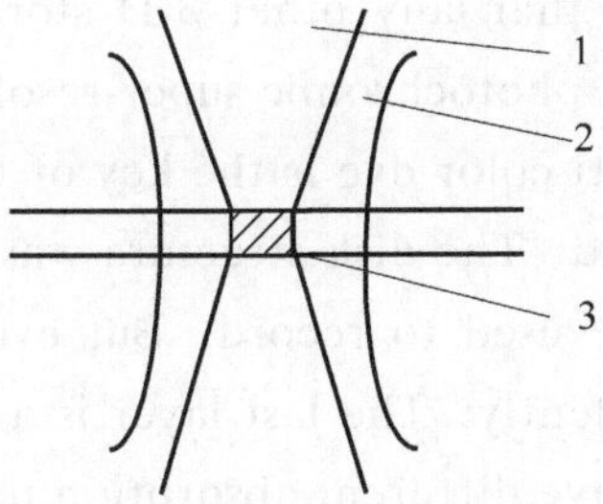

Fig. 3-162 The effective spot of the two wavelength laser beams where 1—read/write beam with shorter wavelength λ_1, 2—write beam with longer wavelength λ_2, 3—the effective spot for recording.

With the development of dyes, the sensitive wavelength can be selected in a narrow range. Besides, photochromic super-resolution mask is developed to reduce the size of the recording spots.

Longitudinal coding technology makes parallel storage available. They make multi-colors 3-dimension optical storage a practical scheme. Here, a novel scheme of 3-dimension is provided.

The optical system is shown in Figure 3-163. The dyes with different sensitive wavelengths are coated layer by layer. The thickness of the dye layers is about 500nm, which is within the focus depth. A reflective layer is coating on the tope dye layer so that the laser beam can focus on it. The laser beams of different wavelengths are modulated independently and converged by a prism. The achromatic object lens focus different lights on the same layer. A splitting grating splits the reflective laser beam into the corresponding photo diodes. Like that, longitudinal data is written and read concurrently. With effective longitudinal coding technology, the data transfer rate increases obviously. Besides, a photochromic super-resolution mask is coated between the dye layer and the substrate to gain smaller effective spot. The reflective laser beam that has indifferent wavelength containing information data is splited by a reflective grating and the different color light enters the corresponding read unit.

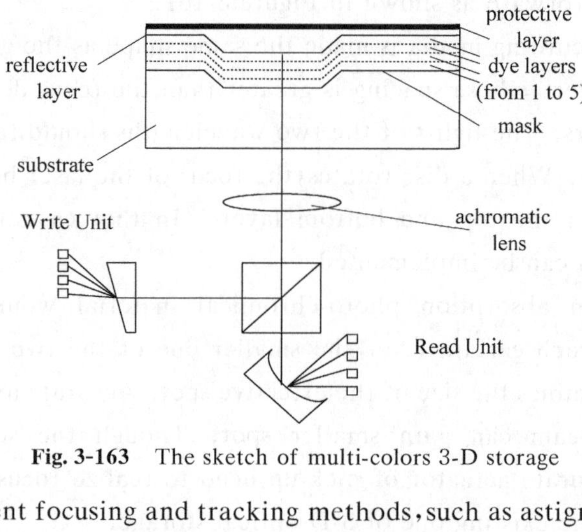

Fig. 3-163 The sketch of multi-colors 3-D storage

In this way, the current focusing and tracking methods, such as astigmatism method as focusing control and push-pull method for tracking control, can be used in the scheme and make it more practical than any other 3-D storage scheme. However, two factors of the scheme, multi-colors dyes and photochromic super-resolution mask, should be considered in detail.

Multi-color dye is the key of the multi-color scheme. Now, certain multi-color dyes have been developed. The disk structure was shown in Figure 3-164(a). There are 5 layers in all. Four of them are used to record. But every layer can store two bits by different level to 4 dye layers independently. The last layer is a mask film. For instance, we use the base on fulgide materials, which have different absorption to wavelength λ. The absorption bands of the 4 dyes are shown in Figure 3-164(b). The absorption peaks are between 450nm and 780nm and the width of the absorption peak of each dye is less than 50nm, with uniform spacing. The 5 dye layers comprise a total recording layer with spin coating technology. There is a separate layer between every two dye layers. The total thickness of the recording layer is controlled within 1μm. Every dye layer of the recording layer is sensitive to different wavelength, 460nm, 550nm, 650nm and 780nm respectively. In this way, the recording intensity will increase dramatically. Besides, another advantage of the multi-color storage is that the data can be read concurrently. In the scheme, 8 bits can be read at each time, which is just a byte with ECC code. The technology, called longitudinal coding, can be adopted. So, not only does the storage intensity increase but also the speed of writing and reading information increases too.

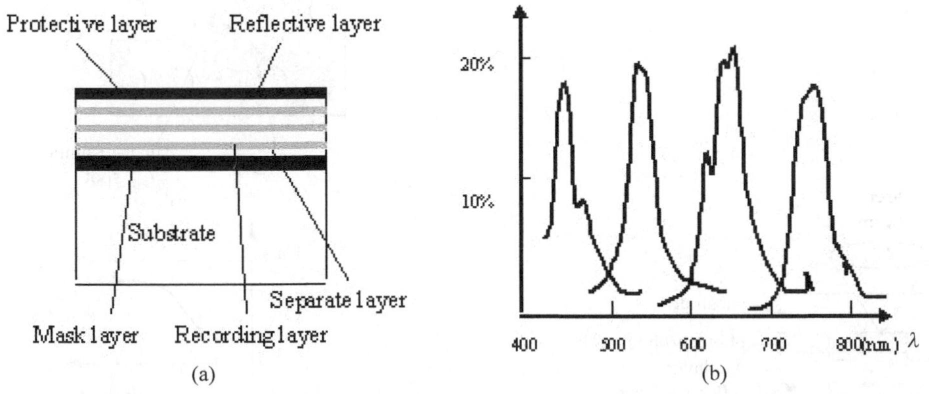

Fig. 3-164 The structure and absorption of multi-color dye layers: (a) Disk structure of multi-layer, (b) Absorption bands of the dyes.

- Photochromic super-resolution mask

A super-resolution mask has been developed to get smaller spots on the disc and increase the storage intensity. At first, MSR (magnetic super-resolution) mask was researched and it could increase storage intensity without changing optical systems. However, MSR only can be used in magnetic media. Now, a new kind of photo-chromic super-resolution mask is being developed. Compared with MSR, PSR has many advantages, such as low cost, wide application and higher intensity. Based on this technology, the 3T run-length is reduced from 0.84μm to 0.48μm and the spacing is reduced from 1.6μm to 0.85μm. The quality of the signal is still good. So, PSR mask can be used to increase the storage intensity.

PSR is adopted in the scheme too. The main reason is to make the spots of indifferent wavelengths the same size. The laser beams of different wavelengths and the same optical system are used in the scheme, so the spots would vary according to different wavelength. In order to gain a same size spot and use the recording material effectively, PSR is a desirable strategy. Obviously, additional benefit is to realize smaller recording spot and increase the storage intensity of course.

The structure of the disc with a PSR mask is shown in Figure 3-165. Compared with common discs, the PSR mask is added between substrate and the dye layer. In Figure 3-166, the principle of PSR mask is shown. When a laser beam scan the colored (initialized) PSR mask, only the zone where a chemical reaction happens can be bleached. Since the effective spot is the intersection of the reading spot and the bleach zone, the effective spot on PSR is smaller than the diffraction spot and therefore achieves the super-resolution.

The bottleneck of traditional 3-dimension optical storage is analyzed. Two practical schemes, multi-layer and multi-color 3-dimension optical storage schemes are provided. The schemes comprise a plenty of advanced technology, such as multi-layer coating, multi-color dyes, photochromic super-resolution mask and effective longitudinal coding. They can adopt both the similar focusing and the tracking way as current optical disc players, so they are far more applicable and have promising utility.

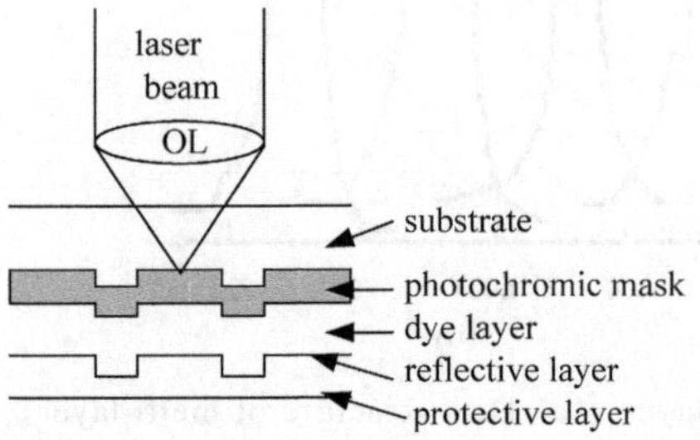

Fig. 3-165 The structure of the disc with a PSR mask.

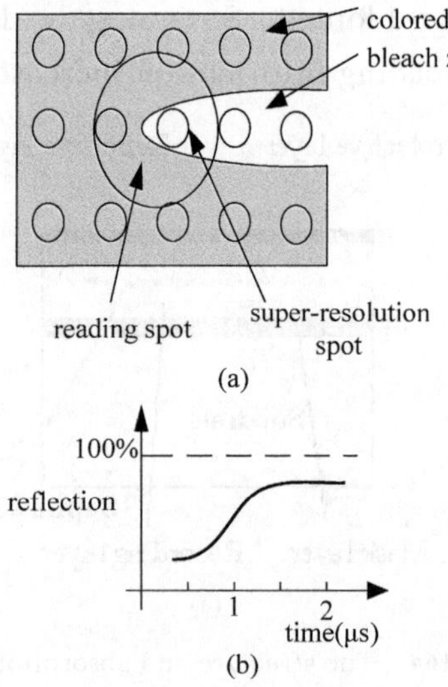

Fig. 3-166 The super-resolution mask working principl.

Chapter 4

Photonic super-resolution memory

4.1 Overview

Super-resolution optical Memory is the important way for increasing density and capacity of optical storage that the key means is adoption of the near field recording (NFR) technology. NFR is based on several basic principles that uses a combination of first surface recording and a flying optical head to achieve its goals. Digital Equipment Corporation developed these concepts in the late 1980s originally. The basic technology behind the Solid Immersion Lens (SIL) was developed and patented by Dr. Gordon Kino at Stanford University in 1992. Quantum Corporation acquired the patents on this technology as part of their acquisition of Digital's storage business in 1994. Co-exclusive patent rights have been granted by Quantum to Tera Stor for both the flying optical head and first surface recording. The primary advantage of NFR technology is achieving the highest areal density, or number of magnetically charged bits, that can fit into a given area. NFR technology is form factor independent and permits both removable and fixed drive applications.

The growing need the transverse spatial resolution of conventional optical lenses is limited by the diffraction nature of the light to $\lambda/(2NA)$, with λ being the wavelength of light and NA being the numerical aperture of the lens. Therefore various remarkable methods, except near-field optical data storage and heat/thermally assisted magnetic recording, including superlens, hyperlens, plasmonic dimple lens, super oscillation, aperiodic metallic waveguide array etc. that have been proposed to achieve resolution beyond the diffraction limit, the immersion technique is still widely used in oil immersion microscopy, solid immersion microscopy, near-field scanning solid immersion microscopy, photolithography, etc. for resolution improvement due to its simplicity. Because the NA of an immersion lens is increased by a factor of the refractive index n, the corresponding resolution also improves n times. Many efforts have been taken to increase the NA by introducing various high index materials. Wu et al. reported an NA of 2.0 using a Gallium Phosphide solid immersion lens (SIL) in the visible band in 1999. Using the NA increasing lens (NAIL) technique, an NA as high as 3.3 was achieved with silicon in the infrared band by Ippolito et al. in 2001. This chapter will introduce various theory and application for data storage.

Near-field optics is from subwave length illumination to nanometric shadowing. Near-field optics uniquely addresses problems of x, y and z resolution by spatially confining the effect of a light source to nanometric domains. The problems in using far-field optics (conventional optical imaging through a lens) to achieve nanometric spatial resolution are formidable. Near-field optics serves a bridging role in biology between optical imaging and scanned probe microscopy. The integration of near-field and scanned probe imaging with far-field optics thus holds promise for solving the so-called inverse problem of optical imaging. The first optical topographic signals with subwavelength resolution were recorded independently by different groups in the world in 1981. These early works were encouraged and stimulated by speculations about the properties of evanescent and confined fields concentrated near the surface of materials. Since this exploratory period, a broad variety of scanning near field optical microscopes (SNOM) have been elaborated and continuously improved. An historical presentation of this pioneering period is detailed in the proceedings of the first near-field optics conference (Pohl and Courjon in 1993). The common feature of all SNOMs is the nanometre-sized detector able to collect or emit photons after coupling with a subwavelength size object deposited on a surface. Depending on the experimental design, this nano-detector can be used to transmit the collected light to an appropriate macro-detector (for example a photomultiplier) located far from the object in 1992. Then many experimental configurations based on this concept of nano-detection provide an increasing amount of optical information about the nanoworld.

Optical resolution can be greatly enhanced by use of a solid immersion lens (SIL). A SIL reduces the wavelength of light inside the optically dense medium of the lens and, as a result, reduces the focused spot size, $s = \lambda / n\sin\theta$, where λ is the wavelength in vacuum, n is the refractive index of the SIL, and θ is the marginal ray angle inside the SIL. Since a SIL acts as a single refracting surface, the paraxial aberrations introduced by the lens have simple geometric relationships. The following sections investigate the 3rd order aberrations for a single refracting surface and their implications for fabrication and usage of SILs. Since Mansfield and Kino introduced a solid immersion lens (SIL) in optical microscopes this technology has been developed for high-density data storage, photolithography, optical microscopes and other applications. The vector diffraction theory established by Richards and Wolf can be used to describe the focal field distribution in a high-NA system designed and demonstrated. The focal spot is asymmetrical when a linearly polarized light is incident on a high-NA SIL system, this asymmetry can be moderately rectified and the so-called super-resolution effect is generated. In addition, some amplitude and phase masks are also used to overcome the disadvantage of the SIL application that asymmetry can be moderately rectified and the so-called super-resolution effect is generated. In addition, some amplitude and phase masks are also used to overcome the disadvantage of the short focal depth in near-field SIL recording. The azimuthal polarization in the focal region of a high-NA lens was calculated by Youngworth and Brown by the vector diffraction theory. When a radially polarized beam illuminates a high NA lens, a tighter and symmetrical spot was obtained. Helseth studied the optical field distribution for the radially polarized beam focused to a solid immersion medium. Recently, Kozawa and Sato calculated the intensity distribution near the focal point of a high-NA. Near the focal point of a high-NA lens illuminated by higher-order radially polarized laser beams. In this chapter will use higher-order radially polarized beams to illuminate a high-NA SIL system and is the radial mode number. The focal point of a high-NA lens illuminated by higher-order radially polarized laser beams.

The adjective mesoscopic is used to define the situations where the sizes are of the order of the incident wavelength λ. For visible light, it corresponds roughly to the length range between 0.1 and 1μm. By nanoscopic, one usually means low-dimensional structures smaller than 100nm. If these structures can be identified with single molecules, the nanoscopic regime also means the molecular range. However, structures smaller than 1nm are commonly viewed as belonging to the atomic range. When λ is much smaller than the size of the scatterers, one speaks of the macroscopic regime. Geometrical optics is a first approximation which describes the scattering of light by macroscopic objects. On a more refined level, Kirchhoff's diffraction theory uses a scalar field to account for phenomena where light displays a wave character on a macroscopic scale. Kirchhoff's theory attributes ideal properties to the scatterers such as a perfect conductivity or a real refractive index. Microscopic systems are objects which are so small when compared to the incident wavelength that the non-retarded approximation becomes applicable. This approximation considers the scatterers as dipoles or a set of dipoles whose susceptibilities may include dissipative effects. For visible wavelengths, this regime corresponds to the atomic range. Near-field optics deals with phenomena involving evanescent electromagnetic wave which becomes significant when the sizes of the objects are of the order of λ or smaller. By object, also mean void structures carved in a surrounding material such as the vacuum gap basic to ATR (attenuated total reflection) experiments or the air gap separating a local probe and a sample surface. In view of the above classification, it is clear that near-field optics is thus concerned with the scattering of electromagnetic waves by meso- and nanoscopic systems. Even in the situations where atomic size structure is involved, near-field optical detection is affected by the nano- and meso-scopic system embedding the atomic size structure. Evanescent waves are important in near-field optics because the typical size of the objects is comparable to λ and the decay of evanescent waves occurs within a range given by λ. The theory of electromagnetic waves describes satisfactorily their interaction with objects which are macroscopic or microscopic relative to the incident wavelength λ. However, the theoretical knowledge about the scattering of electromagnetic waves by mesoscopic systems remains limited. Since many situations involve nanoscopic and mesoscopic systems simultaneously, incomplete information about the mesoscopic range impedes our understanding of nanoscopic systems. Most approximations are not appropriate for studying mesoscopic systems. Unlike macroscopic systems (successfully described by Kirchhoff theory) and microscopic systems (for which retardation is negligible), mesoscopic systems require the detailed solution of the full set of Maxwell equations. A numerical method for solving Maxwell equations is needed because both geometries and dielectric responses of typical mesoscopic systems display a high degree of complexity.

However, numerical methods traditionally used in electrodynamics are not well suited to cope with mesoscopic structures. Cumbersome procedures appear to be uncertain and produce unreliable outputs. The main origin of their problems can be traced to the crucial role played by the evanescent components of the field in the near-field zone close to mesoscopic scatterers. In complete analogy with the tunnel effect for electrons, these evanescent components can lead to optical tunnel effects. In the mesoscopic range, the accurate treatment of evanescent waves requires one to deal carefully with the electromagnetic boundary conditions at each interface and to include realistic dielectric responses. When a radially polarized beam is tightly focused by an aplanatic lens, the electric field in modes can be generally produced by using a polarization-selective optical element or mechanism.

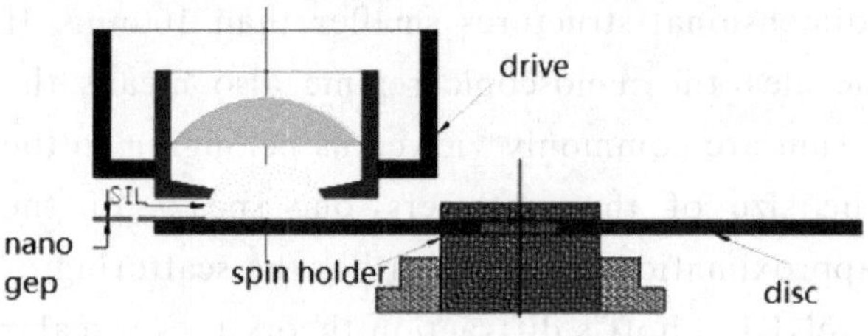

Fig. 4-1 Solid immersion lens.

In addition, some non-sphere solid immersion mirrors are also designed and demonstrated. A new approach to optical disk data storage involves the use of near-field optics in general and the solid immersion lens (SIL) in particular (Kino 1994). The SIL approach requires that a part of the objective lens fly over the surface of the storage medium, as shown in Figure 4-1. The hemispherical glass of refractive index n receives the rays of light at normal incidence to its surface. These rays come to focus at the center of the hemisphere and form a diffraction-limited spot that is smaller by a factor of n compared to what would have been in the absence of the SIL. (This is a well-known fact in microscopy, where oil immersion objectives have been in use for many years.) A typical glass hemisphere having $n = 2$ will reduce the diameter of the focused spot by a factor of 2, thus increasing the recording density fourfold. To ensure that the smaller spot size does indeed increase the resolution of the system, the bottom of the hemisphere must either be in contact with the active layer of the disk or fly extremely closely to it. For a disk spinning at several thousand rpm, it is possible to keep the SIL at a distance of less than 100nm above the disk surface.

The rays of light that are incident at large angles at the bottom of the hemisphere would have been reflected by total internal reflection, except for the fact that light can tunnel through and jump across gaps that are small compared to one wavelength. This tunneling mechanism is known as frustrated total internal reflection, and its presence qualifies the application of SIL in optical data storage as a near-field technique.

The most important application for the increased storage density and data rate afforded by the SIL is the necessity to permanently enclose the disk within the drive, thereby making it non-removable. The possibility of maintaining removability in a SIL system has been suggested, but it remains to be demonstrated in a practical setting. Research and development activities in optical data storage media, Japanese hosts in this subject, as evidenced by the numerous questions that they asked WTEC panelists about the technical and business aspects of the TeraStor approach. Although the Japanese seem to be playing catch-up at this point, it will not be long before they can make flying optical heads using SIL or some such technique. After all, many of the components that TeraStor is using in its experimental drive come from Japan, and it is only a matter of time before the Japanese engineers learn how to put these items together and build a complete system. Sony already has a working system with a variant of the solid immersion lens, one that is not working in the near-field yet, but can provide insight into the subtleties of SIL-based systems. Perhaps what makes the SIL concept so attractive to the Japanese industry is its ability to bridge the gap between optical and magnetic recording technologies. Traditionally, hard disk drives have been the domain of American companies, whereas Japan has been strong in optical recording. The SIL has the potential to marry these two technologies, thus giving the Japanese industry an opening to capture at least a fraction of the hard disk market.

The most important advantages of near-field optical storage are that can achieve high resolution over traditional diffraction limit. The first near-field optical storage experiments by Betzig and Trautman. They used the near-field optical scanning microscope (The NSOM) recoding data on the magneto-optical (MO) thin film, which the minimum recording sites close to 60nm. The design and manufacturing technology of the solid immersion near-field storage has gotten great development in recent years. A lot of achievements of experiment lens of the SIL have been taken on in CIT, IBM, AT & T, NIST at U. S. A, the University of Tokyo, SONY, NIAIR, Matsushita Electric in Japan. South Korea's Samsung Electronics, the company's research institutions have carried out relevant studies also. Japanese National Modern Institute of Science and Technology (AIST) in cooperation with South Korea's Samsung Blue-ray optical head plus super-resolution optical disc, which gets the Super-RENS is shown in Figure 4-2. The track is within 0.32μm and the pitch ($2T$) is about 75nm. The recoding medium is the Co/Pt multilayer magneto-optical film. The minimum character size can reach 10~50nm that could be used for TB level capacity on one disk with single layer. The recoding spot size will be reduced in the further with the near-field optical techniques development. The use of near-field solid immersion lens (SIL) can get the spot size may be under 40nm. The theoretical storage density can be achieved 100GB/in^2 above. The near field solid immersion lens for optical disc storage, its optical system and the structure of the medium are three typical models. The comparison of three types (a-c) near-field solid immersion lens storage systems and Blue-ray disc (BD) storage system (d) are shown in Table 4-1. There are 3~4μm protection layer on the surface of the disc for type (a) and (b). So their efficient numerical aperture (ENA) is 1.45 and 1.65 and the minimum character size is about 87nm and 75nm that can be used for capacity of 75GB and 100GB storage systems with wavelength 405nm laser. The performance size can be down to 35~40nm for the type (c). The capacity could be 1TB even over on a disc. Type (a) and (b) SIL storage systems with protection layer on the surface that the capacity is not too high. But the ability of anti-environmental pollution is well, and possible to produce a multilayer structure which capacity is over 1Tb also. The third option is not protection layer on the surface. So higher resolution can be got and the minimum information symbol size can be reduced to 40nm. However, this program there is no protection layer on the surface that has to be higher environment condition. The third type near-field solid immersion lens optical storage experimental system was developed by the Canadian LGE completed. This system is currently the highest storage density near-field solid immersion lens for optical disc storage. The system uses a laser wavelength of 405nm, numerical aperture of 2.2, the minimum character size

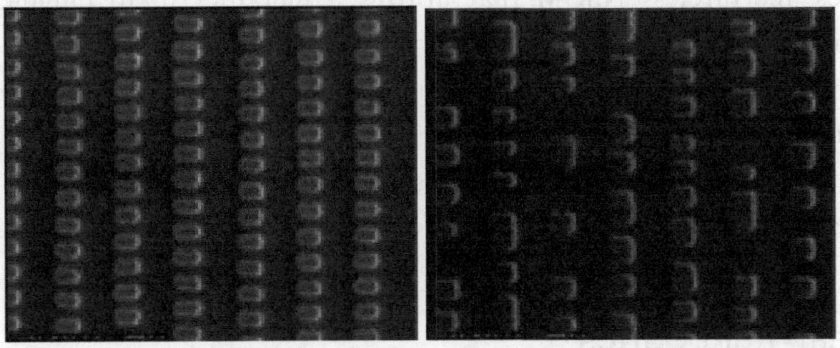

Fig. 4-2 The minimum symbols of Blue-ray Super-RENS of the optical disc is down to 75nm.

53nm. Japan's Sony and Philips developed near-field solid immersion lens optical disk storage system, in general, also reached a similar level. Another important application field of the near-field optical storage technology is the combination with the magnetic storage.

Table 4-1 Comparison of various types of near-field storage of solid immersion lens and far-field effective numerical aperture of the Blue-ray disc storage system where *n* is the refractive index of the SIL and ENA is efficient numerical aperture.

type / sample	(a) half-sphere with protection layer	(b) 60%-sphere with protection layer	(c) 80%-sphere no protection layer	(d) blu-disc with protection layer
protection layer recoding layer isolation layer disc base	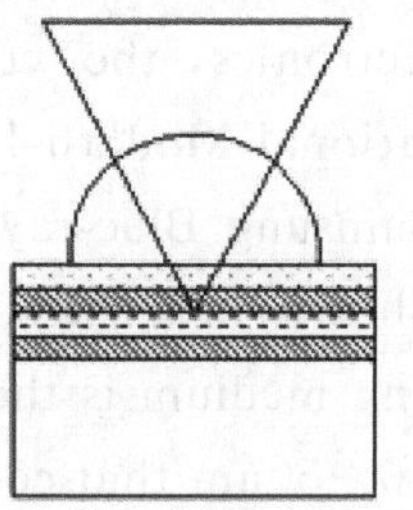 ENA=n x NA	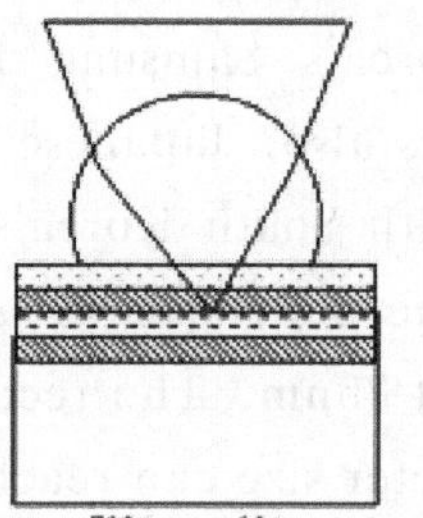 ENA>n x NA	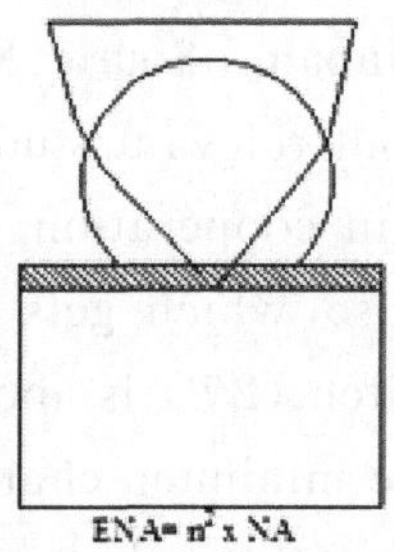 ENA= n^2 x NA	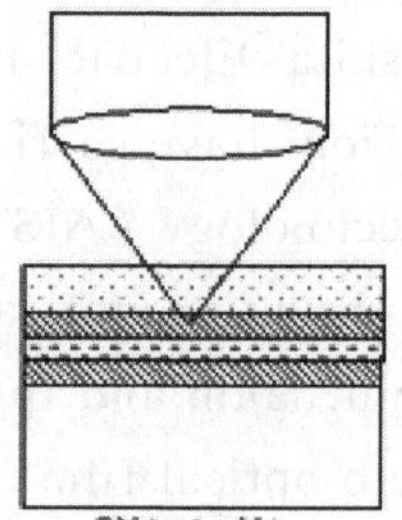 ENA=1 x NA

The co-activity of the magnetic record materials can be reduced as the thermal effect of the small size of light spot. The data write can be completed in the lower magnetic field and can guarantee that has great stability.

4.1.1 Near-field interaction and microscopy

This section will discuss the principles and applications of near-field optics where a near-field geometry is utilized to confine light on nanometer scale. These principles form the basis of near-field scanning optical microscopy (NSOM), which provides a resolution of 100nm, significantly better than the diffraction limit imposed on far-field microscopy. NSOM is emerging as a powerful technique for studying optical interactions in nanodomains as well as for nanoscopic imaging. The applications of NSOM have ranged from single-molecule detection to bioimaging of viruses and bacteria.

After a general description of near-field optics, the section will introduce:

(1) theoretical modeling of near-field nanoscopic interactions;

(2) various approaches used for near-field microscopy, Some illustrative examples of optical interactions and dynamics utilizing NSOM;

(3) discusses spectroscopy of quantum dots and single molecules, as well as studies of nonlinear optical processes in nanoscopic domains;

(4) introduces apertureless NSOM that utilizes a metallic tip to enhance the local field, applications of this approach;

(5) discusses enhancement of optical interactions using a surface plasmon geometry incorporated in an NSOM assembly;

(6) describes time-and space-resolved studies of nanoscale dynamics;

(7) lists some of the commercial manufacturers of near-field microscopes and highlights of the section.

4.1.2 Near-field optics

Near-field optics deals with illumination (and subsequent optical interaction) by light emerging from a subwavelength aperture or scattered by a subwavelength metallic tip or nanoparticle, of an object in the immediate vicinity (or within a fraction of the wavelength of light) of the aperture or scattering source. The light in the near-field contains a large fraction of nonpropagating, evanescent field, which decays exponentially in the far field (far from the aperture or scattering metallic nanostructure). The case where light is passed through a subwavelength aperture as shown in Figure 4-3, or through a tapered fiber (another type of aperture) as shown in Figure 4-4, is also labeled as aperture near-field optics or simply near-field optics.

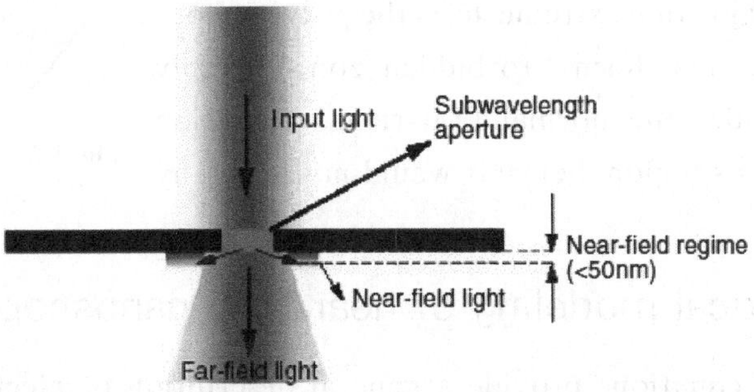

Fig. 4-3 Principle of aperture controlled near-field optics.

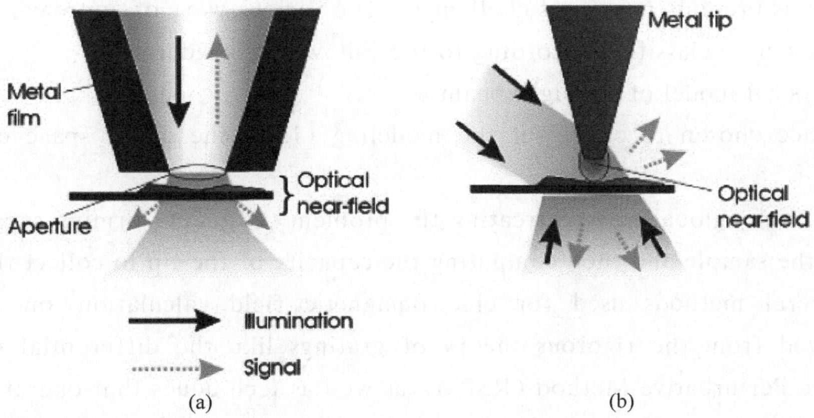

Fig. 4-4 The principles of near-field optics: The aperture-controlled near-field optics using a metal-coated tapered fiber is shown on the left. The schematic on the right-hand side shows apertureless near-field optics utilizing scattering from a metal tip.

Most near-field studies or near-field microscopy utilized near-field optics that generally involves a tapered fiber. An example of apertureless near-field optics is also provided in Figure 4-4, where a sharp metallic tip is used to scatter the radiation. The enhanced electromagnetic field around the metallic tip is strongly confined. This field enhancement near the surface of a metallic nanostructure is discussed in detail later.

In aperture-controlled near-field optics, light is squeezed through an aperature such as an aluminum-coated tapered fiber to confine the light from leaking out. Light then emanates through a tip opening, which is generally anywhere from 50nm to 100nm in diameter, and is incident on a sample within nanometers of the tip. Thus, the sample senses the near-field distribution of the light field. The interesting aspect is that light is confined to a dimension much smaller than its wavelength. Even if one uses IR light of 800nm in wavelength, it can be squeezed down to 50nm or 100nm. Thus, the near-field approach allows one to break the diffraction barrier that limits the focusing in the far field to the dimension of the incident optical wavelength. The field that comes out of the nanoscopic aperture (fiber tip) has some very interesting, unique properties. The field distribution of light emanating from a fiber tip is shown in Figure 4-5. There is a region of propagation in which the wavevector, $\boldsymbol{k}$, of light is real. This light is just the normal far-field-type light which has an oscillating character. Toward the edges, on the extreme right and extreme left, the wavevector, $\boldsymbol{k}$, of light is imaginary. The term "forbidden zone" simply implies that the light under the normal (far-field) condition would not propagate in this region, hence it would not have any field distribution.

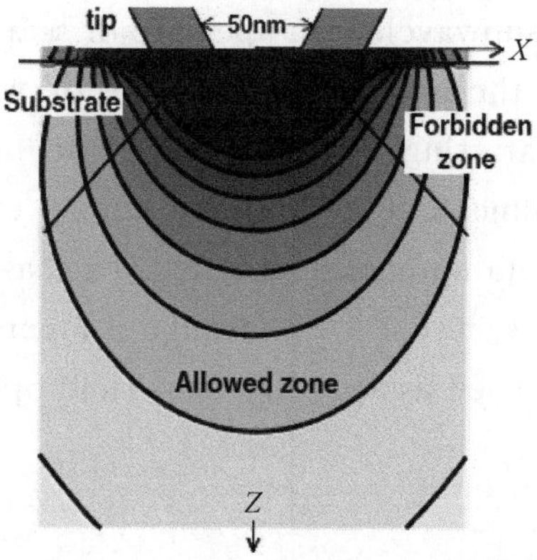

Fig. 4-5 Field distribution of light emanating from a fiber.

4.1.3 Theoretical modeling of near-field nanoscopic interactions

Although Maxwell's equations provide a general description of electromagnetic phenomena, their analytical solutions are limited to relatively simple cases and rigorous treatment of nanoscale optical interactions presents numerous challenges. The various ways of approaching the theory of near-field optics can be classified according to the following considerations:

(1) The physical model of the light beam;

(2) The space chosen to carry out the modeling (i. e., the direct space or Fourier space modeling);

(3) Global or nonglobal way of treating the problem (e. g., performing separate calculation for the field in the sample and then computing the capacity of the tip to collect the field).

Among several methods used for electromagnetic field calculation, one can distinguish techniques derived from the rigorous theory of gratings like the differential method and the Reciprocal-Space Perturbative Method (RSPM), as well as techniques that operate in direct space like the Finite-Difference Time-Domain Method (FDTD) and the Direct-Space Integral Equation Method (DSIEM).

In general, analytical solutions can provide a good theoretical understanding of simple problems, while a purely numerical approach (like that of the FTDT method) can be applied to complex structures. A compromise between a purely analytical and a purely numerical approach is the multiple multipole (MMP) model (Girard and Dereux, 1996). With the MMP model, the system being simulated is divided into homogeneous domains having well-defined dielectric properties. Within individual domains, enumerated by the index i, the electromagnetic field $f(i)(\boldsymbol{r}, \omega_0)$ is expanded as a linear combination of basis functions

$$f^{(i)}(\boldsymbol{r}, \omega_0) \approx \sum_j A_j^{(i)} f_j(\boldsymbol{r}, \omega_0) \tag{4-1}$$

where the basis functions $f_j(\boldsymbol{r}, \omega_0)$ are the analytical solutions for the field within a homogeneous domain. These basic functions satisfy the eigenwave equation for the eigenvalue q_j:

$$-\nabla\times\nabla\times f_j(\boldsymbol{r}, \omega_0) + q_j^2 f_j(\boldsymbol{r}, \omega_0) = 0 \tag{4-2}$$

MMP can use many different sets of basis fields, but fields of multipole character are considered the most useful. The parameters $A_j(i)$ are obtained by numerical matching of the boundary conditions on the interfaces between the domains.

As an example of the use of this technique for investigations of nonlinear optical processes in the near field, we show here investigations of second harmonic generation in a noncentrosymmetric nanocrystal exposed to fundamental light from a near-field scanning tip.

One notes that a consequence of nonlinear optical interaction in the near-field is that the phase-matching conditions do not need to be fulfilled because the domains are much smaller than the coherence length. Starting from Maxwell's equations, the electric fields of the fundamental and the second harmonic (SH) wave can be shown to satisfy the nonlinear coupled vector wave equations

$$\nabla\times\nabla\times\boldsymbol{E}(\boldsymbol{r},\omega_0) - \frac{\omega_0^2}{c^2}\varepsilon(\boldsymbol{r},\omega_0)\boldsymbol{E}(\boldsymbol{r},\omega_0) = 4\pi\frac{\omega_0^2}{c^2}\boldsymbol{P}^2(\boldsymbol{r},\omega_0) \tag{4-3}$$

$$\nabla\times\nabla\times\boldsymbol{E}(\boldsymbol{r},2\omega_0) - \frac{4\omega_0^2}{c^2}\varepsilon(\boldsymbol{r},2\omega_0)\boldsymbol{E}(\boldsymbol{r},2\omega_0) = 4\pi\frac{4\omega_0^2}{c^2}\boldsymbol{P}^2(\boldsymbol{r},2\omega_0) \tag{4-4}$$

where $\varepsilon(\boldsymbol{r},\omega_0)$ and $\boldsymbol{E}(\boldsymbol{r},2\omega_0)$ are linear dielectric functions for the fundamental and the SH waves, respectively. The propagation constant k_z along the z direction is

$$k_z = (k^2 - k_{\parallel}^2)^{1/2} = k_0(1 - n_1^2\sin^2\theta)^{1/2} \tag{4-5}$$

where $k_0 = 2\pi/\lambda$, λ is the wavelength of illumination light in free space; n_1 is the refractive index of the tip, and θ is the incident angle. If $1 - n_1^2\sin^2\theta > 0$ (i.e., k_z is real), the waves will propagate with constant amplitude between the probe and the sample, which corresponds to the "allowed light" in the sample. In the areas where k_z is imaginary, the waves will decay exponentially within distances comparable to the wavelength, thus such waves have evanescent character and produce the "forbidden light" in the sample. From the electrical field distribution of the fundamental wave calculated with the MMP method, we can obtain the electrical field distribution of the SH wave and the different contributions of "allowed light" and "forbidden light." Figure 4-6 shows the three-dimensional perspective view of the optical near-field intensity of the fundamental and the SH wave, respectively. The field intensity of SH wave is orders of magnitude weaker than that of the fundamental wave (FW), and it is highly localized within the area of the probe tip center—that is, about 50nm × 50nm. The fundamental wave is more delocalized compared to the SH wave. Figure 4-7 shows the sectional plot of Figure 4-6 along the x-axis direction, and Figure 4-8 shows the integration of $|E|^2$ for the the field intensity of SHG, $|E|^2(10^{-10})$, SHG over the total solid angle. It is clear that the field intensity close to the probe center comes almost entirely from the allowed light, while a field enhancement appearing at the edge of the tip is due to the field components from the forbidden light. The field intensity decreases very rapidly with the tip-sample distance, and its typical decay length is approximately equal to the tip size about 50nm. Furthermore, the field intensity of the forbidden light, which decays exponentially, exhibits a much larger variation with the probe-sample distance than does the field intensity of the allowed light. Figure 4-8 also indicates that when the probe is very close to the sample surface—that is, $d <$ 50nm—the intensity from the forbidden light dominates. However, when the probe-sample distance is larger than 50nm, the intensity from the allowed light becomes the main contribution to the total field intensity. Because the allowed light only contains the low spatial frequencies of the sample surface, the detection of the forbidden light is essential to investigate details for both linear and nonlinear optical interactions.

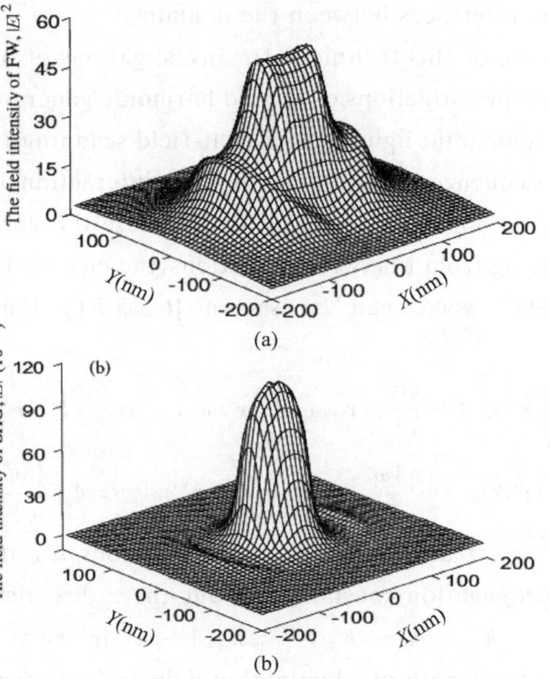

Fig. 4-6 Three-dimensional perspective view of the optical near-field intensity of (a) the fundamental wave and (b) the second harmonic generation. The calculation is performed for *p*-polarization and for the tip-sample distance of 10nm.

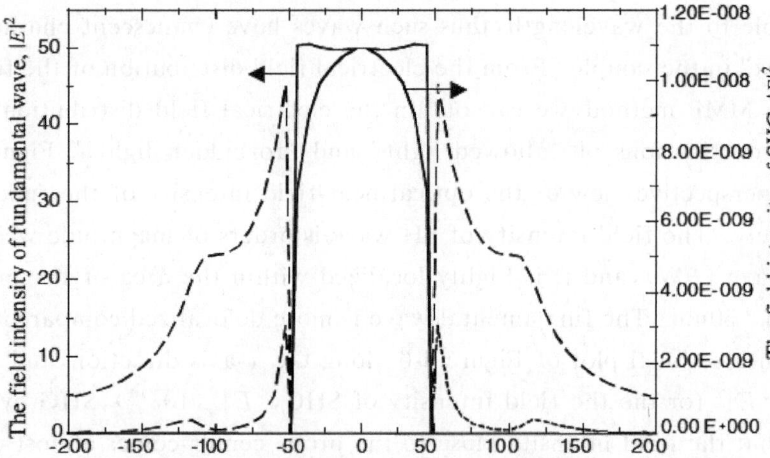

Fig. 4-7 The electric field intensity of the fundamental wave and the second harmonic generation along the sample surface for tip-sample distance of 10nm. The solid curve denotes the field intensity of the allowed light; the dashed curve denotes the field intensity of the forbidden light.

4.1.4 Theoretical modeling of near-field nanoscopic interactions

Near-field optical microscopy utilizes near-field interactions that allow one to achieve a resolution of <100nm, significantly better than that permitted by the diffraction limit. The resolution of conventional (far-field) optical imaging techniques is limited by diffraction of light. The concept of using the near field for imaging was first discussed in 1928 by Synge, who suggested that by combining a subwavelength aperture to illuminate an object, together with a detector very

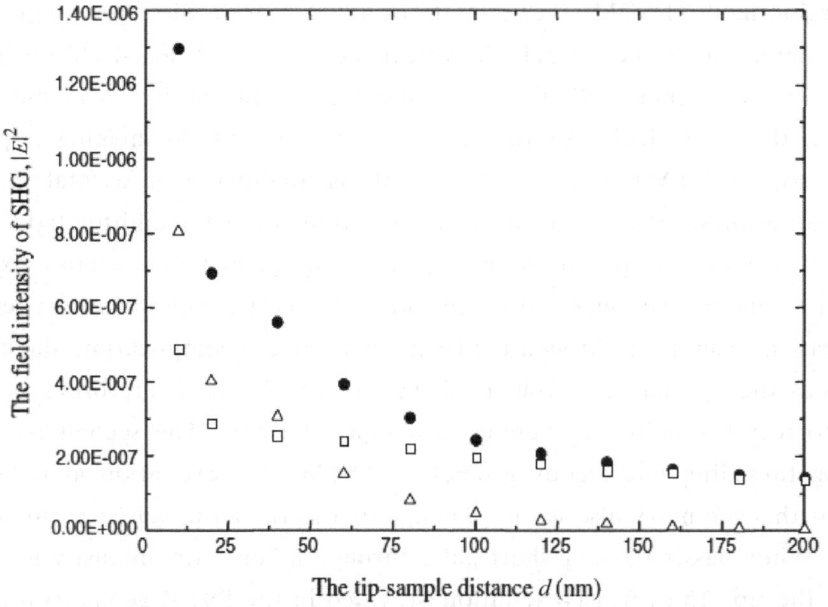

Fig. 4-8 The effect of the tip-sample distance d on the near-field intensity of the SHG from the total field (•), allowed light (□), and forbidden light (△).

close to the sample (one wavelength, or in the "near field"), high resolution could be obtained by a non-diffraction-limited process (see Figure 4-3). The implementation of this principle in practice created the technique of near-field microscopy. Now there are different variations of this technique. One can illuminate the sample in the near field, but collect the signal in the far field or illuminate the sample in the far field while collecting the signal in the near field or do both in the near field. In most methods, the important component is the use of a subwavelength aperture that can be achieved by using a tapered optical fiber with a tip radius small than 100nm.

The most commonly used near-field probe consists of an optical fiber that is tapered and coated on the outside with a reflective aluminum coating. The tip of the fiber is typically about 50nm. Light propagating through this fiber, either for excitation or for collection of emission, produces a resolution determined by the size of the fiber tip and the distance from the sample. The image is collected point to point by scanning either the fiber tip or the sample stage. Hence the technique is called near-field scanning microscopy (NSOM) or scanning near-field microscopy (SNOM). Different modes of near-field microscopy are shown in Figure 4-9.

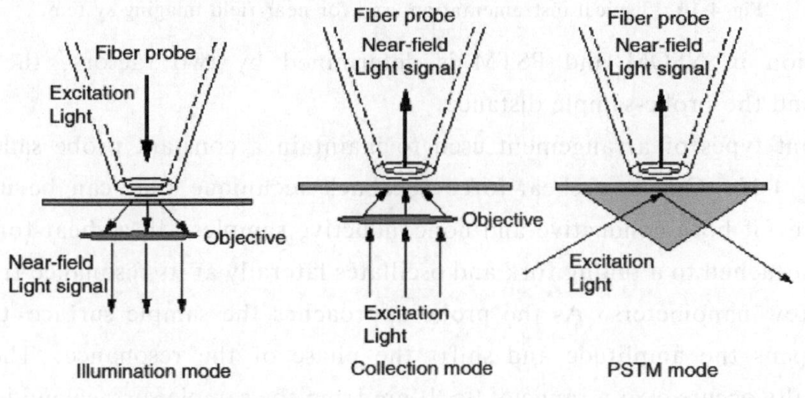

Fig. 4-9 Different modes of near-field microscopy.

In illumination-mode NSOM, the excitation light is transmitted through the probe and illuminates the sample in the near field. A typical setup used for near-field imaging is shown in Figure 4-10. In collection-mode NSOM, the probe collects the optical response (transmitted or emitted light) in the near field. Another mode used in near-field imaging is photon scanning tunneling microscopy (PSTM) in which the sample is illuminated in a total internal reflection geometry using an evanescent wave, as due to photon tunneling; the emitted light is collected by a near-field optical probe. The photon scanning tunneling geometry, or simply exciting from the bottom in the far field, is also much more convenient if one is using ultrashort femtosecond laser pulses. In the case of excitation through the fiber tip, there are complications due to broadening of the short pulses as they propagate through a length of the fiber. Therefore, a pair of gratings is often used to correct for pulse broadening in a tapered fiber. The second reason to choose a photon scanning tunneling microscopy geometry (PSTM) for excitation and the near-field for collection is that this geometry also avoids damage of the fiber tip caused by high peak power of the laser pulse. When passing a very short pulse through a 50nm tip, intensity may be sufficiently high to damage the tip. In contrast, excitation provided in the PSTM geometry, or in the far field from the bottom, minimizes optical damage.

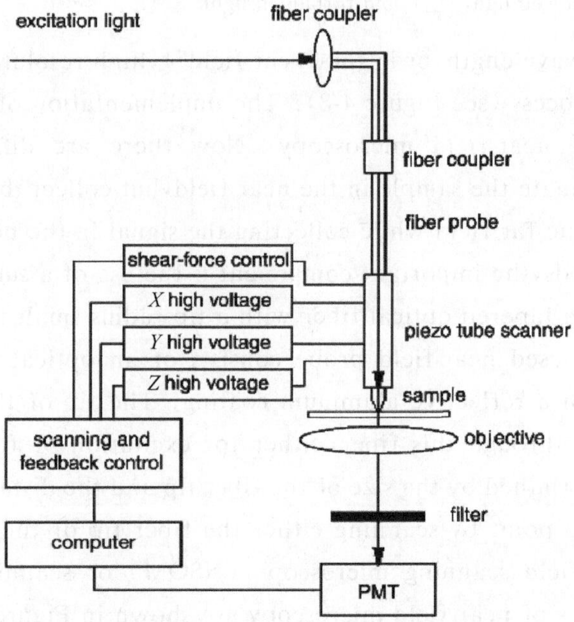

Fig. 4-10 Typical instrumentation used for near-field imaging system.

The resolution in NSOM and PSTM is determined by two factors: the probe aperture (opening) size and the probe-sample distance.

Two different types of arrangement used to maintain a constant probe sample distance are shown in Figure 4-11. One is a shear-force feedback technique that can be used for distance regulation in cases of both conductive and nonconductive samples. In a shear-force feedback, the optical probe is attached to a tuning fork and oscillates laterally at its resonance frequency, with an amplitude of a few nanometers. As the probe approaches the sample surface, the probe-sample interaction dampens the amplitude and shifts the phase of the resonance. The change in the amplitude normally occurs over a range of 0~10nm from the sample surface and is monotonic with the distance, which can be used in a feedback loop for distance regulation. The shear-force feedback can also be used to simultaneously obtain the topographic (AFM) image of the sample, to provide a monitoring reference for NSOM and PSTM. In an alternate arrangement, light reflection from the surface is used while dithering the fiber.

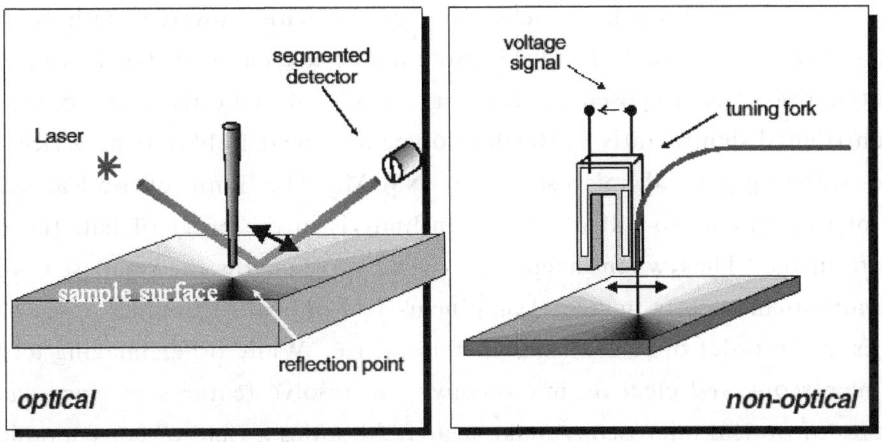

Fig. 4-11 The two types of arrangements used to maintain a constant probe-sample distance.

In regard to the tapered fiber geometry, some of the designs use a straight fiber geometry with the tip at the narrow end. This is shown in the top illustration of Figure 4-11. Some use the cantilever geometry, shown on the bottom of Figure 4-11, where the fiber tip is bent. This cantilever arrangement allows one to use the same probe determines both the spatial resolution and the sensitivity of measurements.

Hence tip fabrication is of major importance. The two methods used for tip fabrication are as follows:

(1) The heating-and-pulling method. Here an optical fiber is locally heated by a CO_2 laser and pulled uniformly on both sides of the heated region.

(2) The chemical etching method. This method utilizes a hydrofluoric acid (HF) solution to etch the glass fiber.

The desired taper angle is achieved by adjusting the composition of a buffered HF solution. It has been shown that the optical transmission efficiency of a double-tapered structure with a large cone angle is two-orders of magnitude greater than that of a single-tapered probe with a small cone angle. Such a double-tapered structure can be obtained by using a multi-step HF etching process.

The principles of near-field optics are extended to many application for digital data storage as super resolution near field structure (Super-RENS), micro-aperture laser, plasmonic near-field recording (PNFR) and meta material immersion lenses (MIL) that will be introduced respectively. For the near-field optical storage technology, experiments conducted with various devices confirmed that improving the resolution requires approaching the tip very close to the sample. Adding to this fact the recent development towards near-field spectroscopy and the exploration of radiation pressure effects, it became clear that a realistic computation should at least take the tip-sample coupling properly into account. Various numerical techniques were then applied in order to include this coupling successfully.

4.2 Principles of near-field optics

Near-field optics is that branch of optics that considers configurations that depend on the passage of light to, from, through, or near an element with subwavelength features, and the coupling of that light to a second element located a subwavelength distance from the first. The barrier of spatial resolution imposed by the very nature of light itself in conventional optical microscopy contributed significantly to the development of near-field optical devices, most notably the near-field scanning optical microscope, or NSOM. The limit of optical resolution in a conventional microscope, the so-called diffraction limit, is in the order of half the wavelength of the light used to image. Thus, when imaging at visible wavelengths the smallest resolvable objects are several hundred nanometers in size. Using near-field optical techniques, researchers currently resolve features in the order of tens of nanometers in size. While other imaging techniques (e.g. atomic force microscopy and electron microscopy) can resolve features of much smaller size, the many advantages of optical microscopy make near-field optics a field of considerable interest.

The early work of developing a near-field optical device was first conceived by Edward Hutchinson Synge in 1928 but was not realized experimentally until the 1950s when several researchers demonstrated the feasibility of sub-wavelength resolution. Published images of sub-wavelength resolution appeared when Ash and Nichols examined gratings with line spacing less than one millimeter using microwaves of 3cm wavelength. In 1982 Dieter Pohl at IBM in Zurich, Switzerland, first obtained sub-wavelength resolution at visible wavelengths using near-field optical techniques.

Microscopic systems are objects which are so small when compared to the incident wavelength that the non-retarded approximation becomes applicable. This approximation considers the scatterers as dipoles or a set of dipoles whose susceptibilities may include dissipative effects. For visible wavelengths, this regime corresponds to the atomic range.

A numerical method for solving Maxwell equations is needed because both geometries and dielectric responses of typical mesoscopic systems display higher degree of complexity. However, numerical methods traditionally used in electrodynamics are not well suited to cope with mesoscopic structures. Cumbersome procedures appear to be uncertain and produce unreliable outputs. The main origin of their problems can be traced to the crucial role played by the evanescent components of the field in the near-field zone close to mesoscopic scatterers. In complete analogy with the tunnel effect for electrons, these evanescent components can lead to optical tunnel effects. In the esoscopic range, the accurate treatment of evanescent waves requires one to deal carefully with the electromagnetic boundary conditions at each interface and to include realistic dielectric responses.

4.2.1 Base theoretical works

The physics of evanescent electromagnetic waves, which is the central concept used in near-field optics, was a poorly developed research area before the mid 1960s. The analysis of the skin depth effect at metallic surfaces by Zenneck and Sommerfeld was probably the first recognition of the existence of evanescent electromagnetic waves. The famous papers of Mie (1908) and Debye (1909) about the scattering of electromagnetic waves by a sphere contained, at least in principle, the solution of the vector wave equation in the near-field zone not only for the microscopic regime but also for the mesoscopic one. However, these works were little exploited in the study of near-field optics. Indeed, for coordinates located in the near-field zone, the formulae of Mie and Debye display convergence problems which are even more difficult to solve in the mesoscopic regime. Much later, Fano (1941) realized that the anomalies observed in the diffraction of light by metal gratings were related to the excitation of evanescent electromagnetic modes bound to the surface.

The classical problem of diffraction by an aperture in a perfectly conducting screen was treated by numerous different approaches (see the review published in this journal by Bouwkamp 1954). In 1944, Bethe proposed a curious model involving magnetic charges in order to extract a solution to the problem of the diffraction of light by a subwavelength aperture. The assessment of the near-field zone at the exit of the aperture was feasible in principle but was ignored because it was not experimentally relevant at that time. Much later, Leviatan (1986) and Roberts (1987, 1989, 1991) applied Bethe's theory to apertures sizes which were significant for near-field microscopes. Such computations confirmed unambiguously the existence of an important exponentially decaying field near the exit of the aperture. In the approximation of a perfectly conducting screen, the generation of this near-field may be understood qualitatively with the help of the Heisenberg uncertainty Near-field optics theories principle. Consider the simpler case of the diffraction of an incident plane wave propagating in the z direction by an aperture of diameter a carved in an infinitely thin screen covering the x-y plane. The Heisenberg uncertainty

$$\Delta s \Delta k \geqslant 1 (D = x;\ y) \tag{4-6}$$

states that passage through a slit implies that the transmitted field acquires a non-zero angular spectrum

$$\Delta ks \geqslant a^{-1} \tag{4-7}$$

In the case of a subwavelength aperture, $k \ll a^{-1}$ and can find

$$k \ll \Delta ks \tag{4-8}$$

The dispersion relation

$$k^2 = \Delta k_x^2 + \Delta k_y^2 + k_z^2 = \varepsilon\omega^2 / c^2 \tag{4-9}$$

where ε is the dielectric function of the background reference medium, shows that k_z^2 may become either positive or negative. The Fourier components of the diffracted field may be classified according to the sign of k_z^2. For positive values, the behaviour e^{ik2z} corresponds to radiative (or propagative) waves along z. These waves reach the far-field. For negative values, the Fourier component takes the evanescent form e^{ik2z}. The set of imaginary values of k_z defines the non-radiative waves existing in the near-field zone. Their exponental decay prevents them from reaching the far-field (Vigoureux et al., 1989, 1992). An important breakthrough was achieved by Levine and Schwinger who established the non-trivial form of the free-space Green dyadic. This dyadic allowed them to formulate a variational method of resolution to the puzzling problem of diffraction by small apertures. As in the case of Bethe's theory, the near-field zone at the exit of the aperture was not assessed at that time. However, the work of Levine and Schwinger inspired the later development of electromagnetic scattering theory.

As discussed above, the assumption of a perfectly conducting screen allows one to reproduce the generation of the near-field as a result of the Heisenberg's uncertainty and can account for polarization effects. However, this assumption is much too restrictive when considering near-field optical phenomena since it hides resonance phenomena.

Indeed, according to their frequency dependent dielectric properties, localized eigenmodes characterized by evanescent wavefunctions may be sustained by small objects and even by surfaces. The understanding of the physical content of the dielectric function was triggered by the work of Huang (1951). Huang brought to the fore the fact of how the parameters driving the infra-red values of the dielectric constant of a polar crystal are closely related to the coupling of light with vibrational eigenmodes (phonons) of the crystal. Hopfield (1958) and Pekar (1960) introduced a similar idea for the range of visible wavelengths by invoking the coupling of light to the excitons of the crystal. In particular, Hopfield developed the concept of polaritons and was the first to observe their dispersion relations by using Raman spectroscopy (Henry and Hopfield, 1965). Polaritons are the polarization waves of a crystal which are excited by incident light. They are the electromagnetic eigenmodes of condensed matter.

The clarification brought by the concept of polariton allowed one to identify the conditions of existence of evanescent electromagnetic eigenmodes bound to a surface (Ferrell, 1958, Stern and Ferrell, 1960) and to the interfaces of a thin film (Kliewer and Fuchs, 1966). In 1968, Otto invented and explained theoretically the attenuated total reflection spectroscopy (ATR) which allowed him to measure the dispersion relations of interface polaritons (plasmon-polaritons in the original experiment) by a simple reflectivity measurement. The principle of ATR consists in approaching a sample surface in the decay range of the evanescent wave produced by total reflexion on a prism surface. By the optical tunnel effect, the evanescent wave can then excite the non-radiative interface modes of the sample. This results in the frustration of the total reflection. Today's near-field optical microscopes using internal illumination (STOM or PSTM devices) rely on the basic discovery of Otto (Reddick et al., 1989, Courjon et al., 1989, Vigoureux et al., 1989).

The success of the polariton concept encouraged further development. Although the subject of light scattering by small particles had already been thoroughly investigated (Mie, 1908, van de Hulst, 1957), electromagnetic eigenmodes of small spherical particles were revisited from the point of view of their frequency-dependent dielectric function (Englman and Ruppin, 1966, Fuchs and Kliewer, 1968). It appeared that, due to the curvature of the particles, the distinction between radiative and non-radiative modes was not so clear as it was in the case of planar interfaces. Nevertheless, it was rapidly recognized that the coupling of such particles with a planar surface changed dramatically as a function of the distance between them. This coupling can be identified as a true near-field effect since it occurs for separation distances smaller than the wavelength so that the evanescent waves scattered by the particle have not yet decayed.

The related phenomenon of enhanced Raman scattering of molecules adsorbed on metallic surfaces proved to be of electromagnetic origin (Otto, 1984). In the non-retarded approximation, the particle is modelled by a point dipole and the surface plasmon is reduced to the image dipole. The local electric field at the coordinate of the molecule is the sum of the external field and the field due to the image dipole. This approximation reproduces the red-shift observed in the absorption spectrum.

The inclusion of retardation in the case of metal particles approaching a metal surface demonstrates that a hybrid plasmon can show up as the result of the coupling of the particle plasmon to the surface one (Takemori et al., 1987). The absorption spectrum of the sphere is much more red-shifted than expected by the electrostatic approximation. The finite size of the sphere makes possible the excitations of several plasmons related to higher multipole modes of the sphere.

If a dielectric (i.e. non-absorbing) sphere is substituted for the metallic one, the absorption can only be due to the surface plasmon of the metal surface. The incident field excites the resonant but undamped modes of the dielectric sphere which are distributed as evanescent waves around the sphere. As the sphere approaches the surface, a larger number of surface plasmons with shorter wavelengths can be excited. Therefore, absorption occurs in the frequency range located below the highest surface plasmon frequency and above the lowest frequency which can sustain an eigenmode of the sphere. Near-field optical microscopes using or detecting the resonance effects of small particles operate by this principle.

Concurrent study of the optical properties of small particles deposited on surfaces, several works brought to the fore the fact of how the presence of a surface alters the light emission of dipole and multipole sources (Lukosz and Kunz, 1977a, b, Lukosz, 1979, Lukosz and Meier, 1981). The physical effects may also be understood as a consequence of the coupling of the dipole or multipole source to the planar surface. In the presence of a perfect conductor, the emitted intensity varies according to the orientation of the dipole:

A dipole placed vertically radiates more than twice the value emitted in a homogeneous environment; for a dipole placed horizontally, the radiated intensity vanishes as the distance to the surface is reduced to contact. As the dipole is approaching a real (absorbing) metal, a larger fraction of the power emitted by the dipole is dissipated in the metal so that the far-field radiation becomes quenched. When the dipole is facing a dielectric, the power transmitted into the dielectric increases as the distance is reduced. This is due to the conversion of evanescent waves into radiative waves in the medium with a larger index of refraction. The above mechanisms apply when modelling the electromagnetic coupling of molecules to solid surfaces. They constitute the basic principles of the recent trend towards near-field fluorescence microscopy (Betzig and Chichester, 1993, Pedarnig et al., 1995).

4.2.2 Perturbative or self-consistent approach

Understanding the optical tip-sample interaction presented surely one of the most serious challenges for the beginning of near-field optical microscopy research. Empirical steps contributed to the progress in designing tips which provide a good imaging quality. Different tip designs evolved according to the type of experimental set-up. Internal illumination devices (STOM or PSTM) exploit bare and sharply elongated optical fibres (Reddick et al., Courjon et al. and Adam et al. in 1993) while external illumination scanning near-field optical microscopes (SNOM) favour metallized tips with a subwavelength aperture at the apex (Betzig et al. and Lieberman et al., 1990). A few wavelengths away from the apex, the general shape of such tips is usually smooth. However, recent configurations involving tips with a tetrahedral termination (entirely or partly coated with metal) were also successful (Fischer and Köglin, 1995). Metal tips were also found appropriate for the scanning surface plasmon microscope (SSPM) (Specht et al., 1992).

All these developments and their numerous variations were supported exclusively by instrumental intuition since classical optical theories were ineffective in describing the basic features of the tip-sample interaction. Moreover, the early theoretical works on near-field optics did not approach this interaction self-consistently. The pioneering investigations provided insights into the field distribution behind subwavelength apertures but without any sample present. After these exploratory studies, the first methods applied to simulate near-field microscopes images computed the lectromagnetic field diffracted above a non-planar sample surface but ignored the presence of any probe. Such procedures were later improved by the use of ideal probes which did not disturb the near-field above the sample. Indeed, these probes were introduced after the computation of the field above the bare surface basically to model devices which integrate the optical near-field over a finite volume. Such ideal probes are thus not coupled to the sample when solving Maxwell's equations. Neglecting this coupling was assumed to be justified for large values of the tip-to-sample distance and for non-resonant wavelengths. It was hoped that less favourable situations could be handled within the first Born approximation. However, all experiments conducted with various devices confirmed that improving the resolution requires approaching the tip very close to the sample. Adding to this fact the recent development towards near-field spectroscopy and the exploration of radiation pressure effects, it became clear that a realistic computation should at least take the tip-sample coupling properly into account. Various numerical techniques were then applied in order to include this coupling successfully.

Originally developed in the context of radar scattering in aeronautics, this method is a Maxwell's equations solver derived from the finite-element method. This purely numerical 664C Girard and A Dereux scheme was recently applied to near-field optical problems (Kann et al. in 1995). The technique directly solves the time-dependent Maxwell equations. This feature imposes time averaging over a period in order to consider harmonically oscillating fields which accurately model the time-dependence of the laser used in near-field optics. Typically, such a procedure requires a supercomputer in order to assess even relatively simple problems. Up to now, it has been able to reproduce some results found previously by the methods that will be in detail latter.

4.2.3 Theories based on matching boundary conditions

Practically all applications of near-field optics have been running under stationary laser illumination. This experimental mode allows one to restrict the theoretical description to electromagnetic fields which depend harmonically on time. With this $\exp(-i\omega t)$ time-dependence, the Maxwell equations in the absence of any external source read (SI units)

$$\nabla \cdot \varepsilon(r,\omega) E(r,\omega) = 0 \tag{4-10}$$

$$\nabla \cdot B(r,\omega) = 0 \tag{4-11}$$

$$\nabla \times E(r,\omega) = i\omega\mu_0 H(r,\omega) \tag{4-12}$$

$$\nabla \times H(r,\omega) = -i\omega\varepsilon_0 \varepsilon(r,\omega) E(r,\omega) \tag{4-13}$$

This set of equations is the starting point of a macroscopic approach of near-field optical phenomena where the response of matter to exciting electromagnetic fields is described by the dielectric function $\varepsilon(r,\omega)$. Roughly speaking, the dielectric function, which is equal to the square of the complex index of refraction, allows one to model the response of a large number of atoms to an external electric field. It is physically meaningful to address problems where the scatterers' size is large enough to justify the use of such a global property. In visible light near-field optics, it allows one to model the response of esoscopic and nanoscopic objects larger than about 10nm.

A first class of numerical methods follows the traditional approach of matching electromagnetic boundary conditions at interfaces. Such methods are typically based on well-established techniques previously developed for other purposes. Solutions are written as linear expansions of a set of eigenfunctions where the coefficients are the unknowns to be found numerically (Van Labeke and Barchiesi, Barchiesi and Van Labeke, Sentenac and Greffet, Bernsten et al., in 1993).

4.2.4 Expansion in plane waves: grating and diffraction theory

Building on the results of the grating theory, the expansions in Fourier series were proposed by Van Labeke and Barchiesi in order to model near-field phenomena above gratings. It is well suited to the study of periodic dielectric surface profiles. Such samples were frequently used some years ago for testing the resolution limit of near-field optical microscopes. In order to account for non-periodic and well localized scatterers, Van Labeke and Barchiesi later used the expansion in a continuum of plane waves typical of diffraction theory which we summarize below.

The method starts from the totally reflected electromagnetic wave $E_0(r,\omega)$ incident on a glass-air plane interface defined by $z = 0$. The diffracted field $E_d(r,\omega)$ due to the surface corrugations is determined within the Rayleigh hypothesis by assuming the following plane wave expansion

$$E_d(r,\omega) = \iint \mathrm{d}k_x \mathrm{d}k_y F_d(k,\omega) e^{ik \cdot r} \tag{4-14}$$

where $k = (k_x, k_y, k_z)$ represents the cartesian components of the different wave vectors associated with the field diffracted by the corrugated surface and $r = x, y, z$. The two-fold integral runs over k_x and k_y. Since the dispersion relation (4-14) must hold, the diffracted field generally contains both radiative and evanescent waves. Indeed, according to the Heisenberg uncertainties, that the dispersion Δk_s (where $s = x, y$) is directly related to the lateral size of the surface corrugations with respect to the incident wavelength λ. When deal with a sample that displays subwavelength details, the Fourier expansion is mainly composed of evanescent components so that the resulting diffracted field $E_d(r,\omega)$ turns out to be confined around the surface corrugations.

4.2.5 Perturbative diffraction theory

The evaluation of the field amplitudes $F(k,\omega)$ is known as a rather difficult task. It requires the introduction of the Fourier transform of the function describing the surface profile $\Gamma(x,y)$ as shown in Figure 4-12.

$$\Gamma(x,y) = \iint \mathrm{d}k_x \mathrm{d}k_y \gamma(k_x, k_y) e^{i(k_x x + k_y y)} \tag{4-15}$$

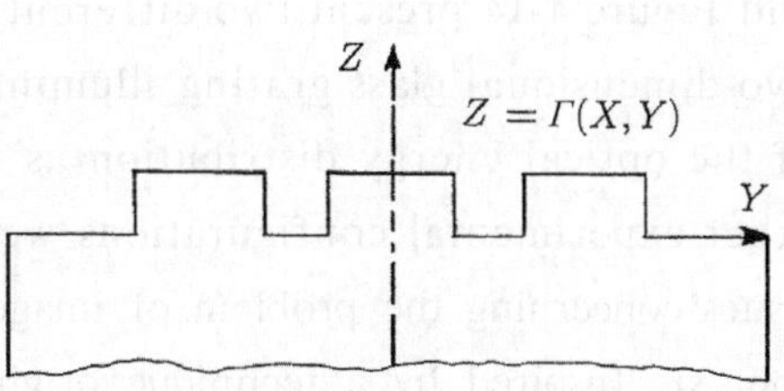

Fig. 4-12 Schematic illustration of a sample limited by an arbitrary surface corrugation function.

The application of the standard boundary conditions at the surface $z = \Gamma(x,y)$ leads to a complex relation between the incident and the diffracted field. This difficulty is reduced by working within the perturbative approximation introduced by Agarwal, Toigo et al. and Elson to study the far-field diffraction and the scattering properties of metallic corrugated surfaces. While applying the boundary conditions, this approximation consists of expanding the exponential function contained in (4-14) as a power series of $k_z\Gamma(x,y)$:

$$e^{i(k_x x + k_y y + k_z \Gamma(x,y))} = e^{i(k_x x + k_y y)}[1 + ik_z\Gamma(x,y) + \cdots] \tag{4-16}$$

For surface corrugations with a weak amplitude, this expansion may be limited to first order. In this case, the diffracted amplitudes are proportional to the Fourier transform of the surface profile $\gamma(k_x; k_y)$ and depend linearly on the zeroth-order field $E_0(r,\omega)$:

$$F_d(k_x,k_y,\omega) \approx i(\varepsilon' - \varepsilon)\gamma(k_x - q_x, k_y - q_y)A(k_x,k_y) \cdot E_0(r,\omega) \quad (4\text{-}17)$$

where q_x and q_y represent the (x, y) components of the incident wave vector. In this linear relation $A(k_x,k_y)$ is the 3×3 transfer matrix defined by

$$A(k_x,k_y) = \frac{\omega^2/c^2}{k_z + k'_z}\mathbf{1} - \frac{1}{k'_z + \varepsilon' k_z}\begin{pmatrix} k_x^2 & k_x k_y & k_x k_z \\ k_y k_x & k_y^2 & k_y k_z \\ k'_z k_x & k'_z k_y & k'_z k_z \end{pmatrix} \quad (4\text{-}18)$$

where $\mathbf{1}$ represents the identity matrix and k'_z is a z component of the wave vectors diffracted inside the sample characterized by the dielectric function ε'.

$$k'^2_z = \frac{\omega^2}{c^2}\varepsilon' - k_x^2 - k_y^2 \quad (4\text{-}19)$$

The total optical field generated near the surface protrusions is thus given by a correction to the result associated with a perfectly flat sample:

$$E(r,\omega) \approx E_0(r,\omega) + i(\varepsilon' - \varepsilon)\iint dk_x dk_y \exp(ik \cdot r) \times \gamma(k_x - q_x, k_y - q_y)A(k_x,k_y) \cdot E_0(r,\omega) \quad (4\text{-}20)$$

At this stage, the field may be determined numerically by using standard fast Fourier transform (FFT) routines. As within any scheme working in reciprocal space, the structural information about the object is thus contained in the Fourier transform $\Gamma(k_x,k_y)$ of the surface profile $\gamma(x,y)$. Consequently, for a given observation distance Z_0, the accuracy of the results will depend on the number of spatial harmonics introduced in the FFT.

The method has a relatively low cost in term of computer time so that it is certainly an interesting tool to help the interpretation of massive amounts of near-field microscope images. Nevertheless, we have to emphasize that such an approximation has a range of validity restricted to surface corrugations of weak amplitude when compared with the incident wavelength. Figure 4-13 and Figure 4-14 present two different numerical calculations based on this method. The system is a two-dimensional glass grating illuminated in total internal configuration. Note a strong variation of the optical energy distribution as a function of the polarization mode. More recently several other experimental configurations were investigated with similar methods. Also, some important issues concerning the problem of image reconstruction (inverse scattering) has been addressed in this context. Inspired by a technique originally developed for antenna design at longer wavelengths and applied the expansion of the solutions on multipolar eigenfunctions to study near-field optical phenomena. The8se multipolar eigenfunctions $F_n(r,\omega)$ do satisfy the vector wave equation for the eigenvalue q_n:

$$-\nabla\times\nabla\times F_n(r,\omega) + q_n^2 F_n(r,\omega) = 0 \quad (4\text{-}21)$$

According to early studies on the representation of electromagnetic fields in terms of scalar fields, they can be constructed from the eigenfunctions $\psi_n(r,\omega)$ of the scalar Helmholtz equation:

$$\nabla^2\psi_n(r,\omega) + q_n^2\psi_n(r,\omega) = 0 \quad (4\text{-}22)$$

larger fraction of the power emitted by the dipole is dissipated in the metal so that since

$$\int dr\psi_n^*(r,\omega)\psi_{n'}(r,\omega) = \delta_{n,n'} \quad (4\text{-}23)$$

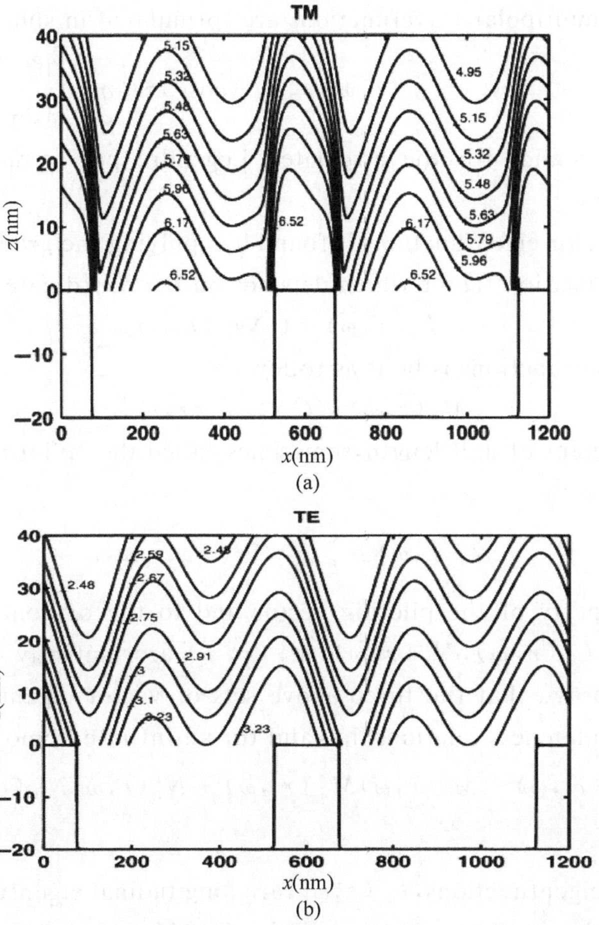

Fig. 4-13 Iso-intensity lines calculated above a lamellar grating. The geometry of the grating is represented by rectangular-shaped surface protrusions. The system is lighted in the TIR configuration with an incident angle $\theta = 45^{\circ}$, a wavelength $\lambda = 632$nm and an optical index $n = 1 : 5$. The calculation has been performed with the perturbative approach: (a) p-polarized illumination mode; (b) s-polarized illumination mode.

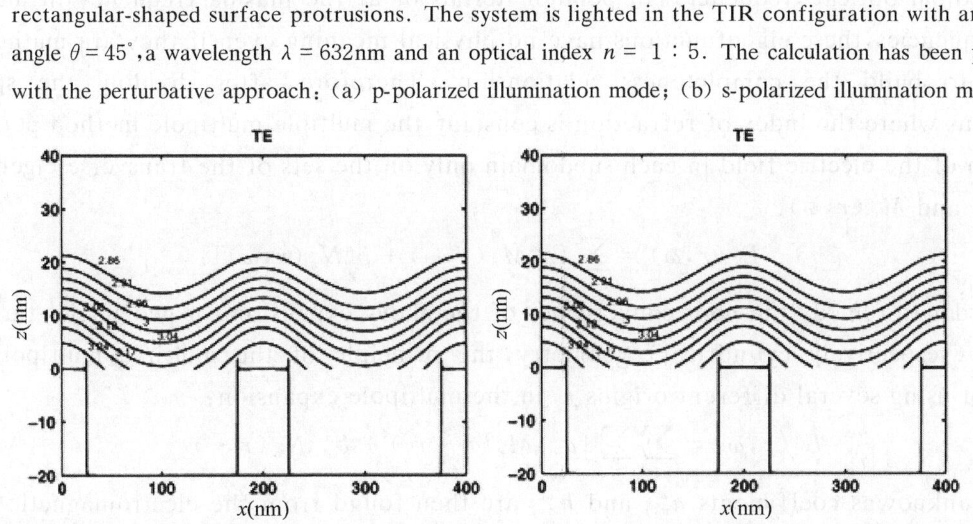

Fig. 4-14 The same as Figure 4-13, but with a smaller lateral extension of the surface structures.

These eigenfunctions form an orthonormal basis set in the Hilbert space. The simplest form is obtained in cartesian coordinates:

$$\psi_n(r, \omega) \equiv \psi_k^*(r, \omega) = \frac{1}{\sqrt{8\pi^3}} \exp(ik \cdot r) \tag{4-24}$$

Three-dimensional multipolar wavefunctions are formulated in spherical coordinates:

$$\psi_n(r,\omega) \equiv \psi_{\sigma,l,m,q_n}(r,\omega) \equiv P_l^m(\cos\theta) z_l(q_n r) \begin{Bmatrix} \cos(m\phi) \\ \sin(m\phi) \end{Bmatrix} \tag{4-25}$$

Where $P_l^m(\cos\theta)$ stands for the associated Legendre polynomials and $z_l(q_n r)$ for the spherical Bessel functions.

A first family of vector eigenfunctions is found by applying the gradient operator to the scalar functions (C is a normalization factor which depends on the coordinate system):

$$L_n(r,\omega) = C\,\nabla\psi_n(r,\omega) \tag{4-26}$$

A second set of eigenfunctions is built as follows

$$M_n(r,\omega) = C\,\nabla\times\psi_n(r,\omega)a \tag{4-27}$$

where a is a constant vector of unit length, sometimes called the "piloting vector". The last group of eigenfunctions is given by

$$N_n(r,\omega) = \frac{C}{k}\nabla\times\nabla\times\psi_n(r,\omega)a \tag{4-28}$$

Thanks to the property of the piloting vector and to the orthonormalization of the scalar eigenfunctions, the sets $L_n(r,\omega)$, $N_n(r,\omega)$, $M_n(r,\omega)$ are mutually orthogonal in the Hilbert sense. One can easily prove that the three above sets of vector eigenfunctions are sufficient to build the following completeness relationship valid for an infinite homogeneous system:

$$\sum_n [L_n(r,\omega)L_n^*(r',\omega) + M_n(r,\omega)M_n^*(r',\omega) + N_n(r,\omega)N_n^*(r',\omega)] = \mathbf{1}\delta(r-r') \tag{4-29}$$

The first family of eigenfunctions, $L_n(r,\omega)$, are longitudinal eigenfunctions which correspond to physical solutions of the wave equation only if $k=0$. This is only possible if $\varepsilon=0$, which occurs at longitudinal optical frequencies in polar materials or at the plasma frequency in metals. For other frequencies, these eigenfunctions have no physical meaning even if they are mathematically required to build the completeness relationship. Therefore, after dividing the space into subdomains where the index of refraction is constant, the multiple multipole method performs the expansion of the electric field in each subdomain only on the sets of the transverse eigenfunctions $N_n(r,\omega)$ and $M_n(r,\omega)$.

$$E^\alpha(r,\omega) = \sum_n [a_n^\alpha M_n(r,\omega) + b_n^\alpha N_n(r,\omega)] \tag{4-30}$$

In order to remedy the poor convergence of the above expansion for geometries far from the spherical (respectively cylindrical) geometry, the principle of the multiple multipole method consists of using several different origins r_j in the multipole expansion:

$$E^\alpha(r,\omega) = \sum_j \sum_n [a_{n,j}^\alpha M_n(r-r_j) + b_{n,j}^\alpha N_n(r-r_j)] \tag{4-31}$$

The unknowns coefficients $a_{n,j}^\alpha$ and $b_{n,j}^\alpha$ are then found from the electromagnetic boundary conditions on the interfaces between adjacent subdomains by a least-squares minimization. This optimization requires one to discretize the curves describing the interfaces. This discretization has some impact on the highest possible degree of a multipole centre. Of course, due to the splitting of the geometry in subdomains and to the existence of several origins r_j, completeness and orthonormalization relationships over the entire composite system are not achieved anymore. Therefore, to avoid mutual dependences, the method relies on semi-empirical rules to fix the separation between the origins r_j. The multipole functions used in the basis set of the multiple multipole methods are rather short range so that they affect their close neighbourhood. The method is thus better suited to account for localized geometries than the expansion in plane waves. It is also well designed to describe complicated structures such as cylindrical waveguides coated with a realistic metal whose dielectric function is complex or the scattering inside sharpened cylindrical tips used in near-field microscopy.

From a mathematical point of view, the idea of spreading a set of multi-pole functions and adjusting the coefficients is similar to quantum mechanical techniques used for computing electronic structures such as the linear combination of atomic orbitals (LCAO). The multipoles in electrodynamics thus play a role similar to the atomic orbitals in quantum mechanics. However, the situation is somewhat clearer in electron physics where the atomic orbitals are centred on each nucleus, whereas a physical meaning is lacking in the mathematical procedure which distributes the coordinates of the multipoles centres in electrodynamics. Figure 4-15 display examples of numerical simulations performed using the MMP methods.

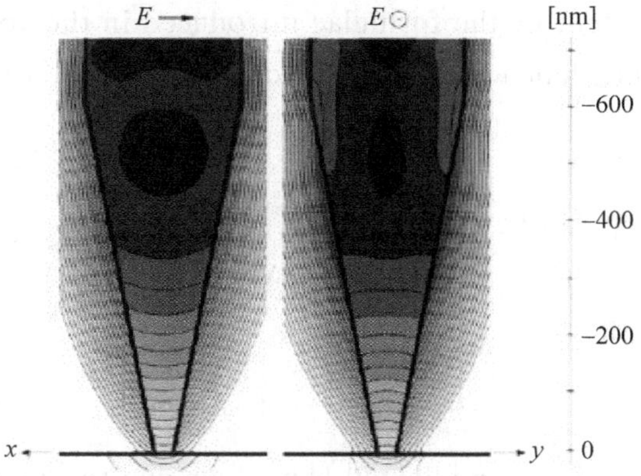

Fig. 4-15 Examples of 3D near-field calculations performed with the MMP method. The physical system is the tip of an aperture SNOM consisting of a cylindrical part and a tapered part. The probe is excited by a waveguide mode of wavelength $\lambda = 488\text{nm}$ and there is a factor three between two successive lines.

4.2.6 Scattering theory

From a mathematical point of view, scattering theory (also known as the field susceptibility or Green dyadic technique) is developed upon the Green function theory applied to the wave equation where a source term is introduced. It simply casts the most general analytical solution of the inhomogeneous wave equation as an integral equation where the kernel is a Green function. Scattering theory has been used for a long time in quantum mechanical problems. In electrodynamics, it has also been exploited extensively to solve engineering problems involving external sources of currents where the solution in the source region is not required. The application to the study of phenomena where the solution in the source region (such as in near-field optics) is of primary importance was hindered by the apparently divergent nature of the Green dyadic in the source region. This divergence is related to depolarization effects. These are well understood today so that an unambiguous renormalization procedure is available. Several variants of lectromagnetic scattering theory where applied successfully to the modelization of near-field optical phenomena. Although the Green function may be expanded in Fourier or multipoles series, most variants preferred an iscretization in the direct space since near-field optical phenomena occur on the subwavelength scale.

- Propagator or Green dyadic

How the Green dyadic (or field susceptibility) method associated with the localized erturbation theory deals efficiently with the resolution of the selfconsistent optical tip-sample interaction. A detailed account of this theory and its numerical implementation was found in 1993(Girard et al., Dereux and Pohl, Martin et al.). A discussion of the convergence and stability of the algorithm was reported in 1994 (Martin et al.). Therefore, this section will focus more on the physical content of the formulae introduced in the above-mentioned references. With the usual $\exp(-i\omega t)$ time-dependence, the vector wave equation issued from Maxwell's equations (SI units),

$$-\nabla\times\nabla\times E(r,\omega)+\frac{\omega^2}{c^2}\varepsilon(r,\omega)E(r,\omega)=0 \tag{4-32}$$

And can be cast as

$$-\nabla\times\nabla\times E(r,\omega)+q^2E(r,\omega)=V(r,\omega)E(r,\omega) \tag{4-33}$$

with

$$q^2=\frac{\omega^2}{c^2}\varepsilon_{\text{ref}} \tag{4-34}$$

Any complicated behaviour due to the anisotropy or to the low symmetry of the geometrical shape of the original dielectric tensor profile $\varepsilon(r,\omega)$ is described as a difference relative to the reference system ε_{ref}:

$$V(r,\omega)=\frac{\omega^2}{c^2}(\mathbf{1}\varepsilon_{\text{ref}}-\varepsilon(r,\omega)) \tag{4-35}$$

As introduced in 1993 (Dereux and Pohl and Girard et al.), the solution of Eq. (4-33) is obtained from the implicit Lippmann-Schwinger equation

$$E(r,\omega)=E_0(r,\omega)+\int_D \mathrm{d}r' G_0(r,r',\omega)V(r',\omega)E(r',\omega) \tag{4-36}$$

In scattering theory, the first term $E_0(r',\omega)$ is referred to as the incident field while the second term is called the scattered field obtained from the integration over the domain D where $V(r',\omega)$ is non-zero. D defines the volume of the scatterer relative to the reference system. Electromagnetic theory traditionally qualifies D as the source region.

To solve the Lippmann-Schwinger equation, we need to know the analytical solution $E_0(r',\omega)$ satisfying

$$-\nabla\times\nabla\times E_0(r,\omega)+q^2E_0(r,\omega)=0 \tag{4-37}$$

and the associated Green dyadic defined by

$$-\nabla\times\nabla\times G_0(r,r',\omega)+q^2G_0(r,r',\omega)=\mathbf{1}\delta(r-r') \tag{4-38}$$

The reference structure ε_{ref} is usually a homogeneous background material or a semi-infinite surface system. For homogeneous media, the analytical form of $G_0(r,r_0,\omega)$ is known from early studies. Its general form may be deduced as follows. To seek a solution (temporarily written as G_0' for a reason which will appear later) of Eq. (4-38) as an expansion over the vector eigenfunctions already defined in the previous section when devising the expansion in multipoles:

$$G_0'(r,r',\omega)=\sum L_n(r,\omega)X_n^*(r',\omega)+M_n(r,\omega)Y_n^*(r',\omega)+N_n(r,\omega)Z_n^*(r',\omega) \tag{4-39}$$

According to the completeness relationship, we can find out the unknowns

$$X_n^*(r',\omega),\quad Y_n^*(r',\omega) \text{ and } Z_n^*(r',\omega)$$

by backsubstitution of Eqs. (4-39) and (4-38), so that

$$G_0'(r,r',\omega) = \sum_n \frac{L_n(r,\omega)L_n^*(r',\omega) + M_n(r,\omega)M_n^*(r',\omega) + N_n(r,\omega)N_n^*(r',\omega)}{q^2 - q_n^2}$$

(4-40)

The above expansion is known as the spectral expansion of the Green dyadic. However, as mentioned in the previous section, the longitudinal wavefunctions are not physical in a homogeneous medium. One might think that discarding the longitudinal wavefunctions out of the spectral expansion would provide the appropriate cure in order to find $G_0(r',r_0,\omega)$. This would be a mistake since it does not take care of the singular behaviour of $G_0(r',r_0,\omega)$ as $r' \to r_0$. Indeed, when introduced in Eq. (4-33), the resulting Green dyadic expansion G_0^t containing only transverse eigenfunctions cannot build up the longitudinal part of $1\delta(r' \to r_0)$ which exists if $r' \to r_0$. Indeed, if

$$G_0^t(r,r',\omega) = \sum_n \frac{M_n(r,\omega)M_n^*(r',\omega) + N_n(r,\omega)N_n^*(r',\omega)}{q^2 - q_n^2}$$

(4-41)

Here have

$$-\nabla \times \nabla \times G_0^t(r,r',\omega) + q^2 G_0^t(r,r',\omega) = D_t(r - r')$$

(4-42)

Where

$$D_t(r - r') = \sum_n [M_n(r,\omega)M_n^*(r',\omega) + N_n(r,\omega)N_n^*(r',\omega)]$$

(4-43)

A longitudinal part must then be added to G_0^1 in order to build G_0:

$$G_0(r,r',\omega) = G_0^t(r,r',\omega) + G_0^1(r,r',\omega)$$

(4-44)

G_0^1 must satisfy

$$-\nabla \times \nabla \times G_0^1(r,r',\omega) + q^2 G_0^1(r,r',\omega) = D_1(r - r')$$

(4-45)

where

$$D_1(r - r') = \sum_n [L_n(r,\omega)L_n^*(r',\omega)]$$

(4-46)

The longitudinal character of G_0^1 allows one to conclude that

$$G_0^1(r,r',\omega) = \frac{1}{q^2}D_1(r - r')$$

(4-47)

So that

$$G_0(r,r',\omega) = G_0^t(r,r',\omega) + G_0^1(r,r',\omega)$$

(4-48)

In Cartesian coordinates, the normalization factor in the vector wavefunctions is given by $C = k^{-1}$. Some algebra leads to

$$G_0(r,r',\omega) = \int dk \left[1 - \frac{1}{q^2}kk\right] \frac{e^{ik\cdot(r-r')}}{8\pi^3(q^2 - k^2)}$$

(4-49)

Integration by the method of residues yields

$$G_0(r,r',\omega) = \left[1 - \frac{1}{q^2}\nabla\nabla\right] g(r,r',\omega)$$

(4-50)

where $g(r,r',\omega)$ is the Green function associated with the scalar Helmholtz equation. The $g(r,r',\omega)$ is given by a spherical wave emitted at r':

$$g(r,r',\omega) = \frac{\exp(iq|r - r'|)}{4\pi|r - r'|}$$

(4-51)

The last two equations are interpreted as follows. The dyadic Green function is not the direct extension of the scalar Green function. Indeed, at large distances (mathematically at infinite distances), one must pay attention to satisfying the Sommerfeld radiation condition which states that the electromagnetic fields are purely transverse. The negative terms inside the brackets of the last two equations take care of discarding any longitudinal components of the field at infinity. For a surface system, the expression of the propagator is somewhat more elaborated.

- The Huygens-Fresnel principles

The implicit character of this Lippmann-Schwinger equation ensures the self-consistency of the solution. This may be understood by revisiting the Huygens-Fresnel principle. The various Kirchhoff theories are themselves a mathematical translation of the Huygens-Fresnel principle. After an abstract discretization of the scatterer into elementary sources, this fundamental principle states that the scattering of light may be explained by the coherent sum of the fields emitted by each elementary source. It is easily checked that this basic feature is embedded into equation (4-31) if assume that we know the value of the field attributed to each elementary source inside D. Moreover, one should notice that Huygens described the field emitted by each elementary source as spherical waves similar to equation (4-46). Relatively to the Lippmann-Schwinger Eq. (4-31), the different look of Kirchhoff integrals originates simply from the use of the Green theorem to transform the volume integrals into surface integrals. One of the basic approximations in classical diffraction theories substitutes the incident field $E_0(r_0', \omega)$ into the values of the field in the source region $E(r_0', \omega)$. This procedure to solve Eq. (4-31) is known as the first Born approximation in scattering theory. The improvement brought by the self-consistent solution is thus related to the accurate determination of the field value in the source region. Here discussed in 1991 (Girard, Bouju Dereux et al.) how the discretization of the Lippmann-Schwinger Eq. (4-31) allowed one to obtain the total field in the source region through the solution of

$$\sum [1\delta(r_i - r_j) - G_0(r_i, r_j, \omega) V(r_j, \omega) w_j] E(r_j, \omega) = E_0(r_i, \omega) \tag{4-52}$$

where w_j is the volume (surface or length according to the dimensionality of the problem) of the mesh. In this linear system of equations, all r_i and r_j belong to the source region so that computation of the diagonal elements $G_0(r_i, r_j, \omega)$ is required. At first sight, this may cause some trouble since $G_0(r_i, r_j, \omega)$ looks singular in Eq. (4-45) when $r_i \to r_j$. A proper renormalization procedure to be explained in the next section is needed to resolve this singularity.

- Renormalization procedure and the depolarizing dyadic

Moreover, this extra source term depends critically on the shape of the mesh used in the discretization procedure. This somewhat confusing feature has hindered the use of the Green dyadic technique in numerical applications. Yaghjian clarified this matter and tabulated the corrective terms for various discretization meshes. Now introduce this correction by discussing further the physical meaning of the self-consistent step which occurs in the numerical scheme. In order to be more intuitive, imagine that the perturbation is made of a single discretization mesh centred around r_i. Eq. (4-52) allows one to find the electric field inside this mesh $E(r', \omega)$ as a function of the applied field $E(r', \omega)$:

$$[\mathbf{1} - G_0(r_i, r_i, \omega) V(r_i, \omega) w_i] E(r_i, \omega) = E_0(r_i, \omega) \tag{4-53}$$

Due to the finite size of the mesh, this evaluation must include the correction arising from the polarization of the mesh. In other words, the self-consistent step Eq. (4-53) must be in agreement with

$$E(r_i,\omega)=E_0(r_i,\omega)-L_\alpha\frac{P(r_i,\omega)}{\varepsilon_0}\tag{4-54}$$

where $\alpha=1,2,3$ according to the dimensionality of the problem and $P(r,\omega)$ is the polarization vector which is related to the dielectric property of the mesh through

$$\frac{P(r_i,\omega)}{\varepsilon_0}=[\varepsilon(r_i,\omega)-\mathbf{1}\varepsilon_{\text{ref}}]E(r_i,\omega)\tag{4-55}$$

The resulting field depends critically on the depolarizing dyadic L_α associated with the shape of the mesh. Here, quote only the results which arise in Cartesian coordinates. When addressing the scattering of electromagnetic waves by multilayers, the shape of the discretization mesh is a very thin plate so that

$$L_1=\begin{pmatrix}0&0&0\\0&0&0\\0&0&1\end{pmatrix}\tag{4-56}$$

In two-dimensional problems, considering infinitely long square rods along the x direction implies the following structure of the depolarizing dyadic:

$$L_2=\begin{bmatrix}0&0&0\\0&\frac{1}{2}&0\\0&0&\frac{1}{2}\end{bmatrix}\tag{4-57}$$

In three-dimensional scattering, the depolarizing dyadic associated with a cubic mesh reads

$$L_3=\begin{bmatrix}\frac{1}{3}&0&0\\0&\frac{1}{3}&0\\0&0&\frac{1}{3}\end{bmatrix}\tag{4-58}$$

Eq. (4-54) evolves then as

$$\left[1-\frac{c^2}{\omega^2}L_\alpha V(r_i,\omega)\right]E(r_i,\omega)=E_0(r_i,\omega)\tag{4-59}$$

which is clearly not included in Eq. (4-53). Indeed, although Eq. (4-54) is related to the shape of the mesh, it does not depend on its size w_i whereas the matrix to be inverted in Eq. (4-53) depends on w_i. Thus, for a vanishing value of the mesh size w_i, Eq. (4-53) does not take any depolarization effect into account. In order to include the depolarization properly, we need to modify the definition of the Green dyadic as follows:

$$G_0(r,r',\omega)=\left[1-\frac{1}{q^2}\nabla\nabla\right]g(r,r',\omega)+L_\alpha\delta(r-r')\frac{c^2}{\omega^2}\tag{4-60}$$

The renormalization procedure outlined above is applied to discretize a large piece of continuous matter. It is based on the bulk dielectric properties of the discretized material which are known to be accurate so as to take particles larger than about 10nm into account. The introduction of smaller particles (clusters, molecules or even atoms) in the computational scheme is possible provided that its polarizability be known from some other source like an experimental measurement or a quantum mechanical calculation. In this case, the corrective delta function term of Eq. (4-55) must not be included (i.e. $L_\alpha = 0$) since the polarizabilities obtained by these means account for the depolarization effect inside the particles.

Point out that the computational procedure introduced in 1993 (Dereux, Pohl and Girard et al.) deliberately avoids any step involving matching boundary conditions. The ability of the method in dealing with scatterers of arbitrary shapes and dielectric responses is of course a fundamental advantage when studying the optical interaction between elongated probe tips and non-planar samples. Not only does this feature enable the handling of arbitrary shaped objects made of continuous matter but, as explained in the preceding paragraph, it also allows one to treat discrete particles in the same framework. This will be a fundamental advantage when addressing the optical interaction of tips with particles whose shapes are intrinsically not sharp. This approach is thus ready to include fuzzy quantum systems (electron gas inside a metallic cluster, molecules or atoms). This is better understood by discussing briefly the relationship between the Green dyadic technique and the concept of field susceptibility.

- The field-susceptibility method-Alternative derivation of the Lippmann-Schwinger equation

Another meaningful equivalent approach may be applied to derive Lippmann-Schwinger equation for optical fields. This alternative derivation is based on the concept of field susceptibility that describes the response of the optical field itself to an infinitesimal fluctuating volume of polarized matter. As illustrated in the fundamental paper of Agarwal, starting from the microscopic Maxwell equations expressed in terms of both charge and current densities is a direct route to derive the classical expression of the free-space field susceptibility. Let us consider a physical system characterized by its time-dependent charge density $\rho(r,t)$ and its current density $j(r,t)$. In the ω-space, the Maxwell equations read

$$\nabla \cdot E(r,\omega) = \frac{\rho(r,\omega)}{\varepsilon_0} \tag{4-61}$$

$$\nabla \cdot B(r,\omega) = 0 \tag{4-62}$$

$$\nabla \times E(r,\omega) = \mathrm{i}\omega\mu_0 H(r,\omega) \tag{4-63}$$

$$\nabla \times H(r,\omega) = -\mathrm{i}\omega\varepsilon_0 E(r,\omega) + j(r,\omega) \tag{4-64}$$

The vectorial wave equation for the electric field is readily obtained by taking the curl of Eq. (4-63). After some straightforward algebraic manipulations, one gets the well known result

$$\Delta E(r,\omega) + q_0^2 E(r,\omega) = \nabla \frac{\rho(r,\omega)}{\varepsilon_0} - \mathrm{i}\omega\mu_0 j(r,\omega) \tag{4-65}$$

where $q_0 = \omega/c$. now express both charge and current densities in terms of the local polarization $P(r,\omega)$ of the material system:

$$\rho(r,\omega) = -\nabla \cdot P(r,\omega) \tag{4-66}$$

And

$$j(r,\omega) = -\mathrm{i}\omega P(r,\omega) \tag{4-67}$$

rewrite the non-homogeneous Eq. (4-60) as follows:

$$\Delta E(r,\omega) + q_0^2 E(r,\omega) = q_0^2 \frac{P(r,\omega)}{\varepsilon_0} + \nabla \left[\nabla \cdot \frac{P(r,\omega)}{\varepsilon_0}\right] \tag{4-68}$$

Let $E_0(r,\omega)$ be the solution of the following homogeneous equation,

$$\Delta E_0(r,\omega) + q_0^2 E_0(r,\omega) = 0 \tag{4-69}$$

the general solution of Eq. (4-68) is the sum of the homogeneous field $E_0(r,\omega)$ plus a particular solution $E_m(r,\omega)$. This particular solution can be derived from the knowledge of the free-space scalar Green function (4-51)

$$E_m(r,\omega) = \int S_0(r,r',\omega) \cdot P(r',\omega)\mathrm{d}r' \tag{4-70}$$

where $S_0(r,r',\omega)$ defines the free-space dyadic field susceptibility

$$S_0(r,r',\omega) = (q_0^2 + \nabla\nabla)g(r,r',\omega) \tag{4-71}$$

At this stage it is interesting to note that the free space Lippmann-Schwinger equation can be deduced from these last relations. First, we write the complete solution of Eq. (4-68) as the sum of both homogeneous and inhomogeneous solutions:

$$E(r,\omega) = E_0(r,\omega) + E_m(r,\omega) \tag{4-72}$$

Second, to introduce the usual constitutive equation for a local medium:

$$\frac{P(r,\omega)}{\varepsilon_0} = \chi(r,\omega) \cdot E(r,\omega) \tag{4-73}$$

Finally, by substituting this expression into Eq. (4-70), and the resulting formula into Eq. (4-72), can find the implicit equation

$$E(r,\omega) = E_0(r,\omega) + \int S_0(r,r',\omega) \cdot \chi(r',\omega) \cdot E(r',\omega)\mathrm{d}r' \tag{4-74}$$

which recovers the result for an arbitrary particle in free space. Indeed, the kernels appearing in Eq. (4-36) and (4-74) are identical,

$$G_0(r,r',\omega)V(r',\omega) = S_0(r,r',\omega)\chi(r',\omega) \tag{4-75}$$

since, in the case of continuous matter where the dielectric susceptibility $X(r',\omega)$ is related to the dielectric function according to

$$\chi(r',\omega) = \varepsilon(\omega) - 1 \tag{4-76}$$

which is consistent with the fact that a simple constant factor is the only difference between the Green dyadic and the field susceptibility:

$$S_0(r,r',\omega) = -q_0^2 G_0(r,r',\omega) \tag{4-77}$$

Now, if introduce a somewhat more complicated surrounding such as, for example, the presence of an extended medium (the surface of a semi-infinite material, a macroscopic sized particle and so on), that just have to replace the free-space dyadic $S_0(r,r',\omega)$ by the following,

$$S(r,r',\omega) = S_0(r,r',\omega) + S_s(r,r',\omega) \tag{4-78}$$

Where the new contribution $S_s(r,r',\omega)$ accounts for the dynamical response of such extended system.

The factorization on the right-hand side of Eq. (4-75) was originally introduced to deal with atoms or molecules adsorbed on a surface. Indeed, if one considers, for example, a system formed of p individual molecules, $X(r,\omega)$ can be expressed as

$$\chi(r',\omega) = \sum_{i=1}^{p} \alpha^{(i)}(\omega)\delta(r' - r_i) + \cdots \tag{4-79}$$

where r_i represents the position vectors of the molecules and $\alpha^i(\omega)$ defines their optical dipolar polarizabilities. However, the most important feature of all variants of scattering theory is the correct description of the self-consistent coupling between all scatterers which raises the numerical implementation to the rank of an accurate predictive procedure. In order to illustrate this versatility, we review the results of numerical applications to the study of optical tip-sample interactions arising in different contexts: optical near-field distributions, near-field spectroscopy and radiation pressure effects.

4.2.7 Near-field distributions

The traditional way of exploiting scattering theory by transforming volume integrals into surface integrals has been applied to near-field optical problems by Carminati et al. and Nieto-Vesperinas and Madrazo in 1995. These authors established an exact numerical method that can account for the interaction of the near-field scattered by a rough surface placed in an interaction with a local probe. Our first numerical applications of the Green function technique will be presented in the framework of the total internal reflection configuration. In this mode, the transparent glass surface bearing the object (surface defect) is illuminated by a monochromatic optical field of frequency ω_0 so that total reflection may occur at the surface of the material. From the experimental point of view, this illuminated mode corresponds to the STOM/PSTM configuration. Consider the three-dimensional (3D) localized surface protrusion schematized in Figure 4-16. Depending on its own dimensions with respect to the incident wavelength, such an object will behave as a more or less efficient obstacle to the propagation of the surface wave. The behaviour of the normalized field intensity defined by

$$I = \frac{|E|^2}{|E_0|^2} \tag{4-80}$$

is described in Figure 4-16.

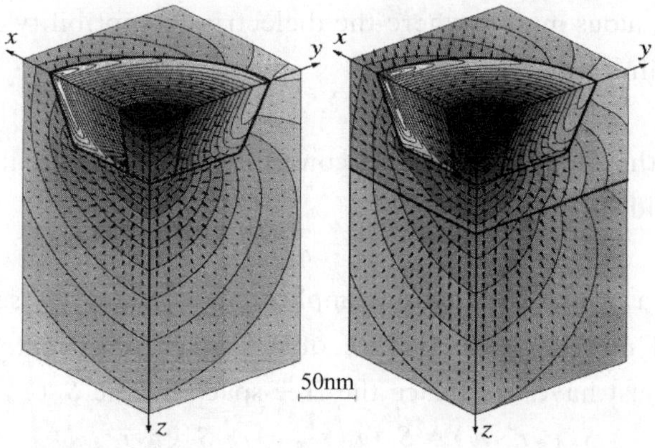

(a) (b)

Fig. 4-16 Contours of constant $|E|^2$ on three perpendicular planes near the aperture of the SNOM probe. The arrows indicate the time-averaged Poynting vector. The polarization is in the plane $y = 0$: The transmission through the probe is increased when a dielectric substrate is approached (b).

In this ratio, $|E|^2$ represents the intensity of the self-consistent field and $|E|_0^2$ the intensity of the incident field (i.e. in the absence of the surface defect). The calculation of I is performed in an observation plane parallel to the reference system, placed a distance $Z_0 = 70\text{nm}$ above the square-shaped surface defect. This observation plane is introduced merely for the convenience of data visualization, since the field can be computed for any arbitrary position outside and inside the system by using the Lippmann-Schwinger Eq. (4-75). The zeroth order solution $E_0(r, \omega)$ is the evanescent field created by total reflection at the surface $Z = 0$. Such an illumination configuration eliminates propagating waves along the z direction. It was seen in Figure 4-17 that the structure of the large scale $2000 \times 2000\text{nm}^2$ calculated image is very complex. It displays a complicated standing field pattern currently observed in the STOM configuration, namely strong confined field effects observed just above the scatterer, scattering along its lateral sides and usual interference patterns due to the interaction between the travelling surface wave and the reflected wave by the surface defect. In particular, this 3D perspective view indicates that at 10nm from the top of the object the enhancement factor of the field intensity reaches. As described in the experimental works of van Hulst and collaborators, for larger objects generally the interference phenomenon dominates and makes the detection of subwavelength features difficult.

Previous have proved with the simulation of Figure 4-18 that, first, the electromagnetic field is more and less confined around surface structures and, second, that the relation between the object profile and the resulting spatial field distribution may be very complex. In fact, the study of the gradual transition between mesoscopic and nanoscopic regimes is of interest for experimentalists working in SNOM since it might allow one to precisely find the fundamental difference between pure topographic signals and artefacts originating from interference and scattering phenomena. In order to get more insight about this important question, consider a second application with a more complex system composed of seven identical square-shaped pads (see Figure 4-17).

The dielectric parameters are the same as those used in the previous application. That present in Figure 4-18 a first simulation by illuminating this system in TM polarization (see Figure 4-17). Each dielectric pad is 100nm high and has a section of $0.25 \times 0.25\mu\text{m}^2$ and the calculation is performed in the plane $Z_0 = 120\text{nm}$. In order to emphasize both interference and scattering effects occurring around the obstacles, used a large computational window $7 \times 7\mu\text{m}^2$.

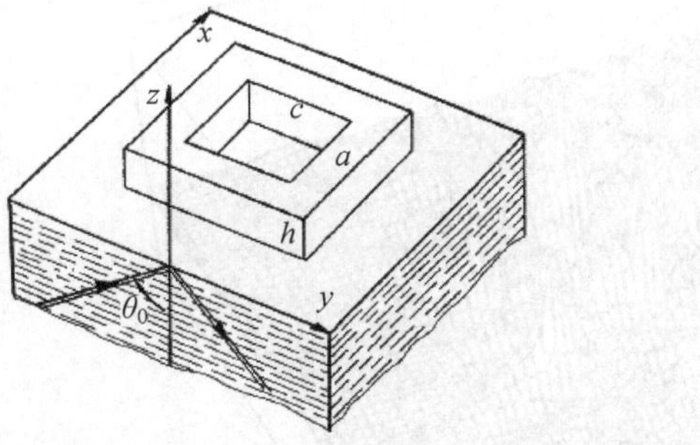

Fig. 4-17 Perspective representation of the square-shaped surface protrusion used in the simulation. The system, of optical index $n = 1.5$, is illuminated in total reflection, θ_0 represents the incident angle and the incident wavelength in vacuum is equal to 620nm. Three geometrical parameters a, c and h have been introduced to define the spatial extension of the protrusion: a represents the external length of the pattern, c the length of the internal side and h its height.

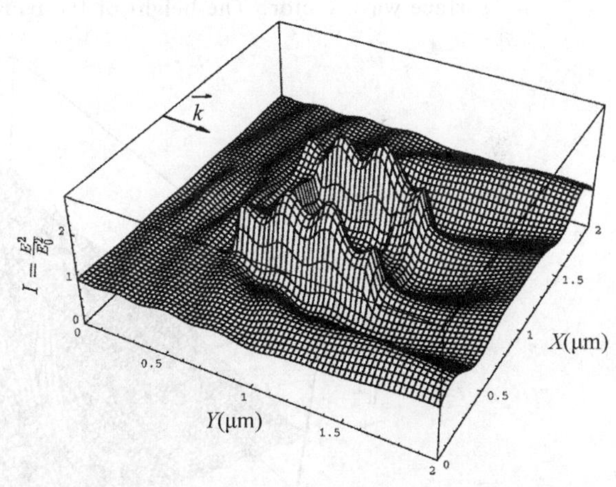

Fig. 4-18 Perspective view of the normalized electric field intensity. The sizes of the object are mesoscopic: $a = 750\text{nm}$, $c = 450\text{nm}$ and $h = 60\text{nm}$. The calculation is performed in the p-polarized mode. For the convenience of the data visualization, the numerical data $I(X, Y)$ have been calculated in the observation plane located at distance $Z_0 = 70\text{nm}$ from the flat surface.

Due to the large spacing between each individual scatterer (1.75μm), the resulting field pattern is a complex mixture of interference phenomena due to multiple reflections between the different pads. As expected, when the number of defects per unit area increases, the standing wave pattern arising from the multiple scattering effects gives rise to the well known "speckle pattern" phenomenon.

Study in Figure 4-21 the evolution of the image upon reduction of the different geometrical parameters P_1, P_2 and H defined in Figure 4-19. The two commonly used polarization modes TE and TM are simultaneously considered in Figure 4-20. Three different typical sizes are successively investigated. In the first example as Figure 4-19(a) start in the mesoscopic range $P_1 = 250$nm, $P_2 = 1750$nm and $h = 100$nm.

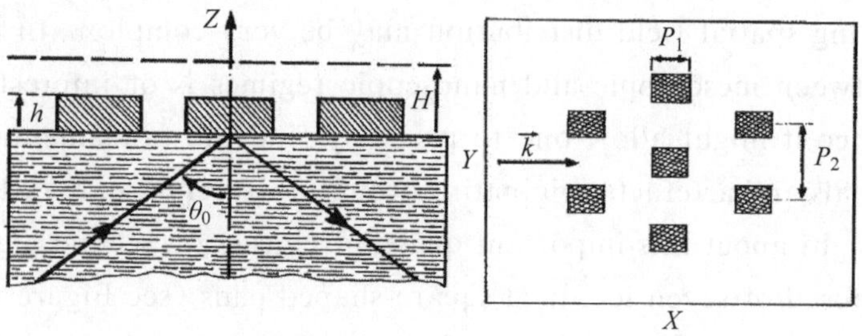

Fig. 4-19 Geometry of a 3D object composed of several identical square-shaped glass protrusions. The dielectric parameters are the same than those used in the previous application (see Figure 4-17). The centre of each pad is located at the nodes of the hexagonal pattern of side P_2. P_1 represents the dimension of each individual protrusion. The system is illuminated in internal reflection configuration and k represents the surface wave vector. The height of the pads is h. (a) side view, (b) top view.

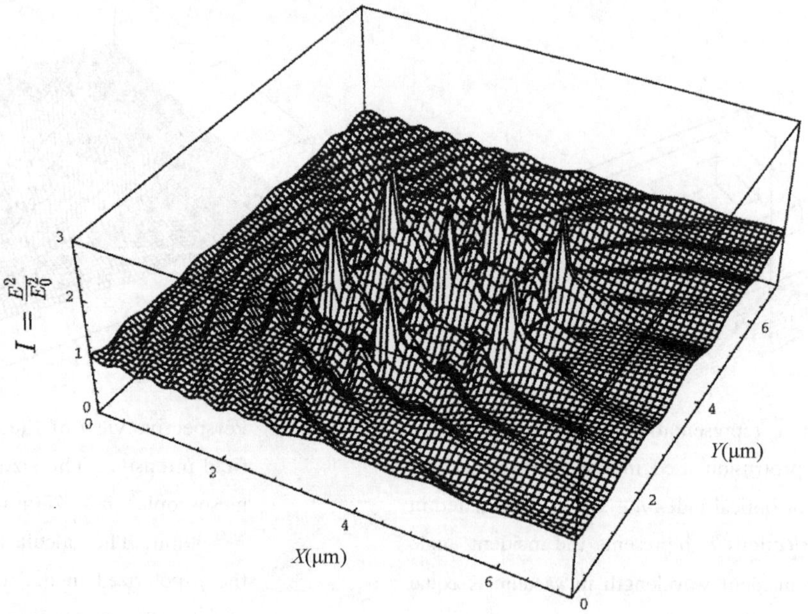

Fig. 4-20 3D perspective view of the normalized electric field intensity. The observation plane is located at a distance $Z_0 = 120$nm above the flat surface. This large scale calculation $7 \times 7\mu m^2$ has been performed in the p-polarized mode. The parameters P_1 and P_2 are equal to 250nm and 1750nm, respectively, and the pads are 100nm high.

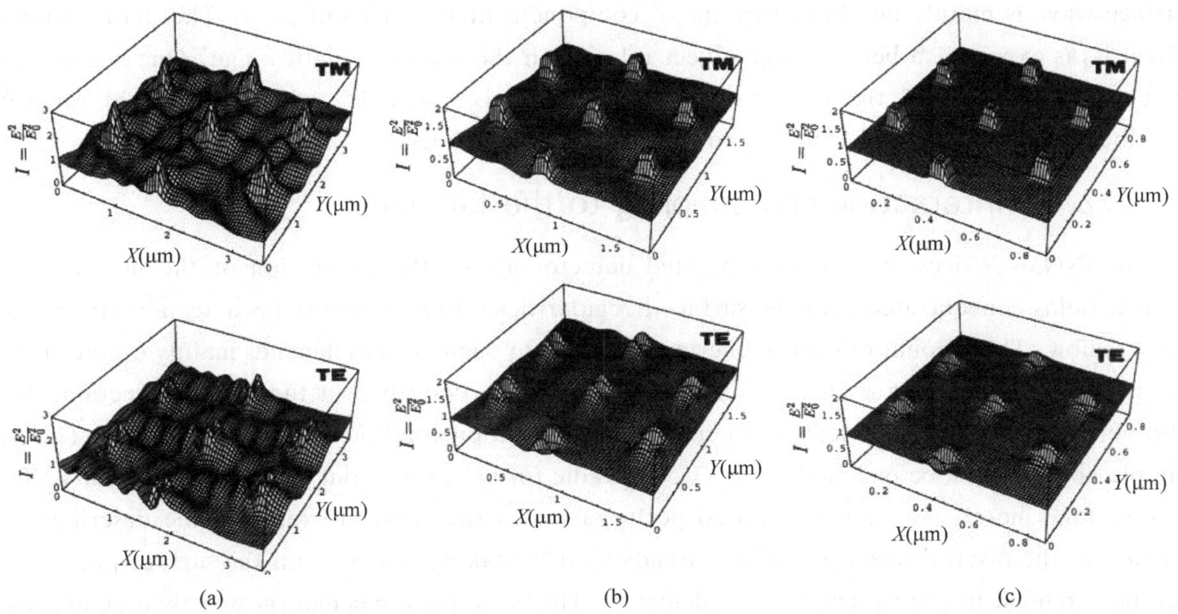

Fig. 4-21 A sequence of 3D maps $I(X,Y)=|E|^2/|E_0|^2$ describing the evolution of the standing field pattern observed around the topographic object described in Figure 4-19. For the same incident wavelength $\lambda=620$nm, the volume occupied by this object is progressively reduced. (a) start in the mesoscopic range with $P_1=250$nm, $P_2=1750$nm and $Z_0=135$nm. The height h of each pad is 100nm, the computational window $3.5\times3.5\mu m^2$ is centred around the structure and the two usual polarization modes have been treated. (b) Same calculation, but after reduction of all the lengths by a factor two: $P_1=125$nm, $P_2=875$nm and $Z_0=67.5$nm and the height h of each pad is 50nm. (c) The reduction factor now reaches four: $P_1=67.5$nm, $P_2=437.5$nm and $Z_0=33.75$nm and the height h of each pad is 25nm.

The evolution of the field pattern raises the following comments.

(1) When the object displays mesoscopic dimensions as Figure 4-18 and Figure 4-19(a), the field distribution is dominated by interference phenomena, so that the field lines do not follow the profile of the square-shaped protrusions.

(2) As the dimensions of the 3D objects enter the subwavelength range as Figure 4-19, the interference pattern around the objects progressively collapses and the field intensity distribution tends to become perfectly symmetrical thereby reproducing the symmetry of the pads. Under such conditions, and in TE polarization, a highly localized field occurs just above the edges located in a perpendicular direction to the incident field E_0.

In fact, when deal with such subwavelength sized objects, the importance of retardation effects decreases dramatically, so that the symmetry of the field distribution is only governed by both the orientation of the incident field and the profile of the object itself. Actually, these features may help us to get more insight into the complex contrast phenomenon observed in the TE mode.

The field distribution is now governed by the depolarization effects which result from the conservation of the normal component of the displacement vector $D(\omega)=\varepsilon(\omega)E(\omega)$ when crossing the surfaces of the dielectric protrusions. Due to the rapid variation of the dielectric constant between air and glass, this conservation imposes a sharp variation of the field near the interfaces perpendicular to E_0.

A completely different behaviour is observed with the TM mode. In this polarization the surface wave is mainly dominated by the Z component of the incident field. The main resulting effect is, as expected, a better image-object relation in the subwavelength range. One can observe that, when the size of the square-shaped protrusions is gradually reduced, the field intensity distribution around the objects tends to reproduce their profiles.

4.2.8 Interaction and coupling to the far-field

In SNOM devices the use of a pointed detector allows the convertion of the non-radiative optical fields concentrated near the surface irregularities into radiative fields detectable in the far-field region. The amount of optical energy converted by such devices depends mainly on the size of the region of interaction with the confined optical fields described in the previous section. It is also very sensitive to the object parameters, the illumination conditions and to tip design. The tip-sample coupling can be included in the Green dyadic formalism without any formal difficulty. This can be done merely by adding a second perturbation in the self-consistent scheme described. In particular, the discretization procedure already used for taking into account the surface protrusions can be extended to the tip-apex of the detector. The conversion mechanism will then be analysed theoretically using the theory described previously. In fact, the knowledge of the effective field distribution inside the perturbation (tip-apex + surface defect) is sufficient to describe the far-field E_{far} crossing a given surface Σ located inside the wave zone of the detector. As described in Girard et al, outside the source region the Lippmann-Schwinger equation may be applied once more to derive the field radiated by the tip extremity. This numerical scheme has been used to study the influence of the detector geometry and of the probing distance on image formation of subwavelength 3D topographic objects.

In order to illustrate the tip-sample coupling phenomenon simultaneously with the concomitant optical non-radiative transfer effect occurring around the contacting zone.

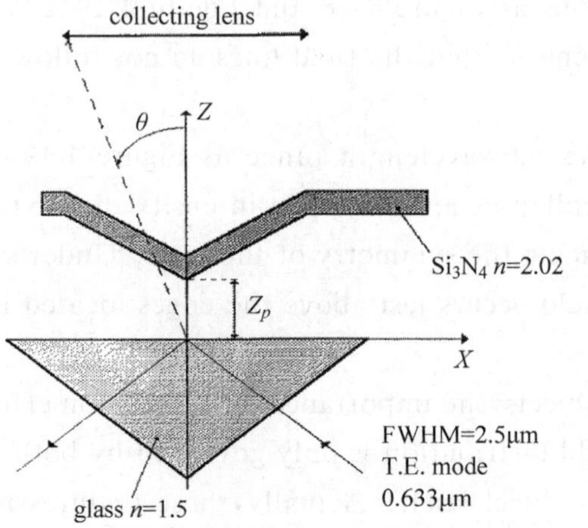

Fig. 4-22 Schematic view of a two-dimensional model of SNOM: a silicon nitride AFM tip placed above a glass prism sample. The incident focused optical wave has a FWHM of 2.5μm, a wavelength of 0.633μm and is in the s-polarized mode.

In Figure 4-22～Figure 4-24 numerical computerized works on a two-dimensional model of a silicon nitride tip facing a glass sample (see Figure 4-24). In fact, during the past few years, a lot of NFO probe designs have been proposed: the nano-apertures, sharpened optical fibres and, more recently, tetrahedral silicon nitride tips supported by cantilevers which are well known in AFM. The main advantage of this last probe is that one can rely on the force measurement to control accurately the tip-sample distance which, otherwise, is badly defined in usual SNOM configurations. A large lens or an optical fibre placed above the cantilever collects the far-field radiated by the SiN tip (see Figure 4-24). A direct consequence of this coupling between the probe and the sample is the strong attenuation of the reflected wave. As expected, other simulations indicate that when the tip is that pushed higher the coupling between the tip and the microprism dramatically decreases (see Figure 4-22). In order to simulate the experimental signal, the far-field intensity has been integrated around the angle θ (see Figure 4-20). This calculation has been performed for various tip positions along the z-axis direction. The calculated approach curve represented in Figure 4-22 is compared to the exponential decay of a pure evanescent wave whose amplitude is fitted to the self-consistent calculation. That such a self-consistent calculation restores the modulations generated by the coupled electromagnetic modes of the tip-sample system.

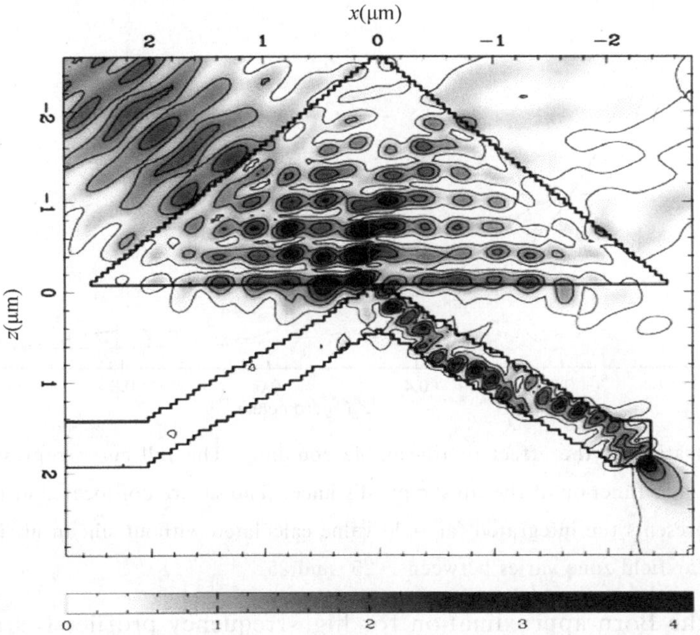

Fig. 4-23 Illustration of the tip-sample optical interaction between a 2D silicon nitride tip and a glass prism. The parameters are described in Figure 4-19. This simulation has been performed with a two-dimensional numerical code built with the Green function technique. In this example the tip touches the surface of the prism.

- Particular conditions for the validity of the Born approximations

As discussed earlier in this review, the Born approximation is generally not expected to work. However, it may be valid in some circumstances which have been cast mathematically by Carminati and Greffet. The Born approximation was found to be valid for low-relief surfaces such that the largest vertical steps h are much smaller than the incident wavelength λ and for small dielectric contrasts such that $\Delta\varepsilon = \sup I\varepsilon(r,\omega) - \varepsilon_{\text{ref}} I \ll 1$.

For a localized defect, these conditions may be detailed as follows,

$$\frac{\pi\Delta\varepsilon h S}{\lambda^2 d} \ll 1 \tag{4-81}$$

where S is the area of the defect and d is the distance of observation of the scattered field. The case of an extended surface profile was also analysed in terms of the Fourier transform of the surface profile. Different validity criteria evolve according to the content of the Fourier spectrum of the surface profile. For low-frequency profiles, the criterion is given by

$$\frac{2\pi\Delta\varepsilon h}{\lambda} \ll 1 \tag{4-82}$$

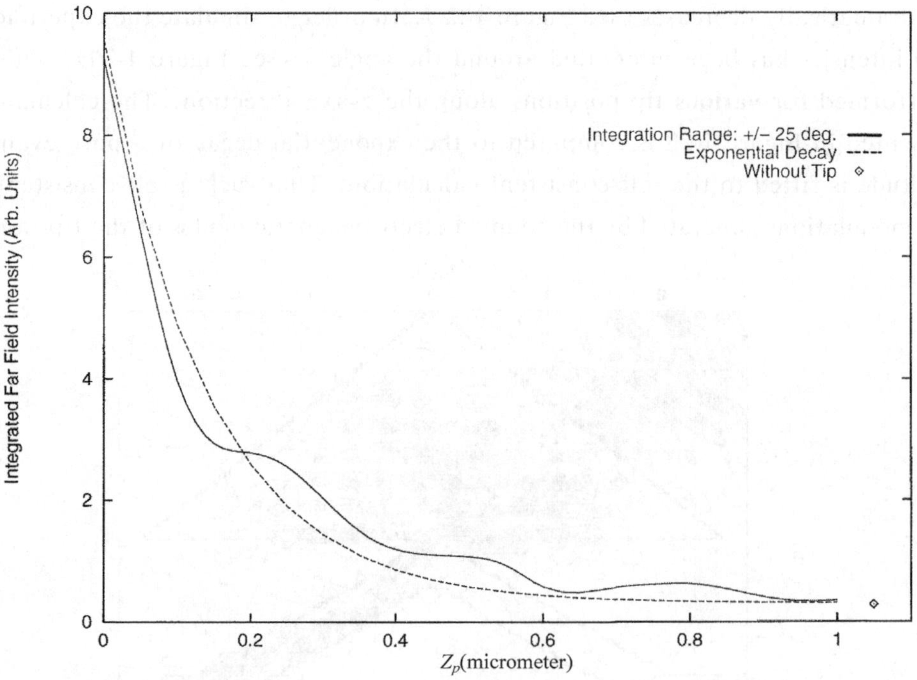

Fig. 4-24 Another illustration of the effect of tip-sample coupling. The full curve represents the integrated far-field intensity as a function of the tip-sample distance. The square dot located at the right-hand corner of the figure represents the integrated far-field value calculated without silicon nitride tip. The integration range in the far-field zone varies between -25° and 25°.

The validity of the Born approximation for high-frequency profiles is strongly dependent on the polarization of the incident wave and on the distance to the plane of observation. In brief, the approximation holds if the observation distance is large and it is also more accurate for TE polarized incident waves.

The dielectric response of metallic clusters with a radius below 15nm cannot be approximated by bulk dielectric function values. For the calculations displayed in this section, we have therefore used a somewhat more elaborate model for the polarizability of the metallic spheres obtained by extending the method of Newns to an electron gas inside a spherical well. A model of multipolar polarizabilities as a function of the size of the metal particle can be described from the knowledge of the interband electronic transitions of the bulk metal. Such a model includes depolarization effects so that, not only $L_3 D_0$, but also the discretization of the metal spheres is no longer required.

Plasmon resonances are very sensitive to slight variations of the dielectric environment so that approaching a glass tip has unavoidable consequences on the detected spectrum. Near-field spectroscopy of metallic nanospheres is thus a typical example of strong tip-sample interactions. Upon approaching the sample, the principal features of the tip-sample interaction are slight red-shifting and broadening of the resonance peak. These behaviours show up simultaneously as a dramatic increase of the intensity which reveals the presence of an enhanced electric near-field (cf. Figure 4-25).

A typical near-field spectroscopic effect shows up as illustrated in Figure 4-26. The isointensity curves account for the detection above a small cluster of nine metallic particles. The

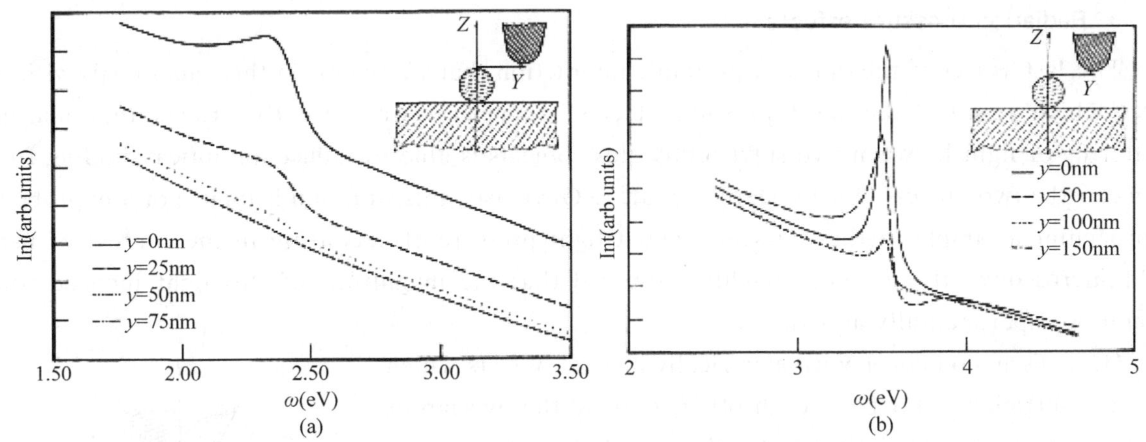

Fig. 4-25 Constant height $Z_p = 11\text{nm}$ iso-intensity curves detected by a probe tip scanning an array of gold spheres deposited on a glass substrate. A silver sphere is introduced at the lower left-hand corner of the array. All spheres have a radius of 5nm. (a) The frequency of the incident field is $\omega = 2\text{eV}$. (b) Image simulated at the resonance frequency of the silver particle $\omega = 3.34\text{eV}$.

sphere located at the lower left is made of silver while the others are made of gold. At an incident frequency which is far from the localized plasmon of a silver particle, the image recorded above the nine spheres follows the square symmetry of the deposited structure. The intensity recorded around the aggregate displays a larger gradient in the direction orthogonal to the incident surface wave vector which is aligned along the y direction. When the incident frequency is tuned to the plasmon frequency of the silver sphere, the intensity is considerably enhanced close to this silver sphere. The resonating particles distort the field above the gold spheres so that the square symmetry of the aggregates is no longer recognizable. This optical resonance effect is interesting to discriminate between different kinds of metallic particles.

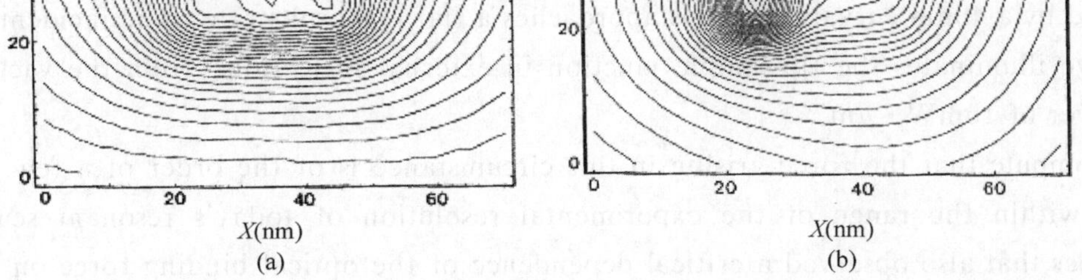

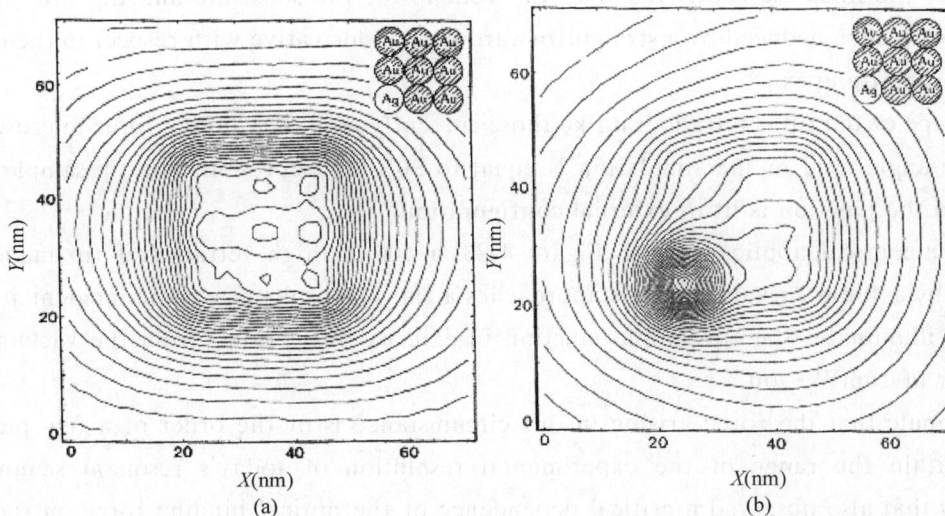

Fig. 4-26 Study of the behaviour of the detected intensity as a function of the incident field frequency. The object is a single metallic particle of 15nm radius located at the origin of the absolute frame. The approach distance is maintained constant: (a) gold sphere; (b) silver sphere.

Pincemin et al. investigated further aspects of the near-field detection of plasmons associated with subsurface particles.

• Radiation pressure effects

The last aspect of the optical tip-sample interaction that we review in this paper deals with the force which results from light pressure effects. Indeed, it was recently proved that multiple scattering of light between two sufficiently close objects is able to induce an optical binding force between the two objects (Burns, Dereux et al.). Obviously, this situation is typical of any probe tip approaching a sample so that questions about light pressure effects arose in the context of near-field microscopy. It was then roughly estimated that the magnitude of this light induced force should be experimentally accessible.

Here discussion copes with electrically neutral objects which are kept sufficiently far from each other to avoid the overlap of electron wavefunctions. When the tip approaches the surface under laser beam illumination, the long-range interaction energy includes the dispersion energy and the optical binding energy induced by the incident light beam (see Figure 4-27).

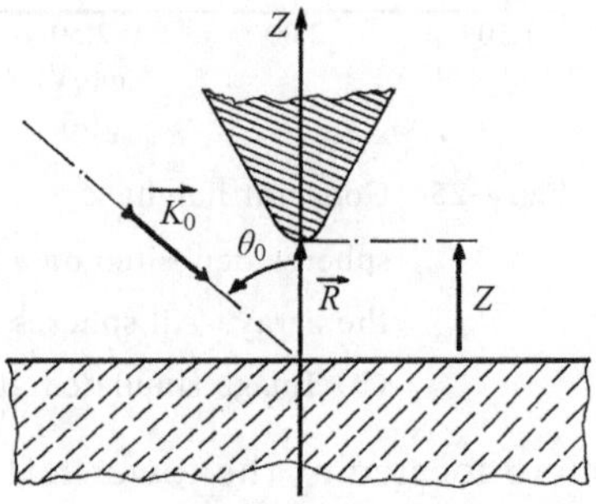

Fig. 4-27 Geometry used in the numerical application.

The van der Waals dispersion energy is due to quantum zero-point fluctuations of the coupled charge densities inside the probe tip and the substrate. Its computation may be achieved within a framework similar to the Lippmann-Schwinger formalism but where the evaluations are performed for imaginary frequencies. The calculation of the optical binding energy induced by the incident beam requires that the spatial and temporal shape of the external field be specified. In the case of harmonically oscillating fields, the optical binding energy is given by the sum of the time-averaged inductive energies experienced by the sample and the probe tip. These inductive energies are obtained by integrating the electric field intensity $|E(r)|^2$ multiplied by the linear susceptibility over the volumes of the substrate and the tip. The optical binding force is then deduced by a straightforward partial derivative with respect to the coordinate Z_p of the probe tip apex.

The shape of the tip is tetrahedral like those currently used in scanning force microscopy. The tip aperture angle used in the simulation is equal to 90°. Z_p represents the tip-sample approach distance and the junction is lit in external configuration.

In the numerical applications of Figure 4-28, a 400nm high tetrahedral tip made of glass terminated by a 40nm curvature radius approaches a flat glass substrate. The incident p-polarized plane wave illuminates the tip-sample junction in external reflection (from the vacuum) with a mean power of $15\text{mW}\cdot\mu\text{m}^{-2}$.

To compute that the force arising in this circumstance is of the order of a few piconewtons which is within the range of the experimental resolution of today's resonant scanning force microscopes that also observed a critical dependence of the optical binding force on the angle of incidence. This variation is understood as follows. If consider the glass surface alone, the electromagnetic field in the vicinity of the surface is reduced to the interference pattern arising from the superposition of the incident and the reflected fields. The pattern is the zeroth-order solution in the description of the multiple scattering in the tip-sample junction. This zeroth-order solution is perturbed by the optical near-field which shows up when approaching the tip.

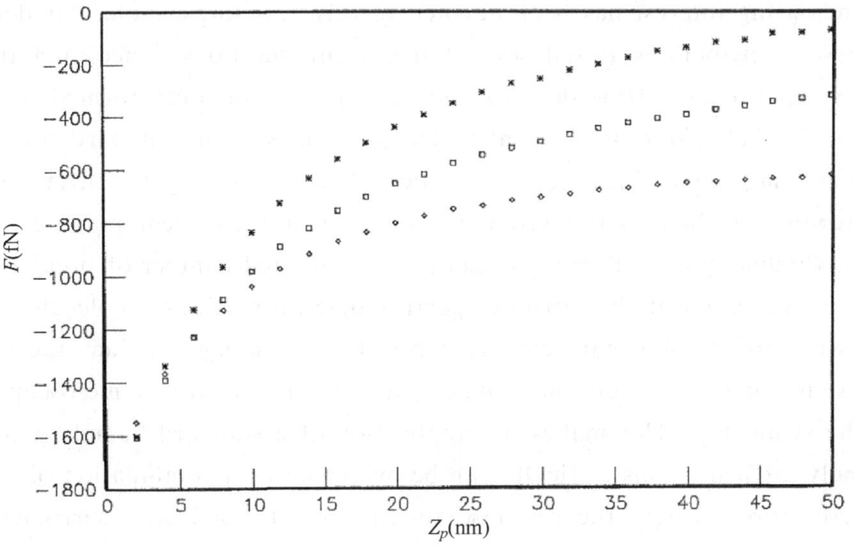

Fig. 4-28 Optically-induced force as a function of the approach distance Z_p and for various angles of incidence. The refraction index of both tip and substrate is equal to 1.5: (∗) $\theta_0 = 56.5^\circ$ (the Brewster angle); (□) $\theta_0 = 50^\circ$; (◇) $\theta_0 = 60^\circ$.

However, the optical field gradients imposed by the zeroth-order solution still dominate. The observation of the optical binding force due only to the multiple scattering between the substrate and the tip is thus better observed in the absence of any reflected field. Such a situation occurs in the vicinity of the Brewster angle (56.5°). Finally, we note that the magnitude of the force increases with the index of refraction of the tip (see Figure 4-29). This last phenomenon also demonstrates the near-field origin of the optical binding force since it was significant for approach distances below 50nm.

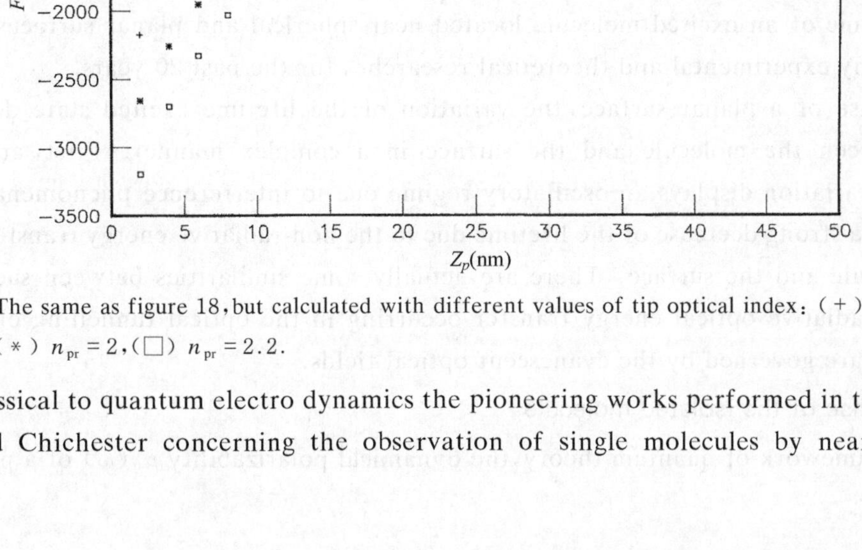

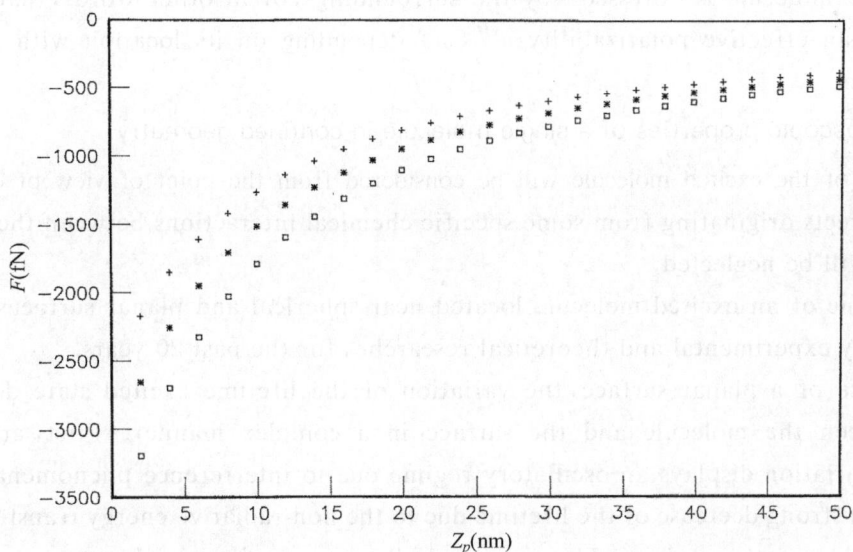

Fig. 4-29 The same as figure 18, but calculated with different values of tip optical index: (+) $n_{pr} = 1.8$, (∗) $n_{pr} = 2$, (□) $n_{pr} = 2.2$.

From classical to quantum electro dynamics the pioneering works performed in the ATT group by Betzig and Chichester concerning the observation of single molecules by near-field optical detection, an increasing interest has been devoted to this new single molecular detection (SMD) technique. These contributions raised several important questions concerning the mechanisms underlying such experiments: How does the macroscopic surrounding formed by the tip-sample junction modify the behaviour of the embedded quantum system? In particular, what kind of relation between image and object may be expected by measuring the lifetime changes of a molecular aggregate attached to the probing tip? From a theoretical point of view, this new experimental methodology is challenging because of the limited number of predictive models able to describe the modification of the intrinsic spectroscopic properties of molecules in the presence of mesoscopic structures dressed with complex optical surroundings. In fact, the presence of any microscopic system (here the molecules) placed in interaction with a mesoscopic environment breaks down the symmetry. This makes the application of a standard boundary conditions based method extremely difficult. This difficulty can be overcome by a combination of microscopic and macroscopic descriptions, where the response function of the dielectric surrounding medium is derived from an appropriate iterative numerical algorithm. To consider a single molecule trapped inside a confined geometry. For example, it may be the junction formed by the tip of a scanning near field optical microscope (SNOM) facing the surface of a sample. When such a low symmetry system is submitted to an external excitation, different kinds of phenomena may be expected.

(1) A highly-confined optical field may be observed in the gap. The magnitude and the shape of the field depends both on the sharpness of the detector and the chemical properties of the junction (metal, semiconductor or dielectric).

(2) The intrinsic polarizability of the molecule is modified. This effect then introduces a significant modification of the lifetime of the excited molecule which is always accompanied by a small shift of the corresponding occupied state. These spectroscopic changes can be understood by saying that the molecule is "dressed" by the surroundings, or in other words, that the molecule responds with an effective polarizability $\alpha^{\mathrm{eff}}(\omega)$ depending on its location with respect to the material system.

- Spectroscopic properties of a single molecule in confined geometry

The decay of the excited molecule will be considered from the point of view of electromagnetic theory. All effects originating from some specific chemical interactions between the molecule and the substrate will be neglected.

The lifetime of an excited molecule located near spherical and planar surfaces has been the subject of many experimental and theoretical researches for the past 20 years.

In the case of a planar surface, the variation of the lifetime excited state depends on the distance between the molecule and the surface in a complex manner. First, at a very large distance, the variation displays an oscillatory regime due to interference phenomena. In the near zone, observe a strong decrease of the lifetime due to the non-radiative energy transfer between the excited molecule and the surface. There are actually some similarities between such phenomena and the non-radiative optical energy transfer occurring in the optical tunnelling effect. Indeed, both of them are governed by the evanescent optical fields.

- Response of the isolated molecule

In the framework of quantum theory, the dynamical polarizability $\alpha_0(\omega)$ of a physical system (atom, molecule, small metallic particle and so on) can be expressed as follows:

$$\alpha_0(\omega) = \frac{1}{\bar{h}}\sum_r\left(\frac{2\omega_{r0}\mu^{0r}\mu^{r0}}{\omega_{r0}^2 - \omega^2 - i\omega\Gamma_{r0}}\right) \tag{4-83}$$

Where μ^{r0} represents the matrix elements of the polarization molecular operator between the fundamental state and the excited states. Moreover, Γ_{r0} represents the natural lifetime widths of the molecule. In what will follow, we will restrict ourselves to the case of a two-level molecular system. In a first stage, this simplication avoids the huge computational difficulties involved in a realistic treatment of the dynamical properties of the molecule Moreover, it provides a comprehensive scheme to analyse the physical mechanisms at the origin of the lifetime changes induced by the presence of highly complex optical systems. This approximation leads to

$$\alpha_0(\omega) = \frac{1}{\bar{h}}\left(\frac{2\omega_0\mu^{01}\mu^{10}}{\omega_0^2 - \omega^2 - i\omega\Gamma_0}\right) \tag{4-84}$$

where μ^{01} is now the matrix element of the polarization operator between the two levels.

Molecule-surrounding medium coupling: towards the concept of effective polarizability, In this subsection introduce the electromagnetic coupling between our molecule and the 3D system described in the inset of Figure 4-30. When this system is perturbed by an optical field $E_0(r, t)$ coming from an external laser source, the field $E(r, t)$ in the gap may be derived from the different numerical schemes described in previous sections. It is important to note that this field is an "observable" since it has already been averaged on the quantum states of the whole primary system (dielectric surrounding medium).

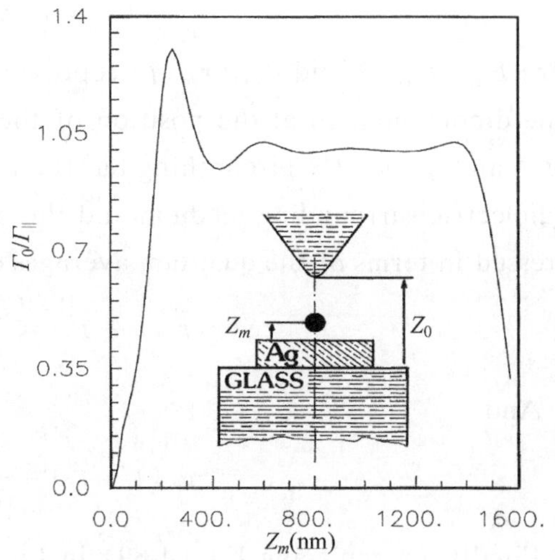

Fig. 4-30 Distance-dependence of the normalized lifetime $\Gamma_0/\Gamma_\parallel$ change of a molecule trapped inside a dielectric-metal-dielectric junction, as a function of the position of the molecule. The geometry used in the simulation is described in the inset and the tip-sample separation is maintained at Z_0 = 1600nm. The glass sample is covered with a square-shaped silver layer 300nm long and 20nm high.

Let to write the interaction Hamiltonian coupling between the primary system and the molecule:

$$H(t) = -(E(r_m, t) + \varepsilon(r_m, t)) \cdot \mu(t) \tag{4-85}$$

where $\mu(t)$ and $\varepsilon(r_m, t)$ are the polarization operator of the molecule and the electric field operator associated with the 3D dielectric surrounding medium, respectively. The vector r_m denotes the position of the molecule. These operators are written here in the interaction representation

$$\varepsilon(r_m, t) = \exp[i\bar{h}^{-1}H_0 t]\varepsilon_m(r)\exp[-i\bar{h}^{-1}H_0 t] \tag{4-86}$$

And

$$\mu(t) = \exp[i\bar{h}^{-1}H_0 t]\mu\exp[-i\bar{h}^{-1}H_0 t] \tag{4-87}$$

where H_0 represents the Hamiltonian for the 3D dielectric surrounding medium. At this stage, since one neglects all chemical interactions between the molecule and its support, one can assume that there is no significant modification of the whole wavefunction $|\psi\rangle$ of the system due to the short-range interaction between the molecule and the 3D dielectric surrounding medium. In this situation it is worthwhile applying the time-dependent Hartree (TDH) approximation in which one assumes that each part of the system moves under the combined effect of the external force and the average displacement of the other system. Within this approximation one can then consider that $|\psi\rangle$ is a tensorial product of the two wavefunctions $|\psi_{\text{mol}}\rangle$ and $|\psi_{\text{sur}}\rangle$ associated with the molecule and the surrounding medium, respectively. A straightforward application of the perturbation theory shows that the linear response of the two variables $\mu(t)$ and $\varepsilon(r,t)$ are given by

$$\begin{aligned} E_{\text{mol}}(r_m,t) &= \langle E(r_m,t)+\varepsilon(r_m,t)\rangle \\ &= E(r_m,t)+\int_{-\infty}^{t} S(r_m,r_m,t-t')\cdot\langle\mu(t')\rangle \mathrm{d}t' \end{aligned} \tag{4-88}$$

And

$$\begin{aligned} \mu_{\text{mol}}(r_m,t) &= \langle\mu(t)\rangle \\ &= \int_{-\infty}^{t} \alpha_0(t-t')\cdot\langle E(r_m,t')+\varepsilon(r_m,t')\rangle \mathrm{d}t' \end{aligned} \tag{4-89}$$

where $E_{\text{mol}}(r_m,t)$ and $\mu_{\text{mol}}(r_m,t)$ represent both the temporal variation of the effective field and of the dipole moment at the position of the molecule. In addition, the dyadic tensors $S(r_m,r_m,t-t')$ and $\alpha_0(t-t')$ are nothing but the temporal representation of the field susceptibility of the 3D dielectric surrounding medium and the polarizability of the molecule. These quantities can be expressed in terms of the quantum averages of the commutators of the operators $\varepsilon(r,t)$ and $\mu(t)$:

$$S(r,r',t-t') = \frac{1}{h}\langle\psi|\,[\varepsilon(r,t),\varepsilon(r',t')]\,|\psi\rangle \tag{4-90}$$

And

$$\alpha_0(t-t') = \frac{1}{h}\langle\psi|\,[\mu(t),\mu(t')]\,|\psi\rangle \tag{4-91}$$

Finally, by replacing Eq. (4-89) in Eq. (4-88) one obtains the time-dependent selfconsistent equation for the molecular electric field:

$$E_{\text{mol}}(r_m,t) = E(r_m,t)+\int_{-\infty}^{t}\mathrm{d}t'\int_{-\infty}^{t'}\mathrm{d}t'' S(r_m,r_m,t-t')\alpha_0(t'-t'')E_{\text{mol}}(r_m,t'') \tag{4-92}$$

Solving this implicit integral equation requires one to pass into the ω-space

$$E_{\text{mol}}(r_m,\omega) = M(r_m,\omega)\cdot E(r_m,\omega) \tag{4-93}$$

where $M(r_m,\omega)$ is a 3×3 matrix defined by

$$M(r_m,\omega) = [I - S(r_m,r_m,\omega)\cdot\alpha_0(\omega)]^{-1} \tag{4-94}$$

Note that from this equation one obtains in a first stage the molecular effective field $E_{\text{mol}}(r_m,\omega)$. The field $E_{\text{mol}}(r_m,\omega)$ generated by the molecule far away from the emitting zone can be described by using once again the Lippmann-Schwinger equation

$$E_{\text{mol}}(r,\omega) = E(r,\omega)+S(r,r_m,\omega)\cdot\alpha_0(\omega)\cdot M(r_m,\omega)\cdot E_{\text{mol}}(r_m,\omega) \tag{4-95}$$

that can be rewritten as

$$E_{\text{mol}}(r,\omega) = E(r,\omega)+S(r,r_m,\omega)\cdot\alpha^{\text{eff}}(\omega)\cdot E_{\text{mol}}(r_m,\omega) \tag{4-96}$$

where $\alpha^{\text{eff}}(\omega)$ defines the effective polarizability of the molecule in the presence of the dielectric surrounding medium:

$$\alpha^{\text{eff}}(r_m,\omega) = \alpha_0(\omega)\cdot M(r_m,\omega) \tag{4-97}$$

Many spectroscopic experiments performed near a solid body require the knowledge of the effective molecular polarizability. In fact, this new response function contains all the dynamical information about the coupling with the dielectric surrounding medium. In other words, the molecule radiates optical energy with a polarizability "dressed" by the dielectric surrounding medium. In the past, several theoretical works have been devoted to its calculation when the molecule interacts with systems of simple symmetry (spheres, cylinders, planes and so on). The symmetry of the tensor $\alpha^{\text{eff}}(r_m, \omega)$ is governed mainly by the symmetry of the molecule-surrounding medium super system even if the molecular polarizability $\alpha_0(\omega)$ is initially isotropic. For example, in the particular case of a single molecule interacting with a perfectly planar surface, the dyadic tensor $\alpha^{\text{eff}}(r_m, \omega)$ belongs to the $C\infty v$ symmetry group, and consequently, may be described with two independent components α_{xx}^{eff} and α_{zz}^{eff}. In the case of a SNOM surrounding medium (corrugated surface + pointed detector), the effective polarizability tensor will be more complex and all components should be taken into account in a realistic calculation.

- Fluorescence lifetime change in complex optical systems

Two important electrodynamic effects are included in the expression of the effective polarizability. The first is a small shift of the excited state, more precisely the frequency ω_0 is shifted towards a lower frequency. This effect can be characterized by the ratio

$$\Omega(r_m) = \omega^{\text{eff}}/\omega_0 \tag{4-98}$$

Under usual conditions this coefficient is weak and, with respect to other effects, it may be chosen close to unity. Moreover, we have to keep in mind the fact that $\Omega(r_m)$ depends on the polarizability component under consideration.

The second effect, much more sensitive to the location of the molecule in the junction, is the lifetime change defined by the ratio

$$\eta(r_m) = \Gamma^{-1}(r_m)/\Gamma_0^{-1} \tag{4-99}$$

This coefficient also depends on all effective polarizability components. Nevertheless, with a good approximation, we may assume that the dressed molecule belongs to the $C\infty v$ group with its main axis in the z direction. This assumption leads to

$$\alpha^{\text{eff}}(r_m, \omega) = \begin{pmatrix} \alpha_{\perp}^{\text{eff}}(r_m, \omega) & 0 & 0 \\ 0 & \alpha_{\perp}^{\text{eff}}(r_m, \omega) & 0 \\ 0 & 0 & \alpha_{\parallel}^{\text{eff}}(r_m, \omega) \end{pmatrix} \tag{4-100}$$

where the two independent components $\alpha_{\perp}^{\text{eff}}(r_m, \omega)$ and $\alpha_{\parallel}^{\text{eff}}(r_m, \omega)$ may then be identified by a two-levels polarizability expression similar to that associated with the isolated molecule see Eq. (4-80). This procedure yields

$$\alpha_{\perp}^{\text{eff}}(r_m, \omega) = \left(\frac{2\omega_{\perp} A_{\perp}}{\omega_{\perp}^2 - \omega^2 - \mathrm{i}\omega\Gamma_{\perp}}\right) = \alpha_0(\omega) M_{xx}(r_m, \omega) \tag{4-101}$$

And

$$\alpha_{\parallel}^{\text{eff}}(r_m, \omega) = \left(\frac{2\omega_{\parallel} A_{\parallel}}{\omega_{\parallel}^2 - \omega^2 - \mathrm{i}\omega\Gamma_{\parallel}}\right) = \alpha_0(\omega) M_{zz}(r_m, \omega) \tag{4-102}$$

For each equation, the three different parameters $\omega_{\perp/\parallel}$, $A_{\perp/\parallel}$ and $\Gamma_{\perp/\parallel}$ may be defined by identification. After some algebra one finds

$$\Gamma_{\perp/\parallel} = \mathrm{Re}\left(\frac{\Gamma_0 M_{xx/zz}(r_m, 0)}{M_{xx/zz}(r_m, \omega_0)}\right) \tag{4-103}$$

In conclusion, one obtains a very compact result, only depending on the dynamical matrix $M(r_m, \omega_0)$ at the resonance frequency of the molecule. Note that the spatial variation of $\Gamma_{\perp/\parallel}$ with respect to the dielectric surrounding medium is contained in the field susceptibility $S(r_m, r_m, \omega)$. In the vicinity of a highly complex system this dyadic may be derived selfconsistently by a recursive sequence of Dyson's equations. Moreover, within the first Born approximation one recovers the well known result

$$\Gamma_{\perp/\parallel} \simeq \Gamma_0 + \frac{2|\mu^{01}|^2}{\hbar}\mathrm{Im}[S_{xx/zz}(r_m, r_m, \omega_0)] \tag{4-104}$$

The molecular parameters used in this simulation are $\Gamma_0 = 2 \times 10^6 \mathrm{s}^{-1}$ and $\alpha_0 = 10\mathrm{A}^3$, and the fluorescing wavelength is $\lambda_0 = 612\mathrm{nm}$. The geometry of the junction is sketched in the inset of Figure 4-30, it consists of a glass support with a thin square silver protrusion, facing a tetrahedral dielectric tip with sharp edges and a 10nm ending curvature radius. When the molecule approaches the metal pad one first observes the usual decay which is then followed by the fluorescence quenching. For intermediate distances, the lifetime variation in the gap region $300n_m \leqslant Z_m \leqslant 1300\mathrm{nm}$, displays standard quasi-periodic oscillations with a period close to the half-fluorescing wavelength λ_0: It may be seen that towards the dielectric tip the decay is less abrupt, and a magnification of the evolution of the coefficient $\Gamma_0/\Gamma_\parallel$ as the molecule approaches the tip extremity indicates that the lifetime drops by about one order of magnitude when the molecule becomes adsorbed on the tip surface. This simulation clearly indicates that the fluorescing molecule behaves as a highly sensitive nano-probe to the external environment. In particular, working in the near-field zone just before the quenching effect occurs should make it possible to increase the lateral SNOM resolution.

The near-field optics is not restricted to the improvement of the resolution of optical microscopy that the principles on which nearfield optics is developed are related to the physics of evanescent electromagnetic waves which have been approached in several different contexts since the mid 1960s. To insist on the fact also that the theoretical description of near-field optical phenomena involves the problem of light diffraction by subwavelength structures which is neglected in most classical textbooks. Since near-field optics is still in rapid development, to give some perspective about trends, for example the study of lifetime effects and their relation to quantum electrodynamics.

4.3 Optical solid immersion lens (OSIL)

Near-field optical data storage with a nano-aperture aims at increasing the optical resolution by scanning the disc with a spot that is much smaller than the diffraction limited spot from an objective lens. There are essentially two methods other than using a SIL for creating sub-wavelength optical spots. The first method uses a metallised tapered optical fibre with a sub-wavelength aperture at the tapered end. Within a subwavelength distance from the tapered aperture the optical resolution is not limited by diffraction as with an imaging lens, but by the size of the sub-wavelength aperture and the distance between the aperture and the optical disc. A major disadvantage of this approach is the extremely small optical efficiency, often much less than 1%, from the light source to the disc and back to the detector. Due to this small efficiency the signal to noise ratio (SNR) of the read-out signal does not allow data transfer rates that are anywhere near what is required to make this technology attractive for data storage. The main application of this technology is in scanning near-field optical microscopy (SNOM). A second method is based on a lens that focuses light onto a sub-wavelength aperture near the optical disc. Again, within the near-field of the aperture, the aperture size and the distance to the optical disc determine the optical resolution. The efficiency of this approach can be larger than that of the tapered optical fibre and may be even further improved by plasmon enhanced anomalous transmission through the aperture.

Significant disadvantages of this method are the required alignment tolerances between the focused spot and the nano-aperture and the fact that a nano-aperture cannot focus through a protective layer of substantial thickness on top of the disc. These difficulties make the nano-aperture approach impractical for optical data storage. Nanoapertures are, however, still being considered for Heat-Assisted Magnetic Recording (HAMR), a high-density hard-disk drive technology. With HAMR the sub-wavelength optical spot is used only for locally heating the magnetic disc. Read-out is done magnetically, not optically. The numerical aperture (NA) of a lens is defined as NA $= n \sin \alpha_m$ with n the refractive index of the medium in which the lens is focusing and α_m the angle of the focused marginal ray with respect to the optical axis, see Figure 4-31(a). Hence, with the refractive index of air essentially equal to unity, the NA of a focusing lens in air cannot exceed unity. In optical disc systems, the objective lens focuses inside the optical disc with refractive index n_d, see Figure 4-31(b). Refraction at the flat disc surface reduces the angle of the marginal ray according to Snell's law and thus the NA of the focusing lens also cannot exceed unity. The Blu-ray Disc system with NA $= 0.85$ has thus approached this upper limit closely. In principle one can achieve an NA greater than unity by filling the space between the objective lens and the optical disc with an immersion liquid, similar to liquid immersion microscopy. The refractive index of the immersion liquid can increase the NA to values greater than unity, resulting in increased optical resolution. However, the use of a liquid between the objective lens and a rotating optical disc does not appear to be practical for a consumer optical disc system, although it has been used for optical disc mastering and in liquid immersion lithography. An alternative to the liquid immersion lens is the solid immersion lens (SIL).

- SIL is typically a truncated spherical lens of a high refractive index n_s

By focusing inside such high refractive index medium the NA can also exceed unity. As discussed in detail, two types of truncated spherical lenses allow aberration-free focusing of a beam inside the lens. The first type is a hemi-spherical SIL placed with its centre of curvature at the focal point of a focusing lens, see Figure 4-31(c). The hemispherical SIL does not refract the focused light and thus NA $= n_s \sin \alpha_m$ which can be greater than unity: $0 \leqslant \text{NA} \leqslant n_s$. The second SIL type is a super-hemispherical SIL that refracts the light towards the optical axis, see Figure 4-31(d). It is shown that such a lens can be aplanatic and that this type of SIL increases the NA of the focusing lens by a factor $2n_s$ hence 2 sin NA $= n_s \alpha_m$ with $0 \leqslant \text{NA} \leqslant n_s$. A SIL thus increases the optical resolution in the focal plane of the lens. However, when a pencil of rays is focused at the flat surface of an NA>1 SIL, a portion of these rays is focused at angles that exceed the critical angle θ_c at which $\sin \theta_c = 1$ with respect to the surface normal, see Figure 4-31(c). Only rays that are

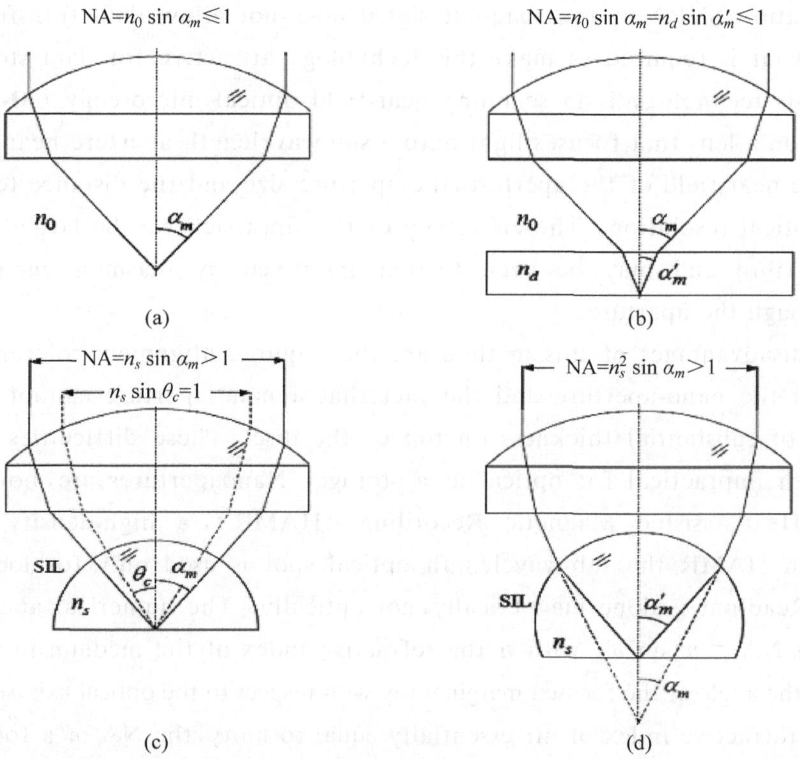

Fig. 4-31 (a)-(d) Lens configurations of (a) conventional objective lens focusing in air, (b) conventional objective lens focusing inside an optical disc, (c) objective lens focusing inside a hemispherical solid immersion lens (SIL), indicated in the figure is the critical angle θ_c inside the SIL, and (d) objective lens focusing inside a super-hemispherical solid immersion lens. Also shown are the equations for the numerical aperture (NA) of the four lens configurations.

focused at angles smaller than θ_c are transmitted through the interface into air. The rays that are focused at angles larger than θ_c in the SIL are reflected back into the SIL by total internal reflection (TIR). The field associated with these rays only contributes to the field in the air by means of evanescent waves. Hence, there can be interaction between the light focused beyond the critical angle in a SIL and an optical disc when the air gap between the SIL and the object is smaller than approximately the 1/e amplitude decay distance of the evanescent waves. This principle is referred to as frustrated total internal reflection (FTIR) or evanescent coupling. To achieve the full optical resolution advantage of a SIL with NA>1, the air gap between the SIL and the optical disc must thus be less than a fraction of the wavelength.

A key-strength of optical data storage is the inherent robustness of the optical discs. The data-layer is read-out through the substrate (CD and DVD) or through a transparent cover-layer (BD) and therefore the data is protected against small scratches, fingerprints and other surface contaminants. This robustness allows the discs to be stored and sold outside the optical disc drive. A cover-layer also functions as a thermally insulating layer. Without a cover-layer the top surface of the optical disc is heated up to several hundred degrees Celsius during recording. As reported this result in instabilities in the optical read-out signals due to evaporation and burning of surface contaminants. Thermal simulations based on a model for optical media showed that a separation of the data storage layer and the air by just 1μm of polymer cover layer already reduced the temperature at the interface to air to ambient temperature. It is preferable that a near-field optical disc system has the same robustness as conventional optical discs by also using discs that are protected by a cover-layer. Prior to the research described in this thesis it was not evident that this would be feasible in practice. In fact, all earlier published experimental results with near-field optical data storage were based on so-called first-surface optical discs in which the data-layer is situated on the surface of the optical disc.

In this research investigated two near-field optical disc systems with a SIL. The first system used first-surface discs with a super-hemispherical NA = 1.9 SIL, see schematic in Figure 4-32(a). At an early stage, this system was similar to the one reported on by Sony. The second system used a hemispherical SIL with a more moderate NA = 1.45 and discs that were protected by a cover-layer. This system served as a first proof-of principle for near-field optical data storage with discs that are protected by a polymer cover-layer. The experiments with this system were part of an investigation into the feasibility of a cover-layer-protected, dual-layer disc system, see schematic in Figure 4-32(b). The advantage of the super-hemispherical SIL system with first-surface discs is that it ultimately allows the largest possible NA and storage density on a single data layer. A disadvantage of an extremely high NA is that it cannot focus through a protective polymer cover-layer. To allow all rays to propagate through the cover-layer, need for its refractive index $n_c >$ NA. Hence the NA = 1.9 system would require a cover-layer with $n_c \approx 2.0$ and such polymers do not exist. All transparent materials with a refractive index greater than 2.0 are inorganic and can only be sputter-deposited or evaporated as a thin layer. Due to their brittleness on plastic disc substrates, their built in stress and long deposition times, the maximum layer-thickness of these inorganic materials is of the order of 100nm. Such materials thus cannot be used as a cover-layer of a few micrometers thickness and neither as a spacer-layer for multilayer discs. This implies that a system with a super-hemispherical SIL with NA≈2.0 will almost certainly consist of a single data-layer without a protective cover-layer.

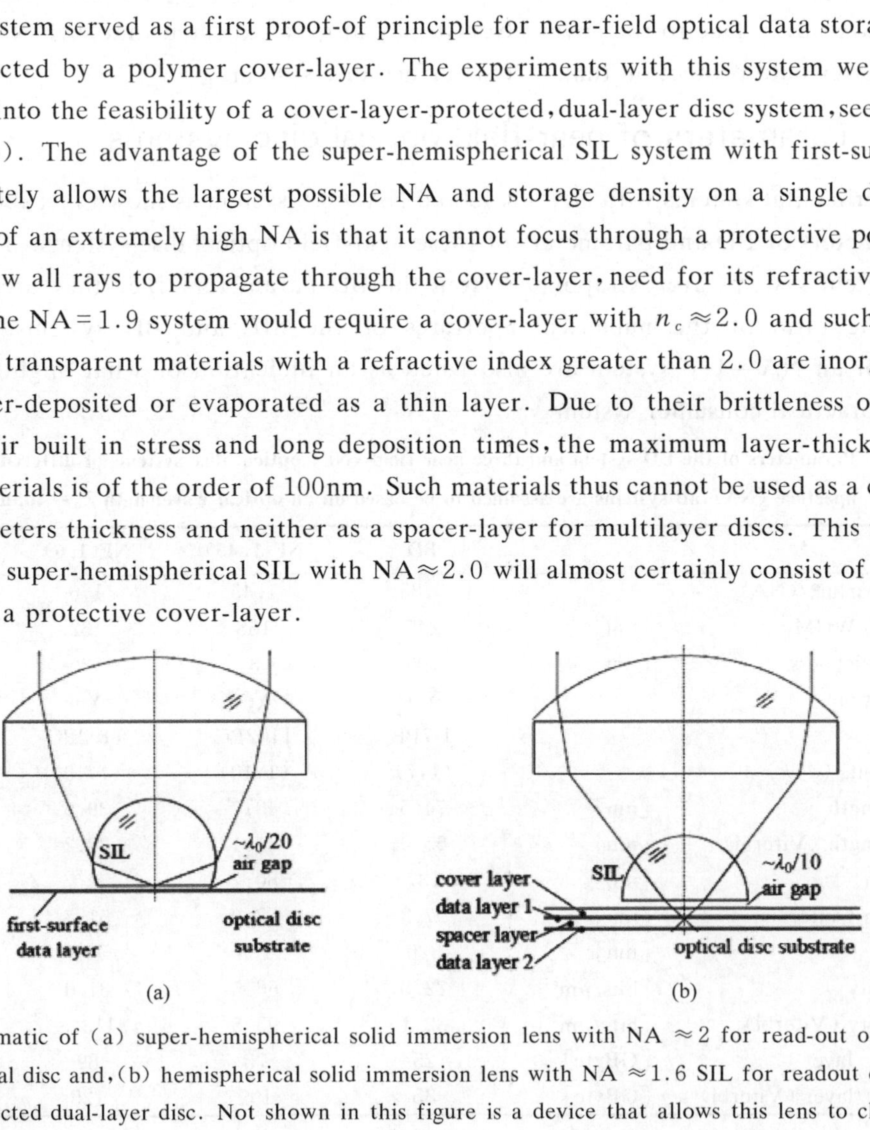

Fig. 4-32 Schematic of (a) super-hemispherical solid immersion lens with NA ≈ 2 for read-out of a first-surface optical disc and, (b) hemispherical solid immersion lens with NA ≈ 1.6 SIL for readout of a cover-layer protected dual-layer disc. Not shown in this figure is a device that allows this lens to change the focus position from data layer 1 to data-layer 2 and vice versa. Such device may, for instance, be an electrically switchable liquid crystal wavefront compensator.

Homogenous polymer materials that are suitable as a cover-layer are now becoming available with a refractive index as high as 1.7 at 405nm wavelength. These materials can be spin coated as layers of several micrometers thickness and are cured with UV radiation. A system with a hemispherical SIL of NA = 1.6 thus seems feasible with cover-layer protected optical discs. Moreover, it should also be possible to use the same polymer material as a spacer-layer in a multilayer disc.

A difficulty with a near-field dual-layer optical disc system is that such system requires an optical device that allows the lens to change focus position from a first data-layer to a second data-layer without a change of air gap and without significant extra aberrations. Such a device may for example be an electrically switchable liquid crystal wavefront compensator. The concept optical design for such a system with NA = 1.5 and with NA = 1.6 that it might even be possible to have four instead of two data-layers in a near-field optical disc. It may thus be expected that a dual-layer NA = 1.6 system is feasible that achieves a larger storage capacity per disc than a single-layer system with NA≈2.0. Due to the cover layer, it may also be expected that such NA = 1.6 system will have a better data protection than a first-surface disc system.

4.3.1 Parameters of near-field optical disc systems

The experimental systems with NA = 1.45 and NA = 1.9 system as shown in Table 4-2 that lists the key-parameters of BD and parameters of these near-field optical disc systems. The parameters for the near-field systems are extrapolated from the BD parameters, i.e. they are multiplied or divided by the ratio of the numerical apertures of the SIL and BD systems. Extrapolated parameters for an NA = 1.6 system are also listed as an indication of what might ultimately be feasible in a practical consumer system.

Table 4-2 Parameters of the BD system and three near-field (NF) optical disc systems of different numerical aperture (NA), all systems are assumed to be based on an optical wavelength $\lambda_0 = 405$nm.

		BD	NF(1.45)	NF(1.6)	NF(1.9)
Numerical Aperture (NA)	—	0.85	1.45	1.6	1.9
Focused spot FWHM	[nm]	245	168	152	128
Cover-layer thickness	[μm]	100	3	3	—
Multilayer option		Yes	Yes	Yes	No
RLL code		1-7PP	1102PC	1102PC	1102PC
Code constraints (d, k)	—	(1,7)	(1,10)	(1,10)	(1,10)
Channel bitlength	[nm]	74.5	43.7	39.6	33.3
Channel bitlength (Viterbi)	[nm]	53.0	31.1	28.2	23.7
User bitlength	[nm]	137	80.2	72.7	61.2
User bitlength (Viterbi)	[nm]	97.3	57.0	51.7	43.5
Track-pitch	[nm]	320	188	170	143
User bit density	[bits/μm^2]	22.8	66.5	81.0	114.2
User bit density (Viterbi)	[bits/μm^2]	32.1	93.5	113.8	160.5
Data Capacity/layer	[GByte]	25	73	89	125
Data Capacity/layer (Viterbi)	[GByte]	35	102	125	175

For all systems it is assumed that they are based on a blue-violet laser wavelength of 405nm. For the channel modulation code of the near-field optical disc systems have assumed the 1102PC run-length-limited (RLL) code with $d = 1$ and $k = 10$ constraints. The 1102PC outperforms the 1-7PP code of BD especially in terms of bit-error rate(bER) at high data densities at which the minimum run-lengths do not exhibit zero-crossings in the CA signal. The better performance of

1102PC compared to 1-7PP is partly due to a smaller repeated minimum transition runlength (RMTR) constraint, known as the r constraint, that limits the maximum number of consecutive minimum run-lengths to r. The 1-7PP code has an $r = 6$ RMTR constraint. The 1102PC code has $r = 2$, leading to a lower bER and improved timing recovery of the bit clock. Both codes have identical code rates of 2/3 and for all systems in the Table an error correction code (ECC) rate of 0.817 was assumed.

Parameters are listed for conventional bit detection with a slicer and for more advanced Viterbi bit detection. Assuming conventional bit detection, the shortest run-length frequency is 1.60[NA/λ], with Viterbi bit detection this frequency is 2.25[NA/λ_0], beyond the cutoff of the MTF.

The NA = 1.5 and 1.6 systems based on hemispherical glass CEL and can be combined with protective polymer cover-layer discs that may contain more than one data-layer. The NA = 1.9 system has to be based on a super-hemispherical SIL and uses single-layer first-surface discs. Here provides sufficient evidence to assume that an NA = 1.6 dual-layer system is feasible with a data storage capacity of approximately two times 125GB or a quarter-Terabyte in total. Depending on the developments with quadruple-layered BD, one might even consider NA = 1.6 systems with four layers resulting in capacities of approximately half a TeraByte. There is sufficient theoretical evidence to show that optics at least allows such a system. The Table further shows that with an NA = 1.9 system it should be possible to achieve a data storage capacity between 125 and 175GB on a CD-size disc. Therefore, as already mentioned in the previous Section, an NA = 1.6 dual-layer optical disc system will likely beat the higher-NA single-layer first-surface optical disc systems in terms of data storage capacity per disc.

4.3.2 Solid immersion lens designs

The lens designs of two types based on a solid immersion lens with a numerical aperture larger than unity are discussed for example. Both types were designed for use with a near-field optical recording system. The first type is an NA = 1.9 lens that focuses through an air gap onto the top surface of an optical disc. The second type is an NA = 1.45 lens that focuses through an air gap and a 3.0μm polymer cover-layer on the data-layer of an optical disc. An overview is presented of high refractive index materials that are transparent for blue light and that have been considered for use as a SIL. It is shown how a phase structure was applied to compensate for the chromatic aberration of the high NA lenses, including the opto-mechanical assembly and the phase structure in the focus and tracking actuator of near-field optical recording set-up. The session is concluded with considerations for dual-layer near-field optical recording.

- The aplanatic solid immersion lens

Aplanatic points: A pencil of focused rays that is refracted at the spherical surface of a truncated sphere of refractive index n_s in a homogenous optical medium of refractive index $n < n_s$ is shown in Figure 4-33. The lens has a thickness L and radius of curvature R. The marginal ray is focused at angle α_m with respect to the optical axis. Refraction at the surface of the sphere causes spherical aberration of the wavefront which strongly depends on the thickness L. In a third order approximation for a pencil of rays, the Seidel spherical aberration coefficient of the wavefront

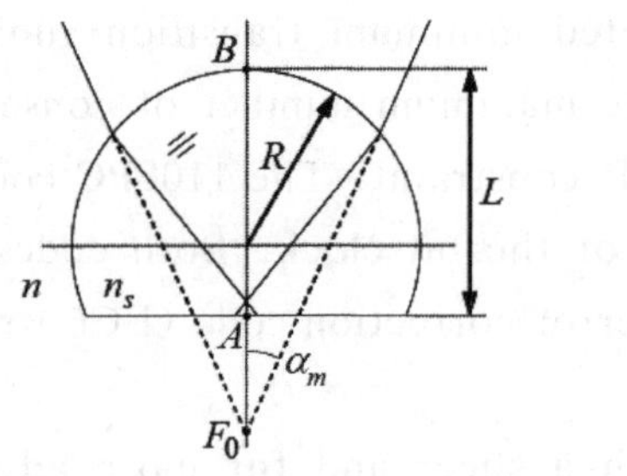

Fig. 4-33 Illustration and notation used for the analysis of the wavefront aberration of a pencil of rays that is focused at the flat surface of a truncated sphere.

aberration in units $[\lambda_0]$ of the wavelength is given by:

$$W_{040} = \frac{1}{8\lambda_0}(\sin\alpha_m)^4\left(\frac{n_sR}{n_sR - L(n_s - n)}\right)^3 \frac{L}{n_s}\left(\frac{L}{R} - 1\right)^2 \times \left\{\frac{L(n_s - n)}{n_s^3 R}\left(\frac{n_sR}{n_sR - L(n_s - n)}\right) + \frac{1}{n_s^2} - \frac{1}{n^2}\right\} \quad (4\text{-}105)$$

Systems with large angular semi-aperture are not within the domain of validity of the approximation that leads to Eq. (4-105) even though the angles of incidence on the first surface are relative small in the neighbourhood of $L = R$. Numerical results with this equation therefore significantly underestimate the amount of spherical aberration for NA >1 SIL systems. Nevertheless, some fundamental insights can be derived from Eq. (4-105) concerning the two types of solid immersion lenses used in this research and the effect of thickness errors of these lenses.

In the Figure 4-34, the Seidel spherical aberration coefficient W_{040} is plotted as a function of the normalised thickness L/R for a system in air with $\lambda_0 = 405\text{nm}$ and with (small) sin $\alpha_m = 0.2$, $n_s = 2.086$ of LaSF_{35} glass and for $R = 0.5\text{mm}$ and $R = 1\text{mm}$. From Eq. (4-105) and the plot, three solutions are found for the lens thickness L at which the focus in the lens is free of spherical aberration. These solutions are $L_0 = 0$, $L_1 = R$ and $L_2 = R(n + n_s)/n_s$. The first two solutions L_0 and L_1 are trivial and correspond to focusing at a point on the sphere and at the centre of curvature of a hemisphere, respectively. The third solution with L_2 is less trivial and corresponds to focusing in a super-hemisphere. From a further third order analysis, that can even by generalised to arbitrary finite angles, it follows that a point imaged near the optical with the spherical aberration free imaging, makes such lenses aplanatic.

Eq. (4-105) implies significantly different thickness tolerances of a hemispherical and super-hemispherical lens. At $L_1 = R$ the derivative of Eq. (4-105) is zero and thus a relatively large tolerance may be expected on the thickness of a hemispherical lens. At $L_2 = R(n + n_s)/n_s$ the curve crosses zero and the derivative of Eq. (4-105) is non-zero and thus a much smaller thickness tolerance is expected for a super-hemispherical lens.

Manufacturing a super-hemispherical solid immersion lens with NA >1 within the thickness tolerance for a well-diffraction limited lens can therefore be a significant challenge. It is not difficult to show that the derivative $\mathrm{d}W_{040}/\mathrm{d}L$ does not depend on R at L_2 and thus the thickness tolerance is independent of the radius of the lens.

Despite the approximate nature of Eq. (4-105), it can easily be shown that the solutions for the zeros at L_0, L_1 and L_2 are also exact for arbitrarily large angular semi-aperture. This is self-evident for the trivial solutions $L_0 = 0$ and $L_1 = R$. The solution $L_2 = R(n + n_s)/n_s$ and the associated position of the virtual focus can be found without the need for approximation by a geometrical construction. For reasons of completeness, this construction as set out by Born and Wolf is included here with some practical remarks about the NA increase with a SIL, the maximum attainable NA and the angle of the marginal ray with respect to the surface normal of the sphere. Figure 4-35 depicts a sphere S of refractive index n_s and radius of curvature R in a homogenous medium of refractive index $n < n_s$. Let S_0 be a sphere with radius $R_0 = n_s/n_R$ and S_1 a sphere with radius $R_1 = n/n_{sR}$ and let both spheres be concentric to the sphere of the lens. Point F_0 is the intersection with sphere S_0 of a line AB along a general ray that intersects the spherical lens surface at point B. The refracted ray intersects the smaller sphere S_1 in point F_1. Both triangles OBF_0 and OF_1B contain angle $F_0OB = F_1OB$ and have equal ratios of the sides that intersect in O:

$$\frac{OB}{OF_0} = \frac{OF_1}{OB} = \frac{n}{n_s} \tag{4-106}$$

which implies that these triangles are similar. This similarity further implies that

$$\frac{\sin\theta_1}{\sin\theta_0} = \frac{OB}{OF_0} = \frac{n_s}{n} \tag{4-107}$$

which is Snell's law with θ_0 the angle of incidence on the sphere and θ_1 the refracted angle. Figure 4-35 proofs that BF_1 is indeed the refracted ray. This shows that all rays that intersect the sphere S and that are virtually focused at F_0 will intersect sphere S_1 at the same point F_1. For the distance VF_1 we obtain $VF_1 = R(n+n_s)/n_s$, or a lens thickness equal to the zero in Eq. (4-101) at $L_2 = R(n+n_s)/n_s$. For OF_0 obtain n_sR/n. The distances VF_1 and OF_0 represent the conditions for aberration free focusing in a super-hemisphere. If assume that OF_0 is the optical axis of our focusing system than F_1 is a point on that axis. It find for the relation between the angle of incidence θ_0 on the sphere S and the angle α with respect to optical axis:

$$\frac{\sin\theta_0}{\sin\alpha} = \frac{F_0O}{BO} = \frac{n_s}{n} \tag{4-108}$$

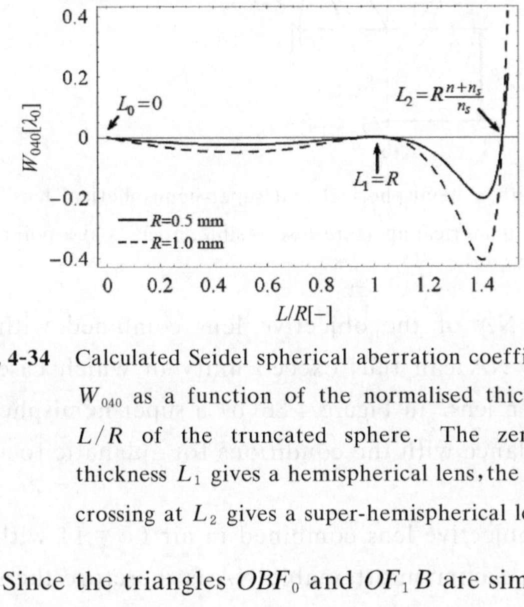

Fig. 4-34 Calculated Seidel spherical aberration coefficient W_{040} as a function of the normalised thickness L/R of the truncated sphere. The zero at thickness L_1 gives a hemispherical lens, the zero-crossing at L_2 gives a super-hemispherical lens.

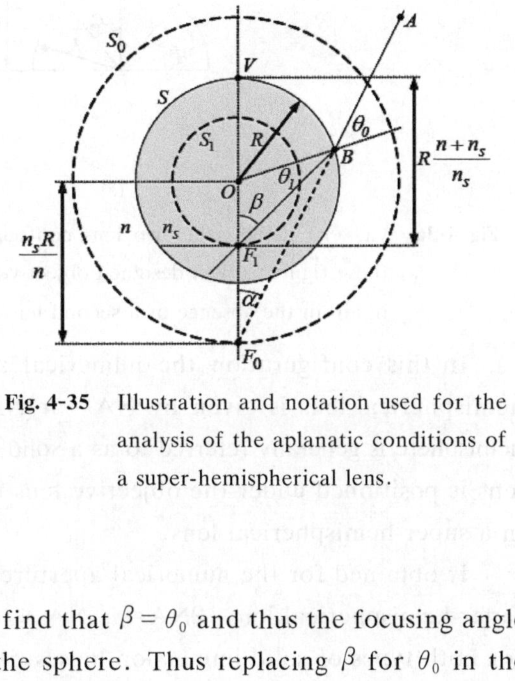

Fig. 4-35 Illustration and notation used for the analysis of the aplanatic conditions of a super-hemispherical lens.

Since the triangles OBF_0 and OF_1B are similar, we find that $\beta = \theta_0$ and thus the focusing angle β in the lens is equal to the angle of incidence θ_0 on the sphere. Thus replacing β for θ_0 in the above equation can obtain

$$\frac{\sin\beta}{\sin\alpha} = \frac{n_s}{n} \tag{4-109}$$

where remark that only rays focused with $\sin\alpha \leqslant n/n_s$ intersect sphere S, which results in an upper limit $\sin\beta = 1$. Thus, a focusing system of angular semiaperture sinam that focuses a pencil of rays at F_0 results in a numerical aperture NA inside the super-hemispherical lens that is given by refractive index for the sphere ($n_s \approx 2$) one can attain very large values for the NA. Since $\beta = \theta_0$, the marginal ray in such a system will intersect the sphere at a large angle with respect to the surface normal.

$$\mathrm{NA} = \frac{n_s^2}{n}\sin\alpha_m \leqslant n_s \tag{4-110}$$

Therefore, unless a wide-angle anti-reflection coating is deposited on the high refractive index sphere, a lot of light will be lost due to reflection at the sphere which would reduce the effective NA. For example, a focusing system with NA = 1.9 with a super-hemispherical lens of refractive index $n_s = 2$ yields a maximum angle of incidence on the sphere of approximately 72° with respect to the surface normal. At this angle the intensity reflectivity for TM-polarised light is only 3% but for TE-polarised light it is as much as 49%. Finally, placed under the objective lens such that its centre of curvature coincides with focal point F_0.

In Figure 4-36(a) a hemispherical lens is Figure 4-36(a)-(b) Lens configurations in which an objective lens with a numerical aperture in air of $\mathrm{NA}_0 = \sin\alpha_m$ focuses in two types of aplanatic solid immersion lenses (a) a hemispherical SIL resulting in a numerical aperture $\mathrm{NA} = n_s\mathrm{NA}_0$ and (b) a super-hemispherical SIL resulting in a numerical aperture $2\mathrm{NA} = n_s\,\mathrm{NA}_0 \leqslant n_s$.

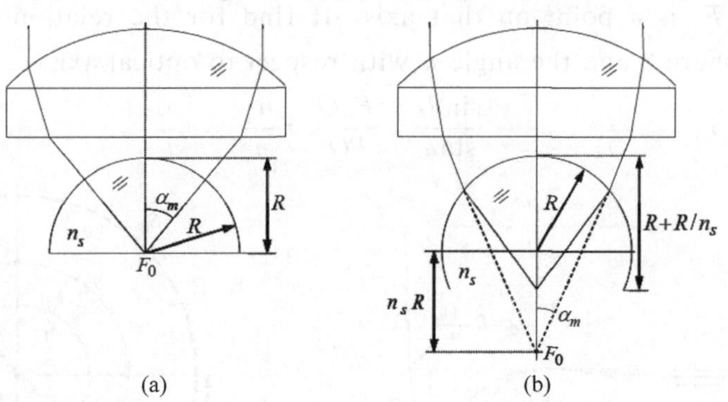

Fig. 4-36 (a)-(b) depicts the two lens configurations with a hemispherical and super-hemispherical lens. In these figures a well-designed objective lens with numerical aperture $\mathrm{NA}_0 = \sin\alpha_m$ focuses at a point F_0 in air, in the absence of a second lens.

In this configuration the numerical aperture NA of the objective lens combined with the hemispherical lens is given by $\mathrm{NA} = n_s^2\mathrm{NA}_0$. The NA can thus exceed unity in which case the hemisphere is generally referred to as a solid immersion lens. In Figure 4-36(b) a super-hemispherical lens is positioned under the objective lens in accordance with the conditions for aplanatic focusing in a super-hemispherical lens.

It obtained for the numerical aperture of the objective lens combined in air ($n = 1$) with the super-hemispherical lens: $2\mathrm{NA} = n_s\mathrm{NA}_0 \leqslant n_s$. The maximum attainable NA is therefore the same for both types of solid immersion lenses, although, in practice, very large NA values near ns are on only attained with the super-hemispherical lens due to the smaller NA_0 that is required.

Sine condition previously, it follows from third order analysis that a point imaged near the optical axis is also free of coma for the solutions $L_0 = 0$, $L_1 = R$ and $L_2 = R(n + n_s)/n_s$. Because of the point-symmetry with respect to O in Figure 4-35, the two dashed circles are perfect conjugate surfaces, thus implying that he super-hemisphere obeys Abbe's sine condition which further implies coma-free imaging in a region near the optical axis. The combination of a well-designed objective lens that focuses inside a super-hemispherical lens, yields again a focusing system that obeys Abbe's sine condition. This is shown with the aid of Figure 4-37. In the figure EP is the entrance pupil of a well-designed objective lens with angular semi-aperture $\sin\alpha_m$. A ray at height h_0 above and parallel to the optical axis defines the focal length f_0 of the objective lens via $h_0 = f_0\sin\alpha$. According to the sine condition each ray parallel to the optical axis in the entrance pupil, intersects its refracted ray AF_0 at a point A on a sphere of radius f_0 centred at F_0. Since in the super-hemisphere $\sin\beta = n_s/n\sin\alpha$, see Eq. (4-105), the same rays that are incident on the entrance pupil intersect their refracted rays BF_1 at point C on a sphere of radius $f_1 = n/n_s f_0$ centred at F_1. Thus find the relations for a ray's height h_0 above the optical axis in the entrance pupil and the angles of the focused rays $h_0 = f_0 \sin\alpha = f_1\sin\beta = n/n_s f_0\sin\beta$, as required by the sine condition.

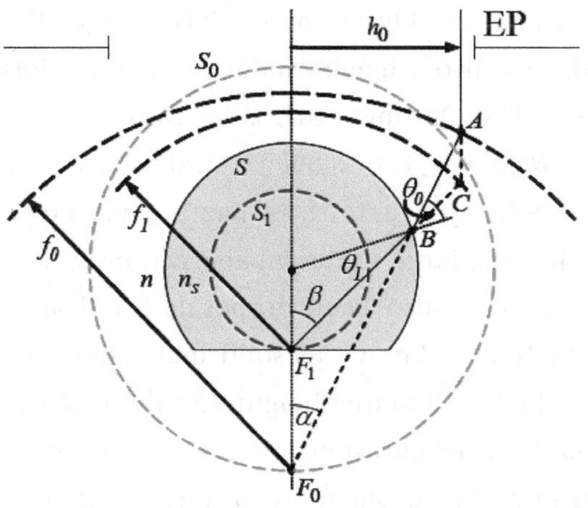

Fig. 4-37 Illustration and notation that shows that the super-hemispherical SIL obeys the sine condition for focusing in an aplanatic super-hemisphere.

Progress in immersion optics The solution for the aplanatic super-hemispherical lens goes back to the 17th century to the French scientist Gilles de Roberval and contemporary Dutch scientist Christiaan Huygens in 1690. While discussing earlier results by Descartes, Huygens remarks that a perfectly imaging refractive surface in his (two-dimensional) problem becomes a perfect circle when the ratio of the geometrical distances F_0V and F_1V (see Figure 4-35) is equal to the ratio of the refractive indices n_s and n. This condition describes the super-hemisphere solution. To (admittedly, mostly Dutch) optical designers the super-hemisphere solution is therefore known as a Huygens aplanat. In other scientific communities, the super-hemisphere is also known as a Weierstrass optic after the 19th century German mathematician. Both names, however, are rarely used in the optical recording research community that sticks to the use of the terms super-hemisphere and sometimes hyper-hemisphere to describe a lens with thickness $L_2 = R(n + n_s)/n_s$. In the first half of the 19th century, Amici introduced the super-hemispherical lens as a solution to increase the numerical aperture of microscope objective lenses. Somewhat later, the Amici objective lens became the basis of immersion microscopy where an index-matching oil between the super-hemispherical lens and a specimen is used to increase the numerical aperture to values larger than unity. The solid immersion lens was first introduced by Mansfield and Kino in 1990 for use in microscopy and they also suggested its use in optical data storage. In 1994 Terris et al. reported first results of near-field optical data storage with a SIL on first-surface magneto-optical media. In 2002 Van Santen and Neijzen reported first results with liquid immersion mastering for optical data storage. Liquid immersion is now also at the heart of nano-lithographic systems such as the water immersion tool from ASML with NA = 1.35 at wavelength $\lambda_0 = 193\text{nm}$.

In 2004 discussed a lens design for a near-field optical recording system with a SIL that focuses through a protective cover-layer of an optical disc, the same year reported results of first prototypes of such lens with NA = 1.45 for a proof of principle experiment. In 2005 showed first read-out results with this lens and in 2006 first recording experiments were reported. The Optimum SIL, which is a super-hemispherical SIL with a thickness that deviates from the Huygens aplanat. The Optimum SIL has a larger thickness tolerance than the aplanatic super-hemispherical SIL, but also a much smaller decenter tolerance for the Optimum SIL with respect to the objective lens. The Optimum SIL is based on the local minimum of Eq. (2-101) in the thickness range $R < L < R(n+n_s)/n_s$. Such a design results in an increase of the numerical aperture of the objective lens NA_0 by a factor less than $2n_s$ but larger than n_s. The reason for the small de-center tolerance is that the large spherical aberration in the local minimum is compensated for in the objective lens. There have also been proposals for designs and manufacturing methods for catadioptric systems with NA>1 known as solid immersion mirrors (SIMs). The main advantages claimed for a SIM are the low building height and the fact that a SIM can be a single optical element, rather than the doublet configurations that are common for SIL systems. Some of these papers report on SIM manufacturing, but no recording results on optical discs have yet been reported.

- High refractive index glasses and crystals

Table 4-3 lists the refractive indices at wavelength $\lambda_0 = 405$nm for a number of transparent high-refractive index glasses and crystals from various suppliers. The crystals ZnS, diamond and ZrO_2 are sometimes used as lens elements for special applications. Until recently, this was not the case with $Bi_4Ge_3O_{12}$ and $KTaO_3$, which were first suggested by Shinoda et al. for use as a high-NA SIL. It is worth noting that a number of older high refractive index glasses have been replaced by more environment-friendly glasses with the same or very similar optical properties. This has for example been the case with $LaSF_9$ which has been replaced by $LaSFN_9$ and the highest refractive index glass $LaSF_{35}$, which has been replaced by N-$LaSF_{35}$. Crystal suppliers are not included in the Table because there are several of them and the crystal names are not unique to one supplier. For the crystals, only the refractive index of the cubic crystallographic structure is listed. Cubic crystals have isotropic refractive index, which is a prerequisite for use as a low-cost SIL. It must also be noted that the pure cubic zirconia crystal is unstable at room temperature and is therefore stabilised with Y_2O_3, MgO or CaO.

Table 4-3 Refractive indices *n* of transparent high-index glasses and cubic crystals at wavelength $\lambda_0 = 405$nm.

Glass type	Supplier	*n*	Crystal	*n*
$LaSFN_9$	Schott	1.898	$Bi_4Ge_3O_{12}$	2.213
SF_{57}	Schott	1.913	ZrO_2	2.220
S-TIH_{53}	Ohara	1.914	$KTaO_3$	2.382
SFL_{57}	Schott	1.915	Diamond	2.462
N-$LaSF_{31}$	Schott	1.918	ZnS	2.545
E73-38	Corning	1.919		
SLAH-58	Ohara	1.921		
$LaSFN_{22}$	Sumita	1.946		
SF_{66}	Schott	2.007		
SLAH-79	Ohara	2.068		
N-$LaSF_{35}$	Schott	2.086		

The value in the Table is for ZrO_2 with 12mol% Y_2O_3. Lens elements are not the main application of the crystals listed in the Table, with the exception perhaps of ZnS. Bismuth germinate $Bi_4Ge_3O_{12}$, for example, is best known as a scintillator material and cubic zirconia is best known as a low-cost substitute for diamond in jewellery. Potassium tantalate, $KTaO_3$, is a ferroelectric material that is used in material science for the semiconductor industry and in non linear optics research. ZnS is the only crystal in the list that is frequently used in lenses. ZnS produced by chemical vapour deposition (CVD) is commercialised under the name Cleartran and is mostly used for infra-red applications. Cleartran, however, is polycrystalline which results in significant scattering of blue light. Besides the crystals listed in the Table, there exist other (mostly infrared) materials with even larger refractive indices. GaP is a relevant example of such material. GaP has a refractive index of 3.295 at wavelength $\lambda_0 = 653$nm (near that of DVD) but is not transparent for blue light. Interestingly, the ratio of the wavelength to refractive index for GaP at $\lambda_0 = 653$nm is almost equal to that of N-LaSF$_{35}$ at $\lambda_0 = 405$nm. This makes GaP an interesting material for use in a SIL since with a given NA_0 objective lens, a hemispherical GaP SIL focuses red light to the same spot size as an N-LaSF$_{35}$ hemispherical SIL with blue light. GaP has indeed been used in solid immersion lenses. Lenses were also manufactured from the other crystals in the Table. In this respect, Shinoda et al. have been most active. In their publications, they reported use of $Bi_4Ge_3O_{12}$ in a super-hemispherical NA=2.05 SIL, diamond in a super-hemispherical NA=2.34 SIL and $KTaO_3$ in a super-hemispherical NA = 2.20 SIL. Nakaoki et al. used $KTaO_3$ in a hemispherical SIL with NA=1.84, which is an extremely large value for a hemispherical SIL with the Lens-grade crystalline exception of $KTaO_3$ material in the Table 4-3 needs to be monocrystalline and should be free of impurities, striae or growth defects that cause stress-birefringence that is very difficult. Another significant difficulty with crystals is their inherent hardness anisotropy which makes it difficult to polish such crystals into nearly perfect spheres. This effect is most severe with polished cubic zirconia and diamond ball lenses that acquired commercially.

4.3.3 Lens design with NA=1.9 for first surface recording

A NA=1.9 lens was designed for use with first-surface optical discs in which the datalayer is on the top surface of the disc, unprotected by a protective cover-layer. The absence of such cover-layer makes this lens design relatively simple since no spherical aberration correction is required. The lens design was optimised for $\lambda_0 = 405$nm using the optical design program Zemax. This program is based on ray-tracing and could therefore not directly correct the lens design for phase effects due to the evanescent coupling or the multiple reflections in the stack of an optical disc. The total RMS value of the phase-only correction for the phasechange medium of our experiments is less than 15mλ for an air gap smaller than 30nm. In the same Section it was shown that the air gap should be well below 30nm. The phase effects due to the air gap and multi-layer structure are therefore small enough that they can safely be ignored in the Zemax design of the NA=1.9 lens. Moreover, a 15mλ RMS limit was used for the tolerancing in the Zemax design. By ignoring the air gap and multi-layer structure in the Zemax design a wavefront aberration is introduced that is comparable to the aberrations due to the mechanical tolerances of the lens design.

- Doublet of a plano-aspherical lens and a super-hemispherical solid immersion lens

The NA = 1.9 doublet lens design consists of an $NA_0 = 0.45$ plano-aspherical lens and an aplanatic super-hemispherical SIL as shown in Figure 4-38. The glass of the SIL is Schott LaSF$_{35}$ with $n_s = 2.086$. Ideally, the assembled doublet lens would result in $2NA = n_s NA_0 = 1.96$. However a small obscuration in the lens holder, designed to centre the SIL, limited the numerical

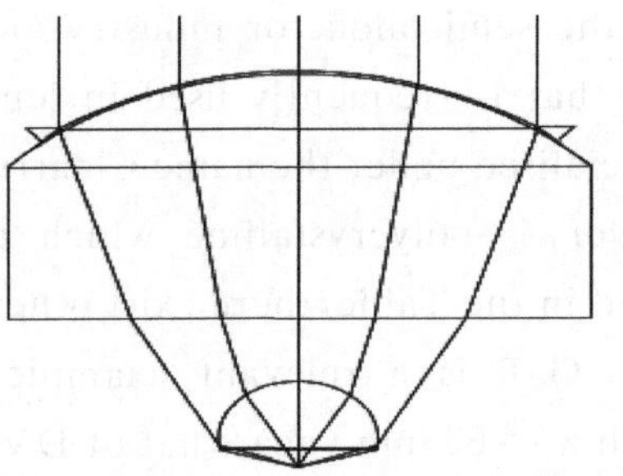

Fig. 4-38 Schematic of the design the NA = 1.9 lens with a $LaSF_{35}$ super-hemispherical solid immersion lens. The lower surface of the SIL-element has a truncated conical shape.

aperture to a value of 1.9. The plano-aspherical lens and the $LaSF_{35}$ spheres were manufactured in Eindhoven, Netherlands. All other grinding and polishing steps and the doublet assembling were done at laboratory. The plano-aspherical lens consists of a plano-spherical glass body on which a UV curable resin is replicated from an aspherical mould to form a thin aspherical layer. With this aspherical layer the lens is designed to focus free of aberration at an on-axis point. The outer dimensions of the lens were designed to fit in a commercial DVD focus and tracking actuator. The free entrance pupil diameter of the plano-aspherical lens is 3.0mm. The mechanical diameter of the planoaspherical lens is 3.65mm and the free working distance without the SIL was 2.084mm. The effective focal length of the focusing system with the SIL is 0.767mm. In the initial design the diameter of the $LaSF_{35}$ spheres was exactly 1mm which results in a thickness of $R(1+n_s)/n_s = 739.7\mu m$ for the SIL in air. The smallest tolerance of the design is on this SIL thickness. Assuming for the wavefront aberration a 15mλ RMS tolerancing limit, the SIL thickness tolerance is only 0.2μm. Taking into account a practical tolerance on the SIL diameter of 1μm and a tolerance in the refractive index of 10^{-4}, that this 0.2μm tolerance cannot be achieved in a single polishing step.

Moreover, for this type of SIL shape and material, the current state of the art polishing techniques are insufficient to achieve a 0.2μm tolerance in a single step. To solve this problem, interferometric wavefront analysis was used in a SIL thickness optimisation and lens assembling process. Given a 15mλ RMS tolerancing limit for the wavefront aberration, the doublet lens design has a field of 0.5° in the entrance pupil and a field diameter of 13.5μm in the image space, where allowed for defocus correction by a small change of the conjugate distance. The decentring tolerance for the optical axes of the two lens elements is also large and is in our design not limited by wavefront aberrations. For a 15mλ RMS tolerancing limit, the decentring tolerance would be 36μm. However at such large decentre, the cone of focused light rotates significantly in the meridional plane resulting in a large asymmetry of the focused cone with respect to the optical axis. At a 36μm decentre of the lens elements, the marginal rays on either side of the optical axis are focused at angles of 60° and 75° with respect to the optical axis, respectively. Moreover, since our lens holder obscures a part of the focused cone of light at this large decentre distance, the decentring tolerance was set at 10μm. The part of the SIL facing the optical disc was designed to be partly conical with a small flat tip of 40μm diameter through which the light is focused onto the disc. This shape was chosen in order to obtain a sufficiently large margin for disc tilt with respect to the flat surface of the SIL at an air gap smaller than 30nm. The conical surface also provides some area to apply a UV curable adhesive during assembling of the SIL in an aluminium lens holder. The spherical surface of the SIL was anti-reflection coated with a single SiO_2 layer which reduced the reflectivity of the marginal ray from 41% without coating to 5% with coating for TE-polarised light. Both sides of the plano-aspherical lens were anti-reflection coated with a reflectivity $<0.5\%$ in the wavelength range $400\text{nm} \leqslant \lambda_0 \leqslant 410\text{nm}$.

- Chromatic aberration correction

The GaN semiconductor laser diode of the same type is also used in Blu-ray Disc recording systems. These lasers emit light of a wavelength in the range $400\text{nm} \leqslant \lambda_0 \leqslant 410\text{nm}$. The wavelength depends on the laser output power and shifts when the output power jumps from a low power level for data reading to a high power level during recording of data. This power-dependent wavelength shift is caused by two phenomena. The dominant effect is a change of the refractive index of the laser cavity due to an increase of the carrier density. A secondary and slower effect is the change of the refractive index of the cavity due to the increased temperature at higher output powers. The main effect of a small wavelength shift with an objective lens is a shift of the position of the focused spot. Since the rise time of the laser pulses is of the order of a nanosecond this wavelength shift is almost instantaneous and much faster than what a closed loop BD focus servo can follow. The bandwidth of such servo systems is typically a few kHz and therefore it takes a few tenths of a millisecond to settle to a new optimum focus position relative to the data-layer in response to a transition from read to write mode. During these few tenths of a millisecond the spot becomes defocused on the data-layer of the disc. The result of this is that the first few thousand marks of a write cycle may be poorly recorded in the data-layer.

Hendriks et al. found that although the wavelength shift is small, of the order of 1nm, it negatively affects the first marks that are recorded on a BD during these few tenths of a millisecond with an NA = 0.85 objective lens without chromatic correction. Such poorly written marks result in a reduced read-out signal modulation with central aperture detection. Without correction for chromatic aberration the NA = 1.9 lens design exhibits a shift of the focused spot position relative to the exit surface of the SIL of 0.24μm per 1nm wavelength shift. This corresponds to a Zernike defocus coefficient $A_{2,0} = 206\text{m}\lambda$ RMS. For spherical aberration we find $A_{4,0} = 11\text{m}\lambda$ RMS, higher orders are negligible. For small wavelength shifts of the order of a few nanometres the chromatism of the NA = 1.9 lens is linear and approximately 3.5 times larger than with the NA = 0.85 objective lens of Hendriks et al. A non-periodic phase structure (NPS) was designed to correct the chromatic aberration of our lens. An NPS is a simple macroscopic phase structure that approximates the required chromatic aberration correction with a number of discontinuous rings, see Figure 4-39. An NPS should not be mistaken for a diffractive optical element. The principle of chromatic aberration correction with an NPS is based on small changes of the optical thicknesses of the rings of the NPS for a small change in the wavelength. An NPS can be mass-produced as a low-cost injection moulded, usually concave, plastic structure. The design of the NPS follows the method and the NPS consists of annular zones that are concentric to a central zone. Each zone j has a step height h_j that is chosen to give rise to a phase step that is exactly an integer m_j multiple of 2π, h_j is given by

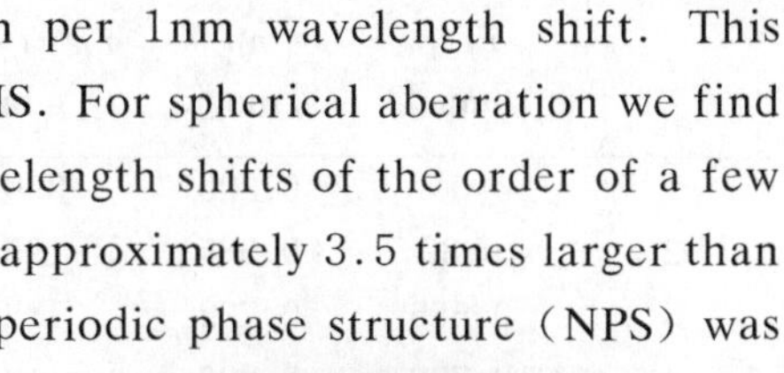

Fig. 4-39 Rendering of the NPS that compensates the chromatic aberration of our NA = 1.9 lens. The diameter of the optical surface is 1.6mm.

$$h_j = \frac{m_j \lambda_0}{n-1} \tag{4-111}$$

Where n is the refractive index of the NPS material at the nominal wavelength λ_0. When the wavelength shifts by $\Delta\lambda$ with respect to λ_0, the refractive index changes by Δn due to dispersion. Consequently, the step heights no longer give rise to phase an integer multiple of 2π. This results in a stepped wavefront that can be designed to approximately compensate the chromatic aberration of the lens. This phase step $\Delta\Phi_j$ for zone j, relative to the central zone, is in lowest order in $\Delta\lambda$ and Δn given by

$$\Delta\Phi_j = -2\pi m_j \left(\frac{\Delta\lambda}{\lambda_0} - \frac{\Delta n}{n-1}\right) \tag{4-112}$$

The zone radii and m_j values are chosen to compensate the chromatic aberration as calculated by Zemax. The nominal wavelength of the recorder set-up from the centre of the spectrum emitted by the laser is just above the laser current threshold, for the nominal wavelength $\lambda_0 = 408.3\text{nm}$. The spectrum is strongly multi-mode with a total width of the spectrum of approximately 0.3nm. The previously mentioned Zernike coefficients for the chromatic aberration are for the lens design that is fine-tuned for exactly $\lambda_0 = 408.3\text{nm}$. The material for the NPS prototypes was polymethyl methacrylate (PMMA), which can be diamond turned on a high-accuracy lathe. At the nominal wavelength, the refractive index of PMMA is $n = 1.546934$ with $\Delta n = 1.546 \times 10^{-4}$ for a 1nm wavelength shift. With 15 annular zones, the residual chromatic aberration for a 1nm wavelength shift is reduced to 50mλ peakto-peak in each zone. Table 4-4 lists the radii and step heights for the NPS design.

Table 4-4 Design of the NPS for the NA=1.9 lens.

j	r_{j-1} [mm]	r_j [mm]	$r_j - r_{j-1}$ [μm]	h_j [μm]	j	r_{j-1} [mm]	r_j [mm]	$r_j - r_{j-1}$ [μm]	h_j [μm]
1	0	0.390	390	0	9	1.159	1.222	63	122.633
2	0.39	0.585	195	15.329	10	1.222	1.280	58	137.962
3	0.585	0.723	138	30.658	11	1.280	1.333	53	153.291
4	0.723	0.835	112	45.987	12	1.333	1.384	51	168.620
5	0.835	0.932	97	61.316	13	1.384	1.432	48	183.949
6	0.932	1.015	83	76.646	14	1.432	1.478	46	199.278
7	1.015	1.091	76	91.975	15	1.478	1.522	44	214.607
8	1.091	1.159	68	107.304	16	1.522	1.600	78	229.940

The total step height is almost 230μm, which is much larger than the NPS. The free pupil diameter was chosen to be somewhat larger than the 3.0mm of the lens. This allowed for some decentre in initial experiments where the NPS was positioned in the recorder near the lens that was mounted in a moving focus and tracking actuator. Ultimately the NPS was mounted together with the lens on the moving body of the focus and tracking actuator. Figure 4-40 shows three calculated cross-sections of chromatic wavefront aberrations for a 1nm wavelength shift with respect to the nominal wavelength. The first cross-section is that of the uncorrected lens which shows the large defocus of $A_{2,0} = 206\text{m}\lambda$ RMS and the small spherical aberration $A_{4,0} = 11\text{m}\lambda$ RMS. The second cross-section is for the NPS without the lens and shows the stepped wavefront. The third cross section is the residual chromatic aberration of the lens with the NPS, i.e. the sum of the previous two cross-sections. The residual chromatic aberration has 50mλ peak-to-peak at the transitions between the zones.

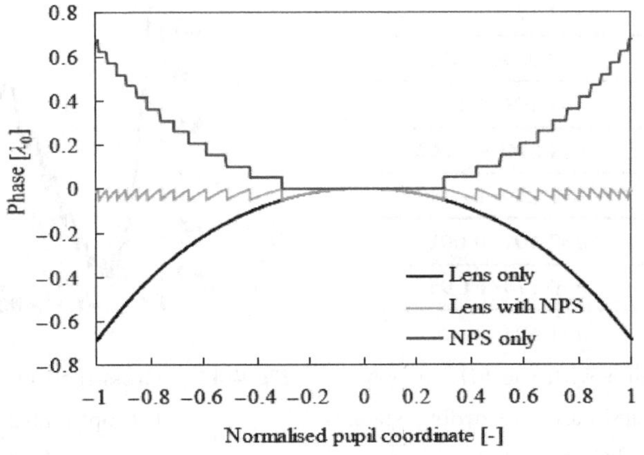

Fig. 4-40 Chromatic wavefront aberration for a wavelength shift of 1nm with respect to the nominal wavelength $\lambda_0 = 408.3$nm. Cross sections are shown for the NA = 1.9 lens only, the lens with the NPS and the NPS only. The pupil coordinate is normalised to unity at a radius corresponding to NA = 1.9.

4.3.4 Air gap dependence of the spot size for practical optical discs

The calculation results of focused spot profiles that are the result of focusing with a SIL in two types of practical near-field optical discs. The maximum air gap height is that can expect in experimental system for a certain tolerance on the broadening of the FWHM of the central peak of the spot relative to the case in which the SIL and the disc are in contact.

- CuSi write-once medium with a 3μm cover-layer and an NA = 1.45 solid immersion lens

The calculation results of the size of a spot that is focused with an NA = 1.45 SIL through an air gap into a write-once optical disc that is protected with a 3.0μm thick polymer cover-layer. The optical parameters of the disc and the focusing system in these calculations are identical to those used in the experiments. The wavelength λ_0 of the light is 405nm, the NA = 1.45 SIL is made of $LaSF_{35}$ glass with $n_{SIL} = 2.086$ and the write-once optical disc consists of a Cu/Si bi-layer that is sandwiched between SiN anti-corrosion layers and $ZnSSiO_2$ layers for optimised recording and read-out performance. An Ag-alloy is used as a mirror and heat sink and the multilayer structure is deposited on a polycarbonate substrate. The structure is schematically shown in Figure 4-41. The recording mechanism of this write-once medium is based on thermal CuSi inter-diffusion at high recording temperatures during high-power laser write pulses. The complex refractive index of the CuSi alloy is different from that of the individual Cu and Si layers and thereby creates optical contrast with the non-written areas.

The NA = 1.45 focusing system is part of the optical geometry as depicted in Figure 4-43(b) illuminated with a circularly polarised plane wave and the system has perfect phaseonly correction. Calculated the spot profiles were at air gap heights of 0nm, 40nm, 80nm, 120nm and 160nm. For each calculation, the undisturbed focus position is chosen such that the position of minimum phase correction in the Si layer coincides with the interface between the Si and the Cu layer. For example, with a 40nm air gap, the corresponding undisturbed focus position is located in the substrate at a distance of 4.855μm from the air-cover interface.

SIL		n=2.086
Air	0..160nm	n=1
Cover	3μm	n=1.58
$ZnSSiO_2$	25nm	n=2.28+i 0.001
SiN	5nm	n=2.10+i 0.001
Si	4nm	n=4.809+i 2.266
Cu	4.5nm	n=1.259+i 2.173
SiN	5nm	n=2.10+i 0.001
Ag	50nm	n=0.174+i 1.95
PC		n=1.62

Fig. 4-41 Optical configuration with the SIL, air gap, cover-layer and CuSi-based recording stack on a polycarbonate (PC) optical disc substrate.

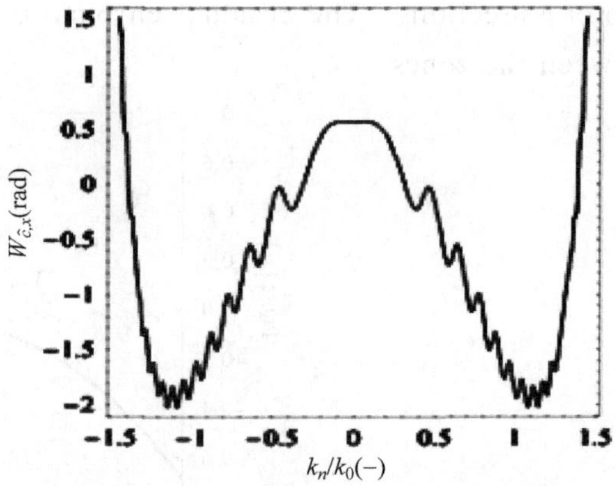

Fig. 4-42 Cross-section of the correction wavefront for optimal focusing through a 40nm air gap into the CuSi-based recording stack of Figure 4-41.

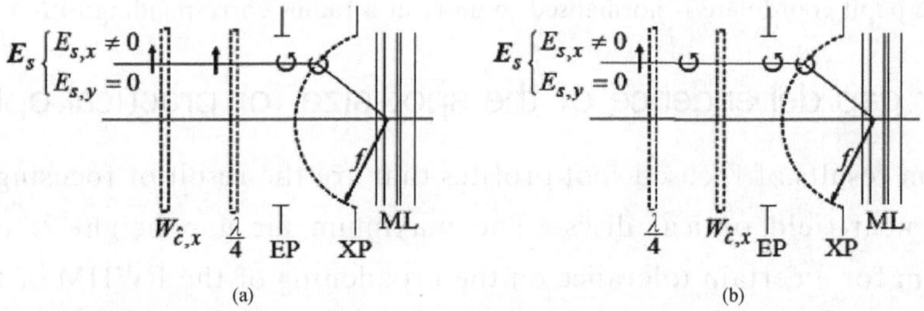

Fig. 4-43 (a)-(b) Two implementations of the simplified phase-only correction for circularly polarised light. EP indicates the entrance pupil, XP the exit pupil, ML the multilayer structure, f the focal length and $W_{\hat{c},x}$ a phase correcting optical element. In (a) a plane wave first passes through the phase correcting element before passing through the quarter-wave plate, (b) the same phase correcting element can be placed between the quarter-wave plate and the entrance pupil or even in the entrance pupil as part of a first aspherical element when either $E_{s,x}$ or $E_{s,y}$ is equal to zero and the fast axis of the quarter-wave plate is oriented at 45° with respect to the x-axis.

Due to the circular polarisation, the applied phase correction has rotation symmetry with respect to the optical axis. Figure 4-43 shows a cross-section through the wavefront of the phase correction for a 40nm air gap. The wavefront can be expanded into a sum of Zernike polynomials with the RMS Zernike coefficients $A_{2,0} = -8.2\text{m}\lambda$, $A_{4,0} = 130.0\text{m}\lambda$, $A_{6,0} = 36.7\text{m}\lambda$, $A_{8,0} = 13.0\text{m}\lambda$ and small higher order coefficients, the total RMS of the wave front is 136.0mλ. The system focuses through a 3.0μm cover-layer which gives rise to the large spherical aberration. We found a value of 40.8mλ per micrometre thickness of the cover-layer, when focusing with the NA = 1.45 lens in a cover-layer with $n = 1.58$ with light of 405nm wavelength. A single pass through the 3.0μm cover-layer therefore results in $A_{4,0} = 122.4\text{m}\lambda$. To conclude the multiple reflections, the air gap and the CuSi stack together add only a small part to the low order spherical aberration coefficients. Multiple reflections in the thick cover-layer are also the cause of the high frequency oscillation in the phase profile of Figure 4-44. A thinner cover-layer results in fewer oscillations. In Figure 4-44(a)-(b) show the computed plots of the normalised (time-averaged) total energy density profiles for the air gaps heights of 0, 40nm, 80nm, 120nm and 160nm. In Figure 4-44(a) the profiles are normalised to unity to show the relative broadening of the central peak with increasing air gap height. Note that, despite the high NA, the field structure has a dark first Airy ring, especially when the SIL and the cover-layer are in contact.

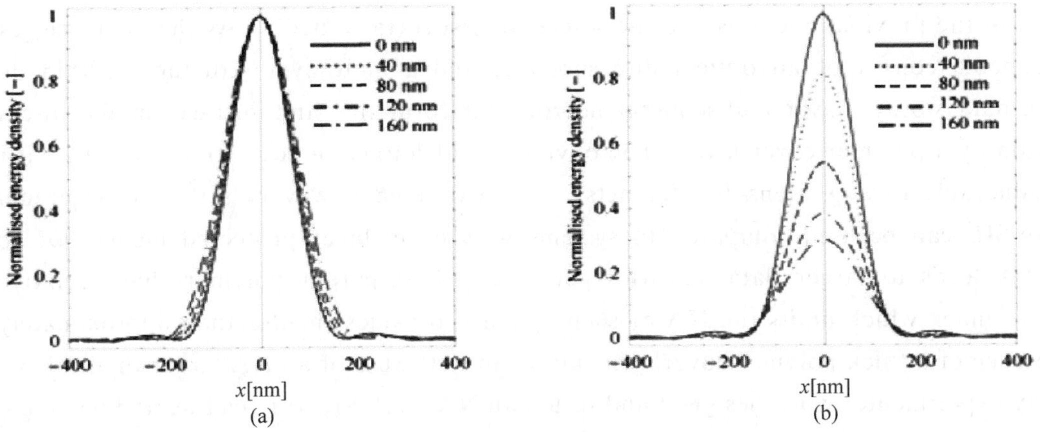

Fig. 4-44 (a)-(b) Computed plots of the normalised (time-averaged) total energy density in the focal region of an NA = 1.45 SIL that focuses through an air gap of 0, 40nm, 80nm, 120nm and 160nm into the CuSi-based recording medium depicted in Figure 4-41. Phase-only correction is applied in the entrance pupil of the lens that is illuminated with a circularly polarised plane wave with wavelength λ_0 = 405nm. The energy density profiles are calculated in the Si layer at the interface between the Si and the Cu layer. In (a) the energy density profiles are normalised to unity to show the relative broadening of the central peak with increasing air gap height and in (b) the spot profiles are normalised on the maximum of the energy density with the SIL in contact with the cover-layer.

The primary reason for this is that the z-component of polarisation is very small due to be large refractive index of the Si. With the SIL in contact with the cover-layer, the FWHM of the central peak in the spot profile is 144nm, equal to the scalar estimate $0.514\lambda_0/\mathrm{NA}$. For air gaps of 40nm, 80nm, 120nm and 160nm the FHWM values are 146nm, 154nm, 164nm and 174nm, respectively. If we accept a maximum increase of the FWHM of 5% compared to the case in which the SIL is in contact with the disc, the air gap should be smaller than 67nm. The broadening of the spot in the Si layer with increasing air gap is to a very large extent caused by a reduction of the amplitude of the plane waves with large kr due to reflections at the various medium transitions and in particular due to the evanescent coupling through the air gap, which becomes weaker for larger air gaps. It is worth noting that while this amplitude reduction is the primary cause of the spot broadening, there is also a small effect due to residual minor phase variations of the field components in k-space. Although we did apply a phase correction on the illumination in the entrance pupil, the phase profiles of the field components in k-space have non-zero variance. The reason for this is the imperfect field correction due to the simplified phase-only correction on only one of two mutually orthogonal field components. Different phase corrections are required for both mutually orthogonal field components to obtain a perfect phase correction. In Figure 4-44(b) normalised the (time-averaged) total energy density on the maximum of the energy density with the SIL in contact with the cover-layer. This Figure shows that intensity of the spot changes much more rapidly with increasing air gap than the spot size at the FWHM. The thermal recording process of the system should thus have power margins that are sufficiently large to be insensitive to small air gap variations that are inevitable in a real consumer product.

• GeSbTe rewriteable first-surface phase-change medium for NA = 1.9 SIL

An alternative to the near-field optical media that are protected by a cover-layer, and that we analysed in the previous Sections, are the so-called first-surface media. As the name suggest, first-surface media consist of an optical disc substrate and a multilayer structure of thin dielectric layers, a data storage layer and a metal mirror that form the first surface on the disc without protection by a polymer cover-layer. The obvious disadvantage of such structure is that it is much more vulnerable to e. g., lens-disc impacts. The advantage however, is that a larger numerical aperture SIL can be used compared to systems with cover-layer protected media. Of course, a larger NA leads to larger data densities per layer. Transparent polymers have relatively low refractive index which limits the NA of such systems to values smaller than approximately 1.65. The absence of a thick polymer coverlayer thus enables the use of an NA larger than this value. In our early experimental work designed and tested an NA = 1.9 SIL for reading and writing of first-surface media such as a Si ROM disc with a pit structure and a disc with a rewriteable phasechange stack. In this Section will analyse the case of the focusing with an NA = 1.9 lens in such a rewriteable phase-change medium. The multilayer structure that we have used for these calculations is taken from the phase-change stack design to use with a SIL. In that paper no refractive indices were mentioned so the refractive indices that are used in these calculations are taken from measured values of these materials as used in experimental Blu-ray Disc media. The structure parameters are shown in Table 4-5. In the Table 4-5, the subscripts "a" and "c" are used to indicate values of the complex refractive index of the GeSbTe phase-change material in the amorphous and crystalline phases, respectively. First, analyse the spot size inside the rewriteable GeSbTe phase-change material as a function of the air gap height.

Table 4-5 The structure parameters of optical configuration with the SIL, air gap, and the rewriteable first-surface GeSbTe phase-change stack on a polycarbonate (PC) optical disc substrate.

SIL	thickness	$n = 2.086$
Air	0.80nm	$n = 1$
SiN	15nm	$n = 2.10 + i0.001$
ZnSSiO$_2$	50nm	$n = 2.28 + i0.001$
GeSbTe	15nm	$n = 2.00 + i3.95$ $n = 2.95 + i3.00$
ZnSSiO$_2$	10nm	$n = 2.28 + i0.001$
SiN	10nm	$n = 2.10 + i0.001$
Ag	50nm	$n = 0.174 + i1.95$
PC	——	$n = 1.62$

In the calculations used optical parameters that are identical to that of our first-surface near field optical recording system. The SIL has NA = 1.9 and has a refractive index $n_{SIL} = 2.086$ equal to that of LaSF$_{35}$ glass at a wavelength of $\lambda_0 = 405$nm that is used in these calculations. The lens is illuminated with a circularly polarised plane wave and applied the simplified phase-only correction and to calculate the spot profiles at air gap heights of 0nm, 20nm, 40nm, 60nm and 80nm. For each calculation, the undisturbed focus position is chosen such that the position of minimum phase correction in the GeSbTe phase-change layer is exactly in the middle of this layer.

In Figure 4-45 (a)-(b) show the computed plots of the normalised (time-averaged) total energy density in the middle of the GeSbTe layer. It was assumed that the GeSbTe was in the crystalline phase. In Figure 4-45 a the energy density profiles are normalised to unity to show the relative broadening of the central peak with increasing air gap height.

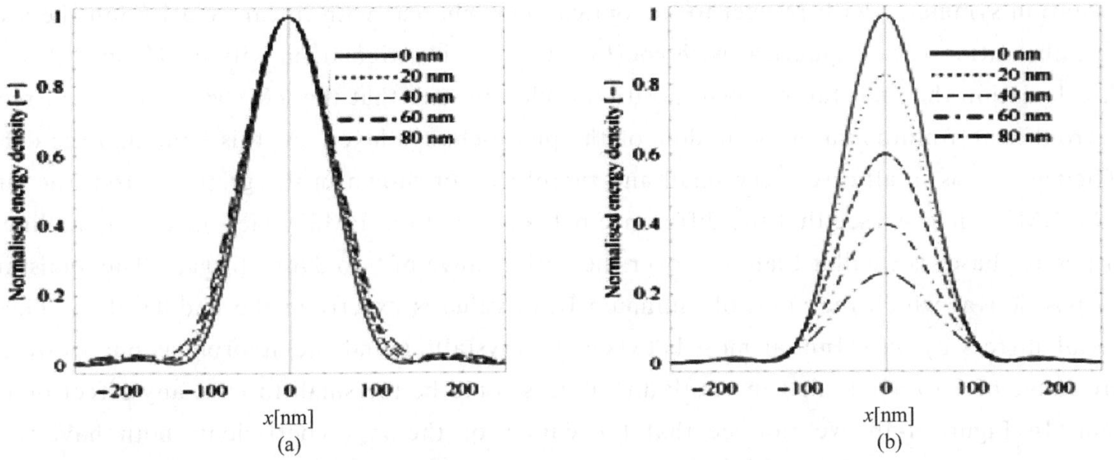

Fig. 4-45 (a)-(b) Computed plots of the normalised (time-averaged) total energy density in the focal region of an NA=1.9 SIL that focuses through an air gap of 0,20nm,40nm,60nm and 80nm into the GeSbTe-based rewriteable first-surface medium depicted in Figure 4-41. Phase-only correction is applied in the entrance pupil of the lens that is illuminated with a circularly polarised plane wave with wavelength $\lambda_0=405$nm. The energy density profiles are calculated in the middle of the GeSbTe phase-change layer in the crystalline phase. In (a) the energy density profiles are normalised to unity to show the relative broadening of the central peak with increasing air gap height and in (b) the spot profiles are normalised on the maximum of the energy density with the SIL in contact with the disc.

The differences of the normalised energy density profiles with those in the amorphous phase are very small and are therefore not shown. When the SIL is in contact with the cover-layer, the FWHM of the central peak in the spot profile is 111nm, very close to to the scalar estimate $0.514\lambda_0/\text{NA}$ of 110nm. For air gaps of 20nm, 40nm, 60nm and 80nm the FHWM values are 116nm, 120nm, 124nm and 128nm, respectively. If accept a maximum increase of the FWHM of 5% compared to the case in which the SIL is in contact with the disc, the air gap should be smaller than 23nm. Clearly, the FWHM increases faster with increasing air gap than the NA=1.45 system of the analysis in the previous Section. In Figure 4-45(b) we normalised the (time-averaged) total energy density on the maximum of the energy density with the SIL in contact with the cover-layer. As expected, this figure shows a more rapid decrease of the peak intensity of the spot with increasing air gap than in Figure 4-45(b) which was calculated for an NA=1.45 system.

Because of the different optical properties of the crystalline and amorphous phases of the phase-change material, we expect different wavefront corrections for optimal focusing in these two phases of the material. If this difference is too large it will be very difficult, if not impossible, to design a single lens that can read high data density marks of both the amorphous and crystalline phases on an optical disc. It is thus interesting to analyse the wavefront corrections that are required for optimum focusing inside the phase-change material for the two phases of this material. Figure 4-46 (a)-(b) shows the dependence on the air gap of the dominant Zernike coefficients and total RMS value of the wavefront correction for focusing with the NA=1.9 SIL exactly in the middle of the GeSbTe layer. In the Figures, solid lines are used for results with the phase-change layer in the crystalline phase and dashed lines for results with the amorphous phase. In Figure 4-46 a results are shown for circularly polarised illumination of the entrance pupil and in Figure 4-47 for linearly polarised illumination. With circularly polarised light, the phase correction has rotation symmetry with respect to the optical axis whereas with linear polarisation the second largest aberration (after spherical with coefficient $A_{4,0}$) is astigmatism with coefficient $A_{2,2}$. For each calculation the undisturbed focus position is chosen such that the RMS value of the correction wavefront is at a minimum in the middle of the phase-change layer. At this minimum the defocus coefficient $A_{2,0}$ is in all cases very small and therefore not shown in the graphs. From the curves for the RMS values we see that the difference between the total RMS values for the crystalline and amorphous phases is smaller than 8mλ over the entire range of 0 to 80nm air gap. The undisturbed focus position, at which the point of minimum RMS value is exactly in the middle of the GeSbTe material, differs by only 4nm at most between the crystalline and the amorphous phases over the entire range of 0 to 80nm air gap. Both differences could be too small to have any effect in a real system. In Figure 4-46, we can see that the curves of the $A_{2,2}$ coefficients both have a zero crossing. This might suggest that at these points the phase correction profile has rotation symmetry with respect to the optical axis, this is however not the case and we find higher order coefficients, most notably $A_{4,2}$, that are nonzero. Finally we remark that only for an air gap larger than approximately 30nm do the RMS values of the correction wavefront become larger than 15mλ, which is a frequently used tolerancing limit for optical recording lens designs. Therefore, if design an NA = 1.9 objective lens with a SIL for air gaps smaller than 30nm, can rely entirely on ray-tracing computer codes such as Zemax to optimise a lens design that works with this disc. For larger air gaps one might consider making a correction to the lens design according to the calculated coefficients. However, as just discussed, for air gaps larger than 30nm the spot size increase is already becoming significant.

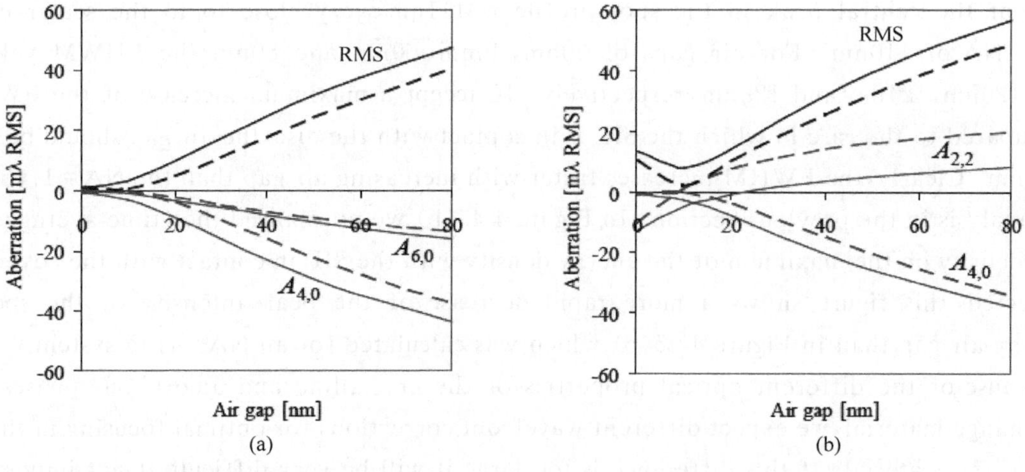

Fig. 4-46 (a)-(b) Dependence on the air gap of low-order Zernike coefficients and total RMS value of the phase-only correction for focusing with an NA = 1.9 SIL in the first-surface rewriteable phase-change medium. (a) Results for circularly polarised illumination of the entrance pupil, (b) results for linearly polarised illumination. Solid lines are used for results with the phase-change layer in the crystalline phase and dashed lines for the amorphous phase.

4.4 Super-resolution near-field structure (S-RENS)

4.4.1 Numerical model for super resolution effect

This was first reported by Bouwhuisl for optical data storage application and is named Super-Resolution (SR). If a nonlinear material is used as a layer in an optical disc, then all diffracted orders would be wider than usual, in such way that it would be possible to have overlap among higher orders within the zeroth order even after the cut-off frequency, breaking the diffraction limit. More recently, a similar technique was demonstrated by Tominaga by using a thin Sb film as nonlinear layer. He was able to observe modulated signal beyond the diffraction limit and the technique was called super resolution near field (S-RENS), becoming a very promising technique for the next generation of ODS. The Figures 4-47(a), 4-47(b) and 4-47(c) show the idea behind the super resolution effect. The SR layer has a refractive index which changes with the laser intensity. If the incident laser intensity is above a certain threshold, then the refractive index of the material abruptly changes, creating a small area with different optical properties. In this way, the laser spot can be smaller than $\lambda/2NA$. As an example of such materials, one can see in the Figure 4-47(a) the refractive index values for an AgInSbTe (AIST) layer below and after a threshold. Figure 4-47(b) shows the formed beam in the layer stack and the Figure and 4-47(c) shows the overlap of the −1st and +1st expanded orders within the zeroth order, illustrating the

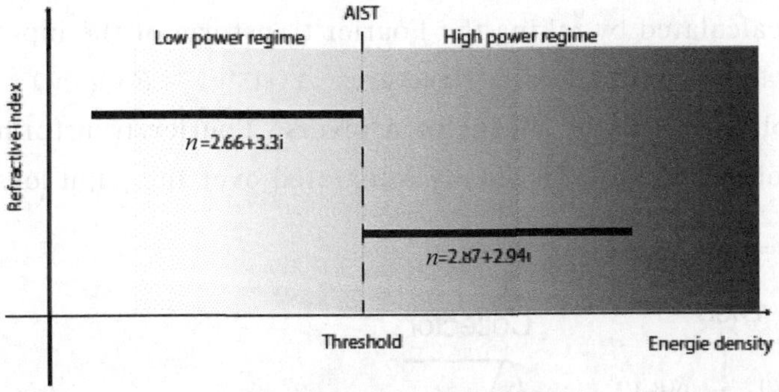

(a) AIST refractive index

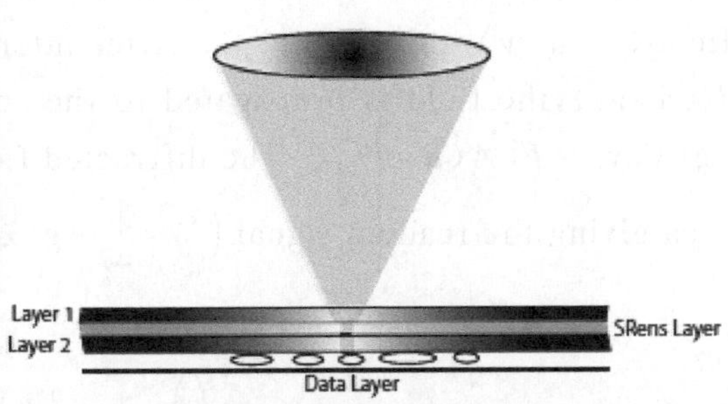

(b) SR effect

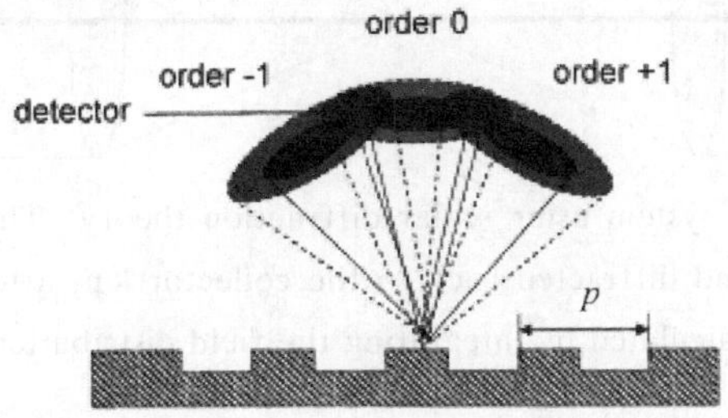

(c) Diffracted orders

Fig. 4-47 A schematic view of super resolution effect on optical discs. Figure 4-47(a) The refractive index values before and after the threshold limit for an AgInSbTe(AIST) based layer. Figure 4-47(b) an aperture is created in the super resolution layer if the laser intensity is above the threshold. In this way, a spot can be obtained smaller than $\lambda/2NA$. Figure 4-47(c) the effect of a nonlinear layer in Fourier space all orders become wider than usual (the inner circle) and eventually fall within the zeroth order resulting in a modulated signal beyond the cutoff frequency.

SR principle. In order to understand the super resolution effect, a computational model is required. A complete computer program based on rigorous vectorial diffraction model is expected to perform calculations of four distinct parts: first, the incident field has to be focused by the lens onto the layer stack. Then the field transmitted throughout all the layers must be calculated taking into account the non-linearity of the super resolution layer as a function of the laser power. After that, a vectorial diffraction theory has to be applied to calculate the diffracted field after the interaction with the data structure.

Finally, the diffracted field must be propagated back to the objective lens where the readout signal is calculated. The whole procedure suggests that a rigorous vectorial diffraction model would require extremely high computational resources and a long processing time. In the present work, a simplified computational model can describe the SR effect. To verify and validate the results, rigorous vectorial calculation is shown for the field immediately after the layer stack a comparison with some experimental results is also done. The present approach, although not rigorous, can be used to get qualitative description of the problem and the main results can be further investigated in a rigorous way.

4.4.2 Numerical approach

- Geometry of the system

The geometry for a readout program based on scalar theory is shown in Figure 4-48. Keep the classical transmission set-up for simplicity. The input field distribution in the objective lens $f(x,y)$ is focused onto the disc surface, which can be calculated by taking the Fourier transform of the input field ($U(u,v)=F[f(x,y)]$). After interaction with the disc structure ($A(u,v)=R(u,v)*U(u,v)$), the field is propagated to the collector lens by means of a inverse Fourier transform ($g(x,y)=F[A(u,v)]$). The diffracted field distribution is finally integrated over the pupil lens area giving the readout signal $\left(I_f=\iint |g(x,y)|^2 \mathrm{d}x\mathrm{d}y\right)$.

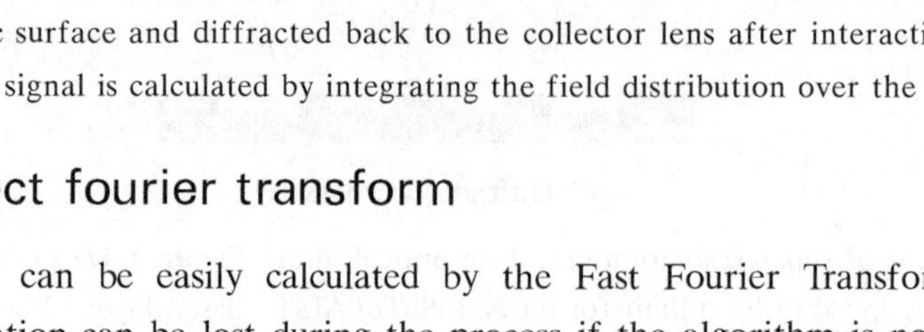

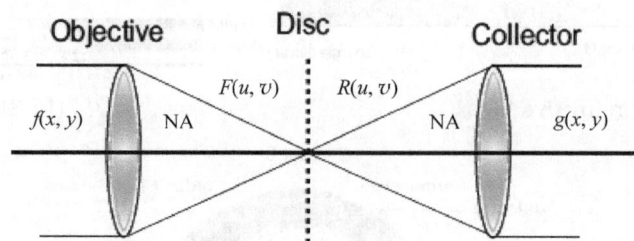

Fig. 4-48 Light path for readout system using scalar diffraction theory. The input field is focused by a lens onto the disc surface and diffracted back to the collector lens after interacting with the disc data. The readout signal is calculated by integrating the field distribution over the collector lens.

4.4.3 Correct fourier transform

Fourier transforms can be easily calculated by the Fast Fourier Transform (FFT) algorithm. However, phase information can be lost during the process if the algorithm is not correctly applied. To make the correct transformation, one need to discretize the function $F(u)=\int_{-\infty}^{+\infty} f(x)\mathrm{e}^{(-2\pi \mathrm{i}ux)}\mathrm{d}x$ in a sample grid and then compare the result with the FFT definition. Using the software package

Matlab, the FFT algorithm in one dimension is defined as

$$F(u(k)) = \sum_{j=1}^{N} f(x(j)) e^{\left[\frac{-2\pi i}{N}(j-1)(k-1)\right]} \tag{4-113}$$

where $j = 1:N, k = 1:N$ and N is the total number of points. Usually, choose $f(x)$ within $x \in [-x_{max}, x_{max}]$ dividing the whole interval in N points. In this case, $\Delta x = 2x_{max}N - 1$ is the smallest resolution in the grid and therefore have $x(j) = -x_{max} + (j+1)\Delta x$. The same holds for the variable $u \in [-u_{max}, u_{max}]$, and $u(k) = -u_{max} + (k-1)\Delta u$, where $\Delta u = 2u_{max}/N - 1$. Substituting $x(j)$ and $u(k)$ in the discretized function, reach, after some manipulations.

$$F(u(k)) = e^{2\pi i u(k) x_{max}} \sum_{j=1}^{N} \{f(x(j)) e^{2\pi i [x(j)+x_{max}] u_{max}}\} e^{-2\pi i [\Delta x \Delta u (j-1)(k-1)]} \Delta x \tag{4-114}$$

We recognize the FFT definition on the second exponential term after the summation if set $\Delta x \Delta u = 1/N$ and define $f((x(j)) = f(x(j))\ e^{2\pi i [x(j)+x_{max}] u_{max}}$. Thus, Eq. (4-114) can be written as

$$F(u(k)) = \text{phase} \sum_{j=1}^{N} f'(x(j)) e^{\frac{-2\pi i}{N}[(j-1)(k-1)]} \Delta x \tag{4-115}$$

The conclusion is that, to perform a correct FFT calculation, one has to use the modified function $f(x(j))$ and add manually the extra phase term. Another alternative is to use the built in function called fftshift together with FFT on the primitive function $f(x(j))$. But the extra phase still has to be added. To summarize, a correct Fourier transformation will be

$$F(u(k)) = \text{phase} * \text{fftshift}(\text{FFT}_{\text{Matlab}}) * \Delta x \tag{4-116}$$

- Adding super resolution effect

It is well known that the refractive index of the SR material shows two values depending on the laser intensity being above or below a given threshold (see Figure 4-47). Based on this fact, we simulated the SR effect in the following way: first, a routine searches in the focused field where the intensity is below and above this threshold (its value is defined in arbitrary units, for simplicity) and then, the layer effect is taken into account by means of $\exp(ikzn)$, where k is the wave vector, z the layer thickness and n the refractive index that can take values either below or above the threshold (see Table 4-6). The result is the modified beams shown in the Figures 4-49(a), 4-49(b) and 4-49(c).

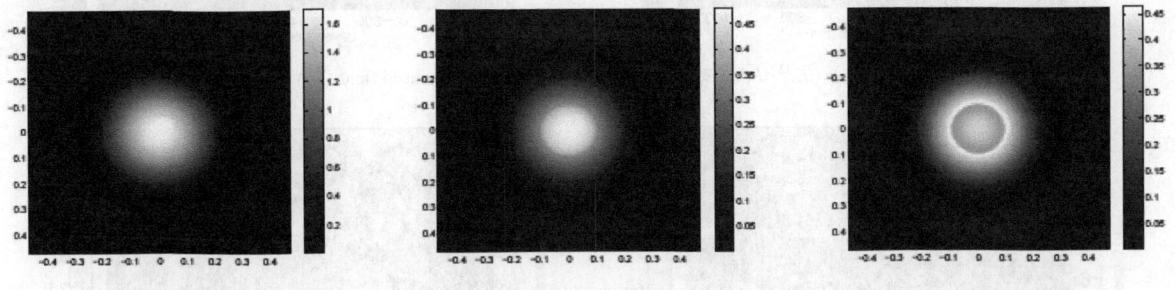

(a) Focused beam (b) SR beam: AIST (c) SR beam: InSb

Fig. 4-49 Focused spot beams. Figure 4-49(a) is the focused beam intensity distribution for $\lambda = 405$nm and NA = 0.85 in air, without any layers. Figure 4-49(b) and Figure 4-49(c) are the focused beam when an AIST or InSb layer is present, respectively. The Table 4-6 gives the refractive index and the layer thickness used in our simulations. The calculated field distribution will be propagated forward after interacting with the disc. This approach only takes into account the changes in the refractive index in the layer, neglecting any possible scattering after the aperture. In this way, this can be seen as a first approximation for the super resolution effect.

Table 4-6 Refractive index of AIST and InSb materials are used as a layer in the simulations.

Material	Refractive index below threshold	Refractive index after threshold	Thickness
AgInSbTe (AIST)	2.66+3.3i	2.87+2.94i	25nm
InSb	3+2.73i	2.1+3.4i	20nm

- Vectorial Calculation on Focal Region

A rigorous approach for the focused beam in the vicinities of the focal region can be found by solving the following diffraction integral.

$$\vec{E}(\vec{r}) = -\frac{\mathrm{i}}{2\pi}\iint_{\Omega} \frac{\vec{a}(k_x, k_y)}{k_z}(x)\mathrm{e}^{\mathrm{i}\vec{k}\cdot\vec{r}}\,\mathrm{d}k_x\mathrm{d}k_y \tag{4-117}$$

where the integration domain Ω is the exit pupil of the system, $\vec{k}$ is the propagating vector and $\vec{a}(k_x, k_y)$ is the vector amplitude of the electrical field. An extension of this theory, which provides the field distribution after a multilayer structure was used to calculate rigorously the field distribution in the focal plane, for the layer depicted in Figure 4-50.

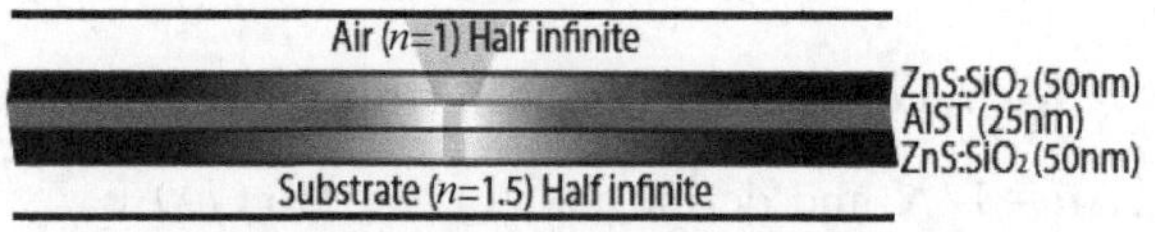

Fig. 4-50 Layer stack used in our simulations. A 25nm AIST (or 20nm InSb) layer sandwiched between two 50nm dielectric layers ($ZnS:SiO_2$, $n=2.2$) on top of a substrate layer ($n=1.5$).

Figure 4-51(a) shows the total field distribution for a focused beam when NA=0.85 and $\lambda=405$nm and no layers are present. Figures 4-51(b), 4-51(c) and 4-51(d) show the components of a circularly polarized electric field incident on the layer stack of the Figure 4-50.

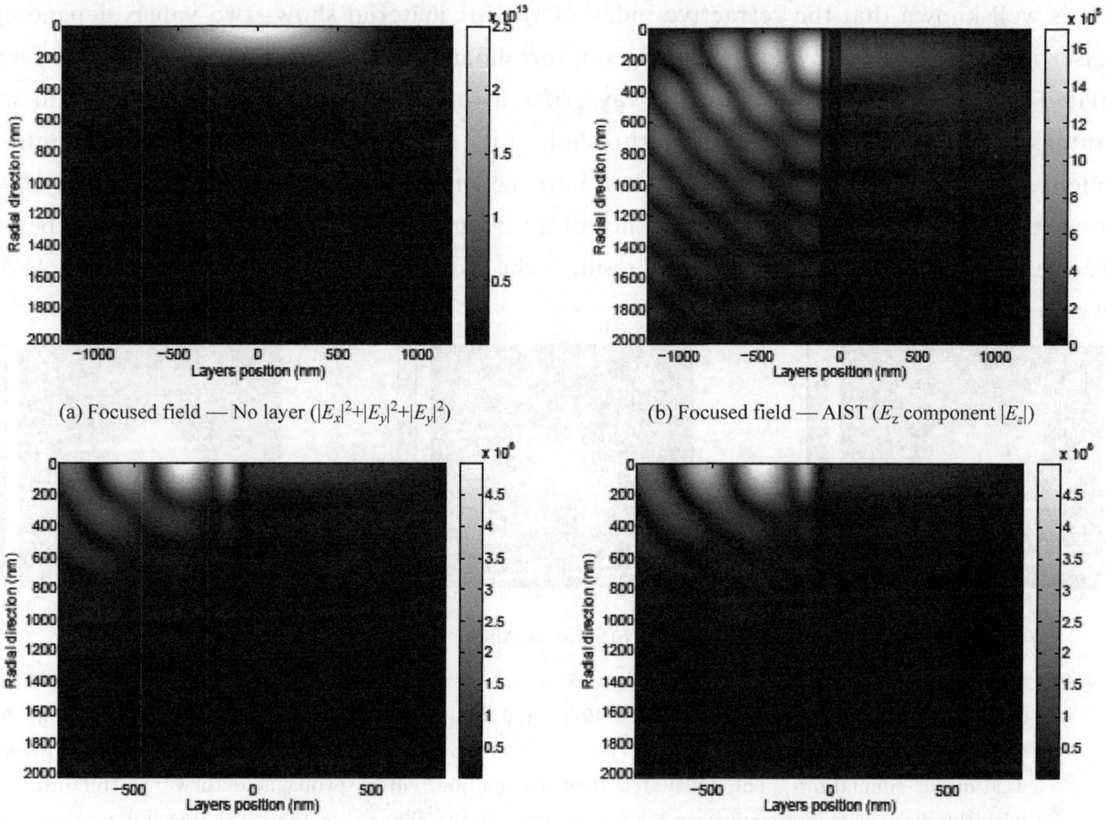

(a) Focused field — No layer ($|E_x|^2+|E_y|^2+|E_y|^2$)

(b) Focused field — AIST (E_z component $|E_z|$)

(c) Focused field—AIST (E_y component $|E_y|$)

(d) Focused field—AIST (E_x component $|E_x|$)

Fig. 4-51 Focused fields distribution.

The simulation of the SR effect in the vectorial case is performed in a different way than in the scalar case. First, the calculation is carried out using the material refractive index below the threshold and after that, a new calculation is carried out using the refractive index value after the threshold. The two results are combined according to the intensity profile of the beam, i. e., the position corresponding to the low intensity sets the below threshold profile, while the high intensity position sets the above threshold profile. Figures 4-52 (a) and 4-52 (b) show the normalized field intensity profile after the super resolution layer. Comparison with scalar results shows that the field profile has a similar shape in both cases, suggesting that both approaches are similar. In this way, a rigorous calculation of the scattered field by the pits structure could use this result as input file, in a first approximation.

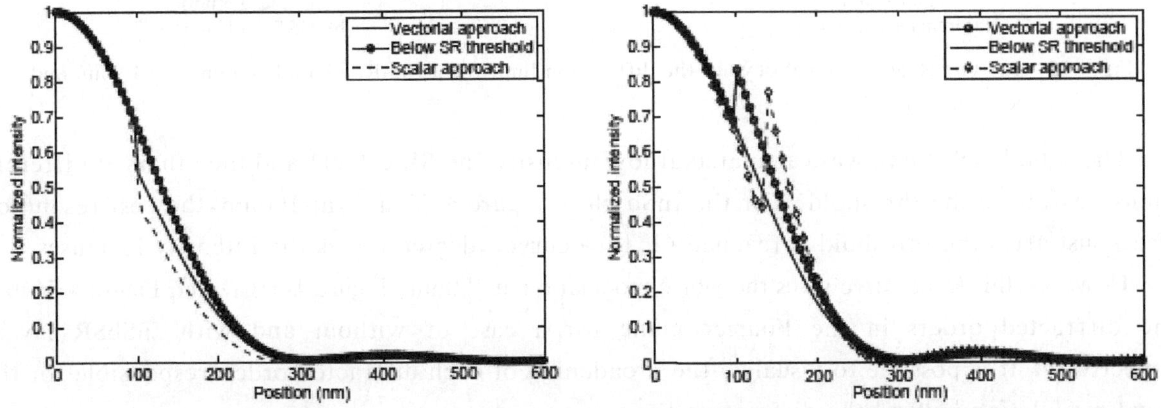

(a) AIST profile (b) InSb profile

Fig. 4-52 Field intensity profiles for AIST and InSb. The refractive indexes and thickness for InSb and AIST are given in the Table 4-6. The parameters for the system were NA = 0.85 and λ = 405nm. The scalar and vectorial field distribution shows a similarity concerning its shape.

4.4.4 Simulation of the readout signal

After the super resolution layer, the light is diffracted according to the Hopkins model. Checked the peak-to-peak signal is a function of the mark length for a periodic structure. The results are plotted in Figure 4-53(a) and Figure 4-53(b) for InS-band AIST-based layer stack. As expected, resolution beyond the diffraction limit (i.e. <120nm) is obtained for both types of layer stack, although the enhancement for AIST is rather weak. As the mark length increases, the resolution for the InSb type disc decreases before taking off again to see Figure 4-53(a). This is not observed for the AIST type media. Its resolution after SR onset is slightly better up to a certain length. For marks large enough, the two regimes coincide. One should notice that the frequency, at which occurs the signal destruction, depends on the laser intensity, i. e. the power threshold for the SR onset. One should notice that this has already been observed experimentally if one agrees to switch the results from InSb- and AIST-based media. This might be due to our model which deals with transmitted light, whereas the reflected light is measured. It is interesting to check the effect of SR layer in the diffracted orders by looking at the Fourier plane. It can be readily verified that the diffracted orders forwarded on the pupil plane, when the super resolution layer is activated, are wider than the diffracted orders when the SR layer is deactivated. Based on that can conclude that the readout beyond the diffraction limit is possible if those expanded orders overlap with the zeroth order, which is consequent upon the onset of the SR layer.

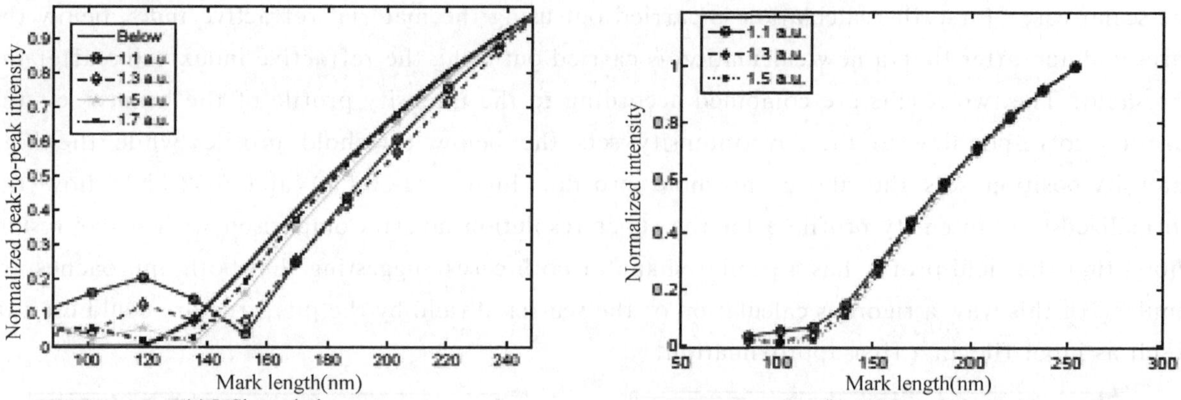

(a) InSb resolution test (b) AIST resolution test

Fig. 4-53 (a)-(b) Readout signal beyond the diffraction limit for a Blu-Ray based system (λ = 405nm and NA = 0.85).

The solid black curve was calculated at low intensity (no SR effect) and the others at different intensities, above the threshold. For the InSb plot (Figure 4-53(a)), at 100nm, the best resolution occurs just after the threshold is reached (1.1u.a curve), decreasing as the intensity increases.

However, this same curve gives the worse modulation at 150nm. Figure 4-54(a) and Figure 4-54(b): The diffracted orders in the Fourier plane for a case of without and with InSbSR layer, respectively. It is possible to visualize the broadening of each diffracted order, responsible for the signal modulation beyond the diffraction limit.

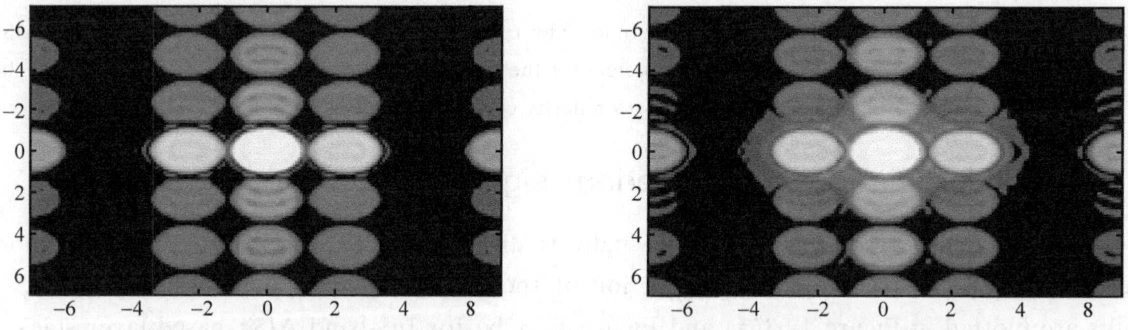

(a) Fourier plane-Without SR (b) Fourier plane-With SR

Fig. 4-54 (a) In the Fourier plane without InSb SR layer and (b) with InSb SR layer.

To get a better understanding of the read-out mechanism of the InSb type disc, the calculation was extended to the case of three isolated marks with variable length (Figure 4-55(a)). The result at 150nm is reported Figure 4-55(b).

At low laser power, i.e. below the SR threshold, the pits are clearly seen with a decrease of reflectivity due to the diffraction. As soon as the SR sets on, the signal is barely modulated, as expected from the Figure 4-53(a). For higher energies, the pits are seen again. Most of the surface exposed to the laser spot is then in the SR state. Thus, the transmitted light is more homogeneous allowing the proper diffractive read-out of the pits. At 120nm, the marks are not visible at low power (Figure 4-55(d)). The SR is needed for the detection. However, only two out of three pits are seen. This was also experimentally observed. When the laser power increases further, the modulation vanishes again. The same behavior is found with the 100nm marks: no modulation below the SR threshold and only two pits out of three are visible after SR onset. But at higher power, instead of a degraded modulation, a signal inversion occurs and the three marks are clearly present. Experiments couldn't confirm this fact, presumably because at too high power the disc degrades very quickly.

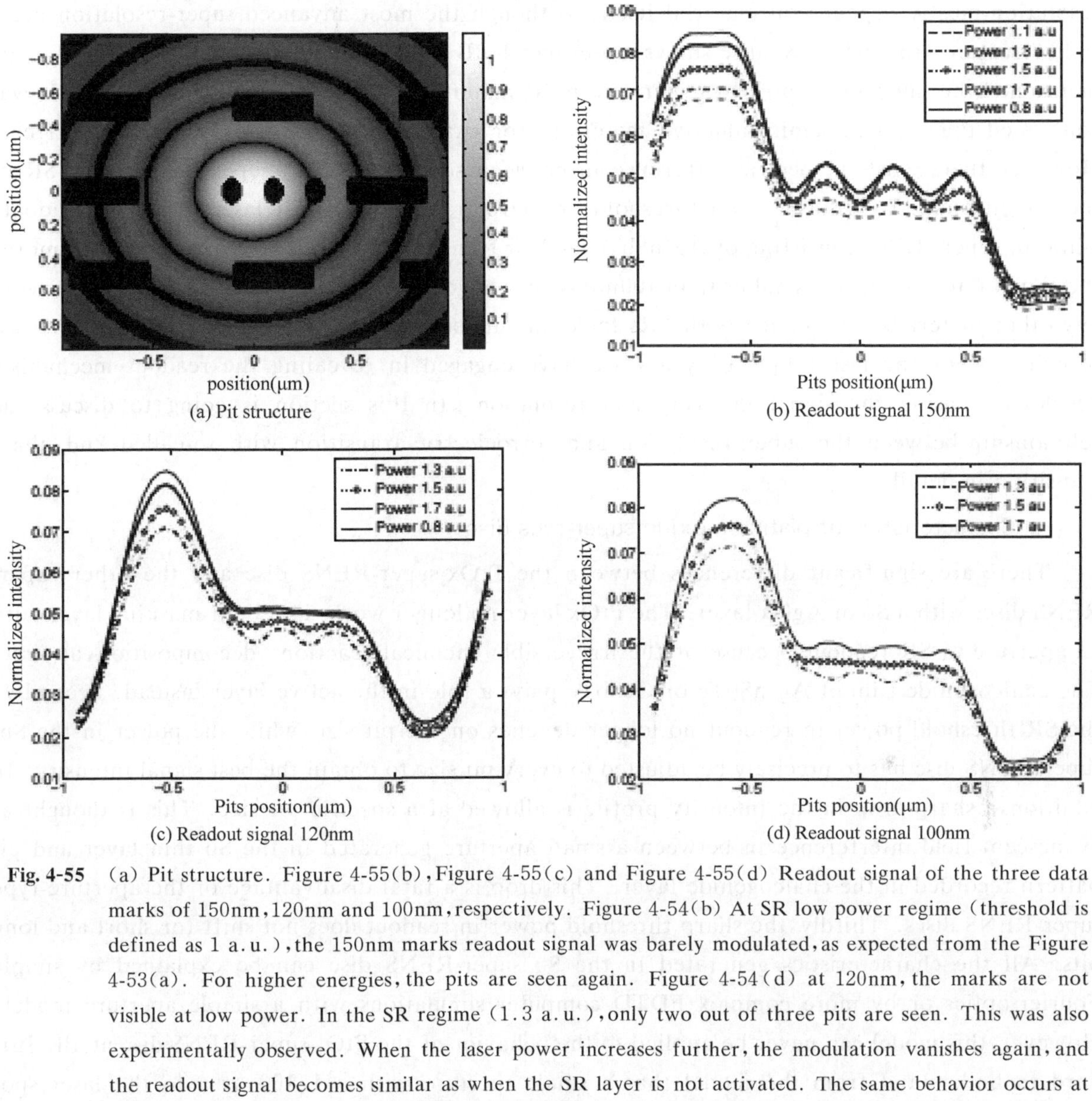

Fig. 4-55 (a) Pit structure. Figure 4-55(b), Figure 4-55(c) and Figure 4-55(d) Readout signal of the three data marks of 150nm, 120nm and 100nm, respectively. Figure 4-54(b) At SR low power regime (threshold is defined as 1 a.u.), the 150nm marks readout signal was barely modulated, as expected from the Figure 4-53(a). For higher energies, the pits are seen again. Figure 4-54(d) at 120nm, the marks are not visible at low power. In the SR regime (1.3 a.u.), only two out of three pits are seen. This was also experimentally observed. When the laser power increases further, the modulation vanishes again, and the readout signal becomes similar as when the SR layer is not activated. The same behavior occurs at 100nm Figure 4-54(d). These results suggest that a laser intensity that is suitable for reading marks at some frequency may not resolve a different frequency.

A simple extension of a scalar readout program was proposed in order to investigate the basics aspects of super resolution effect. Although not rigorous, the model gives results which are qualitatively in agreement with experimental facts. It predicts the signal destruction at specific frequencies and signal inversion for different intensities. The main advantage of such approach is that it is very fast, once FFT algorithm is used. This method could be used to check more global results, followed by vectorial calculations to fully describe the problem.

4.4.5 S-RENS with ferroelectrics of chalcogenides

The former method is clearly based on the theory in physics and optics, and the mark resolution and technical limitation are rigidly determined. In the later type, in contrast, the resolution mostly depends on material itself. Although the most advanced super-resolution near-field structure (super-RENS) disc shows more than 40dB CNR at 50nm pits. Kikukawa et al. first mentioned the importance of heat generated in SR materials by ROM-discs (super-ROM). It was confirmed that almost semiconductive materials, for examples, Si, Ge and W show SR effect. However, there exists a specific difference in between super-ROM and super-RENS. The SR of these semiconductive films has no threshold or sharp transition against laser power in readout, while in super-RENS consisting of AgInSbTe, a clear transition power is confirmed. Furthermore, the signal intensity of pits smaller than 100nm is very huge in comparison of super-RENS discs with any other materials. At glance both SRs look similar, but it would be due to different physical principles. For the last couple of years, we have engaged in revealing the readout mechanism besides increasing the signal intensity and resolution. In this section is going to discuss the relationship between the super resolution and ferroelectric transition with so-called 2nd phase transition in detail.

- Characteristics of platinum-oxide super-rens disc

There are significant differences between the PtOx-super-RENS disc and the other super-RENS discs with a Sb or AgOx layer. The PtOx layer no longer works as a light-masking layer with an aperture in SR readout because of the irreversible chemical reaction: decomposition, at first. The chalcogenide film of AgInSbTe or GeSbTe plays a role in the active layer instead. Secondly, the SR threshold power in readout no longer depends on the pit size, while the power in the Sb-super-RENS disc has to precisely be adjusted to every pit size to obtain the best signal intensity. In addition, a sharp drop in the intensity profile is allowed at a specific pit size. This is thought as evanescent field interference in between a small aperture generated in the Sb thin layer and pit pattern recorded in the chalcogenide layer. This drop is a fatal disadvantage of the aperture-type super-RENS discs. Thirdly, the sharp threshold power in readout does not shift for short and long pits. All the characteristics generated in the Sb super-RENS disc can be explained by simple Fourier optics or by more complex FDTD computer simulations with a simple aperture model. However, this model can never be applied to the behavior of the PtO_x super-RENS disc at all. It is hard to understand from the model that a small aperture with a 1/100 area to the laser spot generates more than 40dB CNR because available photon numbers from the aperture would be very small in comparison to the others bouncing back from the other masked area (99/100). Forth, the SR effect does not depend on laser wavelength in readout. In both using red (635nm wavelength) and blue laser (405nm), it has already been confirmed that the signal enhancement and resolution only depend on the beam spot size on the disc surface. This phenomenon is very attractive for terabyte memory. Once the super-RENS disc and SIL pickup are combined together in one system, ~20nm pits would be recorded and read out with more than a commercial signal level (>40dB) in principle. The last, the strong SR effect has only been observed in dynamic tests, but not clearly seen by static. This means that the strong electrical field in the laser beam spot may assist or induce the SR effect as well as temperature. This is very interesting on the point of view of depending on the electromagnetic field.

In order to explain all the characteristics, a refractive index in the local area (that is, active region) must have a extremely huge value against any other area otherwise the scattered signal photons are almost covered over the major photons bounced back from the masked area. One of physical phenomena to induce the index change is the Kerr effect. This phenomenon is well known as the optical nonlinearity of the 3rd order, in which the index linearly changes with the laser intensity. However, the deviation induced by the effect is not so large in comparison to that observed in the super-RENS disc and the sharp threshold power in the SR cannot be explained. The 2nd order may generate a SHG waves. However, the experimental results by Kim et al. denied it because of the observation of 80nm pit patters with 40dB CNR by a disc drive system with 635nm wavelength and NA 0.60: the theoretical resolution limit by the SHG must be 132nm. Therefore, 2nd and 3rd order optical nonlinearities are little related to the SR of the super-RENS disc.

In turning around to classical physics, refractive index n is expressed by electronic polarizability α with the Clausius-Mossotti equation. In quantum physics, the Eq. (4-118) is further modified by the summation of oscillator strengths attributed to each band transition (4-119).

$$\alpha_{\infty} = \frac{3}{4\pi N_A}\frac{n_{\infty}^2 - 1}{n_{\infty}^2 + 2}V \tag{4-118}$$

$$\alpha_m = \sum_k \frac{2e^2\omega_{mk}\langle m \mid \hat{r} \mid k\rangle\langle k \mid \hat{r} \mid m\rangle E_0\cos(\omega t)}{(\omega_{mk}^2 - \omega^2)\hbar} \tag{4-119}$$

Hence in Eq. (4-118), α_{∞} and n_{∞} are the electronic polarizability and refractive index at the wavelength$\to\infty$. N_A and V are the Avogadro number and volume in a medium. In Eq. (4-119), Eq. (4-118) is modified by the summation of the contributions from each band-transition $m\leftrightarrow k$. Hence, ω, ω_{mk}, e, $e\hat{r}$, E_0 are applied electrical frequency, resonance frequency between the bands, electron charge, electron displacement, and applied electric field, respectively. In solid or liquid, $3V/4\pi N_A$ in Eq. (4-118) may set at a constant ρ, and Eq. (4-118) is further simplified to the

$$n^2 = (2\alpha + \rho)/(\rho - \alpha) \tag{4-120}$$

Hence, ρ is thought as something like a space freedom at around the local position of each atom in the unit cell. When $\rho\sim\alpha$, n^2 may diverse and take a huge value close to the singular point. Ferroelectrics is well known to show such a behavior to temperature with a transition so-called the Curie temperature T_c. It is also known that GeTe holds the ferroelectric characteristics with the 2nd phase-transition at $T_c\sim$352℃. The Raman soft-mode phonons accompanied by the ferroelectric effect were actually observed at around 3.5THz (110cm^{-1}), and the source is attributed to the T_e local-displacement in the unit cell. They discovered relatively very large space deviations at the Sb and the Ge sites Figure 4-50(b), in comparison to the T_e site Figure 4-50(a) in the NaCl-type fcc unit lattice of GeSbTe. It turned out that the lattice transforms into another hexagonal lattice at $\sim$260℃. On the other hand, AIST retains the hexagonal lattice, which is similar to the original Sb lattice with the c-axis expanding from 11.2 to 11.6A at temperature up to 350℃. These results support that AIST show a 2nd phase-transition more anisotropically than the GST system. So far, a large amount of studies have revealed the transition temperatures of optical phase change alloys. However, most of all were only focused on the 1st phase transitions: a transition in between as-deposited amorphous and crystal, and the melting points. None of them has taken care of 2nd phase-transition because of its tiny discontinuity on heat flow in DSC and too small optical change on reflectivity or transmittance in macro-scale.

Because the readout model of the PtO_x super-RENS disc by the effect of the ferroelectric properties of the AIST and GST thin films, the SR effect is only active at a very narrow transition temperature in the ferroelectric catastrophe. Here experimentally determined the relationship between the readout laser power and disc temperature, and clear revealed that the threshold laser power emerging the SR in the super-RENS discs are well agreed with the 2nd phase-transition temperatures. Consider the relationship between the ferroelectrics and the SR effect in much more detail by the Landau theory of the ferroelectrics, that the free energy F_P is decomposed into the power series of the dipole P. Because F_P has to get energy minimums against P, it is only made of the even series of P.

$$F_P = \frac{1}{2}\alpha P^2 + \frac{1}{4}\beta P^4 + \frac{1}{6}\gamma P^6 + \cdots \tag{4-121}$$

Hence, $\alpha = \alpha_0(T - T_0)$ and $\alpha_0 > 0$. Also, put $\beta > 0$. Now, can find the energy minimums of F_P, the first derivative d F_P/dP = 0. Thus, can obtain:

$$\begin{cases} \dfrac{\partial F}{\partial P} = \alpha P + \beta P^3 + \gamma P^5 = E = 0 \\ \dfrac{\partial^2 F}{\partial P^2} = \dfrac{\partial E}{\partial P} = \chi^{-1} \\ 4\pi\varepsilon^{-1} = \dfrac{\partial E}{\partial P} = \alpha_0(T - T_0) + 3\beta P_s^2 = 2\alpha_0(T_0 - T) \end{cases} \tag{4-122}$$

χ in the second equation of (4-118) is a dielectric susceptibility and $P = \chi E$. Here, neglected the higher orders more than the second derivative. As a result, can obtain the famous relationship of the Curie temperature and is $\varepsilon \propto (T_0 - T)^{-1}$.

In optical discs, however, an as-deposited amorphous film must be once crystallized before recording. In the process, the film volume is reduced more than 5% from the original condition. The reduction is in isotropic; that is, the protection layers sandwiching the phase-change film induce a high strain force (Actually, the strain force is not in isotropic but anisotropic) because the crystallization procedure is usually carried out along the tracks and the groove structure may modify or block the strain force across the tracks. From previous experiment with a $ZnSSiO_2$/Sb/ZnS-SiO_2, the strain force is roughly estimated 20～40MPa. As increasing temperature, the crystalline growth is further accelerated with fatting the grain size. Finally, the volume change is balanced in between the thermal expansion and the reduction due to the crystalline growth. Therefore, it easy turns out that Eq. (4-121) is not applicable to the conditions including the strain. Instead Eq. (4-121) is modified including uni-axial strain force (More in general, consider by-axial force) and its coupling term with the dipole.

$$F_P = \frac{1}{2}\alpha P^2 + \frac{1}{4}\beta' P^4 + \frac{1}{2}c(x - x_0)^2 + qxP^2 \tag{4-123}$$

The third term is for the strain and the forth one is for the coupling. Hence, c, x_0 and q are a Young's module, the original position of an atom, in which $\Delta x = (x - x_0)$ means $\sim 1/3\rho$ of Eq. (4-120), and a coupling constant, respectively. Here, in addition to the local minimum of F_P to P, can obtain another minimum by displacement x and $\partial F/\partial x = 0$. As a result, can obtain an attractive relationship between the displacement:

$$\Delta x_s = (x - x_0) \tag{4-124}$$

and

$$P_s, \Delta x_s = -qP_s/c \tag{4-125}$$

The self-distortion now may induce the dipole. Alternatively, the large electrical dipole may induce a very large displacement in the unit cell by the Yield point, resulting in plastic deformation, material flow, or transition to more energetically stable crystalline state. At the transition point, refractive index theoretically has no meaning or no value. This ferroelectric catastrophe probably is the readout mechanism of super-resolution in super-RENS disc.

- Super-resolution model of platinum-oxide super-rens disc

Here, more simplify a PtO_x super-RENS disc structure with a single AIST layer sandwiched by two $ZnS\text{-}SiO_2$ layers. Even after the all film deposition, the AIST layer holds amorphous. In recording, whether the layer is in amorphous or crystal is not a problem because the recording track is almost all crystallized by the pulsed laser beam except for recording longer pits with more than the resolution limit. It means that the track is only crystallized and the layer volume is reduced ~5% as a result. The $ZnS\text{-}SiO_2$ protection layers top and bottom must have a force balance with the crystallized layer, inducing a strong tensile stress to the AIST at this moment. Finally, the disc is deformed by the volume reduction, the disc is bended and large strain forces are induced at the interface between $ZnS\text{-}SiO_2$ and AIST or GST. In readout, as increasing the laser power: temperature, further crystallization with the growth and volume reduction proceeds against the thermal volume expansion by $T_0 \sim 350$℃. During the process, an additional physical phenomenon may occur. Because of the semiconductivity of the crystal, a large amount of carriers are generated in the laser spot area, and diffuse outwards. However, due to the anisotropic crystallization and strain force, a static electrical field is generated along the track direction. The induced static dipole further assists the deformation to the yield point of the first crystalline phase. Carrier generated in the AIST or GST by focusing and increasing the incident laser beam. The strain force may generate anisotropically static field.

At ~350℃, the threshold in the Free-energy, the first crystalline phase cannot endure its unit lattice and probably allow to transit into another stable phase: 2nd phase-transition. Only in a very narrow temperature region in the laser spot at $\alpha \sim \rho$, the refractive index n becomes discontinuous and diverse, resulting in optical super resolution. If this model is true, the transition edge plays a crucial role in the SR, in which the resolution is determined the edge width.

This model suggests that recorded pits smaller than the diffraction limit are read out by the edge, while longer pits than the limit are reproduced by both the edge and far-field diffraction (see Figure 4-56). This model can well explain that the CNRs of small and large pits (but beyond the diffraction limit) are almost constant with more than 40dB by less than 50nm size. Furthermore, the threshold power: temperature only depends on the intrinsic properties of materials and on the strain force field induced by surrounding materials. In addition, it is explained that without the electrical field attributed to the high intensity laser power, this SR is hard to be observed experimentally.

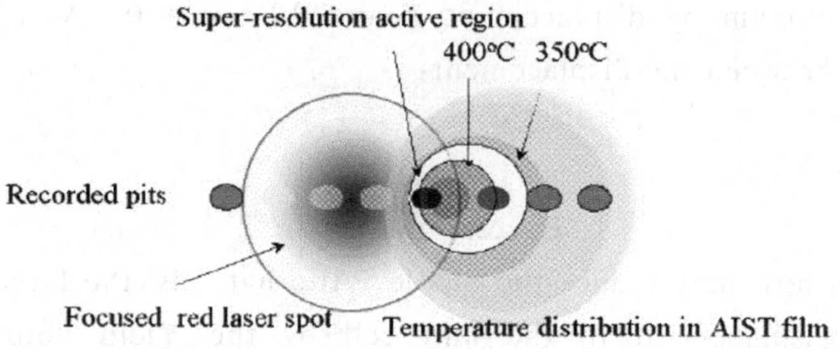

Fig. 4-56 The readout model of PtO_x super-RENS disc by ferroelectric catastrophe.

The SR of the PtO_x-super-RENS disc is probably attributed to the ferroelectric catastrophe induced by the strain force balance with the protective layers. We believe that the chalcogenides, especially GST and AIST give us much more attractive aspects not only for the 1st phase-change application to the optical disc and static memory, but also for future nanotechnological devices based on the 2nd phase-change behavior.

4.5 Micro-aperture laser for NFO data storage

The principles of near-field optics can be used to laser beam writing directly on medium. But the very small aperture laser (VSAL) is an important light source used in the near-field optical storage system. It is necessary for studying the near-field property of VSAL's output light. The optical characters and the intensity distribution in the near-field of the output region of the VSAL have been numerical simulated using two dimension nonlinear FDTD (2D-NL-FDTD) method and the Fox-Li method. Through analyzing the results from the viewpoint of the Optics, the possible application in the near-field optical recording have been discussed, and some curves indicating the near-field optical characters of the output light have been presented. Key words: near-field optical storage 2D-NL-FDTD method Fox-Li method VSAL.

More attention has been paid on the ultra-high optical memory using near field optics for future data storage. A common important issue of near-field optical storage is to realize high-output power source. In the 1999, Partovi suggested a VSAL to be the high-output power light source for near field recording and reported that the VSAL demonstrated more than 10^4 times of increase of output power over coated tapered fibers with comparable aperture diameters. In addition, Shinada etc. fabricated and analyzed the near field characters of the micro-aperture VCSEL. The calculated spot size of output is nearly as small as the aperture width (100nm) when the wavelength is 850nm. It indicated the potentiality of the VSAL as a near-field optical recording tool, so the validate analysis of the optical characters of the output in the VSAL's near field is necessary. In this paper, we carried out a near-field analysis of the VSAL using the 2-D-NL-FDTD method and the Fox-Li method.

4.5.1 Model and numerical methods

Figure 4-57 shows a detail of the VSAL. The typical size of the VSAL is $780\mu m \times 300\mu m \times 150\mu m$. In order to be conveniently analyzed, the VSAL has been simplified to be the physical model showed in the Figure 4-58. It involves two parallel plane mirrors, one of which is a noble metallic film with a micro-aperture.

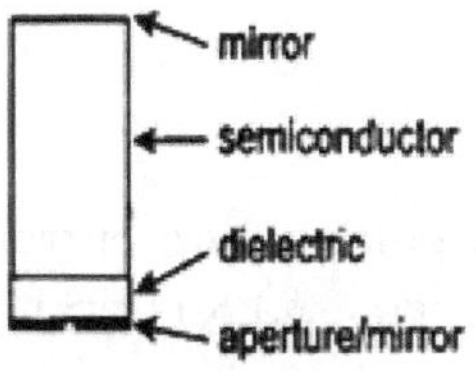

Fig. 4-57 The detail of the VSAL.

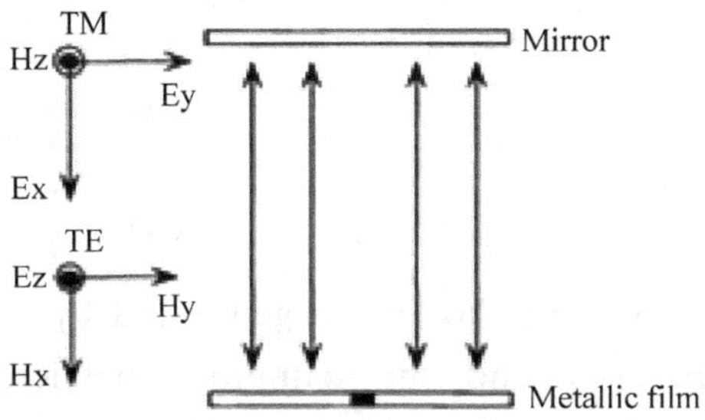

Fig. 4-58 The theoretic model and coordinales.

In the first place, the Fox-Li method has been used to calculating the base mode of the resonant cavity. This method is widely used to numerically obtain the eigenmodes of the side-opened resonators consisting of a pair of plane mirrors. Because of the micro-aperture, the Huygens' integral used in the Fox-Li method is not applicable. Using the angular spectrum method to investigate the near-field diffraction have been reported in 1995. For calculating, first set a field distribution of the pane wave on the metallic film as an initial condition then calculate the distribution on the other mirror by using the angular spectrum method. Moreover to calculate the field distribution on the metallic film in the same manner, can get the distribution of the base mode of the resonator by repeating the round-trip calculation of the field propagation. In the next place, the two dimension nonlinear FDTD method has been used to simulate the light of the base mode propagating through the metallic film. The noble metals (gold, silver, and copper) in the optical regime process complex refractive indices in which the imaginary component is greater than the one, that the requires permittivity to have a negative real component. The boundary condition on the tangential electric field cannot be satisfied with the standard FDTD method. So through combining the full-wave, vector, linear Maxwell equations solver with a Lorentz linear dispersion model, the nonlinear FDTD method has been carried out by Judkins and Ziolkowski. Solving the system of equations yields the following iterative expressions for YE mode. The expressions for TM mode are similar to the ones for TE mode:

$$\begin{cases} H_x(i,j) = H_x(i,j) + \mathrm{CD} \times (E_z(i,j) - E_z(i,j+1)) \\ H_y(i,j) = H_y(i,j) + \mathrm{CD} \times (E_z(i+1,j) - E_z(i,j)) \end{cases}$$

Where

$$z(i,j) = E_z(i,j) + \frac{(c \times \varepsilon_0 \times \Delta t)}{\varepsilon}((H_y(i,j) - H_y(i,j) +$$

$$H_x(i,j-1) - H_x(i,j))/\Delta s - J(i,j))P_z(i,j)$$

$$= P_z(i,j) + \Delta t \times J_z(i,j)$$

And

$$J_z(i,j) = \frac{\dfrac{1}{\Delta t} - \dfrac{\Gamma}{2}}{\dfrac{1}{\Delta t} + \dfrac{\Gamma}{2}} \times J_z(i,j) + \frac{\omega_0^2}{\dfrac{1}{\Delta t} + \dfrac{\Gamma}{2}} \times \left(\frac{\chi_0 \times (E_z(i,j)}{c} - P_z(i,j)\right)$$

$$\begin{cases} CA = \dfrac{1 - 0.5\sigma \times \Delta t}{1 + 0.5\sigma \times \Delta t} \\ CB = \dfrac{\varepsilon_0}{2\varepsilon + \sigma \times \Delta t} \\ CD = \dfrac{1}{2} \end{cases}$$

Where P is the polarization generated by the model, J is the polarization current ω_0 and Γ resonant frequency and the damping coefficient, respectively. The 2d-NL-FDTD algorithm employs a uniform mesh grid in both directions Δs is the unit length of the mesh grid. $\Delta t = \Delta s/2c$ is the unit time in the calculate, where c is the light velocity in the vacuum. In the code, Chebyshev's second-order absorbing boundary conditions are used at all boundary surfaces. The Table 4-7 lists the parameters used in the simulations.

Table 4-7 Parameter values used in the simulations.

Phys. Quantity	Symbol	Value
Wavelength	λ	980nm
Velocity (vacuum)	c	3×10^8 m
Unit Grid	Δs	9.8nm
Unit Time	Δt	1.6×10^{-17} s
Refraction index	n_{Au}	0.175 − j4.91

4.5.2 Numericalresults

In this calculation, the electric field perpendicular and parallel to the sheet have been defined as TE and TM modes, respectively. The coordinate systems are shown in Figure 4-59 and Figure 4-60 shows the intensity profile of base mode of the resonator with a micro-aperture 100nm wide. Because the width of the mirror is nearly 200nm, which is more wider than the aperture of 170nm, and the length of the resonator is nearly 750nm, the influence of the micro-aperture on the mode is negligible. But in our calculating region (163nm × 163nm), the slim difference between the modes with different wide apertures can be shown in the Figure 4-61. Because the 400nm is close to the wavelength (980nm), the change of the mode is obvious. The modes with different aperture have been used in the FDTD simulation as the source.

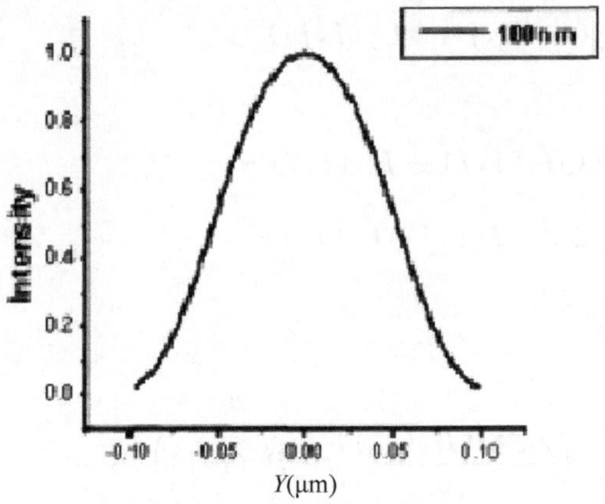

Fig. 4-59 Intensity profile of the base mode of the resonator with aperture 100nm wide.

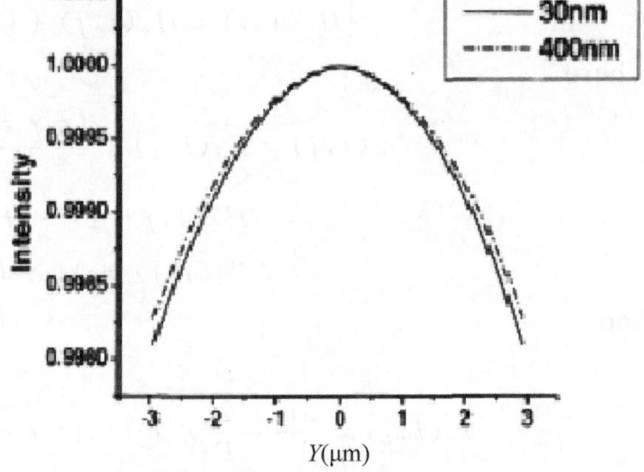

Fig. 4-60 Intensity profiles of the base mode of the resonator with different width aperture.

Figure 4-61 shows the intensity profile of output through a 100nm aperture with different thickness for TM mode at the boundary between Au and air. Two sharp peaks appear at both edges of the aperture. The reason for this result is the surface plasma enhancement of the metallic film for TM mode. The profiles of different thickness are similar each other, the spot size (FWHM) is as small as 118nm, approximately equal to the width of aperture, which is much smaller than the wavelength. At the area away from the center of the curves, the intensity of the thinner film (30nm) is larger than the thicker (50nm) one, because the light can partially penetrate into metal especially when the thickness is much smaller than the wavelength. So the thinner the thickness, the stronger is the background noise. It is possible that this phenomenon can reduce the resolution of the spot. But the affect is negligible for TM mode because of the surface plasma enhancement.

The intensity profile of the same calculation for TE mode is shown in Figure 4-62. One can see that the maximal point of the curves locate in the center of the aperture because of no surface plasma enhancement for TE mode, and the intensity of the thinner (30nm) film is higher than the thicker (50nm) one because of the partially penetrated light. The intensity of TM mode is over 10 times the one of the TE mode, so the intensity of the partially penetrated light is significant compared with the intensity of the spot. It means that the thickness of the metallic film is an important factor to defining the resolution of the outgoing spot for TE mode. Although the output intensity of the thinner film is high, the quality of the spot is not always good. From Figure 4-60, can see that the spot size (FWHM) of the film 50nm thick is 136nm, which is smaller than the one 30nm thick (156nm).

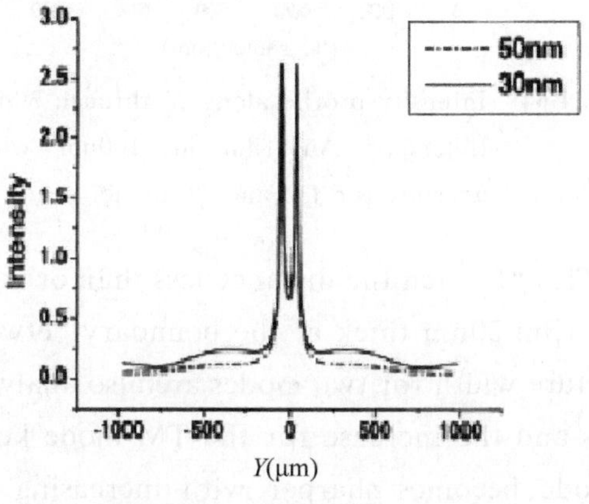

Fig. 4-61 Intensity profiles along Y of the output through the 100nm wide aperture with different thickness for TM mode.

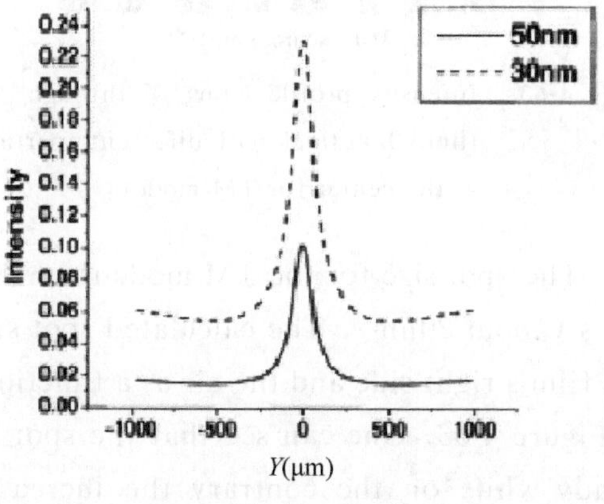

Fig. 4-62 Intensity profiles along Y of the output through the 100mm wide aperture with different thickness for TE mode.

Figure 4-63 shows the calculated intensity distribution across the Au film 50nm thick at the center of the aperture with different widths (30nm, 50nm, 100nm, 200nm and 400nm) for TM modes. The enhancement of the intensity of the electric field at the region of the Au film increases with decreasing the aperture width. The smaller the aperture width, the stronger is the attenuation of the near-field intensity of the output, and the nodes of the intensity profiles shift slightly to the left side of the Au film. It corresponds to the result gotten by SHINADA. The reason has been discussed in his paper. It is obvious that the higher intensity of output decays more rapidly in the near-field region. It means that if to get high powerful output of VSAL, we should realize the precision control of the distance between source and the recording medium.

The trend of the decay of the output in the near-field for TE mode is similar to the TM one. But the important and intuitionistic difference has been indicated through comparing the decay profiles for TM and TE modes. The curves have been shown in Figure 4-64. One can see that the intensity of the TM mode has an obvious enhancement in the Au film, none has the TE mode. In the near-field of the right side of the Au film, the attenuation of the output for TM mode is sharper than the TE one. The intensity of TM mode is over one order the one of the TE mode in the near-field because of the surface plasma enhancement. But these two modes have an approximately equal intensity in the far-field, because the enhancement is only significant in the near-field. Figure 4-65 shows the calculated spot size of the Au film 50nm thick with a 100nm wide aperture as a function of the distance away from the metallic film for TM and TE modes. One can see that the spot size increases and the increase become slower with in creasing the distance.

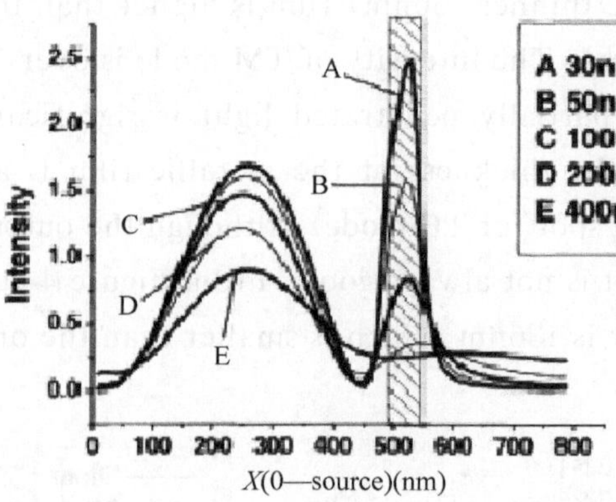

Fig. 4-63 Intensity profile along X through Au film 50nm thick with different aperture at the center for TM mode.

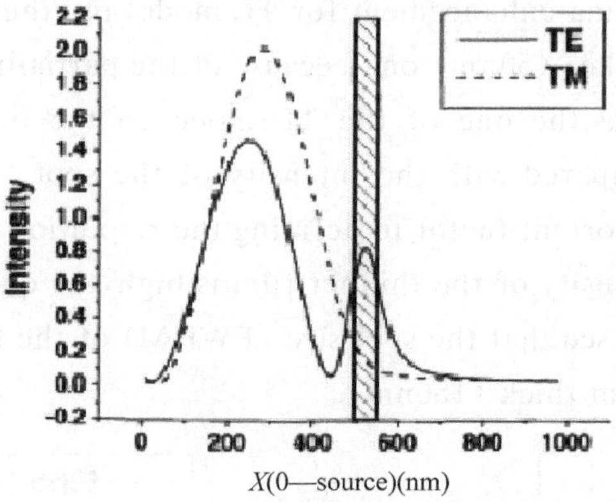

Fig. 4-64 Intensity profile along X through 50nm thilckness Au film ar 100nm wide aperture for TM and TE mode.

The spot size for the TM mode is smaller than the TE one when the distance less than or equal to 3s (about 20nm). The calculated spot size of the Au film 50nm thick at the boundary between the film's right side and the air as a function of the aperture width for two modes are also analyzed in Figure 4-66. One can see that the spot size increases and the increase for the TM mode keeps steady while on the contrary the increase for TE mode becomes sharper with increasing the distance. The spot size of the TM mode is smaller than the TE one when the aperture width is less than 130nm. The minimum spot size is about 32nm at the 30nm wide aperture for TM mode. Therefore, the output of the VSAL can produce the spot beyond the optical diffraction limit. And the TM mode source is more appropriate for near-field optical recording in compare with the TE mode because of its smaller size and higher power.

In conclusion, the Fox-li method has been used to calculate the base mode of the simply resonator of the VSAL, and using the two dimensions NL-FDTD simulation, we have analyzed the near-field optical characters of the output of the VSAL. The intensity profile of the base mode has been shown. The calculated intensity distributions along the transversal and longitudinal directions for TM and TE modes have been presented, respectively. They show that the spot sizes all increase with increasing the distance away from the metallic film, and decrease with decreasing the aperture

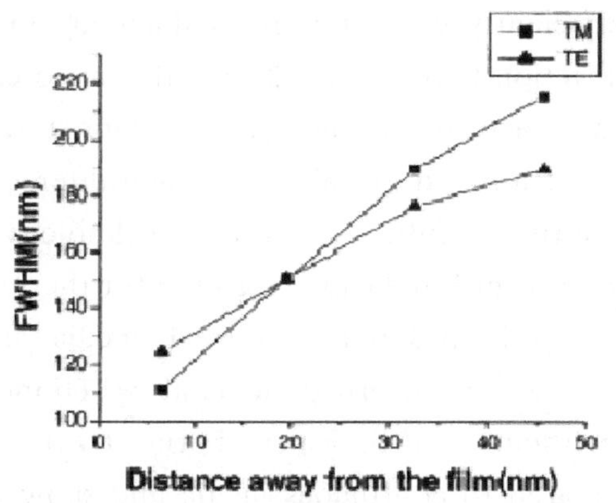

Fig. 4-65 The spot size (FWHM) as a function of the distance away from the film for TM and TE mode (Thickness = 50nm, aperture = 100nm).

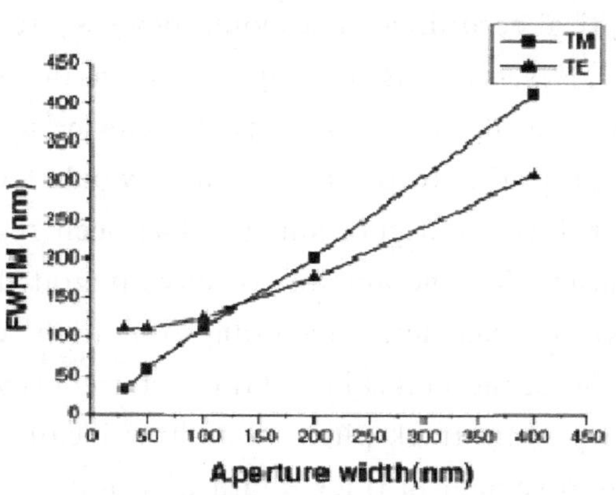

Fig. 4-66 The spot size (FWHM) as a function of the aperture width for TM and TE mode (thickness = 50nm, distance = 6.53nm).

width. But the power of the output for TM mode is higher one order than the YE one, and the decay of the intensity for former also sharper than the latter, because the surface plasma enhancement of the metallic film only appears in the TM mode. The spot size of the TE mode is smaller than the TM one either the aperture width is greater than 110nm or the distance is greater than 20nm with 100nm wide aperture at the right side of the film. But as a near-field recording source, the TM mode source is more suitable because of its higher power of the output.

4.6 Plasmonic near-field recording (PNFR)

The Super-RENS is capable of overcoming the diffraction limit and the issue caused by near field coupling, has been the subject of many studies. Optical resolution was increased by small size aperture generated in a mask layer which blocked a part of the laser spot. However, the method requires a relatively high laser power to generate the aperture in the mask layer not only for recording optical data but also for reading the recorded data. The Plasmonic near-field recording (PNFR) research is ongoing to solve practical issues related to the mask material and readout durability. The structure for an optical data storage medium consisting of a high-transmission metallic nano-aperture array in a dielectric layer can avoid the need to control gap distance for the near field coupling problem. Using the finite differential time domain (FDTD) method, calculate the electric field distribution generated by the nanoaperture array in the proposed data storage medium. The feasibility of the proposed structure is evaluated in terms of its recording power and density for optical data storage.

Both reverse saturable absorption and saturable absorption happened at different range of input power. The results of the optical pump-probe experiments show the temporal dynamic optical response that is closely related to the response time and recording rate of near-field optical recording. The interactions of the localized surface plasmon depended on the properties of the plasmonic materials and their local structures in nanometer scale. The effects of the localized surface plasmon are considered to be the key issue in super-resolution near-field phase-change optical recording. Plasmonic devices are capable of efficiently confining and enhancing optical fields, serving as a bridge between the realm of diffraction-limited optics and the nanoscale. Specifically, a plasmonic device can be used to locally heat a recording medium for data storage. Ideally, the recording medium would consist of individually addressable and non-interacting entities, a configuration that has been regarded as the ultimate future hard-drive technology. A plasmonic nano-antenna is fully integrated into a magnetic recording head and its use for thermally assisted magnetic recording on both continuous and fully-ordered patterned media using nanosecond pulses in a static tester configuration. In the case of patterned media at ~1Tb $inch^{-2}$ with 24nm track pitch, that show ideally written bits without disturbing neighboring tracks. An improvement in track width and optical efficiency compared to continuous media and show that this is largely due to advantageous near-field optical effects. Plasmonic devices are capable of efficiently confining and enhancing optical fields, serving as a bridge between the realm of diffraction-limited optics and the nanoscale. Specifically, a plasmonic device can be used to locally heat a recording medium for data storage. Ideally, the recording medium would consist of individually addressable and non-interacting entities, a configuration that has been regarded as the ultimate future hard-drive technology.

4.6.1 Holographic lithography (HL) application

Dr. Stefan Strauf etc. at Stevens Institute of Technology advanced the HL methodology by using four-beam interference and the concept of a compound lattice to create tunable twin motive shapes into a polymer template, resulting in metallic air gaps down to 7nm, seventy times smaller than the wavelengths of the blue laser light utilized to write the features. Scientists extended the utility of HL to create gaps with results comparable to laborious serial fabrication techniques such as electron beam lithography or focused ion beam milling. Besides being a simpler and more cost-effective production method, the technique does not require a clean room and currently achieves 90% uniformity in the array pattern. Therefore, these innovations provide the foundation for making high-quality, large-scale arrays at a greater speed and lower cost than previously realizable. Measuring the surface enhanced Raman scattering (SERS) effects that result from these arrays and continue to improve the uniformity of the arrays during fabrication.

They using four-beam interference and the concept of a compound lattice to create tunable twin motive shapes into a polymer template, resulting in metallic air gaps down to 7nm, seventy times smaller than the wavelengths of the blue laser light utilized to write the features. The 4-beam holographic lithography can be utilized to create plasmonic nanogap that are 70 times smaller than the laser wavelength (488nm). This was achieved by controlling phase, polarization and laser beam intensity in order to tune the relative spacing of the two sublattices in the interference pattern of a compound-lattice in combination with the nonlinear resist response. Exemplarily, twin and triplet motive features were designed and patterned into polymer in a single exposure step and then transferred into gold nanogap arrays resulting in an average gap size of 22nm and smallest features down to 7nm. These results extend the utility of high-throughput, wafer-scale holographic lithography into the realm of nanoplasmonics.

Nanometric gaps in noble metals can harness surface plasmons, collective excitations of the conduction electrons, for extreme subwavelength localization of electromagnetic energy. Positioning molecules within such metallic nanogaps dramatically enhances light-matter interactions, increasing absorption, emission, and surface-enhanced Raman scattering (SERS). However, the lack of reproducible high-throughput fabrication techniques with nanometric control over the gap size has limited practical applications. The sub-10nm metallic nanogap arrays with precise control of the gap's size, position, shape and orientation. The vertically oriented plasmonic nanogaps are formed between two metal structures by a sacrificial layer of ultrathin alumina grown using atomic layer deposition. The increasing local SERS enhancements of up to 10^9 as the nanogap size decreases to 5nm.

Plasmonic nanogap arrays are essentially uniformly placed metallic nanostructures which feature a tiny air gap between neighbors. By creating strongly confined electrical fields under optical illumination, these tiny air gaps allow scientists to use the arrays in a variety of applications, particularly in the miniaturization of photonic circuits and ultrasensitive sensing. Such sensors could be used to detect the presence of specific proteins or chemicals down to the level of single molecules, or employed in high-resolution microscopy. Nanophotonic circuits, able to transmit huge amounts of information, are considered crucial to bring about the exaflop processing era and a new generation in computing power.

Established fabrication techniques for nanogap arrays have focused on serial methods, which are time-consuming, have a low throughput, and are consequently expensive. Holographic lithography (HL), an optical approach that takes advantage of interference patterns of laser beams to create periodic patterns, had been previously demonstrated to create sub wavelength features.

Plasmonic nanogap arrays are essentially uniformly placed metallic nanostructures which feature a tiny air gap between neighbors. By creating strongly confined electrical fields under optical illumination, these tiny air gaps allow scientists to use the arrays in a variety of applications, particularly in the miniaturization of photonic circuits and ultrasensitive sensing. Such sensors could be used to detect the presence of specific proteins or chemicals down to the level of single molecules, or employed in high-resolution microscopy. Nanophotonic circuits, able to transmit huge amounts of information, are considered crucial to bring about the exaflop processing era and a new generation in computing power.

Established fabrication techniques for nanogap arrays have focused on serial methods, which are time-consuming, have a low throughput, and are consequently expensive. Holographic lithography (HL), an optical approach that takes advantage of interference patterns of laser beams to create periodic patterns, had been previously demonstrated to create sub wavelength features.

Disk-coupled dots-on-pillar antenna (D2PA) structure for surface enhanced Raman scattering. (a) Schematic; (b) top-view scanning electron micrograph (SEM); and (c) cross-sectional SEM of a D2PA structure, which consists of dense 3-D cavity nanoantennas (a metal disk array and a metal backplane on the top and the foot of the SiO_2 pillars respectively) coupled to, through nanogaps, dense plasmonic nanodots on the SiO_2 pillars' sidewall inside the cavity.

Although near-field microscopy has allowed optical imaging with sub-20nm resolution, the optical throughput of this technique is notoriously small. As a result, applications such as optical data storage have been impractical. However, with an optimized near-field transducer design, that optical energy can be transferred efficiently to a lossy metallic medium and yet remain confined in a spot that is much smaller than the diffraction limit. Such a transducer was integrated into a recording head and flown over a magnetic recording medium on a rotating disk. Optical power from a semiconductor laser at a wavelength of 650nm was efficiently coupled by the transducer into the medium to heat a 50nm track above the Curie point in nanoseconds and record data at an areal density of ~1Tb/inch^2. This transducer design should scale to even smaller optical spots.

Further increase the capacity of magnetic HDDs and reduce the bit-cell size by combining two new methods of magnetic recording: plasmonic near-field transducers and bitpatterned magnetic materials Near-field modelling and recording with patterned media. The modelled optical performances of our TAR head on continuous and patterned media. The cobalt media of Figure 4-67 was assumed to be patterned into 16nm-diameter, 20nm high pillars (on a 30nm-thick cobalt underlayer) in a hexagonal lattice with a bit pitch of 28nm. The pattern was transferred to a silicon substrate and a cobalt/palladium medium was deposited on top23 to form a geometry similar to that modelled, with air gaps between 15-20nm-diameter pillars on a 28nm pitch as shown in Figure 4-67(b). Magnetic recording media are made from ferromagnetic materials, such as Fe, Co, Ni and their alloys, and information is stored in localized regions (magnetic domains) that retain a

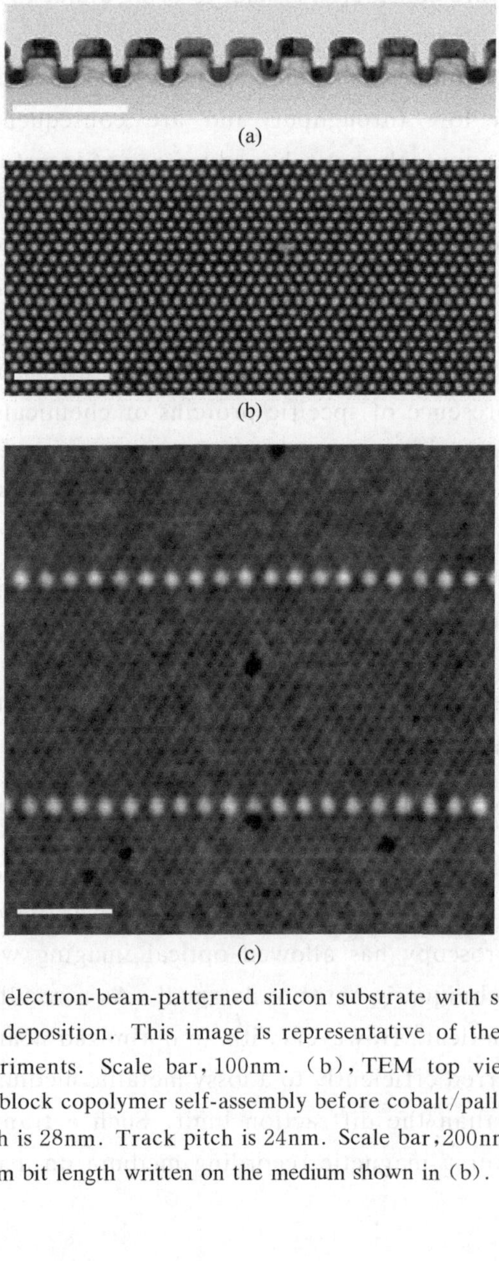

Fig. 4-67 (a), TEM side view of electron-beam-patterned silicon substrate with square lattice at 45nm island pitch after cobalt/palladium deposition. This image is representative of the media geometry for the higher-density recording experiments. Scale bar, 100nm. (b), TEM top view of silicon wafer patterned by electron-beamdirected block copolymer self-assembly before cobalt/palladium deposition. Areal density is 1Tb/inch^2. Island pitch is 28nm. Track pitch is 24nm. Scale bar, 200nm. (c), HR-MFM image of single-tone TAR tracks at 28nm bit length written on the medium shown in (b). Waveguide optical power is 6mW.

spontaneous magnetization that can be aligned in a certain direction by applying an external magnetic field. The main limitation to increasing the areal density of magnetic storage is the magnetic material used to store the data. If the area of the bit cell is reduced, the number of grains within it becomes smaller and smaller until it is no longer possible to distinguish between the grains and the cells. The solution is to reduce the grain size, but below a certain size the grains become unstable and their magnetizations can change arbitrarily.

The orange arrow shows the potential increase of areal density with time.

Longitudinal recording relied on in-plane-oriented magnetic domains on a granular medium.

Perpendicular recording on a granular medium uses smaller, off-plane-oriented magnetic domains. The use of low-magneticanisotropy materials means that this approach is subject to the superparamagnetic limit.

Emperature-assisted perpendicular recording on a granular medium. The use of high-magnetic-anisotropy materials means that this approach is not subject to the superparamagnetic limit, and it should, in principle, be possible to reach areal densities of 100Tb per square inch by using grains with sizes of 2-3nm.

Temperature-assisted perpendicular recording on an artificially patterned medium. Stipe and coworkers/report areal densities of about 1Tb per square inch, and this combined approach has the potential to become the next-generation HDD technology.

- Heat assisted magnetic recording and bit patterned recording

Heat assisted magnetic recording, or HAMR, promises to allow a switch to materials that are more stable than cobalt alloys and can hold magnetization in spots just a few nanometers across. HAMR and bit patterned recording has the potential to allow up to $50Tb/inch^2$. This would be 37.5TB hard 3.5 inch drives. The write speed obtained by the researchers was 250 megabits per second. The researchers only worked with one terabit per square inch densities but believe 10 terabits per square inch is possible. The technique combines two already-established recording methods, thermally-assisted magnetic (TAR) and bit-patterned recording (BPR).

A method for frequency-modulated coding of signals, data recording and storage in the IR and visible regimes used plasmonic nanostructures. The prototype using electron-beam lithography to print plasmonic nanostructures with a certain aspect ratio on a flat glass substrate etched with nano-sized cells. The different nanostructures printed will be effectively frequency-modulated recorded signals. To read this recorded signal, the substrate will be illuminated with a broadband IR or visible light from below. Due to the plasmonic resonance phenomenon, when the plasmonic nanostructure is illuminated with an IR or visible light, which could be broadband, the scattered wave from this structure can be very strong at a given frequency. This resonant frequency can then be read by a NSOM (near-field scanning optical microscope). Using plasmonic nanostructures for optical data storage provides higher storage capacities by using nano-scale-sized structures with various aspect ratios made of plasmonic materials and can lead too smaller unit size for memory in the storage media. The method also provides a technique for optical data storage for N-ary data (i.e. not limited to the current Base 2 system), meaning that the device offers unique and exciting possibilities for future data storage, memory and chip designs. In a $1\mu m 1\mu m$ surface area, with a diameter of 17.5nm, this device will have around 3260 cells, whereas a regular CD has about 4 cells in such an area. Some scientists have noted that nanostructures could offer another thousand-fold increase in data-storage density.

- Photonic crystals and plasmonic structures recorded by multi-exposure of holographic patterns

Different technologies can be used for fabrication of photonic crystals such as: self-assembly of colloidal particles, e-beam lithography (EB), interference lithography (IL) and focused on beam (FIB). Among them, the holographic lithography (HL) is the only technique that is able to fabricate both two-dimensional and three-dimensional photonic crystals, as well as plasmonic structures, in large areas. In this paper we demonstrate the use of the multi-exposure of two-beam interference patterns, with rotation of the sample around different axis, for fabrication of large areas 2D and 3D photonic crystals and plasmonicstructures. Using this technique, achieved aspect ratios of about 4 in 2D photoresist templates recorded in $1cm^2$ glass substrates. In order to generate the 2D photonic band gap layers and plasmonic structures, we combine the use the high aspect ratio photoresist templates with shadow evaporation of appropriated materials, with a further lift-off of the photoresist.

4.6.2 Plasmonic nanostructures

(1) Effects in weakly dissipating plasmonic materials and the properties of different plasmonic structures are investigated last 10 years. In contrast to conventional Rayleigh scattering where only the dipole resonance is pronounced, small plasmonic particles demonstrated the higher order resonances with inverse hierarchy i. e. quadrupole resonance is higher than dipole, etc. There are also dramatic differences between the Rayleigh and anomalous light scatterings in the near fields, which can be seen in the Poynting vector distribution in the Figure 4-68. Various applications of the anomalous scattering in nanotechnologies open new prospects for optical manipulation in field structures at the nanoscale level.

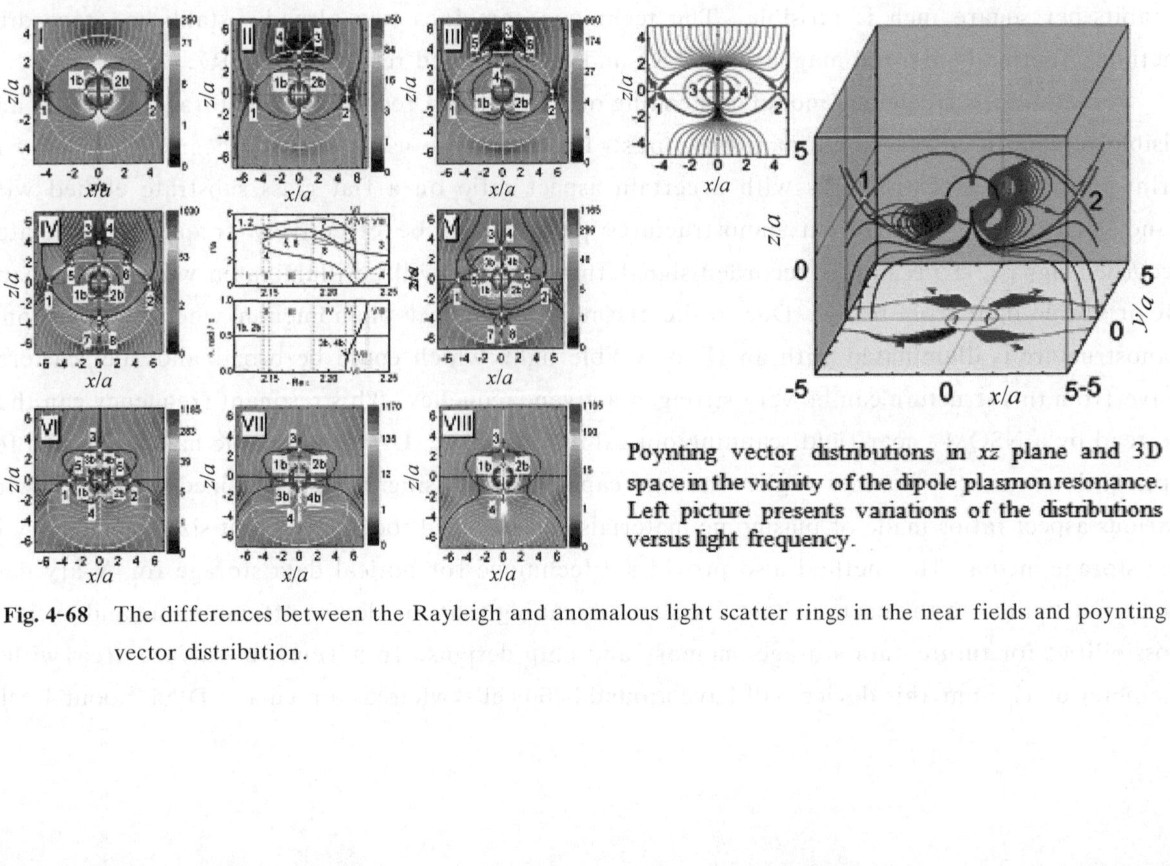
Poynting vector distributions in xz plane and 3D space in the vicinity of the dipole plasmon resonance. Left picture presents variations of the distributions versus light frequency.

Fig. 4-68 The differences between the Rayleigh and anomalous light scatter rings in the near fields and poynting vector distribution.

(2) Fano resonance in plasmonic materials and metamaterials. Since its discovery, the asymmetric Fano resonance has been a characteristic feature of interacting quantum systems. The shape of this resonance is distinctively different from that of conventional symmetric resonance curves as shown in Figure 4-69. Recently, the Fano resonance has been found in plasmonic nanoparticles, photonic crystals, and electro magnetic metamaterials. The steep dispersion of the Fano resonance profile promises applications in sensors, lasing, switching, and nonlinear and slow-light devices.

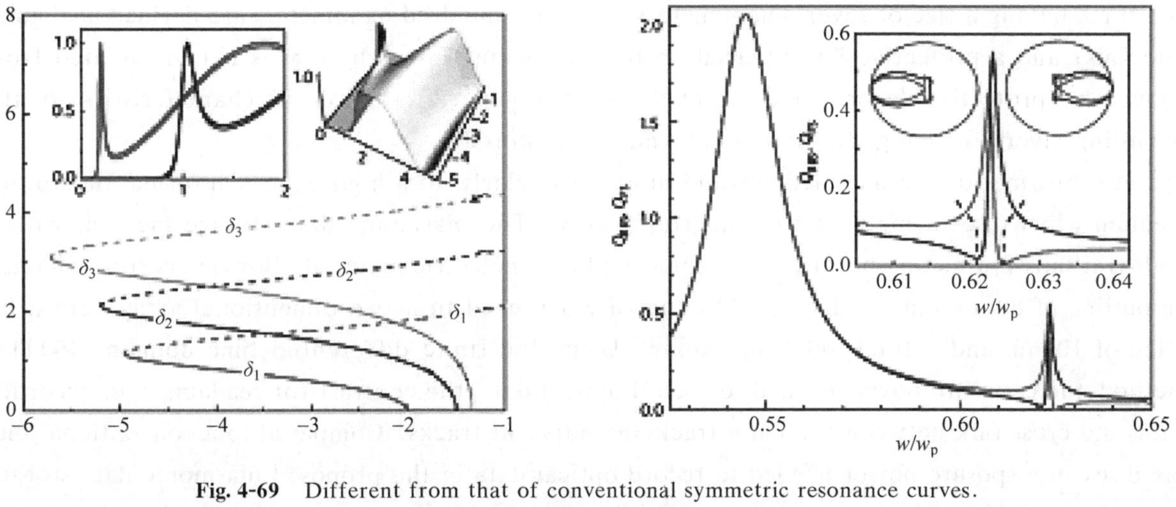

Fig. 4-69 Different from that of conventional symmetric resonance curves.

(3) Huge light scattering from active anisotropic particles as shown in Figure 4-70.

Latest discoveries on nanolasers were based on nanoshell plasmonic structures embedding in active dielectric shell. A novel structure based on active anisotropic materials was suggested, where the radial component of dielectric permittivity is plasmonic and tangential component corresponds to active dielectric. In such structures it is possible to reach huge scattering effects at some specific resonant conditions. This yields a new idea for creation of very efficient nanolasers.

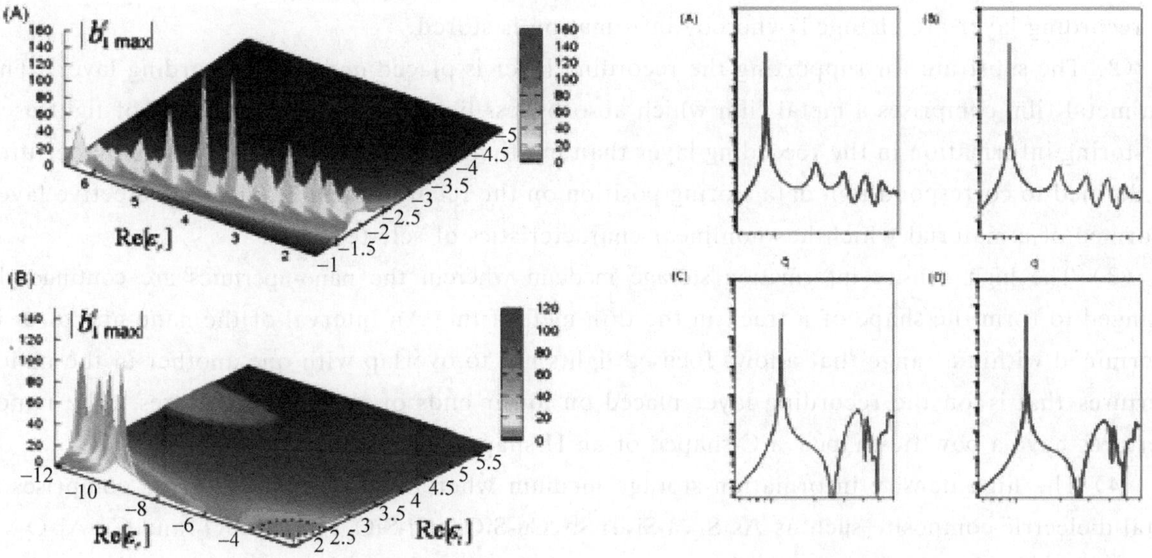

Fig. 4-70 Huge light scattering from active anisotropic particles.

4.6.3 Plasmonic storage medium

The accomplishment of high density by the use of a short wavelength laser beam (a blue laser beam of 405nm) and a high numerical aperture (NA = 0.85) has theoretical and technical limits, and thus, a new way for realizing higher storage capacity is demanded in the art. In one aspect, a high density information storage medium comprises a recording layer in which information is stored; a thin metal film placed on the recording layer and having a structure in which nano-apertures having a size of several nanometers to several hundred nanometers are defined at regular intervals; and a protective layer placed on the thin metal film, wherein, as light irradiated from above the protective layer passes through the nano-apertures, physical characteristics of the recording layer are changed, whereby information is stored.

An information storage medium, and more particularly, to a high density information storage medium which uses a metallic nano-aperture array. The plasmonic data storage medium with a high-transmission metal aperture array embedded in a dielectric material. Bowtie apertures, having an outline of 80nm and a ridge gap of 30nm, are arranged in a two dimensional array with a bit pitch of 100nm and a track pitch of 280nm. Using the finite differential time domain (FDTD) method, the exposure power needed to record optical data, the contrast for readability of recorded data, and cross talk between the main track and adjacent tracks. Compared to a conventional blu-ray disc, the exposure power needed to record optical data in the proposed plasmonic data storage medium is less than a quarter of the conventional threshold power, and the density of the data storage is about 1.8 times larger, the performances as following:

(1) A high density information storage medium comprising: a recording layer in which information is stored; a thin metal film placed on the recording layer and having a structure in which nano-apertures having a size of several nanometers to several hundred nanometers are defined at regular intervals; and a protective layer placed on the thin metal film, wherein, as light irradiated from above the protective layer passes through the nano-apertures, physical characteristics of the recording layer are changed, whereby information is stored.

(2) The substrate for supporting the recording layer is placed under the recording layer. The thin metal film comprises a metal film which absorbs less light in a wavelength band of light used for storing information in the recording layer than in other wavelength bands. The nano-apertures are defined to correspond to a data storing position on the recording layer and the protective layer is formed of a material which has nonlinear characteristics of self-focusing.

(3) The high density information storage medium wherein the nano-apertures are continuously arranged to form the shape of a track in the thin metal film. An interval of the nano-apertures is determined within a range that allows focused lights not to overlap with one another in the nano-apertures, that is, on the recording layer placed on lower ends of the nano-apertures. The nano-apertures have a bow tie-shaped, a C-shaped or an H-shaped transverse section.

(4) The high density information storage medium wherein the protective layer comprises a metal-dielectric composite such as As_2S_3, a-Si, InSb, Cu-SiO_2, Ni-SiO_2, Cu-Ni-SiO_2 and Cu-Al_2O_3, a semiconductor quantum dot dielectric composite material, or a composite material in which a compound of Ⅱ-Ⅳ groups or a compound of Ⅲ-Ⅴ groups is dispersed in glass or resin. A different kind of metal film is additionally formed between the thin metal film and the protective layer. A

lower metal layer is additionally formed between the substrate and the recording layer. The dielectric layer is interposed at least one of between the substrate and the recording layer and between the recording layer and the thin metal film.

A high density information storage medium is a plan view illustrating a compact disk in which a thin metal film defined with nano-apertures according to the embodiment of the present invention is adopted as shown in Figure 4-71. The method for storing information in the high density information storage medium according to the embodiment of the present invention. In Figure 4-71, an information storage medium according to an embodiment of the present invention includes a substrate 1, a recording layer 2, a thin metal film 3 having nano-apertures 4 and a protective layer 5. The substrate 1 can be made of glass, ceramic or resin. In particular, the substrate 1 can be formed of a material which is light in weight and has excellent injection moldability and low birefringence when emitting a laser beam, for example, such as polycarbonate. The recording layer 2 serves as a layer on and from which information is recorded and reproduced. The recording layer 2 can comprise a phase change material layer such as GeSbTe and AgInSbTe or a semiconductor material layer such as Si, Ge and ZnO. Such recording layer 120 undergoes a reversible variation in terms of the index of refraction. The recording layer 2 can be formed as a single layer or multiple layers, and is required to have a thickness capable of reducing the noise level of a reproduced signal and securing sufficient recording sensitivity. Dielectrics 2a and 2b can be formed on the upper and lower surfaces, respectively, of the recording layer 120. The dielectrics 2a and 2b function to physically and chemically protect the recording layer 2 and to prevent recorded information from degrading.

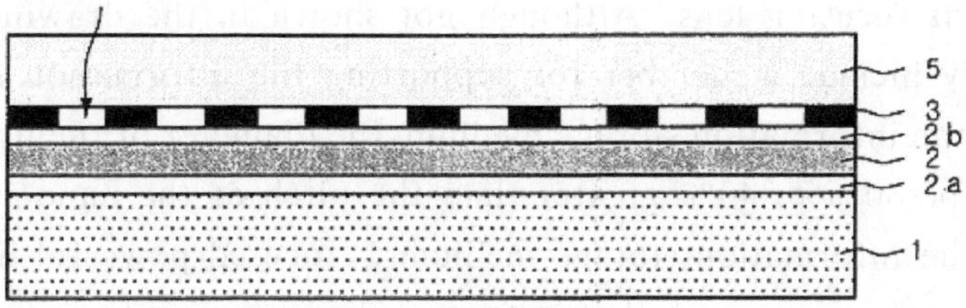

Fig. 4-71 A typical construction of the high density information storage medium.

The thin metal film 3 can be formed of a material which can absorb less light in a wavelength band of light used for the storage and reading of information and can easily produce surface plasmon. In some cases, the thin metal film 3 can be opaque to light incident thereon. For example, when a light source having the wavelength of 400nm is used, an aluminum (Al) film can be employed, and when a light source having the wavelength of 600nm is used, a gold (Au) film can be employed. Here, the thickness of the thin metal film 3 can be determined by considering the transmittance and the thickness of the recording layer 2. For example, in the embodiment of the present invention, the thin metal film 3 can be formed to have a thickness of a few tens of nanometers and several hundreds nanometers. As the nano-apertures 4 having a dimension of several nanometers to several hundred nanometers are designed to have high transmittance of incident light, producing small and bright light spots. The nano-apertures 4 can be designed by considering an information storing position on the recording layer 2. The nano-apertures 4 can be defined to correspond to the information storing position and can be arranged at an interval of about 100 to 200nm so that information can be recorded with high density. At this time, the interval can be determined within a range that allows focused lights not to overlap with one another in the nano-apertures 4. In other words, the interval of the nano-apertures 4 is determined within a range that allows lights reaching the recording layer 2 after passing through the nano-apertures 4 not to overlap with one another. The nano-apertures 4 can have various transverse sectional shapes such as of a bow tie as letter H.

The protective layer 5 is placed on the upper surface of the thin metal film 3. The protective layer 5 can be formed of a nonlinear optical material which has self-focusing effect that the index of refraction changes depending upon the intensity of incident light and of which nonlinear characteristics vary. If the protective layer 5 is formed of the nonlinear optical material, since the flux of light focused on the upper surface of the thin metal film 3 can be decreased, recording density can be increased.

If light is incident on the protective layer 5, self-focusing occurs in the nonlinear optical material, and the numerical aperture of an optical system (a focusing lens not shown) increases, so that the diameter of the flux of focused light can be decreased. The protective layer 150 can comprise a metal-dielectric composite such as As_2S_3, a-Si, InSb, Cu-SiO_2, Ni-SiO_2, Cu-Ni-SiO_2 and Cu-Al_2O_3, a semiconductor quantum dot dielectric composite material, or a composite material in which a compound is dispersed in glass or resin. Information can be stored by focusing laser beams, and the position and the characteristic of information can be read out by the light reflected from the information storage medium 1.

Namely, the information storage medium 1 according to the embodiment can be mounted to a light pick-up device. The light pick-up device can be constituted by an optical system including a light source and a light focusing lens. Although not shown in the drawing, the optical pick-up device can additionally include a member for supporting the information storage medium. If a laser beam focuses to the information storage medium, the diameter of the light flux focused by the focusing lens 220 can be 30% to 40% greater than the pitch of the nano-apertures 4. Here, if a strong laser beam to the information storage medium 1, focused powerful light is applied toward the lower ends of the nano-apertures 4, that is, toward the recording layer 2, by the surface plasma produced around the nano-apertures 4, such that the physical characteristic of the recording layer is changed by the focused light and information is thereby stored. When information is stored using the light focused by the nano-apertures 4, the size of an area in which information is recorded may be a few tens of nanometers much smaller than the pitch (line width) of the nano-apertures 4, and this size can be changed by varying the intensity of the laser beam. That is to say, due to the fact that the thin metal layer 3 having the nano-apertures 4 is placed in the information storage medium 1, light having flux significantly increased when compared to actual incident light can be applied to the recording layer 2, whereby the energy of the laser beam required for storing information can be decreased. According to this, a data storing speed can be increased and the output of the laser beam can be reduced.

Different kind of metal film can be additionally formed between the thin metal film and the protective layer so that the reflectance of the light reflected toward the lower ends of the nano-apertures can be increased. The different kind of metal film can comprise a metal film which is formed of chrome and nickel having excellent conductivity characteristics. By additionally forming the different kind of metal film in this way, the contact characteristic between the materials of the protective layer and the thin metal film can be improved. Further, by the additional formation of the metal film, when stored information is subsequently analyzed, the reflectance of the light reflected from an information storing area can be increased. The metal layer (hereinafter referred to as a lower metal layer) can be additionally interposed between the substrate and the recording layer. The lower metal layer can also be formed of chrome, nickel or aluminum having excellent reflection characteristics. By the additional interposition of the lower metal layer, light which is likely to be absorbed into the substrate can be focused on the recording layer, whereby the focus ability of light can be improved and reflectance of light toward the lower ends of the nano-apertures can be increased.

The thin metal film having nano-apertures is formed in an information storage medium such as a CD, a DVD or a BD, so that, when irradiating light to store and read data, the light can be highly transmitted through the nano-apertures to a recording layer of the information storage medium. According to this, as the diameter of light flux is remarkably decreased by the surface plasmon produced underneath the nano-apertures, data recording density can be significantly increased. Further, even when reading data, a data storing position can be easily detected by analyzing the light reflected through the nano-apertures.

4.6.4 Nanogap control with optical antennas (Metallic nanoantennas)

The optical antennas have been over the past ten years, at the heart of numerous investigations due to their optical properties, like strong enhancement and subwavelength confinement of the electrical field. The coupling between photons and electrons in these nano-objects is the origin of their extraordinary optical properties. In such 3D structures, the electrons oscillate with (spatial) periods smaller than the wavelength of the incident photons, raising plasmon resonances. Such resonances can then give rise to a strongly localized field enhancement. In addition, nanoantennas permit to couple propagative optical waves into evanescent waves and vice versa. They are therefore promising as intermediary between nanoscale optical and electronic devices and optical fields. Their applications include different fields such as nanoscale imaging and spectroscopy, light-emitting devices, photovoltaics, microfluidics, and metamaterials in the infrared.

Recently have demonstrated that single fluorescent molecules can be used as non-perturbative vectorial probes of the local field. Here, expand on such experiments exploiting fluorescence lifetime of single molecules to probe various types of gap nanoantennas. First, studies of the nanoantennas are carried out to evaluate the electric field. Then investigate hybrid systems composed by nanoantennas and randomly positioned fluorescent molecules. Finally, present a fabrication scheme for the controlled placement of fluorescent molecules at well-defined positions with respect to the dimer nanoantenna, which is a more direct route to probe the local field in an a priori determined way.

Now concentrate on gap nanoantennas that can be defined as two symmetric metallic particles (of specific shape and size) separated by a fabricated gap ranging from no gap up to a few tenths of nanometers. In the particular case of gap nanoantennas, the field enhancement and resonance modes have been theoretically predicted and experimentally studied. Numerous works have been reported on luminescence spectroscopy of gold nanoparticles such as nanorods and dimers. Two-photon luminescence (TPL) has also been employed and proven particularly sensitive to local fields near metallic nano-objects. Moreover, TPL spectroscopy permits to measure the plasmon resonances of such nanostructures by evaluating the TPL as a function of excitation wavelengths. Near-field scanning optical microscopy (NSOM) is also extensively used to explore such nano-objects either by locally mapping the field enhancements or by grafting an individual nanoantenna at the apex of an NSOM probe. However, in these studies, the influence of the near-field probe onto the nanostructures can induce a complex coupling response which has to be taken into account. In particular, plasmonics mode detailed mapping of nanoantennas has been achieved with "apertureless" NSOM using Si-tips. Other studies have also looked at the fundamental understanding of resonant processes in such nanostructures by means of Raman imaging or nonlinear effects. In parallel, analytical and numerical models have been developed in an approach to unify the experimental observations. An alternative approach study nanoantennas that is to probe their near fields using single molecules (SM). By doing so, one can access the local mapping of the nanoantenna fields in a nonperturbative way. Furthermore, integrating SM together with metallic nanoantennas is an ultimate step in terms of applications for the technological miniaturization drive. In most cases, the functionality of SM is inherently defined by the electronic or optical properties of the molecules. However, the construction of active molecular-scale devices often requires embedding the molecule in a nanostructured environment that provides a means to address, tailor, or control the molecular functionality. Gap nanoantennas are particularly promising interfaces, because of their strongly localized electromagnetic field modes, both in the spatial and the frequency domain. Consequently, the emerging field of molecular plasmonics holds great promise for applications in areas like sensing, light harvesting and energy conversion, single-photon sources, all-optical components, and electronic-optical interfacing. However, a major task remains in the mutually well-defined positioning of both functional molecules and metallic nanostructures with nanometer-scale precision. Different approaches have been developed relying on random deposition of functional molecules, for example, by spin-coating, bulk overcoating of a metallic structure, or plasmonic micropositioning. In highly symmetric nanoparticle systems, one degree of positional control has been achieved by employing a spacer layer that can be functionalized. The full exploitation of molecular functionality requires localized and designed molecular positioning at arbitrary locations with respect to a tailored plasmonic nanostructure. Recently, encouraging progress in this direction has been made by double e-beam lithography, positioning a single quantum dot on the feed element of an array antenna. The full characterization of antennas needed to realize resonant nanoantenna structures overcoated with low ("single molecule") concentrations of fluorescent molecules. This characterization serves to extract the individual optical response of the nanoantennas in an effort to match the fluorescence spectra of molecules. The fluorescence measurements on nanoantennas over-coated with ultralow concentrations of molecules address measurement of the fluorescence lifetime, to map the local density of states around the nanoantennas, in direct competition with the intrinsic luminescence of the antennas. A fabrication procedure allows for the controlled positioning of molecules with respect to the antenna, together with initial related results.

The nanoantenna samples (dimer and nanorod arrays), explaining the fabrication process and characterizing their spectroscopic response, describe one photon luminescence of dimer antennas. These lay the basis to evaluate the optical response of the nanoantennas, without (dye) molecules but under excitation conditions typical for single-molecule spectroscopy, then discuss the response of randomly scattered single molecules over dimers. Finally, molecular probing will be refined using a well-controlled molecular probing.

The nanoantenna electromagnetic response is known to depend highly on size, geometry, and even chemical treatment. A large panel of sizes, shapes, and compositions has therefore been fabricated at the nanoscale either by bottom-up colloidal chemistry or by top-down nanofabrication techniques with the use of electron beam lithography or ion beam milling. A multitude of shapes is emerging offering a wide range of field pattern and strength. Here, we will report on dimer and nanorod-like structures. The parameters specifying the geometry of the dimer and nanorod-like structures are the aspect ratio (AR), defined as the ratio of length to width, the gap separation distance between the two nanoparticles constituting the gap nanoantenna, and the thickness. The single nanoantennas in these studies have an AR in the range from 1. 0 (dimers) to 4. 5 (nanorods). It kept the width around 100nm and varied the length of the structures. The different fabrication steps are depicted that used conventional e-beam lithography and metal depositions followed by a lift-off process to realize periodic arrays ($100 \times 100\mu m$) of nanoantennas for example. Prior to any investigation, calculated the spectroscopic response (in the excitation wavelength range of 400nm to 700nm) for the given geometry and performed far-field spectroscopic measurements on the arrays of fabricated nanostructures.

Using conventional e-beam lithography and metal depositions followed by a lift-off process to realize periodic arrays ($100 \times 100\mu m$) of nanoantennas that prior to any investigation, calculated the spectroscopic response (in the excitation wavelength range of 400nm to 700nm) for the given geometry and performed far-field spectroscopic measurements on the arrays of fabricated nanostructures. The theoretical curves depict the electric field in the gap of the structures, whereas the data represent the response of the total structure. The two insets give the total electric field distribution for the 450nm long rods at the peaks A (at 580nm) and B (at 500nm).

The measured scattering spectra of dimers, for both excitation s-and p-polarization states, and the calculated spectral response of an individual dimer are displayed. Under p-polarization excitation, the measured peak is close to 635nm, while the calculation presents a peak at around 655nm. This experimental blue shift may be explained both by geometrical variations occurring during the fabrication process and by substrate effects. The numerical work was carried out for a single dimer in free standing space. Under s-polarization excitation (electric field perpendicular to the dimer axis), a very weak peak centered on 599nm is observed. For this wavelength, the simulation does not present any peak as the field distribution is mainly located at the edges of the dimers and not in the gap from where the calculated spectrum is extracted. Finally, secondary peaks are predicted at around 500nm for both polarization states but were not experimentally observed. Neglecting the effect of the substrate as well as retardation effects (more pronounced for shorter wavelengths) in the simulation may explain the discrepancy over these auxiliary peaks.

The nanoscale dimensions of optical antennas bring with them associated characterization challenges. Resonant light scattering is a popular technique both theoretically and experimentally for studying size and shape related optical resonances in antenna structures. Linear antenna structures, such as anorods or nanostrips, are strongly analogous to Fabry-Perot resonators and can support higher-order resonances in addition to the fundamental dipole mode. A direct visualization of field distributions around antennas using near-field scanning microscopy has been demonstrated in the mid-IR for micrometer-sized structures, but the absence of ultrasmall probes capable of resolving details of the order of a nanometer currently impedes the extension of this technique to antennas for the visible. The only technique that currently approaches such level of detail is electron energy loss spectroscopy, which utilizes a tightly focused electron beam to probe the LDOS directly. This nonoptical technique has been used to map energy-resolved Plasmon eigenmodes on single nanoparticles.

An alternative characterization modality uses the nonlinear responses of antennas. TPL is a second-order process especially suited for mapping out intensity hot spots in antennas generating a high degree of field localization, such as the bowtie, half-wave or gap antennas. Figure 4-72 shows field intensities in antenna test structures revealed by far-field TPL measurements using femtosecond laser excitation. As expected, the strongest enhancements arise in the gap region when the incoming light is polarized along the length of the antenna. To increase the spatial resolution in TPL imaging, Bouhelier et al. employed a sharp tip to locally scatter the TPL signal generated by the antenna under study. The tip is raster scanned over the laser-irradiated antenna in close proximity, and the TPL intensity is measured as a function of the tip's scan coordinates. In order not to perturb the antenna's behavior, care must be taken to avoid any strong interaction between the local scatterer and the antenna under study.

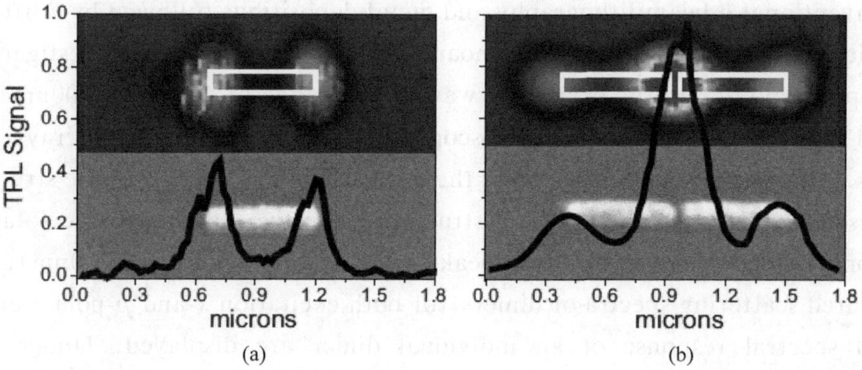

Fig. 4-72 Two-photon excited luminescence as a tool to characterize field intensity distributions in optical antennas. (a) Resonantly excited linear nanostrip antenna showing field enhancement at the ends. (b) TPL from a gap antenna, indicating a strong field enhancement in the gap for incoming light polarized along the long axis.

As an aside, one must note the difference in the role of the feed gap between the radio and optical regimes. The feed gap in a radio antenna is typically impedance matched to a generator (source) and is not a point of high localized energy density. This matching ensures that the two antenna segments do not feel a gap or discontinuity between them. The gap in an optical gap antenna, in contrast, is a region of high local field intensity and dictates the antenna's overall optical response. It is a high impedance point due to the large LDOS and the efforts over the years to engineer the gap for strongest field enhancement are a direct consequence. The mismatched gap can even be turned into an advantage as it provides a means to tune antenna properties, e. g., by loading the gap with various nanoloads. Such a tuning may enable one to go beyond simple field enhancement and might lead to truly impedance-matched energy transfer between a localized source and the antenna.

Theoretical spectra are presented for nanorods for each excitation polarization state, both in the nonretarded and the retarded regime. In the retarded regime, the results are the solutions of the wave equation, whereas in the nonretarded regime, the time dependence has been neglected.

The different nanorods' lengths are highlighted in the colors of the curves, red for the 150nm long, blue for the 350nm, and green for the 450nm. For the *p*-polarization state, a 2-to 4-fold enhancement, defined as the ratio of the electric field in the presence and in the absence of the nanoantennas, is reached, both theoretically (in the retarded regime) and experimentally. The theoretical values obtained in the quasistatic regime are much higher, apart for the case of a nanoantenna with an AR of 1.5. This is to be expected as the time dependence becomes more important for structures with increasing sizes where the distances can be larger than the excitation wavelength. The two models can be directly compared that for the 150nm long rods the quasistatic and retarded responses are similar, whereas for the 450nm long rods, this does not hold anymore. Furthermore, if compare simulations in the retarded regime with the experiment can see that the experimentally observed intensities are mainly higher than the simulated ones. This observation can also be rationalized by the contribution of the additional field imaged on the edges of the structures during the measurements or in the field distributions. However, the theoretical spectra concentrate only on the electric field in the gap.

4.6.5 Plasmonic nano-structures for optical storage

A method of optical data storage that exploits the small dimensions of metallic nano-particles and/or nano-structures can achieve high storage densities. The resonant behavior of these particles (both individual and in small clusters) in the presence of ultraviolet, visible, and near-infrared light may be used to retrieve pre-recorded information by far-field spectroscopic optical detection. In plasmonic data storage, a femtosecond laser pulse is focused to a diffraction-limited spot over a small region of an optical disk containing metallic nano-structures. The digital information stored in each bit-cell modifies the spectrum of the femtosecond light pulse, which is subsequently detected in transmission (or reflection) using an optical spectrum analyzer. The session will present theoretical as well as preliminary experimental results that confirm the potential of plasmonic nano-structures for high-density optical storage applications.

Metallic nano-structures exhibit strong resonances when illuminated with ultraviolet, visible, or near-infrared light in the vicinity of their surface plasmon polariton (SPP) frequencies. These SPP resonance frequencies are sensitive to the geometry and dimensions of the nano-structure, e.g., diameter and depth of a pit or a hole in a metal film, diameter and length of a metallic nano-rod, axial dimensions of an ellipsoidal nano-particle, etc. The resonances are also dependent on the orientation of the nano-structure relative to the polarization state of the incident light. In addition to sensitivity to polarization and wavelength, metallic nano-structures exhibit strong interactions with their environment and with each other; for example, optical transmission through one nano-hole is strongly modulated by the presence of other nano-holes in the neighborhood. A method of optical data storage that exploits the small dimensions of metallic nano-particles can achieve high data densities. It employs the resonant behavior of these particles (both individual and in small clusters) for the purpose of retrieving the stored information using spectroscopic far-field detection. The nanoparticles should be arranged in such a way as to imprint their signature in a unique way on the optical spectrum of the readout laser beam. It should be emphasized at the outset that the large-scale fabrication of such nano-structures in a reliable and cost-effective way is far from trivial for the present-day manufacturing technologies. It is our hope, however, that an exploration of plasmonic nano-structures in the context of optical data storage will bring attention to the unique properties and potential advantages of such structures, thus spurring the development of tools and techniques for their large-scale fabrication.

Conventional methods of optical disk data storage as employed in Compact Disc (CD), Digital Versatile Disc (DVD), and Blue Ray Disc (BD), as well as in the recordable and rewritable versions of these media are well-known. The spectral hole-burning, three-dimensional (3D) optical storage by two-photon point excitation, 3D storage in photochromic and photorefractive materials, and the recently announced five-dimensional optical recording mediated by surface plasmons in gold nano-rods. The plasmonic scheme of optical data storage presented in the sections has some similarities but also major differences with the plasmon-mediated method employed by Zijlstra et al. Compare the two methods and point out their similarities and differences. The principle of plasmonic data storage and demonstrate the feasibility of the concept using numerical simulations and experimental data obtained from nano-apertures milled in thin silver films. Figure 4-61 shows a possible realization of an optical storage medium that incorporates nano-holes and/or nano-slits in a thin metallic film. The data bits are grouped together in small clusters and placed within individual bit-cells, each cell containing several bits of information. As an example, a typical bit-cell may occupy a $0.5 \times 0.5 \mu m^2$ area on the surface of a $0.2 \mu m$-thick silver film, each bit-cell containing ten or more nano-holes whose individual diameters could range from, say, 20nm to 100nm. If, in a given cluster, the presence or absence of a nano-hole of a specific-size is associated with a single information bit (0 or 1), then nano-holes can encode an m-bit sequence within each bit-cell. Transmission of light through a nano-hole (or nano-slit) is a strong function of the aperture diameter and film thickness, as well as the size, shape, and location of the neighboring nano-apertures. For a given state of polarization of the incident beam, certain wavelengths couple strongly to the guided mode through a nano-aperture and reach the opposite side, while other wavelengths are either reflected from the metallic surface or resonantly transmitted through adjacent nanoapertures; see Figure 4-73. It is this property of the nano-holes and nano-slits that provides a mechanism for readout of the stored information. (Although Figure 4-73 shows one track containing nano-holes and an adjacent track containing nano-slits, there is no a priori reason for distinguishing between the two; in other words, it should be possible to mix nano-slits with circular as well as elliptical nano-holes in arbitrary combinations and arrangements.)

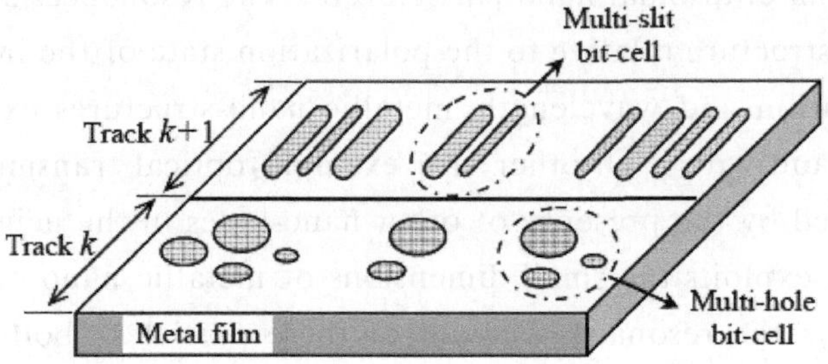

Fig. 4-73 In one realization of the proposed concept, plasmonic features are nano-holes and/or nano-slits in a thin metallic film. A group of such features constitutes a bit-cell, within which several bits of information are encoded in a small (micron-sized) region of the storage medium.

Much like the organization of data on a conventional optical disk, these bit-cells are arranged sequentially along parallel data tracks. Computer simulations indicate that the transmission spectra of nano-hole and/or nano-slit clusters in thin metallic films exhibit multiple resonance peaks, which can be used to identify the structure during readout. Figure 4-74(a) shows a number of transmission spectra computed with the FDTD method for a 250nm-thick silver film suspended in free space. Single, double, and triple nano-holes with various spacings, all filled with a dielectric of refractive index $n_0 = 2.0$, are depicted in this figure. The assumed wavelengths cover the visible range of 400-700nm, although, in principle, the range can be extended to the ultraviolet and nearinfrared wavelengths as well. Similar transmission curves for multiple air-filled (i. e., $n_0 =$ 1.0) nano-slits in a suspended 400nm-thick silver film are shown in Figure 4-74(b).

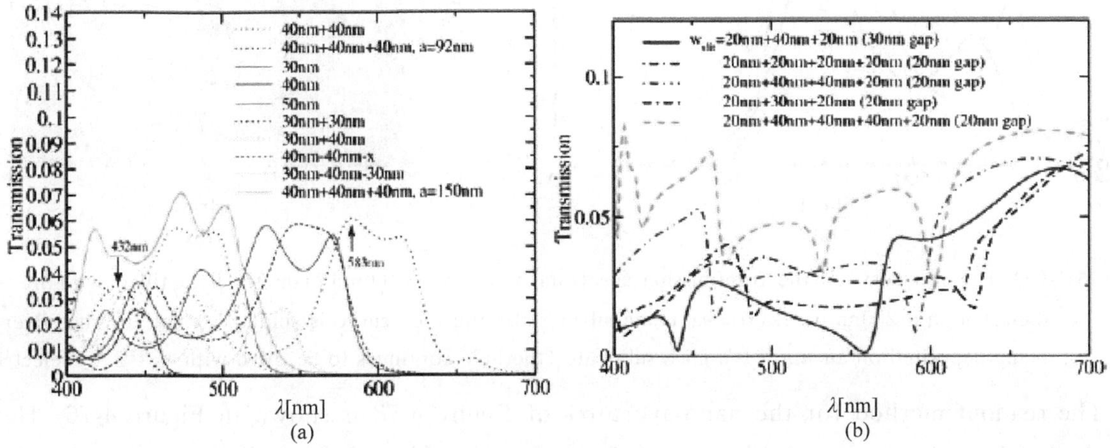

Fig. 4-74 Computed transmissivity versus the vacuum wavelength λ for (a) nano-holes and (b) nano-slits in a silver slab. The Finite Difference Time Domain (FDTD) method has been used to solve Maxwell's equations; transmissivity is defined as the fraction of total incident optical power at each wavelength. The regions on both the incidence and transmission sides of the silver slab are free-space ($n = 1$), and the Drude model is used to simulate the dispersion of the complex dielectric constant $\varepsilon(\omega)$ of silver. (a) The 250nm-thick silver film contains single, double, and triple cylindrical holes of varying diameters, as indicated in the legend by the radii of the holes. All the holes are filled with a transparent dielectric of refractive index $n_0 = 2.0$. The incident beam is a focused Gaussian, linearly-polarized along the y-axis, and having FWHM $= 1\mu$m. The waist of the focused spot is in the xy-plane at a distance of $\Delta z = 55$nm above the top surface of the silver film. Variations of transmissivity with λ are due to resonances within each hole as well as plasmonic coupling between adjacent holes. (b) Multiple air-filled slits ($n_0 = 1$) having widths $W_1 = 20$nm, $W_2 = 30$nm, and $W_3 = 40$nm. Different combinations of these slits are embedded within a 400nm-thick silver film and illuminated with a focused Gaussian beam. Each cluster of slits has a unique transmission spectrum, which could be exploited for identification of the cluster during readout.

Optical properties of the substrate and the surrounding layers, if any, play an important role in determining the details of the transmission/reflection spectra. For example, Figure 4-75(a) shows the transmission spectra of a 100nm-diameter hole (filled with $n_0 = 2.0$ dielectric) with and without a substrate. Presence of a substrate ($n_{sub} = 1.5$) changes the transmissivity of the nano-hole from the red curve to the blue curve. Furthermore, removing the dielectric filling shifts the transmission peak to UV wavelengths, with the result that no visible light is transmitted through an empty 124nm-diameter hole in a 200nm-thick silver film coated on a glass substrate (green curve). In Figure 4-75(b) the profile of E_z, the E-field component of the focused light perpendicular to the silver film's surface, indicates the excitation of surface plasmon polaritons (SPPs) in the vicinity of a pair of 100nm-diameter holes separated in the y-direction by 150nm. Such resonance phenomena, of course, are partly responsible for the peaks and valleys of the transmission spectra of plasmonic nano-structures, with the other relevant factor being the Fabry-Perot-like resonance inside the holes.

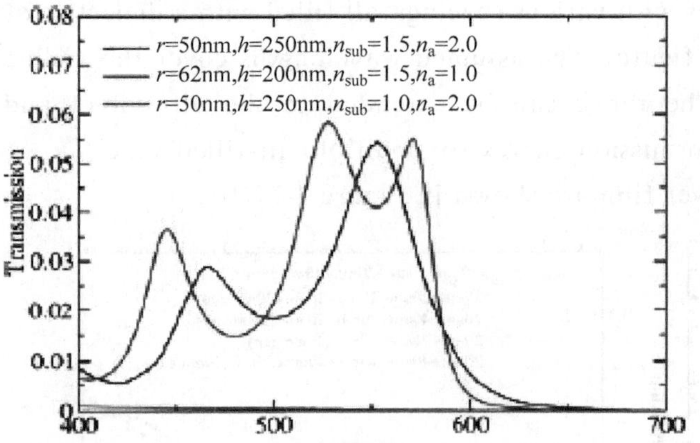

(a)

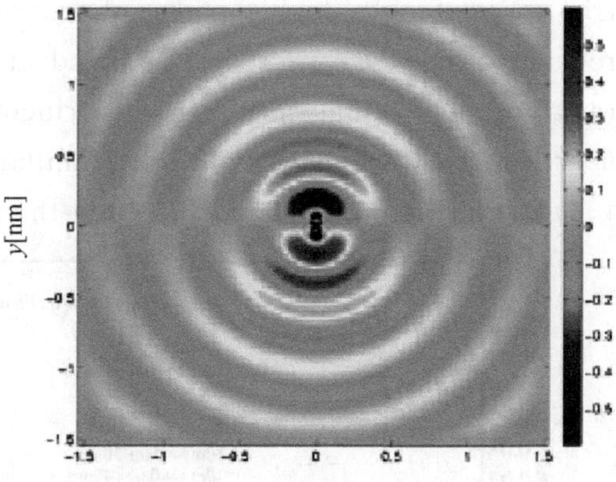

(b)

Fig. 4-75 (a) The red curve is the transmission spectrum of a single, 100nm-diameter hole filled with $n_a = 2.0$ dielectric in a 250nm-thick, free-standing silver slab; the blue curve is similar, except for the silver film being deposited on an $n_{sub} = 1.5$ glass substrate (the hole continues to be filled with $n_a = 2.0$ dielectric).

The readout method for the nano-apertures of Figure 4-73 is shown in Figure 4-76. Here a short pulse from a femtosecond laser is focused on a bit-cell, and the transmitted beam is subsequently sent to a spectrum analyzer.

The green curve at the bottom of the frame corresponds to a 124nm-diameter hole in a 200nm silver film deposited atop an $n_{sub} = 1.5$ substrate. (b) Plot of instantaneous E_z in the xy-plane at $\Delta z = 5$nm below the bottom facet of a silver film containing a pair of nano-holes. The plot reveals the excitation of SPP on both sides of the nano-hole pair along the direction y of incident polarization.

The pulse is short enough ($\sim$10-20fs) that its spectrum covers the entire range of visible frequencies. Each cluster of nano-apertures is thus uniquely identified by its spectral signature, and the entire content of the bit-cell is retrieved upon analyzing the spectrum of the transmitted light. Assuming linear track-velocity of 100m/s and focused spotsize of 0.5μm, the dwell time on each bit-cell is $\sim$5ns, thus requiring a repetition rate of $\sim$200MHz from the femtosecond light source. If one further assumes that a maximum of 10 bits can be stored within a bit-cell, the resulting data-rate will be $\sim$2Gbit/s. Assuming the laser beam delivers an average optical power of 1-2mW to the disk, the integrated power of the focused spot over the xy-plane of the disk (and over the useful spectral bandwidth of the light pulse) will be around 1mJ/s. With a pulse repetition rate of $\sim$200MHz and a linear track velocity of $\sim$100m/s, the focused spot moves a distance of 0.5μm on the disk surface in the time interval between adjacent pulses; this is roughly the diameter of the diffraction-limited spot. The number of photons in each such pulse is $\sim 10^7$, which, even with 1% transmission efficiency through the disk, will result in $\sim 10^5$ photons (per pulse) arriving at the detector. If a 100-element array of photodetectors is needed to monitor the transmitted spectrum, each individual detector will receive $\sim 10^3$ photons per pulse, which should be readily detectable

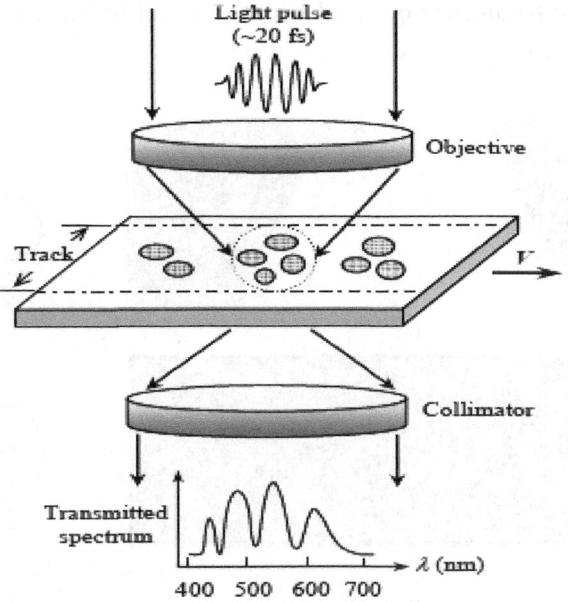

Fig. 4-76 Proposed readout scheme for the plasmonic disk depicted in Figure 4-73. A femtosecond laser pulse is focused by a diffraction-limited objective onto the disk surface. The pulse has a broad spectrum, covering the entire visible range (λ = 400nm to 700nm). The size of the focused spot at the disk surface, ~0.5μm, is comparable to the bit-cell dimensions. Although the nano-holes within a given cell are not individually resolved in a conventional sense, their collective signature, imprinted upon the spectrum of the transmitted light, it can be used to identify the presence or absence of various holes within a cell. With a maximum of 10 nano-holes placed in each cell, the total number of distinct spectral patterns will be 2^{10} = 1024. The spectral patterns can be further optimized by adjusting the nano-holes relative to each other and also relative to the direction of polarization of the incident beam.

with the current silicon photodiode technology. (More efficient and less cumbersome methods of spectral monitoring are also available. For example, the transmitted pulse can be sent through a short length of dispersive optical fiber, which Fourier transforms its spectrum into the time domain. A single fast photodetector can then read the entire spectral content of the pulse as a function of time.)

The energy content of an incident light pulse, given the aforementioned set of parameters, is ~ 5^{-10}pJ. Assuming 25% ~ 50% energy absorption in the metallic film (or other metallic nano-structure), given that a material volume of ~ 10^{-13} cm^3 is typically heated by each pulse, the temperature rise should not exceed 10 ~ 20℃ (silver's heat capacity, for instance, is ~2.4J/cm^3/℃). Therefore, as far as material integrity and longevity are concerned, this level of heating should not pose any serious problems.

Figure 4-77 shows the structure of the samples which fabricated in a silver film in order to verify the predicted optical behavior of nano-holes in a metallic host. A 300nm Si_3N_4 layer was initially grown over the Si substrate. The Si substrate was subsequently etched away through anisotropic etching to create a 200~200μm^2 suspended nitride membrane. This was followed by a 200nm-thick silver film deposition atop the nitride layer. Using the focused ion-beam (FIB) technique, a 10~10μm^2 square of Si_3N_4 was removed to create a freestanding silver layer through which nano-holes with a wide range of diameters and separations were milled. Figure 4-77(b) shows FIB images of single, double, and triple nanoholes (diameter d = 200nm) within the 200nm-thick silver film. Geometric parameters of circular and elliptical hole pairs are defined in Figure 4-77(c) and Figure 4-77(d), respectively.

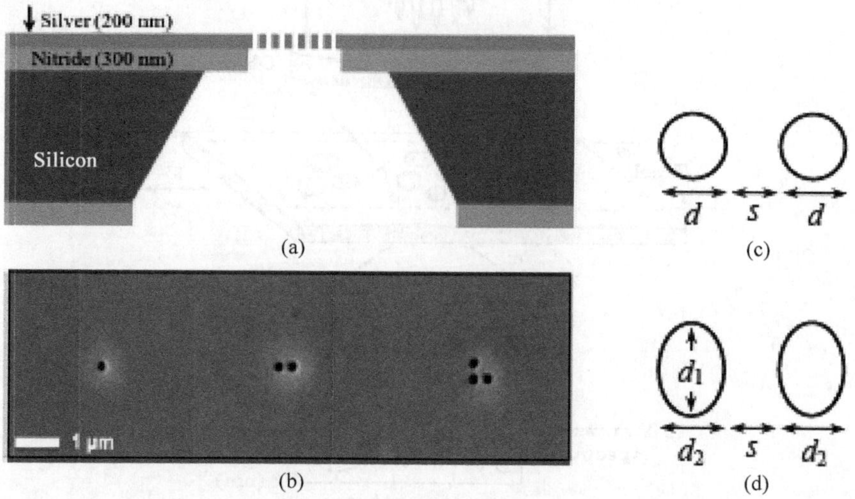

Fig. 4-77 (a) Cross-sectional diagram and (b) FIB images showing nano-holes drilled into a freestanding 200nm-thick silver film on a 300nm-thick silicon nitride membrane. The silicon substrate and the nitride layer are etched away from the region directly beneath the nano-holes. The FIB images show single, double, and triple nano-holes, each having a diameter of 200nm. (c) Circular hole-pairs have diameter d and edge-to-edge separation s. (d) Elliptical hole-pairs have diameters (d_1, d_2) and edge-to-edge separation s.

The measured transmission spectra of single, double, and triple nano-holes for the sample depicted in Figure 4-77 are shown in Figure 4-78. These measurements were carried out with an unpolarized white-light source and a conventional spectro-photometer. The measured structures differ from those simulated in Figure 4-78 in that the nano-holes were not filled with a high-index dielectric material. Nevertheless, the general behavior of transmission spectra in Figure 4-78 agrees qualitatively with that expected from the simulations. In particular, the differences between the spectra of different hole-patterns are large enough to enable reliable readout of the various structures. Similar results for single and double nano-holes in a different sample are shown in Figure 4-79, where a 2μm-diameter aperture was milled in the silver film to improve the calibration procedure.

A quantitative comparison of the measured transmission spectra with FDTD simulation results proved difficult, presumably because the dielectric function $\varepsilon(\omega)$ of bulk silver does not properly represent the optical properties of thin silver films used in our experiments. From these and similar measurements we learn that a single 100nm-diameter hole (blue curve in Figure 4-78(a)) has approximately 0.5% transmission below $\lambda \sim 500$nm. Double-holes of the same diameter, having separations of 75nm, 100nm and 125nm, have nearly identical transmission spectra, with a peak transmissivity of ~0.75% around $\lambda \sim 500$nm. The triplets, having separations of 75nm, 100nm and 125nm, show as much as 0.9%~1.2% transmissivity within the $\lambda \sim 450 \sim 500$nm band. For the 150nm holes depicted in Figure 4-78(b), compared to 100nm holes depicted in Figure 4-78(a), the transmission level is higher across the board, the peak transmission has shifted to slightly longer wavelengths, and transmissivity at longer wavelengths (i. e., $\lambda \sim 550 \sim 650$nm) is substantially higher. Transmission through triple holes is generally greater than that through double holes, which is in turn greater than that through a single hole. Double-hole transmissivity appears to be insensitive to hole separation, whereas the separation distance for triple holes makes a substantial difference in the short-wavelength end of the spectrum (i.e., $\lambda \sim 425 \sim 500$nm).

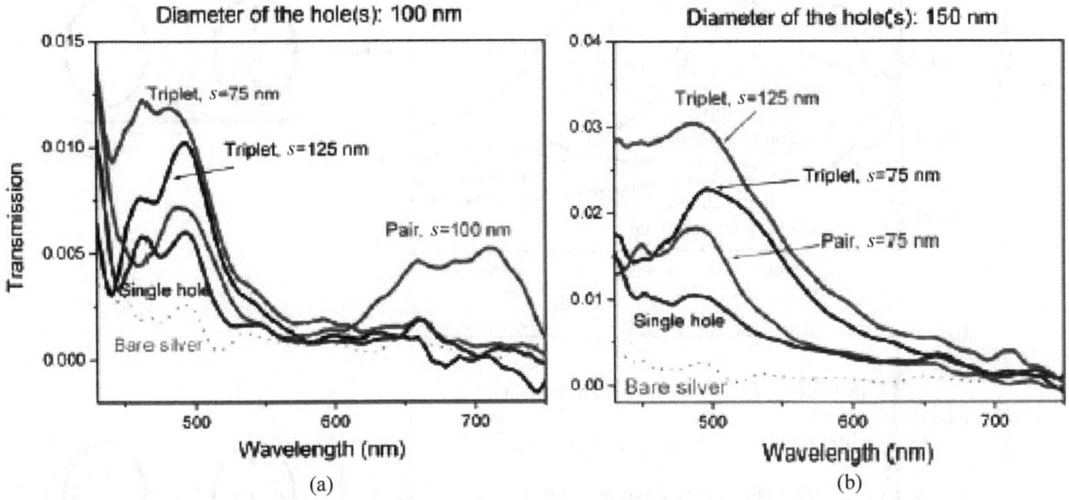

Fig. 4-78 Measured transmission spectra through single, double, and triple circular nano-holes in the suspended 200nm silver film depicted in Figure 4-77. The white-light source used in these measurements was unpolarized, the holes were air-filled, and an incident optical power density of unity /μm^2 was assumed. In the absence of nano-holes, the film's transmissivity is below 0.3% (dotted gray curve). Hole diameters are 100nm in (a) and 150nm in (b).

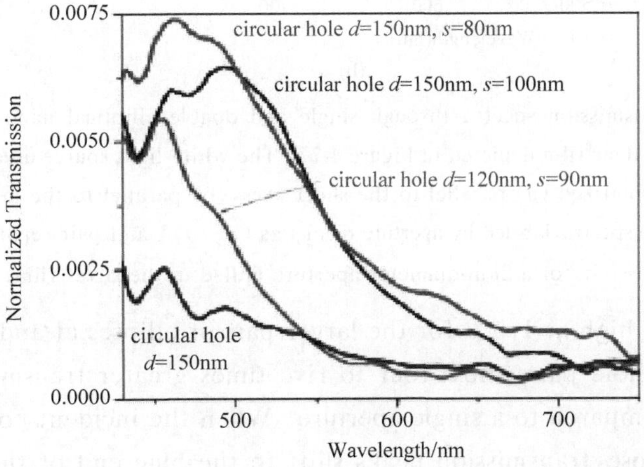

Fig. 4-79 Measured transmission spectra through single and double circular apertures in the suspended 200nm silver film. The white-light source was unpolarized. The spectra are normalized by the transmissivity of a 2μm-diameter aperture milled in the silver film. Blue: single hole, $d = 150$nm. Green: hole pair, $d = 120$nm, $s = 90$nm. Red: hole pair, $d = 150$nm, $s = 60$nm. Black: hole pair, $d = 150$nm, $s = 100$nm.

Figure 4-80 shows the measured transmission spectra for single and double elliptical holes in the suspended, 200nm-thick silver film depicted in Figure 4-80. The transmitted optical power was calibrated with the aid of a 2μm-diameter hole milled in the silver film.

Once again, the spectra are seen to be sufficiently different to provide unique signatures not only for each nano-structure, but also for different orientations of these structures relative to the direction of incident polarization. When the incident polarization is parallel to the short axis,

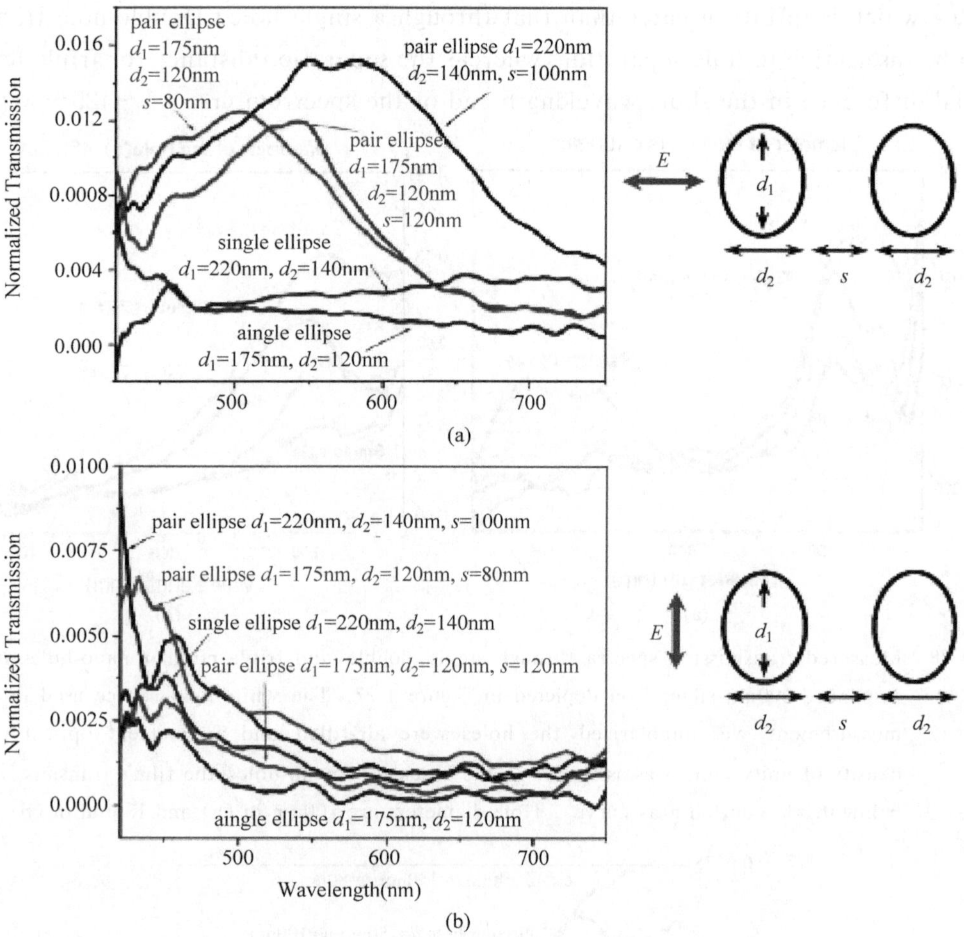

Fig. 4-80 Measured transmission spectra through single and double elliptical nano-holes in the suspended 200nm-thick silver film depicted in Figure 4-77. The white-light source used in these measurements was linearly polarized (a) parallel to the short axes, (b) parallel to the long axes of the elliptical apertures. The spectra, labeled by aperture diameters (d_1, d_2) and pair separation s, are normalized by the transmissivity of a 2μm-diameter aperture milled in the silver film.

transmission could be as high as 1.5% for the larger pair of ellipses at and around $\lambda = 600$nm. In general, these elliptical hole pairs show four to five times greater transmission in the green-red range of wavelengths compared to a single aperture. When the incident polarization is parallel to the long axis of the ellipse, transmission peaks shift to the blue end of the spectrum ($\lambda \sim 400 \sim 500$nm), and the maximum transmission drops to the range between 0.3%-0.6%, depending on aperture size and separation between the apertures. All in all, spectral features in the case of polarization parallel to the long axis are less pronounced compared to those obtained when polarization is parallel to the short axis.

Extending the observation wavelengths to violet and UV (say, $\lambda \sim 300 \sim 400$nm) may be necessary in order to take full advantage of elongated apertures whose long axes are to be aligned (or nearly aligned) with the direction of incident polarization. Alternatively, one may use high-index dielectric fillers to shift the resonance wavelengths toward the red end of the spectrum. In general, a combination of both techniques might be necessary if there is desire to incorporate very small apertures (e.g., $d < 100$nm) within the bit-cells.

Measuring the transmission spectra of some of the circular as well as elliptical hole-pairs using a super-continuum light source (i. e., femtosecond pulsed laser followed by a photonic crystal fiber); the results are shown in Figure 4-81(a). The polarization state of this light source was not determined; therefore, the measured spectra represent a mix of the two linear polarization states used previously in conjunction with the white-light source. Nevertheless, the qualitative features of the spectra obtained with the femtosecond light pulse are in good agreement with those obtained with the white-light source. The presence of the 2μm-diameter calibration aperture on the sample was crucial for the proper normalization of these spectra.

Once again, the pair of large elliptical apertures showed high transmissivity, i. e., 1.0%~1.2%, in the range of $\lambda \sim 500 \sim 650$nm. The smaller apertures exhibited lower transmissivity, and their spectra were oncentrated in the blue-green part of the visible spectrum. Using super-continuum light source obtain the transmission spectra of some samples that were fabricated on a glass substrate. In this case it is covered the nano-holes (milled in the silver film) with a droplet of index-matching fluid. The results are shown in Figure 4-81(b). In these experiments the light pulses were not sufficiently stable, and the results are not quite as reliable as those of Figure 4-81(a). Nevertheless, one can see substantial differences between the spectra of different hole-patterns. Such differences, when monitored with stable femto-second pulses, are expected to provide sufficient information to enable reliable readout of recorded information in a plasmonic data storage system.

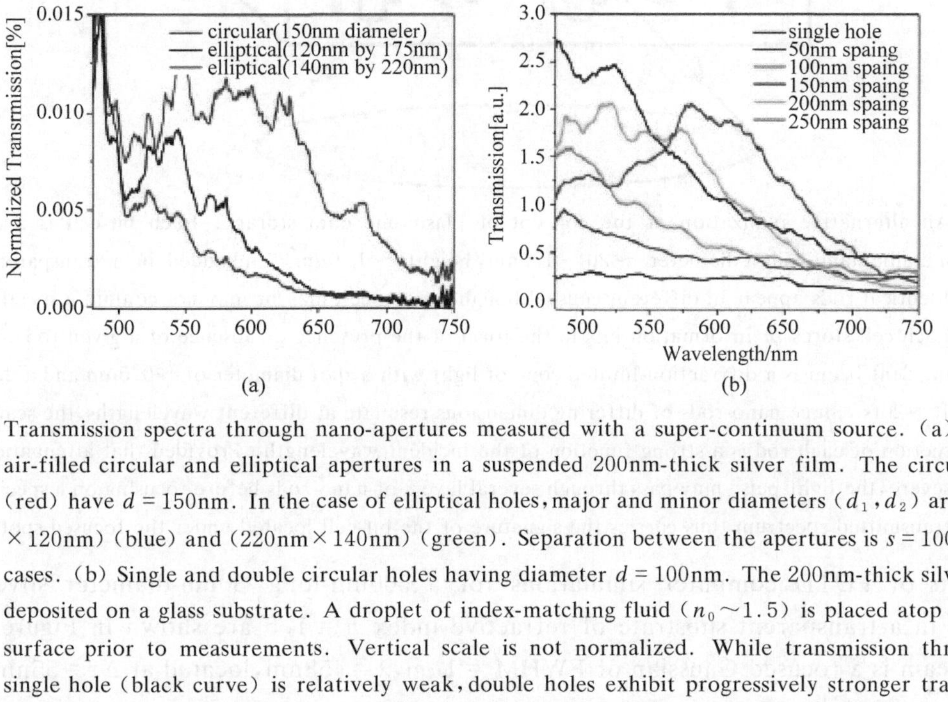

Fig. 4-81 Transmission spectra through nano-apertures measured with a super-continuum source. (a) Pairs of air-filled circular and elliptical apertures in a suspended 200nm-thick silver film. The circular holes (red) have $d = 150$nm. In the case of elliptical holes, major and minor diameters (d_1, d_2) are (175nm ×120nm) (blue) and (220nm×140nm) (green). Separation between the apertures is $s = 100$nm in all cases. (b) Single and double circular holes having diameter $d = 100$nm. The 200nm-thick silver film is deposited on a glass substrate. A droplet of index-matching fluid ($n_0 \sim 1.5$) is placed atop the silver surface prior to measurements. Vertical scale is not normalized. While transmission through the single hole (black curve) is relatively weak, double holes exhibit progressively stronger transmission with increasing hole separation. The spectra of double-holes with separation ⩾150nm extend as far as $\lambda \sim 700$nm. In both (a) and (b), the various spectra are clearly distinguishable from each other, each representing a unique signature for the corresponding hole pattern.

Unless the number of stored bits in individual cells can reach 10 and beyond, it is difficult for the proposed device of Figure 4-73 to surpass the storage capacity of a conventional Blu-Ray disk, i.e., 25GB per layer on a 12cm platter. A variation on the same theme, however, is shown in Figure 4-82, where information is stored in metallic nano-rods embedded in a transparent substrate. Each nano-rod resonates with one or more wavelengths from the UV, visible, and near-IR range, scattering the resonant wavelength(s) out of the main optical path. The transmitted light's spectrum is thus endowed with the collective signature of the cluster of rods embedded within individual bit-cells. This alternative scheme for plasmonic data storage has the advantage that a large fraction of the incident light can pass through each storage layer, thus allowing the stacking of several such layers. The cross-talk among these layers may be negligible so long as the separation between adjacent layers is large enough for the focused cone of laser light to average over a large number of bit-cells in each of the adjacent layers.

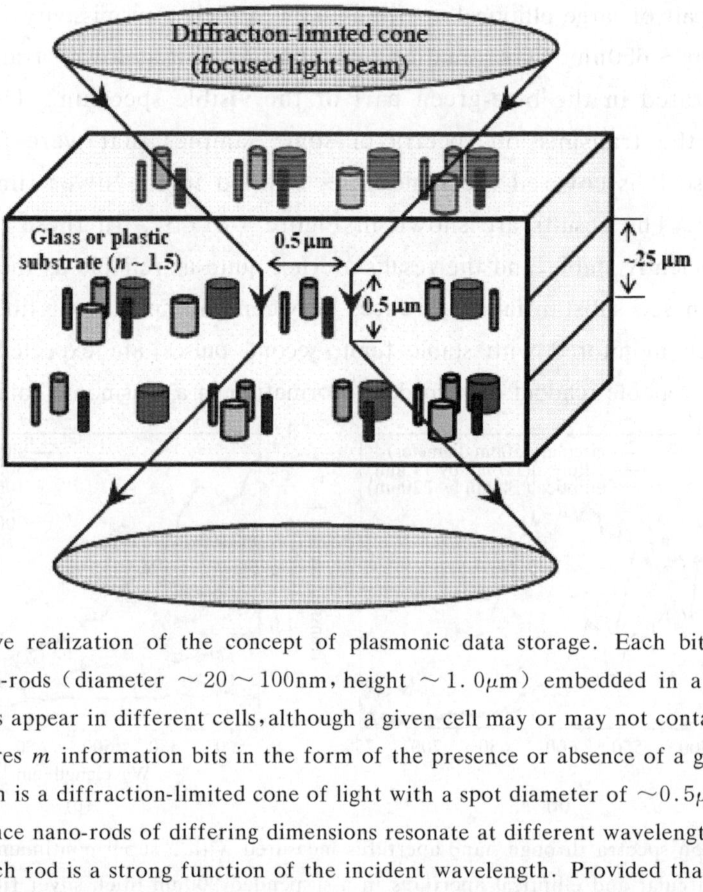

Fig. 4-82 An alternative realization of the concept of plasmonic data storage. Each bit-cell is a collection of metallic nano-rods (diameter ~20~100nm, height ~1.0μm) embedded in a transparent substrate. Identical rods appear in different cells, although a given cell may or may not contain a specific-sized rod. Each cell stores m information bits in the form of the presence or absence of a given rod (0 or 1). The incident beam is a diffraction-limited cone of light with a spot diameter of ~0.5μm and a duration of ~10~20fs. Since nano-rods of differing dimensions resonate at different wavelengths, the scattering cross-section of each rod is a strong function of the incident wavelength. Provided that attenuation is not too severe, the light pulse may pass through several layers of nano-rods before focusing on a specific cell. The transmitted spectrum thus carries the signature of the bit-cell located under the focused spot.

Results of FDTD computer simulations for a 500nm-long, 80nm-diameter silver nanorod embedded in a transparent substrate of refractive index $n = 1.5$ are shown in Figure 4-82. The incident beam is a focused Gaussian of FWHM $= 1\mu$m, $\lambda = 458$nm, located at $\Delta z = 55$nm above the upper surface of the nano-rod and linearly polarized along the y-axis. The incident wavelength is chosen to produce SPP resonance in the nano-rod, thereby producing maximum scattering.

In Figure 4-83 show computed transmission spectra of single, double, and triple nanorods, all having a length of 500nm and embedded in a transparent substrate of refractive index $n = 1.5$. For the smallest rod ($d = 60$nm), the nano-rod is almost invisible at longer wavelengths, but transmission drops to about 90% below $\lambda = 500$nm. As the diameter of the rod increases, the transmissivity minimum shifts to longer wavelengths. For double and triple nano-rods, the transmissivity minima are further shifted to longer wavelengths and the pattern of resonances can be used to uniquely identify the nano-structure.

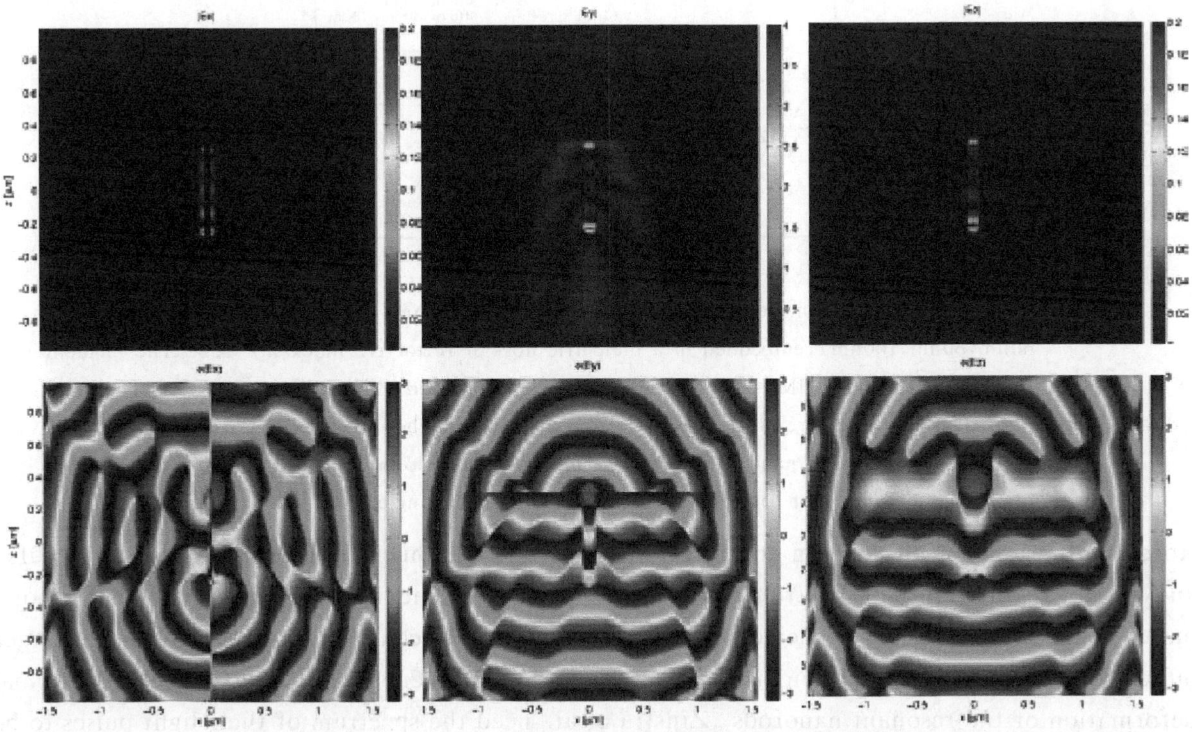

Fig. 4-83 Plots of amplitude and phase in the xz cross-sectional plane for a 500nm-long nanorod (cylinder radius r = 40nm) at the resonance wavelength of $\lambda = 458$nm. The incident beam is a focused Gaussian having FWHM = 1μm, located at Δz = 55nm above the upper surface of the nano-rod and linearly polarized along the y-axis. From left to right: E_x, E_y, E_z components of the electric field. Top row: amplitude; bottom row: phase.

The concept of plasmonic data storage, which exploits optical resonances that occur on the surfaces of metallic films, inside nano-holes and nano-slits milled into such films, and over the length of metallic nano-rods embedded in transparent substrates. Information is encoded into such metallic nano-structures placed on the surface of an optical disk or throughout the entire volume of a 3D storage medium. Based on the results of computer simulations and experimental data from nano-apertures in thin silver films that were presented in this paper, we believe the concept is feasible. It remains to see whether such structures can be fabricated on a large scale, and whether the readout of information can be accomplished at high enough speed and reliability to make the concept worthy of commercialization.

The plasmonic scheme of optical data storage presented in this paper has similarities but also major differences with the plasmon-mediated method employed by Zijlstra et al in. For example, the method provides for read-only storage of information involving prepatterned nano-apertures in metallic films as well as nano-rods or nano-particles of arbitrary shapes embedded in a pre-patterned transparent host. In contrast, the method of Zijlstra et al is a write-once-read-many storage technique based on metallic nano-rods of differing aspectratios all mixed together and applied uniformly over the entire surface of an optical disk. In our method the use of very short femtosecond light pulses (~10~20fs) is essential, as rely on a broad optical spectrum that covers the entire visible range of wavelengths (and beyond) to simultaneously extract all the information

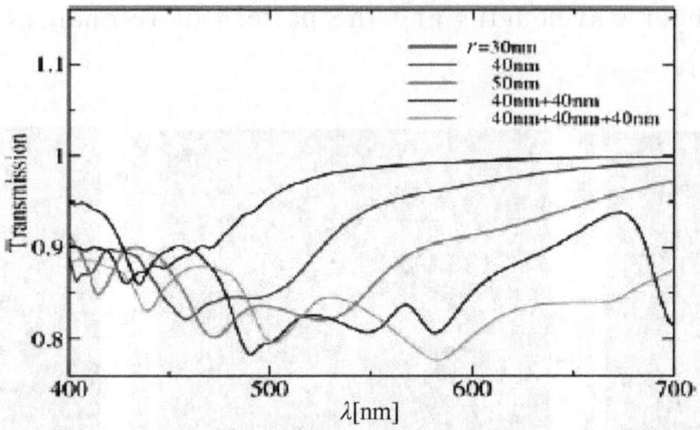

Fig. 4-84 Computed transmission spectra of 500nm-long cylindrical nano-rods of differing diameters (d = 60nm, 80nm, 100nm), embedded in a dielectric host of refractive index $n = 1.5$. The incident focused spot has FWHM = 1.0μm and is linearly polarized along the y-axis. The spectra of individual nano-rods are plotted in black, red and green. The dark-blue curve corresponds to a pair of 80nm-diameter nano-rods, while the light-blue curve represents the transmission spectrum of three 80nm-diameter rods placed at the vertices of a triangle.

stored within a given bit-cell. In contrast, Zijlstra et al. use much longer laser pulses (~100fs) and, do not require a broad spectrum at all. In fact, they even mention that the use of femtosecond pulses in their experiments is incidental and the same results could be obtained by much longer pulses (e.g., nanosecond), so long as the pulse carries enough energy to produce a photo-thermal deformation of the resonant nanorods. Zijlstra et al. need the spectrum of their light pulses to be sufficiently narrow to select specific groups of resonant nano-rods, whereas our laser pulses must have a very broad spectrum to see all the resonances at once. The method of readout proposed by Zijlstra et al. is based on two-photon luminescence, whereas we employ the (linear) plasmonic resonances for readout. Zijlstra et al. use polarization of the light beam to select different orientations of rods having similar aspect ratios within their storage medium, whereas in our scheme the light pulse will always have the same polarization state; we take advantage of the polarization dependence of the resonances by incorporating nano-structures into pre-patterned media that produce different transmission spectra upon illumination by the same light pulse. All in all, we believe that the two methods, rather than being in competition with each other, are complementary, in the sense that one could perhaps use "engineered" super continuum light pulses to write multiple bits simultaneously in the plasmonic media proposed by Zijlstra et al. Similarly, our method of readout could be combined with their method of recording to extract multi-bit information from the same region of the media under a focused laser beam.

4.6.6 The results of FDTD simulations

The results of FDTD simulations for transmission of a focused beam of light through triple nano-holes located at the vertices of an equilateral triangle in a suspended silver film. Figure 4-86 shows profiles of the E-field amplitude (top) and phase (bottom) for a triplet of r = 40nm holes filled with $n_0 = 2.0$ dielectric in a 250nm-thick silver film at λ = 473nm (peak of transmission); the film is suspended in free space ($n = 1$). The holes are at the vertices of an equilateral triangle (side = 150nm) centered at the origin, with one hole centered on the y-axis. The source plane of the incident Gaussian beam, linearly-polarized along the y-axis and having amplitude FWHM =

1μm, is at $\Delta z = 55$nm above the top surface of the film.

Figure 4-85 shows, from left to right, the x, y, z components of the E-field at $\Delta z = 5$nm below the bottom facet of the silver film. The E_z plot shows the location of the electric charge accumulated near the edges of the holes. Figure 4-86 shows the field profiles through a cross section of the silver film; the cross-section is at the yz-plane, with the cut going through the center of one hole and in between the other two holes. One can clearly see in the $|E_y|$ plot the penetration of the incident radiation through the skin depth of the silver film. The E-field inside the hole is strongly resonant, as seen by the strong $|E_y|$ component of the field within the hole, and also by the accumulated charges near the edges of the hole at the top and bottom of the silver film. As can be seen from the light-green curve in Figure 4-74(a), the transmissivity of the triple hole under the conditions shown in Figure 4-85 and Figure 4-86 (i.e., at $\lambda = 473$nm) is ~7%.

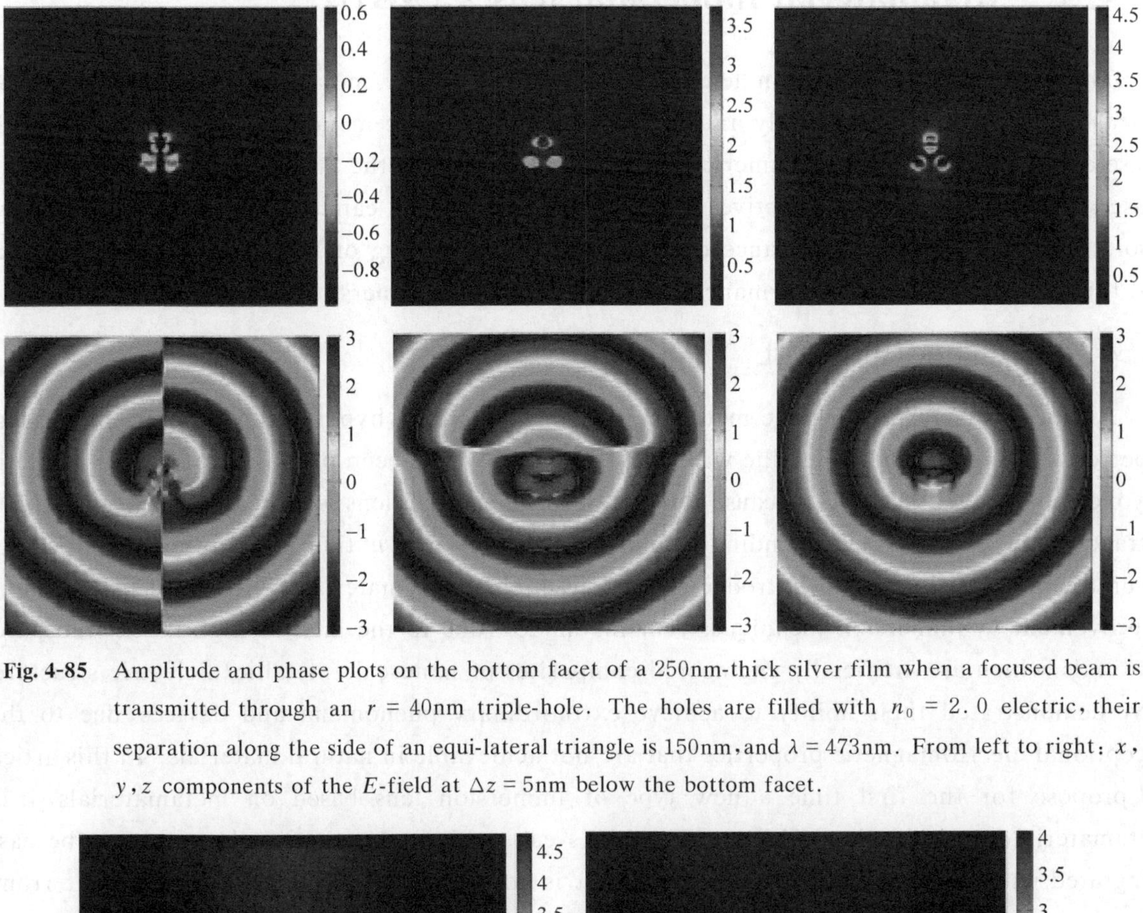

Fig. 4-85 Amplitude and phase plots on the bottom facet of a 250nm-thick silver film when a focused beam is transmitted through an $r = 40$nm triple-hole. The holes are filled with $n_0 = 2.0$ electric, their separation along the side of an equi-lateral triangle is 150nm, and $\lambda = 473$nm. From left to right: x, y, z components of the E-field at $\Delta z = 5$nm below the bottom facet.

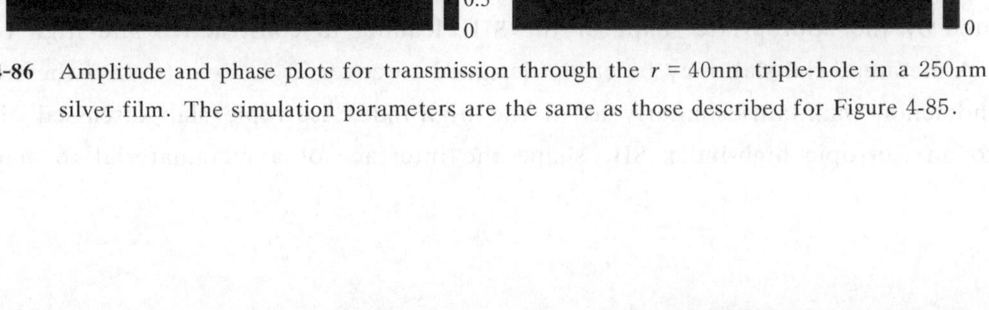

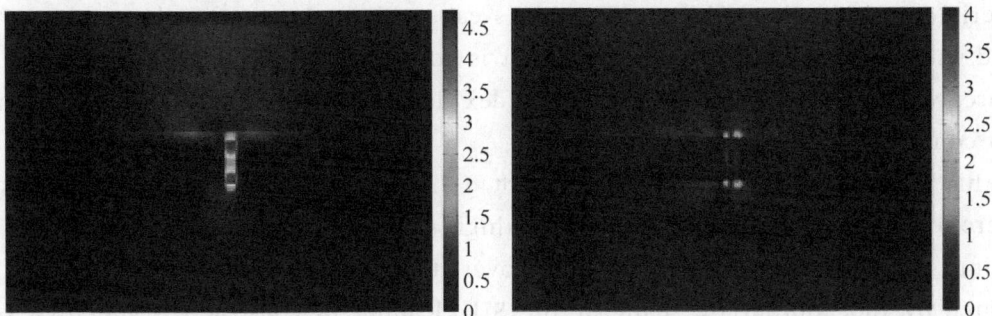

Fig. 4-86 Amplitude and phase plots for transmission through the $r = 40$nm triple-hole in a 250nm thick silver film. The simulation parameters are the same as those described for Figure 4-85.

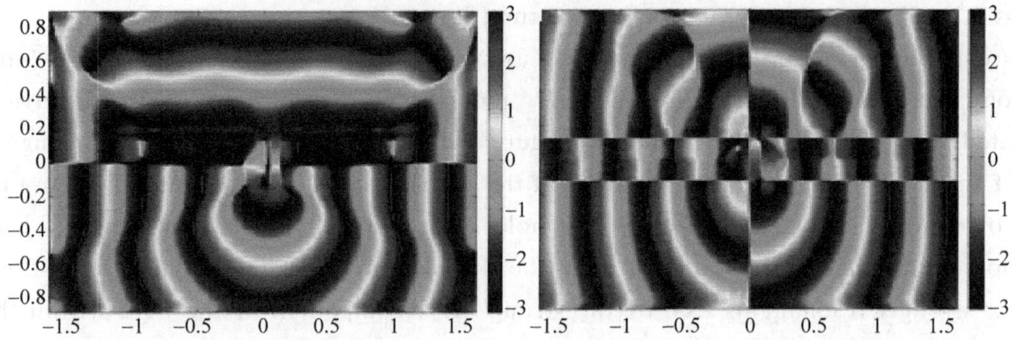

Fig. 4-86 (Continuation diagram)

4.7 Metamaterial immersion lenses (MIL)

The metamaterial immersion lenses by shaping plasmonic metamaterials. The convex and concave shapes for the elliptically and hyperbolically dispersive metamaterials are designed using phase compensation method. Numerical simulations verify that the metamaterial immersion lenses possess exceptionally large effective numerical apertures thus can achieve deep subwavelength resolution focusing. The importance of the losses in modulating of the optical transfer function and thus in enhancing the performance of the metamaterial immersion lenses.

4.7.1 Theory of MIL

Although various remarkable methods, including superlens, hyperlens, plasmonic dimple lens, super oscillation, aperiodic metallic waveguide array etc. have been proposed to achieve resolution beyond the diffraction limit. Because the NA of an immersion lens is increased by a factor of the refractive index n, the corresponding resolution also improves n times. Many efforts have been taken to increase the NA by introducing various high index materials. However, the resolution improvement by immersion techniques remains modest due to the lack of high-index transparent materials, which are artificially engineered composites of nanoscale metallic/dielectric structures, have demonstrated their ability to achieve extraordinary phenomena and devices, due to their exceptional electromagnetic properties that are not achievable in natural materials. In this article, we propose for the first time a new type of immersion lens based on metamaterials, i. e., metamaterial immersion lens (MIL). The MILs can achieve super resolution and can be easily integrated with conventional optical systems. It is commonly known that only the light from a point source within a small cone inside a high-index flat slab can transmit to air. The light outside that light cone will be totally reflected.

For this reason, the shape of an SIL is not a flat but a curved high-index slab, such as a hemisphere or a supersphere. Therefore, all the high k-vector waves supported by the SIL can be coupled into the lens from free space. Moreover, the phases of the incident waves are well compensated by the appropriate shape of the SIL, leading to constructive and high resolution focusing. A metamaterial slab may be easily designed to cover high k-vectors, so the small light cone phenomenon maintains similarly as in the high-index isotropic slab discussed above. In analogy to an isotropic high-index SIL, shape the interface of a metamaterial to achieve the bidirectional coupling between the metamaterial and air with well designed phase compensation. By replacing the isotropic transparent dielectrics with metamaterial, the MIL possesses unprecedented resolving power when compared with conventional SIL.

The idea may be illustrated with a two-dimensional (2D) highly anisotropic metamaterial, which may have either elliptic ($\varepsilon'_x > 0, \varepsilon'_z > 0$) or hyperbolic ($\varepsilon'_x > 0, \varepsilon'_z < 0$) dispersion. Here ε'_x and ε'_z are the real part of the permittivity of the metamaterial in the x and z directions, respectively. The wavefront of a line light source inside such a metamaterial slab may be either convex for elliptic dispersion or concave for hyperbolic dispersion. Accordingly, a metamaterial may be shaped to have either a convex surface for elliptic dispersion or a concave surface for hyperbolic dispersion to achieve the coupling and phase compensation, due to the similarity between the interface shape and the dispersion curve of the metamaterial.

Figure 4-87 shows the dispersion curves (Equifrenquency curves, EFCs) of air and the metamaterials. Let us assume that the principle axes of the metamaterials are always along the horizontal (k_x) and vertical (k_z) directions. When the interface of the metamaterial is along the k_x axis, only the incident light within a small light cone with the transverse k-vectors less than k_0 can transmit into and out of the metamaterial, as shown in Figure 4-87(a). Figure 4-87(b) and Figure 4-87(c) show that the transverse k-vector coverage can be enlarged from k_0 to k_1 when the interface of the metamaterial (k_{xt} axis) has an angle with respect to the material principle axes. Each point on the curved interface of the metamaterial may be considered having an interface not in the x axis direction, thus the transverse k-vector coverage can be extended by a curved interface. As shown in Figure 4-87(b) and Figure 4-87(c), the extended wavevector coverage k_1 can be much larger than k_0 and even the highest achievable k-vectors in natural SIL materials, so super resolution can be achieved. In the following, we demonstrate the MIL concept with both elliptic (referred to as an elliptic MIL) and hyperbolic (referred to as a hyperbolic MIL) dispersions based on phase compensation for focusing in the metamaterials and numerically verify the analysis and designs.

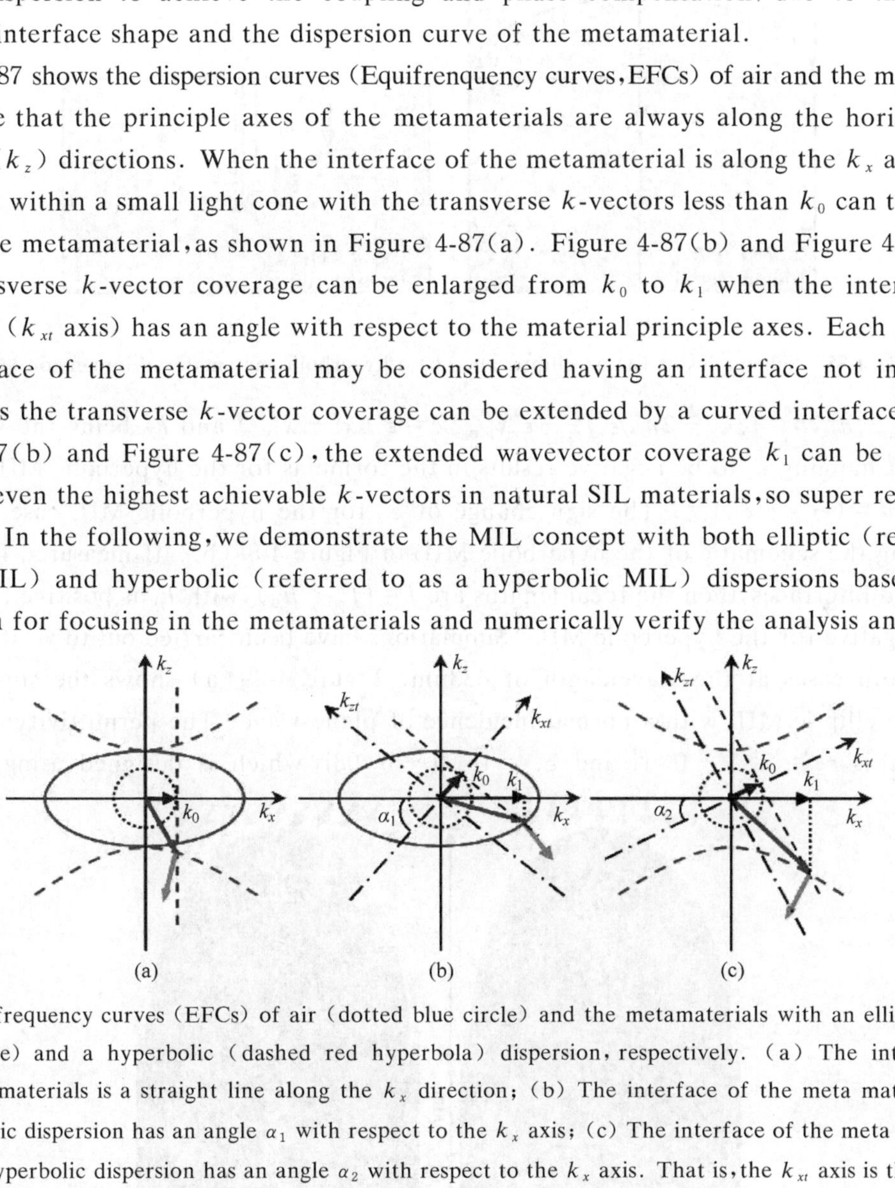

Fig. 4-87 Equifrequency curves (EFCs) of air (dotted blue circle) and the metamaterials with an elliptic (solid red ellipse) and a hyperbolic (dashed red hyperbola) dispersion, respectively. (a) The interface of the metamaterials is a straight line along the k_x direction; (b) The interface of the meta materials with an elliptic dispersion has an angle α_1 with respect to the k_x axis; (c) The interface of the meta materials with an hyperbolic dispersion has an angle α_2 with respect to the k_x axis. That is, the k_{xt} axis is the interface in both (b) and (c).

4.7.2 Simulations and analysis

Figure 4-88(a) shows the schematic of an elliptic MIL. Assuming designing an elliptic MIL with a focal length of f_m and $x \in [-x_{\max}, x_{\max}]$. When a plane wave is incident on the MIL from the top, then refracted by the interface curve and focused to F_m. Phase condition for constructive interference at F_m results in the following quadratic equation

$$k_0(h_0 n + \sqrt{\varepsilon'_x f_m^2 + \varepsilon'_z x_{\max}^2}) = k_0[(h_0 - h)n + \sqrt{\varepsilon'_x (f_m + h)^2 + \varepsilon'_z x^2}] \tag{4-126}$$

Therefore,

$$2h(x) = (-b + b - 4ac)/(2a) \tag{4-127}$$

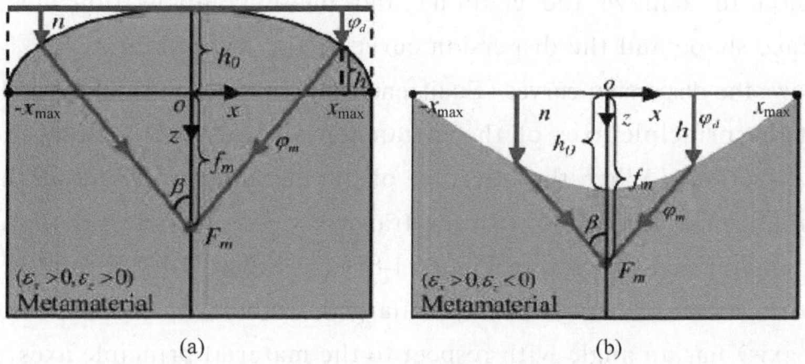

Fig. 4-88 Schematics of (a) an elliptic and (b) a hyperbolic metamaterial immersion lens.

where $a = \varepsilon'_x - n^2$, $b = f_m \varepsilon'_x - 2n\sqrt{\varepsilon'_x f_m^2 + \varepsilon'_z x_{\max}^2}$, $c = \varepsilon'(x^2 - x_{\max}^2)$ and k_0 being the wavevector in free space. Changing ε'_z to be negative results in the formulas for the hyperbolic MIL, with $x_{\max} < f_m$ and $\tan\beta = (-\varepsilon'_x/\varepsilon'_z)^{1/2}$. The sign change of ε'_z for the hyperbolic MIL case can be easily verified using the schematic of the hyperbolic MIL in Figure 4-88(b). If measured from the apex of the curved interfaces, then the focal lengths are $f = (f_m + h_0)$, with h_0 is positive for the elliptic MIL and negative for the hyperbolic MIL. Simulations have been carried out to verify the analysis above for both cases at the wavelength of 633nm. Figure 4-89(a) shows the simulated power profile of an elliptic MIL with a normal incidence of plane wave. The permittivity of the elliptic metamaterial is $\varepsilon_x = 5.1 + 0.1i$ and $\varepsilon_z = 16.0 + 0.08i$, which is designed using an alternate

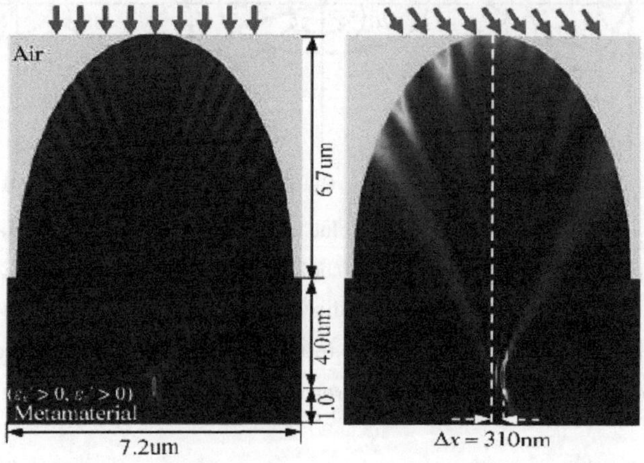

Fig. 4-89 Simulations of an elliptic MIL with (a) normal and (b) oblique incidence of plane wave.

multilayer of silver and Gallium Phosphide with a volume filing factor of silver to the metamaterial of $p = 0.2$.

The multiple layers of the metamaterial lay in the x direction in the MIL. The focal length of the elliptic MIL is $f_m = 3\mu m$ and $x_{max} = 3.4\mu m$, and the calculated $h_0 = 6.7\mu m$. This elliptic MIL achieved a focus with an FWHM (full width at half maximum) of 70nm($\sim\lambda/9$) for the normal plane wave incidence, which is equivalent to an effective NA of 4.5. When the incident light is tilted, the focus will be shifted accordingly. The shift may be estimated by using

$$\Delta x = \mathrm{Re}(f\sqrt{\varepsilon_x/\varepsilon_z}/\sqrt{\varepsilon_z - \sin^2\theta}\sin\theta) \tag{4-128}$$

where θ is the incident angle with respect to the z axis and Re denotes the real part. In practical designing, $|\varepsilon_z'| \gg \sin^2\theta$, so Eq. (4-128) can be reduced to

$$\Delta x \approx \mathrm{Re}(f\sqrt{\varepsilon_x}/\varepsilon_z \sin\theta) \tag{4-129}$$

The simulation in Figure 4-89(b) shows a shift of $\Delta x = 310$nm for $\theta = 14^\circ$, which is comparable to the calculated shift $\Delta x = 331$nm.

The elliptic MIL above has shown its unprecedented high resolution performance over conventional homogeneous SILs, due to its high optical anisotropy. The hyperbolic MIL may behaves differently due to the negative ε_z'. Figure 4-90(a) shows the simulated power profile of a hyperbolic MIL with a normal incidence of plane wave. The permittivity of the elliptic metamaterial is $\varepsilon_x = 8.1 + 0.1i$ and $\varepsilon_z = -12.5 + 0.3i$, which is designed using silver nanowires in air background with a silver volume filing factor of $p = 0.7$. The silver nanowires of the metamaterial are aligned in the z direction. The focal length of the elliptic MIL is $f_m = 2\mu m$ and $x_{max} = 1.29\mu m$, and the calculated $h_0 = -1.23\mu m$. This hyperbolic MIL achieved a focus with an FWHM of 66nm ($\sim\lambda/9.6$) for the normal plane wave incidence, which is equivalent to an effective NA of 4.8. As in the elliptic MIL, when the incident light is tilted, the focus will be shifted. The simulation in Figure 4-90(b) shows a shift of $\Delta x = -120$nm for $\theta = 37.8^\circ$ to the opposite side to the elliptic MIL. This shift is also comparable to the calculated shift $\Delta x = -108$nm using Eq. (4-128).

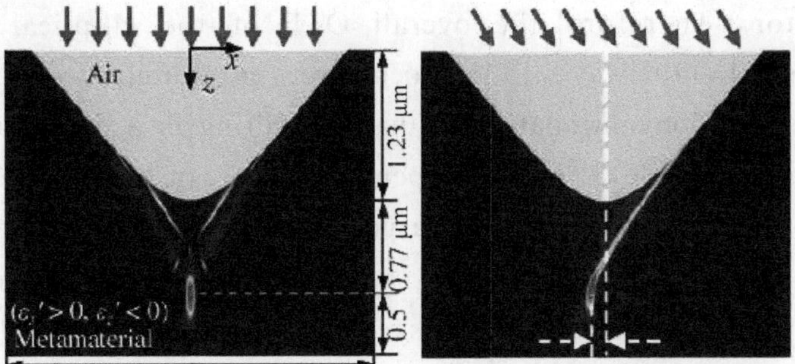

Fig. 4-90 Simulations of a hyperbolic MIL with (a) normal and (b) oblique incidence of plane wave.

4.7.3 Application in the future

As is demonstrated above, the elliptic and hyperbolic MILs can achieve super resolution with convex and concave interfaces, respectively. It is worth emphasizing that the hyperbolic MIL achieves focusing by a concave interface, which is exceptional and not achievable in conventional optics. With the designed curved interfaces, waves with high transverse k-vectors can be coupled into the metamaterial with compensated phase and thus can be focused into deep subwavelength scale. Notice that the concave interface of the hyperbolic MIL in Figure 4-90 is flatter than that of the elliptic MIL in Figure 4-89. This is due to the higher anisotropy, i. e. $|\varepsilon_z'|/\varepsilon_x'$, of the elliptic MIL than the hyperbolic MIL. While both types of MIL can achieve super resolution, the hyperbolic MIL shows anomalous imaging behavior. The opposite displacement of the focus with respect to the incident light is drastically different from the case that as normally been observed in conventional optical lenses. The new phenomenon originates from the negative refraction experienced at the metamaterial/air interface that may open up new possibilities to design lens system with extraordinary functionalities and performances.

There are two major factors affecting the achievable resolution of the MILs:

(1) The intrinsic material properties of the metamaterial. i. e. the k-vector coverage determined by the dispersion relation $k_z^2\varepsilon_x + k_x^2/\varepsilon = k_0^2$ of the metamaterials.

(2) The losses, including propagation and coupling losses, which may strongly modulate the optical transfer function (OTF) of the MILs. The dispersion relation imposes a theoretical cutoff on the coverage of k_x, for an elliptically dispersive metamaterial. On the other hand, the k_x has no limit in the dispersion relation of a hyperbolically dispersive metamaterial.

The losses also play an important role in the performance of the MILs. Different k-vectors may have different loss coefficients and propagation lengths in an MIL, thus the k-vectors attenuate differently. Furthermore, the curved shapes of the MILs cause different refraction (coupling) efficiencies at different positions due to the different angles of incidence. The k-vector spectrum at the focus of an MIL may be significantly modulated by both the propagation and coupling losses and thus substantially different from that of a conventional SIL. The presented elliptic MIL achieved an effective NA of 4.5, which is even a little larger than Re $(\sqrt{\varepsilon\varepsilon_z})$, i. e. 4.0. As shown in Figure 4-89, the low transverse k-vector in the center part propagate longer distances to the focus than the high k-vector on the edge parts, so the low k-vector attenuates more than the high k-vector. Therefore, the overall OTF of the elliptical MIL indicates more transmission for higher k-vector waves to the focus, i. e., more high k-vector contributions to the focus than in the case of a conventional lens, finally resulting in a higher resolution focus and a higher effective NA. In principle, k_x's of a hyperbolic MIL can be infinitely large as there is no cut off in its dispersion relation, so extremely large effective NA's may be obtained. However, in practice, the chosen x_{max} limits the coverage of k_x. In the presented hyperbolic MIL above, the maximum k_x is $4.72k_0$. The loss mechanisms affect the performance of the hyperbolic MIL similarly as in the elliptic MIL, resulting in a higher NA of 4.8.

The metamaterials properties used in the examples above are real values based on the effective media estimation in 2D metal/dielectric multilayer composites and metallic nanowires in dielectric template at the visible wavelength of 633nm. Because the metals are highly dispersive, non-monochromatic illumination may reduce the focusing performance of the MILs. Nonetheless, an MIL for a particular band of frequencies can always be optimized by taking the material dispersion into account. The elliptic and hyperbolic MIL concepts can be easily extended to three-dimensional (3D) devices and other frequency bands, such as ultraviolet, infrared, terahertz, or microwave. The same principle can also be applied to acoustics due to the nature of waves. The presented MILs have achieved super resolution and NAs. The ultimate limitation of MILs is not limited by these examples and can be designed to achieve much higher NAs. A practical usage of the MIL is to make its bottom plane the focal plane so that the resulting near field can be used. As a result, the MIL can find extraordinary applications in lithography, imaging, scanning near-field microscope, sensing, optical storage and heat assisted magnetic recording, etc. As the MILs are typically in microscale, MILs can also be integrated with a flying head for high speed processing as in the flying plasmonic lens, and even an array of MILs can be designed for parallel processing.

4.8 Dynamic pressure air bearing nanogap control

The one of important problems for SIL near-field optical storage is the near-field coupling technology. The light emitted laser beam through the lens to reach the hemispherical or super-hemispherical SIL, because SIL has a very large numerical aperture, the beam is compressed in its underside from the control to achieve near-field coupling of the spot, in order to achieve high-density deposit. SIL program's two core technology based on near-field SIL optical system development and nanogap control.

The flying optical head is as an aerodynamic, mechanical platform for mounting the objective lens and the SIL lens positions is closed distance the surface of the spinning disk. If the distance between the SIL lens and the recording disc has to be much less than the wavelength of the laser, the resolution of the spot within the SIL is maintained across the air gap through evanescent coupling. Using a laser as a reference point, the distance between the bottom of the SIL lens and the recording surface have to be fraction of the wavelength of the laser.

At present, many researchers have designed a variety of SIL-based near-field optical storage nanogap flight systems. However, these systems are suitable principles of near-field optical storage only, so that much work in a static or quasi-static.

Near-field optical storage methods include aperture-type program on super-resolution near-field structure and solid immersion lens (SIL) program. The aperture solution to achieve ultra-high storage density, but its read and write signal is weaker, slower reading and writing speed to limit the development and application of the aperture program. A novel super-resolution near-field structure overcome the near-field pitch control of problem, but the ratio of read to noise signal is too small, unable to meet the actual needs.

In 1999, Sony Corporation laboratory K. Saito, et al. developed a capacitance method for monitoring and control of near-field spacing of the piezoelectric actuator deep sub-micron flight systems. The system focus of the lens and SIL is fixed in a stent to ensure that the SIL and objective lens with high concentricity and the objective lens after working distance constant stent external piezoelectric actuator displacement is controlled by the voltage value, and adjusting the servo movement of the disc to keep the near-field spacing. The system monitoring and control of near-field spacing employe a sensor with a plate capacitor, that the bottom of SIL was made round capacitor with etching processing, and coated with metal film as a plate of the capacitor. The reflective layer of the disc was coated with metallic silver, as another plate of the capacitor. The capacitance access to the LC oscillator circuit, when the near-field distance is changing the capacitance sensitive oscillator of circuit center frequency can measured it and to adjust the near-field spacing soon. The system uses dynamic end jump tiny flotation spindle to reduce the disk axial deflection, and improve the stability of the system. Account local geometrical structure plasmonic flying SIL with high speed and precise gap control physical model of information mark diffraction on optical disk kinematics and dynamics analysis for SIL.

Other most representative design schemes of the near-field optical nanogap control are flight piezoelectric actuator and the gas dynamic pressure support that can keep the spacing of the micro-flying head within 50nm and control accuracy is ±2nm.

The piezoelectric actuator flight system control and high precision have good stability, but the micro-flying head quality is too large, resulting in low dynamic response speed lower data transfer rate, and precision low-voltage piezoelectric actuators are expensive, that this technology is difficult to promote the commercialization application.

OMNERC of Tsinghua University developed and designed an aerodynamic, mechanical flying system for control of position of the objective lens and SIL in 2001. The distance between the bottom of the SIL lens and the recording surface would have to be a fraction of the wavelength of laser. OMNERC was made a laser-based flying head with distance less than 65nm as shown in Figure 4-91. Because the air-bearing slider can control the fly height accurately, that is no need of servo system to focusing between the lens and the media. The areal bit density of NFR is related to fly height directly, as in HDD systems. In future generations, significant increases areal density can be achieved with reducing fly height anywhere close to present HDD technologies.

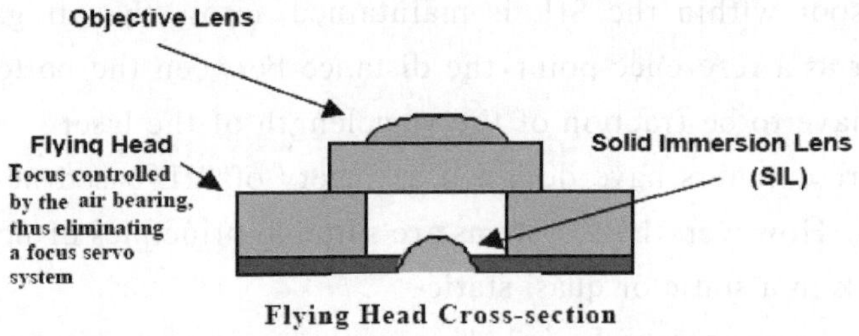

Fig. 4-91 Flying Head With Objective Lens and SIL.

4.8.1 Nanogap flight system design theory model

The gas dynamic pressure support nanogap flight system works is based on the fluid mechanics of micro-flow theory. As shown in Figure 4-92, in the working process of the flight system, optical disc high-speed rotation, due to air molecules viscous to constitute a boundary airflow on under side of the SIL micro flying head air and the surface of the disc to both disengagement. The gas

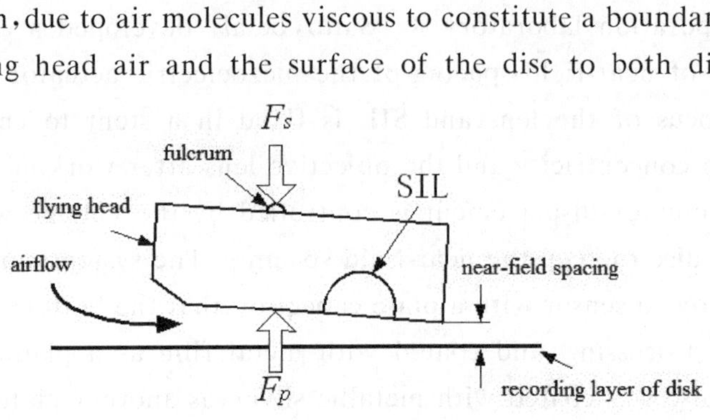

Fig. 4-92 Micro-flight head system.

dynamic pressure supports the micro-flying head that based on the fluid mechanics of micro-flow theory, and the thickness of air dynamic pressure lubrication film is several tens nanometer only. When it maintains equilibrium between the gas buoyancy F_p and weight of micro-flying head added load force F_s, the system reached steady-state operation, and the air film thickness of dynamic pressure becomes a constant, ei built up a near-field spacing between the SIL and the disk.

IBM Research Center developed the flotation micron flight system in 1996, as shown in Figure 4-93. The micro-flying head was composed by glass processing with SIL, its minimum near-field spacing is about 150nm. Due to the quality of the flying head is lighter, that the system has good dynamic response characteristics, shorter seek time and higher data transfer rate. However, due to the flight spacing is too high, and the near-field coupling efficiency is poor, that is leading to the spacing of the objective lens and SIL changing and cannot be accurately focused on the SIL bottom surface.

In order to reduce the flying height, to improve the efficiency of the head/disk interface coupling, have to make intergration SIL system with MEMS that will be introduce in Chapter 10 of this book. The optical system with the integration flying head is shown in Figure 4-94. As the SIL gap is controlled with air bearing, that the optical system is simple than other similar systems.

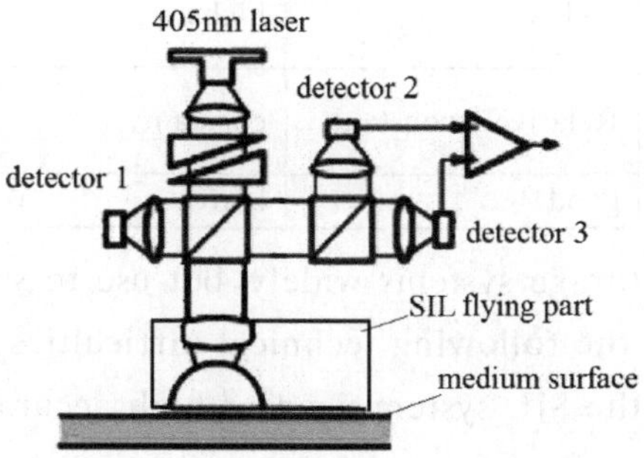

Fig. 4-93 The SIL integrated the SIL micro flying head diagram.

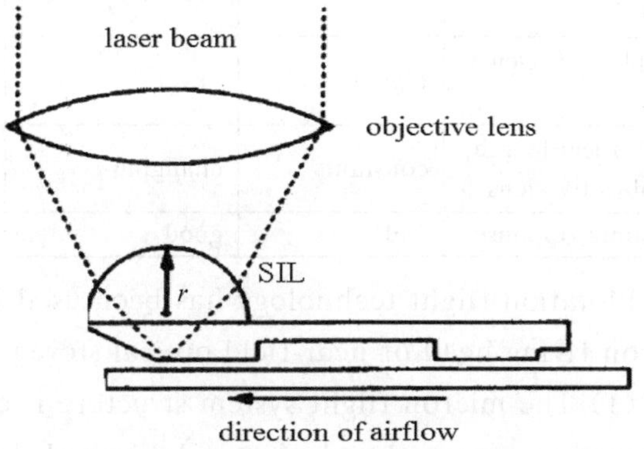

Fig. 4-94 The optical system with micro intergretional SIL flying part in OMNERC.

In order to maintain the beam is focused on the SIL bottom surface accurately and to meet the flight system integration and miniaturization to maintain after the work of the objective lens away from the constant. The Korean company LG Sookyung KIM developed an integrated micro-flying head of deep sub-micron flight systems, as shown in Figure 4-95. The SIL lens was embedded in micro-flying head of the system to ensure the flight system work well. When the flying speed is 2m/s to 10m/s, the near-field spacing is changed from 55nm to 90nm. The near-field spacing is significant change with the increase of flow velocity of the system and the stability of the work need improvement further.

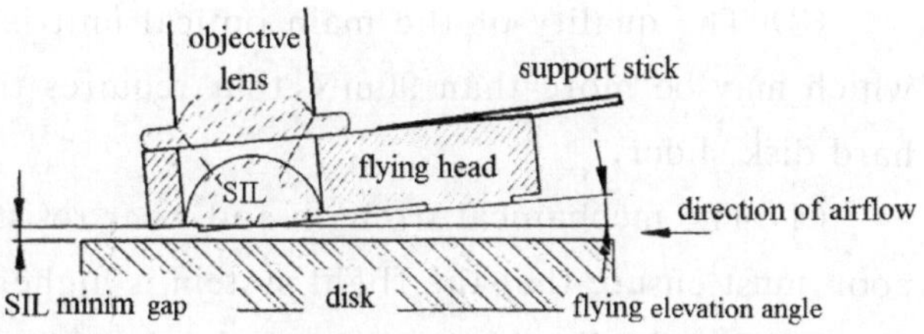

Fig. 4-95 Schematic of integrated micro-flying head.

Comparative analysis of the existing various near-field optical storage with nanogap flight systems are shown in Table 4-8, in which the piezoelectric actuator flight system's control precision, near-field coupling efficiency are higher, but the system structure is complex, expensive and is not suitable for engineering application. The gas dynamic pressure support micron flight system construct compact, anti-disturbance and posture control can be adjusted automatically with main optical unit and the disc. At the same time, the good dynamic response characteristics can improve the data transfer rate. By the way, the technology can be ported to the thermally assisted perpendicular magnetic recording system easily that integration of the near-field optical and magnetic storage technology to achieve higher storage density.

Table 4-8 Comparison of the various near-field optical storage systems with nanogap.

Working model	Piezoelectric actuator	Airbearing			
Countries	Japan	USA	Japan	Japan	Korea
Researcher	K. Saito	B. D. Terries	A. Chekanov	Hirofumi Sukeda	Sookyung KIM
nearfield gap (minimum)	50nm	150nm	100nm	80nm (center of SIL)	55nm (center of SIL)
structure	SIL and objective lens fixed	SIL on flying head only	SIL on flying head only	SIL on flying head, objective lens on actuator	SIL and objective lens on flying head together
Linear velocity of disk	2m/s	1.25m/s	5.0m/s	2.0m/s	2～10m/s
Coupling efficiency SIL	higher	lower	lower	higher	higher
Back focal length of objective lens	constant	changing	changing	Relatively constant	constant
dynamic response	bad	good	good	good	better

Flotation flight technology has been used in magnetic storage systems widely, but use to sub-micron flying head of near-field optical storage system exist the following technical difficulties:

(1) The micron flight system structure is complex, and the SIL system need of high accuracy flight posture control and accurately limited size.

(2) For optical disc storage, the working radius of flight systems is 20～60mm, the boundary flow velocity changing is very significant.

(3) The quality of the main optical unit is resulting loading force of the micro-flying head which may be more than 90mN, that requires the carrying capacity is more than four times the hard disk slider.

(4) The mechanical strength and wear resistance of the SIL precision optical components are poor, must ensure that the flight system is higher stability and cannot contact with disc.

(5) The quality of the micro-flying head has to be ensured accurately and get good high-speed dynamic response characteristics at the same time.

For these reasons, magnetic storage flotation flight design cannot be directly applied to the micron flight system of near-field optical storages. The existing micron flight systems of near-field optical storage are existing following deficiencies:

(1) The disc speed is very low (boundary air flow rate is very low), which greatly reduces the system data transfer rate.

(2) The center of the SIL near-field is changing with the change of flow velocity, so the stability of the system is poor.

(3) The bearing capacity of current micron flight systems is smaller, so that cannot carry the

main optical unit.

(4) The system did not consider the dynamic response characteristics, the system starts and stops to anti-pollution capability and processing error sensitivity.

Therefore, the current research work on micron flight system is far from the practical application, there are still a lot of key issues to be resolved still.

4.8.2 Lubrication model on surface interface of optical head/disc

The flight system of near-field optical head/disk spacing is several tens nanometers only, the classical continuum theory is no longer application. Considering the lubrication model of the gas rarefaction effect on micron head/disk interface, according to the characteristics of the flight system, make up a system dynamics equation. For micro-flying head and the underside of complex graphics, track irregular borders and near-field spacing mutations provide a numerical simulation algorithm based on the finite volume method, and laid out the theoretical foundation for micron flight system design.

- Micron head/disk interface lubrication theory

The micro-flying head of micron flight system is shown in Figure 4-96, that the underside of the micro-flying head is complex three-dimensional morphology, which is composed of along length direction X, width direction that parallel disk Y and along the disk perpendicular to the direction Z-axis that boundary f airflow of micro-flying head of into the side to enter and out from export side. Closed to the center of underside of the micro-flying head is the inside edge, the length side length of the side away from the center is lateral side. The angle of micro-flying head to the X-axis is the pitch angle, to the Y-axis is roll angle.

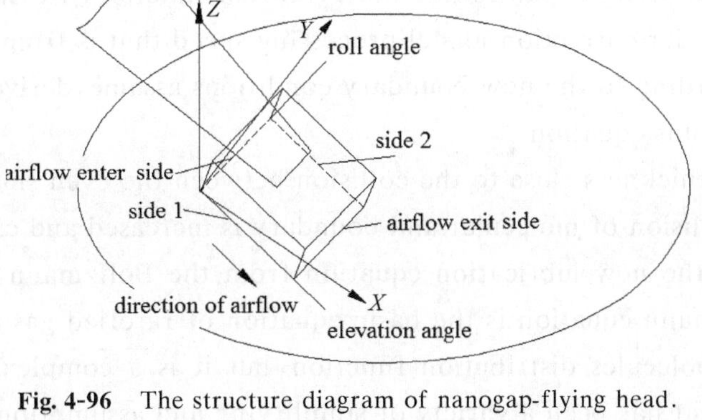

Fig. 4-96 The structure diagram of nanogap-flying head.

According to the Navier-Stokes fluid equation of motion of the continuous fluid, boundary characteristics of the lubricating layer flow and gas to satisfy the equation state can be deduced the general gas hydrodynamic lubrication basic equation, i. e. Generalized Reynolds (Reynolds) equation:

$$\frac{\partial}{\partial x}\left[ph^3\frac{\partial p}{\partial x}\right]+\frac{\partial}{\partial y}\left[ph^3\frac{\partial p}{\partial y}\right]=6U\mu\frac{\partial}{\partial x}[ph]+6V\mu\frac{\partial}{\partial y}[ph]+12\mu\frac{\partial}{\partial t}[ph] \tag{4-130}$$

Where p is the gas buoyancy distribution function, h is the gas film thickness, μ is the air viscosity, U and V are the linear velocity of the boundary flow along the X and Y direction, t is time.

The Reynolds equation based on the following assumptions:

(1) The gas flow is laminar flow, film thickness and the scale of the lubrication area is small

compared to pressure, density and viscosity is constant along the gas film thickness, compared with the viscous force, lubricating fluid inertia and body forces can be neglected.

(2) Fluid executes with Newton's law of viscosity, no relative slip between fluid and solid at the interface with the lubrication region, except direction of the velocity gradient other velocity gradient is negligible.

However, with the reduction of the lubricating film thickness, the collision frequency of the gas molecules with the boundary increases, the lubrication gas rarefaction effect is more significant, so the affect of free course of gas molecules operating characteristics must be considered. According to the rarefied gas lubrication theory, can adopt the Knudsen number K_n generally gas lubrication into the slipstream areas, transition areas and molecular free flow. The near-field spacing of the flight system is close to or even less than the molecular average course, the gas cannot longer be regarded as continuous flow, the lubrication of gas rarefaction effect (Rarefaction effect) is very significant, and cannot longer assume the boundary of the gas molecules are adsorbed on the solid interface, so that have to consider the impact of the slip velocity of the interface molecules.

For account the particularity of micron film lubrication and dynamic pressure accurately, there are a lot of various forms of modified Reynolds equation from various countries. A. Burgdorfer, from the basic equation of the Navier-Stockes assumption of an order slip boundary conditions, obtained an order of modified Reynolds equation. Tseng's Experimental study of a rarefied gas effects, but their research is confined to the sub-micron scale. The Y. T. Hisa, and G. A. Domoto taking into account the size of micro-flying head, and an order slip modified Reynolds equation to promote into account the edge of the air leak in order slip modified Reynolds equation. Mitsuya made the slip model, but the pressure flow to the shear flow and the physical background are not clear, he used the Taylor series expansion of the 1.5-order modified Reynolds equation only. Lin Wu considers the 1.5-order correction model processing speed that is from the mathematical point of departure only, according to the new boundary conditions assume, derived new 1 order and new 2 order modified Reynolds equation.

With the gas film thickness close to the collision between the even smaller than the molecular average free course, collision of molecular and boundary is increased and cannot be ignored. Many scholars have deduced the new lubrication equation from the Boltzmann equation of the kinetic theory of gases. Boltzmann equation is the basic equation of rarefied gas dynamics for describing the speed of the gas molecules distribution function, but it is a complex integral equation with statistical mechanics, that has been a variety of simplifying and assumptions. R. F. Gans according to gas kinematics, isotropic of velocity distribution of the gas molecules and assume of flow velocity on direction Z can be negligible, using BGK form truncated approximation simplified the Boltzmann equation obtained the similar Burgdorfer first order slip modified Reynolds equation. Gans correction derived from the Boltzmann equation, so it can be applicable to any of K_n. Fukui and Kancko consider that the Gans second order approximation has larger error, to renew the BGK model of Boltzmann equation and get generalized gas molecules lubrication equation F-K and give the corresponding data sheet.

In summary, there are two methods for correction of the generalized Reynolds equation:

(1) From the direct correction of the slip velocity, that is through the direct introduction of

the slip velocity enter the universal Reynolds equation, the reaction gas rarefaction effect the impact of this correction method is generally only applies to the case of K_n smaller.

(2) Boltzmann equation was simplified, acording to the impact of molecular motion and the micron dynamic pressure gas film lubrication Reynolds equation, and the correction method is suitable for all K_n.

- Micron head/disk interface lubrication model

Micron head/disk lubrication model of the near-field optical flight system can be obtained from modified Reynolds equation. When micro-flying head length direction is the X axis, the width direction is Y axis, the intersection of airflow enter side and side 1 (see Figure 4-96) is the coordinate origin, it can be written the generalized Reynolds equation with amendments through the dimensionless unified expression:

$$\frac{\partial}{\partial X}\left[QPH^3\frac{\partial P}{\partial X}-\Lambda_x PH\right]+\frac{\partial}{\partial Y}\left[QPH^3\frac{\partial P}{\partial Y}-\Lambda_y PH\right]=\sigma\frac{\partial}{\partial T}[PH] \tag{4-131}$$

Here $X=x/L$, $Y=y/L$, $H=h/h_0$, $P=p/p_a$, L is the length of the micro-flying head, h_0 is spacing of near-field of micro flying head.

Q is Poiseuille coefficient, it is a function of K_n and PH to reaction lubrication of interfacial slip flow. The Q has different correction model with different expressions $K_n=\lambda_a/h_0$, λ_a is the molecular mean free path of air molecular mean free path is about 63.5nm.

$\Lambda_x=6\mu UL/(p_a h_0^2)$, and $\Lambda_x=6\mu VL/(p_a h_0^2)$ is bearing of X and Y direction, μ is air viscosity; $\sigma=12\mu\omega L^2/(p_a h_0^2)$ is squeeze number is the angular frequency of disc rotation;

$T=\omega t$, t is time.

For different correction model, the Poiseuille coefficient $\hat{Q}$ and slip velocity Q_p expression, such as Table 4-9, $a=(2-\alpha)/\alpha$ is the interface correction factor, α is mediation coefficient. It is the film temperature and the interface surface properties. $D=\sqrt{\pi}/2K_n$ is amended Knudsen number.

Table 4-9 $\hat{Q}$ and Q_p with different correction model.

different correction model	$\hat{Q}$	Q_p	Press change
1 step correction	$\hat{Q}=1+6a\frac{K_n}{PH}$	$Q_p=D/6+\sqrt{\pi}a/2$	yes
2 step correction	$\hat{Q}=1+6a\frac{K_n}{PH}+6a\left(\frac{K_n}{PH}\right)^2$	$Q_p=D/6+\sqrt{\pi}a/2+\pi/4D$	no
New 1 step correction	$\hat{Q}=1+4a\frac{K_n}{PH}$	$Q_p=D/6+\sqrt{\pi}a/3$	no
New 2 step correction	$\hat{Q}=1+4a\frac{K_n}{PH}+3\left(\frac{K_n}{PH}\right)^2$	$Q_p=D/6+\sqrt{\pi}a/3+\pi/8D$	no
1.5 step correction	$\hat{Q}=1+6a\frac{K_n}{PH}+\frac{8}{3}\left(\frac{K_n}{PH}\right)^2$	$Q_p=D/6+\sqrt{\pi}a/2+\pi/9D$	yes
Press grads	$\hat{Q}=1+6a\frac{K_n}{PH}+12a\left(\frac{K_n}{PH}\right)^2$	$Q_p=D/6+\sqrt{\pi}a/2+\pi/2D$	no
F-K correction	$\hat{Q}=f\left(\frac{K_n}{PH}\right)$	$Q_p=f(D)$	yes

According to the expression of slip velocity in Table 4-9 to make the slip velocity curve corresponding to the different correction model, as shown in Figure 4-97 (a) corresponds to $0.1 < D < 0.5$, Figure 4-97(b) Equivalent of $0.5 < D < 1$.

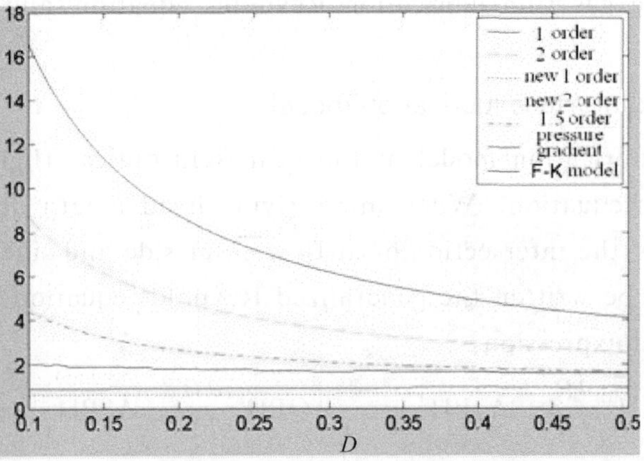

(a) $0.1<D<0.5$ interface slip velocity curve

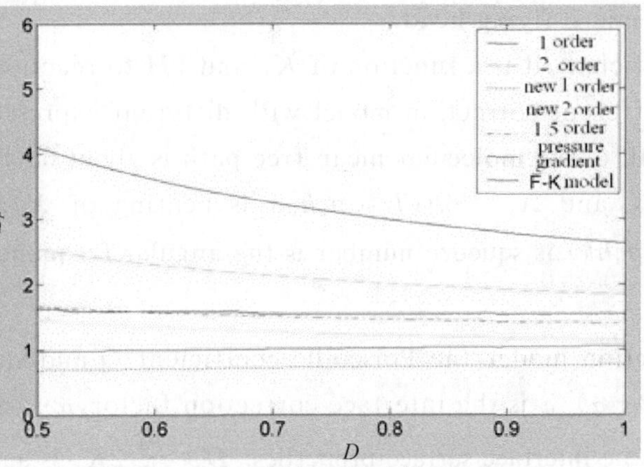

(b) $0.5<D<1$ when the interfacial slip velocity curve

Fig. 4-97 Correction Model interface slip velocity curve.

From the above chart shows:

A new 1 order correction model and first-order correction model slip velocity is smaller; pressure gradient correction model and the order modified model of the interface slip velocity; 1.5-order correction model, the new order correction model and the correction of FK model corresponding to the slip velocity between the middle. When $0.1 < D < 0.5$, the slip velocity of the various correction model was significantly different. When $0.5 < D < 1$, the difference of the slip velocity of the various correction model is relatively small, and the FK correction coincides to the slip velocity curves of 1.5-order correction almost. When the D is increasing continually, the the slip velocity of the various correction model is going to be 0.

According to curves of correction model of slip velocity, comparative analysis above modified Reynolds equation shows that:

(1) D increases (gas film thickness of dynamic pressure increases), effect of rarefied gas decreases. D decreases (hydrodynamic gas film thickness is reduced), difference slip velocity interface of various modified Reynolds equations is significant increased.

(2) As the F-K correction model is derived from the Boltzmann equation, the calculating

precision of dynamic bearing capacity of pressure gas film is believed highest generally. But the calculation of the contact pressure correction exists the singular wrong point, it cannot be used for calculation of nanoscale dynamic pressure of the gas film bearing capacity. In addition, F-K correction model of slip velocity and its computation of the model solution are very complicated.

(3) The theoretical analysis and experiment show that, when $D>0.1$ the pressure gradient correction models and 2-order correction model lower estimated the dynamic pressure of the carrying capacity of the gas film. But the new 1 order correction model and first-order correction model overestimated the carrying capacity of hydrodynamic gas film.

(4) When $0.1<D<0.5$, the 1.5-order correction model with 2 order correction model bearing capacity calculations is very close, but lower than the F-K correction model.

(5) When $0.5<D<1$, F-K correction model and the carrying capacity of the 1.5-order correction model calculation results are very similar, but the 1.5-order correction model has a lower computational complexity.

(6) When $D<0.1$, the dynamic pressure of film thickness is nanoscale, and the rarefied gas effect is very significant, the slip velocity is large. The theoretical and experimental studies show that the calculation model of bearing capacity of the pressure gradient has higher accuracy.

Based on the above analysis, according to the dynamic pressure of the gas film thickness to select the corresponding modified Reynolds equation to establish the model of the head/disk interface lubrication as:

(1) When hydrodynamic gas film thickness is less than 7nm that selects the pressure gradient modified Reynolds equation to establish the head/disk interface lubrication model. It corresponds to the thermally assisted perpendicular magnetic recording near-field flight systems.

(2) When hydrodynamic gas film thickness is greater than 7nm and less than 30nm to select the F-K head/disk interface lubrication model.

(3) When hydrodynamic gas film thickness is greater than 30nm, considering the gas film bearing capacity calculation accuracy and computational complexity of the numerical solution to select the 1.5-order modified Reynolds equation to establish the head/disk interface lubrication model.

- Numerical solution of the head/disk interface lubrication model

In the case of certain operating parameters of the flight system, using the modified Reynolds equation to establish the head/disk interface lubrication model consists of two main variables: the normalized pressure P and the near-field spacing H. Therefore, the operating characteristics of the system will be determined by the lubrication model. Modified Reynolds equation is nonlinear first order partial differential equations, generally unable to find the specific volume. So solving the lubrication model uses numerical methods, including finite element method, finite difference method and finite volume method. The advantage of the finite element method is that can solve pressure distribution of any morphology of underside of micro-flying head. The drawback is the lack of sufficient dissipation to occupying huge memory of computer and longer computing time. Finite difference method just can overcome the limitations of the finite element method and fast calculation. But its physical background is not sufficiency, that the difference equation sometimes cannot maintain the strict conservation of nature, and existence of conservation errors. In addition, the underside of orbital altitude is very small and the boundary points of micro-flying head have some orbits, so the calculation of finite difference method of solution is very unstable.

The physical concept of the finite volume method is more clearly, calculation process are strictly to meet the law of conservation of energy, so it has better convergence and solution accuracy. The mechanical laboratory of University of California Berkeley developed more simulation software. The software features are: Using the control volume method to non-linear Reynolds equation discretize, and using of adaptive meshing and multi-grid parallel algorithm improve its computational efficiency. The accuracy of calculate of the software is high and efficiency is faster, that has been application to industry and manufacturers of hardware widely. So selection the finite volume method numerical solution to modified Reynolds equation applied to calculate the flight system of near-field spacing and gas buoyancy distribution.

• Discretized form of modified Reynolds equation

When the steady-state of the flight system, cannot regard the squeeze effect of time Eq. (4-130) can be rewritten as

$$\frac{\partial}{\partial X}\left[\Lambda_x \mathrm{PH} - Q\mathrm{PH}^3 \frac{\partial P}{\partial X}\right] + \frac{\partial}{\partial Y}\left[\Lambda_y \mathrm{PH} - Q\mathrm{PH}^3 \frac{\partial P}{\partial Y}\right] = 0 \tag{4-132}$$

Eq. (4-132) is integration in control volume can get:

$$\iint_{\Omega}\left[\frac{\partial}{\partial X}\left[\Lambda_x \mathrm{PH} - Q\mathrm{PH}^3 \frac{\partial P}{\partial X}\right] + \frac{\partial}{\partial Y}\left[\Lambda_y \mathrm{PH} - Q\mathrm{PH}^3 \frac{\partial P}{\partial Y}\right]\mathrm{d}\Omega = 0 \tag{4-133}$$

When the integrand Eq. (4-133) is written in divergence form:

$$\oint_S \left[\left[\Lambda_x \mathrm{PH} - Q\mathrm{PH}^3 \frac{\partial P}{\partial X}\right] n_x + \left[\Lambda_y \mathrm{PH} - Q\mathrm{PH}^3 \frac{\partial P}{\partial Y}\right] n_y\right] \mathrm{d}s = 0 \tag{4-134}$$

Where S is the control volume boundary ds is the line element length, n_x and n_y are boundary normal vectors of X direction and Y direction respectively. Eq. (4-134) showed that when micro-flying head is steady-state, the sum of flow in control volume on X direction and Y direction is 0.

The modified Reynolds equation reflects the nature of the flow conservation on the underside of the micro-flying head. According to the physical meaning of modified Reynolds equation, j_x and j_y are defined unit length of the flow speed on X direction and Y direction of the micro-flying head, therefore the Reynolds equation can be written:

$$\begin{cases} j_x = \Lambda_x \mathrm{PH} - Q\mathrm{PH}^3 \dfrac{\partial P}{\partial X} \\ j_y = \Lambda_y \mathrm{PH} - Q\mathrm{PH}^3 \dfrac{\partial P}{\partial Y} \end{cases} \tag{4-135}$$

Consider the feasibility of the numerical calculation, the underside of the micro flying head was rectangular grid, as shown in Figure 4-98. Point A, point B, point C and point D is the four grid nodes on the underside of the micro-flying head, A', B', C', D' is the control element of volume, which size is ΔX, ΔY, δ_x are δ_y.

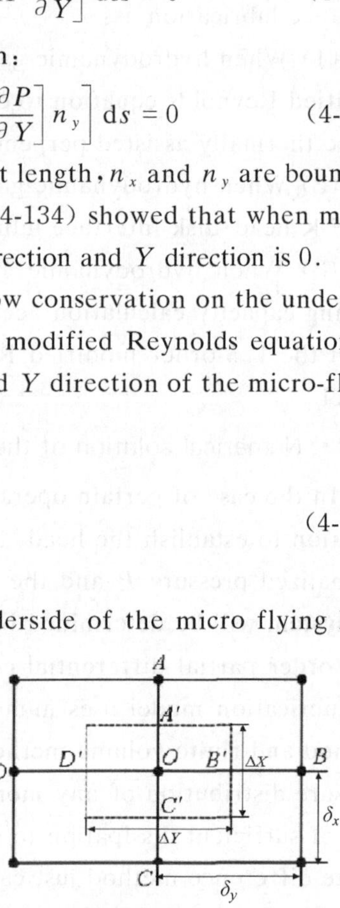

Fig. 4-98 Micro-flying head and the meshing on its underside.

When the flight system is working at steady-state, according to Eq. (4-134) shows that any control gas flow within the volume element on the underside of the micro-flying head is 0, therefore, for A', B', C', D' control volume element:

$$J_{A'} - J_{C'} + J_{B'} - J_{D'} = 0 \tag{4-136}$$

Where: J is fluxes, the subscript indicates the corresponding flux interface.

According to Eq. (4-136), control volume on the flux interface can be expressed as

$$J_{A'} = j_y \big|_{A'} \Delta x = (\Lambda_y \mathrm{PH}) \big|_{A'} \Delta x - \left(\mathrm{QPH}^3 \frac{\partial P}{\partial Y}\right)\bigg|_{A'} \Delta x \tag{4-137}$$

And its difference scheme is

$$J_{A'} = \frac{1}{2}(\Lambda_y H)\big|_{A'}(P_A + P_O)\Delta x - (\mathrm{QPH}^3)\big|_{A'} \frac{(P_A - P_O)}{\delta y\big|_{A'}}\Delta x \tag{4-138}$$

Similarly, the control volume flux of other interface can be expressed as

$$J_{B'} = \frac{1}{2}(\Lambda_x H)\big|_{B'}(P_B + P_O)\Delta y - (\mathrm{QPH}^3)\big|_{B'} \frac{(P_B - P_O)}{\delta x\big|_{B'}}\Delta y \tag{4-139}$$

$$J_{C'} = \frac{1}{2}(\Lambda_y H)\big|_{C'}(P_C + P_O)\Delta x - (\mathrm{QPH}^3)\big|_{C'} \frac{(P_O - P_C)}{\delta y\big|_{C'}}\Delta x \tag{4-140}$$

$$J_{D'} = \frac{1}{2}(\Lambda_x H)\big|_{D'}(P_D + P_O)\Delta y - (\mathrm{QPH}^3)\big|_{D'} \frac{(P_O - P_D)}{\delta x\big|_{D'}}\Delta y \tag{4-141}$$

Take Eq. (4-138) to equation (4-141) into (4-136) can get the discrete format of the modified Reynolds equation:

$$a_o P_o = a_A P_A + a_B P_B + a_C P_C + a_D P_D \tag{4-142}$$

Where the coefficient of pressure of each points is follows:

$$a_A = \frac{\mathrm{QPH}^3}{\delta y}\bigg|_{A'} \Delta x - \frac{1}{2}(\Lambda_y H)\bigg|_{A'} \Delta x \tag{4-143}$$

$$a_B = \frac{\mathrm{QPH}^3}{\delta x}\bigg|_{B'} \Delta y - \frac{1}{2}(\Lambda_x H)\bigg|_{B'} \Delta y \tag{4-144}$$

$$a_C = \frac{1}{2}(\Lambda_y H)\bigg|_{C'} \Delta x + \frac{\mathrm{QPH}^3}{\delta y}\bigg|_{C'} \Delta x \tag{4-145}$$

$$a_D = \frac{1}{2}(\Lambda_x H)\bigg|_{D'} \Delta y + \frac{\mathrm{QPH}^3}{\delta x}\bigg|_{D'} \Delta y \tag{4-146}$$

$$a_O = \left[\frac{\mathrm{QPH}^3}{\delta y}\bigg|_{A'} \Delta x + \frac{\mathrm{QPH}^3}{\delta x}\bigg|_{B'} \Delta y + \frac{\mathrm{QPH}^3}{\delta y}\bigg|_{C'} \Delta x + \frac{\mathrm{QPH}^3}{\delta x}\bigg|_{D'} \Delta y\right] + \frac{1}{2}\left[(\Lambda_y H)\big|_{A'}\Delta x + (\Lambda_x H)\big|_{B'}\Delta y - (\Lambda_y H)\big|_{C'}\Delta x - (\Lambda_x H)\big|_{D'}\Delta y\right] \tag{4-147}$$

- Optimization of the discrete modified Reynolds equation

For discretization scheme of the modified Reynolds equation, the Peclet number P is $P = \Lambda H\delta / \mathrm{QPH}^3$, according to the convergence conditions of the Reynolds equation, when $P_{A'}$, $P_{B'}$ is greater than 2 or $P_{C'}$, $P_{D'}$ less than-2, the coefficient of the discrete format of the Reynolds equation is negative, and the numerical solution is divergence. In addition, if the coefficient a_O is less than 0 could lead to its iterative solution unstable.

Based on the stability and convergence of the numerical solution of optimization methods, the coefficients of the modified Reynolds equation discretization scheme can be rewritten:

$$a'_O P_O = a'_A P_A + a'_B P_B + a'_C P_C + a'_D P_D + b(P_O) \tag{4-148}$$

Definition $A(|P|) = \max(0, 1 - 0.5|P|)$, the coefficient of Eq. (4-148) will be:

$$a'_A = \frac{(QPH^3)}{\delta y}\bigg|_{A'} \Delta x(A(|P_{A'}|)) + \max(-(\Lambda_y H)|_{A'}\Delta x, 0) \tag{4-149}$$

$$a'_B = \frac{(QPH^3)}{\delta x}\bigg|_{B'} \Delta y(A(|P_{B'}|)) + \max(-(\Lambda_x H)|_{B'}\Delta y, 0) \tag{4-150}$$

$$a'_C = \frac{(QPH^3)}{\delta y}\bigg|_{C'} \Delta x(A(|P_{C'}|)) + \max((\Lambda_y H)|_{C'}\Delta x, 0) \tag{4-151}$$

$$a'_D = \frac{(QPH^3)}{\delta x}\bigg|_{D'} \Delta y(A(|P_{D'}|)) + \max((\Lambda_x H)|_{D'}\Delta y, 0) \tag{4-152}$$

$$a_{O'} = a_{A'} + a_{B'} + a_{C'} + a_{D'} + \max(0, (\Lambda_y H)|_{A'}\Delta x + (\Lambda_x H)|_{B'}\Delta y - (\Lambda_y H)|_{C'}\Delta x - (\Lambda_x H)|_{D'}\Delta y) \tag{4-153}$$

$$b(P_O) = \max(0, (\Lambda_y H)|_{C'}\Delta x + (\Lambda_x H)|_{D'}\Delta y - (\Lambda_y H)|_{A'}\Delta x - (\Lambda_x H)|_{B'}\Delta y)P_O \tag{4-154}$$

4.8.3 Solving discrete modified Reynolds equations

To improve the convergence of numerical solution, using line scan method for solving the discrete modified Reynolds equation is shown in Figure 4-99, assuming the first column $i-1$ and $i+1$ column pressure value are known, solving the i column pressure values of all nodes. According to Eq. (4-154), for any three adjacent nodes in column i, the discrete modified Reynolds equation can be rewritten as

$$(a'_{i,j} - b)P_{i,j} - a'_{i,j+1}P_{i,j+1} - a'_{i,j-1}P_{i,j-1} = a'_{i+1,j}P_{i+1,j} + a'_{i-1,j}P_{i-1,j} \tag{4-155}$$

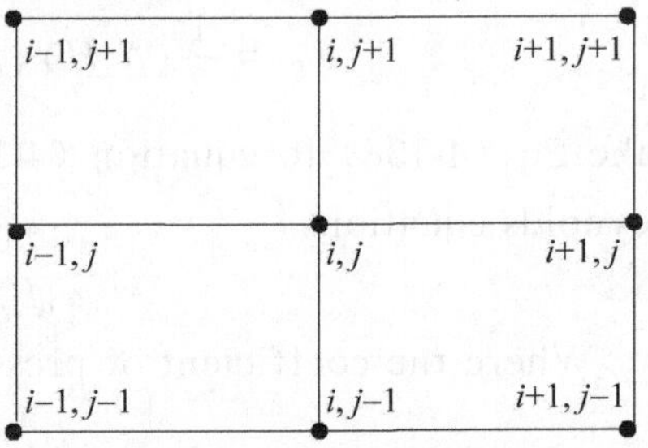

Fig. 4-99 line scanning method for solving the schematic.

Order

$$\begin{cases} A_{i,j} = (a'_{i,j} - b) \\ A_{i,j-1} = -a'_{i,j-1} \\ A_{i,j+1} = -a'_{i,j+1} \\ d_i = a'_{i+1,j}P_{i+1,j} + a'_{i-1,j}P_{i-1,j} \end{cases} \tag{4-156}$$

Then Eq. (4-155) is transformed into:

$$A_{i,j-1}P_{i,j-1} + A_{i,j}P_{i,j} + A_{i,j+1}P_{i,j+1} = d_{i,j} \tag{4-157}$$

According to Eq. (4-157), for all nodes on the i-th column:

$$\begin{bmatrix} A_{i,1} & A_{i,2} & A_{i,3} & & \\ \ddots & \ddots & \ddots & & \\ & A_{i,j-1} & A_{i,j} & A_{i,j+1} & \\ & \ddots & \ddots & \ddots & \\ & & A_{i,n-2} & A_{i,n-1} & A_{i,n} \end{bmatrix} \begin{bmatrix} P_{i,2} \\ \cdots \\ P_{i,j} \\ \cdots \\ P_{i,n-1} \end{bmatrix} = \begin{bmatrix} d_{i,2} \\ \cdots \\ d_{i,j} \\ \cdots \\ d_{i,n-1} \end{bmatrix} \tag{4-158}$$

As the first node and n-th node of column i are boundary point. According to boundary conditions, these two pressure values are standard atmospheric pressure, that take it into Eq. (4-158) can solving other the pressure value by Thomas algorithm.

Solving process is that using the n-th iterative calculation of the pressure values obtained coefficient matrix of Eq. (4-147), and then using the Thomas algorithm to find i-th column pressure values of all nodes ($n+1$). In calculation process, along the direction of air flow, row by row strike the pressure value of each column node, can ultimately the pressure distribution on the entire underside of micro-flying head ultimately. In order to prevent accumulated errors in the iterative process, the solving by X direction and the Y direction has to be alternating, until the pressure distribution values on the underside of the micro-flying head meet the convergence conditions. As iteration direction of the line scanning and the gas flow direction are same, that conform the physical principles of the system work, so the algorithm has good convergence.

- Key issues for solving modified discrete Reynolds equation

In the numerical calculation process, meshing on underside micro-flying head will affect the stability and convergence of the calculation directly. Because the complex morphology on the underside of the micro-flying head, the track boundary is irregular, there will be unit of control volume was crossed by railless border as shown in Figure 4-100. This mesh will result mutations of near-field spacing within a control volume.

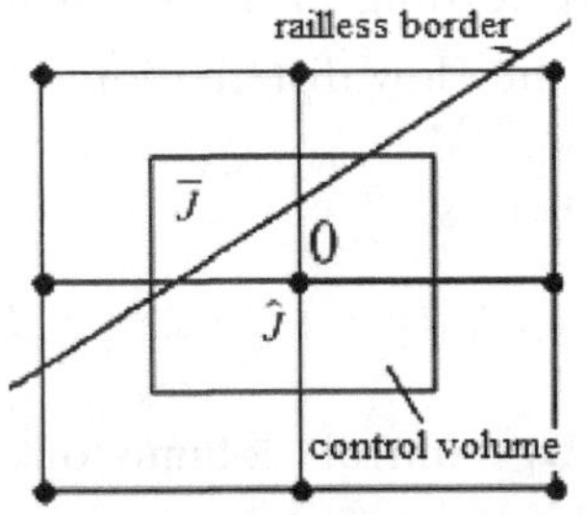

Fig. 4-100 The appearance of railless border is crossing control volume.

To improve mutations of control volume spacing, the mass flux of different near-field spacing was weighted that get the average mass flux of entire control volume. The average mass flux of control volume on X and Y direction can be expressed as

$$\begin{cases} J_x = \xi \bar{J}_x + (1-\xi) J_x \\ J_y = \eta \bar{J}_y + (1-\eta) J_y \end{cases} \tag{4-159}$$

Where J and $\bar{J}$ is near-field spacing corresponds to the mass flux, ξ and η are the weight coefficients that the weight coefficient value is determined by ratio of the different near-field spacing to corresponding side length.

Vary due to the morphology on underside of the micro-flying head is different. According to topographic characteristics on underside, manual meshing can improve the feasibility of the numerical solution. In order to ensure the accuracy of numerical solution, the mesh is fine sufficiently, and cannot allow three different near-field-spacing values at the same time. For complex-shaped of rail design, using average crude sub-grid calculated initial values in the first, then manual re-meshing calculated the pressure distribution to improve the accuracy. General iterative calculation is compared before and after iterations to calculate the difference of the results to determine whether the convergence of the iteration. But in solution process of the pressure distribution, the iterative convergence is relatively slow and convergence condition is prone to miscarriage. Therefore according to the calculation of residual convergence condition and Eq. (4-154) define residuals of the calculation to pressure of nodes:

$$R_O = a'_O P_O - a'_A P_A - a'_B P_B - a'_C P_C - a'_D P_D - b(P_O) \tag{4-160}$$

According to maximum value of entire computational domain residuals set up the convergence condition:

$$\max\left(\sum R_p\right) < \varepsilon \tag{4-161}$$

When the entire computational domain residuals and its maximum value are small enough to meet convergence condition, that indicate iterative end.

4.8.4 Stream function on the underside of micro-flying head

Near-field spacing of the flight system has been close to or less than the molecular mean free path. The boundary air mixture of high-speed dust particle flow between the underside of the micro-flying head and disk constitute three-body friction. High-speed rotation of the disc, the consequence of the friction is disastrous often. Therefore, it is need of study the flow field on underside of the micro-flying head to determine the various parts and volume of flow dust particles on the underside to influence the micro-flying head.

According to the physical meaning of the modified Reynolds equation, the reaction function Ψ of the flow distribution on the underside of the micro-flying head is

$$\begin{cases} j_x = \dfrac{\partial \Psi}{\partial Y} \\ j_y = -\dfrac{\partial \Psi}{\partial X} \end{cases} \tag{4-162}$$

By control volume on underside of micro-flying head mesh to calculate the flow field distribution function, the equation (4-162) can be discretized as

$$\begin{cases} \Psi_{i,j+1} = \Psi_{i,j} + J_{i+1,j+1} \\ \Psi_{i+1,j} = \Psi_{i,j} - J_{i+1,j+1} \end{cases} \tag{4-163}$$

Based on calculation pressure distribution of on underside of the micro-flying head, using equation (4-135) calculates j, then according to equation (4-163) calculates the value of the distribution function of the flow field of each node. According to above numerical algorithm, can program steady-state operating characteristics numerical solver of flight system with FORTRAN language as following three modules:

(1) A program initialization: read into the system operating parameters, mesh, combined with underside surface function of micro-flying head to calculate spacing values of the near field node and the bearing capacity.

(2) Calculation of the pressure distribution: Solve two iterations of inner layer outer layer iteration with the last iteration of the pressure value of the coefficient matrix and vector iteration into a tridiagonal matrix equation group to solve the pressure value of the node.

(3) Calculated the results and lay out the pressure distribution and flow distribution on the underside of micro-flying head.

4.8.5 Dynamic characteristics of micron flight systems

- Kinetic equation of the flight system

According to structure and characteristics of micron flight system, the optical unit is simplified to a spring-damper system with X, Y and Z coordinate system and shown in Figure 4-101. The micro-flying head can move along the Z direction, pitch up around X direction and scroll around Y direction as a freedom body with three directions.

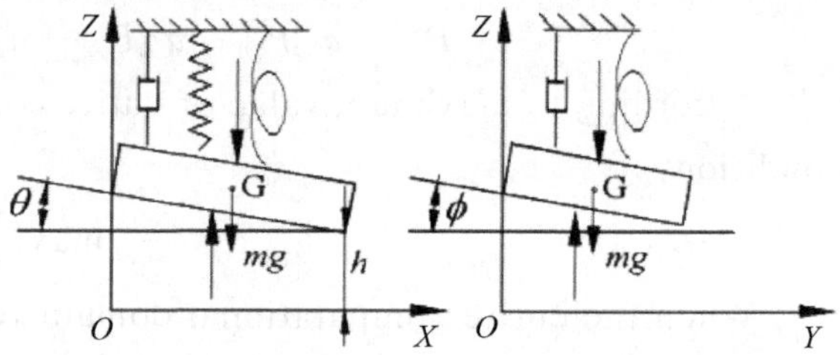

Fig. 4-101 Stress Analysis of micro-flying head.

According to role of the load and air buoyancy, the kinetic equation of the system is

$$\begin{cases} m\ddot{h} + c_h\dot{h} + k_h h = F_a + F_s + G \\ I_\theta\ddot{\theta} + c_\theta\dot{\theta} + k_\theta\theta = M_{a\theta} + M_{s\theta} + G_{s\theta} \\ I_\phi\ddot{\phi} + c_\phi\dot{\phi} + k_\phi\phi = M_{a\phi} + M_{s\phi} + G_{s\phi} \end{cases} \tag{4-164}$$

Where m, I_θ and I_ϕ respectively for the quality, pitch up and rolling moment of inertia of the micro-flying head; h, θ and ϕ for Z displacement, pitch and roll angle; k_h, k_θ and k_ϕ are collectivity stiffness of bracket of micro-flying head; $c_h c_\theta$ and c_ϕ are collectivity damp of bracket; k_h, $M_{s\theta}$ and $M_{s\phi}$ are overall stiffness; F_s, $M_{s\theta}$ and $M_{s\phi}$ are load force, pitch moment and rolling moment of micro-flying head; G, $G_{s\theta}$ and $G_{s\phi}$ are gravity, pitch moment and rolling moment of gravity action; F_a, $M_{a\theta}$ and $M_{a\phi}$ for applied load force, pitching moment and rolling moment of the micro-flying head ie air buoyancy and its pitching moment and rolling moment that can be expressed as

$$\begin{aligned} F_a &= \iint_A (p - p_a)\mathrm{d}A \\ M_{a\theta} &= \iint_A (p - p_a)(x - x_f)\mathrm{d}A \\ M_{a\phi} &= \iint_A (p - p_a)(x - x_f)\mathrm{d}A \end{aligned} \tag{4-165}$$

where x_f and y_f are coordinates of the loading point.

For solving the dynamic characteristics of the micro-flying head is same to steady-state. But have to consider squeeze impact to the modified Reynolds equation. According to Eq. (4-136) can get the physical equations on the underside of the micro-flying head to control the volume as

$$\iint_\Omega \sigma\frac{\partial}{\partial T}(\mathrm{PH})\mathrm{d}\Omega + \oint_S \left[\left[\Lambda_x \mathrm{PH} - \hat{Q}\mathrm{PH}^3\frac{\partial P}{\partial X}\right]n_x + \left[\Lambda_y \mathrm{PH} - \hat{Q}\mathrm{PH}^3\frac{\partial P}{\partial Y}\right]n_y\right]\mathrm{d}s = 0 \tag{4-166}$$

Physical equation of $A'B'C'D'$ four points of control volume element on the underside of micro-flying head is

$$J_{A'} - J_{C'} + J_{B'} - J_{D'} + \sigma\frac{\partial}{\partial T}(\mathrm{PH})\Delta x\Delta y = 0 \tag{4-167}$$

Based dynamic characteristics of the time ΔT, the extrusion of Eq. (4-167) can spread out with Taylor formula and ignore the higher order available:

$$\sigma\frac{\partial}{\partial T}(\mathrm{PH})\Delta x\Delta y = \sigma\frac{(\mathrm{PH})_O^{n+1} - (\mathrm{PH})_O^n}{\Delta T}\Delta x\Delta y \tag{4-168}$$

Formula: the superscript of PH corresponds to the number of iterations of the time domain.

According to calculated results of the previous steady state take Eq. (4-154) and Eq. (4-168) into Eq. (4-167), can get the discrete form of dynamic modified Reynolds equation with $n+1$ time step as

$$\tau_O P_O^{n+1} = \tau_A P_A^{n+1} + \tau_B P_B^{n+1} + \tau_C P_C^{n+1} + \tau_D P_D^{n+1} + \vartheta(P_O^{n+1}) \tag{4-169}$$

Where

$$\tau_O = a'_O + \sigma\frac{\Delta x\Delta y H^{n+1}}{\Delta T}$$

$$\tau_A = a'_A,\quad \tau_B = a'_B,\quad \tau_C = a'_C,\quad \tau_D = a'_D$$

$$\vartheta(P_O^{n+1}) = b(P_O^{n+1}) + \sigma \frac{\Delta x \Delta y H^n}{\Delta T}$$

For solving the dynamic characteristics of the flight system have to make simultaneous equations with micron head/disk interface lubrication model and the kinetic equations, and it consists of two iterations for numerical solution:

(1) The inner iteration: The inner iteration used to solve the pressure distribution on underside of micro-flying head within a time-step. According to Eq. (4-168) establish solving pressure equations on a column of nodes with line scan method, and then using steady-state simulation program to solving.

(2) The outer layer iteration: Outer layer iteration is used to solve posture of the micro-flying head in different time. The third-order Runge-Kutta is utilized to calculate the current flight posture, and then recalculation of the near field spacing of the node, bearing number and squeeze to solve the pressure distribution function in next time step with the enter layer iteration.

4.8.6 Near-field optical dynamic flight experiment system

The deep submicron flight system decided the near-field coupling efficiency of the SIL that is the key technology of the near-field optical storage. According to the requirements of the near-field optical storage to decide the specifications, road map of system design micro-flying head and adaptive suspension, simulation of steady-state operating on gas dynamic pressure characteristics, analysis the negative pressure micro-flight head, near-field coupling efficiency, bearing stiffness, near-field spacing stability and anti-pollution ability to meet the requirements of the near-field optical storage. It is need of a dynamic flight experiment system.

- design specifications of the flight system

The overall structure of the micron flight experiment system is shown in Figure 4-102, including micro-flying head, adaptive suspension, bracket of main optical unit, base, recording disc, driver and other components. The SIL and the objective lens are set in micro-flying head. The underside of the micro-flying head is made of complex three-dimensional graphics to obtain dynamic pressure of airflow under the action of the boundary between the micro-flying head and the disk gas film to formation flight spacing. Therefore, micro-flying head design directly determines the operating characteristics of the near-field spacing and bearing capacity of the flight system. In order to maintain near-field pitch stability, the adaptive suspension must be able to output the proper displacement, so that the micro-flying head can free adjust the flight attitude, real-time tracking the movement to keeping the near-field spacing stability. Therefore, the adaptive suspension system is attached to the bracket of the main optical unit which connected to the base that easy to installed and positioning of the flight system in the near-field optical storage drive.

The interface coupling efficiency of near-field optical storage head-disk is determined by the near-field spacing between the SIL and the disk mainly. Assumptions the refractive index of the SIL is 2.0 and the laser wavelength is 650nm, based on vector diffraction theory the light intensity distribution curve is in Figure 4-103. Due to the evanescent field account a large proportion so the coupling intensity on the medium will decrease with the near-field spacing h increasing. When h is larger than 100nm, the coupling intensity sharp attenuate. Therefore, in order to ensure the

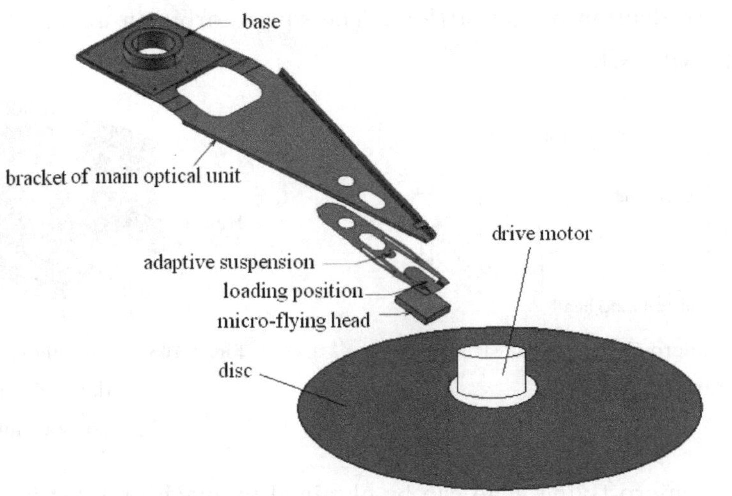

Fig. 4-102 Micron flight system architecture diagram.

coupling of flight, the near-field spacing has to be less than 100nm and the lesser the better. But if it is less than 25nm, the probability of occurrence of the SIL to contact medium will substantial increase. Therefore, under the premise of the head plate near-field coupling efficiency and the system stability, the flight height of the system is greater than 25nm.

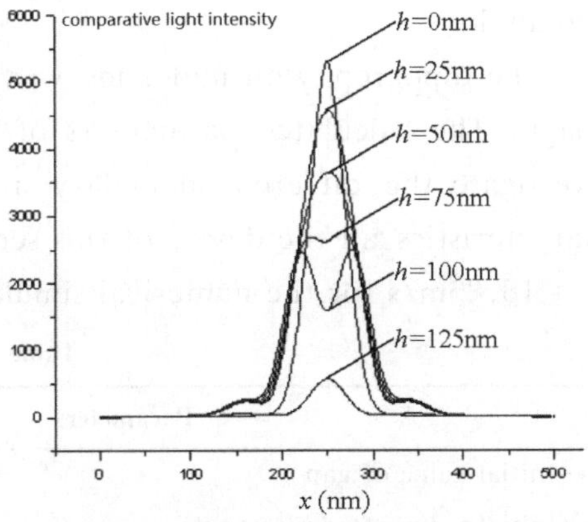

Fig. 4-103 Light intensity changes with the near-field spacing.

The micron flight system seek range is 20~60mm in application. As disc rotation of constant speed, the airflow velocity is great changing on different radius of disc. Taking into account read and write the stability of the optical signal require near-field distance change is less than 10nm in different radius.

4.9 Micro positive pressure nano-gap flying head design

Some flying head, SIL uses air bearing only so it is easy. But high speed disc rotation caused the disc partial swing amplitude of about 5μm that is far over the error limit of $\pm 0.1\mu$m between the SIL and objective lens. In this design, the SIL and the objective lens is integrated to the micro-flying head as shown in Figure 4-104, that the overall size of the micro-flying head is 2.05mm × 1.6mm × 0.8mm. Due to the lens and SIL is fixed when high-speed rotation of disc, even if the flying head real-time tracking disk table movement, objective working distance can remain constant and accurately focus on the SIL bottom surface, thereby overcoming the impact of defocus.

4.9.1 Positive pressure micro-flying head design

The design of morphology diagram on the underside of the positive pressure micro-flying head is given in Figure 4-105. The micro-flying head has a center flow channel with wide 1.175mm and deep 40μm, where both sides of the four straight tracks, airflow export side open square slot embedded with the SIL, the slope surface at the end of the air flow import side of straight track processing for the introduction of the airflow. The square orbit on underside of the micro-flying head is optical unite with SIL.

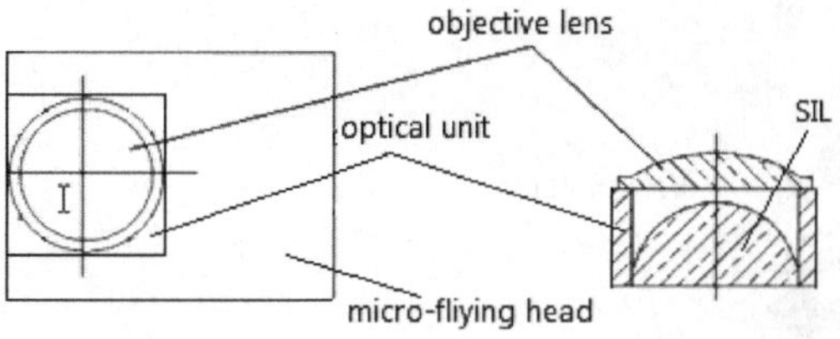

Fig. 4-104 The micro-flying head structure diagram.

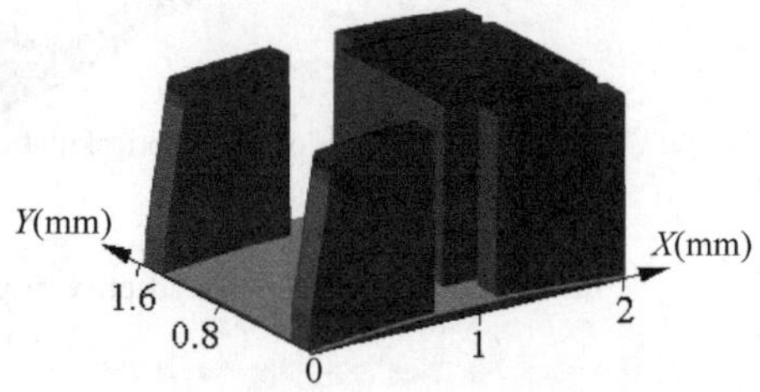

Fig. 4-105 The morphology diagram on the underside of the positive-pressure micro-flying head.

Positive-pressure micro-flying head can be obtained by machining methods, in order to ensure a straight track mechanical strength required thickness greater than 0.2mm with minimum airflow into angle.

The topography on underside was divided into 94 × 94 equipartition grid for calculation and design. The calculated parameters of numerical solution are shown in Table 4-10. In order to investigate the different boundary air flow rate corresponding to the steady-state operating characteristics, all the design of this section were taken to the airflow velocity $V_1 = 9.425$m/s and $V_2 = 18.85$m/s for the numerical simulation.

Table 4-10 Calculation Parameters.

Parameters	value
Initial value of gap	30nm
Initial value of admire angle	70μrad
Initial value of roll angle	0μrad
Bearing capacity	94mN
Pitch angle between airflow velocity and main line of slide	0°
Atmospheric pressure	0.101Mpa
Viscosity of air	1.806×10^{-5}
Airflow line speed	9.425m/s and 18.85m/s
Load center coordinates	$X = 0.5$, $Y = 0.4$
Modified Reynolds equation convergence precision	1×10^{-9}
Bearing capacity error	1×10^{-3}

The simulation results of positive pressure steady state of micro-flying head are shown in Table 4-11. Corresponding to the two air velocity, the SIL near-field spacing is less than 100nm, the carrying capacity is approximately 94mN. Micro-flying head put in order only by the positive pressure, boundary air flow velocity affects positive pressure and near-field spacing significantly. When air flow velocity increases is double, the pitch angle doubles also, and the near-field spacing of SIL increases 28.06nm. The micro-flying head of positive pressure cannot eliminate the affect of gas velocity changing, its steady-state to different radius of disc is no good.

Table 4-11 positive-pressure micro-flying head posture simulation.

parameters	characteristics simulation results	
Boundary air line speed	$V_1 = 9.425$m/s	$V_2 = 18.85$m/s
Minimum gap h_{min}	21.63nm	45.26nm
Maximum gap h_{max}	41.27nm	86.33nm
Center gap of SIL	38.08nm	66.14nm
positive-pressure of flying head	94.08mN	94.08mN
negative-pressure of flying head	2.16×10^{-4}mN	1.08×10^{-4}mN
Bearing capacity	94.08mN	94.08mN
Change of distance between objective lens and SIL	0μm	0μm
pitch angle	19.58μrad	39.71μrad
scroll angle	−0.85μrad	−1.34μrad

The pressure distribution on underside of positive pressure micro-flying head calculated results are shown in Figure 4-106. The surface air on four straight tracks and SIL underside supports micro-flying head with positive pressure. Due to the air bearing surface rail is along the $Y = 0.8$mm symmetry distribution, the micro-flying head will not roll and the roll angle of the system is almost zero. The center of coordinate of micro-flying head is coincided to load point coordinates almost.

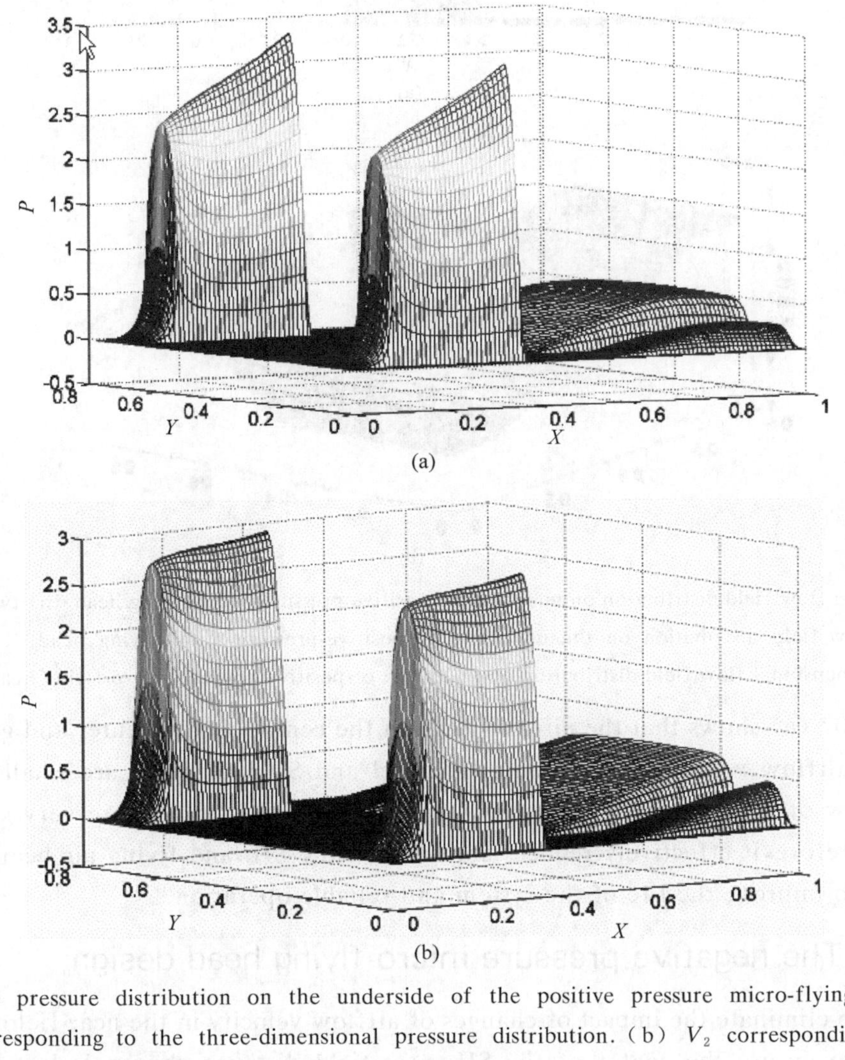

Fig. 4-106 The pressure distribution on the underside of the positive pressure micro-flying head. (a) V_1 corresponding to the three-dimensional pressure distribution. (b) V_2 corresponding to the three-dimensional pressure distribution.

Even the near-field spacing is smaller than the molecular free path, high-speed dust particles will damage to the surface and SIL. Therefore, it is need of the calculation of the flow field distribution on underside of micro-flying head based on numerical simulation algorithm. Due to the two kinks of airflow velocity of the positive pressure of micro flight head is very similar, so only take the V_1 to analysis. As shown in Figure 4-107, in two-dimensional flow field (Figure 4-107 (a)) the flow line towards is direction of air flow, the density of the flow line is air velocity, in three-dimensional flow field (Figure 4-107 (b)) the height difference indicates value of flow.

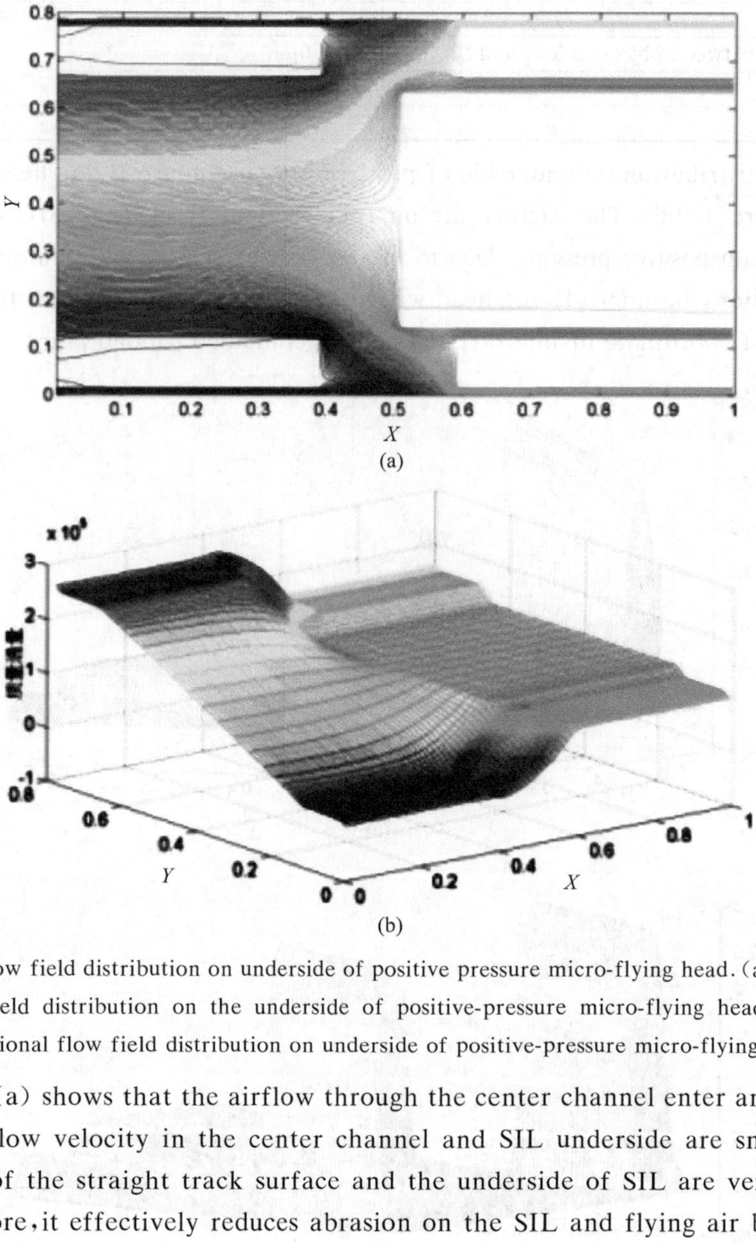

Fig. 4-107 The flow field distribution on underside of positive pressure micro-flying head. (a) two-dimensional flow field distribution on the underside of positive-pressure micro-flying head. (b) the three-dimensional flow field distribution on underside of positive-pressure micro-flying head.

Figure 4-107 (a) shows that the airflow through the center channel enter and go out from the rail gap. The airflow velocity in the center channel and SIL underside are small. At the same time, the airflow of the straight track surface and the underside of SIL are very small as Figure 4-107(b). Therefore, it effectively reduces abrasion on the SIL and flying air bearing surface by dust particles to improve the life of the system and reliable operation.

4.9.2 The negative pressure micro-flying head design

In order to eliminate the impact of changes of airflow velocity in the near-field spacing of the flight system, to ensure the center of the SIL near-field distance change is less than 10nm on different radius of disc. It improved the defect with high sensitivity to flow velocity of positive pressure micro-flying head, and given a design of coexistence of positive and negative pressure as shown in Figure 4-108. The negative pressure micro-flying head employs three-tier structure on the underside with certain vacuum condition. It is spherical crown surface along the length direction and arched ball along the width direction to constitute a class sphere overlay on the underside to prevent the head/disk contact in working process.

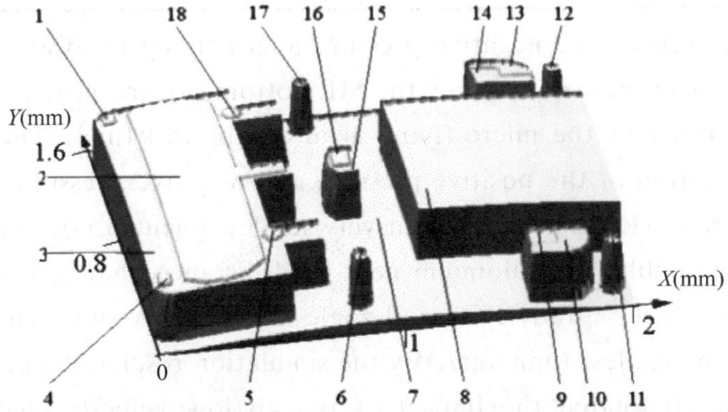

Fig. 4-108 Vacuum micro-flying head and the morphology diagram on the underside.

The underside of the vacuum micro-flying head is etched different depth, which top is the air bearing surface including 1, 2, 4, 8, 10 ~ 13, 1618 track (see Figure 4-108), the middle is light ventilation slots including the 3, 9, 14, 15 tracks. The bottom of the sunken part is the open ventilation slots 7. The height difference between air bearing surface and shallow ventilation slot is 140nm, the difference in height to the main ventilation slot is 2.3μm. For protection of the underside of the micro-flying head from a variety of physical effects and chemical effects of corrosion, the cushion surface is deposited 3nm DLC (diamond-like carbon, DLC) film, which is a transparent insulating passivation film with very high hardness and can improve greatly the tribological properties of the head/disk interface. The underside of the SIL base 8 is a square track of side length of 0.86mm, which four vertex coordinates are (2.032, 1.23), (2.032, 0.37), (1.172, 0.37) and (1.172, 1.23) as in Figure 4-109. In order to obtain a higher near-field coupling efficiency, on SIL underside of the micro-flying head was designed of the structural of air bearing surface track that help to reduce the near-field spacing of the flight system.

The track boundaries of vacuum design irregular and track etch depth is difference, to obtain high precision numerical simulation results that utilized 65 × 65 grid equipartition initial calculation, its initial calculations results are shown in Table 4-12. Based on the initial value, in order to improve the accuracy is need of numerical simulation again. In order to improve the convergence of the iteration, the micro-flying head underside can be meshed of 161 × 161, and calculation the pressure distribution according to initial calculations.

Table 4-12 Steady-state characteristics of initial calculations.

parameters	calculation results	
Linear velocity of airflow	V_1 = 9.425m/s	V_2 = 18.85m/s
initial value of minimum spacing	37.74nm	46.55nm
initial value of pitch angle	49.74μrad	37.32μrad
initial value of the roll angle	0.82μrad	0.47μrad

continued

parameters	calculation results	
Bearing capacity	94mN	94mN
Atmospheric pressure	0.101Mpa	0.101Mpa
Air viscosity	1.806×10^{-5}	1.806×10^{-5}
Reynolds equation convergence accuracy	1×10^{-7}	1×10^{-7}
Bearing capacity error	1×10^{-3}	1×10^{-3}

Final simulation results of the negative pressure micro-flying head are shown in Table 4-13, corresponding to two different air flow rate, the SIL bottom surface near-field spacing is less than 100nm, the bearing capacity of the micro-flying head can be to 94mN. The vacuum micro-flying head works with joint action of the positive pressure and negative pressure to ensure the carrying capacity constant and near-field spacing is relatively stable. Vacuum micro-flying head design, the boundary flow velocity doubles the minimum near-field distance change is less than 10nm, flying pitch angle changes less than 12μrad, flying roll angle is essentially the same, the SIL center point of near-field distance change less than 5nm. By the simulation results, the negative pressure micro-flying head effectively eliminated the impact of the airflow velocity change of the near-field spacing with good stability.

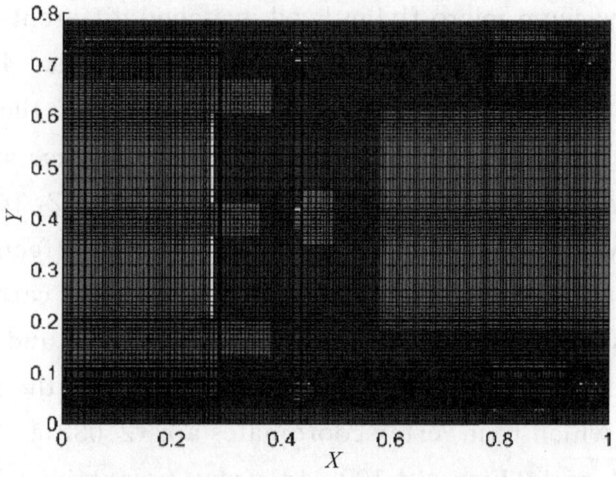

Fig. 4-109 negative pressure flying head underside of meshing.

Table 4-13 The simulation results of the negative pressure type micro-flying head (Steady-state characteristics of the technical indicators calculation results) of the micro-flying head.

Parameters	calculation results of steady-state	
Boundary air line speed	$V_1=9.425$m/s	$V_2=18.85$m/s
Minimum near-field spacing h_{min}	42.12nm	51.07nm
maximum near-field spacing h_{max}	76.55nm	75.38nm
Gap on center of SIL	55.29nm	60.13nm
positive pressure	109.62mN	122.47mN
Negative pressure	15.53mN	28.41mN
bearing capacity	94.09mN	94.06mN
Changes in the objective lens and SIL spacing	0μm	0μm
pitch angle	60.74μrad	48.32μrad
roll angle	0.82μrad	0.47μrad

Negative pressure micro-flying head and the pressure distribution calculation results on the underside are shown in Figure 4-110. The high-speed rotation of the disc, boundary airflow first introduced from the enter side and to bearing surface along shallow ventilation slot track. Due to height of the air bearing surface and shallow ventilation slots is same, the boundary airflow is greatly compressed to form positive pressure on orbit of enter side. When the airflow into the open ventilation slot, due to the height difference between ventilation slots and the air bearing surface, gas into a relatively open area and the rapid expansion rapidly, the main ventilation slot pressure is less than atmospheric pressure to take shape negative pressure effect. The tracks are located in the open ventilation slot center with mediating role to air flow which make two orbits airflow outflow, reducing airflow into the SIL bottom surface and dust particles. Meanwhile the structure also reduces the area of open ventilation slots, increased the carrying capacity of the micro-flying head. The two tracks on the pitch angle of the micro-flying head have some impact also. Airflow after the main ventilation slot flow to SIL underside 8 (see Figure 4-108), that is compressed again to form a positive pressure on out side and enter side to support micro-flying head. The micro-flying head will be balance by positive pressure and negative pressure force and force of load, that the flight system is in steady state operation.

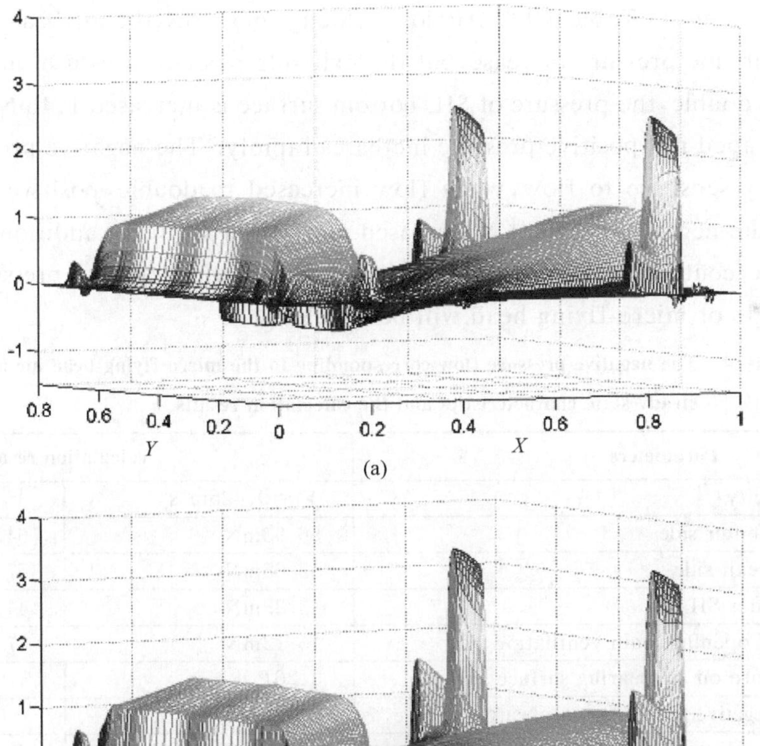

(a)

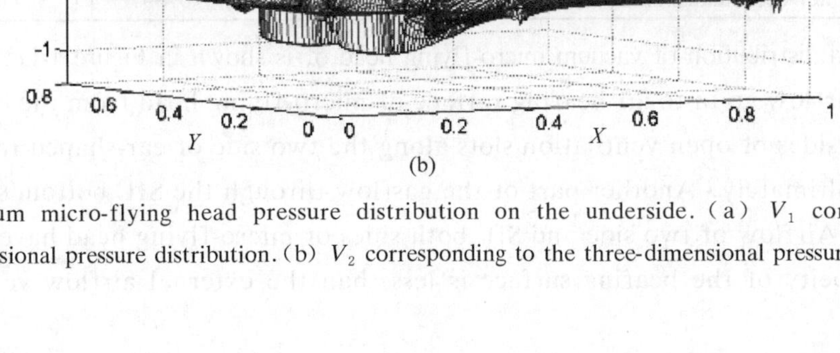

(b)

Fig. 4-110 Vacuum micro-flying head pressure distribution on the underside. (a) V_1 corresponding three-dimensional pressure distribution. (b) V_2 corresponding to the three-dimensional pressure distribution.

Structural design of this co-existence of positive and negative pressure is greatly improved the bearing stiffness and adaptability of the micro-flying head, effectively eliminating the airflow change the course of the near-field spacing, and enhanced the dynamic stability of the flight system. The pressure on both sides of SIL 9 and 10 (see Figure 4-108) and on the track of he 13 and 14 are great, that the positive pressure of two groups improved the rolling stiffness of the micro-flying head, effectively prevent the micro flying head roll in work, and enhance the micro-flying head antidisarrangement. Track 1, 46, 11, 12, 17, 18 and other column-shaped orbit constitute the landing point of the micro-flying head unit. When event of head/disk contact, the collision will be the first occurred a berthing point to avoid scratches SIL and track 2 and 3. Compared Figure 4-110 (a) and Figure 4-110 (b) shows that the gas velocity increases the bearing stiffness of airgas film increase also, that the pressure each enter side increases, and the pressure change on underside of SIL will be gentle, negative pressure on the main ventilation slot of micro-flying head is larger, to ensure the carrying capacity is constant, that the near-field spacing of the system is relatively stable.

Examined in detail the flying characteristics of the negative pressure micro-flying head, in the same carrying capacity to comparation of bearing force with two airflow velocity (V_1 and V_2) analysis results is shown in Table 4-14. Airflow velocity increases, the air bearing surface pressure increases, the enter side pressure increase, but the exit side positive pressure increment is smaller. Flow increased to double, the pressure of SIL bottom surface is increased 1.4mN only, the SIL both sides of the ear-shaped rail positive pressure increases rapidly. The negative pressure of the micro-flying head is very sensitive to flow, when flow increased to double, positive pressure increased 11.7% only, but the negative pressure is increased more than 80%. In addition, as the velocity of flow increases, the center spacing of the positive pressure and negative pressure is significantly reduced, pitch angle of micro-flying head will be smaller.

Table 4-14 The negative pressure flow corresponding to the micro-flying head air buoyancy steady-state characteristics and the calculation results.

Parameters	calculation results	
boundary flow velocity	$V_1 = 9.425$m/s	$V_2 = 18.85$m/s
positive pressure at enter side	56.93mN	64.74mN
positive pressure at exit side	52.65mN	57.66mN
positive pressure under SIL	43.39mN	44.79mN
negative pressure of opening main ventilation slot	15.12mN	27.67mN
great value of pressure on air bearing surface	$3.26P_a$	$4.04P_a$
maximum value of negative pressure on air bearing surface	$-0.39P_a$	$-0.63P_a$
pressure center coordinates on air bearing surface	$X = 0.480$, $Y = 0.394$	$X = 0.472$, $Y = 0.394$
negative pressure center coordinates on air bearing surface	$X = 0.360$, $Y = 0.397$	$X = 0.379$, $Y = 0.397$

The flow field distribution of vacuum micro-flying head of is shown in Figure 4-111. Figure 4-111 (a) shows that the airflow go in to air bearing surface of micro-flying head from the enter side, part of flow to the both sides of open ventilation slots along the two side of ear-shaped rails go out the air bearing surface ultimately. Another part of the gasflow through the SIL bottom surface go out air bearing surface. Airflow of two side and SIL both sides of micro-flying head have higher velocity, but the flow velocity of the bearing surface is less than the external airflow velocity of bearing surface, which the airflow velocity on SIL underside is minimum as shown in Figure 4-111(b), the middle of the corresponding surface of the track and the main ventilation slot is very flat, and airflow is small, But on both sides of surface the air bearing surface and SIL both sides the airflow is lager. The boundary flow is modulated by graphics on underside of the negative pressure micro-flying head, it along both sides of the air bearing surface go out. The gas flow of SIL underside is relatively small due to the adsorption of the negative pressure, a lot of dust is sucked to the open ventilation slot and along both sides go out so it can not reach the SIL bottom surface, meanwhile existence of the DLC protective film on SIL bottom, the negative pressure micro-flying head design has better anti-pollution ability.

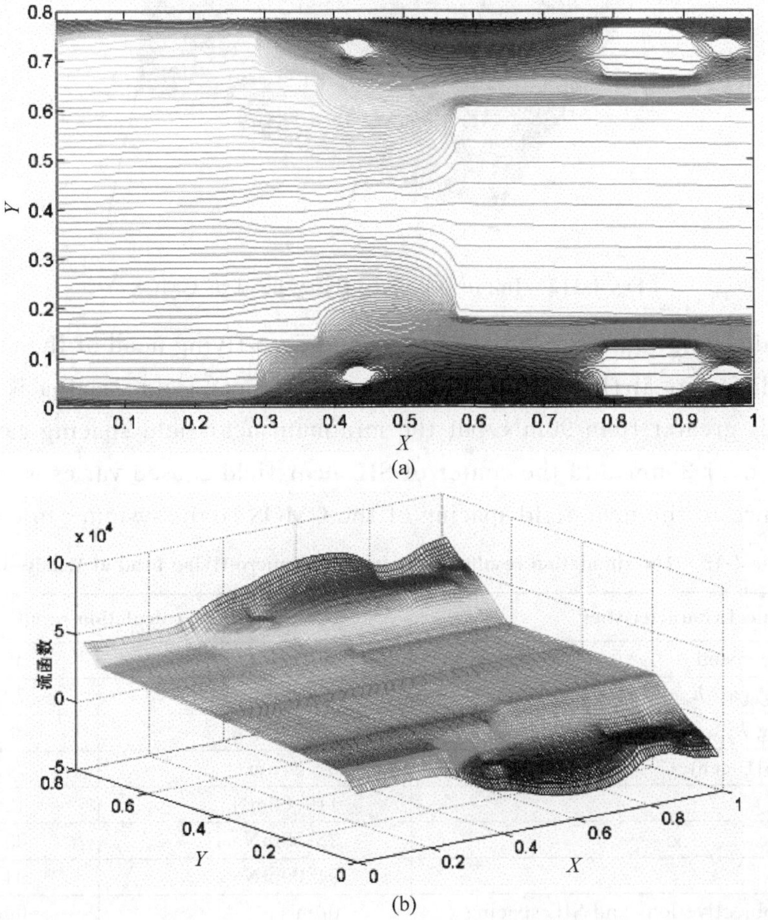

Fig. 4-111 Vacuum on underside of the micro-flying head flow distribution. (a) The negative pressure type micro-flying head and the underside of two-dimensional flow field distribution. (b) the three-dimensional flow field distribution of the underside of the negative pressure type micro-flying head.

4.9.3 Reform design of the slider from magnetic storage

Flotation flight technology has been widely used in magnetic storage systems that technology can be useful for the near-field optical storage flight systems design. The section introduces tow kinds of reformed design of micro-flying head by magnetic storage flotation flight. Because it can be manufacture by CMOS process technology in IC industry, so call it CMOS flight system.

The first improving design is shown in Figure 4-112, that the CMOS underside with two etching depth, the top of the air bearing surface 1, 2, 4, 7 track in Figure 4-112. The middle is shallow ventilation slot track 3, sunken part is main ventilation slot 5, which square orbit is the SIL bottom surface. The height difference of the air bearing surface and shallow ventilation slots is 180nm, and to the main ventilation slot the difference is 1.8μm. The underside of the micro-flying head is coated DLC film with thickness 3nm.

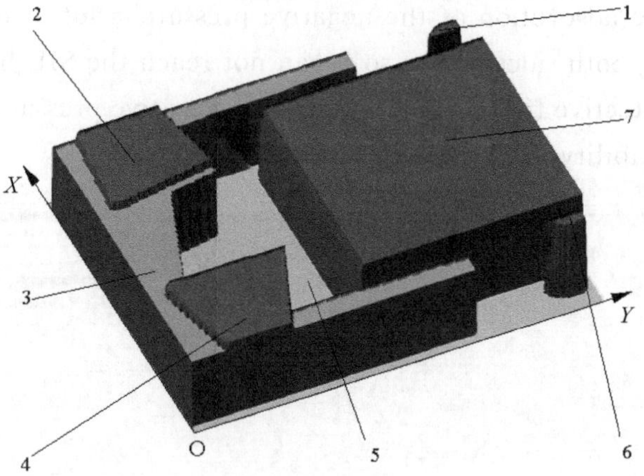

Fig. 4-112 Improved micro-flying head of CMOS.

The numerical simulation results of the improved micro-flying head of the CMOS are shown in Table 4-15. For different airflow velocity, the system of near-field spacing is less than 100nm, carrying capacity is greater than 90mN, but the minimum near-field spacing caused changes with airflow velocity is over 25nm, and the center of SIL near-field caused varies is more than 20nm, ie the airflow influence to the near-field spacing of the CMOS flight system caused larger.

Table 4-15 The simulation results of the improved micro-flying head at steady-state.

technical characteristics	calculation results	
Boundary air line speed	$V_1 = 9.425$m/s	$V_2 = 18.85$m/s
Minimum spacing gap h_{min}	45.75nm	72.28nm
maximum spacing h_{max}	72.19nm	88.87nm
Spacing gap on SIL center	52.27nm	74.17nm
positive pressure	119.68mN	129.86mN
Negative pressure	25.59mN	35.84mN
bearing capacity	94.09mN	94.02mN
Changes of the objective lens and SIL spacing	0μm	0μm
pitch angle	40.43μrad	30.15μrad
roll angle	0.61μrad	0.03μrad

The pressure distribution on the underside of the CMOS is shown in Figure 4-113, the boundary airflow enters guide rail 3 in the first, and form a positive pressure on the enter side of the guide rail 2 and 4. After airflow is into the main ventilation slot, volume expansion causes negative pressure. When airflow reach the SIL bottom surface can form positive pressure on exite side. Comparation in Figure 4-113 (a) and Figure 4-113 (b) shows the flow velocity increases, the positive pressure of the enter side is the same essentially, The pressure distribution of SIL underside is more flat, the pressure on exit side of the micro-flying head is the increases, resulting pitch angle of the CMOS is small, and minimum near-field spacing and the center of the SIL near-field spacing significantly larger.

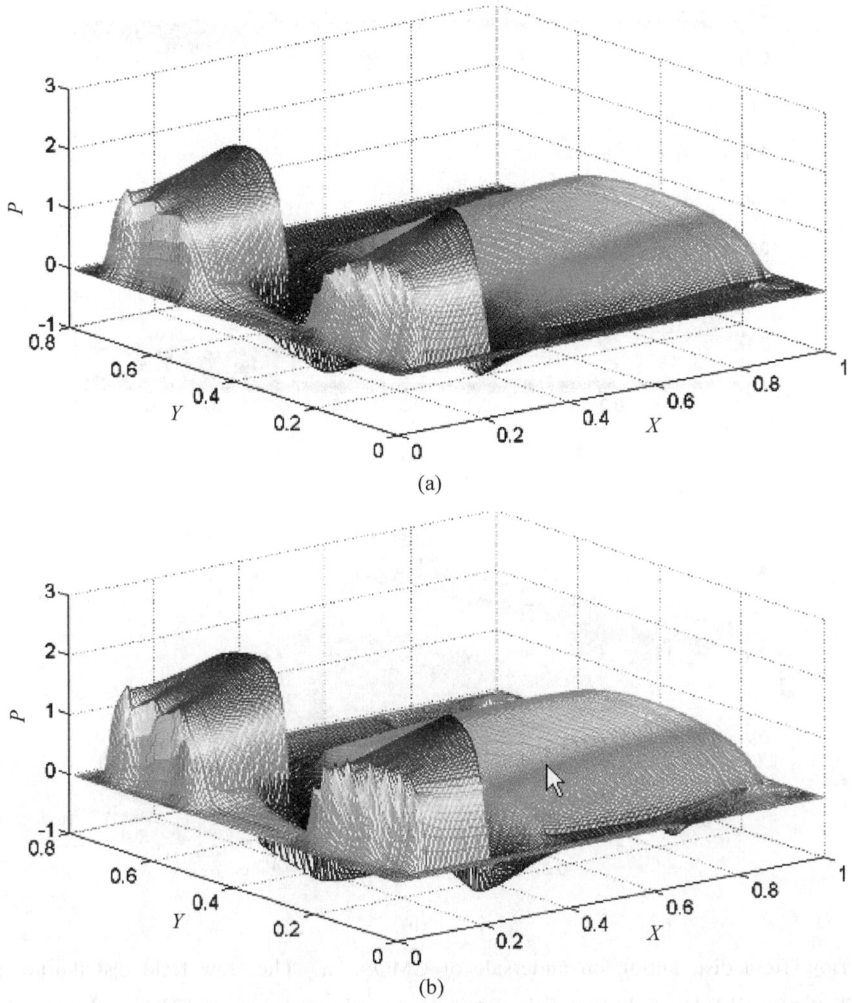

Fig. 4-113 The pressure distribution on underside of the CMOS. (a) The V_1 corresponding to the pressure distribution. (b) V_2, corresponding to the pressure distribution.

The flow field distribution on the underside of the CNOS is the shown in Figure 4-114. The airflow from the import side to closed the main ventilation slot along SIL underside go out SIL underside. The airflow velocity and flow was relatively large. Because CMOS used a closed main ventilation slot structure, resulting in airflow only out of the air bearing surface along the SIL on both sides and the underside of SIL. So dust particles on the SIL bottom surface wear more serious.

Another improved micro-flying head CMOS-I design is shown in Figure 4-115. On the underside there are tracks of 1~3,5,7~9, shallow ventilation slot track 4, the center of the main ventilation slot 6, underside of the square orbit 9 which is the underside of SIL.

When the airflow speed is 18.85m/s, the micro-flying head of the largest near-field spacing is more than 100nm. Corresponding to different airflow, The smallest micro-flying spacing change of CMOS-I is more than 27.13nm, i. e. the stability of the CMOS-I flight system is not good when airflow velocity changing.

The pressure distribution on underside of CMOS-I is shown in Figure 4-116, the boundary airflow first enter the guide rails 4, then on guide rail 2, 7 and SIL underside 9 form positive pressure to support micro-flying head, and on the main ventilation slot 6 form the negative pressure

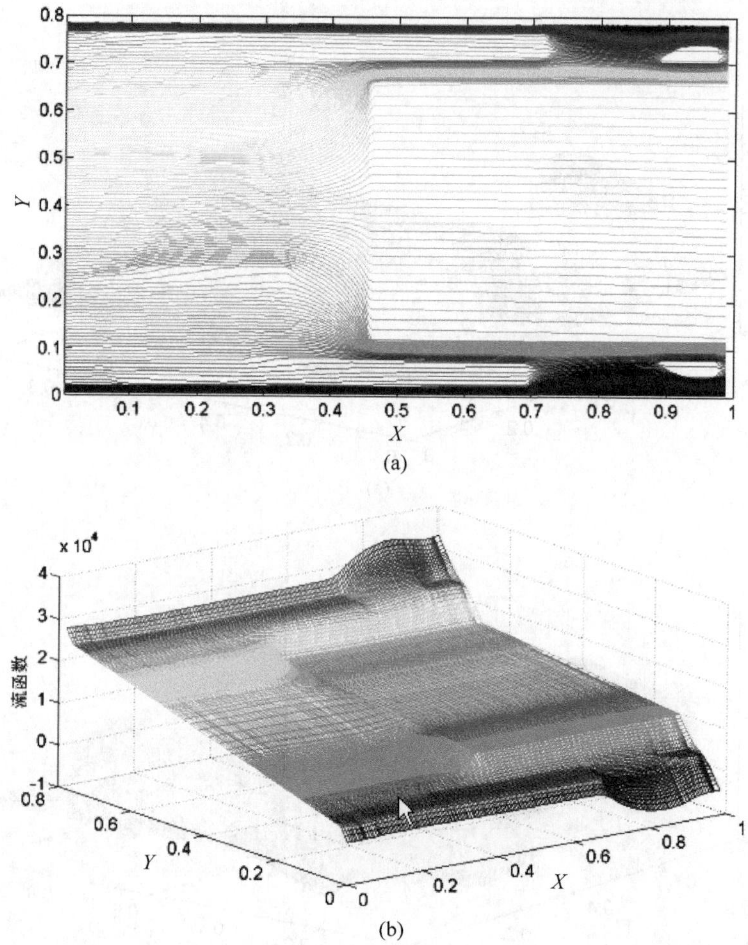

Fig. 4-114 The flow field distribution on underside of CMOS. (a) The flow field distribution on underside of CMOS at plan. (b) The field flow field distribution on the underside of CNOS with three-dimensional map.

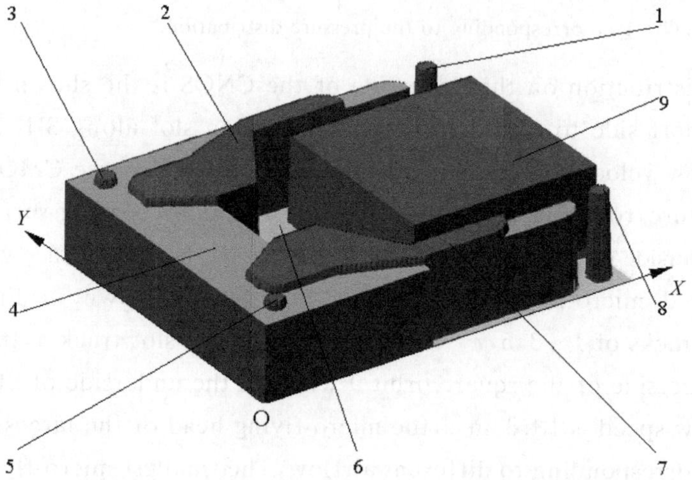

Fig. 4-115 Improved micro-flying head of CMOS-I.

to attract the micro-flying head. Corresponding to a larger airflow velocity, dynamic pressure bearing stiffness of the gas film is larger. As the area of bottom surface of the SIL is greater than the area of enter side of the track, so the airflow change caused the greater pressure increment on bottom surface of the SIL, resulting in micro-flying head pitch angle smaller and near-field spacing increases.

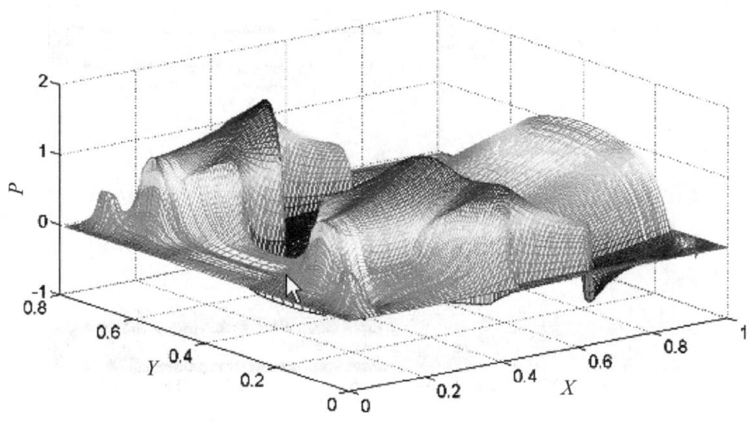

(a)

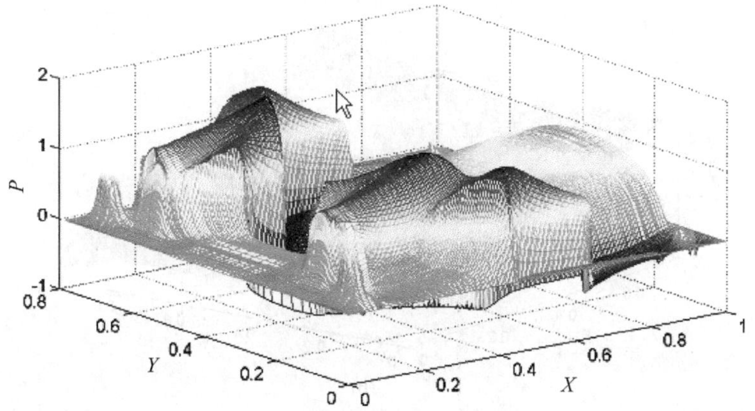

(b)

Fig. 4-116 The pressure distribution on underside of CMOS-I. (a) V_1 corresponding to the pressure distribution (b) V_2 corresponding to the pressure distribution.

The airflow field distribution on underside of the CMOS-I is shown in Figure 4-117. The CMOS-I design is also a closed main ventilation slot, so the airflow velocity and flow on underside of SIL are larger, therefore its anti pollution ability in not good also.

4.9.4 Comparative analysis of the micro-flying head design

By comparing the simulation results of the steady-state characteristics of the four kinds of micro-flying head design analysis shows that:

(1) To a certainty carrying load, when the airflow velocity on the boundary is changing, the positive pressure and negative pressure of the micro-flying head are synchronized changed that ensured the system near-field spacing to be relatively stable. The flight posture change of the positive-pressure is great, but negative pressure micro-flight head was improved greatly and the flow insensitivity is better than positive pressure micro-flying head.

(2) Negative pressure micro flying head design layout complements ear stiffness track on both sides of SIL, open ventilation slots, center air flow adjust orbit around the cushion surface of the main pressure unit and protective column orbit, so the design has better disturbance rejection and higher reliability.

(3) The flow and flow velocity of the dust particles on SIL bottom surface of negative pressure micro-flying head are minimum, meanwhile the negative pressure is adsorption the dust

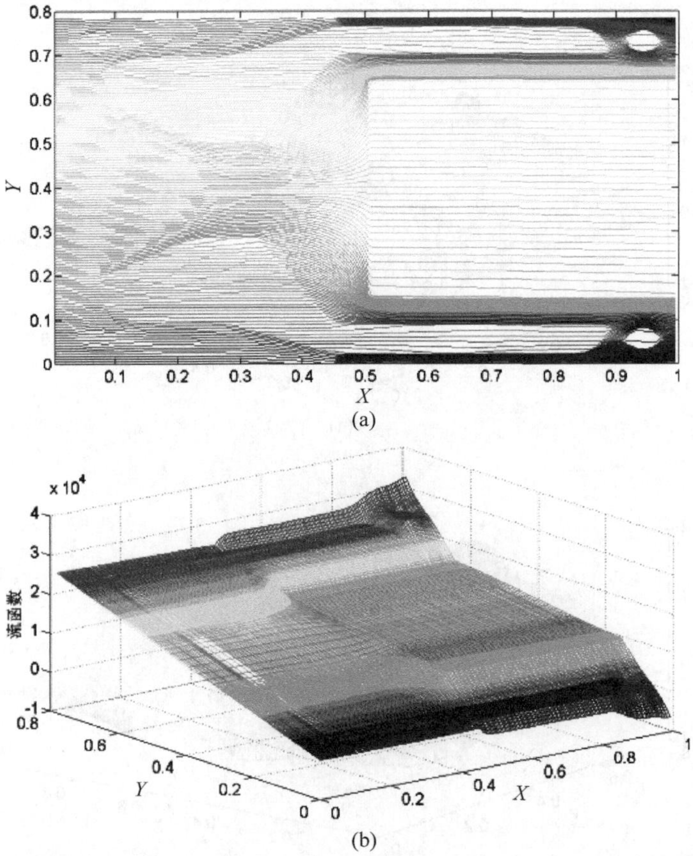

Fig. 4-117 The flow field distribution on underside of CMOS-I. (a) The two-dimensional flow field distribution on underside of CMOS-I (b) The three-dimensional flow field distribution on underside of CMOS-I.

particles, so the design also has the ability of anti-pollution. Closed main ventilation slot micro-flying head dust particle flow velocity is higher, so it is subject to severe wear.

(4) The positive pressure micro flying head can be used machining methods to produce and lower cost. But the negative pressure micro-flying head has to use microfabrication technology to production and cost higher.

(5) Calculation shows that the negative pressure micro-flying head is most sensitive to flow changing.

According to the technical requirements of the flight system, in order to ensure flight systems with stable near-field spacing, good reliability and strong anti-disturbance capacity, selection of negative pressure micro-flying head design is better.

4.9.5 Adaptive suspension design

In micro-flying head working, need of free adjust the flight, to maintain relatively stable near-field space, adaptive suspension bonded micro-flying head that must be able to output the corresponding displacement. Displacement of posture adjustment of micro-flying head is small, that requests the adaptive suspension has higher sensitivity, resolution and reliability.

The structural parameters of flexible hinge of adaptive suspension design affect to the rotational stiffness and pitch stiffness significantly. If flexible hinge stiffness is too large, the increment of the gas buoyancy of the flight system is not sufficient to promote the flexible hinge and cannt do micro-displacement output. The contrary, if the flexible hinge stiffness is too small which is less than external loading and gravity, it cannt obtain security guarantees. A typical design of the adaptive suspension is shown in Figure 4-118, that the micro-flying head is bonded in the center of shrapnel, the suspension has two groups of flexible hinge structure, which flexible hinge 1 provides micro-flying head adjustment and the posture of the pitch rotation deformation, and the flexible hinge 2 provides rolling rotation deformation of micro-flying head and a small deformation of the micro-flying head by vertical translational.

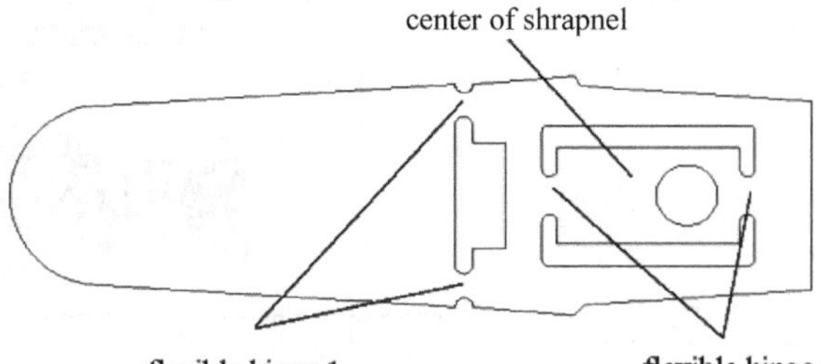

Fig. 4-118 Adaptive suspension diagram.

The adaptive suspension is made of aluminum 6061-T6, which tensile modulus is 69GPa, density is 2710kg/m^3. The overall size of the adaptive suspension is 12mm × 3mm × 0.1mm, the radius of the flexible hinge R_1 and R_2 are 0.1mm, the thickness is 1mm. According to calculation the pitch of direction of rotational stiffness is 3.64uN-m/deg, the rotational stiffness of scroll direction is 2.77uN-m/deg, vertical stiffness is 27N/m, the resistance of the vertical direction is 0.0035Nm/rad, the resistance of the tilt direction is 4.27×10^{-8}Nms/rad, the resistance of the rolling direction is 3.06×10^{-8}Nms/rad.

The micron flying head/disk spacing determined the near-field coupling efficiency of near-field optical storage, flight attitude angle, friction of system starts and stops process and the flight system stability. In addition, the vibration deformation of the system in the seeking influenced its seeking accuracy. Therefore it is need to build a high accuracy overall flight test system as a special testing equipment to test the flight characteristics of the negative pressure micro-flying head and measure the different stop conditions of flight system, the friction between head/disc and the system resonance characteristics etc.

4.10 Nano-gap flight experimental and testing

4.10.1 Main special testing equipment

The overall test device of flight systems of the near-field optical storage in deep submicron is including micron flight systems, precision work bench module, test modules of start and stop characteristics and test module of the resonance characteristics etc., that the block diagram is shown in Figure 4-119.

The system is consists of micro-flight, adaptive suspension, the main optical unit bracket, optical-electron unit, vortex gap sensor, disc spindle motor and stepper driver etc. The micro-flying head is bonding in the center of shrapnel on adaptive suspension, adaptive suspension is welding to consolidation with the bracket of the main optical unit, the optical-electron unit bracket connected with the base through the center hole of the base and column rack-mounted position. Recording disc is installed on the axis of a spindle motor with air bearing. A high precision DC motor is installed outside of the X, Y, Z precision stage to drive the disc for seeking smoothly. Z direction movement of precision stage controls the distance between the head/disk that movement is

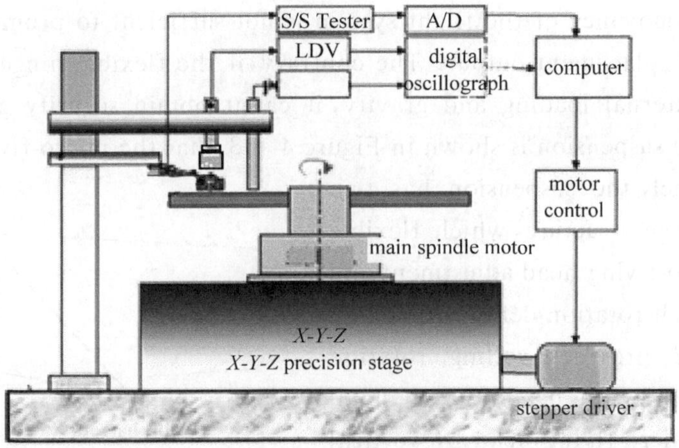

Fig. 4-119 The constructure of the testing for positive and negative pressure micron flying head and near-field optical storage recording experiment system.

composed of a precision worm vice which reduction ratio is 50, the screw pitch is 0.5mm. The drive motor used CoolMusle CM1 series P-17L30 stepper motor, its minimum step angle is 0.2°. So the resolution of Z direction movement of the precision stage can be to 5nm. The test module of start and stop characteristics is used to measure the process of friction between the head/disk when the flight system starts or stops. The module with a stat/stop tester with sensitive strain sensor, its maximum testing range of friction is 200mN, the measurement resolution is less than 3mN, measurement band width is 3.2kHz and measured temperature stability is 0.5mN/℃. The resonance characteristics testing module are used to measurement the resonance characteristics of the flight, it si a laser Doppler vibrometer (LDV) to measurement lateral resonance of micro-flying head with the data processing module analysis to analysis excitation frequency response characteristic and stent-induced vibration to effect to seek accuracy of micro-flying head which corresponding to speed 1mm/s, maximum peak output is 100mm/s with resolution of 0.3μm/s, the maximum band is 50kHz and the linearity error with 0.5%. In addition, the system is used a CCD and a stereomicroscope with magnification of 40× to real-time testing process. All the test results sent to the IPC processing as shown in Figure 4-120. The partial enlarged photo of the adaptive suspension and main optical unit of micro-flying head testing system is shown in Figure 4-121.

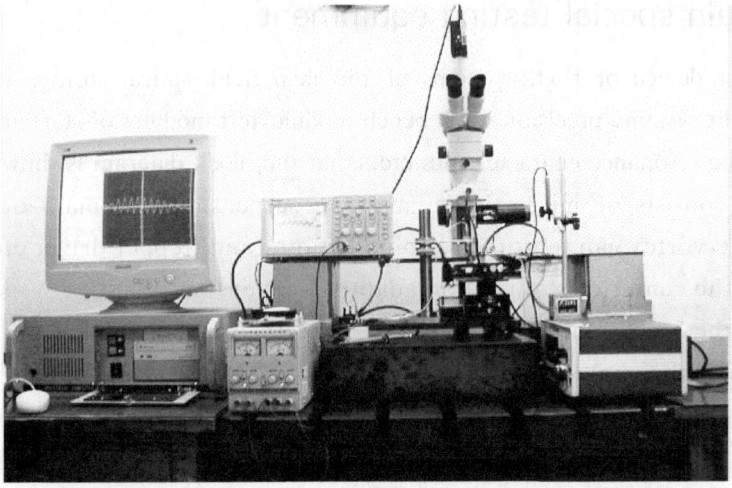

Fig. 4-120 The micron flying head and near-field optical storage recording experiment system.

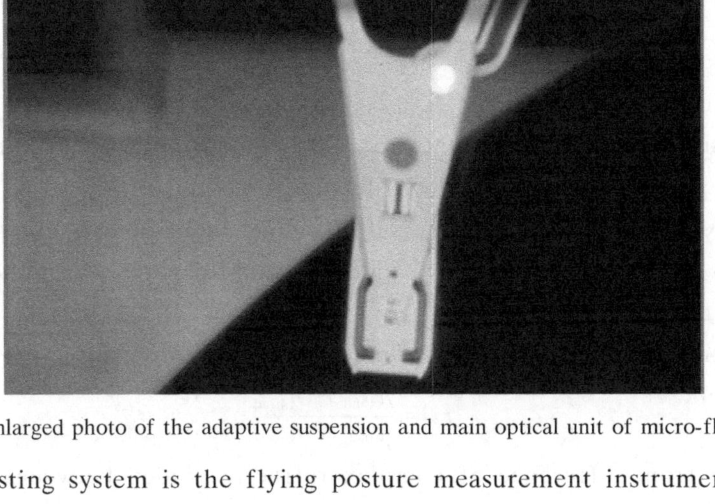

Fig. 4-121 The partial enlarged photo of the adaptive suspension and main optical unit of micro-flying head testing system.

Another key testing system is the flying posture measurement instrument of micron flying head which testing characteristics are including the near-field spacing, flying pitch angle and flight roll angle, the measurement system is shown in Figure 4-122. It has high resolution, high repetition accuracy and high automation with the test accuracy is 1nm, maximum range of the near-field spacing is 600μm and three colored light source of the wavelength 460.36nm, 550.72nm and 650.67nm.

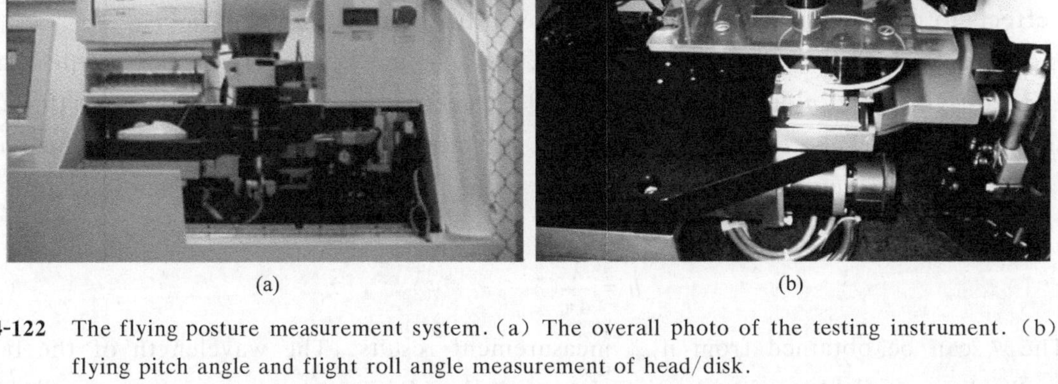

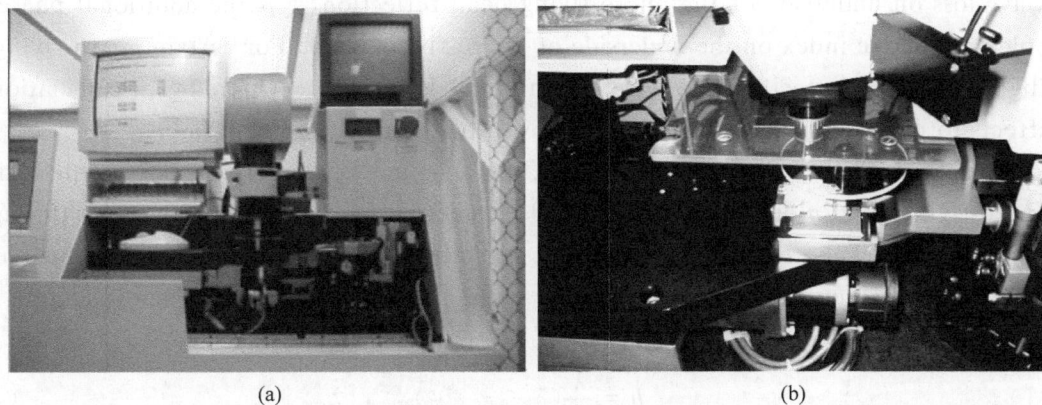

(a) (b)

Fig. 4-122 The flying posture measurement system. (a) The overall photo of the testing instrument. (b) The flying pitch angle and flight roll angle measurement of head/disk.

4.10.2 The near-field spacing testing

The near-field spacing testing is application of multi-wavelength interference measurement principles that is measuring the interference signal of reflected light beams from surface of the recording disc and underside of micro-flying head as shown in Figure 4-123, which using a parallel glass plate replaces recording disc for convenient measurement. The laser beam is vertically incident, so the angle of incidence and refraction angles are 0°. Based on thin film interference theory, the relative intensity I of multi-beam interference in the testing process can

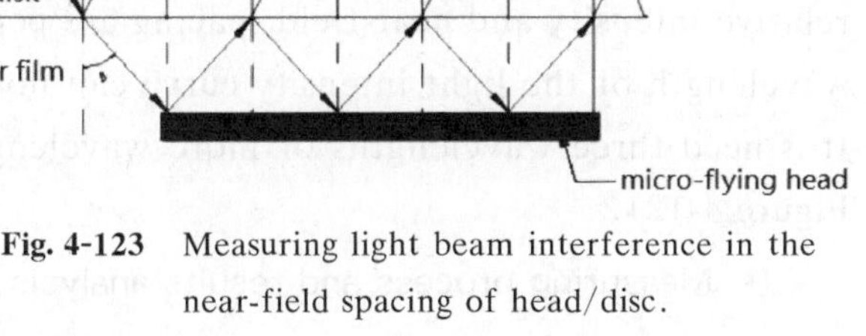

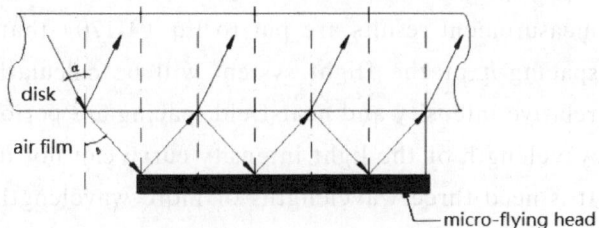

Fig. 4-123 Measuring light beam interference in the near-field spacing of head/disc.

be written:

$$I = \frac{|r_1|^2 + |r_2|^2 + 2|r_1 r_2|\sin\delta}{1 + |r_1^2 r_2^2| + 2|r_1 r_2|\sin\delta} \tag{4-170}$$

Where: relative light intensity $I = I_{\text{interference}}/I_0$, $I_{\text{interference}}$ is the relative light intensity, I_0 is incident light intensity, in the measurement proces it is a constant. r_1 is the reflection coefficient on underside of the glass plate to the airflow film, and $r_1 = (n_{\text{disk}} - n_{\text{fluid}})/(n_{\text{disk}} + n_{\text{fluid}})n_{\text{disk}}$, where n is refractive index of glass plate, the n_{fluid} is d refractive index of dynamic pressure gas film, for air $n_{\text{fluid}} = 1$, n_{head} is the complex refractive index of micro-flying head underside which can be measurements with the ellipsometer.

r_2 is the reflection coefficient of the underside of the micro-flying head:

$$r_2 = \frac{n_{\text{fluid}} - n_{\text{head}}}{n_{\text{fluid}} + n_{\text{head}}} \tag{4-171}$$

δ is the phase from difference of complex refractive index between near-field head/disc spacing of micro-flying head as

$$\delta = \frac{4\pi}{\lambda}h + \pi + \varphi \tag{4-172}$$

Where: λ is the wavelength of the measuring beam, h is the near-field spacing of head/disc, π is half-wave loss on underside of the micro-flying head reflection, φ is the additional phase from the complex refractive index on the underside of micro-flying head. For certain micro-flying head of the flight system and specific laser wavelength, the φ is constant. The relevant calculation of φ and it effects to near-field spacing measurement will be discussed in detail later.

The refractive index of the glass plate $n = 2.48$, the complex refractive index on the underside of the micro-flying head n can be measured by ellipsometer, the light intensity can get through the measurement with the photoelectric detector, so the interference phase δ can be calculated with Eq.(4-170). If interference phase δ was inserted to the Eq.(4-171), the near-field spacing h of the flight system will be calculation as

$$h = \frac{\lambda}{4\pi}(\delta - \pi - \varphi) \tag{4-173}$$

The φ can be obtained from n_{head} measurement results. The wavelength of the beam is known. So the near-field spacing can calculate with Eq.(4-172).

The measuring process is shown in Figure 4-124. First adjust the speed of disc, near-field spacing between the micro-flying head and the disk is continuous change, then measure relative intensity between the near-field spacing with curve to be an initial intensity I_0 curve. Then set the actual parameters of the flight system to measure the interference intensity $I_{\text{interference}}$. These measurement results are put to Eq. (4-170) that can get the phase value δ. Then the near-field spacing h of the flight system will be calculation with Eq. (4-171) and Eq. (4-172). But the relative intensity and near-field spacing are periodic function of wavelength, so only according to a wavelength of the light intensity curve can not determine the near-field spacing value accurately. It is need three wavelengths or more wavelengths to improve measurement accuracy as shown in Figure 4-124.

- Measuring process and results analysis of near-field spacing

Measuring the steady-state near-field spacing, the disc speed is 3600rpm, the micro-flying head

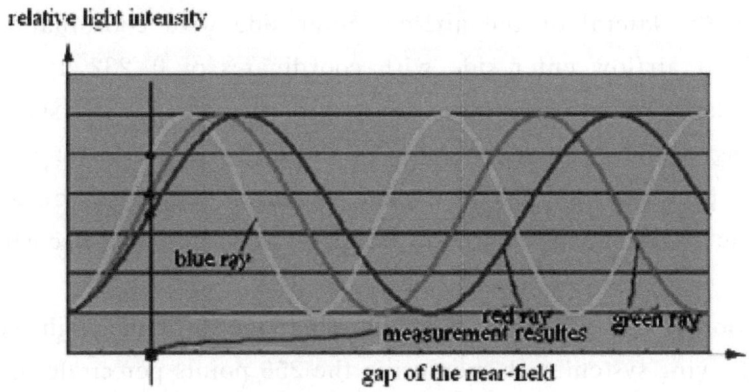

Fig. 4-124 The measuring principles of three wavelengths or multi wavelengths.

is on radius of the disc is 35mm, corresponding to the boundary air flow is 13.2m/s. The load force with the optical unit is 94mN, to ensure the system has a good start and stop characteristics, the flight system static pitch angle is 2°, the static roll angle is 0°. Four point of measurement on underside surface of micro-flying head near-field spacing is shown in Figure 4-125, the test photos is shown in Figure 4-126 and using the average of the complex refractive index to be the complex refractive index on underside of micro-flying head.

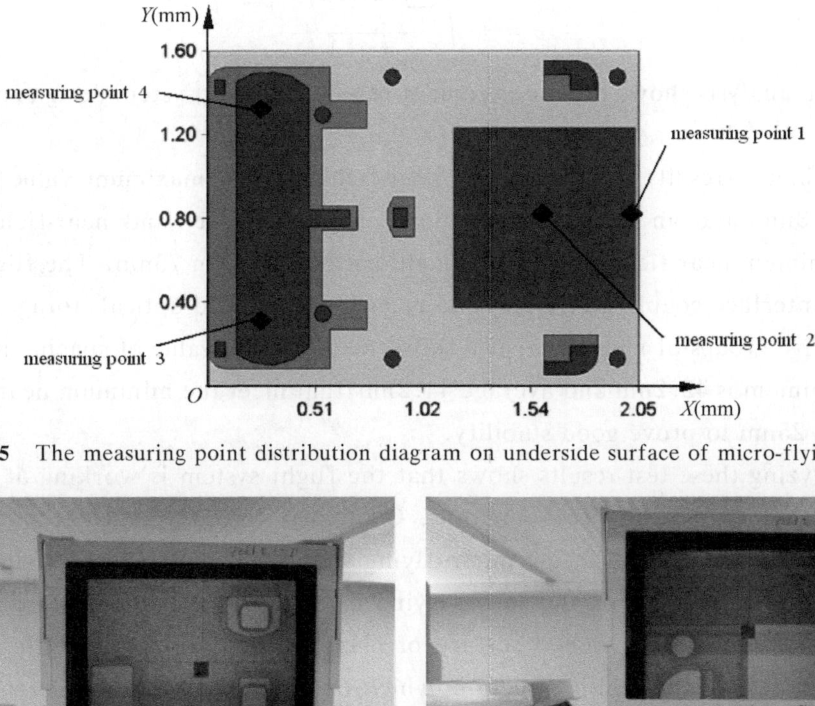

Fig. 4-125 The measuring point distribution diagram on underside surface of micro-flying head.

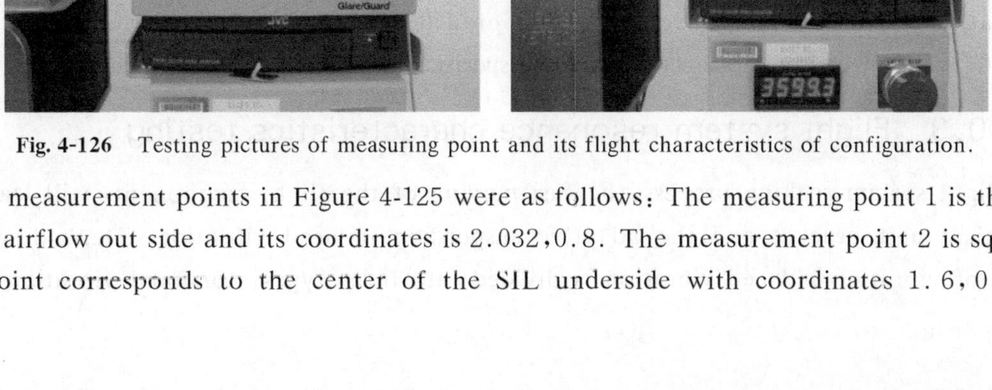

Fig. 4-126 Testing pictures of measuring point and its flight characteristics of configuration.

The measurement points in Figure 4-125 were as follows: The measuring point 1 is the center point of airflow out side and its coordinates is 2.032, 0.8. The measurement point 2 is square rail center point corresponds to the center of the SIL underside with coordinates 1.6, 0.8. The measuring point 3 is the lateral of the airflow enter side with coordinates 0.232, 0.2. The measuring point 4 is at airflow enter side with coordinates of 0.232, 1.4. Among them, the measurement point 2 of near-field spacing is distance of the near field system, which determines the near-field coupling efficiency of the system. In addition, the simulation results shows that the minimum near-field spacing of the flight system is the center point of airflow out side, so measuring point 1 the center point of airflow out side can know the occurrence probability of head/disk contact.

A random selection of five groups of the flight components of the flight system measured the near-field spacing of flying system with measuring the 256 points per circle of disc spins. Sum of measurement results of 8 circles are average to calculate the average value of near-field spacing as

$$\bar{h} = \frac{\sum_{i=1}^{n} h_i}{n} \tag{4-174}$$

Where $\bar{h}$ is the i-th near-field spacing measurement value, n is the number of measurement and $n = 1 \sim 2048$, can get the check sum variance is

$$\sigma = \pm \sqrt{\frac{\sum_{i=1}^{n} (\bar{x} - x_i)^2}{n - 1}} \tag{4-175}$$

Comparative analysis shows the measurement results with the system design specifications can be summary:

(1) Measurement results show that the near-field spacing maximum value is 73.9nm, the minimum is 69.3nm, and an average is 71.6nm. Light intensity and near-field spacing curve achieves the minimum near field working of qualification less than 75nm. The flight system has a high head/disk interface coupling efficiency to meet the near-field optical storage.

(2) In the five groups of measurement results, the maximum value of spacing measurements os 45.8nm, the minimum is 42.2nm and average 44.2nm, that meet the minimum near-field spacing is greater than the 25nm to prove good stability.

(3) By analyzing these test results shows that the flight system is working at the equilibrium position with a slight vibration of about 5nm. Combined with the calculation of the dynamic parameters shows that selection vacuum micro-flying system has better bearing stiffness.

(4) Design and processing of the micro-flying head shows that on underside of the micro-flying head is superimposed an arched surface of class sphere along the length direction of the spherical cap surface and the width direction, which overall height of sphere crown the is 20nm and the overall height of arched surface is 10nm as shown in Figure 4-127. Therefore experiments shows that the flight pitch angle average is 93.56μrad(0.00536°), flight roll angle average is 1.46μrad ($8.37 \times 10^{-5\circ}$), to meet requirements of the design specifications and application of the system.

4.10.3 Flight system resonance characteristics testing

In flight system seeking process, the fast-moving of the flight components will lead to the main optical unit bracket and the adaptive suspension induced vibration. Depending on the direction of vibration induced vibration is divided into three types: torsional vibration, bending

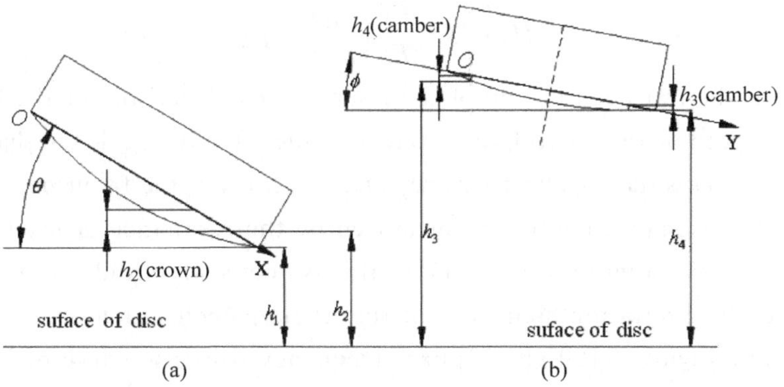

Fig. 4-127 The flight posture schematic of micro flying head, along the direction of the sphere crown of 20nm and along the direction with arched surface of 10nm. (a) The flight posture along sphere crown direction (b) The flight posture along the direction of the arched surface.

and rocking. The bending vibration is driven by micro-flying head movement along the vertical direction mainly, its effect to characteristics of system seeking is very small, and the calculation results shows that its natural frequency of the system in the vertical direction is very high, so induced vibration influence to near-field spacing is smaller also.

The main optical unit support and adaptive suspension to reverse the process will release roll to drive the lateral movement of the micro-flying head, therefore it affects seeking accuracy of the system directly. According to reduce induced vibration to effect the system seek to require that the flight system has a high resonance frequency, to avoid the generation of induced vibration.

In micron flight head testing and experiment system, to adopt the resonance characteristics test module analysis induced vibration to affect the accuracy of the system seek, as shown in Figure 4-128. The computer sends a swept sine wave, after amplifier exciter to drive flight components lateral vibration. The Doppler laser vibration meter (LDVM) can detect the lateral movement of the micro-flight head. Testing results is send to computer for analysis and processing, after the Fourier transform of the two signals measured, can find the transfer function of the resonance characteristics of the flight system as:

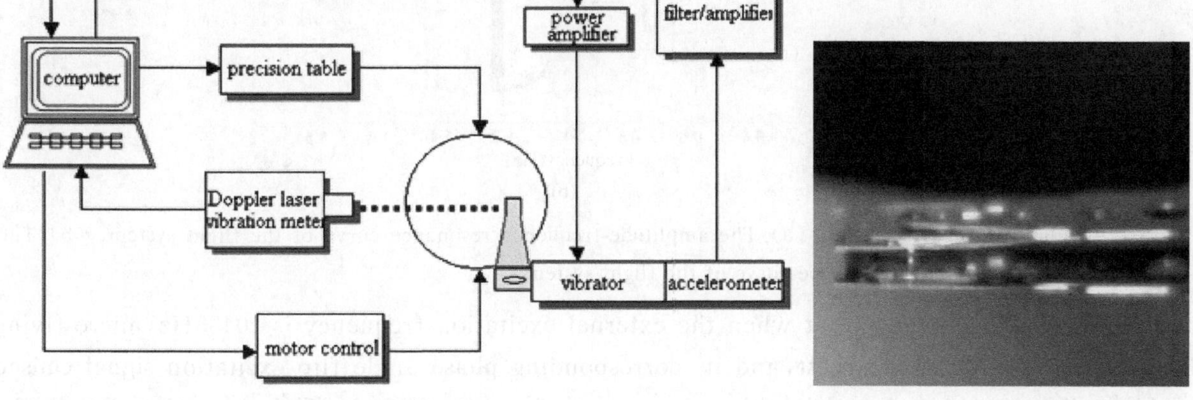

Fig. 4-128 The flight system resonance characteristics testing. (a) The block diagram of the resonance characteristics test module (b) the photo of resonance characteristics testing.

$$H_R(f) = \frac{F(v_{\text{head}})}{F(a_{\text{base}})} \times \mathrm{j}\omega \tag{4-176}$$

Where: $H_R(f)$ is the transfer function of the resonance characteristics of the flight system, v_{head} is the lateral movement speed of the micro-flying head from the DLVM, a_{base} is the shake movement signal from acceleration sensor, ω is the angular frequency and $\omega = 2\pi f$, f is the frequency.

The system flight resonance curves are measured by the resonance characteristics test module as shown in Figure 4-129. Figure 4-129 (a) is the system's amplitude-frequency characteristic curve, which reflects the excitation signals at different frequencies under horizontal amplitude of the micro-flying head. Figure 4-129 (b) is phase frequency response curve of the flying system, it reflects the difference of the excitation signal and the lateral vibration of the micro-flying head.

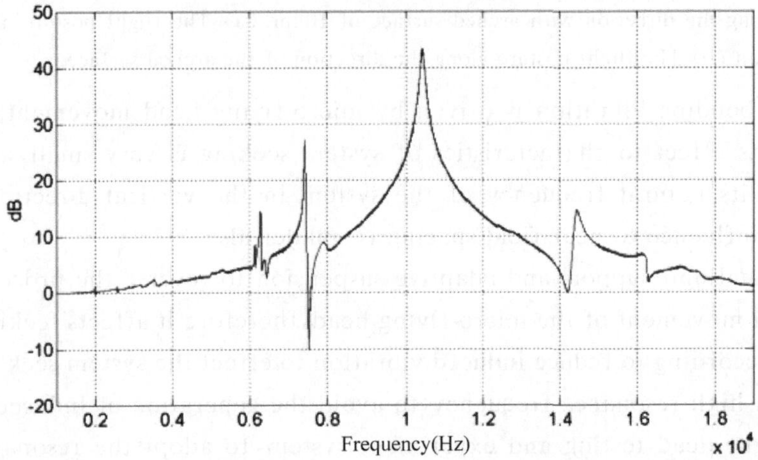

(a)

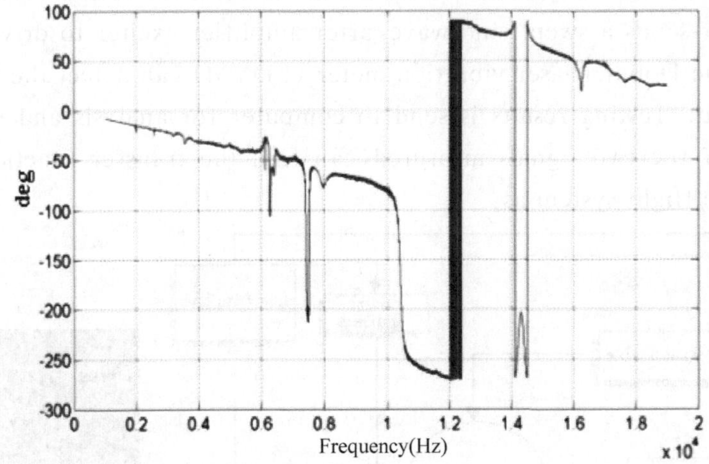

(b)

Fig. 4-129 Flight resonance curves: (a) The amplitude-frequency resonance curve of the flight system. (b) The phase-frequency resonance curve of the flight system.

Figure 4-129(a) shows that when the external excitation frequency is 10435Hz, micro-flying head lateral amplitude is largest, and its corresponding phase angle flip excitation signal caused rocking vibration larger also, ie this is resonance frequency of the flying system. When the excitation frequency is 7428Hz, the micro-flying head also produced larger horizontal vibration. Figure 4-129 (b) shows micro-flying head will be more lateral horizontal vibration obviously, when its frequency is 6128Hz and 14189Hz.

Above analysis results show that when excitation frequency is greater than 6kHz that can be ability to lead lateral vibration of the micro-flying head only. But seek bandwidth of the flight system is 2kHz or less, therefore seek movement triggered system-induced vibration is very small.

4.10.4 Flying start/stop characteristics testing

Start and stop of the flight system are corresponded to the formation and disappearance of gas film of dynamic pressure between the head/disk. When the gas film of dynamic pressure disappears will lead to the head/disk contact easily, resulting in micro-flying head of friction, wear and tear. Therefore, must select the appropriate start and stop, optimized start and stop parameters to achieve a smooth start and stop of the system.

In the micron flight testing system, using the start and stop the tester module (SSTM) measures start and stop characteristics the flight system, as shown in Figure 4-130 (a). The SSTM measures head/disk interface friction indeed, that the sensor of SSTM is attached to the bracket of main optical unit micro-flight head as shown in Figure 4-130 (b). The sensor of SSTM will get signal from vertical direction swing when micro-flying head start or stop lead to friction due to head/disk contact. The vertical direction swing (friction) signal of flight systems are amplification and filtering, the relationship between output voltage and the friction is 50mV/mN.

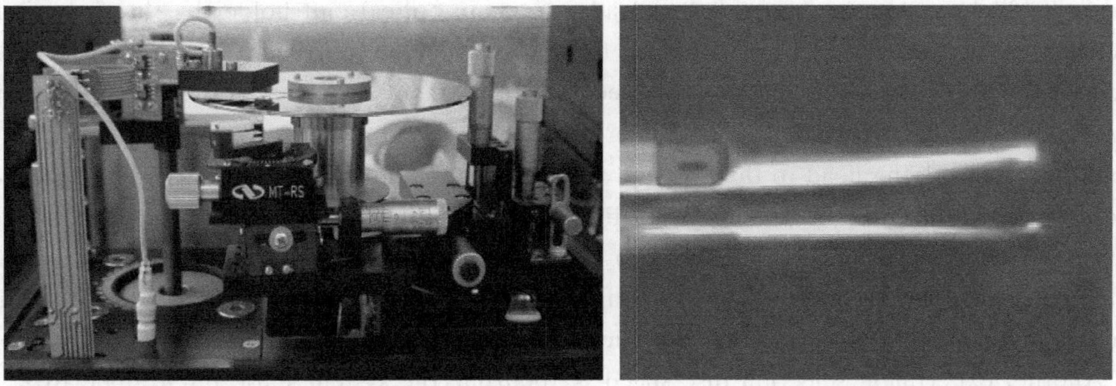

(a) (b)

Fig. 4-130 Dynamic start and stop characteristics testing. (a) SSTM testing module. (b) Flight system working with SSTM sensor.

In testing process, the micro-flying head starts from static state, the disc is from rest gradually accelerate to rated speed, dynamic pressure bearing capacity of the gas film is unable to meet the system requirements in beginning, micro-flying head contacts on the disc may exist serious friction and wear. When the disc speed reaches rated speed, the head to disk begins to form hydrodynamic gas film the micro-flying head, the near-field spacing of the flight system is gradually increasing, the flight system is in steady state operation and the friction of head to disk reaches zero. Similar to the take-off process, the system in the landing process (stop), the disc speed slow down gradually, the carrying capacity of the dynamic pressure gas membrane decreased, when the bearing capacity of the system cannot support the micro-flying head to continue the flight, the micro-flying head docked on the start-stop zone of disc and to appear friction, as shown in Figure 4-131. In the testing process from start to stop of micro-flying head the flight system, the disc accelerates from rest to 3600rpm with 4 seconds, after take-off of the micro-flying head to the steady state for 2 seconds, and then from 3600rpm after 4 seconds to return to the quiescent state. When the disc was accelerated to 1100rpm, the micro flying head take off. When the disc slows down to 1000rpm, the micro flying head is landing. The whole process, micro-flying head by friction is greater than 40mN with serious friction and wear, but after the take-off of the micro-flying head, the friction between the head/disk is approximately 0.

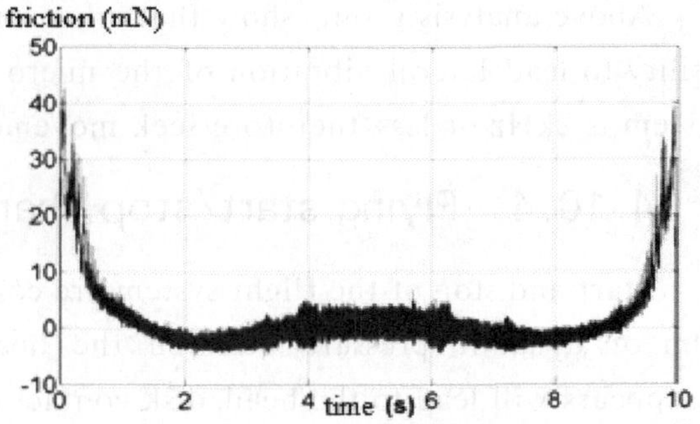

Fig. 4-131 The friction curve of the micro-flying head from start to and stop.

Due to the micro-flying head contact to disc will seriously wear the underside of the micro-flying head, so have to experiment of dynamically flight start and stop. The start and stop experimental parameters are: the dynamic loading process lasts about 0.2 seconds, the dynamic take off process lasted about 0.2 seconds and micro-flying head keeps flying for 2 seconds. During the experiment, the flight system of the first group from the stop parameters increase to the disc speed of 2000rpm, the static pitch angle and static roll angle are 0°, the dynamic loading speed is 20m/s. The friction of start and stop process is shown in Figure 4-132. By the experimental results, the flight system in the 0～0.2 seconds, the main optical unit holder by the slope surface slide the disk, because the hydrodynamic gas film is building, the coefficient friction between head-disk is very small. The strain sensor measured friction is about 5mN, that is the micro-flying head into the disk, static pitch angle is 0°, the micro-flying head load generates greater suction instantly to pull the micro-flying head to the disc and to cause the friction of 15mN. 0.2 seconds later, the head-disk forms the dynamic pressure gas film, micro-flying head gradually into a stable state, the friction between the head-disk decreases to about 0. Flight system in 2.2 seconds, it began to dynamic loading, due to positive pressure and negative pressure release to lack of coordination, resulting in head-disk contact, so that the friction is about 14mN.

Another experiment selection disc speed is 1000rpm, the system static pitch angle to 2° and static roll angle to 0°, the dynamic loading speed is 50m/s, the friction of start and stop process is shown in Figure 4-133.

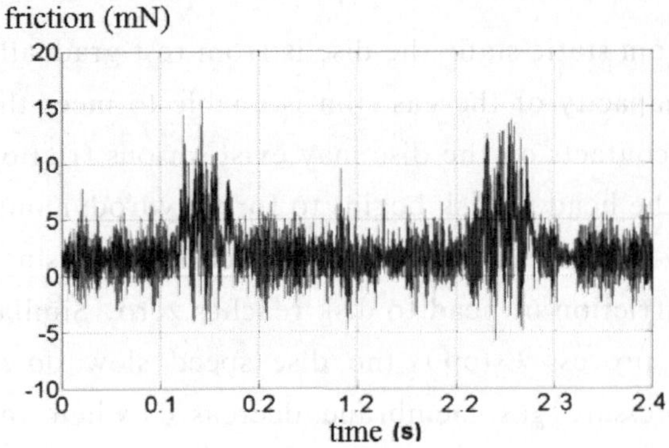

Fig. 4-132 The friction curve of flying head dynamic loading and out the disc with speed of 2000rpm and the static pitch angle and static roll angle are 0°.

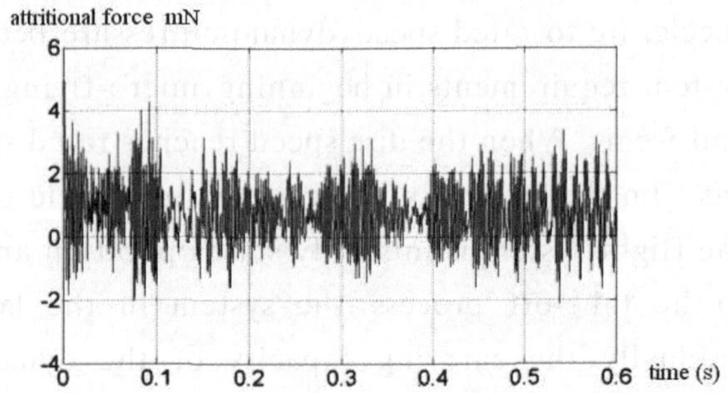

Fig. 4-133 The friction curve of flying head dynamic loading and out the disc with speed of 1000rpm and static pitch angle is 2°, static roll angle is 0°.

The experimental results show that the flight system in the 0~0.2 seconds, micro-flying head enter to the disc to completes the dynamic loading in 0.4~0.6 seconds. The micro-flying head is drived by the bracket of the main optical unit to decline to disc to complete dynamic loading and rise up from the disc to complete dynamic set out. The whole process, friction of the micro-flying head to disk is very small. This is mainly due to the process of dynamic loading as static pitch has an angle of 2° the unit bracket front end and the slope surface is certain buoyancy. While the micro-flying head into the disk or rise up, due to the system is stable state, no head/disk contact and friction between the head/disk is near zero. From throughout the process of start and stop and the measurement results of friction shows that under the conditions, the flight system can implement start and stop smoothly.

Chapter 5

Nanophotonic memory

5.1 Nanophotonics and quantum memories

As society has to deal with the rapid increase of information and the need to store, display, and disseminate it. Hence, increased processing speed, increased bandwidth (more channels to transmit information), high-density storage, and highresolution, flexible thin displays are going to demand new technological innovations. In addition we get more accustomed to wireless communications, coupling of photonics to RF/microwave will play a major role in future information technology. Nanophotonics can be expected to create a major impact in all these areas. Photonic crystal-based integrated photonic circuits, as well as hybrid nanocomposite-based display devices and RF/photonic links, are some specific examples of opportunities. A major challenge lies in formulating processing methods that meet the needs of reliable device performance, together with batch processing and cost effectiveness.

The fundamental aspects of light-matter interactions at a scale significantly smaller than the wavelength of light are most fascinating, and create challenges in the quest of knowledge. This chapter brings the perspective that scientific discoveries lead to technological inventions that can produce commercializable technologies for data memory. Nanophotonics applications already exist, and more will be developed in the years ahead. There is little doubt that nanotechnology in general, and nanophotonics in particular, will play an important role in the development of products and their utilities in the decades ahead. However, any predictions made today are likely to be outdated rather quickly. While there will be the occasional nanophotonics breakthrough giving rise to exciting, new, and highly visible products, most will be made in relative obscurity, and yet will make significant advancements in technologies that will ultimately benefit society.

There are several measures of a new technology's economic impact. A quick search of the Internet clearly shows a rapidly growing number of companies being formed in the area of nanophotonics. Many are still at the research and development stage, but some have matured to the point where products are being offered. One must be careful to distinguish between the technology and the industry it enables. Readers should obviously use caution when encountering market projections because many of them are self-serving and/or exaggerated. Some technological areas are well established, and the commercial potential of nanophotonics can perhaps be best appreciated by looking at the current market sizes and extrapolating what new nanotechnology can offer in the way of improved performance. Other technologies are very young, and it is much more difficult to predict the future market size. This chapter is an attempt to examine some of the most visible and important examples of how nanophotonics is being, and will be used to create nanophotonics for data memory. A general scope of the nanotechnology is cover with lasers and photonics, the areas that together. The examples of materials included are nanoparticles, photonic crystals, fluorescent quantum dots, and nanobarcodes covers quantum-confined lasers. Most semiconductor lasers being commercialized are based on quantum-confined struck. Here, the major application, miniaturization of microprocessors, is analyzed in detail and provides a subjective future outlook for nanophotonics, identifying some examples of future application. First look at the general scope of the science and technology in the areas combine to define nanophotonics. These are nanotechnology, lasers and photonics. Despite the downturn of several technology sectors, nanotechnology continues to attract investments from venture capital groups. There are several sources for information on the market and projections in nanotechnology. Sales figures for lasers are compiled annually by Laser Focus World, and they are reported in January and February issues each year for representational products based on nanophotonics and nanotechnology respectively. More generally, there are some figures and projections for the optoelectronic industry. The greater photonics marketplace is reviewed less quantitatively by Photonics Devices, wherein they discuss important growing industry developments.

5.1.1 Nanophotonics

Nanophotonics and optoelectronics are terms that apply to many areas, including generating, modifying, steering, amplifying and detecting light. Clearly, these enabling technologies are applied to numerous fields of commercial interest, which include CD players, telecom equipment, medicine, manufacturing and other integration optical devices that the technology will trend for the specific case of optical solid state memory. While it is not possible to identify all or most of the specific market opportunities at this time, nanophotonics can be expected to impact further development of all components of photonics described. One may ask why the R&D innovations of today will important to industry tomorrow. Nanophotonics products are smaller, and smaller devices generally are preferred. However, if no other performance feature is added, then one must compare the value of the smaller product against the costs of developing and manufacturing it. In the following sections examine for the various areas of nanophotonics, covered in this book.

Titanium dioxide, TiO_2, and zinc oxide, ZnO, nanoparticles are used as UV blockers. Sunscreen lotions have zinc oxide or titania nanoparticles as UV blockers. While these nanoparticles absorb UV, they are transparent in the visible wavelength and are too small, relative to the wavelength of light, to scatter light. In addition to scattering less, they also absorb UV light more efficiently. Titania nanoparticles have also been investigated for more efficient solar cell applications. There are a number of start-up companies using this nanoparticle concept to develop next-generation solar cells, but they are still in the developmental stage. Nanoparticles also show promise for drug delivery as well as for optical diagnostics and light-activated therapies. Photonic crystals and holey fibers show promise for optical communications. One-dimensional photonic crystals containing liquid crystal nanodroplets can be used for producing dynamically switchable gratings. Dynamically switchable gratings have already been introduced in the marketplace. Photonic crystal components have yet to find a place in the commercial market, but photonic crystal fibers are now being commercialized (Blazephotonics; Crystal Fibres A/S). There are still many practical challenges to be met. Here, a detailed analysis of some selected examples is presented.

The merits of quantum dots for fluorescence based imaging have been discussed. Quantum Dot Corporation, located in Hayward California, is a company commercializing quantum dot-based fluorescent markers. The chief advantages of using these products are the much narrower emission spectra, and their robustness toward photobleaching. Another known fluorescent quantum dots application is security marking as security pigments which contain nanoparticles dispersed in an ink. The resulting dispersion is colorless and completely transparent.

- Photonic Crystals

The property of photonic crystals to steer light at sharp angles has led to great expectations that a large variety of compact integrated optical devices, unattainable with classical waveguide technologies, can be manufactured. Current work is targeting materials and manufacturing methods that are compatible with semiconductor production equipment and methodology. The most popular material is silicon-on-insulator (SOI), where the hole-patterns are etched in a thin layer of silicon on top of an insulating silica substrate; this material has good compatibility with established manufacturing practices and is being pioneered by such companies as Galian, Luxtera, and Clarendon Photonics. Among other approaches being investigated, Nanoopto is developing a proprietary molding process, imprinting the circuit design into a polymer resist layer; the pattern is then created in the silicon/silica layers with anisotropic reactive-ion etching. Neophotonics is creating photonic structures in polymers, while Micro Managed Photons utilizes patterns in thin gold films on a glass substrate. In spite of the all the work being performed on the component level, many industry experts believe that the availability of these products is still some way off. Jaymin Amin, the director of optical systems development at Corning, cautiously states: "I don't think that see photonic-crystal components in the next five years".

- Photonic crystal fibers

There are widespread R&D efforts to realize some of the laboratory demonstrations of photonic crystal fiber applications into practical products. Using modified fiber drawing procedures, fiber manufacturing methods were developed and optimized throughout the 1990s to the point where PCFs have become commercially available at the writing of this book. One of the earliest companies to do so is Blaze Photonics, which is a UK-based company with a substantial number of fiber types, with zero group velocity dispersions at a specified wavelength, these fibers are available in lengths up to many kilometers. Clearly, the price of these fibers will drop with product acceptance and scale-up. Products available already include hollow core fibers (for visible, 800nm, 1060nm, and 1550nm), high nonlinearity fibers (for supercontinuum generation), polarization maintaining fibers, and endlessly single mode fibers. Quantum well, quantum wire, and quantum dot (Q-dot) lasers are nanostructured devices that have been used to produce efficient semiconductor lasers. As most semiconductor lasers, currently sold, utilize quantum are well structures. Hence, there is already a well-established market for this area of nanophotonics, and it is growing, as miniaturization of optical devices continues to be an area of demand. There are certain advantages of Q-dot lasers, which continue to be an area of commercial interest of already lists Q-dot lasers among products. Another laser, being commercialized and finding important application in chemical and biological sensing, is a Quantum Cascade laser (QCL). Alpes Lasers (Switzerland) and Applied Optoelectronics, Inc. (United States) are manufacturers of the Quantum Cascade lasers. Distributed Feedback single-mode QCLs from Alpes Lasers, operating in pulsed mode, are marketed for detection of contaminations in semiconductors, foods, medical diagnostics, and for explosive detection. A promising development in QCL is room-temperature operation at 9.1μm for communication applications. It is believed that this much longer wavelength could greatly enhance transmission through atmosphere in free-space communication.

5.1.2 Nanolithography

This, again, is an area that has generated interest because of advancements in the techniques as described. A major application of nanolithography is in semiconductor industries. The semiconductor industry has distinguished itself by the rapid introduction of technological improvements in its products. Exponentially decreasing the minimum feature sizes used to fabricate integrated circuits (according to the famous Moore's Law, the number of components per chip doubles every 18 months) has resulted in decreasing the cost per function, which translates into improvements of productivity and quality of life as evidenced through the proliferation of computers, electronic communication, and consumer electronics. The typical feature size on a DRAM was 5μm in the 1960s; it is now a quarter micron, and the technology aims at reaching 25nm by 2012. A very thorough resource for further reading on this topic is the International Technology Roadmap for Semiconductors (ITRS) documents. Optical lithography for the patterning of silicon CMOS devices continues to push forward, or downward, creating smaller and smaller transistors. While fabrication facilities that produce silicon chips with the "enormous" minimum MOSFET gate length of 1.6μm still exist, the cutting edge microprocessors are into the deepsubmicron sizes. The newest processor is fabricated using a 25nm technology; that is, the smallest gate length of the MOSFET is no larger than 90nm (actual gate lengths for this process are under 70nm). The original Pentium 4 was fabricated with a 0.18μm technology. Research into smaller technology sizes is ongoing, with an operational 90nm technology process, according to some, expected in late 2003. Beyond this, 30nm gate length devices have been demonstrated using standard optical lithographical techniques (193nm wavelength). However, in order to have production-level processes at 20nm technology, new lithographic techniques must be used. Current thinking is that extreme ultraviolet lithography (EUVL) with wavelengths on the order of 13nm is needed for this step. EUVL is currently the focus of much research. After the end of this decade, it is unclear what the next technology will be for smaller devices, although there are many possibilities such as molecular electronics. In addition, wafer sizes are now as large as 300mm in diameter, requiring mask development and UV light sources that can produce a uniform intensity across the entire wafer. A nanolithography technique, being advertised at the time of writing this book, is dip-pen lithography system based on a fully functional commercial scanning probe microscope (SPM) system, environmental chamber, pens, inkwells, substrates, substrate holders, and accessories for DPN experiments. Another manufacturer of nanolithography tools is Molecular Imprints Inc.,

Austin, TX, which produces a series of systems, based on the unique Step and Flash Imprint Lithography technology called S-FIL. The Imprio 100 provides sub-50nm lithography. Nanonex Inc., Monmouth Junction, NJ offers nanonimprint lithography tools, resists and masks.

The scientific quest for knowledge, along with society's insatiable thirst for compact, energy-efficient, and multimodal technologies, ensures a bright future for nanophotonics. Like in most technology areas, market-driven inventions will create economic opportunities. However, recognizing that nanophotonics is an emerging area, new scientific discoveries will also play a dominant role in the development of new technologies. Not all discoveries and resulting laboratory demonstrations of technology result in commercialized products. The competitive edge of a particular technology, performance reliability, production scalability, and cost-effectiveness are some important measures to be met by a commercially viable product. While research scientists are good at producing innovations, most are not well-suited to transitioning an innovation to commercial opportunities. This is where the university-industry-investment partnership can play a vital role in transitioning scientific discoveries to commercial products.

Even though there is inherent risk in making predictions about future developments and in identifying areas of future growths, it may still be useful to provide perspectives for the future outlook. Four major thrust areas are projected as those that could significantly benefit from breakthroughs in nanophotonics. Therefore, these thrusts are presented with some selected examples of economic opportunities. A nanophotonic approach utilizing inorganic, organic hybrid nanostructures and nanocomposites can produce broadband harvesting of solar energy while using flexible low-cost, large-area roll-to-roll plastic solar panels and solar tents. Other power conversion sources can involve rare-earth-doped nanoparticle up-converters and quantum cutters. These photon converters can be utilized to harvest solar photons at the edges of solar spectrum, specifically in the IR and in the deep UV. A major direction for basic research, needed to mature this technology, is an understanding and subsequent control of dynamic processes at the nanoscale. Much emphasis has been placed on the nanoscopic structure-function relationship. However, the dynamics in nanostructures is equally important in controlling many photonic functions, such as in the case of photon conversion. Another major area of opportunity is the utilization of quantum-cutter nanoparticles for lighting applications. There is a strong push to produce mercury-free, efficient lighting sources. Efficient quantum cutters, which can even be spray-coated, will provide a means to realize this goal. Some other applications of photon up-converting nanoparticles are for display and security marking.

There is an ever-increasing need to enhance the capability of sensor technology for health, structural, environmental monitoring and information memory. One area of great concern is new strains of microbial organisms and the spread of infectious diseases that require rapid detection and identification. This requires point detection as well as environmental monitoring. Another area of major concern, worldwide, is the threat of chemical and biological terrorism. The detection here is not only for the danger posed to health, through chemical and biological agents, but also for structural damage (to bridges, monuments, etc., through explosives). Nanophotonics-based sensors utilizing nanostructured multiple probes provide the ability for simultaneous detection of many threats, as well as the ability for remote sensing where necessary. A useful future approach can utilize nanoscale optoelectronics with hybrid detection methods involving both photonics and electronics. A major impetus for growth of nanotechnology, including nanophotonics, is provided by national priority funding from government in many countries. In the nanotechnology sector, most venture capital funding has gone to the area of nanomaterials. Optical nanomaterials have well-established markets, which are mostly focused on low-technology applications, such as in sunscreen lotions and optical coatings. Fluorescent quantum dots and nanobarcodes are other examples of optical nanostructures being currently commercialized. Another nanophotonic material, recently introduced in the market, is photonic crystal fibers. Most semiconductor lasers utilize quantum well structures. More recently, quantum dot lasers have also been introduced in the market. Quantum cascade lasers are yet another example of quantum-confined lasers, being marketed for detection of contaminants in semiconductors, foods, and medical diagnostics and for explosive detection. A major application of nanolithography is in semiconductor industries for producing smaller microprocessors with more capability. Dip-pen lithography and imprint lithography are some new nanolithographic techniques being commercialized.

- Optical nanofiber sensors

These sensors utilize tapered optical fibers which are used for near field microscopy. The tapered fibers have the tip diameters ranging between 20nm and 100nm. These tapered fibers are also referred to as nanofibers. Like in near-field microscopy, these fibers are metal coated on the wall to confine light. The sensing biorecognition probe, which binds the analyte to be detected, is immobilized at the tip opening (the distal end of the nanofiber). The first optical nanosensors were demonstrated by Kopelman's group for intracellular chemical sensing. Since then, several reports of measurements of pH, various ions, and other chemicals have appeared. An example is a nanobiosensor reported by Vo-Dinh and coworkers. In this work the fiber tip was silanized to allow for covalent attachment of antibody, using a reaction involving carbonyl dimidazole. The antibody employed in this sensor probe recognizes benzo pyene groups as a specific antigen which can detect benzo pyrene tetrol (BPT), a DNA adduct of benzo pyrene found in cells treated with this chemical carcinogen. This provides a convenient and simple method to rapidly detect cells that have be malignantly transformed with this chemical.

5.1.3 Optical nanoscopy for data storage

An optical nanoscopy can records raw data images from living cells and tissues with low levels of light. This advance has been facilitated by the generation of reversibly switchable enhanced green fluorescent protein (RSEGFP), a fluorescent protein that can be reversibly photoswitched more than a thousand times. Distributions of functional RSEGFP-fusion proteins in living bacteria and mammalian cells are imaged at 40nm resolution. Dendritic spines in living brain slices are super-resolved with about a million times lower light intensities than before. The reversible switching also enables all-optical writing of features with subdiffraction size and spacings, which can be used for data storage. In the fluorescence microscope, diffraction prevents (excitation) light being focused more sharply than 1/(2NA), with 1 being the wavelength of light and NA the numerical aperture of the lens. Thus, as they are illuminated together, features residing any closer together than this distance also fluoresce together and appear in the image as a single blur. The diffraction resolution barrier can be overcome by forcing such nearby features to fluoresce sequentially, but this strategy clearly requires a mechanism for keeping fluorophores that are exposed to excitation light non-fluorescent. In stimulated emission depletion (STED) microscopy, this is accomplished by the so-called STED beam, which turns the fluorescence capability of fluorophores off by a photon-induced de-excitation. Because at least a single de-exciting photon must be available within the lifetime of the fluorescent molecular state, the intensity of the focal STED beam must exceed the threshold $I_s = C\tau^{-1}$ with C accounting for the probability of a STED beam photon to interact with the fluorophore. The STED beam, usually formed as a doughnut overlaid with the excitation beam, features a central point of zero intensity at which the fluorophores can still assume the fluorescent state. As this point can be positioned with arbitrary precision in space, the coordinate of the emitting (on-state) fluorophores is known at any instant: it is the position of zero intensity and its immediate vicinity, where the STED beam is still weaker than I_s. The diameter of this area is given by $d = \lambda/[2N_x(1 + I_m/I_s)^{1/2}]$, with I_m (typically $\gg I_s$) denoting the intensity at the doughnut crest. Hence, features that are (just slightly) more apart than $d \approx 1/(2\text{NA})$ cannot fluoresce at the same time even when simultaneously illuminated by excitation light. Scanning the beams across the sample and recording the fluorescence yields images of subdiffraction resolution d automatically and irrespective of the fluorophore concentration in the sample. De-excitation by stimulated emission is the most basic and general mechanism for modulating the fluorescence ability of a molecule.

However, by requiring light intensities $I_s = 1 \sim 10\text{mW} \cdot \text{cm}^{-2}$, attaining high resolutions by this mechanism necessitates large I_m values. For example, $d \approx 40\text{nm}$ typically entails $I_m = 100 \sim 500\text{mW} \cdot \text{cm}^{-2}$. Although intensities of this order have been demonstrated to be live cell compatible, all-optical nanoscopy methods operating at fundamentally lower light levels are highly in demand, because it allows larger fields of view and can avoid photodamage. A route to low light level operation is to replace STED with a fluorescence switching mechanism having a lower threshold I_s. Following the equation for I_s, this can be realized by exploiting transitions between fluorophore states of longer lifetime $\tau \approx 1\mu\text{s}$. Hence, it has been suggested that fluorescence can be switched by transferring the fluorophores transiently to a generic metastable dark (triplet) state of $\tau < 10^{-3}\text{ms}$. A more attractive option is to use fluorophores that can be explicitly "photoswitched", for example by photoisomerization. Hence 2003 it was proposed to implement a STED-like microscope with STED being replaced by a reversible on-off switch as encountered in organic photochromic fluorophores and reversibly photoswitchable fluorescent proteins (RSFPs). In fact, this strategy is more general because any reversible transition between a signalling and a non-signalling state can be used for breaking the diffraction barrier. Therefore, all concepts switch the fluorescence capability of molecules at sample coordinates predefined by patterns of light have been generalized under the name RESOLFT, which stands for reversible saturable optical (fluorescence) transition between two states. A photoswitch is a perfect saturable transition. Concomitantly, the concept was extended to subdiffraction writing and data storage, in which case the on-state is a reactive state from which the molecule be made permanent whereas the off state serves as a temporary "mask" defining the structure to be written.

Super-resolution by switching RSFPs was shown in 2005 with a tetrameric protein with low fluorescence quantum yield. Moreover, when translating the light pattern across the sample, the proteins faded after a few cycles, implying that features that had been turned off could not be turned on again in order to be read out. Biological imaging therefore remained unviable. Other studies using a variant of the RSFP called dronpa faced the same challenge. As a rule of thumb, an m-fold resolution improvement along a certain direction requires, m switching cycles, meaning that $m = 10$ along the x-and y-axes entails $m^2 = 100$ cycles, whereas 1,000 cycles are required for x, y and z. Thus, for RESOLFT super-resolution, the number of switching cycles afforded by the fluorophore assumes a vital role. Because they are able to generate an image with a single on-off cycle, the super-resolution concepts called (F)PALM and STORM, which have emerged in the interim and successfully harnessed the switching between metastable states for gaining subdiffraction resolution. However, these methods rely on the imaging and computation-aided localization of individual fluorophores amidst the scattering and autofluorescence background common in (living) cells and tissues. Moreover, rapid localization of a sufficiently large number of fluorophores requires the excitation light to be intense. In contrast, a RESOLFT approach is able to instantly record the emission from all fluorophores attached to the nanosized feature of interes and can be easily combined with confocal microscopy for three-dimensional imaging and background suppression. Because all RSFPs, conventional fluorescent proteins and photochromic rhodamines seemed unsuitable as shown in Figure 5-1, an all-optical nanoscopy approach operating at low light levels appeared unviable. Similarly, although STED/RESOLFT-inspired optical writing with photochromic compounds has been shown to yield structures $\approx 1/(2NA)$, writing such structures with spacings $\approx 1/(2NA)$ remained challenging, that the impediment being the requirement of many on-off cycles before the structure is made permanent. The RSFP enabling both low-light-level all-optical nanoscopy of living cells and tissues, and far-field optical writing and reading of patterns of subdiffraction size and density.

- Generating a reversibly switchable GFP

All fluorescent proteins have a similar fold, namely the 11-stranded β-barrel with a central helix containing the chromophore, which is typically in a cis-configuration. Light-driven switching of RSFPs generally involves an isomerization of the chromophore, frequently coupled with a change of its protonation state. Started from EGFP and identified, using its X-ray structure, amino acid residues the exchange of which was expected to facilitate isomerization. Expressed numerous EGFP variants in Escherichia coli and screened for colonies expressing an RSFP with an automated microscope. The alternated site-directed and error-prone mutagenesis while maintaining the key amino acids of EGFP concomitantly introduced to ensure that the protein remained a monomer. The amino acid exchange was sufficient to make EGFP reversibly switchable, but the resulting on-off contrast was low. Although it makes the protein switchable and avoided the mutation because it seemed to reduce the number of cycles.

After analysing of ~30,000 clones, identified EGFP as shown in Figure 5-2 that could be reversibly switched on $\lambda = 405$nm and off at 491nm, and named it reversibly switchable EGFP (rsEGFP). At equilibrium, rsEGFP adopts a bright on-state (fluorescence quantum yield $\Phi FL = 0.36$, extinction coefficient $\varepsilon = 47,000 M^{-1}cm^{-1}$. In the on-state, rsEGFP exhibits a single absorption band peaking at 491nm (see Figure 5-1(a)), corresponding to the ionized state of the phenolic hydroxyl of the chromophore. The pKa of the chromophore is 6.5 (see Figure 5-3). Absorption at 490nm yields fluorescence peaking at 510nm and, in a competing process, switches rsEGFP off (see Figures 5-1(a)～(c)). Prolonged irradiation of a pH 7.5 solution of purified rsEGFP at 490 nm reduces the rsEGFP fluorescence to 1%～2% of its initial value. The off-state exhibits a single absorption band at 396 nm, corresponding to the neutral state of the chromophore (see Figure 5-1(b)). Excitation at this band switches the protein back to the onstate. At room temperature rsEGFP converts spontaneously from the off-into the on-state with a half-time of ～23min (see Figure 5-1(d)). Compared the properties of rsEGFP with that of the well-known RSFP dronpa. With the proteinsembedded in a 12.5% polyacrylamide (PAA) layer and using light of

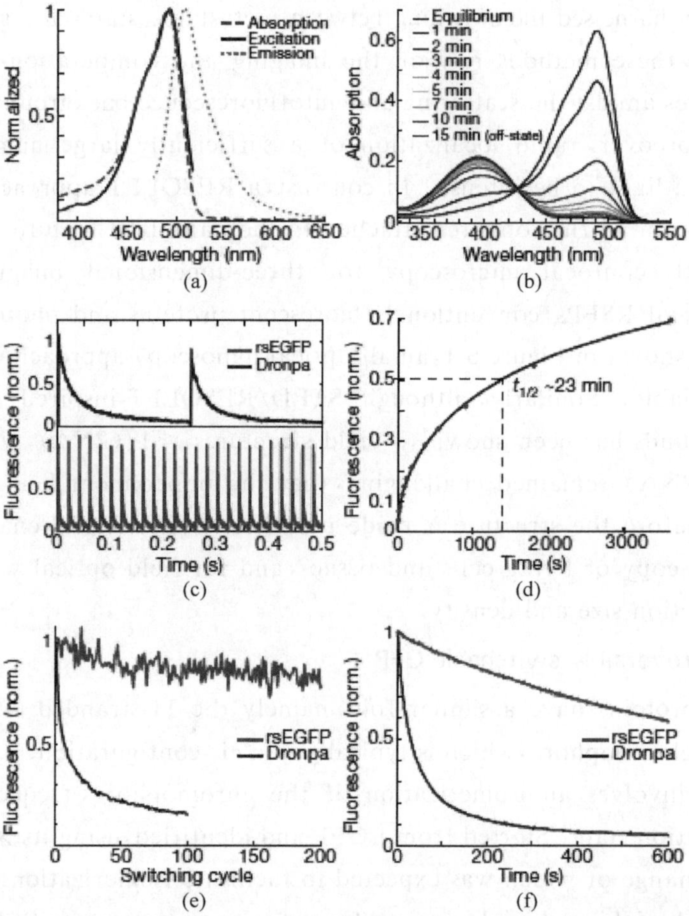

Fig. 5-1 Properties of rsEGFP. (a) Absorption (red dashed line), excitation (solid black line) and fluorescence (dotted green line) spectrum of rsEGFP in the fluorescent equilibrium state at pH7.5. (b) Absorption spectra obtained at different time points during irradiation with 488nm light. (c) Switching curves of dronpa (blue) and rsEGFP (red) immobilized in PAA using the same intensities. Switching was performed by alternating irradiation at 405nm(20mW · mm^{-2}) and at 491nm(60mW · mm^{-2}). The duration of off-switching at 491nm was chosen such that the fluorescence reached a minimum; irradiation with 405nm was chosen so that the proteins were fully switched. (d) Relaxation of rsEGFP embedded in PAA from the off-state into the fluorescent equilibrium state at 22℃. The black line is a stretched exponential fit with a stretching factor of ～0.6 accounting for inhomogeneous spectral broadening or the involvement of multiple dark states. (e) Fluorescence per switching cycle normalized to the initial fluorescence, with the same light intensities and switching durations. (f) Photobleaching: rsEGFP and dronpa embedded in the PAA layer were kept in their on-states by continuous irradiation at 405nm (100mW · mm^{-2}), while fluorescence was probed by irradiation at 491nm (30W · mm^{-2}).

491nm(6W · mm^{-2}) and 405nm(20W · mm^{-2}), a complete on-off cycle took 250ms for dronpa and 20ms for rsEGFP (see Figure 5-1(c)). Dronpa went through 10 cycles before its fluorescence was reduced to 50%, whereas rsEGFP went through 1,200 cycles under the same conditions (see Figure 5-1(e)). To compare bleaching, dronpa and rsEGFP were kept in the on-state by continuous irradiation at 405nm(10W · mm^{-2}) while fluorescence was generated by irradiation at 491nm (30W · mm^{-2}). Whereas dronpa fluorescence was reduced to 50% within $t_{1/2}$ = 30s, for rsEGFP measured $t_{1/2}$ = 800s (see Figure 5-1(f)). The rsEGFP chromophore maturated with a half-time of 3h at 37℃ (see Figure 5-4). The protein behaved as a monomer in vitro, could be fused to various proteins, including a-tubulin and histone H2B and was repeatedly switchable in living cells, tissues and far-field optical writing and reading of patterns of subdiffraction size and density.

5.1.4 Rewritable data storage

To analyse whether immobilized rsEGFP could be used for repeated short-term data storage and coated a microscope slide with $a<1\mu m$ thin layer of rsEGFP (~0.03mm) in PAA. Switching and reading by illumination at 405nm and 491nm in a scanning confocal set-up provided an on-off contrast of ~50 : 1. Wetranslated the text of 25 Grimm's fairy stories into 7-bit binary ASCII code ("0" : off; "1" : on) and wrote and read the ~270,000 letters into a $17\mu m \times 17\mu m$ region in 6,596 frames, each comprising 41 letters (287 bits) (see Figure 5-2). Individual bits were ~0.5μm in diameter with 1mm centre-to-centre spacing, corresponding to a DVD storage density. Discriminating "0" from "1" by a simple threshold entailed 7 bit errors within the entire data set. After ~6, 600read/write cycles in the same region, the average fluorescence of the "1" was reduced by 35%. Hence, the same rsEGFP layer can be used for about 15, 000read/ write processes.

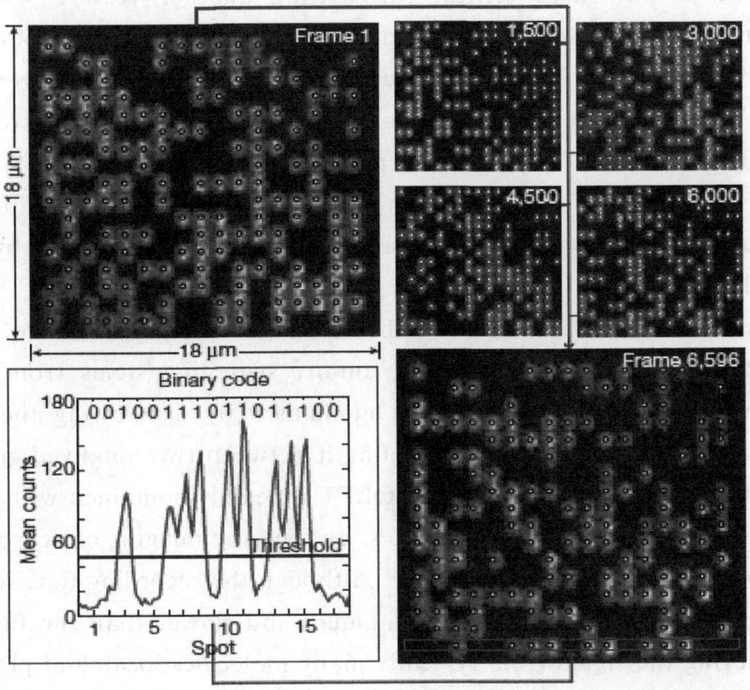

Fig. 5-2 Rewritable data storage. The text of 25 Grimm's fairy stories (ASCII code; 1.9 Mbits) consecutively written and read on a $17\mu m \times 17\mu m$ area of a PAA layer containing rsEGFP, with bits written as spots (representative frames shown). The white dots mark spots that were recognized as set bits.

A scanning confocal set-up with a 405nm (ultraviolet) beam for switching the rsEGFP on 491nm (blue) beam for eliciting fluorescence, and a doughnut-shaped 491nm beam for off-switching. That fused rsEGFP to the amino-terminus of the bacterial actin homologue MreB42 and expressed the fusion protein in E. coli bacteria. Living bacteria on agar-coated slides were recorded by first irradiating each pixel for 100ms with ultraviolet light ($10\text{W}\cdot\text{mm}^{-2}$), thus activating most of the rsEGFP in the focal volume. Then the doughnut-shaped blue beam ($I_m < 10\text{W}\cdot\text{mm}^{-2}$) was applied for 10~20ms to switch all the rsEGFP molecules off, except those located within $d/2$ distance from the doughnut centre. Lastly the rsEGFP fluorescence was read out for 1~2ms by the 491nm beam ($\sim 10\text{W}\cdot\text{mm}^{-2}$). The sequence was repeated for each sample pixel.

The double-helical ytoskeletal structure of rsEGFP-MreB is more clearly revealed by RESOLFT than by its confocal counterpart (see Figure 5-3(a)). The RESOLFT image of a typical filament showed a fullwidth half-maximum (FWHM) of ~70 nm. Because this value seemed to be determined by the thickness of the filament itself, a more accurate upper limit for the resolution d is obtained by imaging the finer keratin-19-rsEGFP intermediate filament network in living mammalian cells (see Figure 5-3(b), (c)). Line profiles from recorded data gave $d < 40\text{nm}$ corresponding to a 5~6 fold all-optical resolution improvement over confocal microscopy (see Figure 5-3(c)).

For investigating subdiffraction resolution writing, an rsEGFP layer was prepared as previously outlined. The writing entailed (1) an ultraviolet beam ($405\text{nm}, 1\text{kW/cm}^2$) applied for 100ms to switch rsEGFP on, (2) a 2ms break for equilibration, (3) a doughnut-shaped blue beam ($491\text{nm}, 0.5\text{kW/cm}^2$) lasting 20ms confining the on-state within $d/2$ around the doughnut centre, and (4) an, 2ms 532nm beam (900kW/cm^2) for transferring on-state rsEGFP to a permanent off (bleached) state (see Figure 5-4(a)). Lastly, the rsEGFP molecules located outside this region were switched back on, which is critical for writing another feature within subdiffraction proximity. To write nine patterns of 333bit fields in an rsEGFP layer, with 250nm centre-to-centre separation between individual bits (see Figure 5-4(b)), both in the conventional and in the RESOLFT mode. Whereas conventional writing and/or confocal reading blurred the data, the bits were fully discernible when both writing and reading were performed by RESOLFT. To write and read the data down to distances of 200nm between the individual bits (see Figure 5-4(b)). Hence this scheme allowed storing and reading out bits that 4 times more densely than by regular focusing. The structures could be read 5~10 times.

Multiphotoninduced optical damage can therefore be virtually excluded. The fundamental reduction in optical intensity required for the on-off switching stems from the fact that the fluorescence capability of the molecule is not modulated by disallowing the population of its nanosecond fluorescent state, but rather by toggling it between two longlived ground states, one in which the fluorophore remains dark when RESOLFT is readily combined with confocal imaging, which increases its use in scattering living samples. In fact, the imaging of neuronal spines in living organotypical brain slices testifies this potential. Although the recording time reported here is still of the order of most other super-resolution techniques and slower than the fastest optical STED recordings, by gathering the signal from typically many molecules located at predefined positions, RESOLFT has all the prerequisites for fast imaging. Scanning with arrays of doughnuts or zero-intensity lines (so-called structured illumination) and detection by a camera will reduce the number

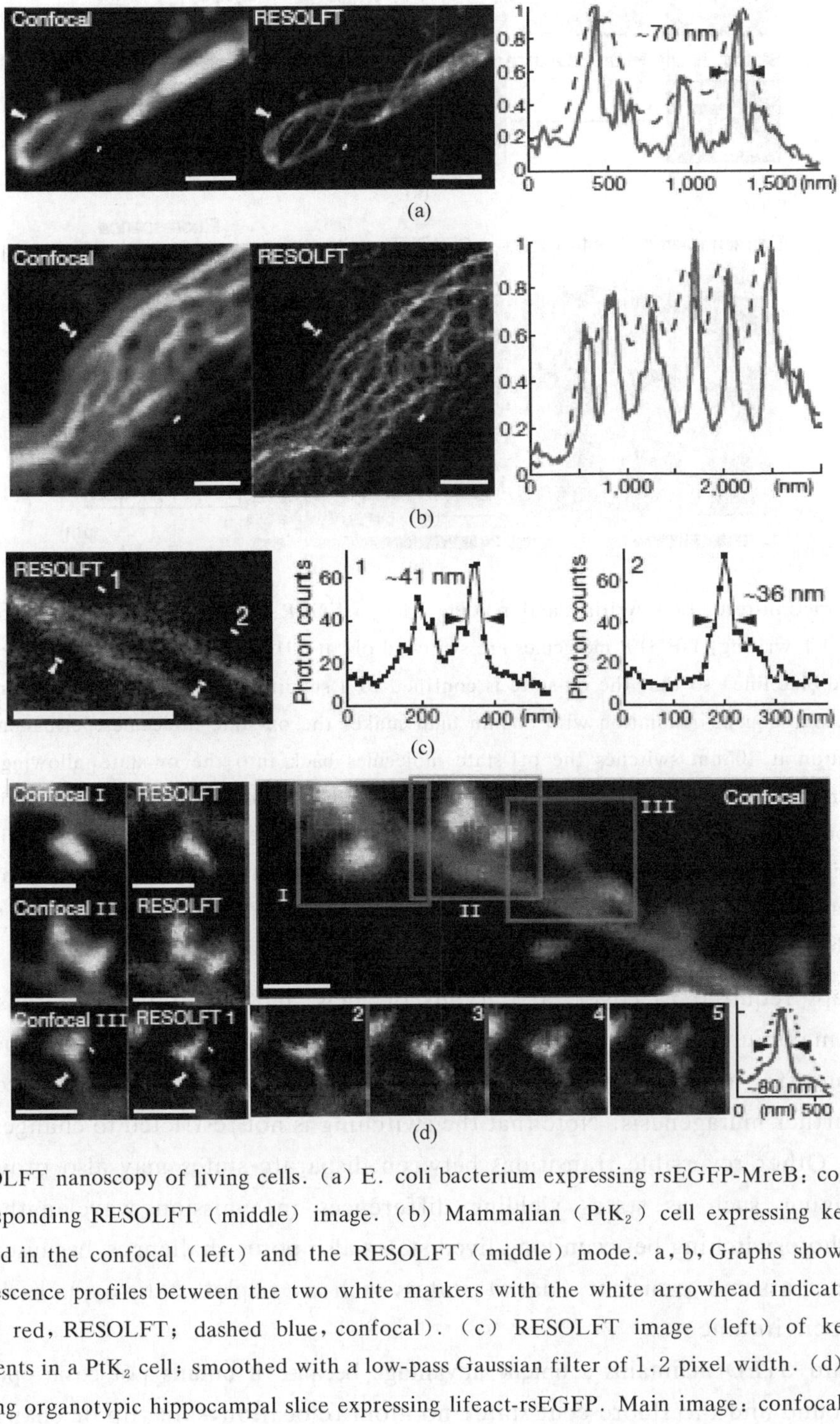

Fig. 5-3 RESOLFT nanoscopy of living cells. (a) E. coli bacterium expressing rsEGFP-MreB: confocal (left) and corresponding RESOLFT (middle) image. (b) Mammalian (PtK_2) cell expressing keratin-19-rsEGFP imaged in the confocal (left) and the RESOLFT (middle) mode. a, b, Graphs show the normalized fluorescence profiles between the two white markers with the white arrowhead indicating the direction (solid red, RESOLFT; dashed blue, confocal). (c) RESOLFT image (left) of keratin-19-rsEGFP filaments in a PtK_2 cell; smoothed with a low-pass Gaussian filter of 1.2 pixel width. (d) Dendrite within a living organotypic hippocampal slice expressing lifeact-rsEGFP. Main image: confocal overview. Ⅰ~Ⅲ: three spines, as indicated on the main image, each imaged in the confocal (left) and the RESOLFT mode (right). Spine Ⅲ was repeatedly imaged in the RESOLFT mode within 5min, demonstrating the changes over time. Graph: normalized profile across a spine neck as imaged in the RESOLFT (solid red) or the confocal mode (dashed blue) between the two white markers.

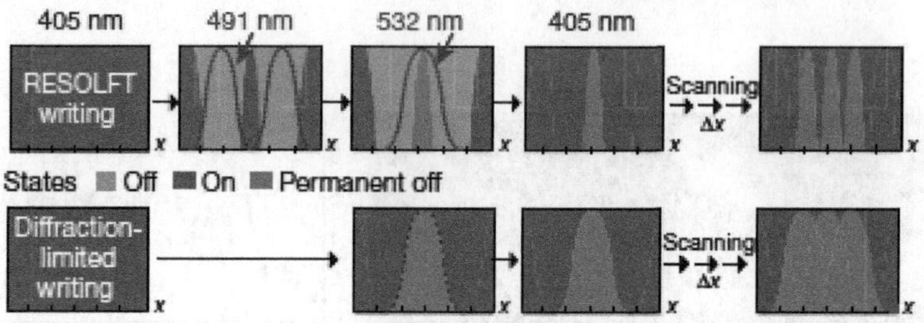

(a)

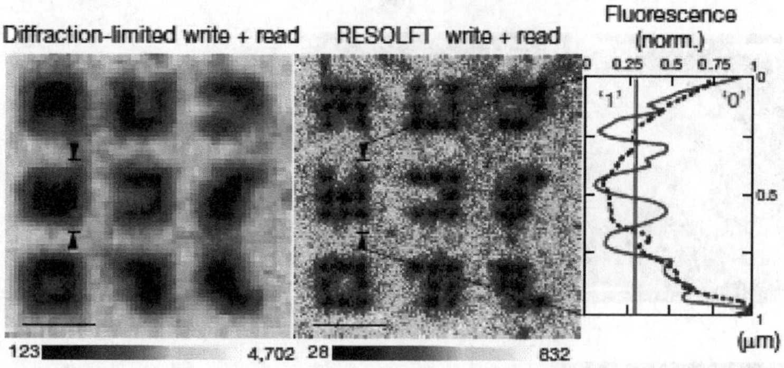

(b)

Fig. 5-4 Subdiffraction-resolution writing and reading using rsEGFP and visible light. (a) Top, schematic of RESOLFT writing: rsEGFP molecules are switched off at 491nm using a doughnut-shaped focal intensity (dashed blue line) so that the on-state is confined to a subdiffraction-sized region around the doughnut centre. Subsequent irradiation with 532nm light makes the on-state molecules permanent by bleaching. Irradiation at 405nm switches the off-state molecules back into the on-state, allowing the writing of another feature in subdiffraction proximity. Bottom, schematic of diffraction-limited writing. (b) Conventional (left) and subdiffraction RESOLFT (middle) joined writing and reading in a layer of immobilized rsEGFP. The outlines of the corresponding 333bit patterns were identical. The distance between two bleached spots was 250nm in each case. Right, normalized line profiles of the fluorescence signal between the two arrows (solid red, RESOLFT; dashed blue, confocal).

of scanning steps required to cover large fields of view and facilitate low-intensity video-rate imaging. The maximum recording speed is determined by the time it takes to establish the disparity of (on-off) states in space, that is, by the switching kinetics, which probably can be improved by further mutagenesis. Note that the switching is not restricted to changes in brightness (on-off) only. Other reversible transitions between disparate states may also prove suitable for RESOLFT imaging, such as states yielding differences in emission wavelengths, lifetime or polarization. Photoswitching between long-lived states also poses challenges, because in the process the molecule can assume transient (dark) states, such as triplet states, which depend on the molecular microenvironment.

In this regard, STED maintains a unique advantage because it entails just basic optical transitions between the ground and the fluorescent state; no atom relocation, spin flip or change in chemical bond is required to switch the fluorescence capability of the molecule just light. Therefore, switching fluorescence by STED is nearly universal and instantaneous.

The switching stamina of rsEGFP also enabled writing and reading of patterns of both subdiffraction size and spacing d, which has so far been difficult for direct far-field optical writing. In our study, the smallest obtainable structure size was co-determined by the fact that the 532nm light moderately bleached the state proteins too, thus reducing the writing contrast. However, this initial demonstration should spur on new advancements in this field, because current nanowriting efforts are dominated by concepts that resort to much shorter wavelengths of electromagnetic radiation at which focusing becomes exceedingly difficult. In fact, RESOLFT and related concepts are unique for creating materials that are nanostructured in three dimensions. To maximize the resolution along the optical axis (z), RESOLFT imaging and writing can also be combined with 4Pi microscopy, in which case three-dimensional resolution of, 10nm should become possible at ultralow light levels. The resolution demonstrated here is similar or even exceeds the resolution attained until now by STED in living cells. Although in both methods the resolution can be continually increased by increasing I_m/I_s, in STED microscopy this strategy will reach practical limits due to the intensities required. Using a threshold intensity I_s that is lower by many orders of magnitude, switching between long-lived states overcomes these limits and, as we have demonstrated here, offers a pathway to lens-based optical imaging and writing at molecular dimensions.

Site-directed mutagenesis was performed with the QuikChange Site Directed Mutagenesis Kit (Stratagene) or a multiplesite approach using several degenerative primers. The proteins were expressed from the high-copy expression vector PQE31 (Qiagen) and expressed in E. coli. A modified Semliki Forest Virus containing the pSCALifeactrsEGFP vector construct was injected into the slice cultures using a patch pipette. Imaging was performed within 16 ~ 48h after incubation. A layer containing immobilized rsEGFP was prepared by mixing 24.5ml purified proteins (0.09mM) with 17.5ml Tris-HCl pH 7.5, 30ml acrylamide (Rotiphorese Gel 30, Roth), 0.75ml 10% ammonium persulfate and 1ml 10% TEMED.

About 10ml of this solution was placed on a glass slide and a cover slip was pressed onto the sample to attain a thin layer. Custom MATLAB programs allowed automated generation of the voltages and signals for moving the sample and for generating the desired laser pulses. Images were also taken using the software Imspector. RESOLFT set-up that implemented a home-built confocal microscope with a normally focused beam for generating fluorescence plus a doughnut-shaped beam for switching rsEGFP off (both at 491nm wavelength). The beams were circularly polarized, superimposed in the focal plane and applied sequentially. The 405nm beam for switching rsEGFP on was also circularly polarized. The fluorescence emitted between 500 ~ 560nm was imaged on the opening of a multimode fibre and detected by a counting avalanche photodiode. The same set-up was used for writing, which was most specific at 532nm. The quantum memory, especially the photon-echo quantum memory in solid state system is a very interesting research area for information storage in the future. Many applications of quantum communication crucially depend on reversible transfer of quantum states between light and matter. Motivated by rapid recent developments in theory and experiment, review research related to quantum memory based on a photon-echo approach in solid state material with emphasis on use in a quantum repeater. After introducing quantum communication, the quantum repeater concept, and properties of a quantum memory required to be useful in a quantum repeater, the historical development from spin echoes, discovered in 1950, to photon-echo quantum memory. A simple theoretical description of the ideal protocol, and comment on the impact of a non-ideal realization on its quantum nature will be described in this chapter. Extensively discuss rare-earth-ion doped crystals and glasses as material candidates, elaborate on traditional photon-echo experiments as a test-bed for quantum state storage, and describe the current state-of-the-art of photon-echo quantum memory and take a brief outlook on current research in this chapter also.

Ideal quantum data storage must be capable of coherently storing multiple quantum states of light for on-demand recall with memory fidelity beyond the classical limit. Arriving at this goal is a challenge for experimentalists and extensive research efforts have been dedicated to the development of such a quantum memory for the last decade. The motivation for this activity is the promise of revolutionary quantum information technologies. Quantum key distribution (QKD) is already a proven technique for the secure distribution of cryptographic keys, but practical implementations are limited to distances on the order of 100km by the transmission losses in optical fibre or the atmosphere. Quantum computing based on optical processes has been shown to work in principles, but this technology is limited in scale by the probabilistic nature of the optical quantum gates. A practical optical quantum memory could overcome the current limits to QKD and optical quantum computing. Many protocols have been proposed to realise such a storage device, these include electromagnetically induced transparency (EIT), off-resonant Raman interactions, controlled reversible inhomogeneous broadening, atomic frequency combs (AFCs) and spin-polarization. Of these techniques, the most impressive efficiencies so far attained are 43% using EIT and 35% using AFC.

If quantum repeaters are to be built successfully, then a quantum memory that is able to store a qubit for a period sufficient to allow several rounds of communication between the nearby nodes (typically several milliseconds) is required. Furthermore, it should either be possible to perform a BSM between two stored qubits or to trigger the release of photons carrying the qubits with a jitter small enough to achieve this, and all of this at wavelengths and bandwidths compatible with existing fibre-optic networks. Today, the best quantum memory by far is a simple fibre loop (though it does not have all the specifications mentioned above). Storing qubits in some atoms, either in traps or in some solid-state devices, is a huge challenge. But the potential applications, both for fundamental experiments (for example, long-distance loophole-free Bell tests) and for a worldwide quantum web, motivates many physicists. Moreover, it is probable that the successful techniques will also find applications in other types of quantum-information processors. At present there is an increasing number of groups working towards quantum memories from a range of different perspectives. The different approaches have so far been motivated by the degree of freedom chosen to encode the quantum state. That have already seen some progress: continuous-variable systems in atomic vapour; atomic ensembles; polarization of atom-photon systems; others are using nitrogen-vacancy centres in diamonds as well as rare-earth ions in fibres and crystals. Indeed this last case is interesting, as most proposals have focused on storing a single mode, or single quantum state, whereas the rare-earth systems offer the possibility of storing several modes and many quantum states, which could have significant practical implications. These and many more approaches are now being actively pursued within national and international collaborative programmes around the world.

Quantum bits (or qubits) of information can be transmitted using photons and put to use in a number of applications, including cryptography. These schemes rely on the fact that photons can travel relatively long distances without interacting with their environment. This means that photon qubits are able, for example, to remain in entangled states with other qubits-something that is crucial for many quantum-information schemes. However, the quantum state of a photon will be gradually changed (or degraded) due to scattering as it travels hundreds of kilometres in a medium such as air or an optical fibre. As a result, researchers are keen on developing quantum repeaters, which take in the degraded signal, store it briefly, and then re-emit a fresh signal. By this way, can build up entanglement over much longer distances.

A quantum memory, which stores and re-emits photons, is the critical component of a quantum repeater. Those made so far in laboratories must be maintained at extremely cold temperatures or under vacuum conditions. They also only tend to work over very narrow wavelength ranges of light and store the qubit for very short periods of time. Walmsley and his colleagues argue that it isn't feasible to use such finicky systems in intercontinental quantum communication-these links will need to cross oceans and other remote areas, where it's difficult to send a repair person to fix a broken cryogenic or vacuum system. Moreover, they should also absorb a broad range of frequencies of light and store data for periods much longer than the length of a signal pulse that enabling step for building big networks. The broad range of frequencies means the memory can handle larger volumes of data, while a long storage time makes it easier to accumulate multiple photons with desired quantum states. A quantum memory for photons that works at room temperature has been created by physicists in the UK. The breakthrough could help researchers to develop a quantum repeater device that allow quantum information to be transmitted over long distances. Working towards this goal, made a cloud of caesium atoms into a quantum memory that operates at an easy-to-achieve temperature of about 62C. Unlike previous quantum memories, the photons stored and re-emitted do not have to be tuned to a frequency that caesium electrons would like to absorb. Instead, a pulse from an infrared control laser converts the photon into a "spin wave", encoding it in the spins of the caesium electrons and nuclei.

5.1.5 Paint it black

Walmsley compares the cloud of caesium atoms to a pane of glass-transparent, so it allows the light through. The first laser paints the glass black in a sense, allowing it to absorb all the light that reaches it. However, instead of becoming dissipating as heat and as it would in the darkened glass, the light that passed into the caesium cloud is stored in the spin wave. Up to 4μs later, a second laser pulse converts the spin wave back into a photon and makes the caesium transparent to light again. The researchers say that the caesium's 30% efficiency in absorbing and re-emitting photons could increase with more energetic pulses from the control laser, while the storage time could be improved with better shielding from stray magnetic fields, which disturb the spins in the caesium atoms. Even at 30% efficiency, Ben Buchler of the Australian National University in Canberra calls the device "a big deal" because it absorbs a wide band of photon frequencies. Due to Heisenberg's uncertainty principle, the ultra-short single-photon pulses from today's sources don't have well defined energies, so an immediately useful quantum memory must be able to absorb a wide range of frequencies-which Buchler says high-efficiency memories can't yet do.

- Noise problem

Background noise, or extra photons generated in the caesium clouds that are unrelated to the signal photons, was a major concern for room-temperature memories. "People thought that if you started using room-temperature gases in storage mode, you'd just have a lot of noise," says Walmsley. Temperatures near absolute zero suppress these extra photons other memories. But because the control and signal pulses in the Oxford team's set-up are far from caesium's favoured frequencies, the cloud was less susceptible to photon-producing excitations and the noise level remained small even at room temperature.

- Towards high-speed optical quantum memories

Entanglement is the fundamental characteristic of quantum physics. Large experimental efforts are devoted to harness entanglement between various physical systems. In particular, entanglement between light and material systems is interesting due to their prospective roles as stationary qubits in future quantum information technologies, such as quantum repeaters and quantum networks. The first demonstration of entanglement between a photon at telecommunication wavelength and a single collective atomic excitation can be stored in a crystal. One photon from an energytime entangled pair is mapped onto a crystal and then released into a well-dened spatial mode after a predetermined storage time. The other photon is at telecommunication wavelength and is sent directly through a 50m fiber link to an analyzer. Successful transfer of entanglement to the crystal and back is proven by a violation of the Clauser-Horne-Shimony-Holt (CHSH) inequality by almost three standard deviations ($S = 2.64 - 0.23$). These results represent an important step towards quantum communication technologies based on solid-state devices. In particular, our resources pave the way for building efficient multiplexed quantum repeaters for long-distance quantum networks.

Quantum information science incorporates quantum principles into information processing and communication. Amongst the most spectacular discoveries and conjectures, that quantum cryptography could enable information-theoretic secure communication through public channels, and quantum computing would efficiently solve certain computational problems that are believed to be intractable by conventional computing. Furthermore quantum dynamics becomes efficiently simulatable on a quantum computer. The prototypical model of quantum information processing represents information as strings of qubits, and processing is effected by unitary quantum gates. The qubit is a single-particle state in a two-dimensional Hilbert space. If the particle is a single photon, then the qubit can be encoded in several ways. For example, in polarization encoding the logical zero state can correspond to a single photon being left-circularly polarized and | to right-circularly polarized. Other examples include path, photon-number, and time-bin encodings. A general qubit state can be expressed as a superposition and general states of quantum information are superpositions of strings of qubits. Quantum memory needs to store qubit strings or parts thereof faithfully and to release them on demand. Storage of a quantum state need not be perfect. Faulttolerant quantum error correction can be employed to make an imperfect memory sufficient as long as the fidelity of the memory "gate" exceeds a particular performance threshold. Next study the specific requirements for optical quantum memory to be effective for quantum information tasks.

The reversible transfer of quantum states of light in and out of matter constitutes an important building block for future applications of quantum communication: it allows synchronizing quantum information1, and enables one to build quantum repeaters and quantum networks. Much effort has been devoted worldwide over the past years to develop memories suitable for the storage of quantum states. Of central importance to this task is the preservation of entanglement, a quantum mechanical phenomenon whose counter-intuitive properties have occupied philosophers, physicists and computer scientists since the early days of quantum physics. The reversible transfer of photonphoton entanglement into entanglement between a photon and collective atomic excitation in a solid-state device can be used for information storage. Employ a thulium-doped lithium niobate waveguide in conjunction with a photon-echo quantum memory protocol, and increase the spectral acceptance from the current maximum of 100MHz to 5GHz. The entanglement-preserving nature of our storage device is assessed by comparing the amount of entanglement contained in the detected photon pairs before and after the reversible transfer, showing, within statistical error, a perfect mapping process. Integrated, broadband quantum memory complements the family of robust, integrated lithium niobate devices. It renders frequency matching of light with matter interfaces in advanced applications of quantum communication trivial and institutes several key properties in the quest to unleash the full potential of quantum communication.

• Performance criteria

In general quantum memory stores a pure or mixed state represented by density matrix ρ and outputs a state ρ' which should be close to ρ. The ultimate performance criterion for quantum memory is the worst-case fidelity with respect to the set of input states where the fidelity for a specific state is given by $F(\rho) = \mathrm{Tr}\sqrt{\sqrt{\rho'}\,\rho\,\sqrt{\rho'}}$. There exists a threshold worst-case fidelity beyond which fault-tolerant quantum error correction methods can overcome memory imperfection. The fidelity of quantum-optical memory for an arbitrary set of input states can be determined by subjecting it to complete quantum process tomography.

However, this procedure is relatively bulky, so in practical experimental implementations, other performance criteria are used. For example the term fidelity sometimes refers to state overlap (possibly after post-selection), which is the square of $F(\rho)$ for the case that ρ is a pure state. Average fidelity is often used, where the average is taken over all input states with respect to an assumed prior distribution. Another popular criterion is efficiency η, which is the ratio between the energies of the stored and retrieved pulses. Efficiency, while easy to determine experimentally, does not account for possible detrimental effects such as contamination of the retrieved state by the excess noise from the storage medium. In continuous-variable implementations, memory can be characterized by the transfer coefficient and conditional variance. These quantities are convenient for characterizing quantum memory for single-mode fields provided that input and retrieved states are Gaussian.

The multimode capacity of quantum memory determines the number of optical modes that can be stored in the memory cell with the requisite performance threshold or better. The multimode capacity strongly depends on the memory mechanism. Quantum memory needs to be able to store the state long enough to perform the task at hand so storage time is another essential memory performance criterion. For many applications, an appropriate figure of merit would be the delay-bandwidth product, i.e. the ratio between storage time and duration of the stored pulse.

• Applications

In optical quantum computation the role of quantum memory is to store quantum bits so that operations can be timed appropriately. Many qubits are being processed in parallel with each other at each step in time, and these processing steps must be synchronized. Quantum communication suffers from imperfect transmission channels, resulting, for example, in quantum key distribution being possible only over finite distances. The quantum repeater solves this problem and permits quantum communication over arbitrary distances with a polynomial cost function. A necessary component of the quantum repeater is quantum memory, which, similar to quantum computation, allows synchronization between entangled resources distributed over adjacent sections of the transmission link.

In addition, quantum memory for light finds applications in precision measurements based on quantum interference of atomic ensembles. By transferring quantum properties of an optical state to the atoms one can reduce the quantum noise level of the observable measured, thereby improving the precision of magnetometry, clocks, and spectroscopy. Finally, quantum optical memory can be used as a component of single-photon sources. If a single-photon detector is placed in one of the emission channels of nondegenerate spontaneous parametric down-conversion, a detection event indicates emission of a photon pair, and thus the presence of a single photon in the other channel. Such a heralded photon is emitted at an arbitrary time, In the following, review recent theoretical and experimental work related to different approaches to quantum memory.

• Realizations

(1) Optical delay lines and cavities

The simplest approach to storage of light is an optical delay line, e.g. an optical fibre. This approach has been used to synchronize photons with the occurrence of certain events. However, the storage half-time, i.e. the time after which half the photons are lost, is, in the case of 1.5μm wavelength and telecommunication fibres, limited to around 70μs, corresponding to a fibre length of ~15km. The half-time decreases at other wavelengths due to increased loss. Furthermore, the storage time in an optical delay is fixed once a delay length is chosen, contrary to the requirement of variable, on-demand out-put as desired in most applications.

Alternatively, light can be stored in a high-Q cavity. The light effectively cycles back and forth between the reflecting boundaries, and can be injected into and retrieved from the cavity using electro-optical or non-linear optical means, or by quantum state transfer with passing atoms. For example, light has been stored for over a nanosecond in wavelength-scale photonic crystal cavities with a tunable Q factor which allows control over the storage time. Dynamic control of Q can achieved by adiabatically tuning the frequency of the stored light, which has yielded output pulses as short as 0.06ns, much shorter than the storage time. Unfortunately, storage of light in cavities suffers from the trade-off between a short cycle time and a long storage time, limiting the efficiency or the delay-bandwidth product. Therefore, where as optical delay lines and nanocavities could be appropriate for obtaining on demand single photons from heralded sources that may not be suitable for quantum memory or quantum repeaters.

(2) Electromagnetically-induced transparency

Electromagnetically-induced transparency (EIT) is a nonlinear optical phenomenon observed in atoms with an energy-level structure. Two optical fields couple the excited level to their respective ground levels: the weak signal field which may carry the quantum information load and the strong control field which is used to steer the atomic system. If the control field is absent, the signal field, which interacts with the resonant two-level system, undergoes partial or complete absorption. In the presence of the control field, the absorption of the signal is greatly reduced whenever the frequency difference of the two optical fields is close to the frequency of the Raman transition between the ground states of the Λ system (the condition known as the two photon resonance) (see Figure 5-5(b)). Transparency is observed when the fields are detuned from the two-photon resonance by no more than

$$W_{\mathrm{EIT}} \sim \frac{4}{\sqrt{\alpha L}} \frac{\Omega^2}{W_{\mathrm{line}}} \tag{5-1}$$

for α the absorption coefficient of the medium in the absence of EIT, L its length, Ω the Rabi frequency of the control field and $\hbar W_{\mathrm{line}}$ the uncertainty of the excited level energy associated with homogeneous or inhomogeneous broadening. The width of the EIT window is thus proportional to the intensity of the control field and can be much narrower than the W_{line}. EIT is largely insensitive to the detuning Δ of the optical fields from their respective individual transitions. It can thus exist in the presence of inhomogeneous broadening as long as both optical transitions experience the same frequency shift. A good example is provided by atoms in a warm gas whose transitions are broadened by the Doppler-effect associated with atomic motion. If the ground levels are of similar energy and the two optical fields are co-propagating, the Doppler shifts cancel and EIT is observed. Counterintuitively, according to Eq. (5-1), EIT window in the presence of inhomogeneous broadening is much narrower than in the case of pure homogeneous broadening.

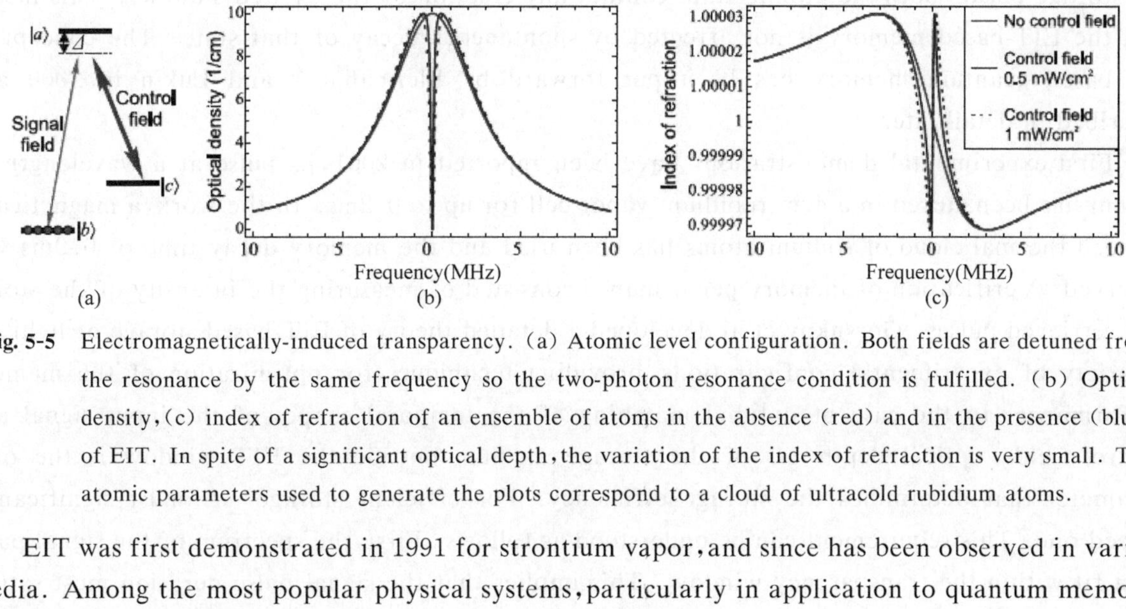

Fig. 5-5 Electromagnetically-induced transparency. (a) Atomic level configuration. Both fields are detuned from the resonance by the same frequency so the two-photon resonance condition is fulfilled. (b) Optical density, (c) index of refraction of an ensemble of atoms in the absence (red) and in the presence (blue) of EIT. In spite of a significant optical depth, the variation of the index of refraction is very small. The atomic parameters used to generate the plots correspond to a cloud of ultracold rubidium atoms.

EIT was first demonstrated in 1991 for strontium vapor, and since has been observed in various media. Among the most popular physical systems, particularly in application to quantum memory, are ensembles of alkali atoms and rare-earth doped solids.

5.1.6 Slow light and memory

According to the Kramers-Kronig relations, an anomaly in the absorption spectrum always comes together with an anomaly in dispersion. The group velocity can in theory be arbitrarily reduced by lowering the intensity of the control field. Experimentally, slowdown by up to seven orders of magnitude have been observed.

Slow light is a "trademark" property of EIT and has a variety of applications, for example buffering of optical communication traffic. It is also the basis for the quantum memory application, which functions as follows(see Figure 5-6(a)). A light pulse, resonant with the EIT window, enters the EIT medium and slows down. The slowdown entails spatial compression, so the pulse, whose initial spatial extent by far exceeds L, will fit inside the medium. Once it is inside, we adiabatically reduce the control field intensity and bring the group velocity down to zero, thereby "collapsing" the EIT window and storing the pulse in the medium. When the pulse needs to be retrieved, the control field is turned back on. The pulse then resumes its propagation and leaves the EIT medium. Before the signal has entered the EIT medium, all atoms are optically pumped by the control field into the ground level $|b\rangle$ so the initial atomic state is

$$|\psi_0\rangle = |b_1 \cdots b_N\rangle \tag{5-2}$$

After the signal pulse has entered the medium and been stored, its quantum state is transferred into a collectiveexcitation of the atoms in the EIT medium. For example, if the signal state is a single photon, the atomic state becomes (neglecting normalization)

$$|\psi_1\rangle_A = \sum_j \psi_j e^{i\Delta k z_j} |b_1 \cdots c_j \cdots b_N\rangle \tag{5-3}$$

for N, the number of atoms in the ensemble, z_j the position of atom j along field propagation and Δk the difference in the wavevectors of the control and signal fields. In other words, one of the atoms is transferred into the other ground state, c_i, but it is not known which atom it is.

In the ideal case of absent ground state decoherence, neither during the transfer, nor during the storage period does the atomic state contain any fraction of the excited state a_i. This means that the EIT-based memory is not affected by spontaneous decay of that state. The concept of EIT-based quantum memory has been put forward by Fleischhauer and Lukin in 2000 and described in detail later.

First experimental demonstrations have been reported in 2001, μs pulse at a wavelength of 795nm has been stored in a 4cm rubidium vapor cell for up to 0.2ms. In the work, a magnetically trapped thermal cloud of sodium atoms has been used and the memory decay time of 0.9ms was observed. Verification of memory performance consisted of measuring the intensity of the stored and retrieved pulses. Gorshkov et al developed a detailed theory of EIT-based storage of light for a variety of experimental configurations, providing techniques for optimization of the memory performance. In the case of optimal matching of the temporal shapes of the input signal and control fields, optical depth αL of the storage medium outside the EIT window is the only parameter that determines the storage efficiency. For efficient storage, αL must significantly exceed one. This requirement can be understood as follows. First, the spectrum of the signal pulse must fit within the transparency window. This implies that the signal pulse duration must satisfy $\tau \gg 1/W_{\text{EIT}}$, where W_{EIT} is given by Eq. (5-1). Second, the pulse must fit geometrically within the EIT medium: the spatial extent of the signal pulse, compressed due to the slow light effect and given by $c\tau/ng \gg 1$, must not exceed L. These bounds can be satisfied at the same time if the slowdown is sufficient, which translates into a demand for a high contrast of the EIT window, i.e. high αL. Achieving high optical density is a challenge in many optical arrangements including magneto-optical traps and solid state systems. In vapor cells, higher atomic density will increase αL, but also will degrade the EIT due to increased ground state decoherence and competing processes such as four-wave mixing and stimulated Raman scattering.

The findings of Gorshkov et al. were verified in an experiment in warm rubidium vapor. For moderate optical densities ($\alpha L \leqslant 25$), the experimental results showed excellent agreement with a three-level theoretical model without any free parameters (see Figure 5-6(b)) For higher optical densities, four-wave mixing effects come into play, leading to an additional idler optical mode being generated. Although the associated parametric gain may lead to better compression of the pulse, it also brings about additional quantum noise that degrades the storage fidelity. EIT-based memory can be implemented in solid media, with the advantage of significantly longer storage times. Following an initial observations of EIT as well as ultraslow and stored light in praseodymium doped Y_2SiO_5 crystal, Longdell et al. stored light in a similar crystal with a decay time of 2.3 seconds. A disadvantage of this crystal in application to light storage is a relatively low optical density, which is due to the inhomogeneous broadening associated with the difference in ionic radii of Y^{3+} and Pr^{3+}. Attempts to increase the dopant concentration only result in a broader line without increasing the optical density. Recently, EIT has been observed in another praseodymium doped crystal, $La_2(WO_4)_3$, which exhibits an inhomogeneous broadening that is 15 times smaller than $Pr^{3+}:Y_2SiO_5$ but at the cost of significantly increased homogeneous decay. Now review a few experiments in which quantum states of light have been stored, and the retrieved pulses were demonstrated to retain some of the nonclassical properties. In 2005, single photons, generated using the DLCZ method (see Figure 5-6), have been stored a cold atom cloud and in a vapor cell and retrieved.

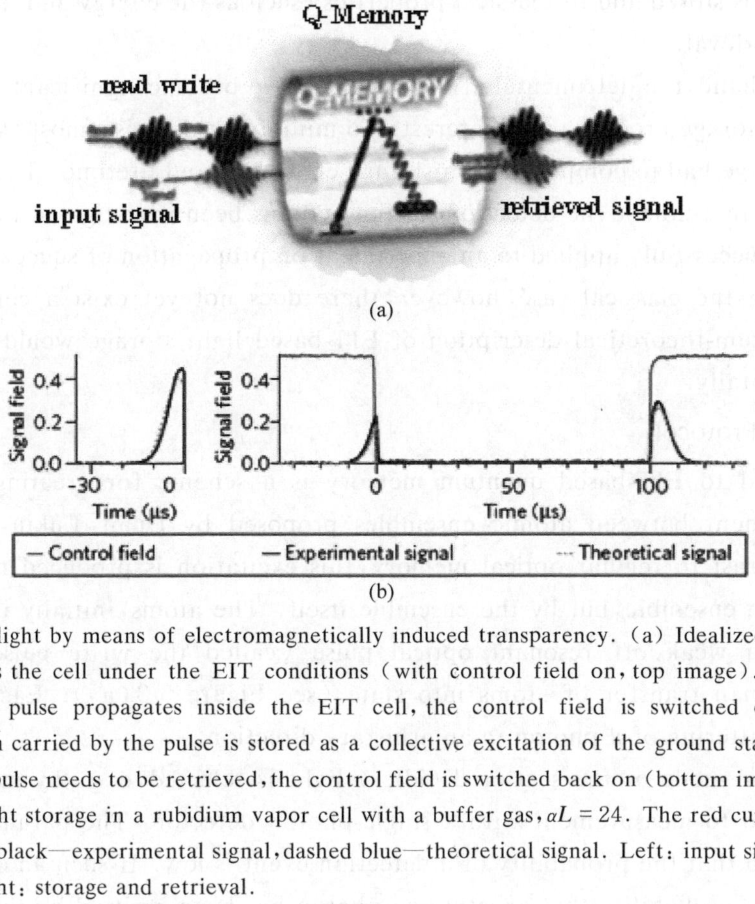

Fig. 5-6 Storage of light by means of electromagnetically induced transparency. (a) Idealized picture. The signal pulse enters the cell under the EIT conditions (with control field on, top image). While the spatially compressed pulse propagates inside the EIT cell, the control field is switched off, so the quantum information carried by the pulse is stored as a collective excitation of the ground states (middle image). When the pulse needs to be retrieved, the control field is switched back on (bottom image). (b) Optimized classical light storage in a rubidium vapor cell with a buffer gas, $\alpha L = 24$. The red curve shows the control field, solid black—experimental signal, dashed blue—theoretical signal. Left: input signal pulse of optimal shape. Right: storage and retrieval.

Because of the finite optical depth, the signal pulse does not entirely fit into the cell, resulting in a fraction of the pulse leaking through the cell before the control field is turned of 0.5μs later. Sub-Poissonian statistics of the retrieved light have been verified using a Hanbury Brown & Twiss detection scheme. In an important step towards applications, a dual-rail single-photon qubit has been stored in 2008. A single photon has been split into two spatially separate optical modes, each of which has been stored in a cloud of ultracold optical atoms. Upon retrieval, the modes were recombined and subjected to an interference measurement, which demonstrated that not only nonclassical photon statistics, but also the phase relation between the two stored modes has been preserved. This experiment constitutes the first mapping of an optical entangled entity in and out of quantum memory. In 2004, propagation of another quantum information primitive, squeezed vacuum, has been observed under EIT conditions, which followed by experimental demonstrations of storage of squeezed vacuum in 2008 Quadrature noise of the retrieved pulses was measured by means of homodyne detection and observed to be below the shot noise level, demonstrating that some of the initial squeezing has been preserved through the storage procedure.

EIT-based storage of quantum light suffers from background noise in the retrieved signal. This noise is likely to originate from the repopulation of the state associated, for example, with the atomic drift into and out of the interaction area. In the presence of the control field, the atoms are pumped from c_i into the excited state a_i, and then spontaneously decay back into the ground state emitting thermal photons that contaminate the signal mode. This effect is negligible when a macroscopic pulse is stored and its classical properties (such as the energy and the pulse shape) are measured upon retrieval.

On the other hand, the detrimental effects of the noise become significant when the quantum properties of the storage process are of interest. To minimize this noise, most experiments on non-classical light storage had to compromise on storage efficiency and lifetime. The background noise has been observed in a homodyne detection setting and has been investigated theoretically. These predictions were successfully applied to an experiment on propagation of squeezed light under EIT conditions. Unlike the classical case, however, there does not yet exist a comprehensive study where a full quantum-theoretical description of EIT-based light storage would be developed and verified experimentally.

- The DLCZ Protocol

Closely related to EIT-based quantum memory is a scheme for creating long-lived, long-distance entanglement between atomic ensembles proposed by Duan, Lukin, Cirac and Zoller (DLCZ). In contrast to regular optical memory, this excitation is produced not by an external photon entering an ensemble, but by the ensemble itself. The atoms, initially in the state b_i, are illuminated with a weak off resonant optical pulse (called the write pulse), resulting in a probability of Raman transfer of atoms into state (see Figure 5-3(a)). Each such transfer is associated with scattering of a photon in an arbitrary direction.

A single spatial mode of the scattered (idler) light is selected, e. g. by means of an optical fiber, and subjected to measurement with a single-photon detector. The parameters of the write pulse are chosen so that the probability of a detection event is low. If such an event does occur, it indicates with high probability that exactly one photon has been emitted by the atomic ensemble into the detection mode.

Spatial filtering erases the information about the location of the atom that has emitted the photon. As a result, the detection event projects the atoms onto a collective superposition of the type. The state of the atomic ensemble becomes equivalent to that as if a single photon has been stored therein using the EIT technique. Therefore one can apply a classical (read) field on the |ci-|ai transition, which will play the role of the EIT control field, leading to retrieval of a signal photon from the ensemble and transfer of the atoms back into the state (Eq. (5-2)) (see Figure 5-7(b)).

In its many aspects, the scheme resembles heralded preparation of a single photon from a biphoton generated via parametric down-conversion. A fundamental difference is that the heralded atomic excitation is longlived and can be retrieved at an arbitrary time, which enables application in a quantum repeater. The protocol can also be viewed as a deterministic single-photon "pistol": once it is "loaded" with an idler detection event, it can "shoot" the signal photon on demand. Figure 5-7(c) illustrates preparation of a single link of longdistance entanglement between two remote atomic ensembles. The ensembles are simultaneously illuminated with write pulses and the spatial modes in which the idler photon is to be detected are mixed on a beam splitter.

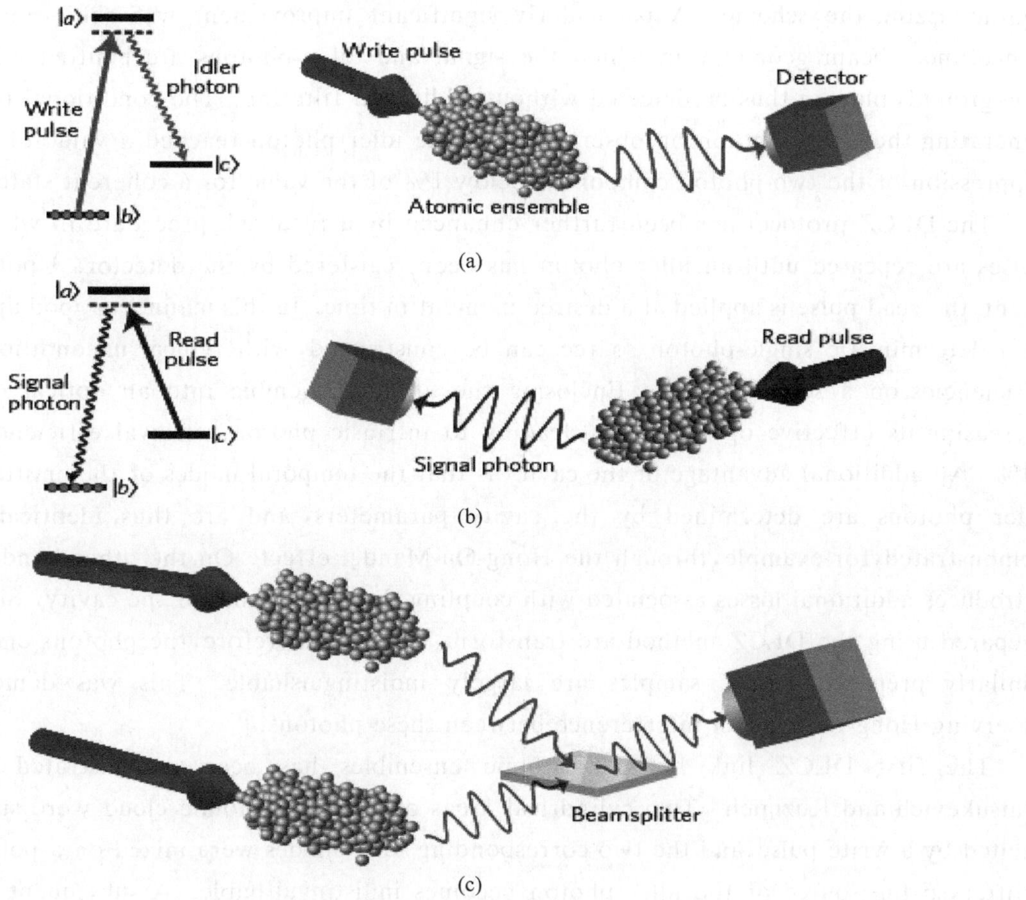

Fig. 5-7 The Duan-Lukin-Cirac-Zoller protocol.

Now, if a photon has been detected in one of the beam splitter outputs, it is impossible to tell which of the two ensembles has emitted the photon. As a result, the state of the two ensembles becomes an entangled superposition:

$$\Psi = \frac{1}{\sqrt{2}} \mid \psi_0 \rangle \mid \psi_1 \rangle + e^{i\phi} \mid \psi_1 \rangle \mid \psi_0 \rangle) \tag{5-4}$$

where phase ψ depends on the lengths of the optical links between the ensembles and the detection apparatus. The DLCZ protocol does not constitute quantum memory for light in the strict sense: being stored is not an external,arbitrary state of light but a heralded excitation. Nevertheless,the scheme fully replaces the "orthodox" memory as far as the quantum repeater application is concerned. Furthermore,it is more convenient in that it requires no additional nonclassical light sources: the role of these is played directly by the memory cells.

There exists a vast body of experimental work on the DLCZ protocol. The first implementations were reported in 2003 by two groups. Kuzmich et al. worked with a cesium MOT and observed nonclassically correlated idler and signal photons for storage times of about 400ns. Van der Wal et al. observed nonclassical correlations between the idler and signal light intensities for a read and write pulse separated by a few hundred nanoseconds.

Regular photodiodes were used rather than photon counters,and both generated pulses were acroscopic. These initial observations were followed by extensive research aimed at refining and characterizing the scheme. A particularly significant improvement was achieved thanks to a noncollinear beam geometry in which the signal and idler photons are emitted against a dark background,and can thus be detected without additional filtering. The conditional probability of generating the signal photon on observation of the idler photon reached a value of 50%,with a suppression of the two-photon component below 1% of the value for a coherent state.

The DLCZ protocol has been further enhanced by a feedback procedure in which the write pulses are repeated until an idler photon has been registered by the detector. Upon a detection event,the read pulse is applied at a desired moment in time. In this manner,a good approximation of a deterministic single-photon source can be constructed, with typical unconditional quantum efficiencies on a scale of 10%. Enclosing the atomic ensemble into an optical cavity allows increasing its effective optical depth, leading to intrinsic photon retrieval efficiencies of up to 84%. An additional advantage of the cavity is that the temporal modes of the emitted signal and idler photons are determined by the cavity parameters, and are thus identical, as can be demonstrated,for example,through the Hong-Ou-Mandel effect. On the other hand,this scheme introduces additional losses associated with coupling the photons out of the cavity. Signal photons prepared using the DLCZ method are transform limited. Therefore,the photons emitted by two similarly prepared DLCZ samples are largely indistinguishable. This was demonstrated by observing Hong-Ou-Mandel interference between these photons.

The first DLCZ link between atomic ensembles has been demonstrated in 2004 by Matsukevich and Kuzmich. Two cylindrical areas of the same atomic cloud were simultaneously excited by a write pulse,and the two corresponding idler modes were mixed on a polarizing beam splitter,so the source of the idler photon becomes indistinguishable. A subsequent pair of read pulses was followed by a polarization measurement of the signal photon,which exhibited Bell-type correlations with the idler.

This work was criticized by van Enk and Kimbl for the post selected character of the measurement. An experiment in a similar setting proving the presence of entanglement without resorting to postselection was later reported by the same group.

After propagating through the beam splitter (see Figure 5-7(c)), the idler photon is in an entangled state with the collective excitations in both atomic samples. This entanglement was used in 2008 by Chen et al. to implement quantum teleportation. A Bell-state measurement was performed on the idler photon and a coherent state of arbitrary polarization, teleporting the polarization state of the coherent state onto the atomic excitations. The excitations were then converted into optical form in order to measure the teleportation fidelity.

On application of the write pulse, the idler photon is emitted by a single DLCZ sample in an arbitrary direction and, generally, with an arbitrary polarization. This feature was utilized by a number of groups to demonstrate entanglement between various degrees of freedom of the optical and atomic excitations: polarization, angular momentum, and direction. By the same principle, entanglement of a frequency-encoded optical qubit with a cold mixture of ^{85}Rb and ^{87}Rb isotopes has been demonstrated. A photon entangled with the atomic ensemble can then be stored in an EIT-based memory cell, leading to entanglement of two remote atomic qubits. Such entanglement can also be produced by a Bell measurement on a pair of polarization encoded optical qubits obtained from two remote samples. The advantage of this approach in comparison with the classic DLCZ protocol is that the polarization encoded qubits are much less sensitive to fluctuations of optical phases in communication links.

Entanglement swapping between two DLCZ links has been demonstrated by Chou et al. in 2007. After preparing entanglement in two pairs of nodes, simultaneous read pulses were applied to two neighboring nodes in different pairs. The generated signal modes were mixed on a beam splitter and subjected to a photon number measurements. Detection of a single signal photon projects the two remaining nodes onto an entangled state, which has been verified by reading out the signal photon from these nodes. The DLCZ scheme is well suited for quantum repeater applications, but does not directly enable storage of arbitrary quantum information from outside the system. However, the scheme could be used as quantum memory for arbitrary qubits by means of quantum teleportation. As the entanglement is generated by spontaneous emission, such an optical quantum memory is only useful in a post-selected way.

5.1.7 Photon-echo quantum memory

- Principles

Similar to EIT-based storage, photon-echo quantum memory relies on the transfer of the quantum state carried by an optical pulse into collective atomic excitation. However, in contrast to EIT, it takes advantage of the inhomogeneous broadening. After absorption of a signal photon at $t=0$, the state is given (in un-normalized form) by

$$|\psi_1\rangle = \sum_j \psi_j e^{-i\delta_j t} e^{ikz_j} \,|b_1 \cdots a_j \cdots b_N\rangle \tag{5-5}$$

where k denotes the wave vector of the signal field, δ_j is the detuning of the transition of atom j with respect to the light carrier frequency, and all other variables are as in Eq. (5-3). Although the atomic dipoles are initially phase aligned with k, this alignment rapidly decays because δ_j is different for each atom.

All photon-echo quantum memory protocols employ a procedure that rephases the atomic dipoles some time later, thereby recreating collective atomic coherence. In other words, the phases of all atoms become equal at some moment t_e. This triggers re-emission of the absorbed signal. The initial distribution of spectral detunings δ_j allows for a classification of photon-echo quantum memory into two categories, which describe now.

- Controlled reversible inhomogeneous broadening (CRIB)

If the spectral distribution is continuous, the requirement of atom-independent phase evolution $\int_0^{t_e} \delta_j(t)\mathrm{d}t$ can only be achieved if, some time t' after absorption of the light, the resonance frequency of all atoms is actively changed from δ_{j1} to δ_{j2} so that:

$$\delta_{j1} t' + \delta_{j2}(t_e - t') = \text{const} \tag{5-6}$$

for all j. This approach to storage can be traced back to 1964, when the well-known spin echo was extended to the optical domain. The two-pulse photon echo was then developed into timeariable storage of data pulses using three-pulse photon echoes. However, conventional photonecho does not allow efficient storage and recall of data encoded into few-photon pulses of light with high fidelity, due to an inherent amplification process. Yet, it recently inspired a quantum memory protocol that is now generally referred to as controlled reversible inhomogeneous broadening (CRIB) Alternatively, the term Gradient echo memory (GEM) is used. First proposed in 2001 for storage in atomic vapor, CRIB has meanwhile been adapted for solid-state storage of microwave photons and optical photons.

The original proposal for CRIB is based on a hidden time-reversal symmetry in the Maxwell-Bloch equations that describe the evolution of the atom-light system during absorption and re-emission. Reversing the evolution of the atom-light system requires that the detuning of all atoms is inverted, i.e. $\delta_{j2} = -\delta_{j1}$ (see Figure 5-8). Additionally, a mode-matching (or phase-matching) operation has to be performed, which consists of applying a phase shift e^{-2ikz_j} to all atoms. This results in mapping the forward-traveling collective atomic coherence created during absorption of the forward-traveling light onto a backward-propagating coherence, which can lead to light emission in the backward direction. Another condition for perfect time reversal is that the optical depth of the medium must be sufficiently large to guarantee absorption of the incoming light. If the conditions of atomic inversion, mode matching and large optical depth are not satisfied, symmetry arguments do not suffice to predict the memory performance.

Of particular interest are the cases where the light is not completely absorbed, due to limited optical depth, and where the mode-matching operation is not implemented, resulting in the light being re-emitted in forward direction. To find the performance, distinguish between different types of inhomogeneous broadening. Two types have been analyzed. In transverse broadening, the atomic absorption line is equally broadened for each position z. Longitudinal broadening refers to the case where the atomic absorption line for each position z is narrow, and the resonance frequency varies monotonically throughout the medium:

$$\delta_j = \chi z_j \tag{5-7}$$

Assuming transverse broadening and limited optical depth αL, the efficiency for recall in the backward direction is given by

$$\varepsilon_b^{(t)} = (1 - \exp\{-\alpha L\})^2 \tag{5-8}$$

For re-emission in the forward direction, is

$$\varepsilon_f^{(t)} = (\alpha L)^2 \exp\{-\alpha L\} \tag{5-9}$$

In this case the maximum efficiency of:

$$\varepsilon_f^{(t)} = 54\% \tag{5-10}$$

Which is obtained for $\alpha L = 2$.

Note that the recalled pulse is time-reversed, resulting in an exchange of the leading and trailing end, regardless of the direction of recall. Longitudinal broadening yields

$$\varepsilon_b^{(l)} = \varepsilon_f^{(t)} = (1 - \exp\{-(\alpha L)_{\text{eff}}\})^2 \text{ where } (\alpha L)_{\text{eff}} \propto \chi^{-1} \tag{5-11}$$

- Characterizes the effective optical depth of the medium

It is interesting that the efficiency for forward recall can reach unity, despite the violation of time reversal: while the output pulse is a timenverted image of the input signal, it is re-emitted in forward direction. The recalled pulse features a frequency chirp, i. e. the fidelity of the retrieved mode deviates from unity.

Depending on the storage medium, the required change of detunings δ_j can be achieved in different ways. For atomic vapor where the inhomogeneous broadening is due to atomic motion, the atoms have to be forced to emit light in the backward direction using the above mentioned mode-matching operation, thereby inverting the Doppler shifts. In solids, the detunings of individual atoms depend on crystal defects and strain. Control over the detunings can be achieved by "tailoring" a narrow absorption line in the naturally broadened transition by optical pumping, followed by controlled and reversible broadening through position-dependent Stark or Zeeman shifts, as shown in STEP 1 in Figure 5-8.

In order to achieve a large delay-bandwidth product when working with a two-level system, we need, on the one hand, a narrow initial line, which determines the storage time. On the other hand, we need a large broadened line, which determines the bandwidth of the pulse to be stored. The required large artificial broadening compromises the optical depth, thereby im-pacting on the storage efficiency. This can be alleviated by working with broader initial lines and rapidly mapping optical coherence (between levels $|b\rangle$ and $|a\rangle$ i in Figure 5-9(a)) onto long-lived ground state coherence (between levels $|b\rangle$ and $|c\rangle$). This transfer can be accomplished using π pulses, or a direct Raman transfer using additional, off resonant control fields connecting levels $|a\rangle$ and $|b\rangle$. If the control fields are counter-propagating, this procedure also implements the mode-matching operation forcing the retrieved signal to propagate backwards. Photon-echo quantum memory with Raman transfer is sometimes referred to as Raman echo quantum memory (REQM). Beyond the possibility to work with relatively broad optical absorption lines, this approach may relax material requirements as the storage bandwidth depends not only on the controlled inhomogeneous broadening of the optical transition, but also on the Rabi frequency of the Raman control fields. Interestingly, reversible mapping of quantum states between light and atomic ensembles can be obtained for fields of arbitrary strengths, i. e. beyond the usual weak field, linear approximation. CRIB was first demonstrated in 2006 using Europium doped Y_2SiO_5 crystal, a reversible external electric field that generated longitudinal broadening, and macroscopic optical pulses recalled in the forward direction. As in all photon-echo based storage, the crystal was cooled to around 4K. Due to limited optical depth, the size of the recalled pulses was six orders of magnitude smaller than the one of the input pulses. Shortly after, the same group demonstrated that amplitude as well as phase information can be stored. The efficiency was similarly small. Since then, the memory performance has been improved tremendously, and a record efficiency of 66% was recently reported using a Praseodymium-doped Y_2SiO_5 crystal and the configuration mentioned above.

The Europium and Praseodymium-doped crystals employed in these experiments feature a favorable level structure and radiative lifetimes for optical pumping, but have somewhat inconvenient transition frequencies around 580nm and 606nm, respectively. This requires working with frequency-stabilized dye lasers. Furthermore, the spectral width of the light to be stored is limited to a few MHz, due to small atomic level spacing in the ground and excited state multiplets. Recently, CRIB was implemented with telecommunication photons in an Erbium doped Y_2SiO_5 crystal. This material features a more convenient transition at 1536nm where standard diode lasers can be employed, but does not provide the same ease for tailoring the initial absorption line as Europium or Praseodymium doped Y_2SiO_5. The recall efficiency for weak coherent laser pulses was below 1%. CRIB based storage was recently combined with a direct Raman transfer. The experiments relied on macroscopic signal pulses and storage in Rubidium vapor. The first investigation established the feasibility and resulted in a recall efficiency of 1%. In the second study, the efficiency could be increased up to 41%. Exploiting the condition that pulse emission can only take place if all dipoles oscillate in phase and the Raman coupling beam is switched on, the authors also demonstrated that the order and the moments of recall of four stored pulses can be chosen at will. Hence, the approach could work as an optical random access memory for quantum information encoded into time-bin qubits. Furthermore, beam splitting of input pulses was observed. The latter has also been demonstrated using EIT systems, traditional stimulated photon echoes, and photonecho quantum memory based on atomic frequency combs.

- Atomic frequency combs (AFC)

In the protocol based on atomic frequency combs (AFC), the distribution of atoms over detuning δ is de-scribed by a periodic, comb-like structure with absorption lines spaced by multiples of Δ (see Figure 5-8). Repetitive rephasing occurs at times $2\pi/\Delta$ when the phases accumulated by atomic dipoles in different "teeth" differ by multiples of 2π. To inhibit re-emission after one fixed cycle time, and allow for long-term storage with on-demand readout, the excited optical coherence can be transferred temporarily to coherence between other atomic levels, e.g. ground state spin-levels, where the comb structure is not present. This condition is well satisfied in rare-earth-ion doped crystals.

The AFC approach originated from the discovery in the late seventies that photon echoes can be stimulated from accumulated frequency gratings. The quantum memory protocol, proposed in 2008, is expected to enable time-variable storage of quantum states with unit efficiency and fidelity. Assuming transverse broadening, the efficiency of the AFC protocol can reach 100% for recall in backward direction, and 54% for recall in forward direction. Compared to CRIB, AFC has the advantage of making better use of the available optical depth as fewer atoms need to be removed through optical pumping. Another advantage is the unlimited multi-mode capacity, rovided that the natural broadened absorption line is sufficiently large. For instance, calculations suggest the possibility to store 100 temporal modes with an efficiency of 90% in Europium Y_2SiO_5. For comparison, the multimode capacity in CRIB scales linearly with the absorption depth, and in EIT it is proportional to the square root thereof.

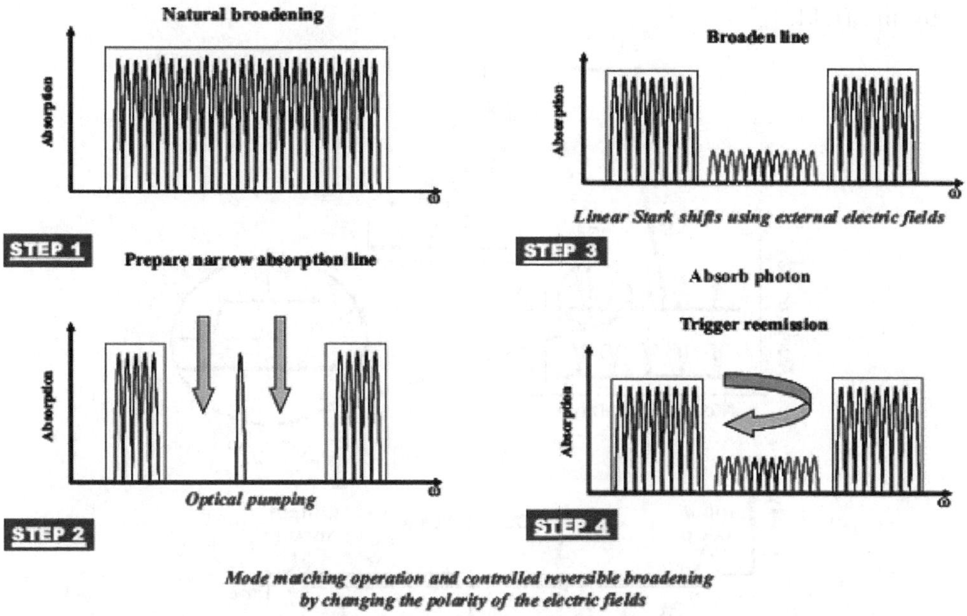

Fig. 5-8 CRIB-based quantum memory in solid state devices featuring optical centers with permanent electric dipole moments. In STEP 1, a narrow absorption line is created from an ensemble of absorbers with broad, inhomogeneously broadened absorption line. This is achieved by means of an optical pumping (or spectral hole burning) procedure, which transfers population to auxiliary atomic levels. In STEP 2, the line is broadened in a controlled and reversible way using DC Stark shifts and position dependent external electric fields. This procedure goes along with a reduction of optical depth. Then, the signal to be stored is directed into the medium and absorbed (STEP 3). In STEP 4, re-emission is activated through the application of a mode (or phase) matching operation. This leads to backwards emission of the signal in a time-reversed version.

AFC quantum memory with re-emission after 250ns, pre-determined by the spacing in the frequency comb, was reported in 2008. The experiment relied on a Neodymium doped YVO_4 crystal, and recall in forward direction. The readout efficiency for weak coherent input states of 20ns duration was 0.5%, and a capacity of up to four temporal modes could be demonstrated. The post-selected storage fidelity for time-bin qubits in various states, defined by average state overlap discussed in sec. IA, exceeded 97%. A shortcoming in this experiment, namely the predetermined emission timing, was overcome in 2009. For on-demand recall, the initially excited optical coherence was temporally transferred to ground state coherence using two π pulses with variable relative delay. 450ns long macroscopic optical pulses could be stored in a Praseodymium-doped Y_2SiO_5 crystal for up to 20μs with around 1% efficiency.

In the same year, AFC-based storage was also demonstrated in a Thulium YAG crystal. Better tailoring of the comb structure resulted in a storage efficiency of 9.1%, i. e. an improvement by almost one order of magnitude. Off-resonant Faraday interaction Consider an optical wave propagating through an ensemble of two-level atoms. When its detuning Δ from the atomic transition is sufficiently large ($\Delta \gg W_{\text{line}}$), the real part of the susceptibility is inversely proportional to Δ, where as its imaginary part behaves as Δ^{-2} (see Figure 5-9(b), (c)). Therefore an off-resonant wave will not excite any atoms, resulting in negligible absorption, but may experience a significant phase shift. The quantum (see Figure 5-9) phase of the atoms will in turn be affected by the field.

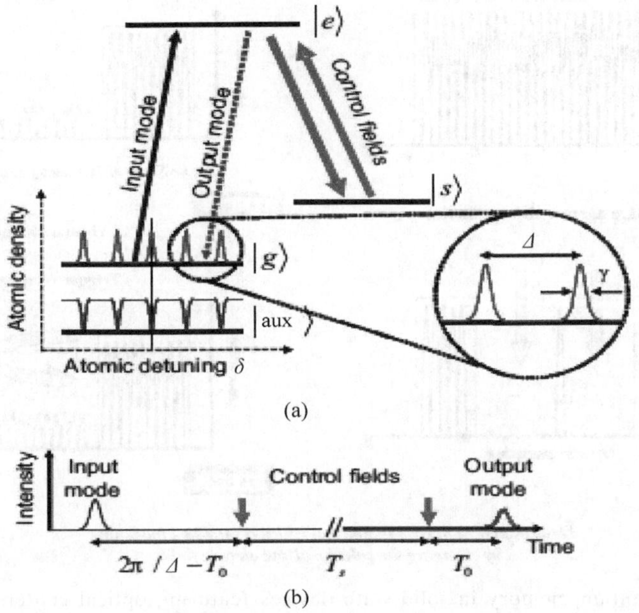

Fig. 5-9 Quantum memory based on atomic frequency combs (AFC). (a) An inhomogeneously broadened absorption line is tailored into an atomic frequency comb using frequency-selective optical pumping to the level. The peaks in the comb are characterized by width and separation Δ. (b) The collective dipole moment created by absorption of the input light rapidly dephases and, due to the discrete structure of the absorption profile tailored in (a), rephases after a time $2\pi/\Delta$. This results in the re-emission of the input light field.

This mutual effect of light and atoms can be utilized to construct an elegant implementation of quantumoptical memory. Consider a signal wave with macroscopic linear polarization along the y axis. The quantum information to be stored is encoded in the microscopic Stokes parameters $\hat{S}_2$ and $\hat{S}_3$ of the wave. The Stokes parameter S_2 is interpreted as the angle of polarization with respect to the y axis, whereas S_3 is proportional to the collective spin of the photons. In order to implement storage, this field propagates, along the z direction, through an off-resonant atomic gas. As a consequence of the angular momentum uncertainty relations, the projections y and $\hat{J}_z$ of the atomic collective are uncertain. In the following, we treat them as quantum operators. The interaction between the light and atoms will lead to the following effects. First, the collective spin of the photons causes the atoms to rotate around the z axis by a microscopic angle θ. Can obtain:

$$\begin{cases} \hat{J}_{y,\text{out}} = \hat{J}_{y,\text{in}} \cos\theta + \hat{J}_{x,\text{in}} \sin\theta \\ \hat{J}_{x,\text{out}} = \hat{J}_{x,\text{in}} \cos\theta + \hat{J}_{y,\text{in}} \sin\theta \\ \hat{J}_{z,\text{out}} = \hat{J}_{z,\text{in}} \end{cases} \tag{5-12}$$

For a small θ, can assume $\cos\theta \approx 1$, $\sin\theta \approx 0$. Furthermore, we can treat the macroscopic quantity $\hat{J}_x$ as a classical number and rewrite Eq. (5-12) as

$$\begin{cases} \hat{J}_{y,\text{out}} = \hat{J}_{y,\text{in}} + \alpha \hat{S}_{3,\text{in}} \\ \hat{J}_{z,\text{out}} = \hat{J}_{z,\text{in}} \end{cases} \tag{5-13}$$

where α is some proportionality coefficient.

On the other hand, the optical wave's polarization will experience Faraday rotation due to the z component of the angular momentum of the atoms:

$$\hat{S}_{2,\mathrm{out}} = \hat{S}_{2,\mathrm{in}} + \beta \hat{J}_{z,\mathrm{in}}$$
$$\hat{S}_{3,\mathrm{out}} = \hat{S}_{3,\mathrm{in}} \tag{5-14}$$

Here assumed again that the rotation angle is small and treated the large x polarization component as a classical number. Eqs. (5-13) and (5-14) describe mutual interaction between the light and the atoms. Can see that the information about the Stokes parameter $\hat{S}_3$ of the optical state is imprinted on the optical state. However, this does not yet constitute memory, because the atoms do not receive any information about $\hat{S}_3$; furthermore, the stored information is compromised by the uncertain values of $\hat{J}_y$ and $\hat{J}_z$.

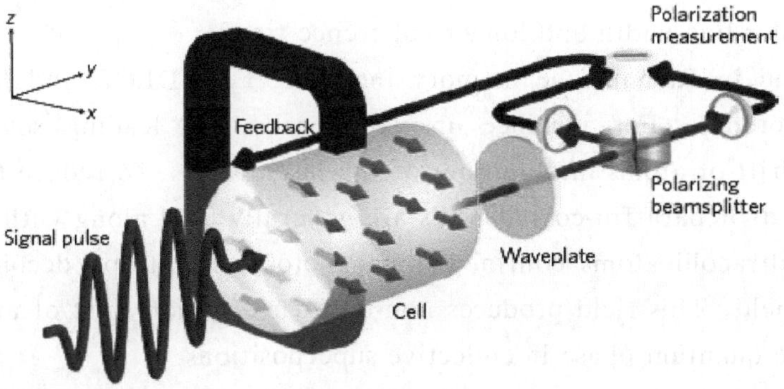

Fig. 5-10 Quantum memory based on off-resonant Faraday interaction beten light and atoms. The quantum information carried by the light is encoded in its polarization. After propagation through the cell, the field is subjected to a polarimetric measurement, whose result is then fed back to the atoms by applying a magnetic field pulse of a known magnitude and duration.

In order to complete the memory protocol, to perform a polarimetric measurement of the pulse emerging from the atomic ensemble, thereby determining its Stokes parameter S_2 out. Then perform a feedback operation on the atoms, displacing its angular momentum $\hat{J}_z$ by the measured quantity as follows:

$$\hat{J}'_z = \hat{J}_{z,\mathrm{out}} - S_{2,\mathrm{out}}/\beta = -\hat{S}_{2,\mathrm{in}}/\beta \tag{5-15}$$

This displacement is performed, for example, by applying a magnetic field and causing the angular momentum to precess around the field direction. Now both components of the optical polarization have been transferred to the atomic angular momentum. Although the $\hat{J}_y$ component is still "contaminated" by its initial noise $\hat{J}_y$, in, this imperfection can be eliminated by initially preparing the atoms in the spin squeezed state, so the uncertainty of $\hat{J}_y$, in is reduced. Following initial theoretical papers in which off-resonant Faraday interaction between light and matter have been proposed as a tool for quantum information applications, a theoretical proposal for quantum memory has been developed in 2003 by Kuzmich and Polzik and further elaborated. In fact, these references propose a scheme in which the light passes through the atomic ensemble twice, in two different directions, which allows elimination of both measurement and feedback used in the scheme described above. Experimentally, the scheme with feedback has been implemented in 2004 by Julsgaard et al. using experimental tools developed earlier by the same group for the purpose of entangling two atomic ensembles. The atoms were prepared without initial spin squeezing. However, the apparatus was shown to beat the classical benchmark for coherent states, which was historically the first demonstration of quantum properties of an optical memory. A common feature in all approaches to quantum memory via atom-light interaction is storage of information in atomic coherence. All methods are thus prone to decoherence, which limits the storage time. In the case of CRIB and AFC, the storage time is also affected by the width of the tailored absorption lines (see e. g. STEP 1 in Figure 5-8) as long as information is stored in coherence between the ground and the optically excited state. This width is fundamentally limited by the intrinsic homogeneous linewidth of the optical transition, and practically limited by laser line fluc-tuations and power broadening during optical pumping. Decoherence can be reduced by temporally mapping the optical coherence onto coherence between ground states, which are generally associated with smaller line-width and longer coherence time.

The storage time in Raman-type memory including EIT, DLCZ and REQM is limited by ground state decoherence. For instance, in vapor cells, the leading source of ground-state decoherence is the drift of atoms into and out of the laser beam. To reduce this effect, cells with inert buffer gases and/or paraffin-coated walls are generally used along with geometrically wider optical modes. In ultracold atoms confined in magneto-optical traps, decoherence often comes from the magnetic field. This field produces non-uniform Zeeman shift of atomic ground levels, leading to loss of the quantum phase in collective superpositions.

Significant improvement of the memory lifetime in atomic ensembles has recently been reported in two DLCZ experiments: The effect of the magnetic field has been minimized by using the atomic "clock" states as the ground states of the system, i. e. such magnetic sublevels whose two-photon detuning is minimally affected by the Zeeman effect. Residual atomic motion has been eliminated by transferring the atoms from a MOT into an optical lattice of a sufficiently small period, instead used a collinear geometry for the four optical fields involved, in which case the dephasing due to atomic motion is greatly reduced. Storage times of 7ms and 1ms, respectively, have been reported in these experiments. Very recently, EIT-based storage of light has been demonstrated in an atomic Mott insulator-a state of a collective of atoms filling a three-dimensional optical lattice, with one atom per lattice site. Absence of mechanical motion and uniformity of the magnetic field resulted in a storage lifetime of 238ms, the current record for atomic media. The residual ground state coherence decay is likely due to heating in the optical lattice and atomic tunneling. Even longer lived coherence can be achieved in rare earth-ion doped crystals. Depending on the crystal and dopant, decoherence mechanisms vary. For instance, the dominant mechanism in Praseodymium-doped Y_2SiO_5 crystals is random Zeeman shifting of Pr^{3+} ions due to fluctuating magnetic fields from Yttrium nuclei. Fortunately, this effect can be significantly reduced by operating in a uniform magnetic field of certain magnitude and direction. Using this approach, coherence times up to 82ms have been reported. This time was further increased to 30s by adding dynamic decoherence control.

Quantum memory for light constitutes a promising, rapidly developing research topic that builds on decades of research into atomic spectroscopy, quantum optics and material science.

Recent results show efficient storage with high fidelity, long lifetime with on-demand recall, and high multi-mode capacities. Yet, these performance characteristics have to date been demonstrated using different storage media and protocols, and more effort is required for combining these benchmarks in a single setting. Once developed, such a device will be invaluable for quantum communication and cryptography as well as optical quantum computation.

5.2 Analysis of a quantum memory for photons

The implementation of quantum memories for photons is an important goal in quantum information processing. It would allow the realization of on-demand single-photon sources based on heralded sources and provide a basic ingredient for quantum repeaters. Several different approaches to the realization of such memories have been proposed, using both single absorbers in high-finesse cavities and dense atomic ensembles. The latter proposals include the use of off-resonant interactions, of electromagnetically induced transparency (EIT), and of nonstandard photon echoes. On the experimental side, storage and retrieval of classical light has been realized using EIT and photon echoes. Coherent states of light have been stored in an atomic ensemble with higher than classical fidelity using off-resonant interactions. Recently the storage and retrieval of single photons have been reported using EIT in atomic ensembles. In the present work we are following the approach that originated, where it was shown that highly efficient photon echoes can be generated in a gas of atoms exploiting the fact that the Doppler shift changes sign if the propagation direction of the light is reversed.

A first adaptation of the effect to solid state systems was suggested in using nuclear magnetic resonance. An attractive experimental realization of the same principle using controlled reversible inhomogeneous broadening (CRIB). These proposals combine spectral hole burning techniques and the use of controllable Stark shifts from electric field gradients. First proof-of-principle experimental demonstrations of the CRIB approach have recently been performed. Here perform a detailed theoretical analysis of the CRIB quantum memory protocol for finite optical depth and general atomic distributions. In section II give the equations of motion for the full system atoms plus light and recall the principle of a memory based on CRIB. In above section derive the general solution of the equations of motion under the condition of weak excitation. To discuss both the complete memory protocol, where the output field is emitted in the backward direction, and a simplified protocol, which does not use any optical control fields, leading to forward emission. In section 4 study the dependence of the memory efficiency on the optical depth of the medium. It approaches one for the complete protocol for sufficiently large optical depth, whereas it can be over 50% for the simplified protocol, for which the efficiency is limited by reabsorption.

Next after section will show that the width of the initial atomic distribution determines the possible storage time. To furthermore analyze what is the optimal broadening for a given initial distribution and pulse width. The effect of shapes of the atomic distribution and light pulse.

5.2.1 Principles

A light pulse propagates through a medium composed of two-level atoms. It resonantly couples the two atomic states, the ground state $|g_i$ and the excited state $|e\rangle$. To look at the behavior of the positive frequency part of the slowly time-varying envelope $E(z,t)$ of the light field decomposed in forward and backward modes

$$E(z,t) = E_{\mathrm{f}}(z,t)\mathrm{e}^{\mathrm{i}\omega_0 z/c} + E_{\mathrm{b}}(z,t)\mathrm{e}^{-\mathrm{i}\omega_0 z/c} \tag{5-16}$$

in a one-dimensional light propagation model. This one-dimensional model is well adapted to the propagation either in an optical wave guide or in a bulk medium under the condition that $A/\lambda\ell \gg 1$, A being the area of the light beam, λ the wavelength and ℓ the propagation length.

The central frequency ω_0 of the light pulse is detuned from the atomic transition frequency ω_{eg} by $\Delta := \omega_0 - \omega_{eg}$. The atomic transitions undergo an inhomogeneousbroadening that we consider uniform in space such that the density of atoms associated to the detuning Δ is $\rho(\Delta)$. To describe the properties of the atomic ensemble, can define a mean field per atoms, slowly varying in time

$$\sigma_{ij}(z,t;\Delta) := \frac{1}{N(\Delta,z)}\sum_{n=1}^{N(\Delta,z)} | i\rangle_{nn}\langle j | \tag{5-17}$$

i,j standing for e or g. In the above sum, the atom index n runs over all atoms

$$N(\Delta,z) := \rho(\Delta)\delta z\delta\Delta/L \tag{5-18}$$

detuning within the interval

$$[\Delta - \delta\Delta/2, \Delta + \delta\Delta/2] \tag{5-19}$$

and position within the interval

$$[z - \delta z/2, z + \delta z/2] \tag{5-20}$$

L denotes the length of the medium. The Hamiltonian describing the system atoms plus light is given in the rotating wave approximation by

$$H = \int_{-\infty}^{+\infty} \mathrm{d}\Delta \frac{\rho(\Delta)}{L}\int_0^L \mathrm{d}z[\Delta\sigma_{ee}(z,t;\Delta) - \wp E(z,t)\sigma_{eg}(z,t;\Delta) + \mathrm{H.c.}] \tag{5-21}$$

Where $\wp$ being the dipole moment of the transition $|e\rangle - |g\rangle$. In analogy with the light field, the positive frequency part of the operator associated to the atomic coherence σ_{ge} is decomposed into two counter-propagating contributions

$$\sigma_{ge}(z,t;\Delta) = \sigma_{\mathrm{f}}(z,t;\Delta)\mathrm{e}^{\mathrm{i}\omega_0 z/c} + \sigma_{\mathrm{b}}(z,t;\Delta)\mathrm{e}^{-\mathrm{i}\omega_0 z/c} \tag{5-22}$$

Under the approximation that most of the population stays in the ground state $\sigma_{ge} \approx 1$, which is well justified for single or few photons light storage, the evolution of the atomic coherence is given by the following Heisenberg-Langevin equations

$$\frac{\partial}{\partial t}\sigma_{\mathrm{f}}(z,t;\Delta) = -\mathrm{i}\Delta\sigma_{\mathrm{f}}(z,t;\Delta) + \mathrm{i}\wp E_{\mathrm{f}}(z,t) \tag{5-23}$$

$$\frac{\partial}{\partial t}\sigma_{\mathrm{b}}(z,t;\Delta) = -\mathrm{i}\Delta\sigma_{\mathrm{b}}(z,t;\Delta) + \mathrm{i}\wp E_{\mathrm{b}}(z,t) \tag{5-24}$$

Here neglect the homogeneous decoherence, and will however take into account inhomogeneous dephasing in the following. Using the slowly varying approximation, the evolution of the forward and backward components of the light pulse is given by

$$\left(\frac{\partial}{\partial t} + c\frac{\partial}{\partial z}\right)E_{\mathrm{f}}(z,t) = \mathrm{i}\beta\int_{-\infty}^{+\infty} \mathrm{d}\Delta G(\Delta)\sigma_{\mathrm{f}}(z,t;\Delta) \tag{5-25}$$

$$\left(\frac{\partial}{\partial t} - c\frac{\partial}{\partial z}\right)E_{\mathrm{b}}(z,t) = \mathrm{i}\beta\int_{-\infty}^{+\infty} \mathrm{d}\Delta G(\Delta)\sigma_{\mathrm{b}}(z,t;\Delta) \tag{5-26}$$

where β is defined by

$$\beta := g_0^2 N\wp \text{ with } g_0 := \sqrt{\omega_0/(2\varepsilon_0 V)} \tag{5-27}$$

$$N := \int \mathrm{d}\Delta \rho(\Delta) \tag{5-28}$$

is the number of atoms in the quantization volume V such that

$$\rho(\Delta) = N \times G(\Delta) \tag{5-29}$$

where $G(\Delta)$ is the normalized spectral atomic distribution.

A complete description of the system is given by the set of Eqs. (5-23) ~ (5-26). These equations of motion describe a situation in which the presence of the laser field introduces changes in the medium by inducing a polarization of the atomic ensemble. Reciprocally, the atomic polarization can serve as a source for the optical field. Owing to the consideration of weak excitations, the coupled equations are linear and of first order offering the possibility to solve them analytically. Furthermore, the linearity of the equations of motion implies the equivalence between classical and single-photon dynamics. In the following, the quantities E_b and E_f can be interpreted either as classical fields or as single-photon wave-functions. Analogously, σ_b and σ_f can describe either a classical atomic polarization or the wave-function of a single atomic excitation created by a single photon.

The principle of a memory based on reversible absorption can be deduced from the symmetry analysis of the equations of motion. The forward components of the light operator and of the atomic coherence evolve following the set of Eqs. (5-30) ~ (5-31). If the following transformations are performed

$$\Delta \to -\Delta \tag{5-30}$$

$$E_b \to -E_b \tag{5-31}$$

the equations for the backward components (5-20) ~ (5-21) define a time reversal evolution of the system compared to the equations for the forward components. This analysis reveals the essence of a quantum memory based on CRIB: a light pulse propagating in the forward direction is completely absorbed by the atomic ensemble. What is required to retrieve the pulse as a time reversed copy of itself is to reverse the detuning between the spectral components of the light pulse and the atomic transition frequency. At the same time, one has to apply a phase matching operation such that the wave function of the single atomic excitation propagates in the backward direction. Now describe how one can construct an optical memory according to the principle discussed above. The following scenario closely follows the ideas.

(1) One first chooses the medium. The relevant atomic transition should have a frequency in the optical domain. It should also have low decoherence rate compared to the total duration of the pulse in order to keep the phase and amplitude informations during the optical absorption. Several experimental realizations are discussed using atomic gases at room temperature or iondoped solid-state materials cooled to cryogenic temperature. Motivated by the work done in our group, here consider a medium based on ion-doped solid materials.

(2) A narrow atomic absorption line is prepared. In ion-doped materials, the absorption transition is broadened inhomogeneously due to the environment in the host which is different for each ion. A narrow absorption line can be prepared using optical pumping techniques. The width of the prepared line is limited ultimately by the homogeneous line-width defined from the decoherence rate of the atomic ensemble.

In practice however, the bandwidth of the absorption line might be limited by the bandwidth of the laser used for the optical pumping. To denote the initial detuning by Δ_0. The selected inhomogeneous atomic distribution, called $G_0(\Delta_0)$ with bandwidth γ_0 is represented in Figure 5-11 as dotted line.

(3) The initial distribution $G_0(\Delta_0)$ associated to the detunings Δ_0 is then broadened to a larger bandwidth using reversible and controlled inhomogeneous broadening. Every detuning Δ_0 is broadened to $\Delta_0 + \Delta'$, where Δ' is distributed according to $G'(\Delta')$ (dashed line in Figure 5-7). In ion-doped materials, a controlled inhomogeneous broadening of the initial absorption line can be obtained by applying an electric field gradient that shifts the transition frequency of the ions by the Stark effect such that the detuning Δ' is given by the induced Stark shift. The final distribution $G(\Delta)$ (full line in Figure 5-11), taking into account the broadening of all the detunings Δ_0, is given by the convolution of the initial distribution $G_0(\Delta_0)$ and of the distribution $G'(\Delta')$. By reversing the electric field, the Stark shift is reversed and consequently, each detuning is reversed such that Eq. (5-30) is replaced by

$$\Delta_0 + \Delta' \rightarrow \Delta_0 - \Delta' \tag{5-32}$$

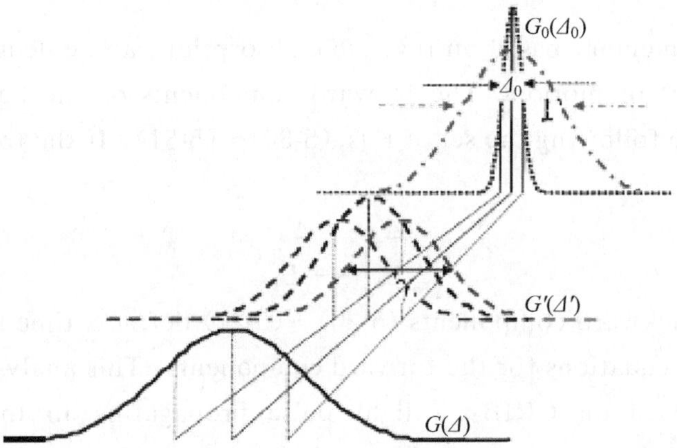

Fig. 5-11 (Color online) Schematic representation of the spectral atomic distribution. The initial distribution $G_0(\Delta_0)$ with characteristic bandwidth 0 is represented as black dotted line. Three spectral components of the initial distribution are considered (blue, black and red vertical lines) and each of them is broadened according to a distribution $G'(\Delta')$ (blue, black and red dashed lines) with bandwidth. The final broadened distribution, called $G(\Delta)$ is the convolution of the initial distribution $G_0(\Delta_0)$ and of the broadened distribution $G'(\Delta')$ associated to a single initial absorption line (Δ_0). The pulse shape with bandwidth γ is represented as green dashed dotted line.

Here consider a controlled inhomogeneous broadening which is independent of the position z. This corresponds to a rare-earth doped fiber or to a crystal in which the electric field gradient is applied in a direction transverse to the propagating pulse.

(4) The equations of motion associated to the forward components are connected to the ones associated to the backward components by applying a position-dependent phase $e^{-i\omega_0 z/c}$ to the atomic ensemble when the light excitation is reduced to zero. This transforms the forward component of into the backward component σ according to the Eq. (5-19). The phase shift $e^{-i\omega_0 z/c}$ can be realized by using counter propagating π-pulses to transfer the atomic coherence back and forth to an auxiliary ground state. This procedure can also offer longer storage duration by choosing a ground state with long coherence time. If the phase shift is not applied, the evolution of

the system is given by the forward set of equations. The pulse reemitted by the medium propagates in the forward direction and is partially reabsorbed, limiting the memory efficiency. On the other hand, this simplified protocol does not require the use of additional control laser fields. This provides an advantage for few photons quantum memory by avoiding the light noise generated by auxiliary laser fields.

In the next section, study the properties of the optical memory for a finite optical depth when the atomic ensemble reemits the light pulse in forward and backward directions. We also look at the impact of the initial distribution bandwidth and we establish the optimal broadening of the initial distribution with respect to the memory efficiency for a given pulse.

5.2.2 General solution

To solve the equations of motion by generalizing the technique, io clearly distinguish the contribution of the initial distribution of atoms $G_0(\Delta_0)$ and of the broadened distribution $G'(\Delta')$ associated to a single absorption line, we introduce the atomic operator

$$\sigma_{ij}(z,t;\Delta',\Delta_0) := \frac{1}{N(\Delta'+\Delta_0,z)} \sum_{n=1}^{N(\Delta'+\Delta_0,z)} |i\rangle_{nn}\langle j| \tag{5-33}$$

where

$$N(\Delta'+\Delta_0,z) \tag{5-34}$$

is the number of atoms with initial detuning around Δ_0 broadened to $\Delta'+\Delta_0$ and with position around z. When the inhomogeneous broadening is reversed, the detuning is changed from $\Delta'+\Delta_0$ to $-\Delta'+\Delta_0$ such that the atomic properties are described by the operator $\sigma_{ij}(z,t;-\Delta',\Delta_0)$. As in the previous section, the atomic operator is decomposed into two counter-propagating contributions. It is interested in the evolution of an incoming light pulse propagating in the forward direction through the atomic ensemble

$$\left(\frac{\partial}{\partial t}+c\frac{\partial}{\partial z}\right)E_{\mathrm{f}}^{\mathrm{in}}(z,t) = \mathrm{i}\beta\int_{-\infty}^{+\infty}\mathrm{d}\Delta_0\mathrm{d}\Delta' G(\Delta_0)G'(\Delta')\sigma_{\mathrm{f}}(z,t;\Delta',\Delta_0) \tag{5-35}$$

$$\frac{\partial}{\partial t}\sigma_{\mathrm{f}}(z,t;\Delta',\Delta_0) = -\mathrm{i}(\Delta_0+\Delta')\sigma_{\mathrm{f}}(z,t;\Delta',\Delta_0)+\mathrm{i}\wp E_{\mathrm{f}}^{\mathrm{in}}(z,t) \tag{5-36}$$

Initially, there is no atomic excitation such that

$$\sigma_{\mathrm{f}}(z,t\to-\infty;\Delta',\Delta_0)=0 \tag{5-37}$$

Taking into account this initial condition, Eq. (5-37) can be solved as

$$\sigma_{\mathrm{f}}(z,t;\Delta',\Delta_0) = \mathrm{i}\wp\int_{-\infty}^{t}\mathrm{d}s\mathrm{e}^{-\mathrm{i}(\Delta'+\Delta_0)(t-s)}E_{\mathrm{f}}^{\mathrm{in}}(z,s) \tag{5-38}$$

It is interested in the regime for which the storage duration T is longer than the pulse duration τ. Thus consider that at the position $z=0$, the incoming light pulse $E_{\mathrm{in}}(z=0)$ is centered around $-T/2$. The inhomogeneous broadening is reversed at time $t=0$ after a time sufficiently long with respect to the pulse duration τ. The Fourier transform of the incoming light pulse is defined by

$$\widetilde{E}_{\mathrm{f}}^{\mathrm{in}}(z,\omega) := \int_{-\infty}^{0}\mathrm{d}t\mathrm{e}^{\mathrm{i}\omega t}E_{\mathrm{f}}^{\mathrm{in}}(z,t) \tag{5-39}$$

The upper bound is equal to 0 since the incoming light pulse is not defined anymore at positive times. Furthermore, consider the regime in which the broadened distribution bandwidth γ is larger than the inverse of the storage duration $1/T$. These considerations allow us to introduce the Fourier transform of the incoming light pulse in Eq. (5-25) in which of is replaced by its expression

$$\left(\frac{\partial}{\partial z}-\frac{\mathrm{i}\omega}{c}+\eta H(\omega)\right)\widetilde{E}_{\mathrm{f}}^{\mathrm{in}}(z,\omega)=0 \tag{5-40}$$

$H(\omega)$ is defined by

$$H(\omega):=\int_{0}^{+\infty}\mathrm{d}x\left(\mathrm{e}^{\mathrm{i}\omega x}\times\int_{-\infty}^{+\infty}\mathrm{d}\Delta_0\mathrm{d}\Delta' G_0(\Delta_0)G'(\Delta')\mathrm{e}^{-\mathrm{i}(\Delta'+\Delta_0)x}\right) \tag{5-41}$$

and

$$\eta:=g_0^2 N\wp^2/c \tag{5-42}$$

Then the solution of Eq. (5-28) is given by

$$\widetilde{E}_{\mathrm{f}}^{\mathrm{in}}(z,\omega)=\widetilde{E}_{\mathrm{f}}^{\mathrm{in}}(0,\omega)\mathrm{e}^{\mathrm{i}\omega z/c}\mathrm{e}^{-\eta H(\omega)z} \tag{5-43}$$

at any position inside the medium $0\ll z\ll L$. The solutions (5-40) and (5-43) describes the absorption of the incoming light pulse by the atomic ensemble. The absorption coefficient $\alpha(\omega)$ defined from the relation

$$|\widetilde{E}_{\mathrm{f}}^{\mathrm{in}}(z,\omega)|^2=\mathrm{e}^{-\alpha(\omega)z}|\widetilde{E}_{\mathrm{f}}^{\mathrm{in}}(0,\omega)|^2 \tag{5-44}$$

is given by

$$\alpha(\omega):=\frac{2g_0^2N\wp^2\mathrm{Re}(H(\omega))}{c} \tag{5-45}$$

It is thus proportional to the coupling between the light pulse and the atomic ensemble and depends on the atomic distribution via the function $H(\omega)$. In what follows, we give the expressions of the outgoing light pulse reemitted by the atomic ensemble by distinguishing the pulses reemitted in the controlled backward and forward directions.

- Complete protocol: Emission in backward direction

The light pulse is partially absorbed such that a part of the light excitation is transfered into the atomic ensemble. At time $t=0$, the inhomogeneous broadening is reversed meaning that the detuning $\Delta_0+\Delta'$ changes to $\Delta_0-\Delta'$. Furthermore, the phase shift $\mathrm{e}^{2\mathrm{i}\omega_0 z/c}$ is applied to the atomic ensemble such that the system evolves backward in space. We assume that at $t=0$ the nonabsorbed part of the light has left the medium so that the field is zero, i.e. the excitation is purely atomic. The system is described by the following set of equations

$$\left(\frac{\partial}{\partial t}-c\frac{\partial}{\partial z}\right)E_{\mathrm{b}}^{\mathrm{out}}(z,t)=\mathrm{i}\beta\int_{-\infty}^{+\infty}\mathrm{d}\Delta_0\mathrm{d}\Delta' G_0(\Delta_0)G(-\Delta')\sigma_{\mathrm{b}}(z,t;-\Delta',\Delta_0) \tag{5-46}$$

$$\frac{\partial}{\partial t}\sigma_{\mathrm{b}}(z,t;-\Delta',\Delta_0)=-\mathrm{i}(\Delta_0-\Delta')\sigma_{\mathrm{b}}(z,t;-\Delta',\Delta_0)+\mathrm{i}\wp E_{\mathrm{b}}^{\mathrm{out}}(z,t) \tag{5-47}$$

with the initial condition

$$\begin{cases}E_{\mathrm{b}}^{\mathrm{out}}(z,t=0)=0\\ \sigma_{\mathrm{b}}(z,t=0;-\Delta',\Delta_0)=\mathrm{i}\wp\int_{-\infty}^{0}\mathrm{d}s\,\mathrm{e}^{\mathrm{i}(\Delta'+\Delta_0)s}E_{\mathrm{f}}^{\mathrm{in}}(z,s)\end{cases} \tag{5-48}$$

The solution of Eq. (5-47) is given by

$$\sigma_{\mathrm{b}}(z,t;-\Delta',\Delta_0)=\sigma_{\mathrm{b}}(z,0;-\Delta',\Delta_0)+\int_0^t\mathrm{d}s[\mathrm{e}^{\mathrm{i}(\Delta'-\Delta_0)(t-s)}\times$$
$$(\mathrm{i}\wp E_{\mathrm{b}}^{\mathrm{out}}(z,s)+\mathrm{i}(\Delta'-\Delta_0)\sigma_{\mathrm{b}}(z,0;-\Delta',\Delta_0))] \tag{5-49}$$

Plugging the previous expression into Eq. (5-48), can get the following propagation equation

$$\left(\frac{\partial}{\partial z}+\frac{\mathrm{i}\omega}{c}-\eta F(\omega)\right)\widetilde{E}_{\mathrm{b}}^{\mathrm{out}}(z,\omega)=\eta\int_{-\infty}^{+\infty}\mathrm{d}\Delta_0 G_0(\Delta_0)J(\omega;\Delta_0)\widetilde{E}_{\mathrm{f}}^{\mathrm{in}}(z,-\omega+2\Delta_0) \tag{5-50}$$

for the Fourier transform of the outgoing pulse

$$\widetilde{E}_{b}^{out}(z,\omega) := \int_{0}^{+\infty} dt e^{i\omega t} E_{b}^{out}(z,t) \tag{5-51}$$

The lower bound is equal to 0 since the outgoing light pulse, centered around $T/2$, is not defined at negative times. It have introduced the functions

$$J(\omega;\Delta_0) := \int_{-\infty}^{+\infty} dx \int_{-\infty}^{+\infty} d\Delta' G'(-\Delta') e^{i(\Delta'-\Delta_0)x} e^{i\omega x} \tag{5-52}$$

and

$$F(\omega) := \int_{0}^{+\infty} dx \left(e^{i\omega x} \times \int_{-\infty}^{+\infty} d\Delta_0 d\Delta' G_0(\Delta_0) G'(-\Delta') e^{i(\Delta'-\Delta_0)x} \right) \tag{5-53}$$

The solution of Eq. (5-32) establishes the connection between the output and the input light pulses when they propagate in opposed directions

$$\begin{aligned}\widetilde{E}_{b}^{out}(z,\omega) = &-\eta \int_{-\infty}^{+\infty} d\Delta_0 G_0(\Delta_0) e^{-i\omega_0 z/c} \times \\ &\left[e^{2i\Delta_0 L/c} e^{-\eta L H(-\omega+2\Delta_0)} e^{\eta(z-L)F(\omega)} - e^{2i\Delta_0 z/c} e^{-\eta z H(-\omega+2\Delta_0)} \right] \times \\ &\frac{J(\omega;\Delta_0)}{2i\Delta_0/c - \eta H(-\omega+2\Delta_0) - \eta F(\omega)} \widetilde{E}_{f}^{in}(0, -\omega+2\Delta_0)\end{aligned} \tag{5-54}$$

In the next subsection, we study the simplified protocol in which the position-dependent phase is not applied to the atomic ensemble. We find the expression of the output pulse when it propagates in the forward direction.

- Protocol without phase shift: Emission in forward direction

If at time $t=0$, the additional phase $e^{2i\omega_0 z/c}$ is not applied to the atomic ensemble, the system evolves forward in space and its evolution is given by the set of Eq. (5-55) in which Δ' has to be changed to $-\ddot{A}'$. As above, the light pulse is partially absorbed and the absorbed light excitation is transfered to the atoms. The initial condition is thus

$$\begin{cases} E_{f}^{out}(z, t=0) = 0 \\ \sigma_f(z, t=0; -\Delta', \Delta_0) = i\wp \int_{-\infty}^{0} ds e^{i(\Delta'+\Delta_0)s} E_{f}^{in}(z,s) \end{cases} \tag{5-55}$$

The equations of motion are resolved as previously. The expression of the outgoing light pulse propagating in the forward direction is given by

$$\begin{aligned}\widetilde{E}_{f}^{out}(z,\omega) = &-\eta z \int_{-\infty}^{+\infty} d\Delta_0 G_0(\Delta_0) J(\omega;\Delta_0) \times \\ &\text{sinhc}\left(\eta z \frac{F(\omega) - H(-\omega+2\Delta_0)}{2} - iz\frac{\omega-\Delta_0}{c} \right) \times \\ &\exp\left(iz\frac{\Delta_0}{c} - \eta z \frac{F(\omega) + H(-\omega+2\Delta_0)}{2} \right) \times \\ &\widetilde{E}_{f}^{in}(0, -\omega+2\Delta_0)\end{aligned} \tag{5-56}$$

The sinhc denotes the hyperbolic sinus cardinal function ($\text{sinhc}(x) = \sinh(x)/x$).

The Eqs. (5-34) and (5-35) establish the expression of the light pulse reemitted by the atomic ensemble in both backward and forward directions as a function of the input light pulse characteristics and of the properties of the atomic medium. They are valid for a general initial distribution broadened to an arbitrary distribution. In the next sections, they are analyzed in order to deduce some characteristics of the memory for specific atomic distributions.

- Finite optical depth

In the framework of a practical realization, it is necessary to know the properties of the memory for a finite optical depth. This topic has also been addressed for the CRIB protocol and for the slow-light based memories. The analysis is done using a simple model in which the initial distribution is reduced to a single narrow absorption line $G_0(\Delta_0)=\delta(\Delta_0)$. The initial absorption line is broadened to be constant on the interval $[-\gamma/2, \gamma/2]$ and null elsewhere such that it satisfies the condition

$$\int_{-\infty}^{+\infty} d\Delta' G'(\Delta') = \int_{-\infty}^{+\infty} d\Delta G(\Delta) = 1 \tag{5-57}$$

In the regime where the spectral pulse bandwidth is smaller than the inhomogeneous broadening $\Gamma \ll \gamma$, here have

$$H(\omega) \approx F(\omega) \approx \bar{J}(\omega;0)/2 \approx \pi/\gamma \tag{5-58}$$

and the outgoing pulse propagating in the backward direction (5-59) takes the following form:

$$E_b^{out}(0,t) = -(1-e^{-\alpha L})E_f^{in}(0,-t) \tag{5-59}$$

at the position $z=0$. The absorption coefficient is given by

$$\alpha = 2\pi g_0^2 N \wp^2/\gamma c \tag{5-60}$$

The memory efficiency, defined by

$$\text{Eff} := \frac{\int d\omega\, |\tilde{E}^{out}(\omega)|^2}{\int d\omega\, |\tilde{E}^{in}(\omega)|^2} \tag{5-61}$$

corresponds to the probability to retrieve an absorbed photon. The memory efficiency depends on the optical depth αL and as can be seen in Figure 5-12, it reaches 100% for large enough optical depth. This limit of a large optical depth corresponds to the result. Under similar conditions, when no phase shift is applied, the expression of the output light pulse propagating in the forward direction (5-38) is given at the position $z=L$ by

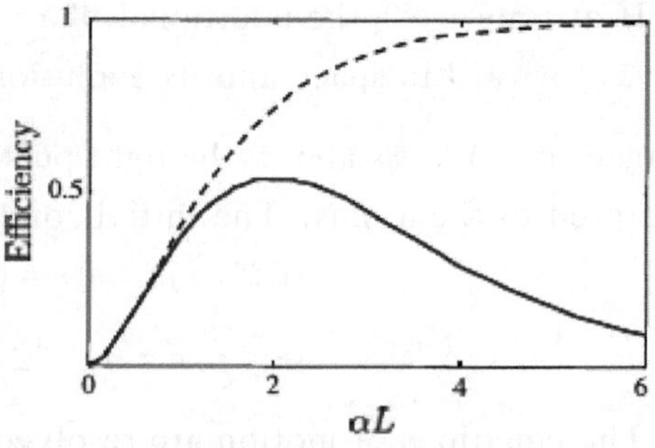

Fig. 5-12 (Color online) Efficiency of the light pulse reemission in forward (full blue line) and backward (dashed black line) direction as a function of the optical depth for a constant broadening. The efficiencies vary as $(\alpha L)^2 e^{-\alpha L}$ for the forward protocol and as $(1-e^{-\alpha L})^2$ for the backward protocol.

$$\tilde{E}_f^{out}(L,\omega) = -\alpha L e^{-\alpha L/2} \frac{\sin(\omega L/c)}{\omega L/c} \tilde{E}_f^{in}(0,-\omega) \tag{5-62}$$

The sinus cardinal function induces a distortion of the light for large pulse bandwidth Γ with respect to c/L. For a pulse bandwidth smaller than c/L, the expression for the forward pulse is reduced to

$$E_f^{out}(L,t) = -\alpha L e^{-\alpha L/2} E_f^{in}(0,-t) \tag{5-63}$$

In this regime, the interaction with the atomic ensemble does not induce distortion neither in the forward direction nor in the backward direction. For small optical depth, the memory efficiency varies quadratically as a function of the optical depth for both forward and backward protocols. However, in the forward direction, the pulse is reabsorbed by the atoms when the optical depth increases and the memory efficiency reaches a maximum of 54% at the optical depth

$\alpha L = 2$. Here we have supposed that the optical depth αL is constant for the whole spectral width of the pulse. If αL is lower for certain spectral components of the pulses, the memory efficiency is lower.

5.3 Atomic distribution and memory efficiency

In previous section has studied the dependence of the memory efficiency on the optical depth. For an initial distribution with characteristic bandwidth $G_0(\Delta_0) \approx \gamma_0$ smaller than the spectral pulse bandwidth Γ and broadened to a distribution bandwidth $G(\Delta) \approx \gamma$, the optical depth can be written as

$$\alpha L \approx \frac{g_0^2 \wp^2 \rho_0(0) L}{c} \frac{\gamma_0}{\gamma} = \alpha_0 L \frac{\gamma_0}{\gamma} \tag{5-64}$$

where $\rho_0(\Delta_0) = NG_0(\Delta_0)$ is the initial density of atoms. The $\alpha_0 L$ corresponds to the initial optical depth in the absence of broadening. For a given medium, the optical depth αL varies with the ratio of the initial distribution bandwidth on the broadened distribution bandwidth. In a first subsection, we show that the initial distribution bandwidth also limits the storage duration. Thus prove the existence of a trade-off between the memory efficiency and the storage duration. In a second subsection, look at the optimal broadening for a given pulse and for a fixed initial distribution.

5.3.1 Memory efficiency versus storage duration

To analyze the role of the initial distribution bandwidth, consider the simple case in which the broadened distribution is constant in a given interval and null elsewhere

$$G'(\Delta') = \frac{1}{\gamma}\theta\left(\Delta' + \frac{\gamma}{2}\right)\theta\left(\frac{\gamma}{2} - \Delta'\right) \tag{5-65}$$

(θ being the Heaviside function) such that in the regime $\Gamma < \gamma$ and $\gamma_0 < \gamma$, we have $J(\omega, \Delta_0)/2 \approx H(\omega) \approx F(\omega) \approx \pi/\gamma$. To discuss the results obtained from the backward process but the conclusions are applicable for the forward process as well. Under the condition $\gamma_0 \ll c/L$, the output light pulse emitted in the backward direction (5-38) is given by the following expression:

$$E_b^{out}(0,t) = -(1 - e^{-\alpha L}) E_f^{in}(0, -t) \times \int_{-\infty}^{+\infty} d\Delta_0 G_0(\Delta_0) e^{-2i\Delta_0 t} \tag{5-66}$$

at the position $z = 0$. Since the output pulse is centered around $T/2$, we are interested in times t around $T/2$. Comparing Eqs. (5-39) and (5-46), it clearly appears that the output pulse is multiplied by the Fourier transform of the initial distribution. For an initial distribution bandwidth γ_0, the storage duration T is thus limited by $1/\gamma_0$. This has already been observed experimentally. The bandwidth of the initial distribution thus constitutes a limitation for the storage duration $T\gamma_0 \ll 1$. Since the initial distribution width also affects the optical depth (see Eq. (5-44)) and thus the memory efficiency, there exits a trade-off between a long storage duration and a large memory efficiency.

Note that this is true for a two-level system. If other long-living levels are available, the storage time can be increased significantly by transferring the excitation to a second long-living state, e.g. a ground state hyperfine level B. Optimal broadening. Now consider that the initial distribution has been chosen. Can define the effective width of the initial distribution

$$\nu := \frac{g_0^2 N \wp^2 L}{c} \approx \frac{g_0^2 \wp^2 \rho_0(0) \gamma_0 L}{c} = \alpha_0 L \gamma_0 \tag{5-67}$$

which depends on the medium properties together with the initial distribution and thus has a fixed value. The optical depth after broadening is given by

$$\alpha L \approx \nu/\gamma \tag{5-68}$$

For a given pulse shape, we look for the optimal width of the broadened distribution with respect to the memory efficiency. First fix the shape of the atomic distributions to be the same as the Lorentzian shape of the light pulse

$$\widetilde{E}_{\mathrm{f}}^{\mathrm{in}}(0,\omega)=\frac{\Gamma}{2\pi}\frac{1}{\Gamma^2/4+\omega^2} \tag{5-69}$$

$$G_0(\Delta_0)=\frac{\gamma_0}{2\pi}\frac{1}{\gamma_0^2/4+\Delta_0^2} \tag{5-70}$$

$$G'(\Delta')=\frac{\gamma}{2\pi}\frac{1}{\gamma^2/4+\Delta'^2} \tag{5-71}$$

For a given value of the effective width of the initial distribution ν, calculate the memory efficiency by varying the broadened distribution width with respect to the pulse bandwidth. In the limit $\gamma_0 \to 0, \rho_0(0) \to +\infty$ such that ν has a finite value, the functions $J(\omega;0)$, $F(\omega)$ and $H(\omega)$ take the simple forms

$$\begin{cases} J(\omega;0)=\dfrac{\gamma}{\gamma^2/4+\omega^2} \\ H(\omega)=F(\omega)=\dfrac{\gamma/2+\mathrm{i}\omega}{\gamma^2/4+\omega^2} \end{cases} \tag{5-72}$$

The output pulse (5-37)~(5-38) is given by

$$\widetilde{E}_{\mathrm{b}}^{\mathrm{out}}(0,\omega)=-\frac{\Gamma}{2\pi(\Gamma^2/4+\omega^2)}\left(1-\exp\left(-\frac{\nu\gamma}{\gamma^2/4+\omega^2}\right)\right) \tag{5-73}$$

$$\widetilde{E}_{\mathrm{f}}^{\mathrm{out}}(L,\omega)=-\frac{\Gamma}{2\pi(\Gamma^2/4+\omega^2)}\frac{\nu\gamma}{\gamma^2/4+\omega^2}\times\exp\left(-\frac{\nu\gamma}{2(\gamma^2/4+\omega^2)}\right)\mathrm{sinc}\left(\frac{\nu\omega}{\omega^2+\gamma^2}\right) \tag{5-74}$$

when propagating in the backward and forward (with the hypothesis Γ smaller than c/L) directions respectively. In Figure 5-13 (top), take $\nu=2\Gamma$ and we plot the efficiency as a function of the broadened distribution bandwidth γ. For large γ, the optical depth (5-40) which is inversely proportional to γ, is small. As shown in Figure 5-13 (top), the memory efficiency for the backward protocol is close to the efficiency of the forward protocol. For smaller values of γ, the efficiency increases and reaches a maximum for the forward process caused by the reabsorption while the efficiency of the backward protocol continues to grow until a value close to 100%.

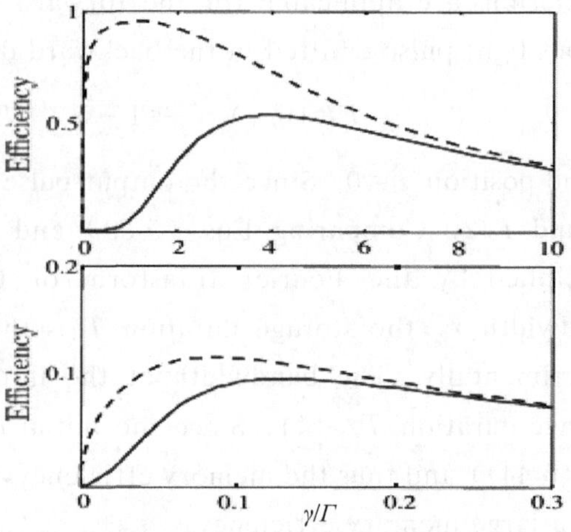

Fig. 5-13 (Color online) Efficiency of the light pulse reemission for forward (full blue line) and backward (dashed black line) protocol as a function of the broadened distribution bandwidth γ (in units of the pulse bandwidth Γ) for effective widths of the initial distribution $\nu=2\Gamma$ (top) and $\nu=0.05\Gamma$ (bottom).

When γ tends to zero, the bandwidth of the broadened distribution is too small to absorb the light pulse and the memory efficiency tends to zero for both backward and forward processes. In both cases, however, the optimum efficiency occurs when the broadened distribution bandwidth and the light bandwidth are of the same order. Taking a smaller value for $\nu = 0.05\Gamma$, we plot in Figure 5-13 (bottom) the emission efficiency versus the distribution bandwidth. The efficiency curves exhibit similar shapes as in the large ν case except that their maxima reach smaller values. These maxima are obtained for a narrower broadened distribution bandwidth than the pulse bandwidth.

5.3.2 Analysis of results

To insure a high efficiency of the CRIB protocol, the optical depth after broadening has to satisfy $\alpha L > 1$. From Eq. (5-48), deduce that the effective width of the initial distribution has to be at least of the order of the broadened distribution bandwidth: $\nu > \gamma$. Thus, for a pulse bandwidth $\Gamma \approx \nu$, one can broaden the initial distribution until $\gamma \approx \nu$ and thus achieve high efficiency. If $\nu \gg \nu$, it is more advantageous to choose a smaller broadening than the pulse width $\gamma \approx \nu \ll \Gamma$. In this case, however, one must expect strong distortion of the output pulse due to filtering effects.

Here investigate how the shape of the broadened distribution influences the memory efficiency. To keep the Lorentzian shape of the light pulse but we take a constant distribution in a given interval

$$\tilde{E}_{\mathrm{f}}^{\mathrm{in}}(0, \omega) = \frac{\Gamma}{2\pi} \frac{1}{\Gamma^2/4 + \omega^2} \tag{5-75}$$

$$G_0(\Delta_0) = \frac{1}{\gamma_0} \theta(\Delta_0 + \gamma_0/2) \theta_0(\gamma_0/2 - \Delta_0) \tag{5-76}$$

$$G'(\Delta') = \frac{1}{\gamma} \theta(\Delta' + \gamma/2) \theta(\gamma/2 - \Delta') \tag{5-77}$$

For a fixed value of the effective width of the initial distribution ν, calculate the memory efficiency and compare it to the efficiency obtained for the Lorentzian shapes of Eqs. (5-69)~(5-70).

In the limit $\gamma_0 \to 0$, the functions J, F and H take the following forms:

$$J(\omega; 0) = \frac{2\pi}{\gamma} \theta(\omega + \gamma/2) \theta(\gamma/2 - \omega) \tag{5-78}$$

$$F(\omega) = H(-\omega)^* = \frac{J(\omega)}{2} + \frac{\mathrm{i}}{\gamma} \log \frac{\gamma + 2\omega}{\gamma - 2\omega} \tag{5-79}$$

such that the light pulse generated in the backward direction (5-38) is given by

$$\tilde{E}_{\mathrm{b}}^{\mathrm{out}}(0, \omega) = -\frac{1}{2\pi}(1 - \mathrm{e}^{-\frac{2\pi\nu}{\gamma}}) \frac{1}{\Gamma^2/4 + \omega^2} \tag{5-80}$$

for $\omega \in [-\gamma/2, \gamma/2]$ and 0 elsewhere. For $\Gamma \ll c/L$, the pulse generated in the forward direction is of the form

$$\tilde{E}_{\mathrm{f}}^{\mathrm{out}}(L, \omega) \approx -\frac{\nu}{\gamma} \mathrm{e}^{-\frac{\pi\nu}{\gamma}} \frac{\Gamma}{\Gamma^2/4 + \omega^2} \tag{5-81}$$

for $\omega \in [-\gamma/2, \gamma/2]$. From these expressions, plot the memory efficiency, to compare it to the efficiency we found for identical broadened distribution and light pulse shapes. Globally, the relative shapes of the broadened distribution and of the light pulse does not modify the memory efficiency. Locally, for small distribution bandwidths, the memory efficiency is a bit larger for identical shapes. However, when the distribution bandwidth increases, the efficiency is higher when these shapes are different.

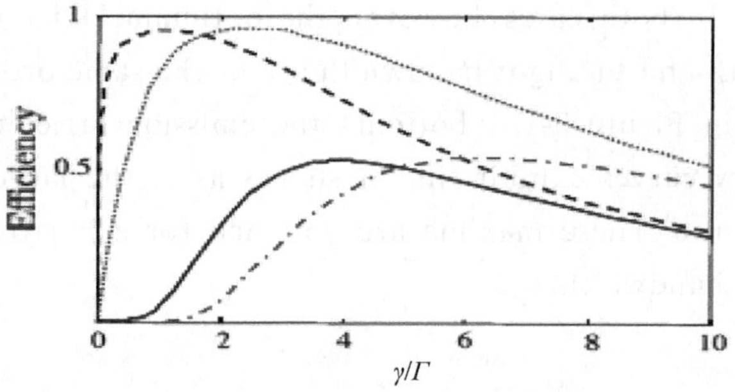

Fig. 5-14 (Color online) Comparison of memory efficiencies for a spectral distribution of atoms and light pulse with identical shapes (forward: full blue line, backward: dashed black line) and with different shapes (forward: dashed-dotted red line, backward: dotted red line) as a function of the inhomogeneous bandwidth (in units of the pulse bandwidth). The effective width of the initial distribution ν is taken equal to 2Γ.

To have obtained the explicit solution of the equations of motion for the system atoms plus light in the weak excitation regime, making it possible to gain insight into the dependence of the memory efficiency on the optical depth, and on the width and shape of the atomic spectral distributions. Furthermore, introduced a simplified CRIB-based memory protocol, which does not require the use of any additional laser field, and showed that its efficiency is limited to 54%. For both the complete and the simplified CRIB protocols, the interaction with the atoms does not induce distortion of the stored light pulse. To have shown that the storage duration is limited by the initial spectral distribution of the atoms. This distribution also defines the absorption of the medium which determines the memory efficiency. There thus exists a trade-off between the storage duration and the memory efficiency. For a given pulse shape, the optimal broadening of the initial distribution is of the order of the effective bandwidth of the initial atomic distribution, defined as the product of the atomic density and the width of the initial atomic distribution. All so shown that the shape of the broadened distribution does not dramatically change the memory efficiency.

5.3.3 Control and releasing of photon

The capable of controllably storing and releasing a photon, are a crucial component for quantum computers and quantum communications. So far, quantum memories have operated with bandwidths that limit data rates to MHz. The coherent storage and retrieval of sub-nanosecond low intensity light pulses with spectral bandwidths exceeding 1GHz in cesium vapor. The novel memory interaction takes place via a far resonant two-photon transition in which the memory bandwidth is dynamically generated by a strong control field. This allows for an increase in data rates by a factor of almost 1000 compared to existing quantum memories. The memory works with a total efficiency of 15% and its coherence is demonstrated by directly interfering the stored and retrieved pulses. Coherence times in hot atomic vapors are on the order of microseconds the expected storage time limit for this memory.

Photons are ideal carriers of quantum information, they have a very large potential information capacity, and do not interact with one another, so that information coded in them is robust. Recent developments in sources, detectors, gates and protocols have laid the ground for the construction of large-scale photonic quantum computers with unique capabilities, as well as intercontinental quantum networks that are immune to eaves-dropping. However, the elects of photon loss and the inherently probabilistic character of some of these functions demand the ability to store photons. The photonic networks will only produce the desired results rarely is overcome if photon storage is possible, since this allows complex protocols to be orchestrated by holding the output of successful computations until all operations have been correctly executed. Quantum memories are therefore an active area of research, with great interest focused on reversibly mapping photons into collective atomic excitations.

Key characteristics for quantum memories are long storage time, high memory efficiency, the ability to store multiple modes (i. e. multiple distinct photons) and high band-width. High bandwidth allows the storage of temporally short photons, enabling quantum information to be processed at a higher "clock rate", increasing the number of computational cycles that can be completed before decoherence sets in. This can be difficult to achieve with atomic memories, since photons must be stored in long-lived atomic states with narrow linewidths.

Previously implemented memory protocols include electromagnetically induced trans-parency (EIT), controlled reversible inhomogeneous broadening (CRIB) and atomic frequency combs (AFC). EIT based memories utilize the extreme dispersion of an induced transparency window to modify the group velocity, and controllably stop, store and retrieve light pulses. CRIB is a photon echo technique that uses an inhomogeneous broadening of the atomic resonance. Reversing this broadening during the readout process causes the atomic spins to rephase and collectively reemit the original signal. In the AFC protocol an artificially created atomic frequency comb absorbs the incident signal, and the periodic structure of the absorption spectrum results in a subsequent rephasing and re-emission of the stored signal. These protocols are resonant of light storage has also been implemented via four-wave mixing, stimulated Brillouin scattering and via the gradient echo memory (GEM) protocol. For all these protocols typical storage times range from μs to ms and achieved efficiencies from 1% to 15%, although two experiments have reported higher values for either storage time or efficiency.

The reported bandwidths range from kHz to MHz. In this letter we present the experimental demonstration of a coherent, efficient and broadband Raman memory for light. In a Raman memory, the bandwidth is generated dynamically by ancillary write/read pulses, which dress the narrow atomic resonances to produce a broad virtual state to which the signal field couples. The resonant nature of the scheme confers some appealing features.

Such light-matter entanglement operations are primitives for the construction of quantum repeaters and for enabling coherent logical information storage. In current experiment, used signal pulses containing several thousand photons. However, because the memory interaction is linear and coherent, the Raman protocol is a genuine quantum memory that would also work in the single photon regime.

In the experiment a strong write pulse and a weak signal pulse, both broadband, are spatially and temporally overlapped and sent together into a cesium vapor cell where the Raman interaction with the storage medium takes place (see Figure 5-15). The signal pulse is mapped via a two-photon transition with the write pulse into a collective atomic excitation called a spin wave. At a later time a strong read pulse is sent into the vapor cell and converts the spin wave into an optical output signal that is measured on a fast detector.

The $F=3,4$ 'clock states' of the ground-level hyperfine manifold serve as the states $|1\rangle$ and $|3\rangle$, which are connected to the excited state 1 (the $6^2P_{3/2}$ manifold) via the D_2 line at 852nm (supplementary information). The cesium is heated to 62.5℃, so that the resonant optical depth

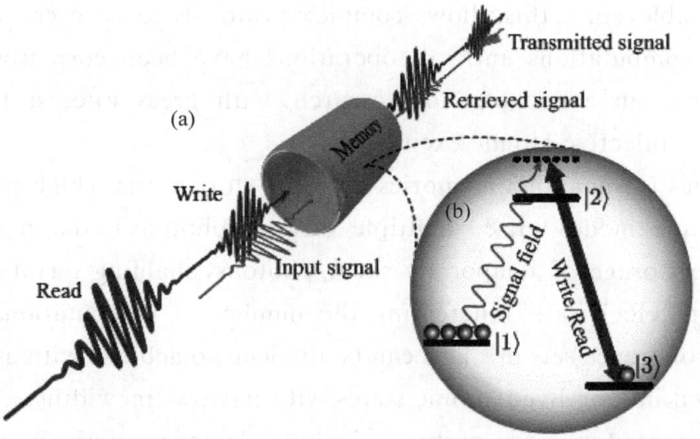

Fig. 5-15 (a) Raman memory. The signal is directed into the memory along with a bright write pulse and is stored. If the storage is partial, any unstored signal is transmitted through the memory. (b) The Λ-level structure of the atoms in the memory. The atoms are prepared in the ground state $|1\rangle$ by optical pumping. The signal is tuned into two-photon resonance with the write field; both are detuned from the excited state 1. Absorption of a signal photon transfers an atom from $|1\rangle$ into the storage state $|1\rangle$ via Raman scattering stimulated by the write field. Upon retrieval the interaction is reversed.

$d \approx 1800$ associated with this transition is high. Experimental data for the storage and retrieval processes are displayed in Figure 5-15. The storage of a signal pulse takes place at time $t=0$ and the retrieval of the stored information is carried out 12.5ns later. The storage and retrieval efficiencies depend on the write and subsequent read pulse extracts the stored excitation, which emerges along with the transmitted read pulse.

5.3.4 Energy control

- Read pulse energy

If read pulse energy is zero, 100% of the incident signal field is transmitted this contrasts with resonant storage protocols, in which the memory becomes absorbing when "inactive". Increasing the write and read pulse energy decreases the transmitted fraction of the incident signal and increases the retrieved signal. The inset in Figure 5-16 clearly shows the short pulse duration of the retrieved signal. The measurement is limited by the response time of the detector, which is 1ns, corresponding to a bandwidth of 1GHz. In practice, our Raman memory should be operating at the full control field bandwidth of 1.5GHz. The time-bandwidth product N of a memory quantise the number of distinct time bins available for computational operations in a hypothetical quantum processor using the memory. The storage time of the memory is limited to several hundred ns.

Precise measurement of the pulse shapes requires greater temporal resolution than the cur-rent detector provides, so to apply the theory assume Gaussian temporal profiles for all pulses; the timing and duration of the signal pulse are then adjusted to account for the dispersive effect of the etalons used to spectrally filter the signal. The observations stand in good agreement with the theory.

The retrieval efficiency $\eta_{\text{ret}}/\eta_{\text{tot}} = \eta_{\text{store}}$ is significantly larger than the storage effect as Figure 5-17.

A broadband quantum-capable memory by coherently storing and retrieving signal pulses with bandwidths greater than 1GHz. This is an increase of almost a factor of 1000 compared to existing quantum memories. Storage efficiencies up to 30% and retrieval efficiencies as high as 50% were

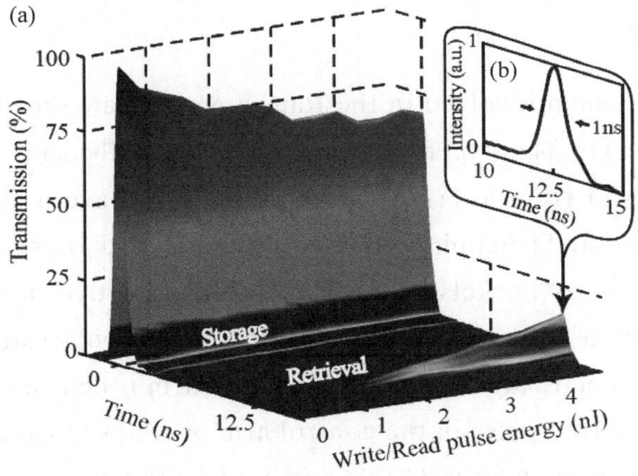

Fig. 5-16 (a) Storage ($t = 0$ns) and retrieval ($t = 12.5$ns) of light pulses vs. write/read pulse energy. With no write pulse present (i.e. 0 nJ), there is 100% transmission; with the highest write/read pulse energies (4.8nJ) this drops to 70%, indicating that 30% of the incident signal is stored. At $t = 12.5$ns 50% of the stored information is retrieved giving a total memory efficiency of 15%. (b) Zoom of retrieved signal field showing the measured full width at half maximum (FWHM) temporal duration of 1ns, limited by the detector response time.

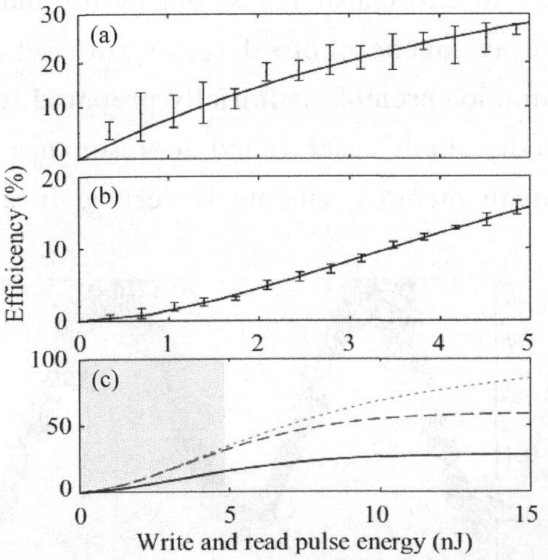

Fig. 5-17 Dependence of memory efficiency on write/read pulse energy. (a) Storage efficiency. (b) Total efficiency. Dots and error bars indicate experimental data solid lines represent predicted theory. (c) Theoretical prediction for total efficiency extrapolated to higher pulse energies. Solid line: Efficiency for current experimental configuration with forward readout. Dashed line: Optimal efficiency using forward retrieval, limited by re-absorption. Dotted line: Optimal efficiency using phase matched backward retrieval. The shaded area denotes the range of pulse energies accessible with the present experiment, efficiency of store, since the total efficiency η_{tot} exceeds store.

observed. If increasing the power of the control pulses should allow further improvements. The excellent coherence of the memory was directly verified by interfering the stored and the retrieved pulses. Such high-speed memories will form the basis of fast, controllable and robust photonic quantum information processors in the near-future.

5.3.5 Methods

The atomic states of cesium involved in the Raman protocol are shown in part (a) of Figure 5-18 (supplementary information). The experimental layout is described in part (b). The read, write and signal pulses are derived from a Ti:Sapph oscillator and have a FWHM of 300ps(1.5GHz bandwidth). The fundamental Ti:Sapph laser frequency is tuned 18.4GHz to the blue of the 1-1 transition in Figure 5-18(a). A Pockels cell selects two consecutive pulses, separated by 12.5ns. The laser beam is split into a strong control arm with vertical polarization (1) and a very weak signal arm with horizontal polarization ($). The control arm is delayed by 12.5ns with respect to the signal arm such that the first pulse in the control arm overlaps in time with the last pulse in the signal arm. An electro-optic modulator (EOM) is used in the signal arm to generate sidebands 9.2GHz shifted from the fundamental laser frequency. After spectral filtering with Fabry Perot etalons only the 9.2GHz red-shifted sideband corresponding to the $|1\rangle$-$|2\rangle$ transition is transmitted and used as the signal field. The control and the signal beam, spectrally separated by 9.2GHz, are recombined and made collinear. They are focused with a beam waist of 350μm into the 7cm long vapor cell filled with cesium and 20 torr of neon buffer gas. Polarization-and spectral filtering are used after the cell to extinguish the strong write and read pulses and transmit the signal field only. A high-speed avalanche photo detector with a bandwidth of 1GHz detects the very weak signal pulse. The atomic ensemble is initially prepared in the ground state 1 by optical pumping using an external cavity diode laser tuned to resonance with the 1-1 transition. The underlying theory for the Raman memory scheme is derived in and briefly summarized in the supplementary section.

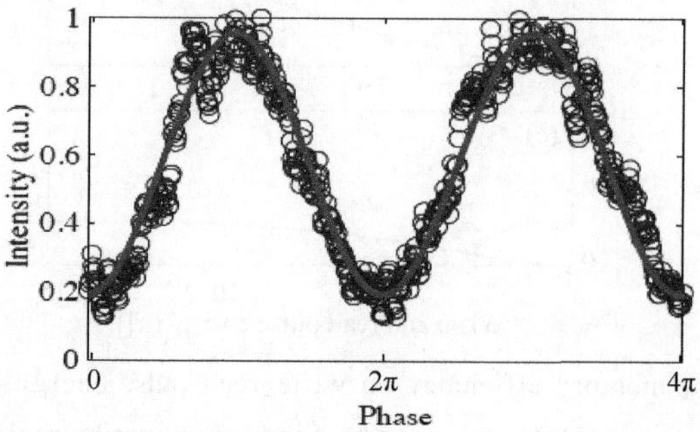

Fig. 5-18 Raw interference and fit of stored and retrieved signal. Circles indicate experimental data, the red curve is a least squares fit. A linear scan of the path length difference in the interferometer results in sinusoidal oscillations of the total intensity. The high visibility of 86% (normalized for interferometer instability) of the interference demonstrates the coherence of the memory, and matches with theoretical calculations, suggesting that the memory interaction is perfectly coherent. Because we retrieve the stored signal after just 12.5ns (the round-trip time of our oscillator), the efficiency we observe is not limited by decoherence, which is only significant over much longer timescales. Instead, it is a direct probe of the intrinsic efficiency of the Raman memory interaction. In addition, it is easy in this configuration to delay a copy of the signal pulse and interfere it directly with the retrieved pulse, to demonstrate the coherence of the interaction. The fringe visibility of 86.5% indicates that the memory is highly coherent. In fact, the theoretical model (supplementary information) predicts a distortion of the retrieved field due to dispersion and Stark shifts (these can be eliminated with backward retrieval), yielding a maximum visibility of 83%. This suggests that the memory interaction is perfectly coherent.

To investigate the coherence properties of the memory, a copy of the incident signal field is attenuated, delayed and overlapped with the retrieved signal in a Mach-Zehnder configuration. Correct for imperfections in the interferometer by interfering the signal and a replica as a benchmark; this yields a visibility of 86.5% (67% uncorrected) for the memory.

The storage and retrieval efficiencies are calculated by solving the semi-classical linearized Maxwell-Bloch equations for the system. The signal field, with amplitude A, propagates through an ensemble of Λ-level atoms in the presence of the write field, whose temporal profile is described by the time-dependent Rabi frequency $\Omega(\tau)$. Provided that the detuning Δ is the dominant frequency in the problem, the optical polarization can be adiabatically eliminated, yielding an explicit expression for the spin wave amplitude B at the end of the storage interaction. The spin wave can be expressed as

$$B_{\text{mem}}(z) = \int_{-\infty}^{\infty} f(\tau) J_0[2C\sqrt{(1-\omega(\tau))z}\,] A_{\text{in}}(\tau)\mathrm{d}\tau \tag{5-82}$$

where A_{in} is the amplitude of the incident signal field to be stored, J_0 is a Bessel function, and the number the

$$C^2 = \mathrm{d}\gamma W/\Delta^2 \tag{5-83}$$

Raman memory coupling, with d the resonant optical depth and the homogeneous linewidth of the excited state 1. Here introduced the dimensionless integrated Rabi frequency

$$\omega(\tau) = \frac{1}{W}\int_{-\infty}^{\tau} |\Omega(\tau')|^2 \mathrm{d}\tau' \tag{5-84}$$

and the normalized Stark-shifted Rabi frequency

$$f(\tau) = C e^{\mathrm{i}W\omega(\tau)/\Delta}\Omega(\tau)/\sqrt{W} \tag{5-85}$$

The constant W, proportional to the control pulse energy, is defined so that $\omega(\infty)=1$, while the longitudinal coordinate z is normalized so that $z=1$ represents the exit face of the ensemble.

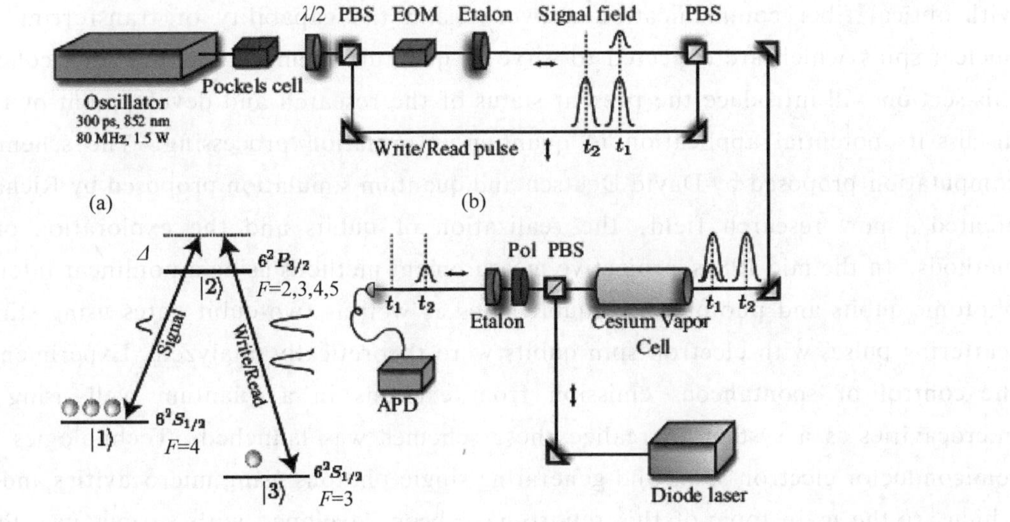

Fig. 5-19 (a) Λ-level scheme for Raman memory. (b) Experimental setup. A cesium vapor cell is optically prepared with a diode laser. In time bin t_1 an incoming signal pulse is mapped by a strong write pulse into a spin wave excitation in the atomic cesium ensemble.

The spin wave can be expressed as

$$N_{\text{mem}} = \int_0^1 |B_{\text{mem}}(z)|^2 \tag{5-86}$$

The storage efficiency $\eta_{\text{store}} = N_{\text{mem}}/N_{\text{in}}$ is the ratio of the number of final spin wave excitations to the number of incident signal photons. If upon retrieval:

$$N_{\text{in}} = \int_{-\infty}^{\infty} |A_{\text{in}}(\tau)|^2 \mathrm{d}\tau \tag{5-87}$$

And

$$N_{\text{out}} = \int_{-\infty}^{\infty} |A_{\text{out}}(\tau)|^2 \mathrm{d}\tau \tag{5-88}$$

photons are recovered from the memory, the total efficiency $\eta_{\text{tot}} = N_{\text{out}}/N_{\text{in}}$ can be computed using the formula:

$$A_{\text{out}}(\tau) = f^*(\tau)\int_0^1 J_0[2C\sqrt{\omega(\tau)(1-z)}]B_{\text{mem}}(z)\mathrm{d}z \tag{5-89}$$

For the retrieved signal field. In general, the field A_{out} is different to A_{in} physically these differences originate in the dispersion associated with propagation through the ensemble, and the Stark shift due to the time-varying control field. The visibility of interference between the incident and retrieved signals is therefore smaller than 1, even for a perfectly coherent memory. If optimal storage and backward retrieval is possible, these distortions can, however, be eliminated.

5.4 Photonic quantum controlle memory function

The scheme of directly controlling electron spins trapped in semiconductor quantum dots or donor impurities as qubits using optical pulses has various advantages, such as the achievements of local excitation and fast operation, low power consumption, easy implementation of an interface with optical fiber communication networks, and the capability of transferring information to nuclear spins, which are expected to serve as quantum memories with a long coherence time. In this section will introduce the present status of the research and development of this scheme and discuss its potential application to quantum information processing. The schemes of quantum computation proposed by David Deutsch and quantum simulation proposed by Richard P. Feynman created a new research field: the realization of qubits and the exploration of their control methods. In the mid 1980s, projective measurement methods using a nonlinear interferometer with photonic qubits and photonic two-qubit gates as well as two-qubit gates using stimulated Raman scattering pulses with electron spin qubits were theoretically analyzed. Experimental research on the control of spontaneous emission from excitons in a quantum well using semiconductor microcavities as a system to realize these schemes was launched. Technologies for controlling semiconductor electron spins and generating single photons using microcavities and optical pulses, which are the main topic of this report, have been developed with various new technologies and new schemes over the last quartercentury.

The goal of this research field is to realize a large-scale quantum information processing system that can be constructed by devices with a finite yield and is fault-tolerant to malfunction and noise (decoherence). The targets of this research that have been particularly focused on are the realization of quantum repeaters that enable longdistance or multi-party quantum communication network by repeating the creation, purification, and swapping of quantum entanglement and measurement based quantum computation, in which an entangled quantum state, called the topological cluster state or surface code, is created in quantum memories, and then quantum computation is performed with repeated projective measurement of the entangled state. To realize such a quantum information processing system, the following qubit technologies must be realized at a high fidelity (the degree of closeness between the target and actually produced quantum states).

(1) Initialization and projective measurement.

(2) Single-qubit control.

(3) Two-qubit control.

(4) Creation and purification of entangled quantum states (implementation of interface for quantum communication channel).

(5) Quantum memory function (long coherence time).

(6) Semiconductor quantum dots controlled by optical pulses have the following unique characteristics as artificial atoms.

The oscillator strength of excitons in semiconductor quantum dots is greater than that of free electron-hole pairs without electron correlation ($f_{\text{exciton}} = f_{\text{e-h}} \times a^2/a_{\text{B}}^{*2}$). Here a is the quantum dot radius and a_{B}^{*2} is the exciton Bohr radius. A quantum dot comprises many electrons in a filled band, and excitons can combine the oscillator strengths of all these electrons, forming a large electric dipole. Such a large oscillator strength of excitons enables the fast control and projective measurement of electron spins using low-energy ultrashort optical pulses.

Semiconductor quantum dots can be grown at fixed positions on a two-dimensional square lattice, which can be sandwiched between two distributed-feedback reflectors placed at an interval of half a wavelength in a monolithic planar microcavity. Excitons in quantum dots can be effectively coupled to the cavity mode of a planar microcavity.

Entangled states of electron spins can be simultaneously created between multiple pairs of adjacent electron spins by two-qubit operation in this cavity mode. The emission wavelength of excitons in quantum dots can be easily taylored to the 1.5μm band, which is used for optical fiber communication channels. Therefore, it is possible to implement an interface function that can transfer semiconductor spin qubits and 1.5μm-band photon qubits via planar microcavities without wavelength conversion. In this section will explain the present status of device technologies based on the control of quantum dot-planar cavity systems using optical pulses and discuss their potential application to quantum information processing systems.

5.4.1 Electron spins in quantum

Dots and Photons in Planar Microcavities Figure 5-20(a) and Figure 5-20(b) show two typical configurations of quantum dots. In Figure 5-20(a), InGaAs quantum dots are selfassembled at random positions on a flat GaAs substrate. For use in devices, these quantum dots are confined in pillar microcavities (see Figure 5-20(c)), which are fabricated from planar microcavities composed of two distributed-feedback reflectors. A single electron is injected and trapped in a quantum dot from the nearby donor impurity doped layer.

The quantum dot-microcavity system has disadvantages, for example, the positions of quantum dots as well as the exciton emission wavelengths are difficult to control and a complicated optical system is required to couple quantum dots with each other. In Figure 5-20(b), InGaAs quantum dots are arrayed on a square lattice. These quantum dots can be placed in a planar microcavity composed of two distributed-feedback reflectors in practical devices, as shown in Figure 5-20(d). In this system, a homogeneous cavity-mode frequency can be easily achieved in the plane of the planar microcavity; therefore, single-qubit gates and two-qubit gates between an arbitrary pair of adjacent quantum dots can be implemented on the basis of the principle described later, by appropriately selecting the irradiation position, angle of incidence and wavelength of optical pulses.

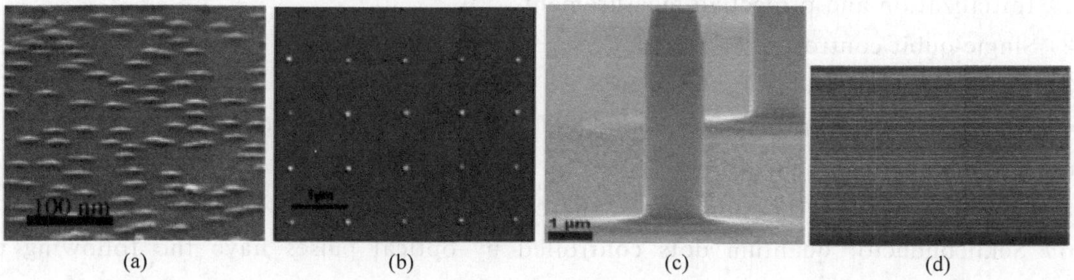

Fig. 5-20 (Color online) (a), (c) Fabricated device in which self-assembled InGaAs quantum dots are confined in pillar microcavities. (b), (d) Fabricated device in which InGaAs quantum dot array on two-dimensional square lattice is confined in planar microcavity.

Zeeman splitting of electron spins and charged excitons (trions) in quantum dots Figure 5-20(a) shows the energy states of a quantum dot doped with a single electron when a DC magnetic field is applied. The direction of the DC magnetic field is perpendicular to the growth direction of the quantum dot (Voigt configuration). The energy level of the electron is split into two, with spin directions parallel and antiparallel to the DC magnetic field (Zeeman splitting). The energy level is given as $\varepsilon_e = \mu_e g_e B_{ext}$, where μ_e is the Bohr magneton, g_e is the g-factor of the electron spin, and B_{ext} is the intensity of the DC magnetic field. The g-factor of an electron spin in vacuum is 2.0013. however an electron spin in an InGaAs quantum dot has a different g-factor because it undergoes strong spin-orbit interactions. For the electron spin in the InGaAs quantum dot shown in Figure 5-20 Zeeman sublevels of this electron spin constitute a qubit. When an electron-hole pair is injected into a quantum dot in which a single electron is already trapped, a trion state, is created Figure 5-20(a). The total spin of the electrons of the trion is zero (singlet state). The energy level of the trion is determined by the Zeeman splitting of the heavy hole spin and is given by $\delta_h = \mu_e g_h B_{ext}$. The g-factor of the hole was $g_h = -0.30$ for the InGaAs quantum dot in Figure 5-20. Upon the Zeeman splitting of the electron and trion, the spectral line of the quantum dot is split into four when the DC magnetic field is increased, as shown in Figure 5-20(b).

Each spectral line is linearly polarized in the quantum-dot plane (x, y) following the spin (polarization) selection rule (see Figure 5-20(b)). Along with the energy levels of two trions, the Zeeman level of the electron spin used as the qubit forms two Λ-type three-level systems. Λ-type three-level systems are the key to realizing fast optical control of electron spins.

5.4.2 Enhancement of excitonic spontaneous emission

The coupling of a single quantum dot with a cavity in resonance can be confirmed by observing photon antibunching, as shown in Figure 5-21(a). Figure 5-21(b) shows changes in the photoluminescence (PL) spectra obtained when the excitonic emission wavelength is varied with respect to the cavity-mode wavelength by changing the temperature of this device. When the two wavelengths coincide (i.e., they are in resonance), vacuum Rabi splitting is observed in the PL spectra, indicating that this system is in a moderately strong coupling regime. For this device, the cavity-mode volume was approximately 20 $(\lambda/n)^3$, the cavity Q-value was approximately 1.5×10^4, and the enhancement factor of the spontaneous emission rate (so-called the Purcell factor, γ/γ_0) was approximately 200. Here, $\gamma=(15\text{ps})^{-1}$ and $\gamma_0=(3\text{ns})^{-1}$ are the spontaneous emission rates when the microcavity is resonant and off-resonant, respectively. The fidelity of quantum information processing using the various types of optical pulses described later depends on how large the Purcell factor is, rather than whether the quantum dot-microcavity system is in a strong or weak coupling regime. Two factors contribute to the enhanced excitonic spontaneous emission of the quantum dot in Figure 5-22 (with a Purcell factor of approximately 200). One is the enhancement of the oscillator strength of excitons in accordance with the relation $f_{\text{exc}}=f_{\text{e-h}}\times a^2/a_{\text{B}}^{*2}$. This is partly why the excitonic emission lifetime was 15ps (confirmed in an experiment using a streak camera) with a cavity, while the emission lifetime of free electron-hole pairs without electron correlation was 3ns. The other is the effect of the cavity enhanced vacuum field strength [cavity quantum electrodynamics (QED) effect]. This is also responsible for the excitonic emission lifetime of 15ps when a cavity was used (also confirmed in an experiment using a streak camera). As the quantum dot radius a increases and ka approaches 1 ($k=2\pi n/\lambda$ is the wavenumber in a semiconductor), an excitonic emission pattern becomes focused in the direction perpendicular to the quantum-dot plane. A reduction in the solid angle of the radiation pattern indicates that the number of electromagnetic-field modes suitable for emission is decreasing, which implies that the enhancement factor of oscillator strength, i.e., $a^2=a_{\text{B}}^{*2}$, is cancelled by the decreased state density of the field. Under the extreme condition of $k_a\gg1$, the ratio of the spontaneous emission rate of excitons (γ_{exc}) to that of free electron-hole pairs ($\gamma_{\text{e-h}}$), i.e., $\gamma_{\text{exc}}/\gamma_{\text{e-h}}$, becomes saturated at a constant value given by $6\lambda/(\pi^2 n a_{\text{B}}^{*2})$ For GaAs quantum dots, the saturated decay rate is independent of a and is estimated to be $\gamma_{\text{exc}}\approx(20\text{ps})^{-1}$, as shown in Figure 5-22 This theoretical prediction has been experimentally verified. Because the radiation pattern becomes focused in the direction perpendicular to the quantum-dot plane, as mentioned above, when the quantum dots are embedded in a planar microcavity, the spontaneous emission rate further increases owing to the effect that the planar microcavity realizes the increase of the vacuum-field density in the perpendicular direction, as

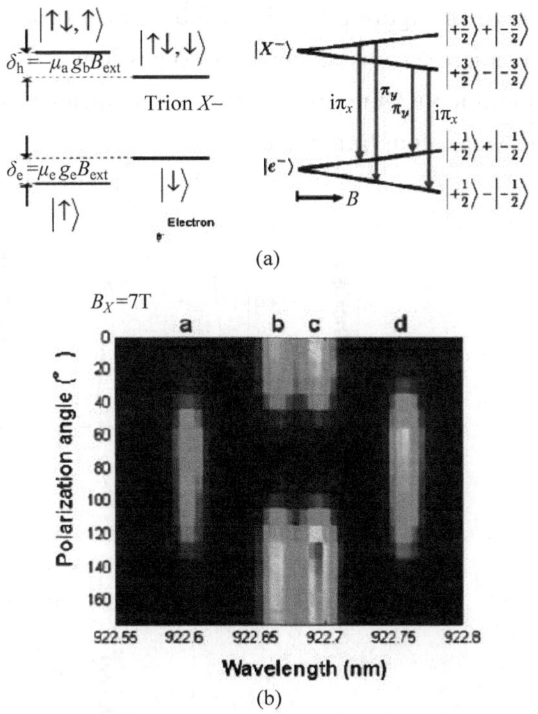

Fig. 5-21 (Color online) (a) Zeeman splitting and spin (polarization) selection rule for ground state (electron) and excited state (trion) in InGaAs quantum dot doped with an electron. (b) Experimental result demonstrating that four split spectral lines follow the polarization selection rule (orthogonal linear polarization).

shown in Figure 5-22. Figure 5-22 also shows the photon lifetime of the planar microcavity. When the mirror transmittance becomes lower than a certain value, the photon lifetime exceeds the spontaneous emission lifetime, indicating that the system enters the strong coupling regime from the weak coupling regime.

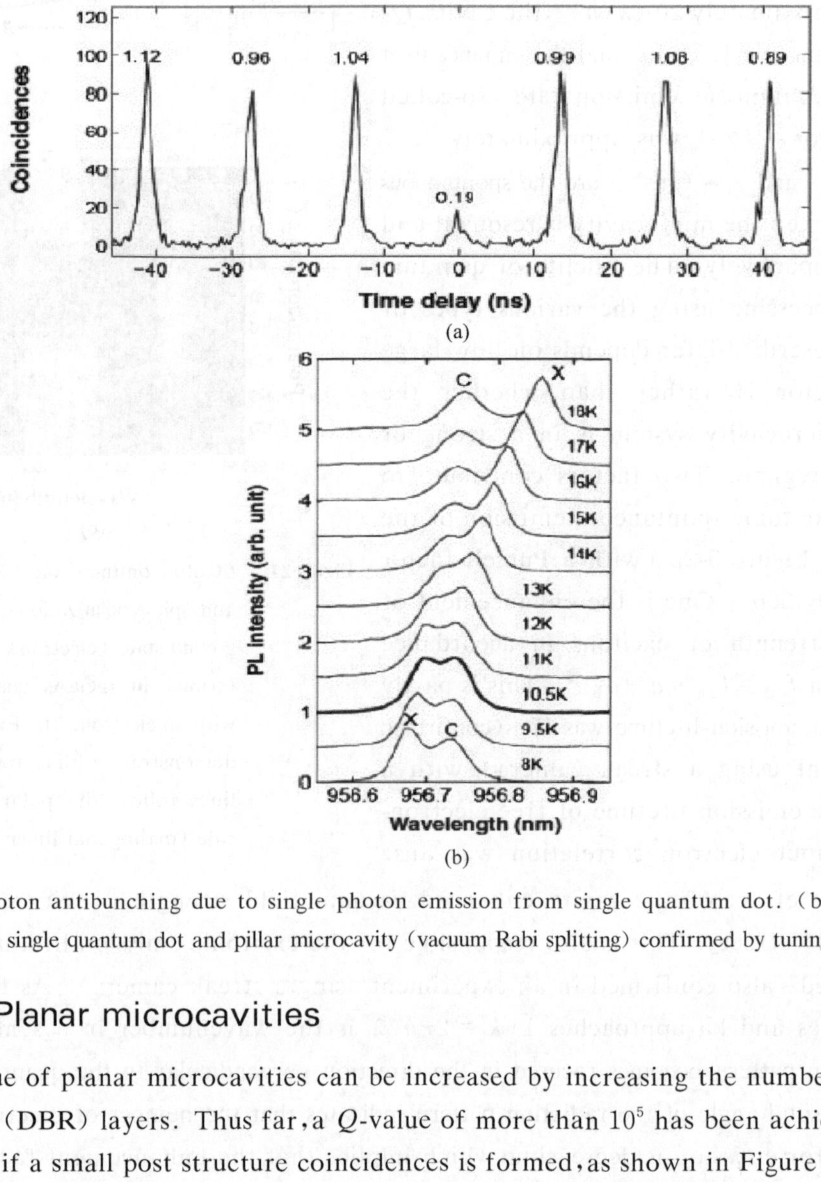

Fig. 5-22 (a) Photon antibunching due to single photon emission from single quantum dot. (b) Strong coupling between single quantum dot and pillar microcavity (vacuum Rabi splitting) confirmed by tuning of temperatures.

5.4.3 Planar microcavities

The Q-value of planar microcavities can be increased by increasing the number of distributed Bragg reflector (DBR) layers. Thus far, a Q-value of more than 10^5 has been achieved but the Q-value decreases if a small post structure coincidences is formed, as shown in Figure 5-23. Although planar microcavities do not have the function of lateral optical confinement, they form a circular spatial mode with a finite size because of the finite photon lifetime. The mode

$$r_{\mathrm{m}} = \frac{\sqrt{\lambda_0 L}(R_1 R_2)^{1/4}}{\sqrt{\pi n [1-(R_1 R_2)^{1/2}]}}$$

where λ_0 is the wavelength in vacuum, L is the effective cavity length including the depth of cavity field penetrationto DBRs, R_1 and R_2 are the reflectances of two DBRs, and n is the effective refractive index of the microcavity. Since the Q-value of planar microcavities is given as

$$Q = \frac{\lambda_0}{\Delta\lambda_{1/2}} = \frac{2\pi L n (R_1 R_2)^{1/4}}{\lambda_0[(1-R_1 R_2)^{1/2}]}$$

there is a one-to-one relationship between Q and γ_m, i.e., $Q=(2\pi^2 n^2/\lambda_0^2)\gamma_m^2$.

When $Q=10^3$, 10^4 and 10^5, $\gamma_m=2$, 6.3 and $20\lambda_m$, respectively. Therefore, when quantum dots are arrayed at intervals of $2\lambda_m$ in a 5cm×5cm planar microcavity with $Q=10^3$, approximately 10^9 qubits composed of a microcavity-quantum dot pair can be independently formed on a single wafer. As described later, such a planar structure is also significantly advantageous when used for realizing two-qubit gates using optical pulses.

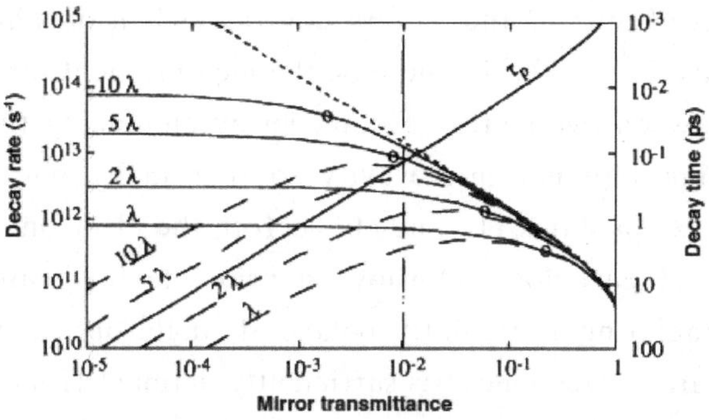

Fig. 5-23 Spontaneous emission decay rate of excitons in quantum dots and decay time of photons in microcavity vs mirror transmittance of halfwavelengthplanar microcavity. The diameter of quantum dots is used as a parameter. The dotted line represents the case of inhomogeneous excitonic broadening. (1) Decay rate. Mirror transmittance. Decay time.

Initialization and Measurement of Electron Spin Qubits Figure 5-24(a) shows the principle of optical pumping forinitialization of a qubit. A Λ-type three-level artificial atom is formed, in which two Zeeman sub-levels of an electronspin are $|0\rangle$ and $|1\rangle$ and the lowest energy level of a trion is $|e\rangle$. Upon the irradiation of narrow-band laser light tuned to the transition frequency between states $|1\rangle$ and $|e\rangle$, the electron spin state is finally initialized to the ground state $|0\rangle$ after several cycles of absorption and spontaneous emission of photons, because the spontaneous emission rate for $|0\rangle$ is equal to that for $|1\rangle$. Figure 5-24(b) shows an experimental result demonstrating the principle of such optical pumping. Assuming, as above, that the electron spin state $|1\rangle$ is excited to state $|e\rangle$ upon the absorption of pumping light and then decays to the ground state $|0\rangle$, a photon with an energy corresponding to the transition frequency between states $|0\rangle$ and $|e\rangle$ should be emitted. In Figure 5-24 (b), it is shown that the count ate of such photons decreases with increasing optical pumping time (corresponding to a process in which the electron spin state is initialized to state $|0\rangle$. The fidelity of the initialization to state $|0\rangle$ achieved during 13ns of optical pumping was 7%. In the initialization through a conventional thermal equilibrium process with phonon reservoirs, however, not only are a strong magnetic field and ultralow temperature required to satisfy $\delta_e=\mu_e g_e B_{ext} \gg k_B T$, but also, a time sufficiently longer than the spin relaxation time of electron spins T_1 (≥1ms) is required for the initialization to be completed.

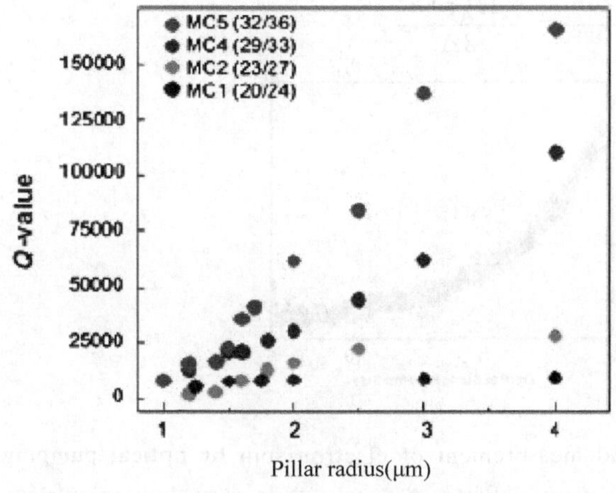

Fig. 5-24 (Color online) Q-value vs radius of pillar microcavity. The number of DBR layers is used as a parameter.

This principle of optical pumping can be used to the measurement of an electron spin. When a photon with an energy corresponding to the transition frequency between states $|0\rangle$ and $|e\rangle$ is counted upon the irradiation of pumping light, the initial state of the electron spin is determined to have been $|0\rangle$ and $|1\rangle$ In contrast, when no photon is counted, the initial state is determined to have been $|0\rangle$. To increase the measurement sensitivity, it is effective to achieve a cyclic transition between two states $|1\rangle$ and jei by enhancing the spontaneous emission between these two states using a planar microcavity so that many photons can be emitted at the transition frequency between states $|1\rangle$ and $|0\rangle$, before the electron spin state is finally initialized to state $|0\rangle$.

Figure 5-25(a) shows a principle for realizing a single-qubit gate for an electron spin by irradiating a circularly polarized single optical pulse. The center frequency of the optical pulse (carrier frequency) is sufficiently detuned from the transition frequencies between states $|0\rangle$ and $|e\rangle$ and between states $|1\rangle$ and $|e\rangle$. The coherent Rabi frequencies in these transitions are denoted as Ω_0 and Ω_1, respectively, and the Zeeman frequency of the electron spin is denoted as δ_e. When the condition $\delta_e \ll \Omega_0, \Omega_1 \ll \Delta_0, \Delta_1$ is satisfied, the electron spin automatically selects upper-and lower-side bands from a wide frequency range in the spectrum of the optical pulse and realizes nonresonant stimulated Raman transition. Here Δ_0 and Δ_1 are the largest and smallest detuning parameters. Such pairs of upper-and lower-side bands are continuously distributed within the spectrum of the optical pulse. When the phase differences between the upper and lower-side bands are constant among all pairs, constructive interference occurs in which the probability amplitude of the nonresonant stimulated Raman scattering is increased, resulting in a highly efficient transition (rotation) of the electron spin between states $|0\rangle$ and $|1\rangle$. The above condition that the phase difference between all upper-and lower-side bands is constant is automatically satisfied when the optical pulse is Fourier transform limited. The time dependent effective Rabi frequency between states $|0\rangle$ and $|1\rangle$ is approximately given by

$$\Omega_{\text{eff}}(t) \simeq \frac{\Omega_0(t)\Omega_1(t)^*}{2\Delta} = \frac{|\Omega(t)|^2}{2\Delta}$$

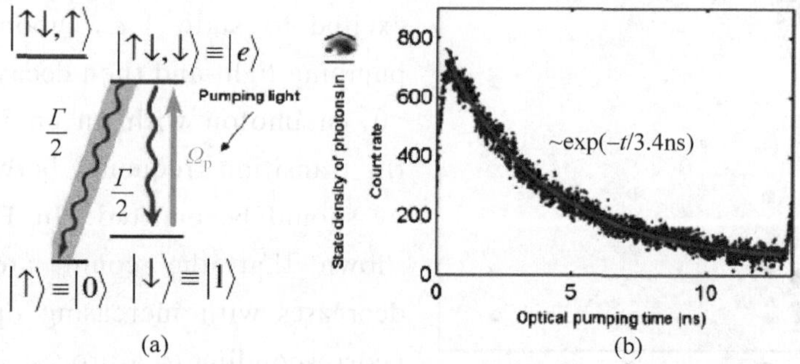

Fig. 5-25 (Color online) (a) Principle of initialization and measurement of electron spin by optical pumping. (b) Count rate of photons emitted upon transition between states $|0\rangle$ and $|e\rangle$ vs optical pumping time.

Here, the approximation $\Delta' \cong \Delta'_0 \cong \Delta_1$ is used. The rotation angle of the electron spin Φ is obtained by time integration of the Rabi frequency as

$$\Phi = \int dt \frac{|\Omega(t)|^2}{2\Delta} \propto \int dt P(t)$$

Because $P(t)$ is the power of the optical pulse at time t, Φ is determined only by the energy of the optical pulse. This enables accurate single-qubit gate operation without the need for accurate control of the width and shape of the optical pulse, which is a significant advantage from a practical viewpoint.

Single-qubit gate operation can be realized at a fidelity of 99.9% for both $\pi/2$ pulses (energy density of $5\mu J/cm^2$) and π pulses (energy density of $14\mu J/cm^2$) under the following circumstance;

the optical pulse width is 100fs, the phase relaxation time of the electron spin is $T_2 = 10\mu s$, the trion emission (spontaneous emission) lifetime is $\tau = 200ps$, and the Zeeman frequency of the electron spin is $\delta_e/2\pi = 100GHz$. This is equivalent to the case in which the electron spin instantaneously rotates when a transverse magnetic field of about $1000T$ is effectively applied by the optical pulses with an ultrashort duration of 100fs.

5.4.4 Clock signals

In conventional experiments on electron spin resonance, the carrier wave phase of a microwave oscillator tuned to the Zeeman frequency provides clock signals to the entire system. In the method based on the control of electron spins by ultrashort optical pulses, the repetition frequency of a mode-locked laser that generates optical pulses is synchronized with the Zeeman frequency of the electron spin, and the arrival time of optical pulses provides clock signals to the entire system. A schematic of this is shown in Figure 5-26. An optical pulse that arrives at $|0\rangle$ rotates the electron spin around the x-axis of a coordinate system rotating with the Larmor frequency of the electron spin. An optical pulse that arrives with a time delay of a quarter-period of the pulse sequence ($t = \pi/2\delta_e$) rotates the electron spin around the y-axis of the rotating coordinate. An optical pulse that arrives with a time delay of half a period ($t = \pi/\delta_e$) rotates the electron spin around the x-axis. The above cases correspond to the phase differences between the upper-and lower-side bands being $0, \pi/2$, and π for optical pulses with the three different arrival times of $t = 0, \pi/2\delta_e$ and π/δ_e, respectively.

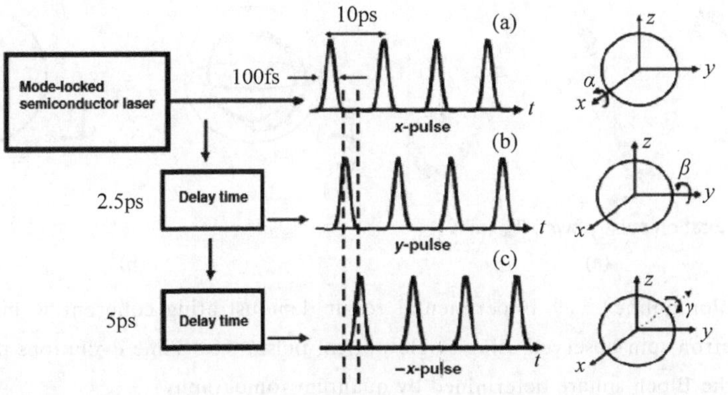

Fig. 5-26 x-pulse, y-pulse, and $-x$-pulse provided by mode-locked semiconductor laser.

According to Euler's theorem, an arbitrary single-qubit gate SU(2) is divided into three rotation operators, $\hat{R}_x(\alpha)$, $\hat{R}_y(\beta)$ and $\hat{R}_x(\gamma)$ therefore, an arbitrary SU(2) can be realized by controlling the areas of three optical pulses with $t = 0, \pi/2\delta_e$ and π/δ_e. Then, an arbitrary single-qubit gate of an electron spin with the Zeeman frequency of $\delta_e/2\pi = 100GHz$ is completed in only 5ps. This indicates that the operating speed can be increased by a factor of $10^3 \sim 10^4$ compared with that achieved with conventional electron spin resonance techniques based on microwaves control pulses.

The most significant advantage of the above scheme is the simultaneous performance of single-qubit gates for multiple individual electron spins arranged at spacing of $2 \sim 3\mu m$. This can be realized because it is easy to construct an optical imaging system in which only one optical pulse is irradiated onto the target cavity mode and electron spin among many simultaneously irradiated optical pulses.

An experiment on a single-qubit gate using a single optical pulse was first carried out for an ensemble of electron spins bound to Si donor impurities in GaAs crystals, and the operating principle was confirmed. A similar experiment was then carried out for an electron spin trapped in a single InGaAs quantum dot, and the following results were obtained. Figure 5-27(a) shows coherent Rabi oscillations observed using a single optical pulse. When the energy level of the single optical pulse increases, coherent Rabi oscillations with rotation angles of up to $\Phi = 12\pi$ are observed for a pulse width of 4ps, an electron spin Zeeman frequency of $\delta_e/2\pi = 26$GHz, and detuning from the trion state of $\Delta = 270$GHz. In Figure 5-27(b), the measured time evolutions of the quantum state of the electron spin are plotted on the Bloch sphere for rotation angles of up to $\Phi = 3\pi$. The electron spin deviates from the south pole at $t = 0$ because the initialization fidelity is approximately 92%.

The deviation of the electron spin quantum state from the target state further increases when the rotation angle increases as $\Phi = \pi, \pi/2$.... The reasons for this are considered to be as follows.

(1) An electron-hole pair produced upon the effective absorption of part of the optical pulse causes the phase relaxation of a virtually excited trion.

(2) Rotation of the electron spin around the z-axis is superposed as shown in Figure 5-27(b) because the pulse width of 4ps is not negligible compared with the Larmor period of the electron spin of 40ps.

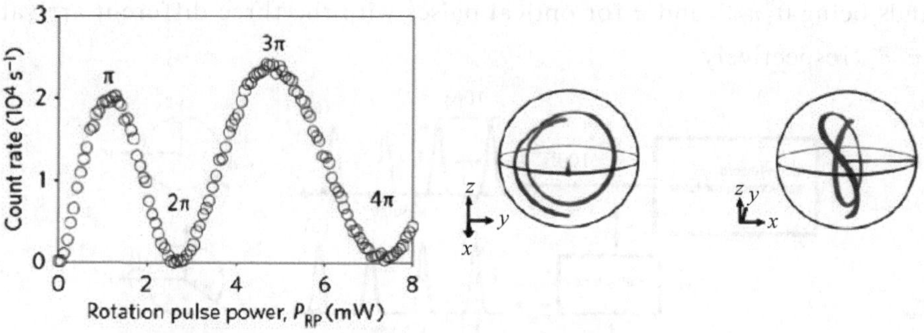

(a) (b)

Fig. 5-27 (Color online) (a) Experimental result demonstrating coherent Rabi oscillations of electron spin observed using single optical pulse. (b) Time evolutions of electron spin in the Bloch sphere determined by quantum tomography.

Figure 5-28 shows experimental results for the Ramsey interference of electron spins obtained using two optical pulses. Figure 5-28(a) shows the probability in which the initial state $|0\rangle$ becomes excited to final state $|1\rangle$ upon the irradiation of two optical pulses (measured by photon count rate) with respect to the rotation angle due to optical pulse irradiation (Φ) and the time difference between the two optical pulses. As shown in Figure 5-28(b), the Ramsey interferometer indicated maximum visibility when $\Phi = \pi/2$, from which the fidelity for the $\pi/2$ pulses was estimated to be 98%～99%. When $\Phi = \pi$, the Ramsey interferometer was expected to output a constant value; however, slight residual oscillations were observed, as shown in Figure 5-28(a). The residual oscillations resulted from the slight deviation of the electron spin from the north pole (only 0.17rad) after the irradiation of π pulses, as shown in Figure 5-27(b).

Two-Qubit Gate Realized by Single Optical Pulses Various schemes for realizing two-qubit gates via a cavity mode has been proposed. The operating principle of such schemes is generalized

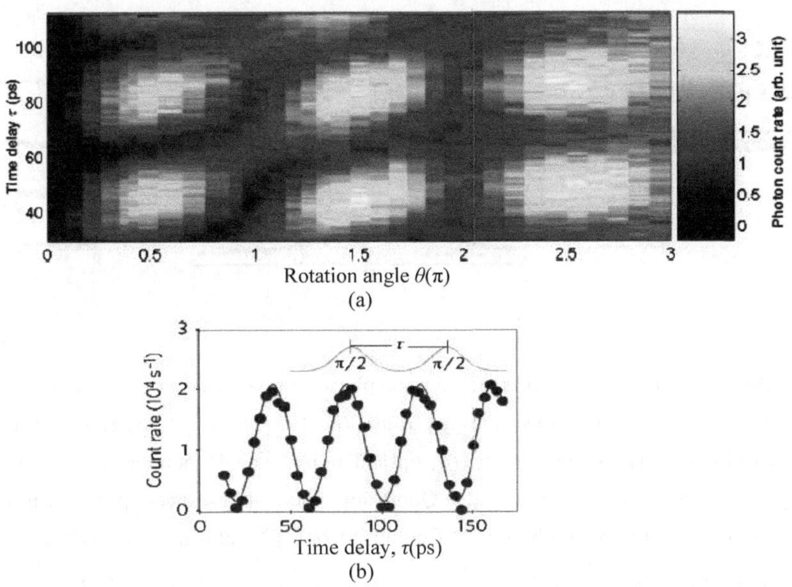

Fig. 5-28 (Color online) Experimental results on Ramsey interference of electron spins obtained using two optical pulses. (a) Dependences of photon count rate and time difference between two pulses on rotation angle due to optical pulse irradiation. (b) Ramsey fringe with respect to time difference between two $\pi/2$ pulses.

as follows: the phase or amplitude of the cavity mode is modulated in accordance with the states of the two qubits, and this modulation causes the time evolution of the two qubits to be modulated. In this case, the cavity mode catalyzes the realization of a two qubit gate. Practically, known examples of this process are either amplitude modulation based on stimulated Raman scattering between qubits via laser light and the cavity mode or phase modulation based on geometrical phases with respect to the time evolution of entanglement in a complex system consisting of the cavity mode and two qubits. Figure 5-29 (a) shows one of the simplest methods of phase modulation. Quantum dots are arrayed on a square lattice in a planar microcavity to form Λ-type three-level artificial atoms doped with a single electron (see Figure 5-29 (b)). Control optical pulses, the spot size of which is set equal to the natural spot size of the planar microcavity, are irradiated to this system so that two adjacent qubits are subjected to the same intensity of the electric field in the spot. Compared with the transition frequency between the electron spin excited state $|1\rangle$ and the trion lowest-energy state $|e\rangle$, the cavity resonance frequency is set in a low-energy region, and the center frequency of the coherent control optical pulse is set in an even lower energy region Figure 5-29 (b). At this time, the cavity resonance frequency shifts by different amounts for different states of the two qubits, as shown in Figure 5-29(c). Therefore, the control optical pulses irradiated from the outside produce different electric fields in the microcavity (see Figure 5-29(d)). When the area enclosed by the phase-space path of the electric field is π in all three states, a universal two-qubit gate called a control phase gate can be realized.

A marked advantage of this two-qubit gate is that a twodimensional cluster state can be created on a square lattice of arbitrary size merely by moving the control optical pulse to four different irradiation positions. If microcavities with a three-dimensional confinement structure, such as pillar microcavities, photonic crystals, and microdisc cavities, are used to realize two-qubit

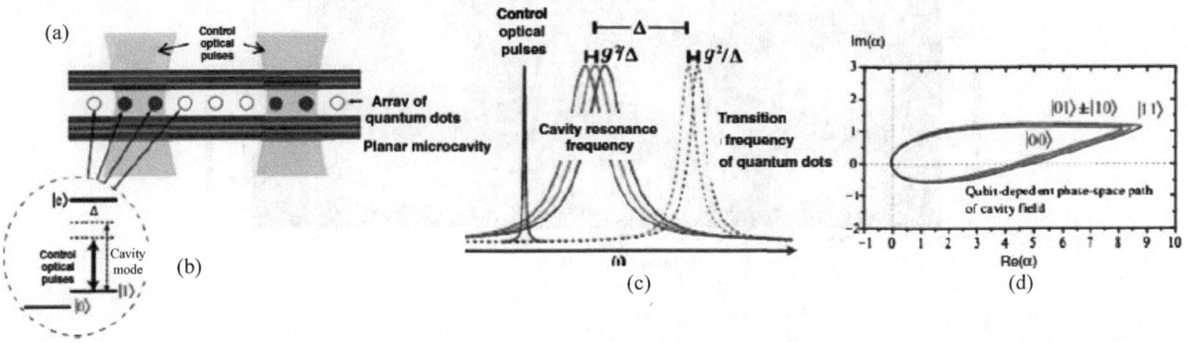

Fig. 5-29 (Color online) (a) Implementation of control phase gate to two adjacent qubits by coherent control optical pulse. (b) Relationships among transition frequency of artificial atom, cavity resonance frequency, and center frequency of control optical pulse. (c) Dependence of shift of cavity resonance frequency on state of two qubits. (d) Qubit-dependent phasespace path with respect to real and imaginary parts of complex amplitude α of coherent optical field inside a cavity.

gates, it is necessary to introduce coupled waveguides serving as quantum communication channels between the microcavities confining the qubits with a high efficiency, which makes a whole system very complicated.

Two-qubit gates have two undesired effects causing decoherence. One effect occurs when a quantum dot absorbs a photon from the control optical pulse and is effectively excited to a trion state. To avoid this decoherence, it is necessary to sufficiently detune the center frequency of the control optical pulse from the transition frequency between states $|1\rangle$ and $|e\rangle$ and to suppress the energy of the control optical pulse to a minimum value. As shown in Fig. 5-29(d), the difference in the phase-space path among the three states, becomes small when the detuning is increased. The area enclosed by each path decreases when the energy of the control optical pulse is decreased. As mentioned above, the difference in area enclosed by each path must be π to realize a control phase gate; therefore, the optimal degree of detuning and pulse energy must be determined while the probability of excitation to the trion state is minimized. The other decoherence effect is caused by that each of the photons in the control optical pulses continuously emitted from a cavity may leak qubit information to external reservoirs through amplitude modulation and phase shift. As shown in Figure 5-29(d), two qubits and the electric field in the cavity are in an entangled quantum state; it is necessary to erase the slight difference in the electric-field paths in the cavity using the intrinsic quantum noise (vacuum-field fluctuation) of the control optical pulse reflected in the cavity (this technique is called quantum erasure). The above complicated optimization problem can be solved by numerical simulation by applying the master equation to coherent control optical pulses and a cavity confining two artificial atoms. Figure 5-30 shows the concurrence

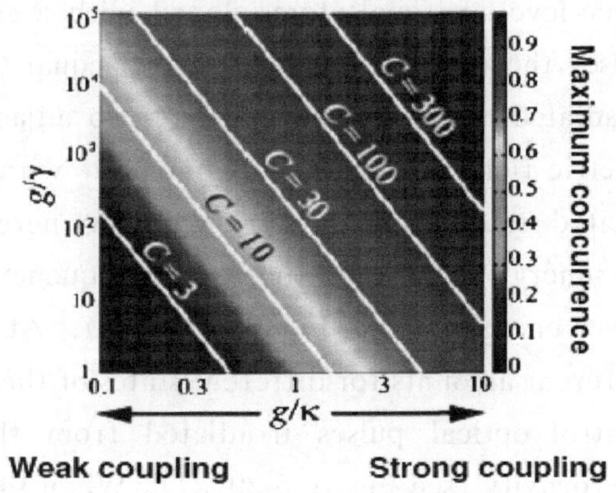

Fig. 5-30 (Color online) Concurrence of entangled quantum states created by two-qubit (control phase) gate. g is the vacuum Rabi frequency, k is the cavity decay rate, and γ is the excitonic spontaneous emission rate.

of entangled quantum states created upon the irradiation of optical pulses as a function of the vacuum Rabi frequency g, the photon decay rate in the cavity, and the spontaneous emission rate of trions γ. Here C is the resulting Purcell factor. A control phase gate is implemented between two electron spins confined in a planar microcavity, indicating that an entangled quantum state can be created at a high fidelity.

5.4.5 Quantum memory and decoherence time

It is necessary to store the information of qubits for a long time in quantum repeaters and quantum computers. In some applications, it is even more advantageous to transfer the information of qubits to nuclear spins with a longer decoherence time than to repeat the gate operation for quantum error correction using many electron spins with a shorter decoherence time. Some donor impurities in semiconductors combine a nuclear-spin memory and an electronspin processor and form a Λ-type three-level system between bound excitons and Zeeman sub-levels of bound electrons.

- Donor impurity atoms, bound electrons, and bound excitons

Figure 5-31(a) shows such a donor impurity, i.e., a ^{19}F atom in a ZnSe crystal (^{19}F: ZnSe). ^{19}F is the only isotope of fluorine that appears in nature and has the simplest nuclear spin of 1/2.

The ^{19}F donor atom captures an unpaired electron at low temperatures to become neutral. A nuclear-spin memory and an electron-spin processor thus coexist in a single ^{19}F system and are coupled with a hyperfine coupling constant of approximately 30MHz; the rapid transfer of qubits within 30ns can be realized using a conventional double-resonance technique. Such neutral donor impurities can capture an exciton by the many-body effect. Figure 5-31(b) shows the PL spectrum of ^{19}F:ZnSe. The strong peak emission of a bound exciton (D_0X) is observed at an energy 6meV lower than that of the peak of a free exciton (X); this energy difference corresponds to the binding energy of a donor bound exciton. A peak PL intensity due to emission by two electron satellite transition, the transition to the bound electron in either 2s or 2p orbital, is also observed at an energy 22meV lower than that of the bound exciton peak.

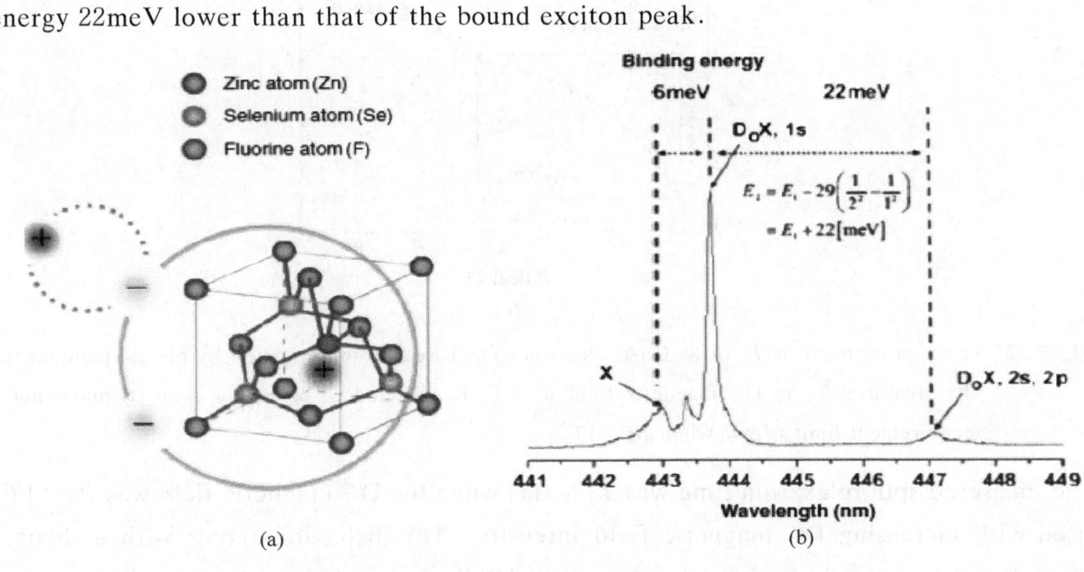

Fig. 5-31 (Color online) (a) Donor impurity (^{19}F:ZnSe), bound electron (D0), and bound exciton (D_0X). (b) PL spectrum of bound exciton in ^{19}F:ZnSe donor impurity.

These spectroscopic characteristics can be theoretically reproduced with high accuracy by Kohn's effective-mass approximation. When a DC magnetic field is applied to this $^{19}F:ZnSe$ system, a Λ-type three-level system similar to that shown in Figure 5-31(a) can be formed. ^{29}Si, another donor impurity, in a GaAs crystal ($^{29}Si:GaAs$) shows qualitatively similar characteristics. Moreover, a ^{31}P donor impurity in a Si crystal ($^{31}P:Si$) also exhibits similar characteristics; however, this particular system is unsuitable for optical control because the host crystal is an indirect transition semiconductor and the quantum efficiency of excitonic emission is much lower than that of Auger nonradiative recombination. To solve this problem, a scheme for increasing the spontaneous emission rate using a microcavity (photonic crystal) has been proposed. Alternatively, an edge state of the quantum Hall effect can be used in place of optical microcavities.

5.4.6 T_1 and T_2 for electron spins

The spin relaxation time (T_1) of the electron spins trapped in a donor impurity is obtained as follows Figure 5-32(a). The electron spin state is initialized to the ground state $|0\rangle$ at a high fidelity by optical pumping in resonance with the transition energy between states $|1\rangle$ and $|e\rangle$. The resonant optical probe pulse is irradiated to the system at time τ after initialization, and the probability of the electron spins leaked to the state $|1\rangle$ is measured. Figure 5-32(b) shows an experimental result for a Si:GaAs donor impurity system.

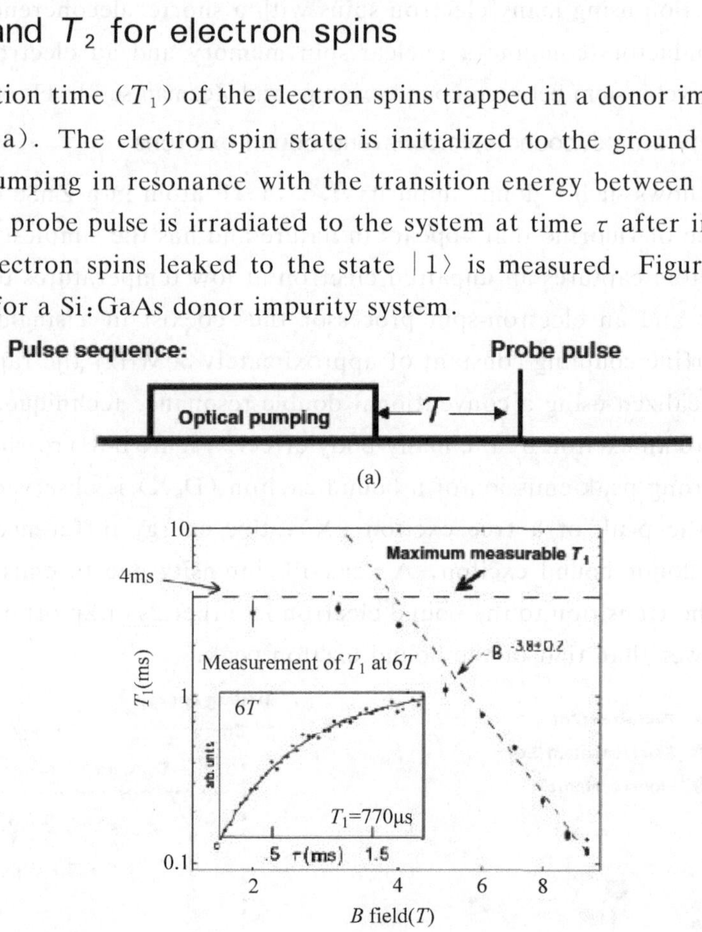

Fig. 5-32 (a) Measurement of T_1 of Si:GaAs electron spins based on initialization by optical pumping. (b) Measured T_1 vs DC magnetic field B_0. T_1 is expected to be longer than its maximum measurement limit (4ms) when $B_0 > 4T$.

The measured spin relaxation time was $T_1 = 4$ms when the DC magnetic field was $B_0 = 4T$ and decreased with increasing DC magnetic field intensity. This behavior agrees with a theoretical estimation based on the spin-lattice relaxation model. The phase relaxation time (T_2') of electron spins in the same system can be determined from a Ramsey interference experiment using the two $\pi/2$ pulses shown in Figure 5-28, and is on the order of $1\sim2$ns. This was also verified in experiments on electromagnetically induced transparency and nonresonant Raman scattering. Such a short T_2' is due to the fluctuation in the longitudinal magnetic field, $\Delta B_0(t)$, formed by the nuclear spins of Ga and As atoms in the host crystal (which is a component parallel to the DC magnetic field). $^{31}P:Si$ and $^{19}F:ZnSe$ donor impurities are more advantageous than $^{29}Si:GaAs$ donor impurities because nuclear spins can be depleted from the host crystals (Si and ZnSe).

It has been reported that $T_2 \approx 10 \sim 100ms$ for a $^{31}P:^{28}Si$ donor impurity system, from which nuclear spins were depleted by isotope engineering. To suppress the effect of $\Delta B_0(t)$, an experiment on Hahn's spin echo was carried out using three optical pulses and the results are shown in Figure 5-33. In the case without a second refocusing pulse (i.e., a conventional Ramsey interference experiment), the interference fringe completely disappeared at $\tau = 26ns$ (see Figure 5-33(b)). In the case with a second refocusing pulse (i.e., Hahn's spin echo experiment), an interference fringe due to Larmor rotation of the electron spins was observed (see Figure 5-33(c)). According to the measurement result for fringe visibility with respect to time delay (see Figure 5-33(d)), the decoherence time after the elimination of the effect of $\Delta B_0(t)$ was estimated to be $T_2 \approx 6.7 \pm 2.5\mu s$. This is similar to T_2 for electron spins trapped in a single InGaAs quantum dot. An even larger T_2 time is expected if a multi-pulse decoupling scheme such as CPMG pulse sequence is employed. The record of T_2 time for an electron spin in GaAs is $\sim 200\mu s$ at the time of writing this article. Therefore, when the running time of a single-qubit gate is assumed to be 5ps, approximately $10^6 \sim 10^7$ single-qubit gates can be completed within T_2 in the above system.

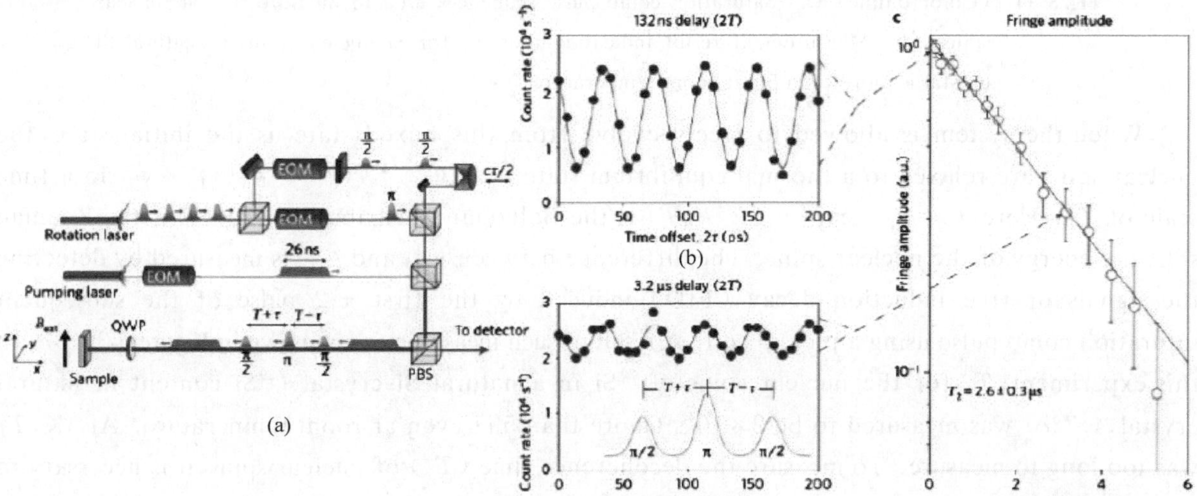

Fig. 5-33 (Color online) (a) Hahn's spin echo experiment using three optical pulses. (b), (c) Interference fringes at $\tau = 132ns$ and $3.2\mu s$, respectively. (d) Measurement result for fringe visibility vs time delay.

5.4.7 T_1 and T_2 for nuclear spins

T_1 for nuclear spins is very long and is normally measured using so-called saturation comb pulses, which are special nuclear magnetic resonance (NMR) pulse sequences Figure 5-34(a). When $\pi/2$ pulse sequences are irradiated to a system with a time interval sufficiently longer than the phase relaxation time (T_2^*) of nuclear spins, a completely mixed state:

$$\hat{I} = 1/2(|\uparrow\rangle\langle\uparrow| + |\downarrow\rangle\langle\downarrow|) \tag{5-90}$$

Which nuclear spins in the ground state and excited state are equally mixed, It is formed in a short time.

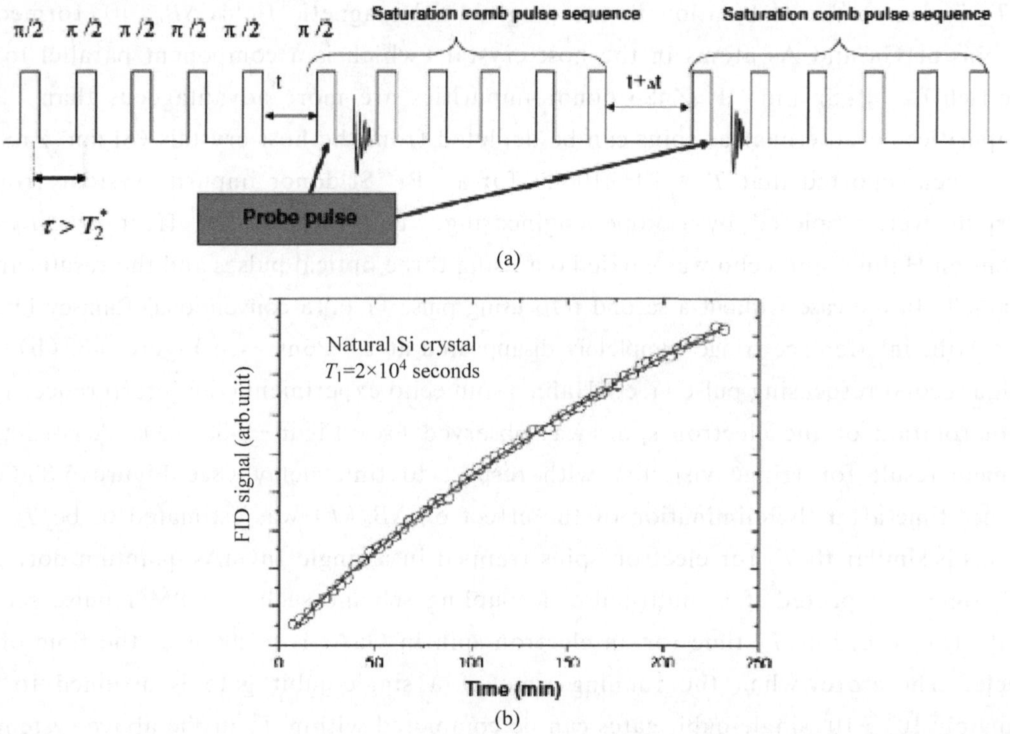

Fig. 5-34 (Color online) (a) Saturation comb pulse sequences used to measure T_1 for nuclear spins. (b) Measurement result indicating that T_1 for Si nuclear spins in natural Si crystal is more than 5h at room temperature.

When the system is allowed to freely evolve from this mixed state as the initial state, the nuclear spins are relaxed to a thermal equilibrium state $\hat{\rho}_{\text{th}} = \rho_{\uparrow} \mid \uparrow\rangle\langle\uparrow\mid + \rho_{\downarrow} \mid \downarrow\rangle\langle\downarrow\mid$ in a time scale of T_1. Here, $\rho_{\uparrow}/\rho_{\downarrow} = \exp(-\Delta E/k_B T)$ is the Boltzmann distribution and ΔE is the Zeeman splitting energy of the nuclear spins. The difference between $\rho\uparrow$ and $\rho\downarrow$ is measured by detecting the signals of free induction decay (FID) induced by the first $\pi/2$ pulse of the subsequent saturation comb pulse using a pickup coil. A result of such measurement is shown in Figure 5-35(b). In this experiment, T_1 for the nuclear spins of ^{29}Si in a natural Si crystal (^{29}Si content in natural crystal, 4.7%) was measured to be 2×10^4 s (more than 5h) even at room temperature. At 4K, T_1 was too long to measure. To measure the decoherence time (T_2) of nuclear spins, it is necessary to not only eliminate the effect of $\Delta B_0(t)$ but also suppress the dipolar interactions between ^{29}Si nuclear spins. In the experiment shown in Figure 5-35, the former is realized using Carr-Purcell-Meiboom-Gill (CPMG) π-pulse sequences, and the latter is realized using MREV-16 $\pi/2$-pulse sequences. These NMR pulse sequences are alternately irradiated, as shown in Figure 5-35. A spin-echo signal is detected as the FID signal after the last $\pi/2$-pulse in each MREV-16 pulse sequence. The results in Figure 5-35 indicate that T_2 for ^{29}Si nuclear spins in a natural Si crystal is 25s at room temperature. Donor nuclear spins (^{31}P:Si, ^{19}F:ZnSe) in semiconductors are expected to have similar T_1 and T_2 times to those shown in Figure 5-34 and Figure 5-35. In conclusion, nuclear-spin memories in semiconductors have a sufficient coherence time for application to quantum information processing systems. Electron-spin processors also have a sufficient coherence time for the transfer of qubits to nuclear spins.

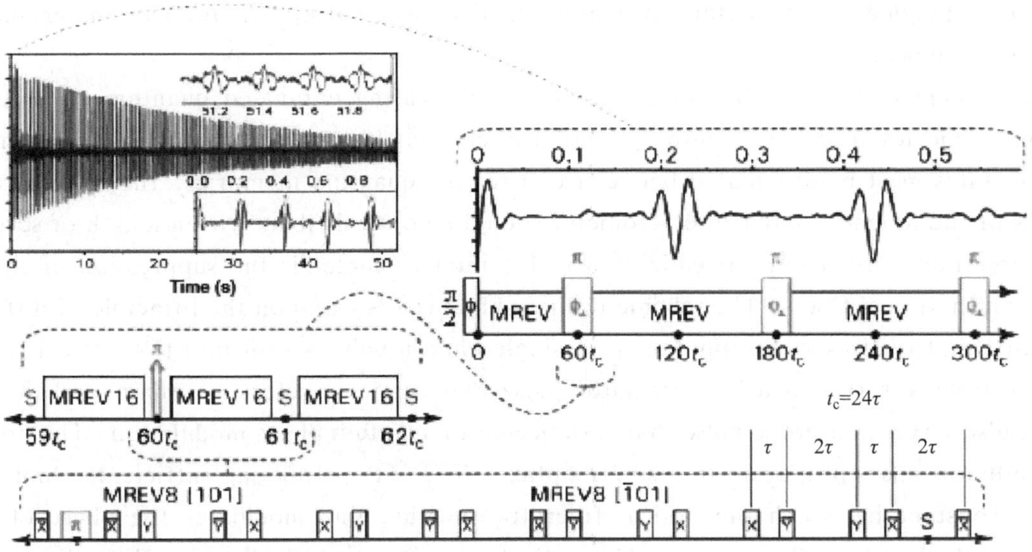

Fig. 5-35 (Color online) Experimental result indicating that T_2 for ^{29}Si nuclear spins in natural Si crystal is 25s at room temperature. CPMG π-pulse sequences and MREV-16 π/2-pulse sequences are alternately irradiated Time (s).

5.5 Single-photon emission and distribution of entangled quantum states

5.5.1 Single-photon interferometer with quantum phase modulators

Consider the case when the phase of the two partial waves of a single photon is in 0 and π linear superposition, i. e:

$$|\Psi_{12}\rangle = 1/\sqrt{2}(|0\rangle + |\pi\rangle)_1 \otimes 1/\sqrt{2}(|0\rangle + |\pi\rangle)_2 \tag{5-91}$$

a Mach-Zehnder interferometer equipped with two quantum phase modulators Figure 5-36(a). Such quantum phase modulators are realized, for example, when a quantum dot or a donor impurity that forms a Λ-type three-level artificial atom is embedded in a microcavity and the ground and excited states of the electron spins are linearly superposed.

When a single-photon pulse is irradiated to the Mach-Zehnder interferometer and detector A captures this photon, the two quantum phase modulators are posteriori found to have been in the state of

$$|\psi_{12}\rangle' = 1/\sqrt{2}(|0\rangle_1 |0\rangle_2 + |\pi\rangle_1 |\pi\rangle_2) \tag{5-92}$$

In contrast, when detector B captures the photon, the two electron spins are posteriori found to have been in the state of

$$|\psi_{12}\rangle'' = 1/\sqrt{2}(|0\rangle_1 |\pi\rangle_2 + |\pi\rangle_1 |0\rangle_2) \tag{5-93}$$

For both detection results, an entangled quantum state is created between the two quantum phase modulators (electron spins). If a coherent pulse is used instead of a single-photon pulse, the quantum states of the two electron spins are individually measured via photons that are lost inside the optical interferometer, resulting in the decoherence of the electron spin state. Therefore, the fidelity of the created entangled quantum state decreases. Single-photon pulses are essential to create the entangled quantum state at a high fidelity in actual optical interferometers in which optical loss is unavoidable.

The scheme in Figure 5-36(a) is effective for creating entangled quantum states between quantum memories in a single chip, but not for creating quantum repeaters because entangled quantum states must be distributed between two remote quantum memories. In this case, the two channels of the Mach-Zehnder interferometer should be optical fibers with a length of several ten kilometers; therefore, a key to realizing a stable interferometer is the suppression of the phase fluctuation in optical fibers. The scheme in Figure 5-36(b) is based on the principle of differential phase shift that realizes such suppression. A single-photon pulse is split into two optical pulses the first optical pulse serves as a localoscillator pulse that carries a reference phase, and the second optical pulse serves as a probe pulse that is subjected to quantum phase modulation. The probe and local oscillator pulses propagate in a single optical fiber with a time lag sufficiently shorter than the time constant by which the optical fiber transmission line modulates the phase of optical pulses; hence, the phase fluctuation in the optical fiber is automatically cancelled. Similar to the experiment using the Mach-Zehnder interferometer, entangled quantum states are distributed when either detector A or detector B counts a single photon, whereas the distribution fails when a single photon is not counted by either detector. In either cases, it is necessary to generate a single-photon pulse with a frequency appropriately detuned from the transition frequency between states $|1\rangle$ and $|e\rangle$ as well as to set two electron spins to the initial linearly superposed state.

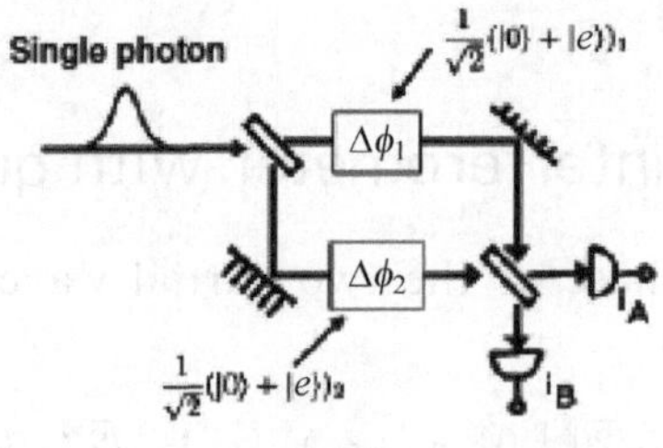

(a)

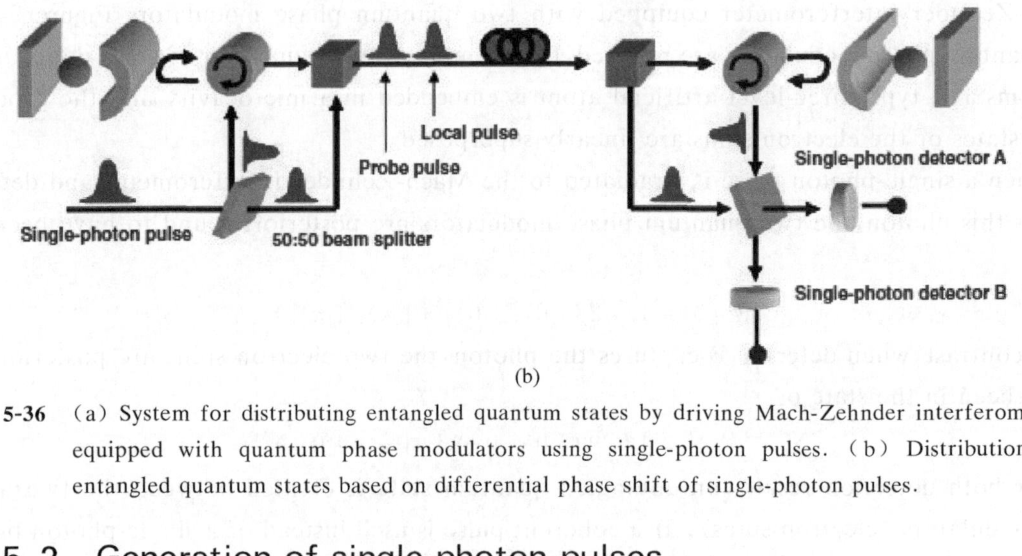

(b)

Fig. 5-36 (a) System for distributing entangled quantum states by driving Mach-Zehnder interferometer equipped with quantum phase modulators using single-photon pulses. (b) Distribution of entangled quantum states based on differential phase shift of single-photon pulses.

5.5.2 Generation of single-photon pulses

Single photons with a wavelength tuned to that of the excitonic transition in quantum dots or donor impurities can be emitted from similar systems. Here, emitted single photons are mutually indistinguishable quantum particles when (1) the start time of the oscillation of single photon electric field $E(t)$ is well controlled (i.e., jitter is negligible) and (2) there is no phase jump in the optical electric field (i.e., the optical pulses are Fourier transform limited).

Figure 5-37 shows the setup and results of an experiment in which such indistinguishable single photons were emitted from a single quantum dot excited by pulsed pumping light. To satisfy the above two conditions (negligible jitter and Fourier-transform-limited pulses), the time width of the single-photon pulse τ_s must be sufficiently longer than the energy relaxation (cooling) time τ_{relax}, which is required for the energy relaxation from the transition level between the excited states (2e-2h) of the quantum dot to that between the ground states (1e-1h) used for emission. For InGaAs quantum dots at $T = 4K$, $\tau_{relax} \approx 10ps$ and $\tau_{phase} \approx 1 \sim 2ns$. A single InGaAs quantum dot was confined in a pillar microcavity, and the indistinguishability of photons emitted from the dot was tested by irradiating two single-photon pulses to a Hong-Ou-Mandel interferometer (see Figure 5-37(a)) for different spontaneous emission lifetimes, which determine τ_s, in the range from 80~380ps. Figure 5-37(b) shows coincident counts when a photon is detected at $t = 0$ and another photon is detected at a different time. Figure 5-37(c) shows the intensities of peak 3 in Figure 5-37(b), for which the arrival times of two single photons are slightly modulated. From the depth of the dip at the center, the quantum-mechanical indistinguishability of two single photons was found to be 72%~81% for a pulse width of 80~380ps. Entangled photon pairs can be generated by such indistinguishable single photons emitted at a certain time using only linear optical devices. Also, experimental demonstrations of violation of Bell inequality and quantum teleportation are ongoing.

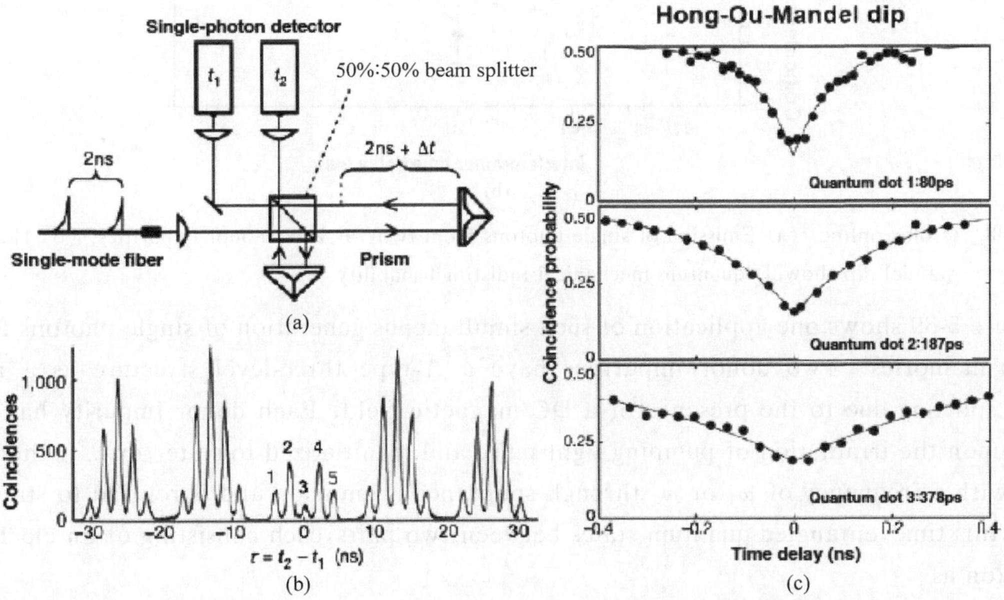

Fig. 5-37 Hong-Ou-Mandel interferometer used to evaluate indistinguishability of two single photons.

- Simultaneous emission of indistinguishable single photons from two quantum memories

In the experimental results in Figure 5-37, consecutive single photons emitted from a single quantum dot were mutually indistinguishable quantum particles. In the experimental results in Figure 5-38, single photons were simultaneously emitted from two ^{19}F:ZnSe donor impurities, and a Hong-Ou-Mandel dip was observed as a result of quantum interference when the two single photons collided at 50% : 50% beam splitter, confirming that the two single photons were identical quantum particles. In this case, the quantum-mechanical indistinguishability of the two single photons was 65%.

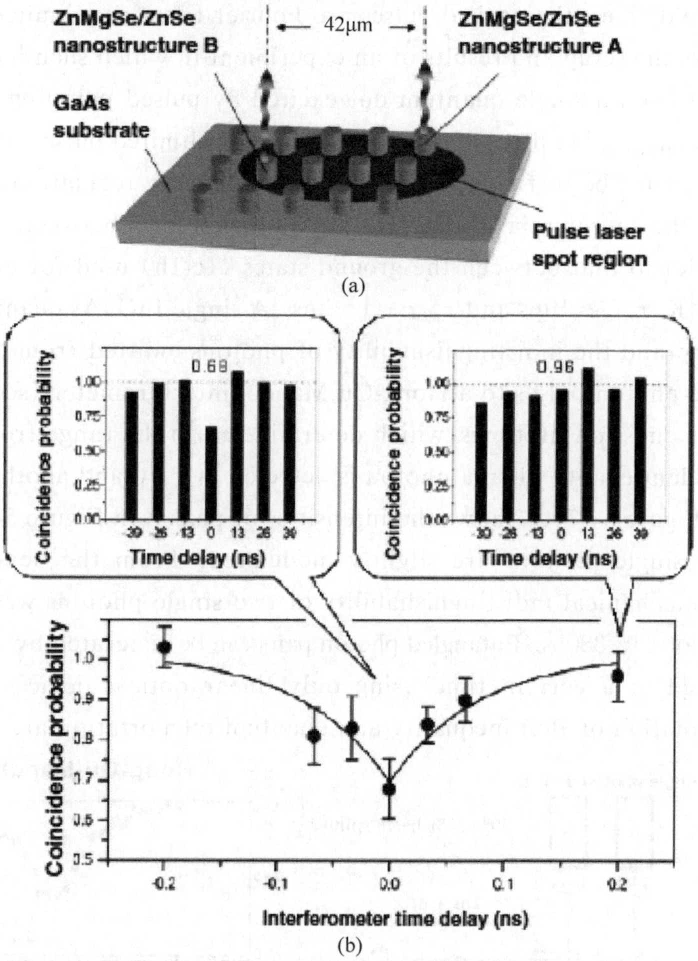

Fig. 5-38 (Color online) (a) Emission of single photons from two ^{19}F:ZnSe donor impurities. (b) Hong-Ou-Mandel dip showing quantum-mechanical indistinguishability.

Figure 5-39 shows one application of such simultaneous generation of single photons from two quantum memories. Two donor impurities have a Λ-type three-level structure as a result of Zeeman splitting due to the presence of a DC magnetic field. Each donor impurity has a bound exciton upon the irradiation of pumping light pulse and is initialized to state $|ex\rangle$. It then emits a photon with a frequency of ω_1 or ω_2 through spontaneous emission and is relaxed to state $|1\rangle$ or $|0\rangle$. At this time, entangled quantum states between two pairs, each consisting of an electron spin and photon as

$$1/\sqrt{2}(|0\rangle_1 |\omega_2\rangle_{f1} + |1\rangle_1 |\omega_1\rangle_{f1}) \tag{5-94}$$

and

$$1/\sqrt{2}(|0\rangle_2 |\omega_2\rangle_{f2} + |1\rangle_2 |\omega_1\rangle_{f2}) \tag{5-95}$$

are created. If detectors D1 and D2 or detectors D3 and D4 simultaneously click when two single-photon pulses f_1 and f_2 are recombined through a 50% : 50% beam splitter, as shown in Figure 5-37, the state of the two donor impurities is projected to the entangled quantum state:

$$1/\sqrt{2}(|\bar{1}\rangle_1 |0\rangle_2 + |\bar{0}\rangle_1 |\bar{1}\rangle_2) \tag{5-96}$$

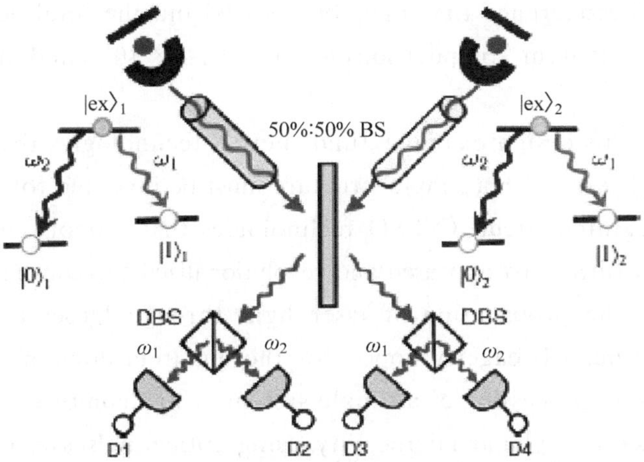

Fig. 5-39 (Color online) Creation of entangled quantum states by coincidence counting of indistinguishable single photons emitted from two quantum memories. Square lattice embedded in a planar microcavity, using optical pulses, as discussed in this section.

In contrast, if detectors D1 and D4 or detectors D2 and D3 simultaneously click, the entangled quantum state of $1/\sqrt{2}(|1\rangle_1|0\rangle_2 - |0\rangle_1|1\rangle_2)$ is selected. This scheme also has the function, although not perfect, of suppressing phase fluctuation in optical fiber transmission lines and is expected to be applied to long-distance quantum repeater systems. With the above-described scheme, experiments on the formation of entangled quantum states using two trapped ions and quantum teleportation have recently been carried out.

Optimum architectures for realizing large-scale quantum computation are expected to continuously evolve using available hardware, which cannot avoid a finite device yield, quantum gate malfunction, and qubit decoherence. Currently, the scheme of quantum computation based on topological surface codes is the most promising one. In particular, the scheme for effectively assembling a three dimensional qubit lattice by combining a two-dimensional spatial resource (qubits) and a one-dimensional temporal resource (the formation and detection of cluster states) proposed by Raussendorf and Harrington can be applied to the systems in which a quantum-dot (donor-impurity) array on a square lattice is embedded in a planar microcavity, as described in this report. This scheme is advantageous because it enables the formation of cluster states simply through two-qubit (control phase) gates between adjacent qubits. The maximum error rate per gate operation in this scheme is approximately ~1% to perform large-scale quantum computation consisting of initialization, gate operation, storage, and projective measurement of qubits.

This is the highest allowable error rate among those for various schemes of fault-tolerant quantum computation reported thus far. An architecture for quantum computers that factorize an integer with $n = 2048$ bits using Shor's algorithm is now being examined on the basis of the above fundamental scheme. The total number of coded logic qubits required for quantum computation is $6n = 12288$ logic bits, which can be supported by $\sim 10^9$ qubits (quantum dots on a two dimensional lattice). This huge number of qubits corresponds to the total number of quantum dots arrayed at intervals of 2μm on a square lattice on a 5×5cm^2 wafer in a planar microcavity. Assuming that the Zeeman frequency of electron spin (system clock frequency) is 10GHz, the percentage of properly functioning quantum dots (donor impurities) is 40%, the gate error rate is 0.2% (i.e., the gate fidelity is 99.8%), the decoherence time of qubits is 1μs, and the total number of Toffoli gates, which are dominant in quantum computation, is $4n^3 = 3.2 \times 10^{11}$, and the computation time is approximately 6 days.

As is obvious from this design example, fundamental technologies that enable the large-scale integration of individual gates with a simple structure must be invented for qubit systems, similar to the very large-scale integrated circuit (VLSI) technologies that support current computers. Forty years ago, low-loss optical fibers first appeared and revolutionalized the conventional concept of optical communication through the propagation of laser light through lenses arranged in underground tunnels. Such a breakthrough is eagerly hoped for the quantum domain. Conducted research on the scheme for the parallel processing of a simple system, which comprises a quantum-dot array on a square lattice embedded in a planar microcavity, using optical pulses, as discussed in this section.

5.6 Single-photon wavepackets and memory in atomic vapor

5.6.1 Electronics and photonics integration

The recent push to for large scale integration of silicon electronics and photonics (the so called electronic-photonic integrated circuit EPIC as show in Figure 5-40) is an embodiment of the decades long trend towards greater functional integration in silicon circuits. It can be instructive to study the history of electronic Systems-on-a-Chip (SoC) where logic, memory, and input-output drivers are monolithically integrated on a single chip. As with the EPICs, SoC integration was driven by the promise of increased speed, lower power consumption, simpler physical interfaces, reduced packaging costs, and increased subsystem reliability. Despite these drivers electronic SoCs faced significant barriers to adoption including: longer design cycles, lower device yields, higher capital investments for manufacturing, incompatible (optimized) processes for the diverse functions, and the challenge of consolidating intellectual property. The same barriers exist for EPICs now that silicon photonics is maturing and discrete device performance is improving. By 2015, the electronic SoC market is expected to reach nearly 40B as the Global Industry Analysts, System-On-a-Chip (SoC): A Global Strategic Business Report in 2011. Overcoming the barriers to SoC integration can be traced back to the development of CAD tools that fostered a "design for manufacture" philosophy and a robust foundry model that lowered the barrier to entry for vendors and supported a robust secondary market in design IP.

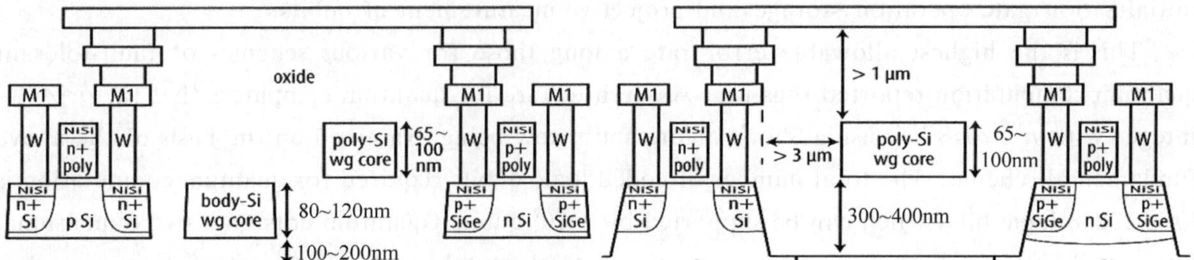

Fig. 5-40 Monolithic EPIC cross-sections integrated within (a) sub-65nm node SOI-CMOS and (b) bulk-CMOS processes.

MIT has been to develop photonic design blocks that can be seamlessly integrated into the electronic foundry environment. With this platform, they hope to one day support EPIC applications from multiprocessor interconnect to coherent-communication receivers which benefit from photonic devices integrated alongside the dense, high-performance transistors only available within state-of-the-art electronic processes. Recently, localized substrate removal technology has enabled photonic device integration through the addition of a single post-fabrication step on designs fabricated in unmodified state-of-the-art CMOS electronic foundries. By sharing all in-foundry processes, EPICs including state-of-the-art transistors can leverage the existing infrastructure and economy of scale provided by the orders of magnitude larger electronics industry. Since process steps are therefore limited to those already used to fabricate the electronic devices, photonic device designs must be tailored for fabrication within this specific set of constraints.

In localized substrate removal monolithic integration, the waveguide cores are implemented in the patternable front-end silicon layers. In the SOI-CMOS process, there are two such available layers: the single-crystalline silicon transistor body layer further referred to as the body-Si layer, and the polycrystalline silicon transistor gate layer further referred to as the poly-Si layer. In the bulk-CMOS process, the transistor body is fabricated directly in the handle wafer leaving only the poly-Si layer available as the waveguide core. By configuring the surrounding material stack-up using available design layers, the EPIC platform cross-sections, shown in Figure 5-41(a) and (b) for the SOI-and bulk-CMOS processes respectively, are available in standard electronic foundries. Using this integration platform, test chips have been produced in various bulk-and SOI-CMOS processes from major semiconductor device manufacturers. Integrated devices such as waveguides and ring resonator filters shown in Figure 5-41 have been demonstrated with zero in-foundry process changes using the existing electronic VLSI design submission data flows and mask-sharing infrastructure. Table 5-1 shows the progression of waveguide loss in scaled, foundry CMOS and deep-submicron DRAM processes. To date, waveguide losses well-below 10dB/cm across the entire infrared spectrum (1250~1550nm) have been demonstrated.

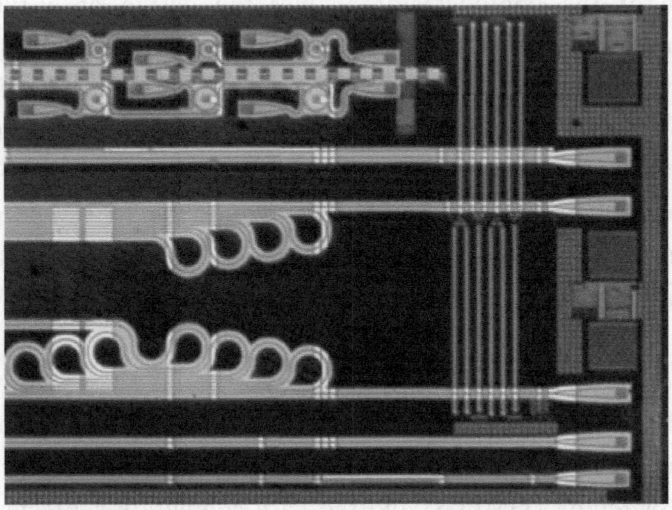

Fig. 5-41 Waveguides and ring resonator filters integrated in a photonic integration test chip produced in 32nm bulk-CMOS process alongside traditional electronic designs.

Table 5-1 Evolution of waveguide loss in foundry CMOS and DRAM photonics.

Technology	Waveguide Material	Loss	Year
65nm CMOS	Polysilicon	55dB/cm	2008
32/28nm CMOS	Polysilicon	55dB/cm	2010/11
90nm CMOS	Polysilicon	50dB/cm	2009
Scaled DRAM	Polysilicon	6dB/cm	2011
45nm CMOS	Silicon	<6dB/cm	Unpublished

Utilizing the existing layers constrains the waveguide core thicknesses to values chosen to be optimal for transistor design at the current process generation. Instead of the ~220nm thickness currently used in silicon photonic projects, the body-Si layer thickness ranges from 80~120nm and the poly-Si layer thickness ranges from 65~100nm depending on process generation. For passive photonic devices, this limits the minimum allowable bend radius for a given operating wavelength. SOI processes offer the designer flexibility to vertically stack the body-Si and poly-Si layers, separated only by a sub-2nm oxide, for tighter bends where necessary.

With the constrained stack-up, the chief task of the photonic designer is therefore to pattern the etch masks of the waveguide core layers. Design rules describing allowed shapes in scaled CMOS processes typically only restrict the geometries of sub-100nm features that are rarely required for photonic devices. Therefore, the designer is largely free to arbitrarily pattern devices using state-of-the-art projection lithography masks addressed on a 1nm grid in a high throughput fabrication environment.

A major benefit of front-end integration is that there are many existing doping and metallization steps present for electronic device formation that are therefore available to the photonic device designer. Since both the body-Si and poly-Si layers must be used to form both polarities of transistors as well as resistors, local interconnect and capacitors, the layers must be doped in a wide variety of densities both n-type and p-type. Recently, our group at MIT demonstrated the first monolithic electronic photonic integration in a sub-100-nm standard SOI process. In this work, the monolithic integration of the photodetector enables the design of a fully-digital, low-energy receiver with high input sensitivity-at 3.5Gb/s. Monolithic EPIC fabrication within scaled CMOS foundries presents a novel photonic device platform subject to specific constraints not present in previous silicon photonic work. The devices presented here are created by making zero-changes to the CMOS foundry process flow, economics will likely require future monolithic EPIC platforms leveraging high-performance transistors to resemble state-of-the-art CMOS processes as closely as possible. Therefore, novel designs based upon standard electronic layer structures will likely reduce the barrier of entry and ultimately the cost of final production devices.

5.6.2 Wavelength switched optical networks

Wavelength Switched Optical Network (WSON), being standardized by the Internet Engineering Task Force (IETF) in April 2011. This document provides a framework for applying Generalized Multi-Protocol Label Switching (GMPLS) and the Path computation Element (PCE) architecture to the control of Wavelength Switched Optical Networks (WSONs). In particular, it examines Routing and Wavelength Assignment (RWA) of optical paths. Wavelength Switched Optical Networks is the application of a GMPLS based control plane and Path Computation Element

(PCE) concepts to an "all optical" network as no electrical switching is part of the WSON scope. Optical-electro-optical conversion is admitted for regeneration and wavelength conversion purposes. Main goal of the WSON work is to reduce the overall network cost, through resource sharing, and to speed up the provisioning and recovery time of optical channels. One of the crucial aspects of WSON is the failure resiliency. Here two options are on the table: Pre-Planned (PP) recovery and On-The-Fly (OTF) recovery. In the PP, path computation is performed before service delivery. Path computation in PP is not time-critical: this allows longer and more accurate computations, important when considering optical impairments, which can be executed in an off-line dedicated computation element, with fully detailed network information. Computation may also be performed together with network design, including the definition of the required hardware including regenerators, if needed. Once computed the PP recovery, paths are reserved, and possibly shared, in the network.

This has two main advantages: recovery resources can not be "stolen" by other primary paths unless explicitly stated and path computation time does not affect recovery time. In the OTF, path computation is time-critical, as its time is added to the restoration time and traffic is being lost. Time-criticality recommends the computation to be performed close to the network, to avoid communication overhead: for this reason, OTF path computation is normally performed in a distributed environment in the network nodes, with summarized information and limited visibility, which may also lead to resource conflict during the signaling phase. However, operators see the OTF as very appealing option. It would be prohibitive, from the hardware cost point of view, to implement PP restoration for more than two simultaneous failures. In this case it can be considered a PP restoration that switches to OTF when all alternative pre-planned paths are exhausted. One of the issues that prevent a real OTF implementation is that path computation cannot involve any resource modification (e. g. adding a piece of hardware where needed): path computation must operate considering the network resources already in place. If regeneration is needed and a regenerator device is not available in the desired location, the path request simply fails. Unfortunately, in large networks, regenerators are often required and these resources need to be pre-planned before being used. No real OTF recovery schemes will work in the current WSON scenario. This section introduces a node configuration set-up called OTN Pit Stop (OTN-PS). The proposed device does not require significant modifications in the conventional Reconfigurable Optical Add Drop Multiplexer (ROADM) architecture because its service is provided by connecting an OTN Switch, or simply an OTN Matrix in some cases, externally to the ROADM. The method and apparatus setup is compatible with centralized or distributed control plane and path computation architectures. The advantages of a set of OTN-PS stations, placed in a realistic WSON landscape, are demonstrated through extensive simulations.

- Equipment setup

The target node is a ROADM based on Wavelength Selective Switch (WSS) technology. The proposed apparatus setup applies to direction-bound or direction-less ROADMs configuration. An OTN Matrix is connected to an Add/Drop section of the ROADM as an external device. Not all the A/D ports of the section shall be connected to the OTN-PS because it's assumed that the proposed device provides a regeneration/wavelength conversion service for a subset of the incoming lightpaths. This is a sensible assumption because, in real WSONs, regeneration is considered a scarce and expensive service and it's avoided whenever possible.

Two possible scenarios are considered and reported in Figure 5-42. In (a), the ROADM is equipped with transponders supporting G. 709 OTN framing on the client side. The connection with the OTN Switch is provided using grey fibers. The OTN Switch functionalities can be performed just by an OTN Matrix (reducing costs). In Figure 5-42(b), the OTN Switch is equipped with DWDM XFPs. Alien wavelengths (colored) are feed directly on the WSS.

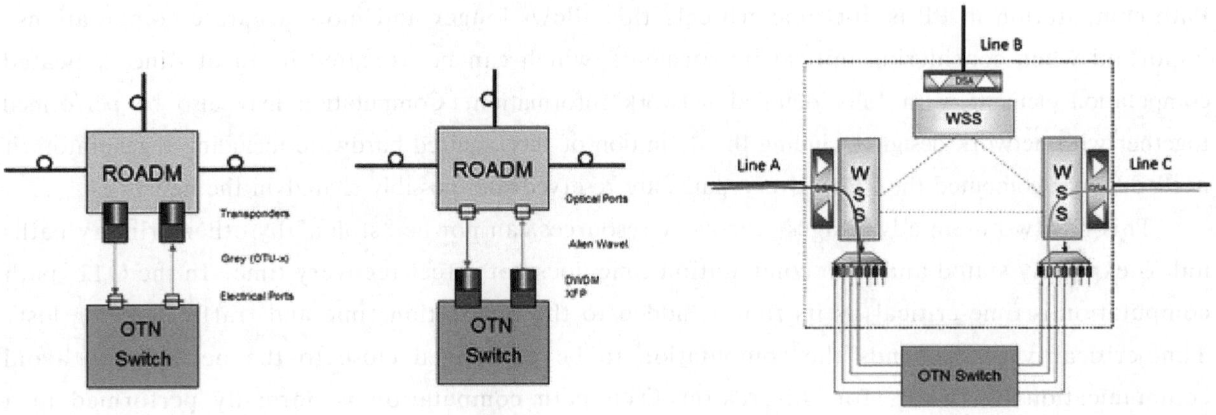

(a) (b) (c)

Fig. 5-42 Equipment configuration.

In order to keep the drawings as simple as possible, (a) and (b) report just one add/drop pair. In a complete setup, multiple transponders, in case (a), or DWDM XFPs, in case (b) ensure the "pit stop" service to multiple lightpaths. If tunable transponders or tunable DWDM XFPs are used, it's possible to provide a reconfigurable wavelength conversion service. This simplifies the wavelength assignment process reducing the blocking probability. In addition, using an OTN-PS supporting different OTU containers, it's possible to apply the method to multi-bit rate WSONs. A typical case is a WSON where some lightpaths has been upgraded from 10Gbit/s (OTU-2) to 40Gbit/s (OTU-3). The two bit rates require a tailored regeneration and thus an OTN-PS having both the OTN layers. An even more appealing scenario consists on the evolution of OTN Switches supporting the ODU-flex container. This could serve also a WSON based on a flexi grid comb. Standards and technologies are not yet mature for this scenario but the principle contained in this method will remain valid.

- Network application

The application landscape of the proposed node setup is a WSON with a high meshing degree. A PCE, centralized or distributed, can leverage on a set of OTN-PS enabled nodes spread all over the network. In the event of failures not recoverable using PP schemes, the PCE can compute OTF recovery paths also with long un-planned detours through spare OTN-PS nodes. Paths that were discarded in the planning phase, for poor signal quality or lack of wavelength continuity, can now be considered by involving an OTN-PS device. For a better understanding a simple example is proposed in Figure 5-43.

In (a) a worker lightpath LSP1 (W) is activated between Node A and Node B using the "orange" wavelength. A pre-planned (PP) alternate path LSP2 is ready for recovery in case of single fault. A couple of faults, Figure 5-43 and Figure 5-44 in (b), affect both LSP1 and LSP2. In

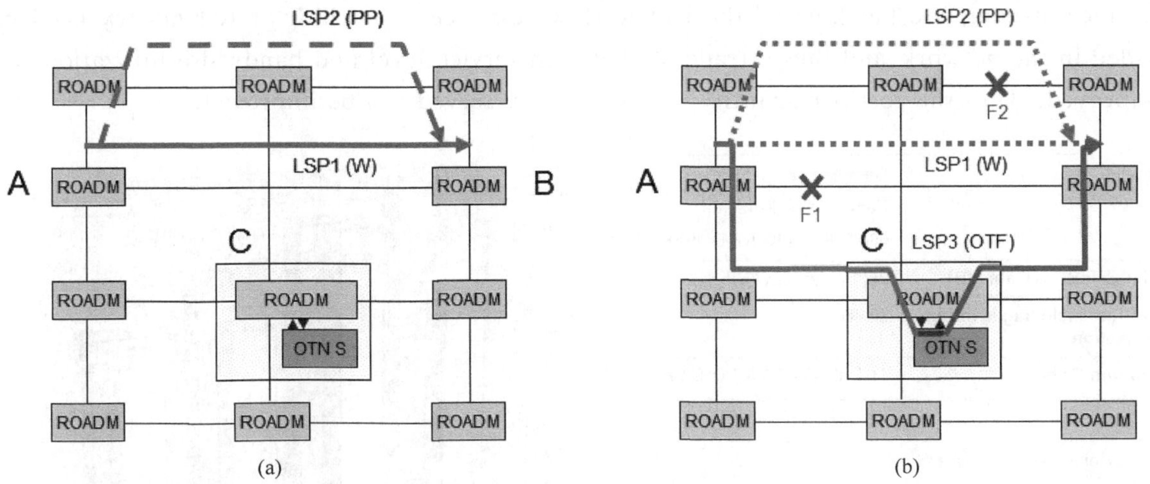

Fig. 5-43 Example of application of the OTN-PS method.

current art of WSON, if no other PP LSPs are available, the traffic is lost. With the proposed node setup, assuming an OTN-PS station in Node C, a new feasible LSP3 path can be activated OTF. Note that the OTN-PS is used here to allow a longer detour (more kms usually require regeneration in between) and to solve the wavelength continuity puzzle from Node A to Node B across this new path (by swapping from "orange" to "pink" in Node C). When one or both the failures are repaired, the traffic could be reverted to the original path(s) and the valuable OTN-PS resources can be released to serve future unpredicted failure events.

The optimal number of OTN-PS enabled nodes with respect to the total number of nodes in the network, and their location shall be defined in the planning phase also considering the desired level of resiliency and blocking probability. Another parameter to be defined during the planning phase is the portion of wavelengths that are sent to the OTN Switch, with respect to the number of wavelengths in the comb. This ratio has an impact on the dimension of the OTN Switch/Matrix, and as e consequence on its cost, but more savings will be achieved at a network level by having the same level of resiliency with a minor required hardware redundancy ("less PP, more OTF").

- Simulations

The behavior of the proposed OTN-PS solution has been tested by means of a custom built C++ planning tool in a demonstration WSON network taken from a publicly accessible network set as shown in Figure 5-44. Eight cases have been considered varying the number of OTN-PS nodes (from 0% to 30% with respect to the total number on nodes, on X axis) and the number of wavelengths potentially served by the OTN-PS device (10% and 20%, on Y axis). On the Z axis is displayed the proportion of optical circuits that have been declared feasible and routed using 40G RZ-DQPSK interfaces. Results hold in an average sense because they depend on the order of traffic population. If the proposed method is not used results indicates $Z = 56\%$. With a 30% of OTN-PS enabled nodes having a 20% of regenerable wavelengths, the best result of $Z = 98\%$ is achieved. Intermediate results are reported in the graph. Results demonstrate that the proposed method can greatly increase the reliability of finding at least one path avoiding the fault, which is increasingly useful as planned paths fill up more of the capacity to increase bandwidth utilization efficiency, or is useful in providing more resilience to multiple faults over a wide area. By reducing the chance of complete blocking of the traffic flow, less free bandwidth or redundancy need be provided in the network, and thus a trade-off between service level and bandwidth utilization can be improved. Thus the cost per quantity of traffic flows served can be improved.

Network	Germany 50 Nodes/Links/Demands 50/88/662 (40G RZ-DQPSK)
NFigure [dB] vs Gain [dB]	8@15, 7@20, 6@25
WSS loss including leveling attenuation	22 dB
Minimum OSNR	17.5 dB @ 40G RZ-DQPSK

X	% of nodes with OTN-PS
Y	% of wavelengths regenerable by OTN-PS
Z	% of demands (optical circuits) that have been declared feasible and routed

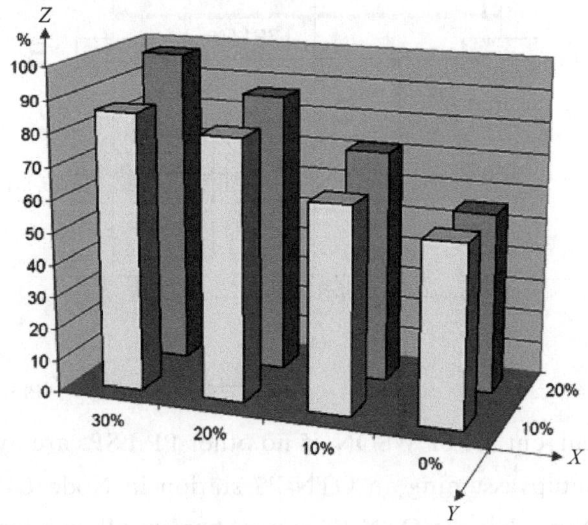

Fig. 5-44 Simulation results for the OTN-PS device.

The proposed OTN-PS solution is the enabler of the OTF approach in WSON: a recovery functionality claimed by operators to reduce the total cost of ownership of their networks. OTN-PS does not require modifications of conventional ROADM architecture because the "pit-stop" service is provided by connecting an OTN Switch or Matrix externally to the equipment. Simulations on 40G optical channels indicate an average reduction of the blocking probability from 44% to 2% if 30% of the network nodes are equipped with OTN-PS, each processing a 20% of the pass-through wavelengths.

5.6.3 Silicon optical phased array

Free-space optical chip-to-chip and board-to-board interconnects offer advantages with respect to packaging and density over waveguide-based approaches due to parallelism and obviation of the need for fiber attachment. The capability to steer the beam is important for such applications and others such as optical scanning, LIDAR, crossconnect switching, and in order to prevent optical misalignment and hence increased insertion loss and crosstalk from thermal or mechanical disturbances. Among the approaches used to accomplish chip-scale free-space beam steering are tunable gain elements, piezo-electric micro-stages, reconfigurable liquid-crystal phase gratings, MEMS microlenses and micromirrors. An approach that avoids mechanical motion can be advantageous in terms of robustness and susceptibility to vibration, while compatibility with standard CMOS silicon processing by fabricating the device in the silicon-oninsulator (SOI) platform is desirable for ease of fabrication and on-chip electronic integration. Furthermore such a platform allows the integration of the free-space beam steerer with tunable optical sources and amplifiers via hybrid integration of Ⅲ-Ⅴ materials with SOI optical waveguides. Optical phased arrays using surface waveguide gratings in SOI have been demonstrated using a combination of wavelength scanning and a single thermo-optic phase tuner for a steering range of $2.3^{\circ}\times 14^{\circ}$, but suffered from the lack of a means to actively eliminate phase errors introduced by fabrication variation and thermal crosstalk. An alternate technique in which a star coupler was integrated with a grating array such that wavelength alone could be used to scan the beam across the far field was also demonstrated, thus eliminating the need for phase tuning altogether at the expense of beam width (4°) and the ability to actively shape the wavefront. Individually phase-tuning the channels in an SOI waveguide optical phased array solves these problems and has been demonstrated for 1D beam steering in a silicon slab. Suppression of side-lobes in such a device is important both to direct a higher fraction of optical power into the central peak (thus enhancing efficiency) and to avoid optical crosstalk between adjacent free-space optical links. Here report a 16-channel independently tuned optical phased array fabricated in SOI for 2D free-space beam steering over a 20°×14° field of view with 1.6°×0.6° beam width, and we investigate the effect of rib waveguide width on far field side-lobe suppression. The emission out coupling angle was determined by wavelength in one axis and by relative phase between emitters in the other axis. Resistive heaters were used to phase-tune individual channels via the thermo-optic effect, and a hill-climber optimization algorithm together with real-time far field feedback from an automated image analyzer was used to minimize phase errors and suppress background peaks. Solution sets were recorded for wavelengths from 1525nm to 1625nm, where a solution was defined by meeting the condition in which beam intensity exceeded the maximum background peak height by 10dB for a field of view chosen so as to exclude the side-lobes. The recorded phase solutions were then used to steer the beam without realtime feedback using a lookup table.

- Fabrication

Rib waveguides were photolithographically defined in 500nm top silicon, 1μm buried oxide SOI. The waveguides had 1μm width and 280±20nm trench depth. The beam was separated into 16 channels spaced at 100μm intervals using 1×2 multi-mode interferometers (MMI), and a separate phase tuning element was fabricated on each channel by e-beam deposition of 72nm/75nm nickel-chrome/gold to form 470μm×4μm resistive heaters adjacent to each waveguide with a 6.5μm separation to avoid metal optical absorption. The top silicon between adjacent channels was etched to the buried oxide so as to minimize thermal crosstalk. The grating array was fabricated with 50% duty cycle, 600nm pitch and 200μm length using e-beam lithography and etched 75nm deep. Within the grating array the waveguide spacing was 3.5μm; this spacing determined the angular separation between the central peak and side-lobes, and hence the angle over which the beam could be swept without introducing side-lobes into the field of view. Waveguide widths within the grating array of 1μm, 2μm, and 3μm were used to evaluate the effect of rib widths on the side-lobe peaks. Schematic diagrams of the device and a scanning electron microscope image of the grating array are shown in Figure 5-45 and Figure 5-46.

- Characterization

A high-numerical-aperture aspheric lens (NA=0.83, effective focal length (EFL)=15mm) was used to collect the optical output and image it into the Fourier plane; two additional lenses (EFL 18cm and 6cm respectively) were used to image the Fourier plane onto an infrared camera for real-time far field imaging following the approach described in. Polarization was aligned along the TE axis using a polarization controller. Beam steering in the longitudinal θ axis (i.e. the axis parallel to the waveguides) was determined by wavelength and measured to be 0.14°/nm with a beam width (FWHM) of 0.6° for the 1μm waveguide array, while steering in the ψ axis (i.e. the axis perpendicular to the waveguides) was determined by relative phase at the emitters.

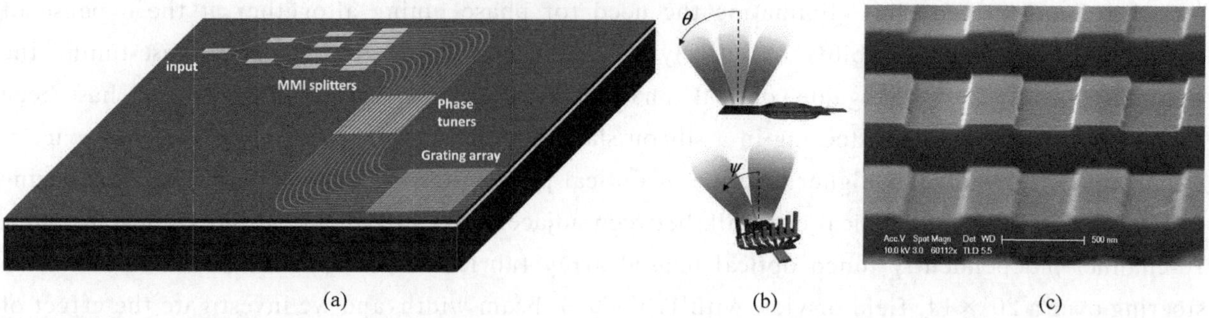

Fig. 5-45 (a) Schematic diagram of the device. (b) Illustration of the longitudinal θ emission angle determined by wavelength (top) and lateral emission angle Ψ determined by phase (bottom). (c) Scanning electron microscope image of the grating array.

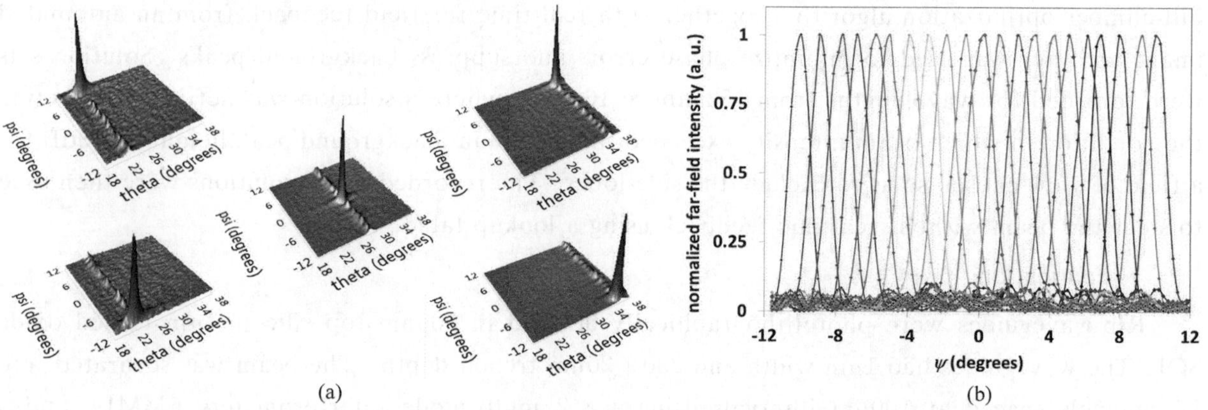

Fig. 5-46 (a) Far field beam profiles for phase and wavelength set to steer the beam to the corners and center of the field of view. The low boundary (left) corresponds to emission at 1625nm, the center corresponds to 1575nm, and the high θ boundary (right) to 1525nm. (b) Cross sections in the ψ axis of the far field beam profile at 1555nm for beam steering at 1° increments with 10dB background suppression.

An optimization algorithm was used to solve for phase solutions with 10dB background suppression within a 20°(ψ axis)×14°(θ axis) field of view at 1° increments. Once solved, these phase settings were stored in a lookup table and used to sweep the beam without real-time feedback. Measured profiles of the beam targeted to the corners and center of the field of view are shown in Figure 5-47(a); cross-sections of the beam in the ψ axis at a wavelength of 1555nm are shown in Figure 5-47(b) for the 1μm waveguide width array. Beam width in ψ was measured to be 1.6°.

With waveguide spacing fixed at 3.5μm, the relative power distribution in the side-lobes versus the central peak was determined by the emitter width, i.e. the rib waveguide width within the grating array. Measured and calculated cross-sections in ψ of the far field profile at 1625nm, where the calculated profile was determined by summing the far field contributions of emitter amplitudes corresponding to a cross-section at the grating etch depth (75nm) of the fundamental mode for each waveguide width. As expected, side-lobe suppression improved for increased rib width, with side-lobe peak heights of 0.6, 0.55, and 0.28 relative to the central peak for rib widths of 1μm, 2μm and 3μm respectively. The 16-channel optical phased array in silicon for 2D free-space beam steering with independently tuned channels. The device exhibited beam steering over a 20°×14°

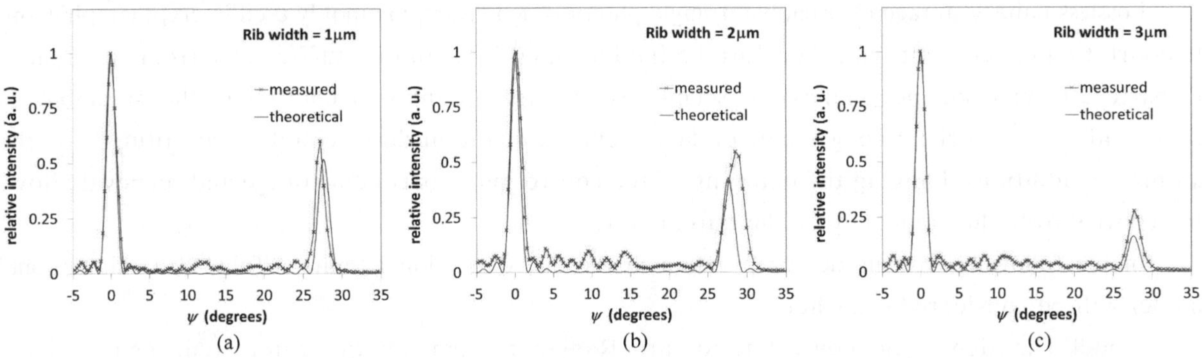

Fig. 5-47 Measured and calculated cross sections of the far field in ψ for 1625nm wavelength showing the first side-lobe for (a) 1mm waveguide width, (b) 2mm waveguide width, and (c) 3mm waveguide width.

field of view with $1.6^{\circ} \times 0.6^{\circ}$ beam width and 10dB background suppression. The effect of waveguide width on side-lobe suppression was investigated and found to decrease side-lobe peak power from 60% to 28% of the central peak height for waveguide widths ranging from 1μm to 3μm. This suggests that the efficiency of the optical phased array can be enhanced by optimizing the waveguide width.

5.6.4 Single-photon wavepackets to atomic memory

The conversion of optical information between photons and atoms forms the basis of a quantum interface that is a critical component of quantum communications networks and distributed quantum computers. Such quantum information processing schemes require the ability both to move quantum information between nodes of a network, and to store the information.

An important class of quantum optical memories is based on the interaction of individual photons with atomic ensembles, in which the information is transferred coherently from a single photon to a collective excitation of the atoms. If active feedback is not used, these memories are generally optically dense, with the density controlled dynamically using an ancillary field.

In this section will analyze a prototypical memory based on the off-resonant Raman interaction of a classical pulsed control field and a broadband signal photon in an atomic medium (see Figure 5-48). The temporal structure of the signal photon is transferred by the control to a long-lived collective atomic excitation, or spin wave. This differs qualitatively from previous narrowband schemes which demonstrated that quadrature squeezing could be transferred from optical fields to a collective atomic spin. We show how proper shaping of the control field allows the mapping of an input wavepacket of arbitrary temporal shape to an output wave packet of a potentially different temporal shape. The dynamics are closely related to those investigated in proposals for entanglement generation via spontaneous and, more recently, stimulated Stokes scattering. However the photon storage process is distinct from these, in that it exhibits an explicit time reversal symmetry; evinced by its fundamental mode structure. Spontaneous emission is suppressed as long as the excited state remains empty.

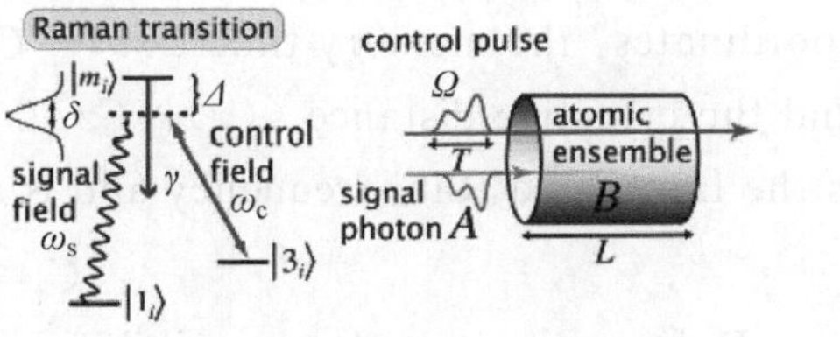

Fig. 5-48 (a) the level structure of the ith atom comprising a quantum memory for broadband photons, with bandwidth δ. (b) a schematic of the read-in process for the quantum memory.

Lossless unitary storage of broadband single photons-such as are commonly used in cryptographic and teleportation experiments can therefore be implemented by detuning sufficiently from resonance. Departure from resonance makes our scheme robust against inhomogeneities in the ensemble, so that solid state absorbers (e.g. semiconductor charge quantum dots) could be substituted for the atoms. In addition, changing the detuning of the control pulse between storage and retrieval allows for control over the frequency of the output state.

In the following we consider propagation in the one dimensional limit; a fully three dimensional model will be considered elsewhere.

Model The signal and control fields are Raman resonant, with center frequencies ω_s, ω_c, respectively. The classical control at time t and position z is represented by the Rabi frequency (τ), where $\tau \equiv t - z/c$ is the local time. The signal and spin wave amplitudes are described by the slowly varying annihilation operators $A(\tau, z)$, and $B(\tau, z)$, respectively. The spin wave is a collective coherence of the form $B(\tau, z) \propto \Sigma_\beta \parallel e^{-i(\omega_s - \omega_c)\tau}$, where the index β runs over all atoms with position z. If the common detuning Δ of the signal and control pulses from single photon resonance is much larger than the signal bandwidth δ, the control Rabi frequency Ω, and the control bandwidth, the excited state $|m_i\rangle$ can be adiabatically eliminated. If the ensemble is prepared in the collective state $|\equiv|$, and if the population of the metastable state $|\equiv|$ is assumed to remain negligible, a linear theory can be used. The Maxwell-Bloch equations, in the slowly varying envelope approximation, are then found to be:

$$[\partial_\tau - i|\Omega(\tau)|^2/\Gamma]B(\tau, z) = -\kappa^*\Omega^*(\tau)A(\tau, z)/\Gamma \tag{5-97}$$

$$[\partial_z - i|\kappa|^2/\Gamma]A(\tau, z) = \kappa\Omega(\tau)B(\tau, z)/\Gamma \tag{5-98}$$

where κ is the signal field coupling and $\Gamma \equiv \Delta - i_\gamma \equiv |\Gamma|e^{-i\theta}$ is the complex detuning, with real phase θ. γ arises from dephasing processes, including spontaneous emission. As have not included the Langevin noise operator which formally accompanies these loss terms, since its contribution vanishes when normally ordered expectation values of A and B are taken. Here neglect the slow decay of the spin wave over the short timescale of the memory interaction. The new set of scaled coordinates: the memory time $\varepsilon(\tau) \equiv C\omega(\tau)/\omega(T)$, where T is the duration of the interaction, and the effective distance $\zeta(z) \equiv Cz/L$, with L the length of the ensemble. Here $\omega(\tau) \equiv R$ which is the integrated Rabi frequency and is a coupling parameter.

$$C \equiv |\kappa|\sqrt{L\omega(T)}/|\Gamma| \tag{5-99}$$

Define dimensionless annihilation operators for the optical field (assuming κ is real for simplicity) and

$$\alpha(\varepsilon, \zeta) \equiv \sqrt{\omega(T)/C}\, e^{-i\chi(\tau, z)} A(\tau, z)/\Omega(\tau) \tag{5-100}$$

for the spin wave, where the exponent is

$$\beta(\varepsilon, \zeta) \equiv \sqrt{L/C}\, e^{-i\chi(\tau, z)} B(\tau, z) \tag{5-101}$$

$$\chi(\tau, z) \equiv [\omega(\tau) + |\kappa|^2 z]/\Gamma \tag{5-102}$$

The first term in χ describes a Stark shift due to the control field; the second represents a modification of the signal group velocity. With these changes, the equations of motion reduce to the simple coupled system $\partial\zeta\alpha = e^{i\theta}\beta; \partial Q\beta = -e^{i\theta}\alpha$.

The solution of these equations then holds for all control pulse shapes and arbitrary inputs. The coupling parameter C sets the size of the region in (ε, ζ)-space over which the memory interaction is driven. For the case considered here, an atomic ensemble, C can be re-written in the form $C \doteq (\pi\alpha_{f}\hbar/m_{e})^{1/2} f\sqrt{NaNc}/(|\Gamma|A)$, where f is the geometric mean of the oscillator strengths for the signal and control transitions, A is the cross-sectional area of the control field, and $Na(Nc)$ is the number of atoms (photons) interacting with (comprising) the control pulse. Here α_{f} is the fine structure constant, and here is the electron mass. The memory read-in and read-out must be unitary to function correctly. We now show that canonical evolution of the field operators α and β is guaranteed by the classical structure of Eqs. (5-103), (5-104) in the dispersive limit $\Delta \gg \gamma$. In this case, the phase θ vanishes and the following continuity relation is satisfied:

$$\partial_{\zeta}\alpha^{\dagger}\alpha + \partial_{\epsilon}\beta^{\dagger}\beta = 0 \tag{5-103}$$

Integration of this expression over a square in (ϵ, ζ)-space yields the flux-excitation conservation condition

$$N_{\alpha}(C) + N_{\beta}(C) = N_{\alpha}(0) + N_{\beta}(0) \tag{5-104}$$

$$\int_{0}^{C}\alpha^{\dagger}(\epsilon,\zeta)\alpha(\epsilon,\zeta)\mathrm{d}\epsilon, \text{and } N_{\beta}(\epsilon) \equiv \int_{0}^{C}\beta^{\dagger}(\epsilon,\zeta)\beta(\epsilon,\zeta)\mathrm{d}\zeta \tag{5-105}$$

where the number operators $N_{\alpha}(\zeta)\equiv$, count the number of signal photons at an effective distance ζ, and the number of excitations of the spin wave at memory time Q, respectively. Eq. (5-105) must hold for arbitrary initial amplitudes $\alpha_{0}(\epsilon)\equiv\alpha(\epsilon,0)\{\alpha_{0}(\epsilon),\beta_{0}(\zeta)\}\rightarrow\{\alpha_{C}(\epsilon),\beta_{C}(\zeta)\}$ and $\beta_{0}(\zeta)\equiv\beta(0,\zeta)$, which fixes the transformation $\{\alpha_{0}(\epsilon),\beta_{0}(\zeta)\}\rightarrow\{\alpha_{C}(\epsilon),\beta_{C}(\zeta)\}$ as unitary (where $\alpha C(\epsilon)\equiv\alpha(\epsilon,C)$ is the signal amplitude at the exit face of the ensemble, and $\beta_{C}(\zeta)\equiv\beta(C,\zeta)$ is the spin wave amplitude at the end of the read-in process. This allows the dynamics to be decomposed into a set of independent transformations between light-field and spin wave modes. In what follows we therefore concentrate on the dispersive limit, and consider the case $\Gamma\rightarrow\Delta;\theta\rightarrow 0$. The solution of the dynamical equations is expressed by the scattering relations,

$$\alpha_{C}(\epsilon) = \int_{0}^{C}[G_{1}(\varepsilon - x, C)\alpha_{0}(x) + G_{0}(C - x, \epsilon)\beta_{0}(x)]\mathrm{d}x \tag{5-106}$$

$$\beta_{C}(\zeta) = \int_{0}^{C}[G_{1}(\zeta - x, C)\beta_{0}(x) - G_{0}(C - x, \zeta)\alpha_{0}(x)]\mathrm{d}x \tag{5-107}$$

The integral kernels are given by $G_{0}(p,q)\equiv J_{0}(2\sqrt{pq})$, and

$$G_{1}(p,q) \equiv \delta(p) - \Theta(p)J_{1}(2\sqrt{pq})\sqrt{q/p} \tag{5-108}$$

with the nth Bessel function of the first kind denoted by J_{n}, and where the Heaviside step function Θ ensures that the convolutions in Eqs. (5-106), (5-107) respect causality.

The integral kernels $G_{0,1}$—as they appear in Eqs. (5-106), (5-107)—share symmetry under reflection about the line $C - x = y$, where y stands for the independent variable: either ε or ζ. This symmetry, along with Eq. (5-107), allows to decompose the kernels using input and output modes related by time reversal (or equivalently space reversal) as follows:

$$G_{0}(C - \epsilon, \zeta) = \sum_{i=1}^{\infty}\phi_{i}(\zeta)\lambda_{i}\phi_{i}(C - \epsilon) \tag{5-109}$$

$$G_{1}(\zeta - \epsilon, C) = \sum_{i=1}^{\infty}\phi_{i}(\zeta)\mu_{i}\phi_{i}(C - \epsilon) \tag{5-110}$$

where $\{\phi_{i}\}$ is a complete orthonormal set of real modefunctions and where the real, positive singular values satisfy the constraint $\lambda_{i}^{2} + \mu_{i}^{2} = 1$. The ensemble begins the read-in process in the state $|0\rangle$, and we are free to replace $\beta_{0}(\zeta)$ by its expectation value; $\beta_{0}(\zeta)\rightarrow 0$. With the above decomposition, Eq. (5-111) then describes a mapping of the optical input mode $\phi_i(C-\varepsilon)$ to the output spin wave mode $\phi_i(\zeta)$, with transfer amplitude-λ_i, for each i. Transforming from memory time ε back to local time τ, the normalized input modes are written:

$$\Phi_i(\tau) \equiv \sqrt{C/\omega(T)}\mathrm{e}^{\mathrm{i}\chi(\tau,0)}\Omega(\tau)\phi_i[C-\varepsilon(\tau)] \tag{5-111}$$

The read-in efficiency can be quantified by evaluating the expectation value of the spin wave number operator $\langle N_\beta(C)\rangle$ at the end of the read-in process. Expanding an incident signal wavepacket $\xi(\tau)$ using the Φ_i, the readin efficiency is expressed as $\langle N_\beta(C)\rangle = \Sigma_i \lambda_i^2 |\xi_i|^2$, with the ith overlap given by

$$\xi_i \equiv \int_0^T \xi^*(\tau)\Phi_i(\tau)\mathrm{d}\tau \tag{5-112}$$

When $\langle N_\beta(C)\rangle = 1$, the read-in works perfectly. The first five transfer amplitudes are plotted as a function of C. These are found using the eigenvalue equation

$$\int_0^C J_0(2\sqrt{xy})\phi_i(y)\mathrm{d}y = \lambda_i\phi_i(x) \tag{5-113}$$

which solve numerically using a 500 by 500 square grid. It is desirable to limit the energy of the Figure 5-49: The five largest singular values of the kernel G_0, plotted as a function of the coupling parameter C.

The intensity $\langle A^\dagger(\tau,z)A(\tau,z)\rangle$ of a Gaussian signal photon |, with wavepacket amplitude $\xi(\tau) \propto \exp\{-2\ln 2[(\tau-\tau_0)/\sigma]^2\}$, as it propagates through an atomic ensemble with $C=2$. Here $\sigma = T/8$, $\tau_0 = 2T/3$. The optimized control field intensity is shown alongside the initial signal field intensity (scaled for clarity). So can find the minimum coupling parameter C which permits complete storage of the signal. For $C \approx 2$ the lowest mode achieves its optimal efficiency $\lambda_1 \approx 1$, but higher modes remain poorly coupled. The efficiency of the memory is therefore maximized by setting:

$$\xi_1 = 1; \xi_i' = 0, \text{so that } \langle N_\beta(C)\rangle = \lambda_1^2 \approx 1 (\text{for } C \geqslant 2) \tag{5-114}$$

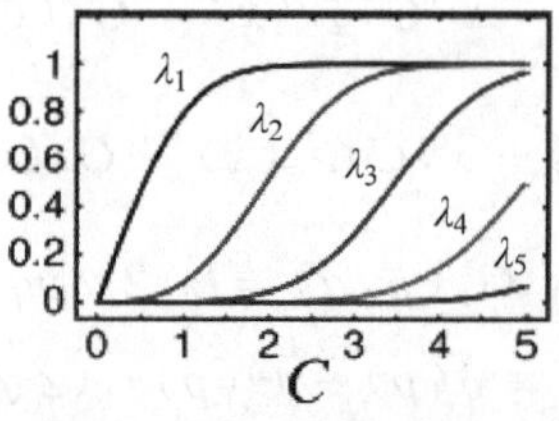

The first five transfer amplitude curves.

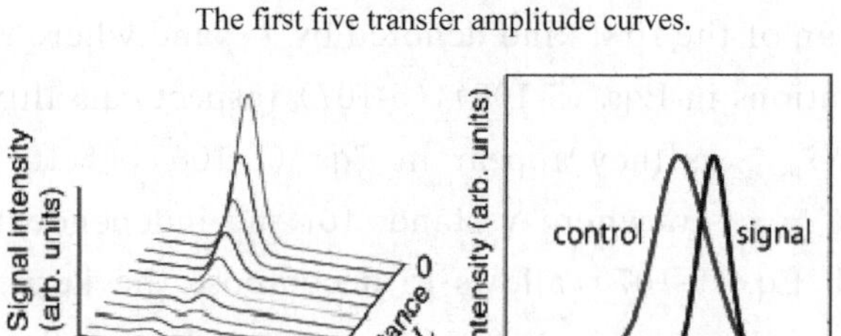

Fig. 5-49 Intensity of a Gaussian signal photon.

To do this, it is necessary to shape the control field so that $\Phi_1(\tau) = \xi(\tau)$. If Gaussian optics ($A \sim cL/\omega_s$) are used to illuminate a region a few cm long in a typical atomic vapour ($f \sim 1$) of modest density ($\sim 1020\mathrm{m}^{-3}$), with 100nJ control pulses, a 1ps photon wavepacket can be stored optimally, with $C=2$. In practice modematching could be achieved through measurement and feedback: the signal and control field sources are locked to a phase reference and operated in pulsed mode. The control pulse profile is characterized, and augmented until the transmission of the signal is minimized. Figure 5-50 shows the result of a simple optimization to find the control pulse shape which modematches the lowest input mode to a Gaussian signal photon, for $C=2$. The photon is absorbed with $\langle N_\beta(C)\rangle \approx 0.96$; the small transmission probability is due to the limitations of our numerical modematching optimization.

Read-out once a properly mode matched photon has been read in to the quantum memory, the ensemble is shown in Figure 5-50.

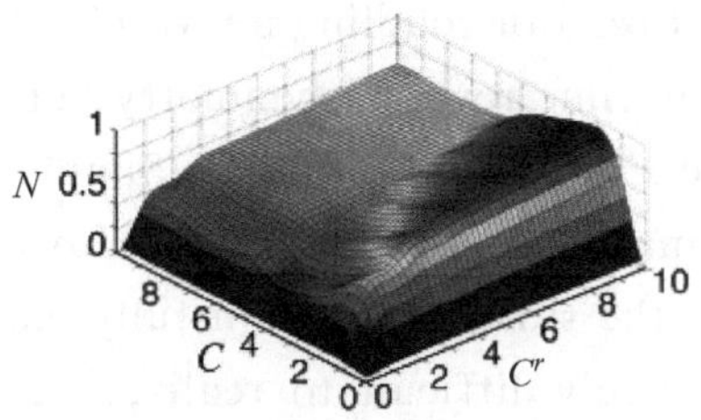

Fig. 5-50 The photon retrieval probability N, for forward readout, plotted as a function of the read-in and read-out coupling parameters C and C'. left in the output mode φ_1 (ζ), with probability amplitude $-\lambda_1$.

Now consider the effect of sending a second control pulse, propagating in the same direction as the initial control pulse, into the ensemble. The center frequency, bandwidth, and intensity of this read-out pulse may differ from that of the first control pulse (herein the read-in pulse). Let use a superscript r to indicate those quantities associated with the read-out.

$$\Psi_i^r(z) \equiv \sqrt{C'/L}\,\mathrm{e}^{\mathrm{i}\chi^r(0,z)}\,\phi_i^r[C'(1-z/L)] \tag{5-115}$$

If neglect decoherence and dephasing of the spin wave over the storage period and set $B^r(0, z)=B(T, z)$.

$$\Psi_i^r(z) \equiv \sqrt{C'/L}\,\mathrm{e}^{\mathrm{i}\chi^r(0,z)}\,\phi_i^r[C'(1-z/L)] \tag{5-116}$$

This provides us with one boundary condition; the second is that the signal field begins in its vacuum state at the start of the read-out process, $\langle N_\alpha^r(0)\rangle = 0$. The efficiency of the read-out depends upon the degree to which the spin wave mode (written in terms of the ordinary spatial variable z) overlaps with the input modes for the read-out process. A measure of the efficiency of the memory is the expectation value of the output photon number operator

$$N \equiv \langle N_\alpha^r(C')\rangle = \lambda_1^2 \sum_i \lambda_i^{r2} \mid f_i \mid^2 \tag{5-117}$$

with the read-out overlaps defined by

$$f_i \equiv \int_0^L \varphi_1^*(z)\Psi_i^r(z)\mathrm{d}z \tag{5-118}$$

The parameter N is the probability of retrieving a photon from the ensemble at read-out, given that a single modematched photon was sent in with the read-in pulse. If the detuning does not change too much, so that

$$\mid \Delta' - \Delta \mid /\Delta \ll \Delta'/(\mid \kappa \mid^2 L) \sim \sqrt{N_c^r/N_a} \tag{5-119}$$

then the phases χ, χ' of the spin wave modes approximately cancel, and then the stored spin wave is phasematched to the readout modes. In Figure 5-50 the variation of N, under this approximation, is plotted as a function of the read-in and read-out coupling parameters C and C'. If $C=C'$, then the lowest read-out mode is just the mirror image of the spin wave mode. For small C, the spin wave mode is monotonic, and relatively flat; f_1 is therefore large. However the transfer amplitudes λ_1 and λ_1^r remain small, so the retrieval probability is low. Increasing λ_1 requires a larger C, but this produces a more asymmetric spin wave mode, and f_1 falls. It is then n necessary to increase C' above C, so that higher modes, with which the spin wave mode overlaps significantly, are efficiently coupled to the optical field. The retrieval probability is maximized along the line $C \approx 2$, which represents the optimal coupling for the read-in process. However, a read-out coupling parameter in excess of 10 is required to achieve $N \geqslant 0.95$. Note also that modematching to the lowest mode at read-in is the best strategy for maximizing the memory efficiency. Modematching to a higher mode, or some combination of modes, simply increases the optimal read-in coupling above $C = 2$.

The time reversal symmetry between the input and output modes makes the read-out for this scheme a nontrivial problem. Simply repeating the read-in process (so that $C = C'$) results in poor performance of the memory. However, the dramatic increase in coupling strength required to extract the stored excitation fully, may make a naïve increase in control pulse energy at read-out prohibitively difficult to realize. An alternative method to boost the coupling is to reduce the bandwidth of the readout pulse, along with its detuning Δ'. The photon recovered from the memory in this way would be frequency shifted (according to the Raman resonance condition), and temporally stretched (since its bandwidth would be diminished as well). Such a memory would act as a "photon transducer", storing broadband photons and converting them to narrowband photons with tunable frequency on demand. Note that switching the propagation direction of the read-out pulse sends $\varphi_i(C_z/L) \to \varphi_i[C(1-z/L)]$, so that the spin wave mode overlaps exactly with the lowest read-out mode with $C' = C$, and we should obtain $N = \lambda_1^4$. Unfortunately the read-out process is no longer phasematched in this situation, and the overlap integrals f_i vanish. However, a solid-state implementation might allow this kind of reverse read-out with the use of quasihasematching, in which the sign of the read-out coupling parameter C' is periodically flipped along the length of the ensemble.

The phasematching problem is obviated in the limit of vanishing Stokes shift, but then the ground state | XXX must be prepared artificially with high purity: if the state | is initially populated, or if selection rules allow residual coupling of the control to the ground state, the memory fidelity at the level of single quanta is greatly reduced. Furthermore, if signal and control are spectrally indistinguishable, another degree of freedom should be used to differentiate between them. Typically, polarization affords discrimination to a part in ~106, but then the control should not contain more than a million photons. A large Stokes shift is therefore desirable, and correct phasematching at read-out is crucial for efficient retrieval. The above considerations demonstrate the importance of propagation effects in the design of an optical quantum memory.

Summary have shown that the off-resonant Raman configuration for a Λ-type ensemble quantum memory can be used to implement deterministic, controllable, and unitary transfer of the temporal structure of broadband single photons to a stationary spin wave, in the adiabatic regime. The dynamics are understood and optimized using a universal mode decomposition, valid for all control pulses and arbitrary input states. The modes are computationally simple to evaluate, and only a few of them are required to approximate the interaction faithfully. The optimal fidelity of the memory depends only upon a single dimensionless parameter (C), which defines an equivalence class for memories with different physical implementations. The authors gratefully acknowledge the support of Hewlett-Packard and the EPSRC (UK) through the QIP IRC.

5.6.5 Solid state light-matter interface at photon

Coherent and reversible mapping of quantum information between light and matter is an important experimental challenge in quantum information science. In particular, it is a decisive milestone for the implementation of quantum networks and quantum repeaters. So far, quantum interfaces between light and atoms have been demonstrated with atomic gases, and with single trapped atoms in cavities. The coherent and reversible mapping of a light field with less than one photon per pulse onto an ensemble of $\sim 10^7$ atoms naturally trapped in a solid. This is achieved by coherently absorbing the light field in a suitably prepared solid state atomic medium. The state of the light is mapped onto collective atomic excitations on an optical transition and stored for a pre-programmed time up of to 1μs before being released in a well defined spatio-temporal mode as a result of a collective interference. The coherence of the process is verified by performing an interference experiment with two stored weak pulses with a variable phase relation. Visibilities of more than 95% are obtained, which demonstrates the high coherence of the mapping process at the single photon level. In addition, experimentally interface allows one to store and retrieve light fields in multiple temporal modes. The results represent the first observation of collective enhancement at the single photon level in a solid and open the way to multimode solid state quantum memories as a promising alternative to atomic gases.

Efficient and reversible mapping of quantum states between light and matter requires strong interactions between photons and atoms. With single quantum systems, this regime can be reached with high finesse optical cavities, which is technically highly demanding. In contrast, light can be efficiently absorbed in ensembles of atoms in free space. Moreover, it is possible to engineer the atomic systems such that the stored light can be retrieved in a well defined spatio-temporal mode due to a collective constructive interference between all the emitters. This collective enhancement is at the heart of protocols for storing photonic quantum states in atomic ensembles, such as schemes based on Electromagnetically-Induced Transparency (EIT), off-resonant Raman interactions and modified photon echoes using Controlled Reversible Inhomogeneous Broadening (CRIB) and Atomic Frequency Combs (AFC).

All previous quantum storage experiments with ensembles have been performed using atomic gases as the storage material. However, some solid state systems have properties that make them very attractive for applications in quantum storage. In particular, rare-earth ion doped solids provide a unique physical system where large ensembles of atoms are naturally trapped in a solid state matrix, which prevents decoherence due to the motion of the atoms and allows the use of trapping free protocols. Moreover, these systems also exhibit excellent coherence properties at low temperature (below 4K), both for the optical and spin transition. Contributed equally to this work, times enable storage of multiple temporal modes in a single QM, which promises significant speed-up in quantum repeater applications. Furthermore, high optical densities can be obtained in rare-earth doped solids, which is required to achieve strong light matter coupling resulting in high efficiency light storage and retrieval. However, despite recent experimental progress, the implementation of a solid state light matter quantum interface has not been reported so far.

The coherent states of light with less than one photon per pulse onto a large number of atoms is in a solid as shown in Figure 5-51. The mapping is done by coherently absorbing the light in an ensemble of inhomogeneously broadened atoms spectrally prepared with a periodic modulation of the absorption profile. The reversible absorption by such a spectral grating is at the heart of the recently proposed multimode quantum memory scheme based on AFC. A proof of principle of an essential primitive of this protocol is the single photon level. Assume an incident weak coherent state of light $|\alpha\rangle_L$ with a mean photon number $n = |\alpha|^2 < 1$. After absorption the photons are stored in a coherent superposition of collective optical excitations de-localized over all the atoms in resonance with the light field. The state of the atoms (not normalized) can be written as

$$|\alpha\rangle_A = |0\rangle_A + \alpha\,|1\rangle_A + O(\alpha^2)$$

$$|0\rangle_A = |g_1 \cdots g_N\rangle \text{ and } {}^4I_{9/2} \rightarrow {}^4F_{3/2} \tag{5-120}$$

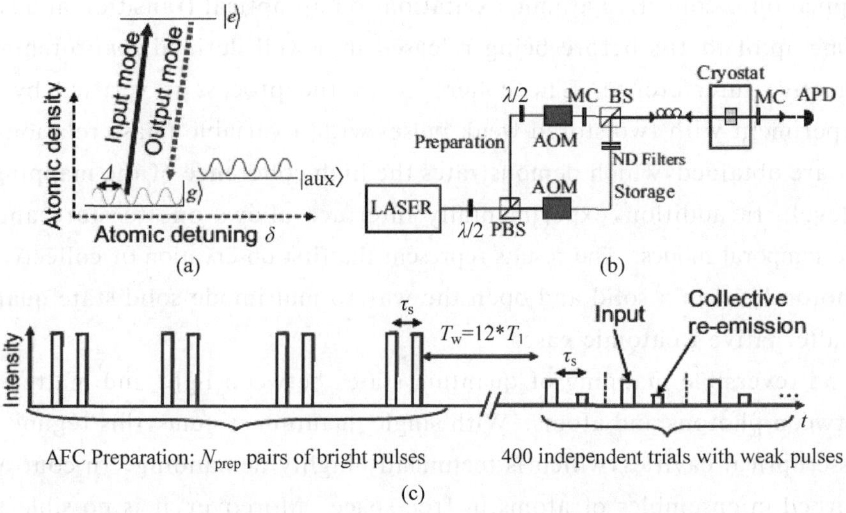

Fig. 5-51 Solid state light matter interface. The light is absorbed on the (${}^4I_{9/2} \rightarrow {}^4F_{3/2}$) transition of Nd^{3+} ions at 880nm.

The inhomogeneous broadening of the optical transition is 2GHz, with a maximal optical depth around 4. The optical relaxation time T_1 of the excited state $|e\rangle$ is equal to 100μs. The ground state is split into two Zeeman levels $|g\rangle$ and $|\text{aux}\rangle$, separated by 3.8GHz through the application of a magnetic field of 300mT. A spectral grating is prepared in $|g\rangle$ by a preparation sequence described in the Methods section. The laser source is a cw external cavity diode laser at 880nm. The laser is split at a variable beam splitter and the pulse sequences for the preparation of the grating and for the pulses to be stored are prepared by independent acousto-optics modulators in different optical paths. The duration of the pulses is about 30ns. These paths are then recombined at a beam splitter (BS) and coupled into a single mode fiber to ensure proper mode matching. The light is then focused onto 1mm long Nd^{3+}: YVO_4 crystal with a beam diameter of 30μm. The crystal is cooled down to 3K by a pulse tube cooler. After the sample, the light passes through a polarizer and is coupled back to a single mode fiber, which is connected to a Silicon Avalanche Photo diode single photon counter. In order to block the preparation light during the storage and to protect the detector from the intense preparation light, two mechanical choppers (MC) are used. The experimental sequence is divided into two parts: the preparation of the spectral grating (see Methods) and the storage of the weak pulses. We wait a time $T_w = 12T_1 = 1200\mu$s between the preparation and the storage sequence in order to avoid fluorescence. During the storage sequence, 400 independent trials are performed at a repetition rate of 200kHz. The entire sequence preparation plus storage is then repeated with a repetition rate of 40Hz.

$$|1\rangle_A = \sum_i c_i e^{i\delta_i t} e^{-ikz_i} |g \cdots e_i \cdots g\rangle \tag{5-121}$$

where z_i is the position of atom i (for simplicity, only consider a single spatial mode defined by the direction of propagation of the input field), k is the wavenumber of the light field, δ_i the detuning of the atom with respect to the laser frequency and the amplitudes ci depend on the frequency and on the spatial position of the particular atom i. This collective state will rapidly diphase since each term acquires an individual phase $e^{i\delta_i t}$ depending on the detuning. However, due to the periodic structure of the absorption $2\pi/\Delta$: profile, the collective state will be re-established after a pre-programmed time here Δ is the period of the spectral grating. This leads to a coherent photon-echo type re-emission in the forward spatial mode. In experiment the light field is stored as a collective excitation on the optical transition, contrary to all previous experiments at the single photon level, where collective excitations of spin states were used. Conceptually, this is an important difference since in our case the light is simply absorbed in the prepared material, without any control field.

The solid state interface is implemented in an ensemble of Neodymium ions (Nd^{3+}) doped into a YVO_4 crystal. The Nd^{3+} ions constitute an ensemble of inhomogeneously broadened atoms having a relevant level structure with two spin ground states $|g\rangle$ and |auxi and one excited state $|e\rangle$. Initially, the two ground states are equally populated for all frequencies over the inhomogeneous broadening. The preparation of the spectral grating is realized by frequency selective optical pumping from $|g\rangle$ to | via the excited state $|e\rangle$ (see Figure 5-52 for an overview of the experiment). This is implemented with a Ramsey type interference using a train of coherent pairs of pulses (see Figure 5-53).

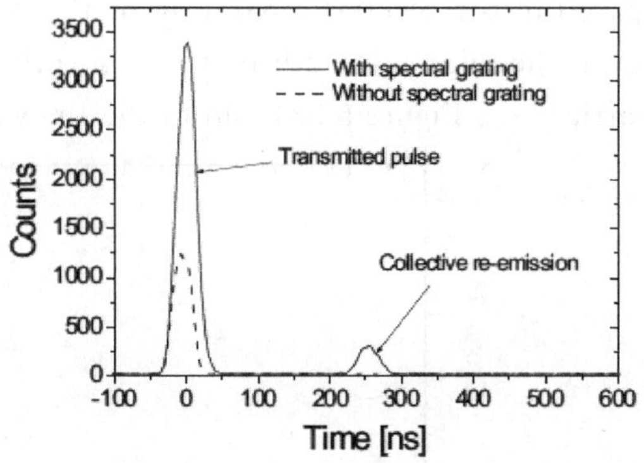

Fig. 5-52 Reversible mapping of a coherent state with $n = 0.5$. The solid line corresponds to the case where a spectral grating is prepared with a periodicity of 4MHz. The peak 250ns after the transmitted input pulse corresponds to the collective retrieval after storage in the solid state medium. The dashed line corresponds to the case where the atomic medium is not prepared with a spectral grating. In that case, we only see the transmitted input pulse (about 2 percent of the incoming pulse is transmitted). The absorption of the input pulse is smaller when the spectral grating is prepared (about 5 percent of the light is transmitted).

To store weak light fields, it is required that there is no population in the excited state, otherwise fluorescence will blur the signal. This is ensured by waiting long enough between the preparation and storage sequences, such that all atoms have returned to the ground states.

As a result of the preparation sequence, a spectral grating is present in $|g\rangle$ before the storage begins, which decays with the population relaxation lifetime T_Z between the spin states ($T_Z = 6$ms in our case).

In the first experiment, demonstrate collective mapping of weak coherent states $|a\rangle L$ on the crystal. An example with $n = 0.5$ is shown in Figure 5-52 (See Methods for the estimation of n). When the sample is prepared with a spectral grating having a periodicity of 4MHz, to observe a strong emission at the expected storage time of 250ns. About 0.5 percent of the incoming light is re-emitted in this signal. This is more than 4 orders of magnitude more than what would be expected from a non collective re-emission, taking into account that we collect a solid angle of 2×10^{-4} and that the optical relaxation time is 3 orders of magnitude longer than the observed signal. This signal thus clearly arises from a collective re-emission, which demonstrates the collective and reversible mapping of a light field with less than one photon onto an large number of atoms in a solid. To further study the mapping process, we record the number of counts in the observed signal for different n ranging from 0.2 to 2.7 (see Figure 5-53(a)). This shows that the mapping is linear and that very low photon numbers can still be mapped and retrieved. Now also investigated the decay of storage efficiency with the storage time (see Figure 5-53(b)). The efficiency of the storage and retrieval and its decay as a function of storage time can be qualitatively understood using the theory of the AFC quantum memory. In order to obtain high storage efficiencies for a given storage time, it is essential to create a spectral grating with narrow absorption peaks as compared to the spectral separation of the peaks, i. e. a high finesse grating. In the experiment, however, the minimal width of the absorption peak is of the order of 1～2MHz, due to material properties (see Figure 5-53) and to the linewidth of our free-running laser.

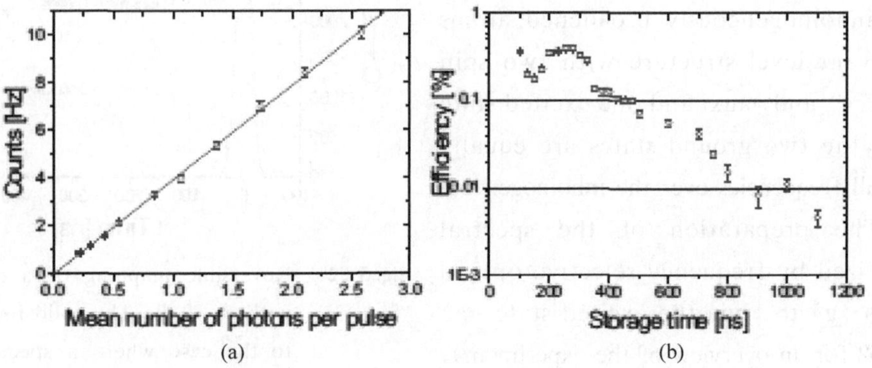

Fig. 5-53 (a) Number of detections in the collective output model as a function of n. (b) Efficiency as a function of the storage time for $n=2.7$. For small storage times (<400ns), an oscillation of the efficiency is clearly visible. This a quantum beat due to the interaction of the electronic spin of the Nd^{3+} ion with the nuclear spin of the surrounding Vanadium ions (called super hyperfine interaction). For longer storage times, the decay is exponential with decay constant of 220ns. This interaction also limits the minimal width of the absorption peaks.

Hence the grating is close to a sinusoidal curve, which limits the storage efficiency and causes the observed decay of the efficiency vs storage time. Moreover, there is still a significant flat absorption background due to imperfect optical pumping in the present experiment. This background can be considered as a loss which strongly limits the observed storage efficiency. However, these are not fundamental limitations of rare-earth materials. The preparation of narrow absorption lines with low absorption background has been demonstrated for several other rare-earth materials, such as Eu: Y_2SiO_5, Pr: Y_2SiO_5 and Tm: YAG. Significantly higher storage efficiencies should be obtained in these materials. Note also that the experiment described here implements a memory device with a fixed storage time. In order to allow for on-demand read-out of the stored field (as required for quantum repeaters) and storage times longer than that given by the spectral grating, the excitations in $|e\rangle$ can be transferred to a third ground state spin level $|s\rangle$. This transfer also enables in principle storage efficiencies close to unity.

So far it is considered the storage of weak light fields in a single temporal mode. However the use of spectral gratings also allows for the storage in multiple temporal modes. The maximal number of modes that can be stored is given by the ratio of the storage time (determined by the spectral grating) to the duration of an individual mode. To illustrate this multimode property store trains of four weak pulses with n from 0.8 to 0.3 during 500ns, as shown in Figure 5-55. It is important to note that the time ordering of the pulses is preserved during the storage, which results in the same storage efficiency for each mode. This is in contrast with:

$$|\psi\rangle_L = |\alpha_t\rangle_L + e^{i\Phi}|\alpha_{t+\tau}\rangle_L \tag{5-122}$$

The spectral grating is prepared for a storage time of 500ns. The first four pulses are the transmitted pulses. After 500ns, we can clearly see the collective re-emission of the four temporal modes. The signal to noise ratio is smaller than for the single mode case (see Figure 5-54), due to the longer storage time of 500ns (see Figure 5-55). For clarity, the output signal part has been magnified by a factor of ten. CRIB based quantum memories where the time ordering is reversed. The number of stored modes for a given storage time can be improved by using shorter preparation pulses, resulting in larger bandwidth of the spectral grating. In this case, shorter pulses can be stored. In the present experiment, the shortest duration of pulses was set to about 20ns (FWHM) by technical limitations. A great advantage of the AFC protocol is that the number of modes that one can store does not depend on the optical depth, contrary to EIT and CRIB. For applications in quantum memories, it is crucial $|\alpha_t\rangle_L$ that the interface conserves the phase of the incoming pulses. To probe the coherence of the mapping process at the single photon level, use a pair of weak pulses separated by a time $\tau = 100$ ns with a fixed relative phase ϕ. This can be seen as a time bin qubit, which can be written as

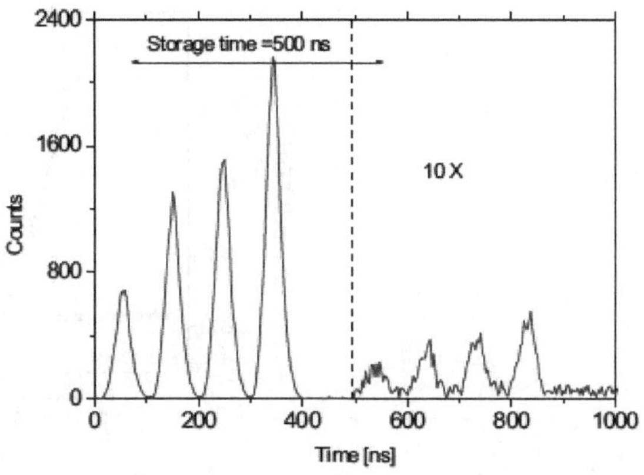

Fig. 5-54 The multimode property store trains of four weak pulses with n from 0.8 to 0.3 during 500ns.

$$|\psi\rangle_L = |\alpha_t\rangle_L + e^{i\Phi}|\alpha_{t+\tau}\rangle_L \tag{5-123}$$

where $|a_t\rangle_L$ represents a weak coherent state at time t. The qubit is stored and thereafter analyzed directly in the memory. This method requires the implementation of partial read-outs at different times. This realizes a projection on a superposition basis, similarly to what can be done with a Mach-Zehnder. If the time τ between the weak pulses matches the time between the two read-outs, the re-emission from the sample can be suppressed or enhanced depending on the phase difference between the two incoming pulses. The visibility of this interference is a measure of the coherence of the mapping process. The two partial read-outs are here achieved by preparing two super-imposed spectral gratings having different periods, corresponding to storage times of 200 and 300ns.

An example of interference for $\bar{n} = 0.85$ is shown in Figure 5-55. Dark-counts subtracted visibilities above 95% have been obtained for various n between 0.4 and 1.7, which demonstrates the high coherence of the storage process, even at the single photon level. From these visibilities, one can infer conditional fidelities above 97% for the storage of single photons. This excellent phase preservation at the single photon level is obtained thanks to the collective enhancement effect and to the almost complete suppression of background noise. In conclusion, we have demonstrated the coherent and reversible mapping of a light field at the single photon level onto a solid. Our results show that the storage of single photons (Fock states) in multiple temporal modes in solids is possible. So can also demonstrated that the quantum coherence of the incident weak light fields is almost perfectly conserved during the storage. Solid state systems can therefore be considered as a promising alternative to atomic gases for photonic quantum storage. This line of research holds promise for the implementation of efficient long distance quantum networks.

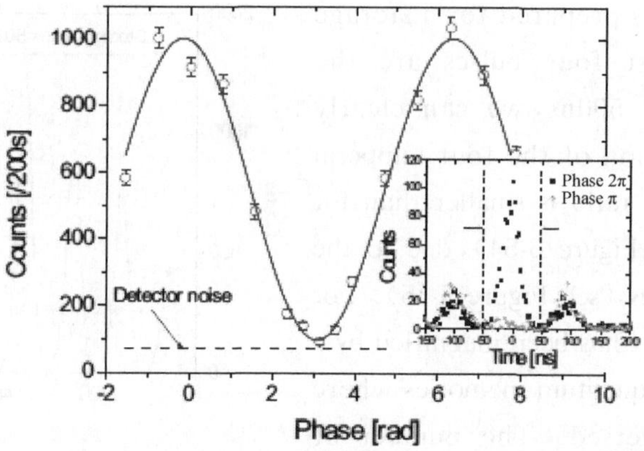

Fig. 5-55 Interference fringe. Time-bin qubits with different phases are stored and analyzed using the interface. The analysis is performed by projecting the time-bin qubit on a fixed superposition basis, which here is achieved by two partial read-outs. The inset shows the histogram of arrival times, where there is a constructive ($\Phi = 2\pi$) and destructive interference ($\Phi = \pi$) in the middle time bin. For this particular interference fringe, obtain a raw visibility of 82%, or 95% when subtracting detector dark counts.

- Methods

The preparation of the spectral grating is realized by a series of pairs of pulses of area $\theta < \pi/2$ resonant with the $|g\rangle \rightarrow |e\rangle$ transition. Each pair of pulses realizes a frequency selective coherent transfer of population from $|g\rangle \rightarrow |e\rangle$. This can be seen as a Ramsey type interference where the two light pulses play the role of beam splitters and the phase shift acquired in the excited state depends on the detuning of the atoms. The periodicity of the created spectral grating is then given by the inverse of the time interval τ_s between the two pulses. The atoms in the excited state can decay to both ground states $|g\rangle$ and $|aux\rangle$ with a relaxation time $T_1 = 100\mu s$. The atoms that decay to $|aux\rangle$ are not affected by the preparation laser and remain in this state for a time of 6ms. The pulse sequence is then repeated 100 times with a time separation between the pairs of 15μs, longer than the optical coherence time $T_2 = 7\mu s$. This allows for the build up of the spectral grating, with population storage in $|aux\rangle$. To estimate the mean number of photons per pulse, we shift the laser out of resonance with the absorbing atoms and record the proportion of detections in the single photon counter (typically between 1% and 20%). By a careful measurement of the detection efficiency ($\eta_D = 0.32$) and of the transmission efficiency from the input face of the cryostat to the detector (typically $\eta_t = 0.2$), one can finally infer the mean number of photons in front of the cryostat, before the sample.

5.6.6 Photon memory in atomic vapor

The study procedures for the optimization of efficiency of photon memory and retrieval based on the dynamic form of electromagnetically induced transparency (EIT) in warm Rb vapor. Here present a detailed analysis of two recently demonstrated optimization protocols: a time-reversal-based iteration procedure, which finds the photon input signal pulse shape for any given control field, and a procedure based on the calculation of a photon control field for any given signal pulse shape. To verify that the two procedures are consistent with each other, and that they both independently achieve the maximum memory efficiency for any given optical depth. Can observe good agreement with theoretical predictions for moderate optical depths, while at higher optical depths the experimental efficiency falls below the theoretically predicted values. We identify possible effects responsible for this reduction in memory efficiency. The ability to store light pulses in matter and then retrieve them while preserving their quantum state is an important step in the realization of quantum networks and certain quantum cryptography protocols. Mapping quantum states of light onto an ensemble of identical radiators (e. g., atoms, ions, solid-state emitters, etc.) offers a promising approach to the practical realization of quantum memory. Recent realizations of storage and retrieval of single-photon wave packets, coherent states, and squeezed vacuum pulses constitute an important step in demonstrating the potential of this method. However, the efficiency and fidelity of the storage must be significantly improved before practical applications become possible.

In this section will present a comprehensive analysis of two recently demonstrated optimization protocols that are based on a recent theoretical proposal. The first protocol iteratively optimizes the input pulse shape for any given control field, while the second protocol uses optimal control fields calculated for any given input pulse shape. Experimentally demonstrate their mutual consistency by showing that both protocols yield the same optimal control-signal pairs and memory efficiencies. That for moderate optical depths, the experimental results presented here are in excellent agreement with a simple three-level theoretical model with no free parameters; discuss the details of the correspondence between the actual atomic system and this simple model. Lastly, study the dependence of memory efficiency on the optical depth that for higher optical depths, the experimental efficiency falls below the theoretically predicted values; we discuss possible effects, such as spinwave decay and four-wave mixing that may limit the experimentally observed memory efficiency.

In this paragraph briefly review the necessary concepts from the theoretical work on which experiments rely. Consider the propagation of a weak signal pulse with envelope $\varepsilon(t)$ and a strong (classical) control field with a Rabi frequency envelope in a resonant Λ-type atomic medium under the conditions of electromagnetically induced transparency (EIT), as shown in Figure 5-56(a). The control field creates a strong coupling between the signal field and a collective atomic spin excitation (spin wave). As a result, the initial pulse gets spatially compressed and slowed down inside the atomic ensemble. The group velocity of the pulse is proportional to the control field intensity:

$$v_g \approx 2|\Omega|^2/(\alpha\gamma) \ll c \tag{5-124}$$

where γ is the decay rate of the optical polarization and α is the absorption coefficient (i. e., unsaturated absorption per unit length), so that αL is the optical depth of an atomic medium of length L. Figure 5-56(b) illustrates schematically the three stages of the light storage process

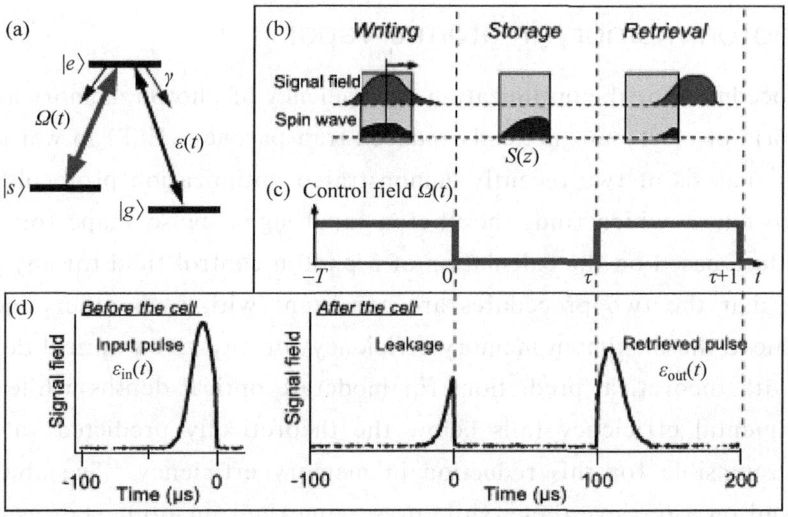

Fig. 5-56 (Color online) (a) The three-level Λ scheme used in theoretical calculations. The schematic (b) and example control (c) and signal (d) fields during light storage. At the writing stage ($t<0$), an input signal pulse $\varepsilon_{in}(t)$ propagates through the atomic medium with low group velocity v_g in the presence of a control field envelope $\Omega(t)$. While compressed inside the cell, the pulse is mapped onto a spin-wave $S(z)$ by turning the control field off at time $t=0$. After a storage period τ, the spin-wave is mapped back into an output signal pulse $\varepsilon_{out}(t)$ using the retrieval control field envelope $\Omega(t)(t>\tau)$.

(writing, storage, and retrieval), while Figure 5-56(c) and (d) show control and signal fields, respectively, during a typical experimental run. At the writing stage, a signal pulse $E_{in}(t)$ is mapped onto the collective spin excitation $S(z)$ by adiabatically reducing the control field to zero. This spin wave is then preserved for some storage time τ (storage stage), during which all optical fields are turned off. Finally, at the retrieval stage, the signal field $E_{out}(t)$ is retrieved by turning the control field back on. In the ideal case, the retrieved signal pulse is identical to the input pulse, provided the same constant control power is used at the writing and the retrieval stages. However, to realize this ideal storage, two conditions must be met. On the one hand, the group velocity vg of the signal pulse inside the medium has to be low enough to spatially compress the whole pulse into the length L of the ensemble and avoid "leaking" the front edge of the pulse past the atoms. This requires $Tv_g \ll L$, where T is the duration of the incoming signal pulse. On the other hand, all spectral components of the incoming pulse must fit inside the EIT transparency window to minimize spontaneous emission losses $1/T \ll \Delta\omega_{EIT}$. The simultaneous satisfaction of both conditions is possible only at very high optical depth $\alpha L \gg 1$. Experimental realization of very high optical depth in atomic ensembles requires high atomic density and/or large sample length. At high atomic density, EIT performance can be degraded by competing processes, such as stimulated Raman scattering and four-wave mixing. Furthermore, spin-exchange collisions and radiation trapping may reduce spin wave lifetime by orders of magnitude, limiting storage time and signal pulse durations. In addition, achieving high optical depth in some experimental arrangements may be challenging, such as in magneto optical traps. Therefore, it is crucial to be able to maximize memory efficiency by balancing the absorptive and leakage losses at moderately large αL via optimal shaping of control and/or signal temporal profiles. To characterize our memory for light, define memory efficiency η as the probability of retrieving an incoming single photon after storage or equivalently, as the energy ratio between initial and retrieved signal pulses:

$$\eta = \frac{\int_{\tau}^{\tau+T} |\varepsilon_{out}(t)|^2 dt}{\int_{-T}^{0} |\varepsilon_{in}(t)|^2 dt} \tag{5-125}$$

The goal of any optimization procedure then is to maximize η under the restrictions and limitations of a given system. In the theoretical treatment of the problem, the propagation of a signal pulse in an idealized three-level Λ system, shown in Figure 5-58(a), is described by three complex, dependent variables, which are functions of time t and position z. These variables are the slowly-varying envelope ε of the signal field, the optical polarization P of the $|g\rangle$-$|e\rangle$ transition, and the spin coherence S. The equations of motion for these variables are

$$(\partial_t + c\partial_z)\varepsilon(z,t) = ig\sqrt{N}P(z,t) \tag{5-126}$$

$$\partial_t P(z,t) = -\gamma P(z,t) + ig\sqrt{N}\varepsilon(z,t) + i\Omega(t - z/c)S(z,t) \tag{5-127}$$

$$\partial_t S(z,t) = -\gamma_s S(z,t) + i\Omega(t - z/c)P(z,t) \tag{5-128}$$

where $g\sqrt{N} = \sqrt{\frac{\gamma\alpha c}{2}}$ is the coupling constant between the atomic ensemble and the signal field, and s are the polarization decay rates for the transitions $|g\rangle$-$|e\rangle$ and $|g\rangle$-$|s\rangle$, respectively. In general, Eqs. (5-126) ~ (5-128) cannot be fully solved analytically, they reveal several important properties of the optimization process. These properties are most evident in the case when spin wave decay rate s is negligible during the processes of writing and retrieval ($\gamma_s T \ll 1$). In this case, the highest achievable memory efficiency depends only on the optical depth αL and the mutual propagation direction of the control fields during the writing and retrieval stages. For each optical depth, there exists a unique spin wave, $S_{opt}(z)$, which provides the maximum memory efficiency. Thus, the focus of the optimization process becomes identifying a matching pair of writing control and signal pulses that maps the signal pulse onto this optimal spin wave. Note that no additional optimization is required with respect to the retrieval control field, because the memory efficiency does not depend on it, provided spin wave decay is negligible during retrieval.

In the experiments, the optimization procedures are tested using weak classical signal pulses rather than quantum fields. Such experimental arrangements greatly improved the experimental simplicity and the accuracy of data analysis. At the same time, the linear equations of motion for classical and quantum signal pulses are identical, which makes the presented results applicable to quantized signal fields, such as, e. g., single photons. It is also important to note that the original theoretical work considered a wide range of interaction processes for storing and retrieving photon wave packets (e. g., EIT, far-off-resonant Raman, and spin echo techniques) under a variety of conditions including ensembles enclosed in a cavity, inhomogeneous broadening, and high-bandwidth nonadiabatic storage ($1/T \sim \alpha L\gamma$). Since the proposed optimization procedures are, to a large degree, common to all interaction schemes and conditions, our results are relevant to a wide range of experimental systems.

- Experiment

The schematic of the experimental apparatus is shown in Figure 5-57. That used an external cavity diode laser (ECDL) tuned near the ^{87}Rb D1 transition ($\lambda = 795$nm) with total available laser power≈45mW. After separating a fraction of original light for a reference beam using a polarizing beam splitter (PBS), the main laser beam passed through an electro-optical modulator (EOM), which modulated its phase at the frequency of the ground-state hyperfine splitting of ^{87}Rb(Δ_{HF} = 6.835GHz) and produced modulation sidebands separated by that frequency. Tuned the zeroth order (carrier frequency) field to the $^{52}S_1/^2 F = 2 \rightarrow ^{52}P_1/^2 F' = 2$ transition. This field was used as the control field during light storage. The +1 modulation sideband played the role of the signal field and was tuned to the $^{52}S_{1/2} F = 1 \rightarrow ^{52}P_{1/2} F' = 2$ transition. To carry out the optimization procedure, had to independently manipulate the amplitudes of the control and the signal fields. Used an acousto-optical modulator (AOM) to adjust the control field intensity. However, since all optical fields traversed the AOM, the intensities of all modulation comb fields were also changed. Thus, accordingly adjusted the r_f power at the EOM input (which controls the strength of the modulation sidebands) to compensate for any changes in the signal field amplitude caused by AOM modulation. To minimize the effects of resonant four-wave mixing, we filtered out the other (−1) first order modulation sideband (detuned by Δ_{HF} to the red from the carrier frequency field) by reflecting the modulation comb off of a temperature-tunable Fabry-Perot etalon (FSR = 20GHz, finesse≈100). The etalon was tuned in resonance with this unwanted modulated sideband, so that most of this field was transmitted. At the same time, the control and signal field frequencies were far from the etalon resonance, and were reflected back with no losses.

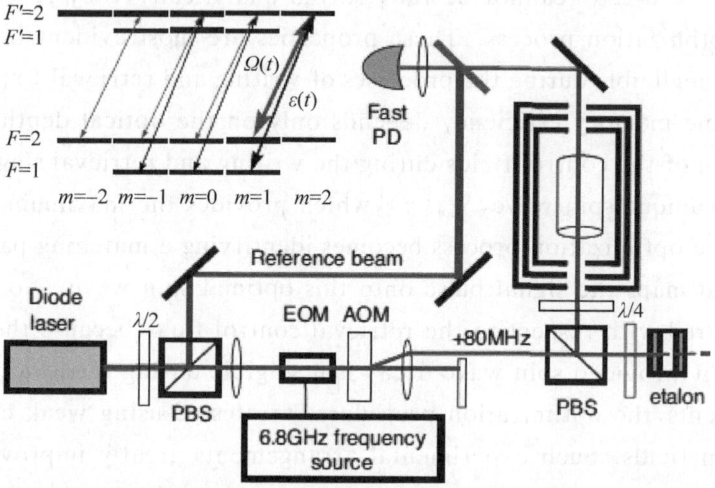

Fig. 5-57 (Color online) Experimental apparatus (see text for abbreviations). Inset: Schematic of the ^{87}Rb D1 line level structure and relevant Λ systems formed by control and signal fields.

Such filtering allowed for suppression of the −1 modulated sideband intensity by a factor of ~10. Typical peak control field and signal field powers were 18mW and 50μW, respectively. The beam was weakly focused to ~5mm diameter and circularly polarized with a quarter-wave plate ($\lambda/4$). A cylindrical Pyrex cell (length and diameter were 75mm and 22mm, respectively) contained isotopically enriched ^{87}Rb and 30 Torr Ne buffer gas, so that the pressure broadened optical transition linewidth was $2\gamma = 2\pi \times 290$MHz. The cell was mounted inside three-layer magnetic shielding to reduce stray magnetic fields. The temperature of the cell was controllably varied between 45℃ and 75℃ using a bifilar resistive heater wound around the innermost magnetic shielding layer. Used relatively short pulses, so that spin decoherence had a negligible effect during writing and retrieval stages and only caused a modest reduction of the efficiency during the storage time $\propto \exp(-2\gamma_s\tau)$. The Rb atom diffusion time out of the laser beam ($\approx$2ms) was long enough to avoid diffusion related effects on EIT dynamics. To extract the spin wave decoherence time by measuring the reduction of the retrieved pulse energy as a function of storage time and fitting it to an exponential decay. Can found the typical decay time to be $1/(2\gamma_s)\approx 500\mu s$, most likely arising from small, uncompensated, remnant magnetic fields.

After the cell, the output laser fields were recombined with the reference beam (at the unshifted laser frequency) at a fast photodetector, and the amplitude of each field was analyzed using a microwave spectrum analyzer. Because of the 80MHz frequency shift introduced by the AOM, the beatnote frequencies of the +1 and −1 modulation sidebands with the reference beam differed by 160MHz, which allowed for independent measurement of the amplitude of each of these fields, as well as of the control field.

To conclude this section, explain the direct cor respondence between the experimental system and the theory based on three-level atoms see Figure 5-58(a), that reviewed later. The goal is to use the structure of the D1 line of ^{87}Rb see inset in Figure 5-58 to identify the optical depth αL and the control field Rabi frequency for the effective three level system. First solve for the ground-state population distribution after control field optical pumping of the Rb D1 line, taking into account Doppler broadening, pressure broadening, and collisional depolarization of the excited state sublevels can find the depolarization to be fast enough. Given this population distribution, we calculate the optical depth αL for the signal field as a function of Rb number density. For example, we find that at 60.5℃ (Rb vapor density of $2.5\times10^{11}\text{cm}^{-3}$) the optical depth is $\alpha L=24.0$.

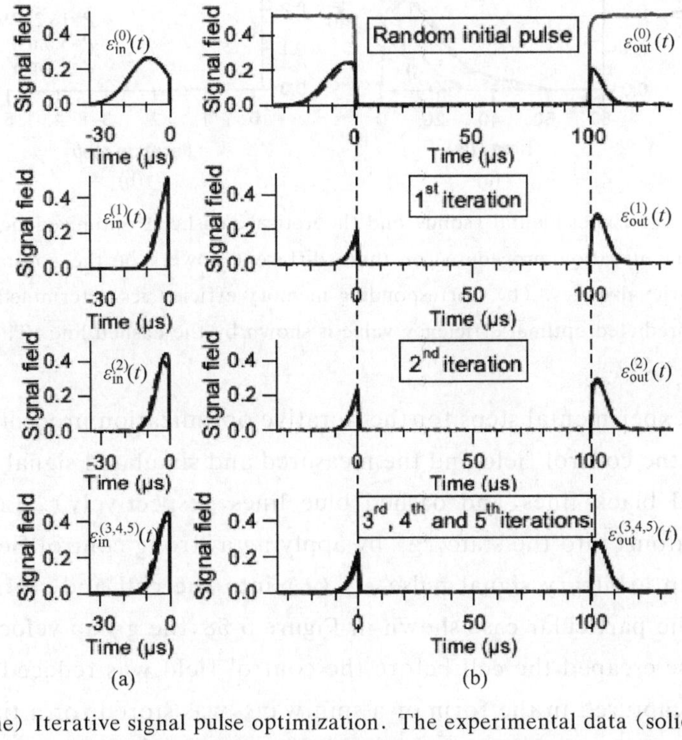

Fig. 5-58 (Color online) Iterative signal pulse optimization. The experimental data (solid black lines) is taken at 60.5℃ ($\alpha L=24$) using 16mW constant control field during writing and retrieval (solid red line in the top panel) with a $\tau=100\mu s$ storage interval. Numerical simulations are shown with blue dashed lines. (a) Input pulses for each iteration. (b) Signal field after the cell, showing leakage of the initial pulse for $t<0$ and the retrieved signal field ε_{out} for $t>100\mu s$.

Approximating a transverse Gaussian laser beam profile with a uniform cylindrical beam of diameter 5mm of the same power, can find that for the control power of 16mW, $\Omega = 2\pi \times 6.13$MHz for example. Since the collisionally broadened optical transition linewidth ($2\gamma = 2\pi \times 290$MHz) is comparable to the width of the Doppler profile, the effects of Doppler broadening are negligible. Note that all the theoretical modeling is done with no free parameters.

- Signal pulse optimization

One approach to the optimization of light storage is based on important time-reversal properties of photon storage that hold even in the presence of irreversible polarization decay. In particular, for co-propagating writing and retrieval control fields, the following is true under optimized conditions (see Figure 5-59): if a signal pulse $E_{in}(t)$ is mapped onto a spin wave using a particular control field (t) and retrieved after some storage time t using the time-reversed control field $(T-t)$, the retrieved signal pulse shape $E_{out}(t)$ is proportional to the time-reversed input signal pulse $E_{in}(T-t)$, but attenuated due to imperfect memory efficiency. Here and throughout the paper, control and signal envelopes are assumed to be real. This symmetry also gives rise to an experimentally realizable iteration procedure, which, for any given writing control field, determines the optimal incoming signal pulse shape. This procedure has been first demonstrated experimentally. The present experiment was performed independently on a different (although similar) experimental setup.

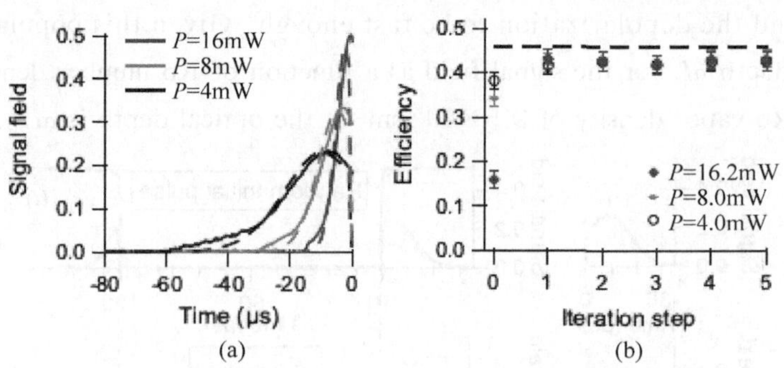

Fig. 5-59 (Color online) (a) Experimental (solid) and theoretical (dashed) optimized signal pulses obtained after five steps of the iteration procedure for three different powers of the constant control fields during writing and retrieval stages. (b) Corresponding memory efficiencies determined for each iteration step. Theoretically predicted optimal efficiency value is shown by the dashed line. The temperature of the cell was 60.5℃ ($\alpha L = 24$).

The sequence of experimental steps for the iterative optimization procedure is shown in Figure 5-58. The plots show the control field and the measured and simulated signal fields (solid red lines in the top panel, solid black lines, and dashed blue lines, respectively). Before each iteration, optically pumped all atoms into the state $|g\rangle$ by applying a strong control field. The optimization sequence by sending an arbitrary signal pulse $\varepsilon_{in}^{(0)}(t)$ into the cell and storing it using a chosen control field(t). In the particular case shown in Figure 5-58, the group velocity was too high, and most of the input pulse escaped the cell before the control field was reduced to zero. However, a fraction of the pulse, captured in the form of a spin wave, was stored for a time period $\tau = 100\mu$s. Then retrieved the excitation using a time-reversed control field $\Omega(t) = \Omega(\tau - t)$ and recorded the output pulse shape $\varepsilon_{out}^{(0)}(t)$. For the sample sequence shown, the control fields at the writing and retrieval stages were constant and identical. This completes the initial (zeroth) iteration step.

The efficiency of light storage at this step was generally low, and the shape of the output pulse was quite different from the time-reverse of the initial pulse. To create the input pulse $\varepsilon_{in}^{(1)}(t)$ for the next iteration step, which digitally timereversed the output $\varepsilon_{out}^{(0)}(t)$ of the zeroth iteration and renormalized it to compensate for energy losses during the zeroth iteration: $\varepsilon_{in}^{(1)}(t) \propto \varepsilon_{out}^{(0)}(\tau - t)$. Then, these steps were repeated iteratively until the rescaled output signal pulse became identical to the time-reversed profile of the input pulse. As expected, the memory efficiency grew with each iteration and converged to 43 ± 2%. To verify that the obtained efficiency is indeed the maximum possible at this optical depth and to confirm the validity of interpretation of the results, compare the experimental data to numerical simulations in Figure 5-60. Using the calculated optical depth and the control Rabi frequency, solve Eqs. (5-126)~(5-128) analytically in the adiabatic limit $T\alpha L\gamma \gg 1$, which holds throughout this section. There is a clear agreement between the calculated and measured lineshapes and amplitudes of the signal pulses. Also, theory and experiment converge to the optimal signal pulse shape in a similar number of iteration steps (see Eqs. (5-126)~(5-128)), and the experimental efficiency (43 ± 2%) converged to a value close to the theoretical limit of 45%.

As in previous study can confirmed that the final memory efficiency and the final signal pulse after a few iteration steps are independent of the initial signal pulse $E_{in}^{(0)}(t)$. Confirmed that the optimization procedure yields the same memory efficiency for different control fields. While constant control fields of three different powers yield different optimal signal pulses Figure 5-60(a), the measured efficiency Figure 5-60(b) converged after a few iteration steps to the same value of 43 ± 2%. With no spin wave decay, the highest achievable memory efficiency for the optical depth $\alpha L = 24$ is 54%. Taking into account spin wave decay during the 100μs storage time by a factor of $\exp[-100\mu s/500\mu s] = 0.82$, the highest expected efficiency is 45% dashed line in Figure 5-60(b), which matches the experimental results reasonably well.

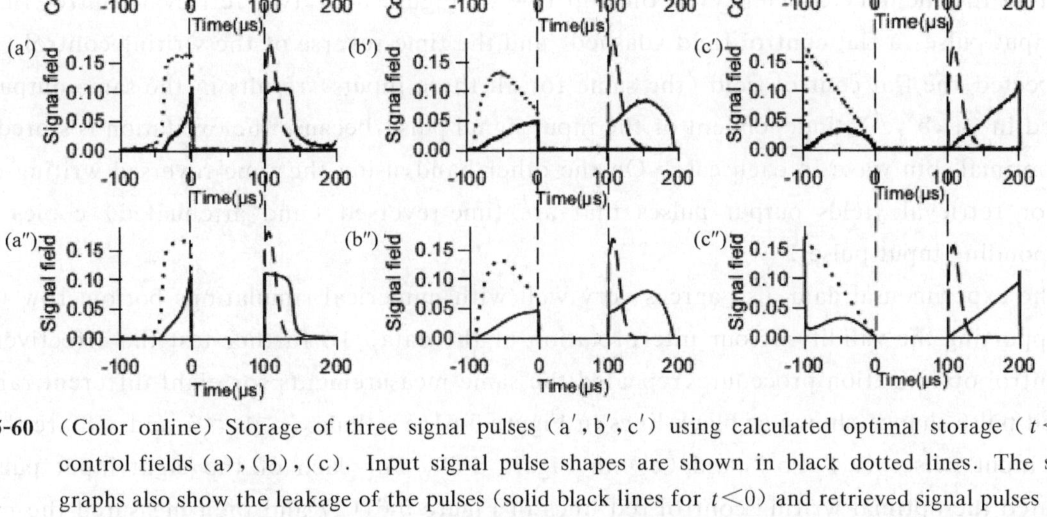

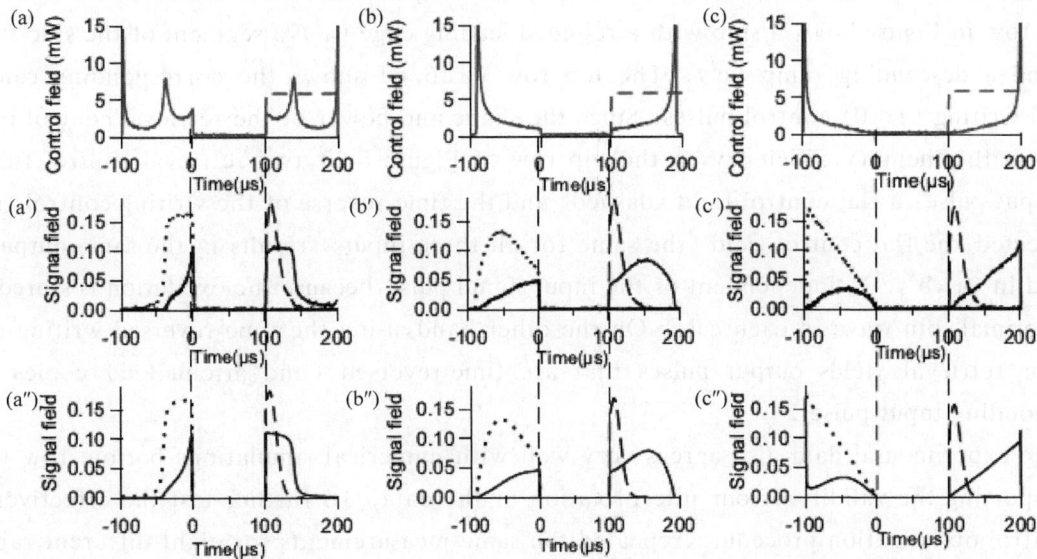

Fig. 5-60 (Color online) Storage of three signal pulses (a′, b′, c′) using calculated optimal storage ($t<0$) control fields (a), (b), (c). Input signal pulse shapes are shown in black dotted lines. The same graphs also show the leakage of the pulses (solid black lines for $t<0$) and retrieved signal pulses ($t>100\mu s$) using flat control fields at the retrieval stage (dashed red lines), or using time-reversed control fields (solid red lines). Graphs (a″, b″, c″) show the results of numerical calculations of (a′, b′, c′). The temperature of the cell was 60.5℃ ($\alpha L = 24$).

- Control pulse optimization

The iterative optimization procedure described in the previous section has an obvious advantage: the optimal signal pulse shape is found directly through experimental measurements without any prior knowledge of the system parameters (e. g., optical depth, control field Rabi frequency, various decoherence rates, etc.). However, in some situations, it is difficult or impossible to shape the input signal pulse (e. g., if it is generated by parametric down-conversion). In these cases, the control field temporal profile must be adjusted in order to optimally store and retrieve a given signal pulse.

To find the optimal writing control field for a given input pulse shape $\varepsilon_{\text{in}}(t)$, maximize η(see Eq. (5-125)) within the three-level model Eqs. (5-126)~(5-128). In this model, for a given optical depth αL and a given retrieval direction (coinciding with the storage direction in the present experiment), there exists an optimal spin wave $S_{\text{opt}}(z)$, which gives the maximum memory efficiency. One way to calculate the control field required to map the input pulse onto this optimal spin wave is to first calculate an artificial "decayless" spin wave mode $S(z)$, which, like $S_{\text{opt}}(z)$, depends only on the optical depth and not on the shape of the incoming pulse. This "decayless" mode $S(z)$ hypothetically allows for unitary reversible storage of an arbitrary signal pulse in a semi-infinite and polarizationdecay-free atomic ensemble, in which the group velocity of the pulse is still given by Eq. (5-125). The unitarity of the mapping establishes a 1-to-1 correspondence between a given input signal pulse shape $\varepsilon_{\text{in}}(t)$ and an optimal writing control field that maps this input pulse onto $S(z)$. The same control field maps this input pulse onto the true optimal spin wave $S_{\text{opt}}(z)$, once polarization decay and the finite length of the medium are taken into account. The details of this construction are described. As an example of control field optimization, we consider the storage of three different initial pulse shapes, shown by dotted black lines in the middle row in Figure 5-60: a step with a rounded leading edge (a′), a segment of the sinc-function (b′), and a descending ramp (c′). The top row (a, b, c) shows the corresponding calculated optimal writing ($t<0$) control pulses. Since the shape and power of the retrieval control pulse do not affect the memory efficiency, in the top row of Figure 5-60, two retrieval control fields for each input pulse: a flat control field (dashed) and the time-reverse of the writing control (solid). As expected, the flat control field (the same for all three inputs) results in the same output pulse [dashed in (a′, b′, c′)] independent of the input signal pulse, because the excitation is stored in the same optimal spin wave in each case. On the other hand, using the time-reversed writing control field for retrieval yields output pulses that are time-reversed (and attenuated) copies of the corresponding input pulses.

The experimental data also agrees very well with numerical simulations bottom row (a″, b″, c″), supporting the validity of our interpretation of the data. To further test the effectiveness of the control optimization procedure, repeated the same measurements for eight different randomly selected pulse shapes, shown as black lines in Figure 5-61(a). Pulses #4, #6, and #8 are the same as the input pulses (a′), (b′), and (c′) in Figure 5-60. For each of the eight input pulses, we calculated the optimal writing control red lines in Figure 5-61(a) and then measured the memory efficiency Figure 5-61(b), retrieving with either a constant control pulse or a time-reversed writing control pulse (open red diamonds and solid black circles, respectively). The measured efficiencies are in good agreement with each other and with the theoretically calculated maximum achievable memory efficiency of 45% (horizontal dashed line) for the given optical depth. By performing these experiments, found that knowledge of accurate values for the experimental parameters, such as optical depth or control field intensity, is critical for calculations of the optimal control field. Even a few percent deviation in their values caused measurable decreases in the output signal pulse amplitude. In the experiment, effective optical depth and control field Rabi frequency were computed accurately directly from measurable experimental quantities with no free parameters. Note that for some other systems, the necessary experimental parameters may be difficult to compute directly with high accuracy; in that case, they can be extracted from the iteration procedure.

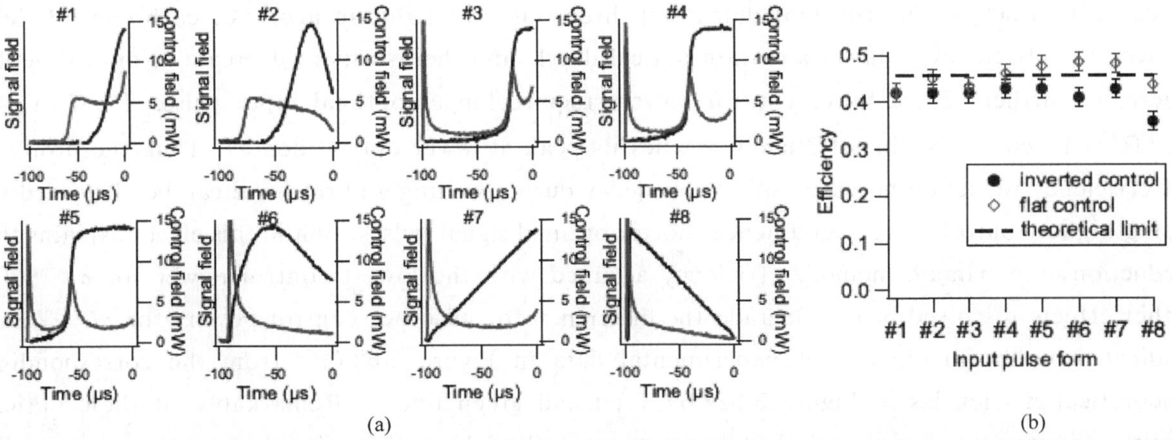

Fig. 5-61 (Color online) (a) Eight randomly selected signal pulse shapes (black lines) and their corresponding optimal control fields (red lines). (b) Memory efficiency for the eight signal pulse shapes using calculated optimized control fields at the writing stage, and flat control fields (open red diamonds) or inverted writing control fields (solid black circles) at the retrieval stage. Theoretically predicted optimal memory efficiency is shown by a dashed line. The temperature of the cell was 60.5℃.

- Dependence of memory efficiency on the optical depth

In the previous two sections, verified at optical depth $\alpha L = 24$, the consistency of the signal and control optimization methods and their agreement with the three-level theory. In this section, we study the dependence of memory efficiency on optical depth. To verify the theoretical prediction that the optimal efficiency depends only on the optical depth of the sample, repeated the iterative signal optimization procedure for several constant control field powers at different temperatures of the Rb cell ranging from 45℃ ($\alpha L = 6$) to 77℃ ($\alpha L = 88$). In Figure 5-62(a), plot the measured efficiencies (markers) along with the maximum achievable efficiency predicated by the theory without spin decay (thin black line) and with spin decay during the storage time (thick black line). This graph allows us to make several important conclusions. First of all, it demonstrates that for relatively low optical depths ($\alpha L \leqslant 25$), the optimized memory efficiency for different control fields is the same, to within the experimental uncertainty, and approximately matches the theoretical value (thick black line).

This confirms that the optimization procedure yields the maximum efficiency achievable for a given optical depth. However, for $\alpha L > 20$, the efficiency obtained with the lowest control field power (black empty circles) dropped below the efficiency obtained for higher control powers. As now show that the most probable reason for such deviation is spin wave decay during writing and retrieval. As the optical depth increases, the duration of the optimal input pulse increases as well, as shown in Figure 5-62(a), following the underlying decrease of group velocity: $T \sim L/v_g \propto \alpha L$. Thus, above a certain value of αL, the duration of the optimal pulse for a given control field becomes comparable with the spin wave lifetime, and the spin wave decoherence during storage and retrieval stages can no longer be ignored. Further increase of the optical depth leads to a reduction of retrieval efficiency, even though the iterative optimization procedure is still valid and produces signal pulses that are stored and retrieved with the highest efficiency possible for a given control field and αL. Figure 5-62(b) shows the calculated maximum achievable efficiencies for different constant control powers as a function of the optical depth, taking into account spin wave decay with a 500μs time constant during all three stages of light storage. For each control field power, the efficiency peaks at a certain optical depth, and then starts to decrease as optical depth increases further. Since lower control powers require longer optimal input pulses $T \sim L/v_g \propto 1/|\Omega|^2$, the corresponding efficiency reaches its peak at lower optical depths. Thus, the problem of efficiency reduction posed by spin wave decay during writing and retrieval can be alleviated by using higher control powers, and hence shorter optimal signal pulses. While this effect explains the reduction of maximum memory efficiency attained with the lowest control power for $\alpha L > 20$, other effects, discussed below, degrade the efficiency for all other control powers for $\alpha L > 25$, as indicated by the divergence of experimental data in Figure 5-62(a) from the corresponding theoretical efficiencies in Figure 5-62(b) (red and green lines). Remarkably, at these optical depths, the iterative signal optimization procedure still yields efficiencies that grow monotonically at each iteration step for the three highest control powers. This suggests that iterative signal optimization may still be yielding the optimum efficiency, although this optimum is lower than what the simple theoretical model predicts.

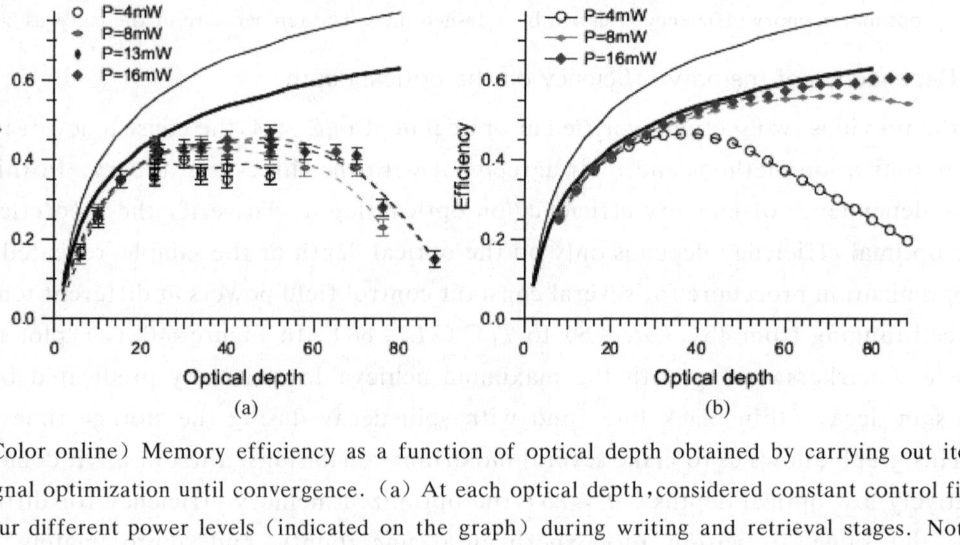

Fig. 5-62 (Color online) Memory efficiency as a function of optical depth obtained by carrying out iterative signal optimization until convergence. (a) At each optical depth, considered constant control fields at four different power levels (indicated on the graph) during writing and retrieval stages. Note that many experimental data points overlap since the converged efficiencies are often the same for different control fields. Dashed lines are to guide the eye. Thin and thick black solid lines show the theoretically predicted maximum efficiency assuming no spin-wave decay and assuming an efficiency reduction by a factor of 0.82 during the 100μs storage period, respectively. (b) Thin and thick black lines are the same as in (a), while the three lines with markers are calculated efficiencies for three different control fields (indicated on the graph) assuming spin wave decay with a 500μs time constant during all three stages of the storage process (writing, storage, retrieval).

To further test the applicability of our optimization procedures at higher optical depths, we complemented the signal-pulse optimization (see Figure 5-63(a), (b)) with the corresponding control field optimization (see Figure 6-63(c), (d)). When stored and retrieved input pulse #4 from Figure 5-63(a) using calculated optimal writing control fields ($t<0$ in Figure 5-63(c)) at different optical depths $\alpha L = 24, 40$, and 50. As expected, the overall control power was higher at higher optical depths to keep the group velocity unchanged: $L/T \sim v_g \propto 2/(\alpha L)$. For each optical depth, used a time-reversed writing control field to retrieve the stored spin wave. This resulted in the output signal pulse shape identical to the time-reversed (and attenuated) copy of the input pulse, as shown in Figure 5-63(d).

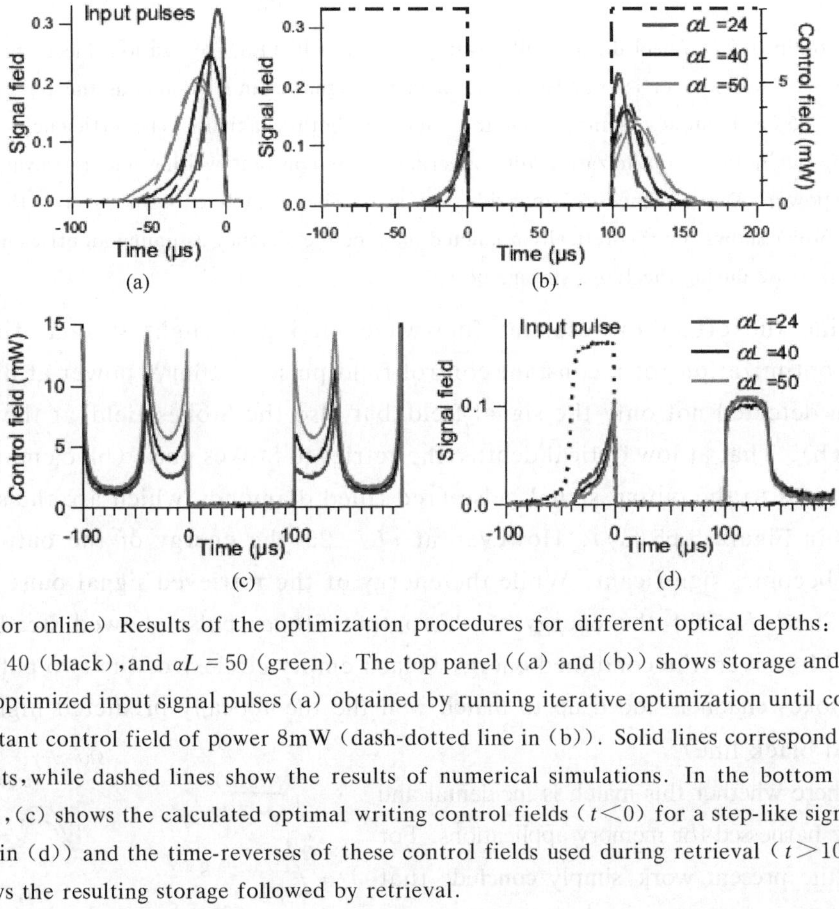

Fig. 5-63 (Color online) Results of the optimization procedures for different optical depths: $\alpha L = 24$ (red), $\alpha L = 40$ (black), and $\alpha L = 50$ (green). The top panel ((a) and (b)) shows storage and retrieval (b) of the optimized input signal pulses (a) obtained by running iterative optimization until convergence for a constant control field of power 8mW (dash-dotted line in (b)). Solid lines correspond to experimental results, while dashed lines show the results of numerical simulations. In the bottom panel ((c) and (d)), (c) shows the calculated optimal writing control fields ($t<0$) for a step-like signal pulse (dotted line in (d)) and the time-reverses of these control fields used during retrieval ($t>100\mu s$), while (d) shows the resulting storage followed by retrieval.

Although the memory efficiency drops below the theoretical value at these high optical depths ($\alpha L = 50$ for the green lines in Figure 5-64(c), (d)), the results suggest that the calculated control field may still be optimal, since it yields the time-reverse of the input signal at the output. To gain insight into what may limit the memory efficiency for $25 < \alpha L < 60$, we investigated the effect of resonant four-wave mixing. Thus far, considered only the ground-state coherence created by the control and signal fields in the one-photon resonant configuration as Figure 5-65(a). However, the strong control field applied to the ground state $|g\rangle$ can also generate an additional Stokes field ES, as shown in Figure 5-65(a). This process is significantly enhanced in EIT media. In particular, it has been shown that a weak signal pulse traversing an atomic ensemble with reduced group velocity generates a complimentary Stokes pulse that travels alongside with a comparably low group velocity.

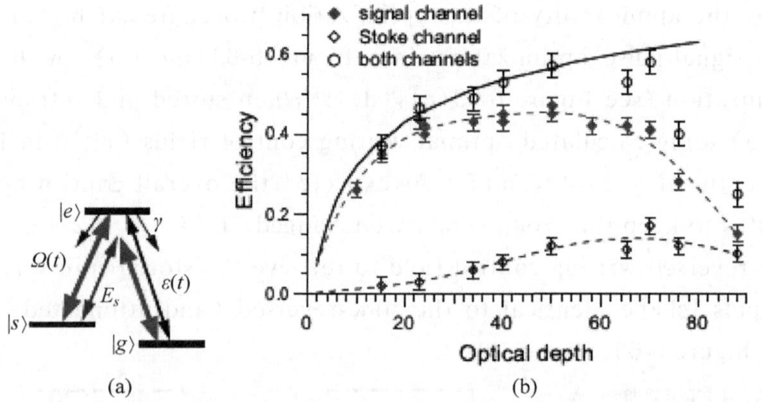

Fig. 5-64 (Color online) (a) Level diagram illustrating Stokes field (ES) generation due to resonant four-wave mixing. (b) Memory efficiency for retrieval in the signal channel (same as the red filled diamonds in Figure 5-65(a)), Stokes channel, and the total for both channels. The efficiencies are obtained by carrying out iterative optimization till convergence for constant writing and retrieval control fields of 16mW power. Dashed lines are to guide the eye. The solid line (same as the thick black line in Figure 5-65) shows the theoretically predicted maximum efficiency assuming an efficiency reduction by a factor of 0.82 during the 100μs storage period.

To determine the effect of resonant four-wave mixing on light storage, first carried out iterative signal optimization for a constant control field pulse of 16mW power at different optical depths, but then detected not only the signal field, but also the Stokes field, at the retrieval stage see Figure 5-65(b). That at low optical depths, the retrieved Stokes pulse (blue empty diamonds) is negligible compared to the output signal pulse (red filled diamonds, which are the same as the red filled diamonds in Figure 5-65(a)). However, at $\alpha L = 25$, the energy of the output pulse in the Stokes channel becomes significant. While the energy of the retrieved signal pulse stayed roughly unchanged for $25 < \alpha L < 60$, the energy of the output Stokes pulse showed steady growth with increasing αL. Moreover, the combined energy (black empty circles) of the two pulses retrieved in the signal and Stokes channels added up to match well the theoretically predicted highest achievable efficiency (solid black line).

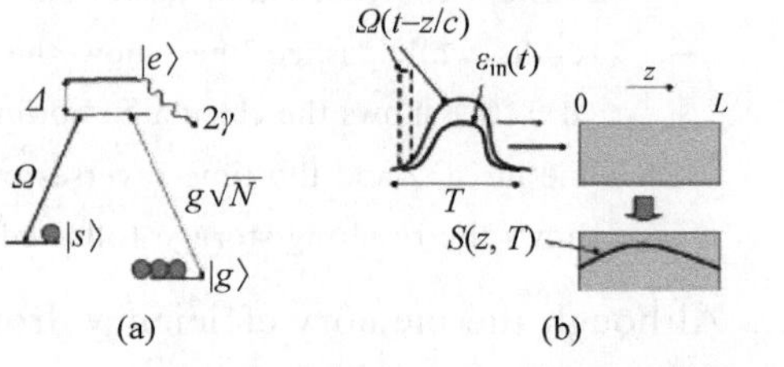

Fig. 5-65 (color online). (a) Λ-type medium coupled to a classical field with Rabi frequency Ω_t and a quantum field with an effective coupling constant $g\sqrt{N}$. (b) Storage setup. The solid Ω curve is the generic control shape for adiabatic storage; the dashed line indicates a π-pulse control field for fast storage. For retrieval, the inverse operation is performed.

Study elsewhere whether this match is incidental and whether it can be harnessed for memory applications. For the purposes of the present work, simply conclude that the effects of four-wave mixing become significant around the same value of αL (~ 25) where experiment starts deviating from theory. Therefore, four-wave mixing may be one of the factors responsible for the low experimental efficiencies at high optical depths. For $\alpha L > 60$, iterative signal optimization still converges, but efficiency does not grow monotonically at each iteration step, which clearly indicates the breakdown of time-reversal-based optimization. In addition, the final efficiency is significantly lower than the theoretical value (Figure 5-65). Many factors, other than four-wave mixing, may be contributing to the breakdown of time-reversal-based optimization and to the rapid decrease of memory efficiency at $\alpha L > 60$. First of all, the absorption of the control field at such high optical depths is significant (measured to be $> 50\%$). In that case, the reabsorption of spontaneous radiation contributes appreciably to spin wave decoherence and can make the spin wave decay rate s grow with αL, reducing the light storage efficiency. Spin-exchange collision rate, which destroys the spin-wave coherence, also becomes significant at high Rb density, reducing spin wave lifetime even further. When studied in detail two quantum memory optimization protocols in warm Rb vapor and demonstrated their consistency for maximizing memory efficiency. To have also observed good agreement between our experimental data and theoretical predictions for relatively low optical depths (< 25), both in terms of the highest memory efficiency and in terms of the optimized pulse shapes. At higher optical depths, however, the experimental efficiency was lower than predicted. That resonant four-wave mixing processes became important at these higher optical depths. The studies will be of importance for enhancing the performance of ensemble-based quantum memories for light.

5.7 Photon storage in atomic media

In quantum networks, states are easily transmitted by photons, but the photonic states need to be stored locally to process the information. Motivated by this and other ideas from quantum information science, techniques to facilitate controlled interactions between single photons and atoms are now being actively explored. A promising approach to a matter-light quantum interface uses classical laser fields to manipulate pulses of light in optically dense media such as atomic gases or impurities embedded in a solid state material. The challenge is to map an incoming signal pulse into a long-lived atomic coherence (referred to as a spin wave), so that it can be later retrieved "on demand" with the highest possible efficiency. Using several different techniques, significant experimental progress towards this goal has been made recently. A central question that emerges from these advances is which approach represents the best possible strategy and how the maximum efficiency can be achieved. In this letter, we present a physical picture that unifies several different approaches to photon storage in Λ-type atomic media and yields the optimal control strategy. This picture is based on two key observations. First, we show that the retrieval efficiency of any given stored spin wave depends only on the optical depth d of the medium.

Physically, this follows from the fact that the branching ratio between collectively enhanced emission into desired modes and spontaneous decay (with a rate 2γ) depends only on d. The second observation is that the optimal storage process is the time reverse of retrieval. This universal picture implies that the maximum the same for all approaches considered and depends only on d. It can be attained by adjusting the control or the shape of the photon wave packet. A generic model for a quantum memory uses the Λ-type level configuration shown in Figure 5-66(a), in which a eak (quantum) signal field with frequency ν is detuned by Δ from the $|g\rangle$-$|e\rangle$ transition. A copropagating (classical) control beam with the same detuning Δ from the $|s\rangle$-$|e\rangle$ transition is used to coherently manipulate the signal propagation and to facilitate the light-atom state mapping. In this system several different approaches to photon storage can be taken. In electromagnetically induced transparency (EIT), resonant fields ($\Delta = 0$) are used to open a spectral transparency window, where the quantum field travels at a reduced group velocity, which is then adiabatically readed to zero. In the Raman configuration, the fields have a large detuning ($|\Delta| \gg \gamma d$) and the photons are absorbed into the stable ground state $|s\rangle$ by stimulated Raman transitions. Finally, in the photon-echo approach, photon storage is achieved by applying a fast resonant π-pulse, which maps excitations from the unstable excited state $|e\rangle$ into the stable ground state $|s\rangle$.

A common problem in all of these techniques is that the pulse should be completely localized inside the medium at the time of the storage. For example, in the EIT configuration, a reduction in group velocity, which compresses the pulse to fit inside the medium, is accompanied by narrowing of the transparency window, which increases spontaneous emission. For retrieval, the inverse operation is performed echo technique, if a photon pulse is very short, its spectral width will be too large to be absorbed by the atoms. To achieve the maximum storage efficiency one thus has to make a compromise between the different sources of errors. Ideal performance is only achieved in the limit of infinite d. In our model, illustrated in Figure 5-66(a), the incoming signal interacts with N atoms in the uniform medium of length $L(z=0$ to $z=L)$ and cross-section area A. The control field is characterized by the slowly varying Rabi frequency $\Omega(t-z/c)$. $P(z,t)=\sqrt{N}\Sigma_i|g\rangle_i\langle s|/N_z$, where the sum is over all N_z atoms in a small region positioned at z, describes the slowly varying collective $|g\rangle$-$|e\rangle$ coherence. All atoms are initially pumped into level $|g\rangle$. As indicated in Figure 5-65(b), the first map a quantum field mode with slowly varying envelope $E_{\text{in}}(t)$ (nonzero on $t\in[0,T]$ and incident in the forward direction at $z=0$) to some slowly varying mode of the collective $|s\rangle$-$|g\rangle$ coherence $S(z,t)=\sqrt{N}\Sigma_i|g\rangle_i\langle s|/N_z$. Then starting at a time $T_r > T$, perform the inverse operation to retrieve S back onto a field mode. As we explain below, the optimal efficiency is achieved by sending the retrieval control pulse in the backward direction; storage followed by forward retrieval is, however, also considered. The goal is to solve the optimal control problem of finding the control fields that will maximize the efficiency of storage followed by retrieval for given optical depth d and input mode $\varepsilon_{\text{in}}(t)$. The efficiency is defined as the ratio of the number of retrieved photons to the number of incident photons. Since the quantum memory operates in the linear regime, an analysis of the interaction process where all variables are treated as complex numbers is sufficient. In this limit the equations of motion read

$$(\partial_t + c\partial_z)E(z,t) = ig\sqrt{N}P(z,t) \tag{5-129}$$

$$\partial_t P(z,t) = -(\gamma + i\Delta)P(z,t) + ig\sqrt{N}E(z,t) + i\Omega(t-z/c)S(z,t) \tag{5-130}$$

$$\partial_t S(z,t) = i\Omega^*(t-z/c)P(z,t) \tag{5-131}$$

Here have neglected the slow decay of S. For storage, the initial conditions are $\varepsilon(0,t)=\varepsilon_{\text{in}}(t)$, $\varepsilon(z,0)=0$, $P(z,0)=0$, and $\varepsilon(z,0)=0$. Being the shape of a mode, $\varepsilon_{\text{in}}(t)$ is normalized according to $(c/L)\int^T \|\varepsilon_{\text{in}}(t)\|^2 dt = 1$, so the storage efficiency is given $\varepsilon_{\text{in}}(c/L)\int_0^T |E_{\text{in}}(t)|^2 dt = 1$, by $\eta_s = (1/L)\int^L \|S(z,T)\|^2 dz$. For the reverse process, i. e. retrieval, the initial conditions are $\varepsilon(0,t)=0$, $\varepsilon(z,T_r)=0$, $P(z,T_r)=0$, and $S(z,T_r)=S(L-z,T)$ for backward retrieval or $S(z,T_r)=S(z,T)$ for forward retrieval. The total efficiency in both cases is η back/forw $=(c/L)\ R\infty T_r|\varepsilon_{\text{out}}(t)|^2 dt$, where $\varepsilon_{\text{out}}(t)\equiv\varepsilon(L,t)$.

It is instructive to first discuss the retrieval process. In a co-moving frame $t' = t - z/c$, using a normalized coordinate $\zeta = z/L$ and a Laplace transformation in space $\zeta \to s$, $\varepsilon(s,t') = i\sqrt{\frac{d\gamma L}{c}}P(s,t')/s$. Therefore, the retrieval efficiency is given by

$$\eta_r = \mathcal{L}^{-1}\left\{\gamma d/(ss')\int_{T_r}^{\infty} dt' P(s,t')[P(s'^*,t')]^*\right\} \tag{5-132}$$

where L^{-1} means that two inverse Laplace transforms ($s \to \zeta$ and $s' \to \zeta'$) are taken and are both evaluated at $\zeta = \zeta' = 1$. To calculate η_r, we insert $E(s,t')$ found from Eq. (5-80) into Eq. (5-81) and use Eqs. (5-81), (5-82) to find

$$\begin{aligned}&\partial_t\{P(s,t')[P(s'^*,t')]^* + S(s,t')[S(s'^*,t')]^*\}\\ &= -\gamma(2 + d/s + d/s')P(s,t')[P(s'^*,t')]^*\end{aligned} \tag{5-133}$$

Eqs. (5-132), (5-133) allow us to express η_r in terms of the initial and final values of the term inside the curly brackets in Eq. (5-133). Assuming $P(s,\infty) = S(s,\infty) = 0$ (i.e. no excitations are left in the atoms) and taking L^{-1}, can get

$$\eta_r = \int_0^1 d\zeta \int_0^1 d\zeta' k_d(\zeta,\zeta') S(\zeta,T_r) S^*(\zeta',T_r) \tag{5-134}$$

$$k_d(\zeta,\zeta') = \frac{d}{2}e^{-d(1-(\zeta+\zeta')/2)} I_0(d\sqrt{(1-\zeta)(1-\zeta')}) \tag{5-135}$$

where I_0 is the zeroth-order modified Bessel function of the first kind. Note that η_r does not depend on Δ and $\Omega(t)$. Physically, this means that a fixed branching ratio exists between the transfer of atomic excitations into the output mode $E_{out}(t)$ and the decay into all other directions.

This ratio only depends on d and $S(\zeta, T_r)$. The efficiency η_r in Eq. (5-85) is an expectation value of a real symmetric operator $k_r(\zeta,\zeta')$ in the state $S(\zeta)$. It is, therefore, maximized when $S(\zeta)$ is the eigenvector (call it $S_d(\zeta)$) with the largest eigenvalue η_r^{max} of the real eigenvalue problem

$$\eta_r S(\zeta) = \int_0^1 d\zeta' k_d(\zeta,\zeta') S(\zeta') \tag{5-136}$$

The $S_d(\zeta)$, start with a trial $S(\zeta)$ and iterate the integral in Eq. (5-136) several times. The resulting optimal spin wave $S_d(1-\zeta)$ is plotted in the inset of Figure 5-66 for $d = 1, 10, 100$, and $d \to \infty$. These shapes represent a compromise attaining the smoothest possible spin wave with the least amount of (backward) propagation. Now discuss storage. To clay backwards from $S_d(1-\zeta)$ into $\varepsilon_{in}^*(T-t)$, for a given d, Δ, and $\varepsilon_{in}(t)$, can find a control (t) that retrieves then the time reverse of this control, $\Omega^*(T-t)$, will give the optimal storage of $\varepsilon_{in}(t)$. To prove this, retrieval transformation as a unitary map $U[\Omega(t)]$ in the Hilbert space H spanned by subspace A of spinwave modes, subspace B of output field

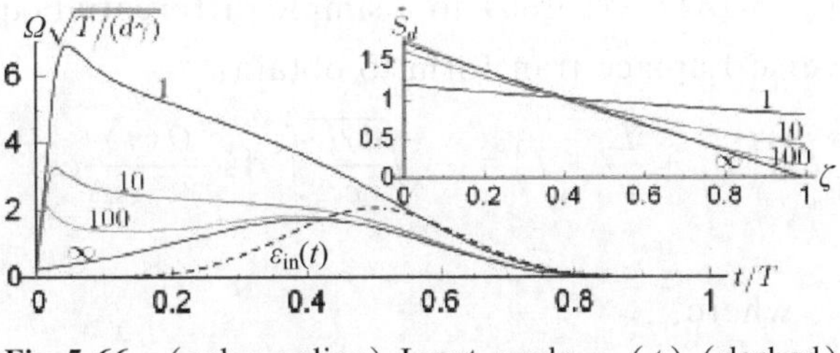

Fig. 5-66 (color online) Input mode $\varepsilon_{in}(t)$ (dashed) and control fields $\Omega(t)$ (in units of) that maximize for this $\varepsilon_{in}(t)$ the efficiency for resonant adiabatic storage (alone or followed by backward retrieval) at $d = 1, 10, 100$, and $d \to \infty$. Inset: Optimal modes $S_d(1-\zeta)$ to retrieve from backwards at $d = 1, 10, 100$, and $d \to \infty$ ($\zeta = z/L$). These are also normalized spin waves $S(\zeta, T)/\sqrt{\eta_s^{max}}$ in adiabatic and fast storage if it is optimized alone or followed by backward retrieval.

modes, as well as a subspace containing (empty) input and reservoir field modes. note that it is essential to include the reservoir modes, since the dynamics is unitary only in the full Hilbert space of the problem.

For a given unit vector $|a\rangle$ in A (a given spin wave), the retrieval efficiency is

$$\eta_r = |\langle b \mid U[\Omega(t)] \mid a\rangle|^2 = |\langle a \mid U^{-1}[\Omega(t)] \mid b\rangle|^2 \tag{5-137}$$

where have used the unitarity of $U[\Omega(t)]$, and where $|b\rangle$ is a normalized projection of $U[\Omega(t)]|a\rangle$ on B, i. e. the mode onto which the spin wave is retrieved. Introducing the time reversal operator T, we find

$$\eta_r = |\langle a \mid TTU^{-1}[\Omega(t)]TT| b\rangle|^2 \tag{5-138}$$

One can show that the time reverse of the inverse propagator

$$TU^{-1}[\Omega(t)]T \text{ is simply } U[\Omega^*(T-t)] \tag{5-139}$$

that have:

$$\eta_r = |\langle a \mid TU[\Omega^*(T-t)]T| b\rangle|^2 \tag{5-140}$$

This immediately tells that the time-reversed control $\Omega^*(T-t)$ will map the time reverse of the retrieved pulse into the complex conjugate of the original spin-wave mode with the same efficiency. The optimal spin waves are, however, real so that complex conjugation plays no role. Furthermore, the storage efficiency cannot exceed $\eta_r^{\max}$ since the time reverse of such storage would then by the same argument give a retrieval efficiency higher than $\eta_r^{\max}$, which is a contradiction. Optimal storage is thus the time reverse of optimal backward retrieval and has the same efficiency $\eta_s^{\max} = \eta_r^{\max}$ (and involves the same optimal spin wave).

To identify the input modes, for which the optimal storage can be achieved, we use Eqs. (5-130) to analytically solve the retrieval problem in two important limits: "adiabatic" and "fast". The "adiabatic" limit, whose two special cases are the Raman and the EIT regimes discussed above, corresponds to a smooth control field, such that P can be adiabatically eliminated in Eq. (5-134).

Using the Laplace transform technique to eliminate E from Eqs. (5-133)~(5-134), reduce Eqs. (5-134)~(5-135) to a simple differential equation on S. To solve it, compute E, and take the inverse Laplace transform to obtain:

$$\mathcal{E}_{\text{out}}\left(T_r + \frac{L}{c} + t\right) = -\sqrt{\frac{d\gamma L}{c}}\int_0^1 \mathrm{d}\zeta \frac{\Omega(t)}{\gamma + i\Delta}\mathrm{e}^{-\frac{\gamma d\zeta + h(t)}{\gamma + i\Delta}} \times I_0\left(2\sqrt{\gamma d\zeta h(t)}/(\gamma + i\Delta)\right) S(1-\zeta, T_r) \tag{5-141}$$

where:

$$h(t) = \int_0^t \mathrm{d}t' \mid \Omega(t') \mid^2 \tag{5-142}$$

It will show that for a given d, Δ, and spin wave $S(\zeta)$, one can always find a control (t) that maps $S(\zeta)$ to any desired normalized output mode $\varepsilon_2(t)$ of duration Tout, so that $\varepsilon_{\text{out}}(T_r + L/c + t) = \sqrt{\eta_r}\varepsilon_2(t)$ (provided we are in the "adiabatic" limit $T_{\text{out}} d\gamma \gg 1$). To replace $\varepsilon_{\text{out}}(T_r + L/c + t)$ in Eq. (5-141) with $\sqrt{\eta_r}\varepsilon_2(t)$, integrate the norm squared of both sides from 0 to t, change variables $t \to h(t)$, and get which allows to solve numerically for the unique $h(t)$. Then $|\Omega(t)| = [h(t)d/\mathrm{d}t]^{1/2}$, while the phase is found by inserting $h(t)$ into Eq. (5-142). Optimal storage controls then follow from the time reversal argument above. Figure 5-66 shows a particular Gaussian-like input mode $E_{\text{in}}(t)$ and the corresponding optimal storage control shapes Ω for the case $\Delta = 0$ and $d = 1, 10, 100$, as well as the limiting shape of the optimal as $d \to \infty$. As have argued, the normalized atomic mode $S(\zeta, T)/\sqrt{\eta_s^{\max}}$, into which $\varepsilon_{\text{in}}(t)$ is stored using these optimal control fields, is precisely $S_d(1-\zeta)$, the optimal mode to retrieve backwards shown in the inset of Figure 5-40. The "fast" limit corresponds to a short and powerful resonant retrieval control satisfying $\Omega \gg d\gamma$ that implements a perfect π-pulse between the optical and spin polarizations, P and S. This retrieval and the corresponding storage technique are similar to the photonecho method. Again using the Laplace transform technique, can find for a perfect π-pulse that enters the medium at time T_r

$$\mathcal{E}_{\text{out}}\left(T_r + \frac{L}{c} + t\right) = -\sqrt{\frac{\gamma d L}{c}}\int_0^1 d\zeta e^{-\gamma t} J_0(2\sqrt{\gamma d \zeta t}) S(1-\zeta, T_r) \tag{5-143}$$

where $J_0(x) = I_0(ix)$. Since the fast retrieval control cannot be shaped, at each d, there is, thus, only one mode (of duration $T \sim 1/(\gamma_d)$) that can be stored optimally. This mode is the time reverse of the output mode in Eq. (5-143) retrieved from the optimal spin wave $\tilde{S}_d$. That time reversal can not only be used to deduce optimal storage from optimal retrieval, but can also be used to find $\tilde{S}_d$ in the first place. In the discussion above, the normalized projection of $U^{-1}|b\rangle$ on A (call it $|a\rangle$) might have a component orthogonal to $|a\rangle$. In this case, the efficiency of U^{-1} as a map from B to A will be $\eta_r' = |\langle a|U^{-1}|b\rangle| > \eta_r$. Now if the normalized projection of $U|a\rangle$ on B is not equal to $|b\rangle$, the map U acting on $|a\rangle$ will similarly have efficiency $\eta_r'' > \eta_r' > \eta_r$.

Therefore, such iterative application of U and U^{-1} converges to the optimal input in A and the corresponding optimal output in B. Indeed, a detailed calculation shows that the search for the optimal spin wave by iterating Eq. (5-143) precisely corresponds to retrieving $S(\zeta)$ with a given control, time-reversing the output, and storing it with the time-reversed control profile.

This time-reversal optimization procedure for finding the optimal $|a\rangle \in A$ can be used to optimize not only retrieval, but also any map including storage followed by retrieval. For storage followed by backward retrieval, this procedure yeilds $S_d(1-\zeta)$ and maximum efficiency $\eta_{\text{back}}^{\max} = (\eta_r^{\max})^2$, since $S_d(1-\zeta)$ optimizes both storage and backward retrieval. Figure 5-67 demonstrates that for resonant adiabatic storage of the field mode in Figure 5-66 followed by backward retrieval, optimal controls result in a much higher efficiency $\eta^{\max}$ back than naive square control pulses on $[0, T]$ with power set by $v_g T = L$ (η_{square} curve), where $v_g = c\Omega^2/(g^2 N)$ is the EIT group velocity. For the case of storage followed by forward retrieval, iterations yield the maximum efficiency $\eta_{\text{forw}}^{\max}$ plotted in Figure 5-67. It is less than $\eta_{\text{back}}^{\max}$ since with backward retrieval, storage and retrieval are each separately optimal, while for forward retrieval a compromise has to be made. From a different perspective, forward retrieval makes it more difficult to minimize propagation since the excitation has to propagate through the entire medium. In conclusion, we have shown that the performance of EIT, Raman, and photon-echo

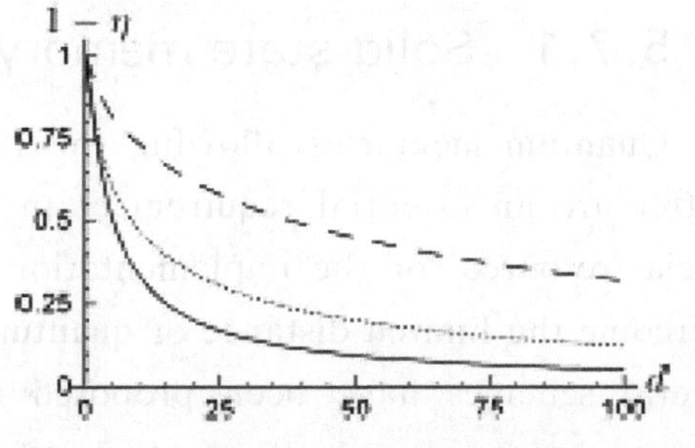

Fig. 5-67 $\eta_{\text{back}}^{\max}$ (solid) and $\eta_{\text{forw}}^{\max}$ (dotted) are maximum total efficiency for storage followed by backward or forward retrieval, respectively. η_{square} (dashed) is the total efficiency for resonant storage of $\varepsilon_{\text{in}}(t)$ from Figure 5-40 followed by backward retrieval, where the storage control field is a naive square pulse.

approaches to a quantum light-matter interface can be understood and optimized within a universal physical picture based on time reversal and a fixed branching ratio between loss and the desired quantum state transfer. For a given optical depth d, the optimal strategy yields a universal maximum efficiency and a universal optimal spin wave, thus, demonstrating a certain degree of equivalence between these three seemingly different approaches. We showed that the optimal storage can be achieved for any smooth input mode with $T\, d\gamma \gg 1$ and any Δ and for a class of resonant input modes satisfying $T\, d\gamma \sim 1$. The presented optimization of the storage and retrieval processes leads to a substantial increase in the memory efficiency. The results described here are of direct relevance to ongoing experimental efforts, where optical depth d is limited by experimental constraints such as density of cold atoms, imperfect optical pumping, or competing nonlinear effects. For example, in two experiments, $d \sim 5$ was used. $H_{\text{back}}^{\text{max}}$ and η_{square} curves in Figure 5-67 indicate that at this d, by properly shaping the control pulses, the efficiency can be increased by more than a factor of 2 . Direct comparison to experiment, however, will require the inclusion of decoherence processes and other imperfections. Discuss some of these imperfections, as well as the details of the present analysis and its extensions to atomic ensembles enclosed in a cavity and to inhomogeneously broadened media. Note that the time-reversal based iterative optimization we suggest is not only a convenient mathematical tool but is also a readily accessible experimental technique for finding the optimal spin wave and optimal input-control pairs: one just has to measure the output mode and generate its time reverse. Indeed, optimization procedure has been recently verified experimentally. To expect this procedure to be applicable to the optimization of other linear quantum maps both within the field of light storage(e. g. light storage using tunable photonic crystals) and outside of it.

5.7.1 Solid-state memory at the single photon level

Quantum memories allowing the reversible transfer of quantum states between light and matter are an essential requirement in quantum information science. They are, for example, a crucial resource for the implementation of quantum repeaters, which are a potential solution to overcome the limited distance of quantum communication schemes due to losses in optical fibers. Several schemes have been proposed to implement photonic quantum memories. Important progress has been made during the last few years, with proof of principle demonstrations in atomic gases, single atoms in a cavity, and solid state systems. For all these experiments the wavelength of the stored light was close to the visible range and thus not suited for direct use in optical telecom fibers. The ability to store and retrieve photons at telecommunication wavelengths (around 1550nm) in a quantum memory would provide an important resource for long distance quantum communication. Such a quantum memory could easily be integrated in the fiber communication network. In combination with a photon pair source, it could provide a narrow band triggered single photon source adapted to long distance transmission. Moreover, quantum memories at telecommunication wavelengths are required for certain efficient quantum repeater architectures. A telecom quantum memory requires an atomic medium with an optical transition in the telecom range, involving a long lived atomic state. The only candidate proposed so far is based on erbium doped solids, which have a transition around 1530nm between the ground state $^4I_{15/2}$ and the excited state $^4I_{13/2}$. These systems have been studied for spectroscopic properties and classical light storage. Photonic quantum storage in these materials is extremely challenging, because of the difficulties in the memory preparation using optical pumping techniques. The experiment of storage and retrieval of weak light fields at the single photon level in an erbium doped solid. Rare-earth doped solids have an inhomogeneously broadened absorption line. Single photons can be mapped onto this optical transition, leading to single collective optical excitations. During the storage, inhomogeneous dephasing takes place, preventing an efficient collective re-emission of the photon. This dephasing can be compensated for using photon echo techniques.

The storage of quantum light (e.g. single photons) is however not possible using traditional photon echo techniques, such as two pulse photon echoes. The main issue is that the application of the strong optical pulse (π-pulse) to induce the rephasing mechanism leads to amplified spontaneous emission and reduce the fidelity of the storage to an unacceptable level. A way to overcome this problem is to induce the rephrasing of the atomic dipoles by generating and reversing an artificial inhomogeneous broadening. This scheme is known as Controlled Reversible Inhomogeneous Broadening (CRIB). In rare-earth doped solids with a permanent dipole moment, this can be done with an electric field gradient using the linear Stark effect. The CRIB scheme was first demonstrated with bright optical pulses, in a $Eu^{3+}:Y_2SiO_5$ crystal at 580nm. The phase of the stored light pulses was shown to be well preserved. It has been dramatically improved in more recent experiments at 606nm in $Pr^{3+}:Y_2SiO_5$. CRIB has also been demonstrated on a spin transition in a rubidium vapor at 780nm. Here report an experiment at telecommunication wavelength. Moreover, we also report the first experimental demonstration of CRIB at the single photon level, opening the road to the quantum regime. In order to realize a CRIB experiment in a rare-earth doped solid, one first has to prepare a narrow absorption line within a large transparency window. The spectrum of this line is then broadened by an electric field gradient using the linear Stark effect to match the bandwidth of the photon to be stored. The incident photon is absorbed by the ions in the broadened line, and mapped into a single collective atomic excitation. During a time t each excited ion i will acquire a phase $\Delta_i t$ due to its shift in the absorption frequency $\omega_i = \omega_0 + \Delta_i$ from the central frequency ω_0. Switching the polarity of the field after a time $t = \tau$ will reverse the broadening ($\omega_i = \omega_0 - \Delta_i$) and after another time τ the ions will be in phase again and re-emit the photon. In order to create the initial narrow absorption line, a population transfer between two ground states (in our case Zeeman states) using optical pumping via the excited state is used. In case of imperfect optical pumping there will be population remaining in the excited state after the preparation sequence. An experimental issue arising when input pulses are at the single photon level is the fluorescence from the remaining excited atoms. If the depletion of this level is slow (as in rare-earth ions, with optical relaxation times T_1 usually in the range of 0.1ms to 10ms), this can lead to a high noise level that may blur the weak echo pulse. The problem is especially important for erbium doped solids, where T_1 is very long ($\approx$11ms in $Er^{3+}:Y_2SiO_5$). In the experiment, the population transfer is enhanced by stimulating ions from the excited state down to the short lived second ground state crystal field level using a second laser at 1545nm (See Figure 5-68). The application of this laser enhances the rate of depletion of the excited state and thus reduces the noise from fluorescence. Together with a suitable waiting time between the preparation and the light storage, it allows the realization of the scheme at the single photon level.

The memory consists of an Y_2SiO_5 crystal doped with erbium (10ppm) cooled to 2.6K in a pulse tube cooler (Oxford Instruments). The crystal has three mutually perpendicular optical-extinction axes labelled D_1, D_2, and b. Its dimensions are 3.5mm×4mm×6mm along these axes. The magnetic field of $B = 1.5$mT used to induce the Zeeman splitting necessary for the memory preparation is provided by a permanent magnet outside the cryostat and is applied in the $D_1 - D_2$ plane at an angle of $\theta = 135^\circ$ with respect to the D_1-axis (Figure 5-68(b)). The light is travelling along b. The electrical field gradient for the Stark broadening is applied with four electrodes placed on the crystal in quadrupole configuration, as shown and decribed in Figure 5-68(b). The induced broadening is proportional to the voltage U applied on the electrodes.

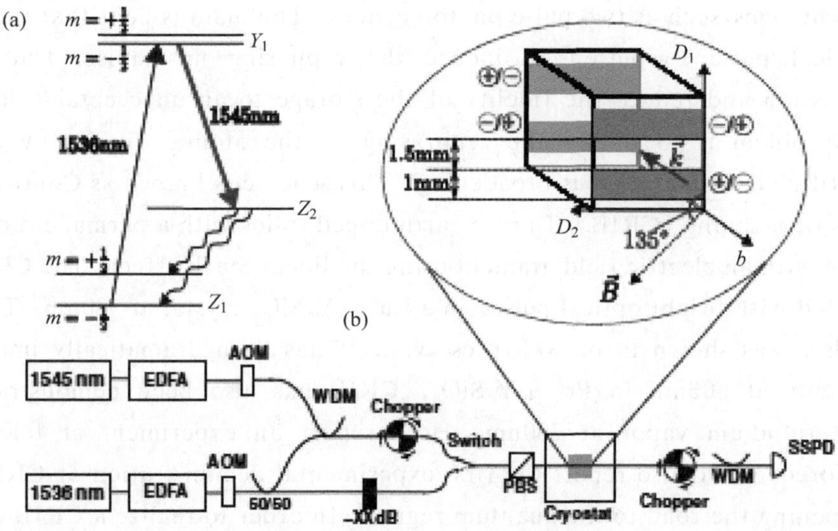

Fig. 5-68 (color online) (a) Level scheme of $Er^{3+}:Y_2SiO_5$. (b) Experimental setup: The pump laser (external cavity diode laser at 1536nm) is split into two paths, one for the preparation pulses and one for the weak pulses to be stored. Pulses are created with acousto-optical modulators (AOMs). In the preparation path, the stimulation laser (DFB laser diode at 1545nm + fiber amplifier) is added using a wavelength division multiplexer (WDM). The pulses to be stored are attenuated to the single photon level with a fiber attenuator. An optical switch allows us to send either of them into the sample. In order to protect the detector (SSPD) and to avoid noise from a leakage of the optical switch, two mechanical choppers are used. The polarization of the light is aligned to maximize the absorption using the polarizing beam splitter (PBS). Inset: illustration of the crystal with electrodes, magnetic field and light propagation directions indicated.

The experiment is divided into two parts: the preparation of the memory and the storage of the weak pulses (see Figure 5-68(b)). Each preparation sequence takes 120ms of optical pumping during which both the pump and the stimulation lasers are sent into the sample. The frequency of the pump laser is repeatedly swept to create a large transparency window into the inhomogeneously broadened absorption line. If the laser is blocked for a short time at the center of each sweep using an acoustic optical modulator (AOM), a narrow absorption feature is left at the center of the pit. The time available to perform the memory protocol is given by the Zeeman lifetime of $T_Z = 130$ms of the material. In order to deplete the excited state the laser at 1545nm is left on for 23.5ms after the using an optical switch and mechanical choppers. The pump pulses. Then the preparation path is closed and the detection path is opened storage sequence begins 86ms after the pump pulse, in order to avoid fluorescence from the excited atoms. It is composed of 8000 independent trials separated by 5μs. In each trial, a weak pulse of duration $\delta\tau$ is stored and retrieved. The initial peak is broadened with an electrical pulse before the absorption of each pulse. The polarity of the field is then inverted at a programable time after the storage, allowing for on demand read-out. The whole sequence is repeated at a rate of 3Hz. The weak output mode is detected using a superconducting single photon detector (SSPD) with an efficiency of 7% and a low dark count rate of 10 ± 5Hz. The incident pulses are weak coherent states of light $|\alpha\rangle_L$ with a mean number of photons $\bar{n} = |\alpha|^2$. To determined n at the input of the cryostat by measuring the number of photons arriving at the SSPD (with the laser out of resonance), compensating for the transmission (16%) and detection efficiency. Now describe the observation of CRIB photon echoes of weak pulses. As a first experiment, we sent pulses with $\bar{n} = 10$ and $\delta\tau = 200$ns into the sample. The polarity of the electric field ($U = \pm 50$V) was reversed directly after each pulse. Figure 5-69 shows a time histogram of the photon counts detected after the crystal. The first peak corresponds to the input photons transmitted through the crystal. The second peak is the CRIB echo. It is clearly visible above the noise floor. Only a small fraction of the incident light is re-emitted in the CRIB echo (about 0.25%). The reasons for this low storage and retrieval efficiency and ways to improve it will be discussed in more detail below. As a consistency check, verified that the echo disappears when the narrow absorption peak is not present (see blue open circles in Figure 5-69). By reversing the electric field gradient at later times, it is possible to choose the retrieval time of the stored light.

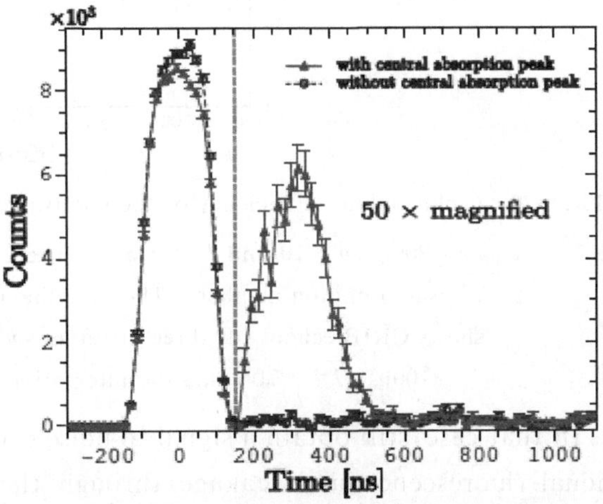

Fig. 5-69 (color online) CRIB measurement with (red triangles, solid line) and without (blue circles, dashed line) absorption peak, with $\bar{n} = 10$ and $\delta\tau = 200$ns. The pulse on the left is the transmitted part of the incident photons. One can clearly see that the absorption is enhanced in the presence of a peak. The electric field ($U = \pm 50$V) was reversed just after the input pulse. Dark counts have been subtracted from the data. Integration time for both curves was 200s.

Figure 5-69 shows the efficiency of the CRIB echo for different storage times. The signal was clearly visible up to a storage time of around 600ns. The decay of the efficiency is due to the finite width of the initial (unbroadened) peak (see below). The solid line is a fit assuming a gaussian shape for the absorption line, giving a decay time of 370ns. Shorter pulses with $\delta_T = 100$ns have also been stored (see the inset of Figure 5-70) with a larger broadening ($U = \pm 70$V), leading to a larger time-bandwidth ratio, with however a reduced storage efficiency. Finally, we gradually lowered n by increasing the attenuation, for input pulses with $\delta_T = 200$ns. The result is shown in Figure 5-70(a). Both the number of photons in the CRIB echo and the signal to noise ratio depend linearly on $\bar{n}$. This means that the efficiency and the noise are independent of n. Figure 5-70(b) shows the result of a measurement with in quantum key distribution terminology-pseudo-single photons ($\bar{n} = 0.6$ photons per pulse).

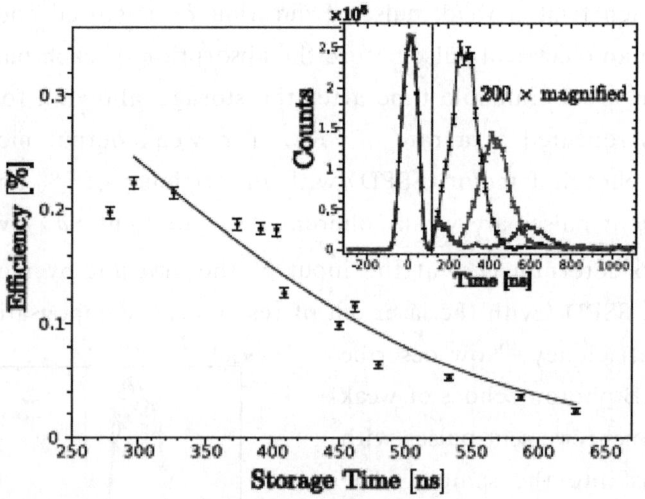

Fig. 5-70 (color online) Efficiency of the CRIB memory as a function of storage time, for input pulses with $\delta\tau = 200\text{ns}$, $\bar{n} = 10$ and $U = \pm 50\text{V}$. The error bars correspond to the statistical uncertainty of the measured photon numbers. The solid line is a gaussian fit (with the first point excluded). The inset shows CRIB echoes for three different switching times of the electrical field. For the inset curves, $\delta\tau = 100\text{ns}$, $U = \pm 70\text{V}$ and the integration time is 500s.

In that case, still obtain a signal to noise ratio of ~3. The remaining noise floor may be due to residual fluorescence and leakage through the AOM creating the pulses to be stored. In the following analyze the efficiency and storage time performances of our memory in more detail. It gives a simplified model for the CRIB memory. In this model, the storage and retrieval efficiency if the echo is emitted in the forward direction is given by

$$\eta_{\text{CRIB}}(t) = d^2 e^{-d} e^{-t^2 \bar{\gamma}^2}$$
$$\bar{\gamma} = 2\pi\gamma \tag{5-144}$$

where is the spectral width (standard deviation) of the initial gaussian absorption peak, and d is the optical depth of the broadened absorption peak. The main assumption here is that the spectral width of the absorption peak is much wider than the spectral bandwidth of the photon to be stored. By fitting the decay curve of Figure 5-71 with Eq. (5-144), we find a full width at half maximum linewidth of the central peak of 1MHz. This corresponds well to the results obtained by a measurement of the transmission spectrum. The minimal width is limited by the linewidth of our unstabilized laser diode and power broadening during the preparation of the peak. The optical coherence time of the transition under our experimental conditions has been measured independently by photon echo spectroscopy. It was found to be $T_2 \approx 2\mu\text{s}$, corresponding to a homogeneous linewidth of 160kHz. Note that the optical coherence in Er^{3+} : Y_2SiO_5 could be drastically increased using lower temperatures and higher magnetic fields.

In the experiment, imperfect optical pumping results in a large absorbing background with optical depth d_0, which acts as a passive loss, such that the experimental storage and retrieval efficiency is given by: $\eta(t) = \eta_{\text{CRIB}}(t)\exp(-d_0)$. The values of d and d_0 can be measured by recording the absorption spectra. This yields an optical depth of the unbroadened peak $d' = 0.5 \pm 0.2$ and an absorbing background of $d_0 = 1.6 \pm 0.1$. A voltage of 50V on the electrodes corresponds to a broadening of a factor of ~3, leading to $d = 0.17 \pm 0.07$.

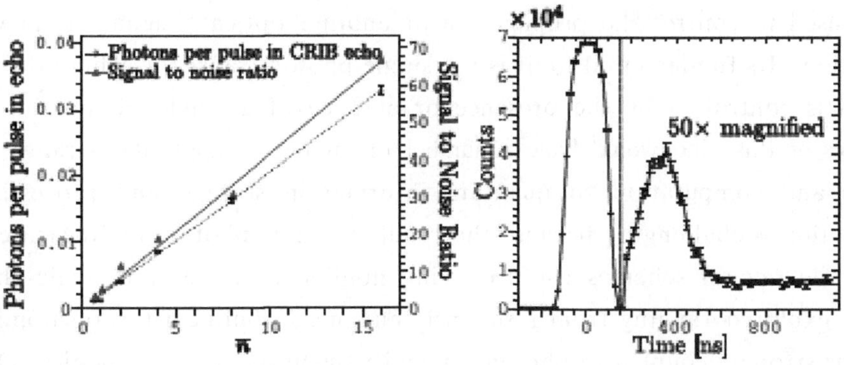

(a) (b)

Fig. 5-71 (color online) (a) Number of photons in the CRIB echo (open circles) and signal to noise ratio (plain triangles) as a function of the number of incident photons $\bar{n}$, for 200ns input pulses. (b) CRIB echo for $\bar{n}=0.6$ (integration time 25000s). Dark counts have been subtracted from the data.

In experiment, the photon bandwidth is of the same order as the broadened peak, so that the assumptions of the simplified model are not fulfilled. In order to have a more accurate description, have solved numerically the Maxwell-Bloch equations with the measured d', using a gaussian initial peak. This gives storage and retrieval efficiencies of order 1.5×10^{-3} (including the passive loss d_0) for a storage time of 300ns, in reasonable agreement with the measured values (see Figure 5-71) This study shows that the main reason of the low storage and retrieval efficiency in the present experiment is the small absorption in the broadened peak and the large absorbing background, due to imperfect optical pumping.

About 80% of the retrieved photons are lost in the absorbing background. The limited optical pumping efficiency is due to the small branching ratio in the Λ system and to the small ratio between the relaxation life times of the optical and the ground state Zeeman transitions. This could be improved in several ways. First technical improvements can be implemented, such as using lower temperatures, higher stimulation laser intensities and spin mixing in the excited state using RF fields, as demonstrated in. Second, the branching ratio and Zeeman life time strongly depend on the applied magnetic field angle and intensity. A full characterization of the optical pumping efficiency with respect to these parameters has not been carried out yet. Finding optimal conditions may lead to significant improvements. It would also be interesting to investigate hyperfine states. Finally, other crystals might be explored, e. g., Y_2O_3, to search for longer Zeeman lifetimes. In summary, have presented a proof-of-principle of quantum memory for photons at telecommunication wavelengths. Pulses of light at the single photon level have been stored and retrieved in an Erbium doped crystal, using the CRIB protocol. Continuing efforts to increase the efficiency and the storage time will be required in order to build a useful device for applications in quantum information science. The experiment is nevertheless an enabling step towards the demonstration of a fiber network compatible quantum light matter interface. It also confirms the feasibility of the CRIB protocol at the single photon level. This work was supported by the Swiss NCCR Quantum Photonics, by the European Commission under the Integrated Project Qubit Applications (QAP) and the ERC-AG Qore.

5.7.2 A single-photon transistor using nano-scale surface plasmons

In analogy with the electronic transistor, a photonic transistor is a device where a small optical "gate" field is used to control the propagation of another optical "signal" field via a nonlinear optical interaction. Its fundamental limit is the single-photon transistor, where the propagation of the signal field is controlled by the presence or absence of a single photon in the gate field. Nonlinear devices of this kind would have a number of interesting applications ranging from optical communication and computation to quantum information storage and processing. However, practical realization is challenging because the requisite single-photon nonlinearities are generally very weak. While several schemes for producing nonlinearrities at the single-photon level are currently being explored, ranging from resonantly enhanced nonlinearities of atomic ensembles to individual atoms strongly coupled to photons in cavity quantum electrodynamics (QED), a robust, practical approach has yet to emerge.

Recently, a new method to achieve strong coupling between light and matter has been proposed and experimentally demonstrated. It makes use of the tight concentration of optical fields associated with guided surface plasmons (SPs) on conducting nanowires to achieve strong interaction with individual optical emitters. In essence, the tight localization of these fields causes the nanowire to act as a very efficient lens that directs the majority of the spontaneously emitted light into the SP modes, resulting in efficient generation of single plasmons (single photons). While this process is essentially a linear optical effect, as it only involves one photon at a time, here show that such a system also allows for the realization of remarkable nonlinear optical phenomena, where individual photons strongly interact with each other. As an example, describe how these nonlinear processes may be exploited to implement a single-photon transistor.

While ideas for developing plasmonic analogues of electronic devices by combining SPs with electronics are already being explored, the process we describe here opens up fundamentally new possibilities, in that it combines the ideas of plasmonics with the tools of quantum optics to achieve unprecedented control over the interactions of individual light quanta.

• Nanowire plasmons: interaction with matter

SPs are propagating electromagnetic modes confined to the surface of a conductordielectric interface. Their unique properties make it possible to confine them to sub-wavelength dimensions, which has led to fascinating new approaches to waveguiding below the diffraction limit, enhanced transmission through sub-wavelength apertures, and sub-wavelength imaging. Large field enhancements associated with Plasmon resonances of metallic nano-particles have also been utilized to detect nearby single molecules via surface-enhanced Raman scattering. Similar properties directly give rise to the strong interaction between single SPs on a conducting nanowire and an individual, proximal optical emitter (see Figure 5-72(a), (b)).

Much like in a single-mode fiber, the SP modes of a conducting nanowire constitute a one-dimensional, single-mode continuum that can be indexed by the wavevectors k along the direction of propagation. Unlike a single-mode fiber, however, the nanowire continues to display good confinement and guiding when its radius is reduced well below the optical wavelength ($R \ll \lambda_0$). Specifically, in this limit, the SPs exhibit strongly reduced wavelengths and small transverse mode areas relative to free-space radiation, which scale like $\lambda_{pl} \propto 1/k \propto R$ and $A_{\text{eff}} \propto R_2$, respectively. The tight confinement results in a large interaction strength between the SP modes and any proximal emitter with a dipole-allowed transition, with a coupling constant that scales like $g \propto 1/$

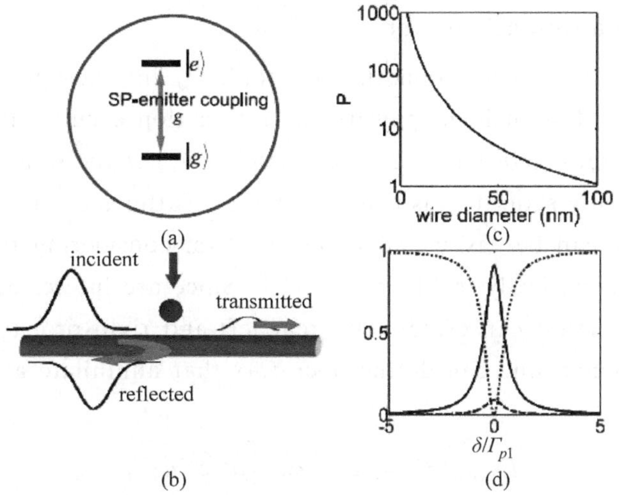

Fig. 5-72 The single surface plasmons with a single emitter. (a) Two-level emitter interacting with the nanowire. States $|g\rangle$ and $|e\rangle$ are coupled via the surface-plasmon modes with a strength g. (b) Schematic diagram of a single incident photon scattered off a near-resonant emitter. The interaction leads to reflected and transmitted fields whose amplitudes can be calculated exactly. (c) The maximum Purcell factor of an emitter positioned near a silver nanowire ($\varepsilon \approx -50+0.6i$) and surrounded by a uniform dielectric ($\varepsilon = 2$), as a function of wire diameter. (d) Probabilities of reflection (solid line), transmission (dotted line) and loss (dashed line) for a single photon incident on a single emitter, as a function of detuning. The Purcell factor for this system is taken to be $P = 20$.

$\sqrt{A_{\mathrm{eff}}}$. The reduction in group velocity also yields an enhancement of the density of states, $D(\omega) \propto 1/R$. Physically, Γ includes contributions both from emission into free space and non-radiative emission via ohmic losses in the conductorl. A relevant figure of merit is an effective Purcell factor $P \equiv \Gamma_{pl}/\Gamma$, which can exceed 10^3 in realistic systems (see Figure 5-72(c)). Furthermore this strong coupling is broadband, as it arises purely from geometrical considerations as opposed to any resonant features of the surface plasmons. This is in direct contrast, for example, to the mechanism by which strong coupling is achieved in cavity QED. Motivated by these considerations, and a general one-dimensional model of an emitter strongly coupled to a set of travelling electromagnetic modes (see Figure 5-72(a), (b)). The first consider a simple two-level configuration for the emitter, consisting of ground and excited states ($|g\rangle$, $|e\rangle$) separated by frequency ω_{eg}. The corresponding Hamiltonian is given by

$$H = \hbar(\omega_{\mathrm{eg}} - i\Gamma'/2)\sigma_{\mathrm{ee}} + \int \mathrm{d}k\, \hbar c\, |k|\, \hat{a}_k^{\dagger}\hat{a}_k - \hbar g \int \mathrm{d}k (\sigma_{\mathrm{eg}} \hat{a}_k \mathrm{e}^{ikz_a} + \mathrm{h.c.}) \tag{5-145}$$

where

$$\sigma_{ij} = |i\rangle\langle j|, \hat{a}_k \tag{5-146}$$

This is the annihilation operator for the mode with wavevector k, g is the emitter-field interaction matrix element. Here have assumed that a linear is persion relation holds over the relevant frequency range, $v_k = c|k|$, where c is the group velocity of SPs on the nanowire, and similarly that g is frequencyindependent. In the spirit of the stochastic wave function or "quantum jump" description of an open system, also included a non-Hermitian term in H due to the decay of state $|e\rangle$ into a reservoir of other radiative and non-radiative modes at a rate Γ' into the other channels. This effective hamiltonian accurately describes the dynamics provided that the thermal energy $k_{\mathrm{B}}T \ll \hbar\omega_{\mathrm{eg}}$, where k_{B} is the Boltzmann constant.

• Single emitter as a saturable mirror

The propagation of SPs can be dramatically altered by interaction with the single two level emitter. In particular, for low incident powers, the interaction occurs with near-unit probability, and each photon can be reflected with very high efficiency. At the same time, for higher powers the emitter response rapidly saturates, as it is not able to scatter more than one photon at a time. The low-power behavior can be easily understood by first considering the scattering of a single photon, as illustrated schematically in Figure 5-72(b). Since are interested only in SP modes near the optical frequency ω_{eg}, that can effectively treat left-and right-propagating SPs as completely separate fields. In particular, one can define operators that annihilate a left (right)-propagating photon at position z,

$$\hat{E}_{L(R)}(z) = (1/\sqrt{2\pi})\int dk e^{ikz}\hat{a}_{L(R),k} \tag{5-147}$$

where operators acting on the left and right branches are assumed to have vanishing commutation relations with the other branch as in Figure 5-73.

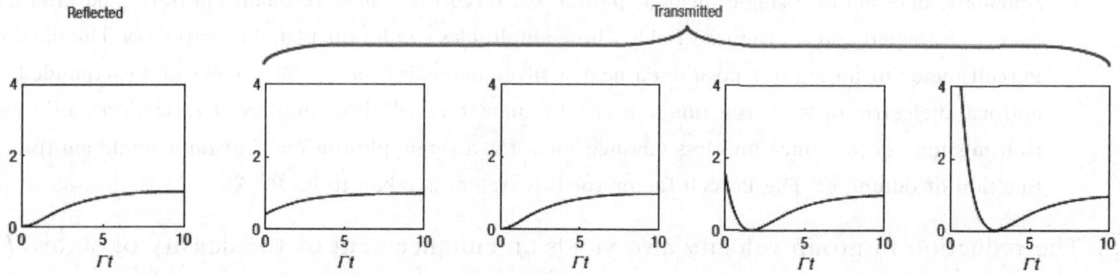

Fig. 5-73 Second-order correlation function $g^{(2)}(t)$ for the reflected and transmitted fields at low incident power ($\Omega c/\Gamma = 0.01$). $g^{(2)}(t)$ for the reflected field is independent of P at low powers. For the transmitted field, going from left to right, the Purcell factors are $P = 0.6, 1, 1.5$ and 2, respectively.

An exact solution to the scattering from the right to left branches in the limit $P \to \infty$ and this approach can be generalized to finite P. In particular, it is possible to solve for the scattering eigenstates of a system containing at most one (either atomic or photonic) excitation, as described in Methods. The reflection coefficient for an incoming photon of wavevector k is

$$r(\delta_k) = -\frac{1}{1 + \Gamma'/\Gamma_{pl} - 2i\delta_k/\Gamma_{pl}} \tag{5-148}$$

where

$$\delta_k \equiv ck - \omega_{eg} \tag{5-149}$$

is the photon detuning, while the transmission coefficient is related to r by $t(\delta_k) = 1 + r(\delta_k)$. Here $\Gamma_{pl} = 4g^2/c$ is the decay rate into the SPs, as obtained by application of Fermi's Golden Rule to the Hamiltonian in Eq. (5-148). On resonance, $r \approx -(1 - 1/P)$, and thus for large Purcell factors the emitter in state $|g\rangle$ acts as a nearly perfect mirror, which simultaneously imparts a π-phase shift upon reflection. The bandwidth $\Delta\omega$ of this process is determined by the total spontaneous emission rate, $\Gamma = \Gamma_{pl} + \Gamma'$, which can be quite large. Furthermore, the probability k of losing the photon to the environment is strongly suppressed for large Purcell factors, $k \equiv 1 - R - T = 2R/P$, where $R(T) \equiv |r|^2(|t|^2)$ is the reflectance (transmittance).

The nonlinear response of the system can be seen by considering the interaction of a single emitter not just with a single photon, but with multi-photon input states. To be specific, we consider the case when the incident field consists of a coherent state, the quantum-mechanical state that most closely corresponds to a classical field (note also similar work where scattering of two-photon states is considered). We assume that the incident field propagates to the right, and that the emitter is initially in the ground state. As shown in the Methods section, by transformation the initial coherent state can be formally mapped to an external Rabi frequency in the hamiltonian, which allows all quantities of interest (for example, field correlation functions) to be calculated exactly. For a narrow bandwidth ($\delta\omega < \Gamma$), resonant ($\delta_k = 0$) input field, the steady-state transmittance and reflectance are found to be

$$\begin{cases} T = \dfrac{1 + 8(1+P)^2(\Omega_c/\Gamma)^2}{(1+P)^2(1+8(\Omega_c/\Gamma)^2)} \\ R = \left(1 + \dfrac{1}{P}\right)^{-2} \dfrac{1}{1+8(\Omega_c/\Gamma)^2} \end{cases} \tag{5-150}$$

At low powers ($\Omega_c/\Gamma < 1$), the emitter has scattering properties identical to the single-photon case, $R \approx (1+1/P)^{-2}$, $T \approx (1+P)^{-1}$, and for large Purcell factors the single emitter again acts as a perfect mirror. At high incident powers ($\Omega_c/\Gamma > 1$), however, the emitter saturates and most of the incoming photons are transmitted past with no effect, $T \to 1$, $R \sim Q((\Gamma/\Omega_c)^2)$.

The significance of these results can be understood by noting that saturation is achieved at a Rabi frequency $\Omega_c \sim \Gamma$ that, in the limit of large P, corresponds to a switching energy of a single quantum within a pulse of duration $1/\Gamma$. While the reflectance and transmittance for this system at low powers resemble those derived for a single photon, the arrival of several photons within the bandwidth ($\delta_\omega < \Gamma$) saturates the atomic response and these photons cannot be efficiently reflected. To be specific, consider the case when the incident field consists of a coherent state, the quantum mechanical state that most closely corresponds to a classical field. A mapping allows the scattering dynamics to be solved exactly. Assume that the incident field propagates to the right and that the emitter is initially in the ground state, such that the initial wave function can be written in the form:

$$|\psi(t \to -\infty)\rangle = D(\{\alpha_k e^{-i\nu_k t}\})\,|\text{vac}\rangle\,|g\rangle \tag{5-151}$$

where the displacement operator

$$D(\{\alpha_k\}) \equiv \exp\left(\int dk \hat{a}^{\dagger}_{R,k}\alpha_k - \alpha^*_k \hat{a}_{R,k}\right) \tag{5-152}$$

creates a multimode coherent state from vacuum. This property of the displacement operator motivates a state transformation given by

$$|\tilde{\psi}\rangle = D(\{\alpha_k e^{-i\nu_k t}\})\,|\psi\rangle \tag{5-153}$$

so that the initial state is transformed into $|\psi(t \to -\infty)\rangle = |\text{vac}\rangle|g\rangle$. An important consequence is that the dynamics of the emitter interacting with the field modes can now be treated under the Wigner-Weisskopf approximation, i. e., interaction with the vacuum modes gives rise to an exponential decay rate from state $|e\rangle$ to $|g\rangle$ at a rate Γ. The evolution of the atomic operators consequently reduces to the usual Langevin-Bloch equations, which enables all properties of the atomic operators and the scattered field to be calculated. For a narrow bandwidth ($\delta_\omega \ll \Gamma$), resonant ($\delta_k = 0$) input field, the steady-state transmittance and reflectance are found as

$$T = \frac{1 + 8(1 + P)^2 (\Omega_c / \Gamma)^2}{(1 + P)^2 (1 + 8(\Omega_c / \Gamma)^2)} \tag{5-154}$$

$$R = \left(1 + \frac{1}{P}\right)^{-2} \frac{1}{1 + 8(\Omega_c / \Gamma)^2} \tag{5-155}$$

At low powers ($\Omega_c / \Gamma \ll 1$), the emitter has scattering properties identical to the single-photon case, $R \approx (1 + 1/P)^{-2}$, $T \approx (1 + P)^{-2}$, and for large Purcell factors the single emitter again acts as a perfect mirror. At high incident powers ($\Omega_c / \Gamma > 1$), however, the single emitter saturates and most of the incoming photons are simply transmitted past with no effect, $T \to 1$, $R \sim O((\Gamma / \Omega_c)^2)$.

5.7.3 Photon correlations

The strongly nonlinear atomic response at the single-photon level leads to dramatic modification of photon statistics that cannot be captured by only considering average intensities. Now consider higher-order correlations of the transmitted and reflected fields. Specifically, focus on the normalized second-order correlation function for the outgoing field: $g^{(2)}_{R,L}(t)$, which for a stationary process is defined as

$$g^{(2)}_{\beta = R,L}(z, t) \equiv \langle \hat{E}^{\dagger}_{\beta}(z, \tau) \hat{E}^{\dagger}_{\beta}(z, \tau + t) \hat{E}_{\beta}(z, \tau + t) \hat{E}_{\beta}(z, \tau) \rangle / \langle \hat{E}^{\dagger}_{\beta}(z, \tau) \hat{E}_{\beta}(z, \tau) \rangle^2 \tag{5-156}$$

where t denotes the difference between the two observation times τ and $\tau + t$. The statistics of the reflected field is identical to the well-known result for resonance fluorescence in three dimensions (see Figure 5-73). This can intuitively be understood because it is a purely scattered field. It follows that the field is strongly anti-bunched, $g^{(2)}(0) = 0$, since the emitter can only absorb and re-emit one photon at a time. The transmitted field, however, has unique properties because it is a sum of the incident and scattered fields. For near-resonant excitation, can find for low powers:

$$g^{(2)}(t) = \mathrm{e}^{-\Gamma t} (P^2 - \mathrm{e}^{\Gamma t/2})^2 + \mathcal{O}(\Omega_c^2 / \Gamma^2) \tag{5-157}$$

while for high powers $g^{(2)}(t)$ approaches unity for all times. The high power result indicates that no change in statistics occurs and is due to saturation of the atomic response. The low-power behavior reflects that of an efficient single-photon switch. In particular, for $P \gg 1$, individual photons have a large reflection probability, but when two photons are incident simultaneously the transition saturates, so that photon pairs have a much larger probability of transmission (for $P \ll 1$ the emitter has little influence and the statistics of the transmitted field are almost unchanged). One also finds a subsequent anti-bunching and perfect vanishing of $g^{(2)}(t)$ at the later time $t_0 = (4\log P)/\Gamma$ for weak input fields. A more detailed understanding of these features can be gained from a quantum jump picture describing the system's evolution following the detection of a photon. Unlike for the reflected field, the picture for the transmitted field is more complicated because one cannot determine whether the detected photon originates from the emitter or from direct transmission of the incident field. For large P, $\hat{E}_T$ is strongly influenced by its atomic component. This is responsible for, e.g., the low transmittance $T \approx (1 + P)^{-2}$ in steady-state, as the field scattered by the emitter destructively interferes with the incoming field. Because multiple incident photons increase the transmission probability, the detection of a photon enhances the conditional probability that another photon is present in the system. In the quantum jump picture this translates into a sudden enhancement of the coherence $\langle \sigma_{ge} \rangle$ by a factor of $1 + P$ over its steady-state value. The destructive interference between the incoming and scattered fields is subsequently lost, and the jump causes a sudden enhancement in the field amplitude $\langle \hat{E}_T \rangle$ while also inducing a π-phase shift relative to its equilibrium value. The initial enhancement in $\langle \hat{E}_T \rangle$ gives rise to bunching. Then, the π-phase shift and subsequent relaxation back to equilibrium causes $\langle \hat{E}_T \rangle$ to pass through zero at time t_0, which yields the subsequent antibunching and reflects the cancellation of the incoming and scattered fields. For $P = 1$, this cancellation happens exactly at $t = 0$ such that $g^{(2)}(0) = 0$.

- Ideal single-photon transistor

While the two-level emitter analyzed previously is capable of acting as a switch that distinguishes between single-and multi-photon fields, a greater degree of coherent control can be gained by considering the interaction of light with a multi-level emitter. For concreteness, consider the three-level configuration. Here, a metastable state $|s\rangle$ is decoupled from the SPs due to, e.g., a different orientation of its associated dipole moment, but is resonantly coupled to $|e\rangle$ via some classical, optical control field with Rabi frequency. States $|g\rangle$, $|e\rangle$ remain coupled via the SP modes as discussed earlier. Using this system, we now describe a process in which a single "gate" photon can completely control the propagation of subsequent "signal" pulses consisting of either individual or multiple photons, whose timing can be arbitrary. In analogy to the electronic counterpart, this corresponds to an ideal single-photon transistor. First describe how one can achieve coherent storage of a single photon, and then how this can be combined with the reflective properties derived above to realize a single-photon transistor. The storage is an important ingredient, as it provides an atomic memory of the gate field and hence allows the gate to interact with the subsequent signal field. The idea behind single-photon storage is to initialize the emitter in $|g\rangle$ and to apply the control field simultaneous with the arrival of a single photon in the SP modes. The control field, if properly chosen (or "impedance-matched"), will result in capture of the incoming single photon while inducing a spin flip from $|g\rangle$ to $|s\rangle$. Generally, one can show by time reversal symmetry that the optimal storage strategy is the time-reversed process of single-photon generation, where the emitter is driven from $|s\rangle$ to $|g\rangle$ by the external field while emitting a single photon, whose wavepacket depends on $\Omega(t)$. By this argument, it is evident that optimal storage is obtained by splitting the incoming pulse and having it incident from both sides of the emitter simultaneously (see Figure 5-74), and that there is a one-to-one correspondence between the incoming pulse shape and the optimal field.

Moreover, the storage efficiency is identical to that of single photon generation and is thus given by $\sim 1 - 1/P$ for large P. This result is also derived explicitly in Supplemental Information, where we solve for the dynamics of this three-level system exactly. Physically, the fidelity of storage is simply determined by the degree to which coupling of the emitter to the SP modes exceeds the coupling to other channels. A detailed analysis reveals that this optimum is achievable for any input pulse of duration $T \gg 1/\Gamma$ and for a certain class of pulses of duration $T \sim 1/\Gamma$. Finally, we note that if no photon impinges upon the emitter, the pulse $\Omega(t)$ has no effect and the emitter remains in state $|g\rangle$ for the entire process. The result is more generally described as a mapping between single SP states and metastable atomic states, $(\alpha|0\rangle+\beta|1\rangle)|g\rangle \rightarrow |0\rangle(\alpha|g\rangle+\beta|s\rangle)$.

The next consider the reflection properties of the emitter when the control field (t) is turned off. If the emitter is in $|g\rangle$, the reflectance and transmittance derived above for the two-level emitter remain valid. On the other hand, if the emitter is in $|s\rangle$, any incident fields will simply be transmitted with no effect since this state is decoupled from the SPs. Therefore, with (t) turned off, the three-level system effectively behaves as a conditional mirror whose properties depend sensitively on its internal state. The techniques of state-dependent conditional reflection and single-photon storage can be combined to create a single-photon transistor, whose operation is illustrated in Figure 5-74. The key principle is to utilize the presence or absence of a photon in an initial "gate" pulse to conditionally flip the internal state of the emitter during the storage process, and to then use this conditional flip to control the flow of subsequent "signal" photons arriving at the emitter. The first step is to implement the storage protocol for the gate pulse, consisting of either zero or one photon, starting with the emitter in $|g\rangle$. The presence (absence) of a photon causes the emitter to flip to (remain in) state $|s\rangle(|g\rangle)$. Next, the interaction of each signal pulse arriving at the emitter depends sensitively on the internal state that results after storage. The storage step and conditional spin flip causes the emitter to be either highly reflecting or completely transparent depending on the gate and the system therefore acts as an efficient switch or transistor for the subsequent signal field.

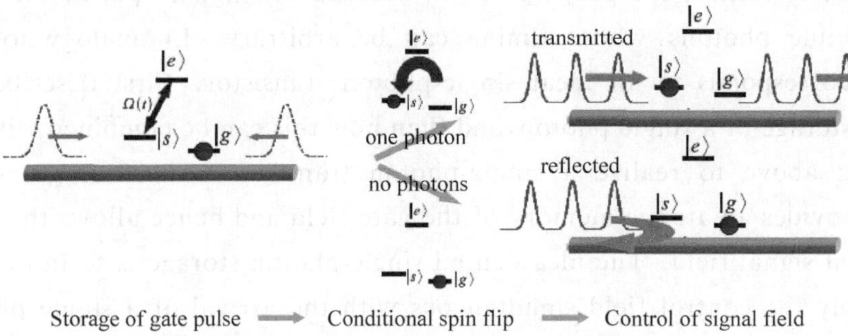

Fig. 5-74 Schematic of transistor operation involving a three-level emitter. In the storage step, a gate pulse consisting of zero or one photon is split equally in counter-propagating directions and coherently stored using an impedance-matched control field (t). The storage results in a spin flip conditioned on the photon number. A subsequent incident signal field is either transmitted or reflected depending on the photon number of the gate pulse, due to the sensitivity of the propagation to the internal state of the emitter.

The ideal operation of the transistor is limited only by the characteristic time over which an undesired spin flip can occur. In particular, if the emitter remains in $|g\rangle$ after storage of the gate pulse, the emitter can eventually be optically pumped to $|s\rangle$ upon the arrival of a sufficiently large number of photons in the signal field.

Finally, to note that there exist other possible realizations of a single-photon transistor as well. The "impedance-matching" condition and the need to split a pulse for optimal storage, for example, can be relaxed using a small ensemble of emitters and photon storage techniques based on electromagnetically induced transparency (EIT). Here, storage also results in a spin flip within the ensemble that sensitively alters the propagation of subsequent photons.

• Integrated systems

Thus far have not dealt with the inevitable losses that SPs experience as they propagate along the nanowire, which could potentially limit their feasibility as long-distance carriers of information and their use in large-scale devices. For the nanowire, one must consider the trade-off between the larger Purcell factors obtainable with smaller diameters and a commensurate increase in dissipation due to the tighter field confinement. However, these limitations are not fundamental, if one can integrate SP devices with low-loss dielectric waveguides and other microphotonic devices. Here, the SPs can be used to achieve strong nonlinear interactions over very short interaction distances, but are rapidly in-and out-coupled to conventional waveguides for long-distance transport. One such integration scheme is illustrated in Figure 5-75, where excitations are transferred to and from the SP modes of a nanowire from an evanescently coupled, phase-matched dielectric waveguide. The losses will be small provided that the distance needed for the SPs to be coupled in and out and interact with the emitter is smaller than the characteristic dissipation length. This can be accomplished by techniques such as optimizing of SP geometries e. g., tapered wires or nanotips and engineering of SP dispersion relations via periodic structures.

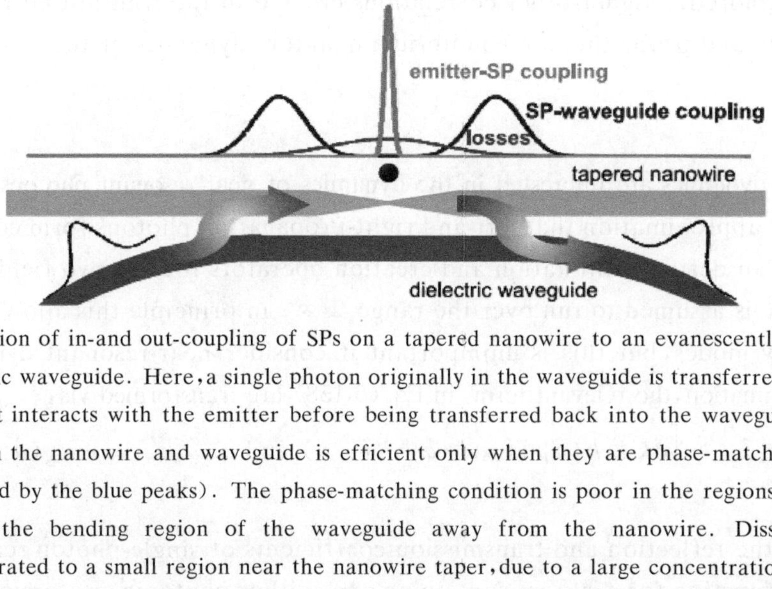

Fig. 5-75 Illustration of in-and out-coupling of SPs on a tapered nanowire to an evanescently coupled, low-loss dielectric waveguide. Here, a single photon originally in the waveguide is transferred to the nanowire, where it interacts with the emitter before being transferred back into the waveguide. The coupling between the nanowire and waveguide is efficient only when they are phase-matched (in the regions indicated by the blue peaks). The phase-matching condition is poor in the regions of the wire taper and in the bending region of the waveguide away from the nanowire. Dissipative losses are concentrated to a small region near the nanowire taper, due to a large concentration of fields here.

Coupling efficiencies exceeding 95%, for example, are predicted using simple systems. Such a conductor/dielectric interface would provide convenient integration with conventional optical elements, enable many nonlinear operations without loss, and make large-scale, integrated photonic devices feasible. Another key feature of nano-scale SPs is that the strong interaction with a single emitter is very robust. In particular, the large coupling occurs over a very large bandwidth and no special tuning of either the emitter or nanowire is required. SPs are thus promising candidates for use with solid-state emitters such as quantum dot nanocrystals or color center, where the spectral properties can vary over individual emitters. Color centers in diamond, for instance, are especially promising because they offer sharp optical lines and three-level internal configurations. At the same time, guided SPs might be used for trapping isolated neutral atoms in the vicinity of suspended wires, thereby creating an effective interface for isolated atomic systems.

Now outline some new directions opened up by this work. That a single emitter near a conducting nanowire provides a strong optical nonlinearity at the level of single photons, which can be exploited to create a single-photon transistor. This can be used for a variety of important applications, such as very efficient single-photon detection, where the large gain in the signal field provides for efficient detection of the gate pulse. This system also finds applications in quantum information science. One can prepare Schrodinger cat states of photons, for example, if the gate pulse contains a superposition of zero and one photon, since this initial pulse can be entangled with the propagation direction of a large number of subsequent signal photons. The controlled-phase gate for photons proposed in for cavity QED is also directly extendable to our plasmonic system. In particular, this scheme relies on the conditional phase shifts acquired as photons are reflected from a resonant cavity containing a single atom, which are analogous to the reflection dynamics derived for single SPs here. In addition, by using SPs it is possible to achieve very large optical depths with just a few emitters, which would make this system effective for realizing EIT-based nonlinear schemes. Furthermore, the present system is an intriguing candidate to observe phenomena associated with strongly interacting, one-dimensional many-body systems. For example, non-perturbative effects such as photon-atom bound states and quantum phase transitions involving photons can be explored. Higher-order correlations created in the transmitted field can become a useful tool to study and probe the non-equilibrium quantum dynamics of these strongly interacting photonic systems.

- Methods

Single-photon dynamics are interested in the dynamics of near-resonant photons with an emitter, that can make the approximation that left-and right-propagating photons form completely separate quantum fields. Can define annihilation and creation operators for the two fields, $\hat{a}_{L(R),k}$, $\hat{a}^{\dagger}_{L(R),k}$, where the index k is assumed to run over the range $\pm\infty$; in principle this allows for the existence of negative-energy modes, but this is unimportant if consider near-resonant dynamics. Under this two-branch approximation, the relevant terms in Eq. (5-158) are transformed via:

$$\int \mathrm{d}k\ \hbar c \mid k \mid \hat{a}^{\dagger}_{k}\hat{a}_{k} \to \int \mathrm{d}k\ \hbar ck(\hat{a}^{\dagger}_{R,k}\hat{a}_{R,k} + \hat{a}^{\dagger}_{L,-k}\hat{a}_{L,-k}) \text{ and } \sigma_{\mathrm{eg}}\hat{a}_{k}\mathrm{e}^{ikz_a} \to \sigma_{\mathrm{eg}}(\hat{a}_{R,k} + \hat{a}_{L,k})\mathrm{e}^{ikz_a} \tag{5-158}$$

To solve for the reflection and transmission coefficients of single-photon scattering, can write the general wave function for a system containing one (either photonic or atomic) excitation in the following way (here a two-level emitter is assumed),

$$\mid \psi_k\rangle = \int \mathrm{d}z\big(\phi_L(z)\hat{E}^{\dagger}_{L}(z) + \phi_R(z)\hat{E}^{\dagger}_{R}(z)\big) \mid g,\mathrm{vac}\rangle + c_e \mid e,\mathrm{vac}\rangle \tag{5-159}$$

The field amplitudes are chosen to correspond to photons of well-defined momenta in the limits $z\to\pm\infty$, e.g., $R(z\to-\infty)\sim \mathrm{e}^{ikza}$, $R(z\to\infty)\sim \mathrm{e}^{ikza}$, and $L(z\to-\infty)\sim \mathrm{e}^{ikza}$ for a photon propagating initially to the right.

5.7.4 Multi-photon dynamics

In the two-branch approximation, the Heisenberg equations of motion for the fields are given by

$$\left(\frac{\partial}{\partial z} + \frac{1}{c}\frac{\partial}{\partial t}\right)\hat{E}_R(z,t) = \frac{\sqrt{2\pi}\,\mathrm{i}g}{c}\sigma_{\mathrm{ge}}(t)\delta(z - z_a) \tag{5-160}$$

which can be formally integrated to give:

$$\hat{E}_R(z,t) = \hat{E}_{R,\text{free}}(z-ct) + \frac{\sqrt{2\pi}\,\mathrm{i}g}{c}\sigma_{\text{ge}}(t-(z-z_a)/c)\theta(z-z_a) \tag{5-161}$$

where $\theta(z)$ is the Heaviside step function. A similar equation holds for $\hat{E}_L$. Assuming that the field initially propagates to the right, $\hat{E}_R(z,t)$ is the field transmitted past the emitter for $z > z_a$, while for $z < z_a$, $\hat{E}_L(z,t)$ is the reflected field. Now discuss how to calculate the transmitted field intensity (a similar method holds for finding the reflected intensity). Under the transformation given by Eq. (5-162), the first-order correlation function for the right-going field is given by

$$G_R^{(1)}(z,t) = \langle(\hat{E}_R^{\dagger}(z,t) + E_c^{*}(z,t))(\hat{E}_R(z,t) + E_c(z,t))\rangle \tag{5-162}$$

which upon evaluating at $z > z_a$ yields the average transmitted intensity. Because the initial photonic state is vacuum following the transformation, $\hat{E}_R$, free has no effect and thus calculation of $G^{(1)}$ reduces to calculating correlations between atomic operators. Techniques for evaluating these correlations are well known using the Langevin-Bloch equations. Calculation of $g^{(2)}(t)$ proceeds in a similar manner by using Eq. (5-102) to express $g^{(2)}(t)$ in terms of two-time atomic correlations, which can be evaluated using the well-known quantum regression theorem.

The system in consideration undergoes a quantum jump following detection of a transmitted photon. Immediately following the detection, the density matrix is given by

$$\rho_{\text{jump}} = \hat{E}_T \rho_{\text{ss}} \hat{E}_T^{\dagger} / \langle \hat{E}_T^{\dagger} \hat{E}_T \rangle_{\text{ss}} \tag{5-163}$$

Where ρ_{ss} is the steady-state density matrix and $\langle\,\rangle_{\text{ss}}$ denotes the average of quantities in steady state. Here $\hat{E}_T$ is the jump operator defined in the "Photon correlations" section and physically corresponds to the transmitted field. In the weak-field limit, it is straightforward to show that

$$\langle \sigma_{\text{ge}} \rangle_{\text{jump}} = (1+P)\langle \sigma_{\text{ge}} \rangle_{\text{ss}} = 2i\Omega_c / \Gamma' \tag{5-164}$$

and

$$\langle \hat{E}_T \rangle_{\text{jump}} / \langle \hat{E}_T \rangle_{\text{ss}} = 1 - P^2 \tag{5-165}$$

Note in particular that for large P, there is an initial enhancement in the transmitted field amplitude as well as a π-phase shift from its equilibrium value.

5.8 Optical dense atomic memory medium

In many papers presented a universal physical picture for describing a wide range of techniques for storage and retrieval of photon wave packets in (Lambda) Λ-type atomic media in free space, including the adiabatic reduction of the photon group velocity, pulse-propagation control via off-resonant Raman techniques, and photon-echo based techniques. This universal picture produced an optimal control strategy for photon storage and retrieval applicable to all approaches and yielded identical maximum efficiencies for all of them. In the present paper, we present the full details of this analysis as well some of its extensions, including the discussion of the effects of non-degeneracy of the two lower levels of the Λ-system.

5.8.1 Λ-type optical dense atomic media

The analysis in the present paper is based on the intuition obtained from the study of photon storage in the cavity model in the preceding paper. They used a universal physical picture to optimize and demonstrate equivalence between a wide range of techniques for storage and retrieval of photon wave packets in Lambda-type atomic media in free space, including the adiabatic reduction of the photon group velocity, pulse-propagation control via off-resonant Raman techniques, and photon-echo-based techniques.

- **Λ**-type atomic media (General Theoretical Physics)

High fidelity storage of a traveling pulse of light into an atomic memory and the subsequent retrieval of the state back onto a light pulse are currently being pursued by a number of laboratories around the world. In 2008 A. V. Gorshkov et al. presented a universal physical picture for describing a wide range of techniques for storage and retrieval of photon wave packets in Λ-type atomic media in free space, including the adiabatic reduction of the photon group velocity, pulse-propagation control via off-resonant Raman techniques, and photon-echo based techniques. This universal picture produced an optimal control strategy for photon storage and retrieval applicable to all approaches and yielded identical maximum efficiencies for all of them. In the present paper, we present the full details of this analysis as well some of its extensions, including the discussion of the effects of non-degeneracy of the two lower levels of the Λ-system. The analysis in the present paper is based on the intuition obtained from the study of photon storage in the cavity model in the preceding paper. They used a universal physical picture to optimize and demonstrate equivalence between a wide range of techniques for storage and retrieval of photon wave packets in Lambda-type atomic media in free space, including the adiabatic reduction of the photon group velocity, pulse-propagation control via off-resonant Raman techniques, and photon-echo-based techniques.

Physics Department of Harvard University and Danish National Research Foundation Centre of Quantum Optics of Denmark presented a universal picture for describing, optimizing, and showing a certain degree of equivalence between a wide range of techniques for photon storage and retrieval in Λ-type atomic media, including the approaches based on electromagnetically induced transparency (EIT), off-resonant Raman processes, and photon echo. In the section, as well as in the preceding and that follows present all the details behind this universal picture and the optimal control shaping that it implies, as well as consider several extensions of this analysis beyond the results. In particular, in the section will present the full details of the optimization in the slightly simpler model where the atoms are placed inside a cavity. Using the intuition gained from the cavity model discussion, we show in the present paper all the details behind the analysis of the free-space model. Here also discuss several extensions of the analysis, such as the inclusion of the decay of coherence between the two lower levels of the system and the effects of nondegeneracy of these two levels. Finally, generalize the treatment to include the effects of inhomogeneous broadening.

For a complete introduction to photon storage in Λ-type atomic media, as it applies to the section, as well as for the full list of references. In the present introduction, only list the two main results. The first important result is the abovementioned proof of a certain degree of equivalence between a variety of different photon storage protocols. In particular, this result means that provided there is a sufficient degree of control over the shape of the incoming photon wave packet and/or over the power and shape of the classical control pulses, all the protocols considered have the same maximum achievable efficiency that depends only on the optical depth d of the medium.

The second important result is a novel time-reversal-based iterative algorithm for optimizing quantum state mappings, a procedure that expect to be applicable beyond the field of photon storage. One of the key features of this optimization algorithm is that it can not only be used as a mathematical tool but also as an experimental technique. In fact, following our theoretical proposal, an experimental demonstration of this technique has already been carried out. Both the experimental results and the theoretical results of the present paper indicate that the suggested optimization with respect to the shape of the incoming photon wave packet and/or the control pulse shape and power will be important for increasing the photon storage efficiencies in current experiments. Although the slightly simpler cavity model discussed is similar enough to the free-space model to provide good intuition for it, the two physical systems have their own advantages and disadvantages, which will discuss in the present paper. One advantage of the free-space model is the fact that it is easier to set up experimentally, which is one of the reasons we study this model in the present paper. Turning to the physics of the two models, the main differences come from the fact that in the cavity model the only spin wave mode accessible is the one that has the excitation distributed uniformly over all the atoms. In contrast, in the free-space model, incoming light can couple to any mode specified by a smooth excitation with position-dependent amplitude and phase. As a consequence of this, the free-space model allows for high efficiency storage of a wider range of input light modes than the cavity model. In contrast, we show in the present paper that high efficiency fast storage in a free-space atomic ensemble with optical depth d is possible for any input light mode of duration T provided $T\gamma \ll 1$ and $Td\gamma \gg 1$. However, the cavity model also has some advantages over the freespace model. The optimal efficiency is therefore higher when the ensemble is enclosed in a cavity, which effectively enhances the optical depth by the cavity finesse to form the cooperativity parameter C. Moreover, if one is forced to retrieve from a spatially uniform spin wave mode (e. g., if the spin wave is generated via spontaneous Raman scattering), the error during retrieval will decrease faster with optical depth in the cavity model ($\sim 1/C$) than in the free-space model ($\sim 1/\sqrt{d}$).

- Model

In this section, only briefly summarize the model and state the equations of motion without derivation. To assume that within the interaction volume the concentration of atoms is uniform in the transverse direction. The atoms have the same Λ-type level configuration as in the cavity case discussed below. They are coupled to a quantum field and a copropagating classical field. To assume that quantum electromagnetic field modes with a single transverse profile are excited. Also assume that both the quantum and the classical field are narrowband fields centered at and $\omega_2 = \omega_{es} - \Delta$ respectively (where ω_{eg} and ω_{es} are atomic transition frequencies). The quantum field is described $\Omega(z, t) = \Omega(t - z/c)$. by a slowly varying operator $\hat{E}(z, t)$, while the classical field is described by the Rabi frequency envelope.

To neglect reabsorption of spontaneously emitted photons. This is a good approximation. That are interested in the storage of single-or few-photon pulses, in which case there will be at most a few spontaneously emitted photons. Although for an optically thick medium they can be reabsorbed and reemitted, the probability of spontaneously emitting into the mode $\hat{E}$ is given by the corresponding far-field solid angle $\sim\lambda^2/A\sim d/N$, where A is the cross section area of both the quantum field mode and the atomic medium (see Appendix A for a discussion of why this choice is not important), $\lambda = 2\pi c/\omega_1$ is the wavelength of the quantum field, and $d\sim\lambda^2 N/A$ is the resonant optical depth of the ensemble.

In most experiments, this probability is very small. Moreover, we will show that for the optimized storage process, the fraction of the incoming photons lost to spontaneous emission will decrease with increasing optical depth. In practice, however, reabsorption of spontaneously emitted photons can cause problems during the optical pumping process, which is used to initialize the sample, and this may require modification of the present model. To treat the problem in a one-dimensional approximation. This is a good approximation provided that the control beam is much wider than the single mode of the quantum field defined by the optics. In this case, the transverse profile of the control field can be considered constant; and, in the paraxial approximation, the equations reduce to one-dimensional equations for a single Hermite-Gaussian quantum field mode.

To define the polarization operator $\hat{P}(z,t) = \sqrt{N}\hat{\sigma}_{ge}(z,t)$ and the spin-wave operator $\hat{S}(z,t) = \sqrt{N}\hat{\sigma}_{gs}(z,t)$. In the dipole and rotating-wave approximations, to first order in $\hat{E}$, and assuming that at all times almost all atoms are in the ground state, the Heisenberg equations of motion read

$$(\partial_t + c\partial_z)\hat{\mathcal{E}} = ig\sqrt{N}\hat{P}n(z)L/N \tag{5-166}$$

$$\partial_t\hat{P} = -(\gamma + i\Delta)\hat{P} + ig\sqrt{N}\hat{\mathcal{E}} + i\Omega\hat{S} + \sqrt{2\gamma}\hat{F}_P \tag{5-167}$$

$$\partial_t\hat{S} = -\gamma_s\hat{S} + i\Omega^*\hat{P} + \sqrt{2\gamma_s}\hat{F}_S \tag{5-168}$$

where introduced the spin-wave decay rate γs, the polarization decay rate γ, and the corresponding Langevin noise operators $\hat{F}_P(z,t)$ and $\hat{F}_S(z,t)$. As in cavity case, collective enhancement results in the increase of the atom-field coupling constant g (assumed to be real for simplicity) by a factor of $\sqrt{N}$ up to $g\sqrt{N}$. As in the cavity discussion that suppose that initially all atoms are in the ground state, i.e., no atomic excitations are present. Also assume that there is only one nonempty mode of the incoming quantum field and that it has an envelope shape $h_0(t)$ nonzero on $[0,T]$. The term "photon storage and retrieval" refers to mapping this mode onto some mode of $\hat{S}$ and, starting at a later time $T_r > T$, retrieving it onto an outgoing field mode. Then precisely as in the cavity case for the purposes of finding the storage efficiency, which is given by the ratio of the numbered of stored excitations to the number of incoming photons

$$\eta_s = \frac{\int_0^L \mathrm{d}z\,\frac{n(z)}{N}\langle\hat{S}^\dagger(z,T)\hat{S}(z,T)\rangle}{\frac{c}{L}\int_0^T \mathrm{d}t\langle\hat{\mathcal{E}}^\dagger(0,t)\hat{\mathcal{E}}(0,t)\rangle} \tag{5-169}$$

In fact, although the resulting equations describe our case of quantized light coupled to the | − | transition, they will also precisely be the equations describing the propagation of a classical probe pulse. Going into a comoving frame $t' = t - z/c$, introducing dimensionless time $t = \gamma t'$ and dimensionless rescaled coordinate

$$\tilde{z} = \int_0^z \mathrm{d}z' n(z')/N$$

absorbing a factor of into the definition of ε and reduce Eqs. (5-166)～(5-168) to:

$$\partial_{\tilde{z}}\,\mathcal{E} = \pi\sqrt{d}P \tag{5-170}$$

$$\partial_{\tilde{t}} P = -(1+\mathrm{i}\tilde{\Delta})P + \mathrm{i}\sqrt{d}\,\mathcal{E} + \mathrm{i}\tilde{\Omega}(\tilde{t})S \tag{5-171}$$

$$\partial_{\tilde{t}} S = \mathrm{i}\tilde{\Omega}^*(\tilde{t})P \tag{5-172}$$

where have identified the optical depth $d = g^2NL/(\gamma c)$ and where $\tilde{\Delta} = \Delta/\gamma$ and $\tilde{\Omega} = \Omega/\gamma$. Moreover, the definition of d that to use here can be related to the intensity attenuation of a resonant probe in our three level system with the control off, in which case the equations give an attenuation of exp $(-2d)$. In Eqs. (5-171)~(5-173) and in the rest of this paper, we neglect the decay γ_s of the spin wave. However, precisely as in the cavity case, nonzero γ_s simply introduces an exponential decay without making the solution or the optimization harder, If note that Eqs. (5-167)~(5-169) are the same as the equations for copropagating fields, generalized to nonzero Δ and γ_s, and taken to first order in ε. During storage, the initial and boundary conditions are (in rescaled variables) $\varepsilon(\tilde{z}=0,\tilde{t}) = \varepsilon_{\text{in}}(\tilde{t})$, $P(\tilde{z},\tilde{t}=0) = 0$, and $S(\tilde{z},\tilde{t}=0) = 0$, where $\varepsilon_{\text{in}}(\tilde{t})$ is nonzero for $\tilde{t}\in[0,\tilde{T}]$ (where $\tilde{T} = T\gamma$) and, being a shape of a mode, is normalized according to:

$$\int_0^{\tilde{T}} \mathrm{d}\tilde{t}\ |\mathcal{E}_{\text{in}}(\tilde{t})|^2 = 1 \tag{5-173}$$

$S(\tilde{z},\tilde{T})$ gives the shape of the spin-wave mode, into which we store, and the storage efficiency is given by

$$\eta_s = \int_0^1 \mathrm{d}\tilde{z}\ |S(\tilde{z},\tilde{T})|^2 \tag{5-174}$$

The remaining initial and boundary conditions during retrieval are $E(\tilde{z}=0,\tilde{t})=0$ and $P(\tilde{z},\tilde{T}_r)=0$. If renormalize the spin wave before doing the retrieval, then the retrieval efficiency will be given by

$$\eta_r = \int_{T_r}^{\infty} \mathrm{d}\tilde{t}\ |\mathcal{E}(1,\tilde{t})|^2 \tag{5-175}$$

If do not renormalize the spin wave before doing the retrieval, this formula will give the total efficiency of storage followed by retrieval $\eta_{\text{tot}} = \eta_s\eta_r$. To solve Eqs. (5-171) ~ (5-173), it is convenient to Laplace transform them in space according to become

$$\mathcal{E} = \pi\frac{\sqrt{d}}{u}P + \frac{\mathcal{E}_{\text{in}}}{u} \tag{5-176}$$

$$\partial_{\tilde{t}} P = -\left(1+\frac{d}{u}+\pi\tilde{\Delta}\right)P + \pi\tilde{\Omega}(t)S + \pi\frac{\sqrt{d}}{u}\mathcal{E}_{\text{in}} \tag{5-177}$$

It is also convenient to reduce Eqs. (5-171)~(5-173) to a single equation:

$$\left[\ddot{S} - \frac{\dot{\tilde{\Omega}}^*}{\tilde{\Omega}^*}\dot{S}\right] + \left(1+\frac{d}{u}+\pi\tilde{\Delta}\right)\dot{S} + |\tilde{\Omega}|^2 S = -\tilde{\Omega}^*\frac{\sqrt{d}}{u}\mathcal{E}_{\text{in}} \tag{5-178}$$

where the overdot stands for the $\tilde{t}$ derivative. As in the cavity case, this second-order differential equation cannot, in general, be fully solved analytically. We can, however, derive several important results regarding the optimal control strategy for storage and retrieval without making any more approximations.

5.8.2 Optimal retrieval

Although Eq. (5-178) cannot, in general, be fully solved analytically, we still make in this and in the following two sections several important statements regarding the optimal strategy for maximizing the storage efficiency, the retrieval efficiency, and the combined (storage followed by retrieval) efficiency without making any more approximations. It is convenient to first consider retrieval, and do so in this section. Although we cannot, in general, analytically solve for the output field $E_{\text{out}}(t)$, we will show now that, the retrieval efficiency is independent of the control shape and the detuning provided no excitations are left in the atoms. Moreover, and will show that the retrieval efficiency is given by a simple formula that depends only on the optical depth and the spin-wave mode. From Eqs. (5-173) and (5-177), it follows that

$$\frac{\mathrm{d}}{\mathrm{d}\tilde{t}}(P(u,\tilde{t})[P(u'^*,\tilde{t})]^* + S(u,\tilde{t})[S(u'^*,\tilde{t})]^*) = -(2 + d/u + d/u')P(u,\tilde{t})[P(u'^*,\tilde{t})]^* \tag{5-179}$$

Using Eqs. (5-176) and (5-179) and assuming $P(u,\infty) = S(u,\infty) = 0$ (i. e., that no excitations are left in the atoms at $\tilde{t} = \infty$), the retrieval efficiency is

$$\begin{aligned}\eta_r &= L^{-1}\left\{\frac{d}{uu'}\int_{\tilde{T}_r}^{\infty} \mathrm{d}\tilde{t}P(u,\tilde{t})[P(u'^*,\tilde{t})]^*\right\} \\ &= L^{-1}\left\{\frac{d}{2uu' + d(u+u')}S(u,\tilde{T}_r)[S(u'^*,\tilde{T}_r)]^*\right\} \\ &= \int_0^1 \mathrm{d}\tilde{z}\int_0^1 \mathrm{d}\tilde{z}' S(1-\tilde{z},\tilde{T}_r)S^*(1-\tilde{z}',\tilde{T}_r)k_r(\tilde{z},\tilde{z}')\end{aligned} \tag{5-180}$$

where L^{-1} means that two inverse Laplace transforms ($u\to\tilde{z}$ and $u'\to\tilde{z}'$) are taken and are both evaluated at $\tilde{z} = \tilde{z}' = 1$ and where the kernel k_r is defined as

$$\begin{aligned}k_r(\tilde{z},\tilde{z}') &= L^{-1}\left\{\frac{d}{2uu' + d(u+u')}\mathrm{e}^{-u(1-\xi)-u'(1-\xi')}\right\} \\ &= \frac{d}{2}\mathrm{e}^{-d\frac{\xi'+\xi}{2}} I_0(d\sqrt{\tilde{z}\tilde{z}'})\end{aligned} \tag{5-181}$$

where I_n is the nth-order modified Bessel function of the first kind. If see that the efficiency is independent of Δ and Ω, which reflects that in Eq. (5-177) (or, equivalently, on the right-hand side of Eq. (5-179)) there is a fixed branching ratio between the decay rates of P. For a given u the rates are (in the original units) γ and $\gamma d/u$ into the undesired modes and the desired mode ε_{out}, respectively, independent of Δ and Ω. In fact, a stronger result than the independence of retrieval efficiency from Δ and Ω can be obtained, the distribution of spontaneous emission loss as a function of position is also independent of the control and detuning.

In contrast to the cavity case in paper I where there was only one spin-wave mode available, in the free-space case, the retrieval efficiency in Eq. (5-180) is different for different spin-wave modes. We can, thus, at each d, optimize retrieval by finding the optimal retrieval spin wave $\tilde{S}_d(\tilde{z})$ (suppress here the argument $\tilde{T}_r$). The expression for the efficiency in the last line of Eq. (5-180) is an expectation value of a real symmetric (and hence Hermitian) operator $k_r(\tilde{z},\tilde{z}')$ in the state $S(1-\tilde{z})$. It is therefore maximized when $S(1-\tilde{z})$ is the eigenvector with the largest eigenvalue of the following eigenvalue problem:

$$\eta_r S(1-\tilde{z}) = \int_0^1 \mathrm{d}\tilde{z}' k_r(\tilde{z},\tilde{z}')S(1-\tilde{z}') \tag{5-182}$$

Since eigenvectors of real symmetric matrices can be chosen real, the resulting optimal spin wave $\tilde{S}_d(\tilde{z})$ can be chosen real. To find it, we start with a trial $S(\tilde{z})$ and iterate the integral in

Eq. (5-182) several times until convergence. In Fig. 5-77, we plot the resulting optimal spin wave $\widetilde{S}_d(\tilde{z})$ for $d=0.1, 1, 10, 100$, as well as its limiting shape ($\widetilde{S}\infty(\tilde{z})=\sqrt{3}\tilde{z}$) as $d \to \infty$. At $d \to 0$, the optimal mode approaches a flat mode. These shapes can be understood by noting that retrieval is essentially an interference effect resembling superradiance, where the emission from all atoms contributes coherently in the forward direction. To get the maximum constructive interference, it is desirable that all atoms carry equal weight and phase in the spin wave. In particular, at low optical depth, this favors the flat spin wave. On the other hand, it is also desirable not to have a sudden change in the spin wave (except near the output end of the ensemble). The argument above essentially shows that excitations can decay through two different paths: by spontaneous emission in all directions or by collective emission into the forward direction. In Eq. (5-182), these two paths are represented by the $-P$ and $\mathrm{i}\sqrt{d}E$ terms, respectively. The latter gives rise to a decay because Eq. (5-171) can be integrated to give a term proportional to:

$$P:\ \mathcal{E}=\mathrm{i}\int \mathrm{d}\tilde{z}\,\sqrt{d}P \tag{5-183}$$

To obtain the largest decay in the forward direction all atoms should ideally be in phase so that the phase of $P(\tilde{z})$ is the same at all $\tilde{z}$. This constructive interference, however, is not homogeneous but builds up through the sample. At $\tilde{z}=0$, have $E=0$, and the spontaneous emission is, thus, the only decay channel, i.e., $d|P(\tilde{z}=0)|2/dt=-2|P(\tilde{z}=0)|2$. To achieve the largest retrieval efficiency, we should therefore put a limited amount of the excitation near $\tilde{z}=0$ and only have a slow build up of the spin wave from $\tilde{z}=0$ to $\tilde{z}=1$. The optimal spin-wave modes in Figure 5-76 represent the optimal version of this slow build up. From the qualitative argument given here, one can estimate the dependence of the optimal retrieval efficiency on the optical depth, consider the emission into a forward mode of cross sectional area A. In the far field, this corresponds to a field occupying a solid angle of λ_2/A, where $\lambda=(2\pi c)/\omega_1$ is the wavelength of the carrier. A single atom will, thus, decay into this mode with a probability $\sim\lambda_2/A$. With N atoms contributing coherently in the forward direction, the emission rate is increased by a factor of N to $\gamma_f \sim \gamma N\lambda_2/A$. The retrieval efficiency can then be found from the rate of desired (γ_f) and undesired (γ) decays as $\eta=\gamma_f/(\gamma_f+\gamma)\sim 1-A/(N\lambda_2)$. By noting that λ_2 is the cross section for resonant absorption of a two-level atom, we recognize $N\lambda_2/A$ as the optical depth d (up to a factor of order 1). The efficiency is then $\eta\sim 1-1/d$, which is in qualitative agreement with the results of the full optimization which gives $1-\eta\approx 2.9/d$.

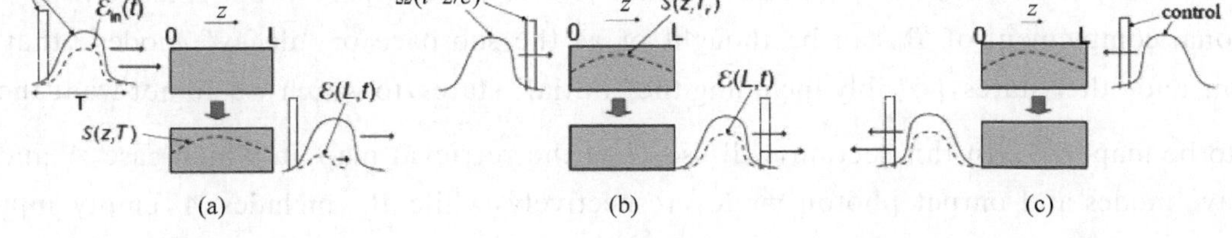

Fig. 5-76 (Color online) (a) Storage, (b) forward retrieval, and (c) backward retrieval setup. The smooth solid curve is the generic control field shape (Ω) for adiabatic storage or retrieval; the dotted square pulse indicates a π-pulse control field for fast storage or retrieval. The dashed line indicates the quantum field E and the spin-wave mode S. During storage, $\varepsilon(L,t)$ is the "leak," whereas it is the retrieved field during retrieval.

- Optimal storage from time reversal

In this section, we prove this result both for the free-space case and for the cavity case (the cavity case differs only in that there is just one spin ave mode involved). In the next section, we generalize these ideas and show that time reversal can used to optimize state mappings generally. Despite the fact that our system contains nonreversible decay γ, time reversal is still an important and meaningful concept: to make time reversal applicable in this situation, expand the system so that not only consider the electric field and the spin wave, but also include all the reservoir modes, into which the excitations may decay. The initial state of the reservoir modes is vacuum. When considering all the modes in the universe that have a closed system described by infinitely many bosonic creation operators $\{\hat{O}_i^\dagger\}$ with commutation relations:

$$[\hat{O}_i, \hat{O}_j^\dagger] = \delta_{ij} \tag{5-184}$$

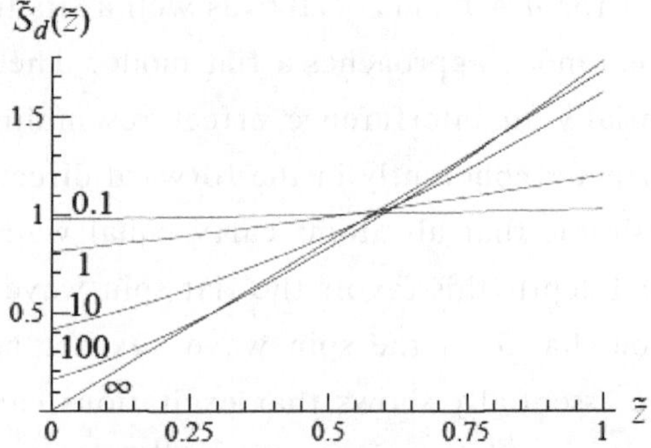

Fig. 5-77 Optimal modes $\widetilde{S}_d(\tilde{z})$ to retrieve from (in the forward direction) at indicated values of d. The flipped versions of these modes $\widetilde{S}_d(1-\tilde{z})$ are the optimal modes for backward retrieval and are also the optimal (normalized) spin waves $S(\tilde{z}, T)$ $p_{\max}$ for adiabatic and fast storage if it is optimized for storage alone or for storage followed by ackward retrieval (max s is the maximum storage efficiency).

The evolution we consider here can be seen as a generalized beam-splitter transformation, and can equivalently be specified as a Heisenberg picture map between the annihilation operators:

$$\hat{O}_{i,\text{out}} = \sum_j U_{ij}[T, 0; \Omega(t)]\hat{O}_{j,\text{in}} \tag{5-185}$$

or as a Schrödinger picture map:

$$\hat{U}[T, 0; \Omega(t)] = \sum_{ij} U_{ij}[T, 0; \Omega(t)] \mid i\rangle\langle j \mid \tag{5-186}$$

in the Hilbert space H with an orthonormal basis of single excitation states $| = \hat{O}_i^\dagger |$. To stress that the mapping depends on the classical control field $\Omega(t)$, here include the argument $\Omega(t)$ in the volution operator $\hat{U}[\tau_2, \tau_1; \Omega(t)]$, which takes the state from time τ_1 to τ_2. The operator must be unitary.

For simplicity of notation, will here use the $\Omega_s^*(T-t) = \Omega_r(t)$. Schrödinger picture. To define two subspaces of H: subspace A of "initial" states and subspace B of "final states." $B_\perp$, the orthogonal complement of B, can be thought of as the subspace of "decay" modes (that is, the reservoir and other states, possibly including the "initial" states, to which we do not want the initial states to be mapped). In this section, will use $\hat{U}$ as the retrieval map, in which case A and B are spin-wave modes and output photon modes, respectively, while $B_\perp$ includes A, empty input field modes, and the reservoir modes, to which the excitations can decay by spontaneous emission. In the cavity solved in the adiabatic limit for the control pulse shape $r(t)$, which retrieves the atomic excitation into a specific mode $e(t)$. then derived the pulse shape $s(t)$, which optimally stores an incoming mode $\varepsilon_{\text{in}}(t)$, and noted that if the incoming mode is the time-reverse of the mode, onto which retrieved, i.e., $\varepsilon'_{\text{in}}(T-t) = e(t)$, then the optimal storage control is the time-reverse of the retrieval control.

Furthermore, in this case the storage and retrieval efficiencies were identical. As we now show, this is not a coincidence, but a very general result. For the free-space case, define the "overlap efficiency" for storing into any given mode $S(z)$ as the expectation value of the number of excitations in this mode $S(z)$. Since the actual mode (call it $S'(z)$), onto which the excitation is stored, may contain components orthogonal to $S(z)$, the overlap efficiency for storing into $S(z)$ is in general less than the (actual) storage efficiency, and is equal to it only if $S(z) = S'(z)$. and will now prove that storing the time reverse of the output of backward retrieval from $S'(z)$ with the time reverse of the retrieval control field gives the overlap storage efficiency into $S(z)$ equal to the retrieval efficiency. To begin the proof, note that the probability to convert under $\hat{U}$ an initial excitation from a state $|a\rangle$ in A into a state $|b\rangle$ in B is just

$$\eta = |\langle b | \hat{U}[T,0;\Omega(t)] | a\rangle|^2 = |\langle a | \hat{U}^{-1}[T,0;\Omega(t)] | b\rangle|^2 \tag{5-187}$$

where, in the last expression, we have used the unitarity of $\hat{U}[T,0;\Omega(t)]$. We now assume that $\hat{U}[T,0;\Omega(t)]$ describes retrieval and that $|a\rangle$ stands $\hat{U}^{-1}[T,0;\Omega(t)]$. for $S^*(z)$, while $|b\rangle$ stands for the output field mode E, onto which S^* is retrieved under $\Omega(t)$. Then η is just the retrieval efficiency from S^*. The last expression then shows that if we could physically realize the operation $\hat{U}^{-1}[T,0;\Omega(t)]$, then it would give the overlap efficiency for storage of E into S^* equal to the retrieval efficiency η. The challenge is, therefore, to physically invert the evolution and realize $\hat{U}^{-1}[T,0;\Omega(t)]$.

As now show, time reversal symmetry allows us to perform this inverse evolution in some cases. To refer the reader to a careful definition of the time reversal operator $\hat{T}$ and for the proof of the following equality:

$$\hat{U}^{-1}[T,0;\Omega(t)] = \hat{T}\hat{U}[T,0;\Omega^*(T-t)]\hat{T} \tag{5-188}$$

where it is implicit that the carrier wave vector of the time-reversed control pulse $\Omega^*(T-t)$ propagates in the direction opposite to the carrier of $\Omega(t)$. Physically, Eq. (5-188) means that we can realize the inverse evolution by time-reversing the initial state, evolving it using a time-reversed control pulse, and finally time-reversing the final state. Then, using Eqs. (5-187) and (5-188), the retrieval efficiency may be rewritten as

$$\eta = |\langle a | \hat{T}\hat{U}[T,0;\Omega^*(T-t)]\hat{T} | b\rangle|^2 \tag{5-189}$$

This means that can retrieve the spin wave S^* backwards onto $\varepsilon(t)$ using $\Omega(t)$, can use $\Omega^*(T-t)$ to store $\varepsilon_{in}(t) = \varepsilon^*(T-t)$ onto S with the overlap storage efficiency equal to the retrieval efficiency η_r. Now will prove that this time-reversed storage is also the optimal solution, i.e., that an overlap efficiency for storage into S greater than η_r is not possible. To begin the proof, let us suppose, on the contrary, that can store $\varepsilon_{in}(t)$ into S with an overlap efficiency $\eta_s > \eta_r$. Applying now the time reversal argument to storage, and find that backward retrieval from S^* with the time-reversed storage control will have efficiency greater than η_r. However, from Eq. (5-183), we know that the retrieval efficiency is independent of the control field and is invariant under the complex conjugation of the spin wave, so have reached a contradiction. Therefore, the maximum overlap efficiency for storage into a given mode S is equal to the backward retrieval efficiency from S^* (and S) and can be achieved by time-reversing backward retrieval from S^*. Finally, the strategy for storing $\varepsilon_{in}(t)$ with the maximum storage efficiency (rather than maximum overlap efficiency into a given mode, as in the previous paragraph) follows immediately from the arguments above: provided can retrieve the (real) optimal backwardretrieval mode $\widetilde{S}_d(L-z)$ backwards into $\varepsilon_{in}^*(T-t)$, the optimal storage of $\varepsilon_{in}(t)$ will be the time reverse of this retrieval and will have the same efficiency as the optimal retrieval efficiency at this d, i.e., the retrieval efficiency from $\widetilde{S}_d$.

While the above argument is very general, it is important to realize its key limitation. The argument shows that it is possible to optimally store a field $\varepsilon_{in}(t)$ provided can optimally retrieve onto $\varepsilon_{in}^*(T-t)$ (i.e., backwardretrieve $\widetilde{S}_d(L-z)$ into $\varepsilon_{in}^*(T-t)$). It may, however, not be possible to optimally retrieve onto $\varepsilon_{in}^*(T-t)$ because it may, for example, be varying too fast. However as show in the next section that time reversal does not only allow one to derive the optimal storage strategy from the optimal retrieval strategy, as did in this section, but also allows one to find the optimal spin wave for retrieval.

- Time reversal as a tool for optimizing quantum state mapping

Now show that time reversal can be used as a general tool for optimizing state mappings. Moreover, and will show that for the photon storage problem considered in this paper, in addition to being a very convenient mathematical tool, the optimization procedure based on time reversal may also be realized experimentally in a straightforward way.

We can found $\widetilde{S}_d(z)$, the optimal spin wave to retrieve from, by starting with a trial spin wave $S_1(z)$ and iterating Eq. (5-185) until convergence. While just used this as a mathematical tool for solving an equation, the iteration procedure actually has a physical interpretation. Suppose that we choose a certain classical control $\Omega(t)$ and retrieve the spin wave $S(z)$ forward onto $\varepsilon(t)$ and then time reverse the control to store $\varepsilon^*(T-t)$ backwards. By the argument in the last section, this will, in general, store into a different mode $S'(z)$ with a higher efficiency (since the actual storage efficiency is, in general, greater than the overlap storage efficiency into a given mode). In this way, can iterate this procedure to compute spin waves with higher and higher forward retrieval efficiencies. In fact, forward retrieval followed by time-reversed backward storage can be expressed as

$$S_2(1-\tilde{z}) = \int_0^1 d\tilde{z}' k_r(\tilde{z}, \tilde{z}') S_1^*(1-\tilde{z}') \tag{5-190}$$

which for real S is identical to the iteration of Eq. (5-185).

Note that the reason why backward storage had to be brought up here (in contrast to the rest of the paper, where storage is always considered in the forward direction) is because Eq. (5-185), which Eq. (5-190) is equivalent to for real S, discusses forward retrieval, whose time-reverse is backward storage.

Since the iterations used to maximize the efficiency in Eq. (5-185) are identical to Eq. (5-190), the physical interpretation of the iterations in Eq. (5-185) is that we retrieve the spin wave and store its time-reverse with the time-reversed control field (i.e., implement the inverse $\hat{U}^{-1}$ of the retrieval map $\hat{U}$ using Eq. (5-188)). Explain below that this procedure of retrieval followed by time-reversed storage can be described mathematically by the operator $\hat{N}\hat{P}_A\hat{U}^{-1}\hat{P}_B\hat{U}$, where $\hat{P}_A$ and $\hat{P}_B$ are the projection operators on the subspaces A and B of spin wave modes and output photon modes, respectively, and where $\hat{N}$ provides renormalization to a unit vector. It is, in fact, generally true that in order to find the unit vector $|a\rangle$ in a given subspace A of "initial" states that maximizes the efficiency $\eta = |\hat{P}_B\hat{U}|^2$ of a given unitary map $\hat{U}$ (where B is a given subspace of "final" states), one can start with any unit vector $|\in A$ and repeatedly apply $\hat{N}\hat{P}_A\hat{U}^{-1}\hat{P}_B\hat{U}$ to it. As an example, let us discuss how this experimental implementation applies to the optimization of backward retrieval. The full Hilbert space is spanned by subspace A of spin-wave modes, subspace B of output field modes, as well as a subspace containing (empty) input and reservoir field modes. Start with a spin-wave mode $|a\rangle$ with a real mode shape $S(z)$, carry out backward retrieval, and measure the outgoing field. We then prepare the time reverse of the measured field shape and store it back into the ensemble using the time-reversed control pulse. The projections $\hat{P}_B$ and $\hat{P}_A$ happen automatically since do not reverse the reservoir modes and the leak. The renormalization can be achieved during the generation of the time-reversed field mode, while the time reversal for the spin wave will be unnecessary since a real spin wave will stay real under retrieval followed by time-reversed storage.

The iteration suggested here is, thus, indeed equivalent to the iteration in Eq. (5-190) with $S(1-\tilde{z})$ replaced with $S(\tilde{z})$ (since Eq. (5-190) optimizes forward retrieval). For single photon states, the measurement of the outgoing field involved in the procedure above will require many runs of the experiment. To circumvent this, one can use the fact that the equations of motion (5-174) ~ (5-176) for the envelope of the quantum field mode are identical to the equations of motion for the classical field propagating under the same conditions. One can, thus, use the optimization procedure with classical light pulses and find optimal pairs of input pulse shapes and control fields, which will give optimal storage into the spin wave $\hat{S}_d(1-\tilde{z})$. However, since the equations of motion for quantum light modes are identical to the classical propagation equations, this data can then be interpreted as optimal pairs of control fields and quantized input photon modes for optimal storage of nonclassical light (such as single photons) into the optimal backward retrieval mode $\hat{S}_d(1-\tilde{z})$. Here will now briefly discuss the application of time reversal ideas to the optimization of the combined process of storage followed by retrieval. For real spin waves, storage and backward retrieval are time reverses of each other since real spin waves are unaltered by complex conjugation. Consequently, the time reversal iteration of storage and backward retrieval optimizes both of them, as well as the combined process of storage followed by backward retrieval. Therefore, for a given input, the storage control field that optimizes storage alone will also be optimal for storage followed by backward retrieval.

In contrast, (forward) storage and forward retrieval are not time reverses of each other, and the entire process of storage followed by forward retrieval has to be optimized as a whole. The general time reversal iteration procedure can still be used in this case with the understanding that the spaces A and B of initial and final states are the right propagating modes to the left of the ensemble (except for later empty input modes during retrieval) and right propagating modes to the right of the ensemble (except for earlier storage leak modes), respectively, while the remaining modes are reservoir modes, spin-wave modes, leak modes, and empty incoming photon modes from the left during retrieval. Since the time-reverse of storage followed by forward retrieval is itself storage followed by forward retrieval except in the opposite direction, the optimization can be carried out physically by starting with a given input field mode, storing it and retrieving it forward with given control pulses, time reversing the output and the control pulses, and iterating the procedure. The optimal control-dependent input field, which the iteration will converge to, will then be stored into a particular optimal spin wave, which itself will be independent of the control used for the iteration. It is important to note that the discussion in this section assumed that the two metastable states are degenerate. If they are not degenerate, a momentum $\Delta_k = \omega_{sg}/c$ will be written onto the spin wave during storage, so that its time reversal will no longer be trivial.

Procedures that are related to ours and that also use time-reversal iterations for optimization are a standard tool in applied optimal control and have been used for a variety of applications in chemistry and atomic physics. In most of these works, time reversal iterations are used as a mathematical tool for computing, for a given initial state $|a\rangle$, the optimal time-dependent control that would result in the final state with the largest projection on the desired subspace B of final states. In fact, this mathematical tool is directly applicable to our problem of shaping the control pulses, as we will discuss elsewhere.

5.8.3 Adiabatic retrieval and storage

• Adiabatic retrieval

Based on the branching ratio and the time reversal arguments, we have found the maximal storage efficiency at each d and have described the optimal storage scenario in the three preceding sections. Since a given input mode can be optimally stored if and only if optimal retrieval can be directed into the time-reverse of this mode, in the following sections, and solve Eq. (5-181) analytically in two important limits to find out, which modes we can retrieve into and store optimally. The first of these two limits, which we will consider in the next five sections, corresponds to smooth input and control fields, such that the term in the square brackets in Eq. (5-181) can be ignored. This "adiabatic limit" corresponds to an adiabatic elimination of the optical polarization P in Eq. (5-180). In this section, to consider the retrieval process. It is instructive to note that in the adiabatic approximation, rescaling variables ε and P by and changing variables $\tilde{t} \to h(\tilde{T}_r, \tilde{t})$, where

$$h(\tilde{t}, \tilde{t}') = \int_{\tilde{t}}^{\tilde{t}'} |\tilde{\Omega}(\tilde{t}'')|^2 d\tilde{t}'' \tag{5-191}$$

makes Eqs. (5-175) ~ (5-176) independent of. This allows one to solve these equations in an-independent form and then obtain the solution for any given by simple rescaling. A special case of this observation has also been made, where the authors treat the Raman limit. However, since Eqs. (5-174) ~ (5-176) are relatively simple, and will avoid causing confusion by using new notation and will solve these equations without eliminating.

To solve for the output field during adiabatic retrieval, to assume for simplicity that retrieval begins at time $\tilde{t} = 0$ rather than at time $\tilde{t} = \tilde{T}_r$ and that the initial spin wave is $S(\tilde{z}, \tilde{t} = 0) = S(\tilde{z})$.

In the adiabatic approximation, Eqs. (5-176) and (5-180) reduce to a linear first order ordinary differential equation on S. Solving this equation for $S(u,\tilde{t})$ in terms of $S(u')$, expressing $\varepsilon(u,\tilde{t})$ in terms of $S(u,\tilde{t})$ using Eqs. (5-179) and (5-180), and taking the inverse Laplace transform $u \to \tilde{z}=1$, as arrive at:

$$\mathcal{E}(1,\tilde{t}) = -\sqrt{d}\tilde{\Omega}(\tilde{t})\int_0^1 \mathrm{d}\tilde{z}\,\frac{1}{1+\mathrm{i}\tilde{\Delta}}\mathrm{e}^{-\frac{h(0,\tilde{t})+d\tilde{z}}{1+\mathrm{i}\tilde{\Delta}}} \times I_0\left(2\frac{\sqrt{h(0,\tilde{t})\mathrm{d}\tilde{z}}}{1+\mathrm{i}\tilde{\Delta}}\right)S(1-\tilde{z}) \tag{5-192}$$

The $\tilde{t}$-dependent and the $\tilde{z}$-dependent phases in the exponent can be interpreted as the ac Stark shift and the change in the index of refraction, respectively. By using the identity

$$\int_0^\infty \mathrm{d}rr\mathrm{e}^{-pr^2} I_0(\lambda r) I_0(\mu r) = \frac{1}{2p}\mathrm{e}^{\frac{\lambda^2+\mu^2}{4p}} I_0\left(\frac{\lambda\mu}{2p}\right) \tag{5-193}$$

for appropriate μ, λ, and p, we find that for a sufficiently large $h(0,\infty)$, the retrieval efficiency (Eq. (5-178) with $\tilde{T}_r$ replaced with 0) is

$$\eta_r = \int_0^1 \mathrm{d}\tilde{z}\int_0^1 \mathrm{d}\tilde{z}' k_r(\tilde{z},\tilde{z}')S(1-\tilde{z})S^*(1-\tilde{z}') \tag{5-194}$$

in agreement with Eq. (5-173). So η_r is independent of detuning and control pulse shape but depends on the spin wave and the optical depth. Thus, the adiabatic approximation does not change the exact value of efficiency and keeps it independent of detuning and classical control field by preserving the branching ratio between desired and undesired state transfers. It is also worth noting that, in contrast to Eq. (5-191), Eq. (5-192) allows us to treat and optimize retrieval even when the energy in the control pulse is limited. However, in the present paper, the treatment of adiabatic retrieval is focused on the case when the control pulse energy is sufficiently large to leave no excitations in the atoms and to ensure the validity of Eq. (5-194) (or, equivalently, Eq. (5-183)).

Although demonstrate in this work that the basic underlying physics and hence the optimal performance are the same for these two photon storage techniques, a more detailed analysis reveals significant differences. It is precisely these differences that obstruct the underlying equivalence between the two protocols. And it is these differences that make this equivalence remarkable. As an example of such a difference, resonant and Raman limits give different dependence on d of the duration Tout of the output pulse. To see this, it is convenient to ignore the decay in Eq. (5-192) (due to the rescaling, this means ignore 1 in $1+\mathrm{i}\tilde{\Delta}$). If do this, can obtain:

$$\mathcal{E}(1,\tilde{t}) = i\sqrt{d}\tilde{\Omega}(\tilde{t})\int_0^1 \mathrm{d}z\,\frac{1}{\tilde{\Delta}}\mathrm{e}^{\mathrm{i}\frac{h(0,\tilde{t})+d\tilde{z}}{\tilde{\Delta}}} J_0\left(2\frac{\sqrt{h(0,t)\mathrm{d}\tilde{z}}}{\tilde{\Delta}}\right) \times S(1-\tilde{z}) \tag{5-195}$$

where $J_0(x)=I_0(ix)$ is the zeroth order Bessel function of the first kind. In the resonant limit ($\mathrm{d}\gamma > |\Delta|$), can find the duration of the output pulse by observing that the $\tilde{\Delta}\to 0$ limit of Eq. (5-195) is

$$\mathcal{E}(1,\tilde{t}) = -\frac{\tilde{\Omega}(\tilde{t})}{\sqrt{d}}S\left(1-\frac{h(0,\tilde{t})}{d}\right) \tag{5-196}$$

with the understanding that $S(\tilde{z})$ vanishes outside of [0, 1]. This is just the ideal lossless and dispersionless group velocity propagation, also to know as EIT. This implies a duration $T_{\text{out}} \sim d\gamma/|\Delta|2$ for the output pulse in the resonant limit, which is consistent with the cavity case analyzed in paper I if one identifies C and d. In the Raman limit ($d\gamma \ll |\Delta|$), the length of the output pulse is simply given by the fall-off of J_0 and is found from $h(0,\tilde{t})d/\tilde{\Delta}^2 \sim 1$ to be Tout $\tilde{\Delta}^2/(d\gamma|\Omega|^2)$. It is worth noting that the appropriate Raman limit condition is $d\gamma \ll |\Delta|$ and not $\gamma \ll |\Delta|$ as one might naively assume by analogy with the single-atom case. It is also important to note that if one is limited by laser power and desires to achieve the smallest possible Tout, the above formulas for Tout imply that EIT retrieval is preferable over Raman.

- Dependence of retrieval error on optical depth *d*

In the cavity case analyzed in paper I, only one spinwave mode is available and the retrieval error is always $1/(1+C)$ ($\approx 1/C$ for $C \gg 1$). In free space, in contrast, the retrieval error depends on the spin-wave mode $S(\tilde{z})$ and as we will explain in this section, scales differently with d depending on the spin wave. Since the retrieval efficiency is independent of Δ, to gain some physical intuition for the error dependence on the spin wave and on d, will focus on the $\Delta = 0$ case, for which the formalism of EIT transparency window has been developed. For $\Delta = 0$ and large d, the integrand in Eq. (5-192) can be approximated with a Gaussian. Then using dimensionless momentum $\tilde{k} = kL$ and defining the Fourier transform of $S(z)$ as

$$S(k) = (2\pi)^{-1}\int_0^1 \mathrm{d}\tilde{z}S(\tilde{z})\exp(-\mathrm{i}k\tilde{z})$$

So Eq. (5-192) can be written as

$$\mathcal{E}(1,\tilde{t}) = -\frac{\tilde{\Omega}(\tilde{t})}{\sqrt{d}}\int_{-\infty}^{\infty} \mathrm{d}\tilde{k}\, \mathrm{e}^{\mathrm{i}\tilde{k}\left(1-\frac{h(0,\tilde{t})}{d}\right)}\, \mathrm{e}^{-h(0,\tilde{t})\frac{\tilde{k}^2}{d^2}}\, S(\tilde{k}) \tag{5-197}$$

In the limit $d \to \infty$, the Gaussian term can be replaced with 1 to yield back the group velocity propagation in Eq. (5-196). Computing the efficiency using Eq. (5-197), can find, after a change of variables $\tilde{t} \to \tau = h(0,\tilde{t})/d$

$$\eta_r = \int_0^{\infty} \mathrm{d}\tau \left| \int_{-\infty}^{\infty} d\tilde{k}\, \mathrm{e}^{\mathrm{i}\tilde{k}(1-\tau)}\, \mathrm{e}^{-\frac{\tilde{k}^2}{d/\tau}} S(\tilde{k}) \right|^2 \tag{5-198}$$

Here will now show that the Gaussian term of width $\Delta\tilde{k}_{\mathrm{EIT}} = pd/\tau$ in the integrand in Eq. (5-199) can be interpreted as the effective momentum-space EIT transparency window for the spin wave. To start by noting that the equivalent of τ in the original units (call it z_{prop}) is equal to $z_{\mathrm{prop}} = L\tau(t) = \int_0^t v_g(t')\mathrm{d}t'$ and, thus, represents the propagation distance ($v_g = c\Omega^2/(g^2N)$ is the EIT group velocity). This interpretation of zprop also follows from the fact that, if one ignores the Gaussian in Eq. (5-198), the spin wave would be evaluated at $\tilde{z} = 1 - \tau$. Thus, in terms of the propagation distance z_{prop}, the width of the momentum-space transparency window can be written, in original units, as

$$\Delta k_{\mathrm{EIT}} = \Delta\tilde{k}_{\mathrm{EIT}}/L = \sqrt{d/\tau}/L = \sqrt{g^2N/(\gamma c z_{\mathrm{prop}})} \tag{5-199}$$

Thus, as the propagation distance zprop decreases, the width Δk_{EIT} of the transparency window gets wider and eventually becomes infinite at the $\tilde{z} = 1$ end of the ensemble, where $z_{\mathrm{prop}} = 0$. The consistency of the expression Δk_{EIT} for the effective momentum-space EIT window with the expression for the frequency-space EIT transparency window $\Delta\omega_{\mathrm{EIT}} = v_g p g^2 N/(\gamma c z_{\mathrm{prop}})$ immediately follows from rescaling by v_g both $\Delta\omega_{\mathrm{EIT}}$ and the dark state polariton bandwidth $\Delta\omega_{\mathrm{p}} = v_g \Delta k_{\mathrm{spin}}$ (where Δk_{spin} is the width of $S(k)$ in the original units).

In fact, the change of variables $t \to \tau$ that led to Eq. (5-198) precisely accomplished this rescaling of the polariton and the EIT window by the group velocity. It is worth noting that this proportionality of both the polariton bandwidth and the frequency-space EIT window width to the group velocity (and hence the existence of the controlindependent effective momentum-space EIT window) is another physical argument for the independence of retrieval efficiency from the control power. An important characterization of the performance of an ensemble-based memory is the scaling of error with optical depth at large optical depth. In the cavity case analyzed, there was only one spin-wavemode available and the retrieval error for it was $1/(1+C)$ ($\approx 1/C$ for $C \gg 1$), where the cooperativity parameter C can be thought of as the effective optical depth enhanced by the cavity. By qualitative arguments, and showed in Sec. III that the retrieval efficiency in free space is $1-\eta \sim 1/d$. A more precise value can be found numerically from the optimal spin wave which gives a maximal retrieval efficiency that scales approximately as $\sim 2.9/d$, i.e., one over the first power of density, precisely as in the cavity case. However, this $1/d$ scaling turns out to be not the only possibility in free space. The scaling of the retrieval error with d can be either $1/d$ or $1/\sqrt{d}$ depending on the presence of steps (i.e., discontinuities in the amplitude or phase of the spin wave). Specifically, numerics show that for a spin wave that does not have steps at any $\tilde{z}<1$, the retrieval error scales as $1/d$, while steps in the phase or amplitude of the spin wave result in a $1/\sqrt{d}$ error. In particular, a step in the amplitude of $S(\tilde{z})$ at position $\tilde{z}$ of height 1 can be found numerically to contribute an error of

$$l^2 \sqrt{2/\pi} \sqrt{1-\tilde{z}} / \sqrt{d} \text{ (at large } d) \tag{5-200}$$

The reason for the importance of steps is that at high d, the effective EIT window is very wide and only the tails of the Fourier transform $S(\tilde{k})$ of the spin wave $S(\tilde{z})$ matter. A step, i.e., a discontinuity, in the function $S(\tilde{z})$ means that its Fourier transform falls off as $S(\tilde{k}) \sim 1/\tilde{k}$. Thus, if assume all frequencies outside of an effective EIT window of width

$$\Delta\tilde{k}_{\mathrm{EIT}} = \sqrt{d/\tau} \sim \sqrt{d/(1-\tilde{z})} \tag{5-201}$$

get absorbed, the error will be proportional to

$$\int_{\Delta\tilde{k}_{\mathrm{EIT}}}^{\infty} d\tilde{k} \mid S(\tilde{k}) \mid^2 \sim \sqrt{(1-\tilde{z})/d} \tag{5-202}$$

precisely as found with numerics. Numerics also show that if a step in $|S(\tilde{z})|$ is not infinitely sharp, at a given d, a feature should be regarded as a step if the slope of $|S(\tilde{z})|^2$ is bigger than $\sim \sqrt{d}$ (and will then contribute a $1/\sqrt{d}$ error). The simple physical reason for this is that only if a feature in $S(\tilde{z})$ is narrower than $1/\sqrt{d}$ will it extend in $\tilde{k}$ space outside the effective EIT window. While we only performed detailed analysis of steps in the amplitude of $S(\tilde{z})$, steps in the phase of $S(\tilde{z})$, as we have already noted, also contribute a $1/\sqrt{d}$ error, and expect that similar dependence on the position and sharpness of such phase steps holds. A useful analytical result on scaling that supports these numerical calculations is the error on retrieval from a flat spin wave $S(\tilde{z})=1$, which can be calculated exactly from Eq. (5-154) to be:

$$1-\eta_r = \mathrm{e}^{-d}(I_0(d)+I_1(d)) \tag{5-203}$$

Using the properties of modified Bessel functions of the first kind, one finds that as $d \to \infty$, the error approaches $1/\sqrt{d}$, which is consistent with the general formula since a flat spin wave has one step at $\tilde{z}<1$, i.e., a step of height 1 at $\tilde{z}=0$. First, in the cavity, the optical depth is enhanced by the value of the cavity finesse from the freespace value of d to form the cooperativity parameter C. Moreover, in terms of d and C the errors during optimal storage in free space and in the cavity scale as $2.9/d$ and $1/C$, respectively. That is, even if one ignores the enhancement due to cavity finesse, the cavity offers a factor of 3 improvement. In addition to that, if one is forced to retrieve from the flat spin wave mode $S(\tilde{z})=1$ (which is the case, for example, if the spin wave is generated via spontaneous Raman scattering), the freespace error is increased from the p optimal and scales as $1/\sqrt{d}$, while in the cavity case the mode $S(\tilde{z})=1$ is, in fact, the only mode coupled to the cavity mode and is, therefore, precisely the one that gives $1/C$ scaling. Finally, because there is only one spin wave mode accessible in the cavity setup, the time reversal based iterative optimization procedure requires only one iteration in the cavity case. On the other hand, the free-space setup described in this paper is much simpler to realize in practice and allows for the storage of multiple pulses in the same ensemble, e.g., time-bin encoded qubits.

These spin waves represent at each d the optimal balance between maximal smoothness (to minimize the momentum space width $\Delta\tilde{k}_{\text{spin}}$ of the spin wave so that it better fits inside the effective EIT $\Delta\tilde{k}_{\text{EIT}}=$ window) and least amount of propagation (to minimize τ and, thus, maximize the width τ of the effective EIT window itself).

5.8.4 Shaping retrieval into an arbitrary mode

Here shown that optimal storage of a given input mode requires the ability to retrieve optimally into the time reverse of this input mode. Thus, by finding the modes can retrieve into, we will also find the modes that can be optimally stored. In this section prove that by adjusting the control during retrieval and can retrieve from any mode $S(\tilde{z})$ into any given normalized mode $\varepsilon_2(\tilde{t})$, provided the mode is sufficiently smooth to satisfy the adiabaticity condition Here know that the retrieval efficiency η_r is independent of the detuning Δ and the control Ω, provided the retrieval is complete. Thus, to find the control that retrieves $S(\tilde{z})$ into any given normalized mode $\varepsilon_2(\tilde{t})$ with detuning $\tilde{\Delta}$, need to solve for $\tilde{\Omega}(\tilde{t})$ with:

$$\sqrt{\eta_r}\,\mathcal{E}_2(\tilde{t})=-\sqrt{d}\tilde{\Omega}(\tilde{t})\int_0^1 \mathrm{d}\tilde{z}\,\frac{1}{1+\mathrm{i}\tilde{\Delta}}\mathrm{e}^{-\frac{h(0,\tilde{t})+d\tilde{z}}{1+\mathrm{i}\tilde{\Delta}}}\times I_0\left(2\frac{h(0,\tilde{t})d\tilde{z}}{1+\mathrm{i}\tilde{\Delta}}\right)S(1-\tilde{z}) \qquad (5\text{-}204)$$

To solve for $\tilde{\Omega}(\tilde{t})$, we integrate the norm squared of both sides from 0 to $\tilde{t}$ and change the integration variable from $\tilde{t}'$ to $h'=h(0,\tilde{t}')$ on the right hand side to obtain

$$\eta_r\int_0^{\tilde{t}}\mathrm{d}\tilde{t}'\,|\,\mathcal{E}_2(\tilde{t}')\,|^2=\int_0^{h(0,\tilde{t})}\mathrm{d}h'\left|\int_0^1\mathrm{d}\tilde{z}\,\frac{\sqrt{d}}{1+\mathrm{i}\tilde{\Delta}}\mathrm{e}^{-\frac{h'+d\tilde{z}}{1+\mathrm{i}\tilde{\Delta}}}\times I_0\left(2\frac{\sqrt{h'd\tilde{z}}}{1+\mathrm{i}\tilde{\Delta}}\right)S(1-\tilde{z})\right|^2 \qquad (5\text{-}205)$$

To solve Eq. (5-204) for $h(0,\tilde{t})$ numerically, we note that both sides of Eq. (5-204) are monotonically increasing functions of $\tilde{t}$ and are equal at $\tilde{t}=0$ and $\tilde{t}=\infty$ (provided $h(0,\infty)$ can be replaced with ∞, which is the case if $d_h(0,\infty)\gg|d+\mathrm{i}\tilde{\Delta}|^2$). $|\tilde{\Omega}(\tilde{t})|$ is then deduced by taking the square root of the derivative of $h(0,\tilde{t})$. The phase of $\tilde{\Omega}$ is found by inserting $|\tilde{\Omega}|$ into Eq. (5-204) and is given by

$$\mathrm{Arg}[\tilde{\Omega}(\tilde{t})] = \pi + \mathrm{Arg}\,[\mathcal{E}_2(\tilde{t})] - \frac{h(0,\tilde{t})}{1+\tilde{\Delta}^2}\tilde{\Delta} -$$

$$\mathrm{Arg}\left[\int_0^1 \mathrm{d}\tilde{z}\,\frac{1}{1+\mathrm{i}\tilde{\Delta}}\mathrm{e}^{-\frac{d\tilde{z}}{1+i\tilde{\Delta}}} \times I_0\left(2\frac{\sqrt{h(0,\tilde{t})d\tilde{z}}}{1+\mathrm{i}\tilde{\Delta}}\right)S(1-\tilde{z})\right] \qquad (5\text{-}206)$$

The second and third terms are the phase of the desired output and the compensation for the Stark shift, respectively. In the resonant limit ($d\gamma \gg |\Delta|$), and can set $\tilde{\Delta}=0$. Then assuming the phase of $S(\tilde{z})$ is independent of $\tilde{z}$, the phase of the optimal control is given (up to a constant) solely by the phase of the desired output. In the Raman limit ($d\gamma \ll |\Delta|$), the integral in the last term is approximately real but at times near the end of the output pulse ($t \approx T_{\mathrm{out}}$) it can change sign and go through zero. At these times, the optimal $|\tilde{\Omega}(\tilde{t})|$ diverges and the phase of $\tilde{\Omega}(\tilde{t})$ changes by π. Numerical simulations show, however, that $|\tilde{\Omega}(\tilde{t})|$ can be truncated at those points without significant loss in the retrieval efficiency and in the precision of $\varepsilon_2(\tilde{t})$ generation. Moreover, these points happen only near the back end of the desired output pulse over a rather short time interval compared to the duration of the desired output pulse. As can therefore often even completely turn off $\tilde{\Omega}(\tilde{t})$ during this short interval without significantly affecting the result, so that the problem of generating large power and π phase shifts can be avoided altogether. However, these potential difficulties in the Raman limit for generating the optimal control (which also has to be chirped according to Eq. (5-206) in order to compensate for the Stark shift) make the resonant (EIT) limit possibly more appealing than the far-off-resonant (Raman) limit. Finally, we note that a divergence in $|\tilde{\Omega}(\tilde{t})|$ can occur at any detuning $\tilde{\Delta}$ even when the above Raman-limit divergences are not present. However, as in the above case of the Raman-limit divergences, the infinite part can be truncated without significantly affecting the efficiency or the precision of $\varepsilon_2(\tilde{t})$ generation. One can confirm this by inserting into the adiabatic solution in Eq. (5-192) a control pulse that is truncated to have a value of $h(0,\infty)$ that is finite but large enough to satisfy $dh(0,\infty) \gg |d+\mathrm{i}\tilde{\Delta}|^2$.

- Adiabatic storage

In principle, the retrieval results of the previous section and the time reversal argument immediately imply that in the adiabatic limit, any input mode $\varepsilon_{\mathrm{in}}(\tilde{t})$ at any detuning $\tilde{\Delta}$ can be stored with the same d-dependent maximum efficiency if one appropriately shapes the control field. However, for completeness and to gain extra physical insight, in this section, we present an independent solution to adiabatic storage. Using the Laplace transform in space to solve retrieval, can find that the adiabatic solution of storage is

$$S(\tilde{z},\tilde{T}) = -\sqrt{d}\int_0^{\tilde{T}} \mathrm{d}\tilde{t}\tilde{\Omega}^*(t)\,\frac{1}{1+\mathrm{i}\tilde{\Delta}}\mathrm{e}^{-\frac{h(\tilde{t},\tilde{T})+d\tilde{z}}{1+\mathrm{i}\tilde{\Delta}}} \times I_0\left(2\frac{\sqrt{h(\tilde{t},\tilde{T})d\tilde{z}}}{1+\mathrm{i}\tilde{\Delta}}\right)\mathcal{E}_{\mathrm{in}}(\tilde{t}) \qquad (5\text{-}207)$$

It is important to note that the retrieval equation (5-203) can be cast in terms of the same-dependent function m as

$$\mathcal{E}_{\mathrm{out}}(\tilde{t}) = \int_0^1 \mathrm{d}\tilde{z}m[\Omega(\tilde{t}'),\tilde{t},\tilde{z}]S(1-\tilde{z}) \qquad (5\text{-}208)$$

$$S(\tilde{z},\tilde{T})=\int_0^{\tilde{T}}\mathrm{d}\tilde{t}m[\Omega^*(\tilde{T}-\tilde{t}'),\tilde{T}-\tilde{t},\tilde{z}]\,\mathcal{E}_{\text{in}}(\tilde{t}) \tag{5-209}$$

However, as we said in the beginning of this section, we will now, for completeness, optimize storage directly without using time reversal and our solution for optimal retrieval. Here would like to solve the following problem: given $\varepsilon_{\text{in}}(\tilde{t})$, Δ and d, we are interested in finding $\tilde{\Omega}(\tilde{t})$ that will give the maximum storage efficiency. To proceed towards this goal, we note that if we ignore decay γ and allow the spin wave to extend beyond $\tilde{z}=1$, can get the "decayless" storage equation

$$s(\tilde{z})=\int_0^{T}\mathrm{d}\tilde{t}q(\tilde{z},\tilde{t})\,\mathcal{E}_{\text{in}}(\tilde{t}) \tag{5-210}$$

where the "decayless" mode $s(\tilde{z})$ is defined for $\tilde{z}$ from 0 to ∞ instead of from 0 to 1 and where

$$q(\tilde{z},\tilde{t})=\mathrm{i}\sqrt{d}\tilde{\Omega}^*(\tilde{t})\frac{1}{\tilde{\Delta}}\mathrm{e}^{\mathrm{i}\frac{h(\tilde{t},\tilde{T})+d\tilde{z}}{\Delta}}J_0\left(2\frac{\sqrt{h(\tilde{t},\tilde{T})d\tilde{z}}}{\tilde{\Delta}}\right) \tag{5-211}$$

Since in Eq. (5-207), both sources of storage loss (the decay γ and the leakage past $\tilde{z}=1$) are eliminated, the transformation between input modes $\varepsilon_{\text{in}}(\tilde{t})$ and decayless modes $s(\tilde{z})$ becomes unitary. Indeed, can show that Eq. (5-206) establishes, for a given $\tilde{\Omega}(\tilde{t})$, a 1-to-1 correspondence between input modes $\varepsilon_{\text{in}}(\tilde{t})$ and decayless modes $s(\tilde{z})$. Moreover, and show in Appendix E that Eq. (4-166) also establishes for a given $s(\tilde{z})$ a 1-to-1 correspondence between input modes $\varepsilon_{\text{in}}(\tilde{t})$ and control fields $\tilde{\Omega}(\tilde{t})$. In particular, this means that we can compute the control field that realizes decayless storage (via Eq. (5-206)) of any given input mode $\varepsilon_{\text{in}}(\tilde{t})$ into any given decayless spin-wave mode $s(\tilde{z})$. A key element in the control shaping procedure just described is the ability to reduce the problem to the unitary mapping by considering the decayless (and leakless) solution. The reason why this shaping is useful and why it, in fact, allows us to solve the actual shaping problem in the presence of decay and leakage is that the spin wave, into which we store in the presence of decay, can be directly determined from the decayless solution: using Eqs. (5-203) and (5-206) and Eq. (5-193) (with appropriate complex values of μ, λ, and p), can find that

$$S(\tilde{z},\tilde{T})=\int_0^{\infty}\mathrm{d}\tilde{z}'\mathrm{d}e^{-d(\tilde{z}+\tilde{z}')}I_0(2d\sqrt{\tilde{z}\tilde{z}'})s(\tilde{z}') \tag{5-212}$$

This means that, remarkably, $S(\tilde{z},\tilde{T})$ (and hence the storage efficiency) depends on $\varepsilon_{\text{in}}(\tilde{t})$ and $\tilde{\Omega}(\tilde{t})$ only through the decayless mode $s(\tilde{z})$, which itself can be computed via unitary evolution in Eq. (5-206). Computing storage efficiency from Eq. (5-208) as a functional of $s(\tilde{z})$ and maximizing it under the constraint that $s(\tilde{z})$ is normalized gives an eigenvalue problem similar to Eq. (5-175) except the upper limit of integration is ∞ and the kernel is different. The mode $S(\tilde{z},\tilde{T})$ used in optimal storage is just the optimal mode for backward retrieval; the optimal storage efficiency and optimal retrieval efficiency are equal; and the optimal storage control for a given input mode is the time-reverse of the control that gives optimal backward retrieval into the time-reverse of that input mode. To give an example of optimal controls, consider a Gaussian-like input mode

$$\mathcal{E}_{\text{in}}(\tilde{t})=A(\mathrm{e}^{-30(\tilde{t}/T-0.5)^2}-\mathrm{e}^{-7.5})/\sqrt{\tilde{T}} \tag{5-213}$$

where for computational convenience we set $\varepsilon_{\text{in}}(0)=\varepsilon_{\text{in}}(\tilde{T})=0$ and where $A\approx 2.09$ is a normalization constant.

The controls are plotted in rescaled units so that the area under the square of the curves shown is equal to $L^{-1}\ RT_0\ dt'v_g(t')$ (where v_g is the EIT group velocity), which is the position (in units of L), at which the front end of the pulse would get stored under ideal decayless propagation. From time reversal and the condition for complete retrieval, it follows that the control pulse energy ($\propto h(0,\tilde{T})$) and hence $L^{-1}\ RT_0\ dt'v_g(t')$ should diverge.

Thus, at any finite d, the optimal plotted in Figure 5-78 should actually diverge at $\tilde{t}=0$. However, the front part of the control pulse affects a negligible piece of $\varepsilon_{in}(\tilde{t})$, so truncating this part (by truncating $s(\tilde{z})$ at some $\tilde{z}$, for example) does not affect the efficiency. Naively, the optimal control should roughly satisfy $L^{-1}RT_0\ dt'v_g(t')=1$: the control (and, thus, the dark state polariton group velocity) should be small enough to avoid excessive leakage; and at the same time it should be as large as possible to have the widest possible EIT transparency window to minimize spontaneous emission losses. For the truncated optimal controls, we see that $L^{-1}RT_0\ dt'v_g(t')$ (i.e., the area under the square of the curves in Figure 5-78 is, in fact, greater than 1. As d decreases, $L^{-1}RT_0\ dt'v_g(t')$ decreases as well, and allows for less and less leakage so that only as $d\to\infty$, $s(\tilde{z})\to\sqrt{3}(1-\tilde{z})$, $L^{-1}R\,T_0\ dt'v_g(t')\to 1$, and no leakage is allowed for. Although optimal storage efficiencies are the same in the Raman and adiabatic limits, the two limits exhibit rather different physical behavior. It is now the dependence on d of the intensity of the optimal control field that can be used to distinguish between the resonant and the Raman regimes.

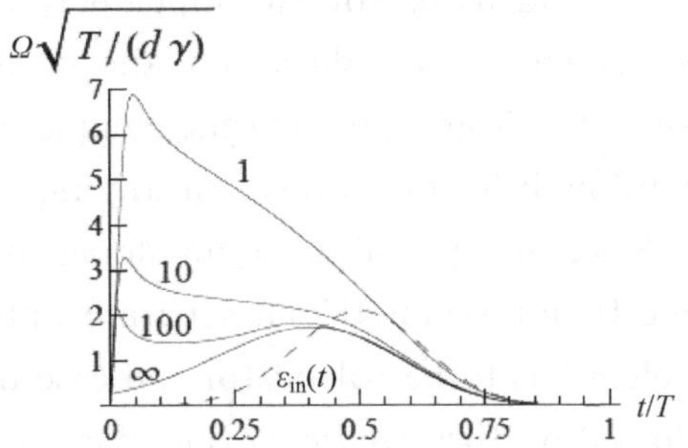

Fig. 5-78 Input mode $\varepsilon_{in}(t)$ (dashed) defined in Eq. (5-209) and control fields $\Omega(t)$ (in units of) that maximize for this $\varepsilon_{in}(t)$ the efficiency of resonant adiabatic storage (alone or followed by backward retrieval) at $d=1,10,100$, and $d\to\infty$.

- Storage followed by retrieval

Moreover, this spatially uniform mode looked the same in the forward and backward directions (assuming negligible metastable state splitting ω_{sg}). Therefore, optimal storage into that mode guaranteed that the combined procedure of storage followed by retrieval was optimal as well and the total efficiency did not depend on the retrieval direction. In contrast, the free-space model allows for storage into a variety of modes, each of which has a different retrieval efficiency that is also dependent on retrieval direction.

Therefore, in free space, we will first discuss the optimization of storage followed by backward retrieval and then the optimization of storage followed by forward retrieval. Since have shown that the optimal spin-wave mode for backward retrieval is also the optimal mode for storage, which optimize storage, are also optimal for storage followed by backward retrieval. Figure 5-79 shows the maximum total efficiency for storage followed by backward retrieval (η_{back}^{max}—solid line), which in the adiabatic limit can be achieved for any input pulse.

For comparison, we also show the total efficiency for storage followed by backward retrieval for a Gaussian-like input mode defined in Eq. (5-209) (assuming the adiabatic limit $Td\gamma\gg 1$) with naive square storage control pulses on $[0,T]$ with power set by $v_gT=L$, where v_g is the EIT group velocity (η_{square}—dashed line). The significant increase in the efficiency up to the input independent optimal efficiency η_{back}^{max} due to the use of optimal storage control pulses instead of naive ones is, of course, not unique to the input pulse of Eq. (5-209) and holds for any input pulse. In fact, since at moderate values of d the naive control pulse is far from satisfying the complete retrieval condition, it is not optimal for any input.

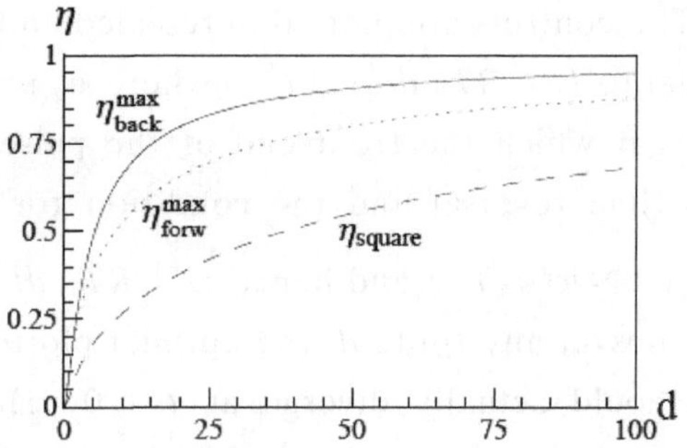

Fig. 5-79 η_{back}^{max} (solid) and η_{forw}^{max} (dotted) are maximum total efficiency for adiabatic (or fast) storage followed by backward or forward retrieval, respectively. η_{square} (dashed) is the total efficiency for resonant storage of $\varepsilon_{in}(t)$ from Eq. (5-209) followed by backward retrieval, where the storage control field is a naive square pulse with the power set by $v_g T = L$.

Since the optimal mode for storage or retrieval alone is not symmetric, a separate optimization problem has to be solved for the case of storage followed by forward retrieval. We show in Eq. (5-206) sets up, for any sufficiently smooth $\varepsilon_{in}(\tilde{t})$, a 1-to-1 correspondence between decayless modes $s(\tilde{z})$ and control fields $\Omega(\tilde{t})$. Moreover, the decayless mode alone determines the total efficiency of storage followed by forward retrieval, which can be found by inserting Eq. (5-208) into Eq. (5-194). Thus, the optimization problem reduces to finding the optimal $s(\tilde{z})$ by the iterative optimization procedure except with a new kernel. Since the reverse process is itself storage followed by forward retrieval (except in the opposite direction) and since the iterations will optimize it as well, the spinwave mode used in optimal storage followed by forward retrieval must be the one that flips under forward retrieval, followed by time reversal and (backward) storage.

Moreover, it follows that the control pulse that and should use for a given $\varepsilon_{in}(\tilde{t})$ is the time-reverse of the control that retrieves the flipped version of the optimal spin-wave mode backwards into $\varepsilon_{in}^{*}(\tilde{T}-\tilde{t})$. Thus, instead of computing the optimal $s(\tilde{z})$, can solve the following eigenvalue problem that finds the optimal mode to store into

$$\lambda S(\tilde{z}) = \int_0^1 \mathrm{d}\tilde{z}' k_r(\tilde{z}, 1-\tilde{z}') S(\tilde{z}') \tag{5-214}$$

This eigenvalue equation is just a simple modification of the retrieval eigenvalue Eq. (5-185): and are now computing the mode that flips under forward retrieval followed by time reversal and storage, while Eq. (5-185) finds the mode that stays the same under backward retrieval followed by time reversal and storage. The total efficiency of storage followed by forward retrieval is then λ^2. However, in contrast to storage followed by backward retrieval, the storage efficiency and the forward retrieval efficiency during the optimal procedure are not each equal to λ; the storage efficiency is greater. The optimal spin-wave modes that result from Eq. (5-200) are shown in Figure 5-80 for the indicated values of d. At small d, the optimal mode is almost flat since at small d the optimal retrieval and storage modes are almost flat and, thus, almost symmetric. As d increases, the optimal mode first bends towards the wedge $\sqrt{3}(1-\tilde{z})$ similarly to the optimal backward retrieval mode. But then above $d \approx 3$, the optimal mode starts shaping towards the parabola $S(z) = (1-4(\tilde{z}-1/2)^2)$, which, as expected, avoids the $1/\sqrt{d}$ error from discontinuities by vanishing at the edges and, simultaneously, maximizes smoothness.

The maximal total efficiency for storage followed by forward retrieval $\eta_{\text{forw}}^{\max}$ is plotted as a dotted curve in Figure 5-80. $H_{\text{forw}}^{\max}$ ($\approx 1 - 19/d$ as $d \to \infty$) is less than $\eta_{\text{back}}^{\max}$ ($\approx 1 - 5.8/d$ as $d \to \infty$) since for optimal backward retrieval, storage and retrieval are each separately optimal, while for forward retrieval a compromise has to be made. From a different perspective, forward retrieval makes it more difficult to minimize propagation since the whole excitation has to propagate through the entire medium.

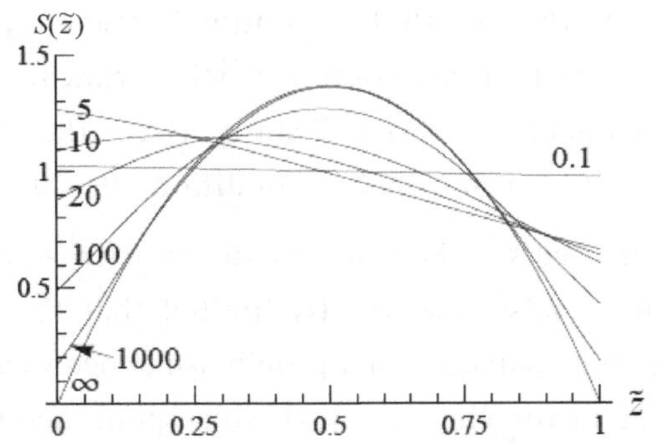

Fig. 5-80 For different values of d, the optimal spin-wave mode to be used during storage followed by forward retrieval.

- Adiabaticity conditions

Before have shown that in the adiabatic limit, any input mode can be stored with the same maximum efficiency. In this section, show that independent of Δ, the necessary and sufficient condition for optimal adiabatic storage of a pulse of duration T to be consistent with the adiabatic approximation is $Td\gamma \gg 1$, except with C replaced with d.

Therefore, we immediately turn to the tasks of verifying numerically that Eq. (5-211) is indeed the correct adiabaticity condition and of investigating the breakdown of adiabaticity for short input pulses. We consider adiabatic storage of a Gaussian-like input mode defined in Eq. (5-209) and shown in Figure 5-81. Can use adiabatic equations to shape the control pulse but then compute the total efficiency of storage followed by backward retrieval numerically from Eqs. (5-194)~(5-196) without making the adiabatic approximation. As $Td\gamma$ decreases to 1 and below, expect the efficiency to be reduced from its optimal value. In Figure 5-81(a), the total efficiency of this procedure is plotted as a function of $Td\gamma$ for $\Delta = 0$ and $d = 1, 10, 100$, and 1000. The dashed lines are the true optimal efficiencies. As expected, when $Td\gamma \leqslant 10$, the efficiency drops. In Figure 5-81(b), when fix $d = 10$ and show how optimal adiabatic storage breaks down at different detunings Δ from 0 to 200γ. However, since the curves for $\Delta = 100\gamma$ and $\Delta = 200\gamma$ almost coincide, $Td\gamma \gg 1$ is still the relevant condition no matter how high Δ is.

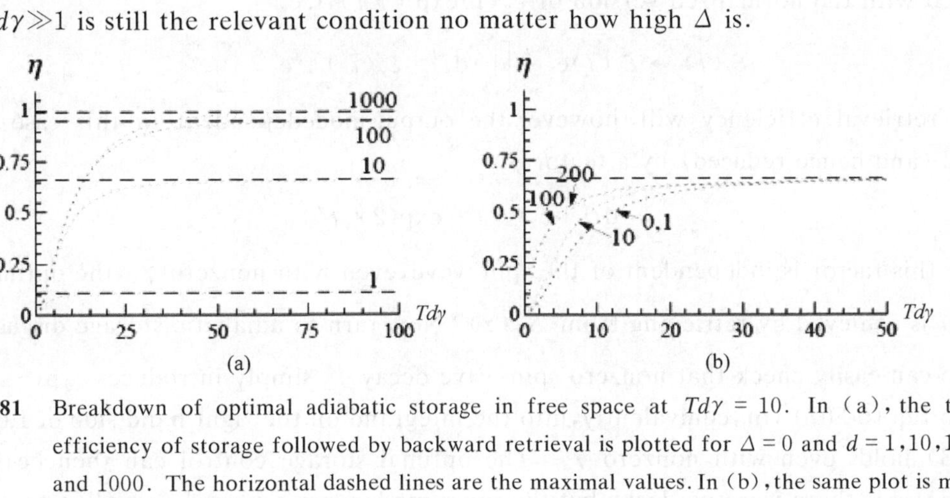

Fig. 5-81 Breakdown of optimal adiabatic storage in free space at $Td\gamma = 10$. In (a), the total efficiency of storage followed by backward retrieval is plotted for $\Delta = 0$ and $d = 1, 10, 100$, and 1000. The horizontal dashed lines are the maximal values. In (b), the same plot is made for $d = 10$ and $\Delta/\gamma = 0, 1, 10, 100, 200$.

As the fact that the optimal efficiency given by the dashed lines in Figure 5-81 is achieved by the dotted curves (obtained with truncated controls) is the proof that truncation of the storage controls does not significantly affect the storage efficiency. The losses associated with truncation are insignificant only if condition $dh\ (0, \infty) \gg |\ d + i\ |\ 2$ is satisfied for the truncated retrieval control pulse (the same condition with ∞ replaced with $\widetilde{T}$ applies to storage). If the energy in the control pulse is so tightly limited that this condition is violated, a separate optimization problem has to be solved. This problem has been considered for the special case of Raman storage in free space in the limit of negligible spontaneous emission loss. In this section, we discuss the effects such a term has on the adiabatic solution and optimal control field shaping discussed. Here will show, in particular, that nonzero γ_s simply introduces exponential decay into the retrieval and storage solutions without making the optimization harder. As will also show that with nonzero spin-wave decay, optimal efficiencies become dependent on the input mode (or, equivalently, on the control fields).

Here first consider adiabatic retrieval discussed. One can easily check that nonzero spinwave decay rate γ_s simply introduces decay described by $\exp(-\widetilde{\gamma}_s\widetilde{t})$ into Eq. (5-192), and, unless we retrieve much faster than $1/\gamma_s$, this makes the retrieval efficiency control dependent. Moreover, if a given fixed control field is not strong enough to accomplish retrieval in a time much faster than $1/\gamma_s$, the problem of finding the optimal retrieval mode for this particular retrieval control will give a different answer from the $\gamma_s = 0$ case. In particular, for forward retrieval, as increase γ_s or, alternatively, decrease the power of the retrieval control, to minimize propagation time at the cost of sacrificing smoothness, the optimal retrieval mode will be more and more concentrated towards the $z = L$ end of the ensemble. As in the $\gamma_s = 0$ case, we can find these optimal modes either by computing the (now-dependent) kernel to replace k_r in Eq. (5-194) and its eigenvector with the largest eigenvalue or, equivalently, by doing the iteration of retrieval, time reversal, and storage. The inclusion of nonzero γ_s also does not prevent from being able to shape retrieval to go into any mode, as described for $\gamma_s = 0$. Here should just shape the control according to Eq. (5-200) as if there were no spin wave decay except the desired output mode $\varepsilon_2(t)$ on the left hand side should be replaced with the normalized version of $\varepsilon_2(t)\exp(\gamma_s t)$, i.e.

$$\mathcal{E}_2(\widetilde{t}) \rightarrow \mathcal{E}_2(\widetilde{t})e^{\widetilde{\gamma}_s\widetilde{t}}\left[\int_0^\infty d\widetilde{t}'\ |\ \mathcal{E}_2(\widetilde{t}')\ |^2 e^{2\widetilde{\gamma}_s\widetilde{t}'}\right]^{-\frac{1}{2}} \tag{5-215}$$

The retrieval efficiency will, however, be output-modedependent in this case: it will be multiplied (and hence reduced) by a factor of

$$\left[\int_0^\infty d\widetilde{t}'\ |\ \mathcal{E}_2(\widetilde{t}')\ |^2 \exp(2\widetilde{\gamma}_s\widetilde{t}')\right]^{-1} \tag{5-216}$$

Since this factor is independent of the spin wave, even with nonzero γ_s, the optimal retrieval into $\varepsilon_2(t)$ is achieved by retrieving from $\widetilde{S}_d(z)$. Now turn to adiabatic storage discussed in Sec. VIB. One can easily check that nonzero spin-wave decay γ_s simply introduces $\exp(-\widetilde{\gamma}_s(\widetilde{T}-\widetilde{t}))$ decay into Eq. (5-207) (or, equivalently, into the integrand on the right hand side of Eq. (5-203)). Eq. (5-208) holds even with nonzero γ_s. The optimal storage control can then be found from Eq. (5-206) as if there were no decay but the input mode were replaced according to

$$\mathcal{E}_{in}(\widetilde{t}) \rightarrow \mathcal{E}_{in}(\widetilde{t})e^{-\widetilde{\gamma}_s(\widetilde{T}-\widetilde{t})}\left[\int_0^T d\widetilde{t}'\ |\ \mathcal{E}_{in}(t')\ |^2 e^{-2\widetilde{\gamma}_s(\widetilde{T}-\widetilde{t}')}\right]^{-\frac{1}{2}} \tag{5-217}$$

However, the optimal storage efficiency will now depend on input pulse duration and shape: it can be multiplied (and hence reduced) by

$$\int_0^{\tilde{T}} \mathrm{d}\tilde{t}' \mid \mathcal{E}_{\text{in}}(\tilde{t}') \mid^2 \exp(-2\tilde{\gamma}_s(\tilde{T}-\tilde{t}')) \tag{5-218}$$

It is also important to note that nonzero γ_s still keeps the general time reversal relationship between storage and retrieval exhibited in Eqs. (5-204) and (5-205). However, for:

$$\mathcal{E}_2(\tilde{t}) = \mathcal{E}_{\text{in}}^*(\tilde{T}-\tilde{t}), \quad \int_0^{\tilde{T}} \mathrm{d}\tilde{t} \mid \mathcal{E}_{\text{in}}(\tilde{t}) \mid^2 \mathrm{e}^{-2\tilde{\gamma}_s(\tilde{T}-\tilde{t})} > \left[\int_0^{\tilde{T}} \mathrm{d}\tilde{t} \mid \mathcal{E}_2(\tilde{t}) \mid^2 \mathrm{e}^{2\tilde{\gamma}_s \tilde{t}}\right]^{-1} \tag{5-219}$$

Which means that with nonzero γ_s, the optimal storage efficiency of a given input mode is greater than the optimal retrieval efficiency into the time-reverse of that mode. Because the two controls involved are not time-reverses of each other, the inequality of the two efficiencies is consistent with the time reversal arguments. Using the fact that nonzero γ_s keeps the general time reversal relationship between storage and retrieval exhibited in Eqs. (5-204) and (5-205), it is not hard to verify that nonzero γ_s still allows one to use time reversal iterations to optimize storage followed by retrieval. In particular, suppose that one is given a storage control field and a (forward or backward) retrieval control field. Then one can find the optimal input mode to be used with these control fields by the following procedure: start with a trial input mode, store and retrieve it with the given pair of controls, time-reverse the whole procedure, and then repeat the full cycle until convergence is reached. Now suppose, on the other hand, one is given an input mode and is asked to choose the optimal storage and retrieval controls. Because of the spin wave decay, it is desirable to read out as fast as possible. As we discuss in the next section, fast readout may be achieved in a time $T \sim 1/\gamma_d$, so that if we assume that $\gamma_s \ll d\gamma$, the spin-wave decay during the retrieval will be negligible. If further assume that the given input mode satisfies the adiabatic limit $Td\gamma \gg 1$, then one should shape the storage control to store into the appropriate optimal spin-wave mode ($\tilde{S}_d(1-\tilde{z})$, depending on the direction of retrieval) as if γ_s were zero and the input were proportional to $E_{\text{in}}(\tilde{t})\exp(-\tilde{\gamma}_s(\tilde{T}-\tilde{t}))$ (see Eq. (5-220)). The total optimal efficiency will now depend on input pulse duration and shape: it will be multiplied (and hence reduced relative to the $\gamma_s = 0$ case) by

$$\int_0^{\tilde{T}} \mathrm{d}\tilde{t} \mid \mathcal{E}_{\text{in}}(\tilde{t}) \mid^2 \exp(-2\tilde{\gamma}_s(\tilde{T}-\tilde{t})) \tag{5-220}$$

Finally, to note that when consider storage followed by retrieval, in order to take into account the spin wave decay during the storage time $[\tilde{T}, \tilde{T}_r]$, one should just multiply the total efficiency by $\exp(-2\tilde{\gamma}_s(\tilde{T}_r - \tilde{T}))$.

- Fast retrieval and storage

As shown that in the adiabatic limit ($Td\gamma \gg 1$, where T is the duration of the incoming pulse), one can optimally store a mode with any smooth shape and any detuning. In this section, solve Eq. (5-191) analytically in the second important limit, the "fast" limit, and demonstrate that this limit allows one to store optimally a certain class of input modes that have duration $T \sim 1/(d\gamma)$. That efficient (but not optimal) fast storage of any smooth pulse is possible as long as $T\gamma \ll 1$ and $T d\gamma \gg 1$.

This gives Rabi oscillations between P and S and allows one to implement a fast storage scheme, in which the input pulse is resonant and the control pulse is a short π pulse at $t = T$, as well as fast retrieval, in which the control is a π pulse at $t = T_r$. During fast retrieval, assuming that the π pulse is perfect and that it enters the medium at $\tilde{t} = 0$ (instead of $\tilde{t} = \tilde{T}_r$), the initial spin wave $S = S(\tilde{z})$ is mapped after the π pulse onto the optical polarization $P = \mathrm{i}S(\tilde{z})$. Then solve Eq. (5-221) for $P(u, \tilde{t})$, express $\varepsilon(u, \tilde{t})$ in terms of $P(u, \tilde{t})$ using Eq. (5-179), and take the inverse Laplace transform $u \to \tilde{z} = 1$ to arrive at

$$\mathcal{E}_{\text{out}}(\tilde{t}) = -\sqrt{d}\int_0^1 \mathrm{d}\tilde{z}\mathrm{e}^{-\tilde{t}} J_0(2\sqrt{d\tilde{t}\tilde{z}})S(1-\tilde{z}) \tag{5-221}$$

When computing the fast retrieval efficiency, one can take the time integral analytically to find that the efficiency is again given by Eq. (5-219). Similarly, the expression in Eq. (5-215) is also a special case of Eq. (5-222) if use:

$$\tilde{\Omega}(\tilde{t}) = (1 + \mathrm{i}\tilde{\Delta})\mathrm{e}^{-\mathrm{i}\tilde{\Delta}\tilde{t}} \tag{5-222}$$

and take the limit $\tilde{t} \to \infty$ (although this violates the approximations made in deriving Eq. (5-193)).

Since the π-pulse control field in fast retrieval is fixed, optimal fast retrieval yields a single possible output mode, that of Eq. (5-215) with the optimal spin wave $S(\tilde{z}) = \tilde{S}_d(\tilde{z})$. By time reversal, the time-reversed version of this input mode (of duration $T \sim 1/(\gamma d)$) is, therefore, the only mode that can be optimally stored using fast storage at this optical depth d. In order to confirm the time reversal argument and for the sake of completeness, one can also compute the optimal input mode for fast storage directly. For an input mode $\varepsilon_{\text{in}}(\tilde{t})$ nonzero for $\tilde{t} \in [0, \tilde{T}]$ and assuming that a perfect π pulse arrives at $\tilde{t} = \tilde{T}$, can find that:

$$S(\tilde{z}, \tilde{T}) = -\sqrt{d}\int_0^{\tilde{T}} \mathrm{d}\tilde{t}\mathrm{e}^{-(\tilde{T}-\tilde{t})} J_0(2\sqrt{d(\tilde{T}-\tilde{t})\tilde{z}})\,\mathcal{E}_{\text{in}}(\tilde{t}) \tag{5-223}$$

One can see that the fast retrieval and storage Eqs. (5-221) and (5-223) obey, as expected, the same general time reversal relationship that we have already verified in the adiabatic limit in Eqs. (5-204) and (5-205). It is worth noting that short exponentially varying pulses, reminiscent of our optimal solution, have been proposed before to achieve efficient photon-echo based storage. The solutions above give an incoming mode $\varepsilon_{\text{in}}(\tilde{t})$ that is optimal for fast storage alone or for fast storage followed by backward retrieval. Similarly, at each d, there is a mode that gives the optimal efficiency for fast storage followed by forward retrieval. Therefore, generating this mode, and hence obtaining high efficiency, may be hard in practice at high values of C. In contrast, in free space, any sufficiently smooth spin wave will have a high retrieval efficiency, and, by time-reversal, the time-reverses of the pulses fast retrieved from these spin waves can also be fast stored with high efficiency. One can, thus, explicitly verify using Eq. (5-217), which allows one to compute these storage efficiencies, that if, in the original units, $T\gamma \ll 1$ but at the same time $Td\gamma \gg 1$, the free-space fast-storage efficiency is close to unity.

5.9 Effects of metastable state nondegeneracy

In the discussion of backward retrieval have so far assumed that the two metastable states $|g\rangle$ and $|s\rangle$ are nearly degenerate. This has meant that during backward retrieval we could simply use the same equations as for forward retrieval but with the spin wave flipped: $S(z) \to S(L-z)$. If $|g\rangle$ and $|s\rangle$ are not degenerate and are split by $\omega_{sg} = c\Delta k$, then during backward retrieval, instead of retrieving from $S(L-z)$, we will have to redefine the slowly varying operators and retrieve from $S(L-z)$ exp $(-2i\_kz)$, which significantly lowers the efficiency unless $\Delta kL \ll \sqrt{d}$. This condition on Δk can be understood based on the concept of the effective EIT window for the Fourier transform of the spin wave. The width of this window is of order (in the original units) $\sim \sqrt{d/L}$. The extra phase just shifts the Fourier transform off center by $2\Delta k$, so that the efficiency will not be significantly affected provided the shift is much smaller than the window width. We have confirmed numerically for $S(\tilde{z})=1$ and for $S(\tilde{z})=\sqrt{3}\,\tilde{z}$ that the ΔkL needed to decrease retrieval efficiency by 50% from its $\Delta kL = 0$ value indeed scales as $\sqrt{d}$ (with proportionality constants$\approx$0.46 and$\approx$0.67, respectively).

There are two ways to understand physically why nondegeneracy of the metastable states ruins the backward retrieval efficiency. The first explanation, also noted, comes from the fact that metastable state nondegeneracy breaks the momentum conservation on backward retrieval. During storage, momentum Δk is written onto the ensemble. Momentum conservation on backward retrieval, however, will require $-\Delta k$ momentum in the spin wave. The second explanation comes from the fact that if $\Delta k = 0$, then backward retrieval of optimal storage is no longer its time reverse. If we had not defined slowly varying operators, the spin wave that we store into our atoms would have had exp $(i\Delta k_z)$ phase written on it. Since time reversal consists of moving in the opposite direction and taking a complex conjugate, backward retrieval will be the time-reverse of storage only if $\Delta k = 0$, in which case complex conjugation is trivial. Thus, if $\Delta k \neq 0$, the optimization of storage does not simultaneously optimize backward retrieval (unless, of course, can apply the desired position-dependent phase to the atoms during the storage time, e. g., by a magnetic field gradient, or alternatively apply a π pulse that flips the two metastable states). We would like now to optimize storage followed by backward retrieval in the presence of nondegeneracy $(\Delta k = 0)$. In order to carry out the optimization, one has to start with an input pulse and a control pulse, do storage, then do backward retrieval with another control pulse. Then one has to time reverse the full process of storage and retrieval, and iterate till one gets convergence to a particular input (and spin wave). Specifically, start with a trial spin wave $S_1(z)$. To find the spin wave in terms of operators that are slowly varying for forward propagation as defined that the optimal storage plus backward retrieval should use, we first rewrite $S_1(z)$ for backward-propagation slowly varying operators (i. e., add the $2k_z$ phase), and then retrieve it backwards, time reverse, and store. Using Eq. (5-224), the iteration get dropping an unimportant constant phase and going to our rescaled units:

$$S_2(\tilde{z}) = \int_0^1 d\tilde{z}' k_r(\tilde{z}, \tilde{z}') e^{-i2\Delta\tilde{k}\tilde{z}'} S_1^*(\tilde{z}') \tag{5-224}$$

where $z\tilde{k} = Lk$. This iteration finds the eigenvector with the largest eigenvalue for the eigenvalue problem:

$$\lambda S(\tilde{z}) = \int_0^1 k_r(\tilde{z}, \tilde{z}') e^{-i2\Delta\tilde{k}z'} S^*(\tilde{z}') \tag{5-225}$$

$|\lambda|2$ will then give the total maximum efficiency of storage followed by backward retrieval. In contrast to the $k_r = 0$ case, the efficiencies of storage and retrieval in the optimal process are not generally equal. It is important to note that since the process we are optimizing followed by its time reverse corresponds to two iterations of Eq. (5-225), after a sufficient number of steps, λ

settles into an oscillation between $|\lambda|\exp(i\alpha)$ and $|\lambda|\exp(-i\alpha)$ for some phase α. The eigenvector will oscillate between two values differing only by an unimportant constant phase, so that either one can be used. While this procedure allows us to find the optimal spin waves, we should, for completeness, also determine, as in the $\Delta k=0$ case, which input fields the optimum can be achieved for. To do this, as before, consider the exactly solvable adiabatic and fast limits. In the adiabatic limit, the argument that retrieval can be shaped into any mode did not require the spin wave $S(z)$ to be real, and it is therefore still applicable. By time reversal, therefore, still achieve the maximum efficiency of storage followed by backward retrieval for any incoming mode of duration T such that $Td\gamma \gg 1$. Similarly, in the fast limit, using fast retrieval and time reversal we can find at each d a pulse shape with $Td\gamma \sim 1$ that gives the maximum efficiency. For completeness, note that one can also generalize to $\Delta k \neq 0$ the method that uses the decayless mode $s(z)$ to shape the optimal storage control. However, since the optimal control is unique, this method will, of course, yield the same control as the method based on retrieval and time reversal. Thus omit here the extension of this method to the $\Delta k \neq 0$ case.

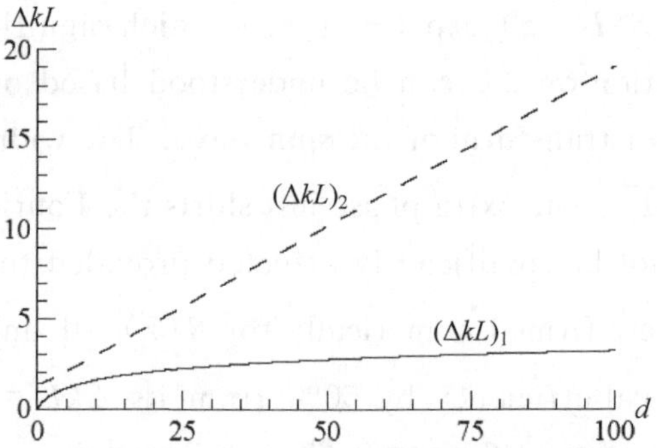

Fig. 5-82 If the two metastable states are not degenerate, the efficiency of storage followed by backward retrieval will be lowered relative to the degenerate case, because the energy difference $\hbar\omega_{sg}$ introduces a momentum difference $\Delta k=\omega_{sg}/c$ between the quantum and classical fields. As a function of d, the figure shows the momentum ΔkL, at which the optimal total efficiency of storage followed by backward retrieval falls to half of the $\Delta kL=0$ value (dashed), and at which it is decreased to the optimal efficiency with forward retrieval (solid).

Thus, demonstrated that we can optimize storage followed by backward retrieval for any given d and Δk. However, as we increase ΔkL, the optimal total efficiency with backward retrieval will drop down to the optimal total efficiency with forward retrieval at some value of (ΔkL). Increasing ΔkL further up to another value (ΔkL) will decrease the optimal total efficiency with backward retrieval to half of its $\Delta kL=0$ value (and then further down to zero). As noted above, without reoptimization, $(\Delta kL)2$ would go as $\sqrt{d}$ but with optimization can see that it is linear in d, i.e., optimization makes the error less severe. ΔkL grows even slower than $\sqrt{d}$. This is not surprising because at $\Delta kL=0$ optimal forward and optimal backward errors both fall off as $1/d$, except with different coefficients and, thus, eventually get very close to each other, so it takes a small ΔkL to make them equal. In Figure 5-83(a) and 5-83(b), can show the magnitude $|S(\tilde{z})|$ and the phase $\mathrm{Arg}[S(\tilde{z})]$, respectively, of the optimal mode (defined for the forward-propagating slowly varying operators as in Eq. (5-224)) at $d=20$ for different values of ΔkL. As we increase ΔkL, the optimal mode becomes concentrated more and more near the back end, i.e., it becomes favorable to effectively decrease the optical depth (i.e., decrease effective L) in order to decrease effective ΔkL. The phase of the optimal mode is approximately linear. At $\Delta kL=0$, $\tilde{k}_0=0$. Interestingly, instead of just growing from 0 linearly with ΔkL, $\tilde{k}_0$ first increases but then above $\Delta kL \sim 7.5$ starts decreasing again.

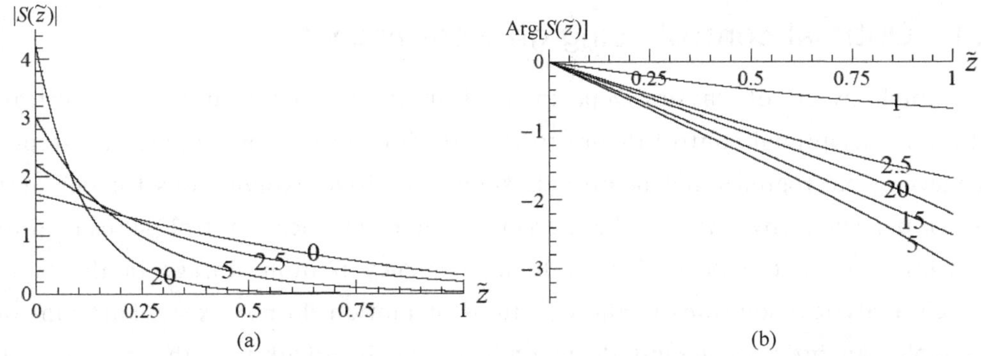

Fig. 5-83 (a) The magnitude and (b) the phase of the optimal mode for storage followed by backward retrieval at $d=20$ for the indicated values of ΔkL. The phase of the optimal mode at $\Delta kL=0$ is 0.

In conclusion, in this section presented a detailed analysis of the storage and retrieval of photons in homogeneously broadened Λ-type atomic media in free space. First of all, the retrieval is essentially an interference effect where the emission from all the atoms interferes constructively in the forward direction. This constructive interference enhances the effective decay rate into the forward direction to $d\gamma$. The branching ratio between the desired forward radiation and the unwanted spontaneous emission is then simply given by the ratio between the various decay rates and is $\eta \sim d\gamma/(\gamma+d\gamma) \sim 1-1/d$, irrespective of the method being used to drive the excitation out of the atoms. Secondly, the storage process is most conveniently viewed as the time reverse of retrieval. In the paragraph used this physical picture to derive the optimal strategy for storage and retrieval and the optimal efficiency that is independent of whether one works in the Raman, EIT, photon-echo, or any other intermediate regime. In particular, showed how to achieve the optimal storage of any smooth input mode at any detuning of duration $T \gg 1/(d\gamma)$ (the adiabatic limit, including Raman and EIT) and of a particular class of resonant input modes of duration $T \sim 1/(d\gamma)$ (the fast or photon-echo limit). This analysis is extendable to other systems. In particular, consider the effects of inhomogeneous broadening on photon storage in Λ-type atomic media. Extensions to other systems, such as the double-Λ system or the tripod system, should also be possible and suggested a novel time reversal based iterative procedure for optimizing quantum state mappings. Moreover, we showed that for the case of photon storage, this procedure is not only a convenient mathematical tool but is also a readily accessible experimental technique for finding optimal spin waves and optimal input-control pairs: one just has to be able to measure the output mode and to generate its time reverse. Following the present work, this procedure has recently been implemented experimentally with classical light. This optimization procedure to be applicable to other systems used for light storage, such as tunable photonic crystals. The presented optimization of the storage and retrieval processes leads to a substantial increase in the memory efficiency whenever reasonable synchronization between the input photon wave packet and the control pulse can be achieved. Therefore, expect this work is important in improving the efficiencies in current experiments, where optical depth is limited by various experimental imperfections such as a limited number of atoms in a trap, competing four-wave mixing processes in a warm vapor cell, or inhomogeneous broadening in solid state samples.

5.9.1 Optimal control using gradient ascent

Alexey Gorshkov etc of Physics Department, Harvard University use the numerical gradient ascent method from optimal control theory to extend efficient photon storage in Λ-type media to previously inaccessible regimes and to provide simple intuitive explanations for our optimization techniques. In particular, by using gradient ascent to shape classical control pulses used to mediate photon storage, open up the possibility of high efficiency photon storage in the non-adiabatic limit, in which analytical solutions to the equations of motion do not exist. This control shaping technique enables an order-of-magnitude increase in the bandwidth of the memory. They also demonstrate that the often discussed connection between time reversal and optimality in photon storage follows naturally from gradient ascent. Finally, discuss the optimization of controlled reversible inhomogeneous broadening. Faithful mapping between quantum states of light (flying qubits) and quantum states of matter (storage and/or memory qubits) is an important outstanding goal in the field of quantum information processing and is being pursued both theoretically and experimentally by a large number of research groups around the world. Photon storage in Λ-type atomic media is a promising avenue for achieving this goal. In a recent theoretical paper, unified a wide range of protocols for photon storage in Λ-type media, including the techniques based on Electromagnetically Induced Transparency (EIT), off-resonant Raman interactions, and photon-echo. Now also demonstrated equivalence between all these protocols and suggested several efficiency optimization procedures, some of which have since been demonstrated experimentally. In the three preceding papers of this series, which will refer to henceforth that presented some details and many extensions of the analysis. It were obtained based on physical arguments and on exact solutions available in certain limits. However, the optimization problems discussed there fall naturally into the framework of optimal control problems, for which powerful numerical optimization methods exist. Thus, in the present paper, we apply these optimal control methods to the problem of photon storage. As a result, we open up the possibility of efficient photon storage in previously inaccessible regimes by increasing the bandwidth of the memory and provide simple intuitive understanding for the optimization methods underlying photon storage. For a comprehensive introduction to photon storage in Λ-type atomic media and for the full list of references. Here summarize only a few important points. In a typical photon storage protocol, an atomic ensemble with Λ-type level structure is assumed to start with all N atoms pumped into the metastable state $|g\rangle$. The incoming quantum light mode is coupled to the $|g\rangle - |e\rangle$ transition with a collectively enhanced coupling constant $g\sqrt{N}$ and is mapped onto the collective coherence (called a spin wave) between the metastable states $|s\rangle$ and $|g\rangle$ using a classical two-photon-resonant control pulse with timedependent Rabi frequency $\Omega(t)$. Ideal mapping of the light mode onto the spin wave and back can be achieved in an ensemble that has infinite resonant optical depth d on the $|g\rangle$-$|e\rangle$ transition. However, despite the existence of proposals for achieving high values, in most current experiments d (or the cooperativity parameter C for ensembles enclosed in a cavity) is limited to $d\sim 10$ due to experimental imperfections such as competing four-wave mixing processes, spatially-varying light shifts, number of atoms in a trap, or inhomogeneous broadening and short interaction lengths.

As a result of the limited optical depth, the experimentally demonstrated efficiencies for the light-matter interface are low, which makes the optimization of photon storage protocols at finite values of d crucial. The optimization as well as in the e knowledge of the shape of the incoming photon mode. Note that such knowledge is not incompatible with storing unknown quantum states because the mode usually acts simply as a carrier while the information is stored in the quantum state of the harmonic oscillator corresponding to this mode. A different type of problem is the storage of an unknown mode or, equivalently, the storage of multiple photonic modes within an ensemble. While we believe that the optimization procedures considered here will probably also be relevant to this situation, we shall not discuss it in more detail here. The main tool used in this paper is a numerical iterative optimization with respect to some set of control parameters, which are updated to yield higher photon storage efficiency at every iteration. Such iterative optimization methods are a standard tool in applied optimal control theory. These methods and their variations are already being used in a variety of applications including laser control of chemical reactions, design of NMR pulse sequences, loading of Bose-Einstein condensates into an optical lattice, atom transport in time-dependent super-lattices, quantum control of the hyperfine spin of an atom, and design of quantum gates. Gradient ascent methods are often more efficient than simple variations of the control parameters using, e. g., genetic algorithms. Moreover, we will show that gradient ascent optimization has the advantage that it can often be understood physically and can provide deeper intuition for the photon storage problem. In particular, used involved physical arguments and exact analytical solutions available in certain limits to derive a time-reversal-based iterative optimization with respect to the shape of the incoming photon mode. In the present paper, we show that these time-reversal iterations and the general and often discussed connection between optimality and time reversal in photon storage naturally follow from the gradient ascent method. The results however, still crucial since they show in certain cases that the solutions obtained via the local gradient ascent method represent global, rather than local, optima. In addition to considering optimization with respect to the shape of the input mode, we consider in the present paper optimization with respect to the storage control field. In particular, we show that shaping the control field via the gradient ascent method allows for efficient storage of pulses that are an order of magnitude shorter than when the control field is optimized in the adiabatic approximation discussed. In other words, this new control shaping method increases the bandwidth of the memory. Finally, we discuss the performance of optimal control pulses in the context of photon storage via controlled reversible inhomogeneous broadening. In particular, assuming one is interested in storing a single known incoming light mode and assuming one can shape control pulses with sufficient precision, we are not able to identify any advantages of CRIB-based photon storage compared to photon storage with optimal control pulses in homogeneously broadened media.

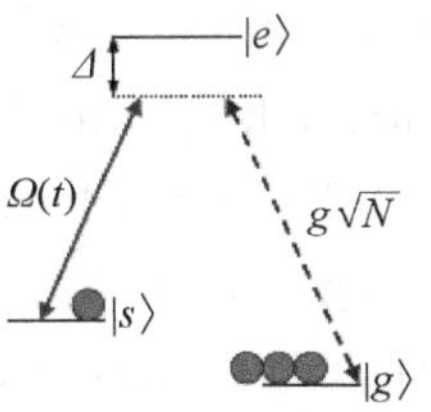

Fig. 5-84 (Color online) Λ-type medium coupled to a quantum field (dashed) with a collectively enhanced coupling constant $g\sqrt{N}$ and a two-photon-resonant classical field (solid) with time-dependent Rabi frequency $\Omega(t)$.

- Optimization with respect to the storage control field

In principle, both the incoming light mode and the classical control pulse may be adjusted to maximize the light storage efficiency. However, it is often easier to vary the classical control pulse. In particular, the photonic state we wish to store may be some non-classical state generated by an experimental setup, where we cannot completely control the shape of the outgoing wave packet. This is, e.g., the case for single photons generated by parametric down conversion or by single nitrogen-vacancy centers in diamond, where the shape of the wave packet will be, respectively, set by the bandwidth of the setup and the exponential decay associated with spontaneous emission. Alternatively, the wave packet may also be distorted in an uncontrollable way by the quantum channel used for transmitting the photonic state.

- Cavity model

The cavity model, in which the atomic ensemble is enclosed in a cavity, is theoretically simpler than the free space model because only one collective atomic mode can be excited. In addition, the cavity setup can yield higher efficiencies in certain cases than the free space model due to the enhancement of the optical depth by the cavity finesse and due to (for certain spin wave modes) better scaling of the error with the optical depth d($1/d$ in the cavity vs. $1/\sqrt{d}$ in free space). Therefore start with the cavity model to get the closest analogy to the free-space regime, Now will discuss only the so-called "bad cavity" limit, in which the cavity mode can be adiabatically eliminated. However, the method of gradient ascent can easily be applied outside of this limit, as well. To simplify the discussion, we first consider the simplest example, in which one stores a given resonant input mode into a homogeneously broadened ensemble enclosed in a cavity and having negligible spin-wave decay rate. It is important to note that, because only one spin-wave mode is accessible in the cavity model, the retrieval efficiency is independent of how the storage is done. This makes it meaningful to optimize storage separately from retrieval the latter does not have to be optimized since its efficiency depends only on the cooperativity parameter.

To adiabatically eliminate the cavity mode and to reduce the equations of motion to the following complex number equations on the time interval $t \in [0, T]$:

$$\dot{P}(t) = -\gamma(1+C)P(t) + \mathrm{i}\Omega(t)S(t) + \mathrm{i}\sqrt{2\gamma C}\,\mathcal{E}_{\mathrm{in}}(t) \tag{5-226}$$

$$\dot{S}(t) = \mathrm{i}\Omega(t)P(t) \tag{5-227}$$

Here the optical polarization $P(t)$ on the $|g\rangle - |e\rangle$ transition and the spin polarization $S(t)$ on the $|g\rangle$-$|s\rangle$ transition satisfy initial conditions $P(0) = 0$ and $S(0) = 0$, respectively, corresponding to the absence of atomic excitations at $t = 0$. In this example, the shape of the incoming mode $\varepsilon_{\mathrm{in}}(t)$ is assumed to be specified, real, and normalized according to $\int_0^T \mathrm{d}t\varepsilon_{\mathrm{in}}^2(t) = 1$. γ is the decay rate of the optical polarization and C is the collectively enhanced cooperativity parameter equal to the optical depth of the atomic ensemble times the cavity finesse. The goal is to find the slowly varying control field Rabi frequency envelope $\Omega(t)$ (assumed to be real) that maximizes the storage efficiency $\eta_s = |S(T)|^2$. To avoid carrying around extra factors of 2, $\Omega(t)$ is defined as half of what is usually called the Rabi frequency: it takes time $\pi/(2\Omega)$ to do a π pulse. For the moment, we suppose that there is no constraint on the energy of the control pulse and return to the possibility of including such a constraint below. It is worth noting that due to their linearity, the equations of motion (and all the results) apply equally well both to classical input fields with pulse shapes proportional to $\varepsilon_{\mathrm{in}}(t)$ and to quantum fields whose excitations are confined to the mode described by $\varepsilon_{in}(t)$. Since the optimization of η_s is constrained by the equations of motion (5-218) and (5-219), we introduce Lagrange multipliers $\bar{P}(t)$ and $\bar{S}(t)$ to ensure that the equations of motion are fulfilled, and turn the problem into an unconstrained maximization of

$$J = S(T)S^*(T) + \int_0^T \mathrm{d}t[\bar{P}^*(-\dot{P} - \gamma(1+C)P + \mathrm{i}\Omega S + \mathrm{i}\sqrt{2\gamma C}\,\mathcal{E}_{in}) + \mathrm{c.c.}] + \int_0^T \mathrm{d}t[\bar{S}^*(-\dot{S} + \mathrm{i}\Omega P) + \mathrm{c.c.}] \tag{5-228}$$

where c.c. stands for the complex conjugate, setting J to be stationary with respect to variations in P and S requires that the Lagrange multipliers (also referred to as the adjoint variables) $\bar{P}$ and $\bar{S}$ satisfy the equations of motion

$$\dot{\bar{P}} = \gamma(1+C)\bar{P} + \mathrm{i}\Omega\bar{S} \tag{5-229}$$

$$\dot{\bar{S}} = \mathrm{i}\Omega\bar{P} \tag{5-230}$$

subject to boundary conditions at time $t = T$

$$\bar{P}(T) = 0 \tag{5-231}$$

$$\bar{S}(T) = S(T) \tag{5-232}$$

These are the same equations as for S and P [Eqs. (5-231) and (5-232)] except that there is no input field and that the decay with rate $(1+C)$ is replaced with growth, which will function as decay for backward evolution. This backward evolution, in fact, corresponds to retrieval with the time-reversed control field and can be implemented experimentally as such. It is satisfying to have obtained this purely mathematical and simple derivation of the often discussed connection between optimality and time reversal in photon storage. Eqs. (5-221)～(5-224) ensure that J is stationary with respect to variations in P and S. To find the optimum it remains to set to zero the functional derivative of J with respect to. This functional derivative is given by

$$\frac{\delta J}{\delta\Omega(t)} = -2\mathrm{Im}[\bar{S}^* P - \bar{P}S^*] \tag{5-233}$$

where "Im" denotes the imaginary part. In general, if one has a real function of several variables, one way to find a local maximum is to pick a random point, compute partial derivatives at that point, move a small step up the gradient, and then iterate. The same procedure can be applied to our optimal control problem. The gradient ascent procedure for finding the optimal storage control pulse $\Omega(t)$ is to take a trial $\Omega(t)$ and then iteratively update $\Omega(t)$ by moving up the gradient in Eq. (5-225) according to:

$$\Omega(t) \to \Omega(t) - \frac{1}{\lambda}\mathrm{Im}[\bar{S}^* P - \bar{P}S^*] \tag{5-234}$$

where $1/\lambda$ regulates the step size. In order to compute the right hand side of Eq. (5-234), one has to evolve the system forward in time from $t = 0$ to $t = T$ using Eqs. (5-218) and (5-219) to obtain $S(t)$ and $P(t)$. Then project the final atomic state described by $S(T)$ and $P(T)$ onto S according to Eqs. (5-223) and (5-224) to obtain $\bar{P}(T)$ and $\bar{S}(T)$. Then evolve $\bar{S}$ and $\bar{P}$ backwards in time from $t = T$ to $t = 0$ according to Eqs. (5-221) and (5-222).

In general, as in any gradient ascent method, the step size $1/\lambda$ in Eq. (5-226) has to be chosen not too big (one should not go up the gradient so quickly as to miss the peak) but not too small (in order to approach the peak relatively quickly). To achieve faster convergence, one could use a different step size $1/\lambda$ for each iteration; but for the problems considered in the present paper, convergence is usually sufficiently fast that do not need to do this (unless the initial guess is too far from the optimum, in which case changing λ a few times helps). Moreover in some optimization problems, $1/\lambda$ has to be chosen such that it depends on the argument of the function we are trying to optimize, i.e., in this case the time t; this is not required for the present problems, and $1/\lambda$ is just taken to be a constant. For example, take $C=1$, $T=10$, and a Gaussian-like input mode

$$\mathcal{E}_{\text{in}}(t) = A(\mathrm{e}^{-30(t/T-0.5)^2} - \mathrm{e}^{-7.5})/\sqrt{T} \tag{5-235}$$

where $A \approx 2.09$ is a normalization constant and where the mode is chosen to vanish at $t=0$ and $t=T$ for computational convenience. Starting with an initial guess $\Omega(t)=\sqrt{\gamma}/T$ and using $\lambda=0.5$, it takes about 45 iterations for the efficiency to converge to within 0.001 of the optimal efficiency of $C/(1+C)=0.5$ for the derivation of this formula. However, if λ is too small (e.g. $\lambda=0.1$), then the step size is too large, and, instead of increasing with each iteration, the efficiency wildly varies and does not converge. Now compare the optimal control field shaping to the adiabatic control field shaping presented in paper I. Here first take $C=10$ and consider the input mode in Eq. (5-227) with $T=50/\gamma$. Following paper I, we calculate the storage control field using the adiabatic equations [Eq. (5-223)], then numerically compute the storage efficiency with this control field, and multiply it by the complete retrieval efficiency $C/(1+C)$ to obtain the total efficiency. Since in the adiabatic limit ($TC\gamma=500 \gg 1$), the resulting total efficiency is equal to the maximum possible efficiency $C^2/(1+C)^2=0.83$. Figure 5-85(a) shows the input mode in Eq. (5-227) (dashed line) and the adiabatic storage control field (dotted line).

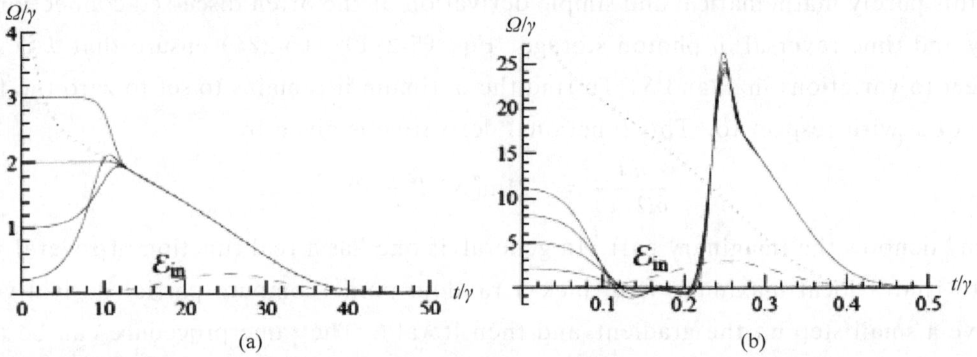

Fig. 5-85 Adiabatic (dotted) and optimal (solid) control fields for the storage of a Gaussian-like input mode $\varepsilon_{\text{in}}(t)$ (dashed) in the cavity model with $C=10$ and $T=50/\gamma$(a) and $T=0.5/\gamma$(b). The four different optimal control pulses correspond to four different initial guesses for the gradient ascent optimization. The adiabatic control field agrees with the optimal one in the adiabatic limit ($TC\gamma \gg 1$) (a) and deviates from it otherwise (b).

The optimal control field shaping using gradient ascent via Eq. (5-226) also yields the maximum possible efficiency $C^2/(1+C)^2=0.83$ independent of the initial guess for $\Omega(t)$. The four solid lines in Fig. 5-85(a) show $\Omega(t)$ resulting from optimal control field shaping for four different initial guesses, $\Omega(t)/=0.2,1,2,$ and 3. The four optimal control fields and the adiabatic control field agree except at small times. The reason for the disagreement is that the dependence of storage efficiency on the front section of the control field is very weak because this section affects only the front part of the excitation, and a large part of this anyway leaks out at the back end of the atomic ensemble. In fact, the dependence is so weak that gradient ascent leaves the front part of the initial guesses almost unperturbed. It is worth noting that, in general, gradient ascent methods are not guaranteed to yield the global optimum, and the iterations may get trapped in a local maximum. However, for our photon storage problem, we know what the global optimum is in some cases. In particular, have shown in the cavity model and in the free space model that, in the adiabatic limit, adiabatic control field shaping yields the global optimum. Since control shaping via gradient ascent agrees with the adiabatic shaping in this limit, have to strong indication that gradient ascent always yields the global optimum also outside of the adiabatic limit. The global optimum is here the (unique) maximum possible efficiency, which, within the numerical error, is achievable for a variety of control fields due to the lack of sensitivity to the control field for small times (see Figure 5-85). Again the four optimal control fields correspond to different initial guesses [$\Omega(t)/\gamma = 2, 5, 8$, and 11]. The adiabatic control field now differs from the optimal one on the entire time interval. The reason is that the adiabatic limit ($TC\gamma \gg 1$) is not satisfied to a sufficient degree ($TC\gamma = 5$), and, as a result, the adiabatic approximation does not work well. Indeed, the efficiency yielded by the adiabatic control (0.49) is much smaller than that yielded by the optimal control (0.81). Since in this regime ($TC\gamma \sim 1$) the optimal control field is turned on abruptly following a time period when it is off [see Figure 5-85(b)], the optimal storage procedure acquires some characteristics of photon-echo type fast storage. In fast storage, the input pulse is first absorbed on the $|e\rangle - |g\rangle$ transition in the absence of the control field, and is then mapped to the $|s\rangle - |g\rangle$ coherence via a control π pulse. This connection is not surprising since fast storage is indeed optimal for certain input modes of duration $T \sim 1/(C\gamma)$. Finally, we note that all the initial guesses for that we tried yielded the same optimal control (up to the unimportant front part) and the same efficiency, which is a signature of the robustness of the optimization protocol and is another strong indication that, for this optimal control problem, gradient ascent yields the global, rather than local, optimum. Having performed the comparison of the control fields generated by adiabatic shaping and by gradient ascent, we turn to the investigation of the dependence on C and on TC of the efficiency achieved by these two methods. In Figure 5-86(a), we compare the efficiency of storage followed by retrieval of the input mode of Eq. (5-227) obtained using the adiabatic control field (dotted lines) and using the control found via gradient ascent (solid lines).

The efficiencies are plotted as a function of $TC\gamma$ for three indicated values of C ($= 1, 10, 100$). Dashed lines correspond to $C^2/(1+C)^2$, the maximum efficiency possible at any given C. We note that the dotted lines have already been shown in Figure 5-86(a). Note that it is impossible to retrieve into a mode much shorter than $1/(C\gamma)$, and hence, by time-reversal, it is impossible to efficiently store such a short mode. Figure 5-86(a) confirms that indeed, when $TC\gamma \ll 1$, even optimal controls cannot give high efficiency. Using gradient ascent instead of adiabatic shaping, one can, however, efficiently store input modes that are about an order of magnitude shorter and, thus, an order of magnitude larger in bandwidth.

It is worth repeating that although the method of gradient ascent is generally not guaranteed to yield the global maximum, the fact that it does give the known global maximum in the limit $TC\gamma \gg 1$ suggests that it probably yields the global maximum at all values of $TC\gamma$. To confirm the robustness and generality of the optimization procedure, we show in Figure 5-86(b) the results of the same optimization as in Figure 5-86(a) but for a square input mode $\varepsilon_{\text{in}}(t) = 1/\sqrt{T}$ instead of

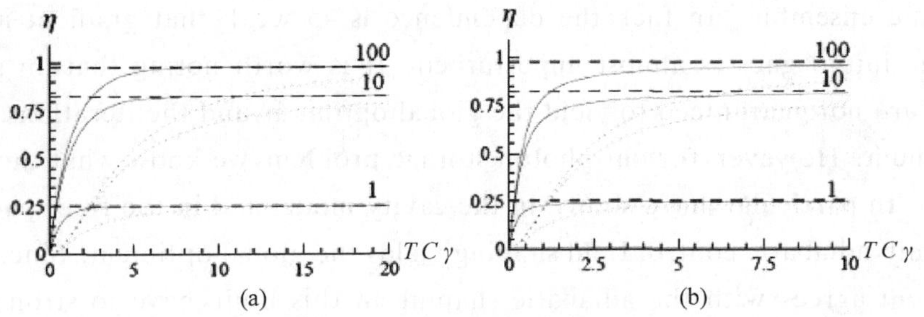

Fig. 5-86 (a) The total efficiency of storage followed by retrieval for the Gaussian-like input mode using adiabatic equations (dotted) and gradient ascent (solid) to shape the storage control field. Results are shown as a function of $TC\gamma$ for the indicated values of $C(=1,10,100)$. The dashed lines are $C^2/(1+C)^2$, the maximum efficiency possible at any given C. (b) Same for $\varepsilon_{in}(t)=1/\sqrt{T}$.

the Gaussian-like input mode of Eq. (5-227). As in Figure 5-86(a), we see that gradient ascent control shaping improves the threshold in the value of $TC\gamma$, where efficiency abruptly drops, by an order of magnitude. This can again be interpreted as an effective increase in the bandwidth of the memory by an order of magnitude. The optimal storage efficiency for the square input pulse falls to half of the maximum at smaller TC than for the Gaussian-like input pulse because the latter has a duration (half-width at half maximum, for example) significantly shorter than T see Eq. (5-227) or Figure 5-85. On the other hand, as $TC\gamma$ is increased, the maximum is approached slower for the square input mode than for the Gaussian-like mode. This is because the high frequency components contributed by the sharp edges of the square pulse are difficult to store. Most experiments have features that go beyond the simple model have just described. Therefore, we generalize this model and the optimization procedure to include the possibility of complex control field envelopes $\Omega(t)$ and input mode envelopes $\varepsilon_{in}(t)$, nonzero single-photon detuning and spin wave decay rate s, and (possibly reversible) inhomogeneous broadening. This model of inhomogeneous broadening is applicable both to Doppler broadening in gases and to the broadening of optical transitions in solid state impurities caused by the differences in the environments of the impurities. For the case of Doppler broadened gases, also allow for the possibility of modeling velocity changing collisions with rate c. Finally, also show how to take into account the possibility that the classical driving fields available in the laboratory are not sufficiently strong to realize the optimal control fields, which may be the case for short input modes and/or large single-photon detuning Δ, both of which require control pulses with large intensities. Although a comprehensive study of optimization for $\Delta \neq 0$ is beyond the scope of the present paper, will now prove that the maximum efficiency for $\Delta \neq 0$ is exactly equal to the maximum efficiency for $\Delta = 0$. Suppose we know the control field $\Omega_0(t)$ that achieves the optimum for a given resonant input $\varepsilon_{in}(t)$.

5.9.2 Free space model

Although the cavity model is theoretically simpler and results, in certain cases, in higher efficiencies than the free space model, the latter is easier to set up experimentally. Moreover, because of the accessibility of a large number of spin wave modes, the free space model can provide higher efficiencies in some other cases and can, in principle, function, as a multi-mode memory. Therefore, we turn in the present section to the analysis of the free space model. To demonstrate how optimization with respect to the control field works in the free space model, we again begin with a simple example of resonant photon storage in a homogeneously broadened atomic ensemble with negligible spin-wave decay. It is important to note that, in contrast to the cavity model, the free space model gives access to many different spin-wave modes, which makes the retrieval efficiency dependent on how storage is carried out. Therefore, optimization of storage alone is not a priori very practical. However, the optimization of storage alone is indeed useful because, in many cases, it also optimizes storage followed by backward retrieval. In order to have slightly simpler mathematical expressions, work in the co-moving frame, although the same argument can be carried out using the original time variable, as well. The complex number equations of motion on the interval $t \in [0, T]$ are then:

$$\partial_{\tilde{z}} \mathcal{E}(\tilde{z}, \tilde{t}) = i\sqrt{d} P(\tilde{z}, \tilde{t}) \tag{5-236}$$

$$\partial_{\tilde{t}} P(\tilde{z}, \tilde{t}) = -P(\tilde{z}, \tilde{t}) + i\sqrt{d}\, \mathcal{E}(\tilde{z}, \tilde{t}) + i\tilde{\Omega}(\tilde{t}) S(\tilde{z}, \tilde{t}) \tag{5-237}$$

$$\partial_{\tilde{t}} S(\tilde{z}, \tilde{t}) = i\tilde{\Omega}(\tilde{t}) P(\tilde{z}, \tilde{t}) \tag{5-238}$$

with initial and boundary conditions

$$\mathcal{E}(0, \tilde{t}) = \mathcal{E}_{\text{in}}(\tilde{t}) \tag{5-239}$$

$$P(\tilde{z}, 0) = 0 \tag{5-240}$$

$$S(\tilde{z}, 0) = 0 \tag{5-241}$$

These equations are written using dimensionless variables, in which (co-moving) time and Rabi frequency are rescaled by $\gamma(\tilde{t} = t\gamma$ and $\tilde{\Omega} = \Omega/\gamma)$ and the position is rescaled by the length L of the ensemble $(\tilde{z} = z/L)$. $\varepsilon(\tilde{z}, \tilde{t})$ describes the slowly varying electric field envelope, the input mode $\varepsilon_{\text{in}}(\tilde{t})$ satisfies the normalization constraint

$$\int_0^{\tilde{T}} |\, \mathcal{E}_{\text{in}}(\tilde{t})\, |^2 \mathrm{d}\tilde{t} = 1 \tag{5-242}$$

d is the resonant optical depth, and $\tilde{\Omega}(\tilde{z})$ and $\varepsilon_{\text{in}}(\tilde{t})$ are for now assumed to be real. [To avoid carrying around extra factors of 2, d is defined as half of what is often referred as the optical depth: the steady-state solution with $\Omega = 0$ gives probe intensity attenuation

$$|\, \mathcal{E}(\tilde{z} = 1)\, |^2 = \mathrm{e}^{-2d}\, |\, \mathcal{E}(\tilde{z} = 0)\, |^2] \tag{5-243}$$

The goal is to maximize the storage efficiency:

$$\eta_s = \int_0^1 \mathrm{d}\tilde{z}\, |\, S(\tilde{z}, \tilde{T})\, |^2 \tag{5-244}$$

with respect to $\tilde{\Omega}(\tilde{t})$. A procedure analogous to that used in the cavity model yields equations of motion (also referred to as the adjoint equations) for the Lagrange multipliers $\bar{\mathcal{E}}(\tilde{z}, \tilde{t})$, $\bar{P}(\tilde{z}, \tilde{t})$, and $\bar{S}(\tilde{z}, \tilde{t})$:

$$\partial_{\tilde{z}} \bar{\mathcal{E}} = i\sqrt{d} \bar{P} \tag{5-245}$$

$$\partial_{\tilde{t}} \bar{P} = \bar{P} + i\sqrt{d}\, \bar{\mathcal{E}} + i\tilde{\Omega}\bar{S} \tag{5-246}$$

$$\partial_{\tilde{t}} \bar{S} = i\tilde{\Omega}\bar{P} \tag{5-247}$$

with initial and boundary condition

$$\bar{P}(\tilde{z}, \tilde{T}) = 0 \tag{5-248}$$

$$\bar{S}(\tilde{z}, \tilde{T}) = S(\tilde{z}, \tilde{T}) \tag{5-249}$$

These equations describe backward retrieval and provide a simple mathematical connection between optimality and time-reversal. In order to move up the gradient, one should update $\Omega(\tilde{t})$ according to

$$\widetilde{\Omega}(\tilde{t}) \to \widetilde{\Omega}(\tilde{t}) - \frac{1}{\lambda}\int_0^1 \mathrm{d}\tilde{z}\,\mathrm{Im}\left[\bar{S}^*(\tilde{z},\tilde{t})P(\tilde{z},\tilde{t}) - \bar{P}(\tilde{z},\tilde{t})S^*(\tilde{z},\tilde{t})\right] \tag{5-250}$$

In the adiabatic limit ($T d\gamma \gg 1$) and for a certain class of input modes of duration $T \sim 1/(d\gamma)$, one can achieve a universally optimal (for a fixed d) storage efficiency that cannot be exceeded even if one chooses a different input mode. In that case obtained control field will also maximize the total efficiency of storage followed by backward retrieval. However, this would not necessarily be the case for a general input mode in the nonadiabatic limit ($Td\gamma < 1$), which is precisely the limit, in which gradient ascent optimization becomes most useful. Moreover, for the case of forward retrieval, the control field that maximizes the storage efficiency does not maximize the total efficiency of storage followed by retrieval even in the adiabatic limit. Thus, describe how to use gradient ascent to maximize (still with respect to the storage control field) the total efficiency of storage followed by retrieval. We assume that $d = 10$ and that $\varepsilon_{\mathrm{in}}(t)$ is the Gaussian-like input mode in Eq. (5-250), shown as a dashed line in Figures 5-87(a) and (b). First consider the case $T = 50/\gamma$ and shape the storage control using adiabatic shaping. Then numerically compute the total efficiency of storage followed by complete backward retrieval using this storage control field (the total efficiency is independent of the retrieval control field provided no excitations are left in the atoms). The adiabatic storage control is shown as a dotted line in Figure 5-87(a). Since for this input mode the adiabatic limit is satisfied ($Td\gamma = 500 \gg 1$), the adiabatic storage control yields an efficiency of 0.66, which is the maximum efficiency possible at this d. For the same reason, the adiabatic control agrees with the control field computed via gradient ascent (solid line), which also yields an efficiency of 0.66.

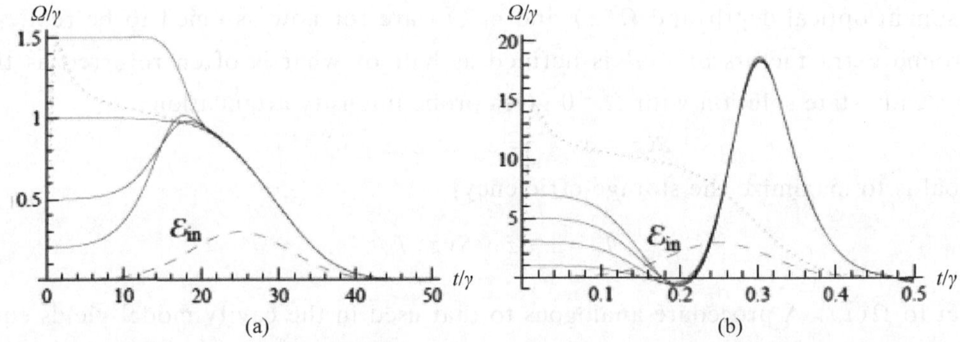

Fig. 5-87 Adiabatic (dotted) and optimal (solid) control fields for the storage followed by backward retrieval of a Gaussian-like input mode $\varepsilon_{\mathrm{in}}(t)$ (dashed) in the free space model with $d = 10$ and $T = 50/\gamma$ (a) and $T = 0.5/\gamma$ (b). Four optimal control pulses were obtained using four different initial guesses for the gradient ascent procedure. The adiabatic control field agrees with the optimal one in the adiabatic limit ($Td\gamma > 1$) (a) and deviates from it otherwise (b).

Figure 5-87(a) shows four solid lines (optimal control fields) corresponding to four initial guesses $\Omega(t)/\gamma = 0.2, 0.5, 1$, and 1.5. As in the cavity model, the difference between the four optimal controls and the adiabatic control is inconsequential. Repeating the calculation for $T = 0.5/\gamma$, can obtain Figure 5-87(b). Since the adiabatic limit ($Td\gamma \gg 1$) is no longer satisfied ($Td\gamma = 5$), the adiabatic approximation does not work and the adiabatic control differs from the optimal control and gives a lower efficiency: 0.24 vs. 0.58.

As in Figure 5-87(a), the four optimal control fields plotted correspond to different initial guesses $\Omega(t)/\gamma = 1, 3, 5$, and 7. As in the cavity discussion, Figure 5-87(b) indicates that, in the regime $Td \sim 1$, where the adiabatic approximation no longer holds, the optimal control field acquires characteristics of the control field used in fast storage. As in the analysis of the cavity model, we now analyze the dependence on d and Td of the efficiency yielded by the adiabatic control shaping and the optimal control shaping. In Figure 5-88, we compare the efficiency of storage followed by complete backward retrieval of the input mode in Eq. (5-227) obtained using the control field shaped using the adiabatic equations (dotted lines) and using gradient ascent (solid lines). The efficiencies are plotted as a function of $T d$ for three indicated values of d (= 1, 10, 100). Horizontal dashed lines represent the maximum efficiency possible at the given d. The dotted lines are the same as in Figure 5-89(a). Similar to the corresponding discussion of the cavity model, Figure 5-88 confirms the predictions that efficient photon storage is not possible for $Td\gamma \leqslant 1$. It also illustrates that optimal control fields open up the possibility of efficient storage of input modes with a bandwidth that is an order of magnitude larger than the bandwidth allowed by the adiabatic storage. In addition, the same reasoning as in the cavity discussion leads to the conclusion that for this problem, gradient ascent most likely yields the global, rather than local, maximum at all values of $Td\gamma$.

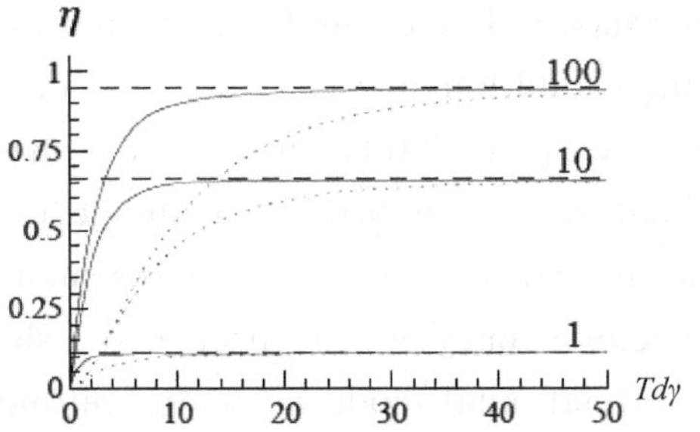

Fig. 5-88 The total efficiency of storage followed by backward retrieval for the Gaussian-like input mode in Eq. (5-244) using adiabatic equations (dotted) and gradient ascent (solid) to shape the storage control field. The results are shown for the indicated values of d (= 1, 10, 100), as a function of $T\,d\gamma$. The dashed lines represent the maximum efficiency possible at the given d.

Various generalizations of the presented procedure can be made. First, the generalization to limited control pulse energy, inhomogeneous broadening, complex Ω and ε_{in}, and nonzero Δ, γ_s, and γ_c can be carried out exactly as in the cavity case. Second, in the case of backward retrieval, if the two metastable states are nondegenerate and have a frequency difference ω_{sg}, one should incorporate an appropriate position-dependent phase shift of the spin wave of the form $\exp(-2\mathrm{i}\Delta kz)$, where $\Delta\tilde{k} = L\omega_{sg}/c$. Finally, another extension can be made for the cases when the total efficiency depends on the retrieval control field (e.g. if s and/or c are nonzero).

In those cases, one can simultaneously optimize with respect to both the storage and the retrieval control fields. However, one may then need to put a limit on the energy in the retrieval control pulse since, for the case of $\gamma_s \neq 0$, for example, the faster one retrieves, the higher is the efficiency, and the optimal retrieval control field may, in principle, end up having unlimited power (e.g. an infinitely short π pulse).

- Optimization with respect to the input field

Although it is usually easier to optimize with respect to the control field, optimization with respect to the input mode can also be carried out in certain systems. For both classical and quantum light, the mode shape can often be controlled by varying the parameters used during the generation of the mode. For example, if the photon mode is created by releasing some generated collective atomic excitation, one can, under certain assumptions, generate any desired mode shape. For the case of classical light, one can also shape the input light pulse simply using an acousto-optical modulator. An important advantage of optimizing with respect to the input mode is that the iterations can be carried out experimentally. In this section, consider the maximization of light storage efficiency with respect to the shape of the input mode. Since one is interested in finding the optimal input mode shape, the optimization has to be carried out subject to the normalization condition:

$$\int_0^T \mathrm{d}t \mid \mathcal{E}_{\text{in}}(t) \mid^2 = 1 \tag{5-251}$$

This condition can be included by adding an extra term with a Lagrange multiplier to the functional J to be optimized. The iterations are then done as follows: one first integrates the storage equations for a trial input mode; then integrates the adjoint equations corresponding to backward retrieval; then updates the trial input mode by adding to it a small correction proportional to the output of backward retrieval ($-\mathrm{i}\bar{P}(t)$ in the cavity model or $\bar{\varepsilon}(0,\tilde{t})$ in the free space model); and finally renormalizes the new input mode to satisfy the normalization condition.

An important feature that distinguishes the optimization with respect to the input mode from the optimization with respect to the control field is the possibility of making finite (not infinitesimal) steps. Standard gradient-ascent improvement (such as via Eqs. (5-251) and (5-252)) is, in principle, infinitesimal due to its reliance on the small parameter $1/\lambda$. Several decades ago, Krotov introduced and developed an important powerful and rapidly converging global improvement method that is not characterized by a small parameter. Largely thanks to the presence of the normalization condition on the input mode, this method can be applied to derive non-infinitesimal quickly converging updates for the problem of optimization of light storage efficiency with respect to the input mode. For the cavity model, this update is given by

$$\mathcal{E}_{\text{in}}(t) \to -i\bar{P}(t) \tag{5-252}$$

followed by a renormalization of $\mathcal{E}_{\text{in}}(t)$, the update is given by

$$\mathcal{E}_{\text{in}}(\tilde{t}) \to \bar{\mathcal{E}}(0,\tilde{t}) \tag{5-253}$$

In these iterations, optimization of light storage with respect to the input field is done by carrying out storage of a trial input mode followed by backward retrieval, and then using the normalized output of backward retrieval as the input mode in the next iteration. The beauty of this update procedure is the possibility of carrying it out experimentally. In fact, the extension of this procedure to the optimization of storage followed by forward retrieval, suggested has already been demonstrated experimentally. In the language of gradient ascent, one can still think of Eqs. (5-252) and (5-253) as steps along the gradient. These steps are, however, finite, not infinitesimal. This allows one to think of time-reversal-based optimization with respect to the input mode as simple intuitive walk up the gradient. Using the terminology of gradient ascent, the optimization with respect to the input field in the cavity model can, surprisingly, be achieved with a single large step up the gradient. Note that the optimization procedure discussed in this section can be easily generalized to include inhomogeneous broadening and (for the case of Doppler broadened gases) the presence of velocity changing collisions. One can show that, even with these features, the iterative optimization procedure still works in exactly the same way by updating the input mode with the output of time-reversed retrieval.

- Optimization with respect to the inhomogeneous profile

Having discussed optimization with respect to the control field and the input mode, we now turn to the optimization with respect to the shape of the inhomogeneous profile. This optimization is most relevant in the context of controlled reversible inhomogeneous broadening (CRIB). The main idea of CRIB is that by introducing inhomogeneous broadening into a homogeneously broadened medium (via Stark or Zeeman shifts, for example) and by optimizing the shape and width of this inhomogeneous profile, one can better match the absorption profile of the medium to the spectrum of the incoming photon mode and, thus, increase the storage efficiency. At the same time, one can minimize the losses caused by dephasing of different frequency classes with respect to each other by using an echo-like process triggered by a reversal of the inhomogeneous profile between the processes of storage and retrieval.

- Cavity model

Although one can, of course, optimize with respect to the inhomogeneous profile in the problem of storage alone (i. e. not followed by retrieval), in the context of CRIB it is more relevant to consider the problem of storage followed by retrieval with the reversed inhomogeneous profile. Moreover, although the approach can be extended to nonzero single-photon detuning and arbitrary control fields, we suppose for simplicity that the input mode $\varepsilon_{in}(t)$ is resonant and that the storage and retrieval control pulses are π pulses.

Now can present the results of gradient ascent optimization with respect to the inhomogeneous profile for a particular example. The dash-dotted line gives the efficiency of fast storage (i. e. storage obtained by applying a control π pulse on the $|g\rangle - |e\rangle$ transition at the end of the input mode at time T) followed by fast retrieval using a homogeneous line. A homogeneous ensemble enclosed in a cavity has only one accessible spin-wave mode and can, therefore, fast-store only one input mode, which has duration $T \sim 1/(C\gamma)$. As a result, the decay at $TC\gamma \gg 1$ of the efficiency represented by the dash-dotted line is dominated by leakage of the input mode into the output mode and not by polarization decay. Now consider introducing reversible inhomogeneous broadening and iteratively optimizing with respect to its shape. As expected, the efficiency grows with each iteration independently of the choice of the number of frequency classes, the choice of Δj, and the initial guess for p_j. The landscape in the control space, however, depends on the number of frequency classes and on Δj. Therefore focus on the case of only two frequency classes with detunings $\pm \Delta I$ and optimize with respect to ΔI [using Eq. (5-235)]. The optimized efficiency is shown with circles in Figure 5-89(a). For $TC\gamma$ less than about 0.75, it is optimal to have $\Delta I = 0$. For larger $TC\gamma$, the optimal ΔI is shown in Figure 5-89(b): at small $TC\gamma$, it scales approximately as $\propto (TC\gamma) - 1$ and then slower. The presence of two frequency classes and hence two accessible spin wave modes instead of one allows us to reduce the leakage error, so that the efficiency [circles in Figure 5-89 (a)] is now limited by polarization decay. To compare the broadeningoptimized efficiency to the homogeneous controloptimized efficiency. Moreover, all inhomogeneous broadening configurations we tried to introduce into the optimized homogeneous protocol converged back to the homogeneous profile. These results suggest that if one wants to store a single mode of known shape using a homogeneously broadened ensemble of Λ-type systems enclosed in a cavity and can shape and time the control field with sufficient precision, it may be better to use optimal homogeneous storage and not to use CRIB. It is, however, worth noting that we have only carried out the simplest optimization of fast storage with CRIB. In particular, the performance of fast storage with CRIB may be further enhanced by optimizing with respect to the time, at which the storage π pulse is applied. Such optimization represents an optimal control problem with a free terminal time and is beyond the scope of the present paper. Although it can be carried out in a straightforward manner by repeating the optimization above systematically for different times of the π-pulse application. It is also important to note that the use of CRIB in the cavity model may allow for implementing a multimode memory in the cavity setup. Unlike the free space model, which allows for the storage of multiple temporal input modes using, e. g., Ramanor EIT-based protocols, the homogeneously broadened cavity model only has a single accessible spin-wave mode.

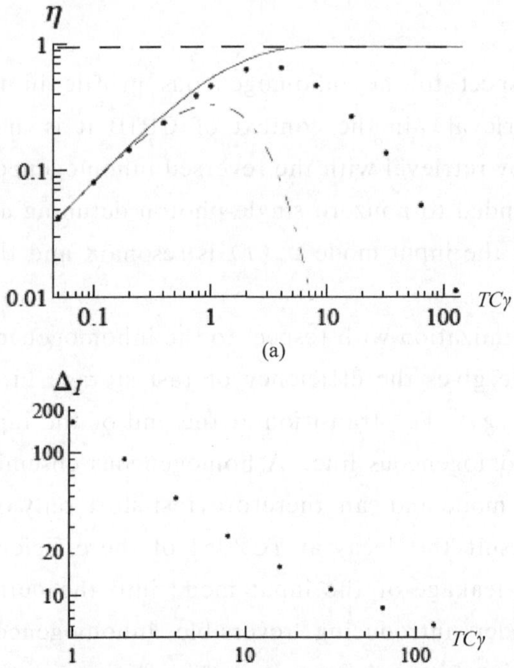

Fig. 5-89 Comparison of the efficiency for storage followed by retrieval in the cavity model with and without controlled reversible inhomogeneous broadening (CRIB).

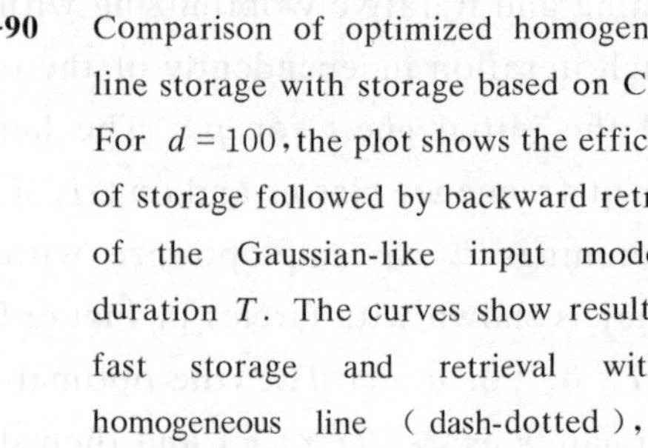

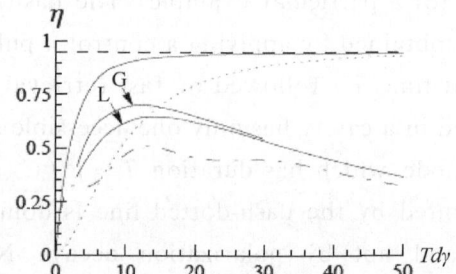

Fig. 5-90 Comparison of optimized homogeneous-line storage with storage based on CRIB. For $d = 100$, the plot shows the efficiency of storage followed by backward retrieval of the Gaussian-like input mode of duration T. The curves show results for fast storage and retrieval with a homogeneous line (dash-dotted), fast storage and retrieval with an optimized reversible Gaussian (G) or Lorentzian (L) inhomogeneous profile.

Therefore, if do not use CRIB or some other inhomogeneous broadening mechanism, it can only store a single input mode B. Free space model discussed the optimization with respect to the inhomogeneous profile in the cavity model, we note that the same procedure can be carried out for the free space model in an analogous manner.

The powerful numerical optimal control method of gradient ascent allows one to obtain simple intuitive understanding and to achieve a significantly improved efficiency and a higher bandwidth in the problem of photon storage in Λ-type atomic ensembles. First showed how to apply gradient ascent to numerically compute optimal control fields even outside of the adiabatic limit both with and without a constraint on the energy in the control pulse. In particular, this opens up the possibility of efficient storage of input modes that are an order of magnitude shorter (and hence an order of magnitude larger in bandwidth) than the shortest modes that can be efficiently stored using adiabatic control field shaping. Second, showed that gradient ascent provides an alternative justification for the often discussed connection between optimality and time-reversal in photon storage, as well as for the iterative time-reversal-based optimization procedure with respect to the input field suggested, discussed in detail and demonstrated mentally. Provided one is interested in storing a single input photon mode of known shape and provided the control pulses can be generated with sufficient precision, we have not, however, been able to identify any advantages of CRIB-based photon storage compared to photon storage with optimal control pulses in homogeneously broadened media. In general, gradient ascent methods do not guarantee the attainment of the global maxima. The global maximum is, however, indeed attained for our problem in the regimes where this maximum is known. This strongly suggests that, for the optimization with respect to the input mode and with respect to the storage control, gradient ascent may indeed be yielding the global optimum. One can optimize simultaneously with respect to various combinations of the control parameters simply by simultaneously updating each of them along the corresponding gradient. One can also include other possible control parameters that are available in a given experimental setup but have not been discussed in the present paper. For example, for the case of photon storage in solid-state systems, one can consider optimizing with respect to the number of atoms put back into the antihole or with respect to a timedependent reversible inhomogeneous profile. Other light storage systems, such as photonic crystals or cavity models where the cavity field cannot be eliminated, are also susceptible to gradient ascent optimization. Therefore, the optimization procedures described in the present paper to allow increased efficiencies and increased bandwidths in many current experiments on quantum memories for light, many of which are narrowband and suffer from low efficiencies. Such improvements would facilitate advances in fields such as quantum communication and quantum computation.

5.9.3 Adjoint equations of motion in the cavity model

Varying J given in Eq. (5-252) with respect to S, S^*, P, and P^*, can obtain:

$$\delta J = S(T)\delta S^*(T) + \int_0^T \mathrm{d}t\bar{P}^*(-\delta\dot{P} - \gamma(1+C)\delta P + \mathrm{i}\Omega\delta S) + \int_0^T \mathrm{d}t\bar{S}^*(-\delta\dot{S} + \mathrm{i}\Omega\delta P) + \text{c.c.} \tag{5-254}$$

where "c. c." means complex conjugate taken of the whole expression after the equal sign. Integrating by parts the terms containing time derivatives, we obtain:

$$\delta J = S(T)\delta S^*(T) - \bar{P}^*(T)\delta P(T) + \int_0^T \mathrm{d}t\dot{\bar{P}}^*\delta P + \int_0^T \mathrm{d}t\bar{P}^*(i\Omega\delta S - \gamma(1+C)\delta P) -$$
$$\bar{S}^*(T)\delta S(T) + \int_0^T \mathrm{d}t\dot{\bar{S}}^*\delta S + \int_0^T \mathrm{d}t\bar{S}^*(i\Omega\delta P) + \text{c.c.} \tag{5-255}$$

Since the initial conditions are fixed, we have here used $\delta S(0) = \delta P(0) = \delta S^*(0) = \delta P^*(0) = 0$ to simplify the expression. The optimum requires that $\delta J = 0$ for any variations δP and δS. Hence we collect the terms multiplying, e. g., $\delta P(T)$ and set the result to zero. Carrying out this procedure for $\delta P(T)$, $\delta S(T)$, and their conjugates, we obtain the boundary conditions (5-254) and (5-255). Collecting terms proportional to δP, δS, and their conjugates, can obtain adjoint equations of motion (5-252) and (5-253).

- Optimal light storage with full pulse shape control

Irina Novikova etc experimentally demonstrate optimal storage and retrieval of light pulses of arbitrary shape in atomic ensembles in 2008. By shaping auxiliary control pulses, we attain efficiencies approaching the fundamental limit and achieve precise retrieval into any predetermined temporal profile. These techniques, demonstrated in warm Rb vapor, are applicable to a wide range of systems and protocols. As an example, here present their potential application to the creation of optical time-bin qubits and to controlled partial retrieval. Quantum memory for light is essential for the implementation of long-distance quantum communication and of linear optical quantum computation. Both applications put forth two important requirements for the quantum memory: (i) the memory efficiency is high (i.e. the probability of losing a photon during storage and retrieval is low) and (ii) the retrieved photonic wavepacket has a well-controlled shape to enable interference with other photons. In this Letter, we report on the first experimental demonstration of this full optimal control over light storage and retrieval: by shaping an auxiliary control field, store an arbitrary incoming signal pulse shape and then retrieve it into any desired wave packet with the maximum efficiency possible for the given memory. While results are obtained in warm Rb vapor using electromagnetically induced transparency (EIT), the presented procedure is universal and applicable to a wide range of systems, including ensembles of cold atoms and solid-state impurities, as well as to other light storage protocols (e. g., the far off-resonant Raman scheme). Consider the propagation of a weak signal pulse in the presence of a strong classical control field in a resonant Λ-type ensemble under EIT conditions, as shown in Figure 5-91(a). An incoming signal pulse propagates with slow group velocity vg, which is uniform throughout the medium and is proportional to the intensity of the control field $v_g \approx 2|\Omega|^2/(\alpha\gamma) \ll c$. Here, Ω is control Rabi frequency, γ is the decay rate of the optical polarization, and α is the absorption coefficient, so that αL is the optical depth of the atomic medium of length L. For quantum memory applications, a signal pulse can be "stored" (i.e. reversibly mapped) onto a collective spin excitation of an ensemble (spin wave) by reducing the control intensity to zero. In the limit of infinitely large optical depth and negligible ground state decoherence, any input pulse can be converted into a spin wave and back with 100% efficiency, satisfying requirement (i). Under the same conditions, any desired output pulse shape can be easily obtained by adjusting the control field power (and hence the group velocity) as the pulse exists the medium, in accordance with requirement (ii). However, most current experimental realizations of ensemble-based quantum memories operate at limited optical depth αL due to various constraints. At finite αL losses limit the maximum achievable memory efficiency to a value below 100%, making efficiency optimization and output-pulse shaping important and nontrivial.

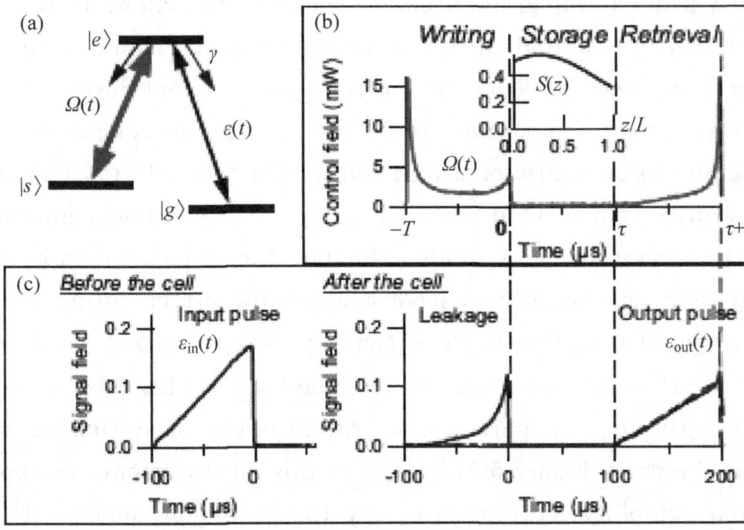

Fig. 5-91 (Color online) (a) Schematic of the three-level Λ interaction scheme. Control (b) and signal (c) fields in pulseshape-preserving storage of a "positive-ramp" pulse using a calculated optimal control field envelope $\Omega(t)$. During the writing stage ($t<0$), the input pulse $\varepsilon_{in}(t)$ is mapped onto the optimal spin-wave $S(z)$ [inset in (b)], while a fraction of the pulse escapes the cell (leakage). After a storage time τ, the spin-wave $S(z)$ is mapped into an output signal pulse $\varepsilon_{out}(t)$ during the retrieval stage. The dashed blue line in (c) shows the target output pulse shape.

Experimentally demonstrate the capability to satisfy both quantum memory requirements in an ensemble with a limited optical depth. Specifically, by adjusting the control field envelopes for several arbitrarily selected input pulse shapes, we demonstrate precise retrieval into any desired output pulse shape with experimental memory efficiency very close to the fundamental limi. This ability to achieve maximum efficiency for any input pulse shape is crucial when optimization with respect to the input wavepacket is not applicable (e.g., if the photons are generated by parametric down-conversion). At the same time, control over the outgoing mode, with precision far beyond the early attempts is essential for experiments based on the interference of photons stored under different experimental conditions (e.g., in atomic ensembles with different optical depths), or stored a different number of times. In addition, control over output pulse duration may also allow one to reduce sensitivity to noise (e.g., jitter). It is important to note that although our experiment used weak classical pulses, the linearity of the corresponding equations of motion ensures direct applicability of our results to quantum states confined to the mode defined by the classical pulse. The experimental setup is described. It phase-modulated the output of an external cavity diode laser to produce modulation sidebands separated by the ground-state hyperfine splitting of ^{87}Rb(HF = 6.835GHz). For this experiment, we tuned the zeroth order (control field) to the $F=2 \rightarrow F'=2$ transition of the ^{87}Rb D1 line, while the +1 modulation sideband played the role of the signal field and was tuned to the $F=1 \rightarrow F'=2$ transition. The amplitudes of the control and signal fields were controlled independently by simultaneously adjusting the phase modulation amplitude (by changing the rf power sent to the electro-optical modulator and the total intensity in the laser beam (using an acousto-optical modulator). Typical peak control field and signal field powers were 18mW and 50μW, respectively. The -1 modulation sideband was suppressed to 10% of its original intensity using a temperature-tunable Fabri-Perot etalon. In the experiment, used a cylindrical Pyrex cell (length and diameter were 75mm and 22mm, respectively) containing isotopically enriched ^{87}Rb and 30 Torr Ne buffer gas, mounted inside three-layer magnetic shielding and maintained at the temperature of 60.5℃, which corresponds to an optical depth of $\alpha L = 24$. The laser beam was circularly polarized and weakly focused to $\approx$5mm diameter inside the cell. We found the typical spin wave decay time to be $1/(2s) = 500\mu s$, most likely arising from small, uncompensated remnant magnetic fields. The duration of pulses used in the experiment was short enough for the spin decoherence to have a negligible effect during writing and retrieval stages and to cause a modest reduction of the efficiency $\propto \exp(-2\gamma_s) = 0.82$ during the storage time $\tau = 100\mu s$. For the theoretical calculations, we model the ^{87}Rb D1 line as a homogeneously broadened Λ-system with no free parameters. An example of optimized light storage with controlled retrieval is shown in Figure 5-91(b,c). In this measurement, we chose the input pulse $\varepsilon_{in}(t)$ to be a "positive ramp". According to theory, the maximum memory efficiency is achieved only if the input pulse is mapped onto a particular optimal spin wave $S(z)$, unique for each αL. The calculated optimal spin wave for $\alpha L = 24$ is shown in the inset in Figure 5-91(b). Then, used the method to calculate the writing control field $\Omega(t)$ $(-T < t < 0)$ that maps the incoming pulse onto the optimal spin wave $S(z)$. To calculate the retrieval control field $\Omega(t)(\tau < t < \tau + T)$ that maps $S(z)$ onto the target output pulse $\varepsilon_{tgt}(t)$, we employ the same writing control calculation together with the following time-reversal symmetry of the optimized light storage. A given input pulse, stored using its optimal writing control field, is retrieved in the time-reversed and attenuated copy of itself $[\varepsilon_{out}(t) \propto \varepsilon_{in}(\tau - t)]$ when the time-reversed control is used for retrieval $[\Omega(t) = \Omega(\tau - t)]$. Thus the control field that retrieves the optimal spin wave $S(z)$ into $\mathcal{E}_{tgt}(t)$ is the time-reversed copy of the control that stores $\mathcal{E}_{tgt}(\tau - t)$ into $S(z)$. The measured output pulse (solid black line in Figure 5-91(c)) matches very well the target shape (dashed blue line in the same figure). This qualitatively demonstrates the effectiveness of the proposed control method. To describe the memory quantitatively, we define memory efficiency η as the probability of retrieving an incoming photon after some storage time equivalently, as the energy ratio between retrieved and initial signal pulses:

$$\eta = \frac{\int_{\tau}^{\tau+T} |\mathcal{E}_{out}(t)|^2 dt}{\int_{-T}^{0} |\mathcal{E}_{in}(t)|^2 dt} \tag{5-256}$$

To characterize the quality of pulse shape generation, we define an overlap integral J^2 as

$$J^2 = \frac{\left| \int_{\tau}^{\tau+T} \mathcal{E}_{out}(t)\, \mathcal{E}_{tgt}(t) dt \right|^2}{\int_{\tau}^{\tau+T} |\mathcal{E}_{out}(t)|^2 dt \int_{\tau}^{\tau+T} |\mathcal{E}_{tgt}(t)|^2 dt} \tag{5-257}$$

The measured memory efficiency for the experiment is 0.42 ± 0.02. This value closely approaches the predicted highest achievable efficiency 0.45 for $\alpha L = 24$, corrected to take into account the spin wave decay during the storage time. The measured value of the overlap integral between the output and the target is $J^2 = 0.987$, which indicates little distortion in the retrieved pulse shape. The definitions of efficiency η and overlap integral J^2 are motivated by quantum information applications. Storage and retrieval of a single photon in a non-ideal passive quantum memory produces a mixed state that is described by a density matrix $\rho = (1-\eta) \parallel + \eta | \phi \rangle \langle \phi |$, where $|\phi\rangle$ is a single photon state with envelope $\varepsilon_{\text{cout}}(t)$, and $|\psi\rangle$ is the vacuum state. Then the fidelity between the target single-photon state $|\psi\rangle$ i with envelope $\varepsilon_{\text{tgt}}(t)$ and the single-photon state $|\phi\rangle$ is given by the overlap integral J^2 [Eq. as(5-257)], while $F = \langle \Psi |_\rho | \Psi \rangle = \eta J^2$ is the fidelity of the output state ρ with respect to the target state $|\Psi\rangle$. The overlap integral J^2 is also an essential parameter for optical quantum computation and communication protocols, since $(1-J^2)/2$ is the coincidence probability in the Hong-Ou-Mandel interference between photons $|\psi\rangle$ and $|\phi\rangle$. One should be cautious in directly using our classical measurements of η and J^2 to predict fidelity for single photon states because single photons may be sensitive to imperfections that do not significantly affect classical pulses. For example, four-wave mixing processes may reduce the fidelity of single-photon storage, although our experiments found these effects to be relatively small at $\alpha L < 25$. Figure 5-92 shows more examples of optimal light storage with full output-pulse-shape control. For this experiment, we stored either of two randomly selected input signal pulse shapes—a Gaussian and a "negative ramp"—and then retrieved them either into their original waveforms (a), (d) or into each other (b), (c). Memory efficiency η and overlap integral J^2 are shown for each graph. Notice that the efficiencies for all four input-output combinations are very similar (0.42 ± 0.02) and agree well with the highest achievable efficiency (0.45) for the given optical depth $\alpha L = 24$. The overlap integrals are also very close to 1, revealing an excellent match

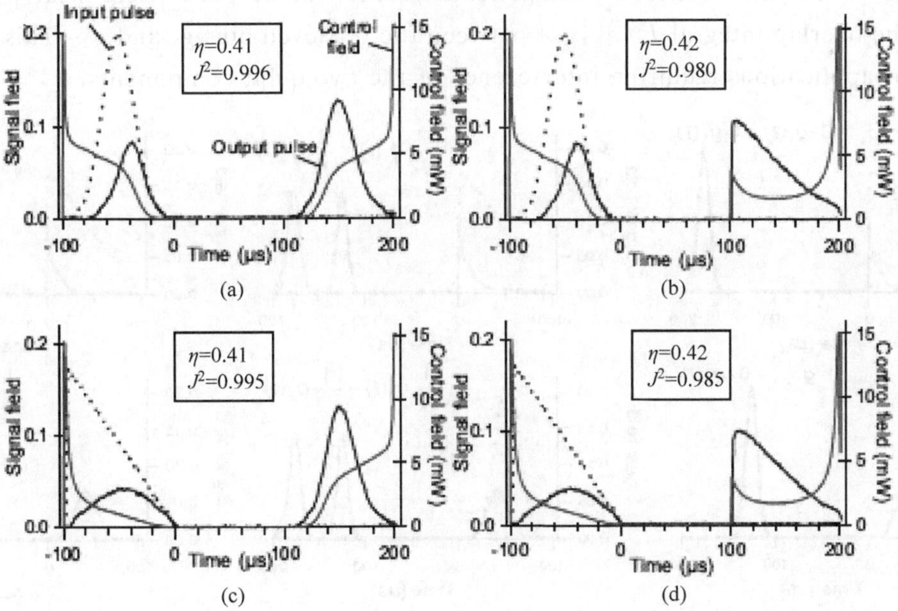

Fig. 5-92 (Color online) An input Gaussian pulse was optimally stored and retrieved either into its original pulse shape (a) or into a ramp pulse shape (b). Similarly, the incoming ramp pulse was optimally stored and retrieved into a Gaussian (c) or into an identical ramp (d). Input and output signal pulses are shown as dotted and solid black lines, respectively, while the optimal control fields are shown in solid red lines.

between the target and the retrieved signal pulse shapes. Note that different input pulses stored using corresponding (different) optimized writing control fields but retrieved using identical control fields [pairs (a),(c) and (b),(d)] had identical output envelopes, very close to the target one. This observation, together with the fact that the measured memory efficiency is close to the fundamental limit, suggests that indeed different initial pulses were mapped onto the same optimal spin wave. This indirectly confirms our control not only over the output signal light field but also over the spin wave. Our full control over the outgoing wavepacket opens up an interesting possibility to convert a single photon into a so-called "time-bin" qubit—a single photon excitation delocalized between two time-resolved wavepackets (bins). The state of the qubit is encoded in the relative amplitude and phase between the two time bins.

Such time-bin qubits are advantageous for quantum communication since they are insensitive to polarization fluctuations and depolarization during propagation through optical fibers. Propose to efficiently convert a single photon with an arbitrary envelope into a time-bin qubit by optimally storing the photon in an atomic ensemble, and then retrieving it into a time-bin output envelope with well-controlled relative amplitude and phase using a customized retrieval control field. To illustrate the proposed wavepacket shaping, in Figure 5-93, we demonstrate storage of two different input pulses (a Gaussian and a positive ramp), followed by retrieval into a time-bin-like output pulse, consisting of two distinct Gaussian wavepackets $g_{1,2}(t)$ with controllable relative amplitude and delay. We obtained the target output independently of what the input pulse shape was. We also attained the same memory efficiency as before (0.41 ± 0.02) for all linear combinations. Also, regardless of the input, the output pulse shapes matched the target envelopes very well, as characterized by the value of the overlap integral close to unity $J^2 = 0.98 \pm 0.01$. Also verified that the envelopes of the two retrieved components of the output pulse were nearly identical by calculating the overlap integral $J^2(g_1, g_2)$ between the retrieved bins g_1 and g_2. This parameter is important for applications requiring interference of the two qubit components.

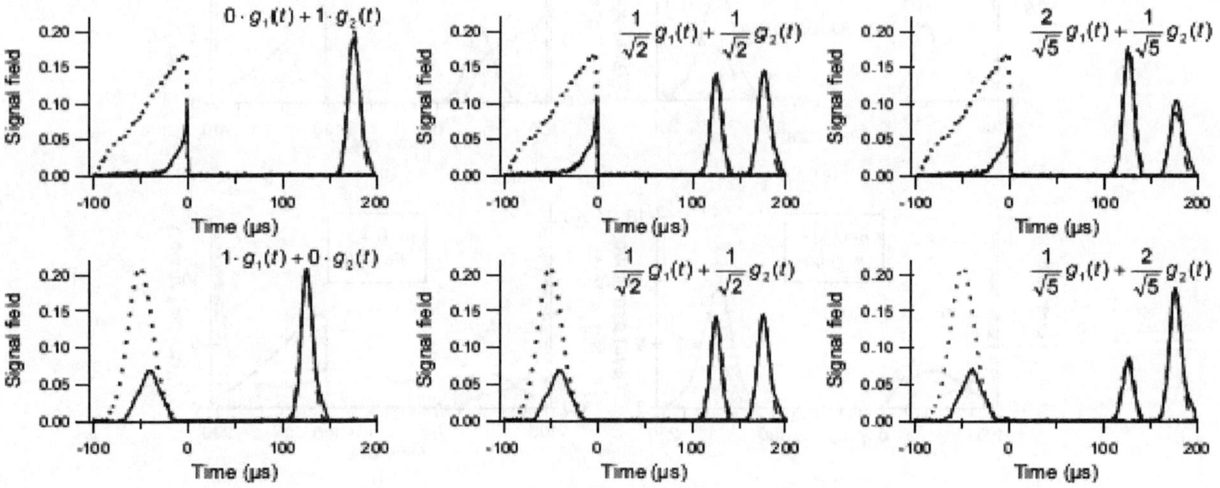

Fig. 5-93 (Color online) Examples of storage of signal input pulses with Gaussian and triangular envelopes, followed by retrieval in a linear combination of two time-resolved Gaussian pulse shapes $g_1(t)$ and $g_2(t)$. Input and output signal fields are shown in dotted and solid black lines, respectively. Dashed blue lines show the target envelopes.

The average value of $J^2(g_1, g_2) = 0.94 \pm 0.02$ was consistently high across the full range of target outputs. The relative ephase of the two qubit components can be adjusted by controlling the phase of the control field during retrieval. The demonstrated control over the amplitude ratio and shape of the two wave packets is essential for achieving high-fidelity time-bin qubit generation. The scheme is also immediately applicable to high-fidelity partial retrieval of the spin wave, which forms the basis for a recent promising quantum communication protocol. To conclude, we have reported the experimental demonstration of optimal storage and retrieval of arbitrarily shaped signal pulses in an atomic vapor at an optical depth $\alpha L = 24$ by using customized writing control fields. Our measured memory efficiency is close to the highest efficiency possible at that optical depth. Demonstrate full precision control over the retrieved signal pulse shapes, achieved by shaping the retrieval control field. A high degree of overlap between the retrieved and target pulse shapes was obtained (overlap integral $J^2 = 0.98 - 0.99$) for all input and target pulse shapes tested in the experiments. Demonstrated the potential application of the presented technique to the creation of optical time-bin qubits and to controlled partial retrieval. Finally, observed excellent agreement between our experimental results and theoretical modeling.

The optimal storage and pulse-shape control presented here are applicable to a wide range of experiments, since the underlying theory applies to other experimentally relevant situations such as ensembles enclosed in a cavity, the off-resonant regime, non-adiabatic storage (i.e., storage of pulses of high bandwidth), and ensembles with inhomogeneous broadening, including Doppler broadening and line broadening in solids. Thus, expect this pulse-shape control to be indispensable for applications in both classical and quantum optical information processing.

5.10 Control field optimization for adiabatic storage

In last section showed how to perform control field optimization in the simplest possible version of the cavity model: a resonant input mode with a real envelope was stored using a control pulse with a real envelope and unlimited power into a homogeneously broadened ensemble with infinite spin-wave lifetime. In this appendix, we show how to optimize the control field in a more general model that includes the possibility of complex control field envelopes $\Omega(t)$ and input mode envelopes $\varepsilon_{in}(t)$, nonzero single-photon detuning Δ and spin wave decay rate γ_s, and (possibly reversible) inhomogeneous broadening such as Doppler broadening in gases or the broadening of optical transitions in solid state impurities. For the case of Doppler broadened gases, we also include velocity changing collisions with rate c. Also show how to take into account possible experimental restrictions on the strength of the classical control fields. The complex number equations describing the generalized model are

$$\dot{P}_j = -[\gamma + i(\Delta + \Delta_j)]P_j - \gamma C\sqrt{p_j}P + i\Omega S_j + i\sqrt{2\gamma C}\sqrt{p_j}\,\mathcal{E}_{in} + \gamma_c(\sqrt{p_j}P - P_j) \quad (5\text{-}258)$$

$$\dot{S}_j = -\gamma_s S_j + i\Omega^* P_j + \gamma_c(\sqrt{p_j}S - S_j) \quad (5\text{-}259)$$

where j labels the frequency class with detuning Δ_j from the center of the line containing a fraction pj of atoms ($\Sigma j\, p_j = 1$) and where the total optical and spin polarizations are respectively.

$$P = \sum_k \sqrt{p_k}P_k \text{ and } S = \sum_k \sqrt{p_k}S_k \quad (5\text{-}260)$$

The terms proportional to γ_c describe completely rethermalizing collisions with rate γ_c. One can, of course, also take γ_c to be different for P and S. For example, if $\gamma_c \ll \gamma$, which is often the case, one can drop the terms proportional to γ_c in Eq. (5-260). In addition to moving atoms from one frequency class to the other, collisions also result in line broadening, which can be taken into account by increasing γ. Assume that the goal is to maximize the efficiency $\gamma_s = |S(T)|^2$ of storage into the symmetric mode $S(T)$ with respect to the control pulse $\Omega(t)$ for a given input mode shape $\varepsilon_{\text{in}}(t)$ satisfying the normalization condition:

$$\int_0^T \mathrm{d}t \, |\mathcal{E}_{\text{in}}(t)|^2 = 1 \tag{5-261}$$

A procedure yields the following equations of motion for the adjoint variables:

$$\dot{\bar{P}}_j = [\gamma - i(\Delta + \Delta_j)]\bar{P}_j + \gamma C \sqrt{p_j}\bar{P} + i\Omega \bar{S}_j - \gamma_c(\sqrt{p_j}\,\bar{P} - \bar{P}_j) \tag{5-262}$$

$$\dot{\bar{S}}_j = \gamma_s \bar{S}_j + i\Omega^* \bar{P}_j - \gamma_c(\sqrt{p_j}\,\bar{S} - \bar{S}_j) \tag{5-263}$$

Where

$$\bar{P} = \sum_k \sqrt{p_k}\bar{P}_k \text{ and } \bar{S} = \sum_k \sqrt{p_k}\,\bar{S}_k \tag{5-264}$$

The corresponding initial conditions for backward propagation are

$$\bar{P}_j(T) = 0 \tag{5-265}$$

$$\bar{S}_j(T) = \sqrt{p_j}S(T) \tag{5-266}$$

After taking an initial guess for $\Omega(t)$ and solving for $P_j, S_j, \bar{P}_j$, and $\bar{S}_j$, one updates $\Omega(t)$ by moving up the gradient:

$$\Omega(t) \rightarrow \Omega(t) + \frac{1}{\lambda} i \sum_j (\bar{S}_j^* P_j - P_j \bar{S}_j^*) \tag{5-267}$$

Short input modes and/or large single-photon detuning Δ require control pulses with large intensities that might not be available in the laboratory. There exist ways to include a bound on the control field amplitude. Alternatively, one may want to consider a slightly simpler optimization problem with a limit on the control pulse energy:

$$\mu'\left(E - \int_0^T |\Omega(t)|^2 \mathrm{d}t\right) \text{ to } J \tag{5-268}$$

$$\int_0^T |\Omega(t)|^2 \mathrm{d}t \leqslant E \tag{5-269}$$

For some E. In order to carry out the optimization subject to this constraint, one should first carry out the optimization without the constraint and see whether the optimal control satisfies the constraint or not. If it does not satisfy the constraint, one has to add a term

$$\mu'\left(E - \int_0^T |\Omega(t)|^2 \mathrm{d}t\right) \text{ to } J \tag{5-270}$$

so that the update becomes

$$\Omega(t) \rightarrow \Omega(t) + \frac{1}{\lambda}\left[i \sum_j (S_j^* P_j - P_j S_j^*) - \mu' \Omega(t)\right] \tag{5-271}$$

where μ' is adjusted to satisfy the constraint. By redefining μ' and λ, this update can be simplified back to Eq. (5-271) followed by a renormalization to satisfy the constraint. Depending on how severe the energy constraint is, one can then sometimes (but not always) further simplify the update by completely replacing $\Delta(t)$ with the gradient followed by a renormalization of $\Delta(t)$, as is done, for example for the problem of laser control of chemical reactions. That these optimization protocols can be trivially extended to the full process of storage followed by retrieval, which, in the presence of inhomogeneous broadening, one might not be able to optimize by optimizing storage and retrieval separately. Similarly, one may include the possibility of reversing the inhomogeneous profile between the processes of storage and retrieval.

- Control field optimization in the free-space model

Generalization to storage followed by retrieval that showed how to use gradient accent to find the control field that maximizes the storage efficiency. However, the obtained storage control field does not always maximize the total efficiency of storage followed by retrieval. Therefore, in this appendix, we consider the maximization of the total efficiency of storage followed by retrieval with respect to the storage control field. While we demonstrate the procedure only for the case of forward retrieval, the treatment of backward retrieval is analogous. Suppose that the control field $\Omega(t)$ consists of a storage control pulse on $t\in[0,T]$ and a retrieval control pulse on $t\in[T_r, T_f]$. To optimize with respect to the former given the latter and the input mode (note that the total efficiency is independent of the retrieval control for sufficiently strong retrieval control pulses, and it is therefore often less important to optimize with respect to the retrieval control pulse). Here $0<T<T_r<T_f$, and the subscripts in T_r and T_f stand for "retrieval" and "final". The time interval $[T, T_r]$ corresponds to the waiting (i.e. storage) time between the processes of storage (which ends at $t=T$) and retrieval (which begins at $t=T_r$). Then forward retrieval that follows after the storage time interval $[T, T_r]$ but on the time interval $t\in[T_r, T_f]$ with initial and boundary conditions:

$$\varepsilon(0,\tilde{t})=0 \tag{5-272}$$

$$P(\tilde{z},\tilde{T}_r)=0 \tag{5-273}$$

$$S(\tilde{z},\tilde{T}_r)=S(\tilde{z},\tilde{T}) \tag{5-274}$$

where $\tilde{T}_r=T_r$ (similarly, $\tilde{T}_f=T_f$). Eq. (5-273) assumes that the polarization has sufficient time to decay before retrieval starts, while Eq. (5-274) assumes that spin-wave decay is negligible during the storage time. The goal is to maximize the total efficiency of storage followed by retrieval,

$$\eta_{\text{tot}}=\int_{\tilde{T}_r}^{\tilde{T}_f}\mathrm{d}\tilde{t}\ |\mathcal{E}(1,\tilde{t})|^2 \tag{5-275}$$

with respect to the storage control field. Constructing J and taking appropriate variations, we obtain initial and boundary conditions for backward propagation:

$$\bar{\mathcal{E}}(1,\tilde{t})=\mathcal{E}(1,\tilde{t})\ \text{for}\ \tilde{t}\in[\tilde{T}_r,\tilde{T}_f] \tag{5-276}$$

$$\bar{P}(\tilde{z},\tilde{T}_f)=0 \tag{5-277}$$

$$\bar{S}(\tilde{z},\tilde{T}_f)=0 \tag{5-278}$$

and

$$\bar{E}(1,\tilde{t})=0\ \text{for}\ \tilde{t}\in[0,\tilde{T}] \tag{5-279}$$

$$\bar{P}(\tilde{z},\tilde{T})=0 \tag{5-280}$$

$$\bar{S}(\tilde{z},\tilde{T})=\bar{S}(\tilde{z},\tilde{T}_r) \tag{5-281}$$

By taking the variational derivative of J on the storage interval, we find that the update is exactly the same as for the optimization of storage alone.

Note that if the retrieval control pulse leaves no atomic excitations, one can obtain the same optimization equations by solving the storage optimization problem but changing the function to be maximized from the number of spin-wave excitations

$$\int_0^1 d\tilde{z} S(\tilde{z},\widetilde{T}) S^*(\tilde{z},\widetilde{T}) \tag{5-282}$$

to the complete retrieval efficiency from $S(\tilde{z},\widetilde{T})$. It is also worth noting that the derivation presented here can trivially be extended to apply to backward (instead of forward) retrieval and to include complex Ω and ε_{in}, (possibly reversible) inhomogeneous broadening, and nonzero Δ, γ_s, and c.

- Optimization with respect to the inhomogeneous profile

The results on the optimization of photon storage with respect to the inhomogeneous broadening without providing the mathematical details. In this appendix, we present these details. The first consider the cavity model, but turn briefly to the free-space model at the end of this appendix. We suppose for simplicity that the input mode $\varepsilon_{in}(t)$ is resonant and that the storage and retrieval control pulses are π pulses at $t = T$ and $t = T_r$, respectively. In order to simplify notation, define $x_j = \sqrt{p_j}$, satisfying the normalization $\Sigma_j x_{2j} = 1$. The storage equation on the interval $t \in [0, T]$ then becomes:

$$\dot{P}_j = -(\gamma + i\Delta_j) P_j - \gamma C x_j P + i\sqrt{2\gamma C} x_j \mathcal{E}_{in} \tag{5-283}$$

with $P = \Sigma_k X_k P_k$ and with the initial condition $P_j(0) = 0$. A π-pulse at $t = T$ mapping P onto S followed by another π-pulse at $t = T_r$ mapping S back onto P result in an overall 2π pulse, so that $P_j(T_r) = -P_j(T)$. Assuming the broadening is reversed at some time between T and T_r, the equations for retrieval on the interval $t \in [T_r, T_f]$ are

$$\dot{P}_j = -(\gamma - i\Delta_j) P_j - \gamma C x_j P \tag{5-284}$$

The total efficiency of storage followed by retrieval is then

$$\eta_{tot} = \int_{T_r}^{T_f} dt \mid \mathcal{E}_{out}(t) \mid^2 = \int_{T_r}^{T_f} dt \mid i\sqrt{2\gamma C} P(t) \mid^2 \tag{5-285}$$

One can show that the equations of motion for the adjoint variables (i. e. the Lagrange multipliers) $\bar{P}_j$ are

$$\dot{\bar{P}}_j = (\gamma + i\Delta_j) \bar{P}_j + \gamma C x_j \bar{P} - 2\gamma C x_j P \tag{5-286}$$

for $t \in [T_r, T_f]$ with $\bar{P}_j(T_f) = 0$ and

$$\dot{\bar{P}}_j = (\gamma - i\Delta_j) \bar{P}_j + \gamma C x_j \bar{P} \tag{5-287}$$

for $t \in [0, T]$ with $\bar{P}_j(T) = -\bar{P}_j(T_r)$, where we defined $\bar{P} = \Sigma_k x_k \bar{P}_k$. The last term in Eq. (5-286) describes an incoming field that is the time-reverse of the retrieved field. Assuming we are optimizing with respect to x_j, the update is

$$x_j \to x_j + \frac{1}{\lambda} A_j \tag{5-288}$$

followed by a rescaling of all x_j by a common factor to ensure the normalization $\Sigma_j x_{2j} = 1$. Here A_j is given by

$$A_j = -\gamma C \mathrm{Re}\left[\int_0^T dt + \int_{T_r}^{T_f} dt\right] (\bar{P}_j^* P + \bar{P}^* P_j) - \sqrt{2\gamma C}\,\mathrm{Im} \int_0^T dt\, \mathcal{E}_{in} \bar{P}_j^* + 2\gamma C \mathrm{Re} \int_{T_r}^{T_f} dt P_j^* P \tag{5-289}$$

where Re denotes the real part. Numerics show that the update can usually be simplified in a way that avoids the search for convenient values of λ and does not lose convergence. Specifically, taking $\lambda \to 0$ in Eq. (5-288), can obtain:

$$x_j \to A_j \tag{5-290}$$

followed by renormalization. By defining a particular functional form for the dependence of x_j on Δ_j, one could also consider optimization with respect to only a few parameters, such as, for example, the width Δ_I and the degree of localization n of the inhomogeneous profile of the form

$$p_j \propto 1/[1+(\Delta_j/\Delta_I)n] \tag{5-291}$$

Equivalently, instead of optimizing with respect to x_j, one can optimize with respect to Δ_j. To illustrate this procedure, we consider a simple optimization procedure with respect to a single parameter, the inhomogeneous width Δ_I. We write $\Delta_j = \Delta_I f_j$ for some fixed dimensionless parameters f_j and consider maximizing the efficiency with respect to Δ_I for fixed x_j and f_j. The equations of motion and the initial conditions for both P_j and $\bar{P}_j$ stay the same as in the optimization with respect to x_j while the update becomes

$$\Delta_I \to \Delta_I + \frac{1}{\lambda}\mathrm{Im}\sum_j\left[\int_0^T \mathrm{d}t - \int_{T_r}^{T_f}\mathrm{d}t\right]\bar{P}_j^* f_j P_j \tag{5-292}$$

By adjusting f_j and x_j, one can choose a particular inhomogeneous profile shape (e.g. Lorentzian, Gaussian, or a square) and optimize with respect to its width. Having discussed the cavity case, now list the corresponding free-space results. In free space, the update of x_j via Eq. (5-291) would use

$$A_j = -\sqrt{d}\,\mathrm{Im}\int_0^1 \mathrm{d}\tilde{z}\left[\int_0^{\tilde{T}}\mathrm{d}\tilde{t} + \int_{\tilde{T}_r}^{\tilde{T}_f}\mathrm{d}\tilde{t}\right](\bar{P}_j^*\,\mathcal{E} + \bar{\mathcal{E}}^*\,P_j) \tag{5-293}$$

Similarly, the update of $\tilde{\Delta}_I = \Delta_I$ would be

$$\tilde{\Delta}_I \to \tilde{\Delta}_I + \frac{1}{\lambda}\mathrm{Im}\sum_j\int_0^1 \mathrm{d}\tilde{z}\left[\int_0^{\tilde{T}}\mathrm{d}\tilde{t} - \int_{\tilde{T}_r}^{\tilde{T}_f}\mathrm{d}\tilde{t}\right]\bar{P}_j^* f_j P_j \tag{5-294}$$

- Details of the model and derivation of the equations of motion

Since the model and the assumptions made are very similar to those presented in the cavity case and review some of them only briefly. The electric field vector operator for the quantum field is given by

$$\hat{E}_1(z) = \varepsilon_1\left(\frac{\hbar\omega_1}{4\pi c\varepsilon_0 A}\right)^{1/2}\int \mathrm{d}\omega \hat{a}_\omega \mathrm{e}^{\mathrm{i}\omega z/c} + \mathrm{h.c.} \tag{5-295}$$

where h.c. stands for Hermitian conjugate and where have a continuum of annihilation operators $\hat{a}_\omega$ for the field modes of different frequencies ω that satisfy the commutation relation $\hat{a}_\omega, \hat{a}^\dagger$

$$[\hat{a}_\omega, \hat{a}_{\omega'}^\dagger] = \delta(\omega - \omega') \tag{5-296}$$

By assumption, the field modes corresponding to $\hat{a}_\omega$ for different ω have the same transverse profile and are nonempty only around $\omega = \omega_1$. In typical experiments, the beam is smaller than the size of the ensemble, and in this case the relevant number of atoms N should only be the number of atoms interacting with the beam. However, the only relevant quantity is the optical depth d, which does not depend on the area A, so that when everything is expressed in terms of d, the precise definition of N and A is irrelevant. The copropagating classical control field vector $E_2(z,t) = Q_2 E_2(t - z/c)\cos(\omega_2(t - z/c))$ is a plane wave with polarization unit vector Q_2 and carrier frequency ω_2 modulated by an envelope $E_2(t - z/c)$, which we assume to be propagating with group velocity equal to the speed of light c since almost all the atoms are assumed to be in the ground state $|g\rangle$ and are, thus, unable to significantly alter the propagation of a strong classical field coupled to the $|s\rangle - |e\rangle$ transition. The Hamiltonian is then modified to

$$\hat{H} = \hat{H}_0 + \hat{V} \tag{5-297}$$

$$\hat{H}_0 = \int d\omega\, \hbar\omega \hat{a}_\omega^\dagger \hat{a}_\omega + \sum_{i=1}^{N} (\hbar\omega_{se}\hat{\sigma}_{ss}^i + \hbar\omega_{ge}\hat{\sigma}_{ee}^i) \tag{5-298}$$

$$\hat{V} = -\hbar\sum_{i=1}^{N}\left(\Omega(t - z_i/c)\hat{\sigma}_{es}^i e^{-i\omega_2(t-z_i/c)+g}\sqrt{\frac{L}{2\pi c}}\int d\omega \hat{a}_\omega e^{i\omega z_i/c}\hat{\sigma}_{eg}^i + \text{h.c}\right) \tag{5-299}$$

Note that in order to avoid carrying extra factors of 2 around, is defined as half of the traditional definition of the Rabi frequency, so that a π pulse, for example, takes time $\pi/2$. Since the position dependence along the ensemble matters, divide the ensemble into thin slices along the length L of the ensemble ($z=0$ to $z=L$) and introduce slowly varying operators

$$\hat{\sigma}_{\mu\mu}(z,t) = \frac{1}{N_z}\sum_{i=1}^{N_z}\hat{\sigma}_{\mu\mu}^i(t) \tag{5-300}$$

$$\hat{\sigma}_{es}(z,t) = \frac{1}{N_z}\sum_{i=1}^{N_z}\hat{\sigma}_{es}^i(t)e^{-i\omega_2(t-z_i/c)} \tag{5-301}$$

$$\hat{\sigma}_{eg}(z,t) = \frac{1}{N_z}\sum_{i=1}^{N_z}\hat{\sigma}_{eg}^i(t)e^{-i\omega_1(t-z_i/c)} \tag{5-302}$$

$$\hat{\sigma}_{sg}(z,t) = \frac{1}{N_z}\sum_{i=1}^{N_z}\hat{\sigma}_{sg}^i(t)e^{-i(\omega_1-\omega_2)(t-z_i/c)} \tag{5-303}$$

$$\hat{\varepsilon}(z,t) = \sqrt{\frac{L}{2\pi c}}e^{i\omega_1(t-z/c)}\int d\omega \hat{a}_\omega(t)e^{i\omega z/c} \tag{5-304}$$

where sums are over all N_z atoms in a slice of atoms positioned at z that is thick enough to contain $N_z \gg 1$ atoms but thin enough that the resulting collective fields can be considered continuous. For these slowly varying operators, the effective Hamiltonian is

$$\hat{H} = \int d\omega\, \hbar\omega\hat{a}_\omega^\dagger\hat{a}_\omega - \hbar\omega_1\frac{1}{L}\int_0^L dz\, \mathcal{E}^\dagger(z,t)\mathcal{E}(z,t) + \int_0^L dz\, \hbar n(z)[\Delta\hat{\sigma}_{ee}(z,t) - (\Omega(t-z/c)\hat{\sigma}_{es}(z,t) + g\hat{\mathcal{E}}(z,t)\hat{\sigma}_{eg}(z,t) + \text{h.c.})] \tag{5-305}$$

and the same-time commutation relations are

$$[\hat{\sigma}_{\mu\nu}(z,t), \hat{\sigma}_{\alpha\beta}(z',t)] = \frac{1}{n(z)}(\delta_{\nu\alpha}\hat{\sigma}_{\mu\beta}(z,t) - \delta_{\mu\beta}\hat{\sigma}_{\alpha\nu}(z,t)) \times \delta(z-z') \tag{5-306}$$

$$[\hat{\mathcal{E}}(z,t), \hat{\mathcal{E}}^\dagger(z',t)] = L\delta(z-z') \tag{5-307}$$

From the generalized Einstein relations, the only nonzero noise correlations are

$$\langle\hat{F}_P(z,t)\hat{F}_P^\dagger(z',t')\rangle = \frac{N}{n(z)}\delta(z-z')\delta(t-t') \tag{5-308}$$

$$\langle\hat{F}_S(z,t)\hat{F}_S^\dagger(z',t')\rangle = \frac{N}{n(z)}\delta(z-z')\delta(t-t') \tag{5-309}$$

The property of our system that guarantees that the incoming noise is vacuum is the absence of decay out of state $|g\rangle$ into states $|e\rangle$ and $|s\rangle$. Under the assumption that almost all atoms are in the ground state at all times, commutation relations (5-306) imply

$$[\hat{S}(z,t), \hat{S}^\dagger(z',t)] = \frac{N}{n(z)}\delta(z-z') \tag{5-310}$$

$$[\hat{P}(z,t), \hat{P}^\dagger(z',t)] = \frac{N}{n(z)}\delta(z-z') \tag{5-311}$$

Equation (5-310) allows us to expand $\hat{S}(z,t)$ in terms of any basis set of spatial modes $\{g_\alpha(z)\}$ satisfying the orthonormality relation and the completeness relation as

$$\hat{S}(z,t) = \sqrt{\frac{N}{n(z)}} \sum_\alpha g_\alpha(z)\hat{c}_\alpha(t) \tag{5-312}$$

where the annihilation operators $\{\hat{c}_\alpha\}$ for the spin-wave modes satisfy

$$[\hat{c}_\alpha(t), \hat{c}_\beta^\dagger(t)] = \delta_{\alpha\beta} \tag{5-313}$$

For the freely propagating input field $\hat{\varepsilon}_{\text{in}}(t) = \hat{\varepsilon}(0,t)$ and output field $\hat{\varepsilon}_{\text{out}}(t) = \hat{\varepsilon}(L,t)$ we have the following commutation relations:

$$[\hat{\mathcal{E}}_{\text{in}}(t), \hat{\mathcal{E}}_{\text{in}}^\dagger(t')] = \frac{L}{c}\delta(t-t') \tag{5-314}$$

$$[\hat{\mathcal{E}}_{\text{out}}(t), \hat{\mathcal{E}}_{\text{out}}^\dagger(t')] = \frac{L}{c}\delta(t-t') \tag{5-315}$$

which differ from their cavity case counterparts in Eq. (5-312) only in normalization. These commutation relations allow us to expand, the input and the output field in terms of any basis set of field (envelope) modes $\{h_\alpha(t)\}$ defined for $t \in [0, \infty)$, satisfying the orthonormality relation

$$\int_0^\infty dt h_\alpha^*(t) h_\beta(t) = \delta_{\alpha\beta} \tag{5-316}$$

and the completeness relation

$$\sum_\alpha h_\alpha^*(t) h_\alpha(t') = \delta(t-t') \tag{5-317}$$

as

$$\hat{\mathcal{E}}_{\text{in}}(t) = \sqrt{\frac{L}{c}} \sum_\alpha h_\alpha(t)\hat{a}_\alpha \tag{5-318}$$

$$\hat{\mathcal{E}}_{\text{out}}(t) = \sqrt{\frac{L}{c}} \sum_\alpha h_\alpha(t)\hat{b}_\alpha \tag{5-319}$$

where annihilation operators $\{\hat{a}_\alpha\}$ and $\{\hat{b}_\alpha\}$ for the input and the output photon modes, respectively, satisfy the usual bosonic commutation relations.

All atoms are initially pumped into the ground state, i.e., no $\hat{P}$ or $\hat{S}$ excitations are present in the atoms. Also assume that the only input field excitations initially present are in the quantum field mode with annihilation operator $\hat{a}_0$ corresponding to an envelope shape $h_0(t)$ nonzero on $[0, T]$.

- Position dependence of loss

In this section, show that for complete retrieval, not only the total efficiency but also the distribution of spontaneous emission loss (or more precisely loss due to polarization decay γ) as a function of position is independent of the control and the detuning. Equations of motion imply that

$$\partial_{\tilde{z}} |\mathcal{E}(\tilde{z},\tilde{t})|^2 + \partial_{\tilde{t}} |P(\tilde{z},\tilde{t})|^2 + \partial_{\tilde{t}} |S(\tilde{z},\tilde{t})|^2 = -2|P(\tilde{z},\tilde{t})|^2 \tag{5-320}$$

Integrating both sides with respect to $\tilde{z}$ from 0 to 1 and with respect to $\tilde{t}$ from $\tilde{T}_r$ to ∞, using the initial conditions $S(\tilde{z}, \tilde{T}_r) = S(\tilde{z})$ $\left(\text{where} \int_0^1 d\tilde{z} \, | S(\tilde{z}) | 2 = 1\right)$ and $P(\tilde{z}, \tilde{T}_r) = 0$, the boundary condition $E(0, \tilde{t}) = 0$, and the complete retrieval condition $S(\tilde{z}, \infty) = P(\tilde{z}, \infty) = 0$, can find that

$$\eta_r = 1 - \int_0^1 d\tilde{z} l(\tilde{z}) \tag{5-321}$$

where the position-dependent loss per unit length is

$$l(\tilde{z}) = 2\int_0^{\infty} \mathrm{d}\tilde{t} \mid P(\tilde{z},\tilde{t}) \mid^2 \tag{5-322}$$

Computing $l(\tilde{z})$, can find

$$l(\tilde{z}) = 2\,\mathcal{L}^{-1}\left\{\left[\int_{\tilde{T}_r}^{\infty} \mathrm{d}\tilde{t}P(u,\tilde{t})[P(u'^*,\tilde{t})]^*\right\}_{u,u'\to\tilde{z}} = \mathcal{L}^{-1}\left\{\frac{2}{2+\frac{d}{u}+\frac{d}{u'}}S(u)[S(u'^*)]^*\right\}_{u,u'\to\tilde{z}} \tag{5-323}$$

where L^{-1} with subscript $u, u'\to\tilde{z}$ means that inverse Laplace transforms are taken with respect to u and u' and are both evaluated at $\tilde{z}$. Therefore, can see that $l(\tilde{z})$ is independent of the detuning and the control. Moreover, the inverse Laplace transforms L^{-1} can be taken analytically to give

$$\begin{aligned} l(\tilde{z}) = {} & | S(\tilde{z}) |^2 - \mathrm{Re}\left[S(\tilde{z})d\int_0^{\tilde{z}} \mathrm{d}\tilde{z}'S^*(\tilde{z}-\tilde{z}')\mathrm{e}^{-\frac{d\tilde{z}'}{2}}\right] + \\ & \int_0^{\tilde{z}} \mathrm{d}\tilde{z}'\int_0^{z} \mathrm{d}\tilde{z}''S(\tilde{z}-\tilde{z}')S^*(\tilde{z}-\tilde{z}'')\frac{d^2}{4}\mathrm{e}^{-\frac{d}{2}(\tilde{z}'+\tilde{z}'')} \times \\ & \left[2I_0(d\sqrt{\tilde{z}'\tilde{z}''}) - \frac{\tilde{z}'+\tilde{z}''}{\sqrt{\tilde{z}'\tilde{z}''}}I_1(d\sqrt{\tilde{z}'\tilde{z}''})\right] \end{aligned} \tag{5-324}$$

5.11 Analysis of photon number in quantum memory

The ability to detect single photons with a high efficiency is a crucial requirement for various quantum information applications. By combining the storage process of a quantum memory for photons with fluorescence-based quantum state measurement, it is, in principle, possible to achieve high-efficiency photon counting in large ensembles of atoms. The large number of atoms can, however, pose significant problems in terms of noise stemming from imperfect initial state preparation and off-resonant fluorescence. Identify and analyse a concrete implementation of a photon number resolving detector based on an ion Coulomb crystal inside a moderately high-finesse optical cavity. The cavity enhancement leads to an effective optical depth of 15 for a finesse of 3000 with only about 1500 ions interacting with the light field. These values allow for essentially noiseless detection with an efficiency larger than 93%. Moderate experimental parameters allow for repetition rates of about 3kHz limited by the time needed for fluorescence collection and re-cooling of the ions between trials. The analysis may lead to the first implementation of a photon number resolving detector in atomic ensembles.

5.11.1 Quantum memory for light

The information carrier of today's communications, a weak pulse of light, is an intrinsically quantum object. As a consequence, complete information about the pulse cannot, even in principle, be perfectly recorded in a classical memory. In the field of quantum information this has led to a long standing challenge: how to achieve a high-fidelity transfer of an independently prepared quantum state of light onto the atomic quantum state. Here propose and experimentally demonstrate a quantum memory based on atomic ensembles to demonstrate for a recording of an externally provided quantum state of light onto the atomic quantum memory with a fidelity up to 70%.

Quantum storage of light is achieved in three steps: an interaction of light with atoms, the subsequent measurement on the transmitted light, and the feedback onto the atoms conditioned on the measurement result. Density of recorded states 33% higher than that for the best classical recording of light on atoms is achieved. A quantum memory lifetime of up to 4ms is demonstrated.

Light is a natural carrier of information in both classical and quantum communications. In classical communications, bits are encoded in large average amplitudes of light pulses which are detected, converted into electric signals, and subsequently stored as charges or magnetization of memory cells. In quantum information processing, information is encoded in quantum states that cannot be accurately recorded by such classical means. Consider a state of light defined by its amplitude and phase, or equivalently by two quadrature phase operators, $\hat{X}_L$ and $\hat{P}_L$, with the canonical commutation relation $[\hat{X}_L \hat{P}_L] = i$. These variables play the same role in quantum mechanics as the classical quadratures X, P do in the decomposition of the electric field of light with the frequency ω as $E \propto X\cos(\omega t) + P\sin(\omega t)$. Other quantum properties of light, such as the photon number

$$\hat{n} = \frac{1}{2}(\hat{X}_L^2 + \hat{P}_L^2 - 1) \tag{5-325}$$

can be expressed in terms of $\hat{X}_L$ and $\hat{P}_L$. The best classical approach to recording a state of light onto atoms would involve homodyne measurements of both observables $\hat{X}_L$ and $\hat{P}_L$ by using, e.g., a beam splitter. The non-commutativity of $\hat{X}_L$ and $\hat{P}_L$ leads to additional quantum noise added during this procedure. The target atomic state has its intrinsic quantum noise (coming from the Heisenberg uncertainty relations). All this extra noise leads to a limited fidelity for the classical recording, e. g., to a maximum fidelity of 50% for coherent states. Thus the challenge of implementing a quantum memory can be formulated as a faithful storing of the simultaneously immeasurable values of $\hat{X}_L$ and $\hat{P}_L$.

A number of quantum information protocols, such as eavesdropping in quantum cryptography, quantum repeaters, and linear optics quantum computing would benefit from a memory meeting the following criteria:

(1) The light pulse to be stored is sent by a third party in a state unknown to the memory party.

(2) The state of light is converted into a quantum state of the memory with a fidelity higher than that of the classical recording.

Several recent experiments have demonstrated entanglement of light and atoms. However, none of these experiments demonstrated the memory obeying the two above criteria, where squeezed light was mapped onto atoms, the atomic state existed only while the light was on, so it was not a memory device. The electromagnetically induced transparency (EIT) approach has led to the demonstration of a classical memory for light. A theoretical proposal for EIT-based quantum memory for light has been published. Other proposals for quantum memory for light with better-than-classical quality of recording have also been published recently.

Quantum state transfer from one species to another is most simply presented if both systems are described by canonical quantum variables $\hat{X}$, $\hat{P}$. All canonical variables have the same commutation relations and the same quantum noise for a given state, providing thus a common frame for the analysis of the state transfer.

In the present work the state of light is stored in the superposition of magnetic sublevels of the ground state of an atomic ensemble. As introduce the operator $\hat{J}$ of the collective magnetic moment (orientation) of a ground state F. All atomic states utilized here are not too far in phase space from the coherent spin state (CSS) for which only one projection has a non-zero mean value, e.g., $\langle \hat{J}_x \rangle = J_x$ whereas the other two projection have minimal quantum uncertainties, $\langle \delta J_y^2 \rangle = \langle \delta J_z^2 \rangle = J_x/2$. For all such states the commutator $[\hat{J}_y, \hat{J}_z] = iJ_x$ can be reduced to the canonical commutator

$$[\hat{X}_A, \hat{P}_A] = i \text{ with } \hat{X}_A = \hat{J}_y/\sqrt{J_x},\quad \hat{P}_A = \hat{J}_z/\sqrt{J_x} \tag{5-326}$$

Hence the y, z-components of the collective atomic angular momentum play the role of canonical variables. Although the memory protocol, in principle, can work with a single atomic ensemble, experimental technical noise is substantially reduced. Combined canonical variables for two ensembles

$$\begin{cases}\hat{X}_A = (\hat{J}_{y1} - \hat{J}_{y2})/\sqrt{2J_x} \\ \hat{P}_A = (\hat{J}_{z1} + \hat{J}_{z2})/\sqrt{2J_x}\end{cases} \tag{5-327}$$

are then introduced where

$$\hat{J}_{x1} = -\hat{J}_{x2} = J_x = FN_{\text{atoms}} \tag{5-328}$$

In the presence of the memory couples to the W-sidebands of light:

$$\hat{X}_L = \frac{1}{\sqrt{T}}\int_0^T (\hat{a}^+(t) + \hat{a}(t))\cos(\Omega t)\,dt,\quad \hat{P}_L = \frac{i}{\sqrt{T}}\int_0^T (\hat{a}^+(t) - \hat{a}(t))\cos(\Omega t)\,dt \tag{5-329}$$

where Ω is the Larmor frequency of spin precession. Quantum storage of light is achieved in three steps: ① an interaction of light with atoms; ② a subsequent measurement of the transmitted light; ③ Feedback onto the atoms conditioned on the measurement result (see Figure 5-94). The off-resonant interaction of light with spin polarized atomic ensembles has been described elsewhere and is summarized in the Methods section. The interaction leads to the equations

$$\begin{cases}\hat{X}_L^{\text{out}} = \hat{X}_L^{\text{in}} + k\hat{P}_A^{\text{in}},\quad \hat{P}_L^{\text{out}} = \hat{P}_L^{\text{in}} \\ \hat{X}_A^{\text{out}} = \hat{X}_A^{\text{in}} + k\hat{P}_L^{\text{in}},\quad \hat{P}_A^{\text{out}} = \hat{P}_A^{\text{in}}\end{cases} \tag{5-330}$$

These equations imply that light and atoms get entangled. The remarkable simplicity of Fq. (5-330) provides a direct link between an input light state, an atomic state, and the output light. Suppose the input light is in a vacuum (or in a coherent) state and atoms are in a CSS with mean values

$$\langle \hat{X}_L \rangle = \langle \hat{X}_A \rangle = \langle \hat{P}_L \rangle = \langle \hat{P}_A \rangle = 0 \tag{5-331}$$

and variances

$$\delta X_L^2 = \delta X_A^2 = \delta P_L^2 = \delta P_A^2 = 1/2 \tag{5-332}$$

The interaction parameter k whose value is crucial for the storage protocol is then readily found as

$$k^2 = 2(\delta X_x^2)^2 - 1$$

For a perfect fidelity of mapping, the initial atomic state must be an entangled spin state with $\delta X_x^2 \to 0$. The pulse to be recorded, combined with the entangling pulse, is sent through, and its

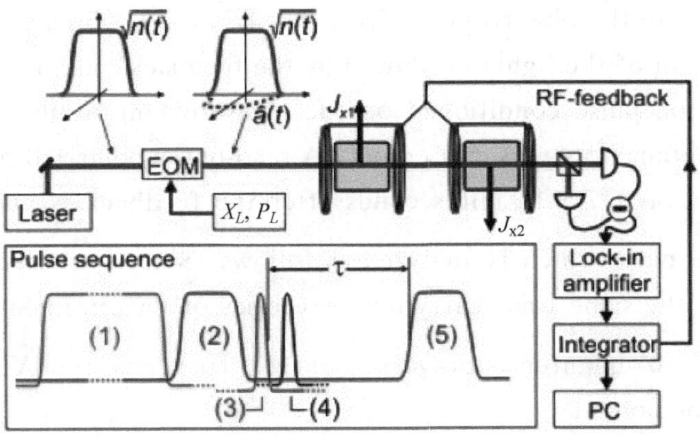

Fig. 5-94 Experimental setup. Atomic memory unit consisting of two Cesium cells inside magnetic shields 1, 2. The path of the recorded and read-out light pulses is shown with arrows. The simplified layout of the experiment. The input state of light with the desired displacements X_L, P_L is generated with the electro-optic modulator (EOM). The inset shows the pulse sequence for the quantum memory recording and read-out. Pulse (1) is the optical pumping (4ms), pulse (2) is the input light pulse $\hat{a}(t)$ overlapped with the strong entangling pulse in orthogonal polarization with the amplitude $\sqrt{n(t)}$. Pulse (3) is the magnetic feedback pulse. Pulse 4 is the magnetic $\pi/2$ pulse used for the read out of one of the atomic operators. Pulse (5) is the read-out optical pulse.

variable $\hat{X}_L^{out}$ is measured. The measurement outcome, $x = \hat{X}_L^{in} + k\hat{P}_A^{in}$. is fed back into the atomic variable $\hat{P}_A$ with a feedback gain g. The result is

$$\hat{P}_A^{mem} = \hat{P}_A^{in} - gx = \hat{P}_A^{in}(1 - kg) - g\hat{X}_L^{in} \tag{5-333}$$

With $g = k = 1$, the mapping of in $\hat{X}_L^{in}$ onto $-\hat{P}_A^{mem}$ is perfect.

The second operator of light is already mapped onto atoms via $\hat{X}_A^{mem} = \hat{X}_A^{in} + \hat{P}_L^{in}$ see Eq. (5-330). For the entangled initial state the mapping is perfect for this component too, $\hat{P}_L^{in} \to \hat{X}_A^{mem}$, leading to the fidelity of the light-to-atoms state transfer $F \to 100\%$. If the initial atomic state is a CSS the mapping is not perfect due to the noisy operator $\hat{X}_A^{in}$. However, $F = 82\%$, still markedly higher than the classical limit, can be achieved. Note that the above discussion holds for an arbitrary single mode input quantum state of light. In the experiment the atomic storage unit consists of two samples of Cesium vapor placed in paraffin coated glass cells placed inside magnetic shields (see Figure 5-94). H is applied along the x-direction with $\Omega = 322$kHz. Optical pumping along H initializes the atoms in the first/second sample in the $F = 4$, $F = \pm 4$m ground state with the orientation above 99%. Hence

$$\hat{J}_{x1} = -\hat{J}_{x2} = J_x = 4N_{atoms} \approx 1.2 \times 10^{12} \tag{5-334}$$

thoroughly check and regularly verify that the initial spin state is close to CSS. The coupling parameter k is varied by adjusting the density of C_s. The input state $\hat{a}(t)$ is encoded in a 1-msec y-polarized pulse. The state is chosen from the set $\{\hat{a}_{input}\}$ of coherent states with the photon number in the range $\{\langle n \rangle = 0, n_{max}\}$ and an arbitrary phase. $\hat{a}(t)$ is generated as W-sidebands by an electro-optical modulator (EOM) and has the same spatial and temporal profile as the strong entangling field. Thus the EOM plays the third party, providing the field to be stored. The pulses are detuned by 700MHz to the blue from the $^6S_{1/2}, F=4, ^6P_{3/2}, F=5$ transition ($\lambda = 852$nm). The polarization measurement of the light is followed by the feedback onto atoms achieved by a 0.2ms radio-frequency magnetic pulse conditioned on the measurement result. Next the experimental verification of the quantum storage is carried out. A read-out expolarized pulse is sent through the samples with the delay of 0.7～10 milliseconds after the feedback is applied. Atomic memory generates a y-polarized pulse which is analyzed as follows. Since both mem $\hat{X}_A^{mem}$ and mem $\hat{P}_A^{mem}$ cannot be measured at the same time, carry out two series of measurements for each input state. Each series consists of 10^4 quantum storage sequences. To verify the $\hat{X}_L^{in} \to -\hat{P}_A^{mem}$ step of the storage, measure the component

$$\hat{X}_L^{read\text{-}out} = \hat{X}_L^{read\text{-}in} + k\hat{P}_A^{mem} \tag{5-335}$$

of the read-out pulse (X_L is a Stokes parameter measured in units of shot noise as discussed in the Methods). For this series

$$\langle \hat{P}_L^{in} \rangle = -4 \text{ and } \langle \hat{X}_L^{in} \rangle = 0 \tag{5-336}$$

corresponding to $\langle \hat{n} \rangle = 8$ photons in the pulse. From this measurement can find the mean

$$\langle \hat{P}_A^{mem} \rangle = \frac{1}{k} \langle X_L^{read\text{-}out} \rangle \tag{5-337}$$

and the variance

$$\sigma_p^2 = \langle (\delta \hat{P}_A^{mem})^2 \rangle = \frac{1}{k^2}\left((\delta X_L^{read\text{-}out})^2 - \frac{1}{2} \right) \tag{5-338}$$

for the quantum state of the memory. Only the knowledge of k and the shot noise level of light is necessary for the determination of the mean values and variances of the atomic canonical variables from the experimental data. Next run another series of storage with the same input state for the verification of the step $\hat{P}_L \to \hat{X}_A^{mem}$.

The $\hat{X}_A^{mem}$ operator does not couple to the read-out pulse in the geometry; therefore, we first apply a $\pi/2$-pulse to atoms converting $\hat{X}_A^{mem} \to \hat{P}_L^{in}$ and then measure $\hat{P}_L^{in}$ with the verifying pulse, then find $\langle \hat{X}_A^{mem} \rangle$ and

$$\sigma_x^2 = \langle (\delta \hat{X}_A^{mem})^2 \rangle \tag{5-339}$$

of the memory state.

The above sequence is repeated for different input states. From

$$\langle \hat{P}_A^{men} \rangle / \langle \hat{X}_L^{in} \rangle \tag{5-340}$$

and

$$\langle \hat{X}_A^{mem} \rangle / \langle \hat{P}_L^{in} \rangle$$

the mapping gains for the two quadratures are determined. For the experimental data presented in Figure 5-95 and Figure 5-96(a), these gains are 0.80 and 0.84 respectively, which is close to the optimal gain for the chosen input set of states. This step would complete the proof of the classical memory performance, because have shown that the *y*-polarized pulse recovered from the memory has the same mean amplitude and mean phase as the input pulse (up to a chosen constant factor).

To prove a quantum memory performance we need in addition to consider the quantum noise

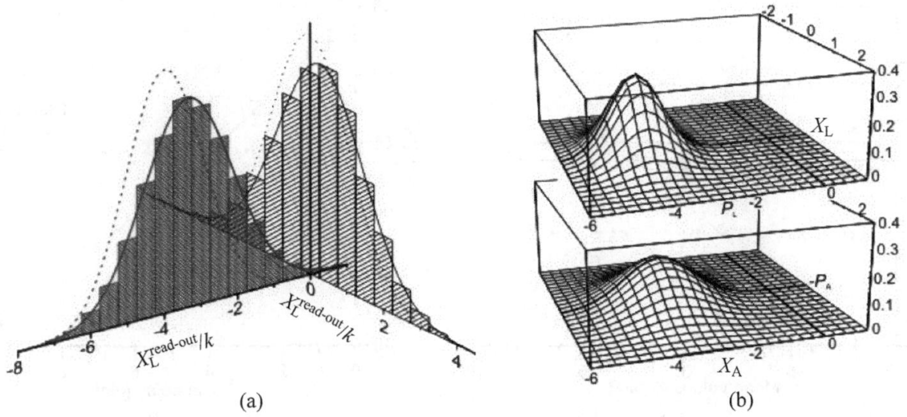

Fig. 5-95 An example of the atomic memory performance. (a) The input state of light in the coherent state with $\langle \hat{X}_L \rangle = 0, \langle \hat{P}_L \rangle = -4$. The results of the read out of this state stored in the atomic memory are shown as histograms of experimental realizations. The left/right histogram shows the results for the $\hat{X}_A / \hat{P}_A$ quadrature read out with/without the $\pi/2$-pulse. Dotted Gaussians represent the distributions for the best possible quantum memory performance (fidelity 100%). (b) The input coherent state of light (upper graph) and the reconstructed state stored in the atomic memory (lower graph) for the input state as in (a). The reconstructed state is obtained from the results presented in (a) after subtracting the noise of the read out pulse.

of the stored state. Towards this end we plot the atomic variances σ_p^2, σ_x^2 for the storage time 0.7msec in Figure 5-96(a). The experimentally obtained variances of the stored state are on average 33% below the best possible variance of the classical recording. Hence the density of stored states 33% higher than that for the best classical recording can be obtained. Thus the goal of quantum storage with less noise than for the classical recording is achieved.

Next the overlap between the input state of light and the state of the atomic memory is determined. An example is shown in Figure 5-95(b). The fidelity F of the quantum recording is then calculated for a given set $\{\hat{a}_{\text{input}}\}$. For example, $F = (66.7 \pm 1.7)\%$ for $\{\hat{a}_{\text{input}}\} = \{n = 0 \rightarrow 8\}$ and $F = (70.0 \pm 2.0)\%$ for $\{\hat{a}_{\text{input}}\} = \{n = 0 \rightarrow 4\}$, respectively for the storage time of 0.7msec. Note that the fidelity of the classical recording can exceed 50% for a limited set $\{\hat{a}_{\text{input}}\}$. The maximum classical fidelity for $\{\hat{a}_{\text{input}}\} = \{n = 0 \rightarrow 8\}$ is 55.4%, and for $\{\hat{a}_{\text{input}}\} = \{n = 0 \rightarrow 4\}$ it is 59.6%—still significantly lower than that for the quantum recording. The main sources of imperfection of our quantum memory are decoherence of the atomic state and reflection off the cell walls. Figure 5-96(b) presents the fidelity of the stored state as a function of the storage time. A simple model provides a good description for the observed fidelity reduction.

The single observable read-out described above can be useful, e. g., in quantum cryptography eavesdropping, where the memory is read after the basis has been publicly announced by Alice and Bob. The present experiment also paves the road towards the proposed quantum cloning of light onto atomic memory. However, other applications require complete state recovery via reverse mapping of the memory state onto light. Proposals for performing this task within our approach have been published. Probably the most intuitively clear protocol for the memory read-out is just to run the storage protocol of the present paper with the reversed roles of light and atoms. The read-out, as the storage, would involve three steps: sending a read-out light pulse through atoms,

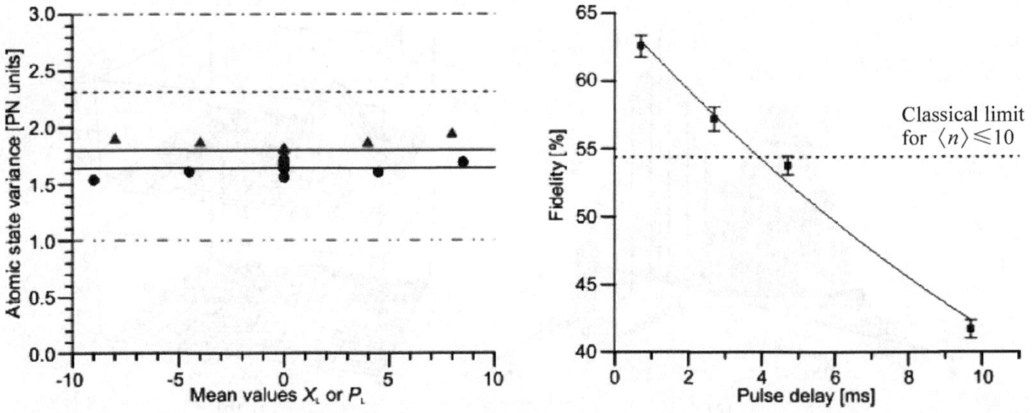

Fig. 5-96 Quantum noise of the stored state and the fidelity of quantum memory as a function of time. (a) Experimental and theoretical (quantum and classical) stored state variances in atomic projection noise (PN) units. Triages and filled circles are the experimental variances for the atomic memory operators, denoted $2\sigma_x^2$ and $2\sigma_p^2$ respectively in the text. Dash-dotted line—the fundamental boundary of three units of noise between quantum and classical mapping for an arbitrary coherent input state. Dashed line—best classical variance for the experimental set of input states with photon numbers between 0 and 8. Double-dot-dashed line—the unity variance corresponding to perfect mapping. (b) Fidelity as a function of storage time for the set of states from 0 to 10 photons. Fidelity higher than the classical limit is maintained for up to 4msec of storage. Error bars (std. dev.) include fitting uncertainty of gains and variances and an additional uncertainty of 2.5% in the projection noise calibration.

measuring the spin projection $\hat{X}_A^{out}$ with an auxiliary light pulse, and applying the feedback conditioned on this measurement to the read-out pulse.

In the present experiment we have demonstrated the memory for a subset of linearly independent coherent states. Due to the linearity of quantum mechanics this demonstration signifies that our method provides faithful mapping for an arbitrary coherent state. Since any arbitrary quantum state can be written as a superposition of coherent states, our approach should in principle work for an arbitrary quantum state, including entangled and single photon (qubit) states.

5.11.2 Methods

Quantum coupling of light to two atomic ensembles in the presence of magnetic field. The off-resonant atom/light interaction is described in terms of Stokes operators for the polarization state of light and the collective spin of atoms. The Stokes operators are defined as one half of the photon number difference between orthogonal polarization modes: S_1-between vertical x-and horizontal y-polarizations, $\hat{S}_2$-between the modes polarized at ± 450 to the vertical axis, and $\hat{S}_3$-between the left-and right-hand circular polarizations. In the experiment a strong entangling x-polarized pulse with the photon flux $n(t)$ is mixed on a polarizing beam splitter with the y-polarized quantum field $\hat{a}(t)$ prior to interaction with atoms. Hence the Stokes operators of the total optical field are

$$\hat{S}_1(t) = S_1(t) = \frac{1}{2}n(t),\quad \hat{S}_2 = \frac{1}{2}\sqrt{n(t)}(\hat{a}^+(t) + \hat{a}(t)),\quad \hat{S}_3 = \frac{i}{2}\sqrt{n(t)}(\hat{a}^+(t) - \hat{a}(t)) \tag{5-341}$$

Note that $\hat{S}_2(t)$ and $\hat{S}_3(t)$ are proportional to the canonical variables for the quantum light mode

$$\hat{X} = \frac{1}{\sqrt{2}}(\hat{a}^+(t) + \hat{a}(t)), \quad \hat{P} = \frac{i}{\sqrt{2}}(\hat{a}^+(t) - \hat{a}(t)) \tag{5-342}$$

Light is transmitted through the atomic samples placed in the bias magnetic field oriented along the x-axis. The magnetic field allows for encoding of the memory at the Larmor frequency Ω, thus dramatically reducing technical noise present at low frequencies. However, in the presence of the Larmor precession, there is an undesired coupling of the single cell variables $\hat{J}_y$ and $\hat{J}_z$ to each other. The introduction of the second cell with the opposite Larmor precession allows us to introduce new two-cell variables $(\hat{J}_{y1} - \hat{J}_{y2})$, $(\hat{J}_{z1} + \hat{J}_{z2})$ that do not couple to each other. Where a similar trick was used, the Stokes parameters of light transmitted through the two cells along the z direction become

$$\hat{S}_2^{\text{out}}(t) = \hat{S}_2^{\text{in}}(t) + aS_1(\cos(\Omega t)[\hat{J}_{z1} + \hat{J}_{z2}] + \sin(\Omega t)[\hat{J}_{y1} + \hat{J}_{y2}]), \quad \hat{S}_3^{\text{out}}(t) = \hat{S}_3^{\text{in}}(t) \tag{5-343}$$

where $\hat{J}_{z,y}$ are the projections in the frame rotating at Ω and $a = \gamma\lambda^2/8\pi\Delta A$, with γ and λ-the natural linewidth and the wavelength of the transition respectively, D-the detuning, and A-the beam cross-section. At the same time, the transverse spin components of the two cells evolve as follows:

$$\frac{\mathrm{d}}{\mathrm{d}t}[\hat{J}_{z1} + \hat{J}_{z2}] = \frac{\mathrm{d}}{\mathrm{d}t}[\hat{J}_{y1} + \hat{J}_{y2}] = 0$$

$$\frac{\mathrm{d}}{\mathrm{d}t}[\hat{J}_{y1} - \hat{J}_{y2}] = 2aJ_x\hat{S}_3^{\text{in}}\cos(\Omega t), \quad \frac{\mathrm{d}}{\mathrm{d}t}[\hat{J}_{z1} - \hat{J}_{z2}] = 2aJ_x\hat{S}_3^{\text{in}}\sin(\Omega t) \tag{5-344}$$

As evident from equation (5-344), in the process of propagation the operator $\hat{S}_3^{\text{in}}$ is recorded onto the operators $\hat{J}_{y1} - \hat{J}_{y2}$ and $\hat{J}_{z1} - \hat{J}_{z2}$ (the "back action" of light on atoms via the dynamic Stark effect caused by light), while the operators $\hat{J}_{y1} + \hat{J}_{y2}$ and $\hat{J}_{z1} + \hat{J}_{z2}$ are left unchanged. The latter are read out onto $\hat{S}_2^{\text{out}}$ via the Faraday rotation (5-343).

Canonical variables are defined for the quantum light mode as

$$\hat{X}_{\mathrm{L}} = \frac{1}{\sqrt{T}}\int_0^T(\hat{a}^+(t) + \hat{a}(t))\cos(\Omega t)\mathrm{d}t, \quad \hat{P}_{\mathrm{L}} = \frac{i}{\sqrt{T}}\int_0^T(\hat{a}^+(t) - \hat{a}(t))\cos(\Omega t)\mathrm{d}t \tag{5-345}$$

the relevant light mode involves the O-side bands. T is the pulse duration, $\hat{a}(t)$ is normalized to the photon flux. $\hat{X}_{\mathrm{L}}$ and $\hat{P}_{\mathrm{L}}$ (i.e., $\hat{S}_2$ and $\hat{S}_3$) are detected by a polarization state analyzer and by lock-in detection of the O component of the photocurrent. Note that the cos (Ωt) component of light couples to the $(\hat{J}_{y1} - \hat{J}_{y2})$, $(\hat{J}_{z1} + \hat{J}_{z2})$ components of atomic storage variables. The equivalent choice of a sin (Ωt) modulation instead would mean the use of $(\hat{J}_{y1} + \hat{J}_{y2})$, $(\hat{J}_{z1} - \hat{J}_{z2})$ for the memory. The atomic canonical variables $\hat{X}_{\mathrm{A}}$, $\hat{P}_{\mathrm{A}}$ are defined in the main section. With the above equations and definitions we straightforwardly derive equation (5-342) under the assumption $\Omega T \gg 1$. Theoretically the dimensionless coupling parameter in equation (5-342) is $k^2 = \left(a^2 J_x \int n(t)\mathrm{d}t\right)/2$.

Experimental calibration of the canonical variances for light and atoms Calculations of the fidelity, the gains, and the variances from the experimental data are based on the experimental calibration of $\langle \delta\hat{X}_{\mathrm{L}}^2 \rangle = \langle \delta\hat{P}_{\mathrm{L}}^2 \rangle$ for the coherent (vacuum) state of light and of $\langle \delta\hat{X}_{\mathrm{A}}^2 \rangle = \langle \delta\hat{P}_{\mathrm{A}}^2 \rangle$ for the coherent spin state (CSS) of atoms. The calibration for light is carried out along the established procedure of determining the shot noise level for measurements of $\hat{S}_2$, $\hat{S}_3$ with the quantum field in a vacuum state. Variances and mean values for light are then measured in units of this shot noise level. The calibration for the atomic CSS variance is carried out with extreme care and has shown excellent reproducibility. As stated in the main text, as soon as the vacuum (shot) noise level for light is established and the atoms are in a CSS, the parameter k^2, important for calculations of atomic variances and fidelity, is easily determined as $k^2 = 2(\delta X_{\mathrm{L}}^{\mathrm{out}})^2$. In the experiment this is equivalent to

$$k^2 = \left((\delta S_2^{\mathrm{out}})^2 - (\delta S_2^{\mathrm{in}})^2\right)/(\delta S_2^{\mathrm{in}})^2 \tag{5-346}$$

Fidelity and the state overlap to calculate the fidelity of the transfer of an input coherent state into an output Gaussian state, first define an overlap function between an input state with mean values x_1, p_1 and the output state with the mean values and variances x_2, p_2, σ_x^2, σ_p^2. Straightforward integration yields

$$O\{x_1, x_2, p_1, p_2\} = 2\exp(-(x_1 - x_2)^2/(1 + 2\sigma_x^2) - (p_1 - p_2)^2/(1 + 2\sigma_p^2))/\sqrt{(1 + 2\sigma_x^2)(1 + 2\sigma_p^2)} \tag{5-347}$$

The fidelity of the transfer for a set of coherent states with mean amplitudes between α_1 and α_2 can then be found as an average overlap

$$F = \pi^{-1}(\alpha_2^2 - \alpha_1^2)^{-1}\int_0^{2\pi} \mathrm{d}\phi \int_{\alpha_1}^{\alpha_2} O\{\alpha\}\alpha \mathrm{d}\alpha \tag{5-348}$$

For classical recording from light onto atoms with the gain g, the overlap between the input coherent state with the mean amplitude $\alpha = \sqrt{(x^2 + p^2)}$ and the output state is given by

$$O\{\alpha\} = (1 + g^2)^{-1}\exp\left(-\frac{1}{2}(1 - g)^2\alpha^2(1 + g^2)^{-1}\right) \tag{5-349}$$

The classical fidelity is then given by

$$F_{\mathrm{class}} = (n_2 - n_1)^{-1}(1 - g)^{-2}\{\exp(-(1 - g)^2 n_1(1 + g^2)^{-1}) - \exp(-(1 - g)^2 n_2(1 + g^2)^{-1})\} \tag{5-350}$$

where have introduced the mean photon number

$$n = \frac{1}{2}\alpha^2 \cdot F_{\mathrm{class}} \to 50\%$$

for arbitrary coherent states when $g \to 1$. If a restricted class of coherent states is chosen as the input, $F_{\mathrm{class}} > 50\%$ can be obtained with a suitable choice of g. For a set of states analyzed in the main text, $\{\hat{a}_{\mathrm{input}}\} = \{n = 0 \to 8\}$, the maximum classical fidelity of 55.4% is achieved with the gain of 0.809.

Calibration of the Atomic Projection Noise level From the measurement of the variance of the Stokes parameter out $\hat{S}_2^{\mathrm{out}}$ as a function of the macroscopic spin size J_x, can determine the contribution of the atomic projection noise to the noise of the transmitted light. The goal here is to measure the light-atoms coupling parameter k that is used for the calculation of the canonical atomic variables in the paper. It is convenient that we do not need to know explicitly the absolute value of the projection noise

$$\langle \hat{J}_{z,y}^2 \rangle = \frac{F}{2} N_{\mathrm{atoms}} \tag{5-351}$$

However, we do need to determine the projection noise contribution to the light noise. In order to determine this contribution we need to (1) extract the linear dependence of $(\hat{\delta}_2^{out})^2$ on J_x and (2) Ensure that the atoms are spin polarized to a high degree. The atomic spin noise is measured for two cells together according to the combined two-cell quantum variables introduced in the main text. As any modulation technique, this approach allows to overcome technical noise by means of lock-in detection at the modulation frequency. In our case we have been able in this way to eliminate technical noise to well below the 10^{-6} level, and thus reach the quantum projection noise limit for up to 3×10^{11} atoms. The atoms are optically pumped with a 4msec pulse preparing a fresh state before each measurement. The Stokes parameter

$$\hat{S}_2 = \frac{1}{\sqrt{2T}}\int_0^T \sqrt{n(t)}(\hat{a}^+(t) + \hat{a}(t))\cos(\Omega t)\mathrm{d}t \tag{5-352}$$

is measured by the lock-in detection. The shot noise of the incoming light $(\hat{\delta}_2^{in})^2$ is measured separately. Repeating the optical pumping and the measurement sequence many times, can obtain the variance of the operator $\hat{\delta}_2^{out}$. By measuring the Faraday rotation angle f of a linearly polarized light propagating along the x direction of the macroscopic spin polarization we obtain the value proportional to the ensemble mean spin J_x. We also determine the degree of optical pumping (spin orientation of the ground state $F=4$) by the magneto-optical resonance method20. We routinely find a degree of optical pumping better than 99%. In the Figure 5-97(a) we plot the atomic contribution to the variance of the transmitted light normalized to the shot noise level:

$$((\delta S_2^{out})^2 - (\delta S_2^{in})^2)/(\delta S_2^{in})^2 \tag{5-353}$$

as a function of Φ. The value of J_x is varied by varying the temperature of the sample. The lower part of the graph shows a nice linear dependence (solid line) which together with a nearly perfect degree of orientation proves that we observe quantum spin noise, i.e., the projection noise of the coherent spin state (CSS) (while classical noise would grow quadratically with J_x). The scattering of the points, especially at high atomic densities, arises from the technical laser noise, as proven by an independent monitoring of this noise.

The above procedure has been carried out on a regular basis to ensure that the contribution of the projection noise is reliably defined. Find that, provided the geometry, detuning, duration, and power of the light beam are carefully reproduced, the excess noise of the laser controlled, and the magnetic shielding of the atoms sufficient, the PNL contribution can be determined with a high level of confidence. As an example, in Figure 5-97(b) we show the PNL calibration 43 days after the data (a) was obtained. The solid line here is the same as in Figure 5-97(a) and it neatly coincides with a linear fit through zero of the lower half of the points. We have thus a reproducible PNL calibration. The procedure described above is quite similar to the determination of the shot noise level of polarized light, a routine well established in the studies of squeezed and entangled light (except, of course, that atoms replace photons in our case). There, similarly to the present work, as soon as light is well polarized, the linear dependence of the noise variance on the photon number (power) signifies that the coherent state noise (shot noise) level is achieved. The PNL is estimated to be stable to within 2.5% and this number is used in the text to calculate the uncertainty of the fidelity $F=(66.7\pm1.6)\%$. However, the PNL uncertainty plays only a minor role here. For example, with a 10% uncertainty in PNL we would get $F=(66.7\pm2.6)\%$. The reason for the weak dependence of the fidelity uncertainty on the PNL uncertainty can be understood as follows: if the PNL is higher than estimated, the variance of the stored state is actually lower (in the PNL units) which leads to a higher fidelity. But at the same time the gain factor is also lower leading to a lower fidelity. The two effects oppose each other and hence the fidelity is a rather slowly varying function of PNL. The parameter k_2 is determined from the linear contribution to the function

$$((\delta S_2^{\mathrm{out}})^2 - (\delta S_2^{\mathrm{in}})^2)/(\delta S_2^{\mathrm{in}})^2 \tag{5-354}$$

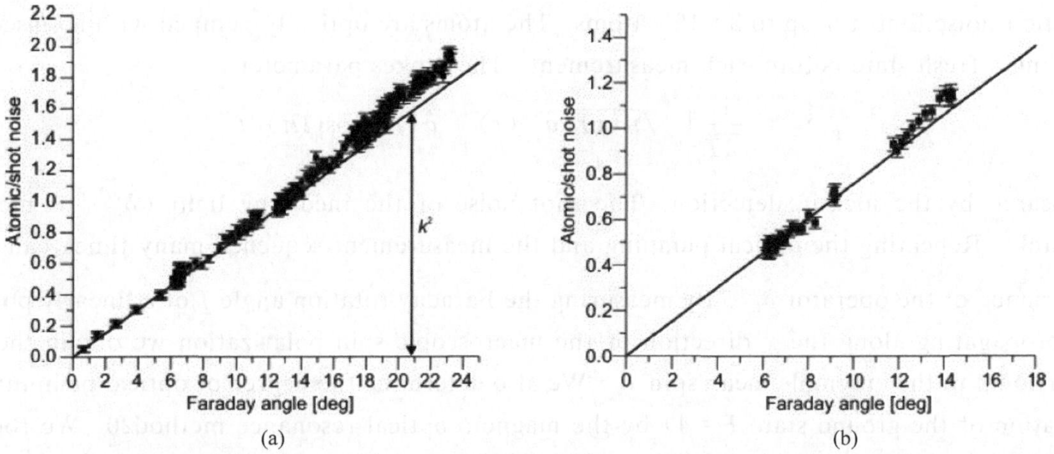

Fig. 5-97 The projection noise calibration. The atomic noise in units of the shot noise of light is plotted as a function of the macroscopic spin size J_x which is proportional to the detected Faraday rotation angle. The error bars are statistical, arising from the fact that the noise variances are obtained from 10.000 cycles of the experiment. An increase in the noise level at high atomic densities seen in part (a) of the figure is due to the classical noise of the lasers. The solid line—the graph of k_2—is the best estimate for the projection noise contribution. The value of k_2 for a particular experimental value of J_x is shown with the arrow. In part (b) the PNL calibration experiment is repeated 43 days later and the same calibration still holds.

as shown in the figure. k_2 is then used to establish the relation between canonical variables of light and canonical variables of memory and to find the variances and mean values of atomic canonical variables, as described in the main text.

Derivation of the quantum feedback relations for an arbitrary quantum state of light Here present a rigorous justification for the feedback relations used in the theoretical part of the paper. In the protocol we make a measurement on the operator of light $\hat{X}_{\mathrm{L}}$ and then displace the atomic ensemble in momentum $\hat{P}_{\mathrm{A}}$ by a quantity proportional to the outcome x. Denote the state of light and atoms after they have interacted as $|\Psi\rangle_{\mathrm{LA}}$. Then, the non-normalized state after the measurement is ${}_{\mathrm{L}}\langle x|\Psi\rangle_{\mathrm{LA}}$. After the displacement the state is $\exp\{-ikx\hat{P}_{\mathrm{A}}\}{}_{\mathrm{L}}\langle x|\Psi\rangle_{\mathrm{LA}}$. Can write this as

$${}_{\mathrm{L}}\langle x \mid \exp\{-ik\hat{X}_{\mathrm{L}}\hat{P}_{\mathrm{A}}\}\Psi\rangle_{\mathrm{LA}} \tag{5-355}$$

by using the fact that $|x\rangle$ is a (generalized) eigenvalue of the operator $\hat{X}_{\mathrm{L}}$, now calculate the density operator obtained by averaging with respect to all outcomes of the measurement (with corresponding probability)

$$\rho = \int_{-\infty}^{\infty} \mathrm{d}x \exp\{-ikx\hat{P}_{\mathrm{A}}\}{}_{\mathrm{L}}\langle x \mid \Psi\rangle_{\mathrm{LALA}}\langle \Psi \mid x\rangle_{\mathrm{L}} \exp\{ikx\hat{P}_{\mathrm{A}}\}$$

$$= \mathrm{Tr_m}(\exp\{-ik\hat{X}_L\hat{P}_A\}\Psi\rangle_{LALA}\langle\Psi \mid \exp\{ik\hat{X}_L\hat{P}_A\}) \tag{5-356}$$

where the trace is taken with respect to the measured mode. The averaged expectation value of any atomic operator $f(\hat{X}_A, \hat{P}_A)$ can be then determined by simply calculating its trace with this density operator. By using the cyclic property of the trace, we can reexpress this quantity as the expectation value of the atomic operator in the Heisenberg picture which is obtained by displacing the atomic momentum operator by the light operator $\hat{X}_L$, i.e. $f(\hat{X}_A, \hat{P}_A + g\hat{X}_L)$. Thus, we can carry out the whole procedure in the Heisenberg picture by performing such a displacement. This is precisely what is done where the outcome of the measurement $x = \hat{X}_L^{in} + k\hat{P}_A^{in}$ is fed back into the atomic variable $\hat{P}_A$ with a feedback gain coefficient g. The result used in the paper is

$$\rho = \int_{-\infty}^{\infty} \mathrm{d}x \exp\{-ikx\hat{P}_A\}_L\langle x \mid \Psi\rangle_{LALA}\langle\Psi \mid x\rangle_L \exp\{ikx\hat{P}_A\}$$

$$= \mathrm{Tr_m}(\exp\{-ik\hat{X}_L\hat{P}_A\}\Psi\rangle_{LALA}\langle\Psi \mid \exp\{ik\hat{X}_L\hat{P}_A\}) \tag{5-357}$$

this analysis is valid for arbitrary input states including mixed states. The improvements of the cooperativity discussed above can probably gain a factor of 4~5. However, many tasks in optical quantum information processing also necessitate quantum memories, whose bandwidth is equally limited. In fact, the kind of photon detector presented here is a quantum memory without retrieval, and it is imaginable that other kinds of quantum memories can be used in a similar way, extending the range of applications for quantum memories significantly.

5.12 Quantum solid memory

Two independent groups of University of Calgary in Canada have demonstrated how a pair of entangled photons can transfer their entanglement to and from a solid—the process that should one day form the backbone of so-called quantum memories or repeaters. These devices would enable quantum communication systems to transmit information over larger distances, with significantly reduced degradation. "While I was sceptical a few years ago that a useful quantum repeater or quantum network could be built, I am now very confident... that this goal can be achieved in the next five to ten years," says Wolfgang Tittel of the University of Calgary, Canada, an author of one of the papers that appear in Nature today. Quantum communication is a means of sending information that is fundamentally secure from eavesdroppers. Two photons must be entangled—that is, have their quantum states inextricably linked—at either end of a channel, over which a "key" for decoding encrypted information can be established. Thanks to the uncertainty principle in quantum mechanics, it is impossible to intercept this key without corrupting it, so the official communicators can always tell if a third party has tried to eavesdrop.

One of the limitations to quantum communication, however, is signal degradation. In conventional information networks, engineers get around this problem by installing repeaters, which record a decaying signal and then re-emit it at its optimum strength. Yet, because it is impossible to record a quantum signal without corrupting it, a quantum repeater must be able to absorb and re-emit the entangled photons without disturbing the entangled state. Quantum memories, a more primitive form of repeaters, have been demonstrated before in single atoms or atomic vapours but not, until now, in the solid state, as is required in a robust communications system. This is the advance made by Tittel's group, which includes members at the University of Paderborn, Germany; and, similarly, by Nicolas Gisin and colleagues at the University of Geneva in Switzerland. Both groups have shown how one photon in an entangled pair can be absorbed by a crystal doped with a rare-earth ion, so that its quantum state becomes stored as an atomic excitation. A fraction of a second later, a new photon is emitted with that entangled state intact.

There are differences between the group's demonstrations. For crystals, Tittel's group used thulium-doped lithium niobate, whereas Gisin's group opted for neodymium-doped yttrium silicate. In addition, a different type of laser set-up has favoured Gisin's group, which reports a maximum storage time of some 200ns at an efficiency of more than 20%; Tittel's group reports a storage time of 7ns at an efficiency of 2%. On the other hand, the quantum memory of Tittel's group functions at a bandwidth of 5GHz—some 40 times greater than Gisin's group—which means, potentially, far more information could be sent in the same time. Val Zwiller, a quantum physicist at the Delft University of Technology in the Netherlands, says the two groups have made "clearly important steps" towards quantum repeaters, but notes several limitations that suggest engineering challenges still lie ahead. One of these is the low efficiency, and the fact that the storage times are not variable, as would be required in a practical device. Another is that the wavelength of the stored photons is not the international standard used in telecommunications, around 1300nm. "The work presented in these two articles still lacks on several fronts," Zwiller concludes.

Members from both groups admit there is some way to go before quantum memories or repeaters can be implemented in practical systems, but believe there are no insurmountable hurdles. "Several solutions are already actively pursued in the world, and the rapid progress that has been achieved recently leads us to believe that these will be significantly improved in the coming years," says Mikael Afzelius, a member of Gisin's group.

5.12.1 Atomic memory

Atomic storage (sometimes called atomic memory) is a nanotechnology approach to computer data storage that works with bits and atoms on the individual level. Like other nanotechnologies, nano-storage deals with microscopic material. An atom is so small that there might be ten million billion in a single grain of sand; optimally, atomic storage would store a bit of data in a single atom. Current data storage methods use millions of atoms to store a bit of data. In 1959, the famous physicist Richard Feynman discussed the potential of atomic storage, explaining that every word ever written up to that point could be stored in a 10 millimeter-wide cubic space, if the words were written with atoms.

Franz Himpsel and colleagues at the University of Wisconsin-Madison created a device that uses 20 atoms to represent a bit of data on a silicone surface. The surface resembles that of a CD (compact disk) but the scale is nanometers rather than micrometers, yielding a storage density a million times higher. Himpsel and colleagues used a scanning tunneling microscope to remove single atoms, and suggest that an extra silicon atom might represent a 1, while a vacant spot represents a 0 (binary language is made up entirely of ones and zeroes). Although the researchers claim their prototype is "proof of (Feynman's) concept," they say that it may yet take decades of work to develop a practical working device that stores bits as single atoms. IBM is working on a different approach to nano-storage in their Millipede project.

- Solid-state photon storage

Quantum communication networks and other quantum information processing will require coherent and efficient transfer of information between light and matter, and the realm of light-matter interfaces is an active area of research. Much of the activity has focused on the mapping of quantum information onto atomic systems. Nicolas Gisin and colleagues at the University of Geneva in Switzerland have now demonstrated the coherent storage and retrieval of information using a solid-state system. The team's quantum memory was an ensemble of roughly 10^7 neodymium ions trapped in a crystal of yttrium vanadium oxide (YVO_4). In such an environment, the resonant frequencies of the rare-earth atoms are inhomogeneously shifted, which broadens the absorption spectrum. That's normally undesirable, but the researchers turned it to their advantage. By optically pumping some of the Nd atoms out of the ground state, they sculpted the spectrum into a series of regularly spaced absorption peaks—an "atomic frequency comb." An incident weak light pulse, with on the order of one photon or less on average, will be uniformly absorbed by the comb and generate a coherent superposition of collective optical excitations, each at a slightly different frequency. The superposition will initially dephase but will get reestablished after a time determined by the comb spacing; once rephased, the atoms will collectively reemit a light pulse that conserves the coherence and phase of the original pulse. Gisin and company achieved storage times of up to a microsecond. Furthermore, they showed that the ensemble can simultaneously store multiple light fields, and they have proposed a means of on-demand retrieval. With such capability, the authors view solid-state systems as a promising contender for quantum storage.

Investigations on improving the quality of reconstructed image of digital holography are carried out. Two processes of recording and numerical reconstruction are taken into account at the same time. Properties of resolution power and contrast of a resolution panel are studied. Some helpful conclusions are obtained. On condition that the aperture size of recording device is larger than an object size, the quality numerical reconstructed image can be improved if the frequency of interference fringe is optimized. In addition, the zero-order and conjugate images are eliminated and the stochastic noises are weakened. The experimental results show that these conclusions are also suitable to general objects.

Optical storage has become one of the most important information storage technologies. The optical discs have become the necessary information media in modern society. And holographic optical data storage (HODS) is one of the most important schemes of super-high density optical storage.

Presents recent technical information and gives an overview of progress in optical memory, neural networks and fractals from the viewpoint of optical information processing. The work introduces holographic optical disks and holographic storage in photorefractive crystal fibre, discusses the optical implementation of neural networks, explains the use of neurochips as artificial retinas, and more.

5.12.2 Stable solid-state source of single photons

The generation of nonclassical light and particularly of single photon states is one of the crucial experimental tasks in quantum optics. The ideal single photon source emits light such that only one of two detectors behind a semitransparent beam splitter registers an event. A specially suited process for that is fluorescence of a single two-level quantum system. Since excitation and subsequent decay to the initial state takes a finite time, only one photon is emitted at a time.

Such two-level systems can be found with good approximation in atoms and ions, where, after initial demonstrations, experiments now concentrate on increasing the yield and the rate of single photons. While manipulation of single trapped atoms or ions still requires significant technical effort, single organic dye molecules as the fluorescent emitter in solvents or polymers allow much simpler setups. Despite recently reported progress, organic molecules still degrade rapidly at room temperature—typically after about 10^9 emissions. Thus, these systems seem impractical for applications requiring single photon sources, i.e., quantum cryptography, where perfect security is given only for single photon pulses. Here, we present an alternative candidate for generating single photons. Single nitrogen-vacancy (NV) centers in diamond combine the robustness of single atoms with the simplicity of experiments with dye molecules. NV centers are one of many well studied luminescent defects in diamond; they are formed by a substitutional nitrogen atom with a vacancy trapped at an adjacent lattice position. Usually, these centers are prepared in type Ib synthetic diamond, where single substitutional nitrogen impurities are homogeneously dispersed. To observe bright luminescence from a sample, additional vacancies are created by electron or neutron irradiation, and allowed to diffuse to the nitrogen atoms by annealing at $900 \pm C$. However, it turned out that already untreated samples of synthetic Ib diamond provide a concentration of NV centers well suited for addressing individual centers.

The high radiative quantum efficiency even at room temperature of close to one as well as a short decay time of the excited state makes them well-suited for single photon generation. The light was focused into the diamond with a relay lens and a standard microscope objective of magnification 60 and a numerical aperture of 0.85 to a spot size of 430nm inside the diamond. The fluorescence light was extracted with the same microscope objective, and focused into a single mode optical fiber. The fiber defines the spatial mode for confocal detection, with the corresponding acceptance area in the diamond sample having a FWHM diameter of 1.6mm. To suppress coupling of pump light into the fluorescence detection setup, we used a combination of a dichroic mirror and a color glass filter.

A piezoelectric two-dimensional scanning unit was used to move the diamond probe transversely to the optical axis, and a motorized linear translation stage to choose the longitudinal position of the focal spot within the diamond. The fluorescence light from the single NV center was analyzed either by a Hanbury-Brown-Twiss configuration or a grating spectrometer.

The inset of Figure 5-98 shows the image of a single NV center. The transverse width of 470nm corresponds to the size of the calculated laser waist in the diamond. In longitudinal direction, we observe a 2.6mm (FWHM) wide acceptance region for fluorescence light detection. Radiation damage in the crystal by exposing diamond samples to fast neutrons and/or heat treatment increased that density, but for investigating single luminescence centers, the untreated samples were sufficient.

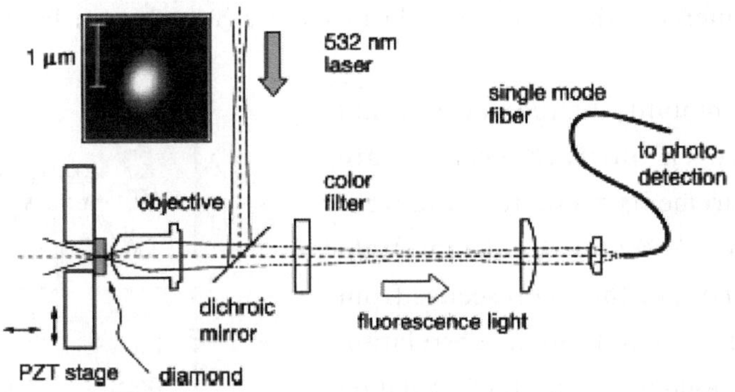

Fig. 5-98 Experimental setup: A frequency doubled diode pumped solid-state laser (532nm) is focused into a type Ib diamond crystal. Fluorescence light is collected with a confocal microscope into a single mode optical fiber, and detected with silicon APDs, shows the fluorescence image of a single NV center.

Spectral analysis allowed us to clearly identify the single NV centers (see Figure 5-99(a)). Even at room temperature the zero phonon line at 637nm is clearly visible, and additional phonon contributions result in the characteristic spectral shape with an overall width of about 120nm. This broad spectral emission is one of the few drawbacks of NV center fluorescence. For applications it

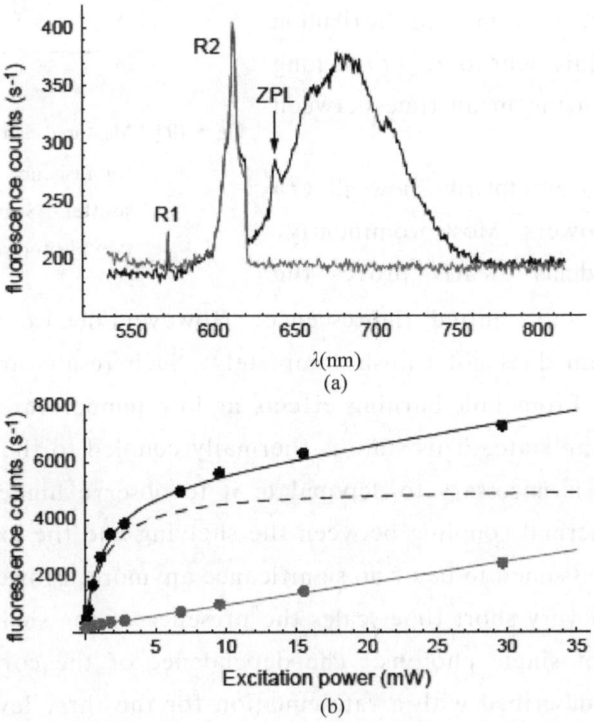

Fig. 5-99 (a) Fluorescence spectrum of a single NV center (black) and reference spectrum from an empty region in type Ib diamond (gray). (b) Fluorescence as a function of the excitation power from a single NV center (black) and from bulk diamond (grey). The background contribution increases linearly with intensity, while the NV contribution (dashed) saturates, with a saturation power of 1. 32mW. (Error bars are smaller than symbols.) bulk diamond at 573nm and between 600nm and 620nm, respectively. For all other measurements, suppressed that light with an additional optical band pass filter. The luminescence from a single NV center placed in the beam waist of the excitation laser shows a clear saturation behavior; the recorded fluorescence on and beside a NV center is shown as a function of pump power.

will be mandatory to increase the spectral yield in narrow wavelength bands, most likely by using microcavities.

Because of the immobility of the vacancy and the substitutional nitrogen atom, NV centers are very stable. All measurements presented here were performed on the same NV center. To demonstrate the nonclassical properties of the fluorescence from single NV centers, the second order correlation function $g_m^{(2)}(t)$ was measured for different excitation powers with two APDs behind a beam splitter (BS, inset of Figure 5-100).

To suppress cross talk due to light created by a detection event of the other detector we inserted a filter (SF) blocking that fluorescence above 750nm. Timing jitter in the APDs and electronics was determined independently to lead to a spreading of t with a FWHM of 1.4ns. A normalized distribution of time differences t is equivalent to $g_m^{(2)}(t)$ as long as t is much smaller than the mean time between detection events.

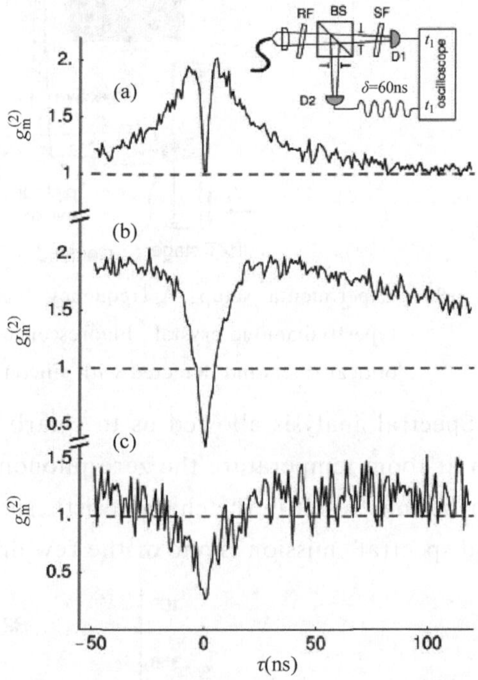

Fig. 5-100 Measured pair correlation function $g_m^{(2)}(t)$: fluorescence light is sent through a beam splitter BS and two filters RF, SF onto two photodiodes D1, D2

Figures 5-100(a)~(c) exemplarily show $g_m^{(2)}(t)$ for different excitation powers. Most prominently, the minimum at zero delay clearly proves the nonclassical character of the emitted fluorescence. However, due to residual background and timing jitter this minimum does not vanish completely. Such results are not compatible with a simple two-level model. From hole-burning effects at low temperatures it is known that there exists a metastable shelving state. This state is thermally coupled to the excited state and at low temperatures, repumping is necessary to depopulate it to observe higher fluorescence rates. At room temperature, the thermal coupling between the shelving and the excited state becomes very strong, and shelving was assumed to be of no significance anymore. However, as becomes apparent from our experiments, on very short time scales the presence of the shelving state still influences the emission robability of single photons. The dependence of the correlation function on the excitation power can be described with a rate equation for the three-level model. Neglecting all coherences, the population dynamics is governed by

$$\begin{pmatrix} \dot{\rho}_1 \\ \dot{\rho}_2 \\ \dot{\rho}_3 \end{pmatrix} = \begin{pmatrix} -k_{12} & k_{21} & 0 \\ k_{12} & -k_{21}-k_{23} & k_{32} \\ 0 & k_{23} & -k_{32} \end{pmatrix} \cdot \begin{pmatrix} \rho_1 \\ \rho_2 \\ \rho_3 \end{pmatrix} \tag{5-358}$$

with the initial condition $\rho_1 = 1, \rho_2 = \rho_3 = 0$ for the system being prepared in the ground state 1 by a fluorescence decay. Here we neglect possible nonradiative transitions from the shelving state 3 to the ground state, which are about 3 orders of magnitude smaller than all other rates. The instantaneous emission probability of a photon is then proportional to $g^{(2)}(t)$, and an analytical expression for the second order correlation function is obtained by normalizing $g^{(2)}(t)$ to $g^{(2)}(\tau)$ resulting in

$$g^{(2)}(\tau) = 1 + c_2 e^{-\tau/\tau_2} + c_3 e^{-\tau/\tau_3} \tag{5-359}$$

where the decay times and coefficients are given by

$$\tau_{2,3} = 2/(A \pm \sqrt{A^2 - 4B})$$

$$c_2 = \frac{(1 - \tau_2 k_{32})}{k_{32}(\tau_2 - \tau_3)}, \quad c_3 = -1 - c_2 \tag{5-360}$$

with

$$A = k_{12} + k_{21} + k_{32} + k_{23}$$

$$B = k_{12}k_{23} + k_{12}k_{32} + k_{21}k_{32}$$

$$\rho_2(t \to \infty) = k_{23}k_{12}/B \tag{5-361}$$

showing a saturation behavior as a function of the pump rate k_{12}.

To link the experimental results with the outcome of the model, it is necessary to correct for background contributions. Their values and the decay times $t_{2,3}$ are shown for a set of pump powers in Figure 5-101. First estimates for the model coefficients in Eq. (5-358) can be deduced from the limiting cases for high or low pump intensities from asymptotical solutions for the coefficients:

$$\begin{cases} k_{32} = 1/(1 + c_3^{(\infty)}\tau_3^{(\infty)}) \\ k_{23} = c_3^{(\infty)} k_{32} \\ k_{21} = \dfrac{1}{\tau_2^{(0)}} - \dfrac{k_{32}}{1 - \tau_2^{(0)} k_{32}} \end{cases} \tag{5-362}$$

where

$$\tau_2^{(0)} = \tau_2(P \to 0), \text{etc.} \tag{5-363}$$

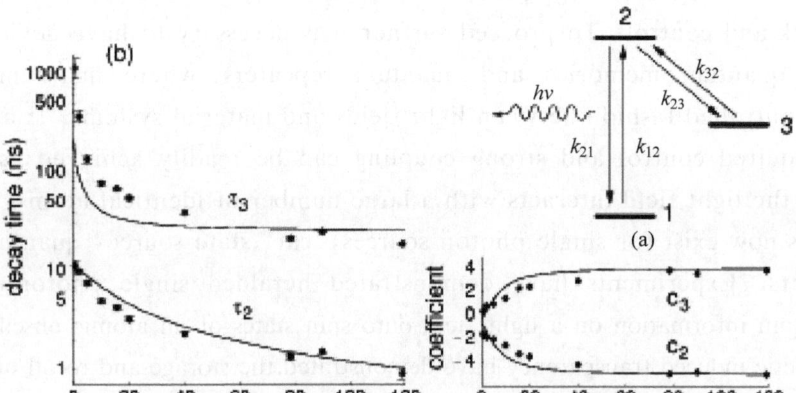

(b) (c)

Fig. 5-101 Three-level model for the fluorescence from the NV center. The excitation is described by a pump rate coefficient k_{12}, fluorescent decay by a coefficient k_{21} and coupling with a shelving state by coefficients k_{23} and k_{32}, respectively.

For the least squares fit of the data the long decay times t_3 should not be considered, since the correlation data are collected over an interval of only 120ns. On the other hand, the short values for t_2 are already

$$\tau_2^{(0)} = 11.7 \pm 0.3\text{ns} \tag{5-364}$$

influenced by the timing jitter of our detectors and should also be disregarded. For small pump power the decay time t_2

$$1/k_{21} = 20.1 \pm 1.6\text{ns}, \quad 1/k_{23} = 31 \pm 2.5\text{ns} \tag{5-365}$$

converges to This is in good agreement with a decay time of 11.6ns measured in a pulsed excitation experiment. The high power limits of $c_{2,3}$ determine the ratio Finally, led to the room temperature rate coefficients, and

$$1/k_{32} = 127 \pm 11\text{ns} \tag{5-366}$$

The fluorescence of NV centers in synthetic Ib diamond shows clear signatures of nonclassical light. The emission spectrum lies in the conveniently detectable red to near infrared region, decay times are short, and the radiative quantum efficiency is close to one. For short times t, can observed strong photon antibunching, making this system a prime candidate for single photon generation by pulsed excitation. For this purpose, the significant influence of a shelving state requires the pulse rate at room temperature to stay below 5MHz in order to avoid population buildup at intermediate times. Foremost, it is the robustness against photobleaching and the simplicity of the all-solid-state setup which distinguishes NV centers from other fluorescing quantum systems. The potential for miniaturizing the setup and the superior stability makes NV centers very attractive both for practical single photon sources and cavity-QED measurements in experimentally simple environments.

5.12.3 Stopped times of light storage

Stopped light with storage times greater than one second using some of the most significant advances in quantum information processing have been made using quantum optics-based techniques. For example, working practical quantum cryptosystems already exist and there have been demonstrations of linear optics quantum computing, quantum teleportation, quantum non-demolition measurements, quantum feedback and control. To proceed further it is necessary to have devices such as single photon sources, quantum memories and quantum repeaters, where quantum information is exchanged in a controlled fashion between light fields and material systems. It has been proposed that both the required control and strong coupling can be readily achieved using an ensemble approach, where the light field interacts with a large number of identical atoms. Such a ensemble based approaches now exist for single photon sources, "cat" state sources, quantum memories and quantum repeaters. Experiments have demonstrated heralded single photon sources and the mapping of quantum information on a light field onto spin states of an atomic ensemble. Experiments using electromagnetic induced transparency have demonstrated the storage and recall of optical pulses.

The quantum systems used for these ensemble based demonstrations have almost exclusively been atomic vapors. An issue with these demonstrations is that even for a laser cooled ensembles, the atomic motion impacts on the devices' performance. Ensembles of solid-state optical centers provide an alternative to atomic systems where the relative motion is zero. In this paper we investigate the use of solid-state system for ensemble based quantum optics and highlight its usefulness by stopping a light pulse using electromagnetically induced transparency (EIT). Unlike an earlier experiment, the current demonstration highlights for the first time two advantages of using optically active solid state centers: a one thousand fold increase in storage time and the ability to operate with a less restrictive beam geometry.

When storing light using EIT characteristics of the field are recorded as a spin wave in the ensemble. The storage time is determined by the coherence times of the hyperfine transitions. In principle coherence times for hyperfine transitions in atomic systems can be very long and many minutes have been measured in ion traps. However, these long coherence times in large ensembles suitable for EIT have not been achieved. Transit time broadening in vapor cells and magnetic inhomogeneity in trapped systems mean that the longest that light has been stored atomic systems is a few milliseconds. In contrast, in earlier work we have demonstrated techniques to obtain hyperfine coherence times of tens of seconds in $Pr:Y_2SiO_5$. Here we show it is possible to utilize these long coherence times to stop light for similar lengths of time.

EIT is sensitive to atomic movement, with the spin wave being scrambled once the atoms have moved significantly compared to the wavevector mismatch between the probe and the coupling beams. To minimize this wavevector mismatch experiments in atomic systems typically operate with the beams co-propagating. Because the probe and coupling beams are close in frequency, in this configuration, the wavevector mismatch is typically less than $1cm^{-1}$. A consequence of this co-propagating operation is that it is difficult to separate the probe and the coupling beam. In a solid-state system, where the optical centers are locked in a crystal lattice, co-propagating operation is not required, in the present work the probe and the coupling beams are counter-propagating. With counter-propagating beams it is easier to separate the probe and coupling beam whilst maintaining optimum overlap. The experimental setup is shown in Figure 5-102. Because of the narrow 2500Hz optical homogeneous linewidth of the ${}^3H_4 \rightarrow {}^1D_2$ transition in $Pr^{3+}:Y_2SiO_5$ a highly frequency stabilized dye laser was required for the experiment not to be limited by laser jitter. The laser used was a modified Coherent 699 dye laser with a linewidth 200Hz over 1 second time scales.

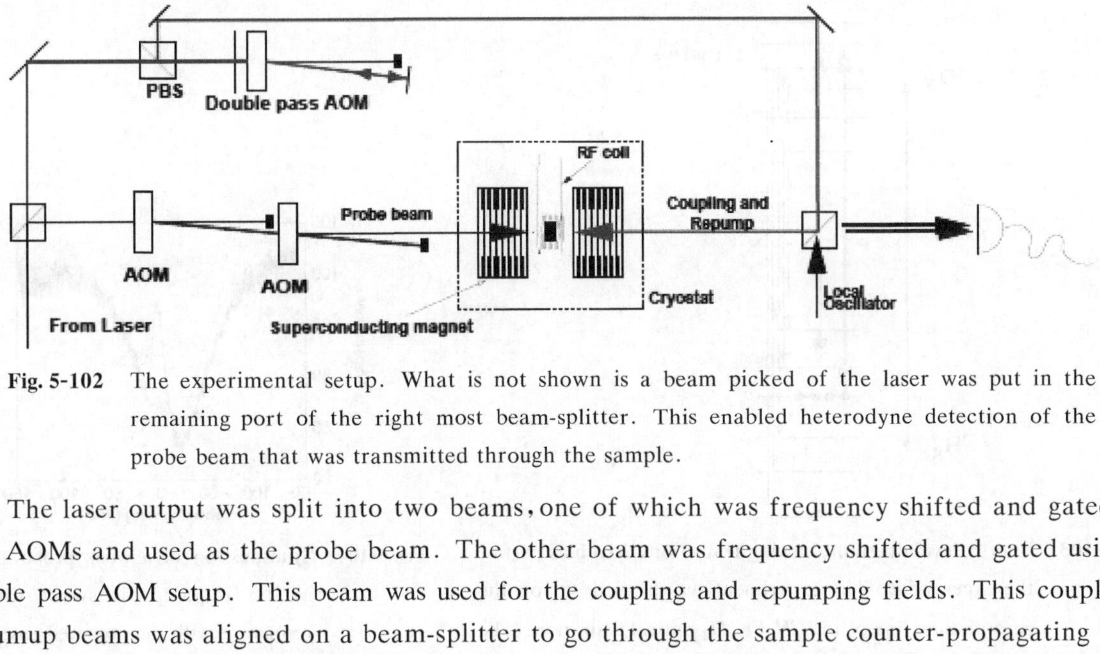

Fig. 5-102 The experimental setup. What is not shown is a beam picked of the laser was put in the remaining port of the right most beam-splitter. This enabled heterodyne detection of the probe beam that was transmitted through the sample.

The laser output was split into two beams, one of which was frequency shifted and gated by two AOMs and used as the probe beam. The other beam was frequency shifted and gated using a double pass AOM setup. This beam was used for the coupling and repumping fields. This coupling/ repumup beams was aligned on a beam-splitter to go through the sample counter-propagating with the probe. The spare port of this right-most beam splitter was used to combine a local oscillator beam with the transmitted probe beam, enabling the heterodyne detection of the signal. The sample used was the same as that used in reference and consisted of 0.05% Praseodymium doped in Y_2SiO_5. It was 4mm thick along the direction of light propagation. The sample was mounted in a bath liquid helium cryostat. Three orthogonal super-conducting magnets were used to apply a DC magnetic field to the sample and a six turn rf coil was used to apply a rf field.

The dominant dephasing mechanism for the hyperfine states of the Pr^{3+} ions is random Zeeman shifting due to fluctuating magnetic fields from the yttrium nuclei. Dramatic increases in coherence times can be achieved by operating at a magnetic field where the transition frequency is insensitive to magnetic field changes to first order. The magnetic field required is 78mT in an orientation described. Once the magnetic field is obtained the remaining fluctuations have reasonably long correlation times.

This situation enables the effective use of dynamic decoherence control (DDC) techniques and coherence times in excess of half a minute have been demonstrated. An energy level diagram showing the transitions driven during the experiments is shown in Figure 5-103. While the optical inhomogeneous line widths is a few GHz. The narrow homogeneous linewidth (of order 1kHz) and long hyperfine population lifetimes (of order 1 minute) enabled the experiment to be carried out on an ensemble with a much smaller range of optical frequencies. At the beginning of each shot a sequence of the five optical frequencies (labelled "R" in Figure 5-103) was applied repeatedly. The repump frequencies were applied sequentially rather that all at once to avoid the possibility of darks states and nonlinear mixing of the different frequencies in the AOM. The gap in time between the repumping and each experimental shot was long enough to ensure that ions had no remaining optical coherence. This repumping procedure prepared an ensemble of ions in the desired hyperfine state and gives a narrow adsorption with an inhomogeneous width of 100kHz when measured by sweeping a week probe in frequency (line given by dots in Figure 5-104). When the coupling beam was applied a narrow transparency was obtained in the absorption of the weak probe (solid trace in Figure 5-104).

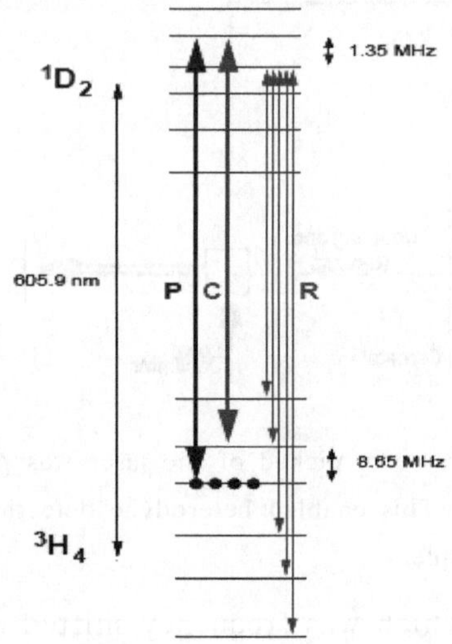

Fig. 5-103 Energy level diagram showing transitions driven as part of the experiment. The experiment was carried out on the zero phonon line of the $^3H_4 \to ^1D_2$ optical transition. The hyperfine levels are shown and these are due to the 5/2 spin of the praseodymium nuclei. In the presence of a magnetic field these are linear combinations of the zero field states. The probe, coupling and repump beams are labelled P, C and R respectively.

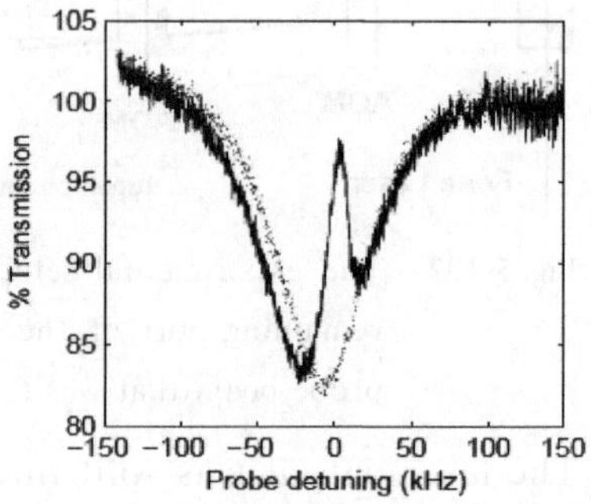

Fig. 5-104 Transmission of a weak probe, 10μW, as its frequency is swept through resonance with the prepared ensemble. The solid line was taken with a 1mW coupling beam on and the dotted line with the coupling off.

The repumping beams were applied after each shot and the 300kHz span shown was swept in 4ms. The transmitted probe beam was detected as a heterodyne beat signal and the bandwidth of the RF detector was comparable to 300kHz, the extra noise at each end of the spectrum came from dividing out this frequency response. For coupling intensities above 1mW the EIT was observed to depend linearly on the amplitude of the coupling beam.

The limiting EIT width at low intensity was 10kHz, corresponding to the hyperfine inhomogeneous linewidth. Below 1mW the EIT transmission decreased with decreasing coupling intensity. It can bee seen from Figure 5-104 that the peak absorption of our ensemble is only about 15% and, as is discussed below, this limits the efficiency of the storing process. The time sequence for the light storage demonstration is shown on the left of Figure 5-105. A 20μs long probe pulse was applied and then the 10mW coupling beam was turned off to transfer the optical coherence onto the spin transition. As in the previous solid state stopped light experiment RF rephasing pulses were used to rephase the inhomogeneous broadening in the spin transition. Although one RF rephasing pulse is enough to rephase the spin-wave it also flips the spin-wave's direction. Therefore when not using co-propagating beams, as is the case here, it is necessary to use an even number of rephasing pulses. The size of the pulse of light recalled as a function of delay can be shown with and without dynamic decoherence control (DDC) and the results are shown in Figure 5-105. The decay constants for the stored signal output were 0.35 seconds without DDC and 2.3 seconds with DDC. These decay rates were comparable to measurements of T_2 made using the same method as Fraval et al. The difference between the present measurements of T_2 and those obtained by Fraval et al. is attributed to not having tuned the magnetic field as carefully as was achieved by Fraval et al. Shown in the inset of Figure 5-105 is the intensity of the output pulse as the intensity of the input probe pulse is varied. At such high powers effects such as self induced transparency (SIT) cannot be ignored. While the effect was linear and scaled to low powers, the efficiency was low, of the order of 1%. This in part can be improved with better timing and shaping of the probe and coupling waveforms. However the main reason for the low efficiency is the low optical absorption at the probe frequency and the accompanying modest group delay.

The sample used for this experiment was only 4mm thick, longer samples as well as multi-pass cells and cavities are simple means to increase the optical absorption. Preliminary measurements on a samples with a range of praseodymium concentrations suggest that at least two or three fold increases in the optical thickness can be achieved by increasing the concentration without significantly increasing the inhomogeneous broadening of the hyperfine transition.

As it is a goal of this line of research to store and retrieve quantum mechanical states it worthwhile to consider the effect that rephasing pulses would have on few photon states stored in the hyperfine coherences. It has been asserted in a theoretical investigation of quantum information storage in the solid state that one would not be able to apply the RF π pulses with sufficient accuracy. It was assumed that the π pulse would have to be applied with an accuracy close to 1 part in N (where N is the number of atoms) in order that the few photon pulse not be swamped by light caused by inaccuracies of the π pulse. However this light will be emitted randomly rather than in the very precise spatio-temporal mode of the output pulse. This should enable the output pulse to be easily separated from the background with very high efficiency.

We have demonstrate stopped light in Pr: Y_2SiO_5 for time scales of several seconds which is

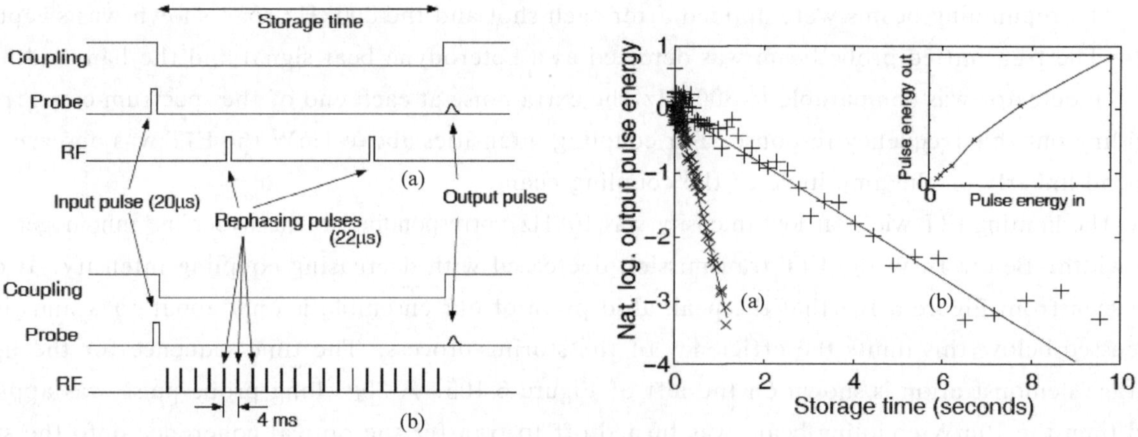

Fig. 5-105 On the left is the time sequence used in the stopped light experiments (a) with simple rephasing of the inhomogeneous broadening of the spin transitions and (b) with "bang-bang" dynamic decoherence control. In (a) two rephasing pulses are used placed 1/4 and 3/4 of the way through the storage time. In (b) N rephasing pulses were used (N even). The first rephasing pulse was applied 2ms after the light was stored, the pulses were separated by 4ms and the last pulse was applied 2ms before the light was recalled. The rephasing pulses lasted 22μs. On the right is the size of the recalled pulse as a function of time. The faster decay was acquired using simple rephasing of the ground state spin coherence (a). The slower decay was acquired using "bangbang" decoherence control (b). The inset shows the energy of the recalled pulse as a function of the energy of the input pulse, the probe pulse length was 20μs and the delay held constant at 100ms.

three orders of magnitude longer than any obtained previously. Based on previous measurements of T_2 it should be possible to extend this storage time by at least one more order of magnitude. For the first time stopped light has been demonstrated in a solid with the coupling and probe beams counter propagating. This configuration is desirable as it allows easy separation of the two beams. However, it is only practical if the atoms movement during the storage time is small compared to the optical wavelength. Even for ultra-cold systems this places significant limits storage time. In a solid where the atoms are locked into position this isn't a problem. The efficiency of the storage process required for a quantum memory should be obtainable by increasing the density of the dopantions and by increasing the interaction length.

5.13 Photon solid-state quantum memories

Quantum memories for light, which allow the reversible transfer of quantum states between light and matter, are central to the development of quantum repeaters, quantum networks, and linear optics quantum computing. Significant progress has been reported in recent years, including the faithful transfer of quantum information from photons in pure and entangled qubit states. However, none of these demonstrations confirm that photons stored in and recalled from quantum memories remain suitable for two-photon interference measurements, such as C-NOT gates and Bell-state measurements, which constitute another key ingredient for all aforementioned applications of quantum information processing. As the interference is always near the theoretical maximum, we conclude that our solid-state quantum memories, in addition to faithfully mapping quantum information, also preserves the entire photonic wavefunction. Hence, demonstrate that our memories are generally suitable for use in advanced applications of quantum information processing that require two-photon interference.

When two indistinguishable single photons impinge on a 50/50 beam-splitter (BS) from different input ports, they bunch and leave together by the same output port. This so-called Hong-Ou-Mandel (HOM) effect is due to destructive interference between the probability amplitudes associated with both input photons being transmitted or both reected. Since no such interference occurs for distinguishable input photons, the interference visibility V provides a convenient way to verify that two photons are indistinguishable in all degrees of freedom, i. e. spatial, temporal, spectral, and polarization modes. The visibility is defined as

$$V = (R_{max} - R_{min})/R_{max} \tag{5-367}$$

where R_{min} and R_{max} denote the rate with which photons are detected in the two output ports in coincidence if the incoming photons are indistinguishable and distinguishable, respectively. Consequently, the HOM effect has been employed to characterize the indistinguishability of photons emitted from a variety of sources, including parametric down-conversion crystals, trapped neutral atoms, trapped ions, quantum dots, organic molecules, nitrogen-vacancy centres in diamond, and atomic vapours. Furthermore, two-photon interference is at the heart of linear optics Bell-state measurements, and, as such, has already enabled experimental quantum dense coding, quantum teleportation, and entanglement swapping. However, to date, the possibility to perform Bell-state measurements with photons that have previously been stored in a quantum memory, as required for advanced applications of quantum information processing, has not yet been established. For these measurements to succeed, photons need to remain indistinguishable in all degrees of freedom, which is more restrictive than the faithful recall of encoded quantum information. Indeed, taking into account that photons may or may not have been stored before the measurement, this criterion amounts to the requirement that a quantum memory preserves a photon's wavefunction during storage. Similar to the case of photon sources, the criterion of indistinguishability is best assessed using HOM interference, provided single-photon detectors are employed. However, with phase incoherent laser pulses obeying Poissonian photon-number statistics, as in our demonstration, the maximally achievable visibility is 50%, irrespective of the mean photon number (see Supplementary Information). Nevertheless, attenuated laser pulses are perfectly suitable for assessing the effect of the quantum memories on the photonic wavefunction. Any reduction of indistinguishability due to storage causes a reduction of visibility, albeit from maximally 50%. This approach extends the characterization of quantum memories using attenuated laser pulses from assessing the preservation of quantum information during storage to assessing the preservation of the entire wavefunction, and from first-to second-order interference.

First deactivate both quantum memories, to examine the interference between directly transmitted pulses, and thereby establish a reference visibility for our experimental setup. To set the mean photon number per pulse before the memories to 0.6, i. e. to the single-photon level. Using the wave plates, we rotate the polarizations of the pulses at the two HOM-BS inputs to be parallel (indistinguishable) or orthogonal (distinguishable). Subsequently, activate memory a while keeping memory b off, and adjust the timing of the pulse preparation so as to interfere a recalled pulse from the active memory with a directly transmitted pulse from the inactive memory. Pulses

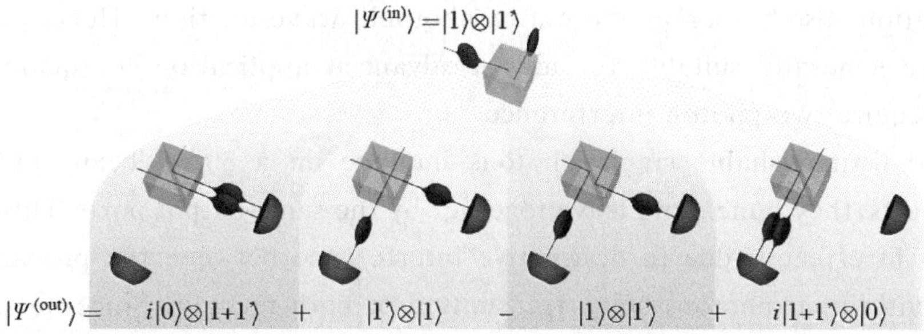

Fig. 5-106 Illustration of HOM-interference in the case of single photons at BS input $|\psi^{(in)}\rangle = |1,1'\rangle$, where the prime on the latter input indicates the possibility to distinguish that input photon from the other in some degree of freedom e. g. by being polarized orthogonally. The four possible paths of the photons are illustrated, together with their corresponding output states. If the input photons are indistinguishable with respect to all degrees of freedom we can ignore the primes in the output states and the paths shown in the two central pictures are identical and, due to the different signs, thus cancel. This leaves in the output state $|\psi^{(out)}\rangle$ only the possibilities in which photons bunch. For distinguishable photons, e. g. having orthogonal polarizations, all paths are distinguishable and all terms remain in $|\psi^{(out)}\rangle$.

are stored for 30ns in memory a, and the mean photon number per pulse at the quantum memory input is 0.6. Taking the limited storage efficiency of $\approx 1.5\%$ and coupling loss into account, this results in 3.4×10^{-4} photons per pulse at the HOM-BS inputs. As before, changing the pulse polarizations from mutually parallel to orthogonal, can find $V = (47.7 \pm 5;4)\%$, which equals our reference value within the measurement.

Uncertainties. As the final step, we activate both memories to test the feasibility of two-photon interference in a quantum-repeater scenario. That in a real-world implementation, memories belonging to different network nodes are not necessarily identical in terms of material properties and environment. This is captured by the setup where the two Ti:Tm:$LiNbO_3$ waveguides feature different optical depths and experience different magnetic fields. To balance the ensuing difference in memory efficiency set the mean photon number per pulse before the less efficient and more efficient memories to 4.6 and 0.6, respectively, so that, as before, the mean photon numbers are 3.4×10^{-4} at both HOM-BS inputs. With the storage time of both memories set to 30ns, get $V = (47.2 \pm 3.4)\%$, in excellent agreement with the values from the previous measurements. The consistently high visibilities, compiled in the first column of Table 5-4, hence conform that our storage devices do not introduce any degradation of photon indistinguishability during the reversible mapping process, and that two-photon interference is feasible with photons recalled from separate quantum memories, even if the memories are different. Now investigate in greater detail the change in coincidence count rates as photons gradually change from being mutually indistinguishable to completely distinguishable w. r. t. each degree of freedom accessible for change in single-mode fibres, i. e. polarization, temporal, and spectral modes. To acquire data more efficiently increase the mean number of photons per pulse at the memory input to between 10 and 50 (referred to as few-photon-level measurements). However, the mean photon number at the HOM-BS remains below one.

Figure 5-108(a) show the coincidence counts rates as a function of the polarization of the recalled pulse for the case of one active memory. The visibilities for all configurations (i.e. zero, one, or two active memories) extracted from fits to the experimental data are listed in column of

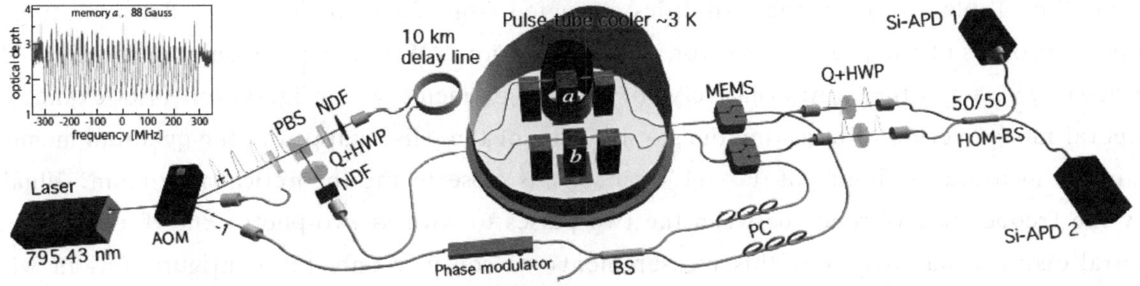

Fig. 5-107 Experimental setup. Light from a 795.43nm wavelength CW laser passes through an acousto-optic modulator (AOM) driven by a sinusoidally varying signal. The first negative refraction order is fibre coupled into a phase modulator and, via a beam-splitter (BS), two polarization controllers (PCs) and two micro-electromechanical switches (MEMS), injected from the back into two Ti: Tm: $LiNbO_3$ waveguides (labelled a and b) cooled to 3K. Waveguide a is placed inside a superconducting solenoid. Using a linear frequency-chirping technique tailor AFCs with 600MHz bandwidth and a few tens of MHz peak spacing, depending on the experiment, into the inhomogeneously broadened absorption spectrum of the thulium ions, as shown for crystal a in the inset. After 3ms memory preparation time and 2ms wait time we store and recall probe pulses during 3ms. The 8ns long probe pulses with $\approx$ 50MHz Fourier-limited bandwidth are derived from the first positive diffraction order of the AOM output at a repetition rate of 2.5~3MHz. Each pulse is divided into two spatial modes by a half-wave plate (HWP) followed by a polarizing beam-splitter (PBS). All pulses are attenuated by neutral-density filters (NDFs) and coupled into optical fibres and injected from the front into the Ti: Tm: $LiNbO_3$ waveguides. After exiting the memories (i.e. either after storage, or after transmission), the pulses pass quarter-and half-wave plates used to control their polarizations at the 50/50BS (HOM-BS) where the two-photon interference occurs. Note that, to avoid first-order interference, pulses passing through memory a propagate through a 10km fibre to delay them w.r.t. the pulses passing through memory b by more than the laser coherence length. Finally, they are detected by two single-photon detectors (actively quenched silicon avalanche photodiodes, Si-APDs) placed at the outputs of the beam-splitter, and coincidence detection events are analyzed with a time-to-digital convertor (TDC) and a computer.

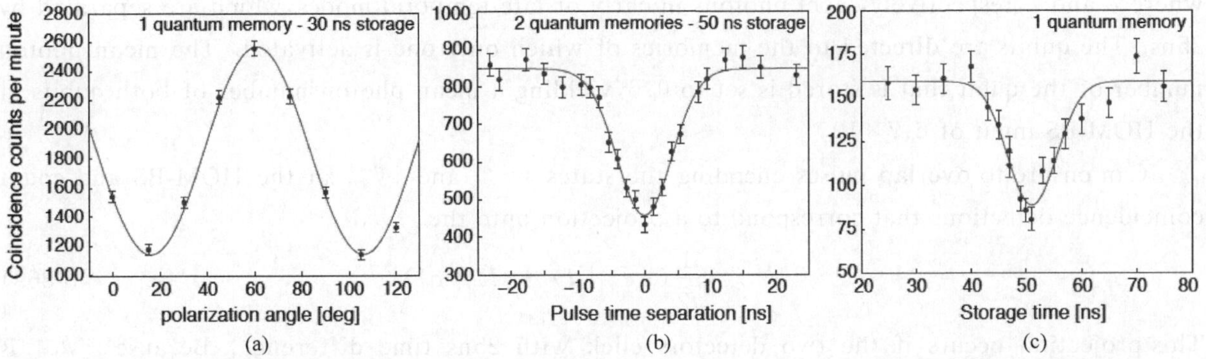

Fig. 5-108 HOM interference plot examples for one or two active memory configurations (as labelled). (a) Varying mutual polarization difference. (b) Varying temporal overlap by changing timing of pulse generation. (c) Varying temporal overlap by changing storage time. The acquisition time per data point is 60s in (a), (b) and 120s in (c).

Table 5-4. They are as in the case of single-photon-level inputs equal to within the experimental uncertainty. Next, in Figure 5-108(b), we depict the coincidence count rates as a function of the temporal overlap (adjusted by the timing of the pulse generation) for the two-memory configuration.

Column 3 of Table 5-4 shows the visibilities extracted from Gaussian fits to the data, reecting the temporal profiles of the probe pulses, for all configurations. Within experimental uncertainty, they are equal to each other. Alternatively, in the single-memory configuration, also change the temporal mode overlap by adjusting the storage time of the pulse mapped to the quantum memory. Again the measured visibility of $V = 44.4 \pm 6.9\%$ is close to the theoretical maximum. Finally, vary the frequency difference between the two pulses to witness two-photon interference w.r.t. spectral distinguishability. For this measurement, we consider only the configurations in which neither, or a just single memory is active. In both cases the visibilities, listed in the last column of Table 5-2, are around 43%. While this is below the visibilities found previously, for reasons discussed in the Supplementary Information, the key observation is that the quantum memory does not affect the visibility.

Table 5-2 Experimental two-photon interference visibilities (%) for different degrees of freedom.

Storage configuration	Single-photon level	Few-photon level		
	Polarization	Polarization	Temporal	Spectral
No-storage	47.9±3.1	51.0±5.6	42.4±2.3	43.7±1.7
Single-storage	47.7±5.4	55.5±4.1	47.6±3.0	42.4±3.5
Double-storage	47.2±3.4	53.1±5.3	46.1±3.2	N.A.

As stated in the introduction, Bell-state measurements (BSM) with photonic qubits recalled from separate quantum memories are key ingredients for advanced applications of quantum communication. To demonstrate this important element, consider the asymmetric (and arguably least favourable) case in which only one of the qubits is stored and recalled. Appropriately driving the AOM in Fig. 5-107, prepare the states $|\psi_1\rangle$ and $|\psi_2\rangle$, which describe time-bin qubits of the form

$$|e\rangle,\ |l\rangle,\quad \frac{1}{\sqrt{2}}(|e\rangle + |l\rangle),\quad \text{or}\quad \frac{1}{\sqrt{2}}(|e\rangle - |l\rangle) \tag{5-368}$$

where e and l, respectively, label photons in early or late temporal modes, which are separated by 25ns. The qubits are directed to the memories of which only one is activated. The mean photon number of the qubit that is stored is set to 0.6, yielding a mean photon number of both qubits at the HOM-BS input of 6.7×10^{-4}.

Can ensure to overlap pulses encoding the states $|\psi_1\rangle$ and $|\psi_2\rangle$ at the HOM-BS and count coincidence detections that correspond to a projection onto the

$$|\psi^-\rangle = \frac{1}{\sqrt{2}}(|e\rangle|l\rangle - |l\rangle|e\rangle) \tag{5-369}$$

This projection occurs if the two detectors click with 25ns time difference. Because $|\psi_1\rangle$ is antisymmetric w.r.t. any basis, the count rate is expected to reach a minimum value $R^{\parallel}$ if the two input pulses are prepared in equal states, and a maximum value $R^{\perp}$ if prepared in orthogonal states. Accordingly, define an error rate that quantifies the deviation of the minimum count rate from its ideal value of zero:

$$e \equiv \frac{R^{\parallel}}{R^{\parallel} + R^{\perp}} \tag{5-370}$$

A subsequent goal is to develop workable quantum repeaters or, more generally, quantum networks, for which longer storage times are additionally needed. Depending on the required value, which may range from hundred micro-seconds to seconds, this may be achieved by storing quantum information in optical coherence, or it may require mapping of optical coherence onto spin states.

5.13.1 Memory operation and properties

A quantum memory is said to be activated when confim the MEMS to allow the optical pumping light to reach the waveguide during the preparation stage and thus tailor an AFC in the inhomogeneously broadend absorption spectrum of thulium ions. If the optical pumping is blocked, the memory is said to be deactivated and light entering the waveguide merely experiences constant attenuation over its entire spectrum. If a memory is activated, an incident photon is mapped onto a collective excitation of thulium ions in the prepared AFC and subsequently re-emitted at a time given by the inverse of the comb tooth spacing, i. e., $t = 1/\Delta$. In all cases, we adjust the mean photon number at the memory inputs so that mean photon numbers are equal at the HOM-BS inputs. This is required for achieving maximum visibility with attenuated laser pulses. The two Ti:Tm:$LiNbO_3$ waveguides are fabricated identically but differ in terms of overall length, yielding optical depths of 2.5 for memory a and 3.2 for memory *b*. As shown in Figure 5-106, memory is placed at the centre of a solenoid in a uniform magnetic field, while memory b is placed outside the solenoid and thus experiences only a much weaker stray field. Therefore it is not possible to achieve the optimal efficiency for both memories at the same time.

- Preparing states for Bell-state measurement

For the Bell-state projection measurement interchangeably prepare the time-bin qubits in either $|e\rangle$ or $|l\rangle$, or in

$$\frac{1}{\sqrt{2}}(|e\rangle + |l\rangle) \text{ and } \frac{1}{\sqrt{2}}(|e\rangle - |l\rangle) \tag{5-371}$$

by setting the relative phase and intensity of the AOM drive signal. Adjusting the timing of the pulse preparation we ensure that qubits in different states overlap at the HOM-BS. Properties of waveguide $LiNbO_3$ crystal and AFC. In the experimental configuration in which the HOM-interference occurs between two pulses recalled from separate quantum memories pointed, in the main text, to the different properties of the two memory devices. In this section wish to elaborate on the differences between the two memories based on their physical dissimilarity and measured optical depth as a function of frequency. Memory waveguide a is 10.4mm long and crystal b is 15.4mm long. The optical depths at 795.43nm are around 2.5 and 3.2 for waveguide a and b, respectively, as shown by the light-grey curves in Figure 5-109(a), (b), corresponding to the case in which the memories are not activated.

In order to spectrally tailor an AFC in Tm:$LiNbO_3$, a magnetic field must be applied along the crystal's c-axis so as to split the ground and excited level multiplets into their two nuclear Zeeman sublevels. However, as one crystal is located at the centre of the setup's solenoid and the other outside the solenoid it is not possible to apply the same B-field at the two crystals. Thus when we activate both memories we generally apply a magnetic field, which provides a reasonable balance in recall efficiencies but is not optimal for either memory. This circumstance is refected by the different shapes of optical-depth profiles of the AFCs shown in red in Figure 5-109(a),(b).

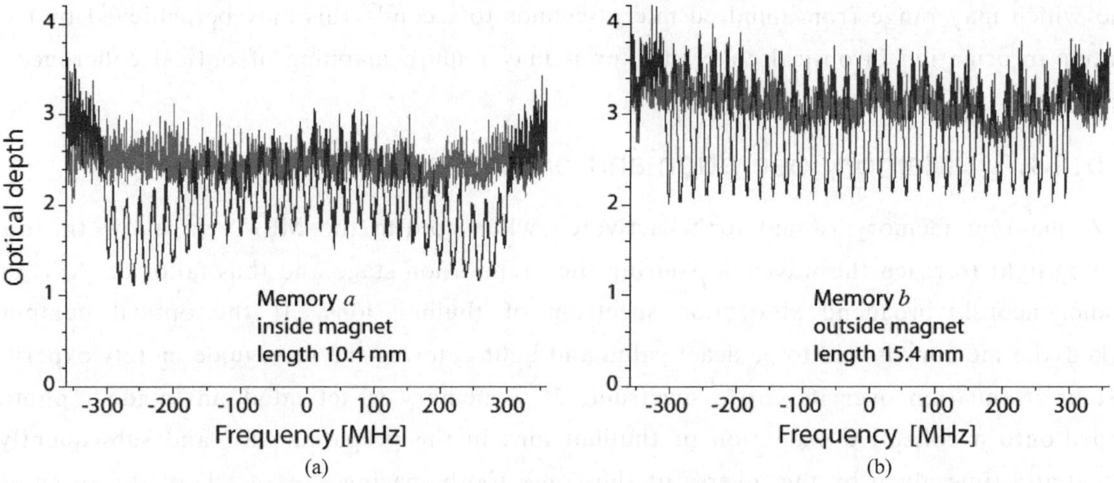

Fig. 5-109 Measured optical depths of our two $Ti:Tm:LiNbO_3$ waveguides as a function of frequency shift of the probing light imparted by the phase-modulator. Light grey traces show optical depths when the memories are inactive, i.e. no AFC is prepared. Dark red traces show the prepared AFCs at a magnetic field of 900 Gauss at the centre of the solenoid.

Two-photon interference in imperfectly prepared memories. In all our demonstrations of the HOM interference consistently observe that the HOM visibility is close to the theoretical maximum for coherent states. Yet, it is important to realize that an improperly configured AFC quantum memory does alter a stored photon's wavefunction, resulting in imperfect HOM interference with a nonstored photon. To support this claim we activate only memory a, whose performance we change by varying the bandwidth of the AFC, and interfere the recalled pulses with pulses directly transmitted through the deactivated memory b. As the AFC bandwidth decreases below that of the probe pulses, the AFC effectively acts as a bandpass filter for the stored photons and we thus expect the recalled pulses to be temporally broadened w.r.t. the original pulse. This is observed in the insert of Figure 5-110, which shows smoothed histograms of photon detection events as a function time. It is worth noting that the small band-width AFC also acts as a bandpass filter for the transmitted pulse by virtue of the different effective optical depths inside and outside the AFC. Thus the broadened transmitted pulse starts to overlap with the echo for the narrow AFC bandwidth traces, as is also observed in the insert of Figure 5-110.

Another consequence of reducing the AFC bandwidth is that the overall efficiency of the quantum memory decreases, which causes an imbalance between the mean photon numbers at the HOM-BS inputs and thus reduces HOM interference visibility. Circumvent the change to the echo efficiency by adapting the mean photon number at the memory input so as to keep the mean photon number of the recalled pulse constant. With this remedial procedure, assess the HOM visibility by changing the HOM-BS inputs from parallel to orthogonal polarizations for a series of different AFC bandwidths. The HOM visibility in Figure 5-110 is steady for bandwidths from around 100MHz and up. However, below 100MHz the visibility begins to drop significantly. The dashed line is a fit of the visibilities to a Gaussian function with full-width at half-maximum (FWHM) of 79 ± 4MHz. Note, that the reason for the visibility being limited to around 40% is solely that, for this measurement, do not go through the usual careful optimization steps.

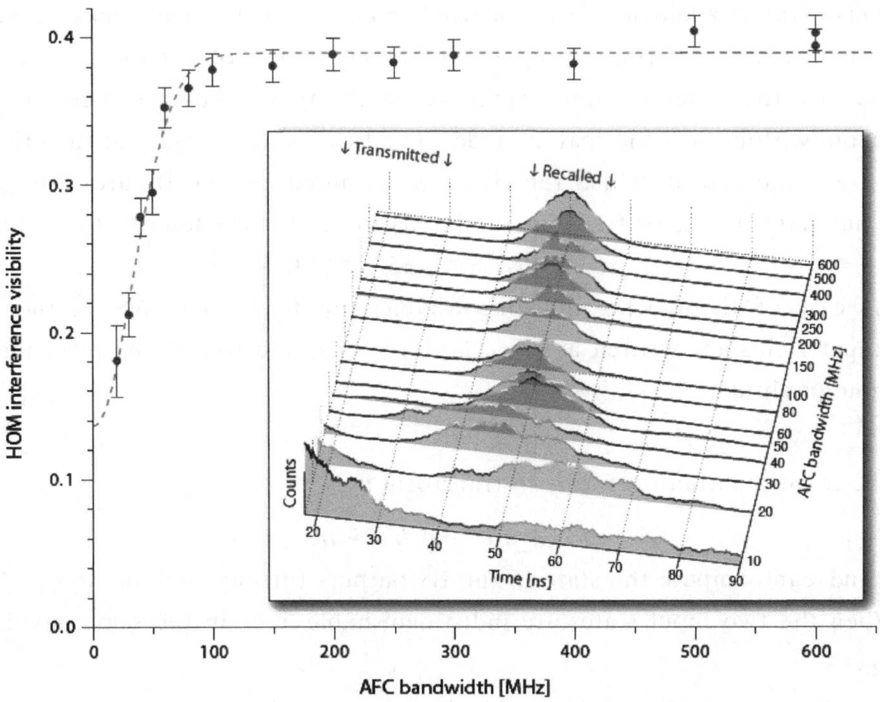

Fig. 5-110 HOM interference visibility if HOM-BS input pulses are recalled from AFCs with varying bandwidths. Insert: Histograms of recalled pulse detection times for different AFC bandwidths clearly showing broadening of recalled (and transmitted) pulses for bandwidths below 100ns.

With these measurements have illustrated how a quantum memory could alter the photonic wavefunction resulting in a reduced HOM interference visibility. A combination of spectral and temporal distortion of the photonic wavefunction is indeed a common type of perturbation by quantum memories. It is particularly worth noting that the gradientecho memory (GEM) quantum memory protocol, though similar to the AFC protocol, imparts a frequency chirp to the recalled pulse. If not corrected, this feature constitutes a perturbation of the wavefunction of the recalled pulse, which may render it unsuitable for applications relying on two-photon interference.

5.13.2 Analytical model of second-order interference in coincidence measurements

In the following theoretical treatment will derive expressions for the coincidence and single-detector counts in terms of probabilities.

By multiplying these probabilities with the average experimental repetition rate can easily calculate the predicted experimental count rates. To a large extent though, we will mainly be interested in relative probabilities or count rates between different settings of the degrees of freedom of pulses. It is reasonably straightforward to derive the rates of detection of photons at the outputs of a BS (note that in this Supplementary Information, the HOM-BS of the main text will be referred to as just BS) In our case coherent states $|\alpha\rangle$ and $|\beta\rangle$, occupy the two spatial input modes of the BS. In the Fock-basis the coherent state can be represented as

$$|\alpha\rangle = \sum_{n=0}^{\infty} e^{-\frac{|\alpha|^2}{2}} \frac{\alpha^n}{\sqrt{n!}} |n\rangle = \sum_{n=0}^{\infty} e^{-\frac{|\alpha|^2}{2}} \frac{\alpha^n}{n!} (\hat{a}^\dagger)^n |0\rangle \tag{5-372}$$

and similarly for. To account for the cases of photons being distinguishable and indistinguishable at the BS must allow for an additional degree of freedom in each of the spatial modes, e.g. polarization, frequency, or time. Thus can write the input state at one of the BS inputs as $\langle \alpha_1, \alpha_2 \rangle \langle \alpha_1 \otimes \alpha_2 \rangle$, where α_1 and α_2 are the coherent state amplitudes in the two orthogonal modes of the auxiliary degree of freedom within the same spatial mode. Treat the coherent state at the other BS input in a similar way. For the case in which the fields at the inputs of the BS are distinguishable with respect to the auxiliary degree of freedom, the inputs to the BS are described as being in the state

$$| \alpha, 0\rangle \, | 0, \beta\rangle \equiv | \alpha, 0\rangle \otimes | 0, \beta\rangle \tag{5-373}$$

whereas in the case of them being indistinguishable (up to a difference in the mean photon number) the input fields are written as | |. The BS is characterized by its refection amplitude r and transmission amplitude

$$t = \sqrt{1 - | r |^2} \tag{5-374}$$

which cause the input creation operators to transform as

$$\hat{a}^\dagger \rightarrow t\hat{c}^\dagger + ir\hat{d}^\dagger \text{ and } \hat{b}^\dagger \rightarrow ir\hat{c}^\dagger + t\hat{d}^\dagger \tag{5-375}$$

With this in hand, can compute the state in the BS outputs for any combination of Fock states at the inputs. When the two input states are indistinguishable, i. e. in the same auxiliary degree of freedom, can get

$$| n, 0\rangle \, | m, 0\rangle \rightarrow \sum_{j=0}^{n} \sum_{k=0}^{m} K_{\parallel}(n, m, j, k) \, | j + k, 0\rangle \, | n + m - j - k, 0\rangle$$

$$K_{\parallel}(n, m, j, k) = t^{m-k+j}(ir)^{n-j+k} \sqrt{\binom{n}{j}\binom{m}{k}\binom{j+k}{j}\binom{n+m-j-k}{n-j}} \tag{5-376}$$

where the binomial coefficient

$$\binom{x}{y} = \frac{x!}{y!(x-y)!} \tag{5-377}$$

For distinguishable input fields the output state is slightly simpler

$$| n, 0\rangle \, | 0, m\rangle \rightarrow \sum_{j=0}^{n} \sum_{k=0}^{m} K_{\perp}(n, m, j, k) \, | j, k\rangle \, | n - j, m - k\rangle$$

$$K_{\perp}(n, m, j, k) = \sum_{j=0}^{n} \sum_{k=0}^{m} t^{m-k+j}(ir)^{n-j+k} \sqrt{\binom{j}{k}\binom{n-j}{m-k}} \tag{5-378}$$

The above calculated output modes impinge on the single photon detectors (SPDs). These may be characterized by the probability of detecting an incident single photon. From this single photon detection probability fit is also possible to deduce the probability of detecting a pulse consisting of multiple photons, keeping in mind that, irrespective of the number of photons, only a single detection event can be generated. To write $p_1(n)$ for the probability for generating one detector event given n incident photns, and it is useful to note that it relates to the probability $p_0(n)$ of detecting nothing as $p_1(n) = 1 - p_0(n)$.

The probability for not detecting n photons is, on the other hand, easily computed as $p_0(n) = (1 - \eta)^n$. Since the two detectors at the BS outputs are independent, the probability $p_{11}(n, m)$ of generating a coincidence event, i. e. having simultaneous detection events in each of the detectors, given n and m photons in one and the other output is simply $p_{11}(n, m) = p_1(n) p_1(m)$. Thus the probability for coincidence detection becomes

$$p_{11}(n,m) = [1-(1-\eta_1)^n][1-(1-\eta_2)^m] \tag{5-379}$$

where η_1 and η_2 are the single photon detection probabilities for detector 1 and 2, respectively. Expressing the coincidence detection probability in terms of Fock states at the BS input have

$$P_{11}^{\|}(n,m) = \sum_{j=0}^{n}\sum_{k=0}^{m} |K_{\|}(n,m,j,k)|^2 p_{11}(j+k, n+m-j-k)$$

$$= \sum_{j=0}^{n}\sum_{k=0}^{m} |K_{\|}(n,m,j,k)|^2 [1-(1-\eta_1)^{j+k}][1-(1-\eta_2)^{n+m-j-k}] \tag{5-380}$$

where $K_{\|}(n,m,j,k)$ should be substituted with the factor from Eq. (5-372). For distinguishable inputs find a similar expression for (n,m) using the factor $(n;m;j;k)$ from Eq. (5-378). It is assumed that the detector at a given spatial output mode is equally sensitive to photons in both auxiliary modes. Now in the position to formulate an expression for the different detection probabilities given a particular set of coherent input fields. The probability to generate a detection event in both detectors, given coherent input fields of amplitudes α and β is

$$P_{11}^{\|(\perp)}(\alpha,\beta) = \sum_{n=0}^{\infty}\sum_{m=0}^{\infty} \mathrm{e}^{-|\alpha|^2-|\beta|^2} \frac{(\alpha^n\beta^m)^2}{n!m!} P_{11}^{\|(\perp)}(n,m) \tag{5-381}$$

(Note that to distinguish the probability in Eq. (5-380), which is applicable to Fock states, from that in Eq. (5-381), which applies to coherent state inputs, we use P to denote the former and P for the latter.) This allows us to derive the visibility of the HOM interference on the two detectors as

$$\nu_{11}(\alpha,\beta,\eta_1,\eta_2,r) = \frac{P_{11}^{\perp}(\alpha,\beta) - P_{11}^{\|}(\alpha,\beta)}{P_{11}^{\perp}(\alpha,\beta)} \tag{5-382}$$

where have spelled out the parameters that affect the value of the visibility. The quantity ν_{11} is referred to as the HOM visibility.

5.13.3 Simplied model for HOM visibility

To gain some intuitive understanding of the way the HOM visibility is affected by the experimental parameters resort to a couple of approximations. Firstly, assume equal mean photon numbers at the inputs of the beam-splitter,

$$|\alpha|^2 = |\beta|^2 \equiv \mu \tag{5-383}$$

the BS ratio to be 50:50, and the detectors to have equal single photon detection probability $\eta_1 = \eta_2 \equiv \eta$. Secondly, since normally work at very low mean photon numbers $\mu < 1$ only the first couple of terms of Eq. (5-389) need to be included. Thus, for the coincidence detection events we get the probabilities

$$P_{11}^{\|} = \eta^2 \frac{\mu^2}{2} \tag{5-384}$$

$$P_{11}^{\perp} = \eta^2\mu^2 \tag{5-385}$$

which results in a HOM visibility of

$$\nu_{11} = \frac{1}{2} \tag{5-386}$$

A key point is that the HOM visibility of 50% is independent of the mean photon number μ. This observation can be explained by noting that in this low order treatment the coincidences in the case of indistinguishable input modes stem mostly from events in which two photons are present at the same input, which occurs with probability $p_0p_2 + p_2p_0$. For distinguishable input modes the coincidences stem from all events that contain two photons at the input, i. e. $p_1 p_1 + p_0 p_2 + p_2 p_0$. Since, according to Eq. (5-372), for coherent input states, all of these probabilities scale in the same way with the mean photon number, their ratio, and thus the visibility of Eq. (5-382), is constant for all mean photon numbers.

Compilation of experimental results for HOM interference at the few-photon level. Here show the plots of coincidence count rates on which the few-photon values in Table 5-4 of the main text are based. To restate that coincidence count rates are proportional to coincidence probabilities by a factor that is given by the average experimental repetition rate. Moreover, when calculating the HOM visibility, only the relative probabilities or count rates in a measurement are important. In the experiments we change the mutual polarization, time separation, or frequency difference of the pulses at the BS in the main text referred to as HOM-BS input as explained in the earlier in the Supplementary Information. Deactivated memories: the data in order of the number of activated memories starting with none is pulses merely pass through attenuated to the BS. In Figure 5-111(a) the coincidence counts as vary the polarization difference of the pulses at the two inputs of the BS. In Figure 5-111(b) display the coincidence counts as step the temporal separation of the pulses at the two inputs of the BS. The count rates for these measurements are generally higher than all the other count rates presented. This is because this data was acquired by looking at coincidences between the transmitted part of the probe pulses in the configuration of two active quantum memories Figure 5-111 (c) shows the coincidence count rates as function of the frequency difference of the two pulses at the BS inputs. The horizontal line and surrounding shaded band shown in Figure 5-111(c), as well as in Figure 5-112(c) give the coincidence counts for completely distinguishable input photons as obtained by making the polarizations orthogonal. As noted earlier in the Supplementary Information, it is necessary to resort to the polarization degree of freedom in order to make the pulses completely distinguishable. The visibility from the fit is noticeably lower than that obtained when change the other degrees of freedom. There are two main reasons for this. The first is that, in order to generate pulses with different frequencies, we drive the AOM at the limits of its bandwidth. This, in turn, necessitates setting the RF drive signal amplitude high whereby the frequency purity of the signal is contaminated by higher-order harmonics. Although it is not expected to change the maximal interference value occurring when the pulses are generated with the same modulation frequency, it will alter the shape of the interference as a function of the pulse frequency difference. Hence, the fitted Gaussian curve, assuming a Fourier limited pulse, may not correctly reproduce the actual frequency dependence of the interference. Indeed, the minimum coincidence factor reducing the observed visibility is related to the need to adjust the AOM drive amplitude to balance the bandwidth limitation.

The limited accuracy with which we are able to estimate the appropriate RF amplitude results in significant scattering of the coincidence counts due to variations in input pulse intensities. Unfortunately, the manifestation of the HOM interference in the single-detector count rates which will be elaborated later in the Supplementary Information means that such a normalization procedure tends to reduce the visibility in the coincidence counts.

One active memory: Next in line are the plots for the case in which only memory is activated, while the other is left inactive. In Figure 5-112 we present the coincidence count rates when changing the same degrees of freedom as in case of both memories being inactive. Additionally, in

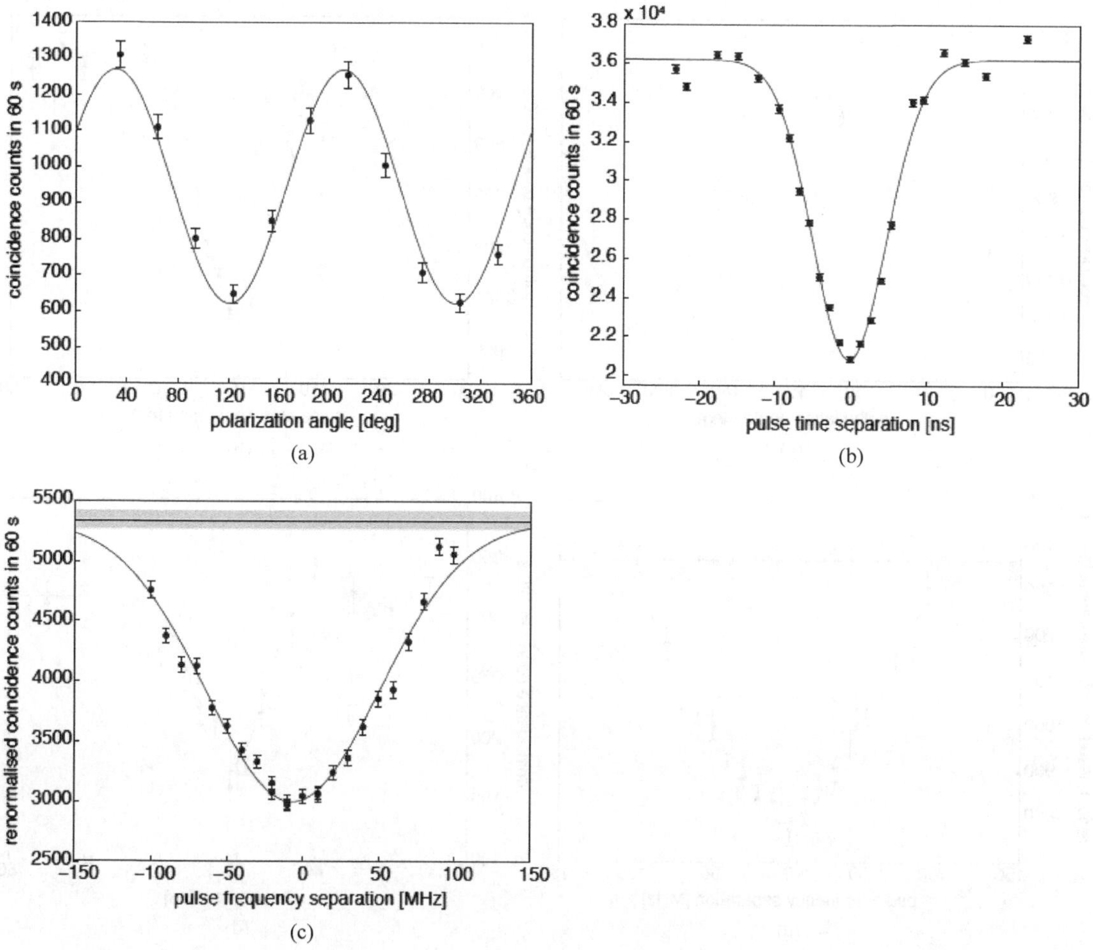

Fig. 5-111 HOM interference manifested in coincidence counts between BS outputs with inactive memories.

Figure 5-112(d), the coincidence count rates when changing the storage time in the quantum memory. Two active memories: Lastly, the plots for the case in which both memories are activated. Due to limitations in current setup it is not possible to simultaneously generate two quantum memories with different storage times, and therefore we do not acquire a storage time scan when both memories are active. Furthermore, skip the characterization with respect to the spectral degree of freedom. The coincidence count data for the remaining two degrees of freedom are plotted in Figure 5-113, which also includes the appropriate fits.

Manifestation of HOM interference in single detector counts, also evaluate the effect of the two-photon interference on the counts registered by a single detector. This is easily done by amending the detection probability to the case of one detection event in one detector and any number of events x (i.e. $x = 0.1$) in the other detector, arrive at

$$p_{1x}(n,m) = 1-(1-\eta_1)^n \tag{5-387}$$

This expression can be inserted into Eq. (5-377) to calculate (n, m), which, through Eq. (5-381), gives

$$P_{1x}^{\parallel(\perp)}(\alpha,\beta)$$

and from which the single-detector visibility ν_{1x} is defined analogous to Eq. (5-382) can formulate a simplified expression by using the same approximations as in the case of coincidence detections:

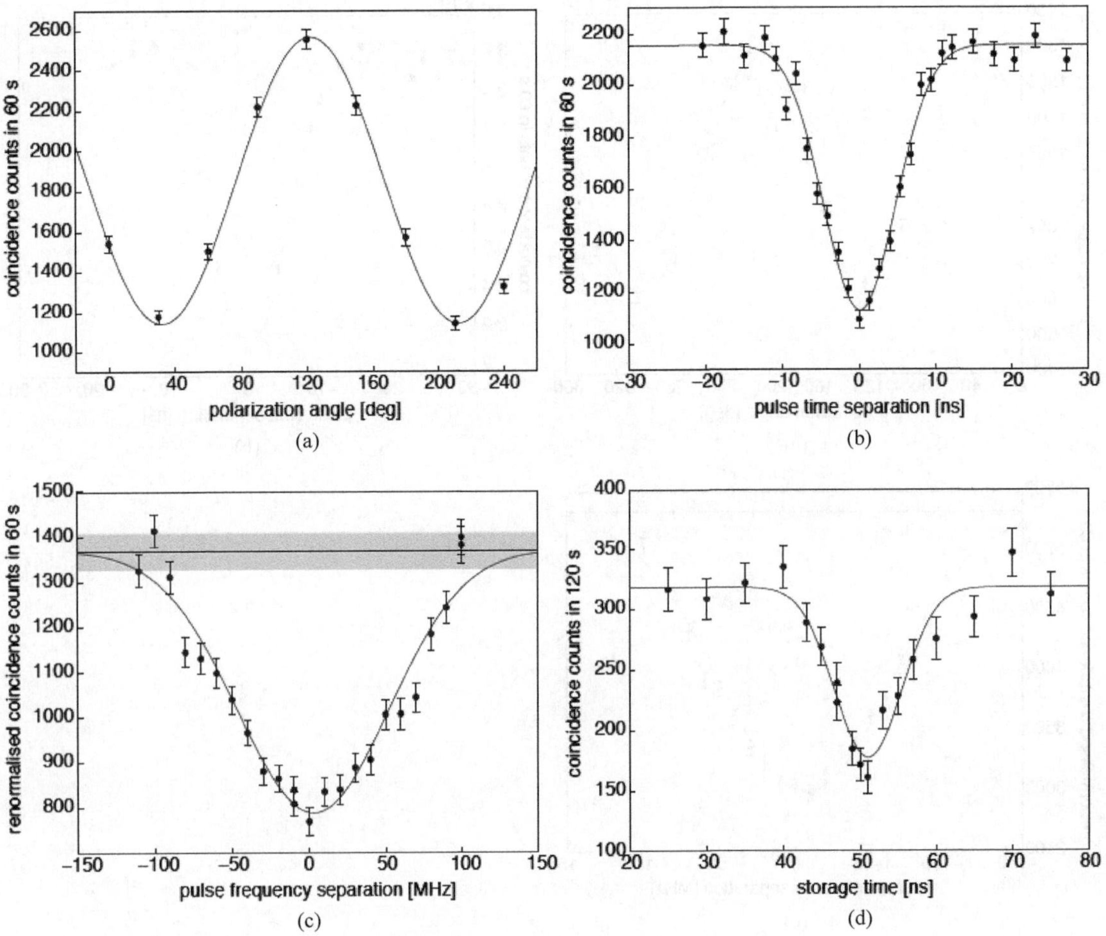

Fig. 5-112 HOM interference manifested in coincidence counts between BS outputs with one active memory.

$$P_{1x}^{\parallel} = \eta\mu + \eta\left(2 - \frac{3\eta}{4}\right)\mu^2 \tag{5-388}$$

$$P_{1x}^{\perp} = \eta\mu + \eta\left(2 - \frac{\eta}{2}\right)\mu^2 \tag{5-389}$$

from which get the single-detector visibility

$$\nu_{1x} = \frac{\eta\mu}{4 + 2(4 - \eta)\mu} \tag{5-390}$$

In the limit of low detector efficiency ν_{1x}, in that case, the probability of detecting two photons impinging on the detector is simply twice that of detecting one. This nulls the limitation that only a single detection event can be generated per pulse. Furthermore, the single-detector visibility also goes to zero for very low mean photon numbers.

In this case it is very unlikely to have two photons either at the same or at different input ports of the BS, hence most of the single-detector counts stem from single photons from either one or the other input of the BS. It is interesting to note that if η is known for a detector, then, from observing the single-detector visibility, it is in principle possible to estimate the mean photon number per pulse μ.

Another important consequence of the manifestation of two-photon interference in the single-

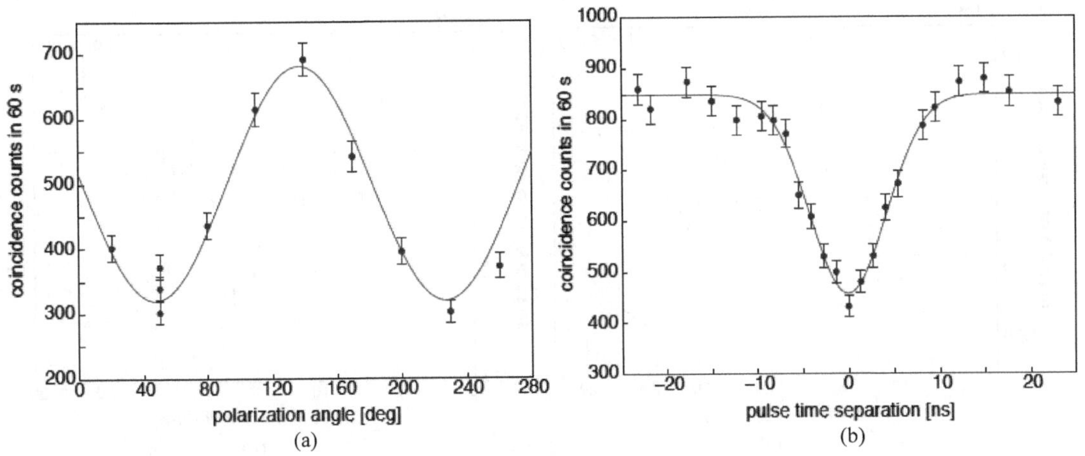

Fig. 5-113 HOM interference manifested in coincidence counts between BS outputs with two active memories.

detector counts is that the single-detector counts cannot generally be used to normalize the coincidence counts w.r.t. compuations in the input pulse intensities. Only for detectors with low detection efficiency or very low mean photon numbers, in which case V_{1x}, is this normalization possible. Experimental results on HOM interference manifested in single-detector counts.

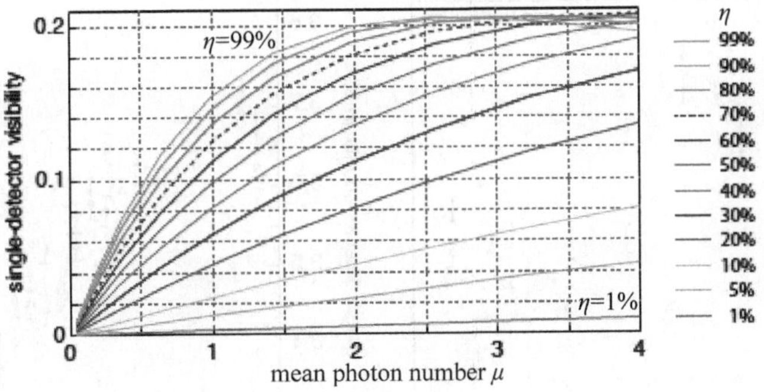

Fig. 5-114 Plots of single-detector visibility as a function of the mean photon number for detectors with a range of single photon detection probabilities η.

First, in Figure 5-115, Now present the single-detector counts corresponding to the coincidence counts depicted in Figure 5-112(a), (b). In the case where we vary the polarization and time separation we see a clear change in the single-detector counts, which, moreover, is evidently correlated with the change in coincidence counts. The count variation due to the two-photon interference is somewhat masked by the single-detector count scatter, which is due to intensity uctuations mainly in the light going through the 10km delay line. To fit the data in Figures 5-115(a) and (b) with a sine and Gaussian function, respectively. For the former we find a mean photon number of $\mu = 0.52$ while from the latter estimate $\mu = 0.54$. From the number of single-detector counts there is some evidence to conclude that the light intensity is about 15% higher. To this should be added about 25% uncertainty for the intensity at the BS w.r.t. the intensity at the detector due to variation in the loss in the fibre mating sleeves.

Finally, the scatter of the counts makes the fits themselves rather uncertain. Nevertheless, the mere fact that the two-photon interference is manifested in the single-detector counts validates the order of magnitude of the mean photon number, as depicted in Figure 5-114. Figure 5-116 depicts the single-detector counts corresponding to the coincidence counts depicted in Figures 5-113(a),

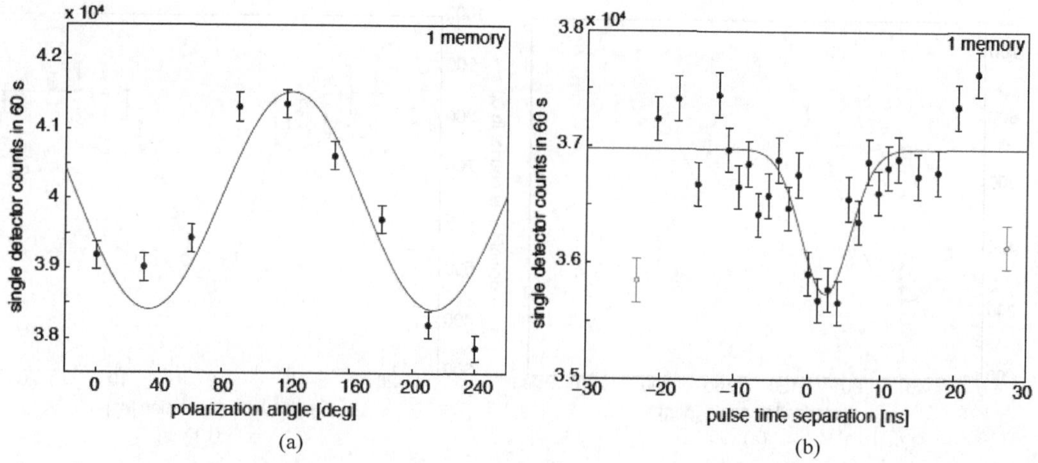

Fig. 5-115 HOM interference manifested in single-detector counts in the case of one active quantum memory when changing (a) polarization and (b) time difference between pulses at BS input.

(b). Again, from fitting the appropriate functions to the polarization and time data yields visibilities around 7%, corresponding to mean photon numbers of around $\mu = 0.5$.

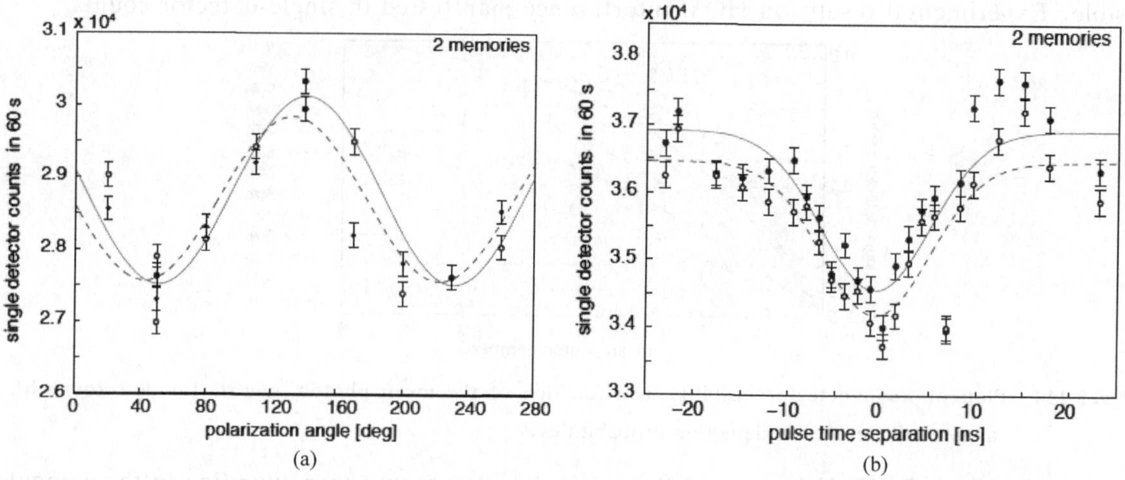

Fig. 5-116 HOM interference manifested in single-detector counts in the case of two active quantum memories when changing (a) polarization and (b) time difference between pulses at BS input.

- Bell-state measurement

In this section derive an analytical expression for the coincidence count rates corresponding to projections onto the $|\psi\rangle$ Bell-state for time-bin qubits detected by the two detectors at the output of the HOM-BS. To that end, will introduce a number of approximations as did previously in order to calculate the HOM interference in the coincidence counts. In the limit of low mean photon numbers, two coherent states impinging onto the two inputs of a 50 : 50BS can be represented in terms of Fock states as

$$
\begin{aligned}
|\psi\rangle_{ab} &= \sqrt{p(1,1)}\,|11\rangle_{a,b} + \sqrt{p(2,0)}\,|20\rangle_{a,b} + \sqrt{p(0,2)}\,|02\rangle_{a,b} \\
&= \left(\sqrt{p(1,1)}\,(\hat{a}^{\dagger} \otimes \hat{b}^{\dagger}) + \frac{1}{\sqrt{2!}}\left[\sqrt{p(2,0)}\,((\hat{a}^{\dagger})^{2} \otimes I) + \sqrt{p(0,2)}\,(I \otimes (\hat{b}^{\dagger})^{2})\right]\right)|00\rangle_{a,b}
\end{aligned}
\tag{5-391}
$$

where the subscripts on the state vector refer to the order of listing the input modes. The factors written as $p(n,m)$ denote the probability of having n and m photons in mode a and b, and are given by

$$p(n,m)=|({}_a\langle n|\otimes{}_b\langle m|)(|\alpha\rangle_a\otimes|\beta\rangle_b)|^2=\frac{e^{-(|\alpha|^2+|\beta|^2)}}{n!m!}(|\alpha|^2)^n(|\beta|^2)^m \tag{5-392}$$

Stemming from the low mean photon number assumption, do not include terms with more than two photons. Assuming that our detectors are noiseless, terms with a total of one or no photons are also left out as they cannot generate any coincidence counts. For a time-bin qubit, the Fock state is created in a superposition of two temporal modes, i.e., an early (e) and a late (l) mode, by the creation operators for the spatial input mode $x^\dagger(x^\dagger=a^\dagger,b^\dagger)$ of the beam-splitter, as

$$(\hat{x}^\dagger)^n|0\rangle_x\rightarrow\left[\cos\left(\frac{\theta_x}{2}\right)\hat{x}_e^\dagger\otimes I+e^{i\phi_x}\sin\left(\frac{\theta_x}{2}\right)I\otimes\hat{x}_l^\dagger\right]^n|00\rangle_{xe,xl} \tag{5-393}$$

where cos $(\theta_x/2)$ and sin $(\theta_x/2)$ are the amplitudes of, and Φ_x is the relative phase between, the two temporal modes composing the time-bin qubit. The subsript xe refers to the early time-bin of the spatial mode x and similarly for xl. Note, that we sometimes simplify the notation for the time-bin qubit states as

$$|e\rangle_x\equiv|10\rangle_{xe,xl}=(\hat{x}_e^\dagger\otimes I)|00\rangle_{xe,xl} \tag{5-394}$$

If insert the expression in Eq. (5-393) in place of the $\hat{a}$ and $\hat{b}$ operators in Eq. (5-391) can get the expression for the wavefunction $|\psi\rangle_{ab}$ for time-bin qubits at the HOM-BS inputs, can split this expression into the various contributions given in Eq. (5-391)

$$\begin{aligned}(\hat{a}^\dagger\otimes\hat{b}^\dagger)|00\rangle_{ab}\rightarrow\frac{1}{2}\Bigg[&\left(ie^{i\phi_b}\cos\left(\frac{\theta_a}{2}\right)\sin\left(\frac{\theta_b}{2}\right)+ie^{i\phi_a}\sin\left(\frac{\theta_a}{2}\right)\cos\left(\frac{\theta_b}{2}\right)\right)(\hat{c}_e^\dagger\hat{c}_l^\dagger+\hat{d}_e^\dagger\hat{d}_l^\dagger)+\\&\left(e^{i\phi b}\cos\left(\frac{\theta_a}{2}\right)\sin\left(\frac{\theta_b}{2}\right)-e^{i\phi_a}\sin\left(\frac{\theta_a}{2}\right)\cos\left(\frac{\theta_b}{2}\right)\right)(\hat{c}_e^\dagger\hat{d}_l^\dagger-\hat{c}_l^\dagger\hat{d}_l^\dagger)+\\&ie^{i(\phi_a+\phi_b)}\sin\left(\frac{\theta_a}{2}\right)\sin\left(\frac{\theta_b}{2}\right)((\hat{c}_l^\dagger)^2+(\hat{d}_l^\dagger)^2)+\\&i\cos\left(\frac{\theta_a}{2}\right)\cos\left(\frac{\theta_b}{2}\right)((\hat{c}_e^\dagger)^2+(\hat{d}_e^\dagger)^2)\Bigg]|0000\rangle_{ce,cl,de,dl}\end{aligned} \tag{5-395}$$

$$\begin{aligned}((\hat{a}^\dagger)^2\otimes I)|00\rangle_{ab}\rightarrow\frac{1}{2}\Bigg[&2e^{i\phi_a}\cos\left(\frac{\theta_a}{2}\right)\sin\left(\frac{\theta_a}{2}\right)(\hat{c}_e^\dagger\hat{c}_l^\dagger-\hat{d}_e^\dagger\hat{d}_l^\dagger)+\\&i2e^{i\phi_a}\cos\left(\frac{\theta_a}{2}\right)\sin\left(\frac{\theta_a}{2}\right)(\hat{c}_e^\dagger\hat{d}_l^\dagger+\hat{c}_l^\dagger\hat{d}_e^\dagger)+\\&\cos^2\left(\frac{\theta_a}{2}\right)((\hat{c}_e^\dagger)^2+i2\hat{c}_e^\dagger\hat{d}_e^\dagger-(\hat{d}_e^\dagger)^2)+\\&e^{i2\phi_a}\sin^2\left(\frac{\theta_a}{2}\right)((\hat{c}_l^\dagger)^2+i2\hat{c}_l^\dagger\hat{d}_l^\dagger-(\hat{d}_l^\dagger)^2)\Bigg]|0000\rangle_{ce,cl,de,dl}\end{aligned} \tag{5-396}$$

and similarly for $(I(\hat{b}^\dagger)^2)|_{ab}$. Again, the subscripts on the state vector refer to the order of listing the temporal and spatial modes, e.g. ce labels the early bin of the spatial output mode c. will look for coincidence detection events that correspond to projections onto the Bell-state

$$|\psi_-\rangle_{cd}=\frac{1}{\sqrt{2}}(\hat{c}_e^\dagger\hat{d}_l^\dagger-\hat{c}_l^\dagger\hat{d}_e^\dagger)|0000\rangle_{ce,cl,de,dl} \tag{5-397}$$

Such projections correspond to a detection event in the early time-bin in one detector followed by a detection event in the late time-bin in the other detector. This projection occurs with a probability

$$P_-(\theta_a,\phi_a,\theta_b,\phi_b)=|{}_{cd}\langle\psi_-|\psi(\theta_a,\phi_a,\theta_b,\phi_b)\rangle_{cd}|^2 \tag{5-398}$$

which can be computed by combining Eq. (5-395) with Eq. (5-391). Assuming equal mean photon numbers at the two inputs $|\alpha|^2 = |\beta|^2\mu$ and averaging over the coherent state phases, i.e. the complex angle between α and β, then get the expression

$$P_-(\theta_a, \phi_a, \theta_b, \phi_b) \propto \frac{\mu^2 e^{-2\mu}}{8}\left[4\sin^2\left(\frac{\theta_a+\theta_b}{2}\right) + \sin^2(\theta_a) + \sin^2(\theta_b) - 2\sin(\theta_a)\sin(\theta_b)(1+\cos(\phi_a - \phi_b))\right] \tag{5-399}$$

With this are able to calculate the probabilities of projection onto | different combinations of qubits at the two BS inputs, i.e. for different choices of the angles θ_x and Φ_x. In turn, this allows to calculate the | Bell-state measurement error rate as

$$e \equiv \frac{P_-^{\parallel}}{P_-^{\parallel} + P_-^{\perp}} \tag{5-400}$$

where is the projection probability when the two input qubit states are identical, i.e. $\psi_a = \psi_b$ and $\theta_a = \theta_b$, while is the projection probability for two orthogonal input qubit states. This is also defined in terms of count rates in Eq. (5-399) in the main text that will treat a number of relevant cases. Expected and observed error rates when $\Phi_a = \Phi_b = 0$. Using the simplified notation this corresponds to the case were the input qubit states are of the form

$$|\psi\rangle = \cos\left(\frac{\theta_z}{2}\right)|e\rangle + \sin\left(\frac{\theta_z}{2}\right)|l\rangle \tag{5-401}$$

When depicted on the Bloch sphere these qubits span the xz-plane. Using Eq. (5-399) compute the projection probability as

$$P_-(\theta_a, 0, \theta_b, 0) \propto \frac{\mu^2 e^{-2\mu}}{8}\left[4\sin^2\left(\frac{\theta_a+\theta_b}{2}\right) + \sin^2(\theta_a) + \sin^2(\theta_b) - 4\sin(\theta_a)\sin(\theta_b)\right] \tag{5-402}$$

interested in the probability for the case in which the input qubits are parallel ($\theta_a = \theta_b$) and for the case in which the input qubit states are orthogonal ($\theta_a = \theta_b - \pi$). Specifically, when we prepare two qubits (one at each input of the BS) in state $|e\rangle$, or two qubits in state $|i\rangle$, we expect = 0. The probability for observing a projection onto $|\psi\rangle$ increases as change θ_a (or θ_b), and reaches a maximum if one qubit is in state $|e\rangle$ and the other one in $|i\rangle$. Hence, using the expression for the error rate above (Eq. (5-400)), can find e(att)e/1 = 0. Now turn to measuring the coincidence rates for all combinations of $|e\rangle$ and $|l\rangle$ input states, and thus extracting and, using 0.6 photons per qubit at the memory input. More precisely, prepare the input qubit state $|e\rangle_a|e\rangle_b$ to measure (1) and then $|e\rangle_a|l\rangle_b$ to measure (1). Subsequently, prepare the input qubit state $|l\rangle_a|l\rangle_b$ to measure (2) and then $|l\rangle_a|e\rangle_b$ to measure (2). These yield the average values

$$P_-^{\parallel} = (P_-^{\parallel(1)} + P_-^{\parallel(2)})/2 \tag{5-403}$$

and

$$P_-^{\perp} = (P_-^{\perp(1)} + P_-^{\perp(2)})/2 \tag{5-404}$$

Expected and observed error rates when $\alpha_a = \alpha_b = \pi/2$. In this case the two input qubits are in equal superpositions of early and late bins, that is of the form

$$|\psi\rangle = \frac{1}{\sqrt{2}}(|e\rangle + e^{i\phi_x}|l\rangle) \tag{5-405}$$

On the Bloch sphere these are qubits that lie in the xy-plane. In this case compute

$$P_{-}(\pi/2,\phi_a,\pi/2,\phi_b) \propto \frac{\mu^2 e^{-2\mu}}{4}(2-\cos(\phi_a-\phi_b)) \tag{5-406}$$

Thus the $|\psi\rangle$ Bell-state projection probability is smallest but nonzero when $\Phi_a - \Phi_b = 0$, i.e. the qubit 16 states are parallel, and largest when the phases differ by π, i.e. the qubit states are orthogonal. Using again 0.6 photons per qubit, measure the coincidence counts for $\Phi_a - \Phi_b = 0$ and π giving us and, respectively. This indicates that either the measurement suffers from imperfections such as detector noise or the modes at the BS are not completely indistinguishable, which in turn could be due imperfectly generated qubit states or imperfect storage of the qubit in the quantum memory. Bounds for attenuated laser pulses stored in quantum and classical memories: ow compare the performance of our Bell-state measurement to a number of relevant bounds assuming always that any imperfections arise from the imperfect storage of the photon in the memory, and will derive bounds to the error rate in the case of one qubit being stored in either a classical memory (CM) or quantum memory (QM). To accommodate this scenario we assume that the memory performs the following operation

$$|\psi\rangle\langle\psi| \to F|\psi\rangle\langle\psi| + (1-F)|\psi^{\perp}\rangle\langle\psi^{\perp}| \tag{5-407}$$

where F denotes the fidelity of the stored state and $|\psi\rangle$ is the state orthogonal to $|\psi\rangle$. For a classical memory FCM = 0.667 whereas for a quantum memory FQM = 1. Doing the replacement

$$P_{-}^{\parallel} \to FP_{-}^{\parallel} + (1-F)P_{-}^{\perp} \tag{5-408}$$

and likewise for can express the error rate expected after imperfect storage of one of the pulses partaking in the Bell-state measurement:

$$e = \frac{FP_{-}^{\parallel} + (1-F)P_{-}^{\perp}}{P_{-}^{\parallel} + P_{-}^{\perp}} \tag{5-409}$$

where in this case the probabilities and refer to those expected without the memory. Since the expected values for and differ between the $e = l$ and $+/-$ bases we treat them separately. Beginning with the $e = l$ basis we use Eq. (5-409) with the values from Eq. (5-412) to derive a bound for the error rate of the Bell-state measurement for one of the two qubits being recalled from a quantum or a classical memory. Here emphasize once more that have assumed that the reduction in error rates is due solely to the memory and thus indicates the fidelity of the memory. However, this is likely not the case as imperfections in the state preparation and detector noise also contribute to the reduction in error rate. Bounds for single photons stored in quantum and classical memories: Although we do not use single photon sources for the experiments reported here, it is interesting to determine how well our results measure up to those that could have been obtained if single photon sources had been employed. In the following will derive the error rate for the Bell-state measurement using qubits encoded into single photons. To this end step back to Eq. (5-391), and note that for single photon sources all probabilities are 0 except for $p(1,1)$, which describes the probability of having a single photon at each BS input. Thus, in the output state we only need to keep the terms from Eq. (5-395), which in turn means that the Bell-state projection probability can be written as

$$P_{-}(\theta_a,\phi_a,\theta_b,\phi_b) \propto \frac{1}{4}\left[\sin^2\left(\frac{\theta_a+\theta_b}{2}\right) + \sin^2\left(\frac{\theta_a-\theta_b}{2}\right) - \sin(\theta_a)\sin(\theta_b)\cos(\phi_a-\phi_b)\right] \tag{5-410}$$

This means that even with a single photon source at ones disposal the error rates that ensured could not have been attained with a classical memory. Experiments at mean photon numbers above one. In this final section we will explore in greater detail the HOM interference dependence on the angle $\Phi_a - \Phi_b$ between a set of equal superposition qubit states

$$|\psi\rangle_x = \frac{1}{\sqrt{2}}(|e\rangle + e^{i\phi_x}|l\rangle) \tag{5-411}$$

which in line with the preceding sections belong to the +/− basis. According to Eq. (5-406) the coincidence count rates vary as function of cos ($\Phi_a - \Phi_b$). In Figure 5-117 show measured coincidence count rates as function of $\Phi_a - \Phi_b$ for a mean photon number per qubit before the memory of around 20.

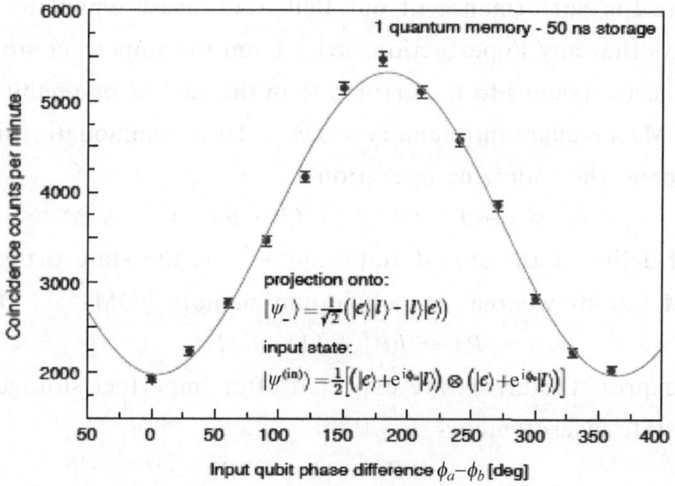

Fig. 5-117 Rate of projection of pairs of time-bin qubits with relative phase $\Phi_a - \Phi_b$ onto $|\psi\rangle$. Each data point was acquired over 60s.

As expected the coincidence detection probability reaches its maximum when two input qubits are orthogonal ($\Phi_a - \Phi_b = \pi$) and when they are identical ($\Phi_a - \Phi_b = 0$) it reaches a minimum. It is natural to define a Bell-state measurement visibility as

$$V = \frac{P_-^{\perp} - P_-^{\parallel}}{P_-^{\perp}} \tag{5-412}$$

- Foundations for nanophotonics

One difference between them is on the length scale. The wavelengths associated with electrons are usually considerably shorter than those of photons. Under most circumstances the electrons will be characterized by relatively larger values of momentum.

This is why electron microscopy (in which the electron energy and momentum are controlled by the value of the accelerating high voltage) provides a significantly improved resolution over optical (photon) microscopy, since the ultimate resolution of a microscope is diffraction-limited to the size of the wavelength. The values of momentum, which may be ascribed to electrons bound in atoms or molecules or the conduction electrons propagating in a solid, are also relatively high compared to those of photons, thus the characteristic lengths are shorter than wavelengths of light. An important consequence derived from this feature is that "size" or "confinement" effects for photons take place at larger size scales than those for electrons.

The propagation of photons as waves is described in form of an electromagnetic disturbance in a medium, which involves an electric field E (corresponding displacement D) and an orthogonal magnetic field H (corresponding displacement B), both being perpendicular to the direction of propagation in a free space. A set of Maxwell's equations describes these electric and magnetic fields. An eigenvalue equation describes a mathematical operation $\hat{O}$ on a function F as

$$\hat{O}F = CF \tag{5-413}$$

to yield a product of a constant C, called eigenvalue, and the same function. Functions fulfilling the above condition are called eigenfunctions of the operator $\hat{O}$. The dielectric constant, and thereby the refractive index, describes the resistance of a medium to the propagation of an electromagnetic wave through it. Therefore, light propagation speed c in a medium is reduced from light propagation speed c_0 in vacuum by the relation

$$c = \frac{c_0}{n} = \frac{c_0}{\varepsilon^{1/2}} \tag{5-414}$$

The eigenvalue equation in Table 5-3 involves the magnetic displacement vector, B.

Table 5-3 Similarities in Characteristics of Photons and Electrons the same relation $\lambda = h/p$, where p is the particle momentum.

Photons	Electrons
Wavelength	
$\lambda = \frac{h}{p} = \frac{c}{v}$	$\lambda = \frac{h}{p} = \frac{h}{mv}$
Eigenvalue (Wave) Equation	
$\left\{\nabla \times \frac{1}{\varepsilon(r)} \nabla \times\right\} B(r) = \left(\frac{\omega}{c}\right)^2 B(r)$	$\hat{H}\psi(r) = -\frac{\hbar^2}{2m}(\nabla \cdot \nabla + V(r))\psi(r) = E\psi$
Free-Space Propagation	
Plane wave	Plane wave:
$E = \left(\frac{1}{2}\right)E^{\circ}(e^{ik \cdot r - \omega t} + e^{-ik \cdot r + \omega t})$	$\Psi = c(e^{ik \cdot r - \omega t} + e^{-ik \cdot r + \omega t})$
k = wavevector, a real quantity	k = wavevector, a real quantity
Interaction Potential in a Medium	
Dielectric constant (refractive index)	Coulomb interactions
Propagation Through a Classically Forbidden Zone	
Photon tunneling (evanescent wave) with wavevector, k, imaginary and hence amplitude decaying exponentially in the forbidden zone	Electron-tunneling with the amplitude (probability) decaying exponentially in the forbidden zone
Localization	
Strong scattering derived from large variations in dielectric constant (e.g., in photonic crystals)	Strong scattering derived from a large variation in Coulomb interactions (e.g., in electronic semiconductor crystals)
Cooperative Effects	
Nonlinear optical interactions	Many-body correlation Superconducting Cooper pairs Biexciton formation

Since the electric field E and the magnetic displacement B are related by the Maxwell's equations, an equivalent equation can be written in terms of E. However, this equation is often solved using B, because of its more suitable mathematical character. (The operator for B has a desirable character called Hermitian). The eigenvalue C in the equations for photon is $(\omega/c)^2$, which gives a set of allowed frequence.

- Photons and electrons

The dielectric constant may be either constant throughout the medium or dependent on spatial location r and the wavevector k. The corresponding wave equation for electrons is the Schrödinger equation, and its time-independent form is often written in the form of an eigenvalue equation shown in Table 5-5 (Levine, 2000). Here $\hat{H}$, called the Hamiltonian operator, consists of the sum of operator forms of the kinetic and the potential energies of an electron and is thus given as

$$\hat{H} = -\frac{\hbar^2}{2m}\left(\frac{\partial^2}{\partial x^2} + \frac{\partial^2}{\partial y^2} + \frac{\partial^2}{\partial z^2}\right) + V(r) = -\frac{\hbar^2}{2m}\nabla^2 + V(r) \tag{5-415}$$

The first term (all in the parentheses) is derived from the kinetic energy; the second term, $V(r)$, is derived from the potential energy of the electron due to its interaction (Coulomb) with the surrounding medium.

Despite the similarities described above, electrons and photons also have important differences. The electrons generate a scalar field while the photons are vector fields (light is polarized). Electrons possess spin, and thus their distribution is described by Fermi-Dirac statistics. For this reason, they are also called fermions.

Photons have no spin, and their distribution is described by Bose-Einstein statistics. For this reason, photons are called bosons. Finally, since electrons bear a charge while the charge of photons is zero, there are principal differences in their interactions with external static electric and magnetic fields.

- Free-Space propagation

In a "free-space" propagation, there is no interaction potential or it is constant in space. For photons, it simply implies that no spatial variation of refractive index n.

In such a case, the propagation of the electromagnetic wave is described by a plane wave for which the electric field described in the complex plane (using a real and an imaginary part) is shown in the Table 5-5. It is a propagating electromagnetic wave with oscillating (sinusoidal) electric field E (and the corresponding magnetic field B). The amplitude of the field is described by E°. The direction of propagation is described by the propagation vector k whose magnitude relates to the momentum as $p = hk$.

The wavevector k has the length equal to

$$k = |k| = \frac{2\pi}{\lambda} \tag{5-416}$$

The positive k describes forward propagation (e. g., left to right), while the negative k describes the propagation in the backward direction (right to left). The corpuscular properties of the electromagnetic wave are described by the photon energy

$$E = h\nu = \hbar\omega = \frac{hc}{\lambda} \tag{5-417}$$

These relationships yield the dispersion relation

$$\omega = c\,|k| \tag{5-418}$$

which describes the dependence of the frequency (or energy) of a photon on its wavevector and is illustrated by the linear dispersion relation shown in Figure 5-118. Similarly, for free-space propagation of an unbound electron, the wavefunction obtained by the solution of the Schrödinger equation is an oscillating (sinusoidal) plane wave, similar to that for a photon, and is characterized by a wavevector k.

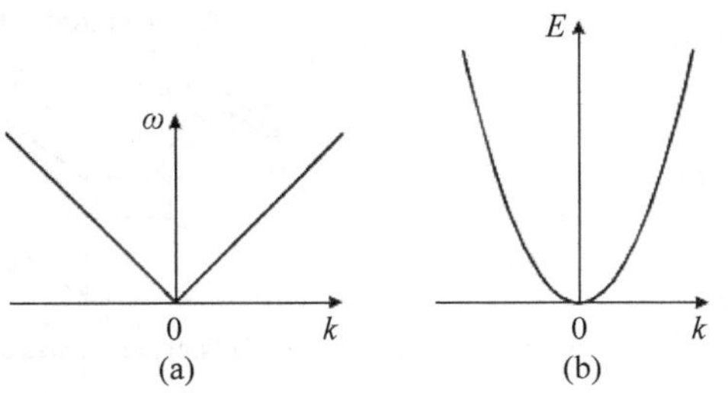

Fig. 5-118 Dispersion relation showing the dependence of energy on the wavevector for a free-space propagation. (a) Dispersion for photons. (b) Dispersion for electrons.

Therefore, the probability density described by the absolute value of the square of the wavefunction is the same everywhere, which again conforms to the free state of the electron. The energy dispersion for a free electron is described by a parabolic relation (quadratic dependence on k) as

$$E = \frac{\hbar^2 k^2}{2m} \tag{5-419}$$

where m is the mass of the electron. A modification of the free electron theory is often used to describe the characteristics of electrons in metals (the so-called Drude model). One often refers to such behavior as delocalization of electron or electronic wavefunction. This dependence of the electronic energy on the wavevector is also represented in Figure 5-118. One can clearly see that, even though analogous descriptions can be used for photons and electrons, the wavevector dependence of energy is different for photons (linear dependence) and electrons (quadratic dependence).

- Confinement of photons and electrons

The propagation of photons and electrons can be dimensionally confined by using areas of varying interaction potential in their propagation path to reflect or backscatter these particles, thus confining their propagation to a particular trajectory or a set of those.

In the case of photons, the confinement can be introduced by trapping light in a region of high refractive index or with high surface reflectivity. This confining region can be a waveguide or a cavity resonator. The examples of various confinements are shown in Figure 5-119. The confinements can be produced in one dimension such as in a plane, as in the case of a planar optical waveguide.

Here, the light propagation is confined in a layer (such as a thin film) of high refractive index, with the condition that the refractive index n_1 of the light-guiding layer is higher than the refractive index n_2 of the surrounding medium, as shown in Figure 5-119.

Thus the envelope of the electric field E (only spatially dependent part), represented in Table 5-5 for free-space propagation, is modified for a waveguide as

$$E = \frac{1}{2} f(x, y) a(z) (e^{i\beta z} + e^{-i\beta z}) \tag{5-420}$$

Equation (5-420) describes guiding in a fiber or a channel waveguide (a rectangular or square guiding channel) which has two-dimensional confinement. The function $a(z)$ is the electric amplitude in the z direction, which (in the absence of losses) is constant. The function $f(x, y)$ represents the electric field distribution in the confinement plane. In the case of a planar waveguide, producing confinement only in the x direction, only the x component shows a spatial distribution limited by the confining potential, while the y component of f is like that for a plane wave (free space) if a plane wave is used to excite the waveguide mode.

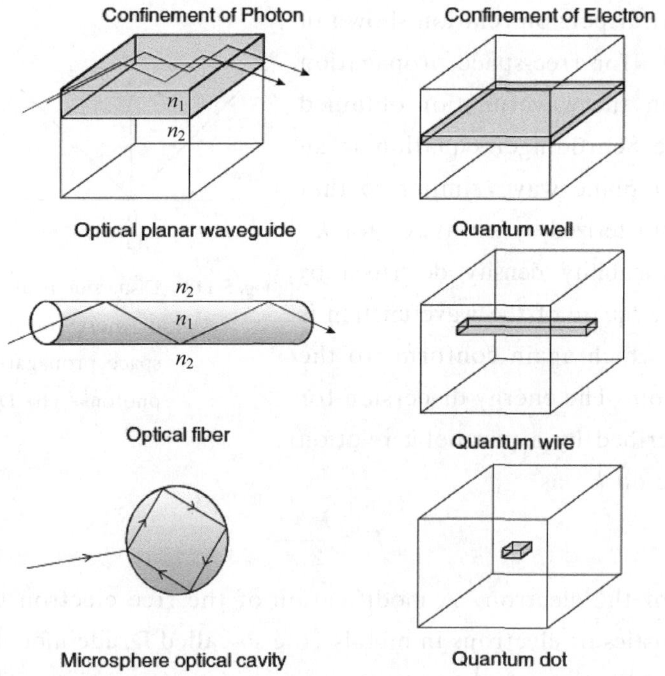

Fig. 5-119 The figure shows the classical optics picture using a ray path to describe light guiding (trapping) due to total internal reflection. In the case of a planar waveguide, the confinement is only in the vertical (x direction). The propagation direction is z. In the case of a fiber or a channel waveguide, the confinement is in the x and y directions. A microsphere is an example of an optical medium confining the light in all dimensions. The light is confined by the refractive index contrast between the guiding medium and the surrounding medium. Thus the contrast n_1/n_2 acts as a scattering potential creating barrier to light propagation.

On the other hand, it will show the characteristic spreading in the y direction if a beam of limited size (e.g., a Gaussian beam) is launched into the waveguide. The field distribution and the corresponding propagation constant are obtained by the solution of the Maxwell's equation and imposing the boundary conditions (defining the boundaries of the waveguide and the refractive index contrast). The solution of the wave equation shows that the confinement produces certain discrete sets of field distributions called eigenmodes, which are labeled by quantum numbers (integer). For a one-dimensional confinement, there is only one quantum number n, which can assume values 0, 1, and so on. Unfortunately, the same letter n, used to represent refractive index, is also used for quantum number. The readers should distinguish them based on the context being described. The examples of the various field distributions in the confining direction x for a planar (or slab) waveguide using a TE polarized light (where the polarization of light is in the plane of the film) is shown in Figure 5-120 for the various

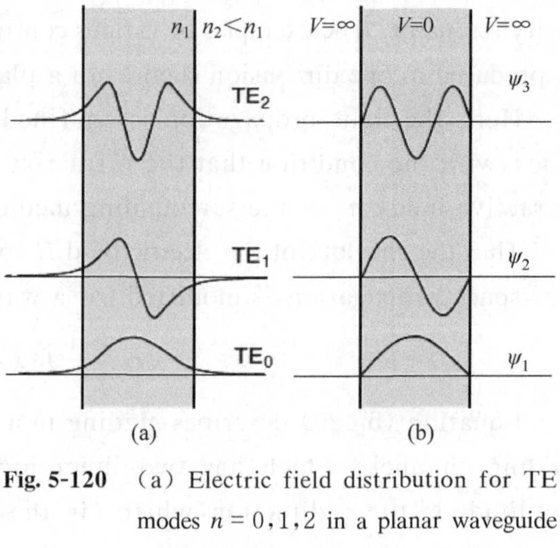

Fig. 5-120 (a) Electric field distribution for TE modes $n = 0, 1, 2$ in a planar waveguide with one-dimensional confinement of photons. (b) Wavefunction ψ for quantum levels $n = 1, 2, 3$ for an electron in a one-dimensional box.

modes (labeled as TE_0, TE_1, TE_2 etc.). From the above description, it is clear that confinement produces quantization—that is, discrete types of the field distributions, labeled by integral sets of quantum numbers used to represent the various eigenmodes.

That confinement of electrons also leads to modification of their wave properties and produces quantization—that is, discrete values for the possible eigenmodes (Merzbacher, 1998; Levine, 2000). The corresponding one-, two-, and three-dimensional confinement of electrons is also exhibited in Figure 5-119 (Kelly, 1995).

Classically, the electron will be completely confined within the potential energy barriers (walls). This is true if the potential barriers are infinite (as shown in the figure). However, for finite potential barriers the wavefunction does enter the region of the barrier and the pattern becomes similar to that for photons in Figure 5-119. The confinement for an electron is thus similar to that for a photon.

However, the length scales are different. To produce confinement effect for photons, the dimensions of the confining regions are in micrometers. But, for electrons, which have a significantly shorter wavelength, the confining dimensions have to be in nanometers to produce a significant quantization effect.

Here, only a general description of electronic confinement is given. A simple example that illustrates confinement effect is demonstrated by the model of an electron in a one-dimensional box, as depicted in Figure 5-121.

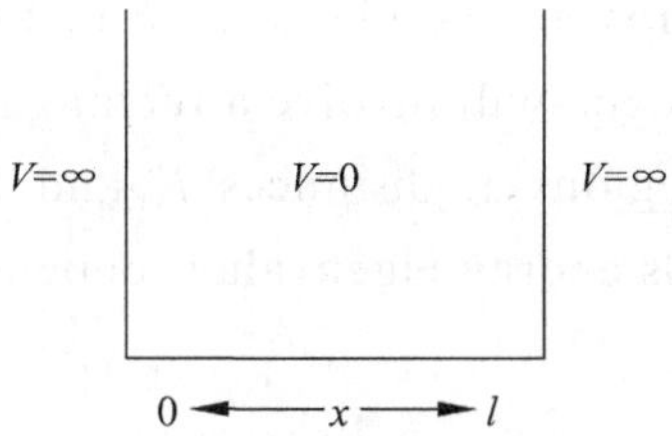

Fig. 5-121 Schematics of a particle in a one-dimensional box.

The electron is trapped (confined) in a box of length l within which the potential energy is zero. The potential energy rises to infinity at the ends of the box and stays at infinity outside the box.

Inside the box, the Schrödinger equation is solved with the following conditions:

$$V(x) = 0$$

$$\Psi(x) = 0 \quad \text{at} \quad x = 0 \quad \text{and} \quad x = l \qquad (5\text{-}421)$$

The solution yields sets of allowed values of E and the corresponding functions $V(x)$, each defining a given energy state of the particle and labeled by a quantum number n which takes an integral value starting from l (Atkins and dePaula, 2002). These values are defined as

$$E_n = \frac{n^2 h^2}{8ml^2} \quad \text{where} \quad n = 1, 2, 3, \cdots \qquad (5\text{-}422)$$

The lowest value of total energy is $E_1 = h_2/8ml^2$. Therefore, total energy E of an electron can never be zero when bound (or confined), even though its potential energy is zero. The discrete energy values are E_1, E_2, and so on, corresponding to quantum numbers $n = 1, 2, 3$, and so on.

The energy levels of an electron confined in a one-dimensional box. The gap between two successive levels describes the effect of quantization (discreteness). If it were zero, we would have a continuous variation of the energy as for a free electron and there would be no quantization.

The gap E between two successive levels E_n and E_{n+1} can be given as

$$\Delta E = (2n + 1)\frac{h^2}{8ml^2} \qquad (5\text{-}423)$$

This equation reveals that the gap between two successive levels decreases as $l2$ when the length of the box increases. Thus the spacing between successive electronic levels decreases as the electron is spread (delocalized) over a longer confining distance, as in the case of the electrons in a conjugated structure. The wavefunction for the various quantum levels is also modified from a plane wave, shown in Table 5-5.

The wavefunctions for the different quantum states n are given as

$$\Psi_n(x) = \left(\frac{2}{l}\right)^{1/2} \sin\left(\frac{n\pi x}{l}\right) = \frac{1}{2i}\left(\frac{2}{l}\right)^{1/2} (e^{ikx} - e^{-ikx}) \tag{5-424}$$

where $k = n/l$.

Figure 5-120 also shows the wavefunction Ψ_n as a function of n for various quantum states $n = 1,2,3$. From a correlation between the field distribution for photons for $n = 0,1,2$ modes of a planar waveguide and the wavefunctions for $n = 1,2,3$ quantum states of electrons, it can be clearly seen that the one-dimensional confinement effects are quite analogous. The probability density $|\Psi_n|^2$ varies with position within the box, and this variation is different for different quantum number states n. Again, this is a modification from the constant probability density for a free electron. For example, for $n = 1$ the maximum probability density is at the center of the box, in contrast to the plane wave picture showing equal probability at every place. The two-dimensional analogue will involve a rectangular box in which the potential barriers in the x and y directions are V regions at distances l_1 and l_2. The solution of a two-dimensional Schrödinger equation now yields energy eigenvalues depending on two quantum numbers n_1 and n_2 as

$$E_{n_1,n_2} = \left(\frac{n_1^2}{l_1^2} + \frac{n_2^2}{l_2^2}\right)\frac{h^2}{8m} \tag{5-425}$$

and the corresponding wavefunction is

$$\Psi_{n_1,n_2}(x,y) = \frac{2}{(l_1,l_2)^{1/2}} \sin\left(\frac{n_1\pi x}{l_1}\right) \sin\left(\frac{n_2\pi y}{l_2}\right) \tag{5-426}$$

with quantum numbers n_1 and n_2 each having allowed values of 1,2,3, and so on.

Similarly, three-dimensional confinement, like in a box of dimension l_1, l_2, and l_3, are characterized by three quantum numbers, n_1, n_2, and n_3, each assuming values 1,2,3, and so on. The eigenvalues En_1, n_2, n_3 and wavefunctions n_1, n_2, n_3 are simple extensions of Eqs. (5-425) and (5-426) to include a third term depending on n_3 and l_3.

- Propagation through a classically forbidden zone: Tunneling

In a classical picture, the photons and electrons are completely confined in the regions of confinement. For photons, it is seen by the ray optics for the propagating wave as shown in Figure 5-122. Similarly, classical physics predicts that, once trapped within the potential energy barriers where the energy E of an electron is less than the potential energy V due to the barrier, the electron will remain completely confined within the walls. However, the wave picture does not predict so. As shown in Figure 5-120, the field distribution of light confined in a waveguide extends beyond the boundaries of the

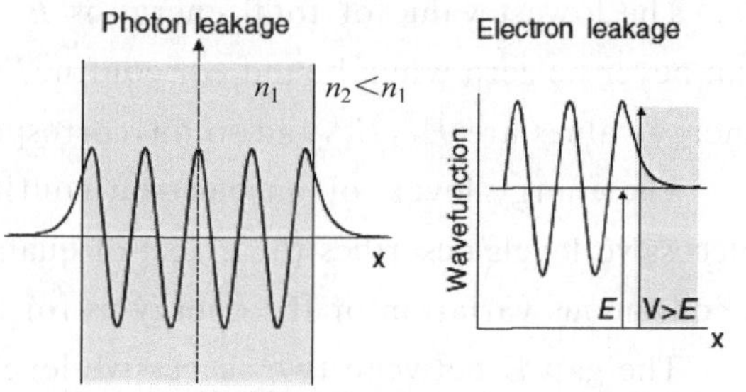

Fig. 5-122 Schematic representation of leakage of photons and electrons into classically energetically.

waveguide. This behavior is also shown in Figure 5-122.

Hence, light can leak into the region outside the waveguide, a classically forbidden region. This light leakage generates an electromagnetic field called evanescent wave (Courjon, 2003). The field distribution in the region outside the waveguide (classically forbidden region) does not behave like a plane wave having the wavevector k, as a real quantity. The electric field amplitude extending into the classically forbidden region decays exponentially with distance x into the medium of lower refractive index, from the boundary of the guiding region, according to the equation

$$E_x = E_0 \exp(-x/d_p) \tag{5-427}$$

E_0 in this equation is the electric field at the boundary of the waveguide. The parameter d_p, also called the penetration depth, is defined as the distance at which the electric field amplitude reduces to 1/e of E_0. Comparing the field E_x of Eq. (5-144)

5.13.4 Forbidden regions

For a plane wave shown in Table 5-4, it can be seen that Eq. (5-427) represents a case where the wavevector k is imaginary. This exponentially decaying wave with an imaginary wavevector k is the evanescent wave. Typically, the penetration depths dp for the visible light are 50～100nm. Thus, this wave can be used for nanophotonics, because optical interactions manifested by the evanescent wave kept localized within nanometer range. Evanescent waves have been used for numerous surface selective excitations.

In Figure 5-122 the represented wavefunction within the box corresponds to a high quantum number n (hence many oscillating cycles). But the wavefunction extending beyond the box into the region of $V>E$ decays exponentially, just like the evanescent wave for confined light.

Electron tunneling is defined as the passage of electrons from one allowed zone ($E>V$) through a classically forbidden zone ($V>E$), called a barrier layer, to another allowed zone ($E>V$) as shown in Figure 5-123 (Merzbacher, 1998). Similarly, photon tunneling is defined as passage of photon through a barrier layer of lower refractive index, also shown in Figure 5-123 (Gonokami et al., 2002; Fillard, 1996). The tunneling probability, often described by T, called the transmission probability, is given as

$$T = a e^{-2kl} \tag{5-428}$$

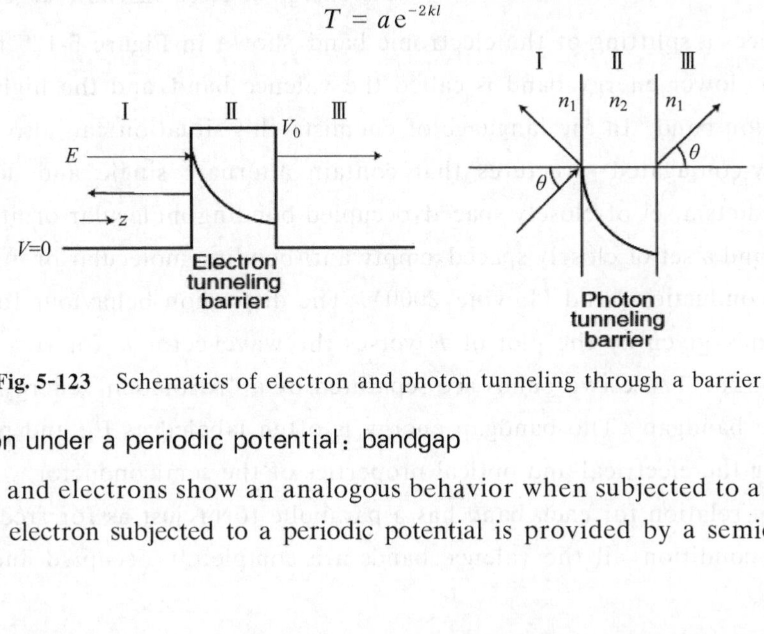

Fig. 5-123 Schematics of electron and photon tunneling through a barrier.

- Localization under a periodic potential: bandgap

Both photons and electrons show an analogous behavior when subjected to a periodic potential. The example of an electron subjected to a periodic potential is provided by a semiconductor crystal, which consists of a periodic arrangement of atoms. The electrons are free to move through a lattice (ordered arrangement) of atoms, but as they move, they experience a strong Coulomb (attractive) interaction by the nucleus of the atom at each lattice site. Here a brief description is presented to draw an analogy between an electronic crystal and a photonic crystal. A photonic crystal represents an ordered arrangement of a dielectric lattice, which represents a periodic variation of the dielectric constant (Joannopoulos et al., 1995). An example presented in Figure 5-123 is that of close-packed, highly uniform colloidal particles such as silica or polystyrene spheres. The refractive index contrast (n_1/n_2), where n_1 is the refractive index of the packing spheres and n_2 is that of the interstitial medium between them (which can be air, a liquid, or more desirably, a very high refractive index material), acts as a periodic potential.

The periodicity in the two cases is of different length scale. In the case of an electronic (semiconductor) crystal, the atomic arrangement (lattice spacings) is on the subnanometer scale. This range of dimensions in the domain of electromagnetic waves corresponds to X-rays, and X-rays can be diffracted on crystal lattices to produce Bragg scattering of X-ray waves. The Bragg equation determining the directions in space at which the diffraction takes place, is given as

$$m\lambda = 2nd\sin\theta \tag{5-429}$$

where d is the lattice spacing and λ is the wavelength of the wave, m is the order of diffraction, n is the refractive index, and θ is the ray incidence angle.

- Photons and electrons: similarities and differences

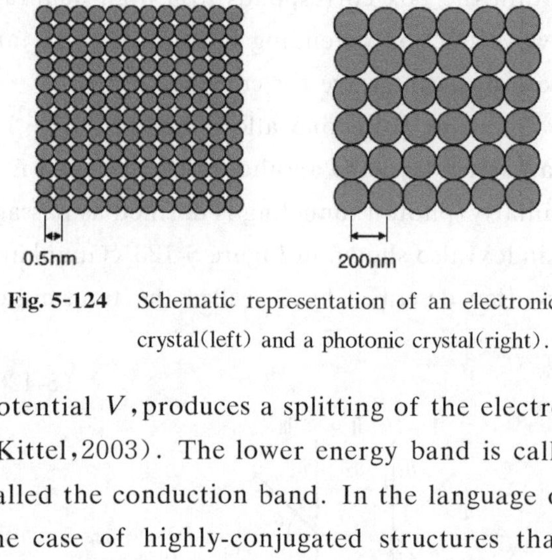

Fig. 5-124 Schematic representation of an electronic crystal(left) and a photonic crystal(right).

In the case of a photonic crystal, the same Bragg scattering produces diffraction of optical waves (Joannopoulos et al., 1995). Use of Eq. (5-429) suggests that the lattice spacing (distance between the centers of the packed spheres) should be, for example, 200nm to produce Bragg scattering of light of wavelength 500nm.

The solution of the Schrödinger equation for the energy of electrons, now subjected to the periodic potential V, produces a splitting of the electronic band, shown in Figure 5-125 for a free electron (Kittel, 2003). The lower energy band is called the valence band, and the higher energy band is called the conduction band. In the language of chemists, this situation can also be described as in the case of highly-conjugated structures that contain alternate single and double bonds. The Hückel theory predicts a set of closely spaced, occupied bonding molecular orbitals, which are like the valence band, and a set of closely spaced, empty anti-bonding molecular orbitals (*) which are equivalent to the conduction band (Levine, 2000). The dispersion behaviour for the valence and the conduction bonds, given by the plot of E versus the wavevector k for two possible cases, are shown in Figure 5-125. These two bands are separated by a "forbidden" energy gap, the width of which is called the bandgap. The bandgap energy is often labeled as Eg and plays an important role in determining the electrical and optical properties of the semiconductor.

The dispersion relation for each band has a parabolic form, just as for free electrons. Under the lowest energy condition, all the valence bands are completely occupied and thus no flow of

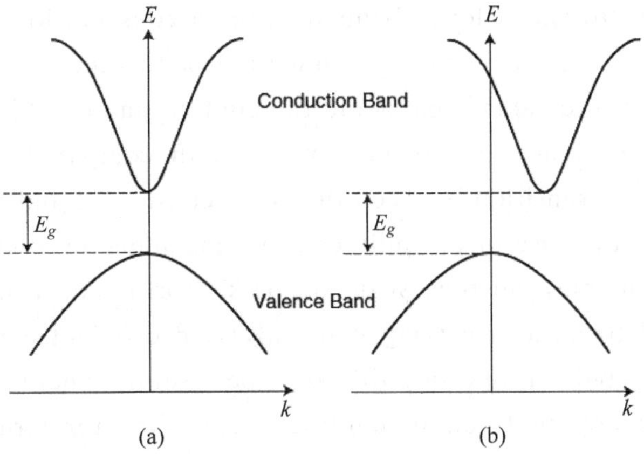

Fig. 5-125 Schematics of electron energy in (a) direct bandgap (e.g., GaAs, InP, CdS) and (b) indirect bandgap (e.g., Si, Ge, GaP) semiconductors.

electrons can occur. The conduction band consequently is empty. If an electron is excited either thermally or optically, or an electron is injected to the conduction band by an impurity (n-doping), this electron can move in the conduction band, producing electronic conduction under an applied electric field. In the case of excitation of an electron to the conduction band, a vacancy is left in the valence band with a net positive charge, which is also treated as a particle of positive charge, called a hole. The hole (positive vacancy) can move through the valence band, providing conduction. The energy, ECB, of an electron near the bottom of the conduction band is given by the relation (Kittel, 2003)

$$E_{\mathrm{CB}} = E_C^0 + \frac{\hbar^2 k^2}{2m_e^*} \tag{5-430}$$

Here E_C^0 is the energy at the bottom of the conduction band, and m_e^* is the effective mass of the electron in the conduction band, which is modified from the real mass of the electron because of the periodic potential experienced by the electron. Similarly, near the top of the valence band, the energy is given as

$$E_{\mathrm{VB}} = E_V^0 + \frac{\hbar^2 k^2}{2m_h^*} \tag{5-431}$$

where $E_V^0 = E_C^0 - E_g$ is the energy at the top of the valence band and m_h^* is the effective mass of a hole in the valence band. Both Eqs. (5-430) and (5-431) predict a parabolic relation between E and k, if m_e^* and m_h^* are assumed to be independent of k. The effective masses m_e^* and m_h^* can be obtained from the curvature of the calculated band structure (E versus k).

Figure 5-126 shows that the two types of cases encountered are as follows: (a) Direct-gap materials for which the top of the valence band and the bottom of the conduction band are at the same value of k. An example is a binary semiconductor GaAs. (b) Indirect-gap materials for which the top of the valence band and the bottom of the conduction band are not at the same value of k. An example of this type of semiconductor is silicon. Hence in the case of an indirect-gap semiconductor, the transition of an electron between the valence band and the conduction band involves a substantial change in the momentum (given by k) of the electron. This has important consequences for optical transitions between these two bands, induced by absorption or emission of photons. For example, emission of a photon (luminescence) leading to transition of an electron from the conduction band to the valence band requires a conservation of momentum. In other words, the momentum of the electron in the conduction band should be the same as the sum of momenta of the electron in the valence band and the emitted photon. Since the photon has a very small momentum ($k \sim 0$) because of its long wavelength compared to that of electrons, the optically induced electronic transition between the conduction band and the valence band requires $k = 0$. Hence, emission in the case of an indirect-gap semiconductor, such as silicon, is essentially forbidden by this selection rule, and thus Si in the bulk form is not a luminescent light emitter. GaAs, on the other hand being a direct-gap material, is an efficient emitter. This discussion of electronic semiconductor bulk property will also be useful, where can see how the bulk semiconductor properties are modified in confined semiconductor structures such as quantum wells, quantum wires, and quantum dots.

Figure 5-126 shows the dispersion curve calculated for a one-dimensional photonic crystal (also known as Bragg stack), which consists of alternating layers of two dielectric media of refractive indices n_1 and n_2.

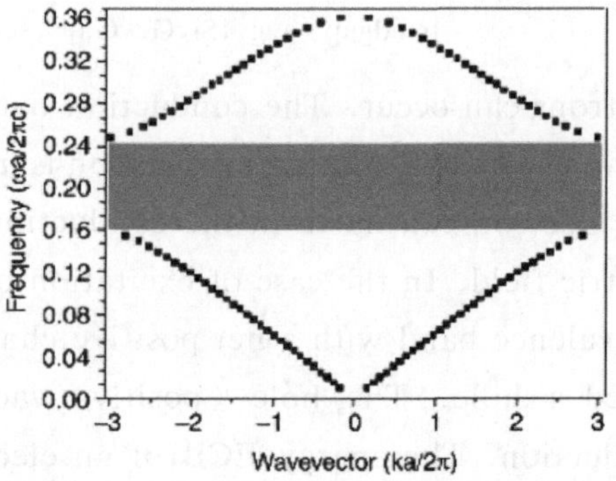

Fig. 5-126 Dispersion curve for a one-dimensional photonic crystal showing the lowest energy bandgap.

Again, a similar type of band splitting is observed for a photonic crystal, and a forbidden frequency region exists between the two bands, similar to that between the valence and the conduction band of an electronic crystal, which is often called the photonic bandgap. Similar to electronic bandgap, no photon frequencies in the gap region of the photonic crystal correspond to the allowed state for this medium.

Hence, the photons in this frequency range cannot propagate through the photonic crystal. Using the Bragg-diffraction model, it can also be seen that the photons of the bandgap frequency meet the Bragg-diffraction conditions. Hence, another way to visualize the photon localization in the photonic bandgap region (not propagating) is that the photons of these frequencies are multiply scattered by the scattering potential produced by the large refractive index contrast n_1/n_2, resulting in their localization. In other words, if photons of frequencies corresponding to the bandgap region are incident on the photonic crystal, they will be reflected from the surface of the crystal and will not enter the crystal. If a photon in the bandgap region is generated inside the crystal by emission, it will not exit the crystal because of its localization (lack of propagation). These properties and their consequences are detailed later.

5.13.5 Cooperative effects for photons and electrons

Cooperative effects refer to the interaction between more than one particle. Although conventionally the cooperative effects for photons and electrons have been described separately, an analogy can be drawn. However, it should be made clear that while electrons can interact directly, photons can interact only through the mediation of a material in which they propagate. In the case of photons, an example of a cooperative effect is the nonlinear optical effect produced in an optically nonlinear medium. In a linear medium, photons propagate as an electromagnetic wave without interacting with each other. As described above, the propagating electromagnetic wave senses the medium response in the form of its dielectric constant or refractive index. For a linear medium, the dielectric constant or the refractive index is related to the linear susceptibility $\chi^{(1)}$ of the medium by the following relation,

$$\varepsilon = n^2 = 1 + \chi^{(1)} \tag{5-432}$$

where $\chi^{(1)}$ is the linear susceptibility, is the coefficient that relates the polarization (charge distortion) of the medium, induced linearly by the electric field E of light to its magnitudes as

$$P = \chi^{(1)} E \tag{5-433}$$

Because $\chi^{(1)}$ relates two vectors, P and E, it actually is a second-rank tensor.

In a strong optical field such as a laser beam, the amplitude of the electric field is so large (comparable with electrical fields corresponding to electronic interactions) that in a highly polarizable nonlinear medium the linear polarization behaviour Eq. (5-434) does not hold. The polarization P in this case also depends on higher powers of electric field, as below:

$$P = \chi^{(1)} E + \chi^{(2)} EE + \chi^{(3)} EEE \cdots \tag{5-434}$$

The higher-order terms in the electric field E produce nonlinear optical interactions, whereby the photons interact with each other. Some of the nonlinear optical interactions discussed in this book are described below. An important manifestation of photon-photon interaction is frequency conversion. The most important examples of such conversion processes are as follows:

(1) Interaction of two photons of frequency $\chi^{(1)}$ by the $\chi^{(2)}$ term to produce a photon of up-converted frequency 2χ. This process is referred to as second harmonic generation (SHG). For example, if the original photon is of wavelength 1.06μm (IR), the new output of $\mu/2$ is at 532nm in the green.

(2) Interaction of two photons of different frequencies $\chi^{(1)}$ and $\chi^{(2)}$, again by the $\chi^{(2)}$ term, to produce a new photon at the sum frequency value $\chi^{(1)} + \chi^{(2)}$ or difference frequency value $\chi^{(1)} - \chi^{(2)}$. This process is called parametric mixing or parametric generation.

(3) Interaction of three photons of frequency $\chi^{(3)}$ through the third-order nonlinear optical susceptibility term. This process is called third harmonic generation (THG).

(4) Simultaneous absorption of two photons (two-photon absorption), to produce an electronic excitation.

These frequency conversion processes are conceptually described by simple energy diagrams in the book Introduction to Biophotonics by this author (Prasad, 2003). Other important types of nonlinear optical interactions produce field dependence of the refractive index of an optically nonlinear medium. These are: Pockels effect, which describes linear dependence of the refractive index on the applied electric field. This linear electro-optic effect can be used to affect the propagation of photons by application of an electric field and thereby produce devices such as electro-optic modulators. Kerr effect (more precisely optical Kerr effect), which describes linear dependence of the refractive index on the intensity of light. This effect is thus all optical, whereby varying the intensity of a controlling intense light beam can affect the propagation of another light beam (signal). This effect provides the basis for all optical signal processing.

An example of a cooperative process for electrons is electron-electron interaction to bind together and produce a Cooper pair in a superconducting medium, as proposed by Bardeen, Cooper, and Schrieffer (BCS) to explain superconductivity (Kittel, 2003). Two electrons, each carrying a negative charge, will be expected to repel each other electrostatically. However, an electron in a cation lattice distorts the lattice around it by so-called electron-phonon (lattice vibration) interactions, creating an area of increased positive charge density around itself, which can attract another electron. In other words, the two electrons are attracted toward each other by electron-phonon interaction, which acts as a spring, and they form what is called a Cooper pair. This pair formation is schematically represented in Figure 5-127.

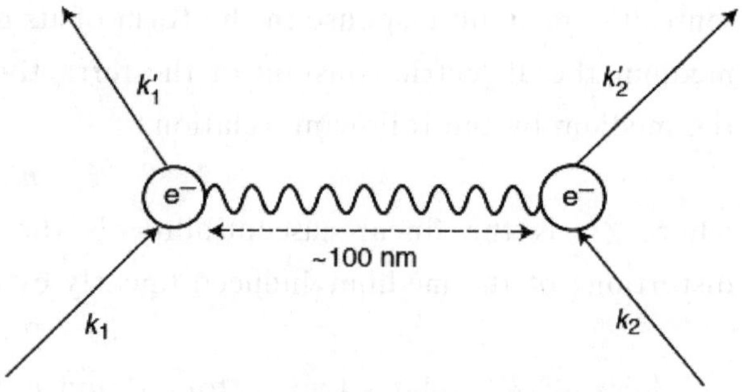

Fig. 5-127 Schematic representation of a phonon-mediated Cooper-pair formation between two electrons.

The binding energy between an electron pair is of the order of milli-electronvolts, sufficient to keep them paired at extremely low temperatures (below a temperature called critical temperature, T_c). The Cooper pair experiences less resistance and leads to superconductivity at lower temperature, where current flows without resistance.

Another example of a cooperative effect is the binding between an electron and a hole to form an exciton, as well as the binding between two excitons to form a bound state called a biexciton (Kittel, 2003). Excitons are formed when an electron and the corresponding hole in the valence band are bound so that they cannot move independently. Thus, the electron and the hole move together as a bound particle called an exciton. In organic insulators, the electron and the hole are tightly bound at the same lattice site (i. e., within a small radius, usually within the same molecule).

Such a tightly bound electron-hole pair is called a Frenkel exciton. In the case of a semiconductor, the electrons in the conduction band and the holes in the valence band are not independent and exhibit coupled behavior, giving rise to an exciton, which may exhibit a larger separation between the electron and the hole (spread over more than one lattice site). This type of exciton is called a Wannier exciton.

The exciton, being composed of a negatively charged electron and a positively charged hole, is a neutral particle and has quantum properties. They are analogous to those of a hydrogen-like atom where an electron and a proton are bound by coulombic interactions. Just like in the case of a hydrogen atom, the energy of an exciton is described by a set of quantized energy levels, which is below the bandgap (E_g) and described by

$$E_n(k) = E_g - \frac{R_y}{n^2} + \frac{\hbar^2 k^2}{2m} \tag{5-435}$$

In the above quotation, R_y is called the exciton Rydberg energy and is defined as

$$R_y = \frac{e^2}{2\varepsilon a_B} \tag{5-436}$$

in which ε is the dielectric constant of the crystal. The term α_B, called the exciton Bohr radius or often simply the Bohr radius of a specific semiconductor, is defined as

$$a_B = \frac{\varepsilon \hbar^2}{\mu e^2} \tag{5-437}$$

In the above equation, μ is the reduced mass of the electron-hole pair defined as

$$\mu^{-1} = m_e^{*-1} + m_h^{*-1} \tag{5-438}$$

The exciton Bohr radius gives an estimate of the size of the exciton (most probable distance of the electron from the hole) in a semiconductor. k is the wavevector for the exciton. For optically generated exciton, k_0. Thus the lowest energy of excitonic transition corresponds to $E_1 = E_g - R_y$ with $n = 1$ in Eq. (5-452) and is below the bandgap E_g. The Rydberg energy R_y is thus an estimate of the exciton binding energy and usually is in the range 1～100meV. A bound exciton will be formed when the thermal energy $k_T < R_y$. If $k_T R_y$, most excitons are ionized and behave like the separated electrons and holes. Under high excitation density, two excitons can bind to form a biexciton (Klingshirn). The formation of biexcitons has been extensively investigated for a number of semiconductors such as CuCl. It has also been a subject of investigation in quantum-confined structures, such as quantum wells, quantum wires, and quantum dots.

5.13.6 Nanoscale optical interactions

The electric field associated with a photon can be confined by using a number of geometries to induce optical interactions on nanoscale. The optical field can be localized on nanoscale both axially and laterally. Figure 5-128 lists the methods that can be used for such a purpose.

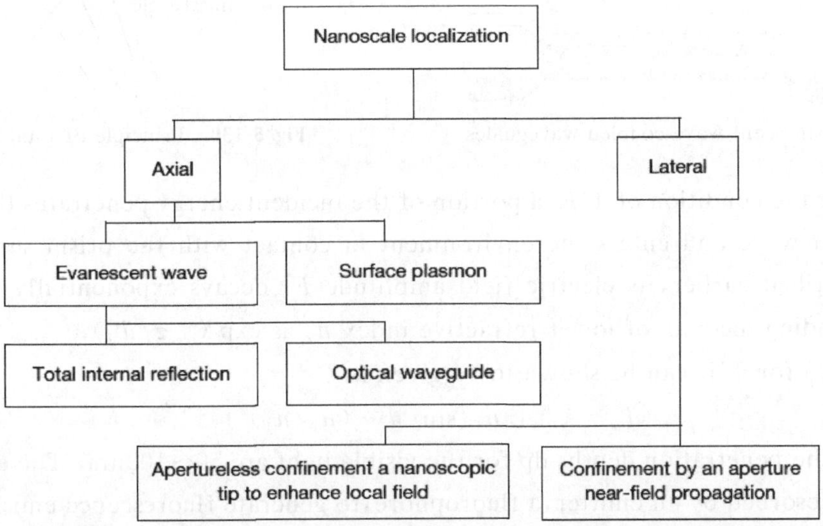

Fig. 5-128 Methods for Nanoscale Localization of Electromagnetic Field.

- Axial nanoscopic localization

Evanescent Wave. The evanescent wave derived from photon tunneling, in the case of a waveguide, has been discussed in the previous section. The evanescent wave from a waveguide surface penetrates the surrounding medium of lower refractive index where it decays exponentially in the axial direction (away from the waveguide). This evanescent field extends about 50～100nm and can be used to induce nanoscale optical interactions. This evanescent wave excitation has been used for fluorescence sensing with high near surface selectivity (Prasad, 2003). Another example of nanoscale optical interaction is coupling of two waveguides by the evanescent wave, which is schematically shown in Figure 5-129. Here, photon launched in one waveguide can tunnel from it to another waveguide. The evanescent wave-coupled waveguides can be used as directional couplers for switching of signal in an optical communication network. Evanescent wave-coupled waveguides have also been proposed for sensor application, where sensing produces a change in photon tunneling from one waveguide channel to another.

Another example of a geometry producing an evanescent wave is provided by total internal reflection involving the propagation of light through a prism of refractive index n_1 to an environment of a lower refractive index n_2. At the interface, the light refracts and partially passes into the second medium at a sufficiently small incidence angle. This process is called total internal reflection (TIR). The critical angle θ_c is given by the equation

$$\theta_c = \sin^{-1}(n_2/n_1) \tag{5-439}$$

As shown in Figure 5-130 for incidence angle θ_c, the light is totally internally reflected back to the prism from the prism/environment interface. The refractive index n_1 of a standard glass prism is about 1.52, while the refractive index n_2 of the surrounding environment, say an aqueous buffer, may be 1.33, yielding a critical angle of 61°.

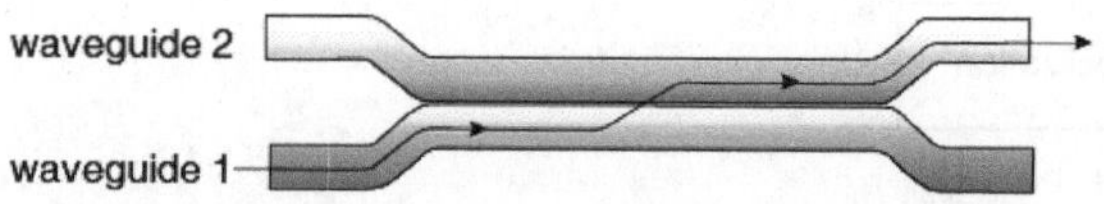

Fig. 5-129 Evanescent wave-coupled waveguides.

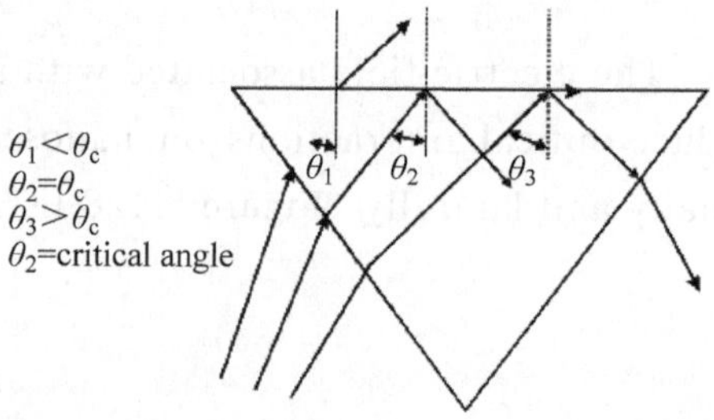

Fig. 5-130 Principle of total internal reflection.

Even under the condition of TIR, a portion of the incident energy penetrates the prism surface as an evanescent wave and enters the environment in contact with the prism surface (Courjon, 2003). As described earlier, its electric field amplitude E_z decays exponentially with distance z into the surrounding medium of lower refractive index n_2 as $\exp(-z/d_p)$.

The term d_p for TIR can be shown to be given as

$$d_p = \lambda/[4\pi n_1\{\sin^2\theta - (n_2/n_1)^2\}^{1/2}] \tag{5-440}$$

Typically, the penetration depths dp for the visible light are 50~100nm. The evanescent wave energy can be absorbed by an emitter, a fluorophore, to generate fluorescence emission that can be used to image fluorescently labeled biological targets (Prasad, 2003). However, because of the short-range exponentially decaying nature of the evanescent field, only the fluorescently labeled biological specimen near the substrate (prism) surface generates fluorescence and can be thus imaged.

The fluorophores, which are further away in the bulk of the cellular medium, are not excited. This feature allows one to use TIR for microscopy and obtain a highquality image of the fluorescently labeled biologic near the surface, with the following advantages:

(1) Very low background fluorescence.

(2) No out-of-focus fluorescence.

(3) Minimal exposure of cells to light in any other planes in the sample, except near the interface Surface Plasmon Resonance (SPR). In principle, the SPR technique provides an extension of evanescent wave interaction, described above, except that a waveguide or a prism is replaced by a metal-dielectric interface. Surface plasmons are electromagnetic waves that propagate along the interface between a metal film and a dielectric material such as organic films.

Since the surface plasmons propagate in a metal film in the frequency and wavevector ranges for which no light propagation is allowed in either of the two media, no direct excitation of surface plasmons is possible. The most commonly used method to generate a surface plasmon wave is attenuated total reflection (ATR).

The Kretschmann configuration of ATR is widely used to excite surface plasmons. This configuration is shown in Figure 5-131. A microscopic slide is coated with a thin film of metal (usually a 40-to 50-nm-thick gold or silver film by vacuum deposition). The microscopic slide is now coupled to a prism through an index-matching fluid or a polymer layer. A p-polarized laser beam (or light from a light-emitting diode) is incident at the prism. The reflection of the laser beam is monitored. At a certain sp, the electromagnetic wave couples to the interface as a surface plasmon. At the same time, an evanescent field propagates away from the interface, extending to about 100nm above and below the metal surface. At this angle the intensity of the reflected light (the ATR signal) drops. This dip in reflectivity is shown by the left-hand curve in Figure 5-132.

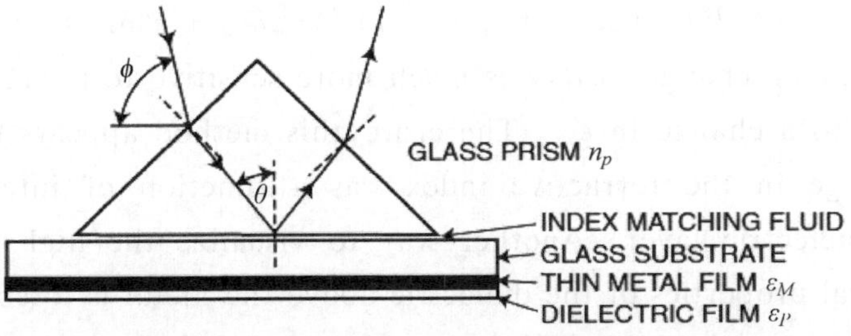

Fig. 5-131 Kretschmann (ATR) geometry used to excite surface plasmons.

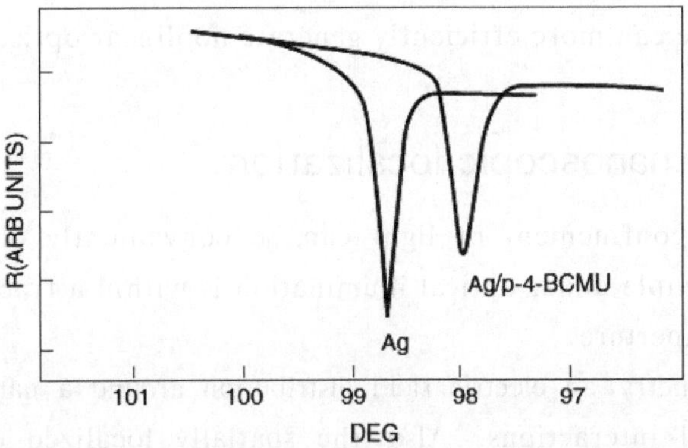

Fig. 5-132 Surface plasmon resonance curves. The left-hand curve is for just the silver film (labeled Ag); the right-hand side shows the curve (labeled Ag/p-4-BCMU) shifted on the deposition of a monolayer Langmuir-Blodgett film of poly-4-BCMU on the silver film.

The angle is determined by the relationship as

$$k_{sp} = kn_p \sin\theta_{sp} \tag{5-441}$$

Where ksp is the wavevector of the surface plasmon, k is the wavevector of the bulk electromagnetic wave, and np is the refractive index of the prism. The surface Plasmon wavevector ksp is given by

$$k_{sp} = (\omega/c)[(\varepsilon_m\varepsilon_d)/(\varepsilon_m + \varepsilon_d)]^{1/2} \tag{5-442}$$

Where ω is the optical frequency, c the speed of light, and ε_m and ε_d are the relative dielectric constants of the metal and the dielectric, respectively, which are of opposite signs. In the case of a bare metal film, ε_d (or square of the refractive index for a dielectric) is the dielectric constant of air and the dip in reflectivity occurs at a certain angle. In the case of metal coated with another dielectric layer (which can be used for photonic processing or sensing), this angle shifts. Figure 5-132 shows as an illustration the shift in the coupling angle on deposition of a monolayer Langmuir-Blodgett film of a polydiacetylene, poly-4-BCMU. The shifted SPR curve is shown on the right-hand side in Figure 5-132.

In this experiment, one can measure the angle for the reflectivity minimum, the minimum value of reflectivity, and the width of the resonance curves. These observables are used to generate a computer fit of the resonance curve using a leastsquares fitting procedure with the Fresnel reflection formulas yielding three parameters: the real and the imaginary parts of the refractive index and the thickness of the dielectric layer.

From the above equations, one can see that the change ε in the surface Plasmon resonance angle (the angle corresponding to minimum reflectivity; for simplicity the subscript sp is dropped) caused by changes ε_m and ε_d in the dielectric constants of the metal and covering film, respectively, is given by

$$\cot\theta\delta\theta = (2\varepsilon_m\varepsilon_d(\varepsilon_m + \varepsilon_d))^{-1}(\varepsilon_m^2\delta\varepsilon_d + \varepsilon_d^2\delta\varepsilon_m) \tag{5-443}$$

Since $|\varepsilon_m| \gg |\varepsilon_d|$, the change in d, ε is much more sensitive to a change in ε_d (i.e., of the dielectric layer) than to a change in ε_m. Therefore, this method appears to be ideally suited to obtain ε_d (or a change in the refractive index) as a function of interactions or structural perturbation in the dielectric layer. Another way to visualize the high sensitivity of SPR to variations in the optical properties of the dielectric above the metal is to consider the strength of the evanescent field in the dielectric, which is an order of magnitude higher than that in a typical evanescent wave source utilizing an optical waveguide as described above. This surface plasmon enhanced evanescent wave can more efficiently generate nonlinear optical processes, which require a much higher intensity.

5.13.7 Lateral nanoscopic localization

A lateral nanoscale confinement of light can be conveniently obtained using a nearfield geometry in which the sample under optical illumination is within a fraction of the wavelength of light from the source or aperture.

In the near-field geometry, an electric field distribution around a nanoscopic structure produces spatially localized optical interactions. Also, the spatially localized electric field distribution contains a significant evanescent character—that is, decaying exponential because of the imaginary wavevector character. A near-field geometry is conveniently realized using a near-field scanning optical microscope, abbreviated as NSOM or sometimes as SNOM (interchanging the terms near-field and scanning). This topic forms the content of the chapter.

In a more commonly used NSOM approach, a submicron size 50nm to 100nm aperture, such as an opening tip of a tapered optical fiber, is used to confine light. For an apertureless NSOM arrangement, a nanoscopic metal tip (such as the ones used in scanning tunneling microscopes, STM) or a nanoparticle (e.g., metallic nanoparticle) is used in close proximity of the sample to enhance the local field. This field enhancement is discussed previously.

- Nanoscale confinement of electronic interactions

This section provides some selected examples of nanoscale electronic interactions, which produce major modifications or new manifestations in the optical properties of a material. Figure

5-133 lists these interactions. Brief discussions of these interactions then follow.

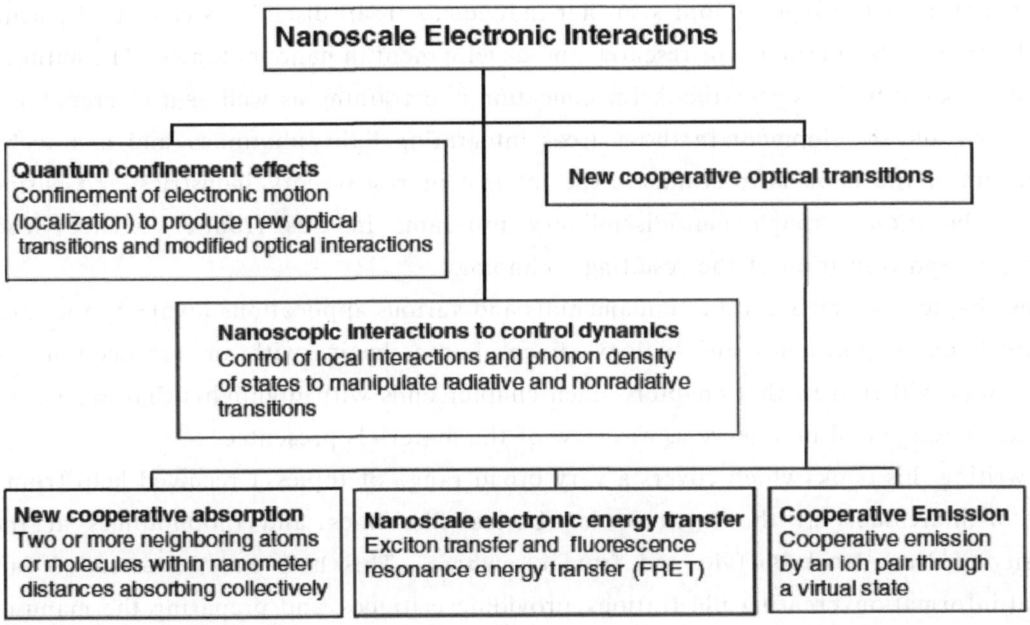

Fig. 5-133 Various Nanoscale Electronic Interactions Producing Important Consequences in the Optical Properties of Materials.

5.13.8 Quantum confinement effects

• Nanoscopic interaction dynamics

Examples of control of nanoscopic interactions, whereby a particular radiative transition (emission at a particular wavelength) is enhanced by local interactions. An example is the use of a nanocrystal host environment with low-frequency phonons (vibrations of lattice) so that multiphonon relaxation of excitation energy in a rare-earth ion is significantly reduced to enhance the emission efficiency.

Because the electronic transitions in a rare-earth ion are very sensitive to nanoscale interactions, only a nanocrystal environment is sufficient to control the nature of electronic interactions. This provides an opportunity to use a glass or a plastic medium containing these nanocrystals for many device applications.

Nanoscale electronic interactions also produce new types of optical transitions and enhanced optical communications between two electronic centers. These interactions are described below.

• Nanophotonics

Nanophotonics, defined by the fusion of nanotechnology and photonics, is an emerging frontier providing challenges for fundamental research and opportunities for new technologies. Nanophotonics has already made its impact in the marketplace. It is a multidisciplinary field, creating opportunities in physics, chemistry, applied sciences, engineering, and biology, as well as in biomedical technology. Nanophotonics has meant different things to different people, in each case being defined with a narrow focus. Several books and reviews exist that cover selective aspects of nanophotonics. However, there is a need for an up-to-date monograph that provides a unified synthesis of this subject. This book fills this need by providing a unifying, multifaceted description of nanophotonics to benefit a multidisciplinary readership. The objective is to provide a basic knowledge of a broad range of topics so that individuals in all disciplines can rapidly acquire the minimal necessary background for research and development in nanophotonics. The author intends this book to serve both as a textbook for education and training as well as a reference book that aids research and development in those areas integrating light, photonics, and nanotechnology. Another aim of the book is to stimulate the interest of researchers, industries, and businesses to foster collaboration through multidisciplinary programs in this frontier science, leading to development and transition of the resulting technology.

This chapter encompasses the fundamentals and various applications involving the integration of nanotechnology, photonics, and biology. Each chapter begins with an introduction describing what a reader will find in that chapter. Each chapter ends with highlights that are basically the take-home message and may serve as a review of the materials presented.

In writing this book, which covers a very broad range of topics, I received help from a large number of individuals at the Institute for Lasers, Photonics, and Biophotonics at the State University of New York-Buffalo and from elsewhere. This help has consisted of furnishing technical information, creating illustrations, providing critiques, and preparing the manuscript. A separate Acknowledgement recognizes these individuals.

- An exciting frontier in nanotechnology

Nanophotonics is an exciting new frontier that has captured the imaginations of people worldwide. It deals with the interaction of light with matter on a nanometer size scale. By adding a new dimension to nanoscale science and technology, nanophotonics provides challenges for fundamental research and creates opportunities for new technologies. The interest in nanoscience is a realization of a famous statement by Feynman that "There's Plenty of Room at the Bottom" (Feynman, 1961). He was pointing out that if one takes a length scale of one micrometer and divides it in nanometer segments, which are a billionth of a meter, one can imagine how many segments and compartments become available to manipulate. Living in an age of "nano-mania." Everything nano is considered to be exciting and worthwhile. Many countries have started Nanotechnology Initiatives. A detailed report for the U. S. National Nanotechnology Initiative has been published by the National Research Council (NRC Report, 2002). While nanotechnology can't claim to provide a better solution for every problem, nanophotonics does create exciting opportunities and enables new technologies. The key fact is that nanophotonics deals with interactions between light and matter at a scale shorter than the wavelength of light itself. This book covers interactions and materials that constitute nanophotonics, and it also describes their applications. Its goal is to present nanophotonics in a way to entice one into this new and exciting area.

Nanophotonics can conceptually be divided into three parts. One way to induce interactions between light and matter on a nanometer size scale is to confine light to nanoscale dimensions that are much smaller than the wavelength of light. The second approach is to confine matter to nanoscale dimensions, thereby limiting interactions between light and matter to nanoscopic dimensions. This defines the field of nanomaterials. The last way is nanoscale confinement of a photoprocess where we induce photochemistry or a light-induced phase change. This approach provides methods for nanofabrication of photonic structures and functional units. Let's look at nanoscale confinement of radiation. There are a number of ways in which one can confine the light to a nanometer size scale. One of them is using near-field optical propagation, which we discuss in detail in section 3 of this book. One example is light squeezed through a metal-coated and tapered optical fiber where the light emanates through a tip opening that is much smaller than the wavelength of light.

The nanoscale confinement of matter to make nanomaterials for photonics involves various ways of confining the dimensions of matter to produce nanostructures. For example, one can utilize nanoparticles that exhibit unique electronic and photonic properties. It is gratifying to find that these nanoparticles are already being used for various applications of nanophotonics such as UV absorbers in sunscreen lotions. Nanoparticles can be made of either inorganic or organic materials.

Nanomers, which are nanometer size oligomers (a small number of repeat units) of monomeric organic structures, are organic analogues of nanoparticles. In contrast, polymers are long chain structures involving a large number of repeat units. These nanomers exhibit size-dependent optical properties. Metallic nanoparticles exhibit unique optical response and enhanced electromagnetic field and constitute the area of "plasmonics." Then there are nanoparticles which up-convert two absorbed IR photons into a photon in the visible UV range; conversely, there are nanoparticles, called quantum cutters, that down-convert an absorbed vacuum UV photon to two photons in the visible range. A hot area of nanomaterials is a photonic crystal that represents a periodic dielectric structure with a repeat unit of the order of wavelength of light. Nanocomposites comprise nanodomains of two or more dissimilar materials that are phase-separated on a nanometer size scale. Each nanodomain in Figure 5-134. Nanophotonics the nanocomposite can impart a particular optical property to the bulk media. Flow of optical energy by energy transfer (optical communications) between different domains can also be controlled.

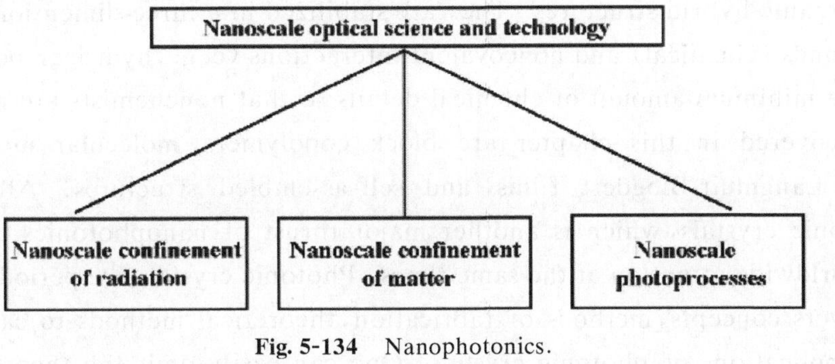

Fig. 5-134 Nanophotonics.

Nanoscale photoprocesses can be used for nanolithography to fabricate nanostructures. These nanostructures can be used to form nanoscale sensors and actuators. A nanoscale optical memory is one of exciting concepts of nanofabrication. An important feature of nanofabrication is that the photoprocesses can be confined to well-defined nanoregions so that structures can be fabricated in a precise geometry and arrangement.

• Near-field interactions

A brief theoretical description of near-field interactions and the various experimental geometries used to effect them are introduced. The theoretical section may be skipped by those more experimentally oriented. Various optical and higher-order nonlinear optical interactions in nanoscopic domains are described. Applications to covers quantum-confined materials whose optical properties are sizedependent.

Described here are semiconductor quantum wells, quantum wires, quantum dots, and their organic analogues. A succinct description of manifestations of quantum confinement effects presented in this chapter should be of significant value to those (e. g., some chemists and life scientists) encountering this topic for the first time. The applications of these materials in semiconductor lasers, described here, exemplify the technological significance of this class of materials.

The topic of metallic nanostructures, now with a new buzzword "plasmonics" describing the subject. Relevant concepts together with potential applications are introduced. Guiding of light through dimensions smaller than the wavelength of light by using plasmonic guiding is described. The applications of metallic nanostructures to chemical and biological sensing is presented.

The deals with nanoscale materials and nanoparticles, for which the electronic energy gap does not change with a change in size. However, the excitation dynamics—that is, emission properties, energy transfer, and cooperative optical transitions in these nanoparticles—are dependent on their nanostructures. Thus nanocontrol of excitation dynamics is introduced. Important processes described are (i) energy up-conversion acting as an optical transformer to convert two IR photons to a visible photon and (ii) quantum cutting, which causes the down-conversion of a vacuum UV photon to two visible photons.

Various methods of fabrications and characterization of nanomaterials. In addition to the traditional semiconductor processing methods such as molecular beam epitaxy (MBE) and metal-organic chemical vapor deposition (MOCVD), the use of nanochemistry, which utilizes wet chemical synthesis approach, is also described. Some characterization techniques introduced are specific to nanomaterials. Nanostructured molecular architectures that include a rich class of nanomaterials often unfamiliar to physicists and engineers. These nanostructures involve organic and inorganic-organic hybrid structures. They are stabilized in a three-dimensional architecture by both covalent bonds (chemical) and noncovalent interactions (e.g., hydrogen bond). This topic is presented with a minimum amount of chemical details so that nonchemists are not overburdened. Nanomaterials covered in this chapter are block copolymers, molecular motors, dendrimers, supramolecules, Langmuir-Blogdett films, and self-assembled structures. Aforementioned the subject of photonic crystals, which is another major thrust of nanophotonics that is receiving a great deal of worldwide attention at the same times. Photonic crystals are periodic nanostructures. The chapter covers concepts, methods of fabrication, theoretical methods to calculate their band structure, and applications of photonic crystals. One can easily omit the theory section and still appreciate the novel features of photonic crystals, which are clearly and concisely described.

A significant emphasis is placed on the nanocomposite materials that incorporate nanodomains of highly dissimilar materials such as inorganic semiconductors or inorganic glasses and plastics. Merits of nanocomposites are discussed together with illustrative examples of applications, such as to energy-efficient broadband solar cells and other optoelectric devices.

Nanolithography, broadly defined, that is used to fabricate nanoscale optical structures. Both optical and nonoptical methods are described, and some illustrative examples of applications are presented. The use of direct twophoton absorption, a nonlinear optical process, provides improved resolution leading to smaller photoproduced nanostructures compared to those produced by linear absorption.

With biomaterials that are emerging as an important class of materials for nanophotonic applications. Bioderived and bioinspired materials are described together with bioassemblies that can be used as templates. Applications discussed are energy-harvesting, low-threshold lasing, and high-density data storage.

The application of nanophotonics for optical diagnostics, as well as for light-guided and light-activated therapy. Use of nanoparticles for bioimaging and sensing, as well as for targeted drug delivery in the form of nanomedicine, is discussed. Current applications of near-field microscopy, nanomaterials, quantum-confined lasers, photonic crystals, and nanolithography are analyzed.

5.13.9 New cooperative transitions

In a collection of ions, atoms, or molecules, two neighboring species can interact to produce new optical absorption bands or allow new multiphoton absorption processes. This produces new optical absorption and emission from a biexcitonic state, whose energy is lower than that of two separate excitons. This difference in energy corresponds to the binding energy of the two excitons. An extension of the biexciton concept to multiexciton or exciton string, produced by binding (condensation) of many excitons, has also been proposed. In the case of a molecular system, an analogy is the formation of various types of aggregates, such as a J-aggregate of dyes. The J-aggregate is a head-to-head alignment of the dipoles of various dyes as schematically represented in Figure 5-135.

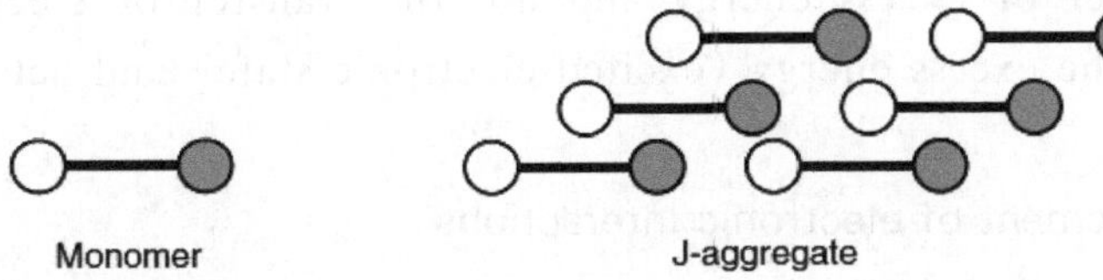

Fig. 5-135 Schematics of a J-aggregate of a linear charge-transfer dye. The two circles represent the two ends of the dipole.

Another type of nanoscale electronic interaction giving rise to new optical transitions manifests when an electron-donating group (or molecule) is in the nearest neighboring proximity within nanoscopic distance of an electron withdrawing group or molecule (electron acceptor). The examples are organometallic structures involving a binding between an inorganic (metallic) ion and many organic groups (ligands). These types of organometallic structures produce novel optical transitions involving metal-to-ligand charge transfer (MLCT) or in some cases the reverse charge transfer induced by light absorption (Prasad, 2003). Another example is an organic donor (D)-acceptor (A) intermolecular complex, which in the excited state produces a charge transfer species D+A−. These charge-transfer complexes display.

Intense visible color derived from new charge-transfer transitions in the visible, even though the components D and A are colorless, thus having no absorption individually in the visible spectral range.

Yet another type of cooperative transition is provided by dimer formation between a species A in the excited electronic state (often labeled as A^*) and another species B in the ground electronic state (Prasad, 2003). This excited-state dimer formation, produced by optical absorption, can be represented as

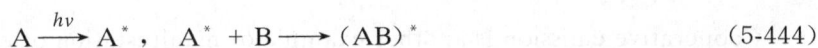

$$A \xrightarrow{hv} A^*, \quad A^* + B \longrightarrow (AB)^* \tag{5-444}$$

If A and B are the same, the resulting excited-state dimer is called an excimer. If A and B are different, the resulting heterodimer is called an exciplex. It should be emphasized that an exciplex does not involve any electron (charge) transfer between A and B. They both are still neutral species in the exciplex state, but bound together by favorable nanoscopic interactions. The optical emission from these excimeric or exciplex state is considerably red-shifted (toward longer wavelength or lower energy), compared to the emission from the monomeric excited form A*. Furthermore, the excimer or exciplex emission is fairly broad and featureless (no structures in the emission band). The excimer and exciplex emission is a very sensitive probe to the nanoscopic structure and orientation surrounding a molecule and has been extensively used to probe local environment and dynamical processes in biology.

Still another example of cooperative transition is shown by rare-earth ion pairs where one ion absorbs energy and transfers it to another ion, which then absorbs another photon to climb to yet another higher electronic level. The emission can then be up-converted in energy compared to the excitation.

5.13.10 Nanoscale electronic energy transfer

The excess electronic energy supplied by an optical transition (or by a chemical reaction as in a chemical laser) can be transferred from one center (ion, atom, or molecule) to another, often on nanoscopic scale, although long-range energy transfer can also be achieved. This electronic energy transfer involves transfer of excess energy and not the transfer of electrons. Hence, in this process, one center has the excess energy (excited electronic state) and acts as an energy donor, which transfers.

- Nanoscale confinement of electronic interactions

Monomer J-aggregate the excitation to an energy acceptor. Consequently, the excited electron in the energy donor returns to the ground state while an electron in the energy acceptor group is promoted to an excited state. The interaction among energetically equivalent centers produces exciton migration either coherently (through many closely spaced levels forming an exciton band) or incoherently by hopping of an electron-hole pair from one center to another. Another type of energy transfer is between two different types of molecules, a process often called fluorescence resonance energy transfer (FRET). This type of transfer, often used with two fluorescent centers within nanometers apart, is detected as fluorescence from the energy acceptor when the energy donor molecule is optically excited to a higher electronic level. FRET is a popular method in bioimaging to probe nanoscale interactions among cellular components, such as to monitor protein-protein interactions (Prasad, 2003). In this case, one protein may be labeled with a fluorescent dye which acts as the energy donor when electronically excited by light. The other protein is labeled by an energy acceptor, which can receive energy when the two proteins are within nanoscopic distances in the range of 1～10nm.

This energy transfer occurs often by dipole-dipole interaction between the energy donor and the energy acceptor. To maximize the FRET process, there should be a significant spectral overlap between the emission spectrum of the donor and the absorption spectrum of the acceptor.

- Cooperative emission

Cooperative emission is another example of manifestation of electronic interactions. Here two neighboring centers within nanoscopic distances, when electronically excited, can emit a photon of higher energy through a virtual state of the pair centers.

This process exhibited by rare-earth ions produces up-converted emission of a higher energy photon than the energy of excitation of individual ions. The interaction is again manifested when the two neighboring ions are separated within nanometers. The interaction between the two ions may be of multipole-multipole or electron exchange type, depending on the nature of electronic excitation in individual ions. It should be pointed out that the emission is not from a real level of the ion pair, but from a virtual level, which is not an allowed electronic level either of the individual ion or of the ion pair.

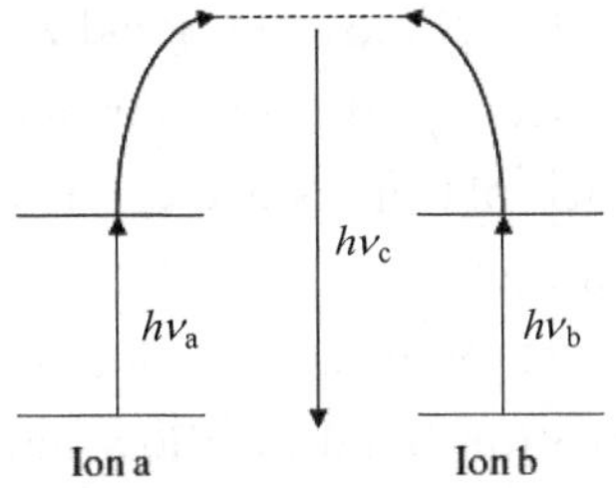

Fig. 5-136 Cooperative emission from an ion pair.

- Both photons and electrons have particle and wave-type simultaneously

(1) The difference is on the length scale, with electrons having wavelengths considerably smaller than that for photons.

(2) Equivalent eigenvalue wave equations describe the propagation of photons and electrons and their allowed energy values in a medium. Similar to the electrostatic interactions introducing resistance to the flow of electrons, the dielectric constant and the related refractive index describe the resistance of a medium to the propagation of photons.

(3) Photons differ from electrons in two ways:

① Photons are vector field (light can be polarized) while the wavefunctions of electrons are scalar; ② photons have no spin and no charge, while electrons possess spin and charge. In a free space, both electrons and photons are described by a propagating plane wave whose amplitude is constant throughout the space, and a propagation vector, k, describes the direction of the propagation; the magnitude of k relates to the momentum.

(4) Confinement of electrons and photons to dimensions comparable to their wavelengths produces quantization where only certain discrete values of energies (for electrons) and field distribution (for photons) are permissible.

(5) Both electrons and photons have finite amplitude in a region, which classically is energetically not allowed for their propagation. For light, this electromagnetic field in the forbidden region is called an evanescent wave, which decays exponentially with the penetration depth.

(6) Like electron tunneling, photon tunneling describes passage of photons from an allowed zone to another through a barrier where its propagation is energetically forbidden.

(7) Electrons face a periodic electrostatic potential (due to nuclear attraction) in an electronic semiconductor crystalline structure, which produces the splitting of the conduction (high-energy) band from the valence band and thus creates a bandgap.

(8) A photonic crystal, analogous to an electronic crystal, describes a periodic dielectric domain (periodic modulation of refractive index), but now with the periodicity at a much longer scale, matched to the wavelength of photons. A photonic bandgap is produced, in analogy to the bandgap for electrons in semiconductors.

- Both photons and electrons exhibit cooperative effects

For photons, the cooperative effects are nonlinear optical effects produced at high field strength (optical intensities). Examples of cooperative effects for electrons are electron-electron interactions in superconductivity and formation of excitons and biexcitons in semiconductors.

(1) Nanoscale localization of optical interactions can be produced axially through the evanescent wave and surface plasmon waves, as well as laterally by using a near-field geometry. Surface plasmons are electromagnetic waves that propagate along interface between a metal film and a dielectric.

(2) Nanoscopic electronic interactions between two neighboring ions produce new optical absorption band or allow new multiphoton absorption processes.

(3) Formation of biexcitons resulting from binding of two excitons in a semiconductor is an example of producing new optical absorption and emission. In molecular systems, the analogous effect is intermolecular interactions producing various types of aggregates (such as J-aggregates) with different spectral characteristics.

(4) Another type of nanoscopic electronic interaction giving rise to new optical transitions (called charge-transfer bands) manifests when an electron-donating specie is within nanoscopic distance from an electron-withdrawing group.

(5) Yet another nanoscopic interaction produces excited-state dimers called excimers (in case of the same molecules) and exciplexes (in case the dimer consists of two different molecules).

(6) Nanoscopic interactions also produce excited-state energy transfer whereby the absorbing molecule (energy donor) returns to ground state by transferring its excitation energy to a different type of molecule of lower-energy excited state (energy acceptor). If the acceptor fluoresces, the process is referred to as fluorescence resonance energy transfer (FRET).

(7) Two neighboring centers within nanoscopic distances, when electronically excited, can emit a photon of higher energy through a virtual state of the pair centers. This process is called cooperative emission.

5.13.11 Quantum dots

For the present we can simply assume that these are nanoparticles of dimensions in the 1nm to 10nm range that exhibit fairly narrow optical transitions that are particle size-dependent. Therefore, particles of a certain size show optical transition (absorption and emission) at a given frequency; but for the increasing size of the particles, the optical resonance shifts to a lower frequency (a longer wavelength). A large number of quantum dots (an ensemble) of different sizes and in different environments, each with its own optical resonance.

This produces an inhomoge-If one looks at a sample in a far field, one samples neous (statistical) broadening of optical transition. One will find a convolution of the distribution of various Q-dot sizes as individual Q-dots cannot be resolved. Nearfield microscopy can allow one to probe domains, which are 50nm in size, whereby it is possible to probe single Q-dots when they are dispersed homogeneously at dilute concentration in a medium (film). Thus, single quantum dot spectroscopy can be achieved. Figure 5-138 shows near-field spectra of single Q-dots obtained using a near-field microscope (website of D. Awschalom). One can image an individual quantum dot region and take a spectrum using near-field excitation. Here the results are shown for two different tip sizes. One of them is 100nm tip size that is shown at the top, the other is 200nm tip size that is in the middle. The bottom spectrum is with 300nm tip size. One can see some structures related to the subsets of Q-dots that are excited when the tip is 100nm. Then going to 200nm and 300nm, this resolution is lost as more than a few Q-dots are simultaneously excited. That reproduced with permission. 100nm tip conducted room temperature photoluminescence study on a single quantum dot from InGaAs quantum dots grown on a GaAs substrate. Their result is shown in Figure 5-138. Because of the spectral resolution obtained by sampling only a single quantum dot no inhomogeneous broadening, that were able to observe, at an appropriate excitation density, emission not only from the lowest level (subband) of the conduction band but also from higher levels. They were able to study the homogeneous line width, determined by the dephasing time of excitation, as a function of the interlevel spacing energy. They found that the line width was larger for a smaller-size quantum dot for which the interlevel spacing is larger. This is predicted by a simple particle in a box model as the length of the box becomes smaller.

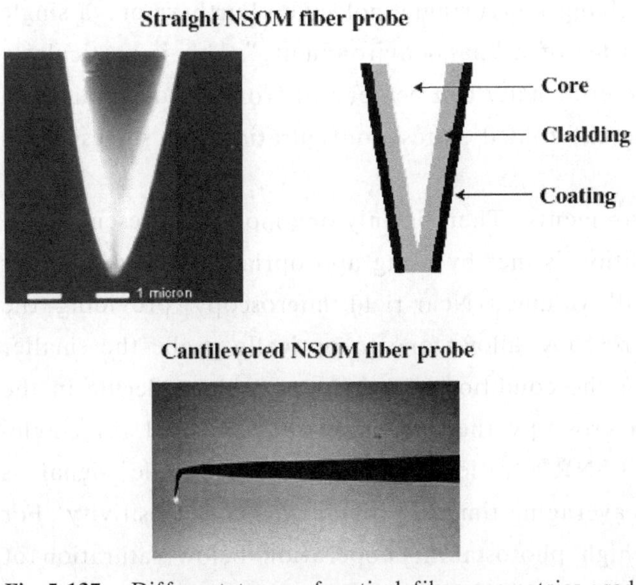

Fig. 5-137 Different types of optical fiber geometries used for NSOM.

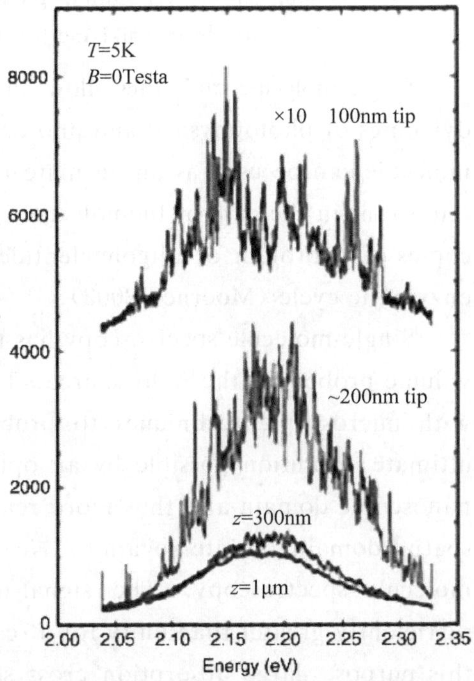

Fig. 5-138 Near-field microscopy/spectroscopy of quantum dots.

- Single-Molecule Spectroscopy

An exciting direction is to push the limit of spatial probing to single molecule detection. A single atom or a molecule represents the ultimate goal of nanoscopic resolution. This means that we are not just looking at individual Q-dots, which are assemblies of hundreds or thousands of atoms, but we can look at individual molecules and atoms.

Single-molecule detection using spectroscopic methods is a highly active field (Moerner, 2002). Single-molecule spectroscopy is a powerful technique to probe individual nanoscale behavior of atoms and molecules in a complex local environment of a condensed phase. The ability to detect a single molecule and study its structure and functions provides the opportunity to elucidate single-molecule properties that are not available in an averaging measurement on an ensemble containing a large number of molecules.

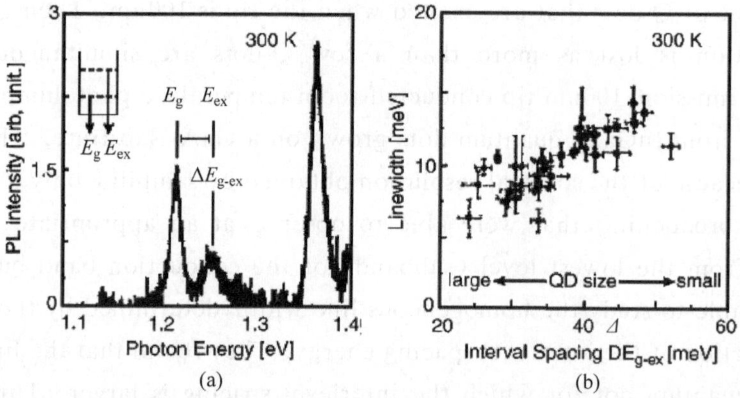

Fig. 5-139 Photoluminescence spectrum of single QD at room temperature (a), and dependence of the homogeneous linewidth of ground-state emission on interval spacing, which is closely related to size of Qd's (b). From Saiki and Narita, reproduced with permission.

Single-molecule study thus allows investigation of hidden heterogeneity and provides information on dynamics of photophysical and photochemical changes in a single molecule. Furthermore, a single molecule can be used as an ultimate local reporter of a "nanoenvironment." This is particularly important in the case of biomolecules where heterogeneity can be derived from various individual copies of a protein or oligonucleotide in different folded states, configurations, or stages of an enzymatic cycle (Moerner, 2002).

Single-molecule spectroscopy has two requirements: There is only one molecule present in the volume probed by the light source. This condition is met by using appropriate dilution together with microscopic techniques to probe a small volume. Near-field microscopy, providing the ultimate resolution possible by an optical microscopy, allows one to optically probe the smaller nanoscopic domain and thus more readily meet the condition of having a single molecule in the spatial domain of interrogation. Near-field microscopy thus has emerged as a tool for single-molecule spectroscopy. The signal-to-noise (SNR) ratio for the single-molecule signal is sufficiently greater than unity for a reasonable averaging time to provide adequate sensitivity. For this purpose, large absorption cross sections, high photostability, operation below saturation of absorption, and (in the case of fluorescence detection) a high fluorescence quantum yield are needed. In addition, the use of a small focal volume also provides an enhancement of SNR. For this reason also, near-field microscopy is desirable. Since the original work of Moerner et al. in the 1990s (Moerner and Kador, 1990; Moerner, 1994; Moerner et al., 1994), this field has seen an explosion of activities and reports. For single-molecule spectroscopy, various spectroscopic techniques such as absorption, fluorescence, and Raman have been used. Fluorescence has been the most widely utilized spectroscopic method for single-molecule detection.

Single-molecule fluorescence detection has been successfully extended to biological systems (Ha et al., 1996, 1999; Dickson et al., 1997). Excellent reviews of the applications of single-molecule detection to bioscience are by Ishii and Yanagida (2001) and Ishijima and Yanagida (2001). Single-molecule detection has been used to study molecular motor functions, DNA transcription, enzymatic reactions, protein dynamics, and cell signaling. The single molecule detection permits one to understand the structure-function relation for individual biomolecules, as opposed to an ensemble average property that is obtained by classical measurements involving a large number of molecules.

Near-field excitation can provide enhancement of the fluorescence to increase SNR for single molecule spectroscopy. Fluorescence probes used for single-molecule detection are fluorescence lifetime, two-photon excitation, polarization anistropy, and fluorescence resonant energy transfer (FRET). Single-molecule fluorescence lifetime measurements have been greatly aided by the use of a technique called time-correlated single-photon detection. Figure 5-140 shows fluorescence NSOM images of single molecules from Barbara's group (Higgins and Barbara, private communication). The variation of fluorescence intensity from one molecular emitter to another may be due to possible variations of the molecule on the substrate. Ambrose et al. (2002) reported an abrupt irreversible photobleaching in Rhodamine-6G dye after repeated excitations of the same molecule. This abrupt bleaching is not observed for an ensemble of molecules which exhibit a gradual decrease of fluorescence. Betzig and Chichester (2003) reported a similar irreversible photobleaching on lipophilic carbocyanine. A measurement of Stark shift in a single molecule can be used as a local sensor of the electric field distribution in nanoscopic domains. For this purpose, Moerner et al. (2004) used a 30-nm-diameter near-field optical probe consisting of an aluminum-coated optical fiber tip. A static electric potential applied to the Al-coated tip produced the Stark shifts of the absorption line of a single molecule.

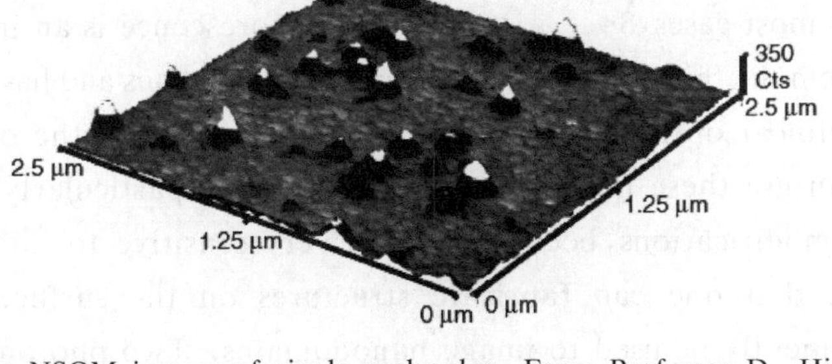

Fig. 5-140 Fluorescence NSOM images of single molecules. From Professor D. Higgins and Professor P. Barbara, unpublished results.

• Nonlinear optical processes

Nonlinear optical processes provide information about nanoscopic-level organization and interactions. The nonlinear optical processes, depend on higher powers of the field strength and involve more than one photon interacting at one time in a medium. For sake of clarity, Figure 5-141 shows some examples of nonlinear optical processes. In the case of second harmonic generation (SHG), two photons of the same frequency, γ, interact with the medium and generate an output at 2ω. This interaction occurs without the absorption of light. Thus, this is a coherent process where a medium simply interacts with light and converts the light, at its fundamental at frequency γ (wavelength $\ddot{e}$), to second harmonic at doubled frequency 2ω (hence wavelength, $\ddot{e}/2$). The SHG process is limited by symmetry requirements to occur only in a noncentrosymmetric medium or at an interface which by nature is asymmetric (two different media on opposite sides).

The second represented process is third harmonic generation (THG). It is derived from third-order nonlinear optical interactions. In this case, the medium interacts with light of fundamental frequency ω, utilizing three photons together to generate an output of a single photon with frequency ω. If λ is in the near infrared at wavelength $1.06\mu m$, the output is at wavelength

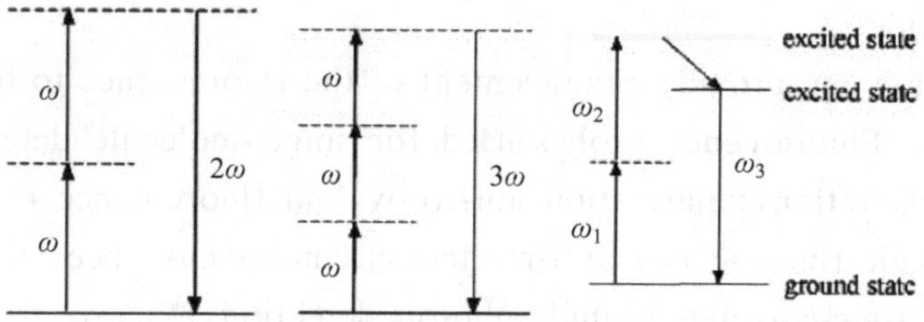

Fig. 5-141 Examples of nonlinear optical processes.

~355nm in the UV. This process can occur in any media because no symmetry restriction on the medium is imposed.

Another process is two-photon excitation (TPE), where the material absorbs two photons simultaneously to reach an excited state. In the case represented in the figure, the two photons have different frequencies, ω_1 and ω_2, to produce nondegenerate two-photon absorption, or they can have the same frequency to produce twophoton absorption at ω_2. In the case represented, $\omega_1 + \omega_2$ generates a real excited state, which can then relax to another lower state and emit a photon of frequency, ω_3. ω_3 is of higher value than the initial photons of frequencies ω_1 or ω_2. The TPE thus produces an up-converted emission process. It is different from SHG, because in SHG the optical output is fixed at 2ω and is a very sharp line at ω_2, related to the line width only of the fundamental at frequency ω. In the TPE, ω_3 is a higher frequency compared to either ω_1 or ω_2, but it is not or $\omega_1 + \omega_2$. In most cases, $3 < (\omega_1 + \omega_2)$. TPE fluorescence is an incoherent process. This process is also derived from the third-order nonlinear optical interactions and has no symmetry restriction.

Some of these nonlinear optical techniques are very sensitive to the orientation of molecules on the surface. One can use these nonlinear optical processes, particularly SHG, to probe surface structures and surface modifications, because SHG is very sensitive to interfacial characteristics. Another application is that one can fabricate structures on the surface. Two-photon excited fluorescence can conveniently be used to image nanodomains. Two-photon excitation can be used to fabricate nanostructures by using photolithographic techniques, as shall be described later.

TPE provides the advantage that this excitation, using two photons, is quadratically dependent on intensity, so it will be significantly more localized near the focal point. Therefore, nanodevices can be fabricated with great precision using nonlinear optical excitations. A number of studies of SHG using near-field optics have been reported.

For the sake of convenience, the work reported here is that performed at our Institute (Shen et al., 2000; Shen et al., 2001a). Figure 5-142 provides an example of SHG in an organic crystal called NPP (structure represented in Figure 5-143) using near-field microscopy. Many different nanocrystals of NPP are formed on the surface of the substrate. The NPP crystallites are ~100nm lengthwise and well-oriented (Shen et al., 2001a). NPP has the appropriate symmetry and structure to generate the second harmonic, so we can use it to map out the local second harmonic domain distribution and discover how these crystalline domains are oriented on the surface.

The polarization of the light is varied from an arbitrary zero degree to further probe the orientation of the nanocrystals. At 90° the crystallites that are bright are not so active in the zero polarization. The top illustration in Figure 5-142 is a shear force topographic image. The topographic image correlates well with the second harmonic image, thus confirming the absence of any artifacts. By analyzing polarization dependence obtained from the study of SHG as a function

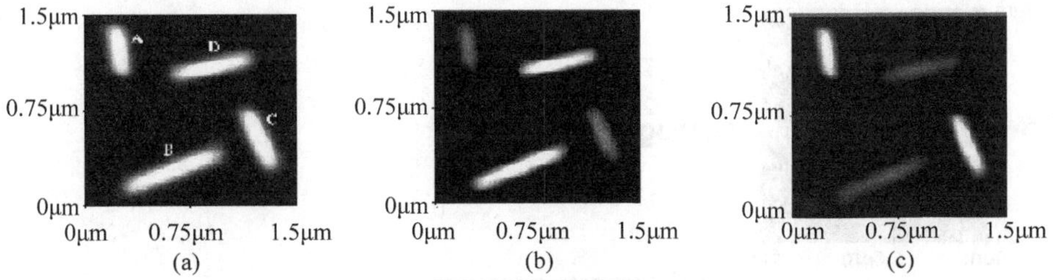

Fig. 5-142 Near-field second harmonic images of NPP nanocrystals in two orthogonal polarizations.

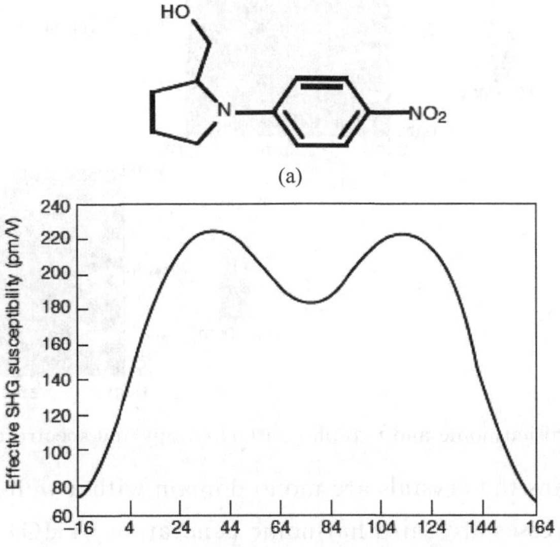

Fig. 5-143 Structure of NPP nanocrystal (a) and Polarization dependence of second harmonic generation in the NPP(b).

of angle of rotation of polarization, one can obtain information on the orientation of the nanocrystals. The experimentally obtained curve is shown in Figure 5-144, which plots SHG as a function of the angle of polarization of light by rotating the polarizer to different angles. The second-order nonlinear optical susceptibility is often described for SHG by a d coefficient that is a tensor. The effective d coefficient, deff, is a measure of the material's figure of merit for the SHG. The deff and its angular distribution is described by Eq. (5-445):

$$d_{\text{eff}} = \{d_{21}^2 \sin^2 2(\theta+\alpha) + [d_{21}\sin^2(\theta+\alpha) + d_{22}\cos^2(\theta+\alpha)]^2\}^{1/2} \tag{5-445}$$

Effective SHG susceptibility (pm/V) A fit of these equations yields $d_{\text{eff}} = 224$ pm/V and provides two tensor components d_{21} and d_{22} of the d coefficient. The other information one gets, by the fit of this anisotropy, is that the crystallographic b axis and the CT axis of this crystal are parallel to the plane of the substrate. Therefore, this study provides a very clear indication of how these nanocrystal domains are oriented on the substrate plane.

This study provided details of structural information on the nanoscopic order of the crystal which are listed in Table 5-4. The result shows that there is a uniform SHG intensity distribution over the entire nanocrystal for each of them, as the nanocrystals do not exhibit bright spots and dark spots. This means that there is a nanoscopic order in these domains. Second, the polarization dependence conforms to this nanoscopic order as demonstrated by the applicability of Eq. (5-445). The fit of deff shows the crystal orientation on the surface. Different locations in the same crystal

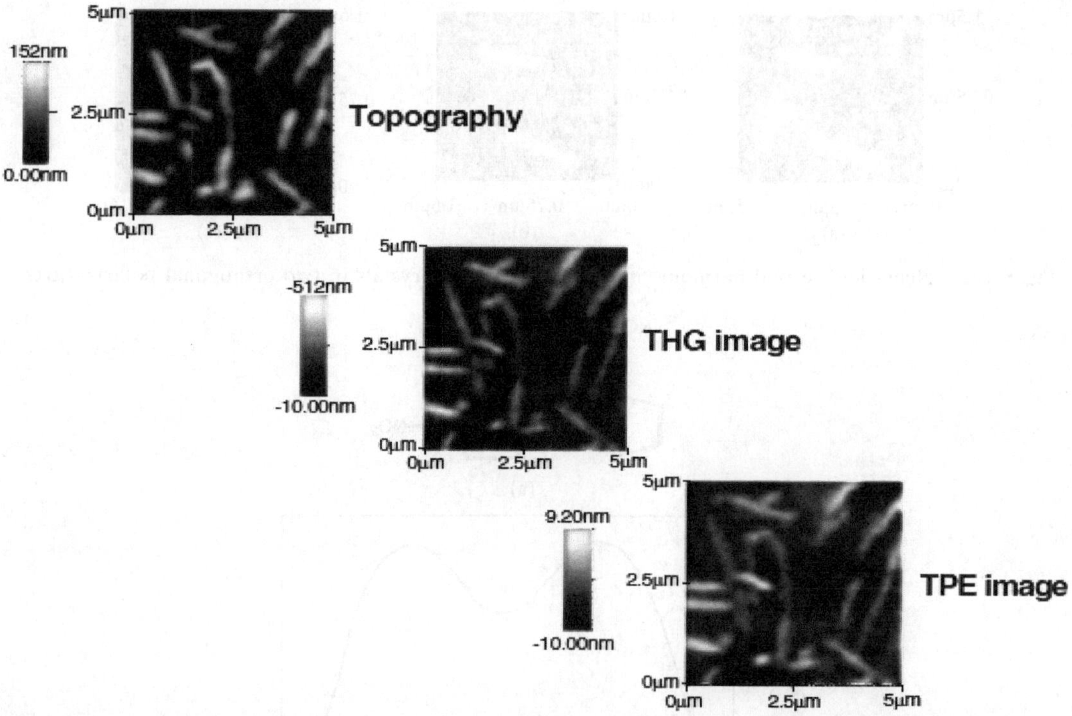

Fig. 5-144 Near-field third harmonic and two-photon microscopy and spectroscopy in DEANST crystals.

yield the same deff. It means the crystals are mono-domain with a well-defined orientation. Other examples of nonlinear processes are third harmonic generation (THG) and two-photon excitation (TPE) (Shen et al., 2001b). As described above, THG involves generation of a beam at a wavelength ~355nm (frequency 3) for the incident beam, called the fundamental beam, of wavelength 1064nm (frequency). TPE, on the other hand, may involve simultaneous absorption of two photons, each of frequency, and generates an up-converted fluorescence of higher frequency. It might appear that THG is a three-photon process and that TPE is a two-photon process. But the theory of nonlinear optical interactions shows that both these processes are derived from third-order nonlinear optical interactions.

Figure 5-144 shows the study of both THG and two-photon fluorescence, observed simultaneously in an organic nanocrystal called DEANST (structure shown in Figure 5-145). The topographic image on the top is shear force generated (as Table 5-4. Nanoscopic Information from Second Harmonic Imaging of Nanocrystals tained by atomic force microscopy, AFM).

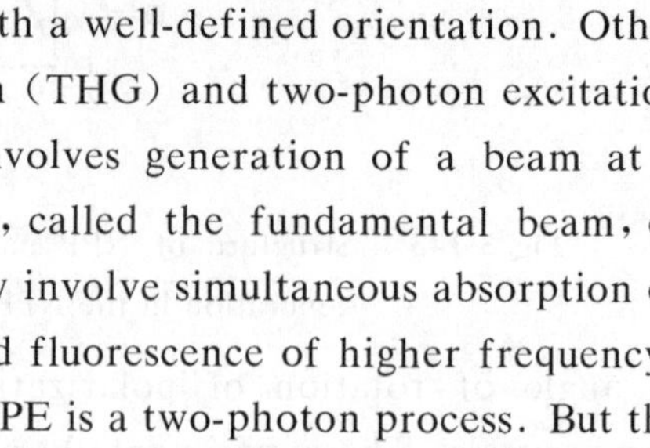

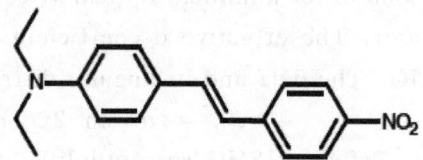

Fig. 5-145 Structure of DEANST.

Table5-4 Nanoscopic Information from Second Harmonic Imaging of Nanocrystals

- Uniform SH intensity distribution→Nanoscopic order
- Polarization dependence→Nanoscopic order
- Applicability of equation for d_{eff}→Crystallographic *b* axis (P_{21}) and the molecular *c* axis parallel to the plane of the substrate
- Locations in the same nanocrystal or different nanocrystals give same d_{eff}→Same symetry and order

The middle image is obtained by monitoring the third harmonic signal. In this case the fundamental light is at 1.064μm. Hence, the third-harmonic output is generated at ~355nm in the

UV. This image correlates fairly well with the topographic image shown on the top. The bottom image is a TPE fluorescence image. Here the fluorescence emission is in the red region with a maximum at ~600nm. All three of the images correlate well. Figure 5-146 shows the spectral distribution of the near-field signal. A very sharp line at ~355nm is due to THG. The line width of the spectroscopic feature is related to the line width of the fundamental at 1.064μm.

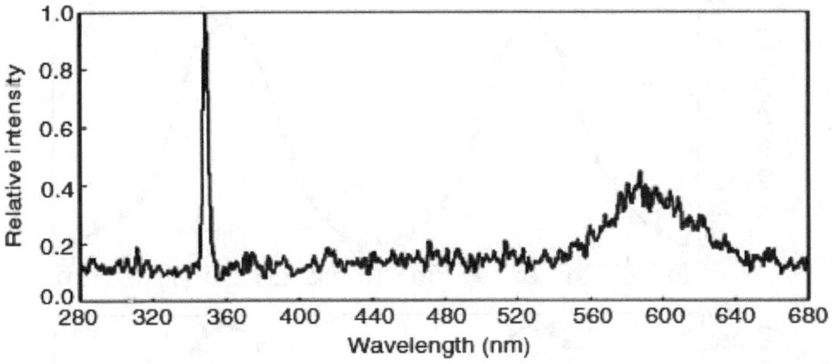

Fig. 5-146 Spectral distribution of THG and TPE near-field signal, in the DEANST nanocrystals.

Then, at around 600nm there is a broad emission peak. This broad curve is due to TPE fluorescence. Hence, the DEANST crystal generates both third harmonic and TPE fluorescence. Thus, both the absolute value (for THG) and the imaginary part (for TPE) of the third-order nonlinear optical interactions are manifested. Both are strong, when a fundamental light of 1.064μm is used. An intensity dependence study shows that TPE appears first, in agreement with the fact that TPE should depend on the square of the input intensity while THG depends on the cube. Detailed spectral study also demonstrates that no SHG is generated, as expected because DEANST crystals are centrosymmetric. It also establishes that THG is generated directly by a $\chi^{(3)}$ process, rather than two coupled (cascaded) second-order processes that first produce 2ω and then sums it with ω. In order to get information on the orientation of the crystal, one can rotate the polarization of light incident on the crystal and can map out the anisotropy. Figure 5-147 shows the results of such a study for both THG and TPE fluorescence. There is a one-to-one correlation between the two curves. Because these angular variations relate to the anisotropy of the third-order nonlinear optical susceptibility, this correlation is not surprising. The observed anisotropy is determined by the relative contributions of various tensor components of $\chi^{(3)}_{eff}$. The fitted result shows that the ratio of the two in-plane diagonal tensor components $\chi^{(3)}_{yyyy}$ component. Furthermore, at a two-photon resonance, the dominant contribution to $\chi^{(3)}_{xxx}$ may be from the imaginary components. Thus even THG may be determined by the imaginary part of $\chi^{(3)}$.

$$\chi^{(3)}_{eff} = \chi^{(3)}_{yyy}\cos^4(\theta+\alpha) + 6\chi^{(3)}_{xxyy}\cos^2(\theta+\alpha)\sin^2(\theta+\alpha) + \chi^{(3)}_{xxx}\sin^4(\theta+\alpha) \quad (5\text{-}446)$$

- Apertureless near-field spectroscopy and microscopy

An emerging approach is the apertureless near-field spectroscopy and microscopy (Novotny et al., 1998; Sanchez et al., 1999; Bouhelier et al., 2003). The use of an aperture such as a tapered fiber opening poses a number of experimental limitations. Some of these are Low light throughput due to the small fiber aperture and the finite skin depth (light penetration) into the aluminum metal coating around the tapered fiber. Absorption of light in the metal coating; this can produce significant heating that can create a problem in imaging, particularly of biological samples. Pulse broadening in the fiber, when using short pulses for nonlinear optical studies. Also, the fiber tip may be damaged by the high peak intensity. The apertureless approach overcomes these limitations, at the same time providing a significantly improved resolution. It has been demonstrated by Novotny, Xie, and Sanchez. Hartschuh et al. that optical images and spectra of nanodomains 25nm can be obtained using the apertureless near-field approach involving a metal tip of end diameter 10nm.

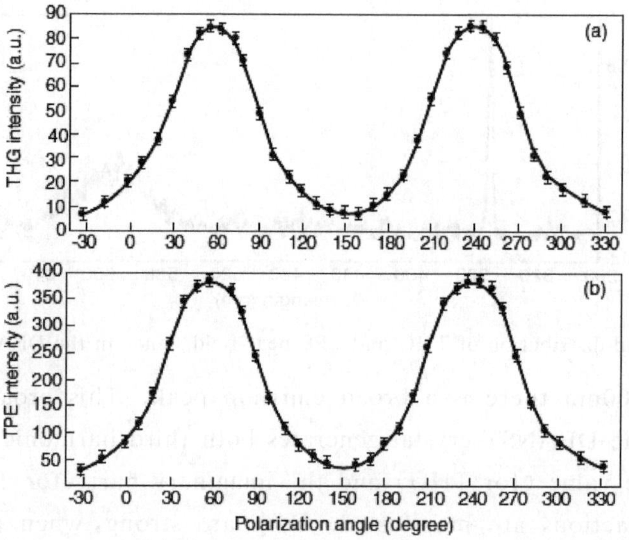

Fig. 5-147 In-plane THG and TPE anisotropy for the DEANST nanocrystals.

- The two approaches used for apertureless NSOM: Scattering type

Scattering type, which involves nanoscopic localization and field enhancement of the electromagnetic radiation by scattering of the light from a metallic nanostructure. An example is provided by Figure 5-148 where the light is scattered by a sharp metallic tip. Scattering and field localization can also be produced by a metallic nanoparticle within nanometers of distance from the sample surface. The localization and enhancement of electromagnetic field by plasmon coupling to a metallic nanoparticle is discussed under "Plasmonics." This principle of obtaining nanoscopic resolution using scattering from a metallic nanoparticle also forms the basis of "plasmonic printing," on "Nanolithography".

Field-enhancing apertureless NSOM, where a metallic tip is used to enhance the field of an incident light in the near field. In this case, the light is incident on the tip as a normal propagating mode (far-field). The strongly enhanced electric field at the metal tip produces nanoscopic localization of optical excitation. This approach offers simplicity and versatility of using light by just focusing on the metallic tip through a high-numerical-aperture lens. Hence it is described here in detail, with examples of some recent studies utilizing this approach. A schematics of "field-enhancing" apertureless NSOM, as used by Novotny, Xie, and co-workers of Novotny et al., and Sanchez et al., is presented in Figure 5-148. This method, as pointed above, combines both the far-field and the near-field techniques. The principle utilized is that if the incident radiation has a polarization component (E field) along the tip axis, a strong surface charge density is induced at the tip end, producing a large field enhancement.

For this purpose, a normal Gaussian beam cannot be used, because it does not provide any polarization component along the tip axis, in the geometry of an on-axis optical illumination as shown in Figure 5-148. Therefore, laser beams of higher order such as Hermite-Gaussian are needed, which provide a longitudinal (along the tip axis) electric field in the focal region. The highly confined fields close to the tip then interact with the neighboring nanodomains. The optical response, whether fluorescence, second harmonic generation, or Raman scattering (at a shifted wavelength, which also collected by the same high N. A. (numerical aperature) lens used for illumination. That a gold tip with an end diameter of 10nm can produce an intensity enhancement factor of over 3000 over the incident intensity. This enhancement is effective even though the incident wavelength is far away from the surface plasmon resonance of the metal (for surface plasmon resonance). As described above, the field enhancement is produced by the high surface charge density at the tip due to the incident light component polarized along the tip. Sanchez et al. showed that with an asymmetrical metal tip (bent shape), even a focused Gaussian beam (TEM00) can be used to produce field enhancement at the tip. That used this geometry with femtosecond pulses from a Ti-sapphire laser to produce efficient two-photon excited emission. This two-photon, near-field microscopy was used to image fragments of photosynthetic membranes as well as J-aggregates for with spatial resolutions of ~20nm. Figure 5-148 shows the near-field, two-photon excited fluorescence image together with the topographic image (Figure 5-148) of Jaggregates of the PIC dye in a polyvinyl sulfate (PVS), coated on a glass substrate. The lower portions of the figure show the simultaneous cross sections taken across the aggregate strands. The emission image cross section has a full width at half-maximum (FWHM) of ~30nm, indicating a superior spatial resolution obtained by this method.

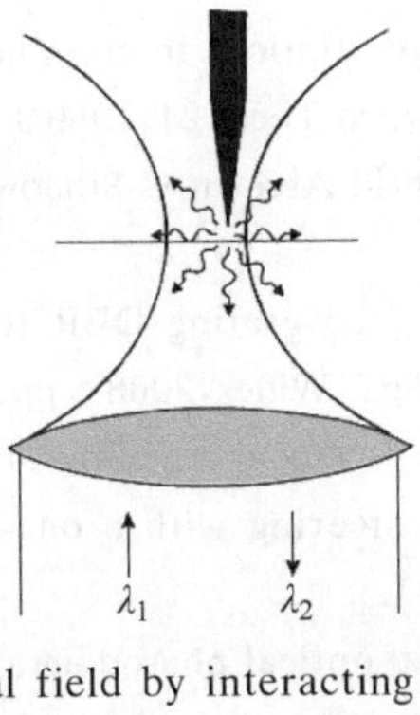

Fig. 5-148 Metallic tip enhancing the local field by interacting with the focused beam at λ_1. The optical response at another wavelength λ_2 is collected by the same objective lens. The second harmonic generation is dominated by excitation fields polarized along the tip axis. Thus the tip can act as a probe for longitudinal (along the tip axis) fields. The high-resolution, near-field Raman microscopy of single-walled carbon nanotubes using a sharp silver tip as a probe (10~15nm in diameter). A spatial resolution of ~25nm and obtained the Raman image with the excitation beam of $\lambda = 633$nm, using the Raman bands at 1596cm^{-1} the so-called G band corresponding to the tangential stretching mode, which is sharp and 2615cm^{-1}, the so-called G band corresponding to an overtone of the disorder-induced mode at 1310cm^{-1}, the second harmonic generation at an infinite tip can be represented with a dipole oscillating at the second harmonic frequency.

References

[1] Novikov, D. V. et al. Plane wave GID topography of defects in lithium niobate after diffusion doping, Nuclear Instruments and Methods in Physics Research B, 97, 342-345(1995).

[2] Quintanilla, M. et al., Optical Materials 30, 1098-1102(2008).

[3] de Riedmatten, H., Afzelius, M., Staudt, M. U., et al. A solid-state lightmatter interface at the single-photon level. Nature 456, 773-777(2008).

[4] Sinclair, N. et al. Spectroscopic investigations of a Ti: Tm: $LiNbO_3$ waveguide for photon-echo quantum memory. J. Lumin. 130(9), 1586-1593(2010).

[5] Afzelius, M., Simon, C., de Riedmatten, et al. Multimode quantum memory based on atomic frequency combs. Phys. Rev. A 79, 052329(2009).

[6] Bonarota, M., Ruggiero, J., Le Gouët, J. L., et al. Efficiency optimization for Atomic Frequency Comb storage. Preprint at http://arxiv.org/abs/0911.4359(2009).

[7] Tittel, W. et al. Photon-echo quantum memory in solid state systems. Laser & Photon. Rev. 4, (2), 244-267(2010).

[8] Thiel, C. W., Sun, Y., Böttger, T., et al. Optical decoherence and persistent spectral hole burning in Tm^{3+}: $LiNbO_3$. J. Lumin, 130(9), 1603-1609(2010).

[9] Afzelius, M. et al. Demonstration of Atomic Frequency Comb Memory for Light with Spin-Wave Storage. Phys. Rev. Lett. 104, 040503(2010).

[10] A. G. Kirk et al. Design rules for highly parallel free-space optical interconnects. IEEE J. Sel. Top. Quant. Electr. 9, 531-547(2003).

[11] C. J. Henderson et al. Free space adaptive optical interconnect at 1.25 Gb/s, with beam steering using a ferroelectric liquid-crystal SLM. J. Lightwave Tech. 24, 1989-1997(2006).

[12] W. Fang et al., "A Continuous Wave Hybrid AlGaInAs-Silicon Evanescent Laser," IEEE Phot. Tech. Lett. 18, 1143-1145(2006).

[13] M. N. Sysak et al., "A hybrid silicon sampled grating DBR tunable laser," in Group IV Photonics, 2008 5th IEEE International Conference on, (Cardiff, Wales, 2008), pp. 55-57.

[14] H. Park et al. A Hybrid AlGaInAs-Silicon Evanescent Amplifier. IEEE Phot. Tech. Lett. 19, 230-232(2007).

[15] K. Van Acoleyen et al., "Off-chip beam steering with a one-dimensional optical phased array on silicon-on-insulator," Opt. Lett. 34, 1477-1479(2009).

[16] K. Van Acoleyen et al., "Two-dimensional optical phased array antenna on silicon-on-Insulator," Opt. Express 18, 13655-13660(2010).

[17] K. Van Acoleyen et al., "Two-Dimensional Dispersive Off-Chip Beam Scanner Fabricated on Silicon-On-Insulator," IEEE Phot. Tech. Lett. 23, 1270-1272(2011).

[18] D. Kwong et al., "1 × 12 Unequally spaced waveguide array for actively tuned optical phased array on a silicon nanomembrane," Appl. Phys. Lett. 99, 051104(2011).

[19] N. Le Thomas et al., "Exploring light propagating in photonic crystals with Fourier optics," J. Opt. Soc. Am. B 24, 2964-2971(2007). OM2J.

[20] Lvovsky, A. I., Sanders, B. C. & Tittel. W. Optical quantum memory. Nature Photonics 13, 706-714(2009).

[21] Sangouard, N., Simon, C., de Riedmatten, H. & Gisin, N. Quantum repeaters based on atomic nsembles and linear optics. Preprint at http://arxiv.org/abs/0906.2699(2009).

[22] Kimble, H. J. The quantum Internet. Nature 453, 1023-1030(2008).

[23] Aspect, A. Talking entanglement. Nature Photonics 3, 486-487(2009).

[24] Afzelius, M., Simon, C., de Riedmatten, H. & Gisin, N. Multimode quantum memory based on atomic frequency combs. Phys. Rev. A 79, 052329(2009).

[25] Usmani, I., Afzelius, M., de Riedmatten, H. & Gisin, N. Mappig multiple photonic qubits into and out of one

solid-state atomic ensemble. Nature Comm. 1: 12, 1-7(2010).

[26] Sohler, W. et al. Integrated Optical Devices in Lithium Niobate. Optics & Photonics News, Jan. 2008, 24-31.

[27] Pan, J.-W, Chen, Z.-B., Zukowski, M., Weinfurter, H. & Zeilinger, A. Multi-photon entanglement and interferometry. Preprint at http://arxiv.org/abs/0805.2853(2008).

[28] Gisin, N., Ribordy, G., Tittel, W. & Zbinden, H. Quantum Cryptography. Rev. Mod. Phys. 74, 145-195(2002).

[29] Tittel, W. & Weihs, G. Photonic Entanglement for Fundamental Tests and Quantum Communication, Quant. Inf. Comp. 1, 3-56(2001).

[30] Julsgaard, B., Sherson, J., Cirac, J. I. J. Fiurášek, J. & Polzik, E. S. Experimental demonstration of quantum memory for light. Nature 432, 482-486(2004).

[31] Chanelière, T. et al. Storage and retrieval of single photons transmitted between remote quantum memories. Nature 438, 833-836(2005).

[32] Eisaman, M. D. et al. Electromagnetically induced transparency with tunable singlephoton pulses. Nature 438, 837-841(2005).

[33] Honda, K. et al. Storage and retrieval of a squeezed vacuum. Phys. Rev. Lett. 100, 093601(2008).

[34] Appel, J., Figueroa, E., Korystov, D., Lobino, M. & Lvovsky, A. Quantum memory for squeezed light. Phys. Rev. Lett. 100, 093602(2008).

[35] de Riedmatten, H., Afzelius, M., Staudt, M. U., Simon, C. & Gisin, N. A solid-state light-matter interface at the single-photon level. Nature 456, 773-777(2008).

[36] Hedges, M. P., Longdell, J. J., Li, Y. & Sellars, M. J. Efficient quantum memory for light, Nature 465, 1052-1056 (2010).

[37] Boozer, A. D. et al. Reversible State Transfer between Light and a Single Trapped Atom. Phys. Rev. Lett. 98, 193601(2007).

[38] Chou, C. W. et al. Measurement-induced entanglement for excitation stored in remote atomic ensembles. Nature 438, 828-832(2005).

[39] Matsukevich, D. N. et al. Entanglement of a Photon and a Collective Atomic Excitation. Phys. Rev. Lett. 95, 040405(2005).

[40] Yuan, Z.-S. et al. Experimental demonstration of a BDCZ quantum repeater node. Nature 454, 1098-1101(2008).

[41] Blinov, B. B., Moehring, D. L., Duan, L.-M. & Monroe, C. Observation of entanglement between a single trapped atom and a single photon. Nature 428, 153-157(2004).

[42] Volz, J., et al. Observation of Entanglement of a Single Photon with a Trapped Atom. Phys. Rev. Lett. 96, 030404(2006).

[43] Wilk, T., Webster, S. C., Kuhn, A., & Rempe, G. Single-Atom Single-Photon Quantum Interface. Science 317, 488-490(2007).

[44] Togan, E., et al. Quantum entanglement between an optical photon and a solid-state spin qubit. Nature 466, 730-735(2010).

[45] Longdell, J., Fraval, E., Sellars, M. & Manson, N. Stopped light with storage times greater than one second using electromagnetically induced transparency in a solid. Phys. Rev. Lett. 95, 063601(2005).

[46] Sinclair, N. et al. Spectroscopic investigations of a Ti: Tm: $LiNbO_3$ waveguide for photon-echo quantum memory. J. Lumin. 130, 1586-1593(2010).

[47] J. B. Altepeter, J. B. Jeffrey, E. R. & Kwiat, P. G. Photonic State Tomography. Advances in Atomic, Molecular, and Optical Physics 52, 105-159(2005).

[48] Plenio M. B. & Virmani, S. An introduction to entanglement measures. Quant. Inf. Comp. 7, 1-51(2007).

[49] Clausen, C. et al. Quantum Storage of Photonic Entanglement in a Crystal. arXiv: 1009.0489(2010).

[50] B. Julsgaard, J. Sherson, J. I. Cirac, J. Fiurášek, and E. S. Polzik, Nature 432, 482(2004).

[51] M. A. Eisaman et al., Nature 438, 837(2005).

[52] S. A. Moiseev, V. F. Tarasov, and B. S. Ham, J. Opt. B.: Quantum Semiclass. Opt. 5, S497(2003).

[53] M. Nilsson, and S. Kröll, Opt. Comm. 247, 393(2005).

[54] A. L. Alexander, J. J. Longdell, M. J. Sellars, and N. B. Manson, Phys. Rev. Lett. 96, 043602(2006).

[55] G. Hétet, J. J. Longdell, A. L. Alexander, P. K. Lam, and M. J. Sellars, quant-phys/0612169. S. R. Hastings-Simon et al., Opt. Comm. 266, 716(2006).

[56] A. V. Gorshkov, A. André, M. Fleischhauer, A. S. Sørensen, and M. D. Lukin, quant-ph/ 0604037.

[57] E. Fraval, M. J. Sellars, and J. J. Longdell Phys. Rev. Lett. 95, 030506(2005).

[58] C. Batten et al., "Building many-core processor-to-DRAM networks with monolithic CMOS silicon photonics," IEEE Micro 29, 8-21(2009).

[59] C. R. Doerr et al., "Monolithic polarization and phase diversity coherent receiver in silicon," J. Lightwave Technol. 28, 520-525(2010).

[60] C. W. Holzwarth et al., "Localized substrate removal technique enabling strong-confinement microphotonics in a bulk CMOS process," in Proc. CLEO/IQEC Conf. Lasers Electro Opt/Intl. Quant. Elec. Conf. (2008).

[61] J. S. Orcutt and R. J. Ram, "Photonic device layout within the foundry CMOS design environment," IEEE Phot. Technol. Lett. 22, 544-546(2010).

[62] J. S. Orcutt et al., "Demonstration of an electronic photonic integrated circuit in a commercial scaled bulk cmos process," in Proc. CLEO/IQEC Conf. Lasers Electro Opt/Intl. Quant. Elec. Conf. (2008).

[63] J. S. Orcutt et al., "Nanophotonic Integration in State-of-the-Art CMOS Foundries," Opt. Exp., 19, 2335-2346 (2011).

[64] J. S. Orcutt et al., "Scaled CMOS photonics," in Proc. Photon. In Switching Conf. (2009).

[65] J. S. Orcutt et al., "Low-loss polysilicon waveguides suitable for integration within a high-volume electronics process,"in Proc. CLEO/IQEC Conf. Lasers Electro Opt/Intl. Quant. Elec. Conf., (2011).

[66] M. Georgas, J. Orcutt, R. J. Ram and V. Stojanovic, " A Monolithically-Integrated Optical Receiver in Standard 45-nm SOI," European Solid-State Circuits Conference, (2011).

[67] Lund, A. P. & Ralph, T. C. Nondeterministic gates for photonic single-rail quantum logic. Phys. Rev. A 66, 032307(2002).

[68] Berry, D. W., Lvovsky, A. I. & Sanders, B. C. Interconvertibility of single-rail optical qubits. Opt. Lett. 31, 107-109(2006).

[69] Marcikic, I. et al. Time-bin entangled qubits for quantum communication created by femtosecond pulses. Phys. Rev. A 66, 062308(2002).

[70] Gottesman, D. Quantum Error Correction and FaultTolerance. in Encyclopedia of Mathematical Physics 4, 196-201. Francoise, J.-P., Naber, G. L. & Tsou, S. T., eds. (Elsevier, Oxford, 2006).

[71] Shor, P. W. Scheme for reducing decoherence in quantum computer memory, Phys. Rev. A 52, R2493-R2496(1995).

[72] Lobino, M. et al. Complete characterization of quantum-optical processes. Science 322, 563-566(2008).

[73] Lobino, M., Kupchak, C., Figueroa E., et al. Memory for Light as a Quantum Process, Phys. Rev. Lett. 102, 203601(2009).

[74] Hammerer, K., Wolf, M. M., Polzik, E. S., et al. Quantum benchmark for storage and transmission of coherent states. Phys. Rev. Lett. 94, 150503(2005).

[75] Hétet, G., Peng, A., Johnsson, M. T., Hope, J. J. & Lam, P. K. Characterization of electromagneticallyinduced-transparency-based continuous-variable quantum memories, Phys. Rev. A 77, 012323(2008).

[76] Afzelius, M., Simon, C., de Riedmatten, H. & Gisin, N. Multimode quantum memory based on atomic frequency combs. Phys. Rev. A 79, 052329(2009).

[77] Nunn, J. et al. Multimode Memories in Atomic Ensembles. Phys. Rev. Lett. 101, 260502(2008).

[78] Raussendorf, R., & Briegel, H. J. A one-way quantum computer, Phys. Rev. Lett. 86, 5188-5191(2001).

[79] P. Kok, Munro, W. J., Nemoto, K., Dowling, J. P. & Milburn, G. J., Linear optical quantum computing with photonic qubits. Rev. Mod. Phys. 79, 135-174(2007).

[80] Briegel, H. J., Dür, W., Cirac, J. I. & Zoller, P. Quantum repeaters: The role of imperfect local operations in quantum communication. Phys. Rev. Lett. 81, 5932-5935(1998).

[81] Sangouard, M., Simon, C., de Riedmatten, H., & Gisin, N. Quantum repeaters based on atomic ensembles and linear optics. arXiv: 0906.2699. Appel, J. et al. Mesoscopic atomic entanglement for precision measurements beyond the standard quantum limit, P Natl. Acad. Sci. USA 106, 10960-10965(2009).

[82] Landry, O., van Houwelingen, J. A. W., Beveratos, A., Zbinden, H. & Gisin, N. Quantum teleportation over the Swisscom telecommunication network. J. Opt. Soc. Am. B 24, 398-403(2007).

[83] Pittman, T. B. & Franson, J. D. Single Photons on pseudodemand from stored parametric down-conversion. Phys. Rev. A 66, 042303(2002).

[84] Pittman, T. B. & Franson, J. D. Cyclical quantum memory for photonic qubits. Phys. Rev. A 66, 062302(2002).

[85] Leung, P. M. & Ralph, T. C. Quantum memory scheme based on optical filters and cavities. Phys. Rev. A 74, 022311(2006).

[86] Maître, X. et al. Quantum Memory with a Single Photon in a Cavity. Phys. Rev. Lett. 79, 769-772(1997).

[87] Tanabe, T., Notomi, M., Kuramochi, E., Shinya, A. & Taniyama, H. Trapping and delaying photons for one nanosecond in an ultrasmall high-Q photonic-crystal nanocavity. Nature Phot. 1, 49-52(2006).

[88] Tanabe, T., Notomi, M., Taniyama, H. & Kuramochi, E. Dynamic Release of Trapped Light from an Ultrahigh-Q Nanocavity via Adiabatic Frequency Tuning. Phys. Rev. Lett. 102, 043907(2009).

[89] Javan, O. Kocharovskaya, H. Lee & M. O. Scully, Narrowing of EIT resonance in a Doppler Broadened Medium, Phys. Rev. A 66, 013805(2002).

[90] Boller, K. J., Imamoglu, A. & Harris, S. E. Observation of electromagnetically induced transparency. Phys. Rev. Lett. 66, 2593-2596(1991).

[91] Kasapi, A., Jain, M., Yin, G. Y. & Harris, S. E. Electromagnetically Induced Transparency: Propagation Dynamics. Phys. Rev. Lett. 74, 2447-2450(1995).

[92] Fleischhauer, M. & Lukin, M. D. Quantum memory for photons: Dark-state polaritons. Phys. Rev. A 65, 022314 (2002).

[93] Budker, D., Kimball, D. F., Rochester, S. M. & Yashchuk, V. V. Nonlinear Magneto-optics and Reduced Group Velocity of Light in Atomic Vapor with Slow Ground State Relaxation, Phys. Rev. Lett. 83, 1767-1770(1999).

[94] Hau, L. V., Harris, S. E., Dutton, Z. & Behroozi, C. H. Light speed reduction to 17 metres per second in an ultracold atomic gas. Nature 397, 594-598(1999)

[95] Burmeister, E. F., Blumenthal, D. J. & Bowers, J. E. A comparison of optical buffering technologies. Optical Switching and Networking 5 10-18(2008)

[96] Fleischhauer, M. & Lukin, M. D. Dark-State Polaritons in Electromagnetically Induced Transparency. Phys. Rev. Lett. 84, 5094-5097(2000).

[97] Lukin, M. D. Colloquium: Trapping and manipulating photon states in atomic ensembles. Rev. Mod. Phys. 75, 457-472(2003).

[98] Fleischhauer, M. Imamoglu, A. & Marangos, J. P. Electromagnetically nduced transparency: Optics in Coherent Media. Rev. Mod. Phys. 77, 633-673(2005).

[99] Phillips, D. F., A. Fleischhauer, A. Mair & Walsworth, R. L. Storage of Light in Atomic Vapor. Phys. Rev. Lett. 86, 783-786(2001)

[100] Liu, C., Dutton, Z., Behroozi, C. H. & Hau, L. V. Observation of coherent optical information storage in an atomic medium using halted light pulses. Nature 409, 490-493(2001).

[101] Gorshkov, A. V., André, A., Fleischhauer, M., Sørensen, A. S. & Lukin, M. D. Optimal storage of photon states in Optically Dense Atomic Media, Phys. Rev. Lett. 98, 123601(2007).

[102] Phillips, N. B. Novikova, I. & Gorshkov, A. V. Optimallight storage in atomic vapor. Phys. Rev. A 78, 023801(2008).

[103] Novikova, I. et al. Optimal Control of Light Pulse Storage and Retrieval Phys. Rev. Lett. 98, 243602(2007).

[104] Novikova, I.. Phillips, N. B. & Gorshkov, A. V. Optimal light storage with full pulse-shape control. Phys. Rev. A 78, 021802(R)(2008).

[105] Camacho, R. M., Vudyasetu, P. K. & Howell, J. C. Four-wave-mixing stopped light in hot atomic rubidium vapour. Nature Phot. 3, 103-106(2009).

[106] Ichimura, K., Yamamoto, K. and Gemma, N. Evidence for electromagnetically induced transparency in a solid medium, Phys. Rev. A 58, 4116-4120(1998).

[107] Turukhin, A. V. et al. Phys. Rev. Lett. 88, 023602(2002).

[108] Longdell, J. J. Fraval, E., Sellars M. & Manson, N. B. Stopped light with storage times greater than one second

using electromagnetically induced transparency in a solid. Phys. Rev. Lett. 95,063601(2005).

[109] Goldner, Ph. et al. Long coherence lifetime and electromagnetically, induced transparency in a highly-spinconcentrated solid. Phys. Rev. A 79,033809(2009).

[110] Chanelière, T. et al. Storage and retrieval of single photons transmitted between remote quantum memories. Nature 438,833-836(2006).

[111] Eisaman, M. D. et al. Electromagnetically induced transparency with tunable single-photon pulses. Nature438, 837-841(2006).

[112] Choi, K. S., Deng, H., Laurat, J. & Kimble,. H. J., Mapping photonic entanglement into and out of a quantum memory. Nature 452,67-71(2008).

[113] Akamatsu, D., Akiba, K. and Kozuma, M. Electromagnetically Induced Transparency with Squeezed Vacuum, Phys. Rev. Lett. 92,203602(2004).

[114] Honda, K. et al. Storage and Retrieval of a Squeezed Vacuum Phys. Rev. Lett. 100,093601(2008).

[115] Arikawa, M. et al. M. Quantum memory of a squeezed vacuum for arbitrary frequency sidebands. arXiv: 0905.2816 Appel, J., Figueroa, E., Korystov, D., Lobino, M. & Lvovsky, A. I., Quantum memory for squeezed light. Phys. Rev. Lett. 100,093602(2008).

[116] Figueroa, E., Vewinger, F., Appel, J. & Lvovsky, A. I. Decoherence of electromagnetically-induced transparency in atomic vapor. Opt. Lett. 31,2625-2627(2006).

[117] Hsu, M. T. L. et al. Quantum Study of Information Delay in Electromagnetically Induced Transparency. Phys. Rev. Lett. 97,183601(2006).

[118] Peng, A. et al. Squeezing and entanglement delay using slow light. Phys. Rev. A 71,033809(2005).

[119] Hétet, G. et al. Erratum: Squeezing and entanglement delay using slow light [Phys. Rev. A 71, 033809 (2005)]. Phys. Rev. A 74,059902(E)(2006).

[120] Figueroa, E., Lobino, M., Korystov, D., Kupchak, C. & Lvovsky, A. I. Propagation of squeezed vacuum under electromagnetically induced transparency, New J. Phys. 11,013044(2009).

[121] Duan, L. M., Lukin, M. D., Cirac, J. I. & Zoller, P. Long-distance quantum communication with atomic ensembles and linear optics. Nature 414,413-418(2001).

[122] Kuzmich, A. A., Bowen, W. P., Boozer, A. D., Boca, A., Chou, C. W., Duan, L. M. & Kimble, H. J. Generation of nonclassical photon pairs for scalable quantum communication with atomic ensembles Nature 423,731-734(2003).

[123] van der Wal, C. H., Eisaman, M. D., André, A., Walsworth, R. L., Phillips, D. F., Zibrov, A. S. & Lukin, M. D. Atomic Memory for Correlated Photon States. Science 301,196-200(2003).

[124] Jiang, W., Han, C., Xue, P., Duan, L. M. & Guo, G. C. Nonclassical photon pairs generated from a roomtemperature atomic ensemble. Phys. Rev. A 69,043819(2004).

[125] Chou, C. W., Polyakov, S. V., Kuzmich, A. & Kimble, H. J. Single-Photon Generation from Stored Excitation in an Atomic Ensemble. Phys. Rev. Lett. 92,213601(2004).

[126] Eisaman, M. D., Childress, L., André, A., Massou, F., Zibrov, A. S. & Lukin, M. D. Shaping Quantum Pulses of Light Via Coherent Atomic Memory. Phys. Rev. Lett. 93,233602(2004).

[127] Polyakov, S. V., Chou, C. W., Felinto, D. & Kimble, H. J. Temporal Dynamics of Photon Pairs Generated by an Atomic Ensemble. Phys. Rev. Lett. 93,263601(2004).

[128] Laurat, J., de Riedmatten, J., Felinto, D., Chou, C. W. Schomburg, E. W. & Kimble, H. J. Efficient retrieval of a single excitation stored in an atomic ensemble. Opt. Express 14,6912-6918(2006).

[129] V. Balić, V., Braje, D. A., Kolchin, P., Yin, G. Y. & Harris, S. E. Generation of Paired Photons with ControllableWaveforms. Phys. Rev. Lett. 94,183601(2005).

[130] D. A. Braje, D. A., Balić, V., Goda, S., Yin G. Y. & Harris, S. E. Frequency Mixing Using Electromagnetically Induced Transparency in Cold Atoms. Phys. Rev. Lett. 93,183601(2004).

[131] Papp, S. B., Choi, K. S., Deng, H., Lougovski, P., van Enk, S. J. & Kimble, H. J. Characterization of Multipartite Entanglement for One Photon Shared Among Four Optical Modes. Science 324,764-768(2009).

[132] Matsukevich, D. N., Chaneliére, T., Jenkins, S. D., Lan, S. Y., Kennedy, T. A. B. & Kuzmich, A. Deterministic

Single Photons via Conditional Quantum Evolution. Phys. Rev. Lett. 97, 013601(2006).

[133] de Riedmatten, H., Laurat, J., Chou, C. W., Schomburg, E. W., Felinto, D. & Kimble, H. J. Direct Measurement of Decoherence for Entanglement between a Photon and Stored Atomic Excitation. Phys. Rev. Lett. 97, 113603(2006).

[134] Chen, S., Chen, Y. A., Strassel, T., Yuan, Z. S., Zhao, B., Schmiedmayer, J. & Pan, J. W. Deterministic and Storable Single-Photon Source Based on a Quantum Memory. Phys. Rev. Lett. 97, 173004(2006).

[135] Black, A. T., Thompson, J. K. & Vuletić, V. OnDemand Superradiant Conversion of Atomic Spin Gratings into Single Photons with High Efficiency Phys. Rev. Lett. 95, 133601(2005).

[136] Thompson, J. K., Simon, J., Loh, H. & Vuletic, V. A High-Brightness Source of Narrowband, Identical-Photon Pairs. Science 313, 74(2006).

[137] Simon, J., Tanji, H., Thompson, J. K. & Vuletić, V. Interfacing Collective Atomic Excitations and Single Photons. Phys. Rev. Lett. 98, 183601(2007).

[138] D. Felinto et al. Conditional control of the quantum states of remote atomic memories for quantum networking. Nature Phys. 2, 844(2006).

[139] Chanelière, T., Matsukevich, D. N., Jenkins, S. D., Lan, S.-Y., Zhao, R., Kennedy, T. A. B. & Kuzmich, A. Quantum Interference of Electromagnetic Fields from Remote Quantum Memories. Phys. Rev. Lett. 98, 113602(2007).

[140] Yuan Z. S., Chen, Y. A., Chen, S., Zhao, B., Koch, M., Strassel, T., Zhao, Y., Zhu, G. J., Schmiedmayer, J. & Pan, J.-W. Synchronized Independent Narrow-Band Single Photons and Efficient Generation of Photonic Entanglement. Phys. Rev. Lett. 98, 180503(2007).

[141] Matsukevich, D. N. & Kuzmich, A. Quantum State Transfer Between Matter and Light. Science 306, 663-666 (2004).

[142] van Enk, S. & Kimble, H. J. Comment on "Quantum State Transfer Between Matter and Light". Science 309, 1187(2005).

[143] Matsukevich, D. N., & Kuzmich, A. Response to Comment on "Quantum State Transfer Between Matter and Light". Science 309, 1187(2005).

[144] Chou, C. W. et al. Measurement-induced entanglement for excitation stored in remote atomic ensembles. Nature 438, 828(2005).

[145] Chen, Y.-A. et al. Memory-build-in quantum teleportation with photonic and atomic qubits. Nature Phys. 4, 103-107(2008).

[146] Matsukevich, D. N. et al.. Entanglement of a Photon and a Collective Atomic Excitation. Phys. Rev. Lett. 95, 040405(2005).

[147] Inoue, R., Kanai, N., Yonehara, T., Miyamoto, Y., Koashi, M. & Kozuma, M. Entanglement of orbital angularmomentum states between an ensemble of cold atoms and a photon Phys. Rev. A 74, 053809(2006).

[148] Chen, S., Chen, Y. A., Zhao, B., Yuan, Z. S., Schmiedmayer, J. & Pan, J. W. Demonstration of a StablePhys. Rev. Lett. 99, 180505(2007).

[149] S.-Y. Lan et al. Dual-Species Matter Qubit Entangled with Light. Phys. Rev. Lett. 98, 123602(2007).

[150] D. N. Matsukevich et al. Entanglement of Remote Atomic Qubits. Phys. Rev. Lett. 96, 030405(2006).

[151] Z.-S. Yuan et al. Experimental demonstration of a BDCZ quantum repeater node. Nature 454, 1098(2008).

[152] Zhao, B., Chen, Z. B., Chen, Y. A., Schmiedmayer, J., & Pan, J.-W. Phys. Rev. Lett. 98, 240502(2007).

[153] Chou, C.-W. et al. Functional Quantum Nodes for Entanglement Distribution over Scalable Quantum Networks. Science 316, 1316(2007).

[154] Kurnit, N., Abella, I. D. & Hartmann, S. R. Observation of a Photon Echo. Phys. Rev. Lett. 13, 567-568(1964).

[155] Elyutin, S. O., Zakharov, S. M. & Manykin, E. A., Theory of formation of photon echo pulses. Sov. Phys. JETP49, 421-431(1979).

[156] Mossberg, T. Time-domain frequency-selective optical data storage. Opt. Lett. 7, 77-79(1982).

[157] Carlson, L. J., Rothberg, L. J., Yodh, A. G., Babbitt, W. R. & Mossberg, T. Storage and time reversal of light pulses using photon echoes. Opt. Lett. 8, 483-485(1983).

[158] Ruggiero, J., Le Gouët, J.-L, Simon, C. & Chanelière, T., Why the two-pulse photon echo is not a good quantum memory protocol. Phys. Rev. A 79, 053851(2009).

[159] Tittel, W. et al. Photon-Echo Quantum Memory in Solid State Systems. Laser & Phot. Rev. DOI 10.1002/lpor.200810056 Hétet, G., Longdell, J. J., Sellars, M. J., Lam, P. K., & Buchler, B. C., Multimodal properties and dynamics of gradient echo memory. Phys. Rev. Lett. 101, 203601(2008).

[160] Moiseev, S. A. & Kröll, S., Complete Reconstruction of the Quantum State of a Single Photon Wave Packet Absorbed by a Doppler-Broadened Transition Phys. Rev. Lett. 87, 173601(2001).

[161] Moiseev, S. A., Tarasov, V. F. & Ham, B. S., Quantum memory photon echo-like techniques in solids. J. Opt. B: Quantum Semiclass. Opt. 5, S497-S502(2003).

[162] Nilsson, N. & Kröll, S. Solid state quantum memory using complete absorption and re-emission of photons by tailored and externally controlled inhomogeneous absorption profiles. Opt. Comm. 247, 393-504(2005).

[163] Alexander, A. L., Longdell, J. J., Sellars, M. J. & Manson, N. B., Photon Echoes Produced by Switching Electric Fields. Phys. Rev. Lett. 96, 043602(2006).

[164] Kraus, B. et al. Quantum memory for nonstationary light fields based on controlled reversible inhomogeneous broadening Phys. Rev. A 73, 020302(2006).

[165] Moiseev, S. A. & Noskov, M. I. The possibilities of the quantum memory realization for short pulses of light in the photon echo technique. Laser Phys. Lett. 1, 303-310(2004).

[166] Sanguard, N., Simon, C., Afzelius, M. & Gisin, N. Analysis of a quantum memory for photons based on controlled reversible inhomogeneous. broadening. Phys. Rev. A 75, 032327(2007).

[167] Longdell, J. J., Hétet, G., Lam, P. K. & Sellars, M. J. Analytic treatment of controlled reversible inhomogeneous broadening quantum memories for light using two-level atoms. Phys. Rev. A 78, 032337(2008).

[168] Hétet, M., Longdell, J. J., Alexander, A. L., Lam, P. K., Sellars, M. Electro-Optic Quantum Memory for Light Using Two-Level Atoms Phys. Rev. Lett. 100, 023601(2008).

[169] Moiseev, S. A. & Arslanov, N. M., Efficiency and fidelity of photon-echo quantum memory in an atomic system with longitudinal inhomogeneous broadening. Phys. Rev. A 78, 023803(2008).

[170] Hétet, G. et al. Photon echoes generated by reversing magnetic field gradients in a rubidium vapor, Opt. Lett. 33, 2323-2325(2008).

[171] Le Gouët, J. L. & Berman, P. R. Raman scheme for adjustable bandwidth quantum memory. Phys. Rev. A 80, 012320(2009)

[172] Hosseini, M. et al. Coherent optical pulse sequencer for quantum applications. Nature 461, 241-245(2009).

[173] Alexander, A. L., Longdell, J. J., Sellars, M. J. & Manson, N. B. Coherent information storage with photon echoes produced by switching electric fields, J. Lumin. 127, 94-97(2007).

[174] Hedges, M. P., Sellars, M. J., Lee, Y.-M. & Longdell, J. J. A Solid State Quantum Memory. Poster presentation at International Conference on Hole Burning, Single Molecule, and Related Spectroscopies: Science Applications(HBSM 2009), Palm Cove, Australia, 22 to 27 June 2009.

[175] Lauritzen, B. et al. Solid state quantum memory for photons at telecommunication wavelengths, arXiv: 0908.2348. Appel, J., Marzlin, K.-P. & Lvovsky, A. I. Raman adiabatic transfer of optical states in multilevel atoms. Phys. Rev. A 73, 013804(2006).

[176] Vewinger, F., Appel, J., Figueroa, E. & Lvovsky, A. I. Adiabatic frequency conversion of quantum optical information in atomic vapor Opt. Lett. 32, 2771-2773(2007).

[177] Campbell, G., Ordog, A. & Lvovsky, A. I. Multimode electromagnetically-induced transparency on a single atomic line. New J. Phys. 11, 103021(2009).

[178] Staudt, M. U. et al. Investigations of optical coherence properties in an erbium-doped silicate fiber for quantum state storage, Opt. Commun. 266, 720-726(2006).

[179] de Riedmatten, H., Afzelius, M., Staudt, M. U., Simon, C. & Gisin, N., A solid-state light-matter interface at the single-photon level, Nature 456, 773-777(2008).

[180] Kuzmich, A., Bigelow, N. P. & Mandel, L. Atomic quantum non-demolition measurements and squeezing. Europhys. Lett. 42, 481-486(1998).

[181] Duan, L.-M., J. I. Cirac, P. Zoller, and E. S. Polzik, 2000, Phys. Rev. Lett. 85, 5643. Kuzmich, A., and E. Polzik, 2000, Phys. Rev. Lett. 85, 5639.

[182] Kuzmich, A. & Polzik, E. S. Quantum Information with Continuous Variables (Kluwer, 2003), pp. 231-265. Muschik, C. A., Hammerer, K., Polzik, E. S. & Cirac, J. I. Efficient quantum memory and entanglement between light and an atomic ensemble using magnetic fields. Phys. Rev. A 73, 062329(2006).

[183] Julsgaard, B., Sherson, J., Cirac, J. I., Fiurášek, J. & Polzik, E. S. Experimental demonstration of quantum memory for light. Nature 432, 482-486(2004).

[184] Julsgaard, B., A. Kozhekin, A. & E. S. Polzik. Experimental long-lived entanglement of two macroscopic objects. Nature 413, 400-403(2001).

[185] Macfarlane, R. M., High-resolution laser spectroscopy of rare-earth doped insulators: a personal perspective, J. Lumin. 100, 1-20(2002).

[186] Sun, Y., Thiel, C. W., Cone, R. L., Equall, R. W. & Hutcheson, R. L. Recent progress in developing new rare earth materials for hole burning and coherent transient applications, J. Lumin. 98, 281-287(2002).

[187] Spectroscopic Properties of Rare Earths in Optical Materials (Springer Series in Materials Science), Guokui Liu and Bernard Jacquier (Editors), Springer Berlin 2005.

[188] Macfarlane, R. M. Optical Stark spectroscopy of solids, J. Lumin. 125, 156-174(2007).

[189] Zhao, B. et al. A millisecond quantum memory for scalable quantum networks. Nature Phys. 5, 95-99(2008).

[190] Zhao, R. et al. Long-lived quantum memory. Nature Phys. 5, 100-104(2008).

[191] Schnorrberger, U. et al. Electromagnetically Induced Transparency and Light Storage in an Atomic Mott Insulator, Phys. Rev. Lett. 103, 033003(2009).

[192] Fraval, E., Sellars, M. J. & Longdell, J. J. Method of Extending Hyperfine Coherence Times in Pr^{3+}: Y_2SiO_5. Phys. Rev. Lett. 92, 077601(2004).

[193] Fraval, E., Sellars, M. J., & Longdell, J. J. Dynamic Decoherence Control of a Solid-State Nuclear Quadrupole Qubit. Phys. Rev. Lett. 95, 030506(2005).

[194] Nicolas Gisin, Grigoire Ribordy, Wolfgang Tittel, and Hugo Zbinden. Quantum cryptography. uantph/0101098, 2001.

[195] L-M. Duan, M. D. Lukin, J. I. Cirac, and P. Zoller. Longdistance quantum communication with atomic ensembles and linear optics. Nature, 414: 413-418, 2001.

[196] H.-J. Briegel, W. Dür, J. I. Cirac, and P. Zoller. Quantum repeaters: The role of imperfect local operations in quantum communication. Phys. Rev. Lett., 81: 5932-5935, 1998.

[197] Jacob Sherson, Brian Julsgaard, and E. S. Polzik. Deterministic atom-light quantum interface. quantph/0601186, 2006.

[198] M. Fleischhauer and M. D. Lukin. Quantum memory for photons: Dark-state polaritons. Am. Phys. Soc., 65: 022814, 2002.

[199] M. D. Lukin. Trapping and manipulating photon states in atomic ensembles. Rev. Mod. Phys., 75(2): 457-472, 2002.

[200] B. Kraus, W. Tittel, N. Gisin, M. Nilsson, S. Kroll, and J. I. Cirac. Quantum memory for non-stationary light fields based on controlled reversible inhomogeneous broadening. quant-ph/0502184, 2005.

[201] E. Kozhekin, K. Mølmer, and E. Polzik. Quantum memory for light. Phys. Rev. A, 62: 033809, 2000.

[202] L. M. Duan, J. I. Cirac, and P. Zoller. Three-dimensional theory for interaction between atomic ensembles and freespace light. Phys. Rev. A, 66: 023818, 2002.

[203] M. G. Raymer. Quantum state entanglement and readout of collective atomic-ensemble modes and optical wave packets by stimulated Raman scattering. J. Mod. Opt. 51(12): 1739-1759, 2004.

[204] Wojciech Wasilewski and M. G. Raymer. Pairwise entanglement and readout of atomic-ensemble and optical wave-packet modes in traveling-wave Raman interactions. Phys. Rev. A, 73: 063816, 2006.

[205] Rupert Ursin, Thomas Jennewein, Markus Aspelmeyer, Rainer Kaltenbaek, Michael Lindenthal, Philip Walther, and Anton Zeilinger. Communications: Quantum teleportation across the Danube. Nature, 430: 849, 2004.

[206] Wang Yao, Ren-Bao Liu, and L. J. Sham. Theory of control of the spin-photon interface for quantum

networks. Phys. Rev. Lett. ,95: 030504,2005.

[207] C. Iaconis and I. A. Walmsley. Spectral phase interferometry for direct electric-field reconstruction of ultrashort optical pulses. Opt. Lett. ,23(10): 792-794,1998.

[208] Shortly after our work was completed, Gorshkov et al. presented a paper, quant-ph/0604037(2006), claiming that the storage of broadband photons on resonance is possible given sufficient atomic density and unlimited control pulse energy.

[209] G. Brassard, N. Lutkenhaus, T. More, and B. Sanders, Phys. Rev. Lett. 85, 1330(2000).

[210] L. M. Duan, M. D. Lukin, J. I. Cirac, and P. Zoller, Nature(London) 414, 413(2001).

[211] C. Kurtsiefer, S. Mayer, P. Zarda, and H. Weinfurter, Phys. Rev. Lett. 85, 290(2000).

[212] Santori et al. , Nature(London) 419, 594(2002).

[213] S. Bracker et al. , Phys. Rev. Lett. 94, 047402(2005).

[214] J. M. Taylor, C. M. Marcus, and M. D. Lukin, Phys. Rev. Lett. 90, 206803(2003).

[215] L. Childress, J. M. Taylor, A. S. Sørensen, and M. D. Lukin, Phys. Rev. A 72, 052330(2005).

[216] F. Jelezko et al. , Phys. Rev. Lett. 93, 130501(2004).

[217] Cabrillo, J. Cirac, P. Garcia-Fernandez, and P. Zoller, Phys. Rev. A 59, 1025(1999).

[218] K. Holman, D. Hudson, J. Ye, and D. Jones, Opt. Lett. 30, 1225(2005).

[219] S. Barrett and P. Kok, Phys. Rev. A 71, 060310(R)(2005).

[220] Simon imon and W. Irvine, Phys. Rev. Lett. 91, 110405(2003).

[221] P. Kok, WJ Munro, K. Nemoto, TC Ralph, J. P. Dowling, and GJ Milburn. Linear optical quantum computing with photonic qubits. Reviews of Modern Physics, 79(1): 135, 174, 2007.

[222] L.-M. Duan, M. D. Lukin, J. I. Cirac, and P. Zoller. Long-distance quantum communication with atomic ensembles and linear optics. Nature, 414(6862): 413, 418, November 2001.

[223] G. Hetet, J. J. Longdell, A. L. Alexander, P. K. Lam, and M. J. Sellars. Electro-Optic Quantum.

[224] Memory for Light Using Two-Level Atoms. Physical Review Letters, 100(2): 023601, 2008.

[225] Hugues de Riedmatten, Mikael Afzelius, Matthias U. Staudt, Christoph Simon, and Nicolas Gisin. A solid-state light-matter interface at the single-photon level. Nature, 456(7223): 773-777, December 2008.

[226] U. Schnorrberger, J. D. Thompson, S. Trotzky, R. Pugatch, N. Davidson, S. Kuhr, and I. Bloch. Electromagnetically Induced Transparency and Light Storage in an Atomic Mott Insulator. Physical Review Letters, 103(3): 033003, July 2009.

[227] D. Boozer, A. Boca, R. Miller, T. E. Northup, and H. J. Kimble. Reversible State Transfer between Light and a Single Trapped Atom. Physical Review Letters, 98(19): 193601, May 2007.

[228] Alexey V. Gorshkov, Axel Andre, Mikhail D. Lukin, and Anders S. Sorensen. Photon storage in Lambda-type optically dense atomic media. II. Free-space model. Physical Review A(Atomic, Molecular, and Optical Physics), 76(3): 033805, 2007.

[229] J. Nunn, I. A. Walmsley, M. G. Raymer, K. Surmacz, F. C. Waldermann, Z. Wang, and D. Jaksch. Mapping broadband single-photon wave packets into an atomic memory. Physical Review A(Atomic, Molecular, and Optical Physics), 75(1): 011401, 2007.

[230] Ryan M. Camacho, Praveen K. Vudyasetu, and John C. Howell. Four-wave-mixing stopped light in hot atomic rubidium vapour. Nat Photon, 3(2): 103, 106, February 2009.

[231] Peter Shor. Polynomial-time algorithms for prime factorization and discrete logarithms on a quantum computer. SIAM J. Sci. Statist. Comput. , 26: 1484, 1997.

[232] N. Sangouard, C. Simon, H. de Riedmatten, and N. Gisin. Quantum repeaters based on atomic ensembles and linear optics. eprint arXiv: 0906.2699, 2009.

[233] S. D. Barrett, P. P. Rohde, and T. M. Stace. Scalable quantum computing with atomic ensembles, 2008.

[234] K. S. Choi, H. Deng, J. Laurat, and H. J. Kimble. Mapping photonic entanglement into and out of a quantum memory. Nature, 452(7183): 67, 71, March 2008.

[235] J. Nunn, K. Reim, K. C. Lee, V. O. Lorenz, B. J. Sussman, I. A. Walmsley, and D. Jaksch. Multimode Memories in Atomic Ensembles. Physical Review Letters, 101(26): 260502, December2008.

[236] Christoph Simon, Hugues de Riedmatten, Mikael Afzelius, Nicolas Sangouard, Hugo Zbinden, and Nicolas Gisin. Quantum repeaters with photon pair sources and multimode memories. Physical Review Letters, 98 (19): 190503, 2007.

[237] M. D. Eisaman, A. Andre, F. Massou, M. Fleischhauer, A. S. Zibrov, and M. D. Lukin. Electromagnetically induced transparency with tunable single-photon pulses. Nature, 438(7069): 837, 841, December 2005.

[238] Stephen E. Harris. Electromagnetically Induced Transparency. Physics Today, 50(7): 36, 1997.

[239] Chien Liu, Zachary Dutton, Cyrus H. Behroozi, and Lene Vestergaard Hau. Observation of coherent optical information storage in an atomic medium using halted light pulses. Nature, 409(6819): 490{493, 2001.

[240] L. Alexander, J. J. Longdell, M. J. Sellars, and N. B. Manson. Photon Echoes Produced by Switching Electric Fields. Physical Review Letters, 96(4): 043602, February 2006.

[241] Kraus, W. Tittel, N. Gisin, M. Nilsson, S. Kroll, and J. I. Cirac. Quantum memory for nonstationary light fields based on controlled reversible inhomogeneous broadening. Physical Review A(Atomic, Molecular, and Optical Physics), 73(2): 020302, February 2006.

[242] M. U. Staudt, S. R. Hastings-Simon, M. Nilsson, M. Afzelius, V. Scarani, R. Ricken, H. Suche, W. Sohler, W. Tittel, and N. Gisin. Fidelity of an Optical Memory Based on Stimulated Photon choes. Physical Review Letters, 98(11): 113601, March 2007.

[243] T. Chaneliere, M. Afzelius, and J L. Le Gouöet. Efficient light storage in a crystal using an Atomic Frequency Comb. arXiv: 0902.2048, February 2009.

[244] Z. Zhu, D. J. Gauthier, and R. W. Boyd. Stored light in an optical fiber via stimulated Brillouin scattering. Science, 318(5857): 1748, 2007.

[245] Mahdi Hosseini, Ben M. Sparkes, Gabriel Hetet, Jevon J. Longdell, Ping Koy Lam, and Ben C. Buchler. Coherent optical pulse sequencer for quantum applications. Nature, 461(7261): 241, 2009.

[246] Irina Novikova, Nathaniel B. Phillips, and Alexey V. Gorshkov. Optimal light storage with full pulse-shape control. Physical Review A(Atomic, Molecular, and Optical Physics), 78(2): 021802, 2008.

[247] J. J. Longdell, E. Fraval, M. J. Sellars, and N. B. Manson. Stopped Light with Storage Times Greater than One Second Using Electromagnetically Induced Transparency in a Solid. Physical Review Letters, 95(6): 063601, 2005.2009

[248] E. Kozhekin, K. Mlmer, and E. Polzik. Quantum memory for light. Physical Review A, 62(3): 033809, 2000.2009

[249] K. Surmacz, J. Nunn, K. Reim, K. C. Lee, V. O. Lorenz, B. Sussman, I. A. Walmsley, and D. Jaksch. Efficient spatially resolved multimode quantum memory. Physical Review A(Atomic, Molecular, and Optical Physics), 78(3): 033806, 2008.

[250] L. S. Pontryagin et. al., Mathematical Theory of Optimal Processes(Gordon & Breach Science Publishers, New York, 1986).

[251] K. Hammerer, A. Sorensen, and E. Polzik, arXiv: 0807.3358(2008).

[252] C. Simon et al., Phys. Rev. Lett. 98, 190503(2007).

[253] N. Sangouard, C. Simon, H. de Riedmatten, and N. Gisin, arXiv: 0906.2699(2009).

[254] Herskind P F, Dantan A, Marler J P, Albert M and Drewsen M 2009 Nature Phys. 5 494.

[255] Albert M, Dantan A and Drewsen M 2011 Nature Photon. 5 633.

[256] N. Sangouard, C. Simon, M. Afzelius, and N. Gisin, Phys. Rev. A 75, 032327(2007).

[257] M. Afzelius, C. Simon, H. de Riedmatten, and N. Gisin, Phys. Rev. A 79, 052329(2009).

[258] K. S. Choi, H. Deng, J. Laurat, and H. J. Kimble, Nature452, 67(2008).

[259] K. Akiba, K. Kashiwagi, M. Arikawa, and M. Kozuma, New Journal of Physics 11, 013049(2009).

[260] K. Honda et al., Phys. Rev. Lett. 100, 093601(2008).

[261] J. Cviklinski et al., Phys. Rev. Lett. 101, 133601(2008).

[262] H. de Riedmatten et al., Nature 456, 773(2008).

[263] N. Sangouard et al., Phys. Rev. A 76, 050301(2007).

[264] N. Sangouard et al., Phys. Rev. A 77, 062301(2008).

[265] T. Böttger, C. W. Thiel, R. L. Cone, and Y. Sun, Phys. Rev. B 79, 115104(2009).

[266] Baldit et al., Phys. Rev. Lett. 95, 143601(2005).

[267] M. U. Staudt et al., Phys. Rev. Lett. 99, 173602(2007).

[268] B. Lauritzen et al., Phys. Rev. A 78, 043402(2008).

[269] J. Ruggiero, J. L. Le Gouët, C. Simon, and T. Chanelière, Phys. Rev. A 79, 053851(2009).

[270] L. Alexander, J. J. Longdell, M. J. Sellars, and N. B. Manson, Phys. Rev. Lett. 96, 043602(2006).

[271] W. Tittel et al., Laser & Photonics Review 1(2009).

[272] Alexander, J. Longdell, M. Sellars, and N. Manson, Journal of Luminescence 127, 94(2007).

[273] Hétet et al., Phys. Rev. Lett. 100, 023601(2008).

[274] S. R. Hastings-Simon et al., Phys. Rev. B 78, 085410(2008).

[275] Gibbs, H. M. Optical bistability: controlling light with light(Academic Press, Inc., Orlando, FL 1985).

[276] Bouwmeester, D., Ekert, A. & Zeilinger, A., Eds. The Physics of Quantum Information(Springer, Berlin, 2000).

[277] Harris, S. E. & Yamamoto, Y. Photon Switching by Quantum Interference. Phys. Rev. Lett. 81, 3611(1998).

[278] Lukin, M. D. Colloquium: Trapping and manipulating photon states in atomc ensembles. Rev. Mod. Phys. 75, 457(2003).

[279] Fleischhauer, M., Imamoglu, A. & Marangos, J. P. Electromagnetically induced transparency: Optics in coherent media. Rev. Mod. Phys. 77, 633(2005).

[280] Vahala, K., Ed. Optical Microcavities(World Scientific, Singapore, 2004).

[281] Miller, R. et al. Trapped atoms in cavity QED: coupling quantized light and matter. J. Phys. 13 B: At. Mol. Opt. Phys. 38, S551(2005).

[282] Duan, L.-M. & Kimble, H. J. Scalable Photonic Quantum Computation through Cavity-Assisted Interactions. Phys. Rev Lett. 92, 127902(2004).

[283] Birnbaum, K. M. et al. Photon blockade in an optical cavity with one trapped atom. Nature 436, 87(2005).

[284] Waks, E. & Vuckovic, J. Dipole Induced Transparency in drop filter cavity-waveguide systems. Phys. Rev. Lett. 96, 153601(2006).

[285] Chang, D. E., Sørensen, A. S., Hemmer, P. R. & Lukin, M. D. Quantum Optics with Surface Plasmons. Phys. Rev. Lett. 97, 053002(2006).

[286] Akimov, A. V. et al., submitted to Nature(2007).

[287] Atwater, H. A. The promise of plasmonics. Scientific American 296, 56(2007).

[288] Maier, S. A. Plasmonics: fundamentals and applications(Springer-Verlag, New York, 2006).

[289] Genet, C. & Ebbesen, T. W. Light in tiny holes. Nature 445, 39(2007).

[290] Klimov, V. V., Ducloy, M. & Letokhov, V. S. A model of an apertureless scanning microscope with a prolate nanospheriod as a tip and an excited molecule as an object. Chem. Phys. Lett. 358, 192(2002).

[291] Smolyaninov, I. I., Elliott, J., Zayats, A. V. & Davis, C. C. Far-Field Optical Microscopy with a Nanometer-Scale Resolution Based on the In-Plane Image Magnification by Surface Plasmon Polaritons. Phys. Rev. Lett. 94, 057401(2005).

[292] Zayats, A. V., Elliott, J., Smolyaninov, I. I. & Davis, C. C. Imaging with short-wavelength surface plasmon polaritons. Appl. Phys. Lett. 86, 151114(2005).

[293] Nie, S. & Emory, S. R. Probing Single Molecules and Single Nanoparticles by Surface-Enhanced.

[294] Raman Scattering. Science 275, 1102(1997).

[295] Chang, D. E., Sørensen, A. S., Hemmer, P. R. & Lukin, M. D. Strong coupling of single emitters to surface plasmons. quant-ph/0603221. Tong, L., Lou, J. & Mazur, E. Single-mode guiding properties of subwavelength-diameter silica and silicon wire waveguides. Opt. Express 12, 1025(2004).

[296] Meystre, P. & Sargent III, M. Elements of Quantum Optics, 3rd ed. (Springer-Verlag, New York, 1999).

[297] Shen, J. T. & Fan, S. Coherent photon transport from spontaneous emission in one-dimensional waveguides. Opt. Lett. 30, 2001(2005).

[298] Gorshkov, A. V., Andre, A., Fleischhauer, M., Sørensen, A. S. & Lukin, M. D. Universal Approach to Optimal Photon Storage in Atomic Media. Phys. Rev. Lett. 98, 123601(2007).

[299] Fleischhauer, M. & Lukin, M. D. Dark-State Polaritons in Electromagnetically Induced Transparency. Phys. Rev. Lett. 84, 5094(2000).

[300] Maier, S. A., Friedman, M. D., Barclay, P. E. & Painter, O. Experimental demonstration of fiber-accessible metal nanoparticle plasmon waveguides for planar energy guiding and sensing. Appl. Phys. Lett. 86, 071103 (2005).

[301] Klimov, V. I. et al. Optical Gain and Stimulated Emission in Nanocrystal Quantum Dots. Science 290, 314 (2000).

[302] Brouri, R., Beveratos, A., Poizat, J.-P. & Grangier, P. Photon antibunching in the fluorescence of individual color centers in diamond. Opt. Lett. 25, 1294(2000).

[303] Tamarat, Ph. et al. Stark Shift Control of Single Optical Centers in Diamond. Phys. Rev. Lett. 97, 083002 (2006).

[304] Lukin, M. D. & Imamoglu, A. Nonlinear Optics and Quantum Entanglement of Ultraslow Single Photons. Phys. Rev. Lett. 84, 1419(2000).

[305] Leclair, A., Lesage, F., Lukyanov, S. & Saleur, H. The Maxwell-Bloch theory in quantum optics and the Kondo model. Phys. Lett. A 235, 203(1997).

[306] Lesage, F. & Saleur, H. Boundary Interaction Changing Operators and Dynamical Correlations in Quantum Impurity Problems. Phys. Rev. Lett. 80, 4370(1998).

[307] Duan, L.-M., Lukin, M. D., Cirac, J. I. & Zoller, P. Longdistance quantum communication with atomic ensembles and linear optics. Nature 414, 413, 418(2001).

[308] Sangouard, N., Simon, C., de Riedmatten, H. & Gisin, N. Quantum repeaters based on atomic ensembles and linear optics(2009).

[309] Kimble, H. J. The quantum internet. Nature 453, 1023, 1030(2008).

[310] Simon, C. et al. Quantum repeaters with photon pair sources and multimode memories. Phys. Rev. Lett. 98, 190503(2007).

[311] Usmani, I., Afzelius, M., de Riedmatten, H. & Gisin, N. Mapping multiple photonic qubits into and out of one solid-state atomic ensemble. Nat Commun 1, 1, 7(2010).

[312] Blinov, B. B., Moehring, D., Duan, L.-M. & Monroe, C. Observation of entanglement between a single trapped ion and a single photon. Nature 428, 153, 7(2004).

[313] Volz, J. et al. Observation of entanglement of a single photon with a trapped atom. Phys. Rev. Lett. 96, 030404 (2006).

[314] Matsukevich, D. N. & Kuzmich, A. Quantum state transfer between matter and light. Science 306, 663, 666 (2004).

[315] Matsukevich, D. N. et al. Entanglement of a photon and a collective atomic excitation. Phys. Rev. Lett. 95, 040405(2005).

[316] de Riedmatten, H. et al. Direct measurement of decoherence for entanglement between a photon and stored atomic excitation. Phys. Rev. Lett. 97, 113603(2006).

[317] Chen, S. et al. Demonstration of a stable atom-photon entanglement source for quantum repeaters. Phys. Rev. Lett. 99, 180505(2007).

[318] Sherson, J. F. et al. Quantum teleportation between light and matter. Nature 443, 557, 560(2006).

[319] Jin, X.-M. et al. Quantum interface between frequencyuncorrelated down-converted entanglement and atomicensemble quantum memory(2010). arXiv: 1004.4691.

[320] Togan, E. et al. Quantum entanglement between an optical photon and a solid-state spin qubit. Nature 466, 730, 734(2010).

[321] Tittel, W. et al. Photon-echo quantum memory in solid state systems. Laser & Photonics Reviews 4, 244, 267 (2010).

[322] Longdell, J. J., Fraval, E., Sellars, M. J. & Manson, N. B. Stopped light with storage times greater than one second using electromagnetically induced transparency in a solid. Phys. Rev. Lett. 95, 063601(2005).

[323] Hedges, M. P., Longdell, J. J., Li, Y. & Sellars, M. J. E_cient quantum memory for light. Nature 465, 1052,